ELECTRONICS FUNDAMENTALS

THIRD
EDITION

ELECTRONICS FUNDAMENTALS

Circuits, Devices, and Applications

Thomas L. Floyd

Prentice Hall
Englewood Cliffs, New Jersey ▪ Columbus, Ohio

Library of Congress Cataloging-in-Publication Data
Floyd, Thomas L.
Electronics fundamentals : circuits, devices, and applications /
Thomas L. Floyd. —3rd ed.
p. cm.
Includes index.
ISBN 0-02-338654-1
1. Electronics. 2. Electronic circuits. 3. Electronic apparatus
and appliances. I. Title.
TK7816.F57 1995
621.381—dc20 94-26234
CIP

Cover art by Todd Yarrington and Thomas Mack
Editor: Dave Garza
Developmental Editor: Carol Hinklin Robison
Production Editor: Rex Davidson
Text Designer: Thomas Mack
Cover Designer: Thomas Mack
Production Buyer: Patricia A. Tonneman
Illustrations: Precision Graphics

This book was set in Times Roman by The Clarinda Company and was printed and bound by R. R. Donnelley & Sons Company. The cover was printed by Lehigh Press, Inc.

Photo credits for chapter openers and part openers: American Stock (Chapter 2); Courtesy of Battelle (Chapter 11); Comstock (Part 2, Chapter 21); Courtesy of DeVry Institutes (Chapters 4, 14, 15, 17, 19); FPG International-The Stock Directory (Chapter 13); Honeywell Home and Building Control, Golden Valley, Minn. (Chapter 7); Pennsylvania College of Technology (Chapter 18); The Stock Market (Part 1, Chapters 10, 20); Tony Stone Images (Chapters 3, 5), Superstock (Part 3); and Tektronix Inc. (Chapters 1, 6, 8, 9, 12, 16).

Printed in the United States of America

10 9 8 7 6 5 4 3 2 1

ISBN 0-02-338654-1

Prentice-Hall International (UK) Limited, *London*
Prentice-Hall of Australia Pty. Limited, *Sydney*
Prentice-Hall Canada Inc., *Toronto*
Prentice-Hall Hispanoamericana, S. A., *Mexico*
Prentice-Hall of India Private Limited, *New Delhi*
Prentice-Hall of Japan, Inc., *Tokyo*
Simon & Schuster Asia Pte. Ltd., *Singapore*
Editora Prentice-Hall do Brasil, Ltda., *Rio de Janeiro*

PREFACE

Electronics Fundamentals: Circuits, Devices, and Applications, Third Edition provides thorough, comprehensive, and practical coverage of basic electrical and electronic concepts and circuits with special emphasis on troubleshooting and applications. This third edition has been carefully reviewed and special efforts have been made to ensure a high level of accuracy and clarity.

This book is divided into three basic parts. Chapters 1 through 7 cover dc topics, Chapters 8 through 15 generally cover ac topics, and Chapters 16 through 21 cover electronic devices and circuits.

New and Improved Features

In addition to the many popular features of the second edition, several new features, as well as improvements to existing features, have been incorporated into this third edition.

- An innovative section at the end of each chapter (except Chapter 1) called Application Assignment
- An Application Assignment preview in each chapter opener (except Chapter 1)
- Distinctive full-color format used throughout
- An introductory message and a list of performance-based objectives at the beginning of each section
- An exercise in each worked example that is related to the example
- More section review questions with a complete answer section at the end of each chapter
- Improved consistency in accuracy (number of significant figures) of numerical results and in the use of metric prefixes
- An increased emphasis on troubleshooting and applications

An improved and expanded ancillary package for this edition includes

- Instructor's Edition
- Test Item File
- Prentice-Hall Test Manager 2.0
- Transparency Masters
- Transparencies
- Lab Manual: *Experiments in Electronics Fundamentals: Circuits, Devices, and Application, Third Edition* by Dave Buchla
- Student Study Guide by Wally McIntyre
- Video library of DC/AC topics

Illustration of Features Within Each Chapter

Chapter Opener As shown in Figure P–1, each chapter begins with a two-page opener. The left page includes a listing of the sections within the chapter and a brief overview

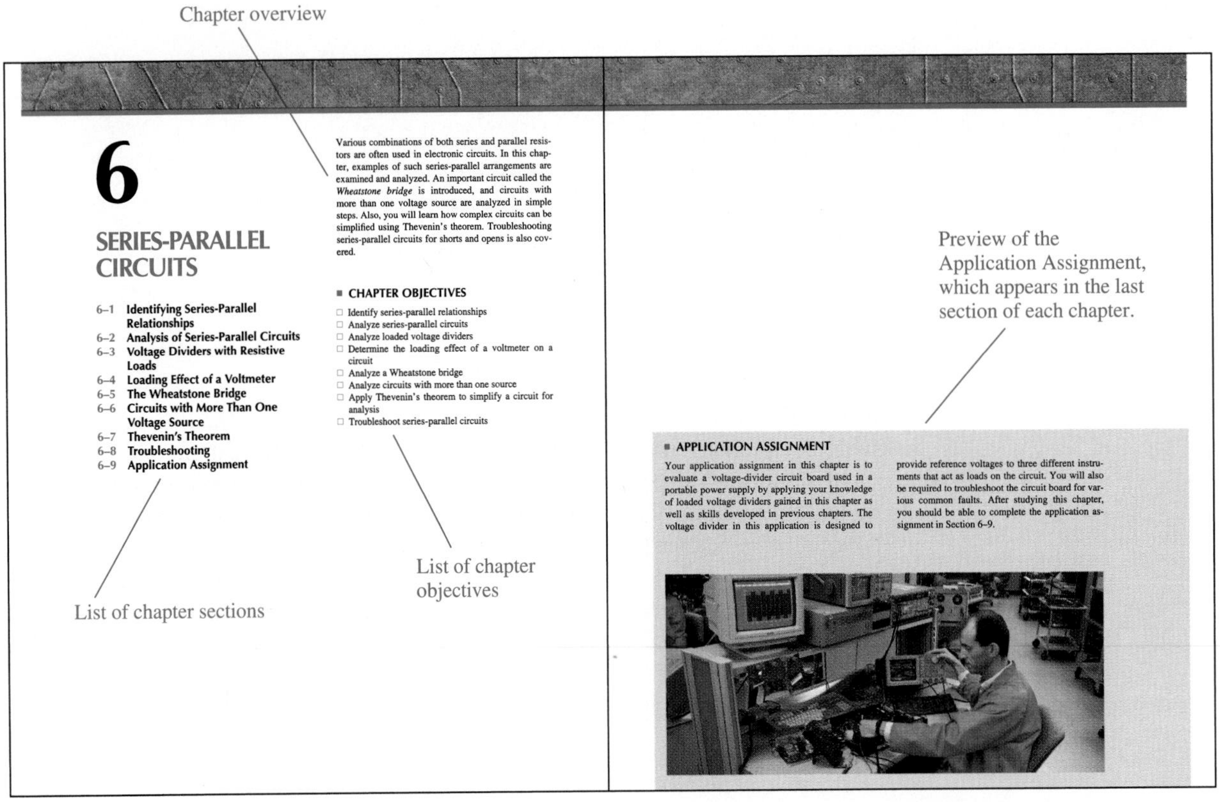

FIGURE P–1
Chapter opener.

and introduction as well as a list of main objectives. The right page presents a preview of the Application Assignment section that is found at the end of the chapter.

Section Opener and Section Review Each section within a chapter begins with a brief introduction that either highlights the material to be covered or provides a general overview and includes a list of performance-based objectives for the section with a main objective and subobjectives. Each section ends with review questions that focus on the main concepts presented in the section. Solutions/answers to these review questions are given at the end of the chapter. Figure P–2 illustrates these two features.

Examples Frequent worked examples help to illustrate and clarify basic concepts. Each example concludes with a related exercise that reinforces or expands on the concept in some way. Some related exercises require the student to repeat the procedure demonstrated in the example but with a different set of values or conditions. Others focus on a more limited part of the example or encourage further thought. Answers to all related exercises are given at the end of the chapter. A typical example problem and related exercise are shown in Figure P–3.

Application Assignment The last section of each chapter (except Chapter 1) provides a practical application of material presented in the chapter. A series of steps generally require the student to compare circuit layouts with a schematic, analyze circuits using concepts or theories learned in the chapter, and evaluate and/or troubleshoot circuits. A typical Application Assignment section is shown in Figure P–4.

Results and answers for the steps in the Application Assignments are provided in the *Instructor's Edition*. Optional worksheets are provided in Dave Buchla's coordinated lab manual.

Review questions end each section.

Introductory statements and a list of performance-based objectives begin each section.

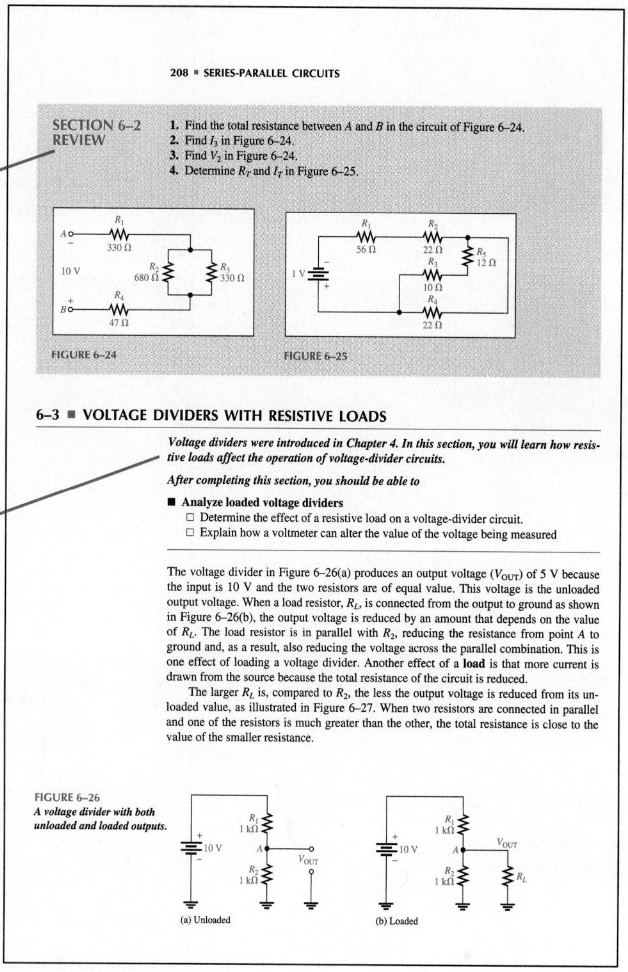

208 ■ SERIES-PARALLEL CIRCUITS

SECTION 6–2 REVIEW

1. Find the total resistance between A and B in the circuit of Figure 6–24.
2. Find I_3 in Figure 6–24.
3. Find V_2 in Figure 6–24.
4. Determine R_T and I_T in Figure 6–25.

FIGURE 6–24

FIGURE 6–25

6–3 ■ VOLTAGE DIVIDERS WITH RESISTIVE LOADS

Voltage dividers were introduced in Chapter 4. In this section, you will learn how resistive loads affect the operation of voltage-divider circuits.

After completing this section, you should be able to

- **Analyze loaded voltage dividers**
  - □ Determine the effect of a resistive load on a voltage-divider circuit.
  - □ Explain how a voltmeter can alter the value of the voltage being measured

The voltage divider in Figure 6–26(a) produces an output voltage (V_{OUT}) of 5 V because the input is 10 V and the two resistors are of equal value. This voltage is the unloaded output voltage. When a load resistor, R_L, is connected from the output to ground as shown in Figure 6–26(b), the output voltage is reduced by an amount that depends on the value of R_L. The load resistor is in parallel with R_2, reducing the resistance from point A to ground and, as a result, also reducing the voltage across the parallel combination. This is one effect of loading a voltage divider. Another effect of a **load** is that more current is drawn from the source because the total resistance of the circuit is reduced.

The larger R_L is, compared to R_2, the less the output voltage is reduced from its unloaded value, as illustrated in Figure 6–27. When two resistors are connected in parallel and one of the resistors is much greater than the other, the total resistance is close to the value of the smaller resistance.

FIGURE 6–26
A voltage divider with both unloaded and loaded outputs.

(a) Unloaded

(b) Loaded

FIGURE P–2
Section review and section opener.

The Application Assignment is a very versatile feature which provides both motivation and practical experiences in the classroom and can be used as

- □ An integral part of the chapter for illustrating how the concepts and theories can be applied in a practical situation. All or selected activities can be assigned and discussed in class or turned in for a grade.
- □ Separate out-of-class assignments to be turned in for extra credit.
- □ In-class activities to promote and stimulate discussion and interaction among students and between students and the instructor.
- □ Examples to help answer the question that most students ask: "Why do I have to learn this?"

Chapter End Matter At the end of most chapters is a summary, a glossary, a formula list, a multiple-choice self-test, and sectionalized basic and advanced problem sets. Also, the answers to section reviews and to related exercises are found at the end of each chapter. Terms that appear boldface in the text are defined in the glossary.

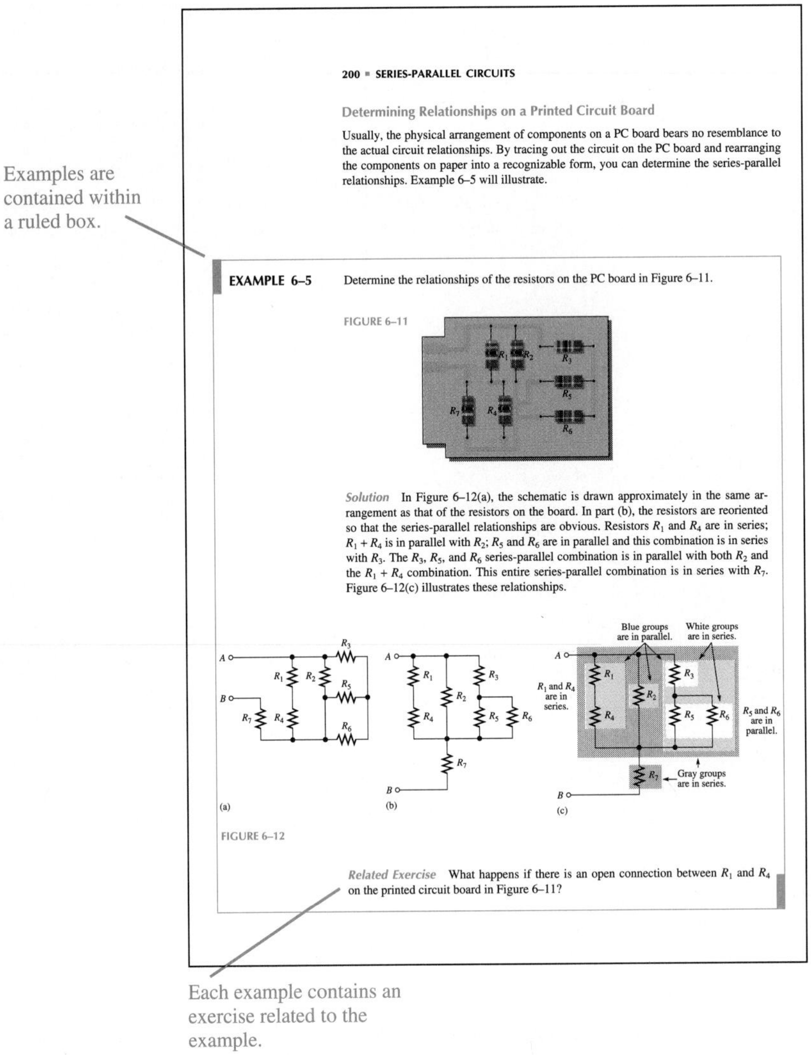
200 ■ SERIES-PARALLEL CIRCUITS

Determining Relationships on a Printed Circuit Board

Usually, the physical arrangement of components on a PC board bears no resemblance to the actual circuit relationships. By tracing out the circuit on the PC board and rearranging the components on paper into a recognizable form, you can determine the series-parallel relationships. Example 6–5 will illustrate.

EXAMPLE 6–5 Determine the relationships of the resistors on the PC board in Figure 6–11.

FIGURE 6–11

Solution In Figure 6–12(a), the schematic is drawn approximately in the same arrangement as that of the resistors on the board. In part (b), the resistors are reoriented so that the series-parallel relationships are obvious. Resistors R_1 and R_4 are in series; $R_1 + R_4$ is in parallel with R_2; R_5 and R_6 are in parallel and this combination is in series with R_3. The R_3, R_5, and R_6 series-parallel combination is in parallel with both R_2 and the $R_1 + R_4$ combination. This entire series-parallel combination is in series with R_7. Figure 6–12(c) illustrates these relationships.

FIGURE 6–12

Related Exercise What happens if there is an open connection between R_1 and R_4 on the printed circuit board in Figure 6–11?

FIGURE P–3

An example problem and related exercise.

Suggestions for Use

Electronics Fundamentals: Circuits, Devices, and Applications, Third Edition can be used to accommodate a variety of scheduling and program requirements. Two suggestions follow.

Option 1 A three-term dc/ac/devices sequence can cover most of the book, although some selectivity may be necessary. Chapters 1 through 7 can be covered in the first term. The topics of capacitors and inductors (Chapters 9 and 10) can be studied at the end of the first term, if desired, by delaying coverage of the ac topics in Sections 9–6, 9–7, 10–6, and 10–7 until the second term. In the third term, Chapters 16 through 21 would be covered.

Option 2 A two-term dc/ac/devices sequence requires a more condensed coverage by combining light treatment or omission of selected topics and a faster-paced coverage of most remaining topics. Since program requirements vary greatly, it is difficult to make specific suggestions for selective coverage.

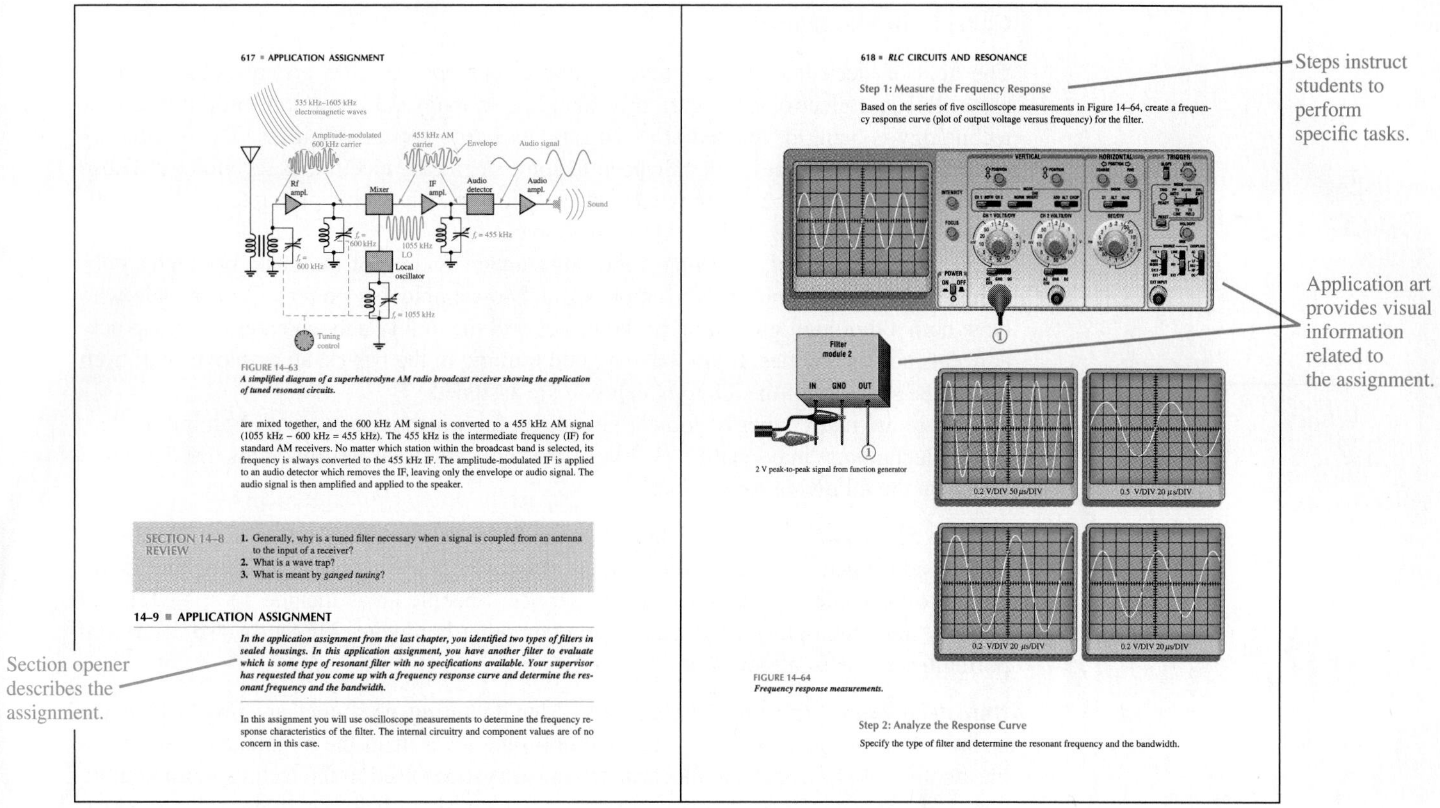

617 ■ APPLICATION ASSIGNMENT

FIGURE 14–63
A simplified diagram of a superheterodyne AM radio broadcast receiver showing the application of tuned resonant circuits.

are mixed together, and the 600 kHz AM signal is converted to a 455 kHz AM signal (1055 kHz – 600 kHz = 455 kHz). The 455 kHz is the intermediate frequency (IF) for standard AM receivers. No matter which station within the broadcast band is selected, its frequency is always converted to the 455 kHz IF. The amplitude-modulated IF is applied to an audio detector which removes the IF, leaving only the envelope or audio signal. The audio signal is then amplified and applied to the speaker.

SECTION 14–8 REVIEW
1. Generally, why is a tuned filter necessary when a signal is coupled from an antenna to the input of a receiver?
2. What is a wave trap?
3. What is meant by *ganged tuning*?

14–9 ■ APPLICATION ASSIGNMENT

In the application assignment from the last chapter, you identified two types of filters in sealed housings. In this application assignment, you have another filter to evaluate which is some type of resonant filter with no specifications available. Your supervisor has requested that you come up with a frequency response curve and determine the resonant frequency and the bandwidth.

In this assignment you will use oscilloscope measurements to determine the frequency response characteristics of the filter. The internal circuitry and component values are of no concern in this case.

618 ■ *RLC* CIRCUITS AND RESONANCE

Step 1: Measure the Frequency Response

Based on the series of five oscilloscope measurements in Figure 14–64, create a frequency response curve (plot of output voltage versus frequency) for the filter.

FIGURE 14–64
Frequency response measurements.

Step 2: Analyze the Response Curve

Specify the type of filter and determine the resonant frequency and the bandwidth.

FIGURE P–4
An Application Assignment section.

To the Student

The material in this preface is intended to help both you and your instructor make the most effective use of this textbook as a teaching and learning tool. Although you should certainly read everything in this preface, this part is especially for you, the student.

I am sure that you realize that knowledge and skills are not obtained easily or without effort. Much hard work is required to properly prepare yourself for any career, and electronics is no exception. You should use this book as more than just a reference. You must really dig in by reading, thinking, and doing. Don't expect every concept or procedure to become immediately clear. Some topics may take several readings, working many problems, and much help from your instructor before you really understand them.

Work through each example step-by-step and then do the related exercise. Answer the review questions at the end of each section. If you don't understand an example or if you can't answer a question, review the section. Check your answers at the end of the chapter. The multiple-choice self-tests at the end of each chapter are a good way to check your overall comprehension and retention of the subjects covered. You should do the self-test before you start the problems. Check your answers to the self-tests at the end of the book.

The problem sets at the end of each chapter provide problems with varying degrees of difficulty. In any technical field, it is important that you work lots of problems. Working through a problem gives you a level of insight and understanding that reading or classroom lectures alone do not provide. Never think that you fully understand a concept or procedure by simply watching or listening to someone else. In the final analysis, you must do it yourself and you must do it to the best of your ability.

Careers in Electronics

The field of electronics is very diverse, and career opportunities are available in many areas. Because electronics is currently found in so many different applications and new technology is being developed at a fast rate, its future appears limitless. There is hardly an area of our lives that is not enhanced to some degree by electronics technology. Those who acquire a sound, basic knowledge of electrical and electronic principles and are willing to continue learning will always be in demand.

The importance of obtaining a thorough understanding of the basic principles contained in this text cannot be overemphasized. Most employers prefer to hire people who have both a thorough grounding in the basics and the ability and eagerness to grasp new concepts and techniques. If you have a good training in the basics, an employer will train you in the specifics of the job to which you are assigned.

There are many types of job classifications for which a person with training in electronics technology may qualify. A few of the most common job functions are discussed briefly in the following paragraphs.

Service Shop Technician Technical personnel in this category are involved in the repair or adjustment of both commercial and consumer electronic equipment that is returned to the dealer or manufacturer for service. Specific areas include TVs, VCRs, CD players, stereo equipment, CB radios, and computer hardware. This area also offers opportunities for self-employment.

Industrial Manufacturing Technician Manufacturing personnel are involved in the testing of electronic products at the assembly-line level or in the maintenance and troubleshooting of electronic and electromechanical systems used in the testing and manufacturing of products. Virtually every type of manufacturing plant, regardless of its product, uses automated equipment that is electronically controlled.

Laboratory Technician These technicians are involved in breadboarding, prototyping, and testing new or modified electronic systems in research and development laboratories. They generally work closely with engineers during the development phase of a product.

Field Service Technician Field service personnel service and repair electronic equipment—for example, computer systems, radar installations, automatic banking equipment, and security systems—at the user's location.

Engineering Assistant/Associate Engineer Personnel in this category work closely with engineers in the implementation of a concept and in the basic design and development of electronic systems. Engineering assistants are frequently involved in a project from its initial design through the early manufacturing stages.

Technical Writer Technical writers compile technical information and then use the information to write and produce manuals and audiovisual materials. A broad knowledge of a particular system and the ability to clearly explain its principles and operation are essential.

Technical Sales Technically trained people are in demand as sales representatives for high-technology products. The ability both to understand technical concepts and to communicate the technical aspects of a product to a potential customer is very valuable. In this area, as in technical writing, competency in expressing yourself orally and in writing is essential. Actually, being able to communicate well is very important in any technical job category because you must be able to record data clearly and explain procedures, conclusions, and actions taken so that others can readily understand what you are doing.

A Look Back

Now, before you begin your study of electric circuits, let's briefly look at the beginnings of electronics and some of the important developments that have led to the electronics technology that we have today. It is always good to have a sense of the history of your career field. The names of many of the early pioneers in electricity and electromagnetics still live on in the names of familiar units and quantities. Names such as Ohm, Ampere, Volta, Farad, Henry, Coulomb, Oersted, and Hertz are some of the better known examples. More widely known names such as Franklin and Edison are also very significant in the history of electricity and electronics because of their tremendous contributions.

The Beginning of Electronics The early experiments in electronics involved electric currents in glass vacuum tubes. One of the first to conduct such experiments was a German named Heinrich Geissler (1814–1879). Geissler removed most of the air from a glass tube and found that the tube glowed when there was an electric current through it. Around 1878, British scientist Sir William Crookes (1832–1919) experimented with tubes similar to those of Geissler. In his experiments, Crookes found that the current in the vacuum tubes seemed to consist of particles.

Thomas Edison (1847–1931), experimenting with the carbon-filament light bulb he had invented, made another important finding. He inserted a small metal plate in the bulb. When the plate was positively charged, there was a current from the filament to the plate. This device was the first thermionic diode. Edison patented it but never used it.

The electron was discovered in the 1890s. The French physicist Jean Baptiste Perrin (1870–1942) demonstrated that the current in a vacuum tube consisted of the movement of negatively charged particles in a given direction. Some of the properties of these particles were measured by Sir Joseph Thomson (1856–1940), a British physicist, in experiments he performed between 1895 and 1897. These negatively charged particles later became known as electrons. The charge on the electron was accurately measured by an American physicist, Robert A. Millikan (1868–1953), in 1909. As a result of these discoveries, electrons could be controlled, and the electronic age was ushered in.

Putting the Electron to Work In 1904 a British scientist, John A. Fleming, constructed a vacuum tube that allowed electrical current in only one direction. The tube was used to detect electromagnetic waves. Called the Fleming valve, it was the forerunner of the more recent vacuum diode tubes. Major progress in electronics, however, awaited the development of a device that could boost, or amplify, a weak electromagnetic wave or radio signal. This device was the audion, patented in 1907 by Lee deForest, an American. It was a triode vacuum tube capable of amplifying small electrical ac signals.

Two other Americans, Harold Arnold and Irving Langmuir, made great improvements in the triode vacuum tube between 1912 and 1914. About the same time, deForest and Edwin Armstrong, an electrical engineer, used the triode tube in an oscillator circuit. In 1914, the triode was incorporated in the telephone system and made the transcontinental telephone network possible. In 1916 Walter Schottky, a German, invented the tetrode tube. The tetrode, along with the pentode (invented in 1926 by Dutch engineer Tellegen), greatly improved the triode. The first television picture tube, called the kinescope, was developed in the 1920s by Vladimir Sworykin, an American researcher.

During World War II, several types of microwave tubes were developed that made possible modern microwave radar and other communications systems. In 1939, the magnetron was invented in Britain by Henry Boot and John Randall. In the same year, the klystron microwave tube was developed by two Americans, Russell Varian and his brother Sigurd Varian. The traveling-wave tube (TWT) was invented in 1943 by Rudolf Komphner, an Austrian-American.

Solid-State Electronics The crystal detectors used in early radios were the forerunners of modern solid-state devices. However, the era of solid-state electronics began with the invention of the transistor in 1947 at Bell Labs. The inventors were Walter Brattain, John Bardeen, and William Shockley.

In the early 1960s, the integrated circuit (IC) was developed. It incorporated many transistors and other components on a single small chip of semiconductor material. Integrated circuit technology has been continuously developed and improved, allowing increasingly more complex circuits to be built on smaller chips.

Around 1965, the first integrated general-purpose operational amplifier was introduced. This low-cost, highly versatile device incorporated nine transistors and twelve resistors in a small package. It proved to have many advantages over comparable discrete component circuits in terms of reliability and performance. Since this introduction, the IC operational amplifier has become a basic building block for a wide variety of linear systems.

Acknowledgments

I wish to express thanks to all the people who had a part in making this revision a reality. In particular, the efforts of David Garza, Carol Robison, Pete Robison, Rex Davidson, Pat Tonneman, and Tom Mack are greatly appreciated. While space does not permit acknowledgment of all those who have submitted suggestions or comments for this new edition, I do wish to thank the following reviewers for their invaluable input: Richard Bacon, Santa Rosa Junior College; Mort Broad, Pinellas Vocational-Technical Institute; Dave Buchla, Yuba College; David Fridenmaker, ITT Technical Institute; Jim Gee, ITT Technical Institute; Jim Hart, DeVry Institute of Technology–Chicago; Carl Jensen, DeVry Institute of Technology–Woodbridge; Osama Maarouf, Microcomputer Technical Institute; Raymond Rich, ITT Technical Institute; David Soldner, ITT Technical Institute; Peter Westray, Los Angeles Valley College; and Jerry Wilson, Lamar University–Beaumont. In addition, I am grateful to Gary Snyder for a superb job of checking the manuscript for accuracy and to Lois Porter for another excellent job of editing the manuscript.

Tom Floyd

CONTENTS

PART ONE ■ DC CIRCUITS

PART TWO ■ AC CIRCUITS

Appendices

PART ONE

DC CIRCUITS

1 Components, Quantities, and Units
2 Voltage, Current, and Resistance
3 Ohm's Law, Energy, and Power
4 Series Circuits
5 Parallel Circuits
6 Series-Parallel Circuits
7 Magnetism and Electromagnetism

1

COMPONENTS, QUANTITIES, AND UNITS

The topics in this chapter present a basic introduction to the field of electronics. An overview of electrical and electronic components and instruments gives you a preview of the types of things you will study throughout this book.

You must be familiar with the units used in electronics and know how to express electrical quantities in various ways using metric prefixes. Scientific notation is an indispensable tool whether you use a computer, a calculator, or do computations the old-fashioned way.

■ CHAPTER OBJECTIVES

- □ Recognize some common electrical components and measuring instruments
- □ State the electrical and magnetic quantities and their units
- □ Use scientific notation (powers of ten)
- □ Use metric prefixes to express large and small quantities
- □ Convert from one metric unit to another

■ APPLICATION ASSIGNMENT

At the beginning of each chapter starting with Chapter 2, you will find an application assignment that relates to that chapter. These assignments present a variety of practical job situations that a technician might encounter in industry.

As you study each chapter, think about how to approach the application assignment. When you have completed each chapter, you should have a sufficient knowledge of the topics covered to enable you to complete the assignment that appears as the last section in each chapter.

1–1 ■ CIRCUIT COMPONENTS AND ELECTRONIC MEASURING INSTRUMENTS

In this book you will study many types of electrical components and measuring instruments. A thorough background in dc and ac circuit fundamentals provides the foundation for understanding electronic devices and circuits. A preview of the basic types of electrical and electronic components and instruments that you will be studying in detail in this and in other courses is provided in this section.

After completing this section, you should be able to

- **Recognize some common electrical components and measuring instruments**
  - ☐ State the function of a resistor
  - ☐ State the function of a capacitor
  - ☐ State the function of an inductor
  - ☐ State the function of a transformer
  - ☐ List the basic types of electronic measuring instruments

Resistors

Resistors resist, or limit, electrical current in a circuit. Several common types of resistors are shown in Figure 1–1.

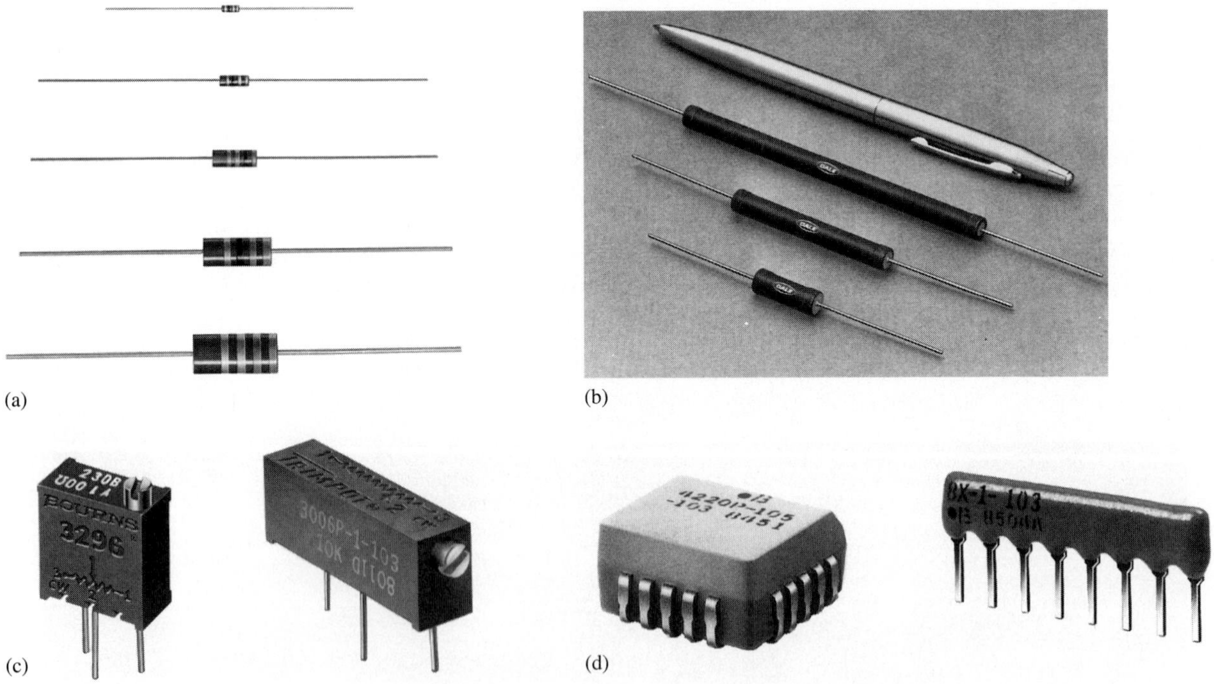

FIGURE 1–1

Typical fixed and variable resistors. (a) Carbon-composition resistors with standard power ratings of 1/8 W, 1/4 W, 1/2 W, 1 W, and 2 W (courtesy of Allen-Bradley Co.). (b) Wirewound resistors (courtesy of Dale Electronics, Inc.). (c) Variable resistors or potentiometers (courtesy of Bourns, Inc.). (d) Resistor networks (courtesy of Bourns, Inc.).

Capacitors

Capacitors store electrical charge; they used to be called *condensers* and are found in a variety of applications. Figure 1–2 shows several typical capacitors.

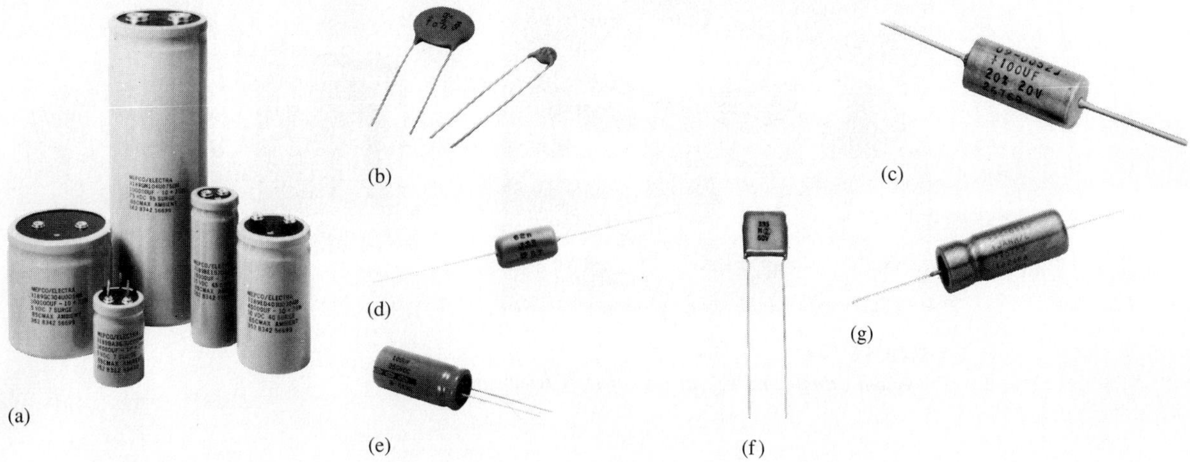

FIGURE 1–2
Typical capacitors. Parts (a, c, d, and g) courtesy of Mepco/Centralab. Part (b) courtesy of Murata Erie North America. Parts (e) and (f) courtesy of Sprague Electric Co.

Inductors

Inductors, also known as *coils,* are used to store energy in an electromagnetic field; they serve many useful functions in an electrical circuit. Figure 1–3 shows several typical inductors.

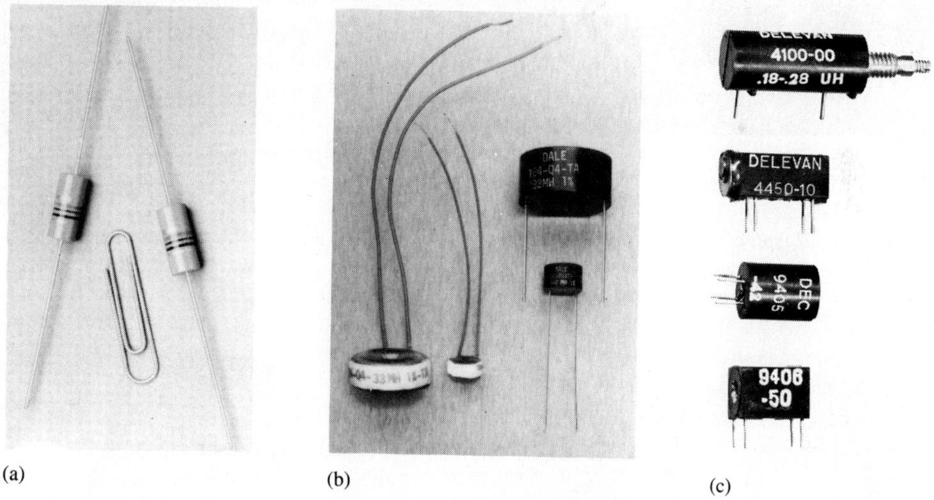

FIGURE 1–3
Typical inductors. Parts (a) and (b) courtesy of Dale Electronics, Inc. Part (c) courtesy of Delevan.

Transformers

Transformers are sometimes used to couple ac voltages from one point in a circuit to another, or to increase or decrease the ac voltage. Several types of transformers are shown in Figure 1–4.

Semiconductor Devices

Several varieties of diodes, transistors, and integrated circuits are shown in Figure 1–5.

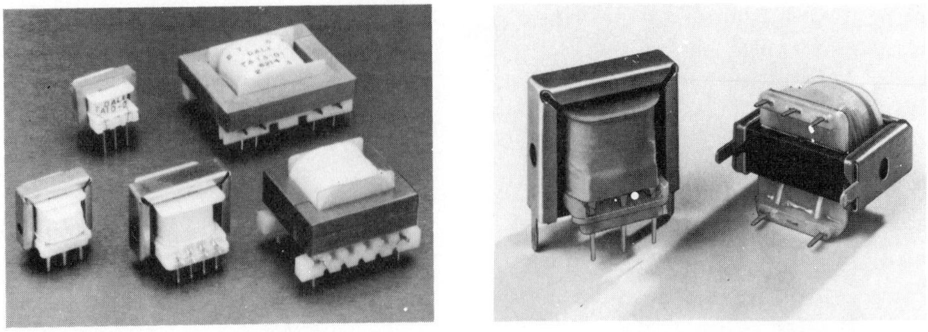

FIGURE 1–4
Typical transformers (courtesy of Dale Electronics, Inc.).

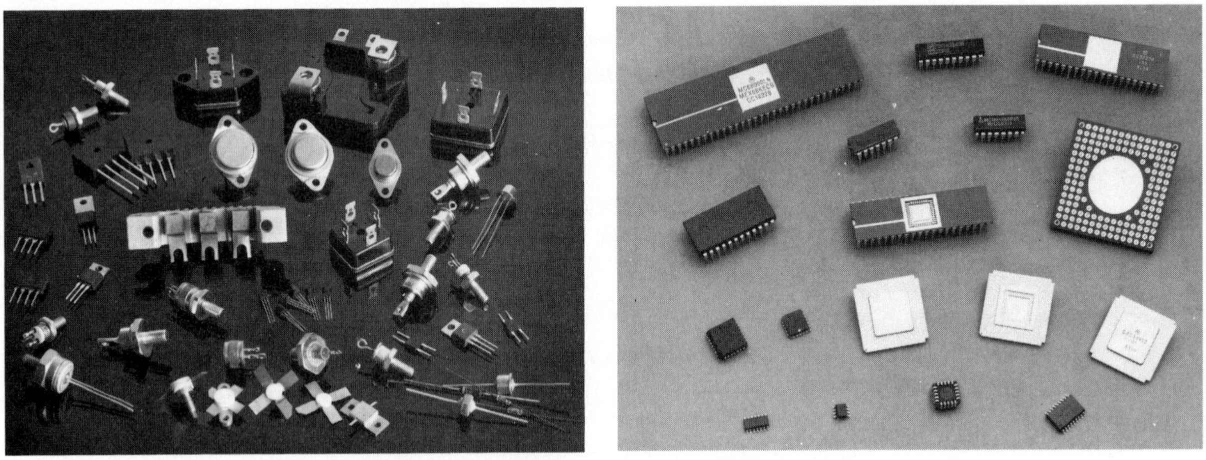

(a) Diodes and transistors

(b) Integrated circuits

FIGURE 1–5
A grouping of typical semiconductor devices (courtesy of Motorola, Inc. Used by permission.).

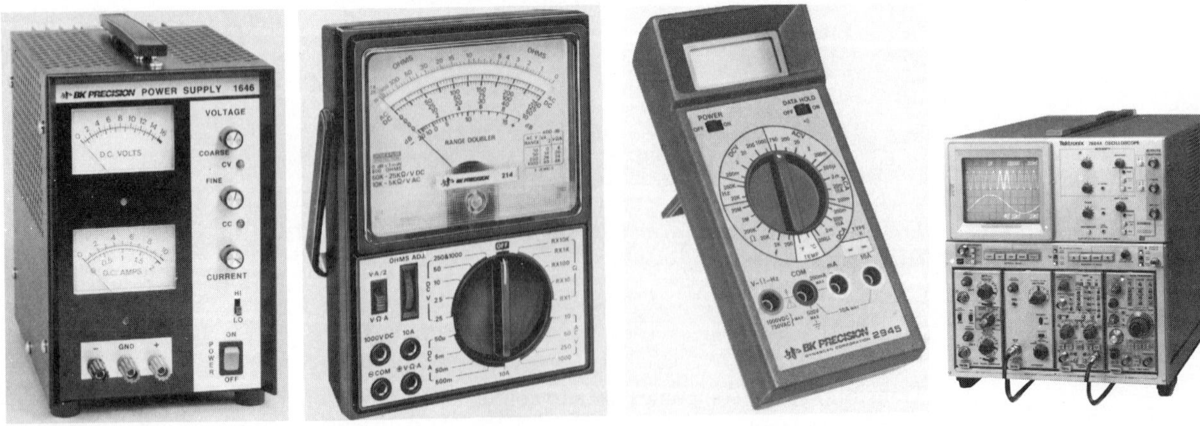

(a)

(b)

(c)

(d)

FIGURE 1–6
Typical instruments (a) DC power supply (courtesy of B & K Precision). (b) Analog multimeter (courtesy of B & K Precision). (c) Digital multimeter (courtesy of B & K Precision). (d) Oscilloscope (courtesy of Tektronix, Inc.).

Electronic Instruments

Figure 1–6 shows a variety of electronic instruments that are discussed throughout the text. Typical instruments include the power supply, for providing voltage and current; the voltmeter, for measuring voltage; the ammeter, for measuring current; the ohmmeter, for measuring resistance; and the oscilloscope for observing and measuring ac voltages.

SECTION 1–1 REVIEW

1. Name three types of common electrical components.
2. What instrument is used for measuring electrical current?
3. What instrument is used for measuring voltage?
4. What instrument is used for measuring resistance?

1–2 ■ ELECTRICAL AND MAGNETIC UNITS

In electronics work, you must deal with measurable quantities. For example, you must be able to express how many volts are measured at a certain test point in a circuit, how much current is flowing through a wire, or how much power a certain amplifier produces. In this section you will learn the units and symbols for many of the electrical and magnetic quantities that are used throughout the book. Definitions of the quantities are presented as they are needed in later chapters.

After completing this section, you should be able to

- **State the electrical and magnetic quantities and their units**
  - ☐ Specify the symbol for each quantity
  - ☐ Specify the symbol for each unit

Symbols are used in electronics to represent both quantities and their units. One symbol is used to represent the name of the quantity, and another is used to represent the unit of measurement of that quantity. For example, P stands for *power,* and W stands for *watt,* which is the unit of power. Table 1–1 lists the most important electrical quantities, along

TABLE 1–1
Electrical quantities and units with SI symbols

QUANTITY	SYMBOL	UNIT	SYMBOL
capacitance	C	farad	F
charge	Q	coulomb	C
conductance	G	siemens	S
current	I	ampere	A
energy	W	joule	J
frequency	f	hertz	Hz
impedance	Z	ohm	Ω
inductance	L	henry	H
power	P	watt	W
reactance	X	ohm	Ω
resistance	R	ohm	Ω
time	t	second	s
voltage	V	volt	V

TABLE 1–2
Magnetic quantities and units with SI symbols

QUANTITY	SYMBOL	UNIT	SYMBOL
flux density	B	tesla	T
magnetic flux	ϕ	weber	Wb
magnetizing force	H	ampere-turns/meter	At/m
magnetomotive force	F_m	ampere-turn	At
permeability	μ	webers/ampere-turns-meter	Wb/Atm
reluctance	$\mathcal{R}$	ampere-turns/weber	At/Wb

with their SI units and symbols. Table 1–2 lists magnetic quantities with their SI units and symbols. The term *SI* is the French abbreviation for *International System (Système International* in French).

SECTION 1–2 REVIEW

1. What does *SI* stand for?
2. Without referring to Table 1–1, list as many electrical quantities as possible, including their symbols, units, and unit symbols.
3. Without referring to Table 1–2, list as many magnetic quantities as possible, including their symbols, units, and unit symbols.

1–3 ■ SCIENTIFIC NOTATION

As an electronics technician, you will encounter both very small and very large quantities. For example, it is common to have electrical current values of only a few thousandths or even a few millionths of an ampere. On the other hand, you will find resistance values of several thousand or several million ohms. This range of values is typical of many other electrical quantities also.

After completing this section, you should be able to

- **Use scientific notation (powers of ten)**
  - ☐ Express any number using a power of ten
  - ☐ Do calculations with powers of ten

Powers of Ten

Scientific notation uses powers of ten, a method that makes it much easier to express large and small numbers and to do calculations involving such numbers.

Table 1–3 lists some powers of ten, both positive and negative. The power of ten is expressed as an exponent of the base 10 in each case. The exponent indicates the number of decimal places to the right or left of the decimal point in the expanded number. If the power is positive, the decimal point is moved to the right. For example,

$$10^4 = 1 \times 10^4 = 1.0000. = 10{,}000.$$

If the power is negative, the decimal point is moved to the left. For example,

$$10^{-4} = 1 \times 10^{-4} = .0001. = 0.0001$$

TABLE 1–3
Some positive and negative powers of ten

$10^6 = 1{,}000{,}000$	$10^{-6} = 0.000001$
$10^5 = 100{,}000$	$10^{-5} = 0.00001$
$10^4 = 10{,}000$	$10^{-4} = 0.0001$
$10^3 = 1{,}000$	$10^{-3} = 0.001$
$10^2 = 100$	$10^{-2} = 0.01$
$10^1 = 10$	$10^{-1} = 0.1$
$10^0 = 1$	

EXAMPLE 1–1

Express each number as a number between 1 and 10 times a positive power of ten.
(a) 200 **(b)** 5000 **(c)** 85,000 **(d)** 3,000,000

Solution In each case, the decimal point is moved an appropriate number of places to the left.
(a) $200 = 2 \times 10^2$ **(b)** $5000 = 5 \times 10^3$
(c) $85{,}000 = 8.5 \times 10^4$ **(d)** $3{,}000{,}000 = 3 \times 10^6$

Related Exercise Express 750,000,000 as a number between 1 and 10 times a positive power of ten.

EXAMPLE 1–2

Express each number as a number between 1 and 10 times a negative power of ten.
(a) 0.2 **(b)** 0.005 **(c)** 0.00063 **(d)** 0.000015

Solution In each case, the decimal point is moved an appropriate number of places to the right.
(a) $0.2 = 2 \times 10^{-1}$ **(b)** $0.005 = 5 \times 10^{-3}$
(c) $0.00063 = 6.3 \times 10^{-4}$ **(d)** $0.000015 = 1.5 \times 10^{-5}$

Related Exercise Express 0.00000093 as a number between 1 and 10 times a negative power of ten.

EXAMPLE 1–3

Express each of the following powers of ten numbers as a regular decimal number:
(a) 1×10^5 **(b)** 2×10^3 **(c)** 3.2×10^{-2} **(d)** 250×10^{-6}

Solution The decimal point is moved to the right or left a number of places indicated by the positive or the negative power of ten respectively.
(a) $1 \times 10^5 = 100{,}000$ **(b)** $2 \times 10^3 = 2000$
(c) $3.2 \times 10^{-2} = 0.032$ **(d)** $250 \times 10^{-6} = 0.000250$

Related Exercise Express 8.2×10^8 as a regular decimal number.

Calculating with Powers of Ten

The advantage of scientific notation is in addition, subtraction, multiplication, and division of very small or very large numbers.

Rules for Addition The rules for adding numbers in powers of ten are as follows:

1. Express the numbers to be added in the same power of ten.
2. Add the numbers without their powers of ten to get the sum.
3. Bring down the common power of ten, which is the power of ten of the sum.

EXAMPLE 1–4 Add 2×10^6 and 5×10^7.

Solution
1. Express both numbers in the same power of ten: $(2 \times 10^6) + (50 \times 10^6)$
2. Add $2 + 50 = 52$.
3. Bring down the common power of ten (10^6) and the sum is 52×10^6.

Related Exercise Add 41×10^3 and 79×10^2.

Rules for Subtraction The rules for subtracting numbers in powers of ten are as follows:

1. Express the numbers to be subtracted in the same power of ten.
2. Subtract the numbers without their powers of ten to get the difference.
3. Bring down the common power of ten, which is the power of ten of the difference.

EXAMPLE 1–5 Subtract 25×10^{-12} from 75×10^{-11}.

Solution
1. Express each number in the same power of ten: $(75 \times 10^{-11}) - (2.5 \times 10^{-11})$
2. Subtract $75 - 2.5 = 72.5$.
3. Bring down the common power of ten (10^{-11}) and the difference is 72.5×10^{-11}.

Related Exercise Subtract 3.5×10^{-6} from 2.2×10^{-5}.

Rules for Multiplication The rules for multiplying numbers in powers of ten are as follows:

1. Multiply the numbers directly without their powers of ten.
2. Add the powers of ten algebraically (the powers do not have to be the same).

EXAMPLE 1–6 Multiply 5×10^{12} by 3×10^{-6}.

Solution Multiply the numbers, and algebraically add the powers:

$$(5 \times 10^{12})(3 \times 10^{-6}) = 15 \times 10^{12+(-6)} = 15 \times 10^{6}$$

Related Exercise Multiply 12×10^{3} by 4×10^{2}.

Rules for Division The rules for dividing numbers in powers of ten are as follows:

1. Divide the numbers directly without their powers of ten.
2. Subtract the power of ten in the denominator from the power of ten in the numerator (the powers do not have to be the same).

EXAMPLE 1–7 Divide 50×10^{8} by 25×10^{3}.

Solution The division problem is written with a numerator and denominator:

$$\frac{50 \times 10^{8}}{25 \times 10^{3}}$$

Divide the numbers and subtract 3 from 8:

$$\frac{50 \times 10^{8}}{25 \times 10^{3}} = 2 \times 10^{8-3} = 2 \times 10^{5}$$

Related Exercise Divide 16×10^{-6} by 8×10^{-10}.

SECTION 1–3 REVIEW

1. Scientific notation uses powers of ten (True or False).
2. Express 100 as a power of ten.
3. Do the following operations:
   (a) $(1 \times 10^{5}) + (2 \times 10^{5})$ **(b)** $(3 \times 10^{6})(2 \times 10^{4})$
   (c) $(8 \times 10^{3}) \div (4 \times 10^{2})$ **(d)** $(25 \times 10^{-6}) - (130 \times 10^{-7})$

1–4 ▪ METRIC PREFIXES

In electrical and electronics applications, certain powers of ten are used more often than others. It is common practice to use metric prefixes to represent these quantities. You can think of the metric prefix as a shorthand way to express a large or small number.

After completing this section, you should be able to

- **Use metric prefixes to express large and small quantities**
  - ☐ List the metric prefixes
  - ☐ Change a power of ten to a metric prefix
  - ☐ Use a calculator for metric prefixed numbers
  - ☐ Use engineering notation

Table 1–4 lists the metric prefix for each of the commonly used powers of ten.

TABLE 1–4
Commonly used metric prefixes and their symbols

METRIC PREFIX	METRIC SYMBOL	POWER OF TEN	VALUE
tera	T	10^{12}	one trillion
giga	G	10^{9}	one billion
mega	M	10^{6}	one million
kilo	k	10^{3}	one thousand
milli	m	10^{-3}	one-thousandth
micro	μ	10^{-6}	one-millionth
nano	n	10^{-9}	one-billionth
pico	p	10^{-12}	one-trillionth

Use of Metric Prefixes

Now some examples will illustrate the use of metric prefixes. The number 2000 can be expressed in scientific notation as 2×10^3. Suppose we wish to represent 2000 W with a metric prefix. Since $2000 = 2 \times 10^3$, the metric prefix *kilo* (k) is used for 10^3. So we can express 2000 W as 2 kW (2 kilowatts).

As another example, 0.015 A can be expressed as 15×10^{-3} A. The metric prefix *milli* (m) is used for 10^{-3}. So 0.015 becomes 15 mA (15 milliamperes).

EXAMPLE 1–8

Express each quantity using a metric prefix.
(a) 50,000 V **(b)** 25,000,000 Ω **(c)** 0.000036 A

Solution
(a) $50{,}000 \text{ V} = 50 \times 10^3 \text{ V} = 50 \text{ kV}$
(b) $25{,}000{,}000\ \Omega = 25 \times 10^6\ \Omega = 25 \text{ M}\Omega$
(c) $0.000036 \text{ A} = 36 \times 10^{-6} \text{ A} = 36\ \mu\text{A}$

Related Exercise Express using metric prefixes:
(a) 56,000,000 Ω **(b)** 0.000470 A

Entering Numbers with Metric Prefixes on the Calculator

To enter a number expressed in scientific notation on the calculator, use the EXP key (the EE key on some calculators). The following example shows how to enter numbers with metric prefixes on a typical scientific calculator.

EXAMPLE 1–9

(a) Enter 3.3 kΩ ($3.3 \times 10^3\ \Omega$) on your calculator.
(b) Enter 450 μA (450×10^{-6} A) on your calculator.

Solution

(a) Step 1: Enter [3] [•] [3]. The display shows 3.3.
Step 2: Press [EXP]. The display shows 3.3 00.
Step 3: Enter [3]. The display shows 3.3 03.

(b) Step 1: Enter [4] [5] [0]. The display shows 450.
Step 2: Press [EXP]. The display shows 450 00.
Step 3: Press [+/−]. Enter [6]. The display shows 450 – 06.

Related Exercise Enter 259 mA on your calculator.

Engineering Notation

Engineering notation is a term that refers to the application of scientific notation in which the powers of ten are limited to multiples of three, such as 10^3, 10^{-3}, 10^6, 10^{-6}, 10^9, 10^{-9}, 10^{12}, and 10^{-12}. The reason for this is that all units used in the fields of engineering and technology, such as ohms, amps, volts, and watts are expressed with prefixes of k (kilo, 10^3), m (milli, 10^{-3}), M (mega, 10^6), μ (micro, 10^{-6}), G (giga, 10^9), n (nano, 10^{-9}), T (tera, 10^{12}), p (pico, 10^{-12}).

SECTION 1–4 REVIEW

1. List the metric prefix for each of the following powers of ten: 10^6, 10^3, 10^{-3}, 10^{-6}, 10^{-9}, and 10^{-12}.
2. Use an appropriate metric prefix to express 0.000001 A.
3. Use an appropriate metric prefix to express 250,000 W.

1–5 ■ METRIC UNIT CONVERSIONS

It is often necessary or convenient to convert a quantity from one metric prefixed unit to another, such as from milliamperes (mA) to microamperes (μA). A metric prefix conversion is accomplished by moving the decimal point in the number an appropriate number of places to the left or to the right, depending on the particular conversion.

After completing this section, you should be able to

- **Convert from one metric unit to another**
  - ☐ Convert between milli, micro, nano, and pico
  - ☐ Convert between kilo and mega

The following basic rules apply to metric unit conversions.

1. When converting from a larger unit to a smaller unit, move the decimal point to the right.

2. When converting from a smaller unit to a larger unit, move the decimal point to the left.
3. Determine the number of places that the decimal point is moved by finding the difference in the powers of ten of the units being converted.

For example, when converting from milliamperes (mA) to microamperes (μA), move the decimal point three places to the right because there is a three-place difference between the two units (mA is 10^{-3} A and μA is 10^{-6} A). The following examples illustrate a few conversions.

EXAMPLE 1–10 Convert 0.15 milliampere (0.15 mA) to microamperes (μA).

Solution Move the decimal point three places to the right.

$$0.15 \text{ mA} = 0.15 \times 10^{-3} \text{ A} = 150 \times 10^{-6} \text{ A} = 150 \ \mu\text{A}$$

Related Exercise Convert 1 mA to microamperes.

EXAMPLE 1–11 Convert 4500 microvolts (4500 μV) to millivolts (mV).

Solution Move the decimal point three places to the left.

$$4500 \ \mu\text{V} = 4500 \times 10^{-6} \text{ V} = 4.5 \times 10^{-3} \text{ V} = 4.5 \text{ mV}$$

Related Exercise Convert 1000 μV to millivolts.

EXAMPLE 1–12 Convert 5000 nanoamperes (5000 nA) to microamperes (μA).

Solution Move the decimal point three places to the left.

$$5000 \text{ nA} = 5000 \times 10^{-9} \text{ A} = 5 \times 10^{-6} \text{ A} = 5 \ \mu\text{A}$$

Related Exercise Convert 893 nA to microamperes.

EXAMPLE 1–13 Convert 47,000 picofarads (47,000 pF) to microfarads (μF).

Solution Move the decimal point six places to the left.

$$47{,}000 \text{ pF} = 47{,}000 \times 10^{-12} \text{ F} = 0.047 \times 10^{-6} \text{ F} = 0.047 \ \mu\text{F}$$

Related Exercise Convert 0.0022 μF to picofarads.

EXAMPLE 1–14 Convert 0.00022 microfarad (0.00022 μF) to picofarads (pF).

Solution Move the decimal point six places to the right.

$$0.00022\ \mu\text{F} = 0.00022 \times 10^{-6}\ \text{F} = 220 \times 10^{-12}\ \text{F} = 220\ \text{pF}$$

Related Exercise Convert 10,000 pF to microfarads.

EXAMPLE 1–15 Convert 1800 kilohms (1800 kΩ) to megohms (MΩ).

Solution Move the decimal point three places to the left.

$$1800\ \text{k}\Omega = 1800 \times 10^{3}\ \Omega = 1.8 \times 10^{6}\ \Omega = 1.8\ \text{M}\Omega$$

Related Exercise Convert 2.2 kΩ to megohms.

When adding (or subtracting) quantities with different metric prefixes, first convert one of the quantities to the same prefix as the other as the next example shows.

EXAMPLE 1–16 Add 15 mA and 8000 μA and express the result in milliamperes.

Solution Convert 8000 μA to 8 mA and add.

$$\begin{aligned} 15\ \text{mA} + 8000\ \mu\text{A} &= 15 \times 10^{-3}\ \text{A} + 8000 \times 10^{-6}\ \text{A} \\ &= 15 \times 10^{-3}\ \text{A} + 8 \times 10^{-3}\ \text{A} = 15\ \text{mA} + 8\ \text{mA} = 23\ \text{mA} \end{aligned}$$

Related Exercise Add 2873 mA to 10,000 μA.

SECTION 1–5 REVIEW

1. Convert 0.01 MV to kilovolts (kV).
2. Convert 250,000 pA to milliamperes (mA).
3. Add 0.05 MW and 75 kW.

▪ SUMMARY

- Resistors limit electrical current.
- Capacitors store electrical charge.
- Inductors store energy electromagnetically.
- Inductors are also known as *coils*.
- Transformers magnetically couple ac voltages.
- Semiconductor devices include diodes, transistors, and integrated circuits.
- Power supplies provide current and voltage.
- Voltmeters measure voltage.
- Ammeters measure current.
- Ohmmeters measure resistance.

■ SELF-TEST

1. Which of the following is not an electrical quantity?
   (a) current **(b)** voltage **(c)** time **(d)** power
2. The unit of current is
   (a) volt **(b)** watt **(c)** ampere **(d)** joule
3. The unit of voltage is
   (a) ohm **(b)** watt **(c)** volt **(d)** farad
4. The unit of resistance is
   (a) ampere **(b)** henry **(c)** hertz **(d)** ohm
5. 15,000 W is the same as
   (a) 15 mW **(b)** 15 kW **(c)** 15 MW **(d)** 15 μW
6. The quantity 4.7×10^3 is the same as
   (a) 470 **(b)** 4700 **(c)** 47,000 **(d)** 0.0047
7. The quantity 56×10^{-3} is the same as
   (a) 0.056 **(b)** 0.560 **(c)** 560 **(d)** 56,000
8. The number 3,300,000 can be expressed as
   (a) 3300×10^3 **(b)** 3.3×10^{-6} **(c)** 3.3×10^6 **(d)** either (a) or (c)
9. Ten milliamperes can be expressed as
   (a) 10 MA **(b)** 10 μA **(c)** 10 kA **(d)** 10 mA
10. Five thousand volts can be expressed as
    (a) 5000 V **(b)** 5 MV **(c)** 5 kV **(d)** either (a) or (c)
11. Twenty million ohms can be expressed as
    (a) 20 mΩ **(b)** 20 MW **(c)** 20 MΩ **(d)** 20 $\mu\Omega$
12. Hertz is the unit of
    (a) power **(b)** inductance **(c)** frequency **(d)** time

■ PROBLEMS

SECTION 1–3 Scientific Notation

1. Express each of the following numbers as a power of ten:
   (a) 3000 **(b)** 75,000 **(c)** 2,000,000
2. Express each number as a power of ten.
   (a) 1/500 **(b)** 1/2000 **(c)** 1/5,000,000
3. Express each of the following numbers in three ways, using 10^3, 10^4, and 10^5:
   (a) 8400 **(b)** 99,000 **(c)** 0.2×10^6
4. Express each of the following numbers in three ways, using 10^{-3}, 10^{-4}, and 10^{-5}:
   (a) 0.0002 **(b)** 0.6 **(c)** 7.8×10^{-2}
5. Express each power of ten in regular decimal form.
   (a) 2.5×10^{-6} **(b)** 50×10^2 **(c)** 3.9×10^{-1}
6. Express each power of ten in regular decimal form.
   (a) 45×10^{-6} **(b)** 8×10^{-9} **(c)** 40×10^{-12}
7. Add the following numbers:
   (a) $(92 \times 10^6) + (3.4 \times 10^7)$ **(b)** $(5 \times 10^3) + (85 \times 10^{-2})$ **(c)** $(560 \times 10^{-8}) + (460 \times 10^{-9})$
8. Perform the following subtractions:
   (a) $(3.2 \times 10^{12}) - (1.1 \times 10^{12})$ **(b)** $(26 \times 10^8) - (1.3 \times 10^9)$ **(c)** $(150 \times 10^{-12}) - (8 \times 10^{-11})$
9. Perform the following multiplications:
   (a) $(5 \times 10^3)(4 \times 10^5)$ **(b)** $(12 \times 10^{12})(3 \times 10^2)$ **(c)** $(2.2 \times 10^{-9})(7 \times 10^{-6})$
10. Divide the following:
    (a) $(10 \times 10^3) \div (2.5 \times 10^2)$ **(b)** $(250 \times 10^{-6}) \div (50 \times 10^{-8})$ **(c)** $(4.2 \times 10^8) \div (2 \times 10^{-5})$
11. Convert each of the following numbers to another number having a multiplier of 10^{-6}:
    (a) 2.37×10^{-3} **(b)** 0.001856×10^{-2} **(c)** 5743.89×10^{-12} **(d)** 100×10^3
12. Perform the following indicated operations:
    (a) $(2.8 \times 10^3)(3 \times 10^2)/2 \times 10^2$ **(b)** $(46)(10^{-3})(10^5)/10^6$
    (c) $(7.35)(0.5 \times 10^{12})/[(2)(10 \times 10^{10})]$ **(d)** $(30)^2(5)^3/10^{-2}$

13. Perform each operation:
(a) $49 \times 10^6/4 \times 10^{-8}$ **(b)** $(3 \times 10^3)^2/1.8 \times 10^3$
(c) $(1.5 \times 10^{-6})^2(4.7 \times 10^6)$ **(d)** $1/[(5 \times 10^{-3})(0.01 \times 10^{-6})]$

SECTION 1–4 Metric Prefixes

14. Express each of the following as a quantity having a metric prefix:
(a) 31×10^{-3} A **(b)** 5.5×10^3 V **(c)** 20×10^{-12} F

15. Express the following using metric prefixes:
(a) 3×10^{-6} F **(b)** 3.3×10^6 Ω **(c)** 350×10^{-9} A

16. Express each quantity as a power of ten:
(a) 5 μA **(b)** 43 mV **(c)** 275 kΩ **(d)** 10 MW

SECTION 1–5 Metric Unit Conversions

17. Perform the indicated conversions:
(a) 5 mA to microamperes **(b)** 3200 μW to milliwatts
(c) 5000 kV to megavolts **(d)** 10 MW to kilowatts

18. Determine the following:
(a) The number of microamperes in 1 milliampere
(b) The number of millivolts in 0.05 kilovolt
(c) The number of megohms in 0.02 kilohm
(d) The number of kilowatts in 155 milliwatts

▪ ANSWERS TO SECTION REVIEWS

Section 1–1

1. Resistors, capacitors, inductors, and transformers
2. Ammeter
3. Voltmeter
4. Ohmmeter

Section 1–2

1. SI is the abbreviation for Système International.
2. Refer to Table 1–1 after you have compiled your list.
3. Refer to Table 1–2 after you have compiled your list.

Section 1–3

1. True **2.** 10^2
3. **(a)** 3×10^5 **(b)** 6×10^{10} **(c)** 2×10^1 **(d)** 12×10^{-6}

Section 1–4

1. Mega (M), kilo (k), milli (m), micro (μ), nano (n), and pico (p)
2. 1 μA (one microampere) **3.** 250 kW (250 kilowatts)

Section 1–5

1. 0.01 MV = 10 kV
2. 250,000 pA = 0.00025 mA
3. 125 kW

▪ ANSWERS TO RELATED EXERCISES FOR EXAMPLES

1–1	7.5×10^8	**1–2**	9.3×10^{-7}
1–3	820,000,000	**1–4**	48.9×10^3
1–5	1.85×10^{-5}	**1–6**	4.8×10^6
1–7	2×10^4	**1–8**	**(a)** 56 MΩ **(b)** 470 μA
1–9	[2] [5] [9] [EXP] [+/−] [3]	**1–10**	1000 μA
1–11	1 mV	**1–12**	0.893 μA
1–13	2200 pF	**1–14**	0.01 μF
1–15	0.0022 MΩ	**1–16**	2883 mA

2

VOLTAGE, CURRENT, AND RESISTANCE

The three basic electrical quantities presented in this chapter are voltage, current, and resistance. No matter what type of electrical or electronic equipment you may work with, these quantities will always be of primary importance.

To help you understand voltage, current, and resistance, the basic structure of the atom is discussed and the concept of charge is introduced. The basic electric circuit is studied, along with techniques for measuring voltage, current, and resistance.

■ CHAPTER OBJECTIVES

- □ Describe the basic structure of an atom
- □ Explain the concept of electrical charge
- □ Define *voltage* and discuss its characteristics
- □ Define *current* and discuss its characteristics
- □ Define *resistance* and discuss its characteristics
- □ Describe a basic electrical circuit
- □ Make basic circuit measurements

■ APPLICATION ASSIGNMENT

As technician for special effects, you are asked to hook up and check out a circuit for use in the special lighting effects for an outdoor play. The requirements are

1. There will be six lamps but only one lamp will be turned on at a time.
2. The sequence in which the lamps are turned on and off will vary.
3. It will be necessary to vary the brightness of each lamp.
4. The lamps must operate from a 12 V battery for portability, and the circuit must be protected by a fuse.

After studying this chapter, you should be able to complete the application assignment in Section 2–8.

2–1 ■ ATOMS

The atomic structure of the materials used in electronics determines, among other things, the ease with which current is established.

After completing this section, you should be able to

- **Describe the basic structure of an atom**
  - ☐ Define *electron*
  - ☐ Define *proton* and *neutron*
  - ☐ Define *atomic weight* and *atomic number*
  - ☐ Define *free electron*
  - ☐ Discuss shells

An **atom** is the smallest particle of an **element** that still retains the characteristics of that element. Different elements have different types of atoms. In fact, every element has a unique atomic structure.

According to the Bohr theory, atoms have a planetary type of structure, consisting of a central nucleus surrounded by orbiting electrons. The nucleus consists of positively charged particles called **protons** and uncharged particles called **neutrons.** The **electrons** are the basic particles of negative charge.

Each type of atom has a certain number of electrons and protons that distinguishes the atom from all other atoms of other elements. For example, the simplest atom is that of hydrogen. It has one proton and one electron, as pictured in Figure 2–1(a). The helium atom, shown in Figure 2–1(b), has two protons and two neutrons in the nucleus, which is orbited by two electrons.

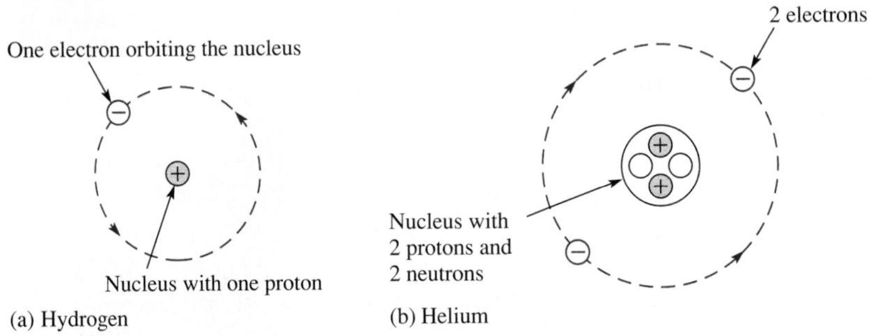

FIGURE 2–1
Hydrogen and helium atoms.

Atomic Number and Weight

All elements are arranged in the periodic table of the elements in order according to their **atomic number,** which is the number of protons in the nucleus. The elements can also be arranged by their **atomic weight,** which is approximately the number of protons and neutrons in the nucleus. For example, hydrogen has an atomic number of one and an atomic weight of one. The atomic number of helium is two, and its atomic weight is four.

In their normal (or neutral) state, all atoms of a given element have the same number of electrons as protons. So the positive charges cancel the negative charges, and the atom has a net charge of zero.

The Copper Atom

Since copper is the most commonly used metal in **electrical** applications, let's examine its atomic structure. The copper atom has 29 electrons in orbit around the nucleus. They do not all occupy the same orbit, however. They move in orbits at varying distances from the nucleus.

The orbits in which the electrons revolve are called **shells.** The number of electrons in each shell follows a predictable pattern according to the formula, $2N^2$, where N is the number of the shell. The first shell of any atom can have up to 2 electrons, the second shell up to 8 electrons, the third shell up to 18 electrons, and the fourth shell up to 32 electrons.

A copper atom is shown in Figure 2–2. Notice that the outermost shell, called the **valence** shell, has only one electron, called the **valence electron.**

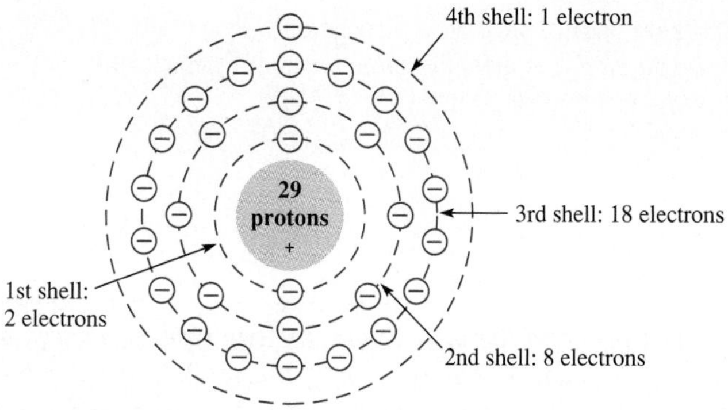

FIGURE 2–2
The copper atom.

Free Electrons

When the valence electron in the outer shell of the copper atom gains sufficient **energy** from the surrounding medium, it can break away from the parent atom and become what is called a **free electron.** Materials with fewer valence electrons in their atoms have more free electrons because the valence electrons are more loosely bound to the atoms. The free electrons in the copper material are capable of moving from one atom to another in the material. In other words, they drift randomly from atom to atom within the copper. As you will see, the free electrons make electrical current possible.

Three categories of materials are used in electronics: conductors, semiconductors, and insulators.

Conductors

Conductors are materials that allow current to flow easily. They have a large number of free electrons and are characterized by one to three valence electrons in their atomic structures. Most metals are good conductors. Silver is the best conductor, and copper is next. Copper is the most widely used conductive material because it is less expensive than silver. Copper wire is commonly used as a conductor in electric circuits.

Semiconductors

Semiconductors are classed below the conductors in their ability to carry current because they have fewer free electrons than do conductors. Semiconductors are characterized by four valence electrons in their atomic structures. However, because of their unique characteristics, certain semiconductor materials are the basis for modern **electronic** devices such as the diode, transistor, and integrated circuit. Silicon and germanium are common semiconductor materials.

Insulators

Insulating materials are poor conductors of electric current. In fact, **insulators** are used to prevent current where it is not wanted. Compared to conductive materials, insulators have very few free electrons and are characterized by more than four valence electrons in their atomic structures.

SECTION 2–1 REVIEW

1. What is the basic particle of negative charge?
2. Define *atom*.
3. What does a typical atom consist of?
4. Do all elements have the same types of atoms?
5. What is a free electron?

2–2 ▪ ELECTRICAL CHARGE

As you learned in the last section, the two types of charge are the positive charge and the negative charge. The electron is the smallest particle that exhibits negative electrical charge. When an excess of electrons exists in a material, there is a net negative electrical charge. When a deficiency of electrons exists, there is a net positive electrical charge.

After completing this section, you should be able to

- **Explain the concept of electrical charge**
  - ☐ Name the unit of charge
  - ☐ Name the types of charge
  - ☐ Describe the forces between charges
  - ☐ Determine the amount of charge on a given number of electrons

The **charge** of an electron and that of a proton are equal in magnitude. Electrical charge is symbolized by Q. Static electricity is the presence of a net positive or negative charge in a material. Everyone has experienced the effects of static electricity from time to time, for example, when attempting to touch a metal surface or another person or when the clothes in a dryer cling together.

Materials with charges of opposite polarity are attracted to each other; materials with charges of the same polarity are repelled, as indicated symbolically in Figure 2–3. A force acts between charges, as evidenced by the attraction or repulsion. This force, called an *electric field,* consists of invisible lines of force as shown in Figure 2–4.

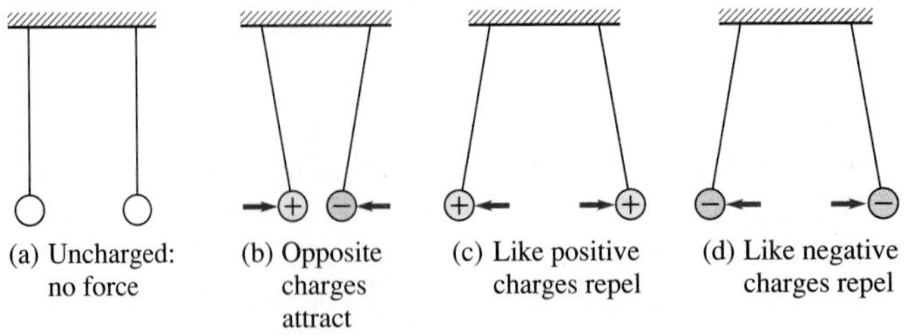

FIGURE 2–3
Attraction and repulsion of electrical charges.

FIGURE 2–4
Electric field between oppositely charged surfaces.

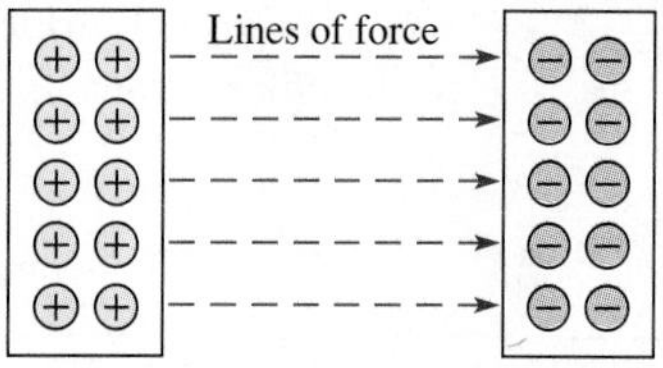

Coulomb: The Unit of Charge

Electrical charge is measured in **coulombs,** abbreviated C.

One coulomb is the total charge possessed by 6.25×10^{18} electrons.

A single electron has a charge of 1.6×10^{-19} C. This unit is named for Charles Coulomb (1736–1806), a French scientist. The total charge, Q, in a given number of electrons is found by the following formula:

$$Q = \frac{\text{number of electrons}}{6.25 \times 10^{18}\ \text{electrons/Coulomb}} \qquad \textbf{(2–1)}$$

Positive and Negative Charge

Consider a neutral atom—that is, one that has the same number of electrons and protons and thus has no net charge. If a valence electron is pulled away from the atom by the application of energy, the atom is left with a net positive charge (more protons than electrons) and becomes a positive ion. If an atom acquires an extra electron in its outer shell, it has a net negative charge and becomes a negative ion.

The amount of energy required to free a valence electron is related to the number of electrons in the outer shell. An atom can have up to eight valence electrons. The more complete the outer shell, the more stable the atom and thus the more energy is required to release an electron. Figure 2–5 illustrates the creation of a positive and a negative ion

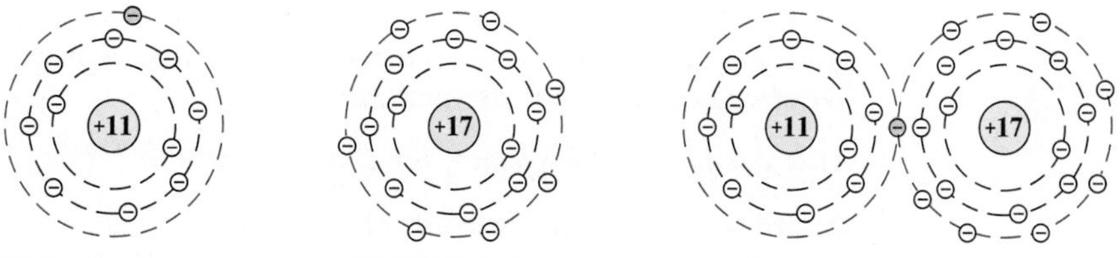

Sodium atom
(11 protons, 11 electrons)

Chlorine atom
(17 protons, 17 electrons)

(a) The sodium atom has a single valence electron.

(b) The atoms combine by sharing the valence electron to form sodium chloride (table salt).

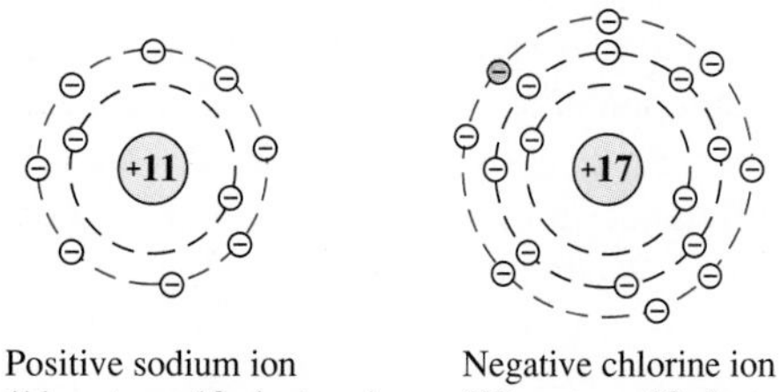

Positive sodium ion
(11 protons, 10 electrons)

Negative chlorine ion
(17 protons, 18 electrons)

(c) When dissolved, the sodium atom gives up the valence electron to become a positive ion, and the chlorine atom retains the extra valence electron to become a negative ion.

FIGURE 2–5
Example of the formation of positive and negative ions.

when sodium chloride dissolves and the sodium atom gives up its single valence electron to the chlorine atom.

EXAMPLE 2–1 How many coulombs of charge do 93.8×10^{16} electrons represent?

Solution

$$Q = \frac{\text{number of electrons}}{6.25 \times 10^{18} \text{ electrons/Coulomb}} = \frac{93.8 \times 10^{16} \text{ electrons}}{6.25 \times 10^{18} \text{ electrons/C}}$$

$$= 15 \times 10^{-2} \text{ C} = 0.15 \text{ C}$$

Related Exercise How many electrons does it take to have 3 C of charge?

SECTION 2–2 REVIEW

1. What is the symbol used for charge?
2. What is the unit of charge, and what is the unit symbol?
3. What are the two types of charge?
4. How much charge, in coulombs, is there in 10×10^{12} electrons?

2–3 ■ VOLTAGE

As you have seen, a force of attraction exists between a positive and a negative charge. A certain amount of energy in the form of work is required to overcome the force and move the charges a given distance apart. All opposite charges possess a certain potential energy because of the separation between them. The difference in potential energy of the charges is the potential difference or voltage. Voltage is the driving force in electric circuits and is what produces current.

After completing this section, you should be able to

- **Define *voltage* and discuss its characteristics**
  - ☐ State the formula for voltage
  - ☐ Name and define the unit of voltage
  - ☐ Describe the basic sources of voltage

Consider a water tank that is supported several feet above the ground. A given amount of energy (work) is required to pump water up to fill the tank. Once the water is stored in the tank, it has a certain potential energy which, if released, can be used to perform work. For example, the water can be allowed to fall down a chute to turn a water wheel.

Potential difference in electrical terms is more commonly called **voltage** (V) and is expressed as energy (W) divided by charge (Q).

$$V = \frac{W}{Q} \qquad \textbf{(2–2)}$$

where W is expressed in **joules** (J) and Q is in coulombs (C).

Volt: The Unit of Voltage

The unit of voltage is the **volt,** symbolized by V.

> **One volt is the potential difference (voltage) between two points when one joule of energy is used to move one coulomb of charge from one point to the other.**

EXAMPLE 2–2 If 50 J of energy are available for every 10 C of charge, what is the voltage?

Solution

$$V = \frac{W}{Q} = \frac{50\text{ J}}{10\text{ C}} = 5\text{ V}$$

Related Exercise How much energy is used to move 50 C from one point in a circuit to another when the voltage between the two points is 12 V?

A Voltage Analogy

You can think of voltage as the pressure difference created by a pump that causes the water to flow through the pipe in a water system.

Sources of Voltage

The Battery A voltage source is a **source** of potential energy that is also called *electromotive force* (emf). The **battery** is one type of voltage source that converts chemical energy into electrical energy. A voltage exists between the electrodes (terminals) of a battery, as shown by a voltaic cell in Figure 2–6. One electrode is positive and the other negative as a result of the separation of charges caused by the chemical action when two different conducting materials are dissolved in the electrolyte.

FIGURE 2–6
A voltaic cell converts chemical energy into electrical energy.

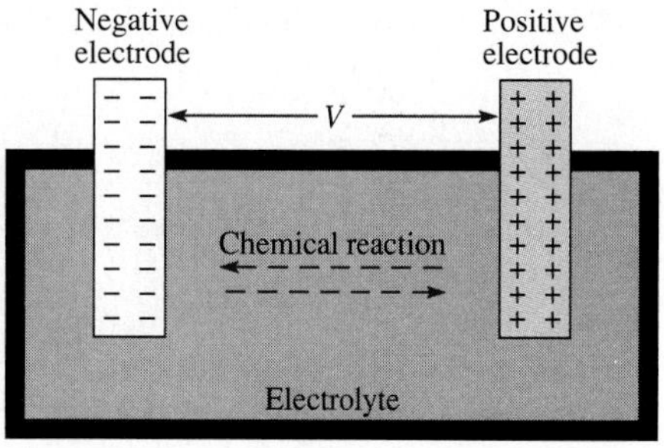

Battery cells are generally classified as *primary cells,* which cannot be recharged, and *secondary cells,* which can be recharged by reversal of the chemical action. Examples of primary cells are carbon-zinc cells and alkaline cells. Examples of secondary cells are lead-acid cells and nickel-cadmium (Ni-Cd) cells. The amount of voltage provided by a cell varies depending on the materials used in the construction of the cell. For example, a carbon-zinc cell (flashlight battery) produces 1.5 V, whereas a lead-acid cell used in automobile batteries produces 2 V. Individual cells are combined to produce higher voltages. A 9 V battery is made up of six 1.5 V cells and a 12 V car battery consists of six 2 V cells. For additional coverage of batteries, see Appendix B.

The Electronic Power Supply These voltage sources convert the ac voltage from the wall outlet to a constant (dc) voltage which is available across two output terminals, as indicated in Figure 2–7(a). Typical commercial **power supplies** are shown in part (b).

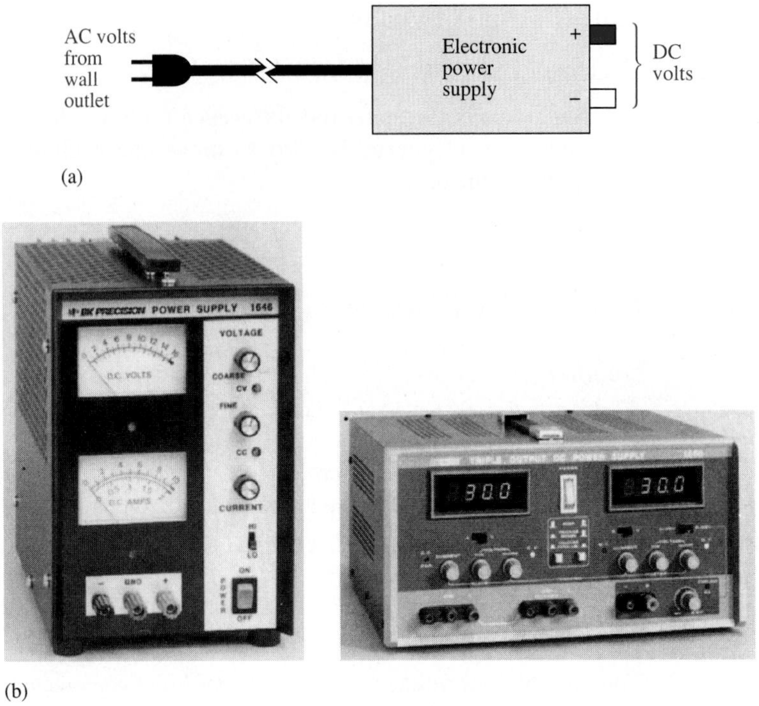

FIGURE 2–7
Electronic power supplies. Part (b) courtesy of B & K Precision.

The Solar Cell The operation of solar cells is based on the principle of *photovoltaic action,* which is the process whereby light energy is converted directly into electrical energy. A basic solar cell consists of two layers of different semiconductive materials joined together to form a junction. When one layer is exposed to light, many electrons acquire enough energy to break away from their parent atoms and cross the junction. This process forms negative ions on one side of the junction and positive ions on the other, and thus a potential difference (voltage) is developed. Basic construction of a solar cell is shown in Figure 2–8.

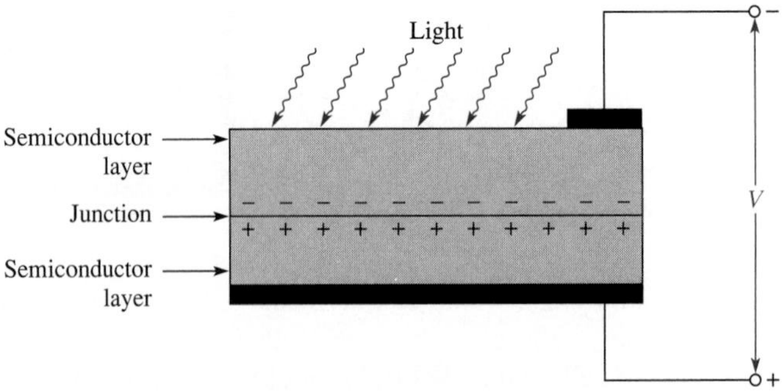

FIGURE 2–8
Basic construction of solar cell.

The Generator **Generators** convert mechanical energy into electrical energy using a principle called *electromagnetic induction* (see Chapter 7). A conductor is rotated through a magnetic field, and a voltage is produced across the conductor. A typical generator is pictured in Figure 2–9.

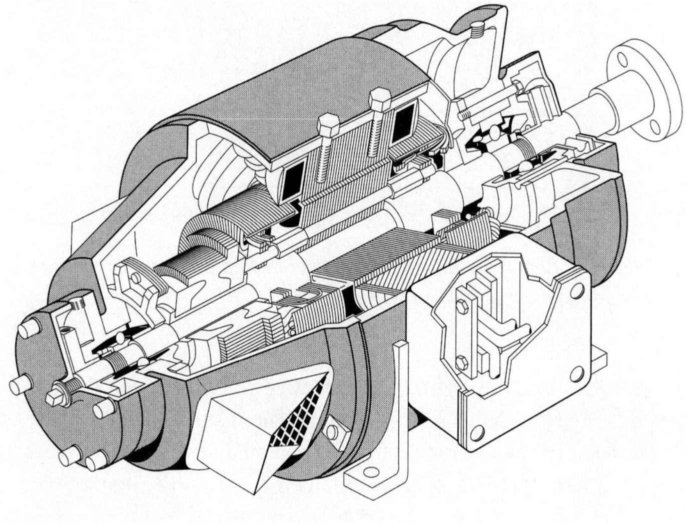

FIGURE 2–9
Cutaway view of dc generator.

SECTION 2–3 REVIEW

1. Define *voltage*.
2. What is the unit of voltage?
3. How much is the voltage of a battery when there are 24 J of energy for 10 C of charge?
4. List four sources of voltage.

2–4 ■ CURRENT

Voltage makes electrons move through a circuit, and this movement of electrons is the current. Current does the work in an electric circuit.

After completing this section, you should be able to

- **Define *current* and discuss its characteristics**
  - ☐ Explain the movement of electrons
  - ☐ State the formula for current
  - ☐ Name and define the unit of current

As you have learned, there are free electrons available in all conductive and semiconductive materials. These electrons drift randomly in all directions, from atom to atom, within the structure of the material, as indicated in Figure 2–10.

FIGURE 2–10
Random motion of free electrons in a material.

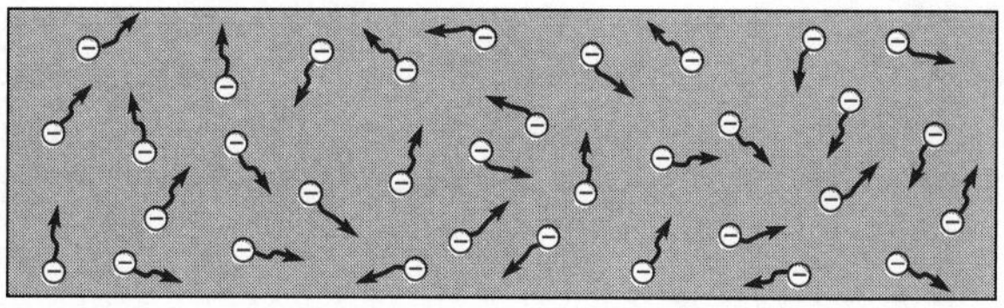

Now, if a voltage is placed across the conductive or semiconductive material, one end becomes positive and the other negative, as indicated in Figure 2–11. The repulsive force between the negative voltage at the left end causes the free electrons (negative charges) to

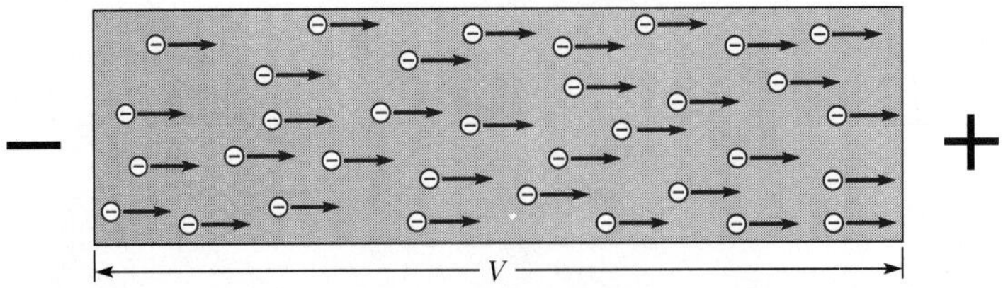

FIGURE 2–11
Electrons flow from negative to positive when a voltage is applied across a conductive or semiconductive material.

move toward the right. The attractive force between the positive voltage at the right end pulls the free electrons to the right. The result is a net movement of the free electrons from the negative end of the material to the positive end, as shown in Figure 2–11.

The movement of the free electrons from the negative end of the material to the positive end is the electrical **current,** symbolized by *I*.

Electrical current is the rate of flow of electrons.

Current is measured by the number of electrons (amount of charge, *Q*) that flow past a point in a unit of time. Current *(I)* equals charge *(Q)* divided by time *(t)*.

$$I = \frac{Q}{t} \qquad \textbf{(2–3)}$$

where *I* is current in amperes, *Q* is the charge of the electrons in coulombs, and *t* is the time in seconds.

Ampere: The Unit of Current

Current is measured in a unit called the **ampere** or *amp* for short, symbolized by A. It is named after André Ampère (1775–1836), a French physicist whose work contributed to the understanding of electrical current and its effects.

One ampere (1 A) is the amount of current when a number of electrons having one coulomb (1 C) of charge move past a given point *(P)* in one second (1 s).

One coulomb is the charge carried by 6.25×10^{18} electrons. See Figure 2–12.

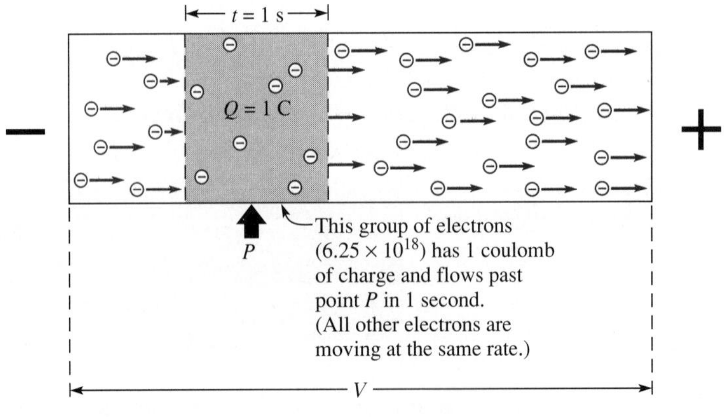

FIGURE 2–12
Illustration of 1 A of current in a material (1 C/s).

EXAMPLE 2–3 Ten coulombs of charge flow past a given point in a wire in 2 s. How many amperes of current are flowing?

Solution

$$I = \frac{Q}{t} = \frac{10\ \text{C}}{2\ \text{s}} = 5\ \text{A}$$

Related Exercise If 8 A of current flow through the filament of a light bulb in 1.5 s, how many coulombs have moved through the filament?

A Current Analogy

You can think of current as the water flowing through the pipe in a water system when pressure (voltage) is applied by the pump (voltage source).

SECTION 2–4 REVIEW

1. Define *current* and state its unit.
2. How many electrons make up one coulomb of charge?
3. What is the current in amperes when 20 C flow past a point in a wire in 4 s?

2–5 ■ RESISTANCE

When there is current in a material, the free electrons move through the material and occasionally collide with atoms. These collisions cause the electrons to lose some of their energy, and thus their movement is restricted. The more collisions, the more the flow of electrons is restricted. This restriction varies and is determined by the type of material. The property of a material that restricts the flow of electrons is called* resistance, *designated with an R.

After completing this section, you should be able to

- **Define *resistance* and discuss its characteristics**
  - ☐ Name and define the unit of resistance
  - ☐ Describe the basic types of resistors
  - ☐ Determine resistance value by color code or labeling

Resistance is the opposition to current.

The schematic symbol for resistance is shown in Figure 2–13.

When current flows through any material that has resistance, heat is produced by the collisions of electrons and atoms. Therefore, wire, which typically has a very small resistance, becomes warm when there is current through it.

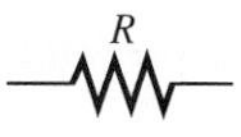

FIGURE 2–13
Resistance/resistor symbol.

Ohm: The Unit of Resistance

Resistance, *R,* is expressed in the unit of **ohms,** named after Georg Simon Ohm and is symbolized by the Greek letter omega (Ω).

> **One ohm (1 Ω) of resistance exists when one ampere (1 A) of current flows in a material when one volt (1 V) is applied across the material.**

Resistors

Components that are specifically designed to have a certain amount of resistance are called **resistors.** The principal applications of resistors are to limit current, divide voltage, and, in certain cases, generate heat. Although there are a variety of different types of resistors that come in many shapes and sizes, they can all be placed in one of two main categories: fixed or variable.

Fixed Resistors The fixed resistor is available with a large selection of resistance values that are set during manufacturing and cannot be changed easily. Fixed resistors are constructed using various methods and materials. Several common types are shown in Figure 2–14.

One common fixed resistor is the carbon-composition type, which is made with a mixture of finely ground carbon, insulating filler, and a resin binder. The ratio of carbon to insulating filler sets the resistance value. The mixture is formed into rods, and lead connections are made. The entire resistor is then encapsulated in an insulated coating for protection. The chip resistor is another type of fixed resistor and is in the category of SMT (surface mount technology) components. It has the advantage of a very small size

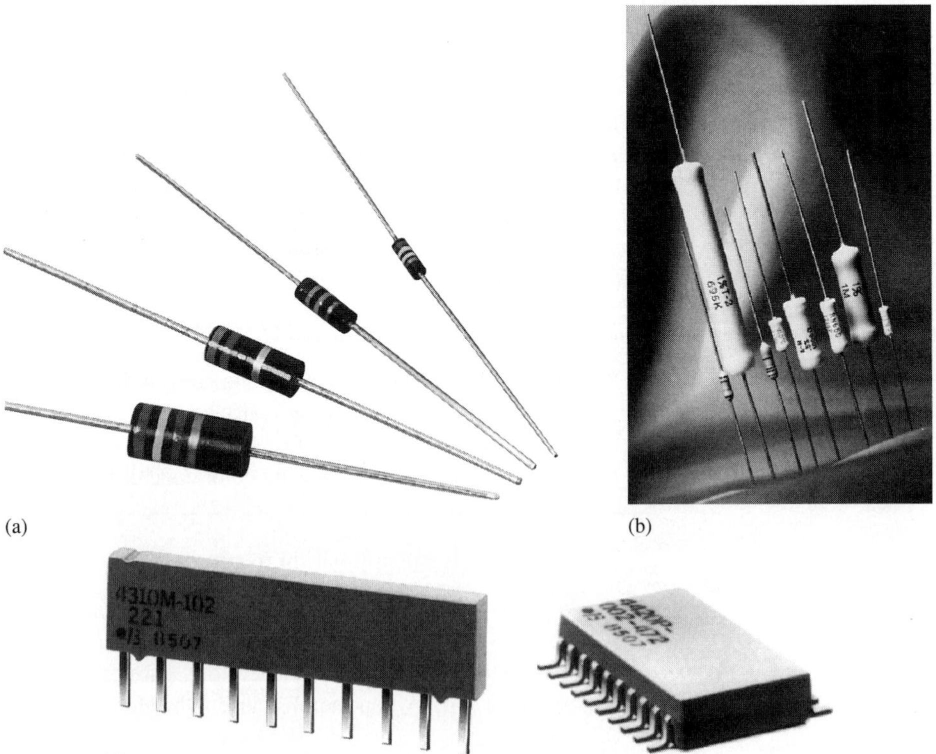

FIGURE 2–14

Typical fixed resistors. Parts (a) and (b) courtesy of Stackpole Carbon Co. (c) Resistor networks (courtesy of Bourns, Inc.).

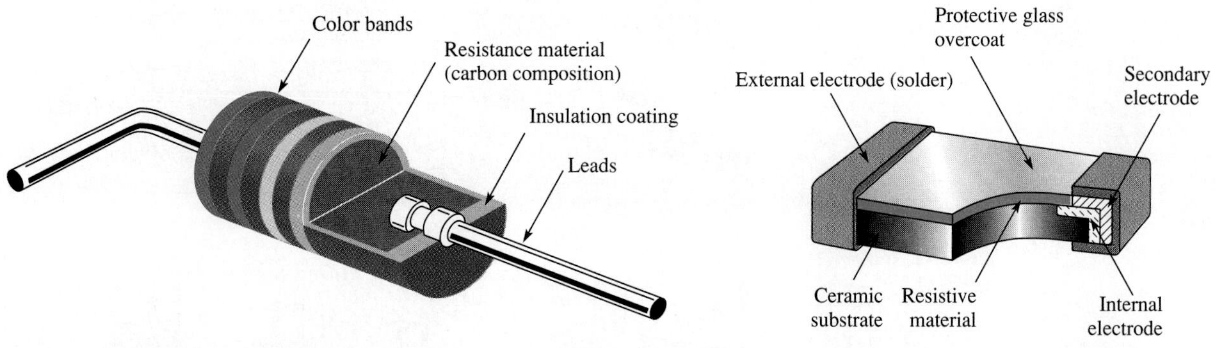

(a) Cutaway view of a carbon-composition resistor

(b) Cutaway view of a chip resistor

FIGURE 2–15
Two types of fixed resistors.

for compact assemblies. Figure 2–15(a) shows the construction of a typical carbon-composition resistor and Figure 2–15(b) shows the construction of a chip resistor.

Other types of fixed resistors include carbon film, metal film, and wirewound. In film resistors, a resistive material is deposited evenly onto a high-grade ceramic rod. The resistive film may be carbon (carbon film) or nickel chromium (metal film). In these types of resistors, the desired resistance value is obtained by removing part of the resistive material in a helical pattern along the rod using a spiraling technique as shown in Figure 2–16(a). Very close **tolerance** can be achieved with this method. Film resistors are also available in the form of resistor networks as shown in Figure 2–16(b).

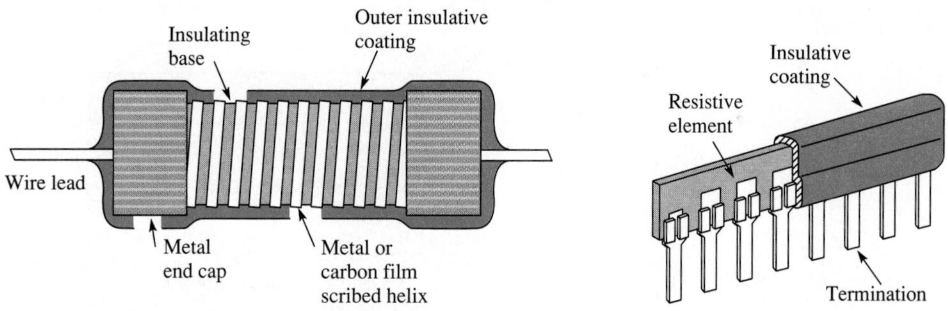

(a) Film resistor showing spiraling technique

(b) Resistor network

FIGURE 2–16
Construction views of typical film resistors.

Wirewound resistors are constructed with resistive wire wound around an insulating rod and then sealed. Normally, wirewound resistors are used because of their relatively high power ratings. Some typical fixed resistors are shown in Figure 2–17.

Resistor Color Codes Many types of fixed resistors with value tolerances of 5%, 10%, or 20% are color coded with four bands to indicate the resistance value and the tolerance. This color-code band system is shown in Figure 2–18, and the color code is listed in Table 2–1.

The color code is read as follows:

1. Beginning at the banded end, the first band is the first digit of the resistance value. If it is not clear which is the banded end, start from the end that does not begin with a gold or silver band.
2. The second band is the second digit.

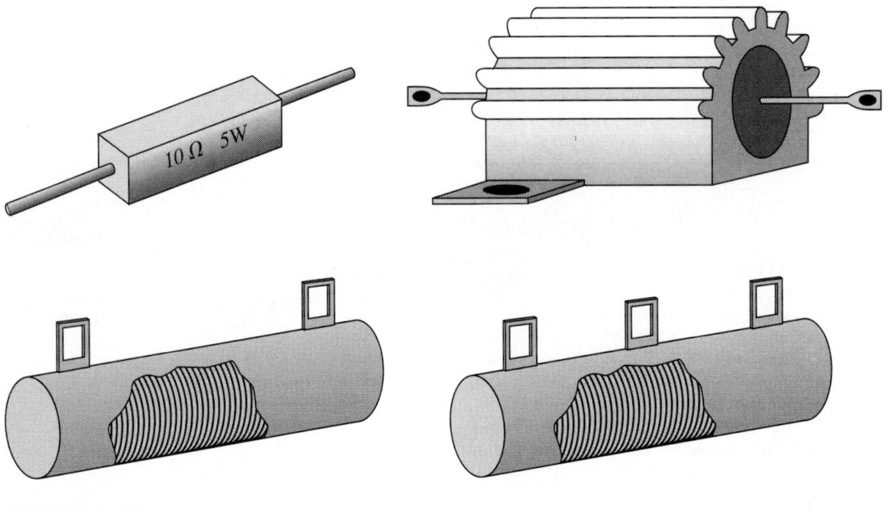

FIGURE 2–17
Typical wirewound power resistors.

FIGURE 2–18
Color-code bands on a resistor.

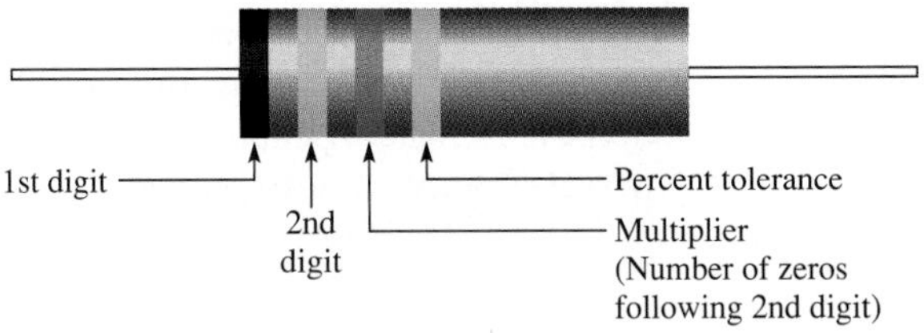

TABLE 2–1
Resistor color code

	DIGIT	COLOR
Resistance value, first three bands: First band—1st digit Second band—2nd digit Third band—number of zeros following the 2nd digit	0	Black
	1	Brown
	2	Red
	3	Orange
	4	Yellow
	5	Green
	6	Blue
	7	Violet
	8	Gray
	9	White
Tolerance, fourth band	5%	Gold
	10%	Silver
	20%	No band

3. The third band is the number of zeros following the second digit, or the *multiplier*.

4. The fourth band indicates the tolerance and is usually gold or silver.

For example, a 5% tolerance means that the *actual* resistance value is within ±5% of the color-coded value. Thus, a 100 Ω resistor with a tolerance of ±5% can have acceptable values as low as 95 Ω and as high as 105 Ω.

For resistance values less than 10 Ω, the third band is either gold or silver. Gold represents a multiplier of 0.1, and silver represents 0.01. For example, a color code of red, violet, gold, and silver represents 2.7 Ω with a tolerance of ±10%.

EXAMPLE 2–4 Find the resistance values in ohms and the percent tolerance for each of the color-coded resistors shown in Figure 2–19.

FIGURE 2–19

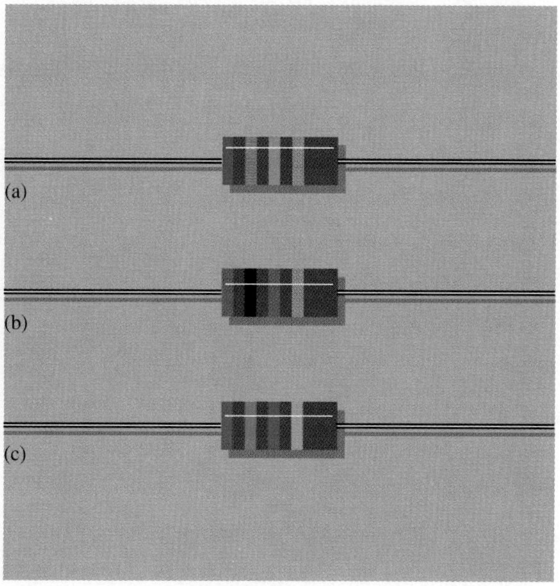

Solution

(a) First band is red = 2, second band is violet = 7, third band is orange = 3 zeros, fourth bank is silver = 10% tolerance.

$$R = 27{,}000\ \Omega \pm 10\%$$

(b) First band is brown = 1, second band is black = 0, third band is brown = 1 zero, fourth band is silver = 10% tolerance.

$$R = 100\ \Omega \pm 10\%$$

(c) First band is green = 5, second band is blue = 6, third band is green = 5 zeros, fourth band is gold = 5% tolerance.

$$R = 5{,}600{,}000\ \Omega \pm 5\%$$

Related Exercise A certain resistor has a yellow first band, a violet second band, a red third band, and no fourth band. Determine the value in ohms.

Certain precision resistors with tolerances of 1% or 2% are color coded with five bands. Beginning at the banded end, the first band is the first digit of the resistance value, the second band is the second digit, the third band is the third digit, the fourth band is the multiplier, and the fifth band indicates the tolerance. Table 2–1 applies, except that gold in the fourth band indicates a multiplier of 0.1 and silver indicates a multiplier of 0.01. Brown in the fifth band indicates 1% tolerance and red indicates 2% tolerance.

Numerical labels are also commonly used on certain types of resistors where the resistance value and tolerance are stamped on the body of the resistor. For example, a common system uses R to designate the decimal point and letters to indicate tolerance as follows:

$$F = \pm 1\%, \quad G = \pm 2\%, \quad J = \pm 5\%, \quad K = \pm 10\%, \quad M = \pm 20\%$$

For values above 100 Ω, three digits are used to indicate resistance value, followed by a fourth digit that specifies the number of zeros. For values less than 100 Ω, R indicates the decimal point.

Some examples are as follows: 6R8M is a 6.8 Ω ±20% resistor; 3301F is a 3300 Ω ±1% resistor; and 2202J is a 22,000 Ω ±5% resistor.

Resistor Reliability Band A fifth band on some color-coded resistors indicates the resistor's reliability in percent of failures per 1000 hours (1000 h) of use. The fifth-band reliability color code is listed in Table 2–2. For example, a brown fifth band means that if a group of like resistors are operated under standard conditions for 1000 h, 1% of the resistors in that group will fail.

TABLE 2–2
Fifth-band reliability color code

COLOR	FAILURES DURING 1000 H OF OPERATION
Brown	1.0%
Red	0.1%
Orange	0.01%
Yellow	0.001%

Variable Resistors Variable resistors are designed so that their resistance values can be changed easily with a manual or an automatic adjustment.

Two basic uses for variable resistors are to divide voltage and to control current. The variable resistor used to divide voltage is called a **potentiometer.** The variable resistor used to control current is called a **rheostat.** Schematic symbols for these types are shown in Figure 2–20. The potentiometer is a three-terminal device, as indicated in part (a). Terminals 1 and 2 have a fixed resistance between them, which is the total resistance. Terminal 3 is connected to a moving contact **(wiper).** You can vary the resistance between 3 and 1 or between 3 and 2 by moving the contact up or down.

Figure 2–20(b) shows the rheostat as a two-terminal variable resistor. Part (c) shows how you can use a potentiometer as a rheostat by connecting terminal 3 to either terminal 1 or terminal 2. Part (d) shows a basic simplified construction diagram of a potentiometer. Some typical potentiometers are pictured in Figure 2–21.

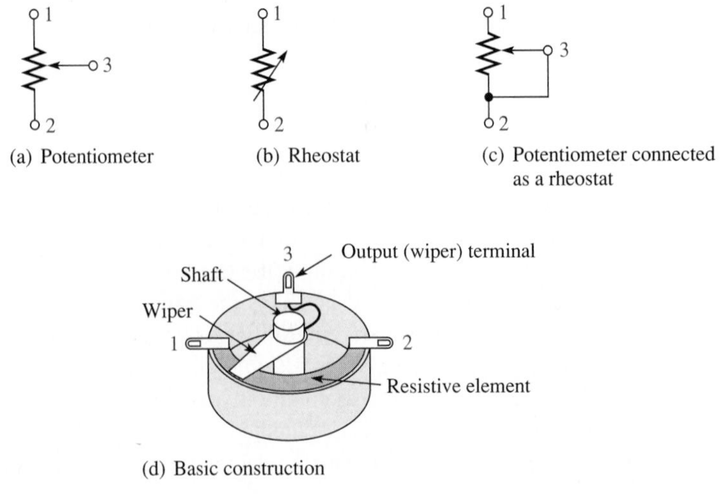

FIGURE 2–20
Potentiometer and rheostat symbols and basic construction of one type of potentiometer.

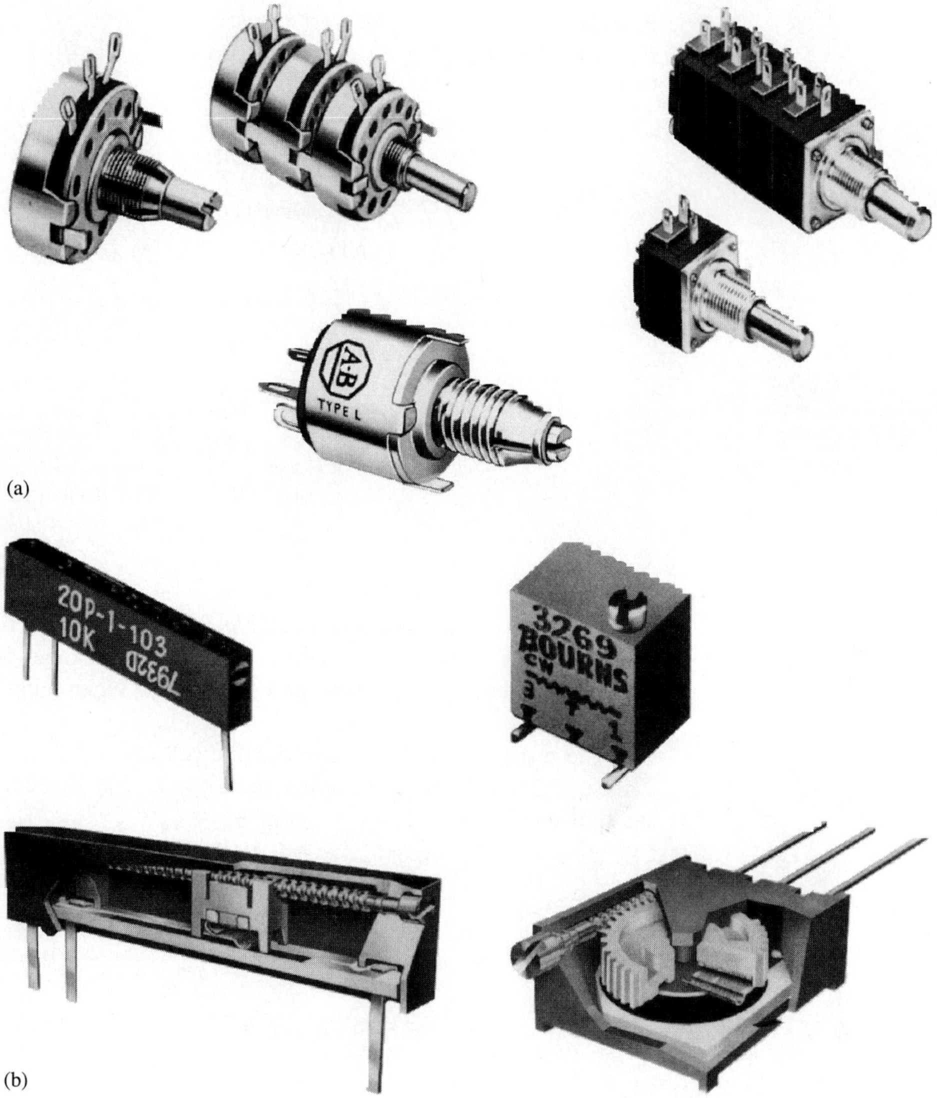

FIGURE 2–21
(a) Typical potentiometers (courtesy of Allen-Bradley Co.). (b) Trimmer potentiometers with construction views (courtesy of Bourns Trimpot).

Potentiometers and rheostats can be classified as linear or tapered, as shown in Figure 2–22, where a potentiometer with a total resistance of 100 Ω is used as an example. As shown in part (a), in a linear potentiometer, the resistance between either terminal and the moving contact varies linearly with the position of the moving contact. For example, one-half of a turn results in one-half the total resistance. Three-quarters of a turn results in three-quarters of the total resistance between the moving contact and one terminal, or one-quarter of the total resistance between the other terminal and the moving contact.

In the **tapered** potentiometer, the resistance varies nonlinearly with the position of the moving contact, so that one-half of a turn does not necessarily result in one-half the total resistance. This concept is illustrated in Figure 2–22(b), where the nonlinear values are arbitrary.

The potentiometer is used as a voltage-control device because when a fixed voltage is applied across the end terminals, a variable voltage is obtained at the wiper contact with respect to either end terminal. The rheostat is used as a current-control device because the current can be changed by changing the wiper position. You will learn the reasons for these functions later.

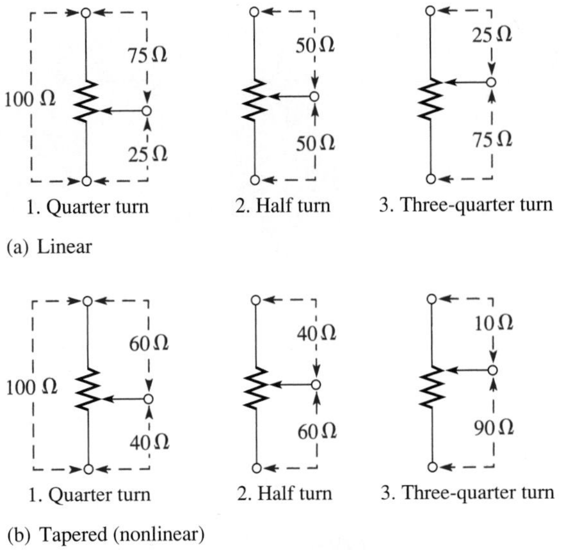

FIGURE 2–22
Examples of (a) linear and (b) tapered potentiometers.

Two Types of Automatically Variable Resistors A **thermistor** is a type of variable resistor that is temperature-sensitive. Its resistance changes inversely with temperature. That is, it has a negative temperature coefficient. When temperature increases, the resistance decreases, and vice versa.

The resistance of a **photoconductive cell** changes with a change in light intensity. This cell also has a negative temperature coefficient. Symbols for both of these devices are shown in Figure 2–23.

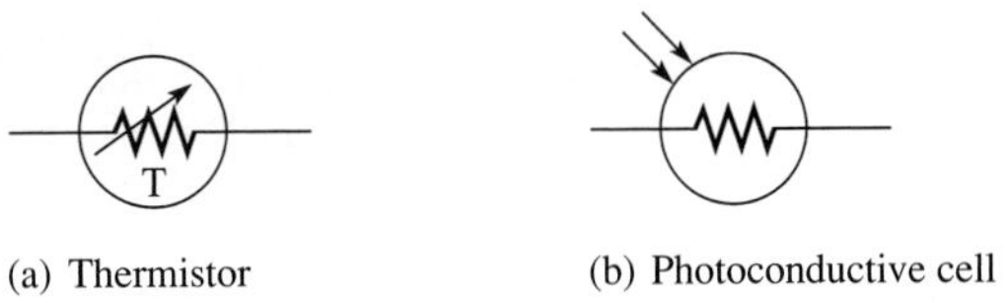

FIGURE 2–23
Resistive devices with sensitivities to temperature and light.

A Resistor Analogy

You can think of a resistor as a partially open valve in a water system that restricts the amount of water flowing through the pipe. If the valve is opened more (less resistance), the water flow (current) increases. If the valve is closed more (more resistance), the water flow (current) decreases.

SECTION 2–5 REVIEW

1. Define *resistance* and name its unit.
2. What are the two main categories of resistors? Briefly explain the difference between them.
3. In the resistor color code, what does each band represent?
4. Determine the resistance and tolerance for each of the following color codes:
   (a) yellow, violet, red, gold **(b)** blue, red, orange, silver
   (c) brown, gray, black, gold
5. What is the basic difference between a rheostat and a potentiometer?
6. What is a thermistor?

2–6 ■ THE ELECTRIC CIRCUIT

A basic electric circuit is an arrangement of physical components which utilize voltage, current, and resistance to perform some useful function.

After completing this section, you should be able to

- **Describe a basic electric circuit**
  - ☐ Relate a schematic to a physical circuit
  - ☐ Define *open circuit* and *closed circuit*
  - ☐ Describe various types of protective devices
  - ☐ Describe various types of switches

Basically, an electric **circuit** consists of a voltage source, a load, and a path for current between the source and the **load.** Figure 2–24 shows an example of a simple electric circuit: a battery connected to a lamp with two wires. The battery is the voltage source, the lamp is the load on the battery because it draws current from the battery, and the two wires provide the current path from the negative terminal of the battery to the lamp and back to the positive terminal of the battery, as shown in part (b). Current flows through the filament of the lamp (which has a resistance), causing it to emit visible light. Current flows through the battery by chemical action. In many practical cases, one terminal of the battery is connected to a *ground* point. For example, in automobiles, the negative battery terminal is connected to the metal chassis of the car. The chassis is the ground for the automobile electrical system. The concept of circuit ground is covered in detail in Chapter 4.

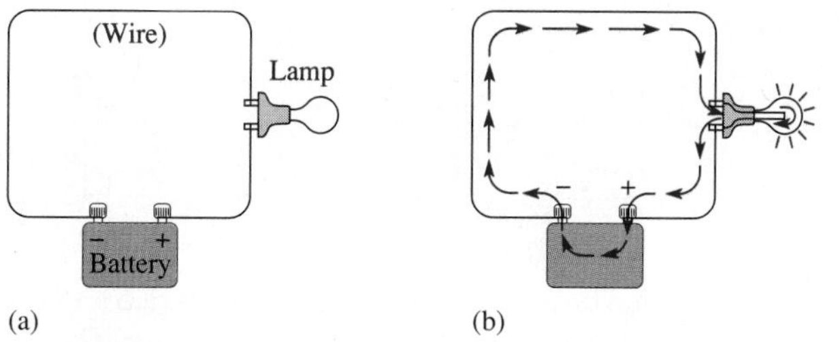

FIGURE 2–24
A simple electric circuit.

The Electrical Schematic

An electric circuit can be represented by a **schematic** using standard symbols for each element, as shown in Figure 2–25 for the simple circuit in Figure 2–24. The purpose of a schematic is to show in an organized manner how the various components in a given circuit are interconnected so that the operation of the circuit can be determined.

FIGURE 2–25
A schematic for the circuit in Figure 2–24.

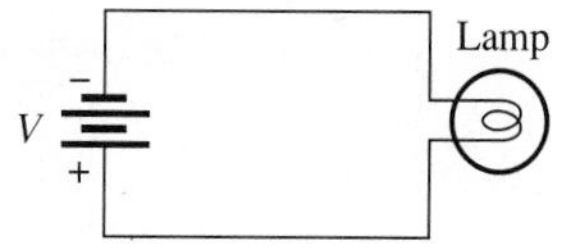

Closed and Open Circuits

The example circuit in Figure 2–24 illustrated a **closed circuit**—that is, a circuit in which the current has a complete path. When the current path is broken so that current cannot flow, the circuit is called an **open circuit.**

Switches **Switches** are commonly used for controlling the opening or closing of circuits by either mechanical or electronic means. For example, a switch is used to turn a lamp on or off as illustrated in Figure 2–26. Each circuit pictorial is shown with its associated schematic. The type of switch indicated is a *single-pole–single-throw* (SPST) toggle switch.

Figure 2–27 shows a somewhat more complicated circuit using a *single-pole–double-throw* (SPDT) type of switch to control the current to two different lamps. When

FIGURE 2–26
Basic closed and open circuits using an SPST switch for control.

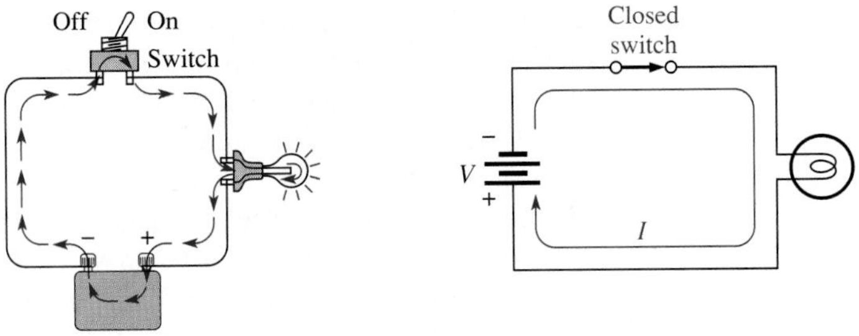

(a) Current flows in a *closed* circuit (switch is ON or in the *closed* position).

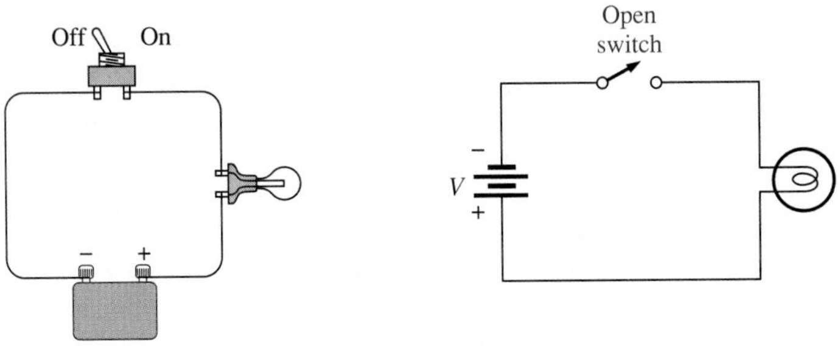

(b) No current flows in an *open* circuit (switch is OFF or in the *open* position).

FIGURE 2–27
An example of an SPDT switch controlling two lamps.

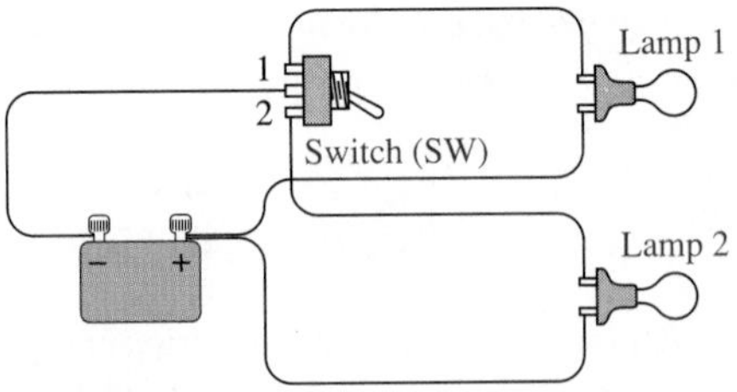

(a) Pictorial

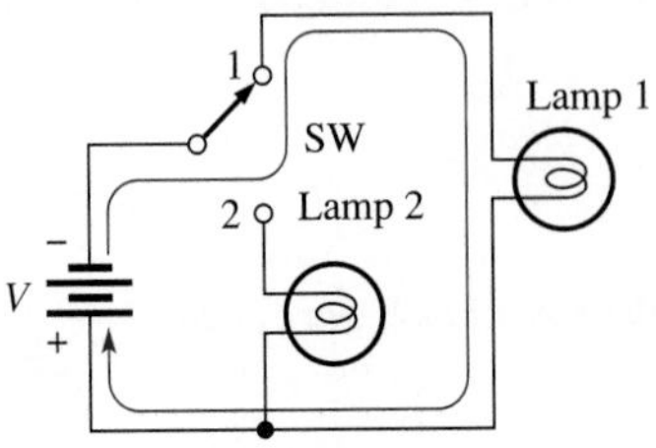

(b) A schematic showing Lamp 1 on and Lamp 2 off

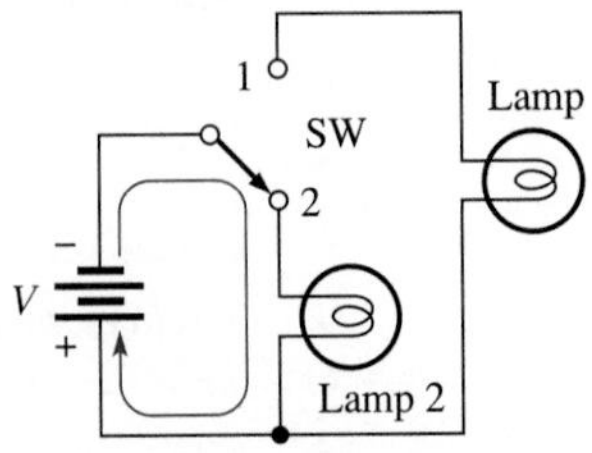

(c) A schematic showing Lamp 2 on and Lamp 1 off

one lamp is on, the other is off, and vice versa, as illustrated by the two schematics that represent each of the switch positions.

The term *pole* refers to the movable arm in a switch, and the term *throw* indicates the number of contacts that are affected (either opened or closed) by a single switch action (a single movement of a pole).

In addition to the SPST and the SPDT switches (symbols are shown in Figure 2–28 (a) and (b)), several other types of switches are important.

- *Double-pole–single-throw (DPST).* The DPST switch permits simultaneous opening or closing of two sets of contacts. The symbol is shown in Figure 2–28(c). The dashed line indicates that the contact arms are mechanically linked so that both move with a single switch action.
- *Double-pole–double-throw (DPDT).* The DPDT switch provides connection from one set of contacts to either of two other sets. The schematic symbol is shown in Figure 2–28(d).
- *Push button (PB).* In the normally open push-button switch (NOPB), shown in Figure 2–28(e), connection is made between two contacts when the button is depressed, and connection is broken when the button is released. In the normally closed push-button switch (NCPB), shown in Figure 2–28(f), connection between the two contacts is broken when the button is depressed.
- *Rotary.* In a rotary switch, a knob is turned to make connection between one contact and any one of several others. A symbol for a simple six-position rotary switch is shown in Figure 2–28(g).

FIGURE 2–28
Switch symbols.

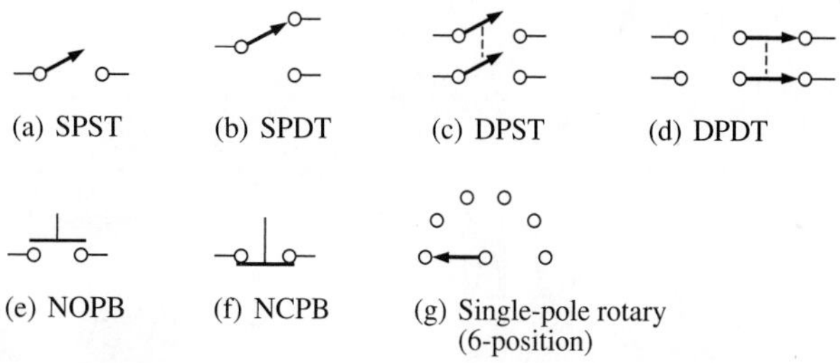

Several varieties of switches are shown in Figure 2–29.

Protective Devices **Fuses** and **circuit breakers** are placed in the current path and are used to deliberately create an open circuit when the current exceeds a specified number of amperes due to a malfunction or other abnormal condition in a circuit. For example, a fuse or circuit breaker with a 20 A rating will open a circuit when the current exceeds 20 A.

The major difference between a fuse and a circuit breaker is that when a fuse is "blown," it must be replaced; but when a circuit breaker opens, it can be reset and reused repeatedly. Both of these devices protect against damage to a circuit due to excess current or prevent a hazardous condition created by the overheating of wires and other components when the current is too great. Because fuses cut off excess current more quickly than circuit breakers, fuses are used whenever delicate electronic equipment needs to be protected. Several typical fuses and circuit breakers, along with their schematic symbols, are shown in Figure 2–30.

Wires

Wires are a common form of conductive material used in electrical circuits. They vary in diameter size and are arranged according to standard gage numbers, called **American Wire Gage (AWG)** sizes. The larger the gage number is, the smaller the wire diameter is. The AWG sizes are listed in Table 2–3 on page 41.

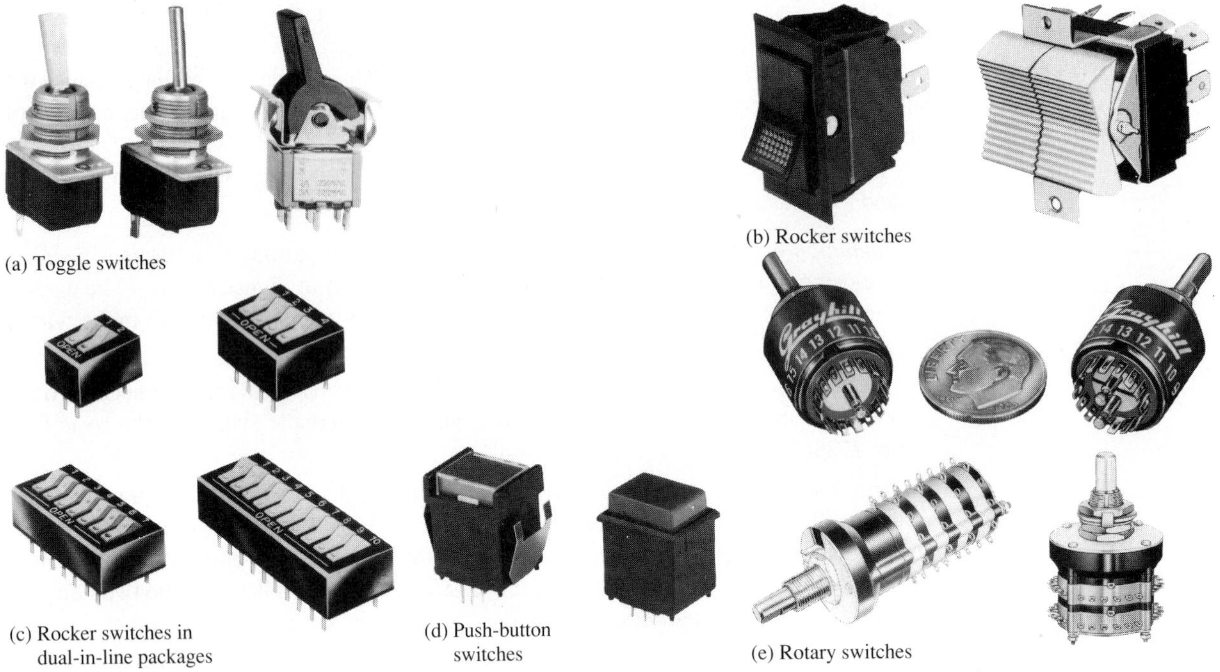

(a) Toggle switches
(b) Rocker switches
(c) Rocker switches in dual-in-line packages
(d) Push-button switches
(e) Rotary switches

FIGURE 2–29
Switches. Parts (a), (b), and (d) courtesy of Eaton Corp. Parts (c) and (e) courtesy of Grayhill, Inc.

(a) Power fuses
(b) Fuses and fuse holders
(c) Fuse symbol
(d) Circuit breakers
(e) Remote control circuit breaker
(f) Circuit breaker symbol

FIGURE 2–30
Fuses and circuit breakers. Parts (a), (b), and (d) courtesy of Bussmann, Cooper Industries. Part (e) courtesy of Eaton Corp.

TABLE 2–3
American Wire Gage (AWG) sizes and resistances for solid round copper

AWG #	AREA (CM)	Ω/1000 FT AT 20°C	AWG #	AREA (CM)	Ω/1000 FT AT 20°C
0000	211,600	0.0490	19	1,288.1	8.051
000	167,810	0.0618	20	1,021.5	10.15
00	133,080	0.0780	21	810.10	12.80
0	105,530	0.0983	22	642.40	16.14
1	83,694	0.1240	23	509.45	20.36
2	66,373	0.1563	24	404.01	25.67
3	52,634	0.1970	25	320.40	32.37
4	41,742	0.2485	26	254.10	40.81
5	33,102	0.3133	27	201.50	51.47
6	26,250	0.3951	28	159.79	64.90
7	20,816	0.4982	29	126.72	81.83
8	16,509	0.6282	30	100.50	103.2
9	13,094	0.7921	31	79.70	130.1
10	10,381	0.9989	32	63.21	164.1
11	8,234.0	1.260	33	50.13	206.9
12	6,529.0	1.588	34	39.75	260.9
13	5,178.4	2.003	35	31.52	329.0
14	4,106.8	2.525	36	25.00	414.8
15	3,256.7	3.184	37	19.83	523.1
16	2,582.9	4.016	38	15.72	659.6
17	2,048.2	5.064	39	12.47	831.8
18	1,624.3	6.385	40	9.89	1049.0

As the table shows, the size of a wire is also specified in terms of its cross-sectional area, as illustrated also in Figure 2–31. The unit of cross-sectional area is the **circular mil** abbreviated CM. One circular mil is the area of a wire with a diameter of 0.001 inch (0.001 in., or 1 mil). Find the cross-sectional area by expressing the diameter in thousandths of an inch (mils) and squaring it, as follows:

$$A = d^2 \qquad \textbf{(2–4)}$$

where A is the cross-sectional area in circular mils and d is the diameter in mils.

FIGURE 2–31
Cross-sectional area of a wire.

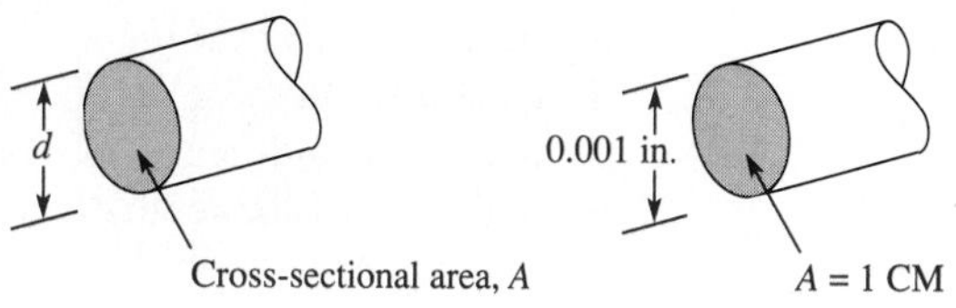

EXAMPLE 2–5 What is the cross-sectional area of a wire with a diameter of 0.005 inch?

Solution

$$d = 0.005 \text{ in.} = 5 \text{ mils}$$
$$A = d^2 = (5 \text{ mils})^2 = 25 \text{ CM}$$

Related Exercise What is the cross-sectional area of a 0.0201 in. diameter wire? What is the AWG # for this wire?

Wire Resistance

Although copper wire conducts electricity extremely well, it still has some resistance, as do all conductors. The resistance of a wire depends on four factors: (1) type of material, (2) length of wire, (3) cross-sectional area, and (4) temperature.

Each type of conductive material has a characteristic called its *resistivity, ρ*. For each material, ρ is a constant value at a given temperature. The formula for the resistance of a wire of length l and cross-sectional area A is

$$R = \frac{\rho l}{A} \tag{2–5}$$

This formula shows that resistance increases with resistivity and length and decreases with cross-sectional area. For resistance to be calculated in ohms, the length must be in feet (ft), the cross-sectional area in circular mils, and the resistivity in CM-Ω/ft.

EXAMPLE 2–6 Find the resistance of a 100 ft length of copper wire with a cross-sectional area of 810.1 CM. The resistivity of copper is 10.4 CM-Ω/ft.

Solution

$$R = \frac{\rho l}{A} = \frac{(10.4 \text{ CM-}\Omega/\text{ft})(100 \text{ ft})}{810.1 \text{ CM}} = 1.284\ \Omega$$

Related Exercise Determine the resistance of a 1000 ft length of #22 AWG copper wire.

Table 2–3 lists the resistance of the various standard wire sizes in ohms per 1000 ft at 20°C. For example, a 1000 ft length of 14-gage copper wire has a resistance of 2.525 Ω. A 1000 ft length of 22-gage wire has a resistance of 16.14 Ω. For a given length, the smaller wire has more resistance. Thus, for a given voltage, larger wires can carry more current than smaller ones.

SECTION 2–6 REVIEW

1. What are the basic elements of an electric circuit?
2. Define *open circuit.*
3. Define *closed circuit.*
4. What is the purpose of a fuse?
5. What is the difference between a fuse and a circuit breaker?
6. Which wire is larger in diameter, AWG #3 or AWG #22?

2–7 ■ BASIC CIRCUIT MEASUREMENTS

An electronics technician cannot function without knowing how to measure voltage, current, and resistance.

After completing this section, you should be able to

- **Make basic circuit measurements**
  - □ Properly place a voltmeter in a circuit
  - □ Properly place an ammeter in a circuit
  - □ Properly connect an ohmmeter to measure resistance
  - □ Set up and read basic meters

Voltage, current, and resistance measurements are commonly required in electronics work. Special types of instruments are used to measure these basic electrical quantities.

The instrument used to measure voltage is a **voltmeter,** the instrument used to measure current is an **ammeter,** and the instrument used to measure resistance is an **ohmmeter.** Commonly, all three instruments are combined into a single instrument known as a **multimeter,** or VOM (volt-ohm-milliammeter), in which you can choose what specific quantity to measure by selecting the switch setting.

Typical multimeters are shown in Figure 2–32. Part (a) shows an analog meter, that is, with a needle pointer, and part (b) shows a digital multimeter (DMM), which provides a digital readout of the measured quantity.

(a) (b)

FIGURE 2–32
Typical multimeters. (a) Analog and (b) digital (courtesy of B & K Precision).

Meter Symbols

Throughout this book, certain symbols will be used in circuits to represent the different meters, as shown in Figure 2–33. You may see any of three types of symbols for volt-

FIGURE 2–33
Meter symbols used in this book.

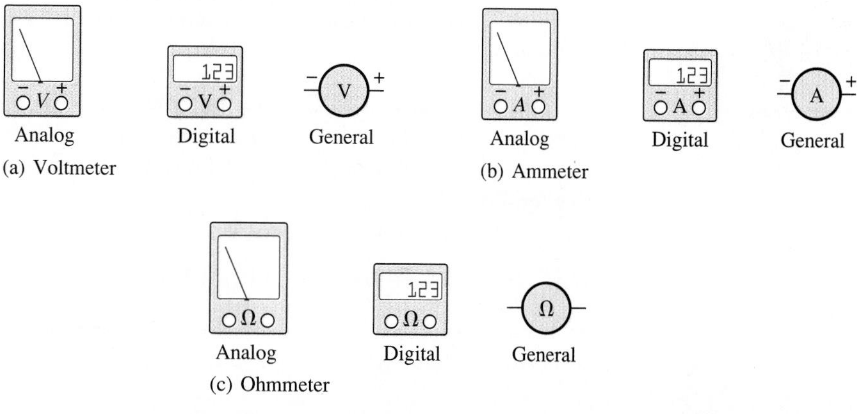

meters, ammeters, and ohmmeters, depending on which symbol most effectively conveys the information required. Although the digital meter is much more widely used than the analog meter in industry, we will use the analog meter symbol in certain situations to illustrate better the operation of a circuit when relative measurements or changes in quantities need to be depicted by the position or movement of the needle. The digital meter symbol is used when fixed values are to be indicated in a circuit. The general schematic symbol is used to indicate placement of meters in a circuit when no values or value changes need to be shown.

How to Measure Current with an Ammeter

Figure 2–34 illustrates how to measure current with an ammeter. Part (a) shows the simple circuit in which the current through the resistor is to be measured. Connect the ammeter in the current path by first opening the circuit, as shown in part (b). Then insert the meter as shown in part (c). Such a connection is a *series* connection. The polarity of the meter must be such that the current flows in at the negative terminal and out at the positive terminal.

FIGURE 2–34
Example of an ammeter connection in a simple circuit.

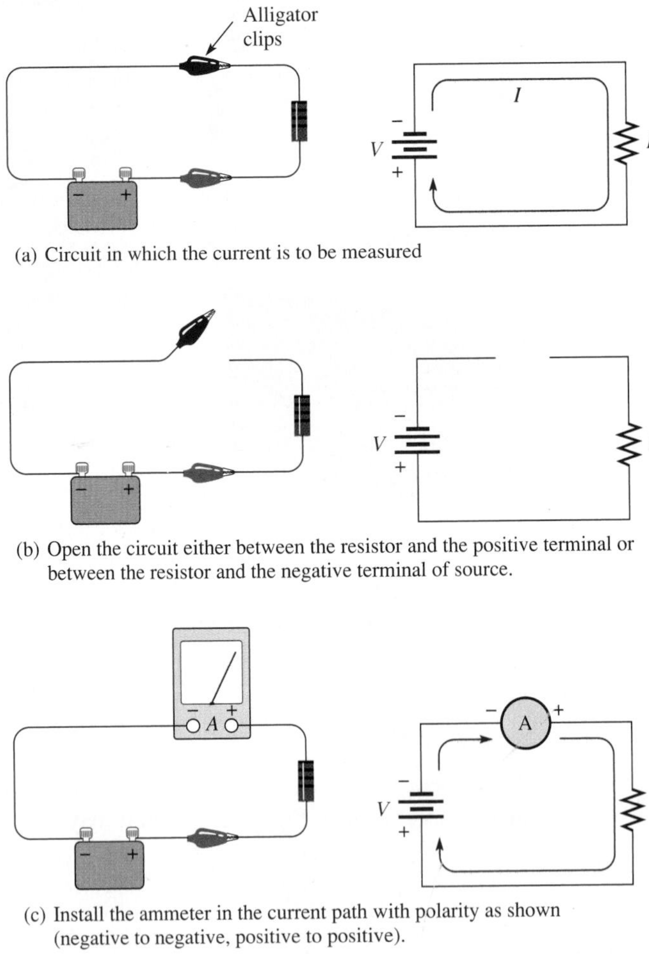

How to Measure Voltage with a Voltmeter

To measure voltage, connect the voltmeter across the component for which the voltage is to be found. Such a connection is a *parallel* connection. The negative terminal of the

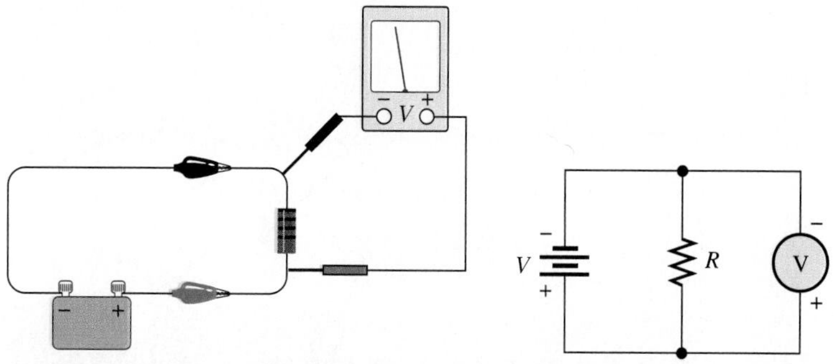

FIGURE 2–35
Example of a voltmeter connection in a simple circuit.

meter must be connected to the negative side of the circuit, and the positive terminal of the meter to the positive side of the circuit. Figure 2–35 shows a voltmeter connected to measure the voltage across the resistor.

How to Measure Resistance with an Ohmmeter

To measure resistance, connect the ohmmeter across the resistor. *The resistor must first be removed or disconnected from the circuit.* This procedure is shown in Figure 2–36.

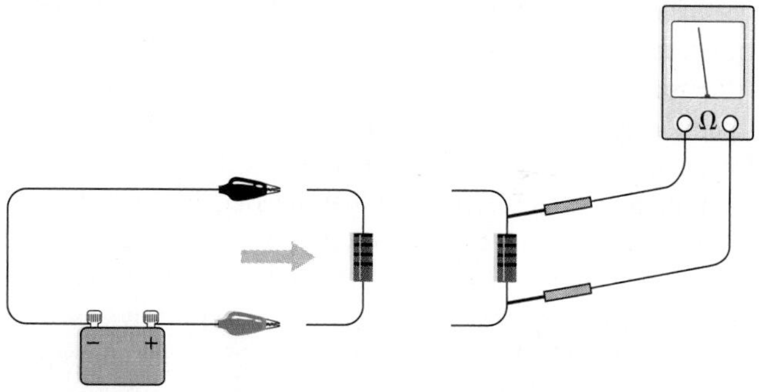

(a) Disconnect the resistor from the circuit to avoid damage to the meter and/or incorrect measurement.

(b) Measure the resistance. (Polarity is not important.)

FIGURE 2–36
Example of using an ohmmeter.

Reading Analog Multimeters

Functions The face of a typical analog meter is represented in Figure 2–37. This particular instrument can be used to measure both direct current (dc) and alternating current (ac) quantities as well as resistance values. It has four selectable functions: dc volts (DC VOLTS), dc milliamperes (DC mA), ac volts (AC VOLTS), and OHMS. Most analog multimeters are similar to this one, although range selections and scales may vary.

Ranges Within each function there are several ranges, as indicated by the brackets around the selector switch. For example, the DC VOLTS function has 0.3 V, 3 V, 12 V, 60 V, 300 V, and 600 V ranges. Thus, dc voltages from 0.3 V full-scale to 600 V full-scale can be measured. On the DC mA function, direct currents from 0.06 mA full-scale to 120 mA full-scale can be measured. On the ohm scale, the range settings are ×1, ×10, ×100, ×1000, and ×100,000.

FIGURE 2–37
A typical analog multimeter.

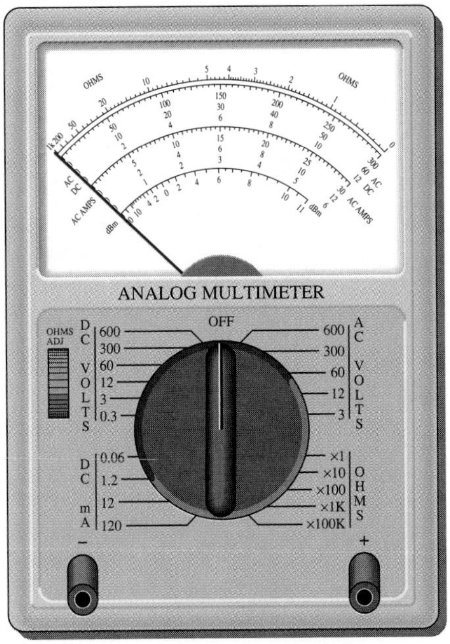

The Ohm Scale Ohms are read on the top scale of the meter. This scale is nonlinear; that is, the values represented by each division (large or small) vary as you go across the scale. In Figure 2–37, notice how the scale becomes more compressed as you go from right to left.

To read the actual value in ohms, multiply the number on the scale as indicated by the pointer by the factor selected by the switch. For example, when the switch is set at ×100 and the pointer is at 20, the reading is $20 \times 100 = 2000\ \Omega$.

As another example, assume that the switch is at ×10 and the pointer is at the seventh small division between the 1 and 2 marks, indicating 17 Ω (1.7×10). Now, if the meter remains connected to the same resistance and the switch setting is changed to ×1, the pointer will move to the second small division between the 15 and 20 marks. This, of course, is also a 17 Ω reading, illustrating that a given resistance value can often be read at more than one switch setting.

The AC-DC Scales The second, third, and fourth scales from the top (labeled "AC" and "DC") are used in conjunction with the DC VOLTS, DC mA, and AC VOLTS functions. The upper ac-dc scale ends at the 300 mark and is used with the range settings that are multiples of three, such as 0.3, 3, and 300. For example, when the switch is at 3 on the DC VOLTS function, the 300 scale has a full-scale value of 3 V. At the range setting of 300, the full-scale value is 300 V, and so on.

The middle ac-dc scale ends at 60. This scale is used in conjunction with range settings that are multiples of 6, such as 0.06, 60, and 600. For example, when the switch is at 60 on the DC VOLTS function, the full-scale value is 60 V.

The lower ac-dc scale ends at 12 and is used in conjunction with switch settings that are multiples of 12, such as 1.2, 12, and 120.

The remaining scales are for ac current and for decibels, which are discussed later in the book.

EXAMPLE 2–7 In Figure 2–38, determine the quantity (voltage, current, or resistance) that is being measured and its value.

FIGURE 2–38

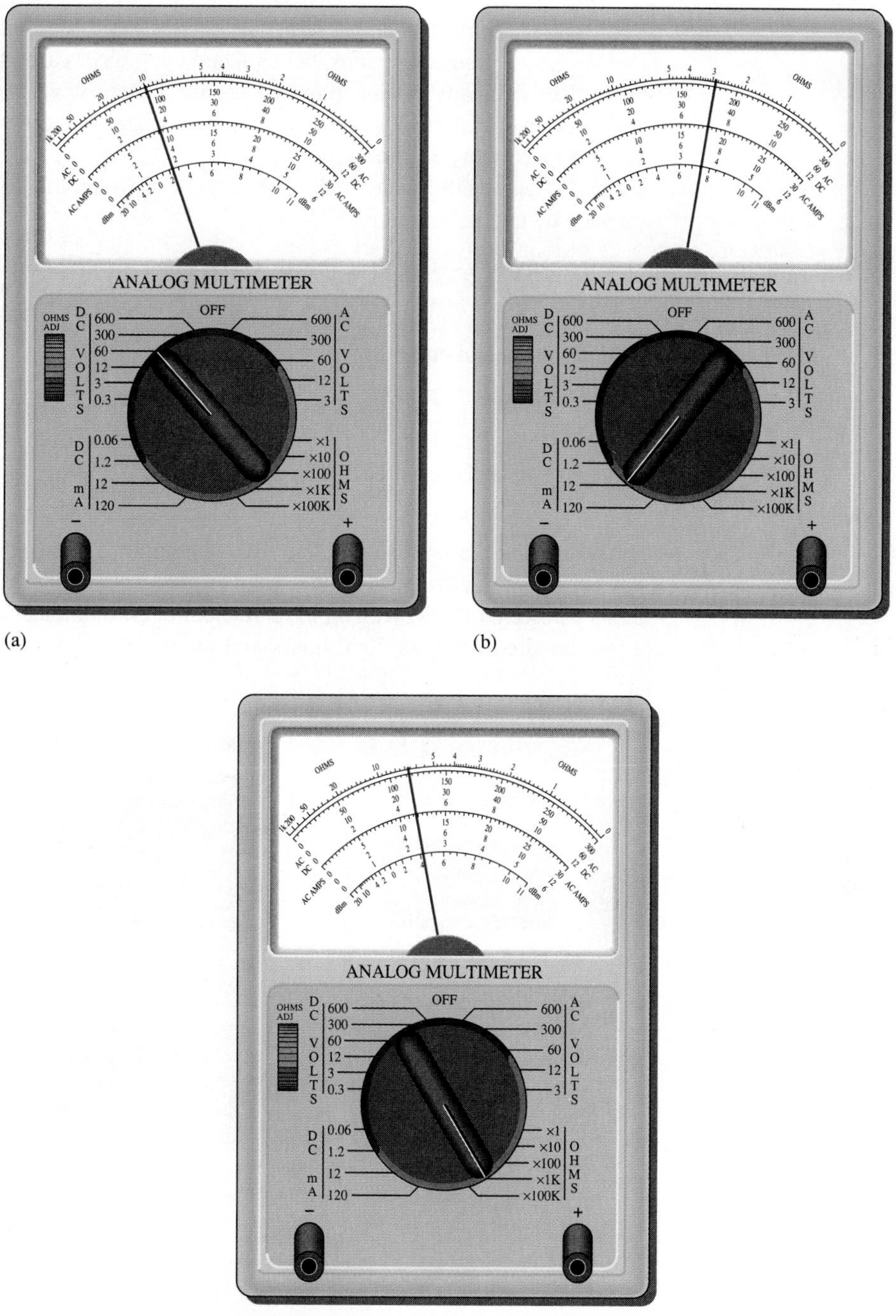

(a) (b) (c)

Solution

(a) The switch in Figure 2–38(a) is set on the DC VOLTS function and the 60 V range. The reading taken from the middle dc scale is 18 V.

(b) The switch in Figure 2–38(b) is set on the DC mA function and the 12 mA range. The reading taken from the upper dc scale is 7.2 mA.

(c) The switch in Figure 2–38(c) is set on the ohm (OHMS) function and the ×1000 range. The reading taken from the ohm scale (top scale) is 7 kΩ.

Related Exercise In Figure 2–38(c) the switch is moved to the ×100 setting. Assuming that the same resistance is being measured, what will the needle do?

Digital Multimeters (DMMs)

DMMs are the most widely used type of electronic measuring instrument. Generally, DMMs provide more functions, better accuracy, greater ease of reading, and greater reliability than do many analog meters. Analog meters have at least one advantage over DMMs, however. They can track short-term variations and trends in a measured quantity that many DMMs are too slow to respond to. Three typical DMMs are shown in Figure 2–39. Many DMMs are autoranging types in which the proper range is automatically selected by internal circuitry.

DMM Functions The basic functions found on most DMMs include the following:

- ☐ Ohms
- ☐ DC voltage and current
- ☐ AC voltage and current

Some DMMs provide special functions such as transistor or diode tests, power measurement, and decibel measurement for audio amplifier tests.

DMM Displays DMMs are available with either LCD (liquid-crystal display) or LED (light-emitting diode) readouts. The LCD is the most commonly used readout in battery-powered instruments because it requires only very small amounts of current. A typical battery-powered DMM with an LCD readout operates on a 9 V battery that will last from a few hundred hours to 2000 hours and more. The disadvantages of LCD readouts are

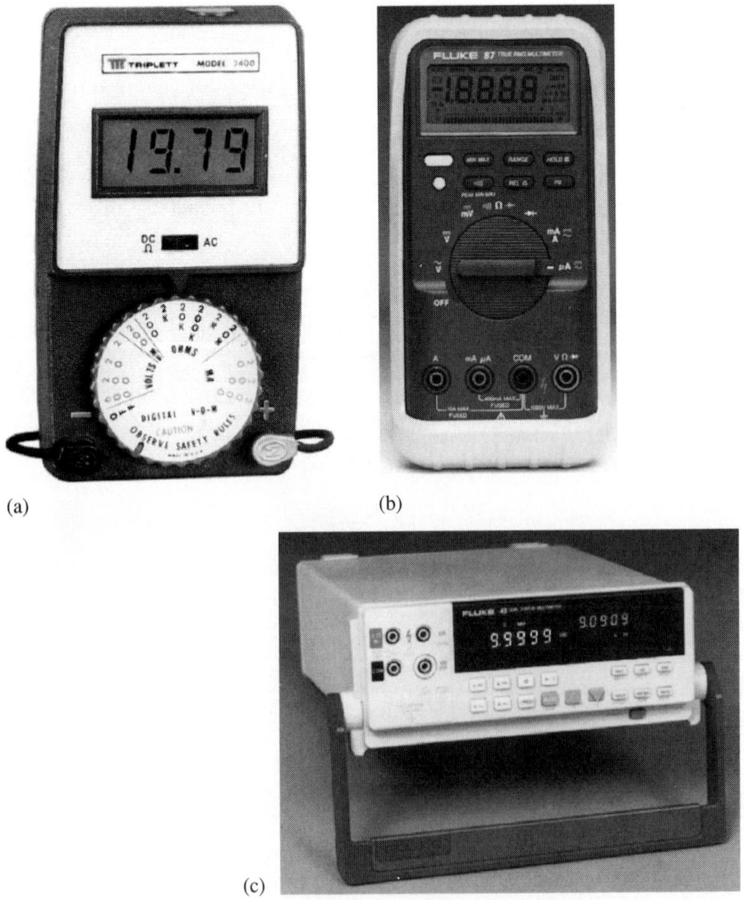

FIGURE 2–39
Typical digital multimeters (DMMs). Part (a) courtesy of Triplett Corp. Parts (b) and (c) courtesy of John Fluke Manufacturing Co., Inc.

that (a) they are difficult or impossible to see in low-light conditions and (b) they are relatively slow to respond to measurement changes. LEDs, on the other hand, can be seen in the dark and respond quickly to changes in measured values. LED displays require much more current than LCDs, and, therefore, battery life is shortened when they are used in portable equipment.

Both LCD and LED DMM displays are in a seven-segment format. Each digit in the display consists of seven separate segments, as shown in Figure 2–40(a). Each of the ten decimal digits is formed by activation of appropriate segments, as illustrated in Figure 2–40(b). In addition to the seven segments, there is also a decimal point.

(a) (b)

FIGURE 2–40
Seven-segment display.

Resolution The **resolution** of a meter is the smallest increment of a quantity that the meter can measure. The smaller the increment, the better the resolution. One factor that determines the resolution of a meter is the number of digits in the display.

Because many meters have 3½ digits in their display, we will use this case for illustration. A 3½-digit multimeter has three digit positions that can indicate from 0 through 9, and one digit position that can indicate only a value of 1. This latter digit, called the *half-digit,* is always the most significant digit in the display. For example, suppose that a DMM is reading 0.999 V, as shown in Figure 2–41(a). If the voltage increases by 0.001 V to 1 V, the display correctly shows 1.000 V, as shown in part (b). The "1" is the half-digit. Thus, with 3½ digits, a variation of 0.001 V, which is the resolution, can be observed.

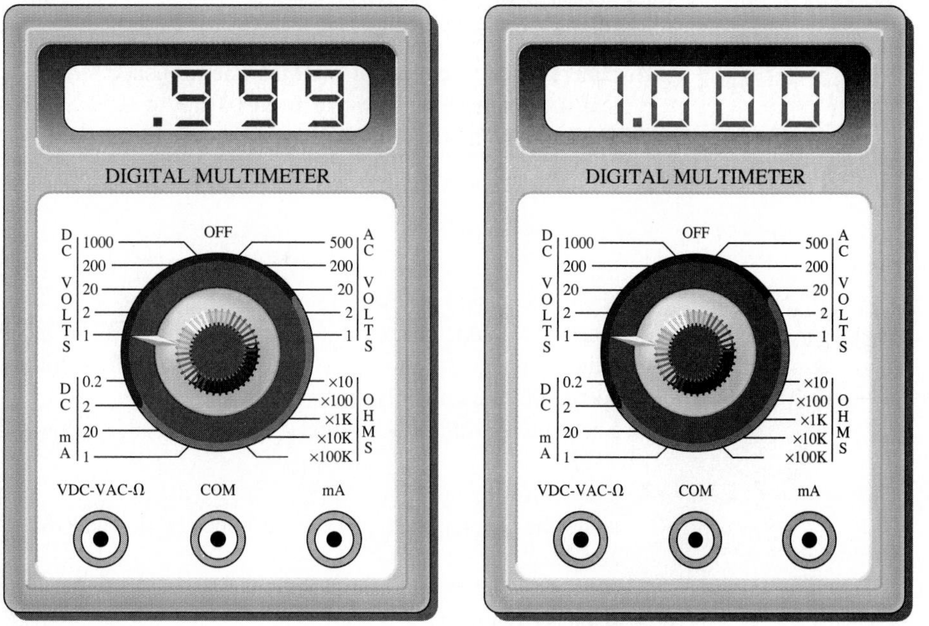

(a) Resolution: 0.001 V (b) Resolution: 0.001 V

FIGURE 2–41
A 3½-digit DMM illustrates how the resolution changes with the number of digits in use.

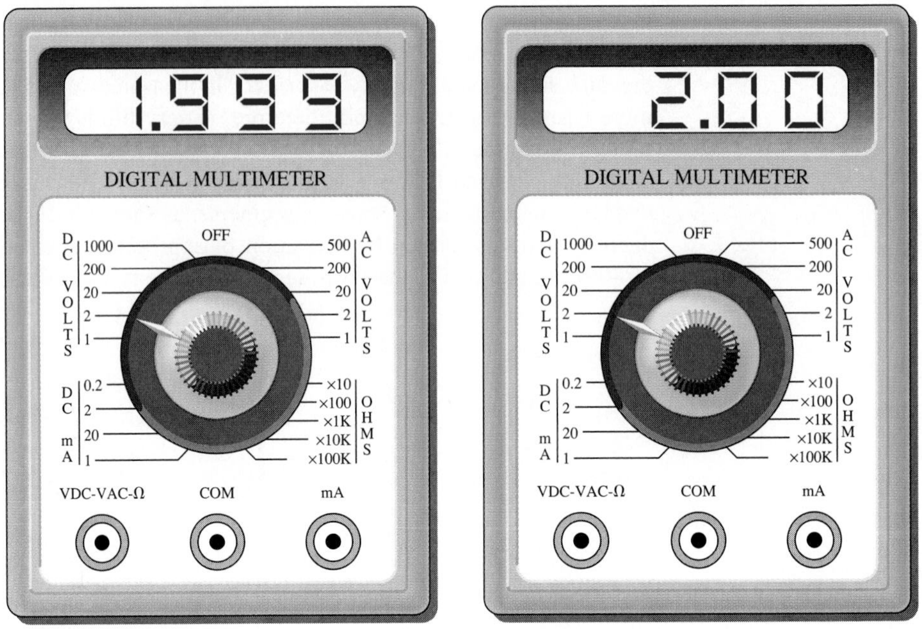

(c) Resolution: 0.001 V

(d) Resolution: 0.01 V

FIGURE 2–41 ***Continued***

Now, suppose that the voltage increases to 1.999 V. This value is indicated on the meter as shown in Figure 2–41(c). If the voltage increases by 0.001 V to 2 V, the half-digit cannot display the "2," so the display shows 2.00. The half-digit is blanked and only three digits are active, as indicated in part (d). With only three digits active, the resolution is 0.01 V rather than 0.001 V as it is with 3½ active digits. The resolution remains 0.01 V up to 19.99 V. The resolution goes to 0.1 V for readings of 20.0 V to 199.9 V. At 200 V, the resolution goes to 1 V, and so on.

The resolution capability of a DMM is also determined by the internal circuitry and the rate at which the measured quantity is sampled. DMMs with displays of 4½ through 8½ digits are also available.

Accuracy The accuracy of a DMM is established strictly by its internal circuitry. For typical meters, accuracies range from 0.01% to 0.5%, with some precision laboratory-grade meters going to 0.002%.

SECTION 2–7 REVIEW

1. Name the meters for measurement of (a) current, (b) voltage, and (c) resistance.
2. Place two ammeters in the circuit of Figure 2–27 to measure the current through either lamp (be sure to observe the polarities). How can the same measurements be accomplished with only one ammeter?
3. Show how to place a voltmeter to measure the voltage across lamp 2 in Figure 2–27.
4. The multimeter in Figure 2–37 is set on the 3 V range to measure dc voltage. Assume the pointer is at 150 on the upper ac-dc scale. What voltage is being measured?
5. How do you set up the meter in Figure 2–37 to measure 275 V dc, and on what scale do you read the voltage?
6. If you expect to measure a resistance in excess of 20 kΩ with the meter in Figure 2–37, where do you set the switch?

7. List two common types of DMM displays, and discuss the advantages and disadvantages of each.
8. Define *resolution* in a DMM.

2–8 ■ APPLICATION ASSIGNMENT

The requirements for a lighting system were stated in the chapter opener. You will go through a step-by-step process of working through this assignment using things that you've learned in this chapter. Recall from the opener that the lighting system will consist of six lamps that are controlled so that only one is turned on at a time and the sequence can be in any order. Also, the system must be fused and operate from a 12 V battery and a method must be provided for varying the brightness of the lamp that is on.

Step 1: The Circuit

Select the circuit from the four configurations in Figure 2–42 that meets the requirements. Explain why the rejected circuits will not meet the requirements. Explain the purpose of each component in your circuit.

FIGURE 2–42

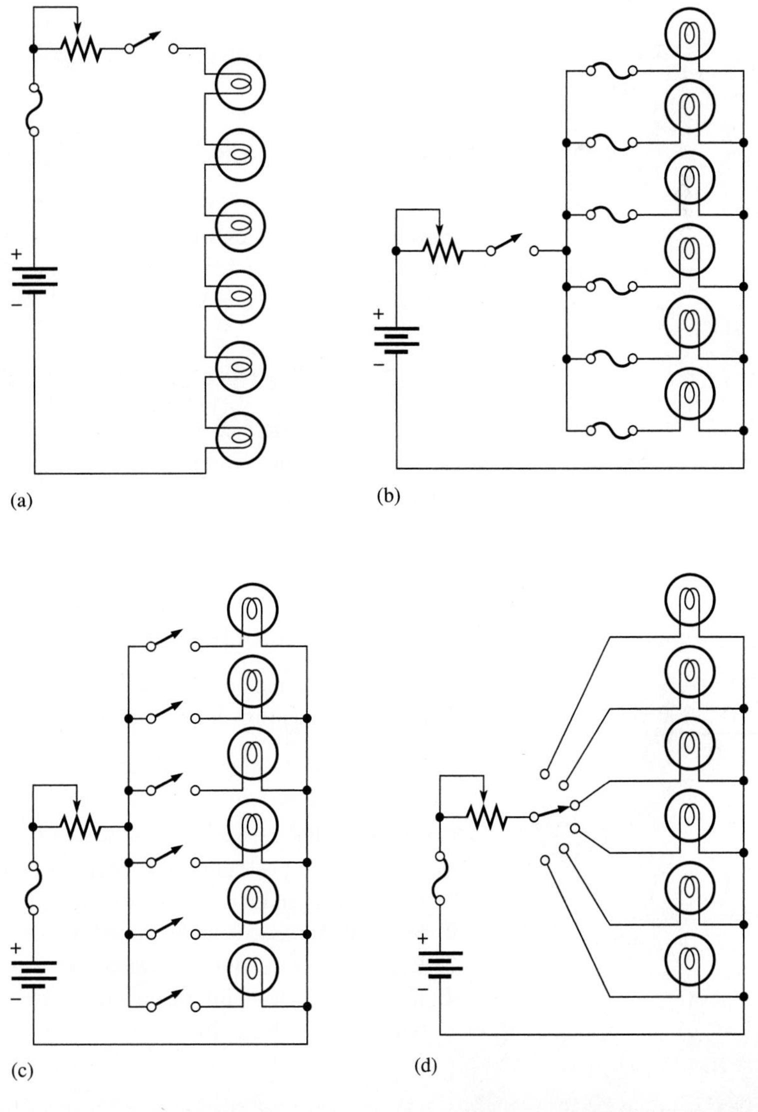

Step 2: Hook Up the Circuit

From the selection of components in Figure 2–43, specify the quantity of each component required for the lighting circuit. Hook up the lighting circuit by developing a point-to-point (From–To) wiring list using component designations and pin numbers.

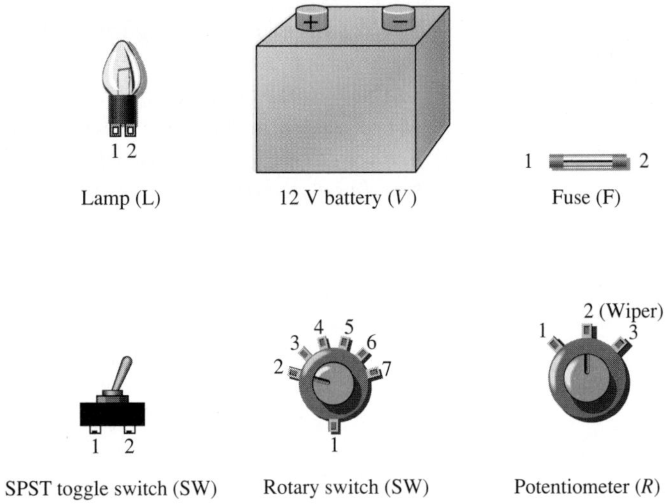

FIGURE 2–43
Component selection.

Step 3: Determine the Current Rating of the Fuse

The fuse protects the circuit by burning open when there is excessive current. The type of lamp to be used draws approximately 5 A when 12 V are applied. Choose a fuse based on its ampere rating that is the most appropriate for your circuit from the following ratings: 1 A, 2 A, 3 A, 4 A, 5 A, 6 A, 7 A, 8 A, 9 A, 10 A, 20 A, and 30 A. Explain your choice.

Step 4: Determine the Capacity of the Battery

A battery can provide a certain amount of current for a certain period of time at its rated voltage. The amount of amps that a battery can deliver to a load for a given period of time at its rated voltage is its *ampere-hour rating*. Determine the minimum ampere-hour rating of the battery if the lighting circuit must operate for four hours before the battery is recharged.

Step 5: Troubleshoot the Circuit

Once you have connected all the components, it is time to check out the operation to verify that everything is functioning properly.

- ☐ Write a brief description of the procedure you would use to thoroughly test the circuit. This is called a *test procedure*.
- ☐ For each of the following cases where the circuit is not working properly, determine the possible fault or faults and specify how you would correct it:
  1. Each lamp can be turned on except one.
  2. None of the lamps can be turned on.
  3. Each lamp is too dim and cannot be brightened by adjusting the rheostat.
  4. Each lamp is too dim; however, the amount of light can be varied with the rheostat, but not to full brightness.

SECTION 2–8 REVIEW

1. If the resistance of the rheostat in your lighting circuit is reduced, does the brightness of the selected lamp increase or decrease?
2. Does removing any one of the lamps affect the operation of any of the other lamps?
3. What is the current from the battery when no lamps are on?
4. What will happen if you try to turn on two lamps at the same time? Explain.

■ SUMMARY

- An atom is the smallest particle of an element that retains the characteristics of that element.
- The electron is the basic particle of negative electrical charge.
- The proton is the basic particle of positive charge.
- An ion is an atom that has gained or lost an electron and is no longer neutral.
- When electrons in the outer orbit of an atom (valence electrons) break away, they become free electrons.
- Free electrons make current possible.
- Like charges repel each other, and opposite charges attract each other.
- Voltage must be applied to a circuit before current can flow.
- Resistance limits the current.
- Basically, an electric circuit consists of a source, a load, and a current path.
- An open circuit is one in which the current path is broken.
- A closed circuit is one which has a complete current path.
- An ammeter is connected in line with the current path.
- A voltmeter is connected across the current path.
- An ohmmeter is connected across a resistor after removal from circuit.
- Figure 2–44 shows the electrical symbols introduced in this chapter.

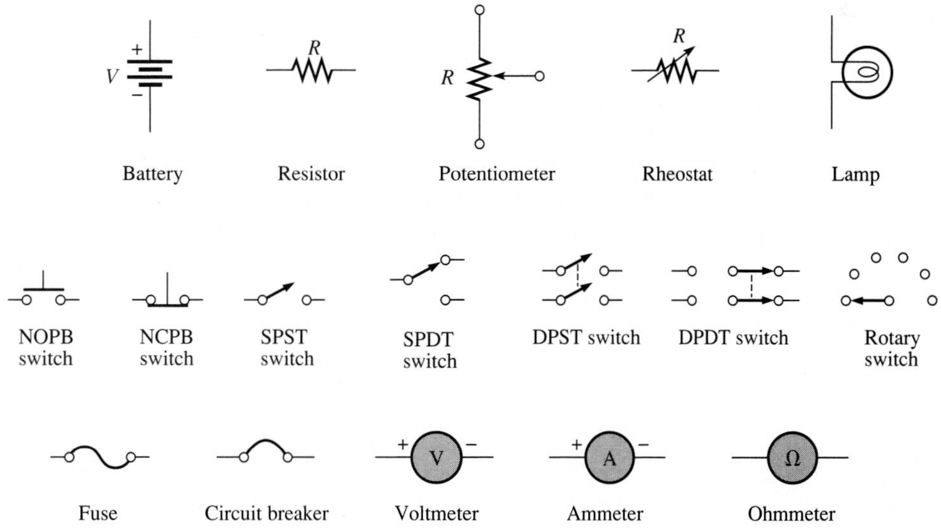

FIGURE 2–44

■ GLOSSARY

Ammeter An electrical instrument used to measure current.

Ampere (A) The unit of electrical current.

Atom The smallest particle of an element possessing the unique characteristics of that element.

Atomic number The number of electrons in a neutral atom.

Atomic weight The number of protons and neutrons in the nucleus of an atom.

AWG American wire gage. A standardization based on wire diameter.

Battery An energy source that uses a chemical reaction to convert chemical energy into electrical energy.

Charge An electrical property of matter that exists because of an excess or a deficiency of electrons. Charge can be either positive or negative.

Circuit An interconnection of electrical components designed to produce a desired result. A basic circuit consists of a source, a load, and an interconnecting current path.

Circuit breaker A resettable protective device used for interrupting excessive current in an electric circuit.

Circular mil (CM) The unit of the cross-sectional area of a wire.

Closed circuit A circuit with a complete current path.

Conductor A material in which electrical current can flow with relative ease. An example is copper.

Coulomb (C) The unit of electrical charge.

Current The rate of flow of electrons.

Electrical Related to the use of electrical voltage and current to achieve desired results.

Electron The basic particle of electrical charge in matter. The electron possesses negative charge.

Electronic Related to the movement and control of free electrons in semiconductors or vacuum devices.

Element One of the unique substances that make up the known universe. Each element is characterized by a unique atomic structure.

Energy The ability to do work.

Free electron A valence electron that has broken away from its parent atom and is free to move from atom to atom within the atomic structure of a material.

Fuse A protective device that burns open when excessive current flows in a circuit.

Generator An energy source that produces electrical signals.

Insulator A material that does not allow current under normal conditions.

Joule (J) The unit of energy.

Load The device in a circuit upon which work is done.

Multimeter An instrument that measures voltage, current, and resistance.

Neutron An atomic particle having no electrical charge.

Ohm (Ω) The unit of resistance.

Ohmmeter An instrument for measuring resistance.

Open circuit A circuit in which there is not a complete current path.

Photoconductive cell A type of variable resistor that is light-sensitive.

Potentiometer A three-terminal variable resistor.

Power supply An electronic instrument that produces voltage, current, and power from the ac power line or batteries in a form suitable for use in various applications to power electronic equipment.

Proton A positively charged atomic particle.

Resistance Opposition to current. The unit is the ohm (Ω).

Resistor An electrical component designed specifically to provide resistance.

Resolution The smallest increment of a quantity that a meter can measure.

Rheostat A two-terminal variable resistor.

Schematic A symbolized diagram of an electrical or electronic circuit.

Semiconductor A material that has a conductance value between that of a conductor and that of an insulator. Silicon and germanium are examples.

Shell The orbit in which an electron revolves.

Source Any device that produces energy.

Switch An electrical device for opening and closing a current path.

Tapered Nonlinear, such as a tapered potentiometer.

Thermistor A type of variable resistor that is temperature-sensitive.

Tolerance The limits of variation in the value of a component.

Valence Related to the outer shell or orbit of an atom.

Valence electron An electron that is present in the outermost shell of an atom.

Volt The unit of voltage or electromotive force.

Voltage The amount of energy available to move a certain number of electrons from one point to another in an electrical circuit.

Voltmeter An instrument used to measure voltage.

Wiper The sliding contact in a potentiometer.

▪ FORMULAS

(2–1)	$Q = \dfrac{\text{number of electrons}}{6.25 \times 10^{18}\ \text{electrons/Coulomb}}$	Charge
(2–2)	$V = \dfrac{W}{Q}$	Voltage equals energy divided by charge.
(2–3)	$I = \dfrac{Q}{t}$	Current equals charge divided by time.
(2–4)	$A = d^2$	Cross-sectional area equals the diameter squared.
(2–5)	$R = \dfrac{\rho l}{A}$	Resistance is resistivity times length divided by cross-sectional area.

▪ SELF-TEST

1. A neutral atom with an atomic number of three has how many electrons?
   (a) 1 **(b)** 3 **(c)** none **(d)** depends on the type of atom
2. Electron orbits are called
   (a) shells **(b)** nuclei **(c)** waves **(d)** valences
3. Materials in which no current can flow are called
   (a) filters **(b)** conductors **(c)** insulators **(d)** semiconductors
4. When placed close together, a positively charged material and a negatively charged material will
   (a) repel **(b)** become neutral **(c)** attract **(d)** exchange charges
5. The charge on a single electron is
   (a) 6.25×10^{-18} C **(b)** 1.6×10^{-19} C **(c)** 1.6×10^{-19} J **(d)** 3.14×10^{-6} C
6. *Potential difference* is another term for
   (a) energy **(b)** voltage **(c)** distance of an electron from the nucleus **(d)** charge
7. The unit of energy is the
   (a) watt **(b)** coulomb **(c)** joule **(d)** volt
8. Which one of the following is not a type of energy source?
   (a) battery **(b)** solar cell **(c)** generator **(d)** potentiometer
9. Which one of the following is not a possible condition in an electric circuit?
   (a) voltage and no current **(b)** current and no voltage
   (c) voltage and current **(d)** no voltage and no current
10. Electrical current is defined as
    (a) free electrons **(b)** the rate of flow of free electrons
    (c) the energy required to move electrons **(d)** the charge on free electrons
11. Current will not flow in a circuit when
    (a) a switch is closed **(b)** a switch is open **(c)** there is no voltage
    (d) both (a) and (c) **(e)** both (b) and (c)
12. The primary purpose of a resistor is to
    (a) increase current **(b)** limit current **(c)** produce heat **(d)** resist current change
13. Potentiometers and rheostats are types of
    (a) voltage sources **(b)** variable resistors **(c)** fixed resistors **(d)** circuit breakers
14. The current in a given circuit is not to exceed 22 A. Which value of fuse is best?
    (a) 10 A **(b)** 25 A **(c)** 20 A **(d)** a fuse is not necessary

■ BASIC PROBLEMS

SECTION 2–2 Electrical Charge

1. How many coulombs of charge do 50×10^{31} electrons possess?
2. How many electrons does it take to make 80 μC of charge?

SECTION 2–3 Voltage

3. Determine the voltage in each of the following cases:
   (a) 10 J/C **(b)** 5 J/2 C **(c)** 100 J/25 C
4. Five hundred joules of energy are used to move 100 C of charge through a resistor. What is the voltage across the resistor?
5. What is the voltage of a battery that uses 800 J of energy to move 40 C of charge through a resistor?
6. How much energy does a 12 V battery in your car use to move 2.5 C through the electrical circuit?

SECTION 2–4 Current

7. Determine the current in each of the following cases:
   (a) 75 C in 1 s **(b)** 10 C in 0.5 s **(c)** 5 C in 2 s
8. Six-tenths coulomb passes a point in 3 s. What is the current in amperes?
9. How long does it take 10 C to flow past a point if the current is 5 A?
10. How many coulombs pass a point in 0.1 s when the current is 1.5 A?

SECTION 2–5 Resistance

11. Figure 2–45(a) shows color-coded resistors. Determine the resistance value and the tolerance of each.
12. Find the minimum and the maximum resistance within the tolerance limits for each resistor in Figure 2–45(a).
13. **(a)** If you need a 270 Ω resistor, what color bands should you look for?
    (b) From the selection of resistors in Figure 2–45(b), choose the following values: 330 Ω, 2.2 kΩ, 56 kΩ, 100 kΩ, and 39 kΩ.

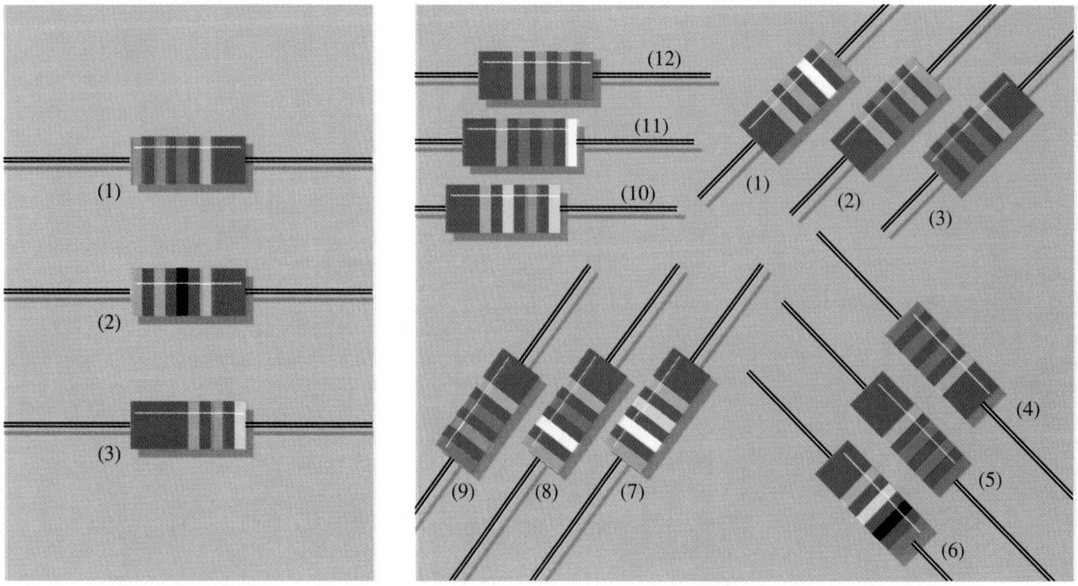

(a) (b)

FIGURE 2–45

14. The adjustable contact of a linear potentiometer is set at the mechanical center of its adjustment. If the total resistance is 1000 Ω, what is the resistance between each end terminal and the adjustable contact?

SECTION 2–6 The Electric Circuit

15. Trace the current path in the lamp circuit of Figure 2–27(a) with the switch in position 2.

16. With the switch in either position, redraw the lamp circuit in Figure 2–27(b) with a fuse connected to protect the circuit against excessive current.

SECTION 2–7 Basic Circuit Measurements

17. Show the placement of an ammeter and a voltmeter to measure the current and the source voltage in Figure 2–46.

18. Show how you would measure the resistance of R_2 in Figure 2–46.

19. In Figure 2–47, how much voltage does each meter indicate when the switch (SW) is in position 1? In position 2?

20. In Figure 2–47, show how to connect an ammeter to measure the current from the voltage source regardless of the switch (SW) position.

21. What are the voltage readings of the multimeters in Figures 2–48(a) and 2–48(b)?

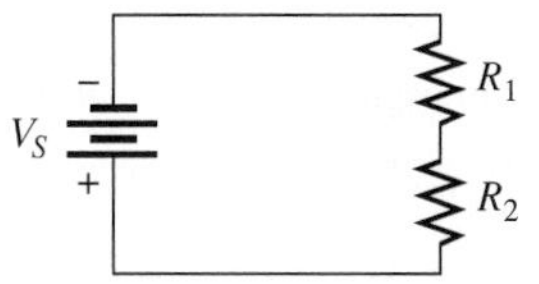

FIGURE 2–46

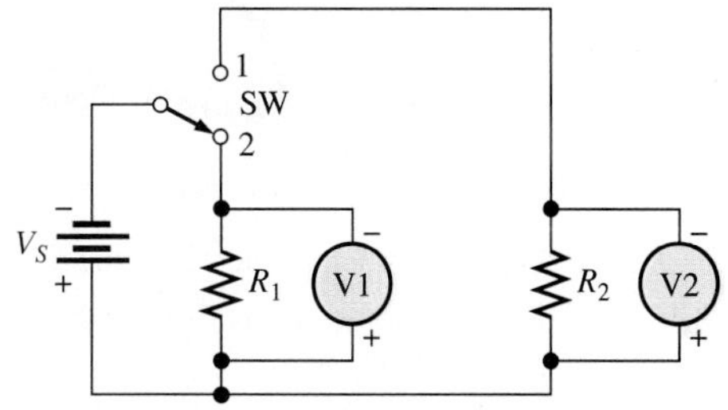

FIGURE 2–47

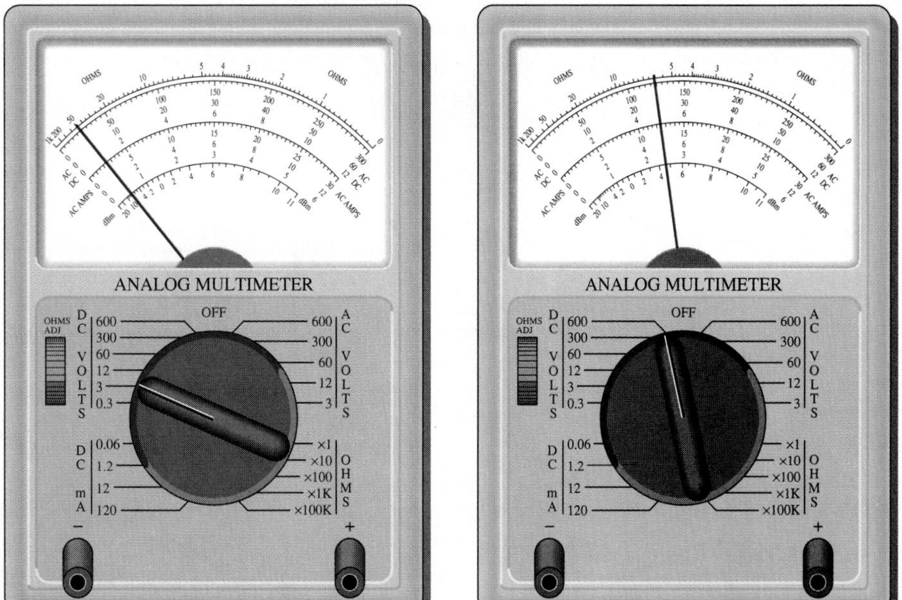

(a) (b)

FIGURE 2–48

FIGURE 2–49

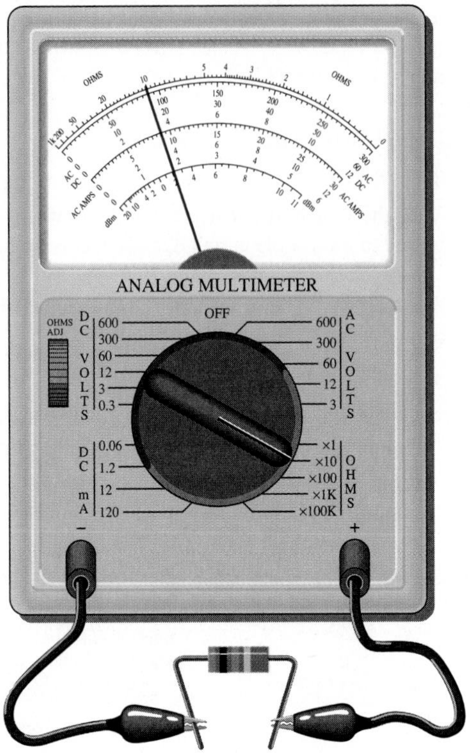

22. How much resistance is the multimeter in Figure 2–49 measuring?
23. Determine the resistance indicated by each of the following ohmmeter readings and range settings:
    (a) pointer at 2, range setting at R ×100
    (b) pointer at 15, range setting at R ×10M
    (c) pointer at 45, range setting at R ×100
24. A multimeter has the following ranges: 1 mA, 10 mA; 100 mA, 100 mV, 1 V, 10 V; R ×1, R ×10, R ×100. Indicate schematically how you would connect the multimeter in Figure 2–50 to measure the following quantities:

    (a) I_1 **(b)** V_1 **(c)** R_1

    In each case indicate the *function* to which you would set the meter and the *range* that you would use.

FIGURE 2–50

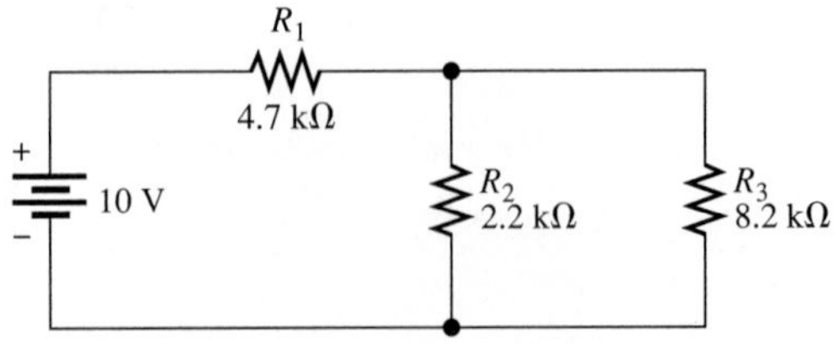

▪ ADVANCED PROBLEMS

25. A resistor with a current of 2 A through it in an amplifier circuit converts 1000 J of electrical energy to heat energy in 15 s. What is the voltage across the resistor?
26. If 574×10^{15} electrons flow through a speaker wire in 250 ms, what is the current in amperes?
27. A 120 V source is to be connected to a 1500 Ω resistive load by two lengths of wire as shown in Figure 2–51. The voltage source is to be located 50 ft from the load. Using Table 2–3, determine the gage number of the *smallest* wire that can be used if the total resistance of the two lengths of wire is not to exceed 6 Ω.

FIGURE 2–51

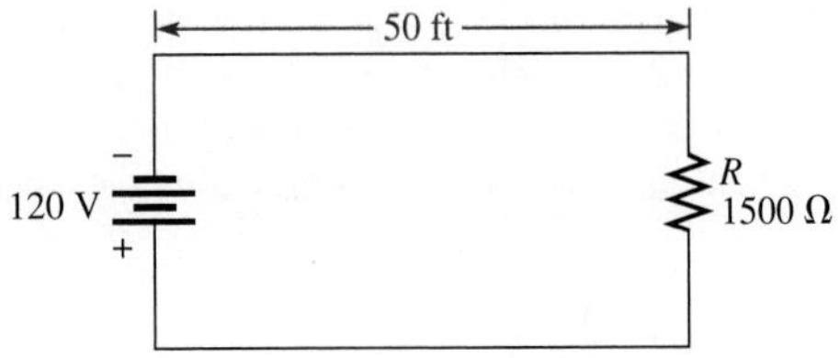

28. Determine the resistance and tolerance of each resistor labeled as follows:
(a) 4R7J **(b)** 5602M **(c)** 1501F

29. There is only one circuit in Figure 2–52 in which it is possible to have all lamps on at the same time. Determine which circuit it is.

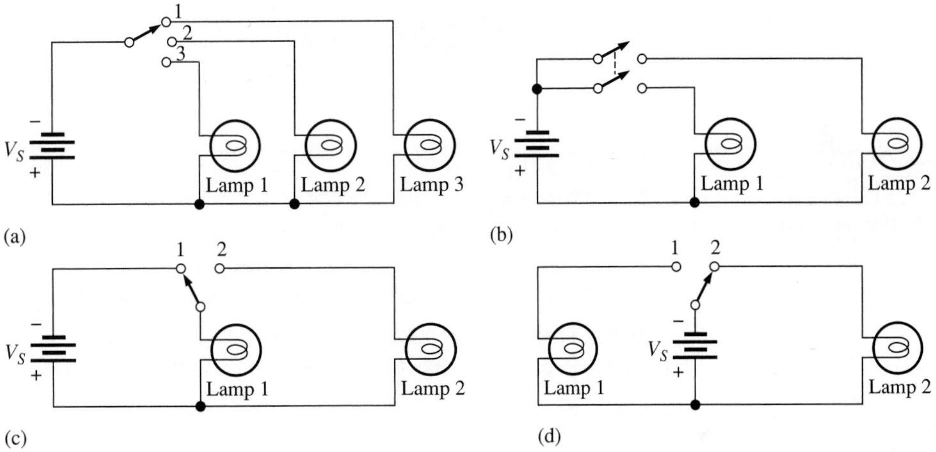

FIGURE 2–52

30. Through which resistor in Figure 2–53 does current always flow, regardless of the position of the switches?

31. In Figure 2–53, show the proper placement of ammeters to measure the current through each resistor and the current out of the battery.

32. Show the proper placement of voltmeters to measure the voltage across each resistor in Figure 2–53.

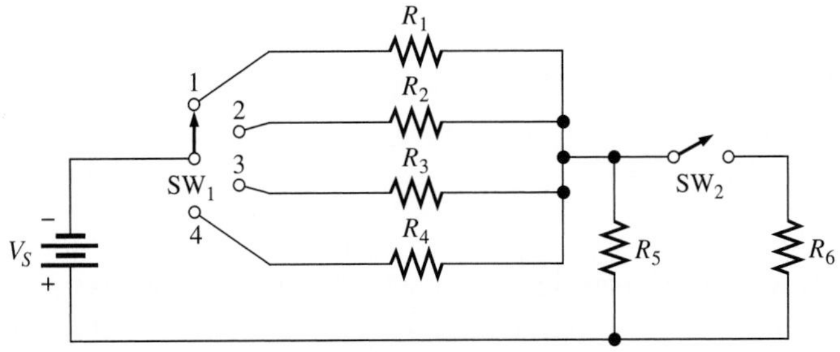

FIGURE 2–53

33. Devise a switch arrangement whereby two voltage sources (V_{S1} and V_{S2}) can be connected simultaneously to either of two resistors (R_1 and R_2) as follows:

V_{S1} connected to R_1 and V_{S2} connected to R_2

or V_{S1} connected to R_2 and V_{S2} connected to R_1

34. The different sections of a stereo system are represented by the blocks in Figure 2–54. Show how a single switch can be used to connect the phonograph, the CD (compact disk) player, the tape deck, the AM tuner, or the FM tuner to the amplifier by a single knob control. Only one section can be connected to the amplifier at any time.

FIGURE 2–54

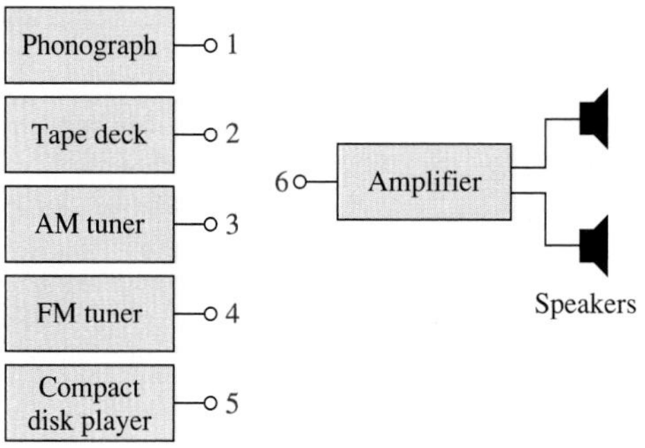

■ ANSWERS TO SECTION REVIEWS

Section 2–1

1. The basic particle of negative charge is the electron.
2. An atom is the smallest particle of an element that retains the unique characteristics of the element.
3. An atom consists of a positively charged nucleus surrounded by orbiting electrons.
4. No, all elements do not have the same type of atoms.
5. A free electron is an outer-shell electron that has drifted away from the parent atom.

Section 2–2

1. Q is the symbol for charge.
2. Coulomb is the unit of charge and C is the symbol for Coulomb.
3. The two types of charge are positive and negative.
4. $Q = \dfrac{10 \times 10^{12} \text{ electrons}}{6.25 \times 10^{18} \text{ electrons/C}} = 1.6 \times 10^{-6}\text{C}.$

Section 2–3

1. Voltage is energy per unit charge.
2. The unit of voltage is the volt.
3. $V = 24 \text{ J}/10 \text{ C} = 2.4 \text{ V}$
4. Sources of voltage are the battery, power supply, solar cell, and generator.

Section 2–4

1. Current is the rate of flow of electrons; the unit of current is the ampere (A).
2. There are 6.25×10^{18} electrons in one coulomb.
3. $I = 20 \text{ C}/4 \text{ s} = 5 \text{ A}$

Section 2–5

1. Resistance is opposition to current and its unit is the ohm (Ω).
2. Two resistor categories are fixed and variable. The value of a fixed resistor cannot be changed, but that of a variable resistor can.
3. *First band:* first digit of resistance value. *Second band:* second digit of resistance value. *Third band:* number of zeros following the 2nd digit. *Fourth band:* tolerance.
4. **(a)** Yellow, violet, red, gold = 4700 Ω ±5%
   (b) Blue, red, orange, silver = 62,000 Ω ±10%
   (c) Brown, gray, black, gold = 18 Ω ±5%
5. A rheostat has two terminals; a potentiometer has three terminals.
6. A thermistor is a temperature-sensitive resistor.

Section 2–6

1. A basic electric circuit consists of source, load, and current path between source and load.
2. An open circuit is one that has no path for current.
3. A closed circuit is one that has a complete path for current.
4. A fuse protects a circuit against excessive current.
5. A fuse must be replaced once blown. A circuit breaker can be reset once tripped.
6. AWG #3 is larger than AWG #22.

Section 2–7

1. **(a)** Ammeter measures current. **(b)** Voltmeter measures voltage. **(c)** Ohmmeter measures resistance.
2. See Figure 2–55.
3. See Figure 2–56.

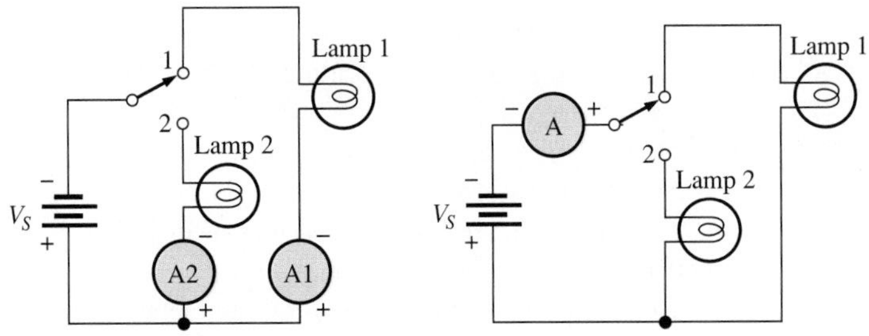

(a) Two ammeters (b) One ammeter

FIGURE 2–55

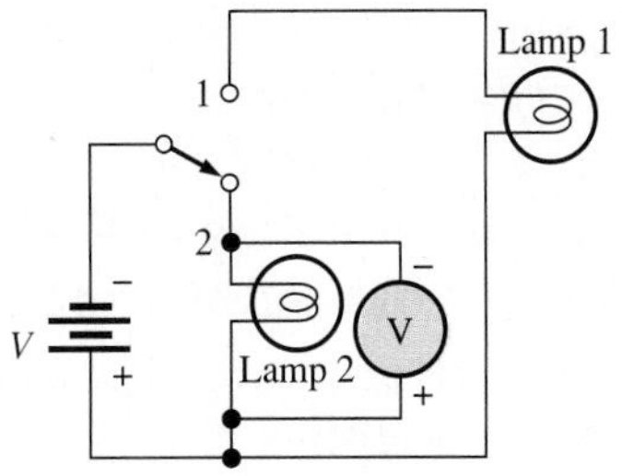

FIGURE 2–56

4. The voltage being measured is 1.5 V.
5. DC volts, 300 setting; 275 V is read on the 300 scale.
6. OHMS ×1000
7. Two types of DMM displays are LED and LCD. The LCD requires little current, but it is difficult to see in low light and is slow to respond. The LED can be seen in the dark, and it responds quickly; however, it requires much more current than does the LCD.
8. Resolution is the smallest increment that the meter can measure.

Section 2–8

1. The brightness increases.
2. No
3. There is no current when all lights are off.
4. The total current would increase, possibly causing the fuse to blow.

■ ANSWERS TO RELATED EXERCISES FOR EXAMPLES

2–1 1.88×10^{19} electrons

2–2 600 J

2–3 12 C

2–4 4.7 kΩ ± 20%

2–5 404.01 CM; #24

2–6 16.14 Ω

2–7 The needle will move left to the “70” mark on the top scale.

3

OHM'S LAW, ENERGY, AND POWER

Georg Simon Ohm (1787–1854) experimentally found that voltage, current, and resistance are all related in a specific way. This basic relationship, known as *Ohm's law,* is one of the most fundamental and important laws in the fields of electricity and electronics. In this chapter, Ohm's law is examined, and its use in practical circuit applications is discussed and demonstrated by numerous examples.

In addition to Ohm's law, the concepts and definitions of energy and power in electric circuits are introduced and the Watt's law power formulas are given.

■ CHAPTER OBJECTIVES

- ☐ Explain Ohm's law
- ☐ Use Ohm's law to determine voltage, current, or resistance
- ☐ Define *energy* and *power*
- ☐ Calculate power in a circuit
- ☐ Properly select resistors based on power rating
- ☐ Explain energy loss and voltage drop
- ☐ Discuss power supplies and their characteristics

■ APPLICATION ASSIGNMENT

As a technician working in the engineering development lab of an electronics company, you are assigned the task of modifying an existing test fixture for use in a new application. The test fixture is a resistor box with an array of switch-selectable resistors of various values. Your job will be to determine and specify the changes to be made in the existing circuit and develop a procedure to test the box once the modifications have been made. To complete the assignment, you will

1. Determine color-coded resistor values.
2. Determine resistor power ratings.
3. Draw the schematic of the existing circuit.
4. Draw the schematic to meet the new requirements.
5. Use Watt's law power formulas to determine required power ratings.
6. Use Ohm's law to determine resistor values to meet specifications.

After studying this chapter, you should be able to complete the application assignment in Section 3–8.

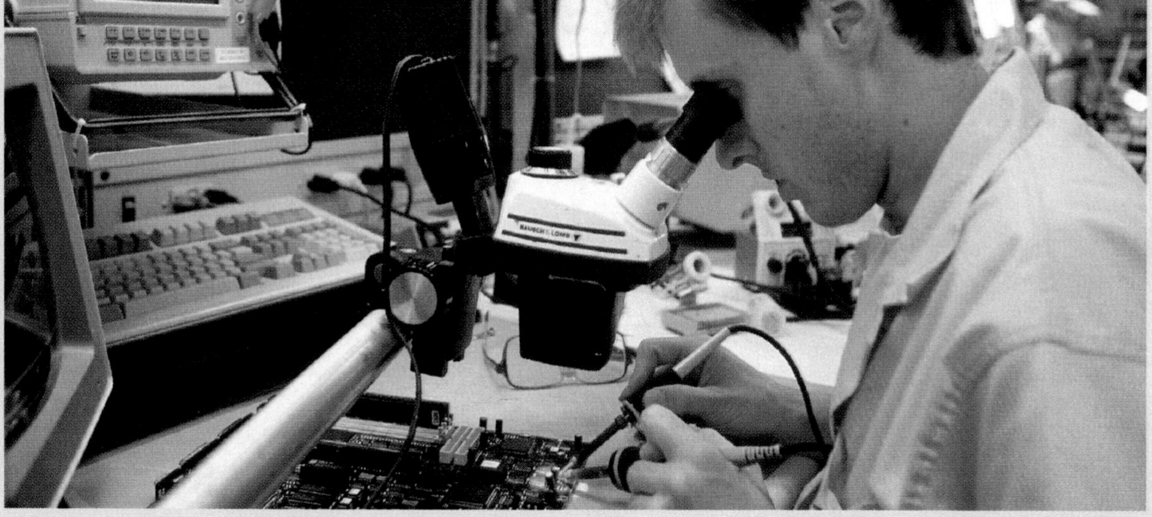

3–1 ■ OHM'S LAW

Ohm's law describes how voltage, current, and resistance in a circuit are related. Ohm's law is used in three equivalent forms depending on which quantity you need to determine. In this section, you will learn each of these forms.

After completing this section, you should be able to

- **Explain Ohm's law**
  - □ Describe how *V*, *I*, and *R* are related
  - □ Express *I* as a function of *V* and *R*
  - □ Express *V* as a function of *I* and *R*
  - □ Express *R* as a function of *V* and *I*

Ohm determined experimentally that if the voltage across a resistor is increased, the current through the resistor will also increase; and, likewise, if the voltage is decreased, the current will decrease. For example, if the voltage is doubled, the current will double. If the voltage is halved, the current will also be halved. This relationship is illustrated in Figure 3–1, with meter indications of voltage and current.

Ohm's law also states that if the voltage is kept constant, less resistance results in more current, and, also, more resistance results in less current. For example, if the resistance is halved, the current doubles. If the resistance is doubled, the current is halved. This concept is illustrated by the meter indications in Figure 3–2, where the resistance is increased and the voltage is held constant.

FIGURE 3–1
Effect on the current of changing the voltage with the resistance at a constant value.

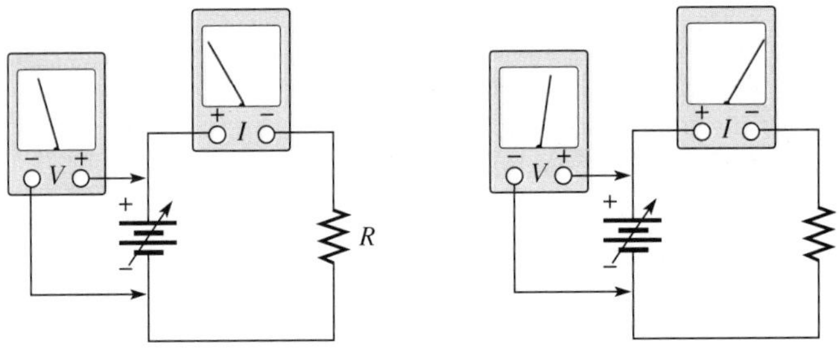

(a) Less *V*, less *I*   (b) More *V*, more *I*

FIGURE 3–2
Effect on the current of changing the resistance with the voltage at a constant value.

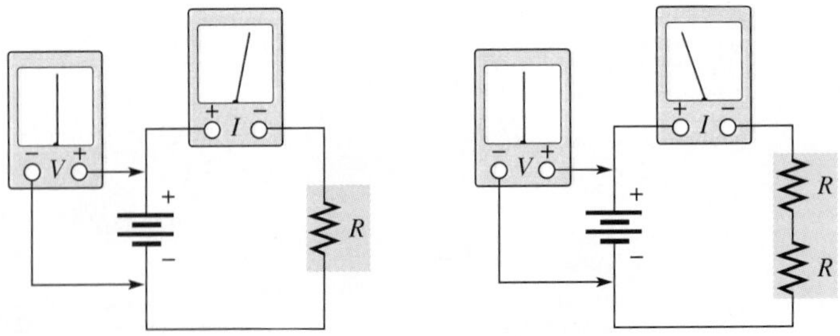

(a) Less *R*, more *I*   (b) More *R*, less *I*

Formula for Current

Ohm's law can be stated as follows:

$$I = \frac{V}{R} \tag{3–1}$$

Equation (3–1) states the relationship indicated by the action in the circuits of Figures 3–1 and 3–2.

For a constant resistance, if the voltage in a circuit is increased, more current will flow; and if the voltage is decreased, less current will flow.

$I = V / R$ — Increase V, I increases

$I = V / R$ — Decrease V, I decreases

R constant

For a constant voltage, if the resistance in a circuit is increased, less current will flow; and if the resistance is decreased, more current will flow.

$I = V / R$ — Increase R, I decreases

$I = V / R$ — Decrease R, I increases

V constant

Using Equation (3–1), you can calculate the current if the values of voltage and resistance are known.

EXAMPLE 3–1 Using the Ohm's law formula in Equation (3–1), verify that the current through a 10 Ω resistor increases when the voltage is increased from 5 V to 20 V.

Solution The following calculations show that the current increases from 0.5 A to 2 A. The calculator sequence for each calculation is also shown.

For $V = 5$ V,

$$I = \frac{V}{R} = \frac{5\text{ V}}{10\ \Omega} = 0.5\text{ A}$$

5 ÷ 1 0 =

For $V = 20$ V,

$$I = \frac{V}{R} = \frac{20\text{ V}}{10\ \Omega} = 2\text{ A}$$

2 0 ÷ 1 0 =

Related Exercise Show that the current decreases when the resistance is increased from 5 Ω to 20 Ω and the voltage is a constant 10 V.

Formula for Voltage

Ohm's law can also be stated another way. By multiplying both sides of Equation (3–1) by R and transposing terms, you can obtain an equivalent form of Ohm's law, as follows:

$$V = IR \tag{3–2}$$

With this equation, you can calculate voltage if the current and resistance are known.

EXAMPLE 3–2 Use the Ohm's law formula in Equation (3–2) to calculate the voltage across a 100 Ω resistor when the current is 2 A.

Solution $V = IR = (2\text{ A})(100\ \Omega) = 200\text{ V}$ 2 × 1 0 0 =

Related Exercise Find the voltage across a 1 kΩ resistor when the current is 1 mA.

Formula for Resistance

There is a third equivalent way to state Ohm's law. By dividing both sides of Equation (3–2) by I and transposing terms, you obtain the following formula:

$$R = \frac{V}{I} \tag{3–3}$$

This form of Ohm's law is used to determine resistance if voltage and current values are known.

The three formulas—Equations (3–1), (3–2), and (3–3)—are all equivalent. They are simply three different ways of expressing Ohm's law.

EXAMPLE 3–3 Use the Ohm's law formula in Equation (3–3) to calculate the resistance in a circuit when the voltage is 12 V and the current is 0.5 A.

Solution $R = \frac{V}{I} = \frac{12\text{ V}}{0.5\text{ A}} = 24\ \Omega$ 1 2 ÷ 0 • 5 =

Related Exercise Find the resistance when the voltage is 9 V and the current is 10 mA.

The Linear Relationship of Current and Voltage

When we say that the current and voltage are **linearly** proportional, we mean that if one is increased or decreased by a certain percentage, the other will increase or decrease by the same percentage, assuming that the resistance is constant in value. For example, if the voltage across a resistor is tripled, the current will triple.

EXAMPLE 3–4 Show that if the voltage in the circuit of Figure 3–3 is increased to three times its present value, the current will triple in value.

FIGURE 3–3

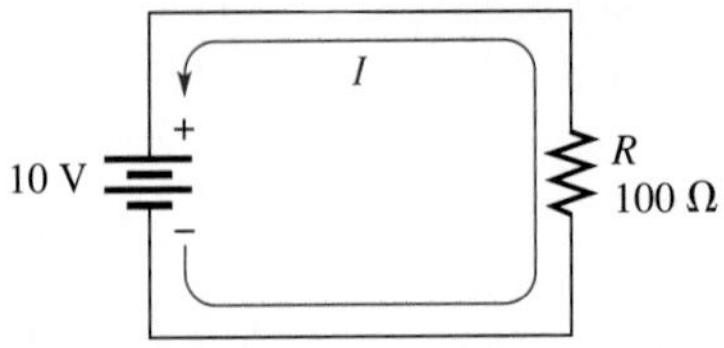

Solution With 10 V, the current is

$$I = \frac{V}{R} = \frac{10\ \text{V}}{100\ \Omega} = 0.1\ \text{A}$$

If the voltage is increased to 30 V, the current will be

$$I = \frac{V}{R} = \frac{30\ \text{V}}{100\ \Omega} = 0.3\ \text{A}$$

The current went from 0.1 A to 0.3 A when the voltage was tripled to 30 V.

Related Exercise Show that if the voltage in Figure 3–3 is quadrupled, the current will also quadruple?

A Graph of Current Versus Voltage

Let's take a constant value of resistance, for example, 10 Ω, and calculate the current for several values of voltage ranging from 10 V to 100 V. The current values obtained are shown in Figure 3–4(a). The graph of the I values versus the V values is shown in Figure 3–4(b). Note that it is a straight line graph. This graph shows that a change in voltage results in a linearly proportional change in current. No matter what value R is, assuming that R is constant, the graph of I versus V will always be a straight line.

FIGURE 3–4
Graph of current versus voltage for R = 10 Ω.

V	I
10 V	1 A
20 V	2 A
30 V	3 A
40 V	4 A
50 V	5 A
60 V	6 A
70 V	7 A
80 V	8 A
90 V	9 A
100 V	10 A

$$I = \frac{V}{10\ \Omega}$$

(a)

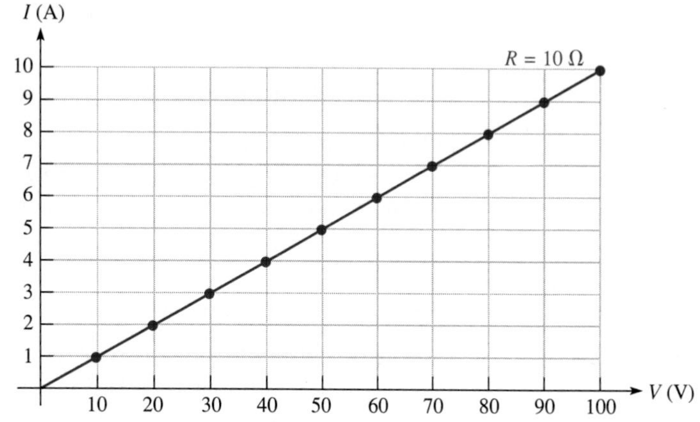

(b)

A Graphic Aid for Ohm's Law

You may find the graphic aid in Figure 3–5 of value for applying any of the three forms of Ohm's law. It is a way to help you remember the formulas.

FIGURE 3–5
A graphic aid for the Ohm's law formulas.

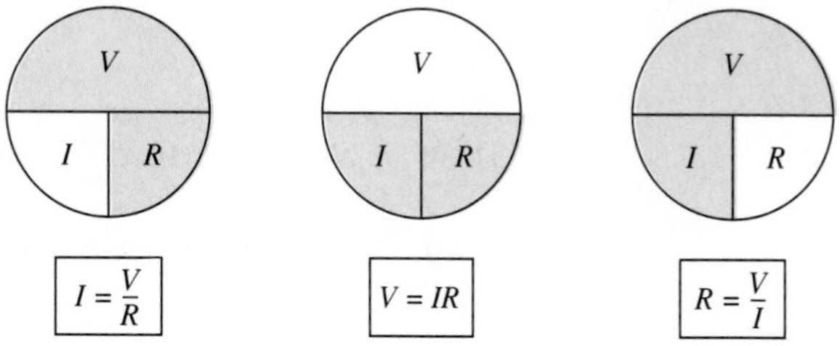

SECTION 3–1 REVIEW

1. Briefly state Ohm's law in words.
2. Write the Ohm's law formula for calculating current.
3. Write the Ohm's law formula for calculating voltage.
4. Write the Ohm's law formula for calculating resistance.
5. If the voltage across a resistor is tripled, does the current increase or decrease? By how much?
6. There is a fixed voltage across a variable resistor, and you measure a current of 10 mA. If you double the resistance, how much current will you measure?
7. Will you observe any change in current in a circuit where both the voltage and the resistance are doubled?

3–2 ■ APPLICATION OF OHM'S LAW

This section provides examples of the application of Ohm's law for calculating voltage, current, and resistance in electric circuits. You will also see how to use quantities expressed with metric prefixes in circuit calculations.

After completing this section, you should be able to

- **Use Ohm's law to determine voltage, current, or resistance**
  - ☐ Use Ohm's law to find current when voltage and resistance are known
  - ☐ Use Ohm's law to find voltage when current and resistance are known
  - ☐ Use Ohm's law to find resistance when voltage and current are known
  - ☐ Use quantities with metric prefixes

How to Determine *I* When *V* and *R* Are Known

In these examples you will learn to determine current values when you know the values of voltage and resistance. In these problems, the formula $I = V/R$ is used. In order to get current in amperes, you must express the value of V in volts and the value of R in ohms.

EXAMPLE 3–5 How many amperes of current are in the circuit of Figure 3–6?

FIGURE 3–6

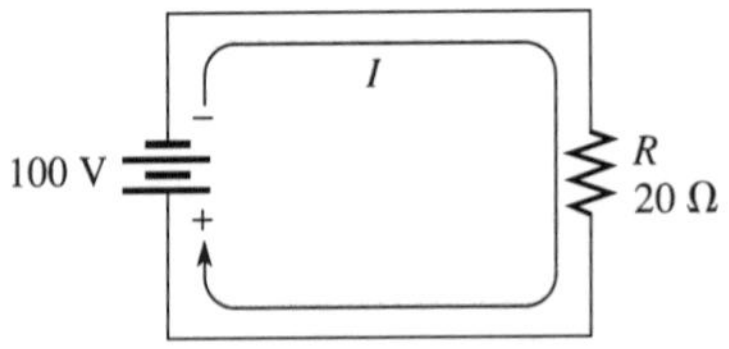

Solution Use the formula $I = V/R$ and substitute 100 V for V and 20 Ω for R:

$$I = \frac{V}{R} = \frac{100\text{ V}}{20\ \Omega} = 5\text{ A}$$

There are 5 A of current in this circuit.

Related Exercise What is the current in Figure 3–6 if the voltage is reduced to 50 V?

In electronics, resistance values of thousands or millions of **ohms** are common. As you learned in Chapter 1, large values of resistance are indicated by the metric system prefixes *kilo* (k) and *mega* (M). Thus, thousands of ohms are expressed in *kilohms* (kΩ), and millions of ohms in *megohms* (MΩ). The following examples illustrate how to use kilohms and megohms when you use Ohm's law to calculate current.

EXAMPLE 3–6 Calculate the current in the circuit of Figure 3–7.

FIGURE 3–7

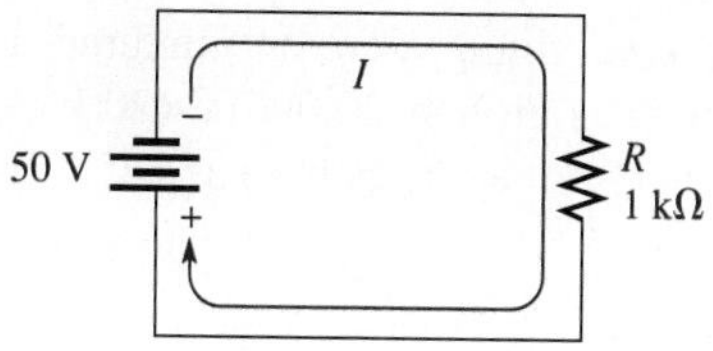

Solution Remember that 1 kΩ is the same as 1×10^3 Ω. Use the formula $I = V/R$ and substitute 50 V for V and 1×10^3 Ω for R:

$$I = \frac{V}{R} = \frac{50\text{ V}}{1\text{ k}\Omega} = \frac{50\text{ V}}{1 \times 10^3\ \Omega} = 50 \times 10^{-3}\text{ A} = 50\text{ mA}$$

The calculator sequence is

5 0 ÷ 1 EXP 3 =

Related Exercise If the resistance in Figure 3–7 is increased to 10 kΩ, what is the current?

In Example 3–6, the current is expressed as 50 mA. Thus, when volts are divided by kilohms, the current will be in milliamperes.

If volts are used when resistance values are in megohms, the current is in microamperes (μA) as Example 3–7 illustrates.

EXAMPLE 3–7 Determine the amount of current in the circuit of Figure 3–8.

FIGURE 3–8

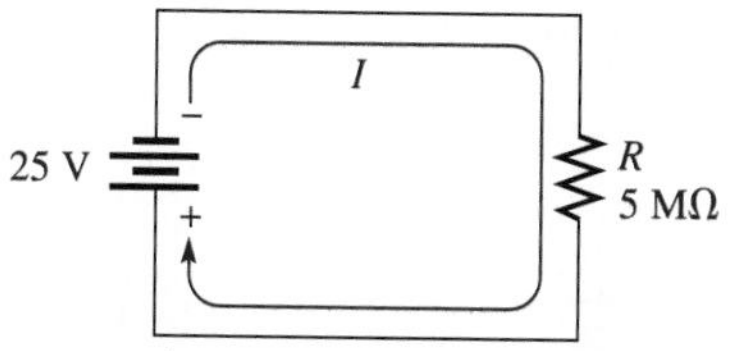

Solution Recall that 5 MΩ equals 5×10^6 Ω. Use the formula $I = V/R$ and substitute 25 V for V and 5×10^6 Ω for R:

$$I = \frac{V}{R} = \frac{25\text{ V}}{5\text{ M}\Omega} = \frac{25\text{ V}}{5 \times 10^6\ \Omega} = 5 \times 10^{-6}\text{ A} = 5\ \mu\text{A}$$

The calculator sequence is

[2] [5] [÷] [5] [EXP] [6] [=]

Related Exercise If the resistance in Figure 3–8 is decreased to 1 MΩ, what is the current?

Small voltages, usually less than 50 V, are common in electronic circuits. Occasionally, however, large voltages are encountered. For example, the high-voltage supply in a television receiver is around 20,000 V (20 kV), and transmission voltages generated by the power companies may be as high as 345,000 V (345 kV).

EXAMPLE 3–8 How much current is there through 100 MΩ when 50 kV are applied?

Solution In this case, divide 50 kV by 100 MΩ to get the current. Substitute 50×10^3 V for 50 kV and $100 \times 10^6\ \Omega$ for 100 MΩ:

$$I = \frac{V}{R} = \frac{50\text{ kV}}{100\text{ M}\Omega} = \frac{50 \times 10^3\text{ V}}{100 \times 10^6\ \Omega} = 0.5 \times 10^{-3}\text{ A} = 0.5\text{ mA}$$

The calculator sequence is

[5] [0] [EXP] [3] [÷] [1] [0] [0] [EXP] [6] [=]

Related Exercise How much current is there through 10 MΩ when 2 kV are applied?

How to Determine *V* When *I* and *R* Are Known

In these examples you will learn to determine voltage values when the current and resistance are known using the formula $V = IR$. To obtain voltage in volts, you must express the value of I in amperes and the value of R in ohms.

EXAMPLE 3–9 In the circuit of Figure 3–9, how much voltage is needed to produce 5 A of current?

FIGURE 3–9

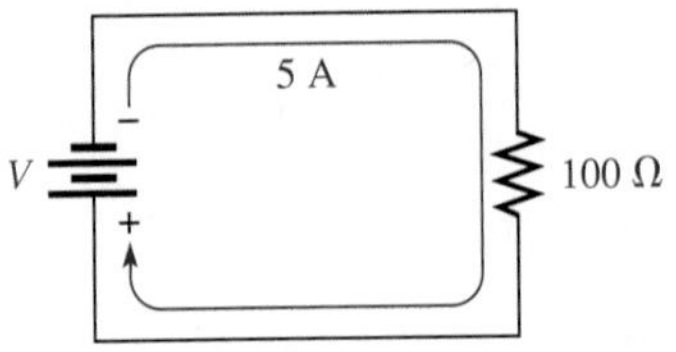

Solution Substitute 5 A for I and 100 Ω for R into the formula $V = IR$:

$$V = IR = (5\text{ A})(100\ \Omega) = 500\text{ V}$$

Thus, 500 V are required to produce 5 A of current through a 100 Ω resistor.

Related Exercise How much voltage is required to produce 8 A in the circuit of Figure 3–9?

EXAMPLE 3–10 How much voltage will be measured across the resistor in Figure 3–10?

FIGURE 3–10

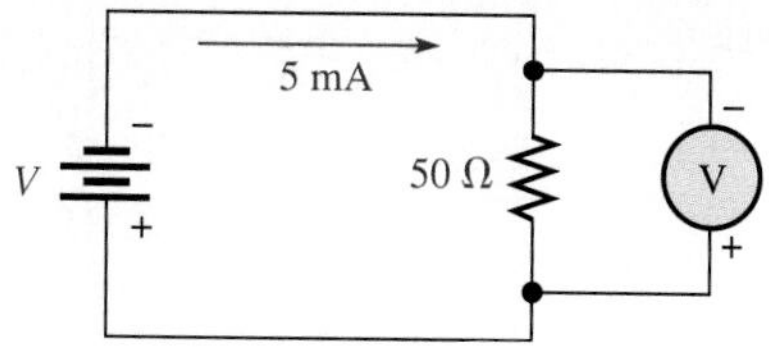

Solution Note that 5 mA equals 5×10^{-3} A. Substitute the values for I and R into the formula $V = IR$:

$$V = IR = (5\text{ mA})(50\ \Omega) = (5 \times 10^{-3}\text{ A})(50\ \Omega) = 250\text{ mV}$$

When milliamperes are multiplied by ohms, the result is millivolts.

The calculator sequence is

[5] [EXP] [+/–] [3] [×] [5] [0] [=]

Related Exercise Change the resistor in Figure 3–10 to 22 Ω and determine the voltage required to produce 10 mA.

EXAMPLE 3–11 The circuit in Figure 3–11 has a current of 10 mA. What is the voltage?

FIGURE 3–11

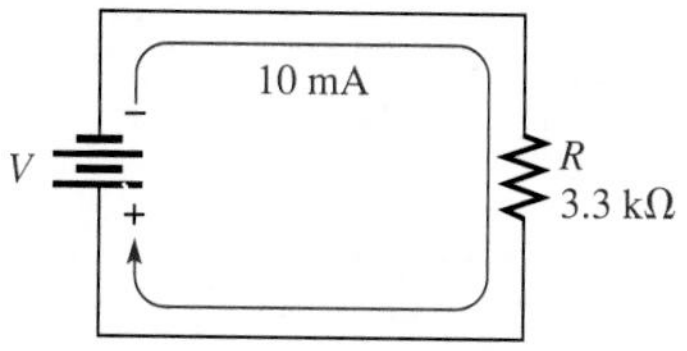

Solution Note that 10 mA equals 10×10^{-3} A and that 3.3 kΩ equals 3.3×10^{3} Ω. Substitute these values into the formula $V = IR$:

$$V = IR = (10\text{ mA})(3.3\text{ k}\Omega) = (10 \times 10^{-3}\text{ A})\ (3.3 \times 10^{3}\ \Omega) = 33\text{ V}$$

When milliamperes and kilohms are multiplied, the result is volts.

The calculator sequence is

[1] [0] [EXP] [+/–] [3] [×] [3] [•] [3] [EXP] [3] [=]

Related Exercise What is the voltage in Figure 3–11 if the current is 5 mA?

How to Determine *R* When *V* and *I* Are Known

In these examples you will learn to determine resistance values when the voltage and current are known using the formula $R = V/I$. To find resistance in ohms, you must express the value of V in volts and the value of I in amperes.

EXAMPLE 3–12 In the circuit of Figure 3–12 how much resistance is needed to draw 3 A of current from the battery?

FIGURE 3–12

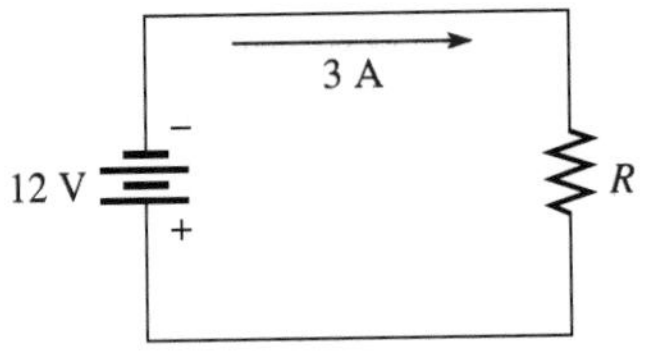

Solution Substitute 12 V for V and 3 A for I into the formula $R = V/I$:

$$R = \frac{V}{I} = \frac{12\text{ V}}{3\text{ A}} = 4\ \Omega$$

Related Exercise How much resistance is required to draw 3 mA from the battery in Figure 3–12?

EXAMPLE 3–13 Suppose that the ammeter in Figure 3–13 indicates 5 mA of current and the voltmeter reads 150 V. What is the value of *R*?

FIGURE 3–13

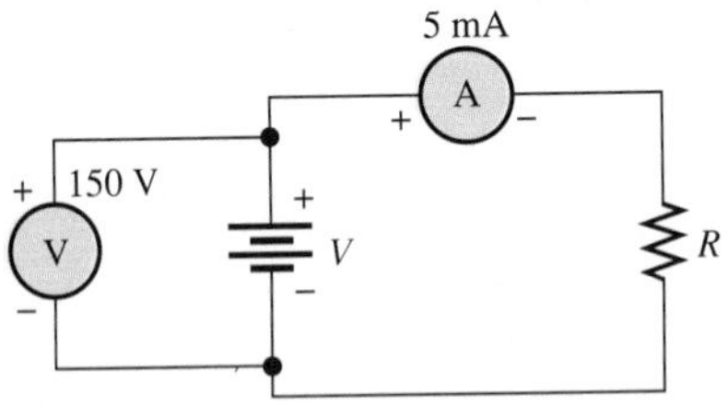

Solution Note that 5 mA equals 5×10^{-3} A. Substitute the voltage and current values into the formula $R = V/I$:

$$R = \frac{V}{I} = \frac{150\text{ V}}{5\text{ mA}} = \frac{150\text{ V}}{5 \times 10^{-3}\text{ A}} = 30 \times 10^{3}\ \Omega = 30\text{ k}\Omega$$

Thus, if volts are divided by milliamperes, the resistance will be in kilohms.

Related Exercise Determine the value of R if $V = 50$ V and $I = 500$ mA in Figure 3–13.

SECTION 3–2 REVIEW

1. $V = 10$ V and $R = 5\ \Omega$. Find I.
2. If a 5 MΩ resistor has 20 kV across it, how much current flows?
3. How much current will 10 kV across 2 kΩ produce?
4. $I = 1$ A and $R = 10\ \Omega$. Find V.
5. What voltage do you need to produce 3 mA of current in a 3 kΩ resistance?
6. A battery produces 2 A of current into a 6 Ω resistive load. What is the battery voltage?
7. $V = 10$ V and $I = 2$ A. Find R.
8. In a stereo amplifier circuit there is a resistor across which you measure 25 V, and your ammeter indicates 50 mA of current through the resistor. What is the resistor's value in kilohms? In ohms?

3–3 ■ ENERGY AND POWER

Energy is the fundamental ability to do work and produce action or results. When there is current through a resistance, energy is released in the form of heat. A common example of this is a light bulb that becomes too hot to touch. The current through the filament that produces light also produces unwanted heat because the filament has resistance. Power is a measure of how fast energy is being used; electrical components must be able to dissipate a certain amount of energy in a given period of time.

After completing this section, you should be able to

- **Define energy and power**
  - Express power in terms of energy
  - State the unit of power
  - State the common units of energy
  - Perform energy and power calculations

Power is the rate at which energy is used.

In other words, **power,** symbolized by P, is a certain amount of **energy** used in a certain length of time, expressed as follows:

$$\text{Power} = \frac{\text{energy}}{\text{time}}$$

$$P = \frac{W}{t} \qquad (3\text{–}4)$$

Energy is measured in **joules** (J), time is measured in seconds (s), and power is measured in **watts** (W). Energy in joules divided by time in seconds gives power in watts. For example, if 50 J of energy are used in 2 s, the power is 50 J/2 s = 25 W. By definition.

One watt is the amount of power when one joule of energy is consumed in one second.

Thus, the number of joules consumed in 1 s is always equal to the number of watts. For example, if 75 J are used in 1 s, the power is 75 W. The following example illustrates use of these units. To avoid confusion between the symbols for energy and watts, remember that an italic W stands for energy and a nonitalic W stands for watts.

EXAMPLE 3–14 An amount of energy equal to 100 J is used in 5 s. What is the power in watts?

Solution

$$P = \frac{\text{energy}}{\text{time}} = \frac{W}{t} = \frac{100\text{ J}}{5\text{ s}} = 20\text{ W}$$

Related Exercise If 100 W of power occurs for 30 s, how much energy in joules is consumed?

Amounts of power much less than one watt are common in certain areas of electronics. As with small current and voltage values, metric prefixes are used to designate small amounts of power. Thus, milliwatts (mW) and microwatts (μW) are commonly found in some applications.

In the electrical utilities field, kilowatts (kW) and megawatts (MW) are common units. Radio and television stations also use large amounts of power to transmit signals.

EXAMPLE 3–15 Express the following powers using appropriate metric prefixes:

(a) 0.045 W **(b)** 0.000012 W **(c)** 3500 W **(d)** 10,000,000 W

Solution

(a) $0.045\text{ W} = 45 \times 10^{-3}\text{ W} = 45\text{ mW}$

(b) $0.000012\text{ W} = 12 \times 10^{-6}\text{ W} = 12\ \mu\text{W}$

(c) $3500\text{ W} = 3.5 \times 10^{3}\text{ W} = 3.5\text{ kW}$

(d) $10{,}000{,}000\text{ W} = 10 \times 10^{6}\text{ W} = 10\text{ MW}$

Related Exercise Express the following amounts of power in watts without metric prefixes:

(a) 1 mW **(b)** 1800 μW **(c)** 3 MW **(d)** 10 kW

The Kilowatt-hour (kWh) Unit of Energy

Since power is the rate of energy usage, power utilized over a period of time represents energy consumption. If we multiply power and time, we have energy, symbolized by W.

$$\text{Energy} = \text{power} \times \text{time}$$

$$W = Pt \qquad \textbf{(3–5)}$$

Earlier, the joule was defined as a unit of energy. However, there is another way to express energy. Since power is expressed in watts and time in seconds, we can use units of energy called the watt-second (Ws), watt-hour (Wh), and kilowatt-hour (kWh).

When you pay your electric bill, you are charged on the basis of the amount of energy you use. Because power companies deal in huge amounts of energy, the most practical unit is the **kilowatt-hour.** *You use a kilowatt-hour of energy when you use the equivalent of 1000 W of power for 1 h.* For example, a 100 W light bulb burning for 10 h uses 1 kWh of energy.

EXAMPLE 3–16 Determine the number of kilowatt-hours for each of the following energy consumptions:

(a) 1400 W for 1 hr **(b)** 2500 W for 2 h **(c)** 100,000 W for 5 h

Solution

(a) 1400 W = 1.4 kW
Energy = power × time = (1.4 kW)(1 h) = 1.4 kWh

(b) 2500 W = 2.5 kW
Energy = (2.5 kW)(2 h) = 5 kWh

(c) 100,000 W = 100 kW
Energy = (100 kW)(5 h) = 500 kWh

Related Exercise How many kilowatt-hours of energy are used by a 250 W light bulb burning for 8 h?

SECTION 3–3 REVIEW

1. Define *power*.
2. Write the formula for power in terms of energy and time.
3. Define *watt*.
4. Express each of the following values of power in the most appropriate units:
   (a) 68,000 W **(b)** 0.005 W **(c)** 0.000025 W
5. If you use 100 W of power for 10 h, how much energy (in kilowatt-hours) have you consumed?
6. Convert 2000 Wh to kilowatt-hours.
7. Convert 360,000 Ws to kilowatt-hours.

3–4 ■ POWER IN AN ELECTRIC CIRCUIT

The loss of energy in an electric circuit is generally an unwanted by-product of current through the resistance in the circuit. In some cases, however, the generation of heat is the primary purpose of a circuit as, for example, in an electric resistive heater. In any case, you must always deal with power in electrical and electronic circuits.

After completing this section, you should be able to

- **Calculate power in a circuit**
  - □ Determine power knowing I and R
  - □ Determine power knowing V and I
  - □ Determine power knowing V and R

When there is current through a resistance, the collisions of the electrons give off heat, resulting in a loss of energy as indicated in Figure 3–14. There is always a certain amount of power in an electric circuit, and it is dependent on the amount of resistance and on the amount of current, expressed as follows:

$$P = I^2R \qquad \textbf{(3–6)}$$

FIGURE 3–14
Power in an electric circuit is seen as heat given off by the resistance.

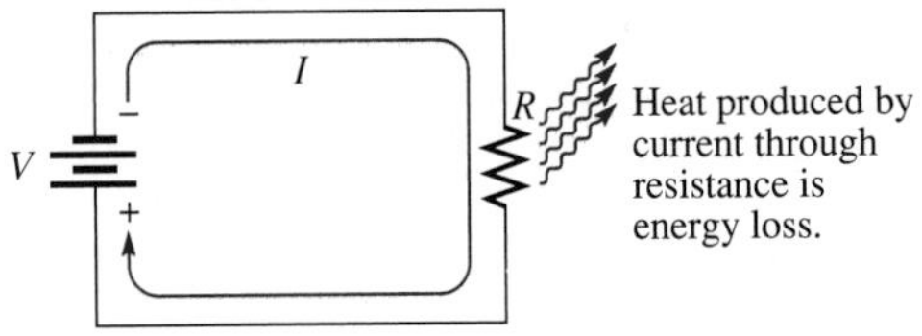

We can produce an equivalent expression for power by substituting V for IR (I^2 is $I \times I$).

$$P = I^2R = (I \times I)R = I(IR) = (IR)I$$

$$P = VI \tag{3–7}$$

We obtain another equivalent expression by substituting V/R for I (Ohm's law) as follows:

$$P = VI = V\left(\frac{V}{R}\right)$$

$$P = \frac{V^2}{R} \tag{3–8}$$

The three power formulas in Equations (3–6), (3–7), and (3–8) are also known as **Watt's law.**

How to Use the Appropriate Power Formula

To calculate the power in a resistance, you can use any one of the three Watt's law power formulas, depending on what information you have. For example, assume that you know the values of current and voltage; in this case, calculate the power with the formula $P = VI$. If you know I and R, use the formula $P = I^2R$. If you know V and R, use the formula $P = V^2/R$.

EXAMPLE 3–17 Calculate the power in each of the three circuits of Figure 3–15.

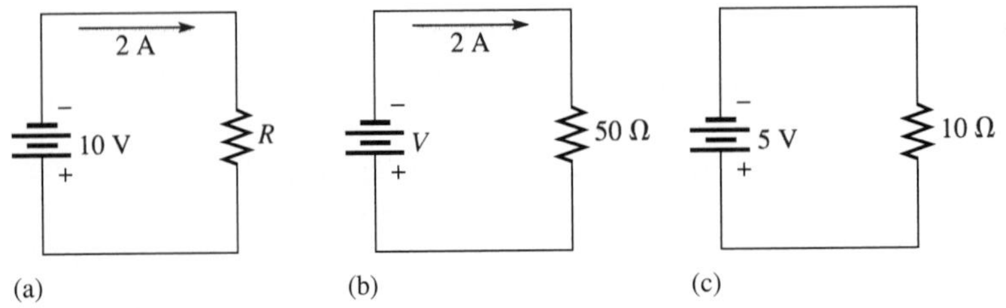

FIGURE 3–15

Solution In circuit (a), V and I are known. The power is determined as follows with calculator sequences also shown:

$$P = VI = (10\ \text{V})(2\ \text{A}) = 20\ \text{W}$$ 1 0 × 2 =

In circuit (b), I and R are known. Therefore,

$$P = I^2R = (2\text{ A})^2(50\ \Omega) = 200\text{ W}$$

[2] [x^2] [×] [5] [0] [=]

In circuit (c), V and R are known. Therefore,

$$P = \frac{V^2}{R} = \frac{(5\text{ V})^2}{10\ \Omega} = 2.5\text{ W}$$

[5] [x^2] [÷] [1] [0] [=]

Related Exercise Determine the power in each circuit of Figure 3–15 for the following changes:
(a) I doubled with V the same **(b)** R doubled with I the same
(c) V halved with R the same

EXAMPLE 3–18

A 100 W light bulb operates on 120 V. How much current does it require?

Solution Use the formula $P = VI$ and solve for I as follows. Transpose the terms to get I on the left of the equation:

$$VI = P$$

Divide both sides of equation by V to get I by itself:

$$\frac{\cancel{V}I}{\cancel{V}} = \frac{P}{V}$$

The Vs cancel on the left, leaving

$$I = \frac{P}{V}$$

Substitute 100 W for P and 120 V for V:

$$I = \frac{P}{V} = \frac{100\text{ W}}{120\text{ V}} = 0.833\text{ A} = 833\text{ mA}$$

Related Exercise A light bulb draws 545 mA from a 110 V source. What is the power?

SECTION 3–4 REVIEW

1. If there are 10 V across a resistor and a current of 3 A through it, what is the power?
2. If a current of 5 A is flowing through a 50 Ω resistor, what is the power?
3. How much power is produced by 20 mA through a 5 kΩ resistor?
4. Five volts are applied to a 10 Ω resistor. What is the power?
5. How much power does a 2 kΩ resistor with 8 V across it produce?
6. What is the resistance of a 75 W bulb that draws 0.5 A?

3–5 ■ THE POWER RATING OF RESISTORS

As you know, a resistor gives off heat when there is current through it. There is a limit to the amount of heat that a resistor can give off, which is specified by its power rating.

After completing this section, you should be able to

- **Properly select resistors based on power rating**
  - □ Define *power rating*
  - □ Explain how physical characteristics of resistors determine their power rating
  - □ Check for resistor failure with an ohmmeter

The **power rating** is the maximum amount of power that a resistor can dissipate without being damaged by excessive heat buildup. The power rating is not related to the ohmic value (resistance) but rather is determined mainly by the physical size and shape of the resistor. The larger the surface area of a resistor, the more power it can dissipate. *The surface area of a cylindrically shaped resistor is equal to the length (l) times the circumference (c),* as indicated in Figure 3–16.

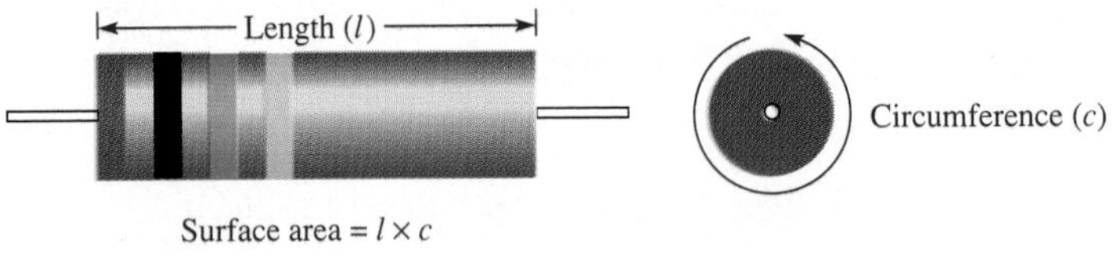

FIGURE 3–16
The power rating of a resistor is directly related to its surface area.

Carbon-composition resistors are available in standard power ratings from ⅛ W to 2 W, as shown in Figure 3–17. Available power ratings for other types of resistors vary. For example, carbon-film and metal-film resistors have ratings up to 10 W, and wire-wound resistors have ratings up to 225 W or greater. Figure 3–18 shows some of these resistors.

FIGURE 3–17
Carbon-composition resistors with standard power ratings of ⅛ W, ¼ W, ½ W, 1 W, and 2 W (courtesy of Allen-Bradley Co.).

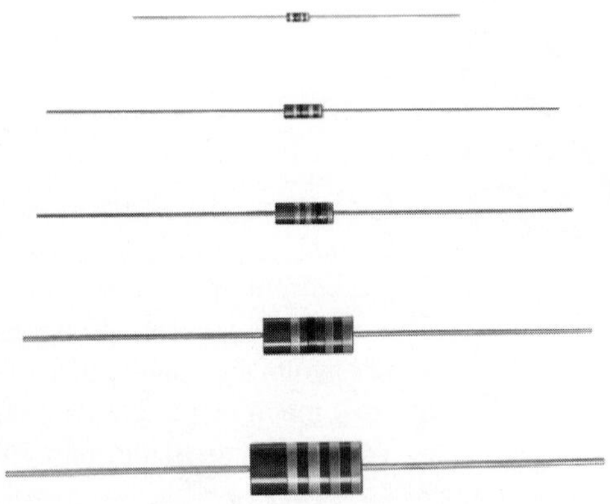

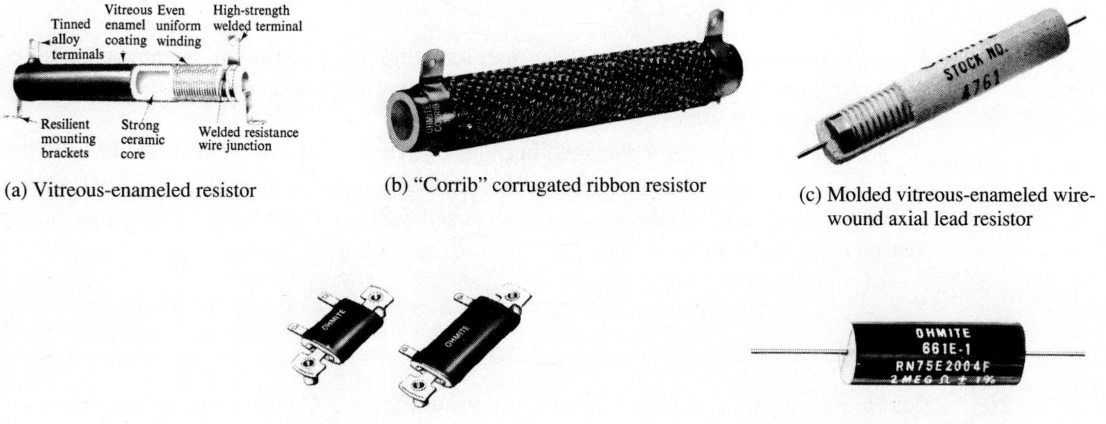

(a) Vitreous-enameled resistor

(b) "Corrib" corrugated ribbon resistor

(c) Molded vitreous-enameled wire-wound axial lead resistor

(d) Thin resistors

(e) Metal-film precision resistors

FIGURE 3–18
Typical resistors with high power ratings (courtesy of Ohmite Manufacturing Co.).

How to Select the Proper Power Rating for an Application

When a resistor is used in a circuit, its power rating must be greater than the maximum power that it will have to handle. For example, if a carbon-composition resistor is to dissipate 0.75 W in a circuit application, its rating should be at least the next higher standard value which is 1 W. *Ideally, a rating that is approximately twice the actual power should be used when possible.*

EXAMPLE 3–19 Choose an adequate power rating for each of the carbon-composition resistors represented in Figure 3–19 ($\frac{1}{8}$ W, $\frac{1}{4}$ W, $\frac{1}{2}$ W, 1 W, or 2 W).

FIGURE 3–19

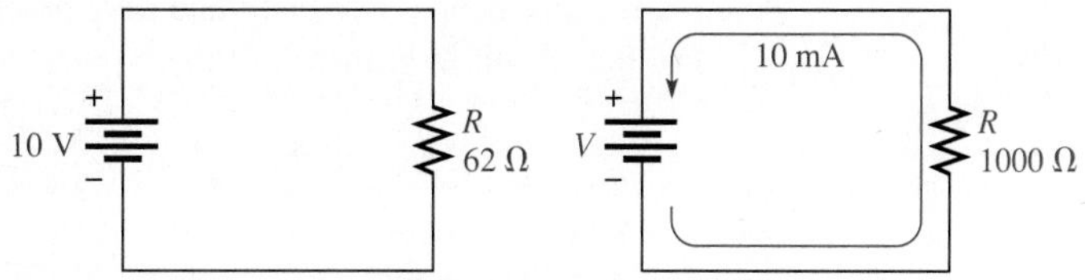

(a) (b)

Solution For the circuit in Figure 3–19(a), the actual power is

$$P = \frac{V^2}{R} = \frac{(10\text{ V})^2}{62\ \Omega} = \frac{100\text{ V}^2}{62\ \Omega} = 1.6\text{ W}$$

Select a resistor with a power rating higher than the actual power. In this case, a 2 W resistor should be used.

For the circuit in Figure 3–19(b), the actual power is

$$P = I^2R = (10\text{ mA})^2(1000\ \Omega) = 0.1\text{ W}$$

A $\frac{1}{8}$ W (0.125 W) resistor should be used in this case.

Related Exercise A certain resistor is required to dissipate 0.25 W ($\frac{1}{4}$ W). What minimum standard rating should be used?

Resistor Failures

When the power in a resistor is greater than its rating, the resistor will become excessively hot. As a result, either the resistor will burn open or its resistance value will be greatly altered.

A resistor that has been damaged because of overheating can often be detected by the charred or altered appearance of its surface. If there is no visual evidence, a resistor that is suspected of being damaged can be checked with an ohmmeter for an open or increased resistance value.

EXAMPLE 3–20 Determine whether the resistor in each circuit of Figure 3–20 has possibly been damaged by overheating.

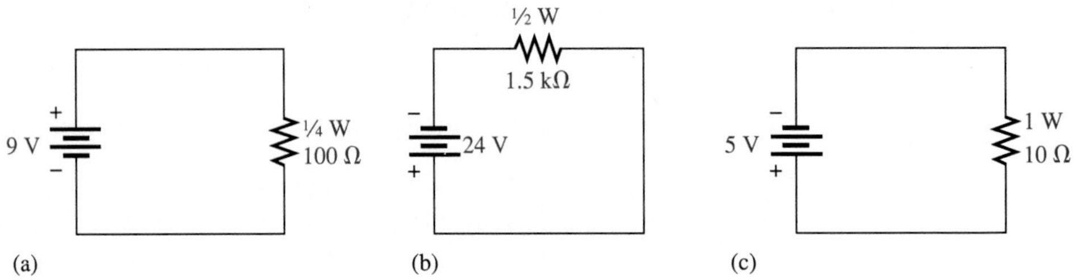

FIGURE 3–20

Solution For the circuit in Figure 3–20(a),

$$P = \frac{V^2}{R} = \frac{(9\text{ V})^2}{100\ \Omega} = 0.81\text{ W}$$

The rating of the resistor is ¼ W (0.25 W), which is insufficient to handle the power. The resistor has been overheated and may be burned out.

For the circuit in Figure 3–20(b),

$$P = \frac{V^2}{R} = \frac{(24\text{ V})^2}{1.5\text{ k}\Omega} = 0.384\text{ W}$$

The rating of the resistor is ½ W (0.5 W), which is sufficient to handle the power.

For the circuit in Figure 3–20(c),

$$P = \frac{V^2}{R} = \frac{(5\text{ V})^2}{10\ \Omega} = 2.5\text{ W}$$

The rating of the resistor is 1 W, which is insufficient to handle the power. The resistor has been overheated and may be burned out.

Related Exercise A ¼ W, 1 kΩ resistor is connected across a 12 V battery. Will it overheat?

How to Check a Resistor with an Ohmmeter

A typical analog multimeter (VOM) and a digital multimeter are shown in Figures 3–21(a) and 3–21(b), respectively. The large switch on the analog meter is called a *range*

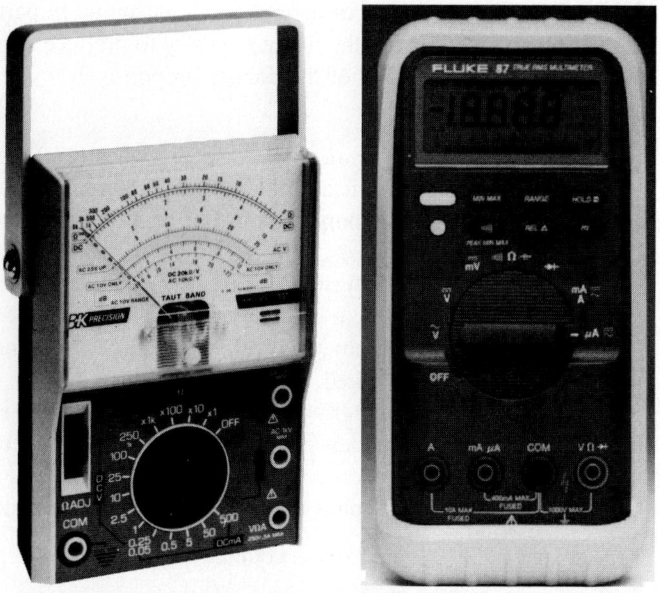

(a) (b)

FIGURE 3–21
Typical portable multimeters. (a) Analog (courtesy B & K Precision). (b) Digital (courtesy John Fluke Manufacturing Co.).

switch. Notice the resistance (OHMS) settings on both meters. For the analog meter in part (a), each setting indicates the amount by which the ohms scale (top scale) on the meter is to be multiplied. For example, if the pointer is at 50 on the ohms scale and the range switch is set at ×100, the resistance being measured is 50 × 100 Ω = 5000 Ω. *If the resistor is open, the pointer will stay at full left scale (∞ means infinite) regardless of the range switch setting.*

For the digital meter in Figure 3–21(b), you use the range switch to select the appropriate setting for the value of resistance being measured. You do not have to multiply to get the correct reading because you have a direct digital readout of the resistance value.

The Basic Analog Ohmmeter A basic ohmmeter, shown in Figure 3–22(a), contains a battery and a variable resistor in series with the meter movement. The range and function selection circuitry is omitted for simplicity. To measure resistance, connect the leads

FIGURE 3–22
Basic analog ohmmeter circuit.

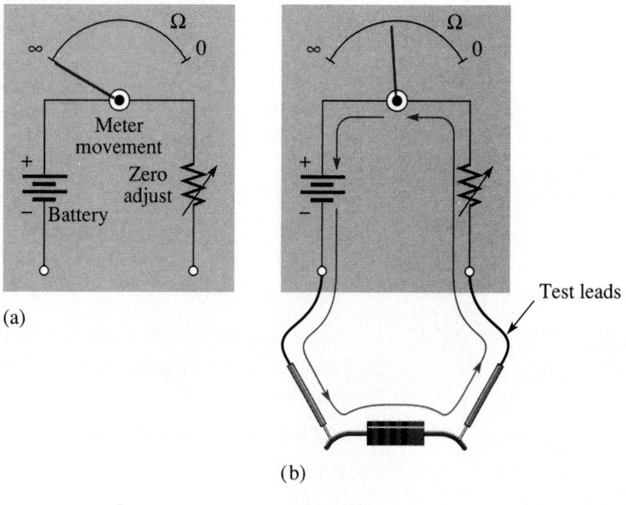

across the external resistor to be measured, as shown in part (b). This connection completes the circuit, allowing the internal battery to produce current through the movement, in turn causing a deflection of the pointer proportional to the value of the external resistance being measured.

Zero Adjustment of an Analog Ohmmeter When the ohmmeter leads are open, as in Figure 3–23(a), the pointer is at full left scale, indicating infinite (∞) resistance (open circuit). When the leads are shorted, as in Figure 3–23(b), the pointer is at full right scale, indicating zero resistance.

The variable resistor adjusts the current so that the pointer is at exactly zero when the leads are shorted. It is used to compensate for changes in the internal battery voltage due to aging. Only analog ohmmeters need to be zeroed and they should be rezeroed each time the ohms scale is changed.

Because of the ohmmeter's internal battery, any voltage source must be disconnected from a circuit before a resistance measurement is made. If the voltage is left on, there is a risk of damaging the meter.

The following steps should be used in measuring resistance:

1. Disconnect the voltage from the circuit.
2. To prevent other parts of the circuit from altering the measurement, disconnect the resistor to be checked from its circuit.
3. Select the proper range switch setting. On the analog meter, it is best to select a setting that produces a reading on the right half of the scale for better accuracy. On the digital meter, set the switch at the lowest setting that is greater than the value you expect to measure.
4. Touch the test leads together, and use the ohms adjust knob to set the pointer at 0 Ω (analog meter only).
5. Connect the meter leads across the resistor. Polarity is not important in resistance measurements. Be careful not to touch the test lead contacts with your fingers because your body resistance will affect the reading.

FIGURE 3–23
Zero adjustment.

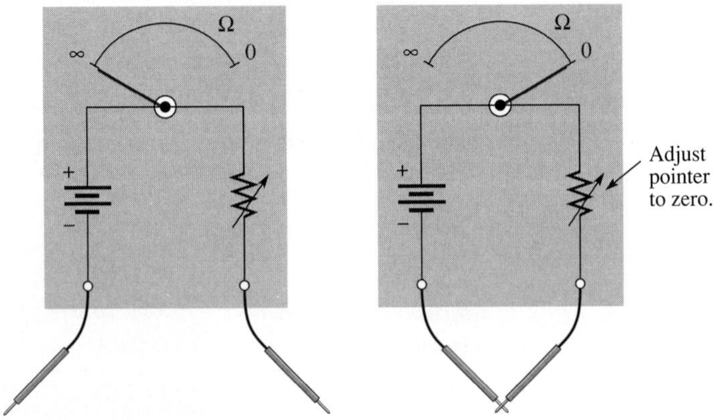

(a) Open (infinite resistance)   (b) Short (zero resistance)

SECTION 3–5 REVIEW

1. Name two important parameters associated with a resistor.
2. How does the physical size of a resistor determine the amount of power that it can handle?
3. List the standard power ratings of carbon-composition resistors.

4. A resistor must handle 0.3 W. What standard size carbon-composition resistor should be used to dissipate the energy properly?
5. Why is an internal battery required in an ohmmeter?
6. If the pointer indicates 8 on the ohmmeter scale and the range switch is set at R × 1k, what resistance is being measured?

3–6 ■ ENERGY LOSS AND VOLTAGE DROP IN RESISTANCE

As you have seen, when there is current through a resistance, energy is dissipated in the form of heat. This heat loss is caused by collisions of the free electrons within the atomic structure of the resistive material. When a collision occurs, heat is given off; and the electron loses some of its acquired energy.

After completing this section, you should be able to

- **Explain energy loss and voltage drop**
  - □ Discuss the cause of energy loss in a circuit
  - □ Define *voltage drop*
  - □ Explain the relationship between energy loss and voltage drop

In Figure 3–24, electrons are flowing out of the negative terminal of the battery. They have acquired energy from the battery and are at their highest energy level at the negative side of the circuit. As the electrons move through the resistor, they lose energy. The electrons emerging from the upper end of the resistor are at a lower energy level than those entering the lower end. The drop in energy level through the resistor creates a potential difference, or **voltage drop,** across the resistor having the polarity shown in Figure 3–24.

The upper end of the resistor in Figure 3–24 is less negative (more positive) than the lower end. When we follow the flow of electrons, this change corresponds to a voltage drop from a more negative potential to a less negative potential.

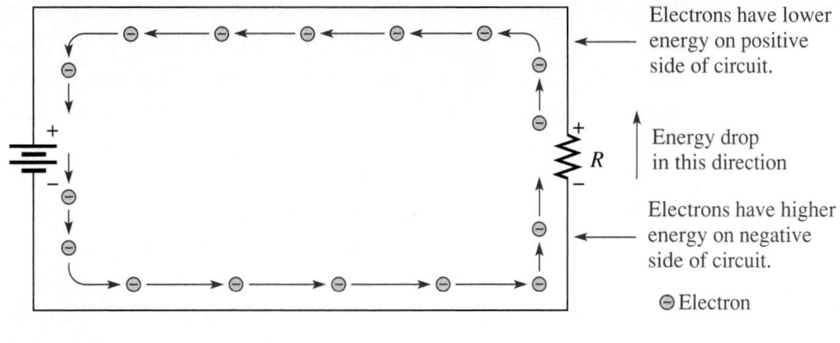

FIGURE 3–24
Illustration of electron flow.

SECTION 3–6 REVIEW

1. What is the basic reason for energy loss in a resistor?
2. What is a voltage drop?
3. What is the polarity of a voltage drop in relation to current direction?

3–7 ■ POWER SUPPLIES

A power supply is a device that provides power to a load. A load is any electrical device or circuit that is connected to the output of the power supply and draws current from the supply.

After completing this section, you should be able to

- **Discuss power supplies and their characteristics**
  - □ Define *ampere-hour rating*
  - □ Discuss power supply efficiency

Figure 3–25 shows a block diagram of a power supply with a loading device connected to it. The load can be anything from a light bulb to a computer. The power supply produces a voltage across its two output terminals and provides current through the load, as indicated in the figure. The product IV_{OUT} is the amount of power produced by the supply and consumed by the load. For a given output voltage (V_{OUT}), more current drawn by the load means more power from the supply.

FIGURE 3–25
Block diagram of power supply and load.

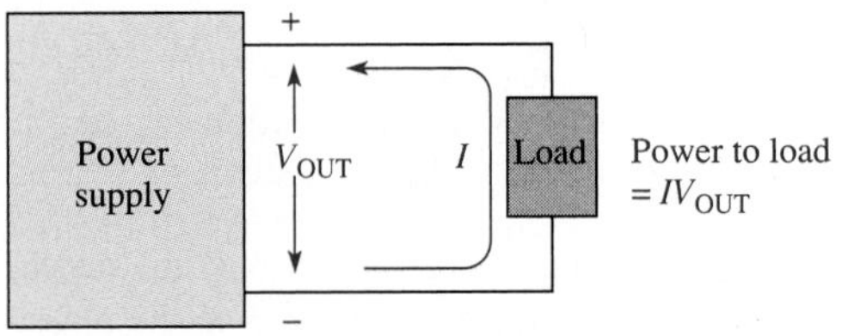

Power supplies range from simple batteries to regulated electronic circuits where an accurate output voltage is automatically maintained. A battery is a dc power supply that converts chemical energy into electrical energy. Electronic power supplies normally convert 110 V ac (alternating current) from a wall outlet into a regulated dc (direct current) or ac voltage.

Ampere-hour Ratings of Batteries

Batteries convert chemical energy into electrical energy. Because of their limited source of chemical energy, batteries have a certain capacity that limits the amount of time over which they can produce a given power level. This capacity is measured in ampere-hours (Ah). The **ampere-hour rating** determines the length of time that a battery can deliver a certain amount of current to a load at the rated voltage.

A rating of one ampere-hour means that a battery can deliver one ampere of current to a load for one hour at the rated voltage output. This same battery can deliver two amperes for one-half hour. The more current the battery is required to deliver, the shorter is the life of the battery. In practice, a battery usually is rated for a specified current level and output voltage. For example, a 12 V automobile battery may be rated for 70 Ah at 3.5 A. This means that it can produce 3.5 A for 20 h.

EXAMPLE 3–21 For how many hours can a battery deliver 2 A if it is rated at 70 Ah?

Solution The ampere-hour rating is the current times the hours.

$$70 \text{ Ah} = (2 \text{ A})(x \text{ h})$$

Solving for x h,

$$x = \frac{70\ \text{Ah}}{2\ \text{A}} = 35\ \text{h}$$

Related Exercise A certain battery delivers 10 A for 6 h. What is its Ah rating?

Power Supply Efficiency

An important characteristic of electron power supplies is efficiency. **Efficiency** is the ratio of the output power to the input power.

$$\text{Efficiency} = \frac{\text{output power}}{\text{input power}}$$

$$\text{Efficiency} = \frac{P_{OUT}}{P_{IN}} \quad \textbf{(3–9)}$$

Efficiency is often expressed as a percentage. For example, if the input power is 100 W and the output power is 50 W, the efficiency is (50 W/100 W) × 100% = 50%.

All power supplies require that power be put into them. For example, an electronic power supply might use the ac power from a wall outlet as its input. Its output may be regulated dc or ac. The output power is *always* less than the input power because some of the total power must be used internally to operate the power supply circuitry. This amount is normally called the *power loss.*

The output power is the input power minus the amount of internal power loss. High efficiency means that little power is lost and there is a higher proportion of output power for a given input power.

EXAMPLE 3–22 A certain power supply unit requires 25 W of input power. It can produce an output power of 20 W. What is its efficiency, and what is the power loss?

Solution

$$\text{Efficiency} = \left(\frac{P_{OUT}}{P_{IN}}\right)100\% = \left(\frac{20\ \text{W}}{25\ \text{W}}\right)100\% = 80\%$$

$$P_{LOSS} = P_{IN} - P_{OUT} = 25\ \text{W} - 20\ \text{W} = 5\ \text{W}$$

Related Exercise A power supply has an efficiency of 92%. If P_{IN} is 50 W, what is P_{OUT}?

SECTION 3–7 REVIEW

1. When a loading device draws an increased amount of current from a power supply, does this change represent a greater or a smaller load on the supply?
2. A power supply produces an output voltage of 10 V. If the supply provides 0.5 A to a load, what is the power output?

3. If a battery has an ampere-hour rating of 100 Ah, how long can it provide 5 A to a load?
4. If the battery in Question 3 is a 12 V device, what is its power output for the specified value of current?
5. An electronic power supply used in the lab operates with an input power of 1 W. It can provide an output power of 750 mW. What is its efficiency?

3–8 ■ APPLICATION ASSIGNMENT

In this assignment, an existing resistor box that is to be used as part of a test setup in the lab is to be checked out and modified. Your task is to modify the circuit so that it will meet the requirements of the new application. You will have to apply your knowledge of Ohm's law and Watt's law in order to complete this assignment.

The specifications are as follows:

1. ***Each resistor is switch selectable and only one resistor is selected at a time.***
2. ***The lowest resistor value is to be 10 Ω.***
3. ***Each successively higher resistance in the switch sequence must be a decade (10 times) increase over the previous one.***
4. ***The maximum resistor value must be 1 MΩ.***
5. ***The maximum voltage across any resistor in the box will be 4 V.***
6. ***Two additional resistors are required, one to limit the current to 10 mA ± 10% and the other to limit the current to 5 mA ± 10% with a 4 V drop.***

Step 1: Inspect the Existing Resistor Box

The existing resistor box is shown in both top and bottom views in Figure 3–26. The switch is a rotary type. The resistors all have a power rating of 1 W.

Step 2: Draw the Existing Schematic

From Figure 3–26, determine the resistor values and draw the schematic for the existing circuit so that you will know what you have to work with. Determine the resistor numbering from the *R* labels on the top view.

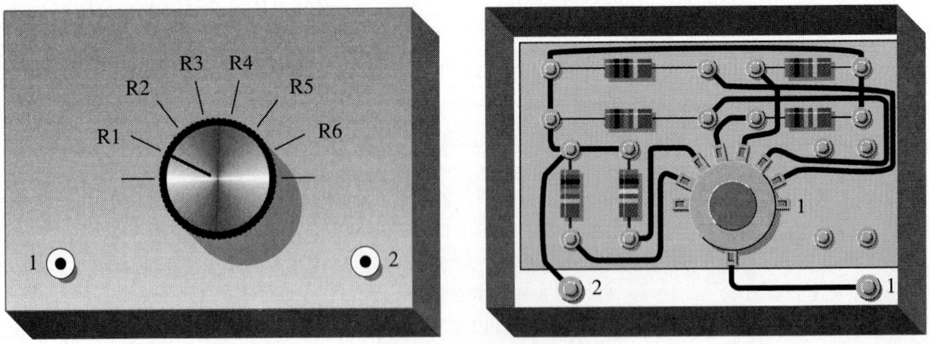

(a) Top view (b) Bottom view

FIGURE 3–26

Step 3: Develop the Schematic for the New Requirements

Draw the schematic for a circuit that will accomplish the following:

1. One resistor at a time is to be connected by the switch between terminals 1 and 2 of the box.
2. Provide switch selectable resistor values beginning with 10 Ω and increasing in decade increments to 1 MΩ.
3. Each of the resistors from Step 1 must be selectable by a sequence of adjacent switch positions in ascending order.
4. In addition to the resistors in Step 1, there must be two switch-selectable resistors, one is in switch position 1 (shown in Figure 3–26, bottom view) and must limit the current to 10 mA ± 10% with a 4 V drop and the other is in switch position 8 and must limit the current to 5 mA ± 10% with a 4 V drop.
5. All the resistors must be standard values with 10% tolerance and have a sufficient power rating for the 4 V operating requirement.

Step 4: Compare the Schematics

Determine the modifications that must be made to the existing box to meet the specifications and develop a detailed list of the changes including resistance values, power ratings, wiring, and new components. You should number each point in the schematic for easy reference.

Step 5: Develop a Test Procedure

After the box has been modified to meet the new specifications, it must be tested to see if it is working properly. Determine how you would test the resistor box and what instruments you would use and then detail your test procedure in a step-by-step format.

Step 6: Troubleshoot the Circuit

When an ohmmeter is connected across terminals 1 and 2 of the resistor box, determine the most likely fault in each of the following cases:

1. The ohmmeter shows an infinitely high resistance when the switch is in position 3.
2. The ohmmeter shows an infinitely high resistance in all switch positions.
3. The ohmmeter shows an incorrect value of resistance when the switch is in position 6.

SECTION 3–8 REVIEW

1. Explain how you applied Watt's law to this application assignment.
2. Explain how you applied Ohm's law to this application assignment.
3. How did you determine the power ratings of the resistors in the existing box?

▪ SUMMARY

- Voltage and current are linearly proportional.
- Ohm's law gives the relationship of voltage, current, and resistance.
- Current is directly proportional to voltage.
- Current is inversely proportional to resistance.
- A kilohm (kΩ) is one thousand ohms.
- A megohm (MΩ) is one million ohms.

- A microampere (μA) is one-millionth of an ampere.
- A milliampere (mA) is one-thousandth of an ampere.
- Use $I = V/R$ when calculating the current.
- Use $V = IR$ when calculating the voltage.
- Use $R = V/I$ when calculating the resistance.
- One watt equals one joule per second.
- Watt is the unit of power, joule is a unit of energy, and second is a unit of time.
- The power rating of a resistor determines the maximum power that it can handle safely.
- Resistors with a larger physical size can dissipate more power in the form of heat than smaller ones.
- A resistor should have a power rating higher than the maximum power that it is expected to handle in the circuit.
- Power rating is not related to resistance value.
- A resistor normally opens when it burns out.
- Energy is equal to power multiplied by time.
- The kilowatt-hour is a unit of energy.
- One kilowatt-hour is one thousand watts used for one hour.
- A power supply is an energy source used to operate electrical and electronic devices.
- A battery is one type of power supply that converts chemical energy into electrical energy.
- An electronic power supply converts commercial energy (ac from the power company) to regulated dc or ac at various voltage levels.
- The output power of a supply is the output voltage times the load current.
- A load is a device that draws current from the power supply.
- The capacity of a battery is measured in ampere-hours (Ah).
- One ampere-hour equals one ampere used for one hour, or any other combination of amperes and hours that has a product of one.
- A power supply with a high efficiency wastes less power than one with a lower efficiency.
- The formula wheel in Figure 3–27 gives the Ohm's law and Watt's law relationships.

FIGURE 3–27

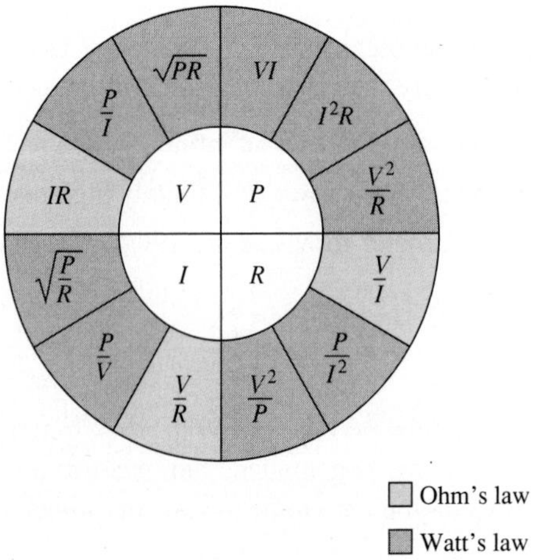

▪ GLOSSARY

Ampere-hour rating A number given in ampere-hours (Ah) determined by multiplying the current (A) times the length of time (h) a battery can deliver that current to a load.

Efficiency The ratio of the output power to the input power.

Energy The fundamental ability to do work and produce action or results..

Joule (J) The unit of energy.

Kilowatt-hour (kWh) A common unit of energy used mainly by utility companies.

Linear Characterized by a straight-line relationship.

Ohm (Ω) The unit of resistance.

Ohm's law A law stating that current is directly proportional to voltage and inversely proportional to resistance.

Power The rate of energy consumption.

Power rating The maximum amount of power that a resistor can dissipate without being damaged by excessive heat buildup.

Voltage drop The potential difference in voltage between two points when there is a drop in energy level due to resistance.

Watt (W) The unit of power.

Watt's law A law that states the relationships of power to current, voltage, and resistance.

■ FORMULAS

(3–1)	$I = \frac{V}{R}$	Ohm's law
(3–2)	$V = IR$	Ohm's law
(3–3)	$R = \frac{V}{I}$	Ohm's law
(3–4)	$P = \frac{W}{t}$	Power equals energy divided by time.
(3–5)	$W = Pt$	Energy equals power multiplied by time.
(3–6)	$P = I^2R$	Power equals current squared times resistance.
(3–7)	$P = VI$	Power equals voltage times current.
(3–8)	$P = \frac{V^2}{R}$	Power equals voltage squared divided by resistance.
(3–9)	$\text{Efficiency} = \frac{P_{OUT}}{P_{IN}}$	Power supply efficiency

■ SELF-TEST

1. Ohm's law states that
(a) current equals voltage times resistance
(b) voltage equals current times resistance
(c) resistance equals current divided by voltage
(d) voltage equals current squared times resistance

2. When the voltage across a resistor is doubled, the current will
(a) triple (b) halve (c) double (d) not change

3. When 10 V are applied across a 20 Ω resistor, the current is
(a) 10 A (b) 0.5 A (c) 200 A (d) 2 A

4. When 10 mA flow through a 1 kΩ resistor, the voltage across the resistor is
(a) 100 V (b) 0.1 V (c) 10 kV (d) 10 V

5. If 20 V are applied across a resistor and there are 6.06 mA of current, the resistance is
(a) 3.3 kΩ (b) 33 kΩ (c) 330 Ω (d) 3.03 kΩ

6. A current of 250 μA through a 4.7 kΩ resistor produces a voltage drop of
(a) 53.2 V (b) 1.175 mV (c) 18.8 V (d) 1.175 V

7. A resistance of 2.2 MΩ is connected across a 1 kV source. The resulting current is approximately
(a) 2.2 mA (b) 0.455 mA (c) 45.5 μA (d) 0.455 A

8. Power can be defined as
(a) energy (b) heat
(c) the rate at which energy is used (d) the time required to use energy

9. For 10 V and 50 mA, the power is
(a) 500 mW (b) 0.5 W (c) 500,000 μW (d) answers (a), (b), and (c)

10. When the current through a 10 kΩ resistor is 10 mA, the power is
(a) 1 W (b) 10 W (c) 100 mW (d) 1000 μW

11. A 2.2 kΩ resistor dissipates 0.5 W. The current is
(a) 15.1 mA (b) 0.227 mA (c) 1.1 mA (d) 4.4 mA

12. A 330 Ω resistor dissipates 2 W. The voltage is
(a) 2.57 V (b) 660 V (c) 6.6 V (d) 25.7 V

13. The power rating of a carbon-composition resistor that is to handle up to 1.1 W should be
(a) 0.25 W (b) 1 W (c) 2 W (d) 5 W

14. A 22 Ω half-watt resistor and a 220 Ω half-watt resistor are connected across a 10 V source. Which one(s) will overheat?
(a) 22 Ω (b) 220 Ω (c) both (d) neither

15. When the needle of an analog ohmmeter indicates infinity, the resistor being measured is
(a) overheated (b) shorted (c) open (d) reversed

BASIC PROBLEMS

SECTION 3–1 Ohm's Law

1. The current in a circuit is 1 A. Determine what the current will be when
(a) the voltage is tripled
(b) the voltage is reduced by 80%
(c) the voltage is increased by 50%

2. The current is a circuit is 100 mA. Determine what the current will be when
(a) the resistance is increased by 100%
(b) the resistance is reduced by 30%
(c) the resistance is quadrupled

3. The current in a circuit is 10 mA. What will the current be if the voltage is tripled and the resistance is doubled?

SECTION 3–2 Application of Ohm's Law

4. Determine the current in each case.
(a) $V = 5$ V, $R = 1$ Ω (b) $V = 15$ V, $R = 10$ Ω
(c) $V = 50$ V, $R = 100$ Ω (d) $V = 30$ V, $R = 15$ kΩ
(e) $V = 250$ V, $R = 5$ MΩ

5. Determine the current in each case.
(a) $V = 9$ V, $R = 2.7$ kΩ (b) $V = 5.5$ V, $R = 10$ kΩ
(c) $V = 40$ V, $R = 68$ kΩ (d) $V = 1$ kV, $R = 2$ kΩ
(e) $V = 66$ kV, $R = 10$ MΩ

6. A 10 Ω resistor is connected across a 12 V battery. How much current is there through the resistor?

7. A resistor is connected across the terminals of a dc voltage source in each part of Figure 3–28. Determine the current in each resistor.

8. Calculate the voltage for each value of I and R.
(a) $I = 2$ A, $R = 18$ Ω (b) $I = 5$ A, $R = 50$ Ω
(c) $I = 2.5$ A, $R = 600$ Ω (d) $I = 0.6$ A, $R = 47$ Ω
(e) $I = 0.1$ A, $R = 500$ Ω

9. Calculate the voltage for each value of I and R.
(a) $I = 1$ mA, $R = 10$ Ω (b) $I = 50$ mA, $R = 33$ Ω
(c) $I = 3$ A, $R = 5$ kΩ (d) $I = 1.6$ mA, $R = 2.2$ kΩ
(e) $I = 250$ μA, $R = 1$ kΩ (f) $I = 500$ mA, $R = 1.5$ MΩ
(g) $I = 850$ μA, $R = 10$ MΩ (h) $I = 75$ μA, $R = 50$ Ω

10. Three amperes of current are measured through a 27 Ω resistor connected across a voltage source. How much voltage does the source produce?

11. Assign a voltage value to each source in the circuits of Figure 3–29 to obtain the indicated amounts of current.

12. Calculate the resistance for each value of V and I.
(a) $V = 10$ V, $I = 2$ A (b) $V = 90$ V, $I = 45$ A
(c) $V = 50$ V, $I = 5$ A (d) $V = 5.5$ V, $I = 10$ A
(e) $V = 150$ V, $I = 0.5$ A

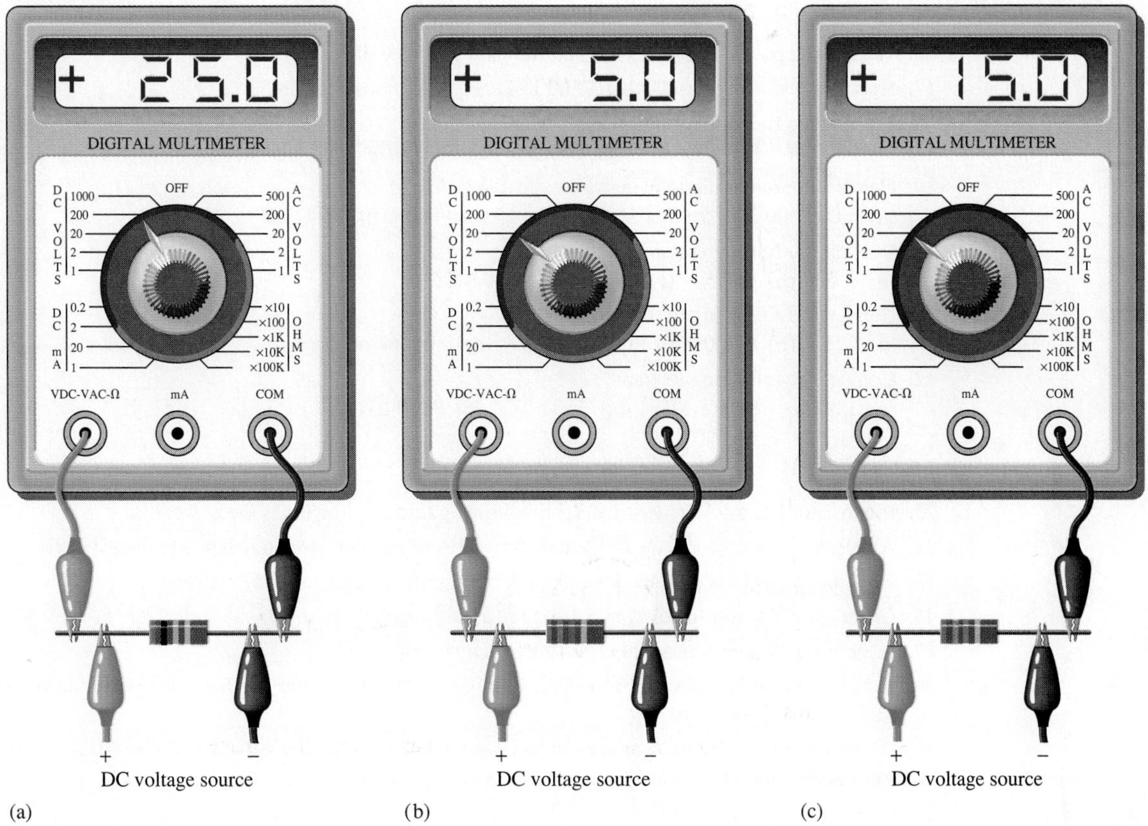

FIGURE 3–28

FIGURE 3–29

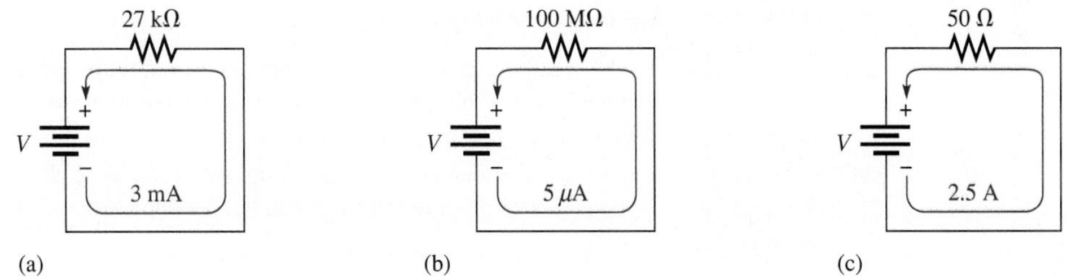

13. Calculate R for each set of V and I values.
(a) $V = 10$ kV, $I = 5$ A **(b)** $V = 7$ V, $I = 2$ mA
(c) $V = 500$ V, $I = 250$ mA **(d)** $V = 50$ V, $I = 500\ \mu$A
(e) $V = 1$ kV, $I = 1$ mA

14. Six volts are applied across a resistor. A current of 2 mA is measured. What is the value of the resistor?

15. Choose the correct value of resistance to get the current values indicated in each circuit of Figure 3–30.

FIGURE 3–30

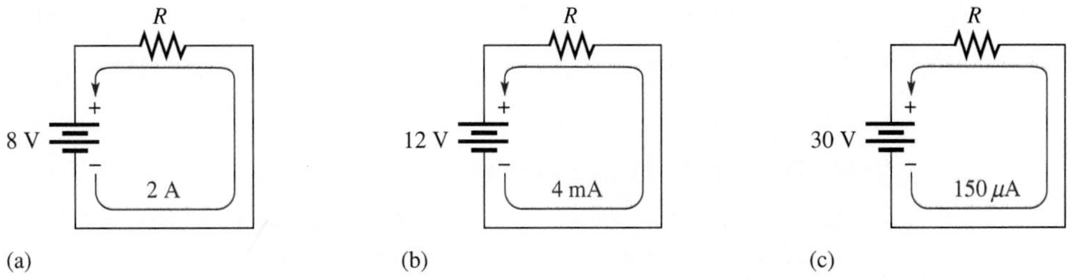

SECTION 3–3 Energy and Power

16. What is the power when energy is consumed at the rate of 350 J/s?

17. How many watts are used when 7500 J of energy are consumed in 5 h?

18. Convert the following to kilowatts:
(a) 1000 W **(b)** 3750 W **(c)** 160 W **(d)** 50,000 W

19. Convert the following to megawatts:
(a) 1,000,000 W **(b)** 3×10^6 W **(c)** 15×10^7 W **(d)** 8700 kW

20. Convert the following to milliwatts:
(a) 1 W **(b)** 0.4 W **(c)** 0.002 W **(d)** 0.0125 W

21. Convert the following to microwatts:
(a) 2 W **(b)** 0.0005 W **(c)** 0.25 mW **(d)** 0.00667 mW

22. Convert the following to watts:
(a) 1.5 kW **(b)** 0.5 MW **(c)** 350 mW **(d)** 9000 μW

SECTION 3–4 Power in an Electric Circuit

23. If a resistor has 5.5 V across it and 3 mA flowing through it, what is the power?

24. An electric heater works on 115 V and draws 3 A of current. How much power does it use?

25. How much power is produced by 500 mA of current through a 4.7 kΩ resistor?

26. Calculate the power handled by a 10 kΩ resistor carrying 100 μA.

27. If there are 60 V across a 600 Ω resistor, what is the power?

28. A 50 Ω resistor is connected across the terminals of a 1.5 V battery. What is the power dissipation in the resistor?

29. If a resistor is to carry 2 A of current and handle 100 W of power, how many ohms must it be? Assume that the voltage can be adjusted to any required value.

30. Convert 5×10^6 Wmin to kWh.

31. Convert 6700 Ws to kWh.

32. How many watt-hours does 50 W used for 12 h equal? How many kilowatt-hours?

SECTION 3–5 The Power Rating of Resistors

33. A 6.8 kΩ resistor has burned out in a circuit. You must replace it with another resistor with the same resistance value. If the resistor carries 10 mA, what should its power rating be? Assume that you have available carbon-composition resistors in all the standard power ratings.

34. A certain type of power resistor comes in the following ratings: 3 W, 5 W, 8 W, 12 W, 20 W. Your particular application requires a resistor that can handle approximately 8 W. Which rating would you use? Why?

SECTION 3–6 Energy Loss and Voltage Drop in Resistance

35. For each circuit in Figure 3–31, assign the proper polarity for the voltage drop across the resistor.

FIGURE 3–31

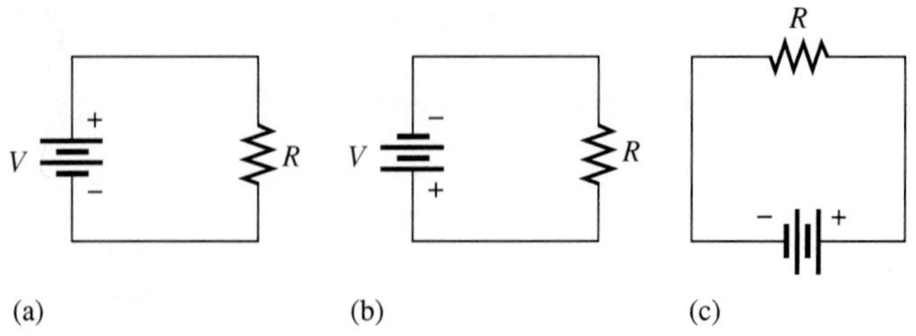

SECTION 3–7 Power Supplies

36. A 50 Ω load consumes 1 W of power. What is the output voltage of the power supply?

37. A battery can provide 1.5 A of current for 24 h. What is its ampere-hour rating?

38. How much continuous current can be drawn from an 80 Ah battery for 10 h?
39. If a battery is rated at 650 mAh, how much current will it provide for 48 h?
40. If the input power is 500 mW and the output power is 400 mW, how much power is lost? What is the efficiency of this power supply?
41. To operate at 85% efficiency, how much output power must a source produce if the input power is 5 W?
42. A certain power supply provides a continuous 2 W to a load. It is operating at 60% efficiency. In a 24 h period, how many kilowatt-hours does the power supply consume?

■ ADVANCED PROBLEMS

43. The filament of a light bulb in the circuit of Figure 3–32(a) has a certain amount of resistance, represented by an equivalent resistance in Figure 3–32(b). If the bulb operates with 120 V and 0.8 A of current, what is the resistance of its filament?

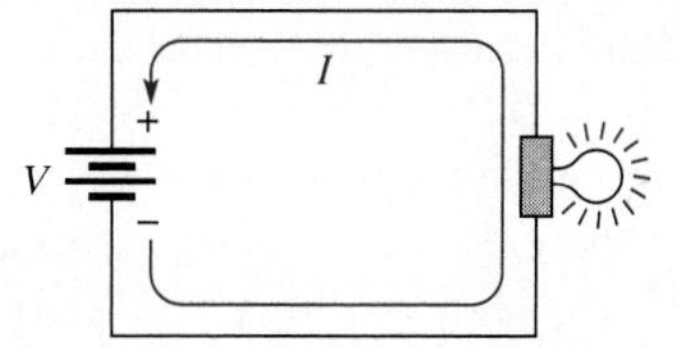

(a)

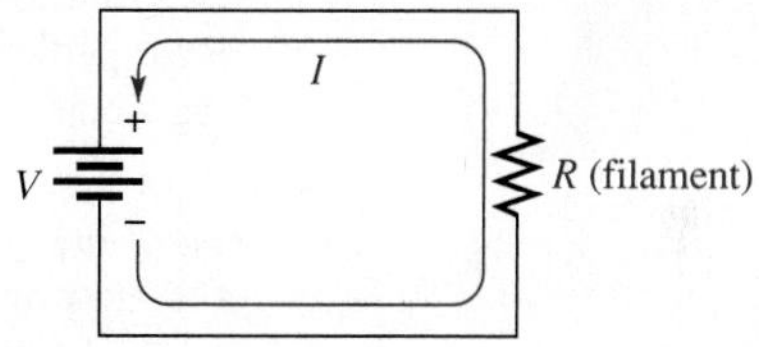

(b)

FIGURE 3–32

44. A certain electrical device has an unknown resistance. You have available a 12 V battery and an ammeter. How would you determine the value of the unknown resistance? Draw the necessary circuit connections.
45. A variable voltage source is connected to the circuit of Figure 3–33. Start at 0 V and increase the voltage in 10 V steps up to 100 V. Determine the current at each voltage value, and plot a graph of V versus I. Is the graph a straight line? What does the graph indicate?
46. In a certain circuit, $V = 1$ V and $I = 5$ mA. Determine the current for each of the following voltages in the same circuit:
(a) $V = 1.5$ V **(b)** $V = 2$ V **(c)** $V = 3$ V **(d)** $V = 4$ V **(e)** $V = 10$ V
47. Figure 3–34 is a graph of voltage versus current for three resistance values. Determine R_1, R_2, and R_3.
48. You are measuring the current in a circuit that is operated on a 10 V battery. The ammeter reads 50 mA. Later, you notice that the current has dropped to 30 mA. Eliminating the possibility of a resistance change, you must conclude that the voltage has changed. How much has the voltage of the battery changed, and what is its new value?
49. If you wish to increase the amount of current in a resistor from 100 mA to 150 mA by changing the 20 V source, by how many volts should you change the source? To what new value should you set it?

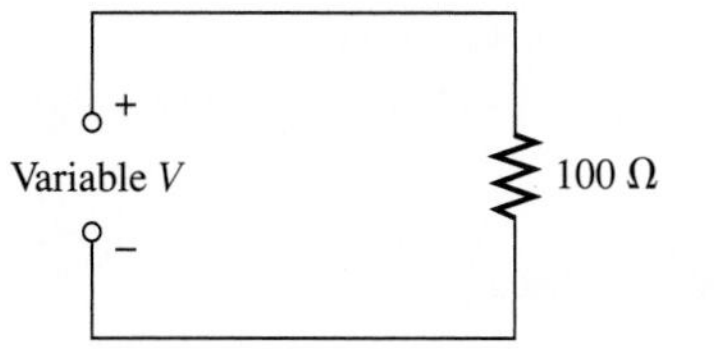

FIGURE 3–33

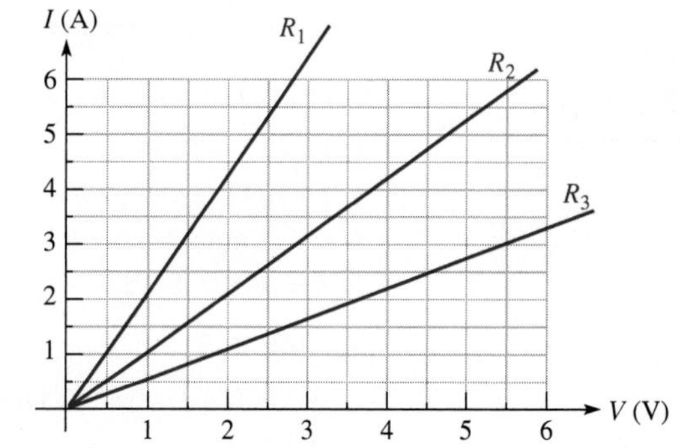

FIGURE 3–34

50. By varying the rheostat (variable resistor) in the circuit of Figure 3–35, you can change the amount of current. The setting of the rheostat is such that the current is 750 mA. What is the ohmic value of this setting? To adjust the current to 1 A, to what resistance value must you set the rheostat?

51. A certain resistor has the following color code: orange, orange, red, gold. Determine the maximum and minimum currents you should expect to measure when a 12 V source is connected across the resistor.

52. A 6 V source is connected to a 100 Ω resistor by two 12 ft lengths of 18-gage copper wire. Refer to Table 2–3 to determine the following:
(a) current **(b)** resistor voltage **(c)** voltage across each length of wire

53. A certain appliance uses 300 W. If it is allowed to run continuously for 30 days, how many kilowatt-hours of energy does it consume?

54. At the end of a 31-day period, your utility bill shows that you have used 1500 kWh. What is your average daily power?

55. A certain type of power resistor comes in the following ratings: 3 W, 5 W, 8 W, 12 W, 20 W. Your particular application requires a resistor that can handle approximately 10 W. Which rating would you use? Why?

56. A 12 V source is connected across a 10 Ω resistor for 2 min.
(a) What is the power?
(b) How much energy is used?
(c) If the resistor remains connected for an additional minute, does the power increase or decrease?

57. The rheostat in Figure 3–36 is used to control the current to a heating element. When the rheostat is adjusted to a value of 8 Ω or less, the heating element can burn out. What is the rated value of the fuse needed to protect the circuit if the voltage across the heating element at the point of maximum current is 100 V?

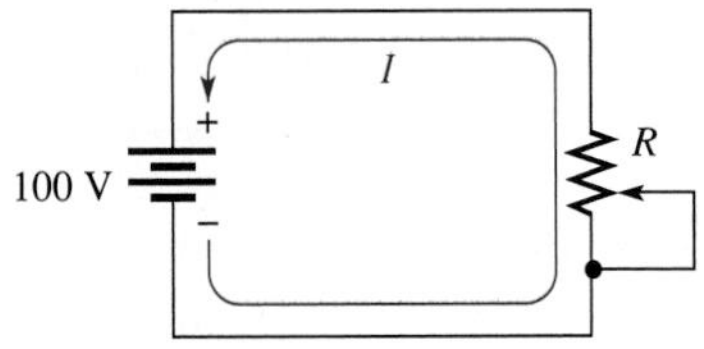

FIGURE 3–35

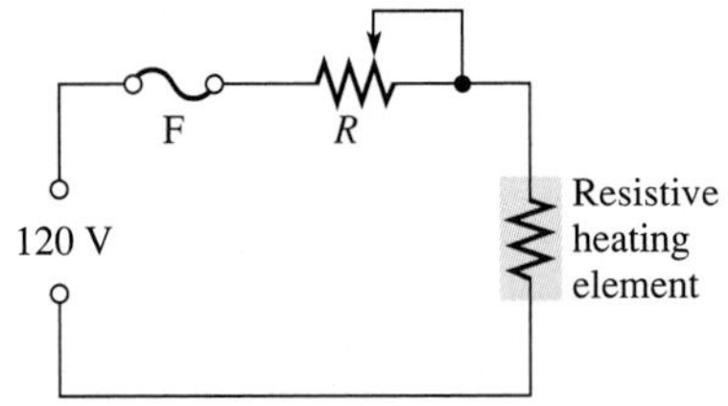

FIGURE 3–36

■ ANSWERS TO SECTION REVIEWS

Section 3–1

1. Ohm's law states that current varies directly with voltage and inversely with resistance.

2. $I = V/R$ **3.** $V = IR$ **4.** $R = V/I$

5. The current increases; three times when V is tripled

6. Doubling R cuts I in half to 5 mA.

7. No change in I if V and R are doubled.

Section 3–2

1. $I = 10$ V/5 Ω = 2 A
2. $I = 20$ kV/5 MΩ = 4 mA
3. $I = 10$ kV/2 kΩ = 5 A
4. $V = (1$ A)(10 Ω) = 10 V
5. $V = (3$ mA)(3 kΩ) = 9 V
6. $V = (2$ A)(6 Ω) = 12 V
7. $R = 10$ V/2 A = 5 Ω
8. $R = 25$ V/50 mA = 0.5 kΩ = 500 Ω

Section 3–3

1. Power is the rate at which energy is used.

2. $P = W/t$

3. Watt is the unit of power. One watt is the power when 1 J of energy is used in 1 s.

4. **(a)** 68,000 W = 68 kW **(b)** 0.005 W = 5 mW **(c)** 0.000025 W = 25 μW
5. (100 W)(10 h) = 1 kWh
6. 2000 Wh = 2 kWh
7. 360,000 Ws = 360 kWs/3600 s/h = 0.1 kWh

Section 3–4

1. $P = (10\text{ V})(3\text{ A}) = 30\text{ W}$
2. $P = (5\text{ A})^2(50\ \Omega) = 1250\text{ W}$
3. $P = (20\text{ mA})^2(5\text{ k}\Omega) = 2\text{ W}$
4. $P = (5\text{ V})^2/10\ \Omega = 2.5\text{ W}$
5. $P = (8\text{ V})^2/2\text{ k}\Omega = 32\text{ mW}$
6. $R = 75\text{ W}/(0.5\text{ A})^2 = 300\ \Omega$

Section 3–5

1. Two resistor parameters are resistance and power rating.
2. A larger resistor surface area dissipates more energy.
3. Standard ratings of carbon-composition resistors are 0.125 W, 0.25 W, 0.5 W, 1 W, 2 W.
4. At least a 0.5 W rating to handle 0.3 W
5. An internal ohmmeter battery is required to produce current through the resistance being measured.
6. $8 \times 1\text{k} = 8\text{ k}\Omega$

Section 3–6

1. Energy loss in a resistor is caused by collisions of free electrons with the atoms in the material.
2. Voltage drop is the potential difference between two points due to energy loss.
3. Voltage drop is negative to positive in the direction of current.

Section 3–7

1. An increased amount of current represents a greater load.
2. $P_{OUT} = (10\text{ V})(0.5\text{ A}) = 5\text{ W}$
3. 100 Ah/5 A = 20 h
4. $P_{OUT} = (12\text{ V})(5\text{ A}) = 60\text{ W}$
5. Efficiency = (750 mW/1000 mW) 100% = 75%

Section 3–8

1. Watt's law was used to find power ratings ($P = V^2/R$).
2. Ohm's law was used to find the two additional resistors ($R = V/I$).
3. Power ratings were determined by the relative sizes of the resistors, given the range of values.

■ ANSWERS TO RELATED EXERCISES FOR EXAMPLES

3–1	$I_1 = 10\text{ V}/5\ \Omega = 2\text{ A}$; $I_2 = 10\text{ V}/20\ \Omega = 0.5\text{ A}$	**3–2**	1 V
3–3	900 Ω	**3–4**	Yes
3–5	2.5 A	**3–6**	5 mA
3–7	25 μA	**3–8**	200 μA
3–9	800 V	**3–10**	220 mV
3–11	16.5 V	**3–12**	4 kΩ
3–13	100 Ω	**3–14**	3000 J
3–15	**(a)** 0.001 W **(b)** 0.0018 W **(c)** 3,000,000 W **(d)** 10,000 W	**3–16**	2 kWh
3–17	**(a)** 40 W **(b)** 400 W **(c)** 0.625 W	**3–18**	60 W
3–19	0.5 W (½ W)	**3–20**	No
3–21	60 Ah	**3–22**	46 W

4

SERIES CIRCUITS

Resistive circuits can be of two basic forms: series or parallel. In this chapter we discuss series circuits. Parallel circuits are covered in Chapter 5, and combinations of series and parallel circuits are examined in Chapter 6. In this chapter you will see how Ohm's law is used in series circuits; and you will study another important law, Kirchhoff's voltage law. Also, several important applications of series circuits are presented.

■ CHAPTER OBJECTIVES

- ☐ Identify a series circuit
- ☐ Determine the current in a series circuit
- ☐ Determine the total series resistance
- ☐ Apply Ohm's law in series circuits
- ☐ Determine the total effect of voltage sources in series
- ☐ Apply Kirchhoff's voltage law
- ☐ Use a series circuit as a voltage divider
- ☐ Determine power in a series circuit
- ☐ Determine and identify ground in a circuit
- ☐ Troubleshoot series circuits

■ APPLICATION ASSIGNMENT

For the application assignment in this chapter, your supervisor has given you a voltage-divider board to evaluate and modify if necessary. You will use it to obtain five different voltage levels from a 12 V battery. The voltage divider will be used to provide positive reference voltages to an electronic circuit in an analog-to-digital converter. Your task will be to check the circuit to see if it provides the required voltages and, if not, you will modify it so that it does. Also, you must make sure that the power ratings of the resistors are adequate for the application and determine how long the battery will last with the voltage divider connected to it. After studying this chapter, you should be able to complete the application assignment in Section 4–11.

4–1 ■ RESISTORS IN SERIES

When connected in series, resistors form a "string" in which there is only one path for current.

After completing this section, you should be able to

- **Identify a series circuit**
  - □ Translate a physical arrangement of resistors into a schematic

Figure 4–1(a) shows two resistors connected in **series** between point A and point B. Part (b) shows three in series, and part (c) shows four in series. Of course, there can be any number of resistors in series.

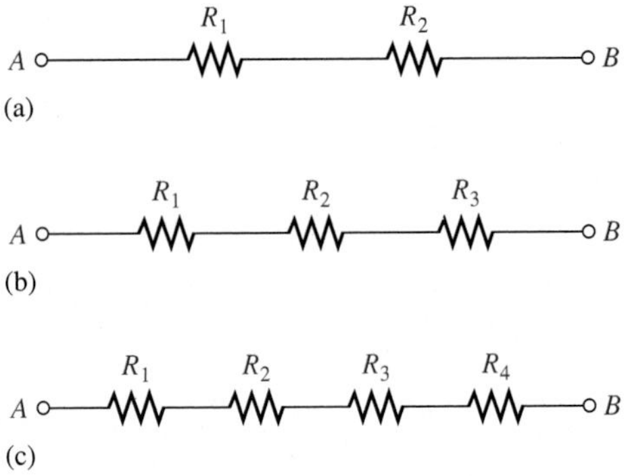

FIGURE 4–1
Resistors in series.

The only way for electrons to get from point A to point B in any of the connections of Figure 4–1 is to go through each of the resistors. The following statement identifies a series circuit:

A series circuit provides only one path for current between two points in a circuit so that the current is the same through each series resistor.

Identifying Series Circuits

In an actual circuit diagram, a series circuit may not always be as easy to identify as those in Figure 4–1. For example, Figure 4–2 shows series resistors drawn in other ways. Remember, if there is only one current path between two points, the resistors between those two points are in series, no matter how they appear in a diagram.

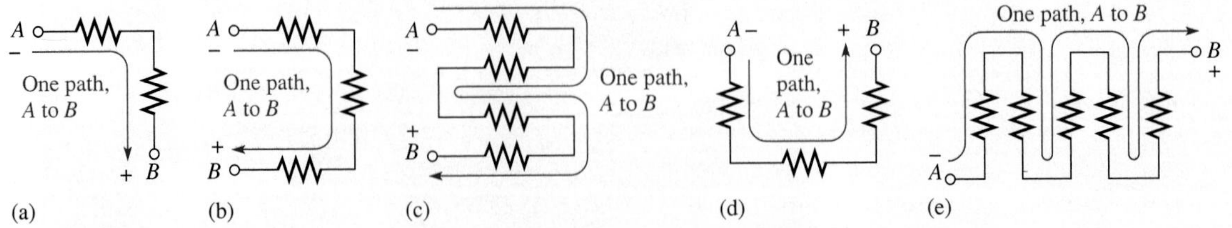

FIGURE 4–2
Some examples of series circuits. Notice that the current must be the same at all points.

EXAMPLE 4–1 Five resistors are positioned on a circuit board as shown in Figure 4–3. Wire them together in series so that, starting from the negative (–) terminal, R_1 is first, R_2 is second, R_3 is third, and so on. Draw a schematic showing this connection.

FIGURE 4–3

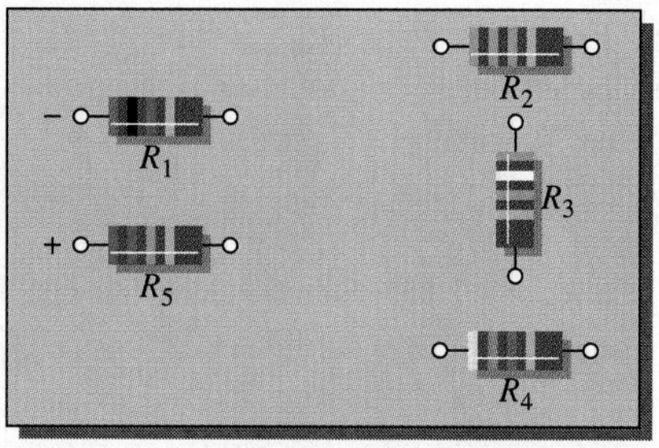

Solution The wires are connected as shown in Figure 4–4(a), which is the assembly diagram. The schematic is shown in Figure 4–4(b). Note that the schematic does not necessarily show the actual physical arrangement of the resistors as does the assembly diagram. The schematic shows how components are connected electrically; the assembly diagram shows how components are arranged physically.

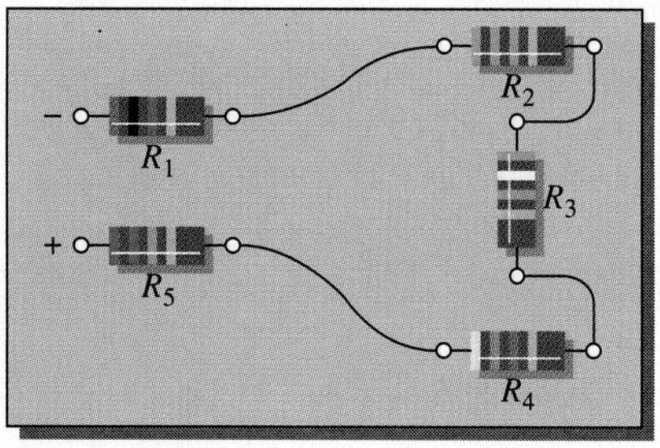

(a) Assembly diagram

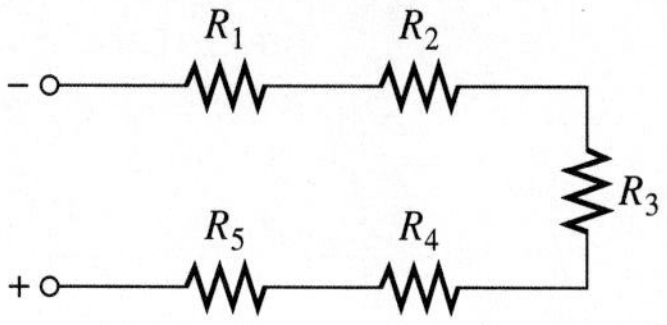

(b) Schematic

FIGURE 4–4

Related Exercise **(a)** Show how you would rewire the circuit board in Figure 4–4(a) so that all the odd-numbered resistors come first followed by the even-numbered ones. **(b)** Determine the value of each resistor.

EXAMPLE 4–2 Describe how the resistors on the printed circuit (PC) board in Figure 4–5 are related electrically. Determine the resistance value of each resistor.

FIGURE 4–5

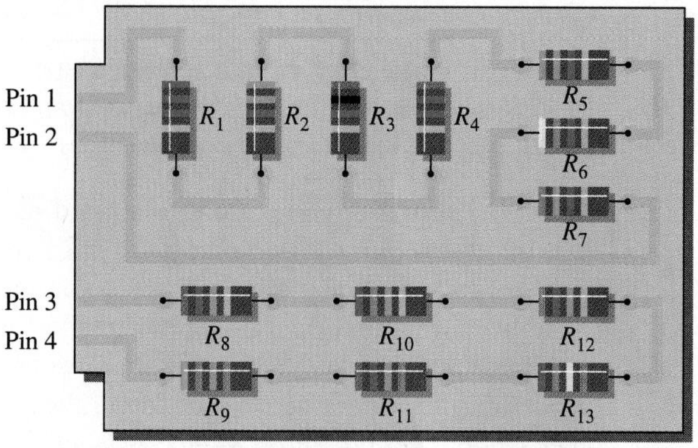

Solution Resistors R_1 through R_7 are in series with each other. This series combination is connected between pins 1 and 2 on the PC board.

Resistors R_8 through R_{13} are in series with each other. This series combination is connected between pins 3 and 4 on the PC board.

The values of the resistors are $R_1 = 2.2\ \text{k}\Omega$, $R_2 = 3.3\ \text{k}\Omega$, $R_3 = 1\ \text{k}\Omega$, $R_4 = 1.2\ \text{k}\Omega$, $R_5 = 3.3\ \text{k}\Omega$, $R_6 = 4.7\ \text{k}\Omega$, $R_7 = 5.6\ \text{k}\Omega$, $R_8 = 12\ \text{k}\Omega$, $R_9 = 68\ \text{k}\Omega$, $R_{10} = 27\ \text{k}\Omega$, $R_{11} = 12\ \text{k}\Omega$, $R_{12} = 82\ \text{k}\Omega$, and $R_{13} = 270\ \text{k}\Omega$.

Related Exercise How is the circuit in Figure 4–5 changed when pin 2 and pin 3 are connected?

SECTION 4–1 REVIEW

1. How are the resistors connected in a series circuit?
2. How can you identify a series circuit?
3. Complete the schematics for the circuits in each part of Figure 4–6 by connecting the resistors in series in numerical order from *A* to *B*.
4. Now connect each group of series resistors in Figure 4–6 in series.

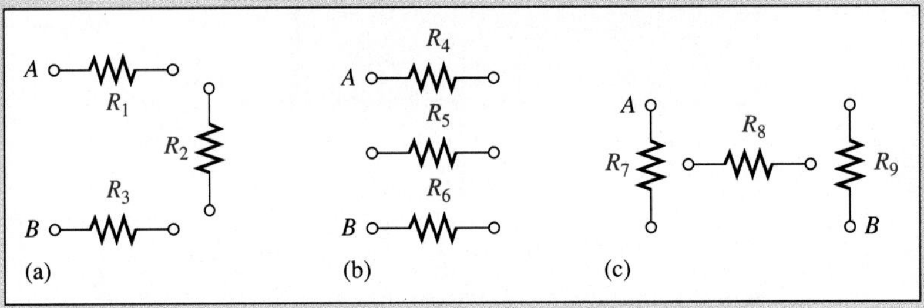

FIGURE 4–6

4–2 ■ CURRENT IN A SERIES CIRCUIT

The same amount of current is through all points in a series circuit. The current through each resistor in a series circuit is the same as the current through all the other resistors that are in series with it.

After completing this section, you should be able to

- **Determine the current in a series circuit**
  - ☐ Show that the current is the same at all points in a series circuit.

Figure 4–7 shows three resistors connected in series to a voltage source. *At any point in this circuit, the current into that point must equal the current out of that point,* as illustrated by the directional arrows. Notice that the current out of each of the resistors must equal the current in, because it is essentially forced through the resistor and out the other end. Since there are no paths other than the single series path into which the current can branch off, the current in each section of the circuit must be the same as the current in all other sections. *Although resistance reduces the current, the reduction is felt at all points equally.*

FIGURE 4–7
Current entering any point in a series circuit is the same as the current leaving that point.

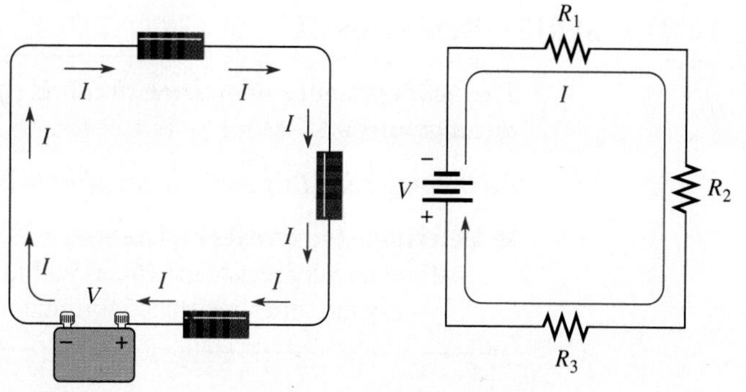

(a) Pictorial (b) Schematic

FIGURE 4–8
Current is the same at all points in a series circuit.

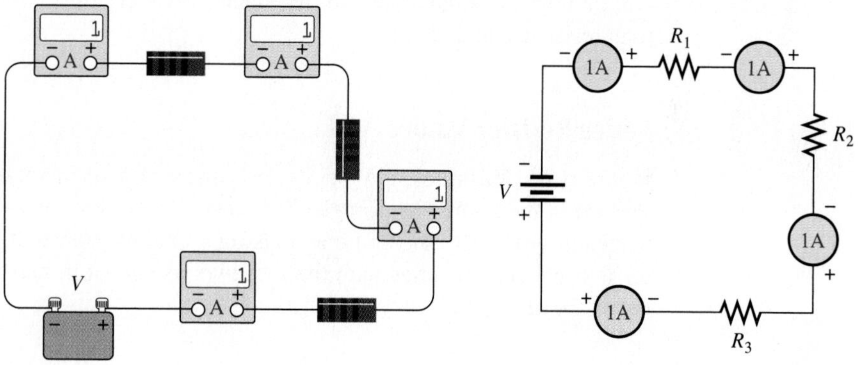

(a) Pictorial (b) Schematic

In Figure 4–8, the battery supplies 1 A of current to the series resistors. One ampere is flowing out of the battery's negative terminal. When ammeters are connected in series at several points as shown, each meter reads 1 A.

SECTION 4–2 REVIEW

1. What statement can you make about the amount of current at any point in a series circuit?
2. In a series circuit with a 10 Ω and a 5 Ω resistor in series, 2 A flow through the 10 Ω resistor. How much current flows through the 5 Ω resistor?
3. A milliammeter is connected between points *A* and *B* in Figure 4–9. It measures 50 mA. If you move the meter and connect it between points *C* and *D*, how much current will it indicate? Between *E* and *F*?
4. In Figure 4–10, how much current does ammeter 1 indicate? Ammeter 2?

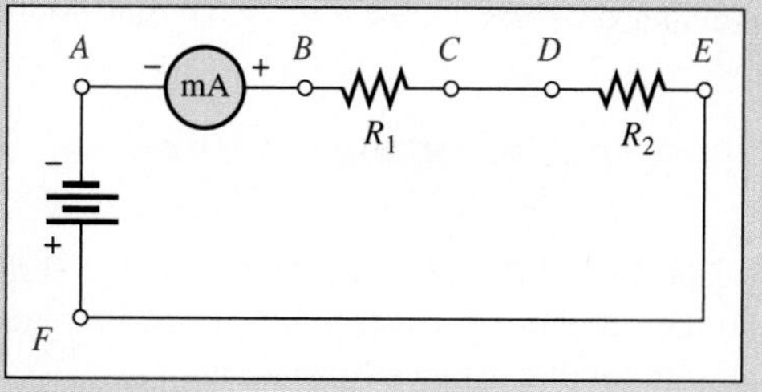

FIGURE 4–9

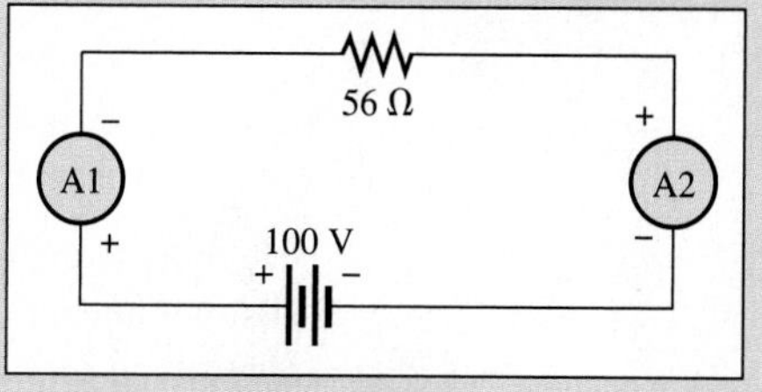

4–3 ■ TOTAL SERIES RESISTANCE

The total resistance of a series circuit is equal to the sum of the resistances of each individual resistor.

After completing this section, you should be able to

- **Determine total series resistance**
  - ☐ Explain why resistance values add in series
  - ☐ Apply the series resistance formula

It is understandable that resistor values add in series because each of the resistors in series offers opposition to the current in direct proportion to its resistance. A greater number of resistors connected in series creates more opposition to current. More opposition to current implies a higher value of resistance. So every time a resistor is added in series, the total resistance increases.

Series Resistor Values Add

Figure 4–11 illustrates how series resistances add to increase the total resistance (R_T). Part (a) of the figure has a single 10 Ω resistor. Part (b) shows another 10 Ω resistor connected in series with the first one, making a total resistance of 20 Ω. If a third 10 Ω resistor is connected in series with the first two, as shown in part (c), the total resistance becomes 30 Ω.

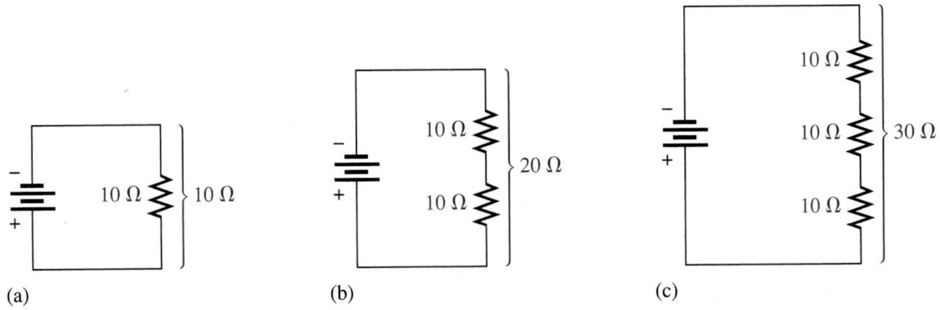

FIGURE 4–11
Total resistance increases with each additional series resistor.

Series Resistance Formula

For any number of individual resistors connected in series, the total resistance is the sum of each of the individual values:

$$R_T = R_1 + R_2 + R_3 + \cdots + R_n \qquad \textbf{(4–1)}$$

where R_T is the total resistance and R_n is the last resistor in the series string (n can be any positive integer equal to the number of resistors in series). For example, if there are four resistors in series ($n = 4$), the total resistance formula is

$$R_T = R_1 + R_2 + R_3 + R_4$$

If there are six resistors in series ($n = 6$), the total resistance formula is

$$R_T = R_1 + R_2 + R_3 + R_4 + R_5 + R_6$$

To illustrate the calculation of total series resistance, let's determine R_T of the circuit of Figure 4–12, where V_S is the source voltage. This circuit has five resistors in series. To find the total resistance, simply add the values:

$$R_T = 56\ \Omega + 100\ \Omega + 27\ \Omega + 10\ \Omega + 5.6\ \Omega = 198.6\ \Omega$$

Note in Figure 4–12 that the order in which the resistances are added does not matter. You can physically change the positions of the resistors in the circuit without affecting the total resistance.

FIGURE 4–12
Example of five resistors in series.

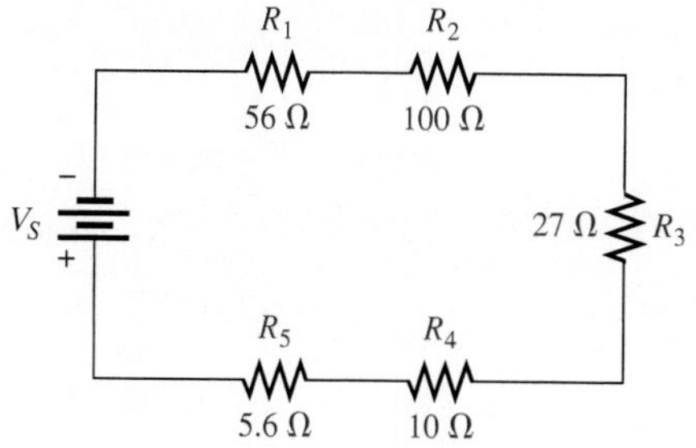

EXAMPLE 4–3 Connect the resistors in Figure 4–13 in series, and determine the total resistance, R_T.

FIGURE 4–13

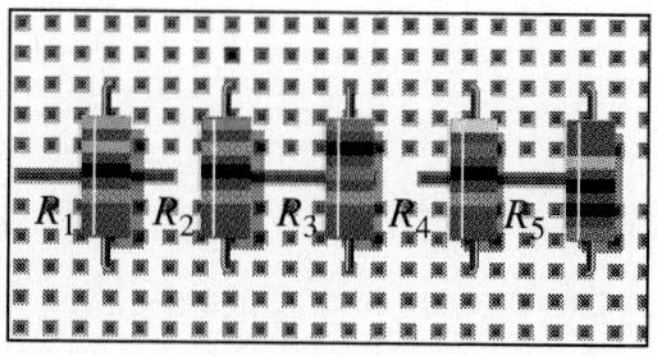

Solution The resistors are connected as shown in Figure 4–14. Find the total resistance by adding all the values:

$$R_T = 33\ \Omega + 68\ \Omega + 100\ \Omega + 47\ \Omega + 10\ \Omega = 258\ \Omega$$

FIGURE 4–14

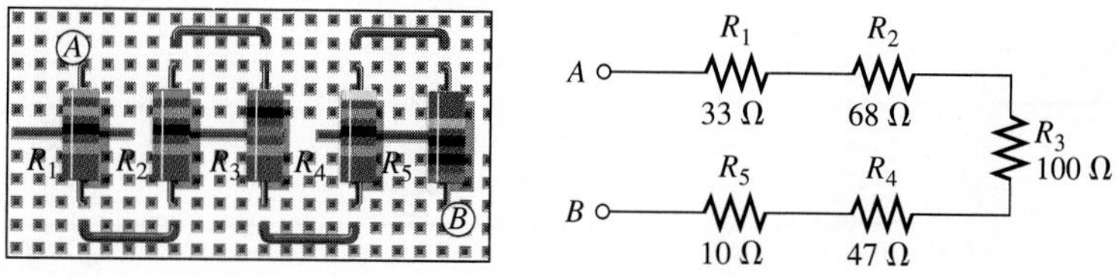

(a) Circuit assembly (b) Schematic

Related Exercise Determine the total resistance in Figure 4–14(a) if the positions of R_2 and R_4 are interchanged.

EXAMPLE 4–4 Calculate R_T for each circuit in Figure 4–15.

FIGURE 4–15

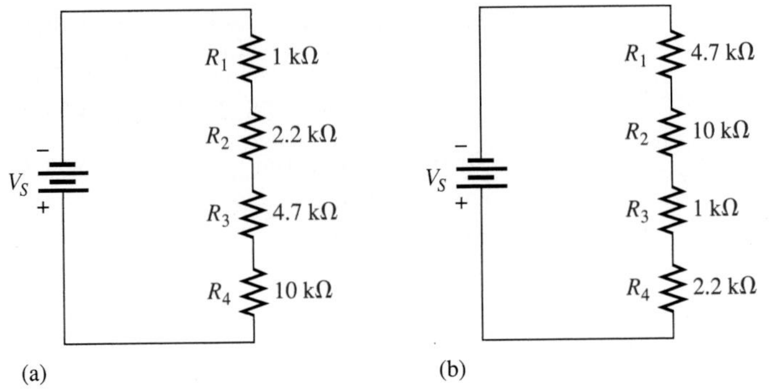

Solution For circuit (a),

$$R_T = 1\text{ k}\Omega + 2.2\text{ k}\Omega + 4.7\text{ k}\Omega + 10\text{ k}\Omega = 17.9\text{ k}\Omega$$

For circuit (b),

$$R_T = 4.7\text{ k}\Omega + 10\text{ k}\Omega + 1\text{ k}\Omega + 2.2\text{ k}\Omega = 17.9\text{ k}\Omega$$

Notice that the total resistance does not depend on the position of the resistors. Both circuits are identical in terms of total resistance.

The calculator sequence for R_T in circuit (a) is

1 EXP 3 + 2 • 2 EXP 3 + 4 • 7 EXP 3 + 1 0 EXP 3 =

Related Exercise What is the total resistance for the following series resistors: 1 kΩ, 2.2 kΩ, 3.3 kΩ, and 5.6 kΩ?

EXAMPLE 4–5 Determine R_4 in the circuit of Figure 4–16.

FIGURE 4–16

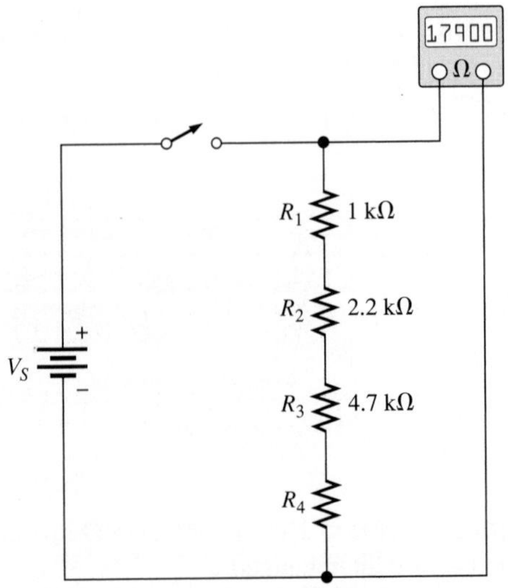

Solution From the ohmmeter reading, $R_T = 17.9\ \text{k}\Omega$.

$$R_T = R_1 + R_2 + R_3 + R_4$$

Solving for R_4 yields

$$R_4 = R_T - (R_1 + R_2 + R_3)$$
$$= 17.9\ \text{k}\Omega - (1\ \text{k}\Omega + 2.2\ \text{k}\Omega + 4.7\ \text{k}\Omega) = 10\ \text{k}\Omega$$

Related Exercise Determine R_4 in Figure 4–16 if the ohmmeter reading is 14,700 Ω.

Equal-Value Series Resistors

When a circuit has more than one resistor of the same value in series, there is a shortcut method to obtain the total resistance: Simply multiply the resistance value of the resistors having the same value by the number of equal-value resistors that are in series. This method is essentially the same as adding the values. For example, five 100 Ω resistors in series have an R_T of 5(100 Ω) = 500 Ω. In general, the formula is expressed as

$$R_T = nR \tag{4–2}$$

where n is the number of equal-value resistors and R is the resistance value.

EXAMPLE 4–6 Find the R_T of eight 22 Ω resistors in series.

Solution Find R_T by adding the values:

$$R_T = 22\ \Omega + 22\ \Omega + 22\ \Omega + 22\ \Omega + 22\ \Omega + 22\ \Omega + 22\ \Omega + 22\ \Omega = 176\ \Omega$$

However, it is much easier to multiply.

$$R_T = 8(22\ \Omega) = 176\ \Omega$$

Related Exercise Find R_T for three 1 kΩ and two 720 Ω resistors in series.

SECTION 4–3 REVIEW

1. Calculate R_T between points A and B for each circuit in Figure 4–17.

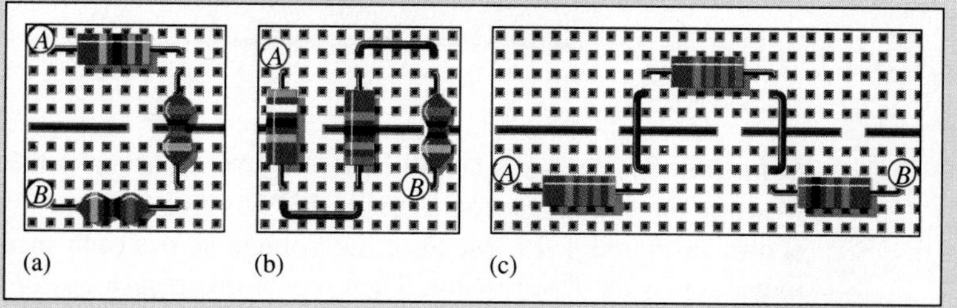

FIGURE 4–17

2. The following resistors are in series: one 100 Ω, two 47 Ω, four 12 Ω, and one 330 Ω. What is the total resistance?
3. Suppose that you have one resistor each of the following values: 1 kΩ, 2.7 kΩ, 3.3 kΩ, and 1.8 kΩ. To get a total resistance of 10 kΩ, you need one more resistor. What should its value be?
4. What is the R_T for twelve 47 Ω resistors in series?

4–4 ■ OHM'S LAW IN SERIES CIRCUITS

The use of Ohm's law and the basic concepts of series circuits are applied in several examples.

After completing this section, you should be able to

- **Apply Ohm's law in series circuits**
  - ☐ Find the current in a series circuit
  - ☐ Find the voltage across each resistor in series

EXAMPLE 4–7

Some resistors are not color coded with bands but have the values stamped on the resistor body. When the portion of the circuit board shown in Figure 4–18 was assembled, someone mounted the resistors with the labels turned down, and there is no documentation showing the resistor values. Assume that there are a voltmeter, ammeter, and power supply immediately available, but no ohmmeter. Without removing the resistors from the board, use Ohm's law to determine the resistance of each one.

FIGURE 4–18

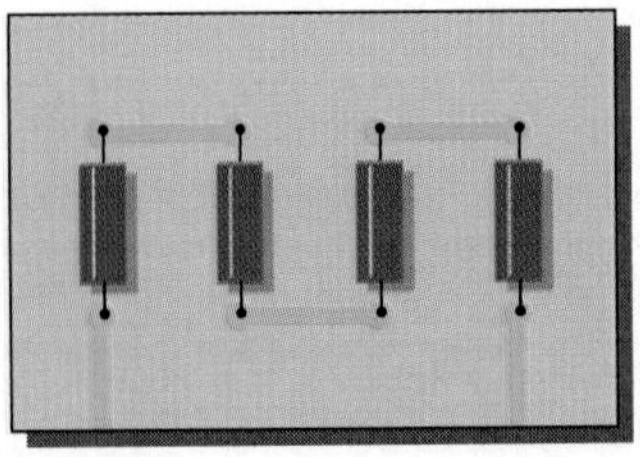

Solution The resistors are all in series, so the current is the same through each one. Measure the current by connecting a 12 V source (arbitrary value) and an ammeter as shown in Figure 4–19. Measure the voltage across each resistor by placing the voltmeter across the first resistor. Then repeat this measurement for the other three resistors. For illustration, the voltage values indicated in color are assumed to be the measured values.

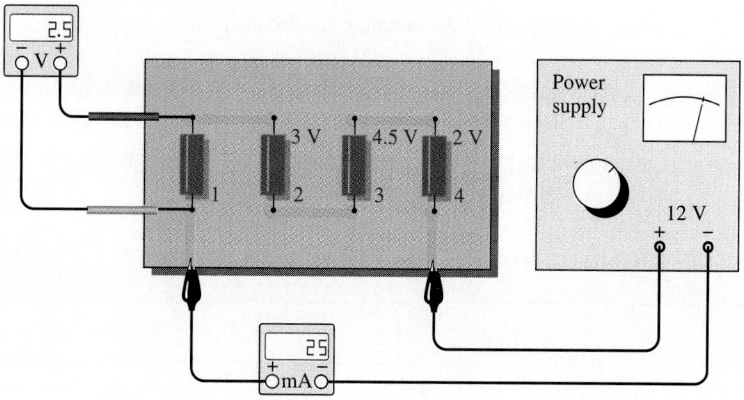

FIGURE 4–19
The voltmeter readings across each resistor are indicated.

Determine the resistance of each resistor by substituting the measured values of current and voltage into the Ohm's law formula as follows:

$$R_1 = \frac{V_1}{I} = \frac{2.5\text{ V}}{25\text{ mA}} = 100\ \Omega$$

$$R_2 = \frac{V_2}{I} = \frac{3\text{ V}}{25\text{ mA}} = 120\ \Omega$$

$$R_3 = \frac{V_3}{I} = \frac{4.5\text{ V}}{25\text{ mA}} = 180\ \Omega$$

$$R_4 = \frac{V_4}{I} = \frac{2\text{ V}}{25\text{ mA}} = 80\ \Omega$$

The calculator sequence for R_1 is

[2] [•] [5] [÷] [2] [5] [EXP] [+/−] [3] [=]

Related Exercise What is an easier way to determine the resistance values?

EXAMPLE 4–8 Find the current in the circuit of Figure 4–20.

FIGURE 4–20

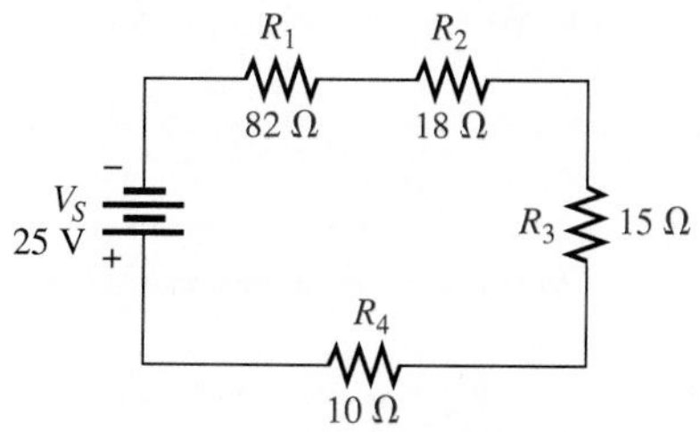

Solution The current is determined by the voltage and the total resistance. First, calculate the total resistance:

$$R_T = R_1 + R_2 + R_3 + R_4$$
$$= 82\ \Omega + 18\ \Omega + 15\ \Omega + 10\ \Omega = 125\ \Omega$$

Next, use Ohm's law to calculate the current:

$$I = \frac{V_S}{R_T} = \frac{25\text{ V}}{125\ \Omega} = 0.2\text{ A} = 200\text{ mA}$$

Remember, the same current flows at all points in the circuit. Thus, each resistor has 200 mA through it.

Related Exercise What is the current in Figure 4–20 if R_4 is changed to 100 Ω?

EXAMPLE 4–9 The current in the circuit of Figure 4–21 is 1 mA. For this amount of current, what must the source voltage V_S be?

FIGURE 4–21

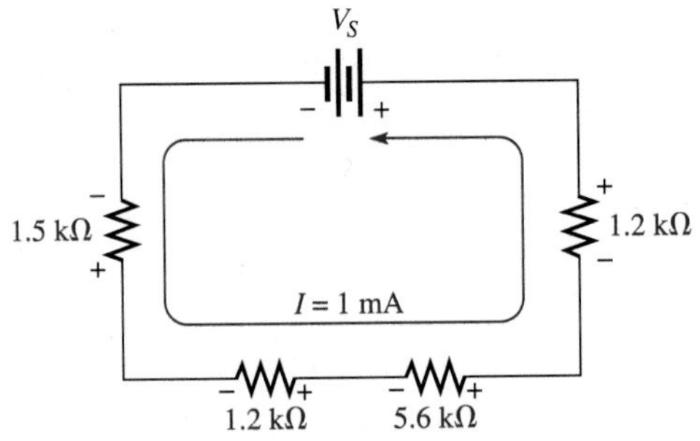

Solution In order to calculate V_S, first determine R_T:

$$R_T = 1.2\text{ k}\Omega + 5.6\text{ k}\Omega + 1.2\text{ k}\Omega + 1.5\text{ k}\Omega = 9.5\text{ k}\Omega$$

Now use Ohm's law to get V_S:

$$V_S = IR_T = (1\text{ mA})(9.5\text{ k}\Omega) = 9.5\text{ V}$$

Related Exercise If the 5.6 kΩ resistor is changed to 3.9 kΩ, what value of V_S is necessary to keep the current at 1 mA?

EXAMPLE 4–10 Calculate the voltage across each resistor in Figure 4–22, and find the value of V_S. To what value can V_S be raised before the 5 mA fuse blows?

FIGURE 4–22

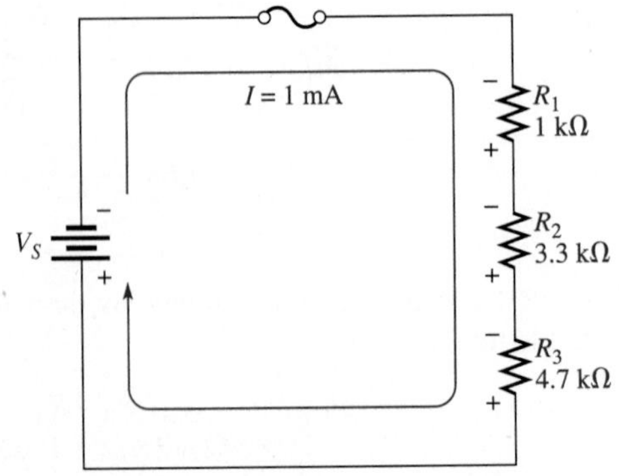

Solution By Ohm's law, the voltage across each resistor is equal to its resistance multiplied by the current through it. Using the Ohm's law formula $V = IR$, determine the voltage across each of the resistors. Keep in mind that the same current flows through each series resistor. The voltage across R_1 is

$$V_1 = IR_1 = (1\text{ mA})(1\text{ k}\Omega) = 1\text{ V}$$

The voltage across R_2 is

$$V_2 = IR_2 = (1\text{ mA})(3.3\text{ k}\Omega) = 3.3\text{ V}$$

The voltage across R_3 is

$$V_3 = IR_3 = (1\text{ mA})(4.7\text{ k}\Omega) = 4.7\text{ V}$$

The source voltage V_S is equal to the current times the total resistance.

$$R_T = 1\text{ k}\Omega + 3.3\text{ k}\Omega + 4.7\text{ k}\Omega = 9\text{ k}\Omega$$

$$V_S = (1\text{ mA})(9\text{ k}\Omega) = 9\text{ V}$$

Notice that if you add the voltage drops of the resistors, they total 9 V, which is the same as the source voltage.

The fuse can handle a maximum current of 5 mA. The maximum value of V_S is therefore

$$V_S = IR_T = (5\text{ mA})(9\text{ k}\Omega) = 45\text{ V}$$

Related Exercise Determine all the voltages in the circuit of Figure 4–22 if R_3 is changed to 2.2 kΩ and I is maintained at 1 mA.

SECTION 4–4 REVIEW

1. A 10 V battery is connected across three 100 Ω resistors in series. What is the current through each resistor?
2. How much voltage is required to produce 5 A through the circuit of Figure 4–23?

FIGURE 4–23

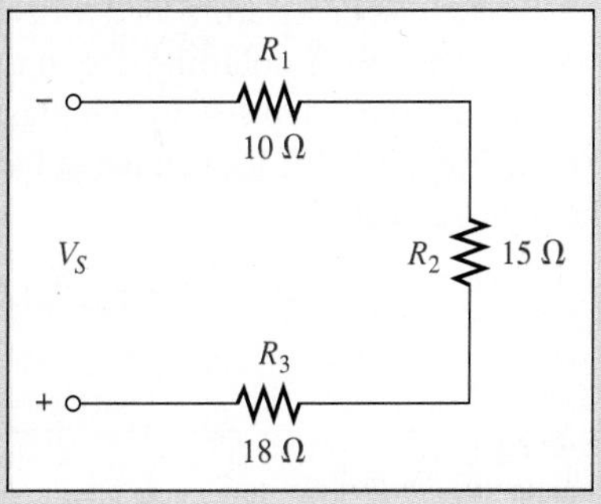

3. How much voltage is dropped across each resistor in Figure 4–23 with 5 A?
4. Four equal-value resistors are connected in series with a 5 V source. The measured current is 4.63 mA. What is the value of each resistor?

4–5 ■ VOLTAGE SOURCES IN SERIES

A voltage source is an energy source that provides a constant voltage to a load. Batteries and power supplies are practical examples. When two or more voltage sources are in series, the total voltage is equal to the algebraic sum of the individual source voltages.

After completing this section, you should be able to

- **Determine the total effect of voltage sources in series**
  - □ Determine the total voltage of series sources with the same polarities
  - □ Determine the total voltage of series sources with opposite polarities

When batteries are placed in a flashlight, they are connected series-aiding to produce a larger voltage, as illustrated in Figure 4–24. In this example, three 1.5 V batteries are placed in series to produce a total voltage (V_{ST}) as follows:

$$V_{ST} = V_{S1} + V_{S2} + V_{S3} = 1.5\text{ V} + 1.5\text{ V} + 1.5\text{ V} = 4.5\text{ V}$$

FIGURE 4–24
Example of series-aiding voltage sources.

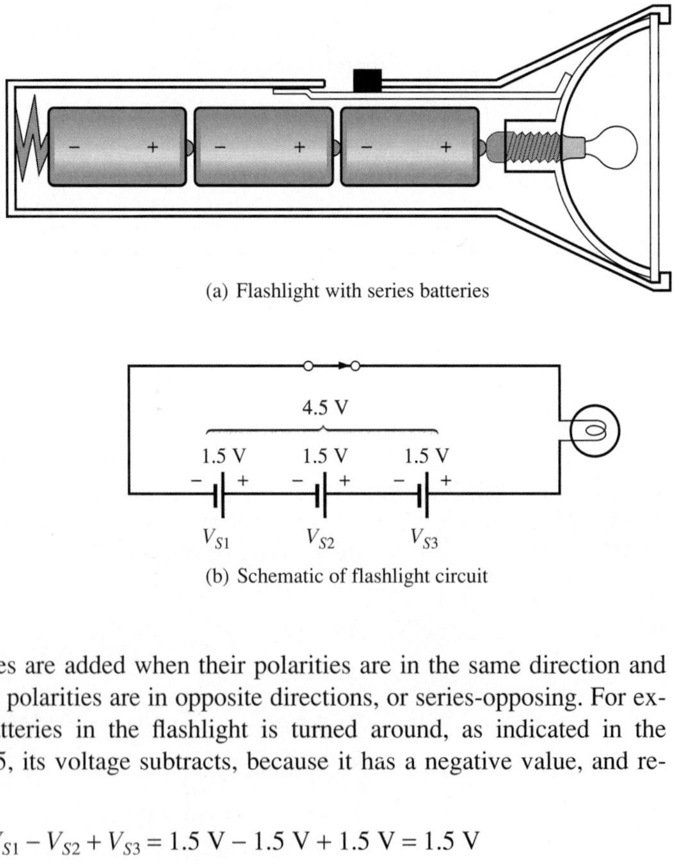

(a) Flashlight with series batteries

(b) Schematic of flashlight circuit

Series voltage sources are added when their polarities are in the same direction and are subtracted when their polarities are in opposite directions, or series-opposing. For example, if one of the batteries in the flashlight is turned around, as indicated in the schematic of Figure 4–25, its voltage subtracts, because it has a negative value, and reduces the total voltage:

$$V_{ST} = V_{S1} - V_{S2} + V_{S3} = 1.5\text{ V} - 1.5\text{ V} + 1.5\text{ V} = 1.5\text{ V}$$

FIGURE 4–25
Opposite polarities subtract.

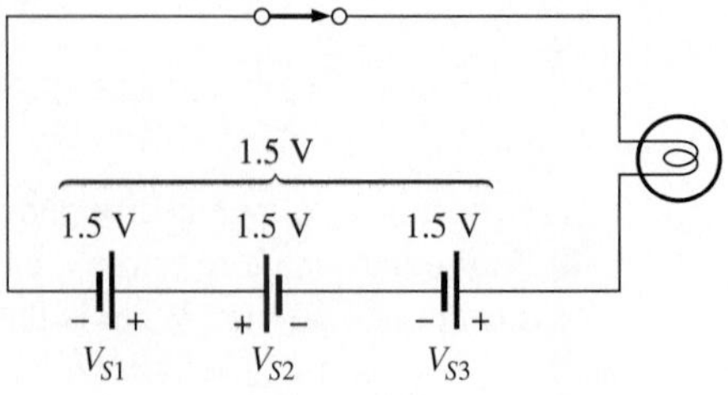

EXAMPLE 4–11 What is the total source voltage (V_{ST}) in Figure 4–26?

FIGURE 4–26

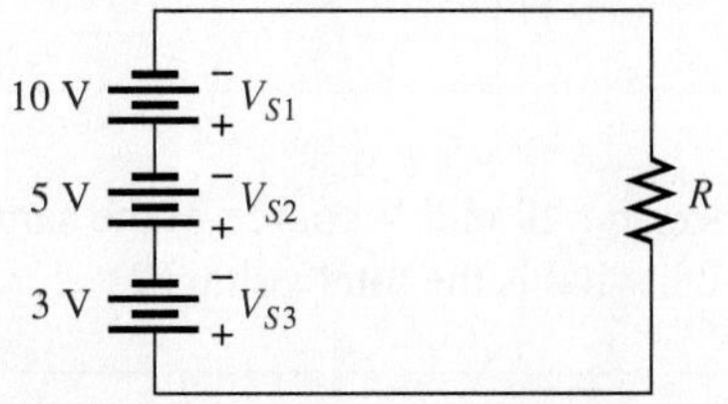

Solution The polarity of each source is the same (the sources are connected in the same direction in the circuit). Therefore, sum the three voltages to get the total:

$$V_{ST} = V_{S1} + V_{S2} + V_{S3} = 10\text{ V} + 5\text{ V} + 3\text{ V} = 18\text{ V}$$

The three individual sources can be replaced by a single equivalent source of 18 V with its polarity as shown in Figure 4–27.

FIGURE 4–27

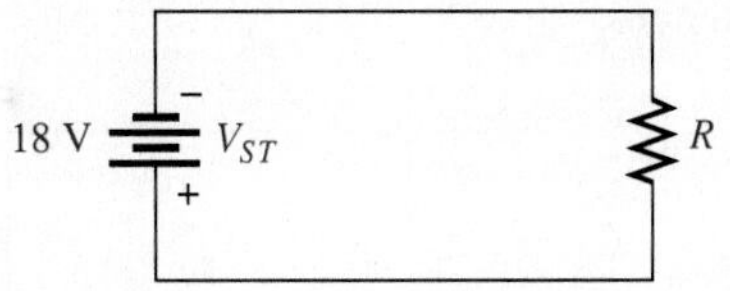

Related Exercise If the battery V_{S3} in Figure 4–26 is reversed, what is the total voltage?

EXAMPLE 4–12 Determine V_{ST} in Figure 4–28.

FIGURE 4–28

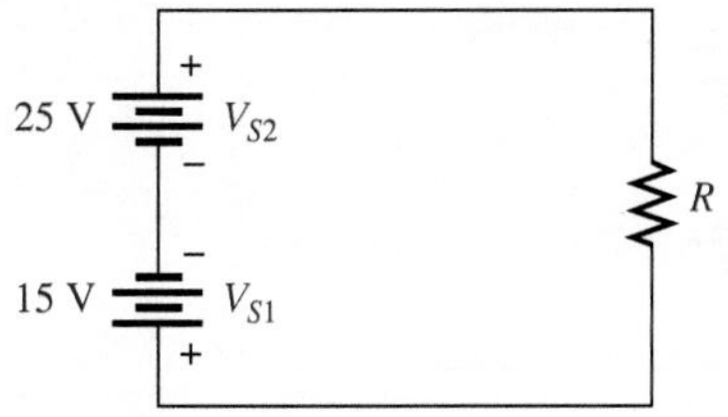

Solution These sources are connected in opposing directions. If you go clockwise around the circuit, you go from plus to minus through V_{S1}, and minus to plus through V_{S2}. The total voltage is the difference of the two source voltages (algebraic sum of oppositely signed values). The total voltage has the same polarity as the larger-value source. Here we will choose V_{S2} to be positive.

$$V_{ST} = V_{S2} - V_{S1} = 25\text{ V} - 15\text{ V} = 10\text{ V}$$

The two sources in Figure 4–28 can be replaced by a 10 V equivalent one with polarity as shown in Figure 4–29.

FIGURE 4–29

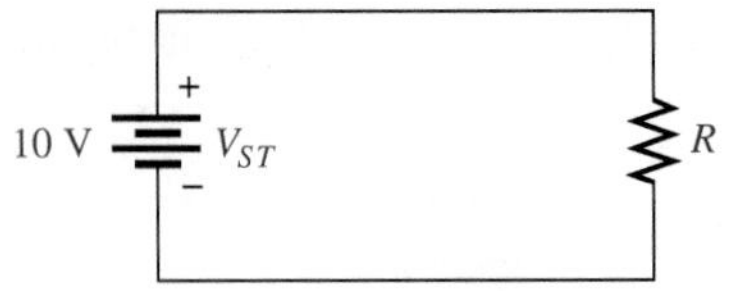

Related Exercise If an 8 V source in the same direction as V_{S1} is connected in series in Figure 4–28, what is the total voltage?

SECTION 4–5 REVIEW

1. How many 12 V batteries must be connected in series to produce 60 V? Sketch a schematic that shows the battery connections.
2. Four 1.5 V flashlight batteries are connected in series positive to negative. What is the total voltage across the bulb?
3. The resistive circuit in Figure 4–30 is used to bias a transistor amplifier. Show how to connect the resistors to two 15 V power supplies in order to get 30 V across the series resistors.

FIGURE 4–30

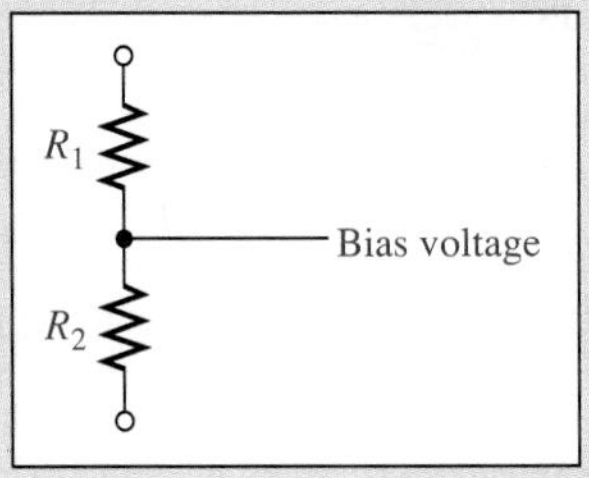

4. Determine the total source voltage in each circuit of Figure 4–31.
5. Sketch the equivalent single source circuit for each circuit of Figure 4–31.

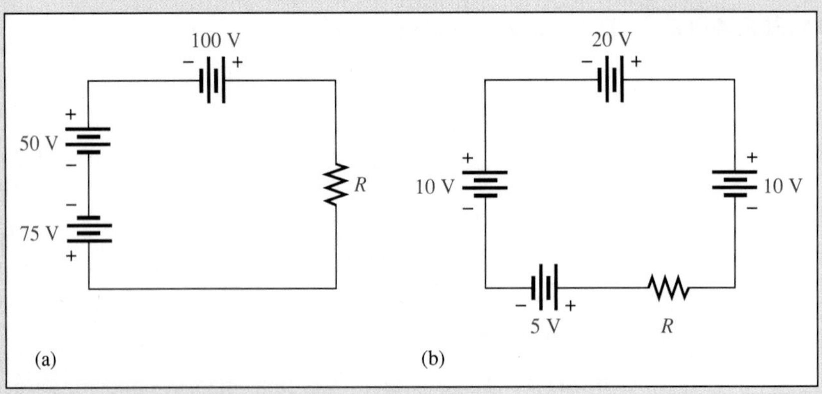

FIGURE 4–31

4–6 ■ KIRCHHOFF'S VOLTAGE LAW

Kirchhoff's voltage law is a fundamental circuit law that states that the algebraic sum of all the voltages around a closed path is zero or, in other words, the sum of the voltage drops equals the total source voltage.

After completing this section, you should be able to

- **Apply Kirchhoff's voltage law**
  - ☐ State Kirchhoff's voltage law
  - ☐ Determine the source voltage by adding the voltage drops
  - ☐ Determine an unknown voltage drop

In an electric circuit, the voltages across the resistors (voltage drops) *always* have polarities opposite to the source voltage polarity. For example, in Figure 4–32, follow a clockwise loop around the circuit and note that the source polarity is plus-to-minus and each voltage drop is minus-to-plus.

Also notice in Figure 4–32 that the current is out of the negative side of the source and through the resistors as indicated by the arrows. The current is into the negative side of each resistor and out the positive side. As the electrons flow through a resistor, they lose energy and are therefore at a lower energy level when they emerge. The lower energy side is less negative (more positive) than the higher energy side. The drop in energy level across a resistor creates a potential difference, or voltage drop, with a minus-to-plus polarity in the direction of the current.

Notice that the voltage from point A to point B in the circuit of Figure 4–32 equals the source voltage, V_S. Also, the voltage from A to B is the sum of the series resistor voltage drops. Therefore, the source voltage is equal to the sum of the three voltage drops.

This discussion is an example of **Kirchhoff's voltage law,** which is generally stated as:

The sum of all the voltage drops around a single closed loop in a circuit is equal to the total source voltage in that loop.

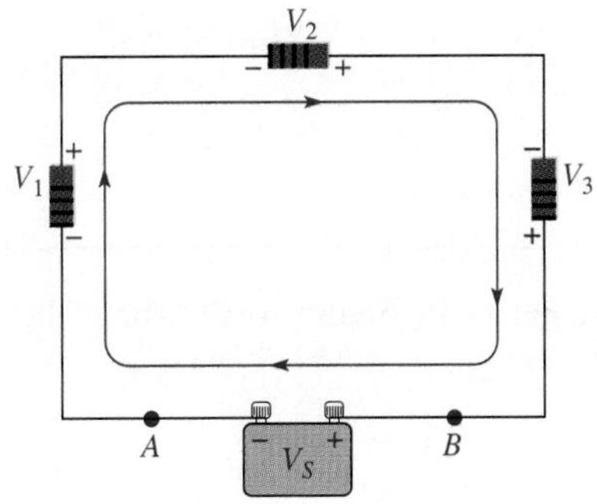

FIGURE 4–32
Kirchhoff's voltage law: The sum of the voltage drops equals the source voltage.

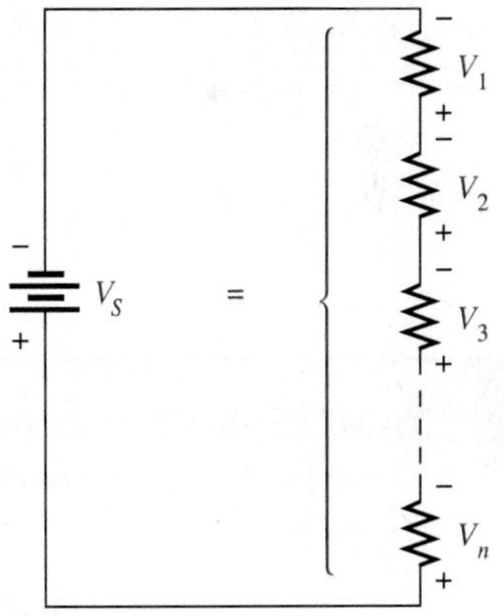

FIGURE 4–33
Sum of n voltage drops equals the source voltage.

The general concept of Kirchhoff's voltage law is illustrated in Figure 4–33 and expressed by Equation (4–3):

$$V_S = V_1 + V_2 + V_3 + \cdots + V_n \qquad \textbf{(4–3)}$$

where n represents the number of voltage drops.

Another Way to State Kirchhoff's Voltage Law

If all the voltage drops around a closed loop are added and then this total is subtracted from the source voltage, the result is zero. This result occurs because the sum of the voltage drops always equals the source voltage.

The algebraic sum of all voltages (both source and drops) around a closed path is zero.

Therefore, another way of expressing Kirchhoff's voltage law is as follows:

$$V_S - V_1 - V_2 - V_3 = 0 \qquad \textbf{(4–4)}$$

You can verify Kirchhoff's voltage law by connecting a circuit and measuring each resistor voltage and the source voltage as illustrated in Figure 4–34. When the resistor voltages are added together, their sum will equal the source voltage. Any number of resistors can be added.

FIGURE 4–34
Experimental verification of Kirchhoff's voltage law.

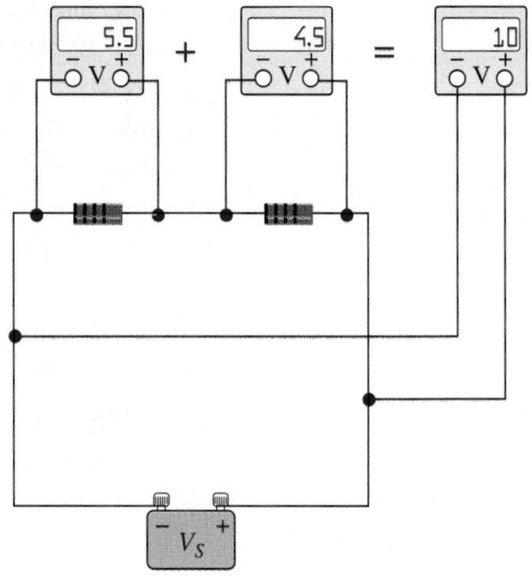

EXAMPLE 4–13 Determine the source voltage V_S in Figure 4–35 where the two voltage drops are given.

FIGURE 4–35

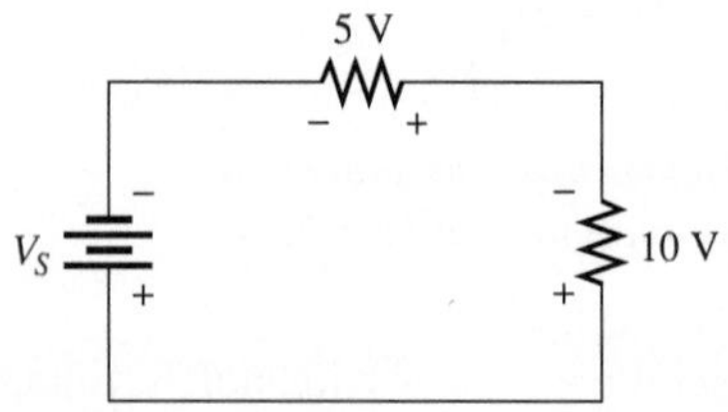

Solution By Kirchhoff's voltage law (Eq. 4–3), the source voltage (applied voltage) must equal the sum of the voltage drops. Adding the voltage drops gives the value of the source voltage:

$$V_S = 5\text{ V} + 10\text{ V} = 15\text{ V}$$

Related Exercise If V_S is increased to 30 V in Figure 4–35, what are the two voltage drops?

EXAMPLE 4–14 Determine the unknown voltage drop, V_3, in Figure 4–36.

FIGURE 4–36

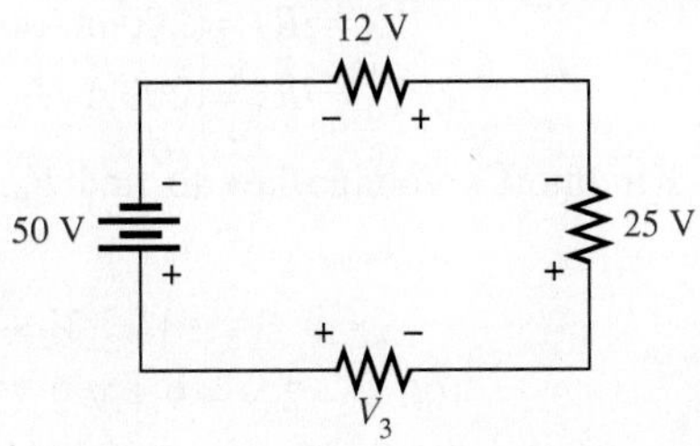

Solution By Kirchhoff's voltage law (Eq. 4–4), the algebraic sum of all the voltages around the circuit is zero (the signs of the voltage drops are opposite the sign of the source).

$$V_S - V_1 - V_2 - V_3 = 0$$

The value of each voltage drop except V_3 is known. Substitute these values into the equation:

$$50\text{ V} - 12\text{ V} - 25\text{ V} - V_3 = 0\text{ V}$$

Next combine the known values:

$$13\text{ V} - V_3 = 0\text{ V}$$

Transpose 13 V to the right side of the equation, and cancel the minus signs:

$$-V_3 = -13\text{ V}$$
$$V_3 = 13\text{ V}$$

The voltage drop across R_3 is 13 V, and its polarity is as shown in Figure 4–36.

Related Exercise Determine V_3 in Figure 4–36 if the source is changed to 25 V.

EXAMPLE 4–15 Find the value of R_4 in Figure 4–37.

FIGURE 4–37

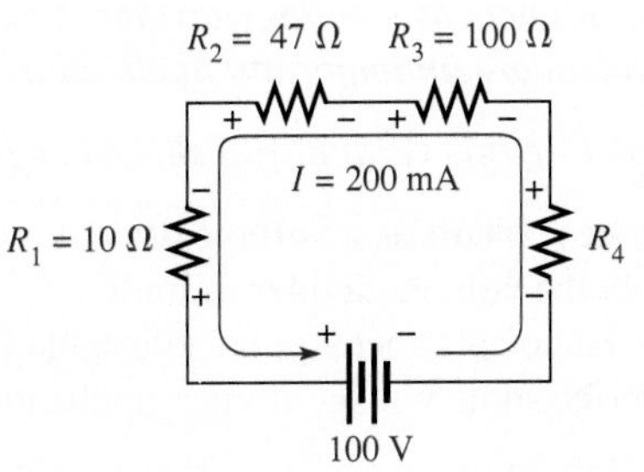

Solution In this problem you must use both Ohm's law and Kirchhoff's voltage law. Follow this procedure carefully.

First, use Ohm's law to find the voltage drop across each of the known resistors:

$$V_1 = IR_1 = (200 \text{ mA})(10\ \Omega) = 2 \text{ V}$$
$$V_2 = IR_2 = (200 \text{ mA})(47\ \Omega) = 9.4 \text{ V}$$
$$V_3 = IR_3 = (200 \text{ mA})(100\ \Omega) = 20 \text{ V}$$

Next, use Kirchhoff's voltage law to find V_4, the voltage drop across the unknown resistor:

$$V_S - V_1 - V_2 - V_3 - V_4 = 0 \text{ V}$$
$$100 \text{ V} - 2 \text{ V} - 9.4 \text{ V} - 20 \text{ V} - V_4 = 0 \text{ V}$$
$$68.6 \text{ V} - V_4 = 0 \text{ V}$$
$$V_4 = 68.6 \text{ V}$$

Now that you know V_4, you can use Ohm's law to calculate R_4 as follows:

$$R_4 = \frac{V_4}{I} = \frac{68.6 \text{ V}}{200 \text{ mA}} = 343\ \Omega$$

R_4 is most likely a 330 Ω resistor because 343 Ω is within a standard tolerance range (+5%) of 330 Ω.

Related Exercise Determine R_4 in Figure 4–37 for $V_S = 150$ V and $I = 200$ mA.

SECTION 4–6 REVIEW

1. State Kirchhoff's voltage law in two ways.
2. A 50 V source is connected to a series resistive circuit. What is the total of the voltage drops in this circuit?
3. Two equal-value resistors are connected in series across a 10 V battery. What is the voltage drop across each resistor?
4. In a series circuit with a 25 V source, there are three resistors. One voltage drop is 5 V, and the other is 10 V. What is the value of the third voltage drop?
5. The individual voltage drops in a series circuit are 1 V, 3 V, 5 V, 7 V, and 8 V. What is the total voltage applied across the series circuit?

4–7 ■ VOLTAGE DIVIDERS

A series circuit acts as a voltage divider. You will learn what this term means and why voltage dividers are an important application of series circuits.

After completing this section, you should be able to

- **Use a series circuit as a voltage divider**
  - ☐ Apply the voltage-divider formula
  - ☐ Use the potentiometer as an adjustable voltage divider
  - ☐ Describe some voltage-divider applications

To illustrate how a series string of resistors acts as a voltage divider, we will examine Figure 4–38(a), where there are two resistors in series. As you already know, there are two voltage drops: one across R_1 and one across R_2. These voltage drops are V_1 and V_2, respectively, as indicated in the diagram.

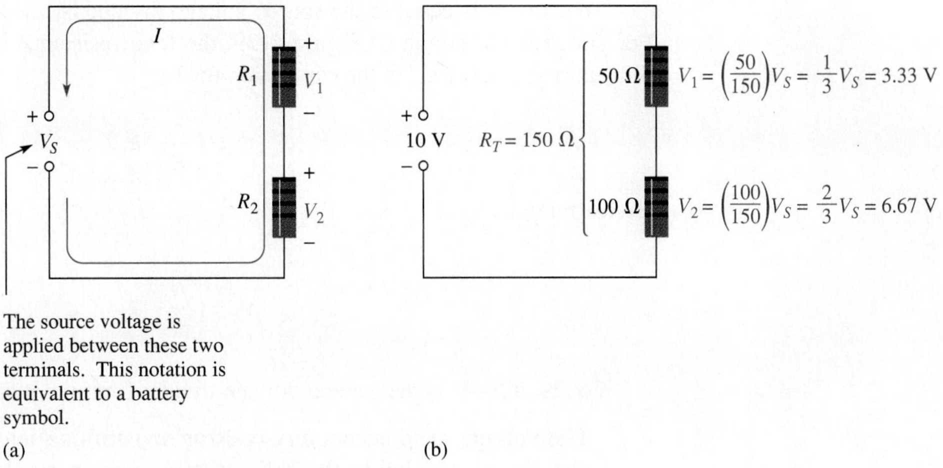

FIGURE 4–38
A two-resistor voltage divider.

Since the same current flows through each resistor, the voltage drops are proportional to the values of the resistors. For example, if the value of R_2 is twice that of R_1, then the value of V_2 is twice that of V_1. In other words, the total voltage drop divides among the series resistors in amounts directly proportional to the resistance values. The smallest resistance has the least voltage, and the largest resistance has the most voltage ($V = IR$). For example, in Figure 4–38(b), if V_S is 10 V, R_1 is 50 Ω, and R_2 is 100 Ω, then V_1 is one-third the total voltage, or 3.33 V, because R_1 is one-third the total resistance. Likewise, V_2 is two-thirds V_S, or 6.67 V, because R_2 is two-thirds the total resistance.

Voltage-Divider Formula

With a few calculations, a formula for determining how the voltages divide among series resistors can be developed. Let's assume that we have several resistors in series as shown in Figure 4–39. This figure shows five resistors, but there can be any number.

Let's call the voltage drop across any one of the resistors V_x, where x represents the number of a particular resistor (1, 2, 3, and so on). By Ohm's law, the voltage drop across any of the resistors in Figure 4–39 can be written as follows:

$$V_x = IR_x$$

where $x = 1, 2, 3, 4,$ or 5.

FIGURE 4–39
Five-resistor voltage divider.

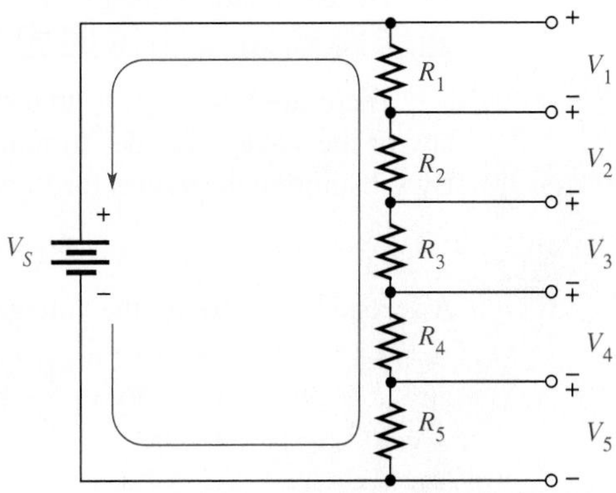

The current is equal to the source voltage divided by the total resistance ($I = V_S/R_T$). For our example circuit of Figure 4–39, the total resistance is $R_1 + R_2 + R_3 + R_4 + R_5$. Substituting V_S/R_T for I in the expression for V_x,

$$V_x = \left(\frac{V_S}{R_T}\right)R_x$$

By rearranging,

$$V_x = \left(\frac{R_x}{R_T}\right)V_S \quad \textbf{(4–5)}$$

Equation (4–5) is the general voltage-divider formula. It can be stated as follows:

The voltage drop across any resistor or combination of resistors in a series circuit is equal to the ratio of that resistance value to the total resistance, multiplied by the source voltage.

The following three examples illustrate use of the voltage-divider formula.

EXAMPLE 4–16

Determine V_1, the voltage across R_1, and V_2, the voltage across R_2, in the voltage divider in Figure 4–40.

FIGURE 4–40

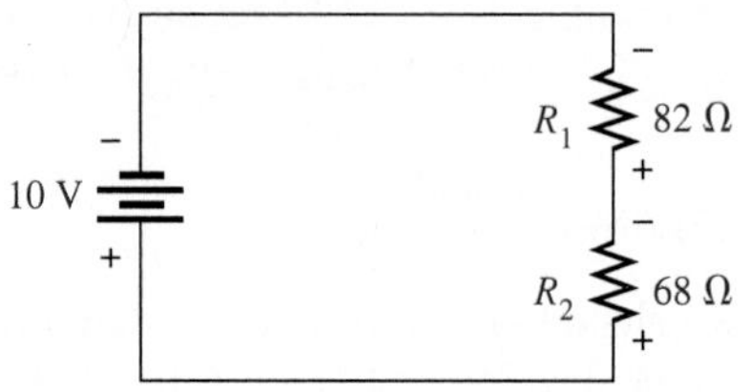

Solution Use the voltage-divider formula, $V_x = (R_x/R_T)V_S$. In this problem you are looking for V_1; so $V_x = V_1$ and $R_x = R_1$. The total resistance is

$$R_T = R_1 + R_2 = 82\ \Omega + 68\ \Omega = 150\ \Omega$$

R_1 is 82 Ω and V_S is 10 V. Substitute these values into the voltage-divider formula:

$$V_1 = \left(\frac{R_1}{R_T}\right)V_S = \left(\frac{82\ \Omega}{150\ \Omega}\right)10\ \text{V} = 5.47\ \text{V}$$

The calculator sequence is

8 2 ÷ 1 5 0 × 1 0 =

There are two ways to find the value of V_2 in this problem: Kirchhoff's voltage law or the voltage-divider formula. First, use Kirchhoff's voltage law, $V_S = V_1 + V_2$. By substituting the values for V_S and V_1, you can solve for V_2 as follows:

$$V_2 = 10\ \text{V} - 5.47\ \text{V} = 4.53\ \text{V}$$

A second way is to use the voltage-divider formula to find V_2.

$$V_2 = \left(\frac{R_2}{R_T}\right)V_S = \left(\frac{68\ \Omega}{150\ \Omega}\right)10\ \text{V} = 4.53\ \text{V}$$

The calculator sequence is
6 8 ÷ 1 5 0 x 1 0 =

Related Exercise Find the voltage drops across R_1 and R_2 if R_2 is changed to 180 Ω in Figure 4–40.

EXAMPLE 4–17 Calculate the voltage drop across each resistor in the voltage divider of Figure 4–41.

FIGURE 4–41

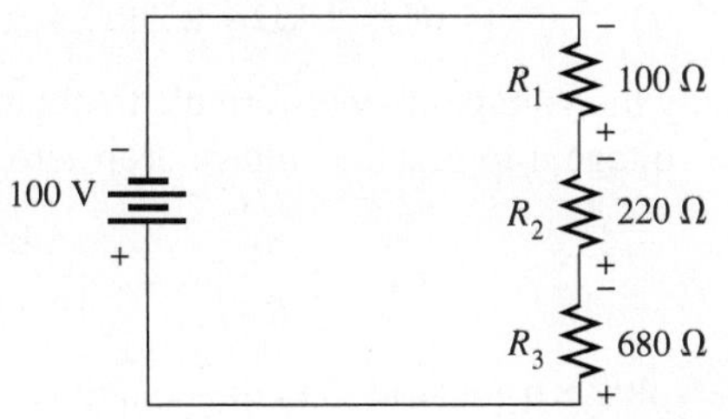

Solution Look at the circuit for a moment and consider the following: The total resistance is 1000 Ω. Ten percent of the total voltage is across R_1 because it is 10% of the total resistance (100 Ω is 10% of 1000 Ω). Likewise, 22% of the total voltage is dropped across R_2 because it is 22% of the total resistance (220 Ω is 22% of 1000 Ω). Finally, R_3 drops 68% of the total voltage (680 Ω is 68% of 1000 Ω).

Because of the convenient values in this problem, it is easy to figure the voltage drops mentally (0.10 × 100 V = 10 V, 0.22 × 100 V = 22 V, and 0.68 × 100 V = 68 V). Such is not always the case, but sometimes a little thinking will produce a result more efficiently and eliminate some calculating.

Although you have already reasoned through this problem, the calculations are

$$V_1 = \left(\frac{R_1}{R_T}\right)V_S = \left(\frac{100\ \Omega}{1000\ \Omega}\right)100\text{ V} = 10\text{ V}$$

$$V_2 = \left(\frac{R_2}{R_T}\right)V_S = \left(\frac{220\ \Omega}{1000\ \Omega}\right)100\text{ V} = 22\text{ V}$$

$$V_3 = \left(\frac{R_3}{R_T}\right)V_S = \left(\frac{680\ \Omega}{1000\ \Omega}\right)100\text{ V} = 68\text{ V}$$

Notice that the sum of the voltage drops is equal to the source voltage, in accordance with Kirchhoff's voltage law. This check is a good way to verify your results.

Related Exercise If R_1 and R_2 are both changed to 680 Ω in Figure 4–41, what are all the voltage drops?

EXAMPLE 4–18 Determine the voltages between the following points in the voltage divider of Figure 4–42: **(a)** A to B **(b)** A to C **(c)** B to C **(d)** B to D **(e)** C to D

FIGURE 4–42

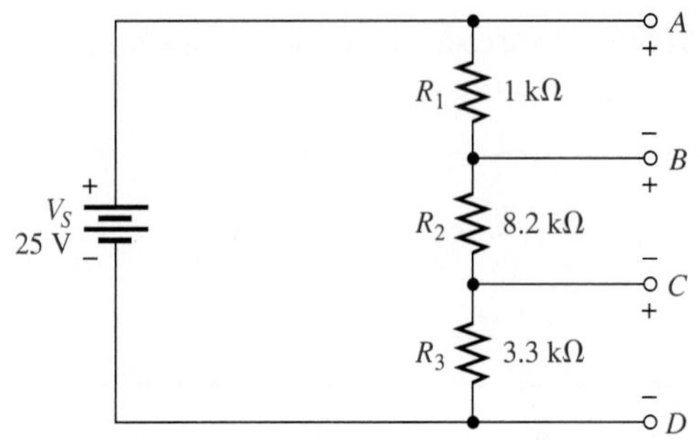

Solution First determine R_T.

$$R_T = 1\ \text{k}\Omega + 8.2\ \text{k}\Omega + 3.3\ \text{k}\Omega = 12.5\ \text{k}\Omega$$

Now apply the voltage-divider formula to obtain each required voltage.

(a) The voltage A to B is the voltage drop across R_1.

$$V_{AB} = \left(\frac{R_1}{R_T}\right)V_S = \left(\frac{1\ \text{k}\Omega}{12.5\ \text{k}\Omega}\right)25\ \text{V} = 2\ \text{V}$$

(b) The voltage from A to C is the combined voltage drop across both R_1 and R_2. In this case, R_x in the general formula given in Equation (4–5) is $R_1 + R_2$.

$$V_{AC} = \left(\frac{R_1 + R_2}{R_T}\right)V_S = \left(\frac{9.2\ \text{k}\Omega}{12.5\ \text{k}\Omega}\right)25\ \text{V} = 18.4\ \text{V}$$

(c) The voltage from B to C is the voltage drop across R_2.

$$V_{BC} = \left(\frac{R_2}{R_T}\right)V_S = \left(\frac{8.2\ \text{k}\Omega}{12.5\ \text{k}\Omega}\right)25\ \text{V} = 16.4\ \text{V}$$

(d) The voltage from B to D is the combined voltage drop across both R_2 and R_3. In this case, R_x in the general formula is $R_2 + R_3$.

$$V_{BD} = \left(\frac{R_2 + R_3}{R_T}\right)V_S = \left(\frac{11.5\ \text{k}\Omega}{12.5\ \text{k}\Omega}\right)25\text{V} = 23\ \text{V}$$

(e) Finally, the voltage from C to D is the voltage drop across R_3.

$$V_{CD} = \left(\frac{R_3}{R_T}\right)V_S = \left(\frac{3.3\ \text{k}\Omega}{12.5\ \text{k}\Omega}\right)25\ \text{V} = 6.6\ \text{V}$$

If you connect this voltage divider in the lab, you can verify each of the calculated voltages by connecting a voltmeter between the appropriate points in each case.

Related Exercise Determine each of the previously calculated voltages if V_S is doubled.

The Potentiometer As an Adjustable Voltage Divider

Recall from Chapter 2 that a potentiometer is a variable resistor with three terminals. A potentiometer connected to a voltage source is shown in Figure 4–43. Notice that the two end terminals are labeled 1 and 2. The adjustable terminal or wiper is labeled 3. The potentiometer acts as a voltage divider, which can be illustrated by separating the total resistance into two parts as shown in Figure 4–43(c). The resistance between terminal 1 and terminal 3 (R_{13})

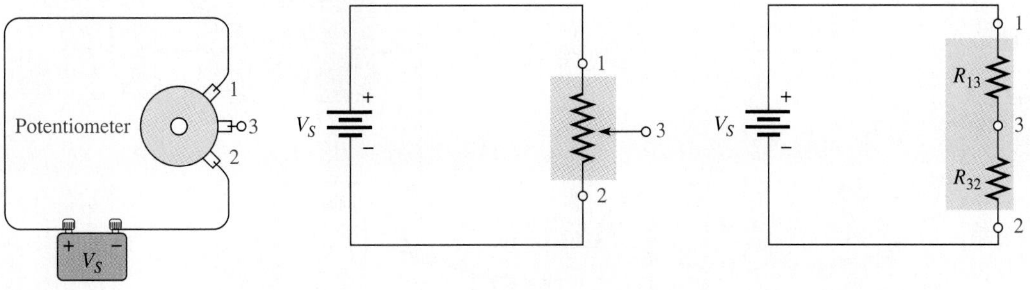

FIGURE 4–43
The potentiometer as a voltage divider.

is one part, and the resistance between terminal 3 and terminal 2 (R_{32}) is the other part. So this potentiometer actually is a two-resistor voltage divider that can be manually adjusted.

Figure 4–44 shows what happens when the wiper terminal (3) is moved. In part (a), the wiper is exactly centered, making the two resistances equal. If we measure the voltage across terminals 3 to 2 as indicated by the voltmeter symbol, we have one-half of the total source voltage. When the wiper is moved up, as in part (b), the resistance between terminals 3 and 2 increases, and the voltage across it increases proportionally. When the wiper is moved down, as in part (c), the resistance between terminals 3 and 2 decreases, and the voltage decreases proportionally.

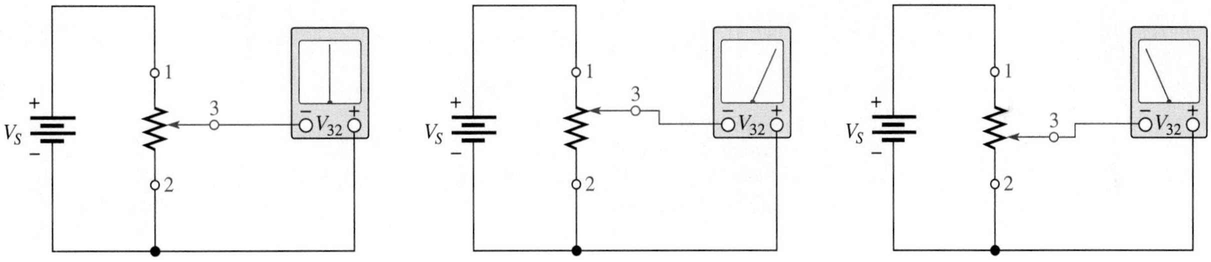

FIGURE 4–44
Adjusting the voltage divider.

Applications of Voltage Dividers

The volume control of radio or TV receivers is a common application of a potentiometer used as a voltage divider. Since the loudness of the sound is dependent on the amount of voltage associated with the audio signal, you can increase or decrease the volume by adjusting the potentiometer, that is, by turning the knob of the volume control on the set. The block diagram in Figure 4–45 shows how a potentiometer can be used for volume control in a typical receiver.

FIGURE 4–45
A voltage divider used for volume control.

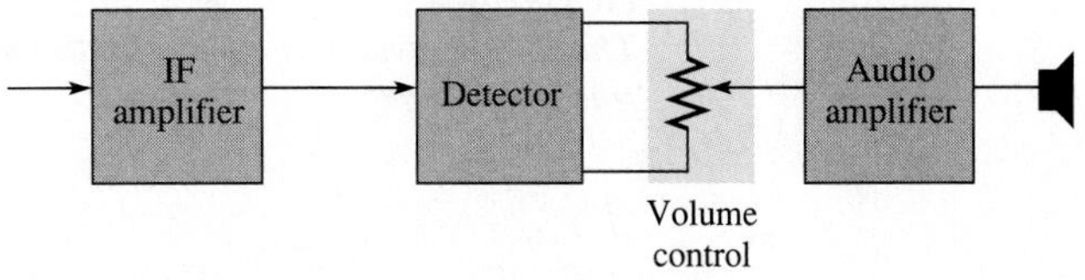

Another application of a voltage divider is illustrated in Figure 4–46, which depicts a potentiometer voltage divider as a fuel-level sensor in an automobile gas tank. As shown in part (a), the float moves up as the tank is filled and moves down as the tank empties.

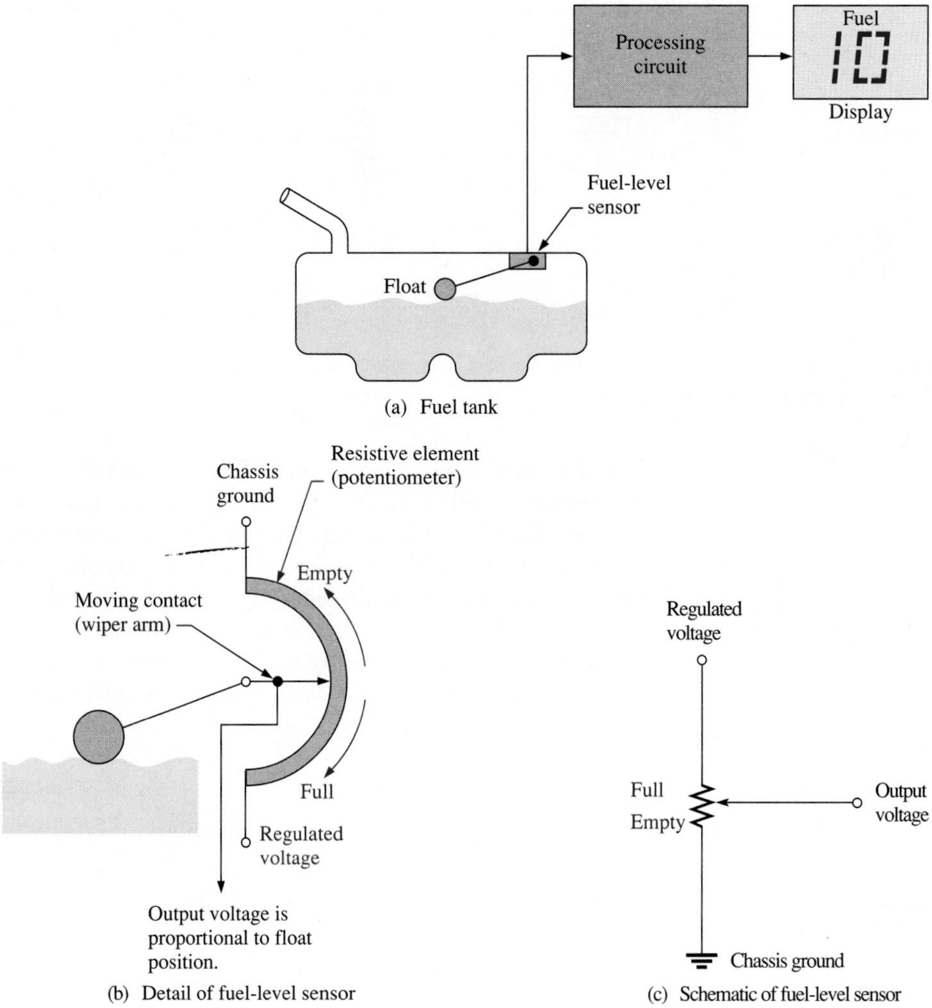

FIGURE 4–46
A potentiometer voltage divider used as an automotive fuel-level sensor.

The float is mechanically linked to the wiper arm of a potentiometer, as shown in part (b). The output voltage varies proportionally with the position of the wiper arm. As the fuel in the tank decreases, the sensor output voltage also decreases. The output voltage goes to the indicator circuitry, which controls the fuel gauge or digital readout to show the fuel level. The schematic of this system is shown in part (c).

Still another application for voltage dividers is in setting the dc operating voltage (bias) in transistor amplifiers. Figure 4–47 shows a voltage divider used for this purpose. You will study transistor amplifiers and biasing later, so it is important that you understand the basics of voltage dividers at this point.

These examples are only three out of many possible applications of voltage dividers.

FIGURE 4–47
The voltage divider as a bias circuit for a transistor amplifier.

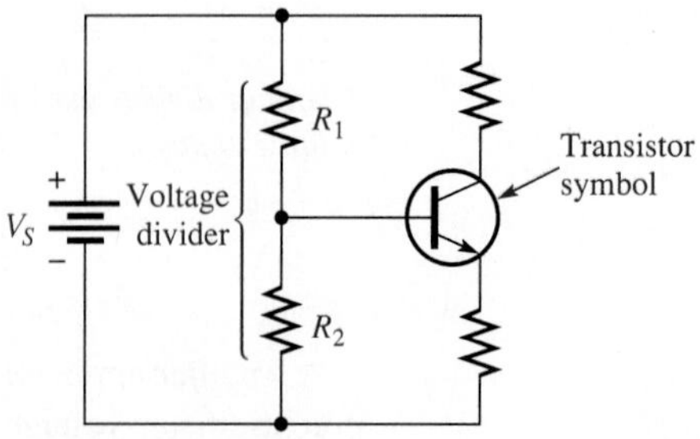

SECTION 4–7 REVIEW

1. What is a voltage divider?
2. How many resistors can there be in a series voltage divider?
3. Write the general voltage-divider formula.
4. If two series resistors of equal value are connected across a 20 V source, how much voltage is there across each resistor?
5. A 56 Ω resistor and an 82 Ω resistor are connected as a voltage divider. The source voltage is 100 V. Sketch the circuit, and determine the voltage across each of the resistors.
6. The circuit of Figure 4–48 is an adjustable voltage divider. If the potentiometer is linear, where would you set the wiper in order to get 5 V from *B* to *A* and 5 V from *C* to *B?*

FIGURE 4–48

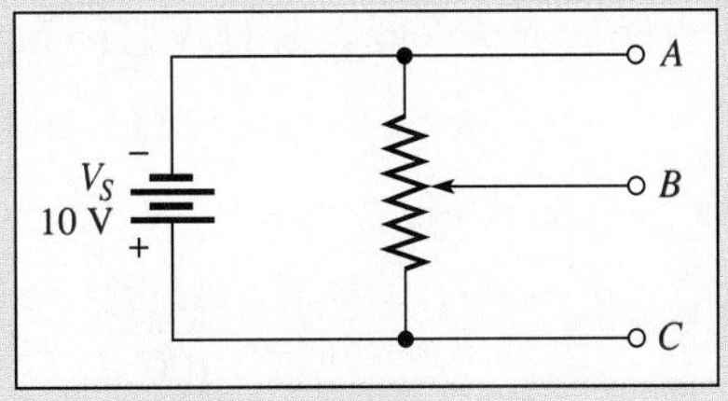

4–8 ▪ POWER IN A SERIES CIRCUIT

The power dissipated by each individual resistor in a series circuit contributes to the total power in the circuit. The individual powers are additive.

After completing this section, you should be able to

- **Determine power in a series circuit**

The total amount of power in a series resistive circuit is equal to the sum of the powers in each resistor in series:

$$P_T = P_1 + P_2 + P_3 + \cdots + P_n \tag{4–6}$$

where n is the number of resistors in series, P_T is the total power, and P_n is the power in the last resistor in series. In other words, the powers are additive.

The power formulas that you learned in Chapter 3 are, of course, directly applicable to series circuits. Since the same current flows through each resistor in series, the following formulas are used to calculate the total power:

$$P_T = V_S I$$
$$P_T = I^2 R_T$$
$$P_T = \frac{V_S^2}{R_T}$$

where V_S is the source voltage across the series circuit and R_T is the total resistance.Example 4–19 illustrates how to calculate total power in a series circuit.

EXAMPLE 4–19 Determine the total amount of power in the series circuit in Figure 4–49.

FIGURE 4–49

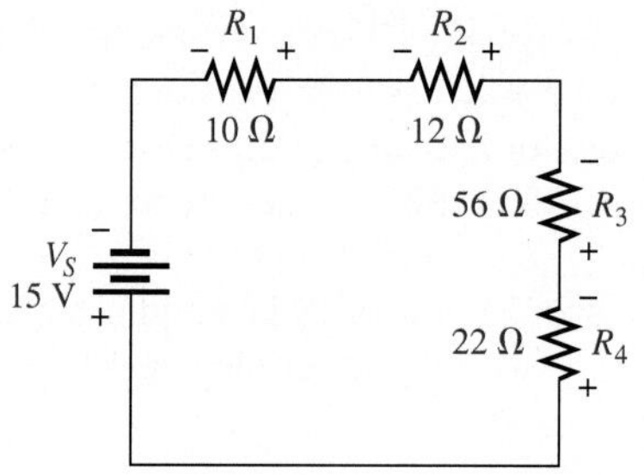

Solution The source voltage is 15 V. The total resistance is

$$R_T = 10\ \Omega + 12\ \Omega + 56\ \Omega + 22\ \Omega = 100\ \Omega$$

The easiest formula to use is $P_T = V_S^2/R_T$, since you know both V_S and R_T.

$$P_T = \frac{V_S^2}{R_T} = \frac{(15\ \text{V})^2}{100\ \Omega} = \frac{225\ \text{V}^2}{100\ \Omega} = 2.25\ \text{W}$$

If you determine the power of each resistor separately and add all these powers, you obtain the same result. Another calculation will illustrate. First, find the current:

$$I = \frac{V_S}{R_T} = \frac{15\ \text{V}}{100\ \Omega} = 0.15\ \text{A}$$

Next, calculate the power for each resistor using $P = I^2R$:

$$P_1 = (0.15\ \text{A})^2(10\ \Omega) = 0.225\ \text{W}$$
$$P_2 = (0.15\ \text{A})^2(12\ \Omega) = 0.270\ \text{W}$$
$$P_3 = (0.15\ \text{A})^2(56\ \Omega) = 1.260\ \text{W}$$
$$P_4 = (0.15\ \text{A})^2(22\ \Omega) = 0.495\ \text{W}$$

Now, add these powers to get the total power:

$$P_T = 0.225\ \text{W} + 0.270\ \text{W} + 1.260\ \text{W} + 0.495\ \text{W} = 2.25\ \text{W}$$

This result shows that the sum of the individual powers is equal to the total power as determined by one of the power formulas.

Related Exercise What is the total power in the circuit of Figure 4–49 if V_S is increased to 30 V?

SECTION 4–8 REVIEW

1. If you know the power in each resistor in a series circuit, how can you find the total power?
2. The resistors in a series circuit have the following powers: 1 W, 2 W, 5 W, and 8 W. What is the total power in the circuit?
3. A circuit has a 100 Ω, a 330 Ω, and a 680 Ω resistor in series. A current of 1 A flows through the circuit. What is the total power?

4–9 ■ CIRCUIT GROUND

Voltage is relative. That is, the voltage at one point in a circuit is always measured relative to another point. For example, if we say that there are +100 V at a certain point in a circuit, we mean that the point is 100 V more positive than some reference point in the circuit. This reference point in a circuit is usually called ground.

After completing this section, you should be able to

- **Determine and identify ground in a circuit**
  - □ Measure voltage with respect to ground
  - □ Define the terms *circuit ground* and *earth ground*

The term *ground* derives from the method used in ac power lines, in which one side of the line is neutralized by connecting it to a water pipe or a metal rod driven into the ground. This method of grounding is called **earth ground.**

In most electronic equipment, a large conductive area on a printed circuit board or the metal chassis that houses the assembly is used as the common or reference point, called the **circuit ground** or *chassis ground,* as illustrated in Figure 4–50. This ground provides a convenient way of connecting all common points within the circuit back to one side of the battery or other energy source. The circuit or chassis ground does not necessarily have to be connected to earth ground. However, in many cases it is earth grounded in order to prevent a shock hazard due to a potential difference between circuit and earth ground.

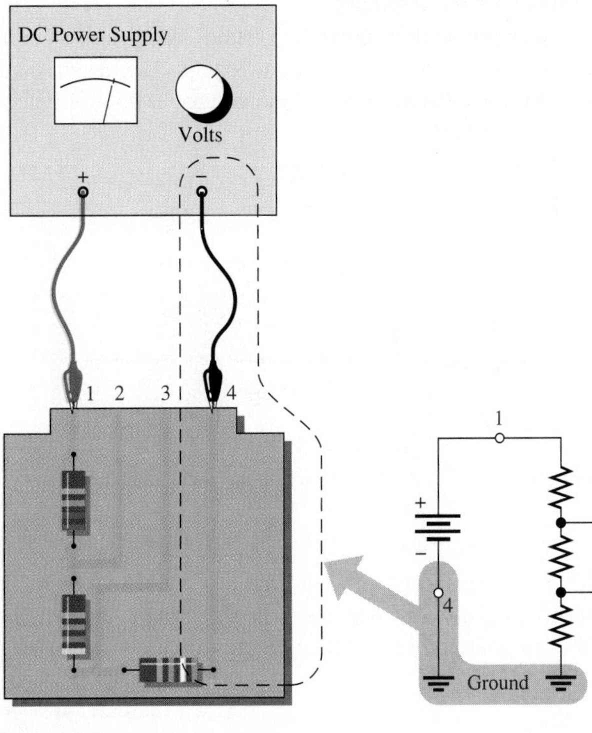

FIGURE 4–50
Simple illustration of circuit ground.

In summary, **ground** is the reference point in electronic circuits. It has a potential of zero volts (0 V) with respect to all other points in the circuit that are referenced to it, as illustrated in Figure 4–51. In part (a), the negative side of the source is grounded, and all voltages indicated are positive with respect to ground. In part (b), the positive side of the source is ground. The voltages at all other points are therefore negative with respect to ground.

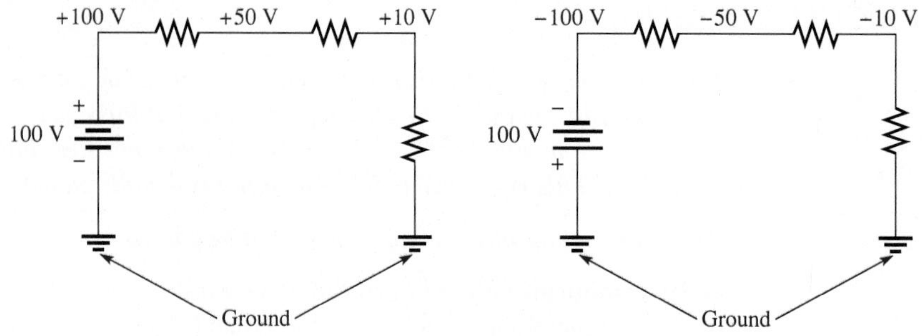

(a) Negative ground

(b) Positive ground

FIGURE 4–51
Examples of negative and positive grounds.

Measurement of Voltages with Respect to Ground

When voltages are measured in a circuit, one meter lead is connected to the circuit ground, and the other to the point at which the voltage is to be measured. In a negative ground circuit, the negative meter terminal is connected to the circuit ground. The positive terminal of the voltmeter is then connected to the positive voltage point. Measurement of positive voltage is illustrated in Figure 4–52, where the meter reads the voltage at point *A* with respect to ground.

For a circuit with a positive ground, the positive voltmeter lead is connected to ground, and the negative lead is connected to the negative voltage point, as indicated in Figure 4–53. Here the meter reads the voltage at point *A* with respect to ground.

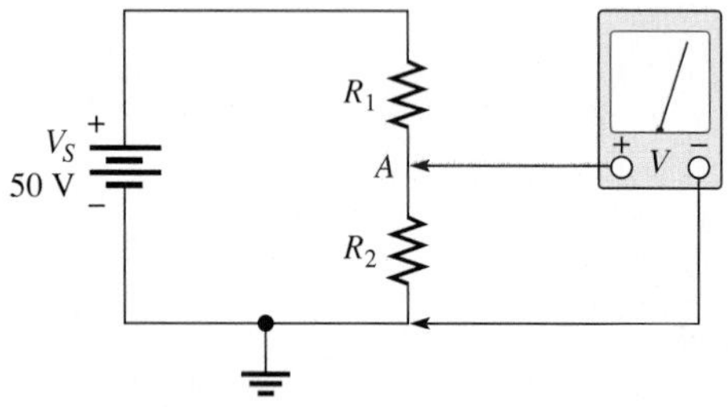

FIGURE 4–52
Measuring a voltage with respect to negative ground.

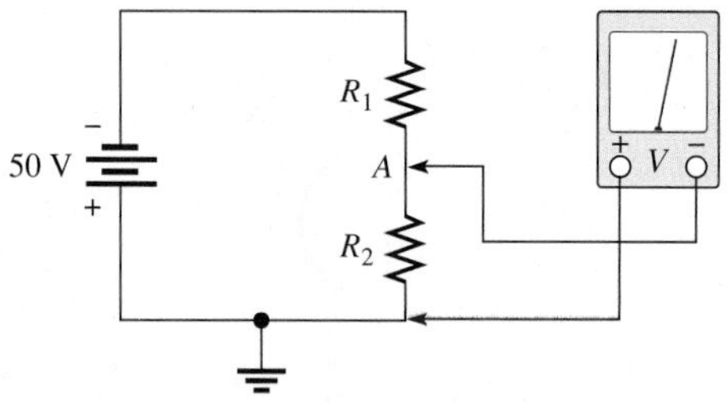

FIGURE 4–53
Measuring a voltage with respect to positive ground.

When voltages must be measured at several points in a circuit, the ground lead can be clipped to ground at one point in the circuit and left there. The other lead is then moved from point to point as the voltages are measured. This method is illustrated in Figure 4–54.

Measurement of Voltage across an Ungrounded Resistor

Voltage can normally be measured across a resistor, as shown in Figure 4–55, even though neither side of the resistor is connected to circuit ground. In some cases, when the meter is not isolated from power line ground, the negative lead of the meter will ground one side of the resistor and alter the operation of the circuit. In this situation, another method must be used, as illustrated in Figure 4–56. The voltages on each side of the resistor are measured with respect to ground. The difference of these two measurements is the voltage drop across the resistor.

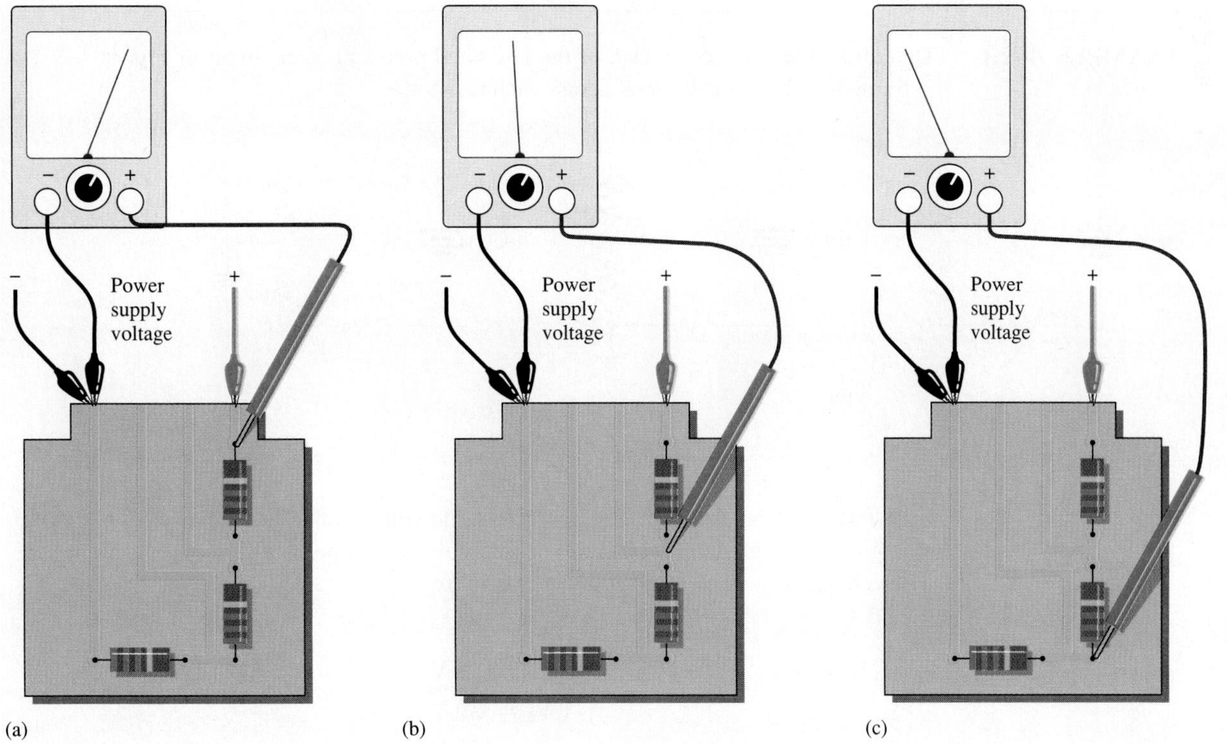

FIGURE 4–54
Measuring voltages at several points in a circuit.

FIGURE 4–55
Measuring voltage across a resistor.

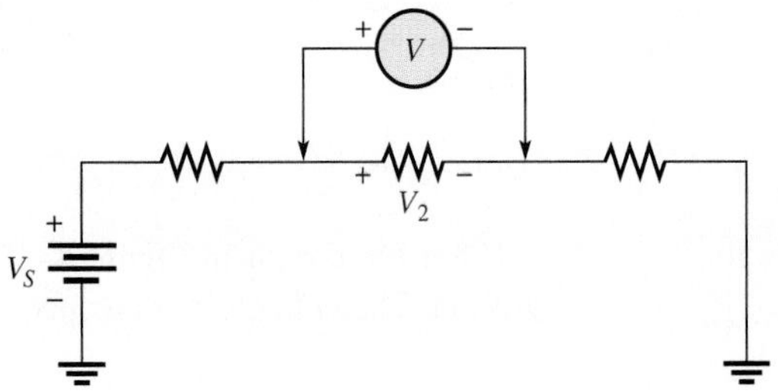

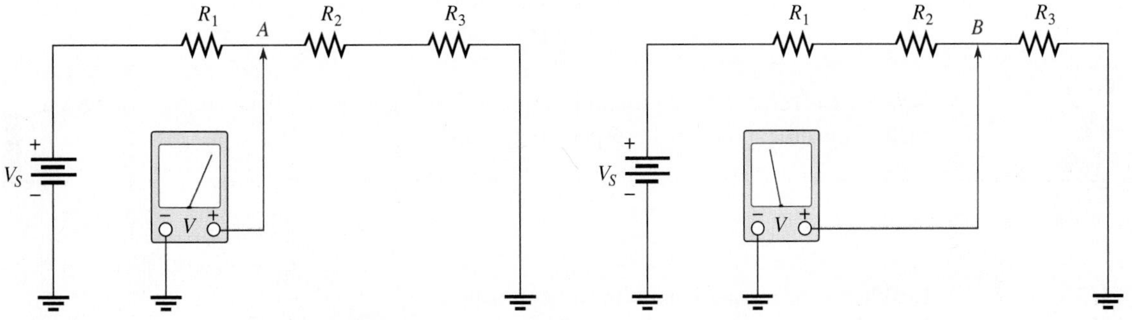

$$V_2 = V_A - V_B$$

FIGURE 4–56
Measuring voltage across a resistor with two separate measurements to ground.

EXAMPLE 4–20 Determine the voltages at each of the indicated points in each circuit of Figure 4–57. Assume that 25 V are dropped across each resistor.

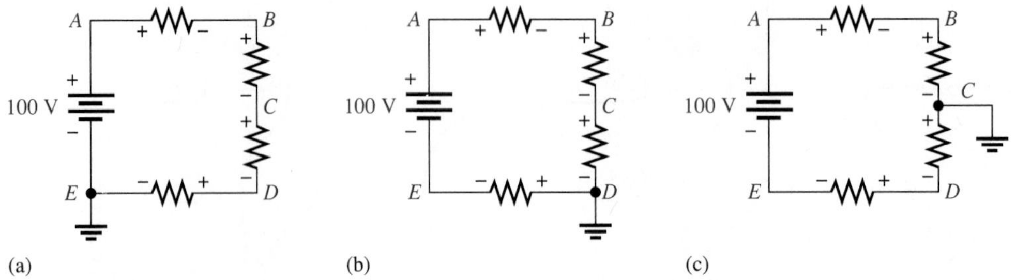

FIGURE 4–57

Solution For the circuit in Figure 4–57(a), the voltage polarities are as shown. Point *E* is ground. Single-letter subscripts denote voltage at a point with respect to ground. The voltages with respect to ground are as follows:

$$V_E = 0\text{ V}$$
$$V_D = +25\text{ V}$$
$$V_C = +50\text{ V}$$
$$V_B = +75\text{ V}$$
$$V_A = +100\text{ V}$$

For the circuit in Figure 4–57(b), the voltage polarities are as shown. Point *D* is ground. The voltages with respect to ground are as follows:

$$V_E = -25\text{ V}$$
$$V_D = 0\text{ V}$$
$$V_C = +25\text{ V}$$
$$V_B = +50\text{ V}$$
$$V_A = +75\text{ V}$$

For the circuit in Figure 4–57(c), the voltage polarities are as shown. Point *C* is ground. The voltages with respect to ground are as follows:

$$V_E = -50\text{ V}$$
$$V_D = -25\text{ V}$$
$$V_C = 0\text{ V}$$
$$V_B = +25\text{ V}$$
$$V_A = +50\text{ V}$$

Related Exercise If the ground is at point *A* in Figure 4–57, what are the voltages at each of the other points with respect to ground?

SECTION 4–9 REVIEW

1. What is the common point in a circuit called?
2. Most voltages in a circuit are referenced to ground (T or F).
3. The housing or chassis is often used as circuit ground (T or F).
4. What is the symbol for ground?
5. What does *earth ground* mean?

4–10 ■ TROUBLESHOOTING

Open resistors or contacts and one point shorted to another are common problems in all circuits including series circuits.

After completing this section, you should be able to

- **Troubleshoot series circuits**
  - □ Check for an open circuit
  - □ Check for a short circuit
  - □ Identify primary causes of shorts

Open Circuit

The most common failure in a series circuit is an open. For example, when a resistor or a lamp burns out, it causes a break in the current path and creates an **open circuit** as illustrated in Figure 4–58.

An open in a series circuit prevents current.

FIGURE 4–58
An open circuit prevents current.

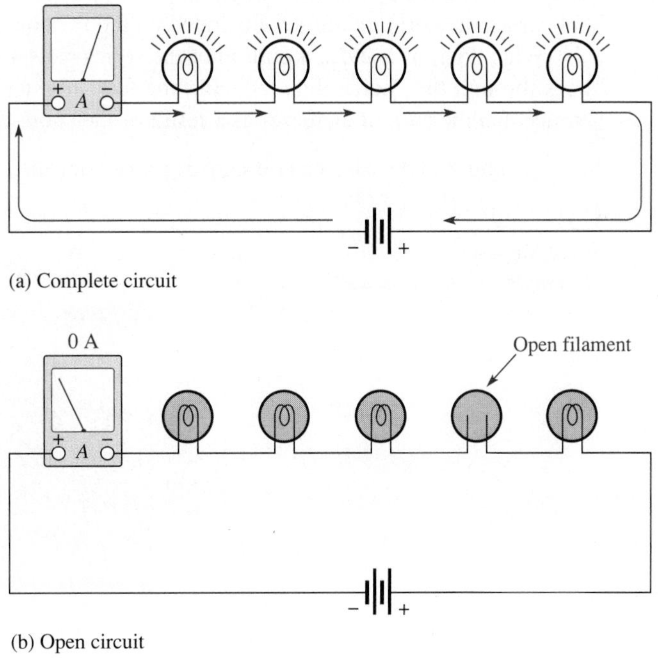

How to Check for an Open Element Sometimes a visual check will reveal a charred resistor or an open lamp filament. However, it is possible for a resistor to open without showing visible signs of damage. In this situation, a voltage check of the series circuit is required. The general procedure is as follows: *Measure the voltage across each resistor in series. The voltage across all of the good resistors will be zero. The voltage across the open resistor will equal the total voltage across the series combination.*

This condition occurs because an open resistor will prevent current through the series circuit. With no current, there can be no voltage drop across any of the good resistors. Since $IR = (0\text{ A})R = 0\text{ V}$, in accordance with Ohm's law, the voltage on each side of a good resistor is the same. The total voltage must then appear across the open resistor in accordance with Kirchhoff's voltage law, as illustrated in Figure 4–59. To fix the circuit, replace the open resistor.

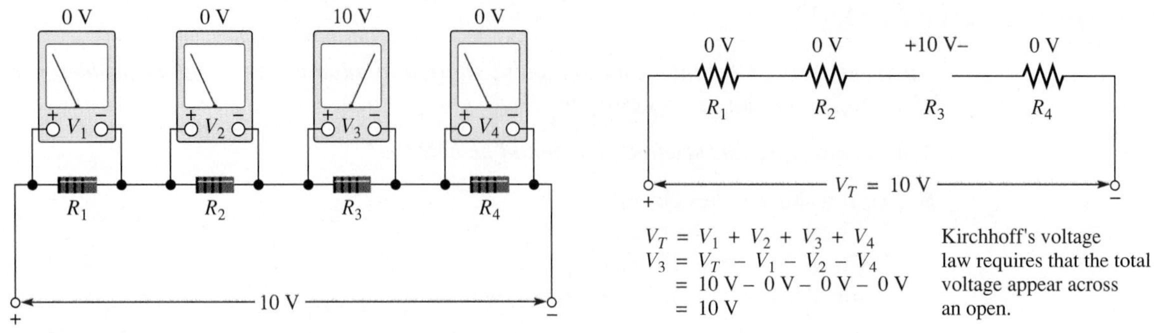

(a) Voltage check shows that R_3 is open.

(b)

FIGURE 4–59
Troubleshooting a series circuit for an open element.

Short Circuit

Sometimes a **short circuit** occurs when two conductors touch or a foreign object such as solder or a wire clipping accidentally connects two sections of a circuit together. This situation is particularly common in circuits with a high component density. Several potential causes of short circuits are illustrated on the PC board in Figure 4–60.

When there is a short, a portion of the series resistance is bypassed (all of the current goes through the short), thus reducing the total resistance as illustrated in Figure 4–61. Notice that the current increases as a result of the short.

A short in a series circuit causes more current.

FIGURE 4–60
Examples of shorts on a PC board.

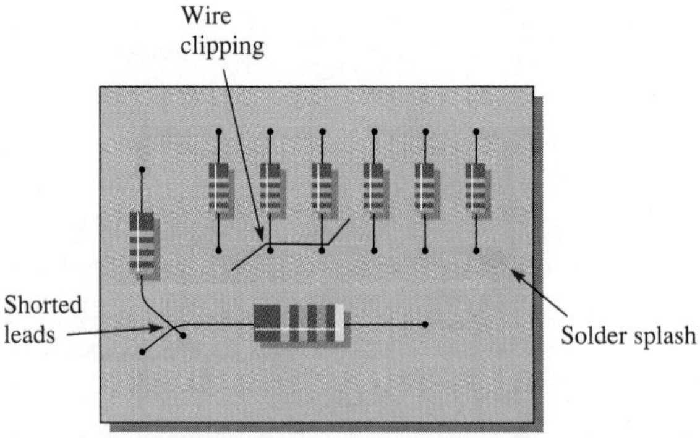

FIGURE 4–61
The effect of a short in a series circuit.

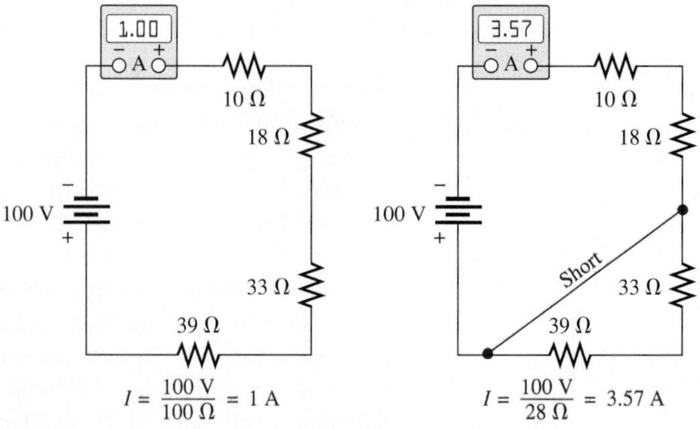

(a) Before short

(b) After short

SECTION 4–10 REVIEW

1. Define *open circuit.*
2. Define *short circuit.*
3. What happens when a series circuit opens?
4. Name two general ways in which an open circuit can occur in practice. Why might a short circuit occur?
5. When a resistor fails, it will normally open (T or F).
6. The total voltage across a string of series resistors is 24 V. If one of the resistors is open, how much voltage is there across it? How much is there across each of the good resistors?

4–11 ■ APPLICATION ASSIGNMENT

For this assignment, your supervisor has given you a voltage-divider board to evaluate and modify if necessary. It will be used to obtain five different voltage levels from a 12 V battery that has a 6.5 Ah rating. The voltage divider is to be used to provide positive reference voltages to an electronic circuit in an analog-to-digital converter. Your task will be to check the circuit to see if it provides the following voltages within a tolerance of ±5% with respect to the negative side of the battery: 10.4 V, 8.0 V, 7.3 V, 6.0 V, and 2.7 V. If the existing circuit does not provide the specified voltages, you will modify it so that it does. Also, you must make sure that the power ratings of the resistors are adequate for the application and determine how long the battery will last with the voltage divider connected to it.

Step 1: Draw the Schematic of the Circuit

Use Figure 4–62 to determine the resistor values and draw the schematic of the voltage-divider circuit so you will know what you are working with. All the resistors on the board are ¼ W.

FIGURE 4–62

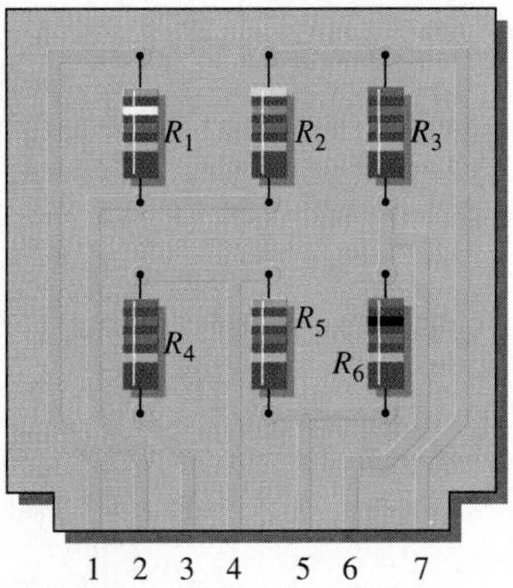

Step 2: Determine the Voltages

Determine each output voltage on the existing circuit board when the positive side of the 12 V battery is connected to pin 3 and the negative side is connected to pin 1. Compare the existing output voltages to the following specifications:

Pin 1: negative terminal of 12 V battery

Pin 2: 2.7 V ± 5%

Pin 3: positive terminal of 12 V battery

Pin 4: 10.4 V ± 5%

Pin 5: 8.0 V ± 5%

Pin 6: 7.3 V + 5%

Pin 7: 6.0 V ± 5%

Step 3: Modify The Existing Circuit (if necessary)

If the output voltages of the existing circuit are not the same as those stated in the specifications of Step 2, make the necessary changes in the circuit to meet the specifications. Draw a schematic of the modified circuit showing resistor values and adequate power ratings.

Step 4: Determine the Life of the Battery

Find the total current drawn from the 12 V battery when the voltage-divider circuit is connected and determine how many days the 6.5 Ah battery will last.

Step 5: Develop a Test Procedure

Determine how you would test the voltage-divider board and what instruments you would use and then detail your test procedure in a step-by-step format.

Step 6: Troubleshoot the Circuit

Determine the most likely fault for each of the following cases (voltages are with respect to the negative battery terminal (pin 1 on the circuit board):

1. No voltage at any of the pins on the circuit board
2. 12 V at pins 3 and 4. All other pins have 0 V.
3. 12 V at all pins except 0 V at pin 1
4. 12 V at pin 6 and 0 V at pin 7
5. 3.3 V at pin 2

SECTION 4–11 REVIEW

1. What is the total power dissipated by the voltage-divider circuit with a 12 V battery?
2. What are the output voltages from the voltage divider if a 6 V battery is used?
3. When the voltage-divider board is connected to the electronic circuit to which it is providing positive reference voltages, which pin on the board should be connected to the ground of the electronic circuit?

■ SUMMARY

- The same amount of current flows at all points in a series circuit.
- The total series resistance is the sum of all resistors in the series circuit.
- The total resistance between any two points in a series circuit is equal to the sum of all resistors connected in series between those two points.
- If all of the resistors in a series circuit are of equal value, the total resistance is the number of resistors multiplied by the resistance value.

- Voltage sources in series add algebraically.
- First statement of Kirchhoff's voltage law: The sum of the voltage drops equals the total source voltage.
- Second statement of Kirchhoff's voltage law: The sum of all the voltages around a closed path is zero.
- The voltage drops in a circuit are always opposite in polarity to the total source voltage.
- Current is out of the negative side of a source and into the positive side.
- Current is into the negative side of each resistor and out of the positive side.
- A voltage divider is a series arrangement of resistors.
- A voltage divider is so named because the voltage drop across any resistor in the series circuit is divided down from the total voltage by an amount proportional to that resistance value in relation to the total resistance.
- A potentiometer can be used as an adjustable voltage divider.
- The total power in a resistive circuit is the sum of all the individual powers of the resistors making up the series circuit.
- All voltages in a circuit are referenced to ground unless otherwise specified.
- Ground is zero volts with respect to all points referenced to it in the circuit.
- *Negative ground* is the term used when the negative side of the source is grounded.
- *Positive ground* is the term used when the positive side of the source is grounded.
- The voltage across an open series element equals the source voltage.

■ GLOSSARY

Circuit ground A method of grounding whereby a large conducive area on a printed circuit board or the metal chassis that houses the assembly is used as the common or reference point; also called *chassis ground.*

Earth ground A method of grounding whereby one side of a power line is neutralized by connecting it to a water pipe or a metal rod driven into the ground.

Ground In electronic circuits, the common or reference point.

Kirchhoff's voltage law A law stating that (1) the sum of the voltage drops around a closed loop equals the source voltage or (2) the sum of all the voltages (drops and sources) around a closed loop is zero.

Open circuit A circuit in which the current path is broken.

Series In an electrical circuit, a relationship of components in which the components are connected such that they provide a single current path between two points.

Short circuit A zero or abnormally low resistance between two points; usually an inadvertant condition.

■ FORMULAS

(4–1)	$R_T = R_1 + R_2 + R_3 + \cdots + R_n$	Total resistance of *n* resistors in series
(4–2)	$R_T = nR$	Total resistance of *n* equal-value resistors in series
(4–3)	$V_S = V_1 + V_2 + V_3 + \cdots V_n$	Kirchhoff's voltage law
(4–4)	$V_S - V_1 - V_2 - V_3 = 0$	Kirchhoff's voltage law
(4–5)	$V_x = \left(\dfrac{R_x}{R_T}\right)V_S$	Voltage-divider formula
(4–6)	$P_T = P_1 + P_2 + P_3 + \cdots + P_n$	Total power

■ SELF-TEST

1. Five resistors are connected in series and there is a current of 2 A into the first resistor. The amount of current out of the second resistor is
(a) 2 A **(b)** 1 A **(c)** 4 A **(d)** 0.4 A

2. To measure the current out of the third resistor in a circuit consisting of four series resistors, an ammeter can be placed
(a) between the third and fourth resistors **(b)** between the second and third resistors
(c) at the positive terminal of the source **(d)** at any point in the circuit

3. When a third resistor is connected in series with two series resistors, the total resistance
   (a) remains the same (b) increases (c) decreases (d) increases by one-third
4. When one of four series resistors is removed from a circuit and the circuit reconnected, the current
   (a) decreases by the amount of current through the removed resistor
   (b) decreases by one-fourth
   (c) quadruples
   (d) increases
5. A series circuit consists of three resistors with values of 100 Ω, 220 Ω, and 330Ω. The total resistance is
   (a) less than 100 Ω (b) the average of the values (c) 550 Ω (d) 650 Ω
6. A 9 V battery is connected across a series combination of 68 Ω, 33 Ω, 100 Ω, and 47 Ω resistors. The amount of current is
   (a) 36.3 mA (b) 27.6 A (c) 22.3 mA (d) 363 mA
7. While putting four 1.5 V batteries in a flashlight, you accidentally put one of them in backward. The light will be
   (a) brighter than normal (b) dimmer than normal (c) off (d) the same
8. If you measure all the voltage drops and the source voltage in a series circuit and add them together, taking into consideration the polarities, you will get a result equal to
   (a) the source voltage (b) the total of the voltage drops
   (c) zero (d) the total of the source voltage and the voltage drops
9. There are six resistors in a given series circuit and each resistor has 5 V dropped across it. The source voltage is
   (a) 5 V (b) 30 V (c) dependent on the resistor values (d) dependent on the current
10. A series circuit consists of a 4.7 kΩ, a 5.6 kΩ, and a 10 kΩ resistor. The resistor that has the most voltage across it is
    (a) the 4.7 kΩ (b) the 5.6 kΩ
    (c) the 10 kΩ (d) impossible to determine from the given information
11. Which of the following series combinations dissipates the most power when connected across a 100 V source?
    (a) one 100 Ω resistor (b) two 100 Ω resistors
    (c) three 100 Ω resistors (d) four 100 Ω resistors
12. The total power in a certain circuit is 10 W. Each of the five equal-value series resistors making up the circuit dissipates
    (a) 10 W (b) 50 W (c) 5 W (d) 2 W
13. When you connect an ammeter in a series resistive circuit and turn on the source voltage, the meter reads zero. You should check for
    (a) a broken wire (b) a shorted resistor (c) an open resistor (d) both (a) and (c)
14. While checking out a series resistive circuit, you find that the current is higher than it should be. You should look for
    (a) an open circuit (b) a short (c) a low resistor value (d) both (b) and (c)

■ BASIC PROBLEMS

SECTION 4–1 Resistors in Series

1. Connect each set of resistors in Figure 4–63 in series between points *A* and *B*.
2. Determine which resistors in Figure 4–64 are in series. Show how to interconnect the pins to put all the resistors in series.

FIGURE 4–63

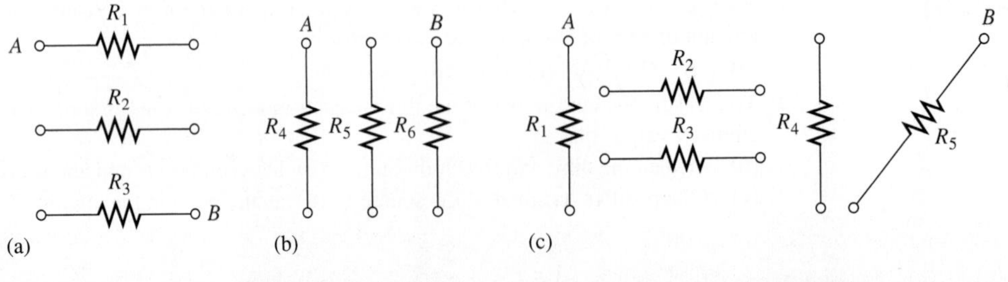

FIGURE 4–64

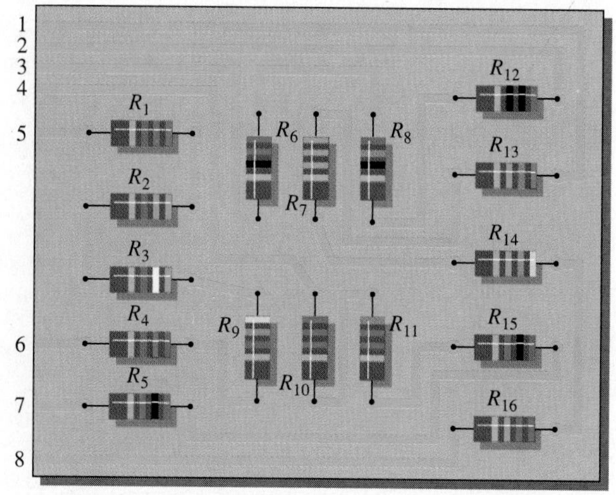

SECTION 4–2 Current in a Series Circuit

3. What is the current through each of four resistors in a series circuit if the source voltage is 12 V and the total resistance is 120 Ω?
4. The current from the source in Figure 4–65 is 5 mA. How much current does each milliammeter in the circuit indicate?

FIGURE 4–65

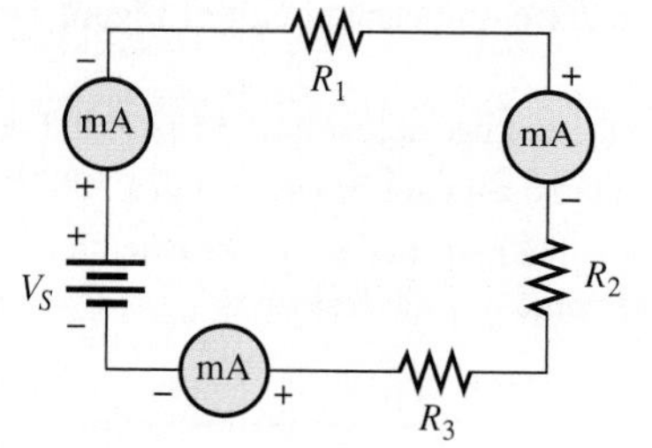

SECTION 4–3 Total Series Resistance

5. An 82 Ω resistor and a 56 Ω resistor are connected in series. What is the total resistance?
6. Find the total resistance of each group of series resistors shown in Figure 4–66.

FIGURE 4–66

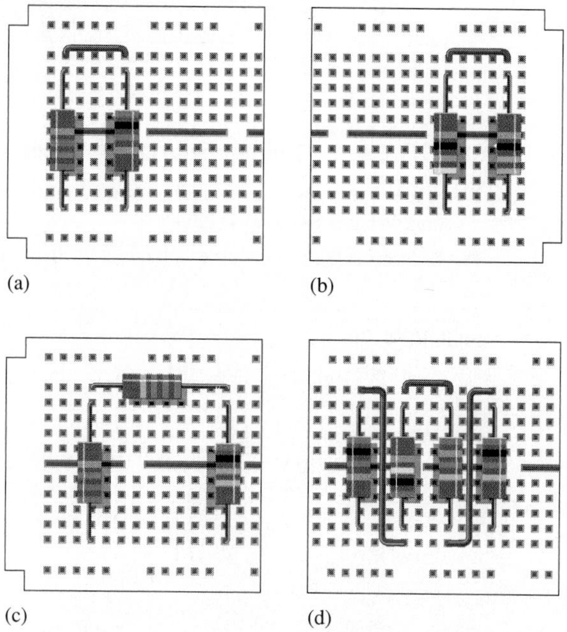

7. Determine R_T for each circuit in Figure 4–67. Show how to measure R_T with an ohmmeter.
8. What is the total resistance of twelve 5.6 kΩ resistors in series?
9. Six 47 Ω resistors, eight 100 Ω resistors, and two 22 Ω resistors are in series. What is the total resistance?
10. The total resistance in Figure 4–68 is 20 kΩ. What is the value of R_5?
11. Determine the resistance between each of the following sets of pins on the PC board in Figure 4–64.
(a) pin 1 and pin 8 (b) pin 2 and pin 3 (c) pin 4 and pin 7 (d) pin 5 and pin 6
12. If all the resistors in Figure 4–64 are connected in series, what is the total resistance?

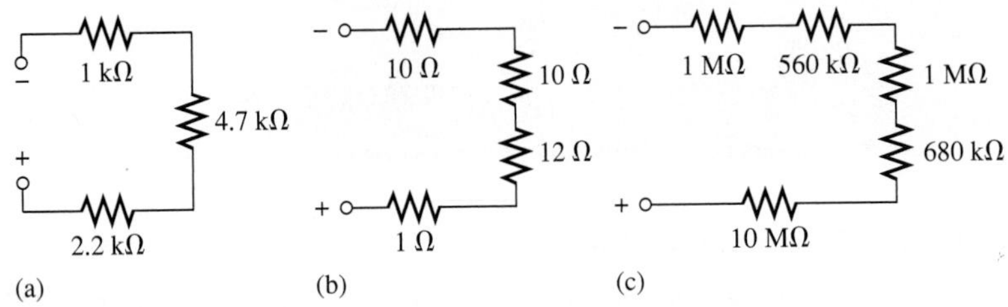

FIGURE 4–67

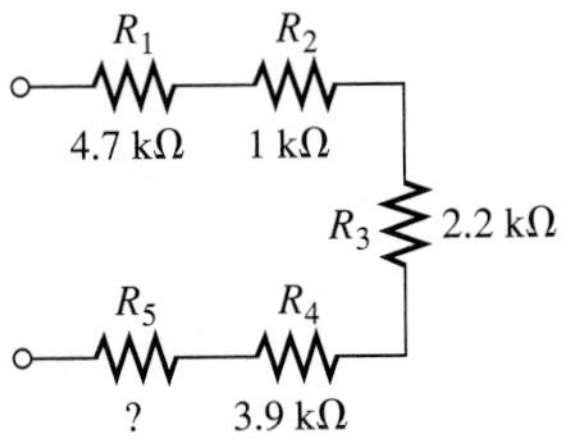

FIGURE 4–68

SECTION 4–4 Ohm's Law in Series Circuits

13. What is the current in each circuit of Figure 4–69? Show how to connect an ammeter in each case.
14. Determine the voltage across each resistor in Figure 4–69.
15. Three 470 Ω resistors are in series with a 500 V source. How much current is there?
16. Four equal-value resistors are in series with a 5 V source, and a current of 1 mA is measured. What is the value of each resistor?

FIGURE 4–69

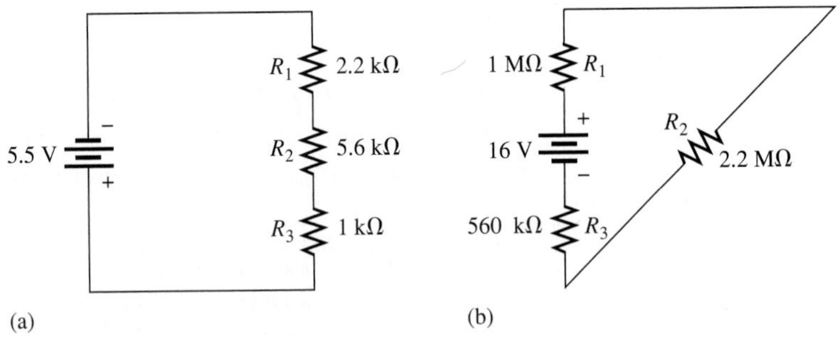

SECTION 4–5 Voltage Sources in Series

17. A 5 V battery and a 9 V battery are connected in series with their polarities in the same direction. What is the total voltage?
18. Determine the total source voltage in each circuit of Figure 4–70.

FIGURE 4–70

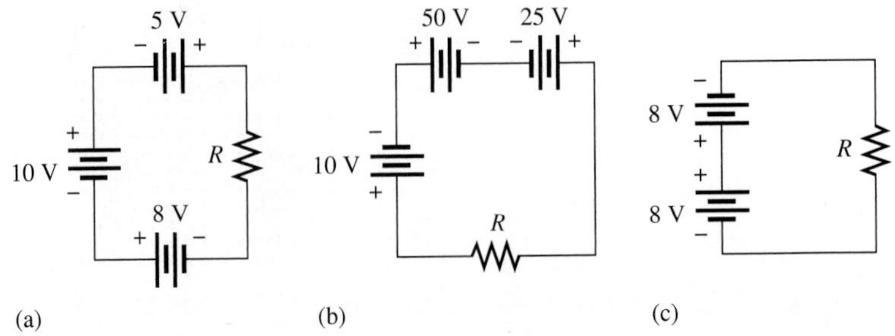

SECTION 4–6 Kirchhoff's Voltage Law

19. The following voltage drops are measured across each of three resistors in series: 5.5 V, 8.2 V, and 12.3 V. What is the value of the source voltage to which these resistors are connected?

20. Five resistors are in series with a 20 V source. The voltage drops across four of the resistors are 1.5 V, 5.5 V, 3 V, and 6 V. How much voltage is across the fifth resistor?

21. Determine the unspecified voltage drop(s) in each circuit of Figure 4–71. Show how to connect a voltmeter to measure each unknown voltage drop.

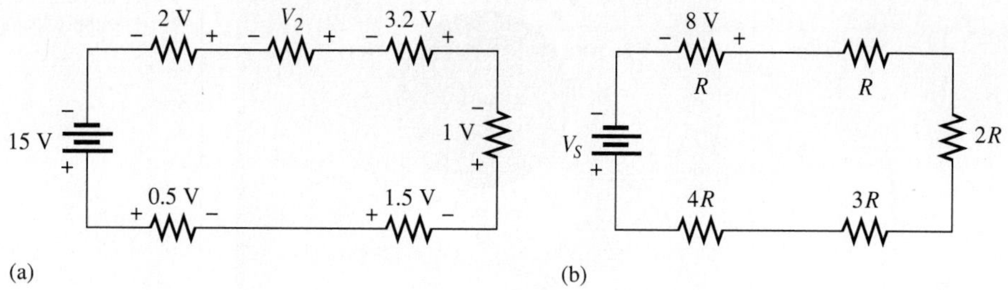

FIGURE 4–71

SECTION 4–7 Voltage Dividers

22. The total resistance of a series circuit is 500 Ω. What percentage of the total voltage appears across a 22 Ω resistor in the series circuit?

23. Find the voltage between *A* and *B* in each voltage divider of Figure 4–72.

24. What is the voltage across each resistor in Figure 4–73? *R* is the lowest value and all others are multiples of that value as indicated.

25. What is the voltage across each resistor in Figure 4–74?

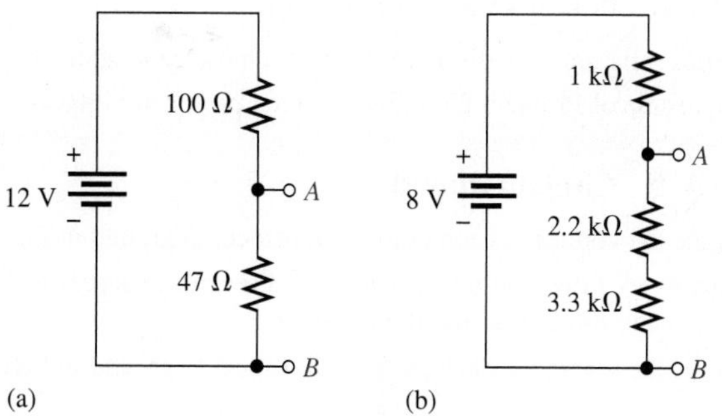

FIGURE 4–72

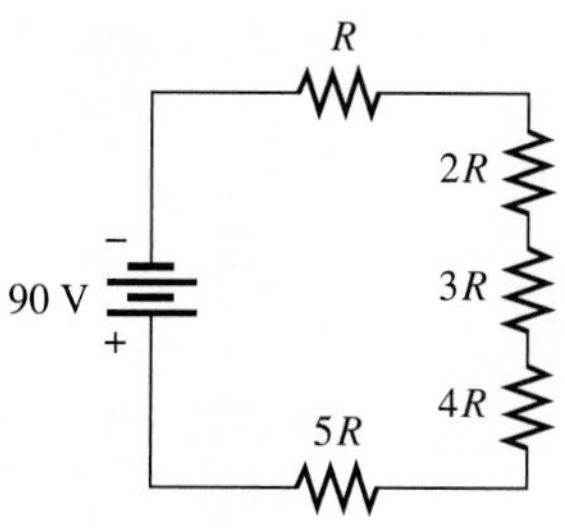

FIGURE 4–73

(a) Meter with leads going to circuit board

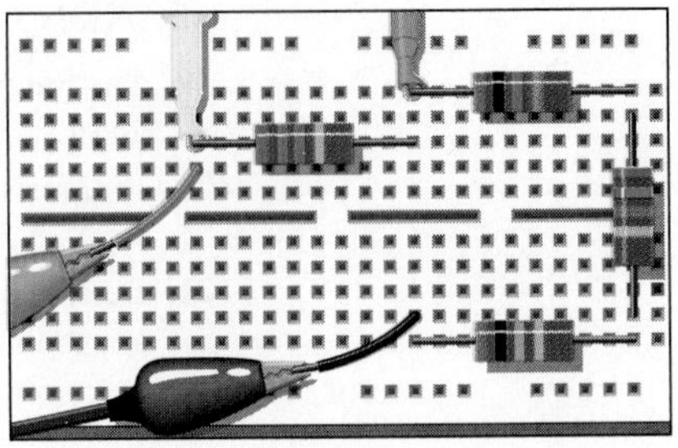

(b) Circuit board with meter leads and power supply leads (red and gray) connected

FIGURE 4–74

SECTION 4–8 Power in a Series Circuit

26. Five series resistors each dissipate 50 mW of power. What is the total power?

27. Use the results of Problem 25 to find the total power in Figure 4–74.

SECTION 4–9 Circuit Ground

28. Determine the voltage at each point with respect to ground in Figure 4–75.

29. In Figure 4–76, how would you determine the voltage across R_2 by measurement, without connecting a meter directly across the resistor?

30. Determine the voltage at each point with respect to ground in Figure 4–76.

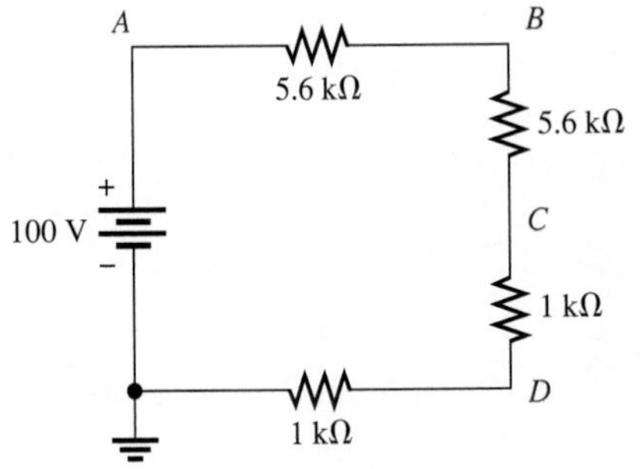

FIGURE 4–75

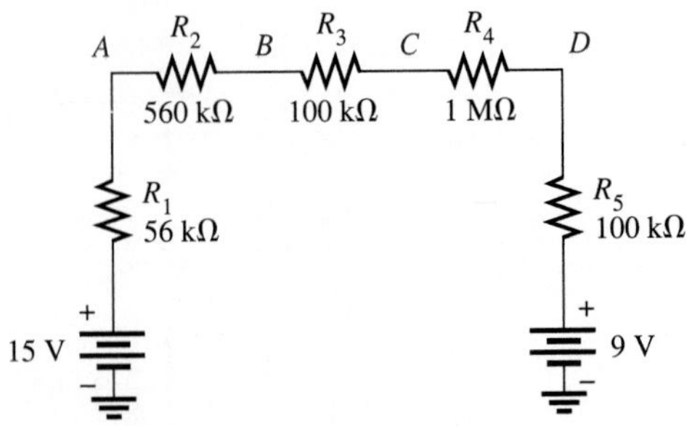

FIGURE 4–76

SECTION 4–10 Troubleshooting

31. By observing the meters in Figure 4–77, determine the types of failures in the circuits and which components have failed.

32. Is the ohmmeter reading in Figure 4–78 correct? If not, what is wrong?

FIGURE 4–77

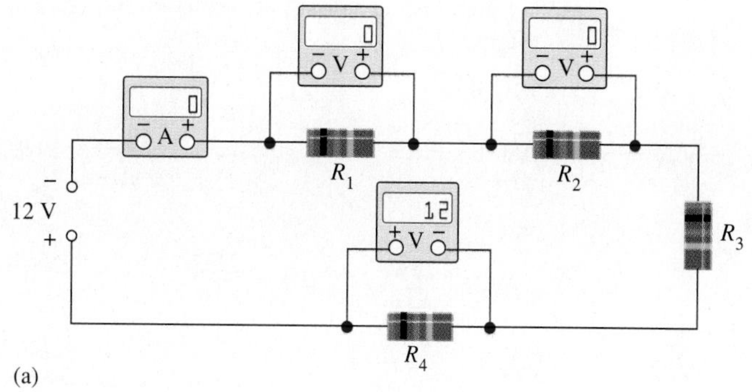

(a)

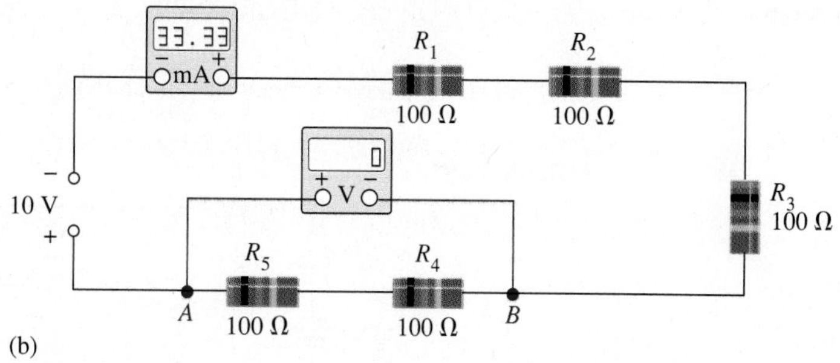

(b)

FIGURE 4–78

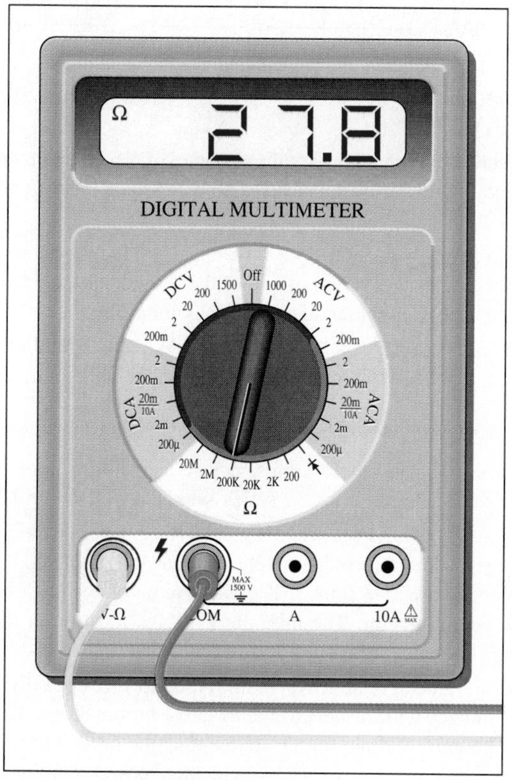

(a) Meter with leads going to circuit board

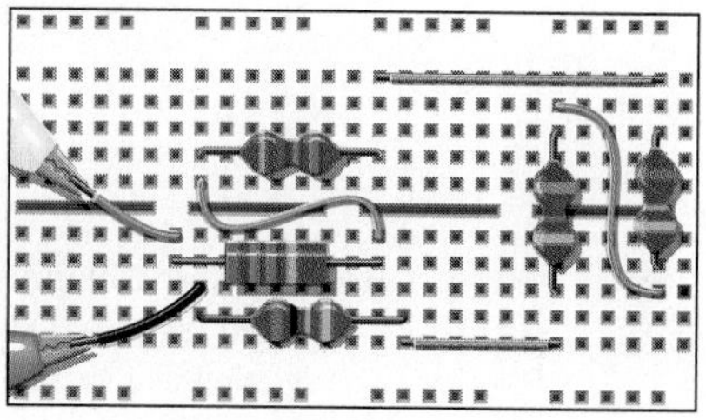

(b) Circuit board with meter leads connected

■ ADVANCED PROBLEMS

33. Determine the unknown resistance (R_3) in the circuit of Figure 4–79.

34. You have the following resistor values available to you in the lab in unlimited quantities: 10 Ω, 100 Ω, 470 Ω, 560 Ω, 680 Ω, 1 kΩ, 2.2 kΩ, and 5.6 kΩ. All of the other standard values are out of stock. A project that you are working on requires an 18 kΩ resistance. What combination of available values can you use to obtain the needed value?

35. Determine the voltage at each point in Figure 4–80 with respect to the negative side of the battery.

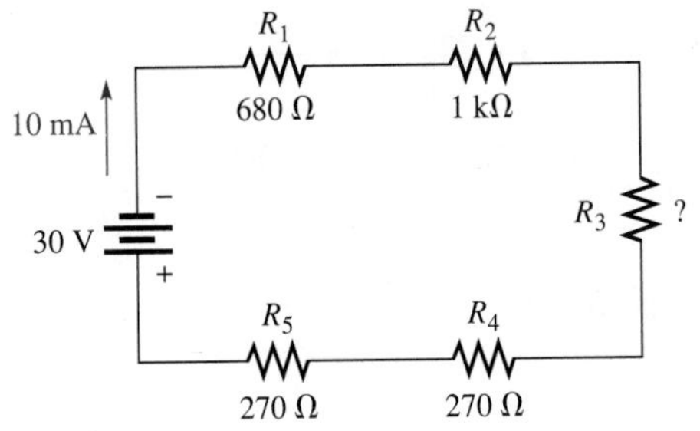

FIGURE 4–79

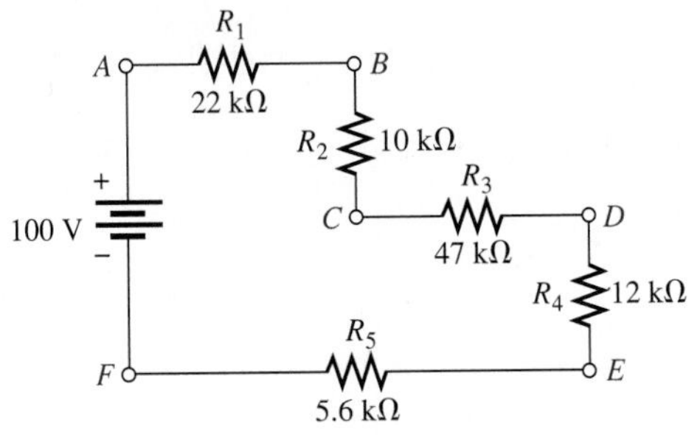

FIGURE 4–80

36. Find all the unknown quantities (in color) in Figure 4–81.

FIGURE 4–81

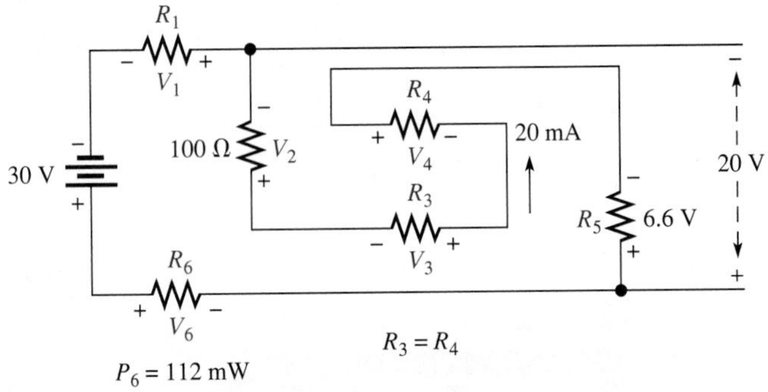

37. There are 250 mA in a series circuit with a total resistance of 1.5 kΩ. The current must be reduced by 25%. Determine how much resistance must be added in order to accomplish this reduction in current.

38. Four ½ W resistors are in series: 47 Ω, 68 Ω, 100 Ω, and 120 Ω. To what maximum value can the current be raised before the power rating of one of the resistors is exceeded? Which resistor will burn out first if the current is increased above the maximum?

39. A certain series circuit is made up of a ⅛ W resistor, a ¼ W resistor, and a ½ W resistor. The total resistance is 2400 Ω. If each of the resistors is operating at its maximum power level, determine the following:
(a) I **(b)** V_S **(c)** the value of each resistor

40. Using 1.5 V batteries, a switch, and three lamps, devise a circuit to apply 4.5 V across one lamp, two lamps in series, or three lamps in series with a single-control switch. Draw the schematic.

41. Develop a variable voltage divider to provide output voltages ranging from a minimum of 10 V to a maximum of 100 V using a 120 V source. The maximum voltage must be at the maximum resistance setting of the potentiometer. The minimum voltage must be at the minimum resistance (zero ohms) setting. The maximum current is to be 10 mA.

42. Using the standard resistor values given in Appendix A, build a voltage divider to provide the following approximate voltages with respect to the negative terminal of a 30 V source: 8.18 V, 14.73 V, and 24.55 V. The current drain on the source must be limited to no more than 1 mA. The number of resistors, their resistance values, and their power ratings must be specified. Draw a schematic showing the circuit with all resistor values indicated.

FIGURE 4–82

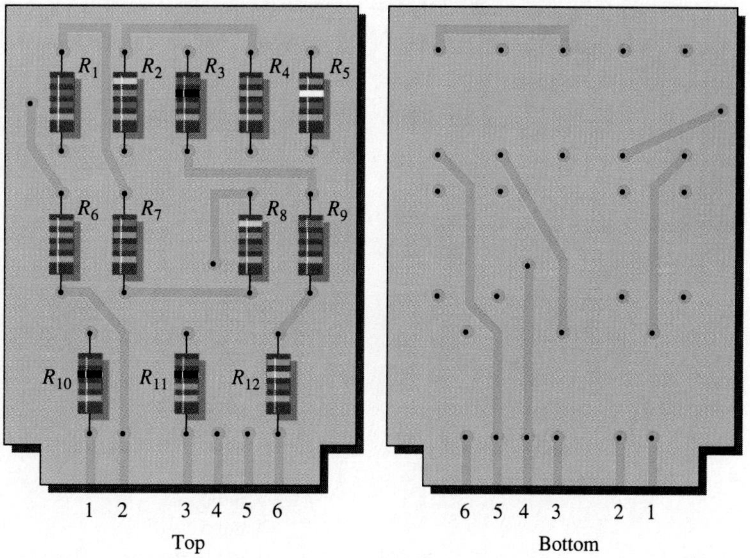

43. On the double-sided PC board in Figure 4–82, identify each group of series resistors and determine its total resistance. Note that many of the interconnections feed through the board from the top side to the bottom side.

44. What is the total resistance from *A* to *B* for each switch position in Figure 4–83?

45. Determine the current measured by the meter in Figure 4–84 for each switch position.

46. Determine the current measured by the meter in Figure 4–85 for each position of the ganged switch.

47. Determine the voltage across each resistor in Figure 4–86 for each switch position.

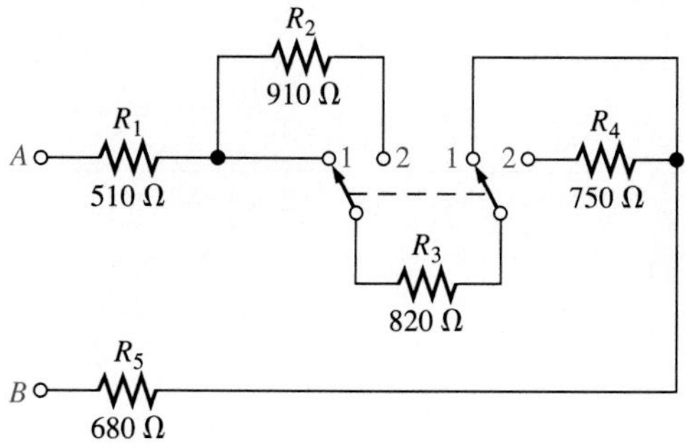

FIGURE 4–83

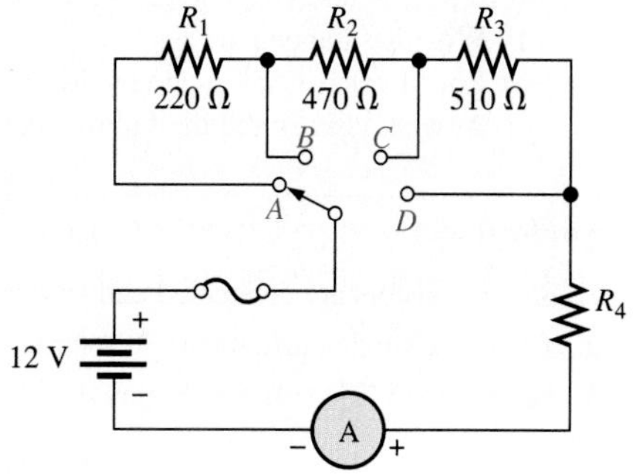

FIGURE 4–84

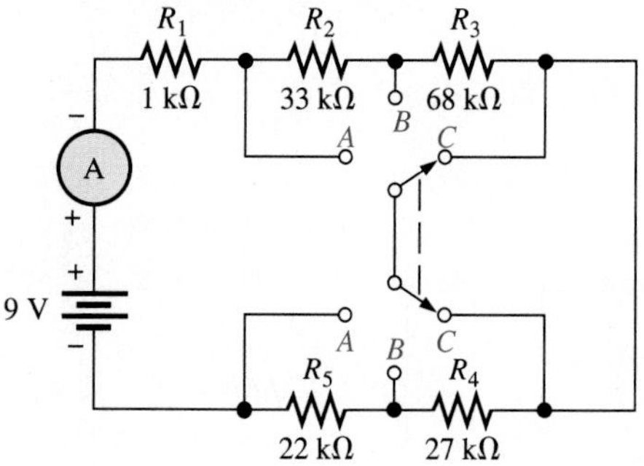

FIGURE 4–85

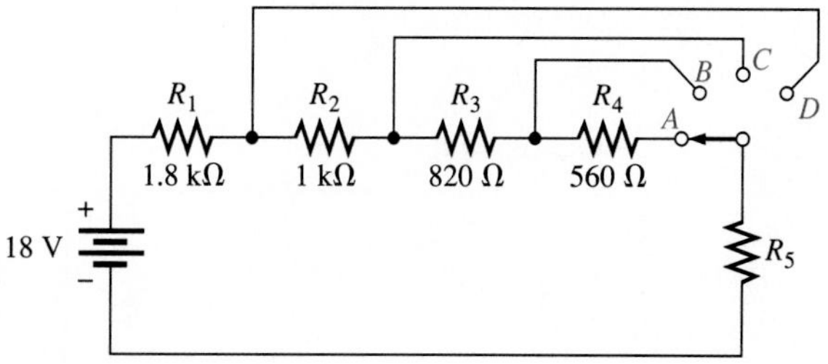

FIGURE 4–86

48. Table 4–1 shows the results of resistance measurements on the PC circuit board in Figure 4–82. Are these results correct? If not, identify the possible problems.

TABLE 4–1

BETWEEN PINS	RESISTANCE
1 and 2	∞
1 and 3	∞
1 and 4	4.23 kΩ
1 and 5	∞
1 and 6	∞
2 and 3	23.6 kΩ
2 and 4	∞
2 and 5	∞
2 and 6	∞
3 and 4	∞
3 and 5	∞
3 and 6	∞
4 and 5	∞
4 and 6	∞
5 and 6	19.9 kΩ

49. You measure 15 kΩ between pins 5 and 6 on the PC board in Figure 4–82. Does this indicate a problem? If so, identify it.

50. In checking out the PC board in Figure 4–82, you measure 17.83 kΩ between pins 1 and 2. Also, you measure 13.6 kΩ between pins 2 and 4. Does this indicate a problem on the PC board? If so, identify the fault.

51. The three groups of series resistors on the PC board in Figure 4–82 are connected in series with each other to form a single series circuit by connecting pin 2 to pin 4 and pin 3 to pin 5. A voltage source is connected across pins 1 and 6 and an ammeter is placed in series. As you increase the source voltage, you observe the corresponding increase in current. Suddenly, the current drops to zero and you smell smoke. All resistors are ½ W.
   (a) What has happened?
   (b) Specifically, what must you do to fix the problem?
   (c) At what voltage did the failure occur?

■ ANSWERS TO SECTION REVIEWS

Section 4–1

1. Series resistors are connected end-to-end in a "string."

2. There is a single current path in a series circuit.

3. See Figure 4–87. **4.** See Figure 4–88.

FIGURE 4–87

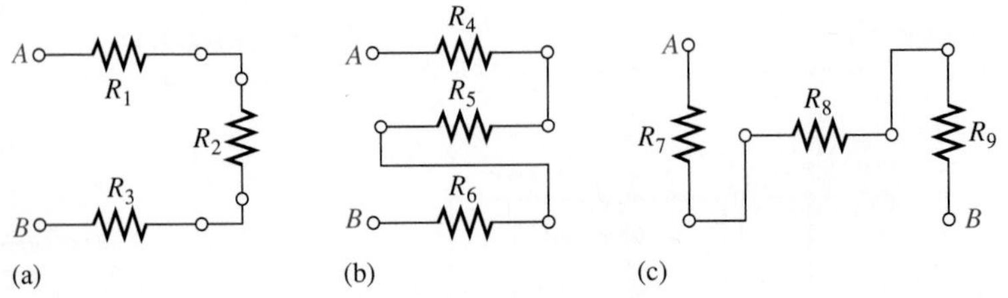

FIGURE 4–88

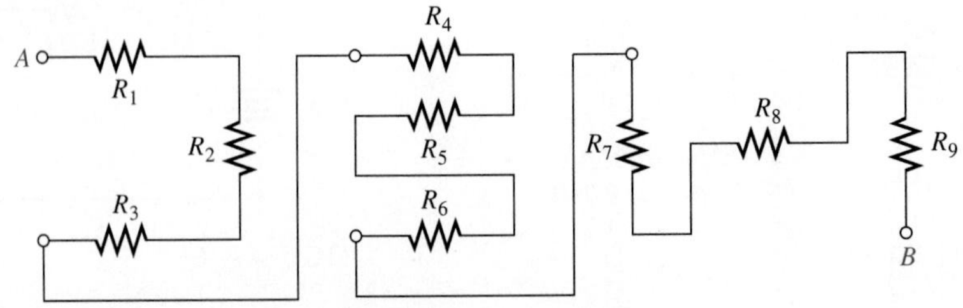

Section 4–2

1. The current is the same at all points in a series circuit.
2. There are 2 A through the 5 Ω resistor.
3. 50 mA between *C* and *D*, 50 mA between *E* and *F*
4. Ammeter 1 indicates 1.79 A; Ammeter 2 indicates 1.79 A.

Section 4–3

1. **(a)** $R_T = 33\ \Omega + 100\ \Omega + 10\ \Omega = 143\ \Omega$
   (b) $R_T = 39\ \Omega + 56\ \Omega + 10\ \Omega = 105\ \Omega$
   (c) $R_T = 820\ \Omega + 2200\ \Omega + 1000\ \Omega = 4020\ \Omega$
2. $R_T = 100\ \Omega + 2(47\ \Omega) + 4(12\ \Omega) + 330\ \Omega = 572\ \Omega$
3. $10\ \text{k}\Omega - 8.8\ \text{k}\Omega = 1.2\ \text{k}\Omega$
4. $R_T = 12(47\ \Omega) = 564\ \Omega$

Section 4–4

1. $I = 10\ \text{V}/300\ \Omega = 0.033\ \text{A} = 33\ \text{mA}$
2. $V = (5\ \text{A})(43\ \Omega) = 215\ \text{V}$
3. $V_1 = (5\ \text{A})(10\ \Omega) = 50\ \text{V}$, $V_2 = (5\ \text{A})(15\ \Omega) = 75\ \text{V}$, $V_3 = (5\ \text{A})(18\ \Omega) = 90\ \text{V}$
4. $R = 125\ \text{V}/4.63\ \text{mA} = 270\ \Omega$

Section 4–5

1. Five (5)(12 V) = 60 V; see Figure 4–89.

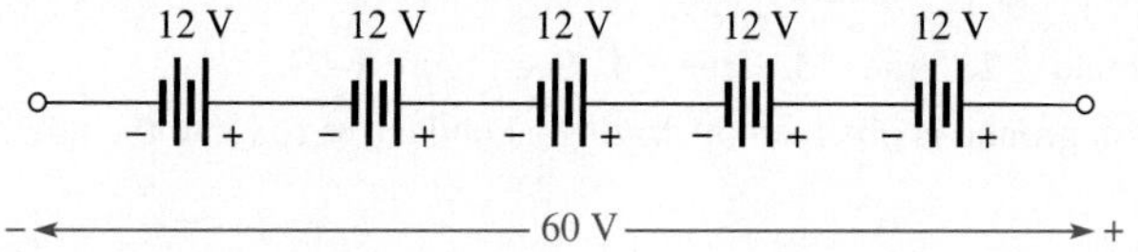

FIGURE 4–89

2. $V_T = (4)(1.5\ \text{V}) = 6.0\ \text{V}$
3. See Figure 4–90.
4. **(a)** $V_{ST} = 100\ \text{V} + 50\ \text{V} - 75\ \text{V} = 75\ \text{V}$
   (b) $V_{ST} = 20\ \text{V} + 10\ \text{V} - 10\ \text{V} - 5\ \text{V} = 15\ \text{V}$
5. See Figure 4–91.

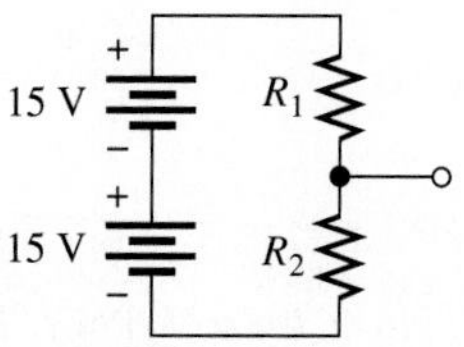

FIGURE 4–90

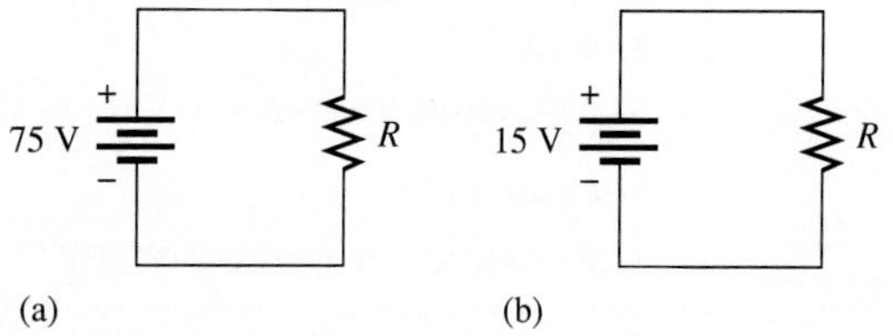

FIGURE 4–91

Section 4–6

1. Kirchhoff's voltage law states:
   (a) The sum of the voltages around a closed path is zero.
   (b) The sum of the voltage drops equals the total source voltage.
2. $V_{RT} = V_{ST} = 50\ \text{V}$
3. $V_R = 10\ \text{V}/2 = 5\ \text{V}$
4. $V_{R3} = 25\ \text{V} - 5\ \text{V} - 10\ \text{V} = 10\ \text{V}$
5. $V_T = 1\ \text{V} + 3\ \text{V} + 5\ \text{V} + 7\ \text{V} + 8\ \text{V} = 24\ \text{V}$

Section 4–7

1. A voltage divider is a series circuit with two or more resistors with an output taken across one or more of the resistors.
2. There can be any number of resistors in a voltage divider.
3. $V_x = (R_x/R_T)\,V_S$ is the general voltage-divider formula.
4. $V_R = 20\text{ V}/2 = 10\text{ V}$
5. $V_{56\Omega} = (56\ \Omega/138\ \Omega)100\text{ V} = 40.6\text{ V}$, $V_{82\Omega} = (82\ \Omega/138\Omega)100\text{ V} = 59.4\text{ V}$. See Figure 4–92.
6. Set the potentiometer at the midpoint.

FIGURE 4–92

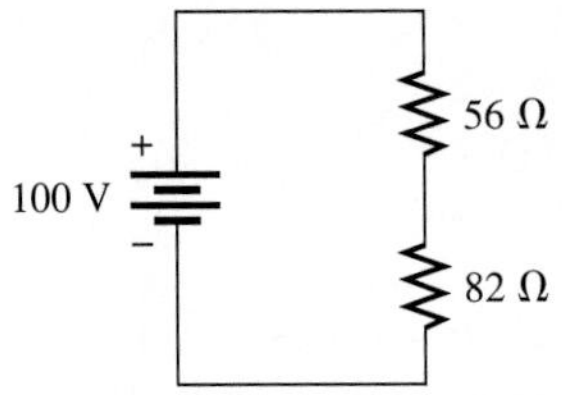

Section 4–8

1. Add the powers in each resistor to get total power.
2. $P_T = 1\text{ W} + 2\text{ W} + 5\text{ W} + 8\text{ W} = 16\text{ W}$
3. $P_T = (1\text{ A})^2(100\ \Omega + 330\ \Omega + 680\ \Omega) = 1110\text{ W}$

Section 4–9

1. Ground **2.** True **3.** True **4.** See Figure 4–93.
5. Earth ground is obtained by driving a conductive rod into the ground.

FIGURE 4–93

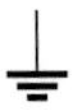

Section 4–10

1. An open is a break in the current path.
2. A short is a zero resistance path that bypasses a portion of a circuit.
3. Current ceases when a series circuit opens.
4. An open can be created by a switch or a component failure. A short can be created by a switch or by wire clippings, solder splashes, etc.
5. True
6. 24 V across the open R; 0 V across the other Rs.

Section 4–11

1. $P_T = (12\text{ V})^2/16.6\text{ k}\Omega = 8.67\text{ mW}$
2. Pin 2: 4.59 V; Pin 6: 2.35 V; Pin 5: 1.99 V; Pin 4: 795 mV; Pin 7: 2.89 V
3. Pin 3 connects to ground.

■ ANSWERS TO RELATED EXERCISES FOR EXAMPLES

4–1 (a) Left end of R_1 to "minus," right end of R_1 to top end of R_3, bottom end of R_3 to right end of R_5, left end of R_5 to left end of R_2, right end of R_2 to right end of R_4, left end of R_4 to "plus".

(b) $R_1 = 1\text{ k}\Omega$, $R_2 = 33\text{ k}\Omega$, $R_3 = 39\text{ k}\Omega$, $R_4 = 470\ \Omega$, $R_5 = 22\text{ k}\Omega$

4–2 The two series circuits are connected in series.

4–3 258 Ω (No change)

4–4 12.1 kΩ

4–5 6.8 kΩ

4–6 4.44 kΩ
4–7 Measure each R with an ohmmeter.
4–8 116 mA
4–9 7.8 V
4–10 $V_1 = 1$ V, $V_2 = 3.3$ V, $V_3 = 2.2$ V, $V_S = 6.5$ V
4–11 12 V
4–12 2 V
4–13 10 V, 20 V
4–14 6.5 V
4–15 593 Ω
4–16 $V_1 = 3.13$ V; $V_2 = 6.87$ V
4–17 33.3 V
4–18 $V_{AB} = 4$ V; $V_{AC} = 36.8$ V; $V_{BC} = 32.8$ V; $V_{BD} = 46$ V; $V_{CD} = 13.2$ V
4–19 9 W
4–20 $V_A = 0$ V; $V_B = -25$ V; $V_C = -50$ V; $V_D = -75$ V; $V_E = -100$ V

5

PARALLEL CIRCUITS

In this chapter, parallel circuits are introduced. Parallel circuits are found in many applications, such as lighting systems, sound systems, appliances, and most electronic equipment.

■ CHAPTER OBJECTIVES

- □ Identify a parallel circuit
- □ Determine the voltage across each parallel branch
- □ Apply Kirchhoff's current law
- □ Determine total parallel resistance
- □ Apply Ohm's law in a parallel circuit
- □ Use a parallel circuit as a current divider
- □ Determine power in a parallel circuit
- □ Describe some basic applications of parallel circuits
- □ Troubleshoot parallel circuits

■ APPLICATION ASSIGNMENT

As a line technician in an electronic instruments company, your job is to troubleshoot any defective instrument that fails the routine line check. In this particular assignment, you must determine the problem(s) with a defective five-range milliammeter so that it can be repaired. The knowledge of parallel circuits and of basic ammeters that you will acquire in this chapter plus your understanding of Ohm's law, current dividers, and the resistor color code will be put to good use. After studying this chapter, you should be able to complete the application assignment in Section 5–10.

5–1 ■ RESISTORS IN PARALLEL

When two or more resistors are individually connected between the same two points, they are in parallel with each other. A parallel circuit provides more than one path for current.

After completing this section, you should be able to

- **Identify a parallel circuit**
  - □ Translate a physical arrangement of resistors into a schematic

Each parallel path in a circuit is called a **branch.** Two resistors connected in parallel are shown in Figure 5–1(a). As shown in part (b), the current out of the source divides when it gets to point A. Part of it goes through R_1 and part through R_2. The two currents come back together at point B. If additional resistors are connected in parallel with the first two, more current paths are provided, as shown in Figure 5–1(c).

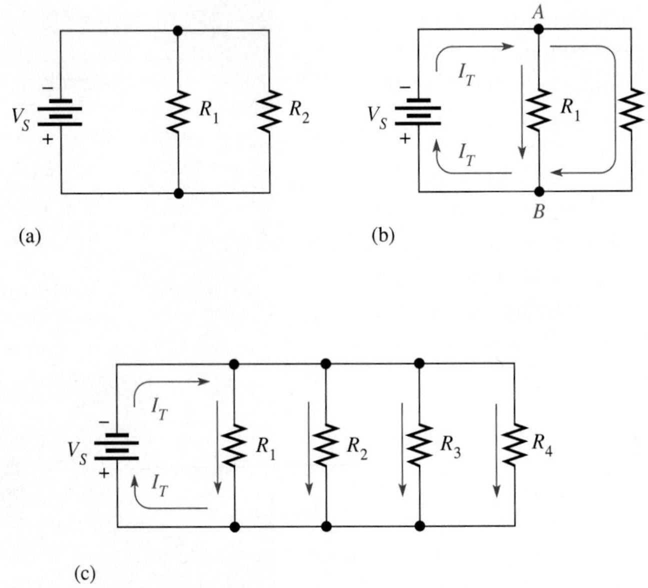

FIGURE 5–1
Resistors in parallel.

Identifying Parallel Circuits

It is obvious that the resistors in Figure 5–1 are connected in **parallel.** However, in actual circuit diagrams, the parallel relationship often is not so clear. It is important that you learn to recognize parallel circuits regardless of how they may be drawn.

A way to identify a parallel resistive circuit is as follows:

If there is more than one resistor connected from point *A* to point *B* and if the voltage across each of the resistors is the same, then the resistors are in parallel.

Figure 5–2 shows parallel resistors drawn in different ways between two points labeled A and B. Notice that in each case, the current "travels" two paths going from A to B, and the voltage across each branch is the same. Although these figures show only two parallel paths, there can be any number of resistors in parallel.

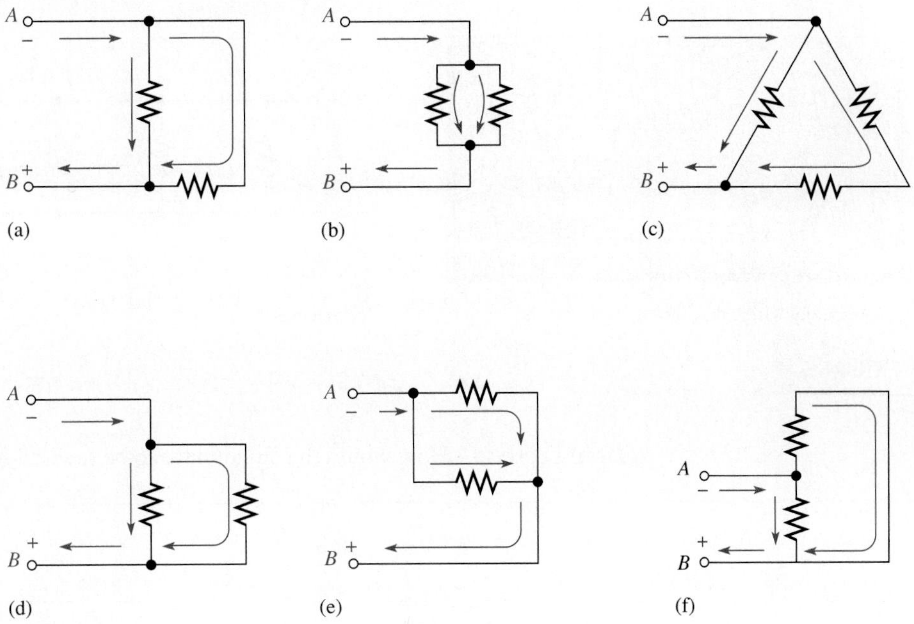

FIGURE 5–2
Examples of circuits with two parallel paths.

EXAMPLE 5–1

Five resistors are positioned on a circuit board as shown in Figure 5–3. Show the wiring required to connect all the resistors in parallel, and draw a schematic showing this circuit with resistance values.

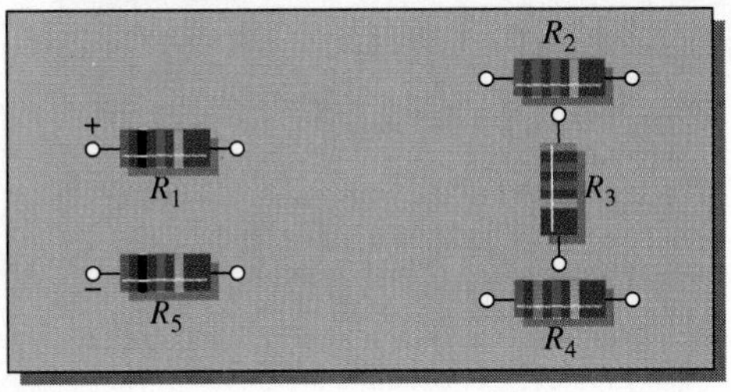

FIGURE 5–3

Solution Wires are connected as shown in the assembly wiring diagram of Figure 5–4(a). The schematic is shown in Figure 5–4(b) with resistance values from the color codes. Again, note that the schematic does not necessarily have to show the actual physical arrangement of the resistors. The schematic shows how components are connected electrically.

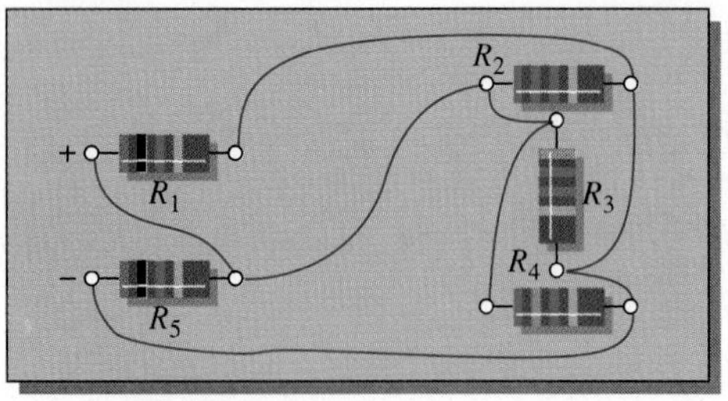

(a) Assembly wiring diagram

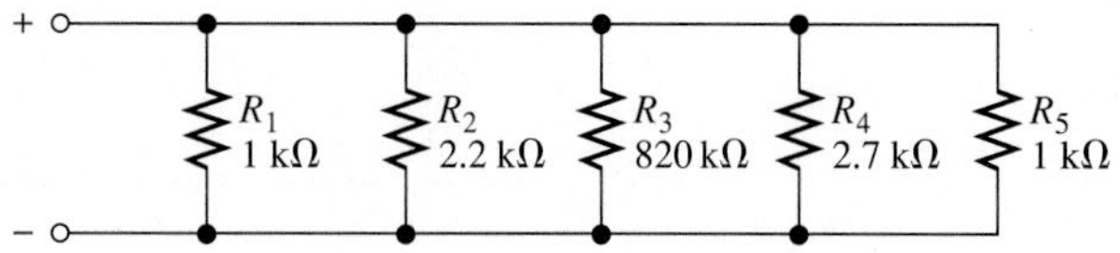

(b) Schematic

FIGURE 5–4

Related Exercise How would the circuit have to be rewired if R_2 is removed?

EXAMPLE 5–2

Determine the parallel groupings in Figure 5–5 and the value of each resistor.

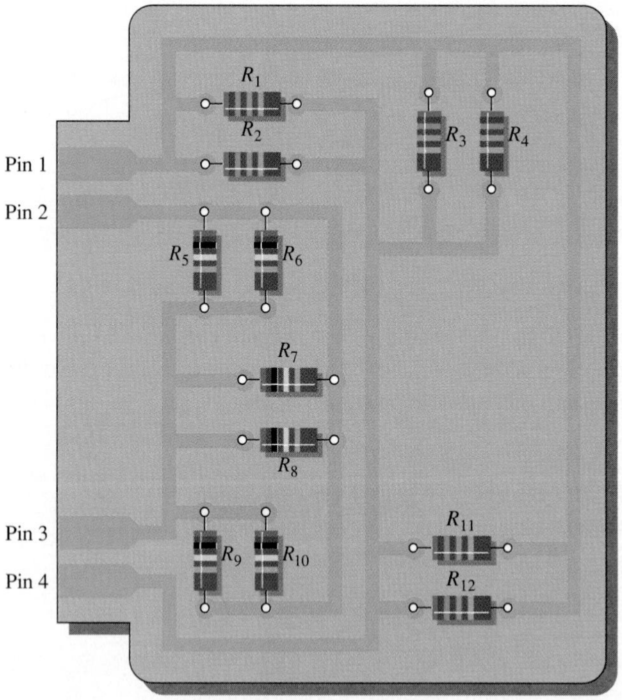

FIGURE 5–5

Solution Resistors R_1 through R_4 and R_{11} and R_{12} are all in parallel. This parallel combination is connected to pins 1 and 4. All resistors in this group are 56 kΩ.

Resistors R_5 through R_{10} are all in parallel. This combination is connected to pins 2 and 3. All resistors in this group are 100 kΩ.

Related Exercise How would you connect all the resistors on the PC board in parallel?

SECTION 5–1 REVIEW

1. How are the resistors connected in a parallel circuit?
2. How do you identify a parallel circuit?
3. Complete the schematics for the circuits in each part of Figure 5–6 by connecting the resistors in parallel between points A and B.
4. Now connect each group of parallel resistors in Figure 5–6 in parallel with each other.

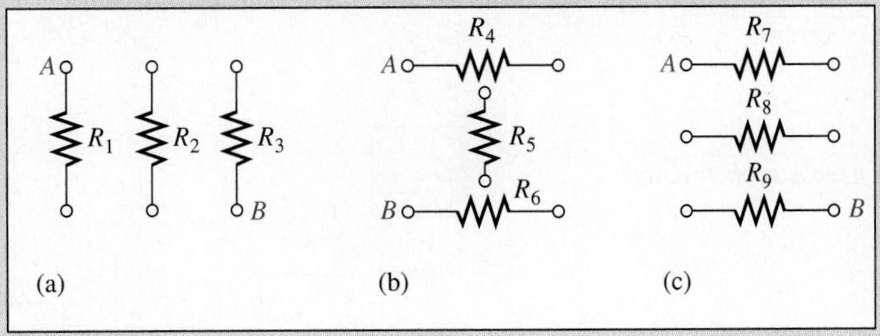

FIGURE 5–6

5–2 ▪ VOLTAGE DROP IN PARALLEL CIRCUITS

As mentioned in the previous section, each current path in a parallel circuit is called a branch. The voltage across any given branch of a parallel circuit is equal to the voltage across each of the other branches in parallel.

After completing this section, you should be able to

- **Determine the voltage across each parallel branch**
  - ☐ Explain why the voltage is the same across all parallel resistors

To illustrate voltage drop in a parallel circuit, let's examine Figure 5–7(a). Points A, B, C, and D along the left side of the parallel circuit are electrically the same point because the voltage is the same along this line. You can think of all of these points as being connected by a single wire to the negative terminal of the battery. The points E, F, G, and H along the right side of the circuit are all at a voltage equal to that of the positive terminal of the

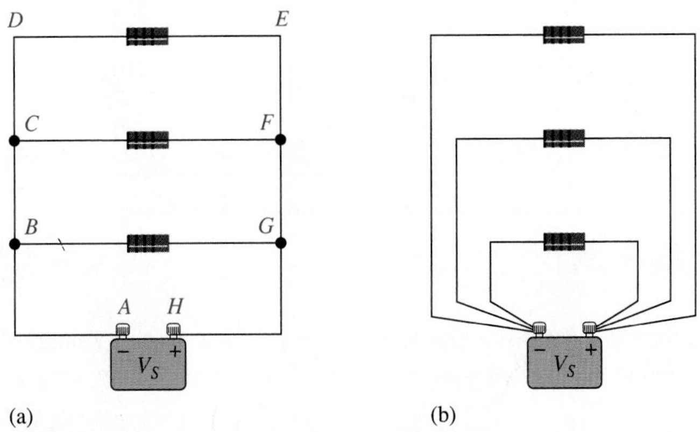

FIGURE 5–7
Voltage across parallel branches is the same.

source. Thus, voltage across each parallel resistor is the same, and each is equal to the source voltage.

Figure 5–7(b) is the same circuit as in part (a), drawn in a slightly different way. Here the left side of each resistor is connected to a single point, which is the negative battery terminal. The right side of each resistor is connected to a single point, which is the positive battery terminal. The resistors are still all in parallel across the source.

In Figure 5–8, a 12 V battery is connected across three parallel resistors. When the voltage is measured across the battery and then across each of the resistors, the readings are the same. As you can see, the same voltage appears across each branch in a parallel circuit.

FIGURE 5–8
The same voltage appears across each resistor in parallel.

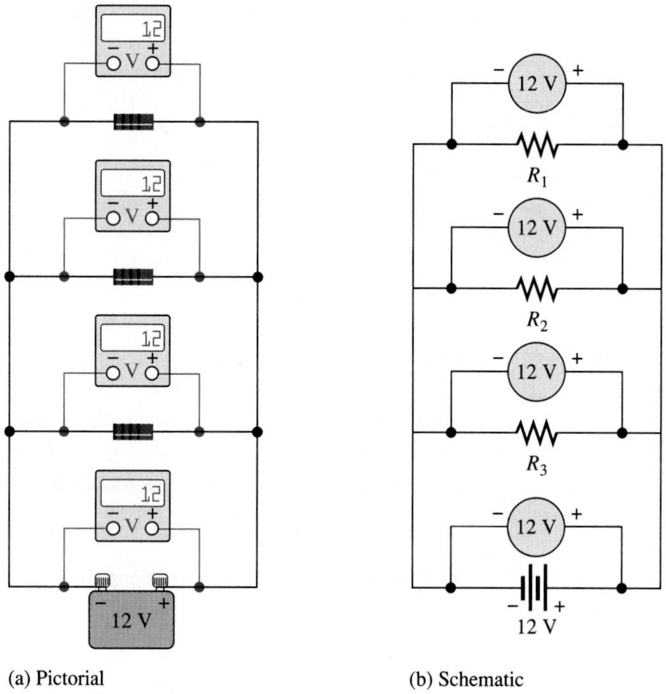

EXAMPLE 5–3

Determine the voltage across each resistor in Figure 5–9.

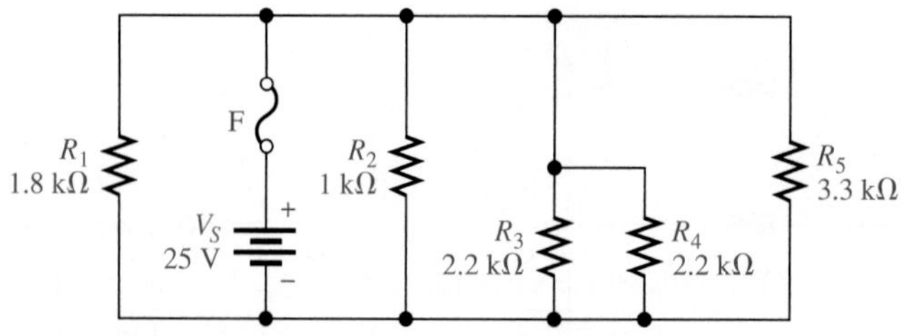

FIGURE 5–9

Solution The five resistors are in parallel; so the voltage drop across each one is equal to the applied source voltage. There is no voltage drop across the fuse.

$$V_{ST} = V_1 = V_2 = V_3 = V_4 = V_5 = 25\text{ V}$$

Related Exercise If R_4 is removed from the circuit, what is the voltage across R_3?

SECTION 5–2 REVIEW

1. A 10 Ω and a 22 Ω resistor are connected in parallel with a 5 V source. What is the voltage across each of the resistors?
2. A voltmeter connected across R_1 in Figure 5–10 measures 118 V. If you move the meter and connect it across R_2, how much voltage will it indicate? What is the source voltage?
3. In Figure 5–11, how much voltage does voltmeter 1 indicate? Voltmeter 2?
4. How are voltages across each branch of a parallel circuit related?

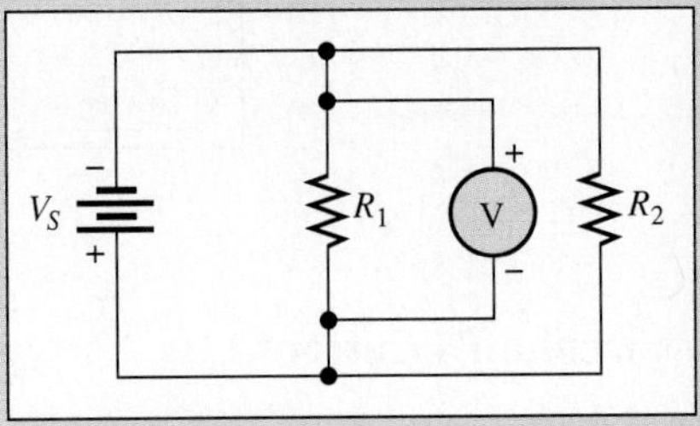

FIGURE 5–10

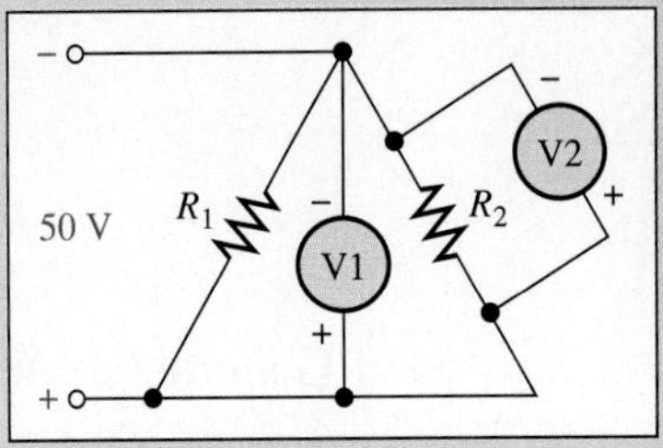

FIGURE 5–11

5–3 ■ KIRCHHOFF'S CURRENT LAW

In the last chapter, you learned Kirchhoff's voltage law that dealt with voltages in a closed series circuit. Now, you will learn Kirchhoff's current law that deals with currents in a parallel circuit.

After completing this section, you should be able to

- **Apply Kirchhoff's current law**
  - ☐ State Kirchhoff's current law
  - ☐ Determine the total current by adding the branch currents
  - ☐ Determine an unknown branch current

Kirchhoff's current law is stated as follows:

> **The sum of the currents into a junction (total current in) is equal to the sum of the currents out of that junction (total current out).**

A **junction** is any point in a circuit where two or more circuit paths come together. In a parallel circuit, a junction is a point where the parallel branches connect together.

For example, in the circuit of Figure 5–12, point A is one junction and point B is another. Let's start at the negative terminal of the source and follow the current. The total current I_T flows from the source and *into* the junction at point A. At this point, the current splits up among the three branches as indicated. Each of the three branch currents (I_1, I_2, and I_3) flows *out of* junction A. Kirchhoff's current law says that the total current into junction A is equal to the total current out of junction A; that is,

$$I_T = I_1 + I_2 + I_3$$

Now, following the currents in Figure 5–12 through the three branches, you see that they come back together at point B. Currents I_1, I_2, and I_3 flow into junction B, and I_T flows out. Kirchhoff's current law formula at this junction is therefore the same as at junction A:

$$I_T = I_1 + I_2 + I_3$$

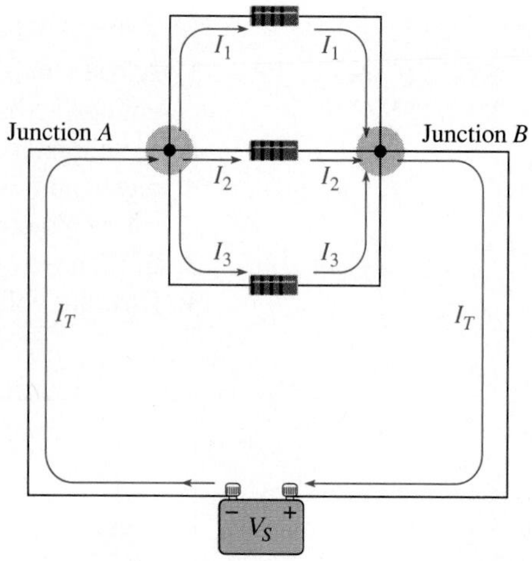

FIGURE 5–12
Kirchhoff's current law: The current into a junction equals the current out of that junction.

General Formula for Kirchhoff's Current Law

The previous discussion used a specific case to illustrate Kirchhoff's current law, often abbreviated KCL. Now let's look at the general case. Figure 5–13 shows a generalized circuit junction where a number of branches are connected to a point in the circuit. Currents $I_{IN(1)}$ through $I_{IN(n)}$ flow into the junction (n can be any number). Currents $I_{OUT(1)}$ through $I_{OUT(m)}$ flow out of the junction (m can be any number but not necessarily equal to n).

By Kirchhoff's current law, the sum of the currents into a junction must equal the sum of the currents out of the junction. With reference to Figure 5–13, the general formula for Kirchhoff's current law is

$$I_{IN(1)} + I_{IN(2)} + \cdots + I_{IN(n)} = I_{OUT(1)} + I_{OUT(2)} + \cdots + I_{OUT(m)} \quad \textbf{(5–1)}$$

If all of the terms on the right side of Equation (5–1) are brought over to the left side, their signs change to negative, and a zero is left on the right side. Kirchhoff's current law can also be stated in this way:

The algebraic sum of all the currents entering and leaving a junction is equal to zero.

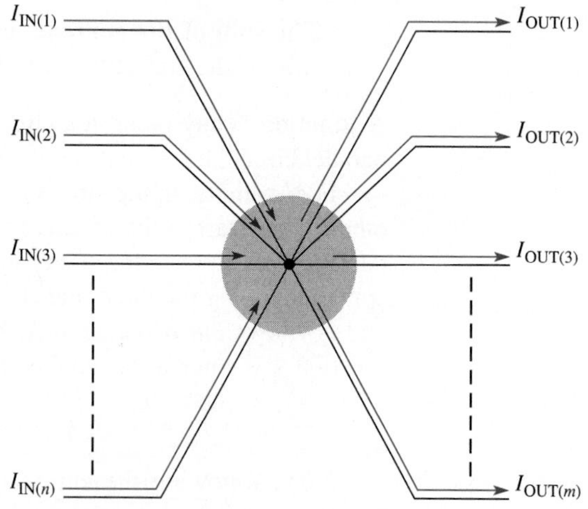

$$I_{IN(1)} + I_{IN(2)} + I_{IN(3)} + \cdots + I_{IN(n)}$$
$$= I_{OUT(1)} + I_{OUT(2)} + I_{OUT(3)} + \cdots I_{OUT(m)}$$

FIGURE 5–13
Generalized circuit junction to illustrate Kirchhoff's current law.

FIGURE 5–14
Experimental verification of Kirchhoff's current law.

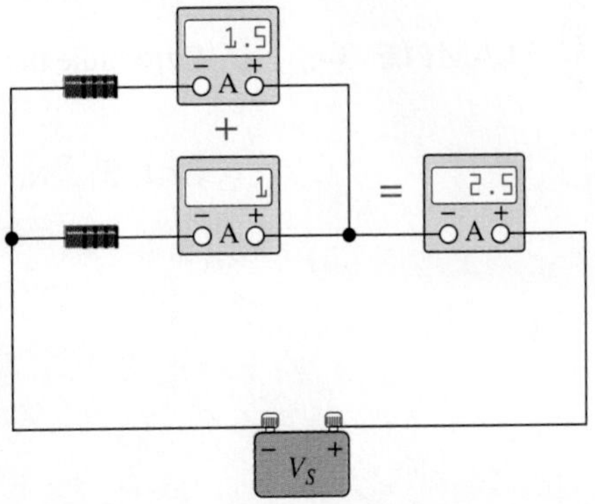

You can verify Kirchhoff's current law by connecting a circuit and measuring each branch current and the total current from the source, as illustrated in Figure 5–14. When the branch currents are added together, their sum will equal the total current. This rule applies for any number of branches.

The following three examples illustrate use of Kirchhoff's current law.

EXAMPLE 5–4

The branch currents in the circuit of Figure 5–15 are known. Determine the total current entering junction A and the total current leaving junction B.

FIGURE 5–15

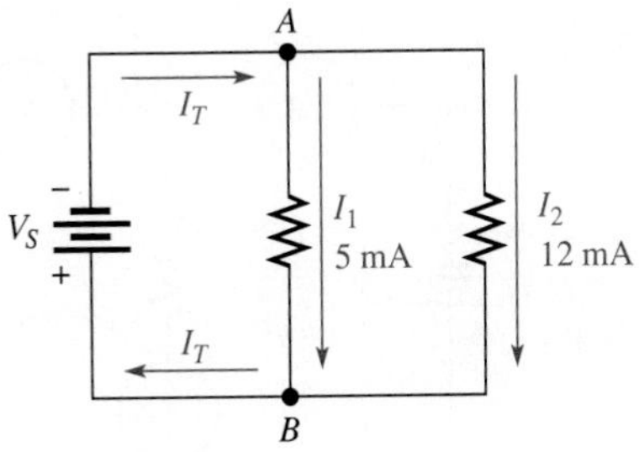

Solution The total current out of junction A is the sum of the two branch currents. So the total current into A is

$$I_T = I_1 + I_2 = 5\text{ mA} + 12\text{ mA} = 17\text{ mA}$$

The total current entering point B is the sum of the two branch currents. So the total current out of B is

$$I_T = I_1 + I_2 = 5\text{ mA} + 12\text{ mA} = 17\text{ mA}$$

The calculator sequence is

[5] [EXP] [+/−] [3] [+] [1] [2] [EXP] [+/−] [3] [=]

Related Exercise If a third resistor is connected in parallel to the circuit of Figure 5–15, and its current is 3 mA, what is the total current into junction A and out of junction B?

EXAMPLE 5–5 Determine the current through R_2 in Figure 5–16.

FIGURE 5–16

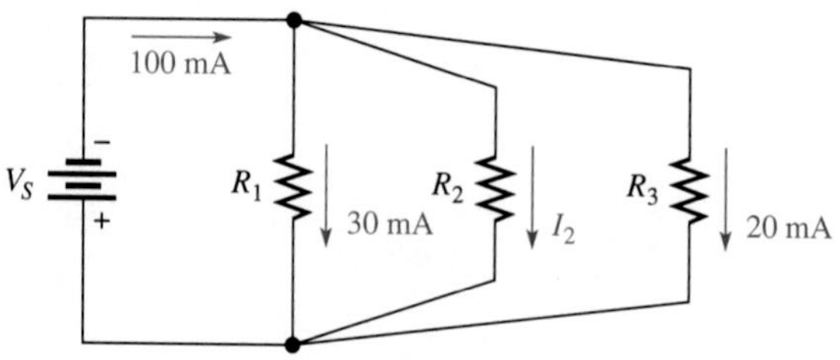

Solution The total current into the junction of the three branches is known, and two of the branch currents are known. The current equation at this junction is

$$I_T = I_1 + I_2 + I_3$$

Solving for I_2,

$$I_2 = I_T - I_1 - I_3 = 100\text{ mA} - 30\text{ mA} - 20\text{ mA} = 50\text{ mA}$$

Related Exercise Determine I_T and I_2 if a fourth branch is added to the circuit in Figure 5–16 and it has 12 mA through it.

EXAMPLE 5–6 Use Kirchhoff's current law to find the current measured by ammeters A1 and A2 in Figure 5–17.

FIGURE 5–17

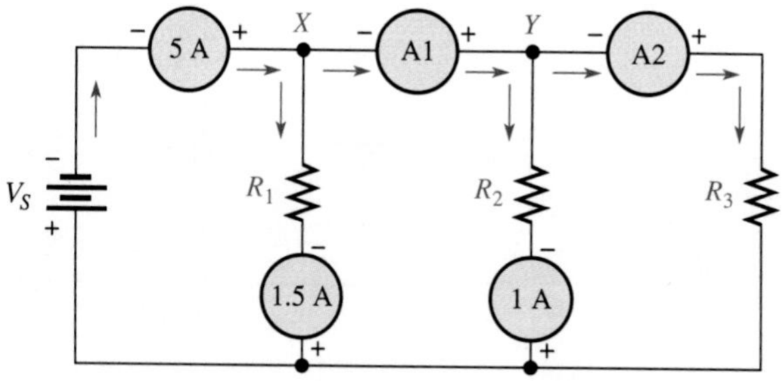

Solution The total current into junction X is 5 A. Two currents are out of junction X: 1.5 A through resistor R_1 and the current through A1. Kirchhoff's current law at junction X is

$$5\text{ A} = 1.5\text{ A} + I_{A1}$$

Solving for I_{A1} yields

$$I_{A1} = 5\text{ A} - 1.5\text{ A} = 3.5\text{ A}$$

The total current into junction Y is $I_{A1} = 3.5$ A. Two currents are out of junction Y: 1 A through resistor R_2 and the current through A2 and R_3. Kirchhoff's current law applied to junction Y gives

$$3.5\text{ A} = 1\text{ A} + I_{A2}$$

Solving for I_{A2} yields

$$I_{A2} = 3.5\text{ A} - 1\text{ A} = 2.5\text{ A}$$

Related Exercise How much current will an ammeter measure when it is placed in the circuit in Figure 5–17 right below R_3? Below the positive battery terminal?

SECTION 5–3 REVIEW

1. State Kirchhoff's current law in two ways.
2. A total current of 2.5 A is into the junction of three parallel branches. What is the sum of all three branch currents?
3. In Figure 5–18, 100 mA and 300 mA are into the junction. What is the amount of current out of the junction?
4. Determine I_1 in the circuit of Figure 5–19.
5. Two branch currents enter a junction, and two branch currents leave the same junction. One of the currents entering the junction is 1 A, and one of the currents leaving the junction is 3 A. The total current entering and leaving the junction is 8 A. Determine the value of the unknown current entering the junction and the value of the unknown current leaving the junction.

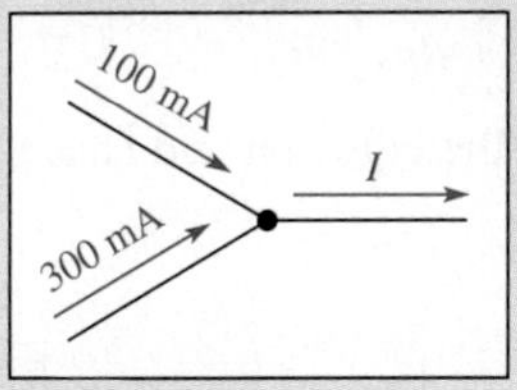

FIGURE 5–18

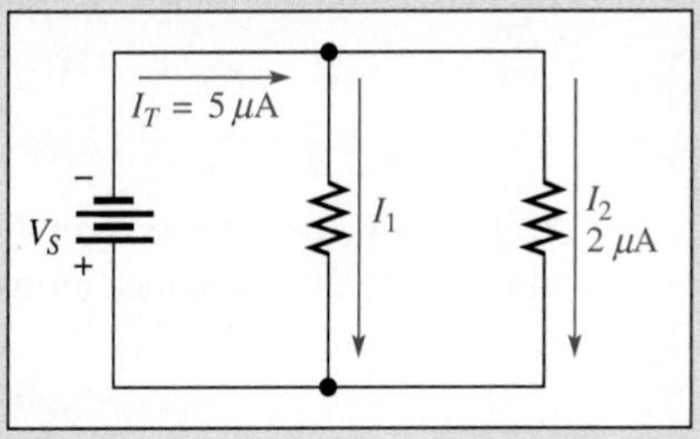

FIGURE 5–19

5–4 ■ TOTAL PARALLEL RESISTANCE

When resistors are connected in parallel, the total resistance of the circuit decreases. The total resistance of a parallel circuit is always less than the value of the smallest resistor. For example, if a 10 Ω resistor and a 100 Ω resistor are connected in parallel, the total resistance is less than 10 Ω.

After completing this section, you should be able to

- **Determine total parallel resistance**
  - ☐ Explain why resistance decreases as resistors are connected in parallel
  - ☐ Apply the parallel-resistance formula

The Number of Current Paths Affects Resistance

As you know, when resistors are connected in parallel, the current has more than one path. The number of current paths is equal to the number of parallel branches.

For example, in Figure 5–20(a), there is only one current path because it is a series circuit. A certain amount of current, I_1, flows through R_1. If resistor R_2 is connected in

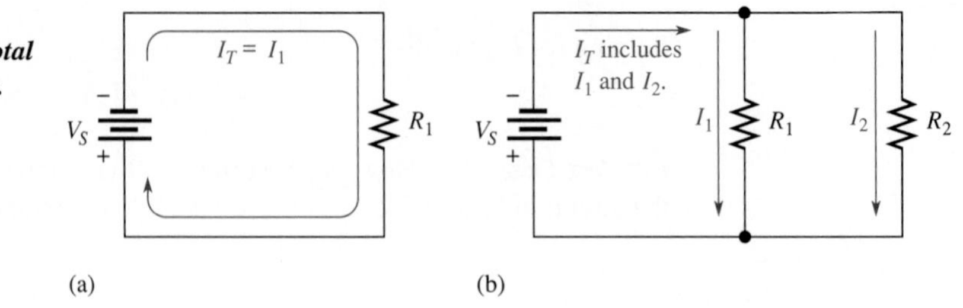

FIGURE 5–20
Adding resistors in parallel reduces total resistance and increases total current.

parallel with R_1, as shown in Figure 5–20(b), an additional amount of current, I_2, flows through R_2. The total current coming from the source has increased with the addition of the parallel branch. An increase in the total current from the source means that the total resistance has decreased, in accordance with Ohm's law. Additional resistors connected in parallel will further reduce the resistance and increase the total current.

Formula for Total Parallel Resistance

The circuit in Figure 5–21 shows a general case of n resistors in parallel (n can be any number). From Kirchhoff's current law, the current equation is

$$I_T = I_1 + I_2 + I_3 + \cdots + I_n$$

Since V_S is the voltage across each of the parallel resistors, by Ohm's law, $I_1 = V_S/R_1$, $I_2 = V_S/R_2$, and so on. By substituting into the current equation,

$$\frac{V_S}{R_T} = \frac{V_S}{R_1} + \frac{V_S}{R_2} + \frac{V_S}{R_3} + \cdots + \frac{V_S}{R_n}$$

We factor V_S out of the right side of the equation and cancel it with V_S on the left side, leaving only the resistance terms.

$$\frac{1}{R_T} = \frac{1}{R_1} + \frac{1}{R_2} + \frac{1}{R_3} + \cdots + \frac{1}{R_n} \qquad \textbf{(5–2)}$$

We get another useful form of Equation (5–2) by taking the reciprocal of (that is, by inverting) both sides of the equation.

$$R_T = \frac{1}{\left(\frac{1}{R_1}\right) + \left(\frac{1}{R_2}\right) + \left(\frac{1}{R_3}\right) + \cdots + \left(\frac{1}{R_n}\right)} \qquad \textbf{(5–3)}$$

Equation (5–3) shows that to find the total parallel resistance, add all the $1/R$ terms and then take the reciprocal of the sum. Example 5–7 shows how to use this formula in a specific case, and Example 5–8 shows a simple method of using the calculator to determine parallel resistances. The reciprocal of resistance ($1/R$) is called *conductance*. Conductance is symbolized by G, and its unit is the siemens (S).

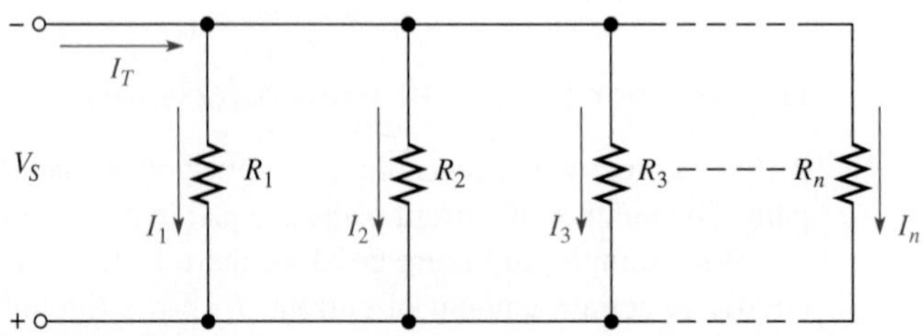

FIGURE 5–21
Circuit with n resistors in parallel.

EXAMPLE 5–7 Calculate the total parallel resistance between points A and B of the circuit in Figure 5–22.

FIGURE 5–22

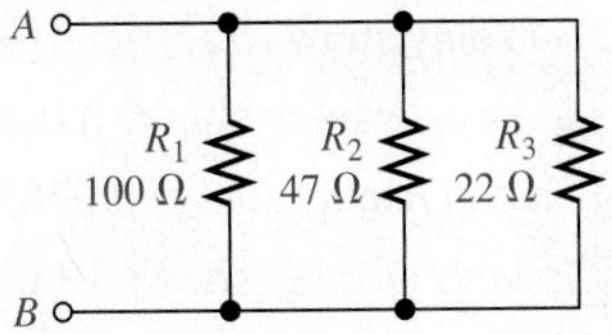

Solution First, find the reciprocal (conductance) of each of the three resistors:

$$\frac{1}{R_1} = \frac{1}{100\ \Omega} = 10\text{ mS}$$

$$\frac{1}{R_2} = \frac{1}{47\ \Omega} = 21.3\text{ mS}$$

$$\frac{1}{R_3} = \frac{1}{22\ \Omega} = 45.5\text{ mS}$$

Next, calculate R_T by adding $1/R_1$, $1/R_2$, and $1/R_3$ and taking the reciprocal of the sum:

$$R_T = \frac{1}{10\text{ mS} + 21.3\text{ mS} + 45.5\text{ mS}} = \frac{1}{76.8\text{ mS}} = 13.0\ \Omega$$

For a quick accuracy check, notice that the value of R_T (13.0 Ω) is smaller than the smallest value in parallel, which is R_3 (22 Ω), as it should be.

Related Exercise If a 33 Ω resistor is connected in parallel in Figure 5–22, what is R_T?

Calculator Solution

The parallel-resistance formula is easily solved on an electronic calculator. The general procedure is to enter the value of R_1 and then take its reciprocal by pressing the [1/x] key ([1/x] is a secondary function on some calculators.). Next press the [+] key; then enter the value of R_2 and take its reciprocal. Repeat this procedure until all of the resistor values have been entered and the reciprocal of each has been added. The final step is to press the [1/x] key to convert $1/R_T$ to R_T. The total parallel resistance is now on the display. This calculator procedure is illustrated in Example 5–8.

EXAMPLE 5–8 Show the steps required for a calculator solution of Example 5–7.

Solution

Step 1: Enter 100. Display shows 100.

Step 2: Press [1/x] key. Display shows 0.01.

Step 3: Press [+] key. Display shows 0.01.

Step 4: Enter 47. Display shows 47.
Step 5: Press [1/x] key. Display shows 0.0212766.
Step 6: Press [+] key. Display shows 0.0312766.
Step 7: Enter 22. Display shows 22.
Step 8: Press [1/x] key. Display shows 0.0454545.
Step 9: Press [=] key. Display shows 0.0767311.
Step 10: Press [1/x] key. Display shows 13.0325183.

The number displayed in Step 10 is the total resistance in ohms. Round it to 13.0 Ω.

Related Exercise Show the additional calculator steps for R_T when a 33 Ω resistor is placed in parallel in the circuit of Example 5–7.

Two Resistors in Parallel

Equation (5–3) is a general formula for finding the total resistance of any number of resistors in parallel. It is often useful to consider only two resistors in parallel because this situation occurs commonly in practice.

Derived from Equation (5–3), the formula for two resistors in parallel is

$$R_T = \frac{R_1 R_2}{R_1 + R_2} \tag{5–4}$$

Equation (5–4) states

The total resistance of two resistors in parallel is equal to the product of the two resistors divided by the sum of the two resistors.

This equation is sometimes referred to as the "product over the sum" formula. Example 5–9 illustrates how to use it.

EXAMPLE 5–9 Calculate the total resistance between the positive and negative terminals of the source of the circuit in Figure 5–23.

FIGURE 5–23

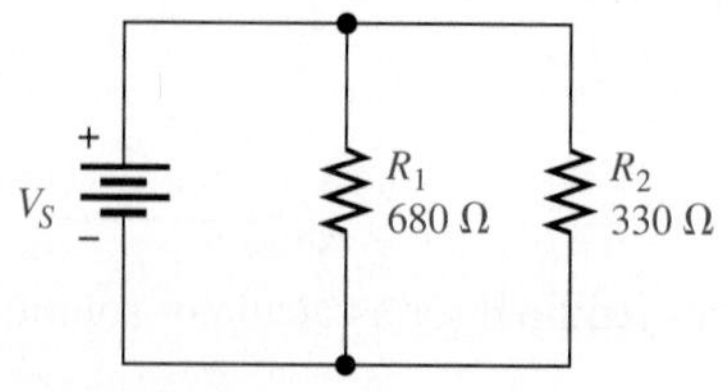

Solution Use Equation (5–4) as follows:

$$R_T = \frac{R_1 R_2}{R_1 + R_2} = \frac{(680\ \Omega)(330\ \Omega)}{680\ \Omega + 330\ \Omega} = \frac{224{,}000\ \Omega^2}{1{,}010\ \Omega} = 222\ \Omega$$

The calculator sequence is

6 8 0 × 3 3 0 ÷ (6 8 0 + 3 3 0) =

Related Exercise Find R_T if a 220 Ω resistor is added in parallel in Figure 5–23.

Resistors of Equal Value in Parallel

Another special case of parallel circuits is the parallel connection of several resistors having the same value. The following is a shortcut method of calculating R_T when this case occurs:

$$R_T = \frac{R}{n} \tag{5–5}$$

Equation (5–5) says that when any number of resistors (n), all having the same resistance (R), are connected in parallel, R_T is equal to the resistance divided by the number of resistors in parallel. Examples 5–10 and 5–11 show how to use this formula.

EXAMPLE 5–10 Find the total resistance between points A and B in Figure 5–24.

FIGURE 5–24

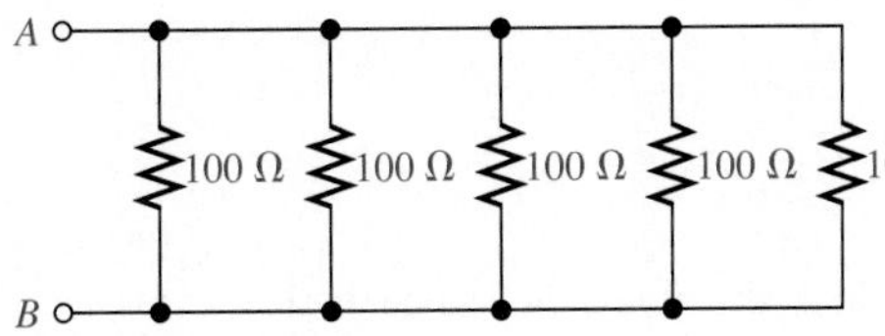

Solution There are five 100 Ω resistors in parallel. Use Equation (5–5) as follows:

$$R_T = \frac{R}{n} = \frac{100\ \Omega}{5} = 20\ \Omega$$

Related Exercise Find R_T for three 100 kΩ resistors in parallel.

EXAMPLE 5–11 A stereo amplifier drives two 8 Ω speakers in parallel from each channel as shown in Figure 5–25. What is the total resistance across the output terminals of each channel of the amplifier?

FIGURE 5–25

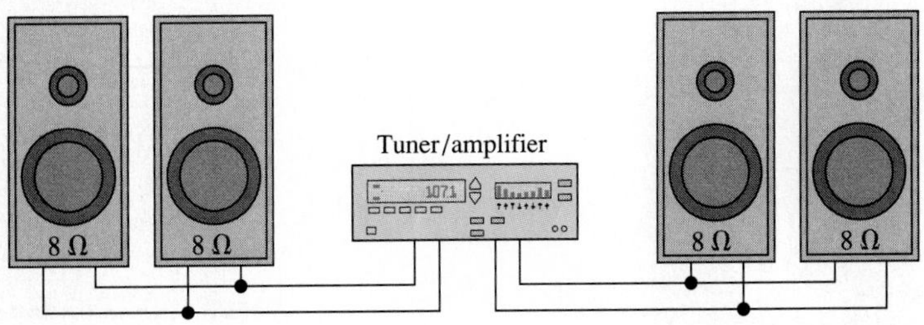

Solution The total parallel resistance across each channel output with the two 8 Ω speakers is

$$R_T = \frac{R}{n} = \frac{8\ \Omega}{2} = 4\ \Omega$$

Related Exercise If two additional 8 Ω speakers are added to each channel, what is the resistance across each channel output?

Notation for Parallel Resistors

Sometimes, for convenience, parallel resistors are designated by two parallel vertical marks. For example, R_1 in parallel with R_2 can be written as $R_1 \parallel R_2$. Also, when several resistors are in parallel with each other, this notation can be used. For example,

$$R_1 \parallel R_2 \parallel R_3 \parallel R_4 \parallel R_5$$

indicates that R_1 through R_5 are all in parallel.

This notation is also used with resistance values. For example,

$$10\ \text{k}\Omega \parallel 5\ \text{k}\Omega$$

means that a 10 kΩ resistor is in parallel with a 5 kΩ resistor.

SECTION 5–4 REVIEW

1. Does the total resistance increase or decrease as more resistors are connected in parallel?
2. The total parallel resistance is always less than ______ .
3. Determine R_T (between pin 1 and pin 4) for the circuit in Figure 5–26.

FIGURE 5–26

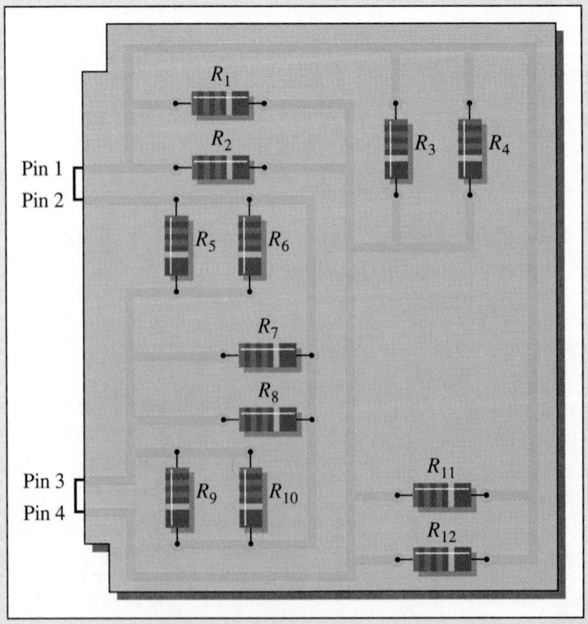

5–5 ■ OHM'S LAW IN PARALLEL CIRCUITS

In this section, you will see how Ohm's law can be applied to parallel circuit analysis. You will learn by several examples how to determine the total current and branch currents in a parallel circuit. You will also learn how to find the voltage across a parallel circuit and the resistance in a parallel circuit.

After completing this section, you should be able to

- **Apply Ohm's law in a parallel circuit**
  - ☐ Find the total current in a parallel circuit
  - ☐ Find each branch current in a parallel circuit
  - ☐ Find the voltage across a parallel circuit
  - ☐ Find the resistance in a parallel circuit

EXAMPLE 5–12 Find the total current produced by the battery in Figure 5–27.

FIGURE 5–27

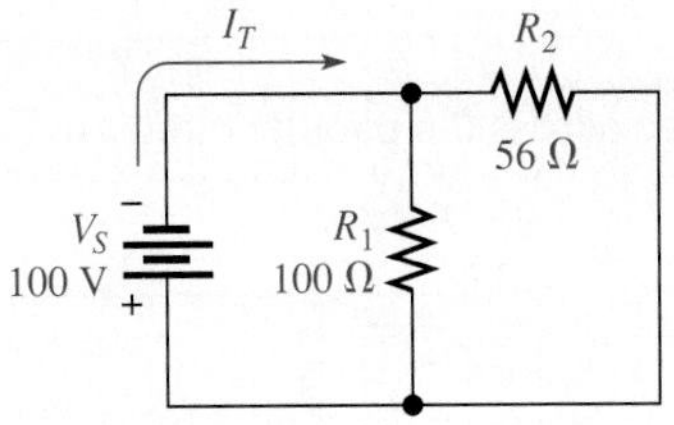

Solution The battery "sees" a total parallel resistance that determines the amount of current generated. First, calculate R_T:

$$R_T = \frac{R_1R_2}{R_1 + R_2} = \frac{(100\ \Omega)(56\ \Omega)}{100\ \Omega + 56\ \Omega} = \frac{5600\ \Omega^2}{156\ \Omega} = 35.9\ \Omega$$

The calculator sequence is

1 0 0 × 5 6 ÷ (1 0 0 + 5 6) =

The battery voltage is 100 V. Use Ohm's law to find I_T:

$$I_T = \frac{100\ \text{V}}{35.9\ \Omega} = 2.79\ \text{A}$$

1 0 0 ÷ 3 5 • 9 =

Related Exercise Find the currents through R_1 and R_2 in Figure 5–27.

EXAMPLE 5–13 Determine the current through each resistor in the parallel circuit of Figure 5–28.

FIGURE 5–28

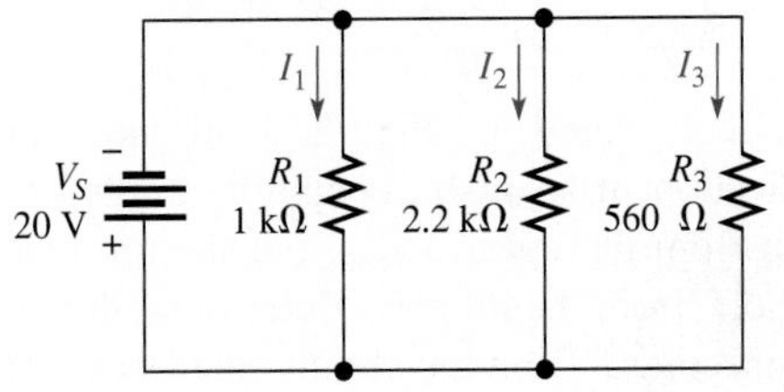

Solution The voltage across each resistor (branch) is equal to the source voltage. That is, the voltage across R_1 is 20 V, the voltage across R_2 is 20 V, and the voltage across R_3 is 20 V. The current through each resistor is determined as follows:

$$I_1 = \frac{V_S}{R_1} = \frac{20\text{ V}}{1\text{ k}\Omega} = 20\text{ mA}$$

$$I_2 = \frac{V_S}{R_2} = \frac{20\text{ V}}{2.2\text{ k}\Omega} = 9.1\text{ mA}$$

$$I_3 = \frac{V_S}{R_3} = \frac{20\text{ V}}{560\ \Omega} = 35.7\text{ mA}$$

Related Exercise If an additional resistor of 910 Ω is connected in parallel to the circuit in Figure 5–28, determine all the branch currents.

EXAMPLE 5–14 Find the voltage across the parallel circuit in Figure 5–29.

FIGURE 5–29

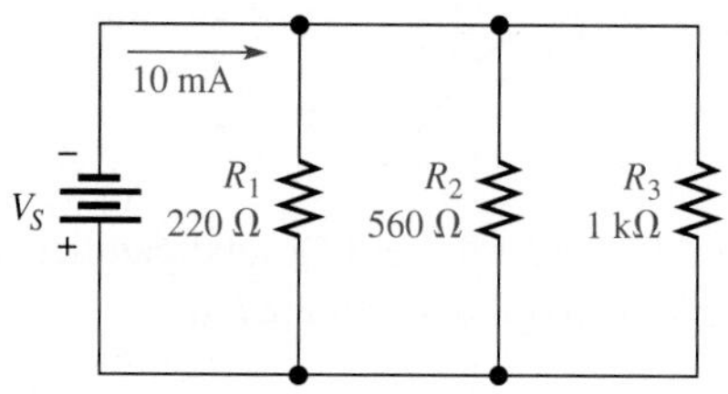

Solution The total current into the parallel circuit is 10 mA. You need to find the total resistance, and then apply Ohm's law to get the voltage. The total resistance is

$$R_T = \frac{1}{\left(\frac{1}{R_1}\right) + \left(\frac{1}{R_2}\right) + \left(\frac{1}{R_3}\right)} = \frac{1}{\left(\frac{1}{220\ \Omega}\right) + \left(\frac{1}{560\ \Omega}\right) + \left(\frac{1}{1\text{ k}\Omega}\right)}$$

$$= \frac{1}{4.55\text{ mS} + 1.79\text{ mS} + 1\text{ mS}} = \frac{1}{7.34\text{ mS}} = 136\ \Omega$$

$$V_S = I_T R_T = (10\text{ mA})(136\ \Omega) = 1.36\text{ V}$$

Related Exercise Find the voltage if R_3 is decreased to 680 Ω in Figure 5–29. Assume I_T remains 10 mA.

EXAMPLE 5–15 The circuit board in Figure 5–30 has three resistors in parallel used for bias modification in an instrumentation amplifier. The values of two of the resistors are known from the color codes, but the top resistor is unmarked (maybe the bands are worn off from handling). Determine the value of the unknown resistor without disconnecting it from the circuit board or without using an ohmmeter.

FIGURE 5–30

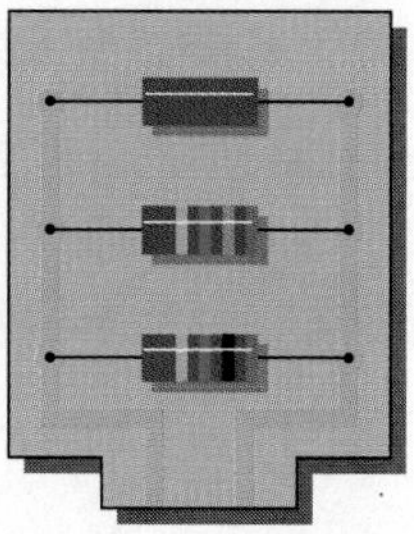

Solution If you can determine the total resistance of the three resistors in parallel, then you can use the parallel-resistance formula to calculate the unknown resistance. You can use Ohm's law to find the total resistance if voltage and total current are known.

In Figure 5–31, a 10 V source (arbitrary value) is connected across the resistors and the total current is measured. Using these measured values, the total resistance is

$$R_T = \frac{V}{I_T} = \frac{10\text{ V}}{20.1\text{ mA}} = 498\ \Omega$$

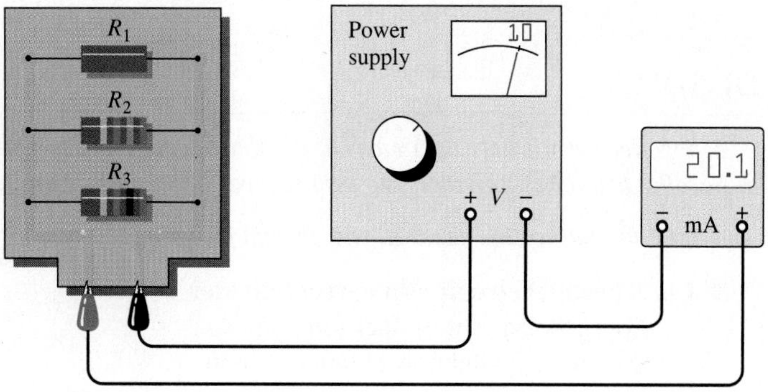

FIGURE 5–31

Use Equation (5–2) to find the unknown resistance as follows:

$$\frac{1}{R_T} = \frac{1}{R_1} + \frac{1}{R_2} + \frac{1}{R_3}$$

$$\frac{1}{R_1} = \frac{1}{R_T} - \frac{1}{R_2} - \frac{1}{R_3} = \frac{1}{498\ \Omega} - \frac{1}{1.8\text{ k}\Omega} - \frac{1}{1\text{ k}\Omega} = 453\ \mu\text{S}$$

$$R_1 = \frac{1}{453\ \mu\text{S}} = 2.2\text{ k}\Omega$$

The calculator sequence for R_1 is

4 9 8 1/x − 1 EXP 3 1/x − 1 • 8 EXP 3 1/x = 1/x

Related Exercise Explain how to determine the value of R_1 using an ohmmeter and without removing R_1 from the circuit.

SECTION 5–5 REVIEW

1. A 10 V battery is connected across three 68 Ω resistors that are in parallel. What is the total current from the battery?
2. How much voltage is required to produce 2 A of current through the circuit of Figure 5–32?

FIGURE 5–32

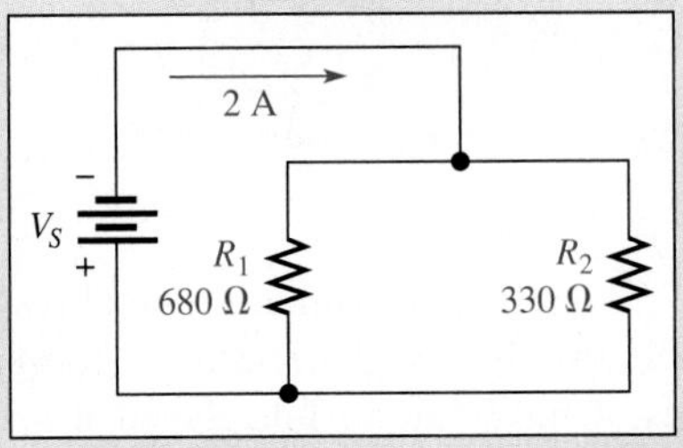

3. How much current is there through each resistor of Figure 5–32?
4. There are four equal-value resistors in parallel with a 12 V source, and 6 mA of current from the source. What is the value of each resistor?
5. A 1 kΩ and a 2.2 kΩ resistor are connected in parallel. A total of 100 mA flows through the parallel combination. How much voltage is dropped across the resistors?

5–6 ■ CURRENT DIVIDERS

A parallel circuit acts as a current divider because the current entering the junction of parallel branches "divides" up into several individual branch currents.

After completing this section, you should be able to

- **Use a parallel circuit as a current divider**
  - □ Apply the current-divider formula
  - □ Determine an unknown branch current

In a parallel circuit, when the total current flows into the junction of the parallel branches, it divides among the branches. Thus, a parallel circuit acts as a current divider. This current divider principle is illustrated in Figure 5–33 for a two-branch parallel circuit in which part of the total current I_T goes through R_1 and part through R_2.

Since the same voltage is across each of the resistors in parallel, the branch currents are inversely proportional to the values of the resistors. For example, if the value of R_2 is twice that of R_1, then the value of I_2 is one-half that of I_1. In other words,

The total current divides among parallel resistors in a manner inversely proportional to the resistance values.

FIGURE 5–33
Total current divides between two branches.

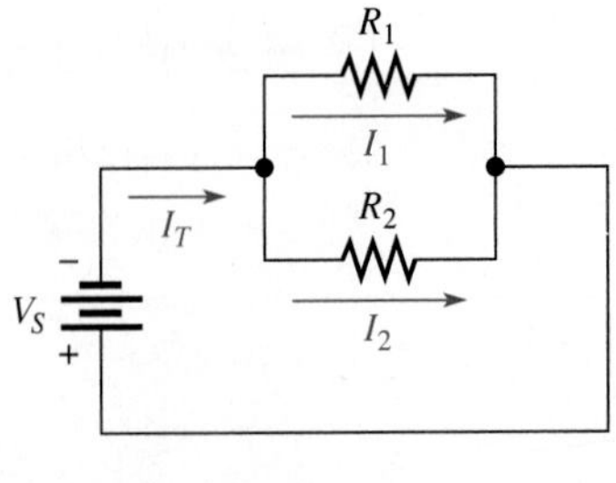

FIGURE 5–34

The branch with the lowest resistance has the most current, and the branch with the highest resistance has the least current.

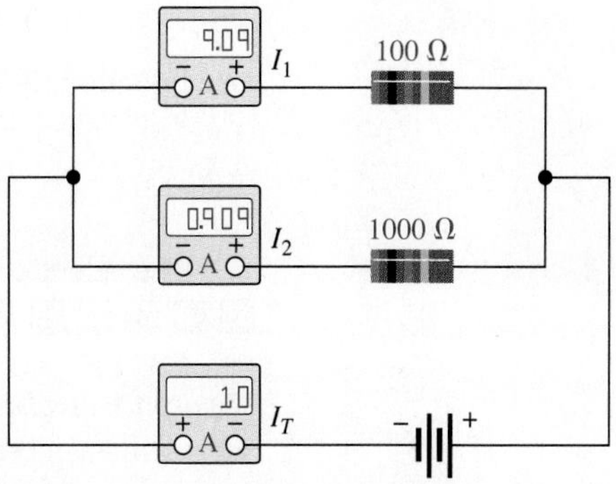

The branches with higher resistance have less current, and the branches with lower resistance have more current, in accordance with Ohm's law. If all the branches have the same resistance, the branch currents are all equal.

Figure 5–34 shows specific values to demonstrate how the currents divide according to the branch resistances. Notice that in this case the resistance of the upper branch is one-tenth the resistance of the lower branch, but the upper branch current is ten times the lower branch current.

Current-Divider Formulas for Two Branches

You already know how to use Ohm's law ($I = V/R$) to determine the current in any parallel branch when the voltage and resistance are known. When the voltage is not known but the total current is, you can find both of the branch currents (I_1 and I_2) by using the following formulas:

$$I_1 = \left(\frac{R_2}{R_1 + R_2}\right)I_T \quad \text{(5–6)}$$

$$I_2 = \left(\frac{R_1}{R_1 + R_2}\right)I_T \quad \text{(5–7)}$$

The current in either branch is equal to the opposite branch resistance divided by the sum of the two resistors and then multiplied by the total current.

EXAMPLE 5–16 Find I_1 and I_2 in Figure 5–35.

FIGURE 5–35

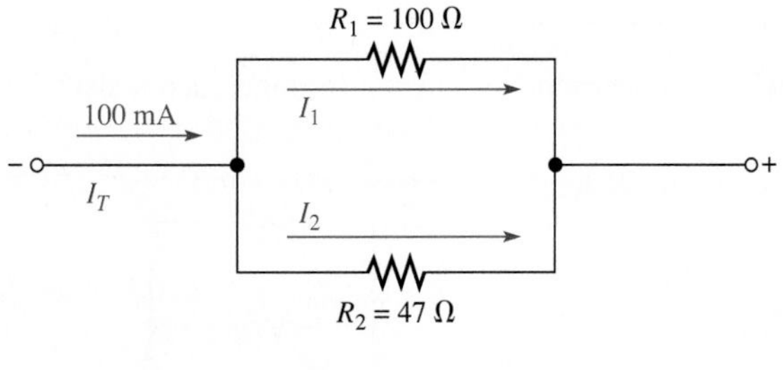

Solution Use Equation (5–6) to determine I_1:

$$I_1 = \left(\frac{R_2}{R_1 + R_2}\right)I_T = \left(\frac{47\ \Omega}{147\ \Omega}\right)100\text{ mA} = 32.0\text{ mA}$$

Use Equation (5–7) to determine I_2:

$$I_2 = \left(\frac{R_1}{R_1 + R_2}\right)I_T = \left(\frac{100\ \Omega}{147\ \Omega}\right)100\ \text{mA} = 68.0\ \text{mA}$$

The calculator sequence for I_2 is

1 0 0 ÷ (1 0 0 + 4 7) × 1 0 0 EXP +/− 3 =

Related Exercise If $R_1 = 56\ \Omega$ and $R_2 = 82\ \Omega$ in Figure 5–35 and I_T stays the same, what will each branch current be?

General Current-Divider Formula for Any Number of Branches

A generalized parallel circuit with n branches is shown in Figure 5–36 (n can represent any number).

FIGURE 5–36
Generalized parallel circuit with n branches.

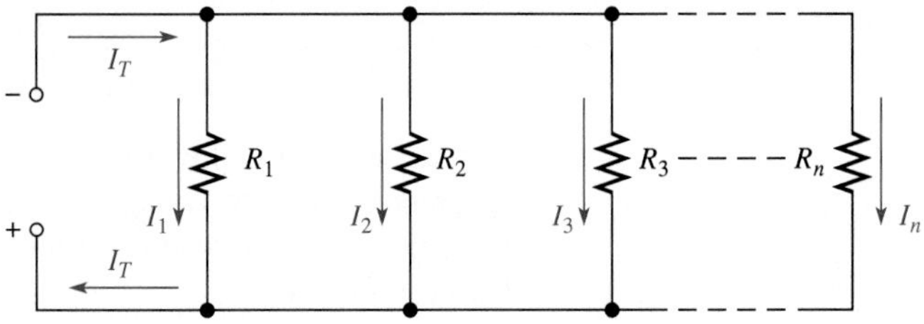

The current in any branch can be determined with the following formula:

$$I_x = \left(\frac{R_T}{R_x}\right)I_T \qquad \textbf{(5–8)}$$

where I_x represents any branch current (I_1, I_2, and so on) and R_x represents any resistance (R_1, R_2, and so on). For example, the formula for current in the second branch in Figure 5–36 is

$$I_2 = \left(\frac{R_T}{R_2}\right)I_T$$

Equation (5–8) can be used for a parallel circuit with any number of branches. Notice that you must determine R_T in order to use Equation (5–8) to find a branch current.

EXAMPLE 5–17 Determine the current through each resistor in the circuit of Figure 5–37.

FIGURE 5–37

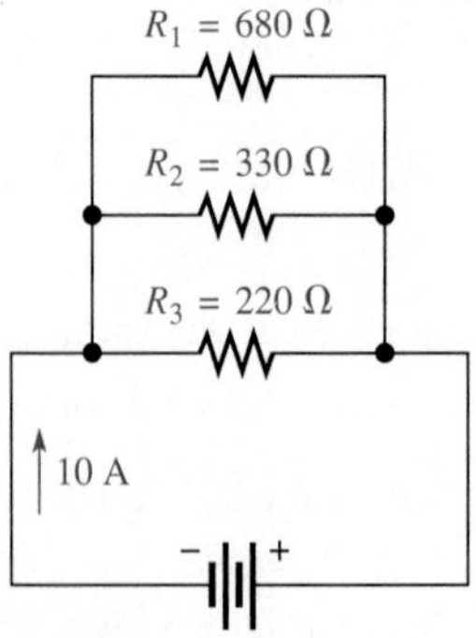

Solution First calculate the total parallel resistance:

$$R_T = \frac{1}{\left(\frac{1}{R_1}\right)+\left(\frac{1}{R_2}\right)+\left(\frac{1}{R_3}\right)} = \frac{1}{\left(\frac{1}{680\ \Omega}\right)+\left(\frac{1}{330\ \Omega}\right)+\left(\frac{1}{220\ \Omega}\right)} = 111\ \Omega$$

The total current is 10 A. Use Equation (5–8) to calculate each branch current:

$$I_1 = \left(\frac{R_T}{R_1}\right)I_T = \left(\frac{111\ \Omega}{680\ \Omega}\right)10\text{ A} = 1.63\text{ A}$$

$$I_2 = \left(\frac{R_T}{R_2}\right)I_T = \left(\frac{111\ \Omega}{330\ \Omega}\right)10\text{ A} = 3.36\text{ A}$$

$$I_3 = \left(\frac{R_T}{R_3}\right)I_T = \left(\frac{111\ \Omega}{220\ \Omega}\right)10\text{ A} = 5.05\text{ A}$$

Related Exercise Determine the current through R_1 and R_2 in Figure 5–37 if R_3 is removed. Assume the source voltage remains the same.

SECTION 5–6 REVIEW

1. A circuit has the following resistors in parallel: 220 Ω, 100 Ω, 68 Ω, 56 Ω, and 22 Ω. Which resistor has the most current through it? The least current?
2. Determine the current through R_3 in Figure 5–38.
3. Find the currents through each resistor in the circuit of Figure 5–39.

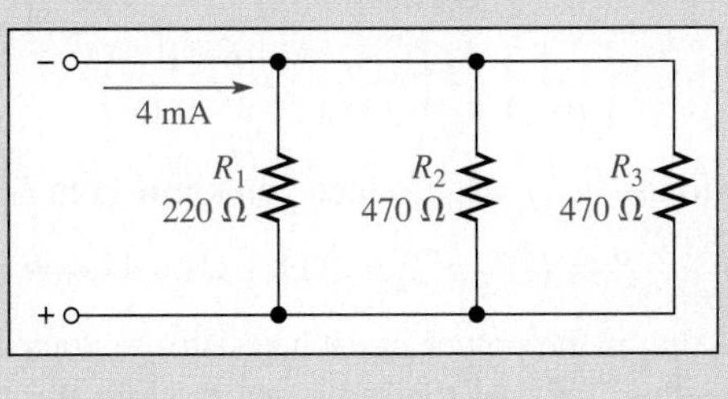

FIGURE 5–38

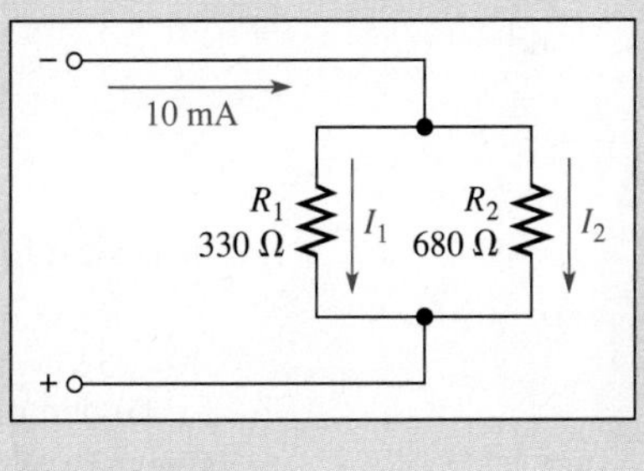

FIGURE 5–39

5–7 ■ POWER IN PARALLEL CIRCUITS

Total power in a parallel circuit is found by adding up the powers of all the individual resistors, the same as you did for series circuits.

After completing this section, you should be able to

■ **Determine power in a parallel circuit**

Equation (5–9) states the formula for finding total power in a concise way for any number of resistors in parallel.

$$P_T = P_1 + P_2 + P_3 + \cdots + P_n \tag{5–9}$$

where P_T is the total power and P_n is the power in the last resistor in parallel. As you can see, the power losses are additive, just as in the series circuit.

The power formulas in Chapter 3 are directly applicable to parallel circuits. The following formulas are used to calculate the total power P_T:

$$P_T = VI_T$$
$$P_T = I_T^2 R_T$$
$$P_T = \frac{V^2}{R_T}$$

where V is the voltage across the parallel circuit, I_T is the total current into the parallel circuit, and R_T is the total resistance of the parallel circuit. Example 5–18 shows how total power can be calculated in a parallel circuit.

EXAMPLE 5–18 Determine the total amount of power in the parallel circuit in Figure 5–40.

FIGURE 5–40

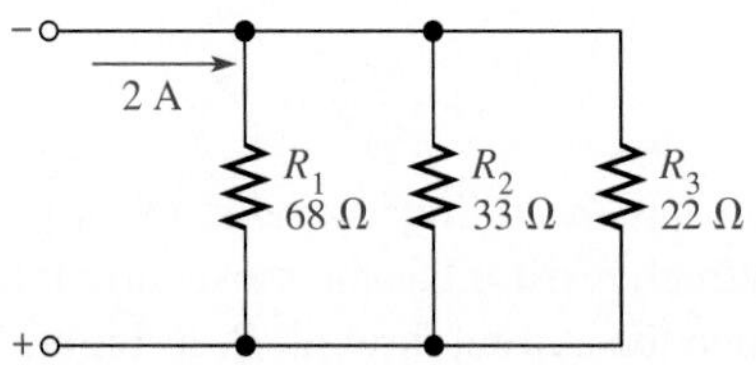

Solution The total current is 2 A. The total resistance is

$$R_T = \frac{1}{\left(\frac{1}{68\ \Omega}\right) + \left(\frac{1}{33\ \Omega}\right) + \left(\frac{1}{22\ \Omega}\right)} = 11.1\ \Omega$$

The easiest formula to use is $P_T = I_T^2 R_T$ since you know both I_T and R_T. Thus,

$$P_T = I_T^2 R_T = (2\text{ A})^2(11.1\ \Omega) = 44.4\text{ W}$$

To demonstrate that if the power in each resistor is determined and if all of these values are added together, you get the same result, work through another calculation. First, find the voltage across each branch of the circuit:

$$V_S = I_T R_T = (2\text{ A})(11.1\ \Omega) = 22.2\text{ V}$$

Remember that the voltage across all branches is the same.

Next, use $P = V^2/R$ to calculate the power for each resistor:

$$P_1 = \frac{(22.2\text{ V})^2}{68\ \Omega} = 7.25\text{ W}$$

$$P_2 = \frac{(22.2\text{ V})^2}{33\ \Omega} = 14.9\text{ W}$$

$$P_3 = \frac{(22.2\text{ V})^2}{22\ \Omega} = 22.4\text{ W}$$

Now, add these powers to get the total power:

$$P_T = 7.25\text{ W} + 14.9\text{ W} + 22.4\text{ W} = 44.6\text{ W}$$

This calculation shows that the sum of the individual powers is equal (approximately) to the total power as determined by one of the power formulas. Rounding to three significant figures accounts for the difference.

Related Exercise Find the total power in Figure 5–40 if the total current is doubled.

EXAMPLE 5–19 The amplifier in one channel of a stereo system as shown in Figure 5–41 drives four speakers as shown. If the maximum voltage to the speakers is 15 V, how much power must the amplifier be able to deliver to the speakers?

FIGURE 5–41

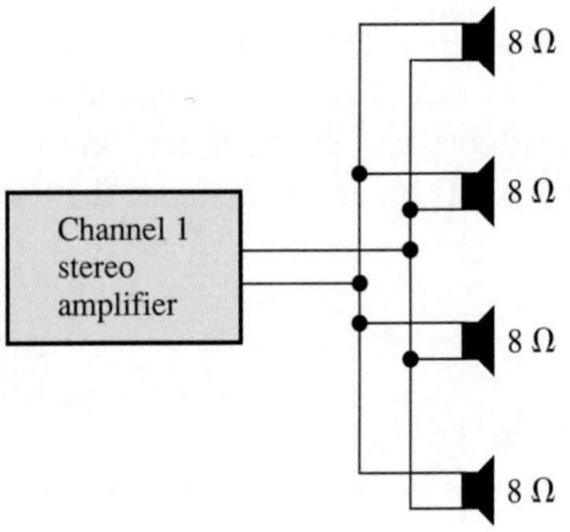

Solution The speakers are connected in parallel to the amplifier output, so the voltage across each is the same. The maximum power to each speaker is

$$P_{max} = \frac{V_{max}^2}{R} = \frac{(15\text{ V})^2}{8\ \Omega} = 28.1\text{ W}$$

The total power that the amplifier must be capable of delivering to the speaker system is four times the power in an individual speaker because the total power is the sum of the individual powers.

$$P_{T(max)} = P_{(max)} + P_{(max)} + P_{(max)} + P_{(max)} = 4P_{max} = 4(28.1\text{ W}) = 112.4\text{ W}$$

Related Exercise If the amplifier can produce a maximum of 18 V, what is the maximum total power to the speakers?

SECTION 5–7 REVIEW

1. If you know the power in each resistor in a parallel circuit, how can you find the total power?
2. The resistors in a parallel circuit dissipate the following powers: 1 W, 2 W, 5 W, and 8 W. What is the total power in the circuit?
3. A circuit has a 1 kΩ, a 2.7 kΩ, and a 3.9 kΩ resistor in parallel. A total current of 1 A flows into the parallel circuit. What is the total power?

5–8 ▪ EXAMPLES OF PARALLEL CIRCUIT APPLICATIONS

Parallel circuits are found in some form in virtually every electronic system. In many of these applications, the parallel relationship of components may not be obvious until you have covered some advanced topics that will come later. For now, we will look at some common and familiar applications of parallel circuits.

After completing this section, you should be able to

- **Describe some basic applications of parallel circuits**
  - ☐ Discuss the lighting system in automobiles
  - ☐ Discuss residential wiring
  - ☐ Describe how current through a heating element is controlled
  - ☐ Explain basically how a multiple-range ammeter works

Automotive

One advantage of a parallel circuit over a series circuit is that when one branch opens, the other branches are not affected. For example, Figure 5–42 shows a simplified diagram of an automobile lighting system. When one headlight on your car goes out, it does not cause the other lights to go out, because they are all in parallel.

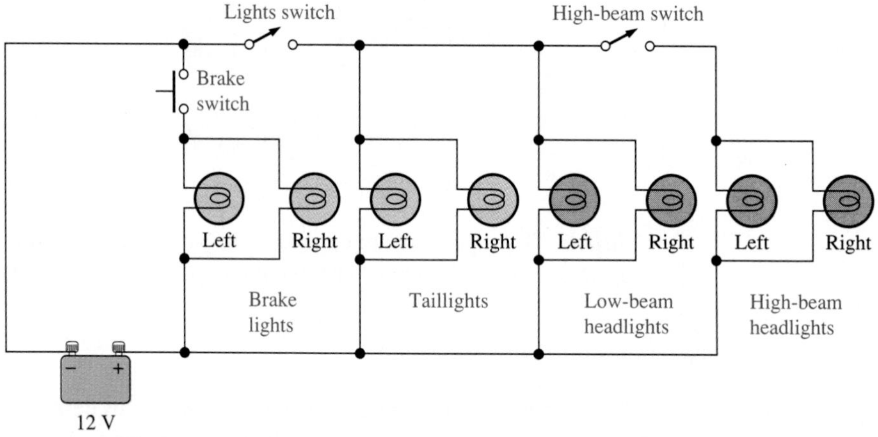

FIGURE 5–42
Simplified diagram of the exterior light system of an automobile. All lights are off with switches in the positions shown.

Notice that the brake lights are switched on independently of the head and taillights. They come on only when the driver closes the brake light switch by depressing the brake pedal.

When the lights switch is closed, both low-beam headlights and both taillights are on. The high-beam headlights are on only when both the lights switch and the high-beam switch are closed. If any one of the lights burns out (opens), there is still current in each of the other lights.

Residential

Another common use of parallel circuits is in residential electrical systems. All the lights and appliances in a home are wired in parallel. Figure 5–43(a) shows a typical room wiring arrangement with two switch-controlled lights and three wall outlets in parallel.

Figure 5–43(b) shows a simplified parallel arrangement of four heating elements on an electric range. The four-position switches in each branch allow the user to control the

FIGURE 5–43
Examples of parallel circuits in residential wiring and appliances.

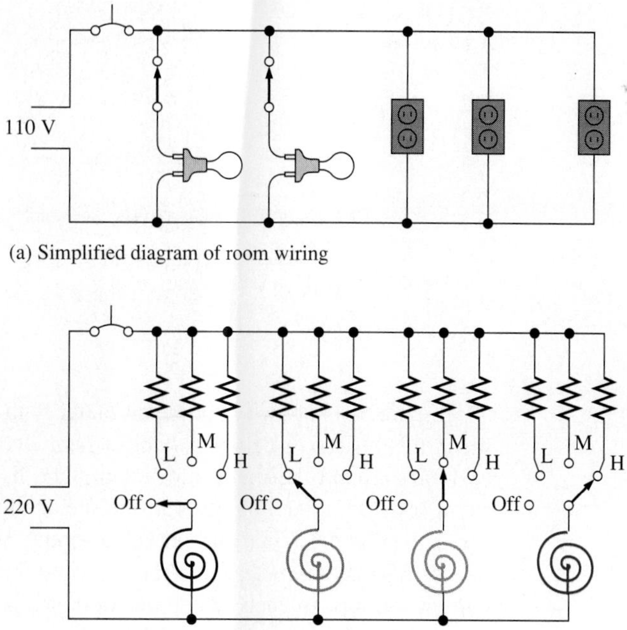

(a) Simplified diagram of room wiring

(b) Simplified diagram of a four-burner range

amount of current through the heating elements by selecting the appropriate limiting resistor. The lowest resistor value (H setting) allows the highest amount of current for maximum heat. The highest resistor value (L setting) allows the least amount of current for minimum heat; M designates the medium settings.

Ammeters

Another example in which parallel circuits are used is the familiar analog (needle-type) ammeter or milliammeter. Parallel circuits are an important part of the operation of the ammeter because they allow the user to select various ranges in order to measure many different current values.

The mechanism in an ammeter that causes the pointer to move in proportion to the current is called the *meter movement,* which is based on a magnetic principle that you will learn later. Right now, all you need to know is that a meter movement has a certain resistance and a maximum current. This maximum current, called the *full-scale deflection* current, causes the pointer to go all the way to the end of the scale. For example, a certain meter movement has a 50 Ω resistance and a full-scale deflection current of 1 mA. A meter with this particular movement can measure currents of 1 mA or less. Currents greater than 1 mA will cause the pointer to "peg" (or stop) at full scale. Figure 5–44 illustrates a 1 mA meter.

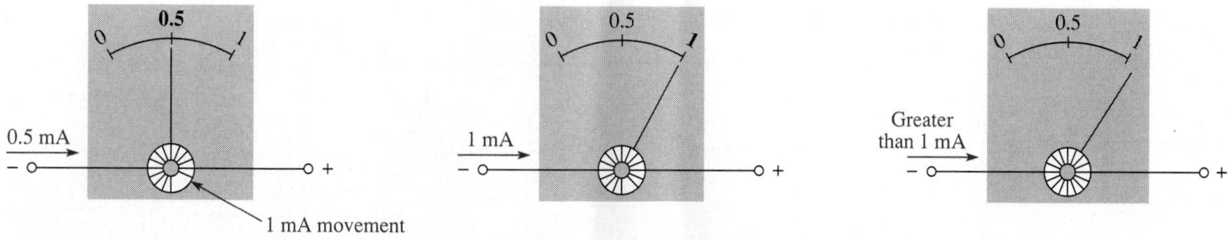

(a) Half-scale deflection

(b) Full-scale deflection

(c) Pegged

FIGURE 5–44
A 1 mA meter.

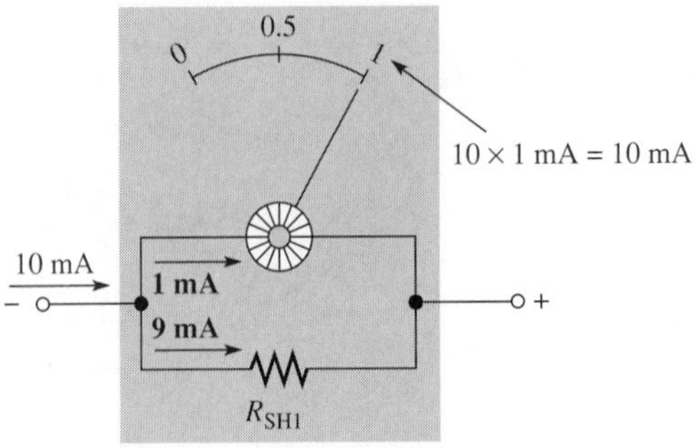

FIGURE 5–45
A 10 mA meter.

Figure 5–45 shows a simple ammeter with a resistor in parallel with the meter movement; this resistor is called a *shunt resistor*. Its purpose is to bypass any current in excess of 1 mA around the meter movement. The figure specifically shows 9 mA through the shunt resistor and 1 mA through the meter movement. Thus, up to 10 mA can be measured. To find the actual current value, simply multiply the reading on the scale by 10.

A practical ammeter has a range switch that permits the selection of several full-scale current settings. In each switch position, a certain amount of current is bypassed through a parallel resistor as determined by the resistance value. In our example, the current through the movement is never greater than 1 mA.

Figure 5–46 illustrates a meter with three ranges: 1 mA, 10 mA, and 100 mA. When the range switch is in the 1 mA position, all of the current coming into the meter goes through the meter movement. In the 10 mA setting, up to 9 mA goes through R_{SH1} and up to 1 mA through the movement. In the 100 mA setting, up to 99 mA goes through R_{SH2} and the movement can still have only 1 mA for full-scale.

For example, in Figure 5–46, if 50 mA of current are being measured, the needle points to the 0.5 mark on the scale; you must multiply 0.5 by 100 to find the current value. In this situation, 0.5 mA flows through the movement (half-scale deflection), and 49.5 mA flows through R_{SH2}.

Effect of the Ammeter on a Circuit As you know, an ammeter is connected in series to measure the current in a circuit. Ideally, the meter should not alter the current that it is intended to measure. In practice, however, the meter unavoidably has some effect on the circuit, because its internal resistance is connected in series with the circuit resistance. However, in most cases, the meter's internal resistance is so small compared to the circuit resistance that it can be neglected.

For example, if the meter has a 50 Ω movement (R_M) and a 100 μA full-scale current (I_M), the voltage dropped across the movement is

$$V_M = I_M R_M = (100\ \mu\text{A})(50\ \Omega) = 5\ \text{mV}$$

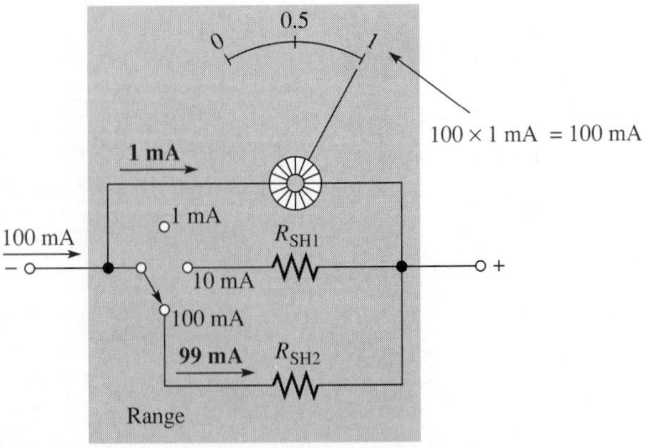

FIGURE 5–46
A milliammeter with three ranges.

The shunt resistance (R_{SH}) for the 10 mA range, for example, is

$$R_{SH} = \frac{V_M}{I_{SH}} = \frac{5\text{ mV}}{9.9\text{ mA}} = 0.505\ \Omega$$

As you can see, the total resistance of the ammeter on the 10 mA range is the resistance of the movement in parallel with the shunt resistance:

$$R_M \parallel R_{SH} = 50\ \Omega \parallel 0.505\ \Omega = 0.5\ \Omega$$

EXAMPLE 5–20 How much does an ammeter with a 100 μA, 50 Ω movement affect the current in the circuit of Figure 5–47?

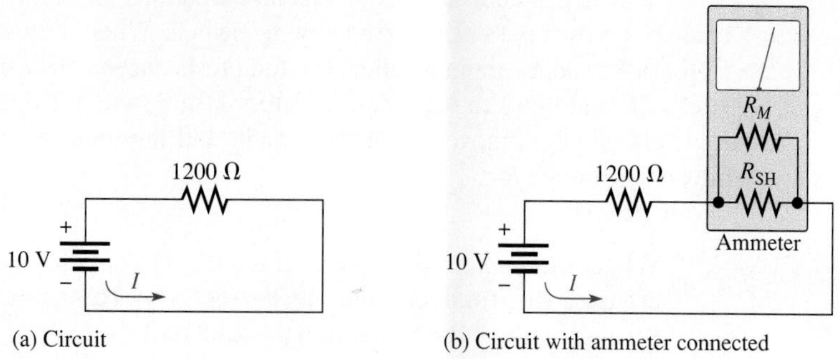

FIGURE 5–47

Solution The true current in the circuit is

$$I = \frac{10\text{ V}}{1200\ \Omega} = 8.3333\text{ mA}$$

The meter is set on the 10 mA range in order to measure this particular amount of current. It was found that the meter's resistance on the 10 mA range is 0.5 Ω. When the meter is connected in the circuit, its resistance is in series with the 1200 Ω resistor. Thus, there is a total of 1200.5 Ω.

The current actually measured by the ammeter is

$$I = \frac{10\text{ V}}{1200.5\ \Omega} = 8.3299\text{ mA}$$

This current differs from the true circuit current by only 0.04%. Therefore, the meter does not significantly alter the current value, a situation which, of course, is necessary because the measuring instrument should not change the quantity that is to be measured accurately.

Related Exercise How much will the measured current differ from the true current if the circuit resistance is 12 kΩ instead of 1200 Ω?

SECTION 5–8 REVIEW

1. For the ammeter in Figure 5–46, what is the maximum resistance that the meter will have when connected in a circuit? What is the maximum current that can be measured at the setting?
2. Do the shunt resistors have resistance values considerably less than or more than that of the movement? Why?

5–9 ▪ TROUBLESHOOTING

In this section, you will see how an open in a parallel branch affects the parallel circuit.

After completing this section, you should be able to

- **Troubleshoot parallel circuits**
  - ☐ Check for an open circuit

Open Branches

Recall that an open circuit is one in which the current path is interrupted and no current flows. In this section we examine what happens when a branch of a parallel circuit opens.

If a switch is connected in a branch of a parallel circuit, as shown in Figure 5–48, an open or a closed path can be made by the switch. When the switch is closed, as in Figure 5–48(a), R_1 and R_2 are in parallel. The total resistance is 50 Ω (two 100 Ω resistors in parallel). Current flows through both resistors. If the switch is opened, as in Figure 5–48(b), R_1 is effectively removed from the circuit, and the total resistance is 100 Ω. Current is now only through R_2.

In general,

When an open circuit occurs in a parallel branch, the total resistance increases, the total current decreases, and the same current continues through each of the remaining parallel paths.

The decrease in total current equals the amount of current that was previously in the open branch. The other branch currents remain the same.

Consider the lamp circuit in Figure 5–49. There are four bulbs in parallel with a 120 V source. In part (a), there is current through each bulb. Now suppose that one of the bulbs burns out, creating an open path as shown in Figure 5–49(b). This light will go out

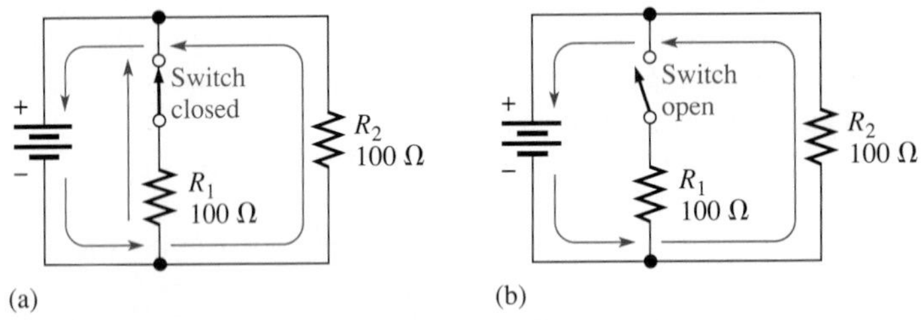

FIGURE 5–48
When the switch opens, total current decreases and current through R_2 remains unchanged.

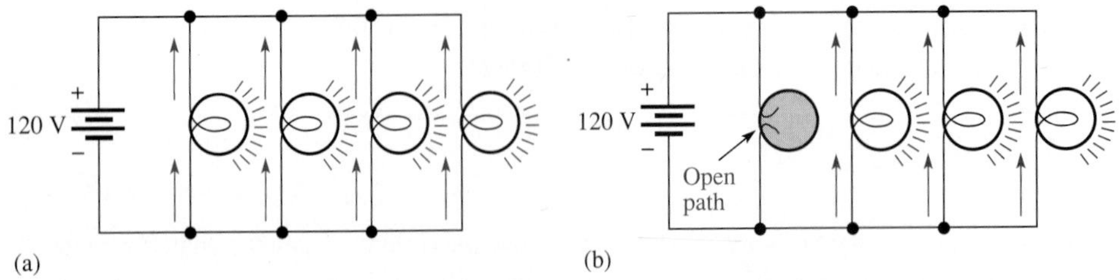

FIGURE 5–49
When a lamp filament opens, total current decreases and the other branch currents remain unchanged.

because there is no current through the open path. Notice, however, that current continues through all the other parallel bulbs, and they continue to glow. The open branch does not change the voltage across the parallel branches; it remains at 120 V.

You can see that a parallel circuit has an advantage over a series connection in lighting systems because if one or more of the parallel bulbs burn out, the others will stay on. In a series circuit, when one bulb goes out, all of the others go out also because the current path is completely interrupted.

When a resistor in a parallel circuit opens, the open resistor cannot be located by measurement of the voltage across the branches, because the same voltage exists across all the branches. Thus, there is no way to tell which resistor is open by simply measuring voltage. The good resistors will always have the same voltage as the open one, as illustrated in Figure 5–50 (note that the middle resistor is open).

If a visual inspection does not reveal the open resistor, it can be located by current measurements. In practice, measuring current is more difficult than measuring voltage because you must insert the ammeter in series to measure the current. Thus, a wire or a printed circuit connection must be cut or disconnected, or one end of a component must be lifted off the circuit board, in order to connect the ammeter in series. This procedure, of course, is not required when voltage measurements are made, because the meter leads are simply connected across a component.

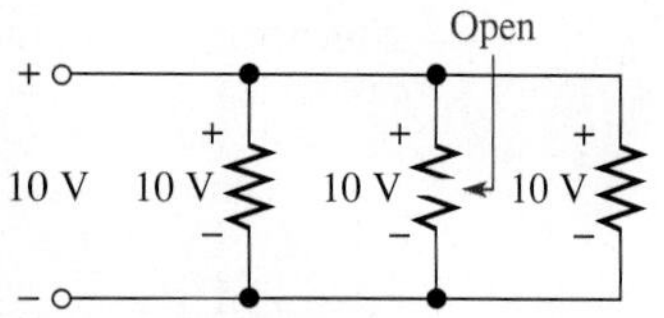

FIGURE 5–50
All parallel branches (open or not) have the same voltage.

Finding an Open Branch by Current Measurement

In a parallel circuit, the total current should be measured. *When a parallel resistor opens, I_T is always less than its normal value.* Once I_T and the voltage across the branches are known, a few calculations will determine the open resistor when all the resistors are of different values.

Consider the two-branch circuit in Figure 5–51(a). If one of the resistors opens, the total current will equal the current in the good resistor. Ohm's law quickly tells you what the current in each resistor should be:

$$I_1 = \frac{50\text{ V}}{560\ \Omega} = 89.3\text{ mA}$$

$$I_2 = \frac{50\text{ V}}{100\ \Omega} = 500\text{ mA}$$

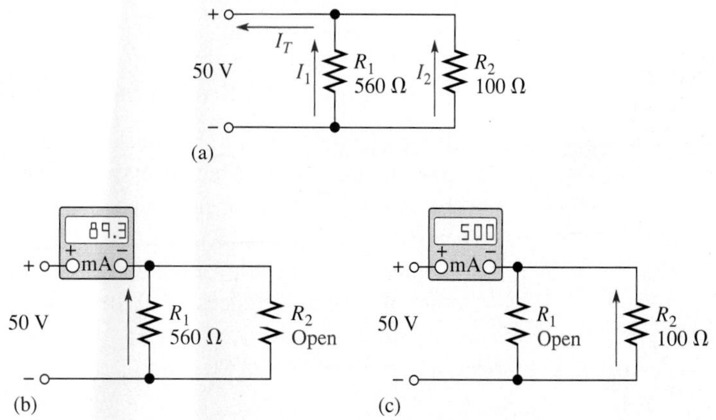

FIGURE 5–51
Finding an open path by current measurement.

If R_2 is open, the total current is 89.3 mA, as indicated in Figure 5–51(b). If R_1 is open, the total current is 500 mA, as indicated in Figure 5–51(c). This procedure can be extended to any number of branches having unequal resistances.

If the parallel resistances are all equal, the current in each branch must be checked until a branch is found with no current. This branch has the open resistor.

EXAMPLE 5–21

In Figure 5–52, there is a total current of 31.09 mA, and the voltage across the parallel branches is 20 V. Is there an open resistor, and, if so, which one is it?

FIGURE 5–52

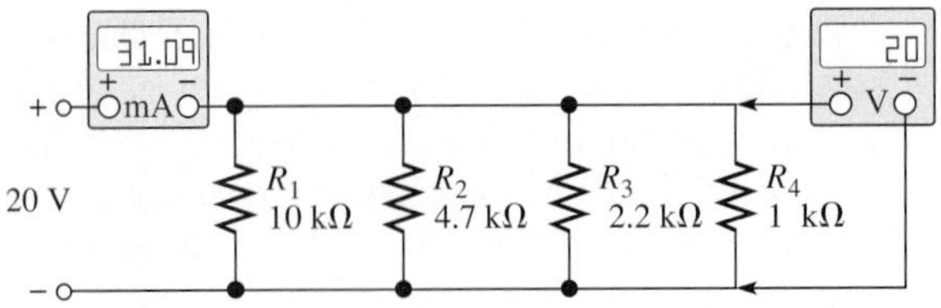

Solution Calculate the current in each branch:

$$I_1 = \frac{V}{R_1} = \frac{20\text{ V}}{10\text{ k}\Omega} = 2\text{ mA}$$

$$I_2 = \frac{V}{R_2} = \frac{20\text{ V}}{4.7\text{ k}\Omega} = 4.26\text{ mA}$$

$$I_3 = \frac{V}{R_3} = \frac{20\text{ V}}{2.2\text{ k}\Omega} = 9.09\text{ mA}$$

$$I_4 = \frac{V}{R_4} = \frac{20\text{ V}}{1\text{ k}\Omega} = 2.0\text{ mA}$$

The total current should be

$$I_T = I_1 + I_2 + I_3 + I_4$$
$$= 2\text{ mA} + 4.26\text{ mA} + 9.09\text{ mA} + 20\text{ mA} = 35.35\text{ mA}$$

The actual measured current is 31.09 mA, as stated, which is 4.26 mA less than normal, indicating that the branch carrying 4.26 mA is open. Thus, R_2 must be open.

Related Exercise What is the total current measured in Figure 5–52 if R_4 and not R_2 is open?

Finding an Open Branch by Resistance Measurement

If the parallel circuit to be checked can be disconnected from its voltage source and from any other circuit to which it may be connected, a measurement of the total resistance can be used to locate an open branch.

Recall that conductance, G, is the reciprocal of resistance ($1/R$) and its unit is the siemens (S). Based on Equation (5–2), the total conductance of a parallel circuit is the sum of the conductances of all the resistors.

$$G_T = G_1 + G_2 + G_3 + \cdots + G_n \qquad \textbf{(5–10)}$$

To locate an open branch, do the following steps:

1. Calculate what the total conductance should be using the individual resistor values:

$$G_{T(\text{calc})} = \frac{1}{R_1} + \frac{1}{R_2} + \frac{1}{R_3} + \cdots + \frac{1}{R_n}$$

2. Measure the total resistance with an ohmmeter and calculate the total measured conductance:

$$G_{T(\text{meas})} = \frac{1}{R_{T(\text{meas})}}$$

3. Subtract the measured total conductance (Step 2) from the calculated total conductance (Step 1). The result is the conductance of the open branch and the resistance is obtained by taking its reciprocal ($R = 1/G$).

$$R_{\text{open}} = \frac{1}{G_{T(\text{calc})} - G_{T(\text{meas})}} \quad \textbf{(5–11)}$$

EXAMPLE 5–22 Check the printed circuit board in Figure 5–53 for open branches.

FIGURE 5–53

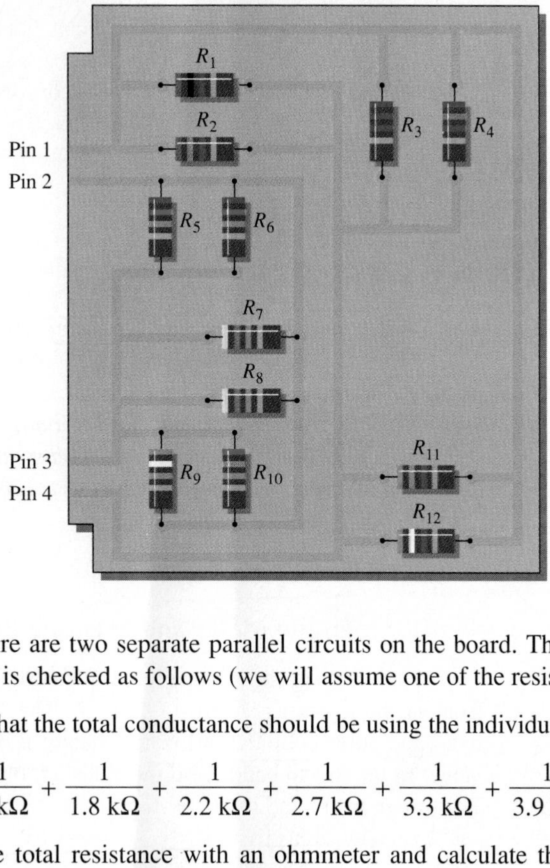

Solution There are two separate parallel circuits on the board. The circuit between pin 1 and pin 4 is checked as follows (we will assume one of the resistors is open):

1. Calculate what the total conductance should be using the individual resistor values:

$$G_{T(\text{calc})} = \frac{1}{1\text{ k}\Omega} + \frac{1}{1.8\text{ k}\Omega} + \frac{1}{2.2\text{ k}\Omega} + \frac{1}{2.7\text{ k}\Omega} + \frac{1}{3.3\text{ k}\Omega} + \frac{1}{3.9\text{ k}\Omega} = 2.94\text{ mS}$$

2. Measure the total resistance with an ohmmeter and calculate the total measured conductance. Assume that your ohmmeter measures 402 Ω.

$$G_{T(\text{meas})} = \frac{1}{402\ \Omega} = 2.49\text{ mS}$$

3. Subtract the measured total conductance (Step 2) from the calculated total conductance (Step 1). The result is the conductance of the open branch and the resistance is obtained by taking its reciprocal.

$$G_{open} = G_{T(calc)} - G_{T(meas)} = 2.94 \text{ mS} - 2.49 \text{ mS} = 0.45 \text{ mS}$$

$$R_{open} = \frac{1}{G_{open}} = \frac{1}{0.45 \text{ mS}} = 2.2 \text{ k}\Omega$$

Resistor R_3 is open and must be replaced.

Related Exercise Your ohmmeter indicates 9.6 kΩ between pin 2 and pin 3 on the PC board in Figure 5–53. Determine if this is correct and, if not, which resistor is open.

SECTION 5–9 REVIEW

1. If a parallel branch opens, what changes can be detected in the circuit's voltage and the currents, assuming that the parallel circuit is across a constant voltage source?
2. What happens to the total resistance if one branch opens?
3. If several light bulbs are connected in parallel and one of the bulbs opens (burns out), will the others continue to glow?
4. One ampere of current flows in each branch of a parallel circuit. If one branch opens, what is the current in each of the remaining branches?
5. A three-branch circuit normally has the following branch currents: 1 A, 2.5 A, and 1.2 A. If the total current measures 3.5 A, which branch is open?

5–10 ▪ APPLICATION ASSIGNMENT

A five-range milliammeter has just come off the assembly line, but has failed the line check. For any current in excess of 1 mA, the needle "pegs" (goes off scale to the right) on all the range settings except 1 mA and 50 mA. In these two ranges, the meter seems to work correctly. As you know, a multiple-range meter circuit is based on shunt resistance placed in parallel with the meter movement to achieve the specified full-scale deflection of the needle. Your task is to troubleshoot the meter and determine what repairs are necessary.

Step 1: Open the Meter and Examine the Circuit

The meter face is shown in Figure 5–54(a). The circuitry is exposed by removing the front of the meter as shown in part (b). The precision resistors are 1% tolerance with a five-band color code. Recall from Chapter 2 that the first three bands indicate resistance value. In the fourth band, gold indicates a multiplier of 0.1 and silver a multiplier of 0.01. The brown fifth band is for 1% tolerance.

Step 2: Compare the Meter Circuit to the Schematic

The meter schematic is shown in Figure 5–55. Carefully check the circuit connections to make sure they agree with the schematic. If there is a wiring problem, specify how to repair it. Resistors R_1, R_2, etc. can be identified by the way they are wired in the circuit. The 100 mA switch position is labeled "1". Resistance values are not shown in the schematic.

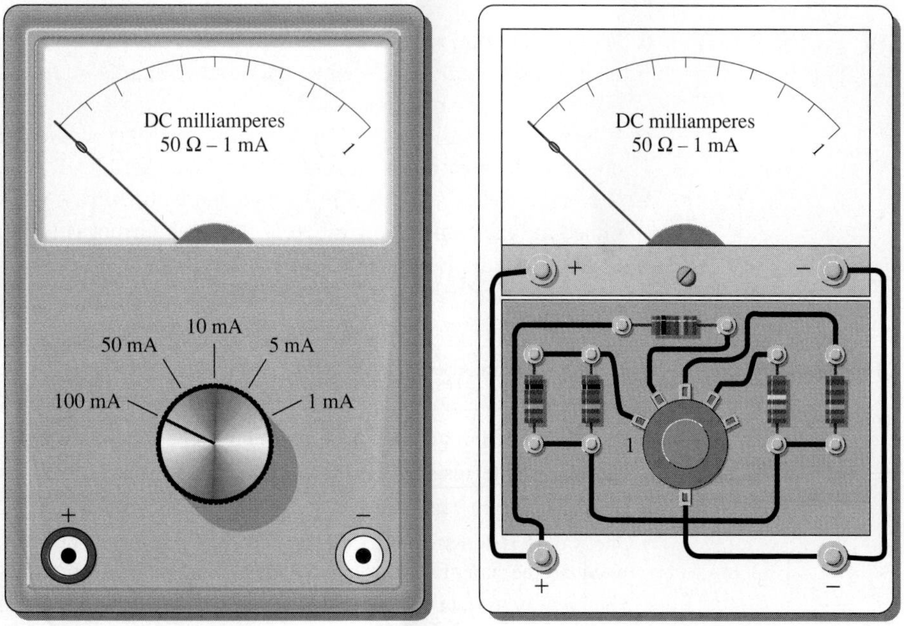

(a) (b)

FIGURE 5–54

FIGURE 5–55

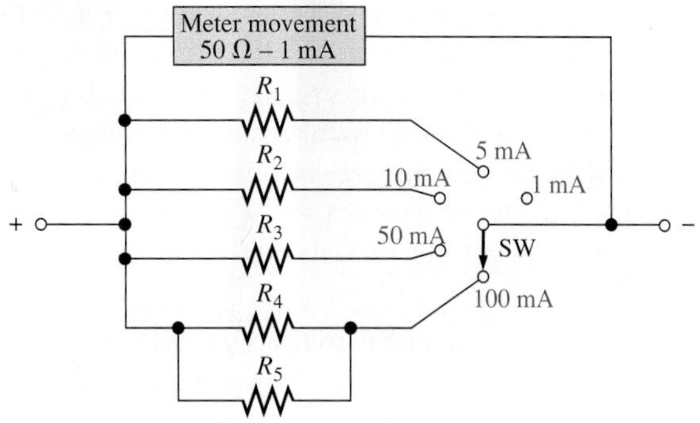

Step 3: Check for Additional Problems

There was a wiring problem in the meter circuit which, let's assume, you have located and corrected. Now, assume that when you connect the meter to the test instrument to check for proper operation you find that the meter works properly on the 1 mA, 5 mA, 10 mA, and 50 mA ranges but the needle still "pegs" on the 100 mA range setting for a current of 50 mA or greater.

Step 4: Analyze the Circuit to Determine the Additional Problem

Since the needle pegs on the 100 mA range, all or most of the current must be going through the meter movement. Recall that a shunt resistor carries most of the current being measured so that the current through the meter movement does not exceed its 1 mA maximum. For this problem to occur, one of the shunt resistors for this range is either not connected, failed open, or has a resistance value that is much too high. Check the circuit in Figure 5–54 again for another wiring error and check the color coded values of R_4 and R_5. For some reason, the schematic does not show the resistor values so you will have to calculate what they should be using your knowledge of Ohm's law and current dividers. Refer to the table of standard resistor values in Appendix A.

SECTION 5–10 REVIEW

1. What is the total resistance between the positive and negative terminals of the milliammeter in Figure 5–54 when the range switch is in the 1 mA position? The 5 mA position? The 10 mA position?
2. When the switch is in the 5 mA position and the needle points to the middle mark on the scale, how much current is being measured?
3. When the meter is on the 5 mA range and is measuring the amount of current determined in Question 2, how much of the total current is through the meter movement and how much through the shunt resistor?

▪ SUMMARY

- Resistors in parallel are connected across the same points.
- A parallel combination has more than one path for current.
- The number of current paths equals the number of resistors in parallel.
- The total parallel resistance is less than the lowest-value resistor.
- The voltages across all branches of a parallel circuit are the same.
- One way to state Kirchhoff's current law: The sum of the currents into a junction (total current in) equals the sum of the currents out of the junction (total current out).
- Another way to state Kirchhoff's current law: The algebraic sum of all the currents entering and leaving a junction is zero.
- A parallel circuit is a current divider, so called because the total current entering the parallel junction divides up into each of the branches.
- If all of the branches of a parallel circuit have equal resistance, the currents through all of the branches are equal.
- The total power in a parallel-resistive circuit is the sum of all of the individual powers of the resistors making up the parallel circuit.
- The total power for a parallel circuit can be calculated with the power formulas using values of total current, total resistance, or total voltage.
- If one of the branches of a parallel circuit opens, the total resistance increases, and therefore the total current decreases.
- If a branch of a parallel circuit opens, there is still current through the remaining branches.

▪ GLOSSARY

Branch One current path in a parallel circuit.

Junction A point at which two or more branches are connected.

Kirchhoff's current law A law stating that the total current into a junction equals the total current out of the junction.

Parallel The relationship in electric circuits in which two or more current paths are connected between the same two points.

▪ FORMULAS

(5–1) $$I_{IN(1)} + I_{IN(2)} + I_{IN(3)} + \cdots + I_{IN(n)} = I_{OUT(1)} + I_{OUT(2)} + I_{OUT(3)} + \cdots + I_{OUT(m)}$$ Kirchhoff's current law

(5–2) $$\frac{1}{R_T} = \frac{1}{R_1} + \frac{1}{R_2} + \frac{1}{R_3} + \cdots + \frac{1}{R_n}$$ Reciprocal for total parallel resistance

(5–3) $$R_T = \frac{1}{\left(\frac{1}{R_1}\right) + \left(\frac{1}{R_2}\right) + \left(\frac{1}{R_3}\right) + \cdots + \left(\frac{1}{R_n}\right)}$$ Total parallel resistance

(5–4) $$R_T = \frac{R_1 R_2}{R_1 + R_2}$$ Special case for two resistors in parallel

(5–5) $$R_T = \frac{R}{n}$$ Special case for n equal-value resistors in parallel

(5–6) $$I_1 = \left(\frac{R_2}{R_1 + R_2}\right) I_T$$ Two-branch current-divider formula

(5–7)	$I_2 = \left(\frac{R_1}{R_1 + R_2}\right)I_T$	Two-branch current-divider formula
(5–8)	$I_x = \left(\frac{R_T}{R_x}\right)I_T$	General current-divider formula
(5–9)	$P_T = P_1 + P_2 + P_3 + \cdots + P_n$	Total power
(5–10)	$G_T = G_1 + G_2 + G_3 + \cdots + G_n$	Total conductance
(5–11)	$R_{open} = \frac{1}{G_{T(calc)} - G_{T(meas)}}$	Open branch resistance

▪ SELF-TEST

1. In a parallel circuit, each resistor has
(a) the same current **(b)** the same voltage **(c)** the same power **(d)** all of the above
2. When a 1.2 kΩ resistor and a 100 Ω resistor are connected in parallel, the total resistance is
(a) greater than 1.2 kΩ **(b)** greater than 100 Ω but less than 1.2 kΩ
(c) less than 100 Ω but greater than 90 Ω **(d)** less than 90 Ω
3. A 330 Ω resistor, a 270 Ω resistor, and a 68 Ω resistor are all in parallel. The total resistance is approximately
(a) 668 Ω **(b)** 47 Ω **(c)** 68 Ω **(d)** 22 Ω
4. Eight resistors are in parallel. The two lowest-value resistors are both 1 kΩ. The total resistance
(a) cannot be determined **(b)** is greater than 1 kΩ
(c) is less than 1 kΩ **(d)** is less than 500 Ω
5. When an additional resistor is connected across an existing parallel circuit, the total resistance
(a) decreases **(b)** increases
(c) remains the same **(d)** increases by the value of the added resistor
6. If one of the resistors in a parallel circuit is removed, the total resistance
(a) decreases by the value of the removed resistor **(b)** remains the same
(c) increases **(d)** doubles
7. The currents into a junction flow along two paths. One current is 5 A and the other is 3 A. The total current out of the junction is
(a) 2 A **(b)** unknown **(c)** 8 A **(d)** the larger of the two
8. The following resistors are in parallel across a voltage source: 390 Ω, 560 Ω, and 820 Ω. The resistor with the least current is
(a) 390 Ω **(b)** 560 Ω
(c) 820 Ω **(d)** impossible to determine without knowing the voltage
9. A sudden decrease in the total current into a parallel circuit may indicate
(a) a short **(b)** an open resistor **(c)** a drop in source voltage **(d)** either (b) or (c)
10. In a four-branch parallel circuit, there are 10 mA of current in each branch. If one of the branches opens, the current in each of the other three branches is
(a) 13.33 mA **(b)** 10 mA **(c)** 0 A **(d)** 30 mA
11. In a certain three-branch parallel circuit, R_1 has 10 mA through it, R_2 has 15 mA through it, and R_3 has 20 mA through it. After measuring a total current of 35 mA, you can say that
(a) R_1 is open **(b)** R_2 is open
(c) R_3 is open **(d)** the circuit is operating properly
12. If there are a total of 100 mA into a parallel circuit consisting of three branches and two of the branch currents are 40 mA and 20 mA, the third branch current is
(a) 60 mA **(b)** 20 mA **(c)** 160 mA **(d)** 40 mA
13. A complete short develops across one of five parallel resistors on a PC board. The most likely result is
(a) the shortest resistor will burn out
(b) one or more of the other resistors will burn out
(c) the fuse in the power supply will blow
(d) the resistance values will be altered
14. The power dissipation in each of four parallel branches is 1 W. The total power dissipation is
(a) 1 W **(b)** 4 W **(c)** 0.25 W **(d)** 16 W

■ BASIC PROBLEMS

SECTION 5–1 Resistors in Parallel

1. Connect the resistors in Figure 5–56 in parallel across the battery.
2. Determine whether or not all the resistors in Figure 5–57 are connected in parallel on the printed circuit board. Sketch the schematic including resistor values.

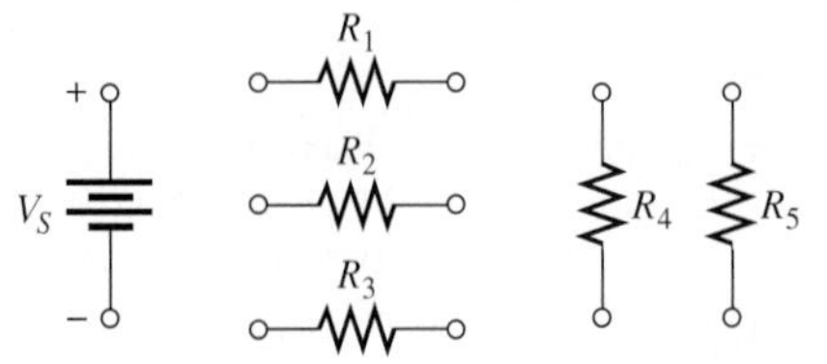

FIGURE 5–56

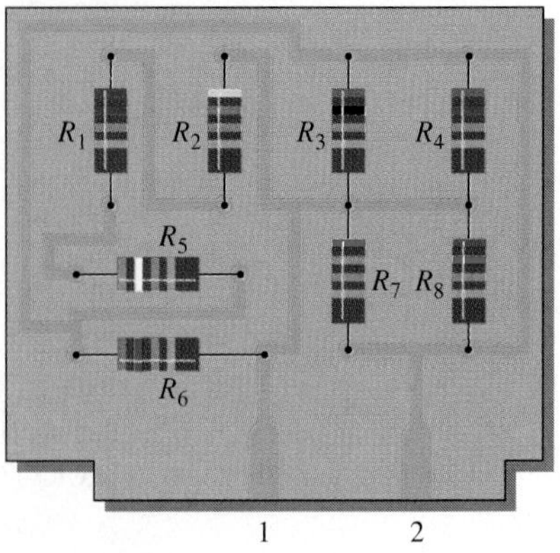

FIGURE 5–57

SECTION 5–2 Voltage Drop in Parallel Circuits

3. Determine the voltage across and the current through each parallel resistor if the total voltage is 12 V and the total resistance is 600 Ω. There are four resistors, all of equal value.
4. The source voltage in Figure 5–58 is 100 V. How much voltage does each of the meters V1–V3 read?

FIGURE 5–58

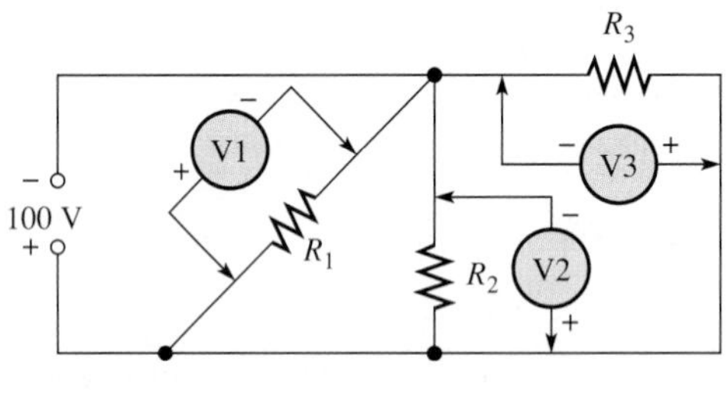

SECTION 5–3 Kirchhoff's Current Law

5. The following currents are measured in the same direction in a three-branch parallel circuit: 250 mA, 300 mA, and 800 mA. What is the value of the current into the junction of these three branches?
6. Five hundred milliamperes (500 mA) flow into five parallel resistors. The currents through four of the resistors are 50 mA, 150 mA, 25 mA, and 100 mA. How much current flows through the fifth resistor?
7. How much current flows through R_2 and R_3 in Figure 5–59 if R_2 and R_3 have the same resistance? Show how to connect ammeters to measure these currents.

FIGURE 5–59

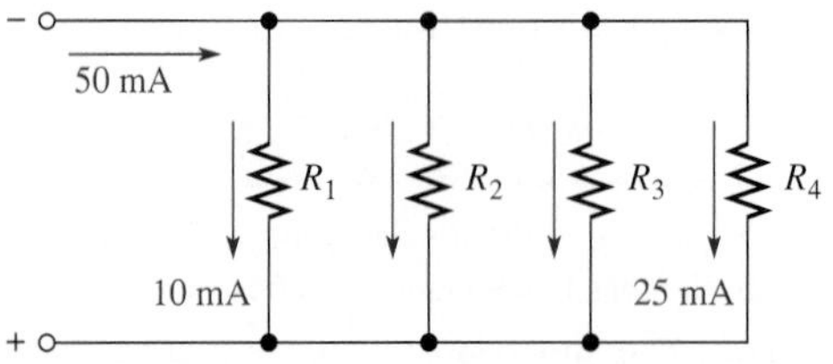

SECTION 5–4 Total Parallel Resistance

8. The following resistors are connected in parallel: 1 MΩ, 2.2 MΩ, 4.7 MΩ, 12 MΩ, and 22 MΩ. Determine the total resistance.
9. Find the total resistance for each group of parallel resistors in Figure 5–60.
10. Calculate R_T for each circuit in Figure 5–61.
11. What is the total resistance of eleven 22 kΩ resistors in parallel?
12. Five 15 Ω, ten 100 Ω, and two 10 Ω resistors are all connected in parallel. What is the total resistance?

FIGURE 5–60

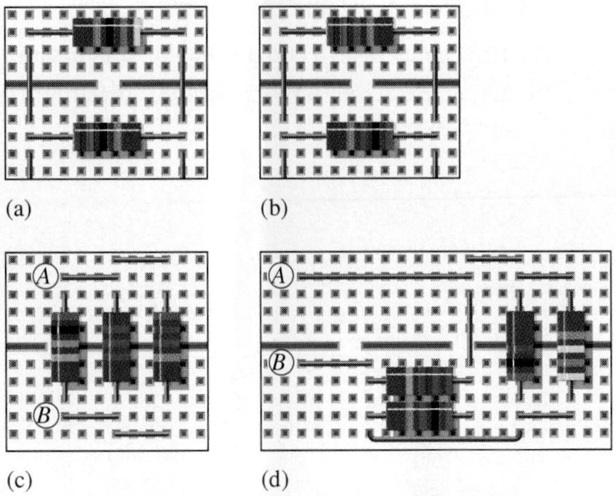

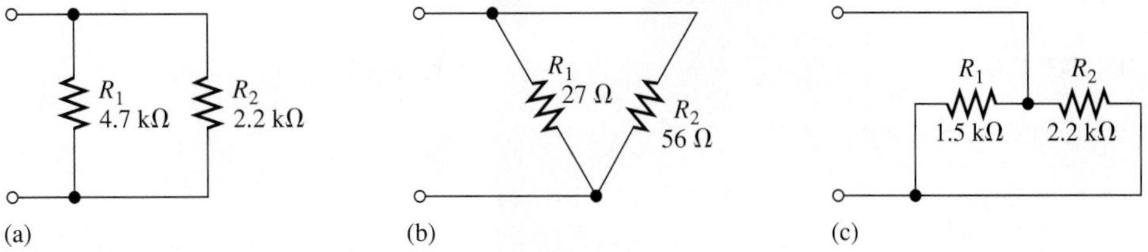

FIGURE 5–61

SECTION 5–5 Ohm's Law in Parallel Circuits

13. What is the total current I_T in each circuit of Figure 5–62?
14. Three 33 Ω resistors are connected in parallel with a 110 V source. How much current is flowing from the source?
15. Which circuit of Figure 5–63 has more total current?
16. Four equal-value resistors are connected in parallel. Five volts are applied across the parallel circuit, and 2.5 mA are measured from the source. What is the value of each resistor?

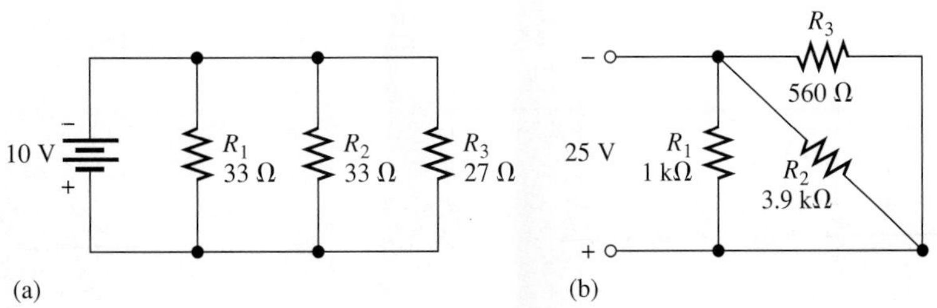

FIGURE 5–62

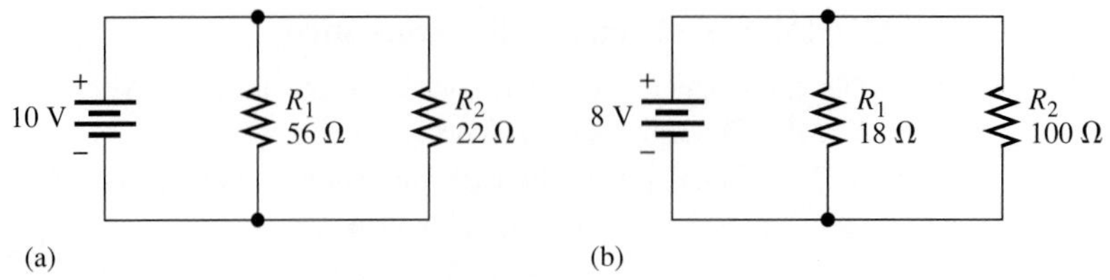

FIGURE 5–63

SECTION 5–6 Current Dividers

17. How much branch current should each meter in Figure 5–64 indicate?

18. Using the current-divider formula, determine the current in each branch of the circuits of Figure 5–65.

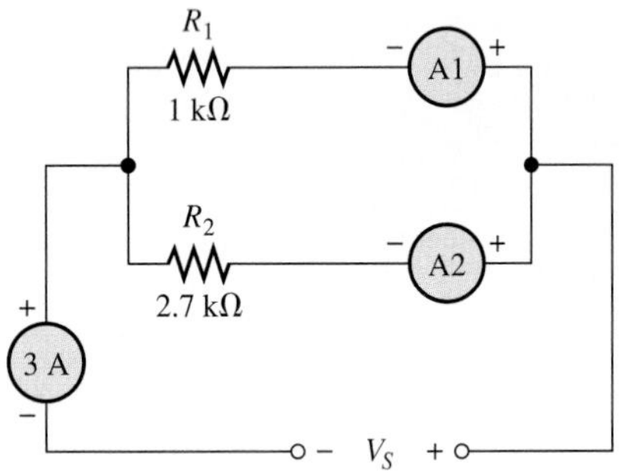

FIGURE 5–64

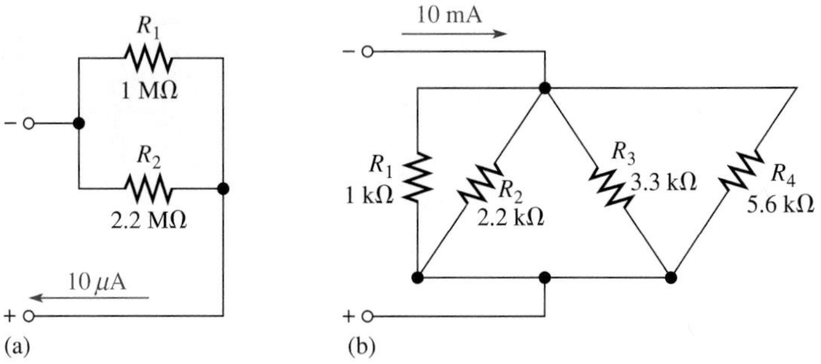

FIGURE 5–65

SECTION 5–7 Power in Parallel Circuits

19. Five parallel resistors each handle 40 mW. What is the total power?

20. Determine the total power in each circuit of Figure 5–65.

21. Six light bulbs are connected in parallel across 110 V. Each bulb is rated at 75 W. How much current is through each bulb, and what is the total current?

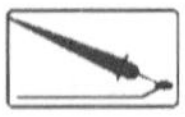

SECTION 5–9 Troubleshooting

22. If one of the bulbs burns out in Problem 21, how much current will flow through each of the remaining bulbs? What will the total current be?

23. In Figure 5–66, the current and voltage measurements are indicated. Has a resistor opened, and, if so, which one?

24. What is wrong with the circuit in Figure 5–67?

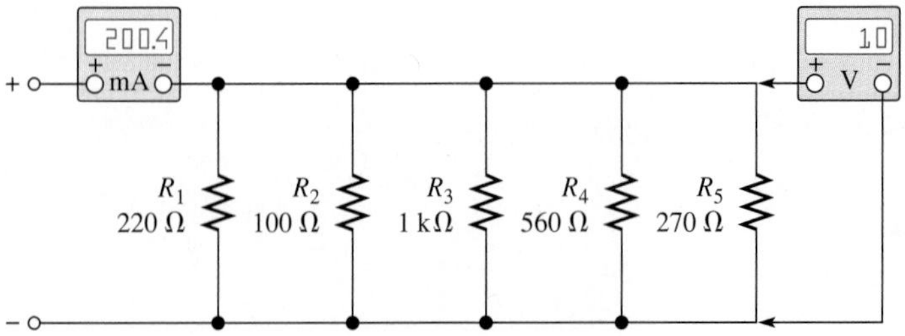

FIGURE 5–66

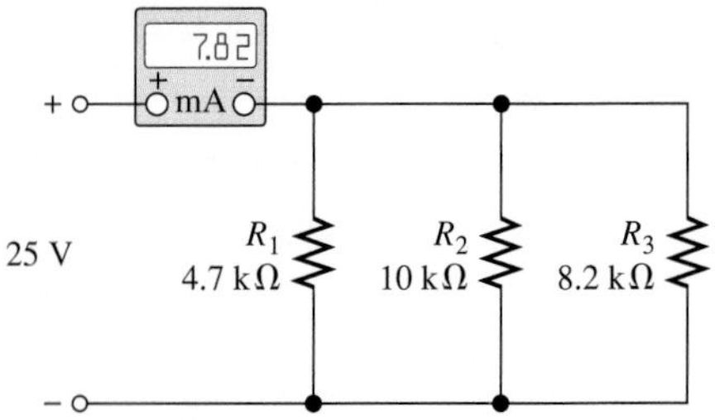

FIGURE 5–67

■ ADVANCED PROBLEMS

25. In the circuit of Figure 5–68, determine resistances R_2, R_3, and R_4.

26. The total resistance of a parallel circuit is 25 Ω. If the total current is 100 mA, how much current flows through a 220 Ω resistor that makes up part of the parallel circuit?

27. What is the current through each resistor in Figure 5–69? R is the lowest-value resistor, and all others are multiples of that value as indicated.

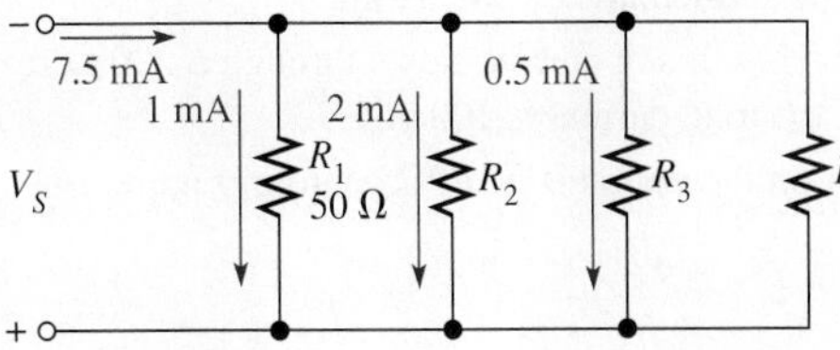

FIGURE 5–68

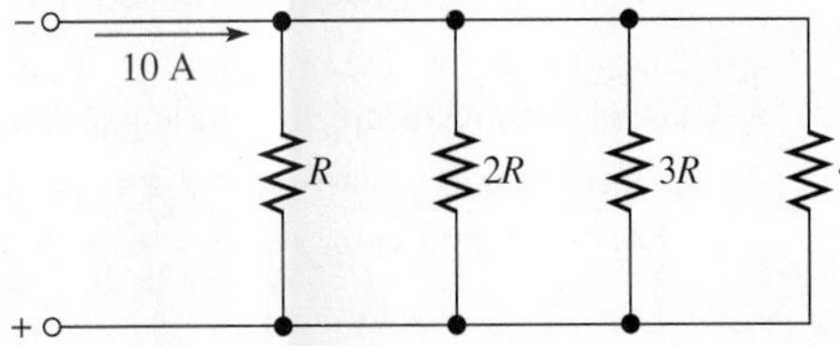

FIGURE 5–69

28. A certain parallel network consists of only ½ W resistors. The total resistance is 1 kΩ, and the total current is 50 mA. If each resistor is operating at one-half its maximum power level, determine the following:

(a) the number of resistors **(b)** the value of each resistor

(c) the current in each branch **(d)** the applied voltage

29. Find the values of the unspecified labeled quantities in each circuit of Figure 5–70.

FIGURE 5–70

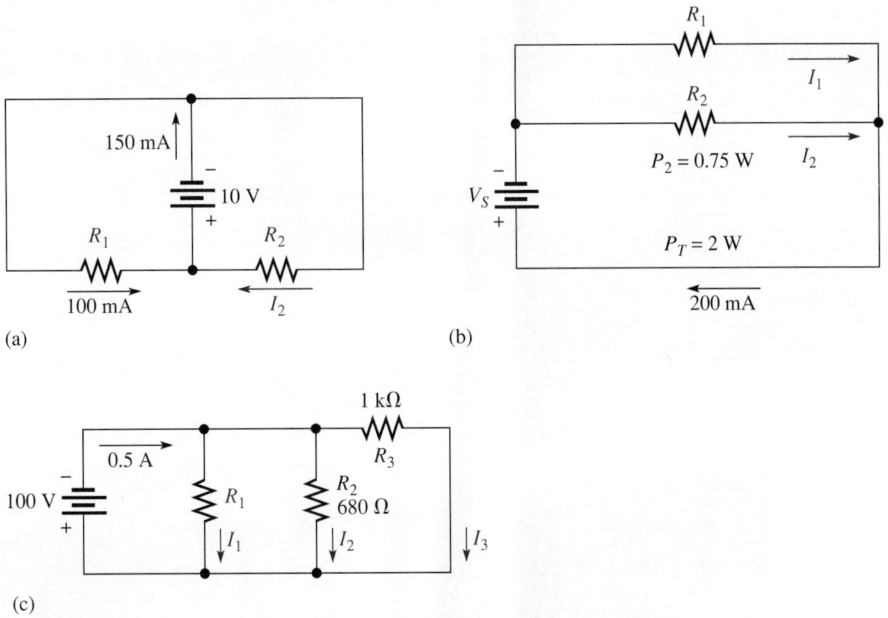

30. What is the total resistance between point A and ground in Figure 5–71 for the following conditions?

(a) SW_1 and SW_2 open **(b)** SW_1 closed, SW_2 open

(c) SW_1 open, SW_2 closed **(d)** SW_1 and SW_2 closed

31. To what minimum value can the 100 Ω rheostat in Figure 5–72 be adjusted before the 0.5 A fuse blows?

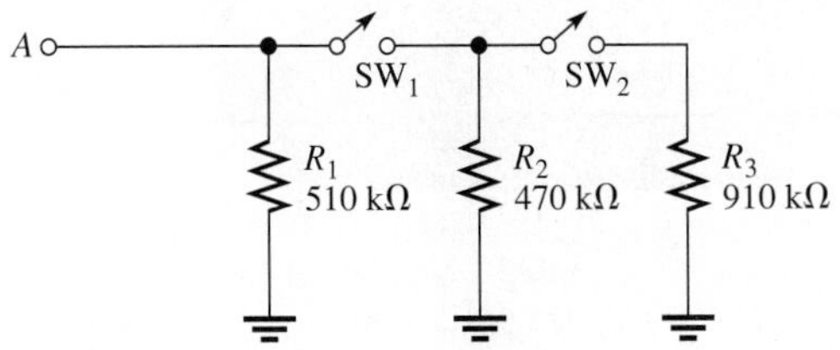

FIGURE 5–71

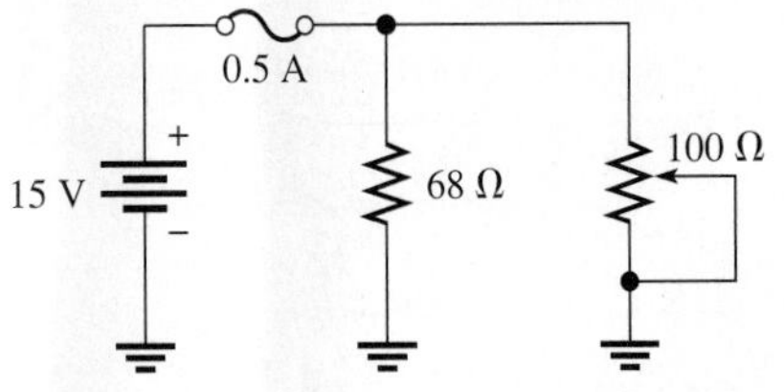

FIGURE 5–72

32. Determine the total current from the source and the current through each resistor for each switch position in Figure 5–73.

33. The electrical circuit in a room has a ceiling lamp that draws 1.25 A and four wall outlets. Two table lamps that each draw 0.833 A are plugged into two outlets, and a TV set that draws 1 A is connected to the third outlet. When all of these items are in use, how much current is in the main line serving the room? If the main line is protected by a 5 A circuit breaker, how much current can be drawn from the fourth outlet? Draw a schematic of this wiring.

34. The total resistance of a parallel circuit is 25 Ω. How much current flows through a 220 Ω resistor that makes up part of the parallel circuit if the total current is 100 mA?

35. Identify which groups of resistors are in parallel on the double-sided PC board in Figure 5–74. Determine the total resistance of each group.

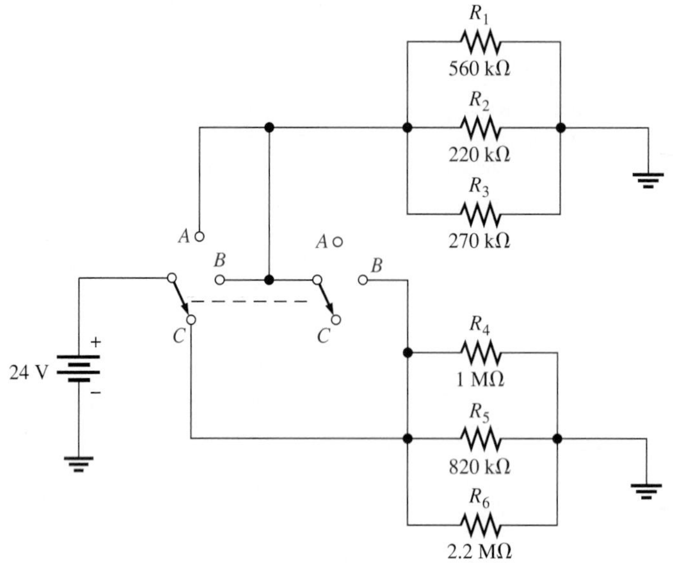

FIGURE 5–73

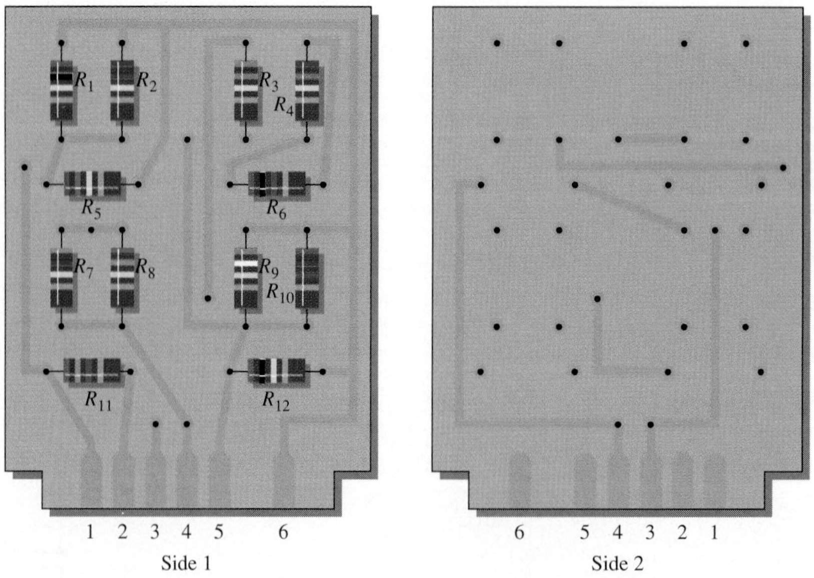

FIGURE 5–74

36. If the total resistance in Figure 5–75 is 200 Ω, what is the value of R_2?

37. Determine the unknown resistances in Figure 5–76.

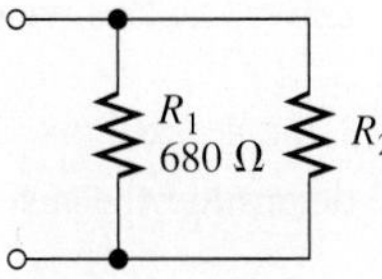

FIGURE 5–75

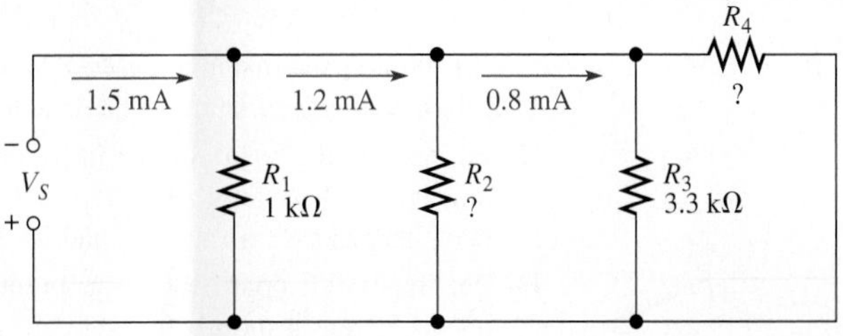

FIGURE 5–76

38. A total of 250 mA flows into a parallel circuit with a total resistance of 1.5 kΩ. The current must be increased by 25%. Determine how much resistance to add in parallel to accomplish this increase in current.

39. Sketch the schematic for the setup in Figure 5–77 and determine what is wrong with the circuit, if 25 V are applied across the red and black leads.

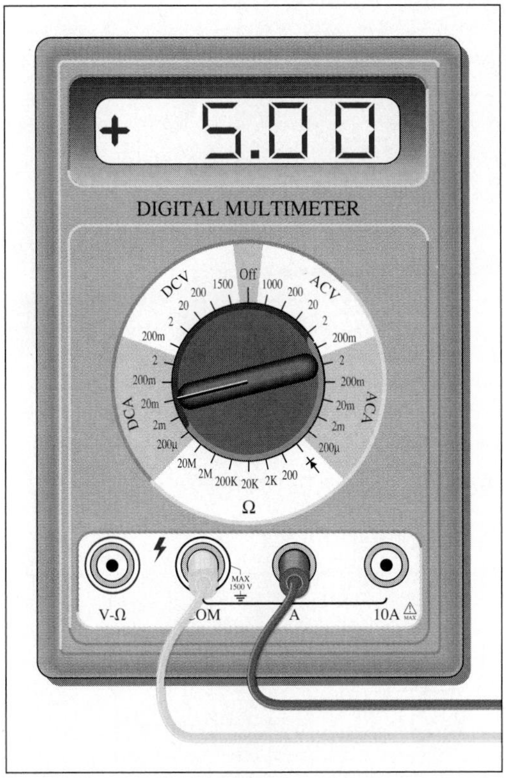

(a) Meter with yellow lead going to circuit board and red lead going to the positive terminal of the 25 V power supply.

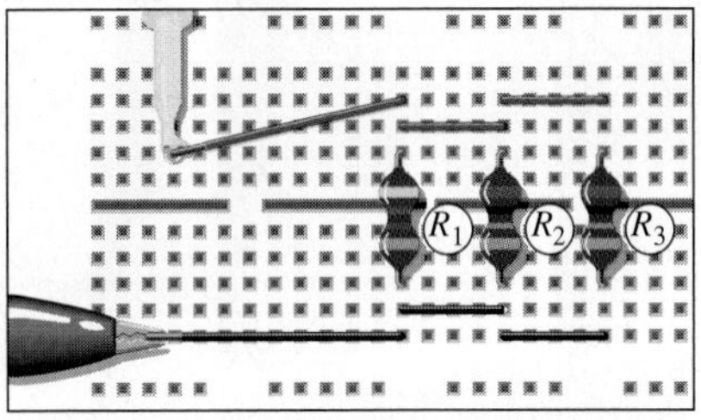

(b) Circuit board with leads connected. Yellow lead is from meter and gray lead is from 25 V power supply ground. The red meter lead goes to +25 V.

FIGURE 5–77

40. Develop a test procedure to check the circuit in Figure 5–78 to make sure that there are no open components. You must do this test without removing any component from the board. List the procedure in a detailed step-by-step format.

41. A certain parallel circuit consists of five ½ W resistors with the following values: 1.8 kΩ, 2.2 kΩ, 3.3 kΩ, 3.9 kΩ, and 4.7 kΩ. As you slowly increase the voltage across the parallel circuit, the total current slowly increases. Suddenly, the total current drops to a lower value.
(a) Excluding a power supply failure, what happened?
(b) What is the maximum voltage you should have applied?
(c) Specifically, what must be done to repair the fault?

42. For the circuit board shown in Figure 5–79, determine the resistance between the following pins if there is a short between pins 2 and 4.
(a) pins 1 and 2 **(b)** pins 2 and 3 **(c)** pins 3 and 4 **(d)** pins 1 and 4

43. For the circuit board shown in Figure 5–79, determine the resistance between the following pins if there is a short between pins 3 and 4.
(a) pins 1 and 2 **(b)** pins 2 and 3 **(c)** pins 2 and 4 **(d)** pins 1 and 4

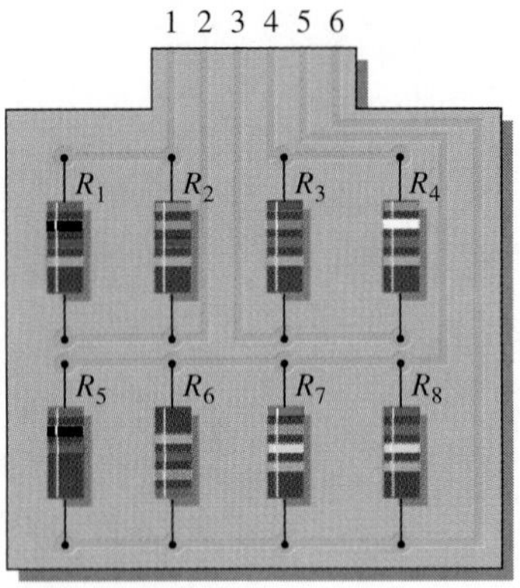

FIGURE 5–78

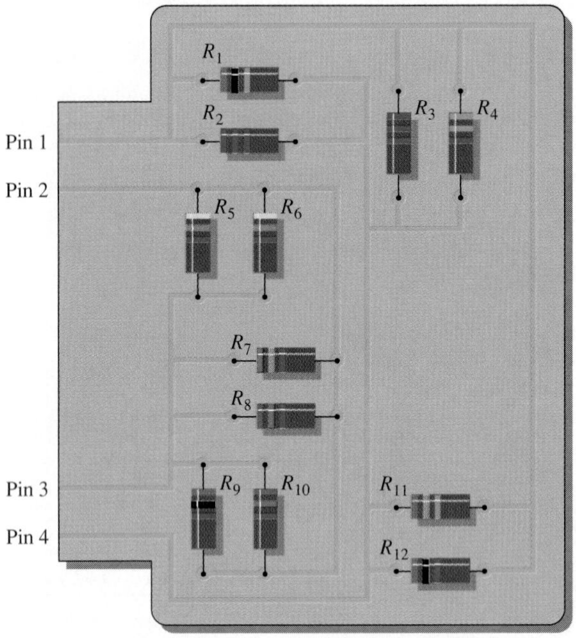

FIGURE 5–79

▪ ANSWERS TO SECTION REVIEWS

Section 5–1

1. Parallel resistors are connected between the same two points.
2. A parallel connection has more than one current path between two given points.
3. See Figure 5–80.
4. See Figure 5–81.

FIGURE 5–80

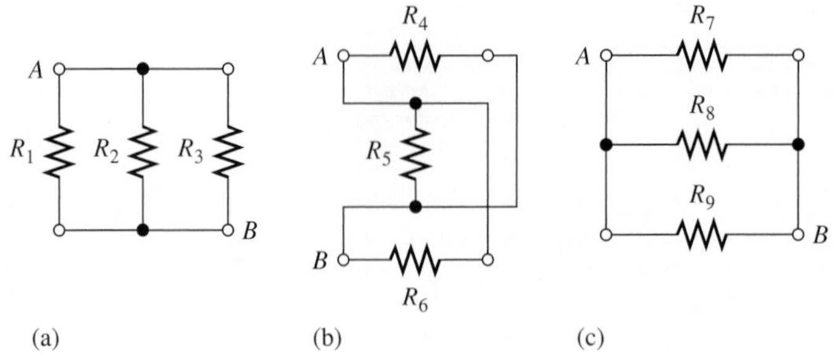

FIGURE 5–81

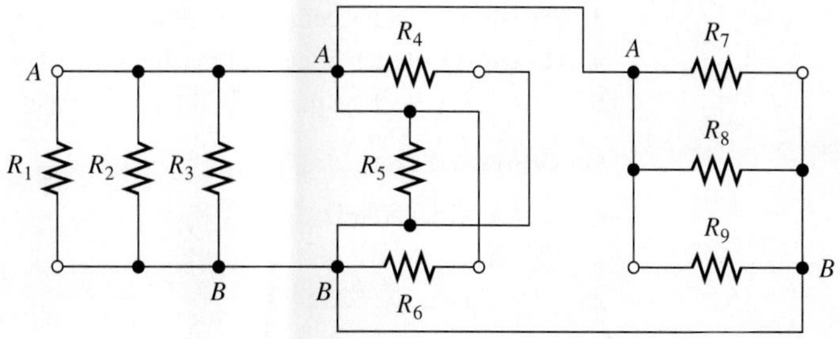

Section 5–2

1. 5 V
2. $V_{R1} = 118$ V; $V_S = 118$ V
3. $V_1 = 50$ V; $V_2 = 50$ V
4. Voltage is the same across all branches.

Section 5–3

1. Kirchhoff's current law states that the algebraic sum of all currents at a junction is zero; the sum of the currents entering a junction equals the sum of the currents leaving that junction.
2. $I_1 + I_2 + I_3 = 2.5$ A
3. 100 mA + 300 mA = 400 mA
4. $I_1 = 5\ \mu\text{A} - 2\ \mu\text{A} = 3\ \mu\text{A}$
5. 8 A − 1 A = 7 A entering, 8 A − 3 A = 5 A leaving

Section 5–4

1. As more resistors are connected in parallel, the total resistance decreases.
2. R_T is always less than the smallest resistance value.
3. $R_T = 2.2\ \text{k}\Omega/12 = 183\ \Omega$

Section 5–5

1. $I_T = 10\ \text{V}/(68\ \Omega/3) = 441$ mA
2. $V_S = (2\ \text{A})(680\ \Omega \parallel 330\ \Omega) = 444$ V
3. $I_1 = (330\ \Omega \parallel 680\ \Omega/330\ \Omega)2\ \text{A} = 653$ mA; $I_2 = (330\ \Omega \parallel 680\ \Omega/680\ \Omega)2\ \text{A} = 1.35$ A
4. $R = 4(12\ \text{V}/6\ \text{mA}) = 8\ \text{k}\Omega$
5. $V = (1\ \text{k}\Omega \parallel 2.2\ \text{k}\Omega)100\ \text{mA} = 68.8$ V

Section 5–6

1. The 22 Ω has most current; the 200 Ω has least current.
2. $I_3 = (R_T/R_3)4\ \text{mA} = (113.6\ \Omega/470\ \Omega)4\ \text{mA} = 967\ \mu\text{A}$
3. $I_2 = (R_T/680\ \Omega)10\ \text{mA} = 3.27$ mA; $I_1 = (R_T/330\ \Omega)10\ \text{mA} = 6.73$ mA

Section 5–7

1. Add the powers in each resistor to get P_T.
2. $P_T = 1\ \text{W} + 2\ \text{W} + 5\ \text{W} + 8\ \text{W} = 16$ W
3. $P_T = (1\ \text{A})^2 R_T = 615$ W

Section 5–8

1. $R_{max} = 50\ \Omega$; $I_{max} = 1$ mA
2. Shunt resisters have less resistance than the movement because the shunt resistors must allow currents much greater than the current through the meter movement.

Section 5–9

1. When a parallel branch opens, there is no change in voltage and the total current decreases.
2. When the total resistance is greater than it should be, a branch is open.

3. Yes, all other bulbs continue to glow.

4. The current in each nonopen branch remains 1 A.

5. 1 A + 2.5 A = 3.5 A; therefore, the 1.2 A branch is open.

Section 5–10

1. 50 Ω; 9.94 Ω; 5.08 Ω

2. 2.5 mA

3. $I_M = 0.5$ mA; $I_{R1} = 2$ mA

■ ANSWERS TO RELATED EXERCISES FOR EXAMPLES

5–1 No rewiring is necessary.

5–2 Connect pin 1 to pin 2 and pin 3 to pin 4.

5–3 25 V

5–4 20 mA

5–5 $I_T = 112$ mA; $I_2 = 50$ mA

5–6 2.5 A; 5 A

5–7 9.34 Ω

5–8 Replace Step 9 with "Press [+] key." Display shows 0.0767311.
Replace Step 10 with "Enter 33." Display shows 33.
Step 11: Press [1/x]. Display shows 0.030303.
Step 12: Press [=] key. Display shows 0.1070342.
Step 13: Press [1/x] key. Display shows 9.3428107.

5–9 111 Ω

5–10 33.3 Ω

5–11 2 Ω

5–12 $I_1 = 1$ A; $I_2 = 1.79$ A

5–13 $I_1 = 20$ mA; $I_2 = 9.1$ mA; $I_3 = 35.7$ mA; $I_4 = 22.0$ mA

5–14 1.28 V

5–15 Use an ohmmeter to measure R_T and then calculate R_1 using the parallel-resistance formula.

5–16 $I_1 = 59.4$ mA; $I_2 = 40.6$ mA

5–17 $I_1 = 1.63$ A; $I_2 = 3.36$ A

5–18 177.6 W

5–19 162 W

5–20 An insignificant amount

5–21 15.4 mA

5–22 Not correct; R_{10}(68 kΩ) is open.

6

SERIES-PARALLEL CIRCUITS

Various combinations of both series and parallel resistors are often used in electronic circuits. In this chapter, examples of such series-parallel arrangements are examined and analyzed. An important circuit called the *Wheatstone bridge* is introduced, and circuits with more than one voltage source are analyzed in simple steps. Also, you will learn how complex circuits can be simplified using Thevenin's theorem. Troubleshooting series-parallel circuits for shorts and opens is also covered.

■ CHAPTER OBJECTIVES

- ☐ Identify series-parallel relationships
- ☐ Analyze series-parallel circuits
- ☐ Analyze loaded voltage dividers
- ☐ Determine the loading effect of a voltmeter on a circuit
- ☐ Analyze a Wheatstone bridge
- ☐ Analyze circuits with more than one source
- ☐ Apply Thevenin's theorem to simplify a circuit for analysis
- ☐ Troubleshoot series-parallel circuits

■ APPLICATION ASSIGNMENT

Your application assignment in this chapter is to evaluate a voltage-divider circuit board used in a portable power supply by applying your knowledge of loaded voltage dividers gained in this chapter as well as skills developed in previous chapters. The voltage divider in this application is designed to provide reference voltages to three different instruments that act as loads on the circuit. You will also be required to troubleshoot the circuit board for various common faults. After studying this chapter, you should be able to complete the application assignment in Section 6–9.

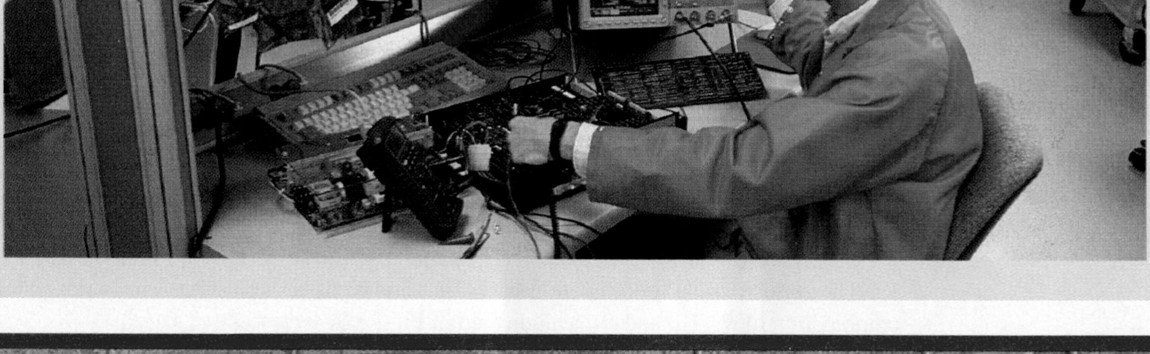

6–1 ■ IDENTIFYING SERIES-PARALLEL RELATIONSHIPS

A series-parallel circuit consists of combinations of both series and parallel current paths. It is important to be able to identify how the components in a circuit are arranged in terms of their series and parallel relationships.

After completing this section, you should be able to

- **Identify series-parallel relationships**
  - ☐ Recognize how each resistor in a given circuit is related to the other resistors
  - ☐ Determine series and parallel relationships on a PC board

Figure 6–1(a) shows a simple series-parallel combination of resistors. Notice that the resistance from point A to point B is R_1. The resistance from point B to point C is R_2 and R_3 in parallel ($R_2 \parallel R_3$). The resistance from point A to point C is R_1 in series with the parallel combination of R_2 and R_3, as indicated in Figure 6–1(b).

FIGURE 6–1
A simple series-parallel circuit.

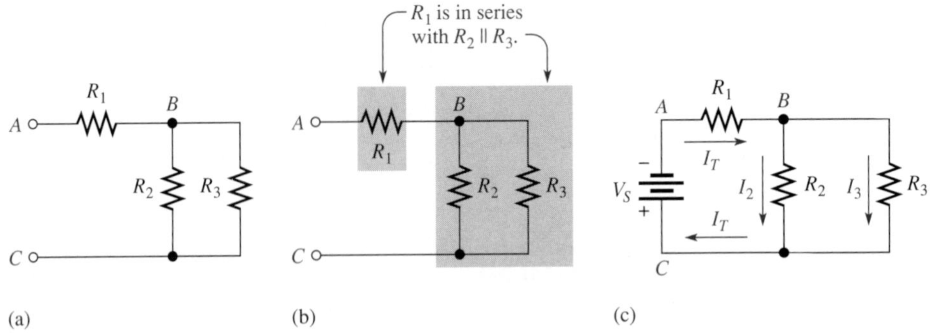

When the circuit of Figure 6–1(a) is connected to a voltage source as shown in part (c), the total current flows through R_1 and divides at point B into the two parallel paths. These two branch currents then recombine, and the total current flows into the positive source terminal as shown.

Now, to illustrate series-parallel relationships, we will increase the complexity of the circuit in Figure 6–1(a) step-by-step. In Figure 6–2(a), another resistor (R_4) is connected in series with R_1. The resistance between points A and B is now $R_1 + R_4$, and this combination is in series with the parallel combination of R_2 and R_3, as illustrated in Figure 6–2(b).

FIGURE 6–2
R_4 is added to the circuit in series with R_1.

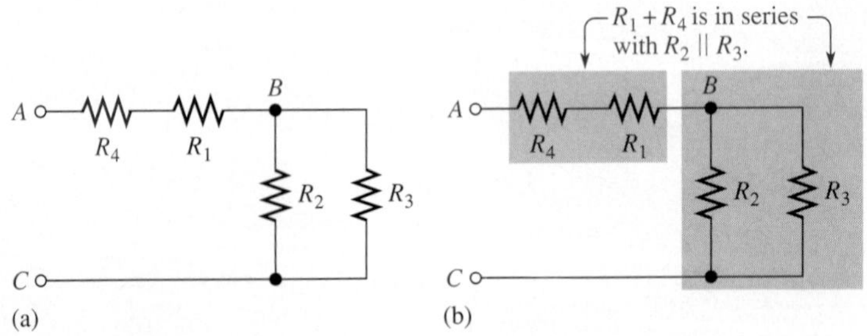

In Figure 6–3(a), R_5 is connected in series with R_2. The series combination of R_2 and R_5 is in parallel with R_3. This entire series-parallel combination is in series with the $R_1 + R_4$ combination, as illustrated in Figure 6–3(b).

In Figure 6–4(a), R_6 is connected in parallel with the series combination of R_1 and R_4. The series-parallel combination of R_1, R_4, and R_6 is in series with the series-parallel combination of R_2, R_3, and R_5, as indicated in Figure 6–4(b).

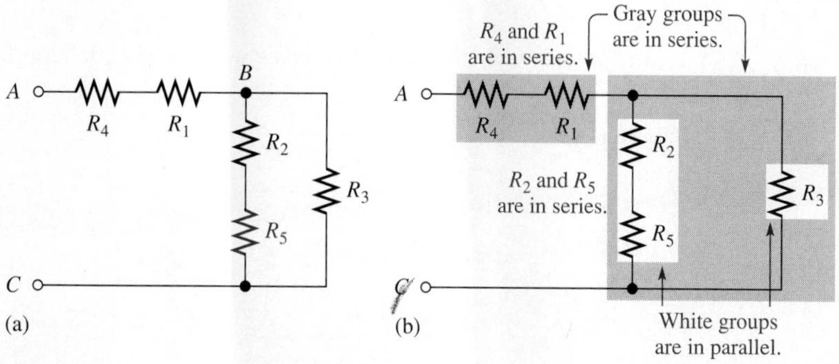

FIGURE 6–3
R_5 is added to the circuit in series with R_2.

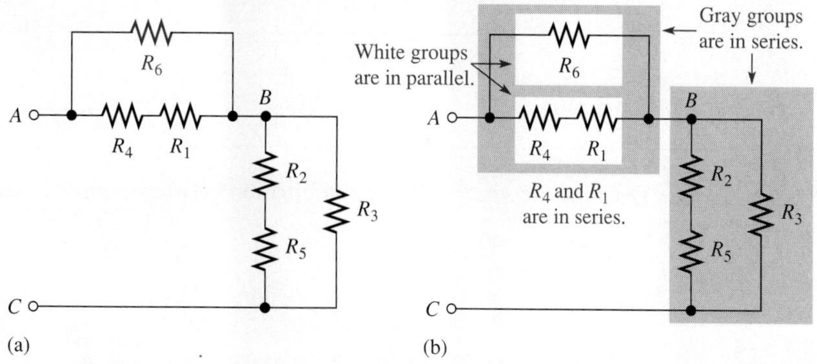

FIGURE 6–4
R_6 is added to the circuit in parallel with the series combination of R_1 and R_4.

EXAMPLE 6–1 Identify the series-parallel relationships in Figure 6–5.

FIGURE 6–5

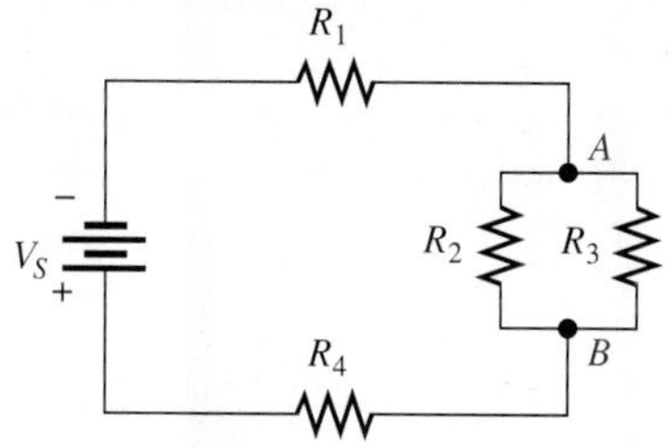

Solution Starting at the negative terminal of the source, follow the current paths. All the current produced by the source must go through R_1, which is in series with the rest of the circuit.

The total current takes two paths when it gets to point A. Part of it flows through R_2, and part of it through R_3. R_2 and R_3 are in parallel with each other, and this parallel combination is in series with R_1.

At point B, the currents through R_2 and R_3 come together again. Thus, the total current flows through R_4. R_4 is in series with R_1 and the parallel combination of R_2 and R_3. The currents are shown in Figure 6–6, where I_T is the total current.

In summary, R_1 and R_4 are in series with the parallel combination of R_2 and R_3.

FIGURE 6–6

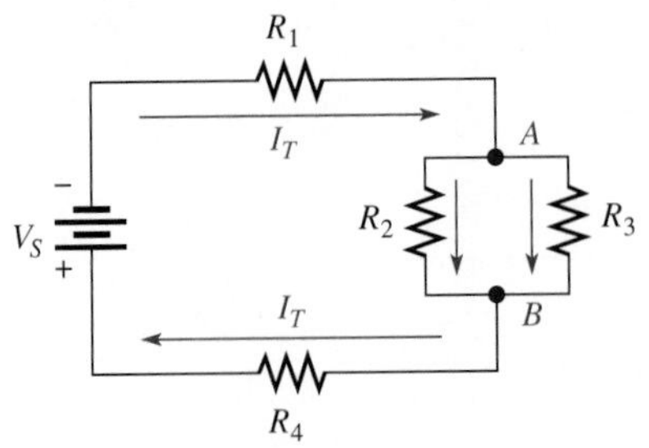

Related Exercise If another resistor, R_5, is connected from point A to the positive side of the source in Figure 6–6, what is its relationship to the other resistors?

EXAMPLE 6–2

Describe the series-parallel combination between points A and D in Figure 6–7.

FIGURE 6–7

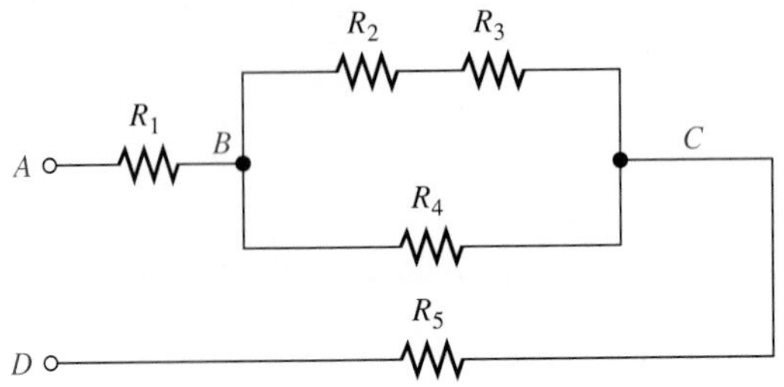

Solution Between points B and C, there are two parallel paths. The lower path consists of R_4, and the upper path consists of a series combination of R_2 and R_3. This parallel combination is in series with both R_1 and R_5.

In summary, R_1 and R_5 are in series with the parallel combination of R_4 and $R_2 + R_3$.

Related Exercise If a resistor is connected from point C to point D in Figure 6–7, describe its parallel relationship.

EXAMPLE 6–3

Describe the total resistance between each pair of points in Figure 6–8.

FIGURE 6–8

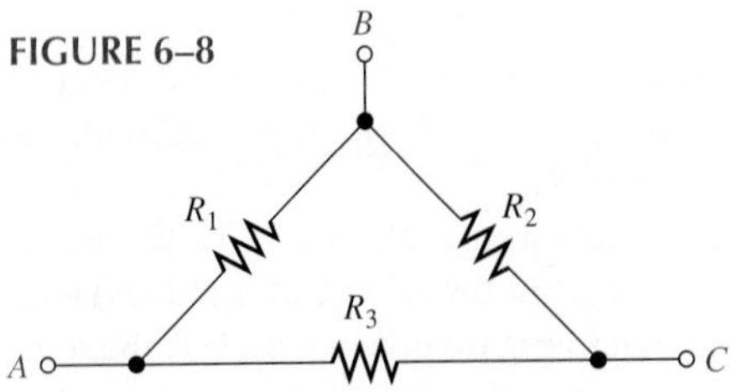

Solution

1. From point A to point B: R_1 is in parallel with the series combination of R_2 and R_3.
2. From point A to point C: R_3 is in parallel with the series combination of R_1 and R_2.
3. From point B to point C: R_2 is in parallel with the series combination of R_1 and R_3.

Related Exercise Describe the total resistance between each point and ground if a resistor, R_4, is connected from point C to ground in Figure 6–8.

Redrawing a Schematic to Determine the Series-Parallel Relationships

Sometimes it is difficult to see the series-parallel relationships on a schematic because of the way in which it is drawn. In such a situation, it helps to redraw the diagram so that the relationships become clear. Example 6–4 will illustrate.

EXAMPLE 6–4 Identify the series-parallel relationships in Figure 6–9.

FIGURE 6–9

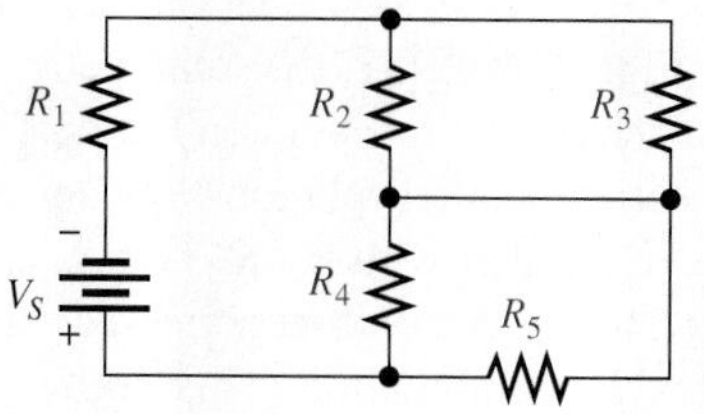

Solution Sometimes it is easier to see a particular circuit arrangement if it is drawn in a different way. The circuit schematic is redrawn in Figure 6–10, which better illustrates the series-parallel relationships. Now you can see that R_2 and R_3 are in parallel with each other and also that R_4 and R_5 are in parallel with each other. Both parallel combinations are in series with each other and with R_1.

FIGURE 6–10

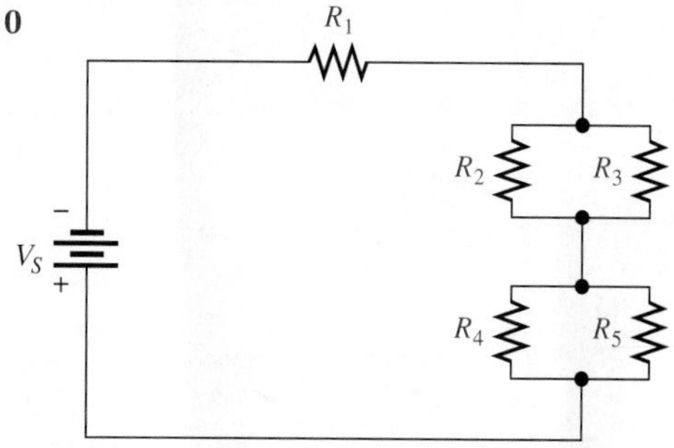

Related Exercise If a resistor is connected from the bottom end of R_3 to the top end of R_5 in Figure 6–10, what effect does it have on the circuit?

Determining Relationships on a Printed Circuit Board

Usually, the physical arrangement of components on a PC board bears no resemblance to the actual circuit relationships. By tracing out the circuit on the PC board and rearranging the components on paper into a recognizable form, you can determine the series-parallel relationships. Example 6–5 will illustrate.

EXAMPLE 6–5

Determine the relationships of the resistors on the PC board in Figure 6–11.

FIGURE 6–11

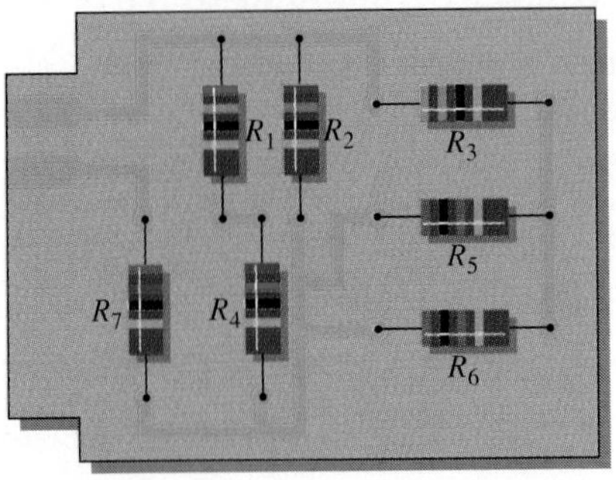

Solution In Figure 6–12(a), the schematic is drawn approximately in the same arrangement as that of the resistors on the board. In part (b), the resistors are reoriented so that the series-parallel relationships are obvious. Resistors R_1 and R_4 are in series; $R_1 + R_4$ is in parallel with R_2; R_5 and R_6 are in parallel and this combination is in series with R_3. The R_3, R_5, and R_6 series-parallel combination is in parallel with both R_2 and the $R_1 + R_4$ combination. This entire series-parallel combination is in series with R_7. Figure 6–12(c) illustrates these relationships.

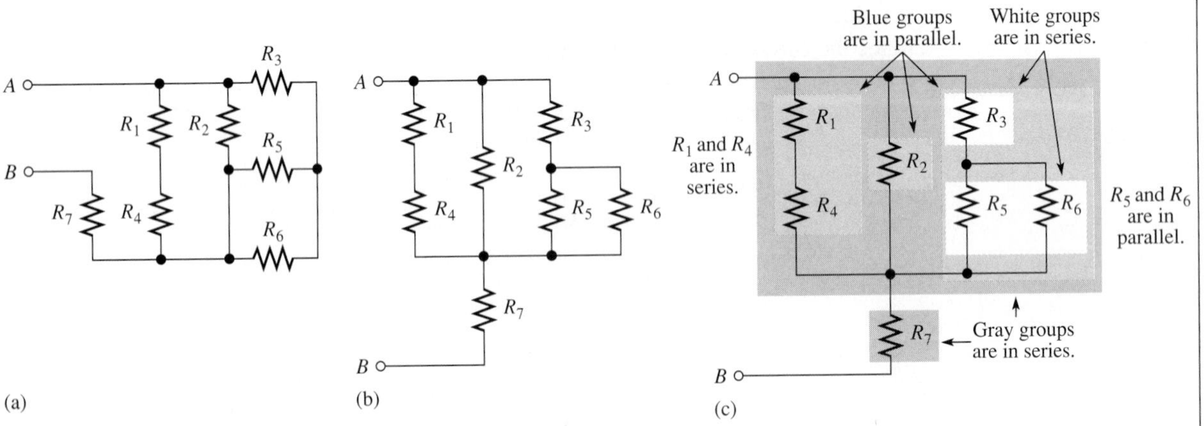

FIGURE 6–12

Related Exercise What happens if there is an open connection between R_1 and R_4 on the printed circuit board in Figure 6–11?

SECTION 6–1 REVIEW

1. A certain series-parallel circuit is described as follows: R_1 and R_2 are in parallel. This parallel combination is in series with another parallel combination of R_3 and R_4. Sketch the circuit.
2. In the circuit of Figure 6–13, describe the series-parallel relationships of the resistors.
3. Which resistors are in parallel in Figure 6–14?

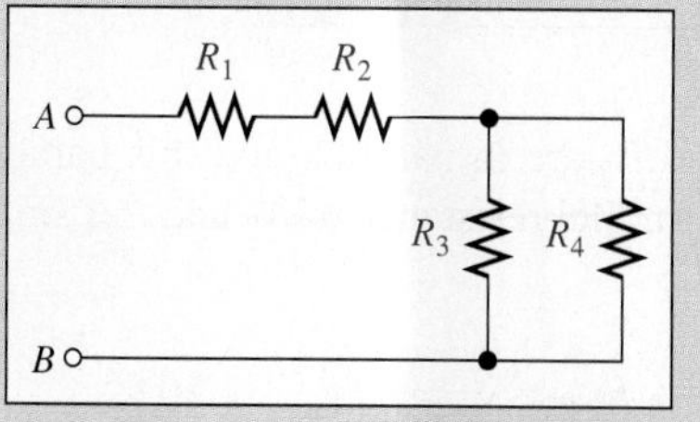

FIGURE 6–13

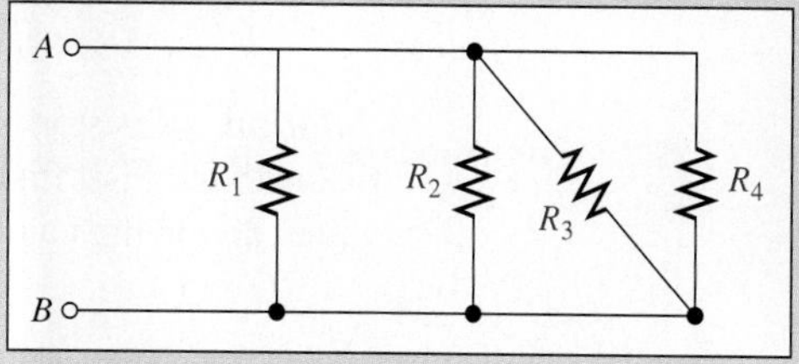

FIGURE 6–14

4. Identify the parallel resistors in Figure 6–15.
5. Are the parallel combinations in Figure 6–15 in series?

FIGURE 6–15

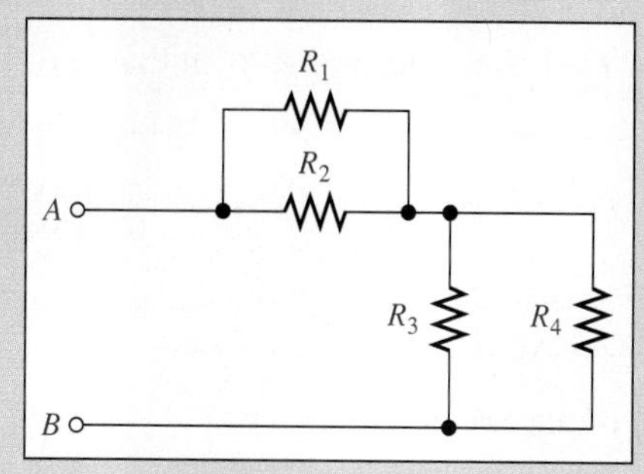

6–2 ■ ANALYSIS OF SERIES-PARALLEL CIRCUITS

Several quantities are important when you have a circuit that is a series-parallel configuration of resistors.

After completing this section, you should be able to

- **Analyze series-parallel circuits**
  - ☐ Determine total resistance
  - ☐ Determine all the currents
  - ☐ Determine all the voltage drops

Total Resistance

In Chapter 4, you learned how to determine total series resistance. In Chapter 5, you learned how to determine total parallel resistance.

To find the total resistance R_T of a series-parallel combination, first identify the series and parallel relationships, and then apply what you have previously learned. The following two examples illustrate the general approach.

EXAMPLE 6–6 Determine R_T between points A and B of the circuit in Figure 6–16.

FIGURE 6–16

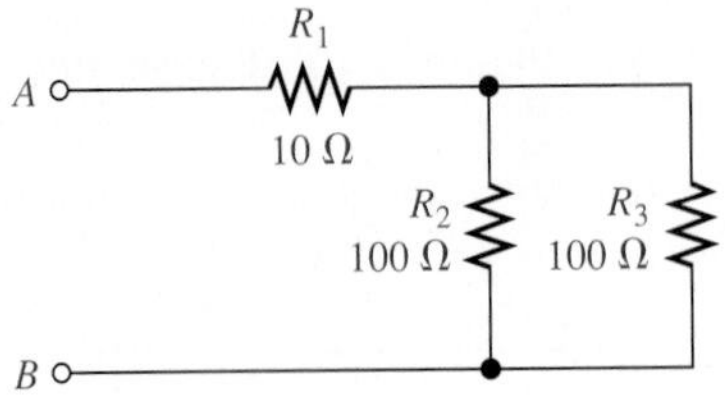

Solution Resistors R_2 and R_3 are in parallel, and this parallel combination is in series with R_1. First find the parallel resistance of R_2 and R_3. Since R_2 and R_3 are equal in value, divide the value by 2:

$$R_{2\text{-}3} = \frac{R}{n} = \frac{100\ \Omega}{2} = 50\ \Omega$$

Now, since R_1 is in series with $R_{2\text{-}3}$, add their values:

$$R_T = R_1 + R_{2\text{-}3} = 10\ \Omega + 50\ \Omega = 60\ \Omega$$

The calculator sequence for R_T is

[1] [0] [0] [÷] [2] [=] [+] [1] [0] [=]

Related Exercise Determine R_T in Figure 6–16 if R_3 is changed to 82 Ω.

EXAMPLE 6–7 Find R_T of the circuit in Figure 6–17.

FIGURE 6–17

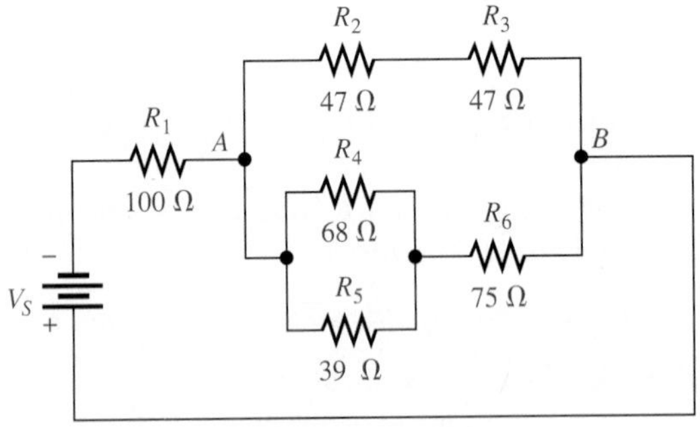

Solution In the upper branch, R_2 is in series with R_3. The series combination is designated $R_{2\text{-}3}$ and is equal to $R_2 + R_3$.

$$R_{2\text{-}3} = R_2 + R_3 = 47\ \Omega + 47\ \Omega = 94\ \Omega$$

In the lower branch, R_4 and R_5 are in parallel with each other. This parallel combination is designated $R_{4\text{-}5}$ and is calculated as follows:

$$R_{4\text{-}5} = \frac{R_4R_5}{R_4 + R_5} = \frac{(68\ \Omega)(39\ \Omega)}{68\ \Omega + 39\ \Omega} = 24.8\ \Omega$$

Also in the lower branch, the parallel combination of R_4 and R_5 is in series with R_6. This series-parallel combination is designated $R_{4\text{–}5\text{–}6}$ and is calculated as follows:

$$R_{4\text{-}5\text{-}6} = R_6 + R_{4\text{–}5} = 75\ \Omega + 24.8\ \Omega = 99.8\ \Omega$$

Figure 6–18 shows the original circuit in a simplified equivalent form.

FIGURE 6–18

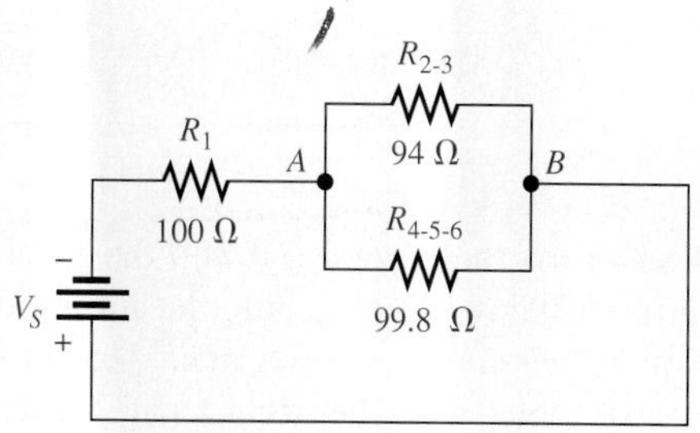

Now you can find the resistance between points A and B. It is $R_{2\text{-}3}$ in parallel with $R_{4\text{-}5\text{-}6}$. The equivalent resistance is calculated as follows:

$$R_{AB} = \frac{1}{\left(\dfrac{1}{94\ \Omega}\right) + \left(\dfrac{1}{99.8\ \Omega}\right)} = 48.4\ \Omega$$

Finally, the total resistance is R_1 in series with R_{AB}.

$$R_T = R_1 + R_{AB} = 100\ \Omega + 48.4\ \Omega = 148.4\ \Omega$$

Related Exercise Determine R_T if a 68 Ω resistor is connected to the circuit in Figure 6–17 from point A to point B.

Total Current

Once the total resistance and the source voltage are known, you can find total current in a circuit by applying Ohm's law. Total current is the total source voltage divided by the total resistance:

$$I_T = \frac{V_S}{R_T}$$

For example, let's find the total current in the circuit of Example 6–7 (Figure 6–17). Assume that the source voltage is 30 V. The calculation is

$$I_T = \frac{V_S}{R_T} = \frac{30\ \text{V}}{148.4\ \Omega} = 202\ \text{mA}$$

Branch Currents

Using the current-divider formula, Kirchhoff's current law, Ohm's law, or combinations of these, you can find the current in any branch of a series-parallel circuit. In some cases it may take repeated application of the formula to find a given current.

EXAMPLE 6–8 Determine the current through R_4 in Figure 6–19 if $V_S = 50$ V.

FIGURE 6–19

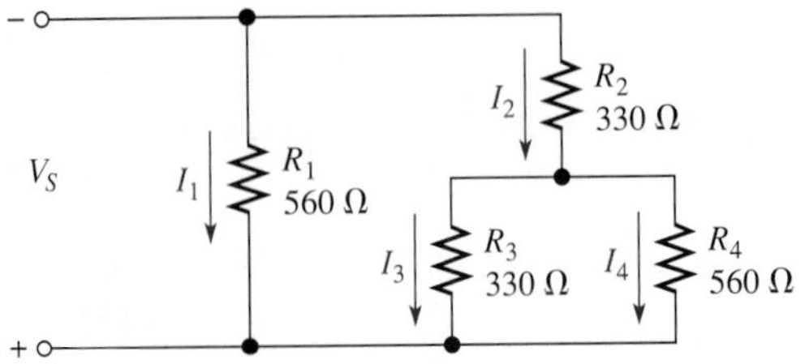

Solution First the current (I_2) into the junction of R_3 and R_4 must be found. Once this current is known, you can use the current-divider formula to find I_4.

Notice that there are two main branches in the circuit. The left-most branch consists of only R_1. The right-most branch has R_2 in series with the parallel combination of R_3 and R_4. The voltage across both of these main branches is the same and equal to 50 V. Find the current (I_2) into the junction of R_3 and R_4 by calculating the resistance of the right-most main branch and then applying Ohm's law as follows:

$$R_{2\text{-}3\text{-}4} = R_2 + \frac{R_3R_4}{R_3 + R_4} = 330\ \Omega + \frac{(330\ \Omega)(560\ \Omega)}{890\ \Omega} = 538\ \Omega$$

$$I_2 = \frac{V_S}{R_{2\text{-}3\text{-}4}} = \frac{50\text{ V}}{538\ \Omega} = 93\text{ mA}$$

Using the current-divider formula, calculate I_4 as follows:

$$I_4 = \left(\frac{R_3}{R_3 + R_4}\right)I_2 = \left(\frac{330\ \Omega}{890\ \Omega}\right)93\text{ mA} = 34.5\text{ mA}$$

You can find the current in R_3 by applying Kirchhoff's law ($I_2 = I_3 + I_4$) and subtracting I_4 from $I_2 (I_3 = I_2 - I_4)$.

You can find the current in R_1 by using Ohm's law ($I_1 = V_S/R_1$).

The calculator sequence for I_4 is

3 3 0 × 5 6 0 ÷ (3 3 0 + 5 6 0) +
3 3 0 = STO 5 0 ÷ RCL = STO 3 3 0 ÷
(3 3 0 + 5 6 0) × RCL =

Related Exercise Find I_1, I_3, and I_T in Figure 6–19.

Voltage Relationships

The circuit in Figure 6–20 illustrates voltage relationships in a series-parallel circuit. Voltmeters are connected to measure each of the resistor voltages, and the readings are indicated.

Some general observations about Figure 6–20 are as follows:

1. V_1 and V_2 are equal because R_1 and R_2 are in parallel. (Recall that voltages across parallel branches are the same.) V_1 and V_2 are the same as the voltage from A to B.
2. V_3 is equal to $V_4 + V_5$ because R_3 is in parallel with the series combination of R_4 and R_5. (V_3 is the same as the voltage from B to C.)
3. V_4 is about one-third of the voltage from B to C because R_4 is about one-third of the resistance $R_4 + R_5$ (by the voltage-divider principle).

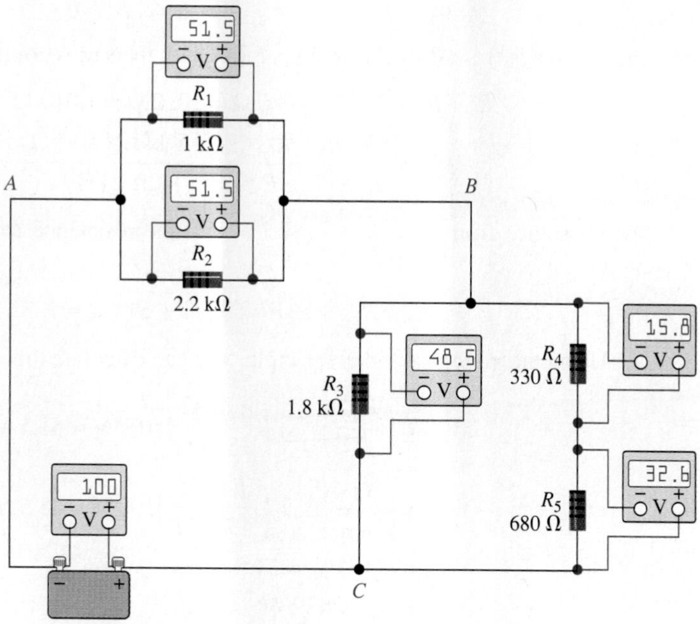

FIGURE 6–20
Illustration of voltage relationships.

4. V_5 is about two-thirds of the voltage from B to C because R_5 is about two-thirds of $R_4 + R_5$.
5. $V_1 + V_3$ equals V_S because, by Kirchhoff's voltage law, the sum of the voltage drops must equal the source voltage.

Example 6–9 will verify the meter readings in Figure 6–20.

EXAMPLE 6–9

Verify that the voltmeter readings in Figure 6–20 are correct. The circuit is redrawn as a schematic in Figure 6–21.

FIGURE 6–21

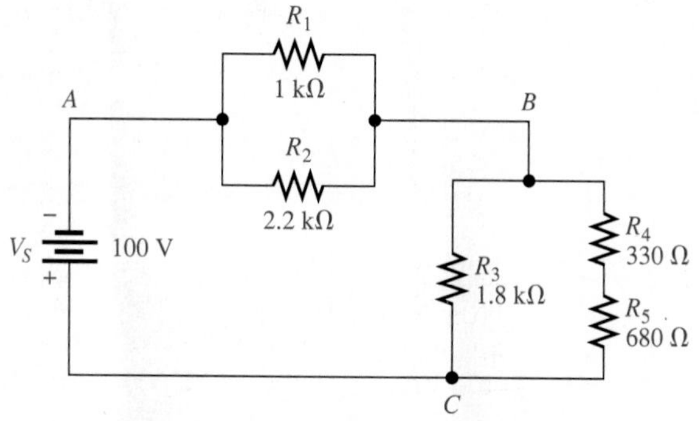

Solution The resistance from A to B is the parallel combination of R_1 and R_2.

$$R_{AB} = \frac{R_1 R_2}{R_1 + R_2} = \frac{(1\ \text{k}\Omega)(2.2\ \text{k}\Omega)}{3.2\ \text{k}\Omega} = 688\ \Omega$$

The resistance from B to C is R_3 in parallel with the series combination of R_4 and R_5.

$$R_4 + R_5 = 330\ \Omega + 680\ \Omega = 1010\ \Omega = 1.01\ \text{k}\Omega$$

$$R_{BC} = \frac{R_3(R_4 + R_5)}{R_3 + R_4 + R_5} = \frac{(1.8\ \text{k}\Omega)(1.01\ \text{k}\Omega)}{2.81\ \text{k}\Omega} = 647\ \Omega$$

The resistance from A to B is in series with the resistance from B to C, so the total circuit resistance is

$$R_T = R_{AB} + R_{BC} = 688\ \Omega + 647\ \Omega = 1335\ \Omega$$

Using the voltage-divider principle, you can calculate the voltages as follows:

$$V_{AB} = \left(\frac{R_{AB}}{R_T}\right)V_S = \left(\frac{688\ \Omega}{1335\ \Omega}\right)100\ \text{V} = 51.5\ \text{V}$$

$$V_{BC} = \left(\frac{R_{BC}}{R_T}\right)V_S = \left(\frac{647\ \Omega}{1335\ \Omega}\right)100\ \text{V} = 48.5\ \text{V}$$

$$V_1 = V_2 = V_{AB} = 51.5\ \text{V}$$

$$V_3 = V_{BC} = 48.5\ \text{V}$$

$$V_4 = \left(\frac{R_4}{R_4 + R_5}\right)V_{BC} = \left(\frac{330\ \Omega}{1010\ \Omega}\right)48.5\ \text{V} = 15.8\ \text{V}$$

$$V_5 = \left(\frac{R_5}{R_4 + R_5}\right)V_{BC} = \left(\frac{680\ \Omega}{1010\ \Omega}\right)48.5\ \text{V} = 32.7\ \text{V}$$

Related Exercise Determine each voltage drop in Figure 6–21 if the source voltage is doubled.

EXAMPLE 6–10 Determine the voltage drop across each resistor in Figure 6–22.

FIGURE 6–22

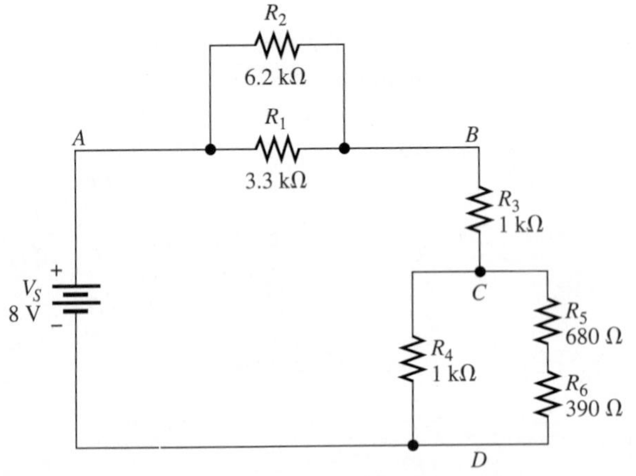

Solution Because you know the total voltage, you can solve this problem using the voltage-divider formula. First reduce each parallel combination to an equivalent resistance. Since R_1 and R_2 are in parallel between points A and B, combine their values.

$$R_{AB} = \frac{R_1R_2}{R_1 + R_2} = \frac{(3.3\ \text{k}\Omega)(6.2\ \text{k}\Omega)}{9.5\ \text{k}\Omega} = 2.15\ \text{k}\Omega$$

Since R_4 is in parallel with the series combination of R_5 and R_6 between points C and D, combine these values.

$$R_{CD} = \frac{R_4(R_5 + R_6)}{R_4 + R_5 + R_6} = \frac{(1\text{ k}\Omega)(1.07\text{ k}\Omega)}{2.07\text{ k}\Omega} = 517\ \Omega$$

The equivalent circuit is drawn in Figure 6–23.

FIGURE 6–23

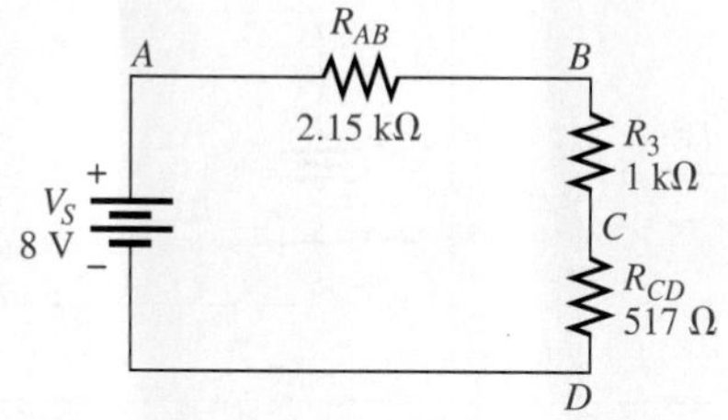

The total circuit resistance is

$$R_T = R_{AB} + R_3 + R_{CD} = 2.15\text{ k}\Omega + 1\text{ k}\Omega + 517\ \Omega = 3.67\text{ k}\Omega$$

Apply the voltage-divider formula to solve for the voltages:

$$V_{AB} = \left(\frac{R_{AB}}{R_T}\right)V_S = \left(\frac{2.15\text{ k}\Omega}{3.67\text{ k}\Omega}\right)8\text{ V} = 4.69\text{ V}$$

$$V_{CD} = \left(\frac{R_{CD}}{R_T}\right)V_S = \left(\frac{517\text{ k}\Omega}{3.67\text{ k}\Omega}\right)8\text{ V} = 1.13\text{ V}$$

$$V_3 = \left(\frac{R_3}{R_T}\right)V_S = \left(\frac{1\text{ k}\Omega}{3.67\text{ k}\Omega}\right)8\text{ V} = 2.18\text{ V}$$

V_{AB} equals the voltage across both R_1 and R_2.

$$V_1 = V_2 = V_{AB} = 4.69\text{ V}$$

V_{CD} is the voltage across R_4 and across the series combination of R_5 and R_6.

$$V_4 = V_{CD} = 1.13\text{ V}$$

Now apply the voltage-divider formula to the series combination of R_5 and R_6 to get V_5 and V_6:

$$V_5 = \left(\frac{R_5}{R_5 + R_6}\right)V_{CD} = \left(\frac{680\ \Omega}{1070\ \Omega}\right)1.13\text{ V} = 718\text{ mV}$$

$$V_6 = \left(\frac{R_6}{R_5 + R_6}\right)V_{CD} = \left(\frac{390\ \Omega}{1070\ \Omega}\right)1.13\text{ V} = 412\text{ mV}$$

Related Exercise Determine the current and power in each resistor in Figure 6–22.

As you have seen in this section, the analysis of series-parallel circuits can be approached in many ways, depending on what information you need and what circuit values you know. The examples in this section do not represent an exhaustive coverage, but they give you an idea of how to approach series-parallel circuit analysis.

If you know Ohm's law, Kirchhoff's laws, the voltage-divider formula, and the current-divider formula, and if you know how to apply these laws, you can solve most resistive circuit analysis problems. The ability to recognize series and parallel combinations is, of course, essential.

SECTION 6–2 REVIEW

1. Find the total resistance between A and B in the circuit of Figure 6–24.
2. Find I_3 in Figure 6–24.
3. Find V_2 in Figure 6–24.
4. Determine R_T and I_T in Figure 6–25.

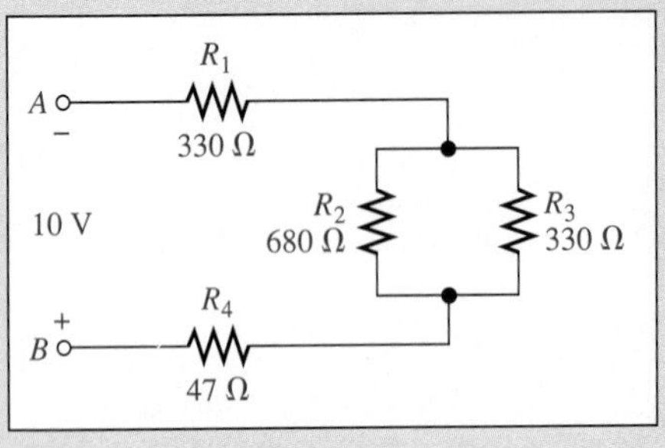

FIGURE 6–24

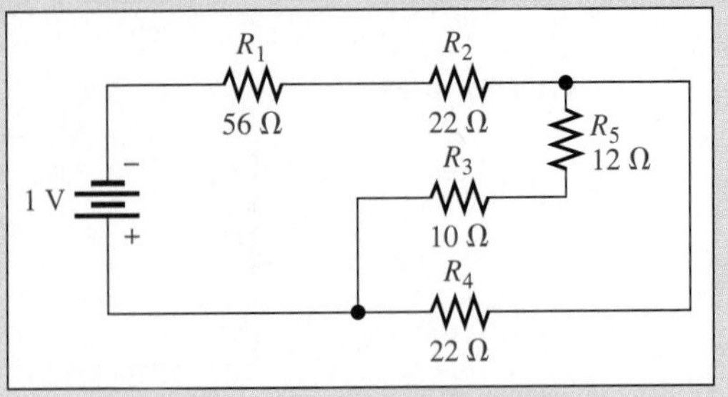

FIGURE 6–25

6–3 ▪ VOLTAGE DIVIDERS WITH RESISTIVE LOADS

Voltage dividers were introduced in Chapter 4. In this section, you will learn how resistive loads affect the operation of voltage-divider circuits.

After completing this section, you should be able to

- **Analyze loaded voltage dividers**
  - □ Determine the effect of a resistive load on a voltage-divider circuit.
  - □ Explain how a voltmeter can alter the value of the voltage being measured

The voltage divider in Figure 6–26(a) produces an output voltage (V_{OUT}) of 5 V because the input is 10 V and the two resistors are of equal value. This voltage is the unloaded output voltage. When a load resistor, R_L, is connected from the output to ground as shown in Figure 6–26(b), the output voltage is reduced by an amount that depends on the value of R_L. The load resistor is in parallel with R_2, reducing the resistance from point A to ground and, as a result, also reducing the voltage across the parallel combination. This is one effect of loading a voltage divider. Another effect of a **load** is that more current is drawn from the source because the total resistance of the circuit is reduced.

The larger R_L is, compared to R_2, the less the output voltage is reduced from its unloaded value, as illustrated in Figure 6–27. When two resistors are connected in parallel and one of the resistors is much greater than the other, the total resistance is close to the value of the smaller resistance.

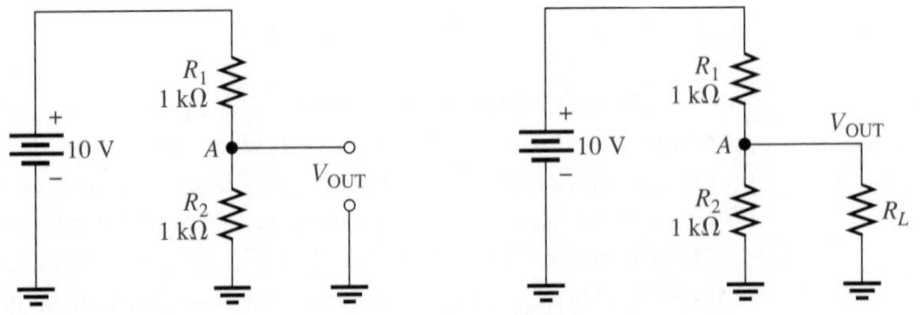

FIGURE 6–26
A voltage divider with both unloaded and loaded outputs.

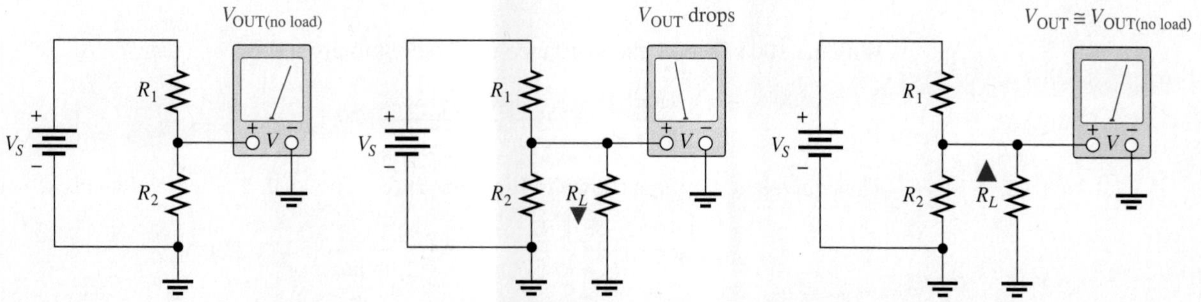

(a) No load (b) R_L not significantly greater than R_2 (c) R_L much greater than R_2

FIGURE 6–27
The effect of a load resistor.

EXAMPLE 6–11

(a) Determine the unloaded output voltage of the voltage divider in Figure 6–28.
(b) Find the loaded output voltages of the voltage divider in Figure 6–28 for the following two values of load resistance: $R_L = 10\ \text{k}\Omega$ and $R_L = 100\ \text{k}\Omega$.

FIGURE 6–28

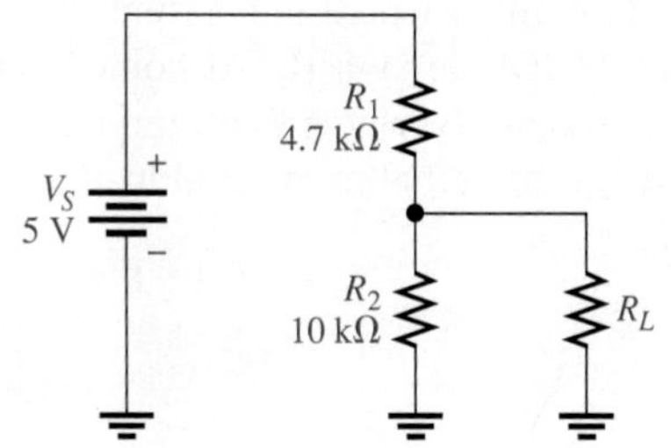

Solution

(a) The unloaded output voltage is

$$V_{OUT(unloaded)} = \left(\frac{R_2}{R_1 + R_2}\right)V_S = \left(\frac{10\ \text{k}\Omega}{14.7\ \text{k}\Omega}\right)5\ \text{V} = 3.40\ \text{V}$$

(b) With the 10 kΩ load resistor connected, R_L is in parallel with R_2, which gives 5 kΩ, as shown by the equivalent circuit in Figure 6–29(a). The loaded output voltage is

$$V_{OUT(loaded)} = \left(\frac{R_2 \parallel R_L}{R_1 + R_2 \parallel R_L}\right)V_S = \left(\frac{5\ \text{k}\Omega}{9.7\ \text{k}\Omega}\right)5\ \text{V} = 2.58\ \text{V}$$

FIGURE 6–29

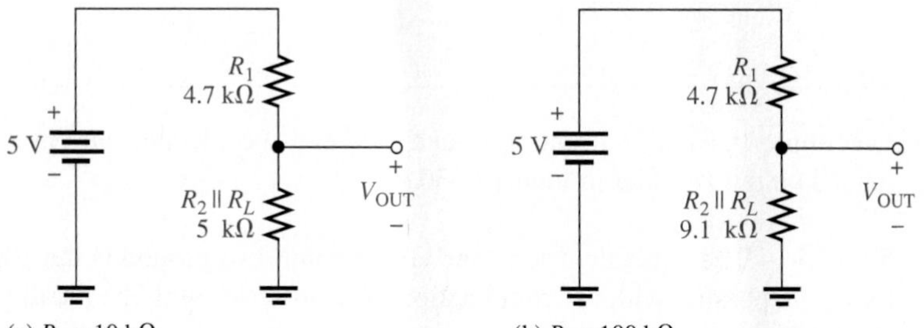

(a) $R_L = 10\ \text{k}\Omega$ (b) $R_L = 100\ \text{k}\Omega$

With the 100 kΩ load, the resistance from output to ground is

$$\frac{R_2 R_L}{R_2 + R_L} = \frac{(10\ \text{k}\Omega)(100\ \text{k}\Omega)}{110\ \text{k}\Omega} = 9.1\ \text{k}\Omega$$

The equivalent circuit is shown in Figure 6–29(b). The loaded output voltage is

$$V_{\text{OUT(loaded)}} = \left(\frac{R_2 \parallel R_L}{R_1 + R_2 \parallel R_L}\right)V_S = \left(\frac{9.1\ \text{k}\Omega}{13.8\ \text{k}\Omega}\right)5\ \text{V} = 3.30\ \text{V}$$

The smaller value of R_L causes a greater reduction in V_{OUT} (3.40 V − 2.58 V = 0.82 V) than does the larger value of R_L (3.40 V − 3.30 V = 0.10 V). This illustrates the loading effect of R_L on the voltage divider.

Related Exercise Determine V_{OUT} in Figure 6–28 for a 1 MΩ load resistance.

Load Current and Bleeder Current

In a loaded voltage-divider circuit, the total current drawn from the source consists of currents through the load resistors and the divider resistors. Figure 6–30 shows a voltage divider with two voltage outputs or two taps. Notice that the total current, I_T, flows through R_1 to point A where the current divides into I_{RL1}, which flows through R_{L1} and into I_2, which flows through R_2. At point B, the current I_2 divides into I_{RL2}, which flows through R_{L2} and into I_3, which flows through R_3. Current I_3 is called the **bleeder current,** which is the current left after the total load current is subtracted from the total current into the circuit:

$$I_{\text{BLEEDER}} = I_T - I_{RL1} - I_{RL2} \quad \textbf{(6–1)}$$

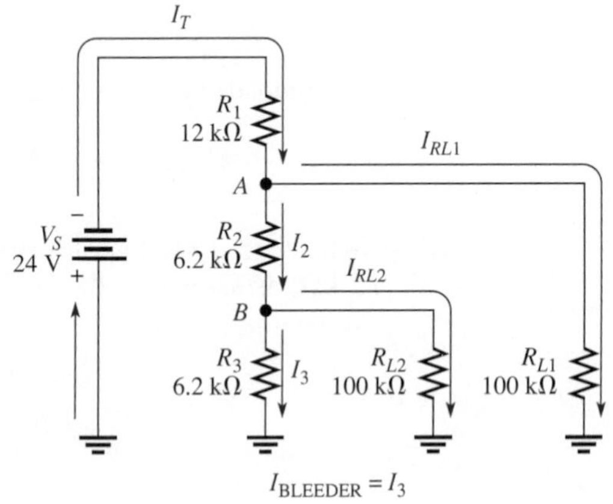

FIGURE 6–30
Currents in a two-tap loaded voltage divider.

EXAMPLE 6–12 Determine the load currents I_{RL1} and I_{RL2} and the bleeder current I_3 in the two-tap loaded voltage divider in Figure 6–30.

Solution The equivalent resistance from point A to ground is the 100 kΩ load resistor R_{L1} in parallel with the combination of R_2 in series with the parallel combination of R_3 and R_{L2}.

Determine the resistance values first. R_3 in parallel with R_{L2} is designated R_B and is calculated as follows:

$$R_B = \frac{R_3 R_{L2}}{R_3 + R_{L2}} = \frac{(6.2\ \text{k}\Omega)(100\ \text{k}\Omega)}{106.2\ \text{k}\Omega} = 5.84\ \text{k}\Omega$$

R_2 in series with R_B is designated $R_{2\text{-}B}$ and is calculated as follows:

$$R_{2\text{-}B} = R_2 + R_B = 6.2\ \text{k}\Omega + 5.84\ \text{k}\Omega = 12.0\ \text{k}\Omega$$

R_{L1} in parallel with $R_{2\text{-}B}$ is designated R_A and is calculated as follows:

$$R_A = \frac{R_{L1} R_{2\text{-}B}}{R_{L1} + R_{2\text{-}B}} = \frac{(100\ \text{k}\Omega)(12.0\ \text{k}\Omega)}{112\ \text{k}\Omega} = 10.7\ \text{k}\Omega$$

R_A is the total resistance from point A to ground. The total resistance for the circuit is

$$R_T = R_A + R_1 = 10.7\ \text{k}\Omega + 12.0\ \text{k}\Omega = 22.7\ \text{k}\Omega$$

The voltage across R_{L1} is determined as follows:

$$V_{RL1} = V_A = \left(\frac{R_A}{R_T}\right)V_S = \left(\frac{10.7\ \text{k}\Omega}{22.7\ \text{k}\Omega}\right)24\ \text{V} = 11.3\ \text{V}$$

The load current through R_{L1} is

$$I_{RL1} = \frac{V_{RL1}}{R_{L1}} = \left(\frac{11.3\ \text{V}}{100\ \text{k}\Omega}\right) = 113\ \mu\text{A}$$

The voltage at point B is determined by the parallel combination of R_{L2} and R_3 (R_B), R_2, and the voltage at point A:

$$V_B = \left(\frac{R_B}{R_2 + R_B}\right)V_A = \left(\frac{5.84\ \text{k}\Omega}{12.0\ \text{k}\Omega}\right)11.3\ \text{V} = 5.50\ \text{V}$$

The load current through R_{L2} is

$$I_{RL2} = \frac{V_{RL2}}{R_{L2}} = \frac{V_B}{R_{L2}} = \frac{5.50\ \text{V}}{100\ \text{k}\Omega} = 55\ \mu\text{A}$$

The bleeder current is

$$I_3 = \frac{V_B}{R_3} = \frac{5.50\ \text{V}}{6.2\ \text{k}\Omega} = 887\ \mu\text{A}$$

Related Exercise How can the bleeder current in Figure 6–30 be reduced without affecting the load currents?

Bipolar Voltage Dividers

An example of a voltage divider that produces both positive and negative voltages from a single source is shown in Figure 6–31. Notice that neither the positive nor the negative terminal of the source is connected to ground. The voltages at points A and B are positive with respect to ground, and the voltages at points C and D are negative with respect to ground.

FIGURE 6–31
A bipolar voltage divider. The positive and negative voltages are with respect to ground.

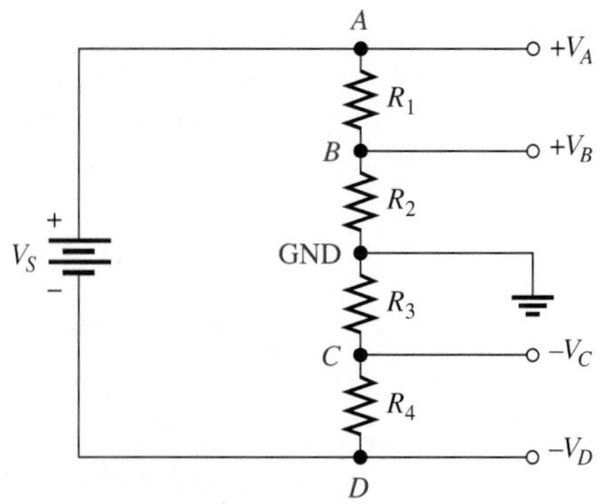

SECTION 6–3 REVIEW

1. A load resistor is connected to an output on a voltage divider. What effect does the load resistor have on the output voltage?
2. A larger-value load resistor will cause the output voltage of a voltage divider to change less than a smaller-value one will (T or F).
3. For the voltage divider in Figure 6–32, determine the unloaded output voltage. Also determine the output voltage with a 10 kΩ load resistor connected from the output to ground.

FIGURE 6–32

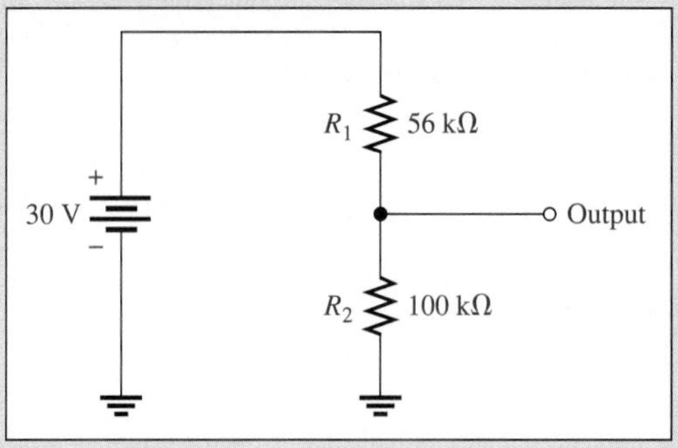

6–4 ■ LOADING EFFECT OF A VOLTMETER

As you have learned, voltmeters must be connected in parallel with a resistor in order to measure the voltage across the resistor. Because of its internal resistance, a voltmeter puts a load on the circuit and will affect, to a certain extent, the voltage that is being measured. Until now, we have ignored the loading effect because the internal resistance of a voltmeter is very high, and normally it has negligible effect on the circuit that is being measured. However, if the internal resistance of the voltmeter is not sufficiently greater than the circuit resistance across which it is connected, the loading effect will cause the measured voltage to be less than its actual value. You should always be aware of this effect.

After completing this section, you should be able to

- **Determine the loading effect of a voltmeter on a circuit**
  - □ Explain why a voltmeter can load a circuit
  - □ Discuss the internal resistance of a voltmeter

Why a Voltmeter Can Load a Circuit

When a voltmeter is connected to a circuit as shown, for example, in Figure 6–33(a), its internal resistance appears in parallel with R_3, as shown in part (b). The resistance from

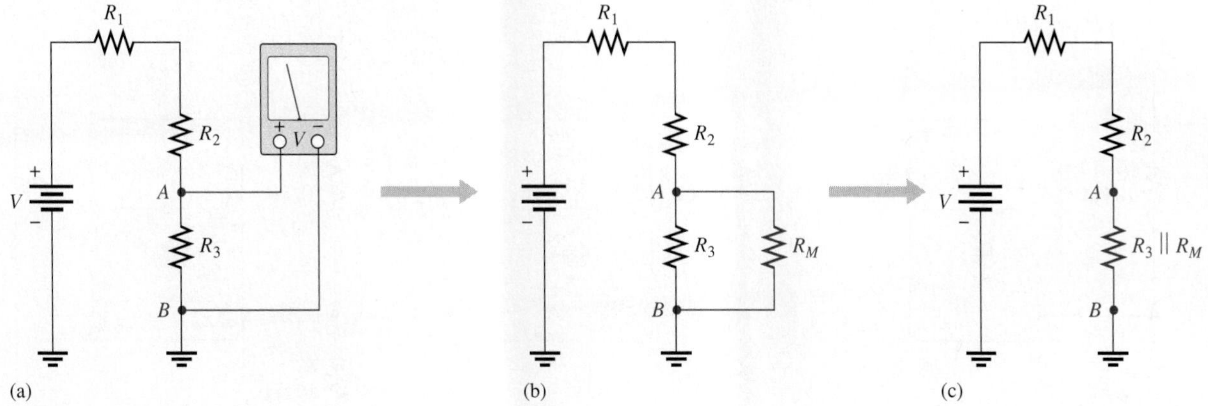

FIGURE 6–33
The loading effect of a voltmeter.

point A to point B is altered by the loading effect of the voltmeter's internal resistance R_M and is equal to $R_3 \parallel R_M$, as indicated in part (c).

If R_M is much greater than R_3, the resistance from A to B changes very little, and the meter indicates the actual voltage. If R_M is not sufficiently greater than R_3, the resistance from A to B is reduced significantly, and the voltage across R_3 is altered by the loading effect of the meter. A good rule of thumb is that *if the meter resistance is at least ten times greater than the resistance across which it is connected, the loading effect can be neglected.*

Internal Resistance of a Voltmeter

Three basic categories of voltmeters are the electromagnetic analog voltmeter, whose internal resistance is determined by its **sensitivity factor;** the electronic analog voltmeter, whose internal resistance is set by an input amplifier and is typically at least 10 MΩ; and the digital voltmeter, whose internal resistance is also typically at least 10 MΩ. The electronic analog voltmeter and the digital voltmeter present fewer loading problems than the electromagnetic type because their internal resistances are much higher.

As mentioned, the internal resistance of an electromagnetic voltmeter is specified by the sensitivity factor. A common sensitivity value is 20,000 Ω/V. The internal resistance depends on the dc voltage range switch setting of the meter and is determined by multiplying the sensitivity by the range setting. For example, if the dc voltage range switch is set at 10 V, the internal resistance is

$$R_M = 20{,}000\ \Omega/\text{V} \times 10\ \text{V} = 200{,}000\ \Omega$$

The lower the range setting, the less the internal resistance and the greater the loading effect of the voltmeter on a circuit. Always use the highest possible range setting that will allow an accurate reading from the meter scale. However, you would not use a 60 V setting to measure 1 V because you could not read 1 V accurately on the 60 V meter scale. The following example shows how the range setting affects the loading.

EXAMPLE 6–13 How much does the electromagnetic voltmeter affect the voltage being measured for each range switch setting indicated in Figure 6–34? Assume the meter has a sensitivity of 20,000 Ω/V.

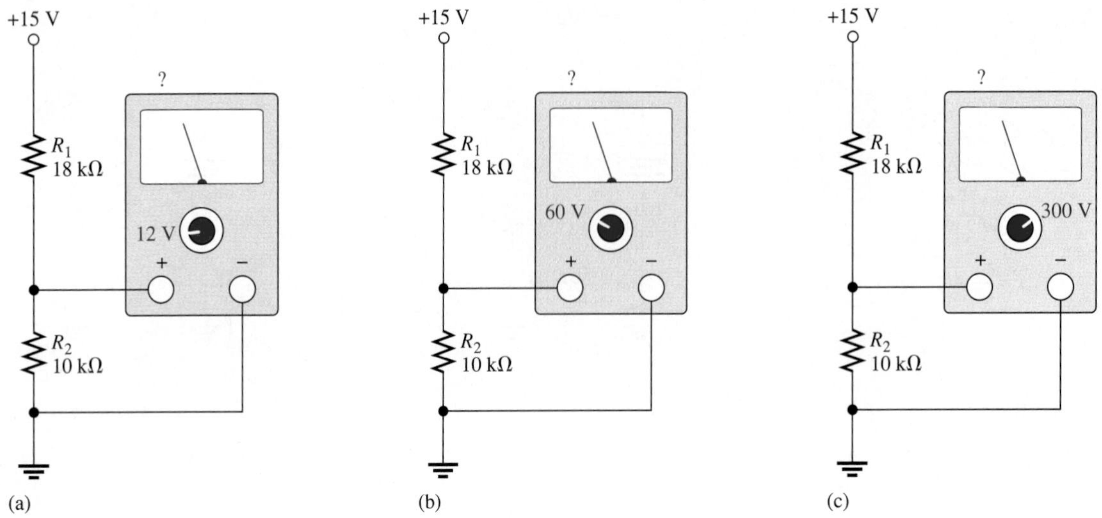

FIGURE 6–34

Solution The unloaded voltage across R_2 in the voltage-divider circuit is

$$V_{R2} = \left(\frac{R_2}{R_1 + R_2}\right)15\text{ V} = \left(\frac{10\text{ k}\Omega}{28\text{ k}\Omega}\right)15\text{ V} = 5.36\text{ V}$$

Refer to Figure 6–34(a). The internal resistance of the meter is

$$R_M = 20{,}000\ \Omega/\text{V} \times 12\text{ V} = 240\text{ k}\Omega$$

The meter's resistance in parallel with R_2 is

$$R_2 \parallel R_M = \left(\frac{R_2 R_M}{R_2 + R_M}\right) = \frac{(10\text{ k}\Omega)(240\text{ k}\Omega)}{250\text{ k}\Omega} = 9.6\text{ k}\Omega$$

The voltage actually measured by the meter is

$$V_{R2} = \left(\frac{R_2 \parallel R_M}{R_1 + R_2 \parallel R_M}\right)V = \left(\frac{9.6\text{ k}\Omega}{27.6\text{ k}\Omega}\right)15\text{ V} = 5.22\text{ V}$$

Refer to Figure 6–34(b). The internal resistance of the meter is

$$R_M = 20{,}000\ \Omega/\text{V} \times 60\text{ V} = 1.2\text{ M}\Omega$$

and

$$R_2 \parallel R_M = \frac{R_2 R_M}{R_2 + R_M} = \frac{(10\text{ M}\Omega)(1.2\text{ M}\Omega)}{1.21\text{ M}\Omega} = 9.92\text{ k}\Omega$$

The voltage actually measured is

$$V_{R2} = \left(\frac{R_2 \parallel R_M}{R_1 + R_2 \parallel R_M}\right)V = \left(\frac{9.92\text{ k}\Omega}{27.9\text{ k}\Omega}\right)15\text{ V} = 5.33\text{ V}$$

Refer to Figure 6–34(c). The internal resistance of the meter is

$$R_M = 20{,}000\ \Omega/\text{V} \times 300\text{ V} = 6\text{ M}\Omega$$

and

$$R_2 \parallel R_M = \frac{R_2 R_M}{R_2 + R_M} = \frac{(10\text{ k}\Omega)(6\text{ M}\Omega)}{6.01\text{ M}\Omega} = 9.98\text{ k}\Omega$$

The voltage actually measured is

$$V_{R2} = \left(\frac{R_2 \parallel R_M}{R_1 + R_2 \parallel R_M}\right)V = \left(\frac{9.98\ \text{k}\Omega}{27.98\ \text{k}\Omega}\right)15\ \text{V} = 5.35\ \text{V}$$

As you can see, the loading decreases as you increase the range setting on the voltmeter because the measured voltage is closer to the unloaded voltage. However, the 12 V setting is the best choice in this case because, although the loading effect is greater, you can read the expected voltage much more accurately than you can on the 60 V range. Of course, using the 300 V range is not a good choice in this case.

Related Exercise Calculate the voltage across R_2 in Figure 6–34 if the range switch is set on 100 V.

SECTION 6–4 REVIEW

1. What is meant when we say a voltmeter loads a circuit?
2. If a voltmeter with a 10 MΩ internal resistance is measuring the voltage across a 1 kΩ resistor, should you normally be concerned about the loading effect?
3. A certain electromagnetic-type voltmeter is set on the 12 V range to measure 2.5 V. If you switch it down to the 3 V range, will there be an increase or a decrease in the measured voltage? What is the advantage of using the 3 V range instead of the 12 V range in this situation?
4. What is the internal resistance of a 30,000 Ω/V voltmeter on its 50 V range setting?

6–5 ■ THE WHEATSTONE BRIDGE

Bridge circuits are widely used in measurement devices and other applications that you will learn later. In this section, you will study the balanced resistive bridge, which can be used to measure unknown resistance values.

After completing this section, you should be able to

- **Analyze a Wheatstone bridge**
  - □ Determine when a bridge is balanced
  - □ Determine an unknown resistance using a balanced bridge
  - □ Describe how a Wheatstone bridge can be used for temperature measurement

The circuit shown in Figure 6–35(a) is known as a *Wheatstone bridge*. Part (b) is the same circuit electrically, but it is drawn in a different way.

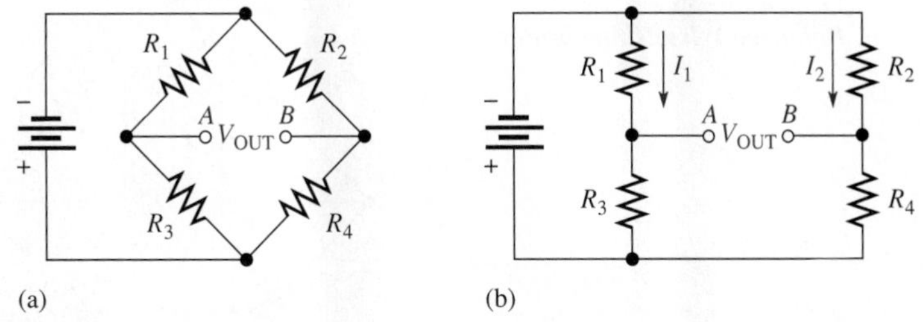

FIGURE 6–35
Wheatstone bridge.

A bridge is said to be balanced when the voltage (V_{OUT}) across the output terminals A and B is zero; that is, $V_A = V_B$. In Figure 6–35(b), if V_A equals V_B, then $V_{R1} = V_{R2}$, since the top sides of both R_1 and R_2 are connected to the same point. Also $V_{R3} = V_{R4}$, since the bottom sides of both R_3 and R_4 connect to the same point. The voltage ratios can be written as

$$\frac{V_1}{V_3} = \frac{V_2}{V_4}$$

Substituting by Ohm's law yields

$$\frac{I_1R_1}{I_1R_3} = \frac{I_2R_2}{I_2R_4}$$

The currents cancel, leaving

$$\frac{R_1}{R_3} = \frac{R_2}{R_4}$$

Solving for R_1,

$$R_1 = R_3\left(\frac{R_2}{R_4}\right)$$

How can you use this formula to determine an unknown resistance? First, make R_3 a variable resistor and call it R_V. Also, set the ratio R_2/R_4 to a known value. If R_V is adjusted until the bridge is balanced, the product of R_V and the ratio R_2/R_4 is equal to R_1, which is the unknown resistor (R_{UNK}). The formula for unknown resistance is stated as follows:

$$R_{UNK} = R_V\left(\frac{R_2}{R_4}\right) \qquad \textbf{(6–2)}$$

The bridge is balanced when the voltage across the output terminals equals zero ($V_A = V_B$). A galvanometer (a meter that measures small currents in either direction and is zero at center scale) is connected between the output terminals. Then R_V is adjusted until the galvanometer shows zero current ($V_A = V_B$), indicating a balanced condition. The setting of R_V multiplied by the ratio R_2/R_4 gives the value of R_{UNK}. Figure 6–36 shows this arrangement. For example, if $R_2/R_4 = 1/10$ and $R_V = 680\ \Omega$, then $R_{UNK} = (680\ \Omega)(1/10) = 68\ \Omega$.

FIGURE 6–36
Balanced Wheatstone bridge.

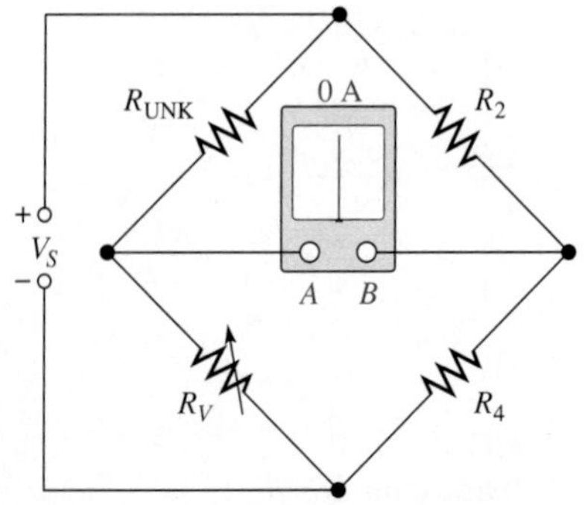

EXAMPLE 6–14 What is R_{UNK} under the balanced bridge conditions shown in Figure 6–37?

FIGURE 6–37

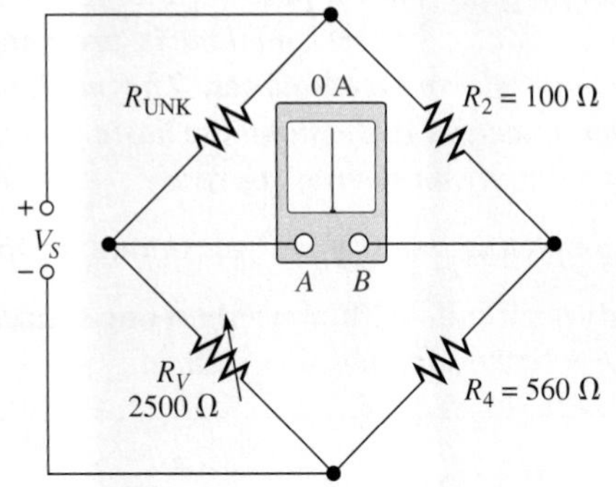

Solution

$$R_{UNK} = R_V\left(\frac{R_2}{R_4}\right) = 2500\ \Omega\left(\frac{100\ \Omega}{560\ \Omega}\right) = 446\ \Omega$$

Related Exercise If R_V must be adjusted to 1.8 kΩ in order to balance the bridge, what is R_{UNK}?

A Bridge Application

Bridge circuits are used for many measurements other than that of unknown resistance. One application of the Wheatstone bridge is in accurate temperature measurement. A temperature-sensitive element such as a thermistor is connected in a Wheatstone bridge as shown in Figure 6–38. An amplifier is connected across the output from *A* to *B* in order to increase the output voltage from the bridge to a usable value. The bridge is calibrated so that it is balanced at a specified reference temperature. As the temperature changes, the resistance of the sensing element changes proportionately, and the bridge becomes unbalanced. As a result, V_{AB} changes and is amplified (increased) and converted to a form for direct temperature readout on a gauge or a digital-type display.

FIGURE 6–38
A simplified circuit for temperature measurement.

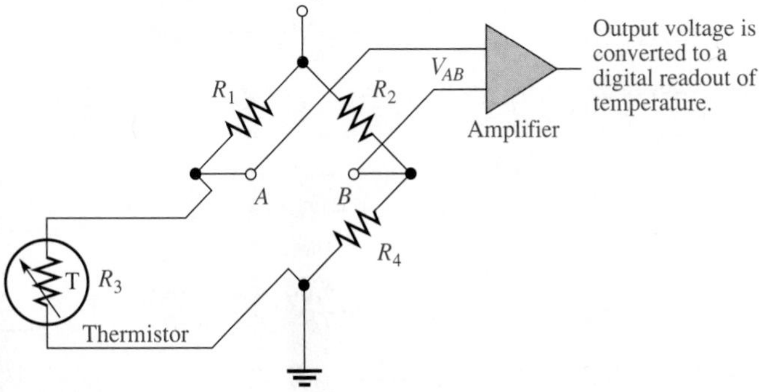

SECTION 6–5 REVIEW

1. Sketch a basic Wheatstone bridge circuit.
2. Under what condition is a bridge balanced?
3. What is the unknown resistance in Figure 6–36 when $R_V = 3.3\ \text{k}\Omega$, $R_2 = 10\ \text{k}\Omega$, and $R_4 = 2.2\ \text{k}\Omega$?

6–6 ▪ CIRCUITS WITH MORE THAN ONE VOLTAGE SOURCE

Some circuits require more than one voltage source. For example, certain types of amplifiers require both a positive and a negative voltage source for proper operation. In this section, a general method is presented for evaluating series-parallel circuits having multiple voltage sources. The method is illustrated with two-source circuits. However, once you understand the basic principle, you can easily extend this method to circuits with any number of sources.

After completing this section, you should be able to

- **Analyze circuits with more than one source**
  - Apply the superposition method

The Superposition Method

The **superposition** method is a way to determine currents and voltages in a circuit that has multiple sources by taking one source at a time. The other sources are replaced by their internal resistances. The ideal voltage source has a zero internal resistance. In this section, all voltage sources will be treated as ideal in order to simplify the coverage.

The steps in applying the superposition method are as follows:

Step 1: Take one voltage source at a time and replace each of the other voltage sources with a short (a short represents zero resistance).

Step 2: Determine the current or voltage that you need just as if there were only that one source in the circuit.

Step 3: Repeat Steps 1 and 2 for each source in turn.

Step 4: To find an actual current or voltage, add or subtract the currents or voltages due to each individual source. If the currents are in the same direction or the voltages of the same polarity, add them. If the currents are in opposite directions or the voltages of opposite polarities, subtract them with the direction of the resulting current or voltage the same as the larger of the original quantities.

An example of the approach to superposition is demonstrated in Figure 6–39 for a series-parallel circuit with two voltage sources. Study the steps in this figure.

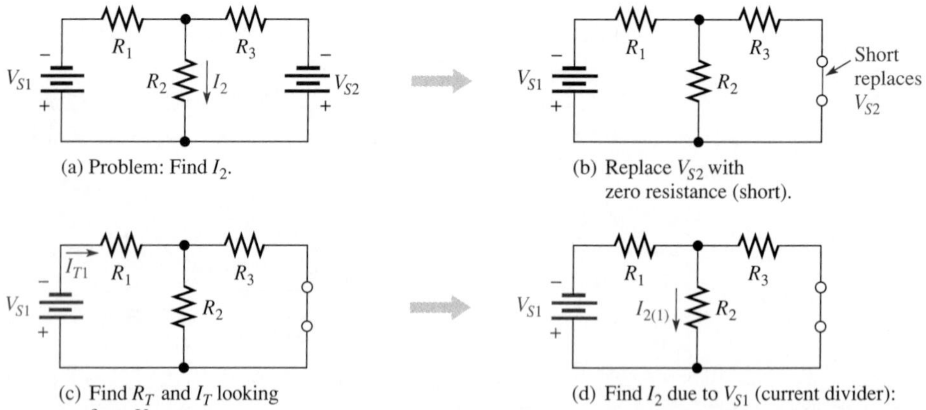

(a) Problem: Find I_2.

(b) Replace V_{S2} with zero resistance (short).

(c) Find R_T and I_T looking from V_{S1}:
$R_{T1} = R_1 + R_2 \parallel R_3$
$I_{T1} = V_{S1}/R_{T1}$

(d) Find I_2 due to V_{S1} (current divider):
$$I_{2(1)} = \left(\frac{R_3}{R_2 + R_3}\right) I_{T1}$$

FIGURE 6–39
Demonstration of the superposition method.

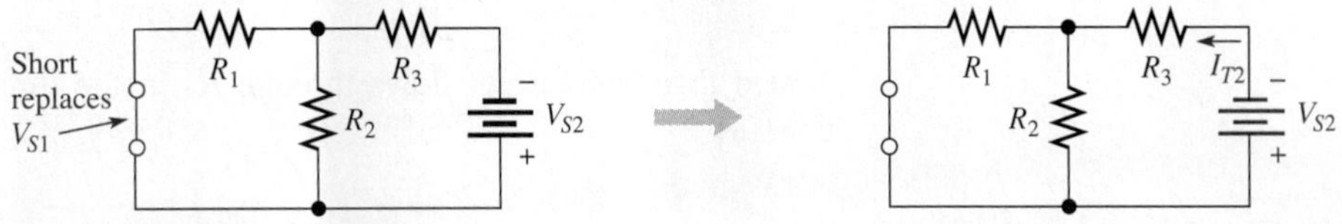

(e) Replace V_{S1} with zero resistance (short).

(f) Find R_T and I_T looking from V_{S2}: $R_{T2} = R_3 + R_1 \parallel R_2$
$I_{T2} = V_{S2}/R_{T2}$

(g) Find I_2 due to V_{S2}:

$$I_{2(2)} = \left(\frac{R_1}{R_1 + R_2}\right) I_{T2}$$

(h) Add $I_{2(1)}$ and $I_{2(2)}$ to get the actual I_2 (they are in same direction): $I_2 = I_{2(1)} + I_{2(2)}$

FIGURE 6–39 ***Continued***

EXAMPLE 6–15

Find the total current through R_3 in Figure 6–40.

FIGURE 6–40

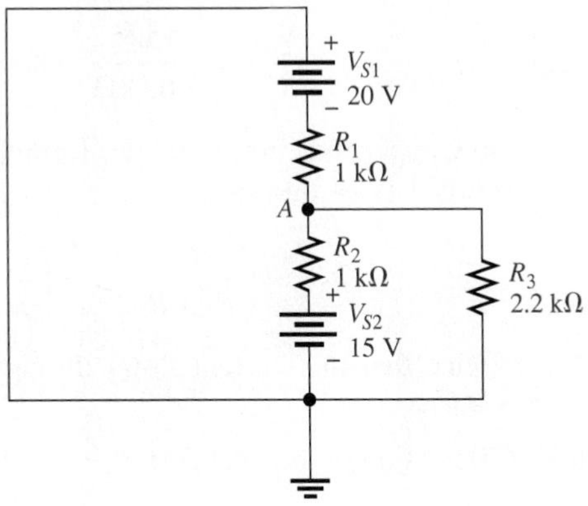

Solution

Step 1: Find the current through R_3 due to source V_{S1} by replacing source V_{S2} with a short, as shown in Figure 6–41(a). Looking from V_{S1},

$$R_{T1} = R_1 + \frac{R_2R_3}{R_2 + R_3} = 1\text{ k}\Omega + \frac{(1\text{ k}\Omega)(2.2\text{ k}\Omega)}{3.2\text{ k}\Omega} = 1.69\text{ k}\Omega$$

$$I_{T1} = \frac{V_{S1}}{R_{T1}} = \frac{20\text{ V}}{1.69\text{ k}\Omega} = 11.8\text{ mA}$$

Now apply the current-divider formula to find the current through R_3 due to source V_{S1} as follows:

$$I_{3(1)} = \left(\frac{R_2}{R_2 + R_3}\right)I_{T1} = \left(\frac{1\text{ k}\Omega}{3.2\text{ k}\Omega}\right)11.8\text{ mA} = 3.69\text{ mA}$$

Notice that this current flows through R_3 in the direction shown in Figure 6–41(a).

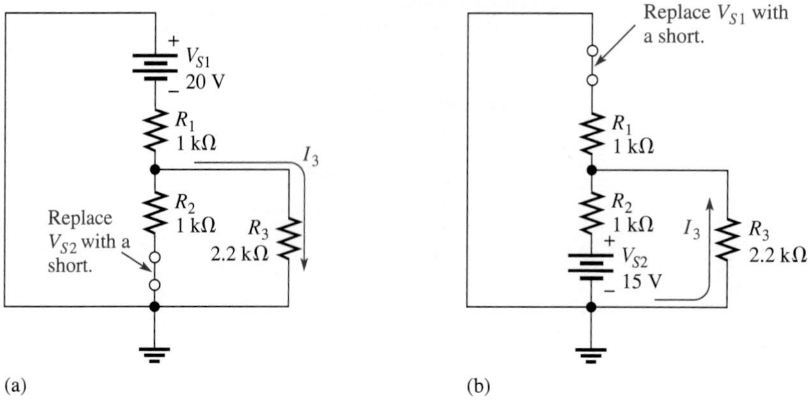

FIGURE 6–41

Step 2: Next find I_3 due to source V_{S2} by replacing source V_{S1} with a short, as shown in Figure 6–41(b). Looking from V_{S2},

$$R_{T2} = R_2 + \left(\frac{R_1 R_3}{R_1 + R_3}\right) = 1\text{ k}\Omega + \frac{(1\text{ k}\Omega)(2.2\text{ k}\Omega)}{3.2\text{ k}\Omega} = 1.69\text{ k}\Omega$$

$$I_{T2} = \frac{V_{S2}}{R_{T2}} = \frac{15\text{ V}}{1.69\text{ k}\Omega} = 8.88\text{ mA}$$

Now apply the current-divider formula to find the current through R_3 due to source V_{S2} as follows:

$$I_{3(2)} = \left(\frac{R_1}{R_1 + R_3}\right)I_T = \left(\frac{1\text{ k}\Omega}{3.2\text{ k}\Omega}\right)8.88\text{ mA} = 2.78\text{ mA}$$

Notice that this current flows through R_3 in the direction shown in Figure 6–41(b).

Step 3: Calculation of the total current through R_3 is as follows:

$$I_{3(tot)} = I_{3(1)} - I_{3(2)} = 3.69\text{ mA} - 2.78\text{ mA} = 0.91\text{ mA} = 910\ \mu\text{A}$$

This current flows through R_3 in the same direction as the larger component current, $I_{3(1)}$.

Related Exercise Determine the voltage at point A in Figure 6–40.

SECTION 6–6 REVIEW

1. State the superposition method.
2. Why is the superposition method useful for analysis of multiple-source linear circuits?
3. Why is a voltage source shorted and a current source opened when the superposition method is applied?

4. Using the superposition method, find the current through R_1 in Figure 6–42.

FIGURE 6–42

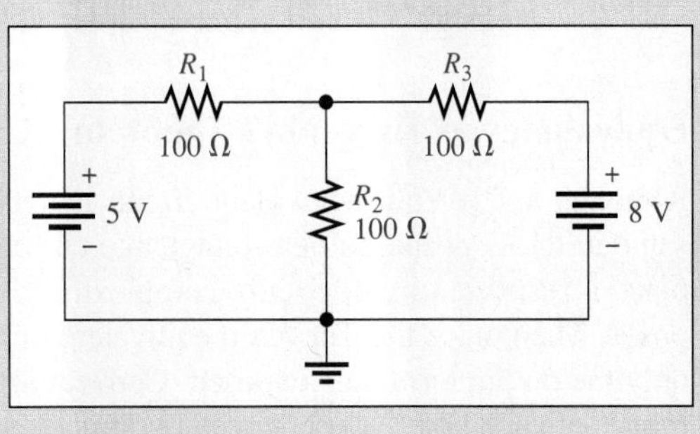

5. If, as a result of applying the superposition method, two currents are in opposing directions through a branch of a circuit, in which direction does the net current flow?

6–7 ■ THEVENIN'S THEOREM

Thevenin's theorem provides a method for simplifying a circuit to a standard equivalent form. In many cases, this theorem can be used to simplify the analysis of complex series-parallel circuits.

After completing this section, you should be able to

- **Apply Thevenin's theorem to simplify a circuit for analysis**
  - □ Describe the form of a Thevenin equivalent circuit
  - □ Obtain the Thevenin equivalent voltage source
  - □ Obtain the Thevenin equivalent resistance
  - □ Explain terminal equivalency in the context of Thevenin's theorem
  - □ Thevenize a portion of a circuit
  - □ Thevenize a bridge circuit

The Thevenin equivalent form of any resistive circuit consists of an equivalent voltage source (V_{TH}) and an equivalent resistance (R_{TH}), arranged as shown in Figure 6–43. The values of the equivalent voltage and resistance depend on the values in the original circuit. Any resistive circuit can be simplified regardless of its complexity.

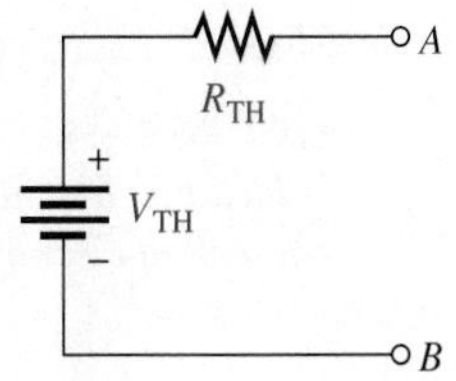

FIGURE 6–43
The general form of a Thevenin equivalent circuit is simply a voltage source in series with a resistance. Any resistive circuit can be reduced to this form.

Thevenin's Equivalent Voltage (V_{TH}) and Equivalent Resistance (R_{TH})

As you have seen, the equivalent voltage, V_{TH}, is one part of the complete Thevenin equivalent circuit. The other part is R_{TH}.

> **V_{TH} is defined to be the open circuit voltage between two specified points in a circuit.**

Any component connected between these two points effectively "sees" V_{TH} in series with R_{TH}. As defined by **Thevenin's theorem,**

> **R_{TH} is the total resistance appearing between two specified points in a given circuit with all sources replaced by their internal resistances.**

Equivalency in Thevenin's Theorem

Although a Thevenin equivalent circuit is not the same as its original circuit, it acts the same in terms of the output voltage and current. For example, as shown in Figure 6–44, place a resistive circuit of any complexity in a box with only the output terminals exposed. Then place the Thevenin equivalent of that circuit in an identical box with, again, only the output terminals exposed. Connect identical load resistors across the output terminals of each box. Next connect a voltmeter and an ammeter to measure the voltage and current for each load as shown in the figure. The measured values will be identical (neglecting tolerance variations), and you will not be able to determine which box contains the original circuit and which contains the Thevenin equivalent. That is, in terms of your observations, both circuits are the same. This condition is sometimes known as **terminal equivalency,** because both circuits look the same from the "viewpoint" of the two output terminals.

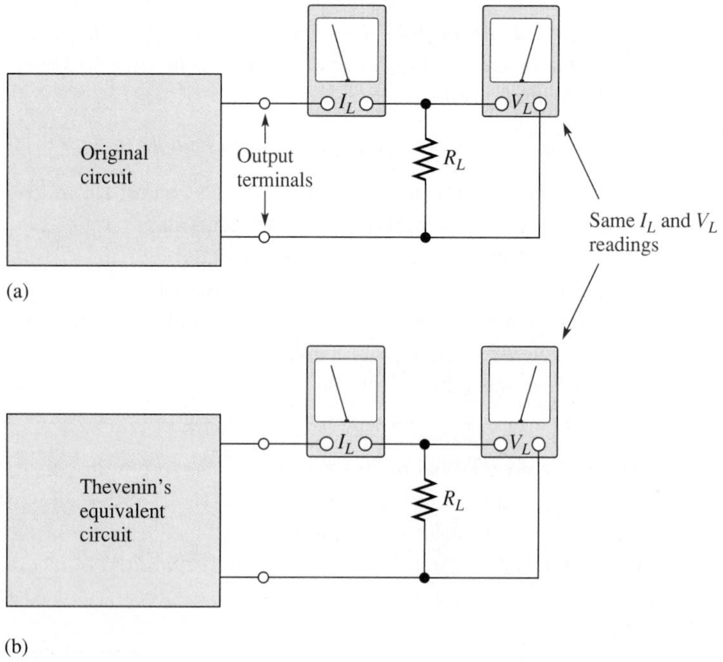

FIGURE 6–44
Which box contains the original circuit and which contains the Thevenin equivalent circuit? You cannot tell by observing the meters.

The Thevenin Equivalent of a Circuit

To find the Thevenin equivalent of any circuit, determine the equivalent voltage, V_{TH}, and the equivalent resistance, R_{TH}. For example, in Figure 6–45, the Thevenin equivalent for the circuit between points A and B is found as follows.

In Figure 6–45(a), the voltage across the designated points A and B is the Thevenin equivalent voltage. In this particular circuit, the voltage from A to B is the same as the voltage across R_2 because there is no current through R_3 and, therefore, no voltage drop across it. V_{TH} is expressed as follows for this particular example:

$$V_{TH} = \left(\frac{R_2}{R_1 + R_2}\right)V_S$$

In Figure 6–45(b), the resistance between points A and B with the source replaced by a short (zero internal resistance) is the Thevenin equivalent resistance. In this particular circuit, the resistance from A to B is R_3 in series with the parallel combination of R_1 and R_2. Therefore, R_{TH} is expressed as follows:

$$R_{TH} = R_3 + \frac{R_1R_2}{R_1 + R_2}$$

The Thevenin equivalent circuit is shown in Figure 6–45(c).

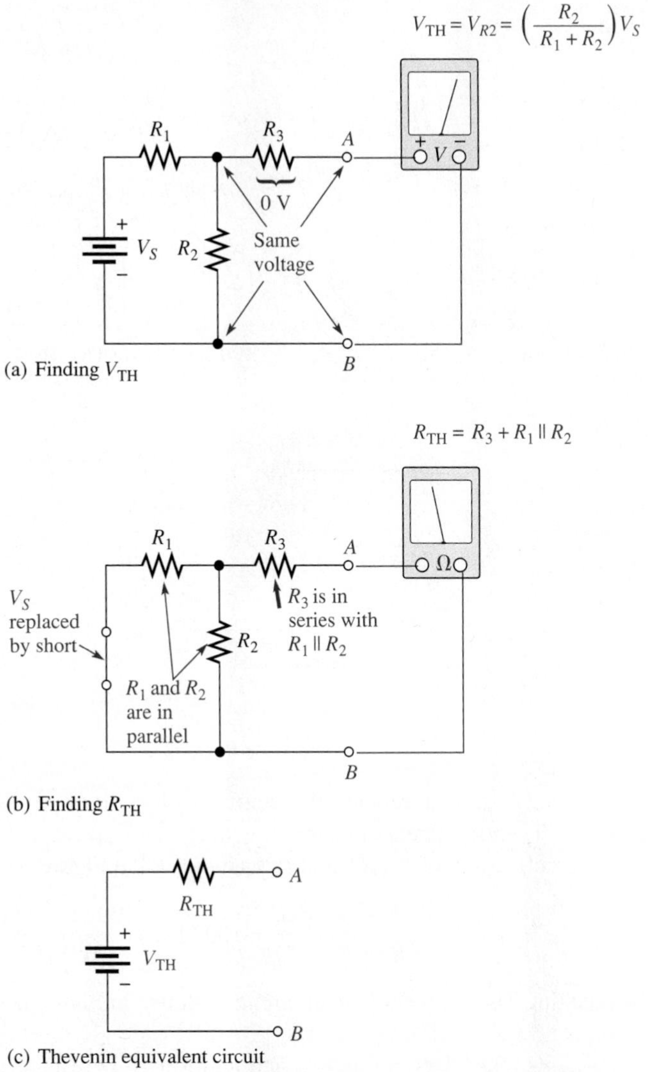

FIGURE 6–45
Example of the simplification of a circuit by Thevenin's theorem.

EXAMPLE 6–16 Find the Thevenin equivalent between the output terminals of the circuit in Figure 6–46.

FIGURE 6–46

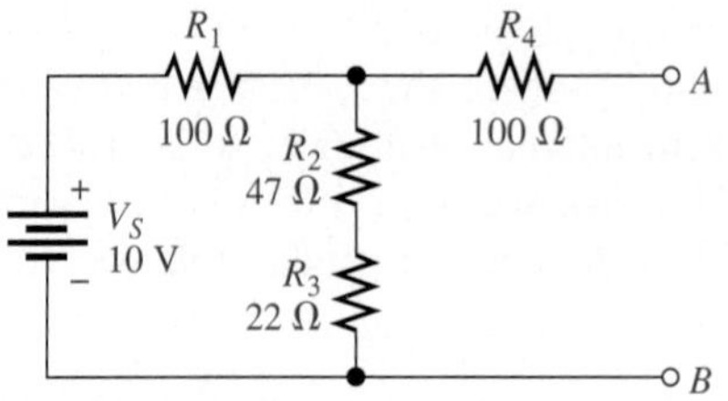

Solution V_{TH} equals the voltage across $R_2 + R_3$ as shown in Figure 6–47(a). Use the voltage-divider principle to find V_{TH}.

$$V_{TH} = \left(\frac{R_2 + R_3}{R_1 + R_2 + R_3}\right)V_S = \left(\frac{69\ \Omega}{169\ \Omega}\right)10\text{ V} = 4.08\text{ V}$$

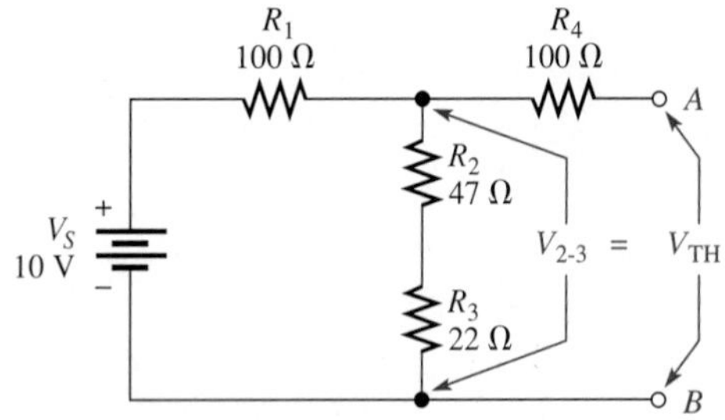

(a) The voltage from *A* to *B* is V_{TH} and equals $V_{2\text{-}3}$.

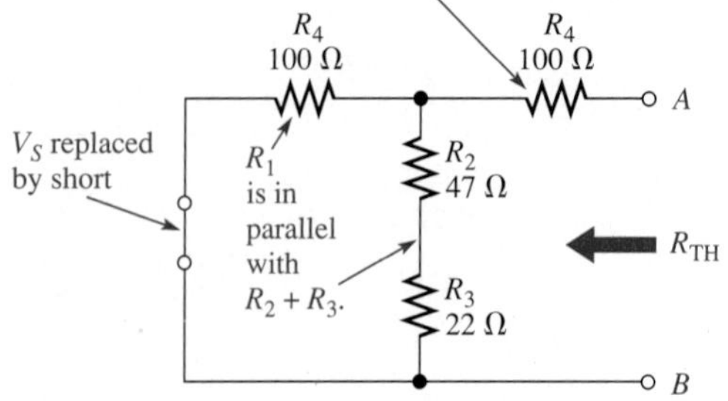

(b) Looking from terminals *A* and *B*, R_4 appears in series with the combination of R_1 in parallel with $(R_2 + R_3)$.

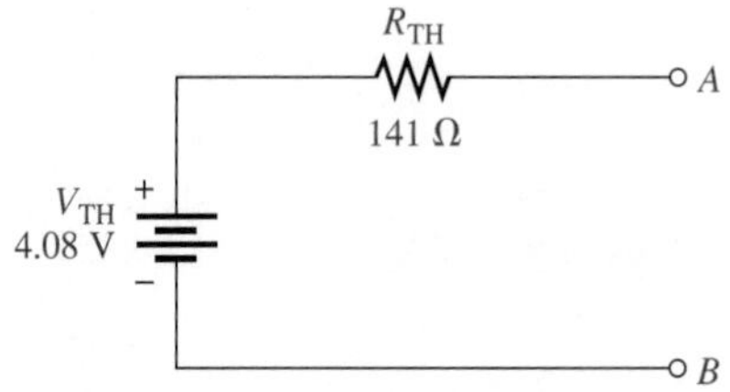

(c) Thevenin equivalent circuit

FIGURE 6–47

To find R_{TH}, first replace the source with a short to simulate a zero internal resistance. Then R_1 appears in parallel with $R_2 + R_3$, and R_4 is in series with the series-parallel combination of R_1, R_2, and R_3 as indicated in Figure 6–47(b).

$$R_{TH} = R_4 + \frac{R_1(R_2 + R_3)}{R_1 + R_2 + R_3} = 100\ \Omega + \frac{(100\ \Omega)(69\ \Omega)}{169\ \Omega} = 141\ \Omega$$

The resulting Thevenin equivalent circuit is shown in Figure 6–47(c).

Related Exercise Determine V_{TH} and R_{TH} if a 56 Ω resistor is connected in parallel across R_2 and R_3 in Figure 6–46.

Thevenin Equivalency Depends on the Viewpoint

The Thevenin equivalent for any circuit depends on the location of the two points from between which the circuit is "viewed." In Figure 6–46, we viewed the circuit from between the two points labeled *A* and *B*. Any given circuit can have more than one Thevenin equivalent, depending on how the viewpoints are designated. For example, if we view the circuit in Figure 6–48 from between points *A* and *C*, we obtain a completely different result than if we view it from between points *A* and *B* or from between points *B* and *C*.

FIGURE 6–48
Thevenin's equivalent depends on viewpoint.

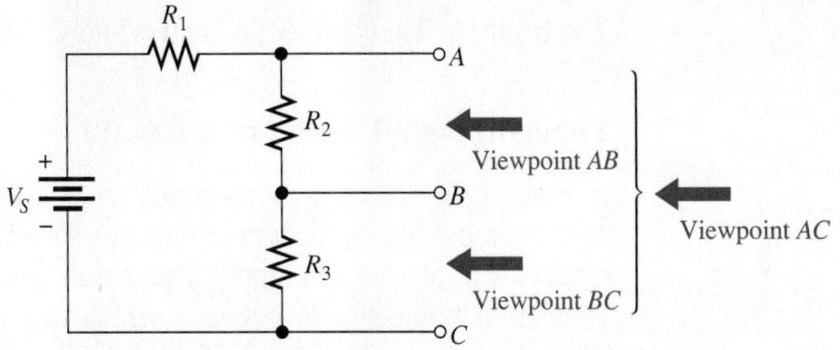

In Figure 6–49(a), when viewed from between points *A* and *C*, V_{TH} is the voltage across $R_2 + R_3$ and can be expressed using the voltage-divider formula as

$$V_{TH} = \left(\frac{R_2 + R_3}{R_1 + R_2 + R_3}\right)V_S$$

Also, as shown in Figure 6–49(b), the resistance between points *A* and *C* is $R_2 + R_3$ in parallel with R_1 (the source is replaced by a short) and can be expressed as

$$R_{TH} = \frac{R_1(R_2 + R_3)}{R_1 + R_2 + R_3}$$

The resulting Thevenin equivalent circuit is shown in Figure 6–49(c).

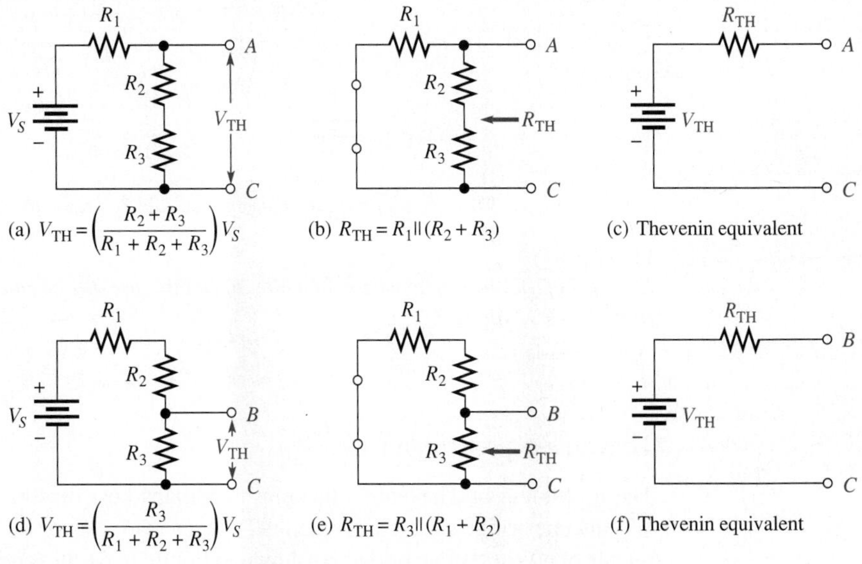

FIGURE 6–49
Example of circuit thevenized from two viewpoints, resulting in two different equivalent circuits. (The V_{TH} and R_{TH} values are different.)

When viewed from between points B and C as indicated in Figure 6–49(d), V_{TH} is the voltage across R_3 and can be expressed as

$$V_{\text{TH}} = \left(\frac{R_3}{R_1 + R_2 + R_3}\right)V_S$$

As shown in Figure 6–49(e), the resistance between points B and C is R_3 in parallel with the series combination of R_1 and R_2.

$$R_{\text{TH}} = \frac{R_3(R_1 + R_2)}{R_1 + R_2 + R_3}$$

The resulting Thevenin equivalent is shown in Figure 6-49(f).

Thevenizing a Portion of a Circuit

In many cases it helps to thevenize only a portion of a circuit. For example, when we need to know the equivalent circuit as viewed by one particular resistor in the circuit, we remove that resistor and apply Thevenin's theorem to the remaining part of the circuit as viewed from the points between which that resistor was connected. Figure 6–50 illustrates the thevenizing of part of a circuit.

Using this type of approach, we can easily find the voltage and current for a specified resistor for any number of resistor values using only Ohm's law. This method eliminates the necessity of reanalyzing the original circuit for each different resistance value.

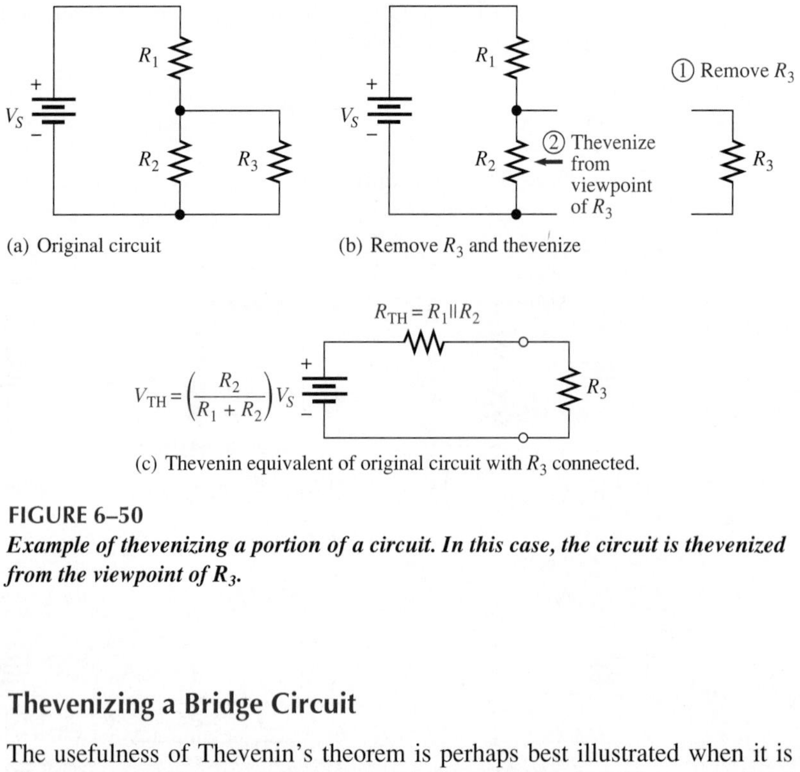

FIGURE 6–50
Example of thevenizing a portion of a circuit. In this case, the circuit is thevenized from the viewpoint of R_3.

Thevenizing a Bridge Circuit

The usefulness of Thevenin's theorem is perhaps best illustrated when it is applied to a Wheatstone bridge circuit. For example, when a load resistor is connected to the output terminals of a Wheatstone bridge, as shown in Figure 6–51, the circuit is very difficult to analyze because it is not a straightforward series-parallel arrangement. If you doubt that this analysis is difficult, try to identify which resistors are in parallel and which are in series.

FIGURE 6–51
Wheatstone bridge with a load resistor is not a series-parallel circuit.

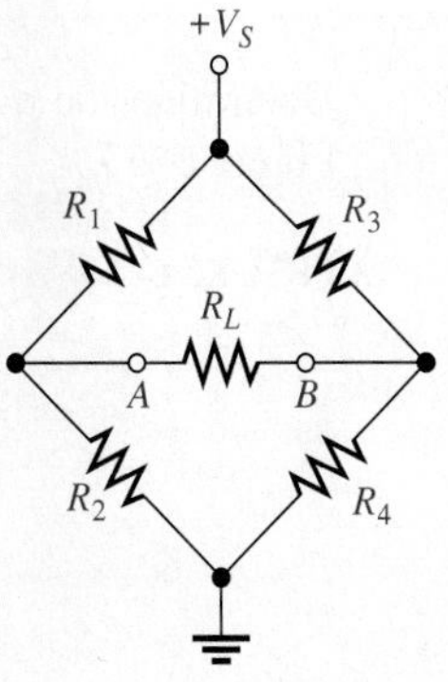

Using Thevenin's theorem, we can simplify the bridge circuit to an equivalent circuit viewed from the load resistor as shown step-by-step in Figure 6–52. Study carefully the steps in this figure. Once the equivalent circuit for the bridge is found, the voltage and current for any value of load resistor can easily be determined.

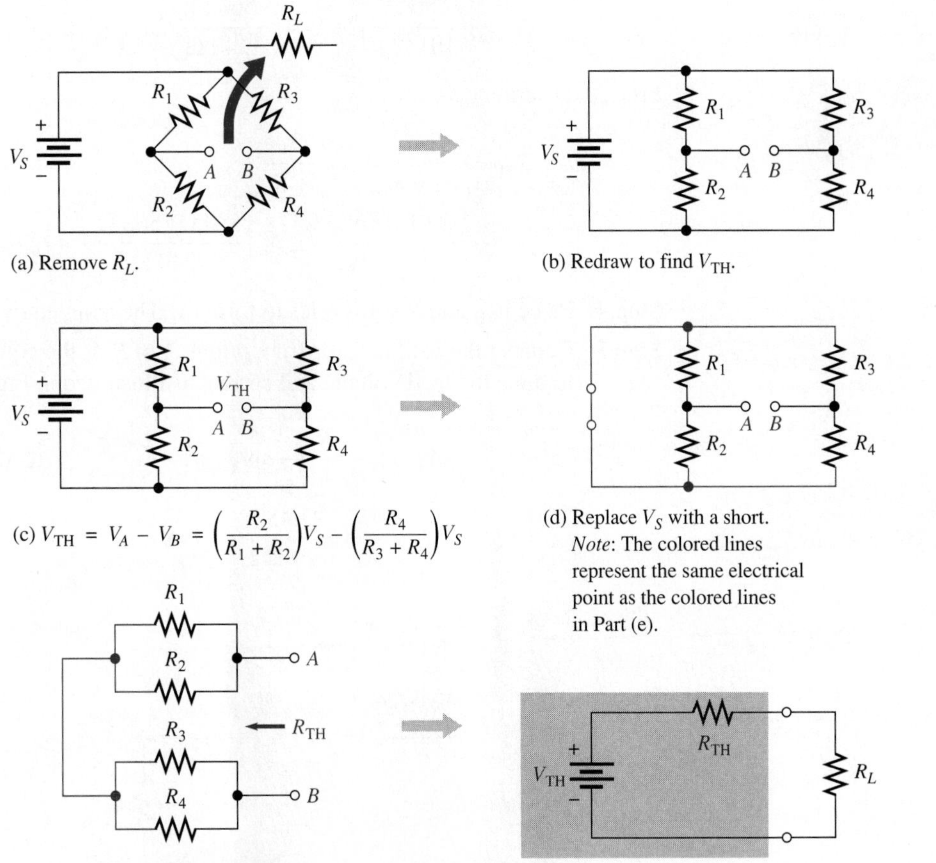

(a) Remove R_L.

(b) Redraw to find V_{TH}.

(c) $V_{TH} = V_A - V_B = \left(\frac{R_2}{R_1 + R_2}\right)V_S - \left(\frac{R_4}{R_3 + R_4}\right)V_S$

(d) Replace V_S with a short. *Note*: The colored lines represent the same electrical point as the colored lines in Part (e).

(e) Redraw to find R_{TH}: $R_{TH} = R_1 \| R_2 + R_3 \| R_4$

(f) Thevenin's equivalent with R_L reconnected

FIGURE 6–52
Simplifying a Wheatstone bridge with Thevenin's theorem.

EXAMPLE 6–17 Determine the voltage and current for the load resistor, R_L, in the bridge circuit of Figure 6–53.

FIGURE 6–53

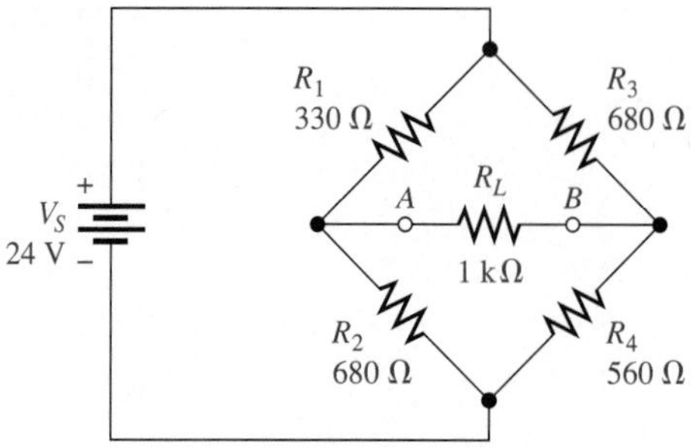

Solution

Step 1: Remove R_L.

Step 2: To Thevenize the bridge as viewed from between points A and B, as was shown in Figure 6–52, first determine V_{TH}.

$$V_{TH} = V_A - V_B = \left(\frac{R_2}{R_1 + R_2}\right)V_S - \left(\frac{R_4}{R_3 + R_4}\right)V_S$$

$$= \left(\frac{680\ \Omega}{1010\ \Omega}\right)24\ \text{V} - \left(\frac{560\ \Omega}{1240\ \Omega}\right)24\ \text{V} = 16.16\ \text{V} - 10.84\ \text{V} = 5.32\ \text{V}$$

Step 3: Determine R_{TH}.

$$R_{TH} = \frac{R_1R_2}{R_1 + R_2} + \frac{R_3R_4}{R_3 + R_4}$$

$$= \frac{(330\ \Omega)(680\ \Omega)}{1010\ \Omega} + \frac{(680\ \Omega)(560\ \Omega)}{1240\ \Omega} = 222\ \Omega + 307\ \Omega = 529\ \Omega$$

Step 4: Place V_{TH} and R_{TH} in series to form the Thevenin equivalent circuit.

Step 5: Connect the load resistor from points A to B of the equivalent circuit, and determine the load voltage and current as illustrated in Figure 6–54.

$$V_L = \left(\frac{R_L}{R_L + R_{TH}}\right)V_{TH} = \left(\frac{1\ \text{k}\Omega}{1.529\ \Omega}\right)5.32\ \text{V} = 3.48\ \text{V}$$

$$I_L = \frac{V_L}{R_L} = \frac{3.48\ \text{V}}{1\ \text{k}\Omega} = 3.48\ \text{mA}$$

FIGURE 6–54

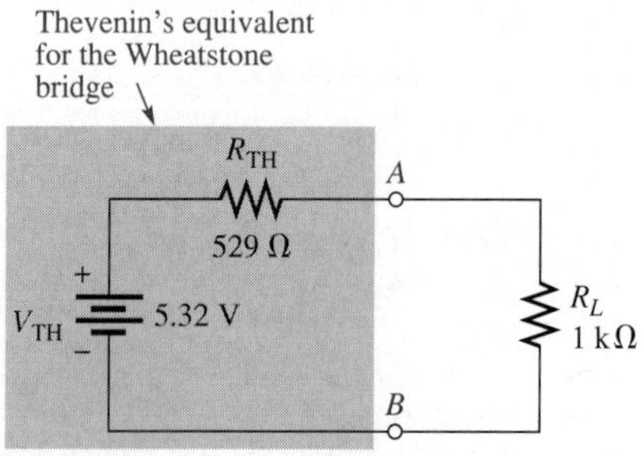

Related Exercise Calculate I_L for $R_1 = 2.2\ \text{k}\Omega$, $R_2 = 3.3\ \text{k}\Omega$, $R_3 = 3.9\ \text{k}\Omega$, and $R_4 = 2.7\ \text{k}\Omega$ in Figure 6–53.

Summary of Thevenin's Theorem

Remember, the Thevenin equivalent circuit is *always* an equivalent voltage source in series with an equivalent resistance regardless of the original circuit that it replaces. The significance of Thevenin's theorem is that the equivalent circuit can replace the original circuit as far as any external load is concerned. Any load resistor connected between the terminals of a Thevenin equivalent circuit will have the same current through it and the same voltage across it as if it were connected to the terminals of the original circuit.

A summary of steps for applying Thevenin's theorem is as follows:

Step 1: Open the two terminals (remove any load) between which you want to find the Thevenin equivalent circuit.

Step 2: Determine the voltage (V_{TH}) across the two open terminals.

Step 3: Determine the resistance (R_{TH}) between the two terminals with all voltage sources shorted and all current sources opened.

Step 4: Connect V_{TH} and R_{TH} in series to produce the complete Thevenin equivalent for the original circuit.

Step 5: Place the load resistor removed in Step 1 across the terminals of the Thevenin equivalent circuit. The load current can now be calculated using only Ohm's law, and it has the same value as the load current in the original circuit.

SECTION 6–7 REVIEW

1. What are the two components of a Thevenin equivalent circuit?
2. Draw the general form of a Thevenin equivalent circuit.
3. How is V_{TH} defined?
4. How is R_{TH} defined?
5. For the original circuit in Figure 6–55, draw the Thevenin equivalent circuit as viewed by R_L.

FIGURE 6–55

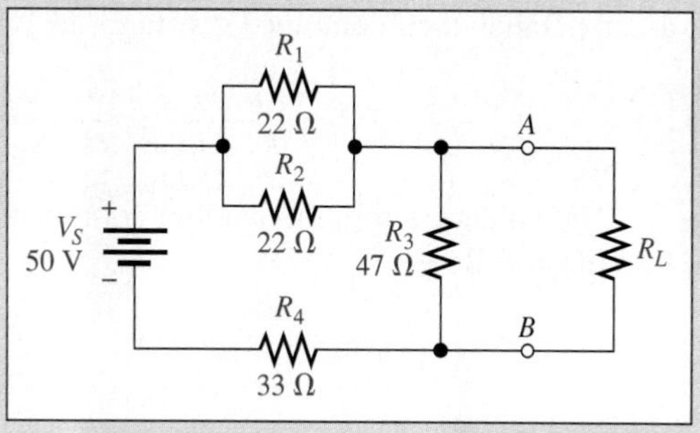

6–8 ■ TROUBLESHOOTING

Troubleshooting is the process of identifying and locating a failure or problem in a circuit. Some troubleshooting techniques have already been discussed in relation to both series circuits and parallel circuits. These methods are now extended to the series-parallel networks.

After completing this section, you should be able to

- **Troubleshoot series-parallel circuits**
  - ☐ Determine the effects of an open circuit
  - ☐ Determine the effects of a short circuit
  - ☐ Locate opens and shorts

Opens and shorts are typical problems that occur in electric circuits. As mentioned in Chapter 4, if a resistor burns out, it will normally produce an open circuit. Bad solder connections, broken wires, and poor contacts can also be causes of open paths. Pieces of foreign material, such as solder splashes, broken insulation on wires, and so on, can often lead to shorts in a circuit. A short is a zero resistance path between two points.

In addition to complete opens or shorts, partial opens or partial shorts can develop in a circuit. A partial open would be a much higher than normal resistance, but not infinitely large. A partial short would be a much lower than normal resistance, but not zero.

The following three examples illustrate troubleshooting series-parallel circuits.

EXAMPLE 6–18

From the indicated voltmeter reading in Figure 6–56, determine if there is a fault. If there is a fault, identify it as either a short or an open.

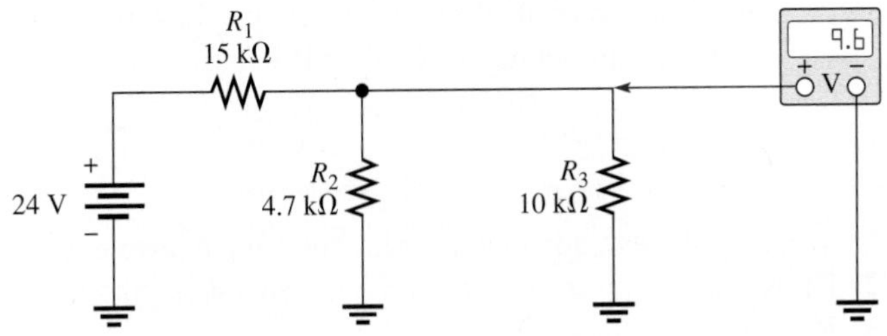

FIGURE 6–56

Solution First determine what the voltmeter should be indicating. Since R_2 and R_3 are in parallel, their combined resistance is

$$R_{2\text{-}3} = \frac{R_2R_3}{R_2 + R_3} = \frac{(4.7\text{ k}\Omega)(10\text{ k}\Omega)}{14.7\text{ k}\Omega} = 3.20\text{ k}\Omega$$

The voltage across the parallel combination is determined by the voltage-divider formula as follows:

$$V_{2\text{-}3} = \left(\frac{R_{2\text{-}3}}{R_1 + R_{2\text{-}3}}\right)V_S = \left(\frac{3.2\text{ k}\Omega}{18.2\text{ k}\Omega}\right)24\text{ V} = 4.22\text{ V}$$

This calculation shows that 4.22 V is the voltage reading that you should get on the meter. But the meter reads 9.6 V instead. This value is incorrect, and, because it is higher than it should be, R_2 or R_3 is probably open. Why? Because if either of these two resistors is open, the resistance across which the meter is connected is larger than expected. A higher resistance will drop a higher voltage in this circuit, which is, in effect, a voltage divider.

Start by assuming that R_2 is open. If it is, the voltage across R_3 is

$$V_3 = \left(\frac{R_3}{R_1 + R_3}\right)V_S = \left(\frac{10\text{ k}\Omega}{25\text{ k}\Omega}\right)24\text{ V} = 9.6\text{ V}$$

This calculation shows that R_2 is open. Replace R_2 with a new resistor.

Related Exercise What would be the voltmeter reading if R_3 were open in Figure 6–56? If R_1 were open?

EXAMPLE 6–19 Suppose that you measure 24 V with the voltmeter in Figure 6–57. Determine if there is a fault, and, if there is, identify it.

FIGURE 6–57

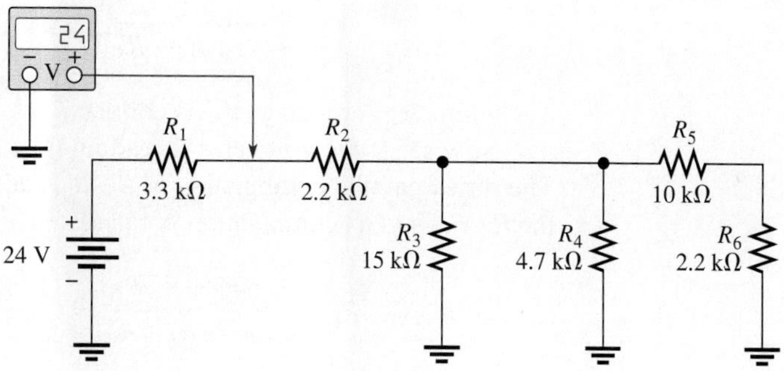

Solution There is no voltage drop across R_1 because both sides of the resistor are at +24 V. Either no current is flowing through R_1 from the source, which tells us that R_2 is open in the circuit, or R_1 is shorted.

If R_1 were open, the meter in Figure 6–57 would not read 24 V. The most logical failure is an open R_2. If it is open, then no current will flow from the source. To verify this, measure across R_2 with the voltmeter as shown in Figure 6-58. If R_2 is open, the meter will indicate 24 V. The right side of R_2 will be at zero volts because no current is flowing through any of the other resistors to cause a voltage drop.

FIGURE 6–58

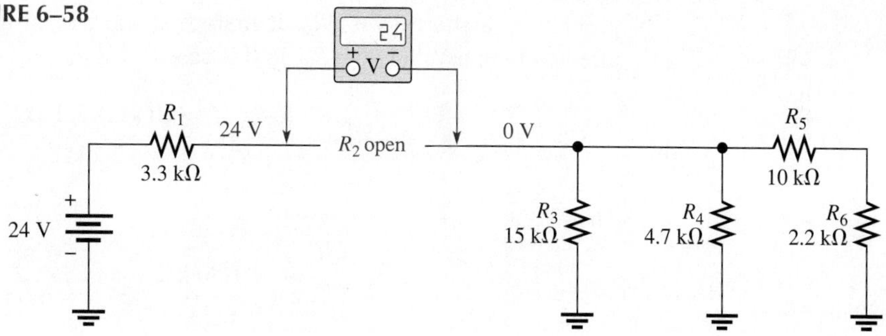

Related Exercise What would be the voltage across an open R_5 in Figure 6–57 assuming no other faults?

EXAMPLE 6–20 The two voltmeters in Figure 6–59 indicate the voltages shown. Determine if there are any opens or shorts in the circuit and, if so, where they are located.

FIGURE 6–59

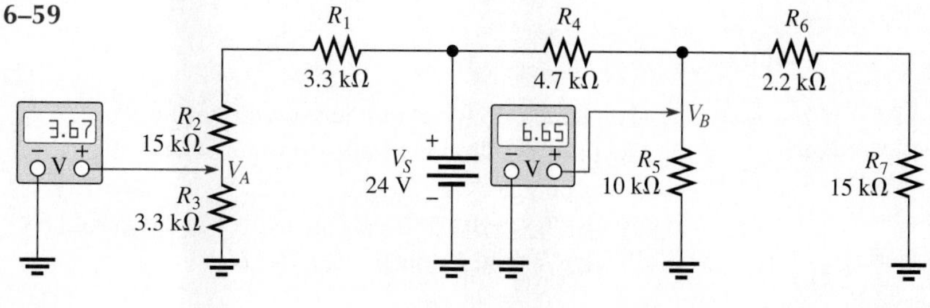

Solution First determine if the voltmeter readings are correct. R_1, R_2, and R_3 act as a voltage divider. The voltage across R_3 (V_A) is calculated as follows:

$$V_A = \left(\frac{R_3}{R_1 + R_2 + R_3}\right)V_S = \left(\frac{3.3\text{ k}\Omega}{21.6\text{ k}\Omega}\right)24\text{ V} = 3.67\text{ V}$$

The voltmeter A reading (V_A) is correct.

Now see if the voltmeter B reading (V_B) is correct. $R_6 + R_7$ is in parallel with R_5. The series-parallel combination of R_5, R_6, and R_7 is in series with R_4. The resistance of the R_5, R_6, and R_7 combination is calculated as follows:

$$R_{5\text{-}6\text{-}7} = \frac{(R_6 + R_7)R_5}{R_5 + R_6 + R_7} = \frac{(17.2\text{ k}\Omega)(10\text{ k}\Omega)}{27.2\text{ k}\Omega} = 6.32\text{ k}\Omega$$

$R_{5\text{-}6\text{-}7}$ and R_4 form a voltage divider, and voltmeter B is measuring the voltage across $R_{5\text{-}6\text{-}7}$. Is it correct? Check as follows:

$$V_B = \left(\frac{R_{5\text{-}6\text{-}7}}{R_4 + R_{5\text{-}6\text{-}7}}\right)V_S = \left(\frac{6.32\text{ k}\Omega}{11\text{ k}\Omega}\right)24\text{ V} = 13.8\text{ V}$$

Thus, the actual measured voltage (6.65 V) at this point is incorrect. Some further thought will help to isolate the problem.

R_4 is not open, because if it were, the meter would read 0 V. If there were a short across it, the meter would read 24 V. Since the actual voltage is much less than it should be, $R_{5\text{-}6\text{-}7}$ must be less than the calculated value of 6.32 kΩ. The most likely problem is a short across R_7. If there is a short from the top of R_7 to ground, R_6 is effectively in parallel with R_5. In this case,

$$R_5 \parallel R_6 = \frac{R_5R_6}{R_5 + R_6} = \frac{(10\text{ k}\Omega)(2.2\text{ k}\Omega)}{12.2\text{ k}\Omega} = 1.80\text{ k}\Omega$$

Then V_B is

$$V_B = \left(\frac{1.80\text{ k}\Omega}{6.5\text{ k}\Omega}\right)24\text{ V} = 6.65\text{ V}$$

This value for V_B agrees with the voltmeter B reading. So there is a short across R_7. If this were an actual circuit, you would try to find the physical cause of the short.

Related Exercise If R_2 in Figure 6–59 were shorted, what would voltmeter A read? What would voltmeter B read?

SECTION 6–8 REVIEW

1. Name two types of common circuit faults.
2. For the following faults in Figure 6–60, determine what voltage would be measured at point *A*:
   (a) no faults **(b)** R_1 open **(c)** short across R_5
   (d) R_3 and R_4 open **(e)** R_2 open

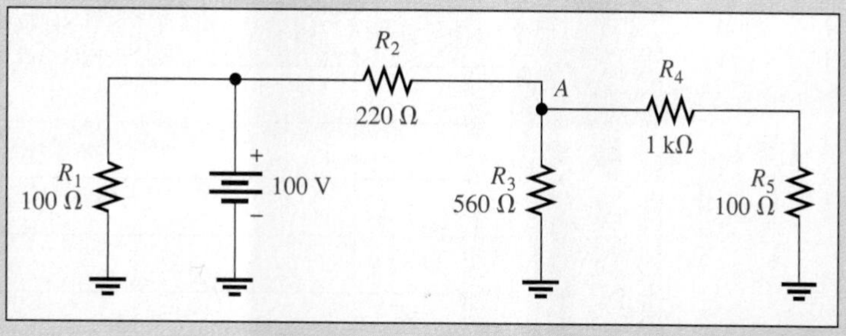

FIGURE 6–60

3. In Figure 6–61 one of the resistors in the circuit is open. Based on the meter reading, determine which is the bad resistor.

FIGURE 6–61

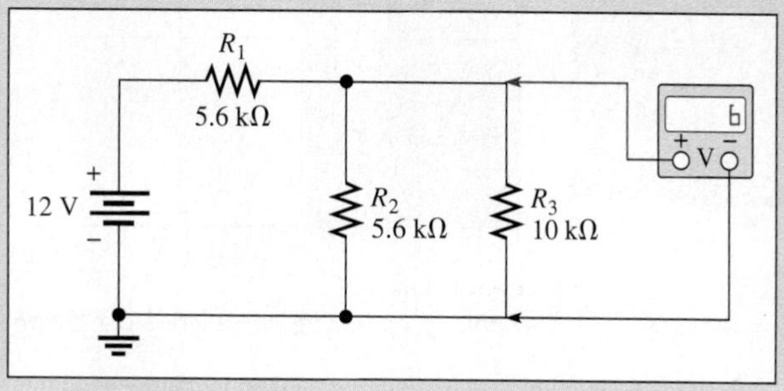

6–9 ■ APPLICATION ASSIGNMENT

A voltage divider with three output voltages has been designed and constructed on a PC board. The voltage divider is to be used as part of a portable power supply unit for supplying up to three different reference voltages to measuring instruments in the field. The power supply unit contains a battery pack combined with a voltage regulator that produces a constant +12 V to the voltage-divider circuit board. In this assignment, you will apply your knowledge of loaded voltage dividers, Kirchhoff's laws and Ohm's law to determine the operating parameters of the voltage divider in terms of voltages and currents for all possible load configurations. You will also troubleshoot the circuit for various malfunctions.

Step 1: Draw the Schematic of the Voltage Divider

Draw the schematic and label the resistor values for the circuit board in Figure 6–62.

Step 2: Connect the 12 V Power Supply

Specify how to connect a 12 V power supply to the circuit board so that all resistors are in series and pin 2 has the highest output voltage.

Step 3: Determine the Unloaded Output Voltages

Calculate each of the output voltages with no loads connected. Add these voltage values to a copy of the table in Figure 6–63.

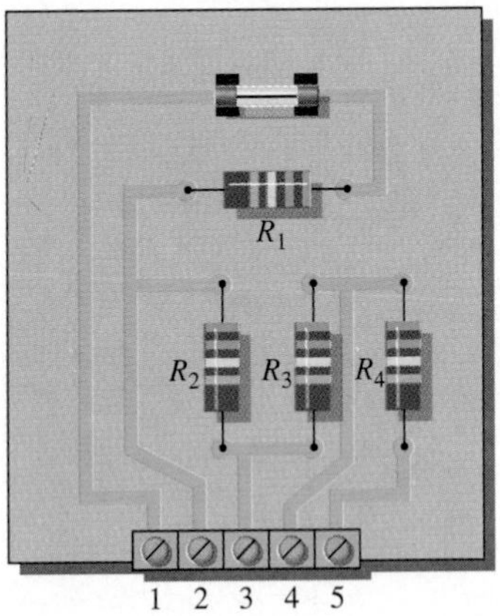

FIGURE 6–62
Voltage-divider circuit board.

10MΩ Load	$V_{OUT\,(2)}$	$V_{OUT\,(3)}$	$V_{OUT\,(4)}$	% Deviation	$I_{LOAD\,(2)}$	$I_{LOAD\,(3)}$	$I_{LOAD\,(4)}$
None							
Pin 2							
Pin 3							
Pin 4							
Pins 2 and 3				2			
				3			
Pins 2 and 4				2			
				4			
Pins 3 and 4				3			
				4			
Pins 2, 3, and 4				2			
				3			
				4			

FIGURE 6–63
Table of operating parameters for the power supply voltage divider.

Step 4: Determine the Loaded Output Voltages

The instruments to be connected to the voltage divider each have a 10 MΩ input resistance. This means that when an instrument is connected to a voltage-divider output there is effectively a 10 MΩ resistor from that output to ground (negative side of source). Determine the output voltage across each load for the following load combinations and add these voltage values to a copy of the table in Figure 6–63.

1. A 10 MΩ load connected to pin 2.
2. A 10 MΩ load connected to pin 3.
3. A 10 MΩ load connected to pin 4.
4. 10 MΩ loads connected to pin 2 and pin 3.
5. 10 MΩ loads connected to pin 2 and pin 4.
6. 10 MΩ loads connected to pin 3 and pin 4.
7. 10 MΩ loads connected to pin 2, pin 3, and pin 4.

Step 5: Determine the Percent Deviation of the Output Voltages

Calculate how much each loaded output voltage deviates from its unloaded value for each of the load configurations listed in Step 4 and express as a percentage using the following formula.

$$\text{Percent deviation} = \left(\frac{V_{OUT(unloaded)} - V_{OUT(loaded)}}{V_{OUT(unloaded)}}\right)100\%$$

Add the values to the table in Figure 6–63.

Step 6: Determine the Load Currents

Calculate the current to each 10 MΩ load for each of the load configurations listed in Step 4. Add these values to the table in Figure 6–63.

Step 7: Specify a Minimum Value for the Fuse

Step 8: Troubleshoot the Circuit Board

The voltage-divider circuit board is connected to a 12 V power supply and to the three instruments to which it provides reference voltages, as shown in Figure 6–64. Voltages at each of the numbered test points are measured with a voltmeter in each of eight different cases. For each case, determine the problem indicated by the voltage measurements.

FIGURE 6–64

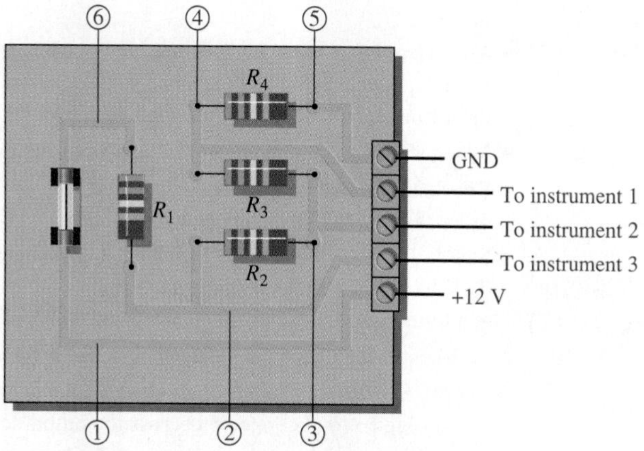

The following voltmeter readings are taken at test points 1 through 6 going from left to right. The readings are in volts.

	①	②	③	④	⑤	⑥
Case 1	0	0	0	0	0	0
Case 2	12	0	0	0	0	0
Case 3	12	0	0	0	0	12
Case 4	12	11.6	0	0	0	12
Case 5	12	11.3	10.9	0	0	12
Case 6	12	11	10.3	10	0	12
Case 7	12	5.9	0	0	0	12
Case 8	12	7.8	3.8	0	0	12

SECTION 6–9 REVIEW

1. If the portable unit covered in this section is to supply reference voltages to all three instruments, how many days can a 100 mAh battery be used?
2. Can ⅛ W resistors be used on the voltage-divider board?
3. If ⅛ W resistors are used, will an output shorted to ground cause any of the resistors to overheat due to excessive power?

■ SUMMARY

- A series-parallel circuit is a combination of both series paths and parallel paths.
- To determine total resistance in a series-parallel circuit, identify the series and parallel relationships, and then apply the formulas for series resistance and parallel resistance from Chapters 4 and 5.
- To find the total current, divide the total voltage by the total resistance.
- To determine branch currents, apply the current-divider formula, Kirchhoff's current law, or Ohm's law. Consider each circuit problem individually to determine the most appropriate method.
- To determine voltage drops across any portion of a series-parallel circuit, use the voltage-divider formula, Kirchhoff's voltage law, or Ohm's law. Consider each circuit problem individually to determine the most appropriate method.
- When a load resistor is connected across a voltage-divider output, the output voltage decreases.
- The load resistor should be large compared to the resistance across which it is connected, in order that the loading effect may be minimized. A *10-times* value is sometimes used as a rule of thumb, but the value depends on the accuracy required for the output voltage.
- To find any current or voltage in a circuit with two or more voltage sources, take the sources one at a time using the superposition method.
- A Wheatstone bridge can be used to measure an unknown resistance.
- A bridge is balanced when the output voltage is zero. The balanced condition produces zero current through a load connected across the output terminals of the bridge.
- Any resistive circuit, no matter how complex, can be replaced by its Thevenin equivalent.
- The Thevenin equivalent circuit is made up of an equivalent resistance (R_{TH}) in series with an equivalent voltage source (V_{TH}).
- Open circuits and short circuits are typical faults.
- Resistors normally open when they burn out.

■ GLOSSARY

Bleeder current The current left after the total load current is subtracted from the total current into the circuit.

Load An element (resistor or other component) connected across the output terminals of a circuit that draws current from the circuit.

Sensitivity factor The ohms-per-volt rating of a voltmeter.

Superposition A method for analyzing circuits with two or more sources by examining the effects of each source by itself and then combining the effects.

Terminal equivalency A condition that occurs when two circuits produce the same load voltage and load current when the same value of load resistance is connected to either circuit.

Thevenin's theorem A circuit theorem that provides for reducing any resistive circuit to a single equivalent voltage source in series with an equivalent resistance.

■ FORMULAS

(6–1)	$I_{BLEEDER} = I_T - I_{RL1} - I_{RL2}$	Bleeder current
(6–2)	$R_{UNK} = R_V\left(\frac{R_2}{R_4}\right)$	Unknown resistance in a Wheatstone bridge

■ SELF-TEST

1. Which of the following statements are true concerning Figure 6–65?
   (a) R_1 and R_2 are in series with R_3, R_4, and R_5
   (b) R_1 and R_2 are in series
   (c) R_3, R_4, and R_5 are in parallel
   (d) The series combination of R_1 and R_2 is in parallel with the series combination of R_3, R_4, and R_5
   (e) both (b) and (d)

FIGURE 6–65

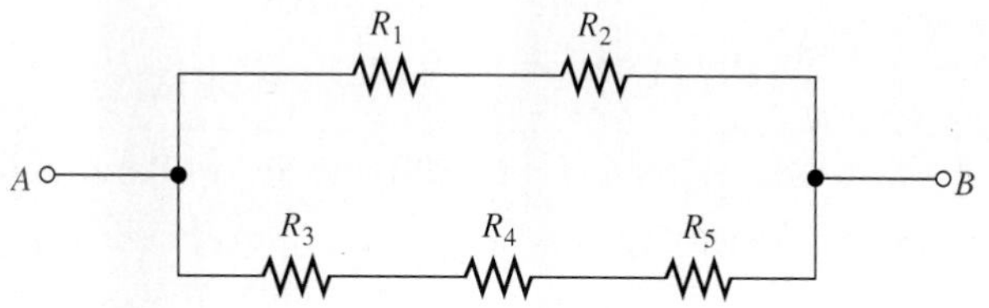

2. The total resistance of Figure 6–65 can be found with which of the following formulas?
   (a) $R_1 + R_2 + R_3 \| R_4 \| R_5$ (b) $R_1 \| R_2 + R_3 \| R_4 \| R_5$
   (c) $(R_1 + R_2) \| (R_3 + R_4 + R_5)$ (d) none of the above
3. If all of the resistors in Figure 6–65 have the same value, when voltage is applied across terminals A and B, the current is
   (a) greatest in R_5 (b) greatest in R_3, R_4, and R_5
   (c) greatest in R_1 and R_2 (d) the same in all the resistors
4. Two 1 kΩ resistors are in series and this series combination is in parallel with a 2.2 kΩ resistor. The voltage across one of the 1 kΩ resistors is 6 V. The voltage across the 2.2 kΩ resistor is
   (a) 6 V (b) 3 V (c) 12 V (d) 13.2 V
5. The parallel combination of a 330 Ω resistor and a 470 Ω resistor is in series with the parallel combination of four 1 kΩ resistors. A 100 V source is connected across the circuit. The resistor with the most current has a value of
   (a) 1 kΩ (b) 330 Ω (c) 470 Ω
6. In the circuit described in Question 5, the resistor(s) with the most voltage has a value of
   (a) 1 kΩ (b) 470 Ω (c) 330 Ω
7. In the circuit described in Question 5, the percentage of the total current through any single 1 kΩ resistor is
   (a) 100% (b) 25% (c) 50% (d) 31.25%
8. The output of a certain voltage divider is 9 V with no load. When a load is connected, the output voltage
   (a) increases (b) decreases (c) remains the same (d) becomes zero
9. A certain voltage divider consists of two 10 kΩ resistors in series. Which of the following load resistors will have the most effect on the output voltage?
   (a) 1 MΩ (b) 20 kΩ (c) 100 kΩ (d) 10 kΩ
10. When a load resistance is connected to the output of a voltage-divider circuit, the current drawn from the source
    (a) decreases (b) increases (c) remains the same (d) is cut off
11. The output voltage of a balanced Wheatstone bridge is
    (a) equal to the source voltage
    (b) equal to zero
    (c) dependent on all of the resistor values in the bridge
    (d) dependent on the value of the unknown resistor
12. The primary method of analyzing a circuit with two or more voltage sources is usually
    (a) Thevenin's theorem (b) Ohm's law
    (c) superposition (d) Kirchhoff's law
13. In a certain two-source circuit, one source acting alone produces 10 mA through a given branch. The other source acting alone produces 8 mA in the opposite direction through the same branch. The total current through the branch is
    (a) 10 mA (b) 8 mA (c) 18 mA (d) 2 mA

14. A Thevenin equivalent circuit consists of
   (a) a voltage source in series with a resistance
   (b) a voltage source in parallel with a resistance
   (c) a current source in parallel with a resistance
   (d) two voltage sources and a resistance
15. You are measuring the voltage at a given point in a circuit that has very high resistance values and the measured voltage is a little lower than it should be. This is possibly because of
   (a) one or more of the resistance values being off
   (b) the loading effect of the voltmeter
   (c) the source voltage is too low
   (d) all the above

■ BASIC PROBLEMS

SECTION 6–1 Identifying Series-Parallel Relationships

1. Identify the series-parallel relationships in Figure 6–66 as seen from the source terminals.

FIGURE 6–66

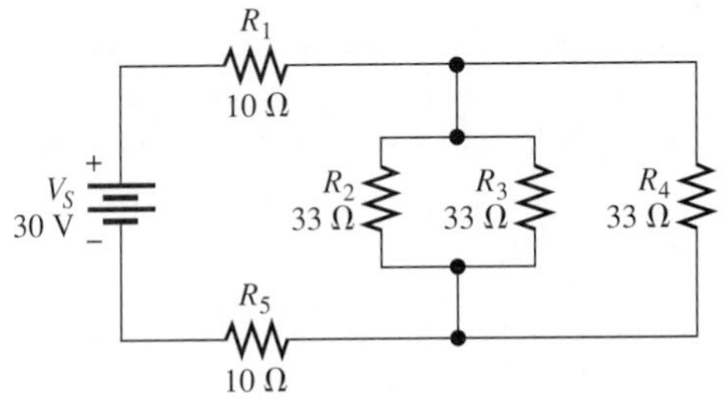

2. Visualize and sketch the following series-parallel combinations:
   (a) R_1 in series with the parallel combination of R_2 and R_3
   (b) R_1 in parallel with the series combination of R_2 and R_3
   (c) R_1 in parallel with a branch containing R_2 in series with a parallel combination of four other resistors
3. Visualize and sketch the following series-parallel circuits:
   (a) a parallel combination of three branches, each containing two series resistors
   (b) a series combination of three parallel circuits, each containing two parallel resistors
4. In each circuit of Figure 6–67 identify the series and parallel relationships of the resistors viewed from the source.

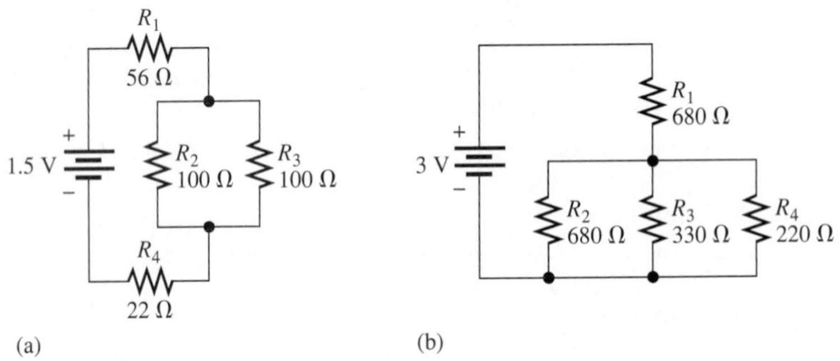

FIGURE 6–67

SECTION 6–2 Analysis of Series-Parallel Circuits

5. A certain circuit is composed of two parallel resistors. The total resistance is 667 Ω. One of the resistors is 1 kΩ. What is the other resistor?
6. For the circuit in Figure 6–68, determine the total resistance between points A and B.

7. Determine the total resistance for each circuit in Figure 6–67.
8. Determine the current through each resistor in Figure 6–66; then calculate each voltage drop.
9. Determine the current through each resistor in both circuits of Figure 6–67; then calculate each voltage drop.
10. In Figure 6–69, find the following:
    (a) total resistance between terminals A and B
    (b) total current drawn from a 6 V source connected from A to B
    (c) current through R_5
    (d) voltage across R_2

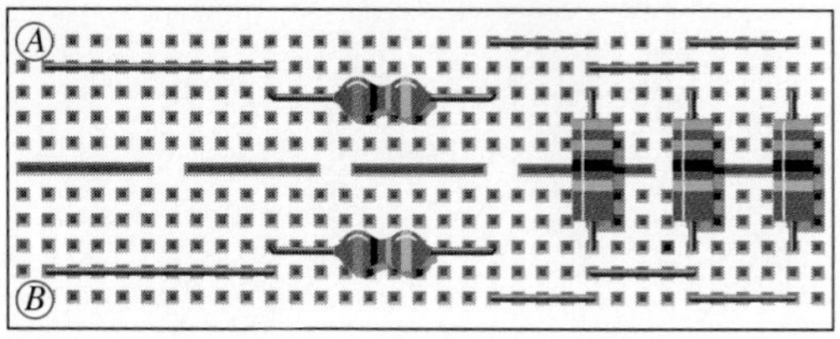

FIGURE 6–68

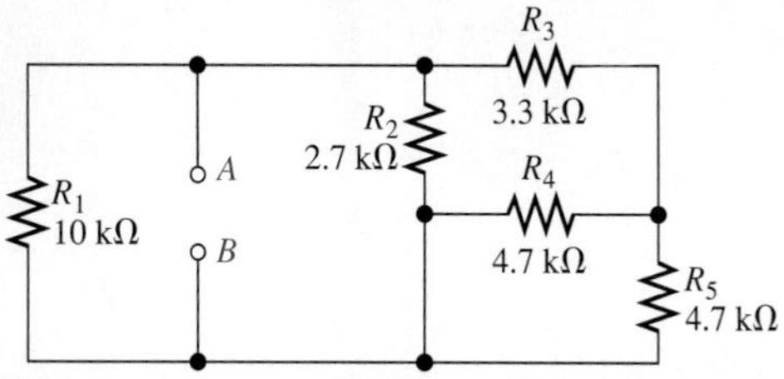

FIGURE 6–69

SECTION 6–3 Voltage Dividers with Resistive Loads

11. A voltage divider consists of two 56 kΩ resistors and a 15 V source. Calculate the unloaded output voltage. What will the output voltage be if a load resistor of 1 MΩ is connected to the output?
12. A 12 V battery output is divided down to obtain two output voltages. Three 3.3 kΩ resistors are used to provide the two outputs with only one output at a time loaded with 10 kΩ. Determine the output voltages in both cases.
13. Which will cause a smaller decrease in output voltage for a given voltage divider, a 10 kΩ load or a 56 kΩ load?
14. In Figure 6–70, determine the current drain on the battery with no load across the output terminals. With a 10 kΩ load, what is the battery current?

FIGURE 6–70

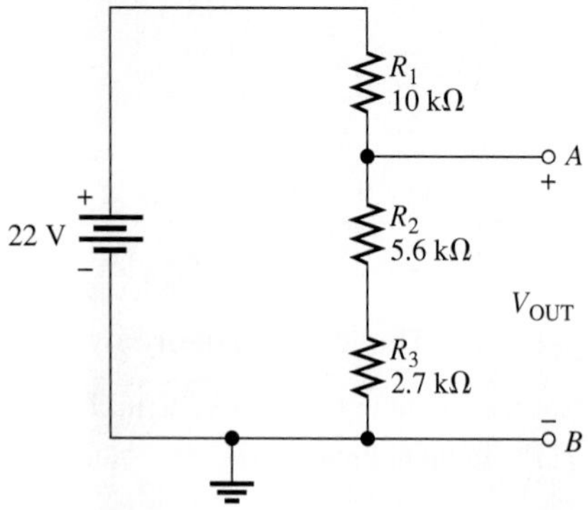

SECTION 6–4 Loading Effect of a Voltmeter

15. On which one of the following voltage range settings will a voltmeter present the minimum load on a circuit?
    (a) 1 V (b) 10 V (c) 100 V (d) 1000 V
16. Determine the internal resistance of a 20,000 Ω/V voltmeter on each of the following range settings.
    (a) 0.5 V (b) 1 V (c) 5 V (d) 50 V (e) 100 V (f) 1000 V

17. The voltmeter described in Problem 16 is used to measure the voltage across R_4 in Figure 6–67(a).
  (a) What range should be used?
  (b) How much less is the voltage measured by the meter than the actual voltage?

18. Repeat Problem 17 if the voltmeter is used to measure the voltage across R_4 in the circuit of Figure 6–67(b).

SECTION 6–5 The Wheatstone Bridge

19. A resistor of unknown value is connected to a Wheatstone bridge circuit. The bridge parameters for a balanced condition are set as follows: $R_V = 18$ kΩ and $R_2/R_4 = 0.02$. What is R_{UNK}?

20. A bridge network is shown in Figure 6–71. To what value must R_V be set in order to balance the bridge?

FIGURE 6–71

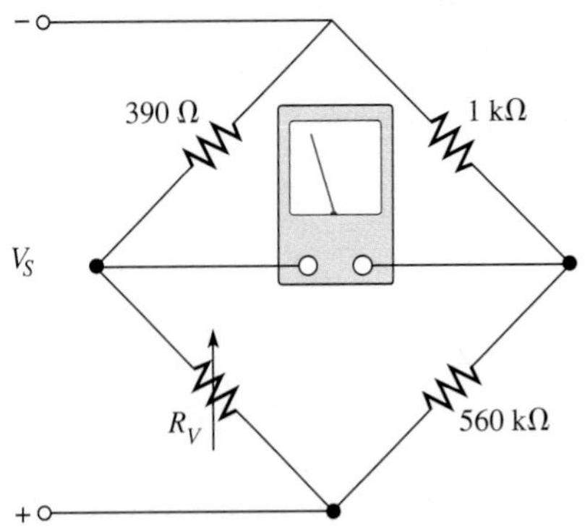

SECTION 6–6 Circuits with More Than One Voltage Source

21. In Figure 6–72, use the superposition method to find the total current in R_3.

22. In Figure 6–72, what is the total current through R_2?

FIGURE 6–72

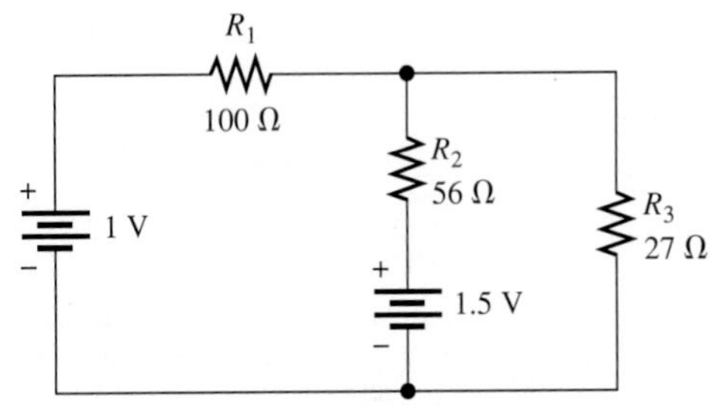

SECTION 6–7 Thevenin's Theorem

23. Reduce the circuit in Figure 6–73 to its Thevenin equivalent as viewed from terminals A and B.

24. For each circuit in Figure 6–74, determine the Thevenin equivalent as seen by R_L.

FIGURE 6–73

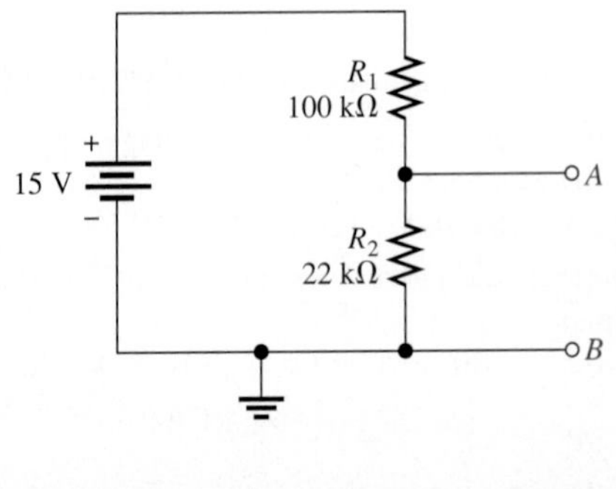

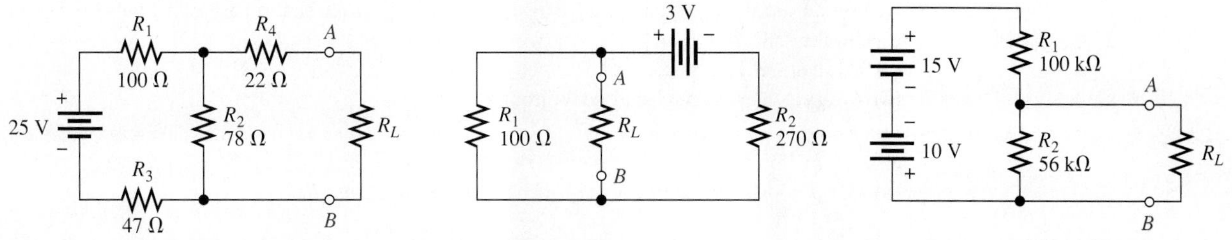

(a) (b) (c)

FIGURE 6–74

SECTION 6–8 Troubleshooting

25. Is the voltmeter reading in Figure 6–75 correct? If not, what is the problem?

26. If R_2 in Figure 6–76 opens, what voltages will be read at points A, B, and C?

27. Check the meter readings in Figure 6–77 and locate any fault that may exist.

FIGURE 6–75

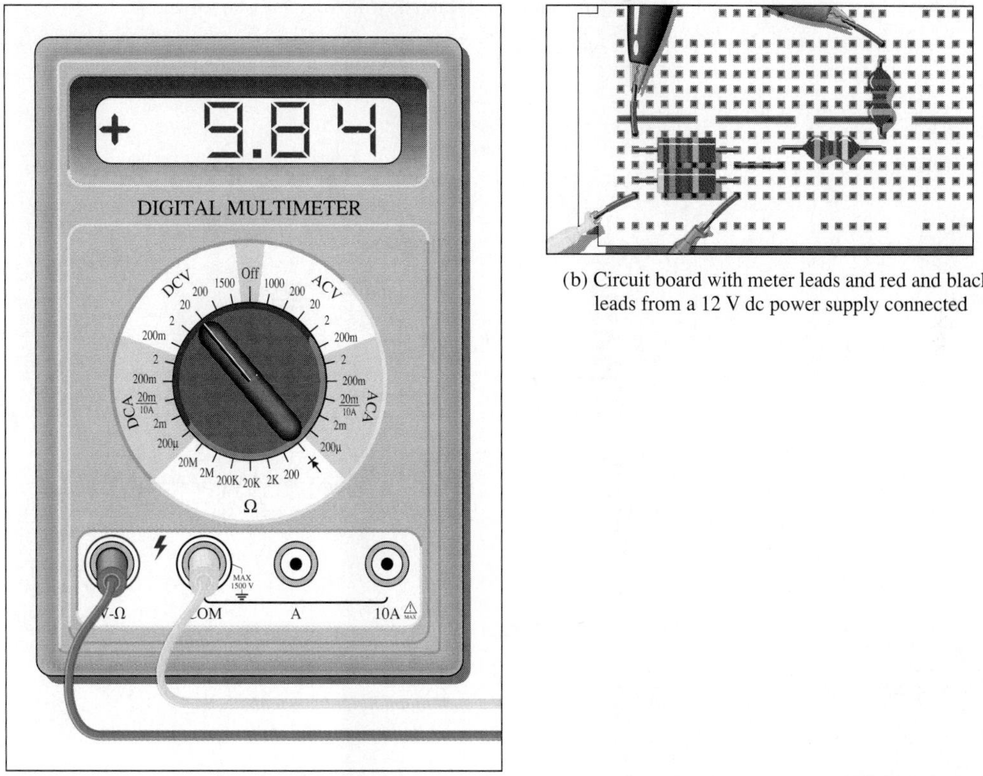

(b) Circuit board with meter leads and red and black leads from a 12 V dc power supply connected

(a) Meter with leads going to circuit board

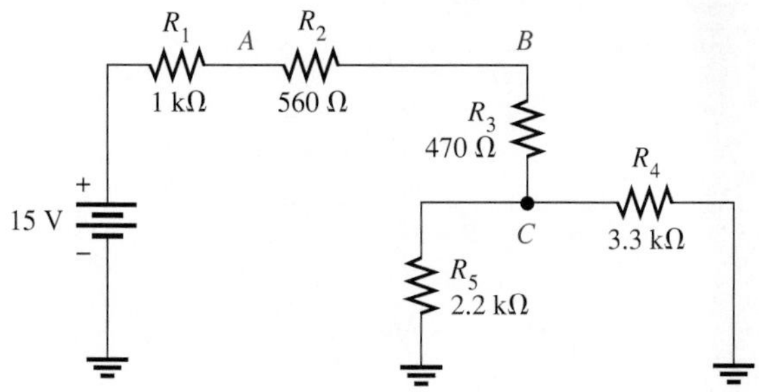

FIGURE 6–76

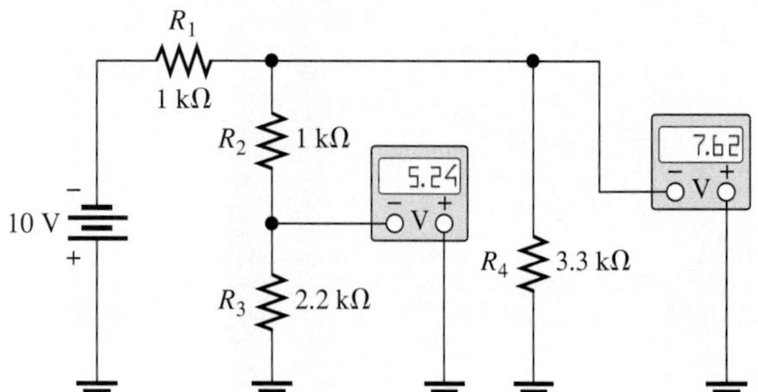

FIGURE 6–77

28. Determine the voltage you would expect to measure across each resistor in Figure 6–76 for each of the following faults:
(a) R_1 open **(b)** R_3 open **(c)** R_4 open
(d) R_5 open **(e)** point C shorted to ground

29. Determine the voltage you would expect to measure across each resistor in Figure 6–77 for each of the following faults:
(a) R_1 open **(b)** R_2 open **(c)** R_3 open **(d)** a short across R_4

■ ADVANCED PROBLEMS

30. In each circuit of Figure 6–78, identify the series and parallel relationships of the resistors viewed from the source.

31. Draw the schematic of the PC board layout in Figure 6–79 showing resistor values and identify the series-parallel relationships. Which resistors, if any, can be removed with no effect on R_T?

32. For the circuit shown in Figure 6–80, calculate the following:
(a) total resistance across the source **(b)** total current from the source
(c) current through the 910 Ω resistor **(d)** voltage from point A to point B

33. Determine the total resistance and the voltage at points A, B, and C in the circuit of Figure 6–81.

34. Determine the total resistance between terminals A and B of the circuit in Figure 6–82. Also calculate the current in each branch with 10 V between A and B.

35. What is the voltage across each resistor in Figure 6–82?

FIGURE 6–78

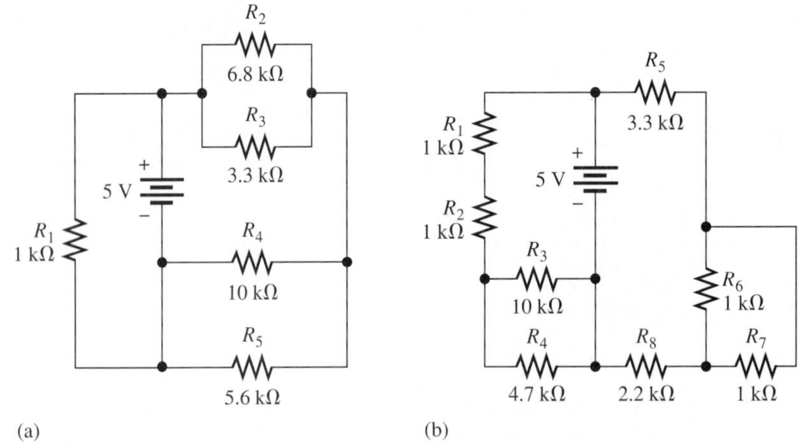

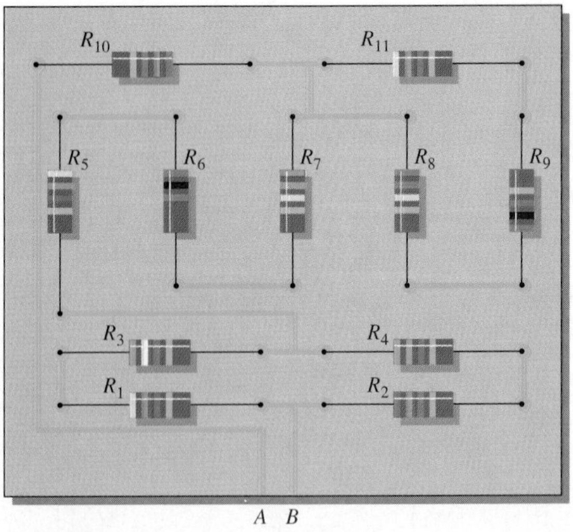

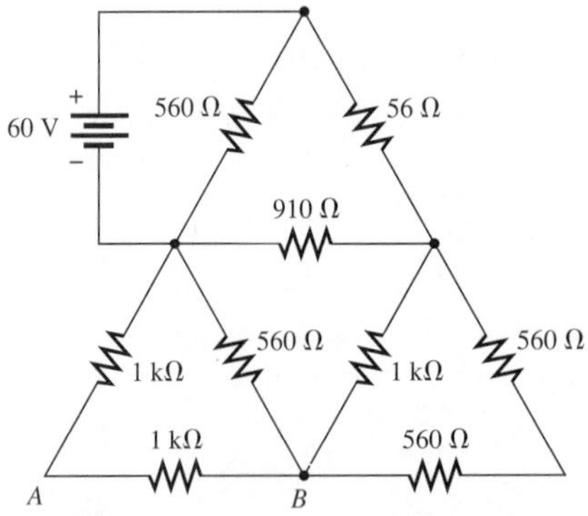

FIGURE 6–79

FIGURE 6–80

FIGURE 6–81

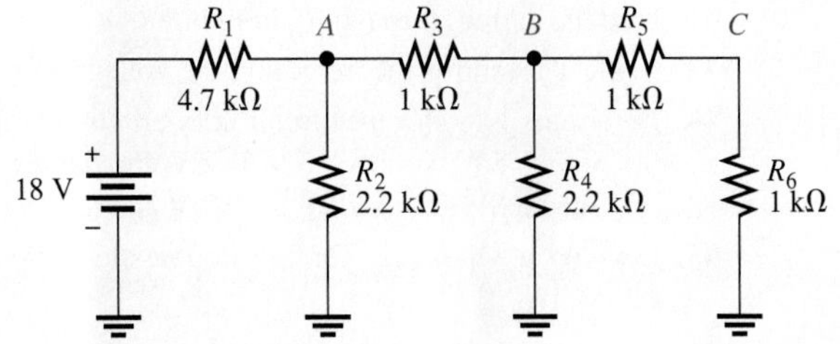

FIGURE 6–82

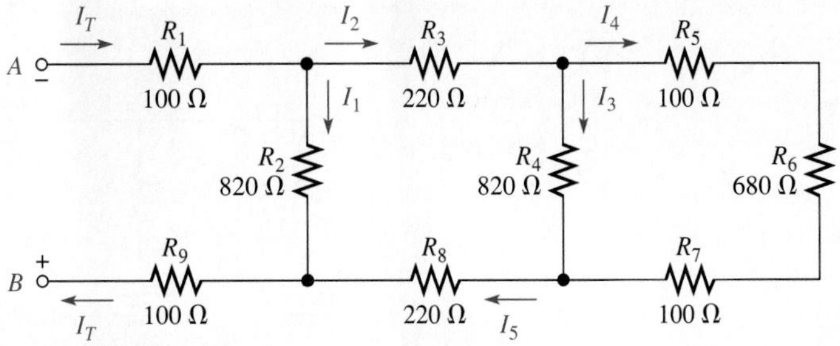

36. Determine the voltage, V_{AB}, in Figure 6–83.

37. Find the value of R_2 in Figure 6–84.

38. Determine the total resistance and the voltage at points A, B, and C in the circuit of Figure 6–85.

39. Develop a voltage divider to provide a 6 V output with no load and a minimum of 5.5 V across a 1 kΩ load. The source voltage is 24 V, and the unloaded current is not to exceed 100 mA.

40. Determine the resistance values for a voltage divider that must meet the following specifications: The current under an unloaded condition is not to exceed 5 mA. The source voltage is to be 10 V. A 5 V output and a 2.5 V output are required. Sketch the circuit. Determine the effect on the output voltages if a 1 kΩ load is connected to each output.

41. Using the superposition method, calculate the current in the right-most branch of Figure 6–86.

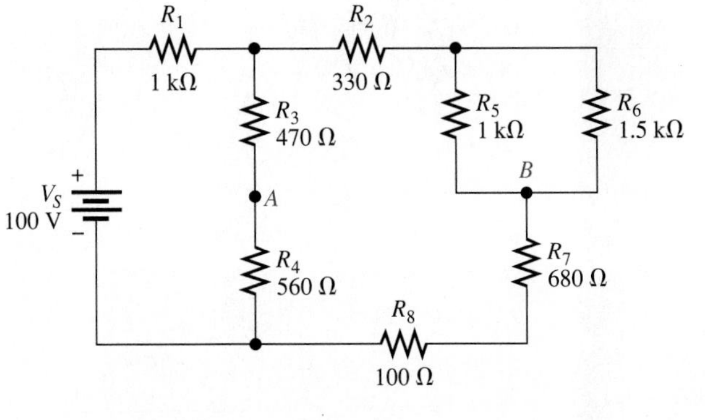

FIGURE 6–83

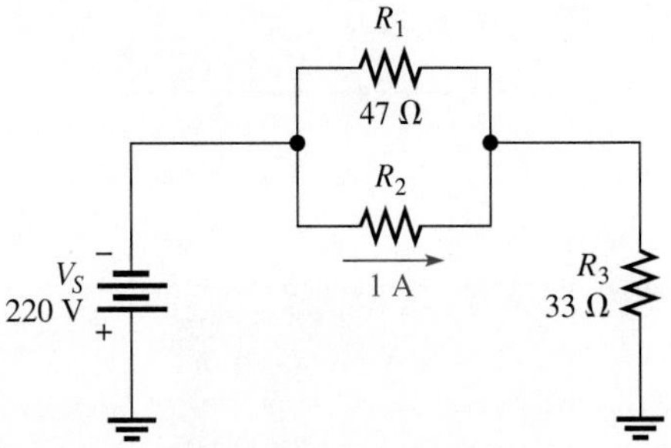

FIGURE 6–84

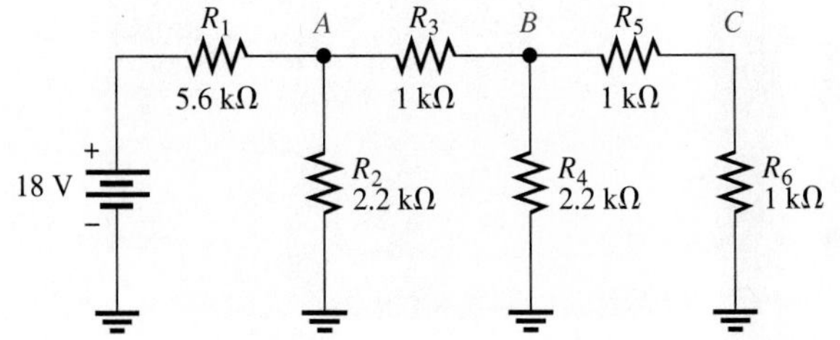

FIGURE 6–85

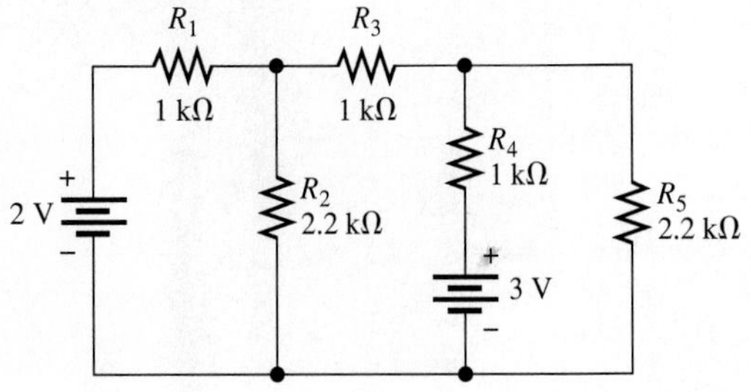

FIGURE 6–86

42. Find the current through R_L in Figure 6–87.

43. Using Thevenin's theorem, find the voltage across R_4 in Figure 6–88.

44. Determine V_{OUT} for the circuit in Figure 6–89 for the following conditions:
(**a**) Switch SW_2 connected to +12 V and the rest to ground
(**b**) Switch SW_1 connected to +12 V and the rest to ground

45. Develop a schematic for the double-sided PC board in Figure 6–90 and label the resistor values.

46. Lay out a PC board for the circuit in Figure 6–78(b). The battery is to be connected external to the board.

FIGURE 6–87

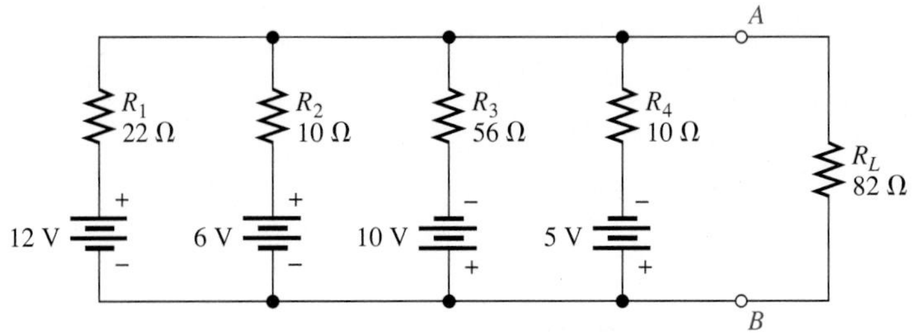

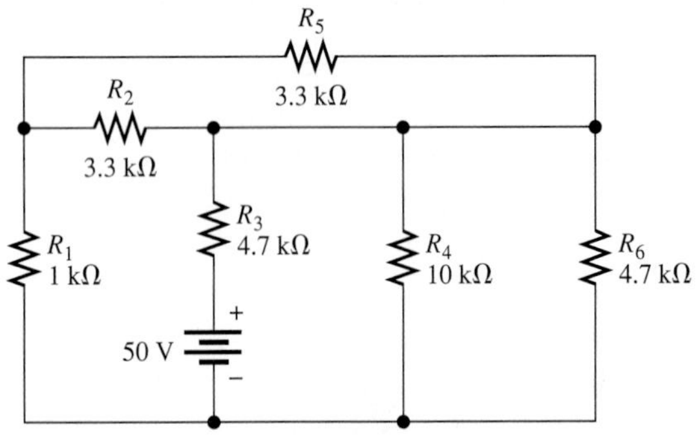

FIGURE 6–88

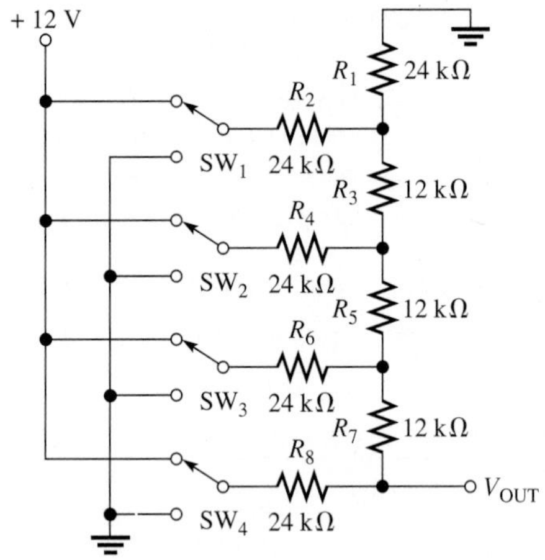

FIGURE 6–89

FIGURE 6–90

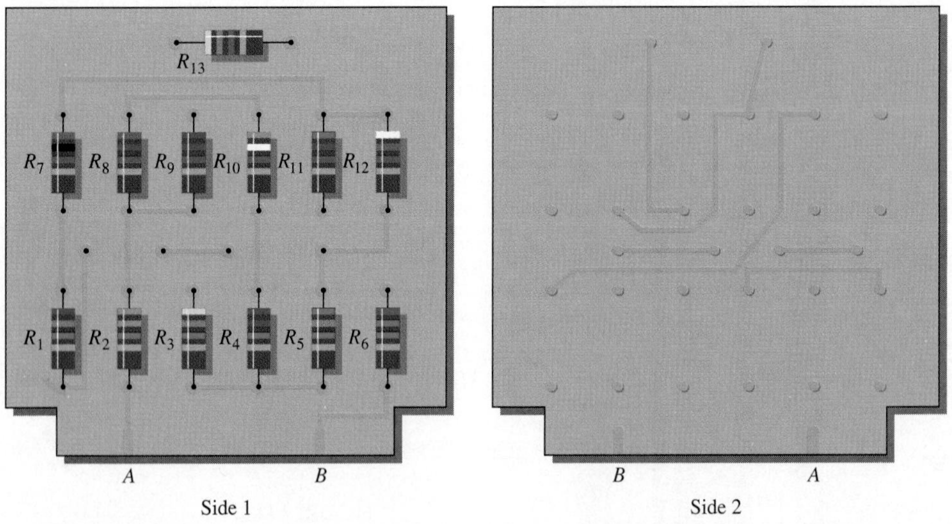

47. The voltage divider in Figure 6–91 has a switched load. Determine the voltage at each tap (V_1, V_2, and V_3) for each position of the switch.

48. Figure 6–92 shows a dc biasing arrangement for a field-effect transistor amplifier. Biasing is a common method for setting up certain dc voltage levels required for proper amplifier operation. Although you may not be familiar with transistor amplifiers at this point, the dc voltages and currents in the circuit can be determined using methods that you already know.
(a) Find V_G and V_S with respect to ground **(b)** Determine I_1, I_2, I_D, and I_S
(c) Find V_{DS} and V_{DG}

49. Look at the voltmeters in Figure 6–93 and determine if there is a fault in the circuit. If there is a fault, identify it.

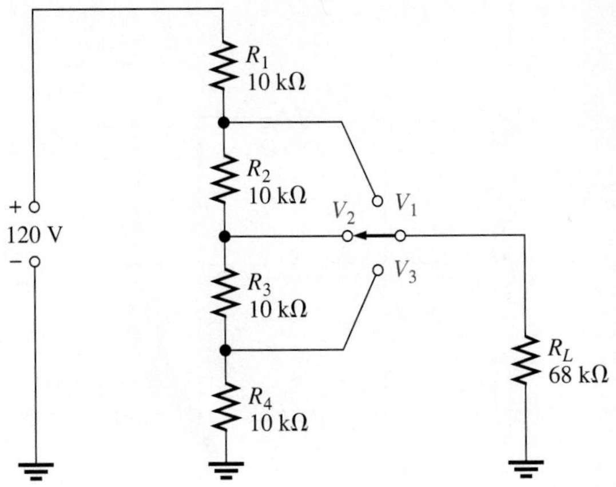

FIGURE 6–91

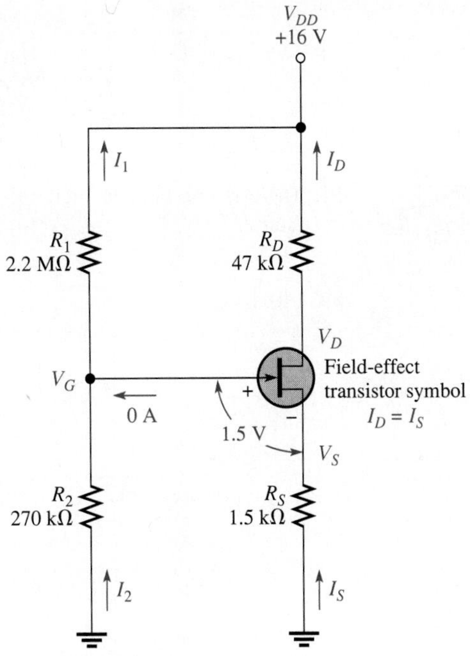

FIGURE 6–92

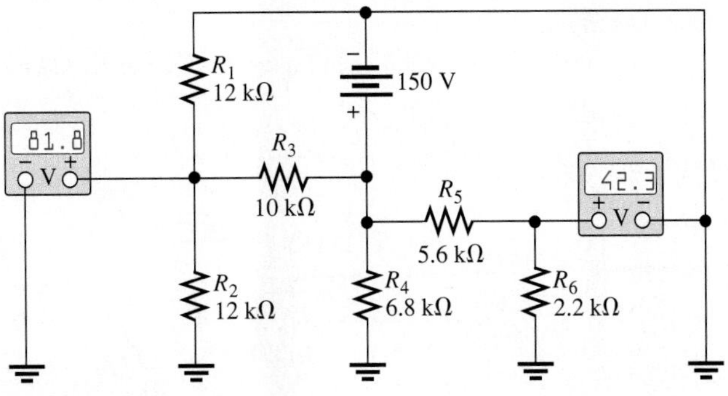

FIGURE 6–93

50. Are the voltmeter readings in Figure 6–94 correct?

FIGURE 6–94

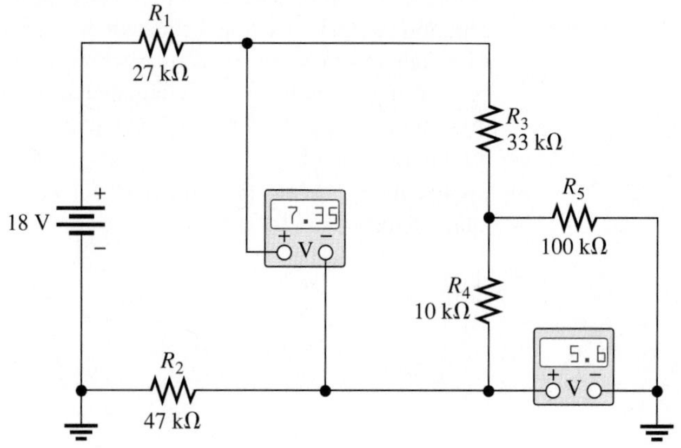

51. There is one fault in Figure 6–95. Based on the voltmeter indications, determine what the fault is.

FIGURE 6–95

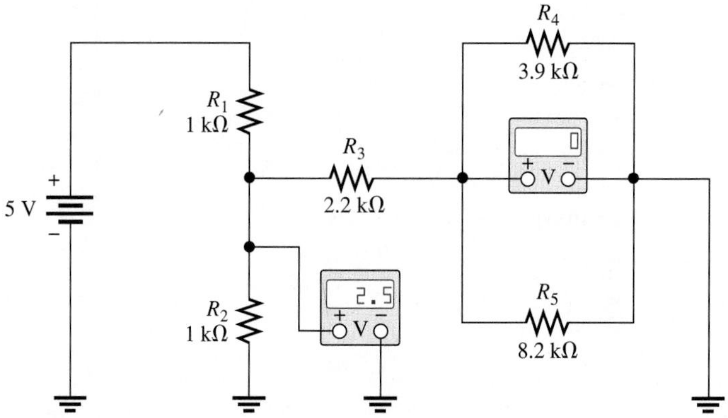

52. Look at the voltmeters in Figure 6–96 and determine if there is a fault in the circuit. If there is a fault, identify it.

53. Determine the voltmeter readings in Figure 6–96 if the 4.7 kΩ resistor opens.

FIGURE 6–96

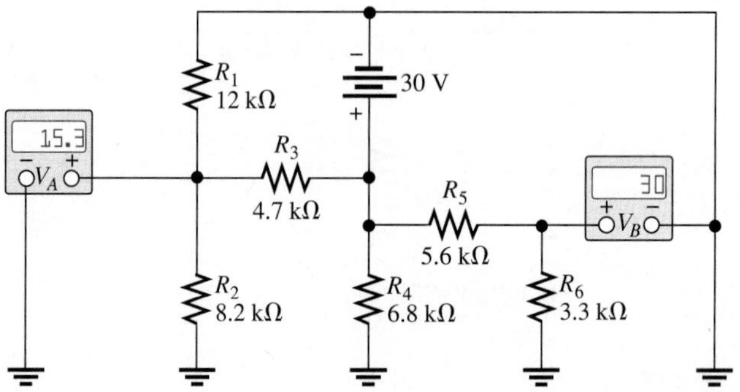

■ ANSWERS TO SECTION REVIEWS

Section 6–1

1. See Figure 6–97.

FIGURE 6–97

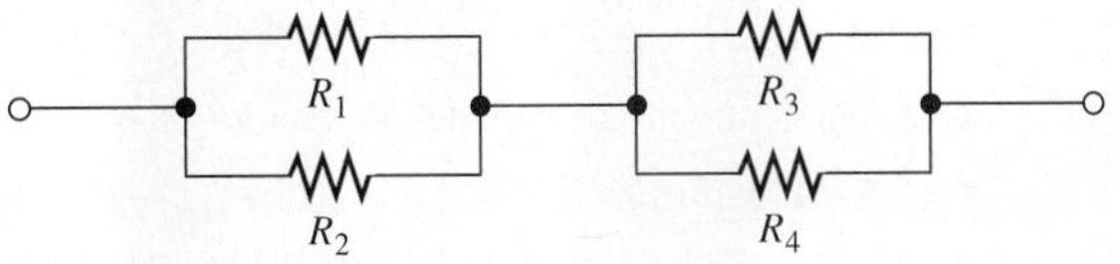

2. R_1 and R_2 are in series with the parallel combination of R_3 and R_4.
3. All resistors are in parallel.
4. R_1 and R_2 are in parallel; R_3 and R_4 are in parallel.
5. Yes, the two parallel combinations are in series with each other.

Section 6–2

1. $R_T = R_1 + R_4 + R_2 \| R_3 = 599\ \Omega$
2. $I_3 = 11.2$ mA
3. $V_2 = I_2R_2 = 3.7$ V
4. $R_T = 89\ \Omega$; $I_T = 11.2$ mA

Section 6–3

1. The load resistor decreases the output voltage.
2. True
3. $V_{OUT(unloaded)} = 19.2$ V, $V_{OUT(loaded)} = 4.19$ V

Section 6–4

1. A voltmeter loads a circuit because the internal resistance of the meter appears in parallel with the circuit resistance across which it is connected, reducing the resistance between those two points of the circuit and drawing current from the circuit.
2. No, because the meter resistance is much larger than 1 kΩ.
3. There is a decrease in measured voltage because R_M is smaller. You can read the scale more accurately on the 3 V range.
4. $R_M = 1.5$ MΩ

Section 6–5

1. See Figure 6–98.
2. A bridge is balanced when the output voltage is zero.
3. $R_{UNK} = 15$ kΩ

FIGURE 6–98

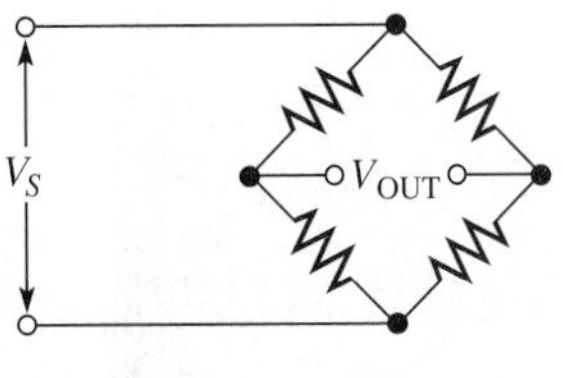

Section 6–6

1. The total current in any branch of a multiple-source linear circuit is equal to the algebraic sum of the currents due to the individual sources acting alone, with the other sources replaced by their internal resistances.
2. The superposition method allows each source to be treated independently.
3. A short simulates the internal resistance of an ideal voltage source; an open simulates the internal resistance of an ideal current source.
4. $I_{R1} = 6.67$ mA
5. The net current is in the direction of the larger current.

Section 6–7

1. A Thevenin circuit includes V_{TH} and R_{TH}.
2. See Figure 6–99.
3. V_{TH} is the open circuit voltage between two points in a circuit.
4. R_{TH} is the resistance as viewed from two points in a circuit, with all sources replaced by their internal resistances.
5. See Figure 6–100.

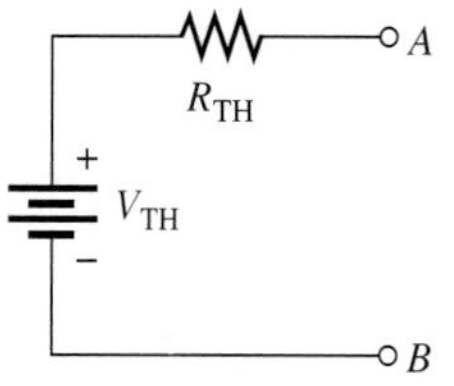

FIGURE 6–99

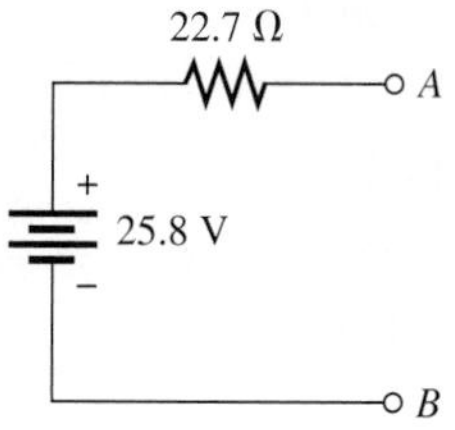

FIGURE 6–100

Section 6–8

1. Opens and shorts are two common faults.
2. **(a)** 62.8 V **(b)** 62.8 V **(c)** 62 V **(d)** 100 V **(e)** 0 V
3. The 10 kΩ resistor is open.

Section 6–9

1. The battery will last 417 days.
2. Yes, ⅛ W resistors can be used.
3. No, none of the resistors will overheat.

▪ ANSWERS TO RELATED EXERCISES FOR EXAMPLES

6–1 The added resistor would be in parallel with the combination of R_4 in series with R_2 and R_3 in parallel.

6–2 The added resistor is in parallel with R_5.

6–3 A to gnd: $R_T = R_3 \parallel (R_1 + R_2) + R_4$
B to gnd: $R_T = R_2 \parallel (R_1 + R_3) + R_4$
C to gnd: $R_T = R_4$

6–4 None

6–5 R_1 and R_4 are effectively removed from the circuit.

6–6 55.1 Ω

6–7 128.3 Ω

6–8 $I_1 = 89.3$ mA; $I_3 = 58.5$ mA; $I_T = 182.3$ mA

6–9 $V_1 = V_2 = 103$ V; $V_3 = 97$ V; $V_4 = 31.6$ V; $V_5 = 65.4$ V

6–10 $I_1 = 1.42$ mA, $P_1 = 6.67$ mW; $I_2 = 756\ \mu$A, $P_2 = 3.55$ mW; $I_3 = 2.18$ mA, $P_3 = 4.75$ mW; $I_4 = 1.13$ mA, $P_4 = 1.28$ mW; $I_5 = 1.06$ mA, $P_5 = 758\ \mu$W; $I_6 = 1.06$ mA, $P_6 = 435\ \mu$W

6–11 3.37 V

6–12 Increase the values of R_1, R_2, and R_3 proportionally.

6–13 5.34 V

6–14 321 Ω

6–15 −2 V

6–16 2.36 V; 124 Ω

6–17 1.17 mA

6–18 5.73 V; 0 V

6–19 9.46 V

6–20 $V_A = 12$ V; $V_B = 13.8$ V

7

MAGNETISM AND ELECTROMAGNETISM

This chapter is somewhat of a departure from the previous six chapters because two new concepts are introduced: magnetism and electromagnetism. The operation of many types of electrical devices is based partially on magnetic or electromagnetic principles.

■ CHAPTER OBJECTIVES

- ☐ Explain the principles of the magnetic field
- ☐ Explain the principles of electromagnetism
- ☐ Describe the principle of operation for several types of electromagnetic devices
- ☐ Explain magnetic hysteresis
- ☐ Discuss the principle of electromagnetic induction
- ☐ Describe some applications of electromagnetic induction

■ APPLICATION ASSIGNMENT

For the application assignment in this chapter, you will apply what you learn in this chapter about relays and other devices to a simple burglar alarm system. You will determine how to connect the components to form a complete system and specify a check-out procedure to make sure the system is working properly. After studying this chapter, you should be able to complete the application assignment in Section 7–7.

7–1 ■ THE MAGNETIC FIELD

A permanent magnet has a magnetic field surrounding it. The magnetic field consists of lines of force that radiate from the north pole to the south pole and back to the north pole through the magnetic material.

After completing this section, you should be able to

- **Explain the principles of the magnetic field**
  - □ Define *magnetic flux*
  - □ Define *magnetic flux density*
  - □ Discuss how materials are magnetized
  - □ Explain how a magnetic switch works

A permanent magnet, such as the bar magnet shown in Figure 7–1, has a magnetic field surrounding it. The magnetic field consists of lines of force, or flux lines, that radiate from the north pole (N) to the south pole (S) and back to the north pole through the magnetic material. For clarity, only a few lines of force are shown in the figure. Imagine, however, that many lines surround the magnet in three dimensions. The lines shrink to the smallest possible size and blend together, although they do not touch. This effectively forms a continuous magnetic field surrounding the magnet.

Attraction and Repulsion of Magnetic Poles

When unlike poles of two permanent magnets are placed close together, an attractive force is produced by the magnetic fields, as indicated in Figure 7–2(a). When two like poles are brought close together, they repel each other, as shown in part (b).

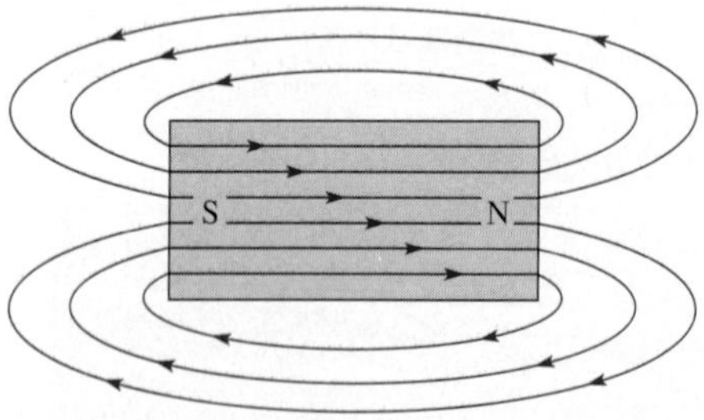

Colored lines represent a few magnetic lines of force in the magnetic field.

FIGURE 7–1
Magnetic lines of force around a bar magnet.

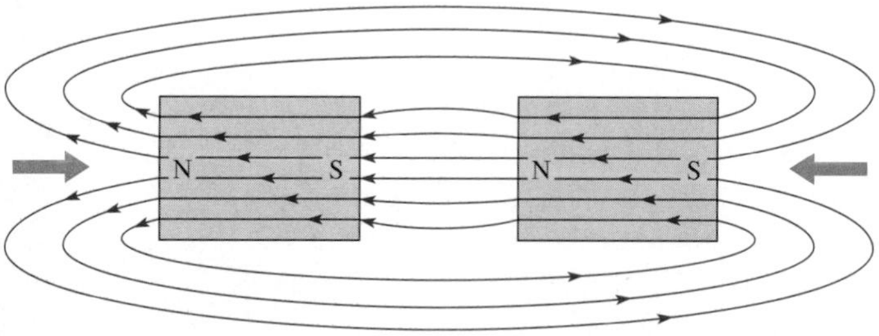

(a) Unlike poles attract.

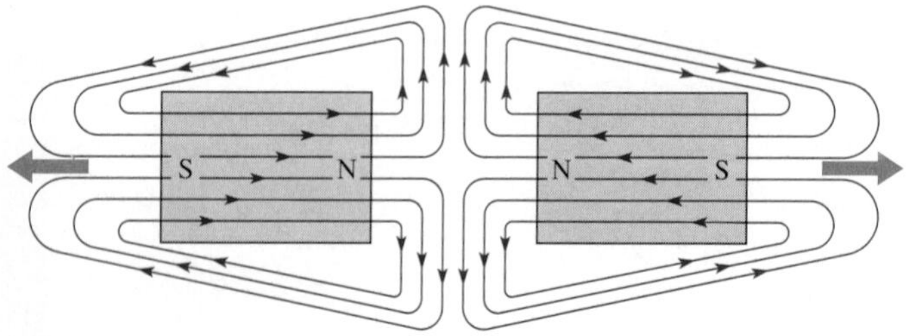

(b) Like poles repel.

FIGURE 7–2
Magnetic attraction and repulsion.

Altering a Magnetic Field

When a nonmagnetic material such as paper, glass, wood, or plastic is placed in a magnetic field, the lines of force are unaltered, as shown in Figure 7–3(a). However, when a magnetic material such as iron is placed in the magnetic field, the lines of force tend to change course and pass through the iron rather than through the surrounding air. They do so because the iron provides a magnetic path that is more easily established than that of air. Figure7–3(b) illustrates this principle.

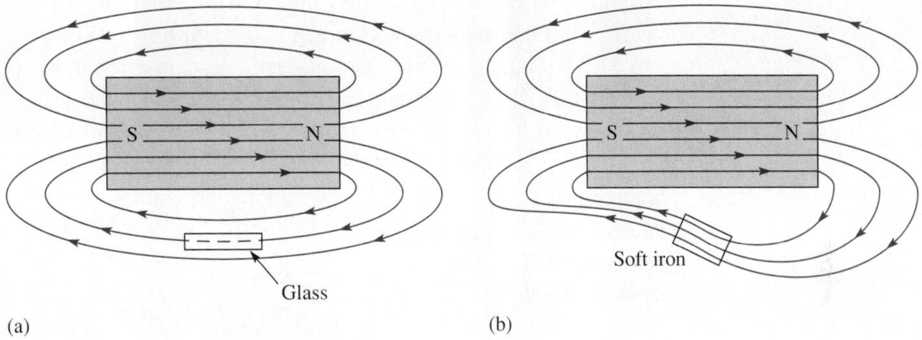

FIGURE 7–3
Effect of (a) nonmagnetic and (b) magnetic materials on a magnetic field.

Magnetic Flux (ϕ)

The group of force lines going from the north pole to the south pole of a magnet is called the **magnetic flux**, symbolized by ϕ (the lowercase Greek letter phi). The number of lines of force in a magnetic field determines the value of the flux. The more lines of force, the greater the flux and the stronger the magnetic field.

The unit of magnetic flux is the **weber** (Wb). One weber equals 10^8 lines. The weber is a very large unit; thus, in most practical situations, the microweber (μWb) is used. One microweber equals 100 lines of magnetic flux.

Magnetic Flux Density (B)

The **magnetic flux density** is the amount of flux per unit area in the magnetic field. Its symbol is B, and its unit is the **tesla** (T). One tesla equals one weber per square meter (Wb/m^2). The following formula expresses the flux density:

$$B = \frac{\phi}{A} \tag{7–1}$$

where ϕ is the flux and A is the cross-sectional area of the magnetic field.

EXAMPLE 7–1 Find the flux density in a magnetic field in which the flux in 0.1 m^2 is 800 μWb.

Solution

$$B = \frac{\phi}{A} = \frac{800\ \mu\text{Wb}}{0.1\ \text{m}^2} = 8000 \times 10^{-6}\ \text{T}$$

Related Exercise Repeat this example for a flux of 1.2 mWb.

How Materials Become Magnetized

Ferromagnetic materials such as iron, nickel, and cobalt become magnetized when placed in the magnetic field of a magnet. We have all seen a permanent magnet pick up paper clips, nails, iron filings, and so on. In these cases, the object becomes magnetized (that is, it actually becomes a magnet itself) under the influence of the permanent magnetic field and becomes attracted to the magnet. When removed from the magnetic field, the object tends to lose its magnetism.

Ferromagnetic materials have minute magnetic domains created within their atomic structure by the orbital motion and spin of electrons. These domains can be viewed as very small bar magnets with north and south poles. When the material is not exposed to an external magnetic field, the magnetic domains are randomly oriented, as shown in Figure 7–4(a). When the material is placed in a magnetic field, the domains align themselves, as shown in part (b). Thus, the object itself effectively becomes a magnet.

FIGURE 7–4
Ferromagnetic domains in (a) an unmagnetized and (b) a magnetized material.

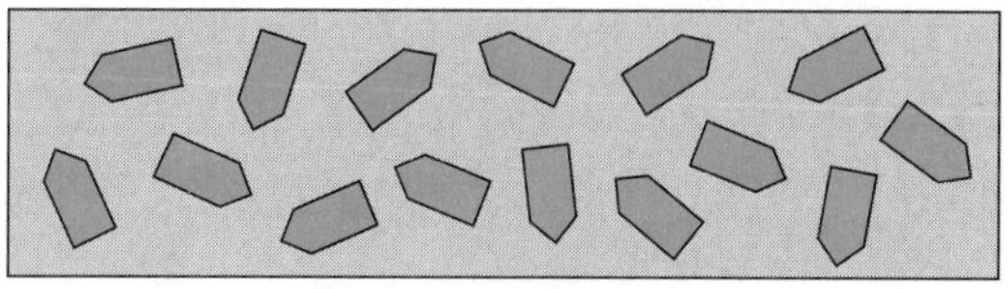

(a) The magnetic domains (N S) are randomly oriented in the unmagnetized material.

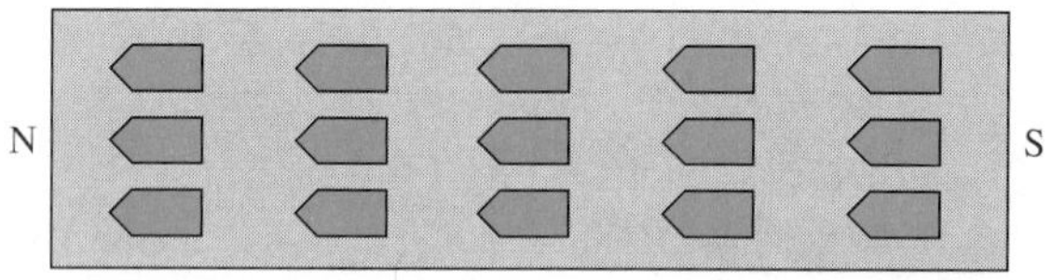

(b) The magnetic domains become aligned when the material is magnetized.

An Application

Permanent magnets have almost endless applications, one of which is presented here as an illustration. Figure 7–5 shows a typical magnetically operated, normally closed (NC) switch. When the magnet is near the switch mechanism, the metallic arm is held in its NC position. When the magnet is moved away, the spring pulls the arm up, breaking the contact as shown in Figure 7–6.

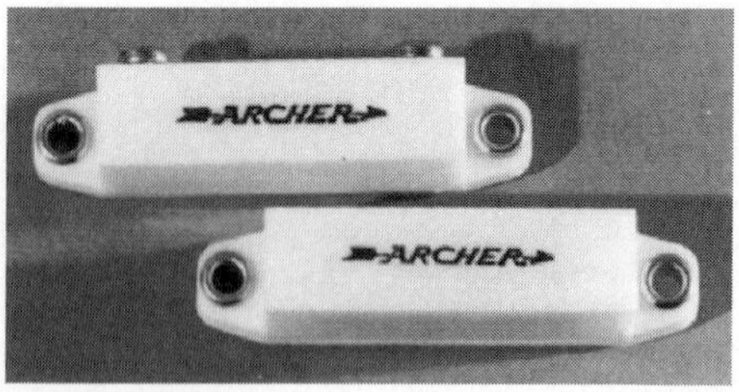

FIGURE 7–5
Magnet and switch set (courtesy of Tandy Corp.).

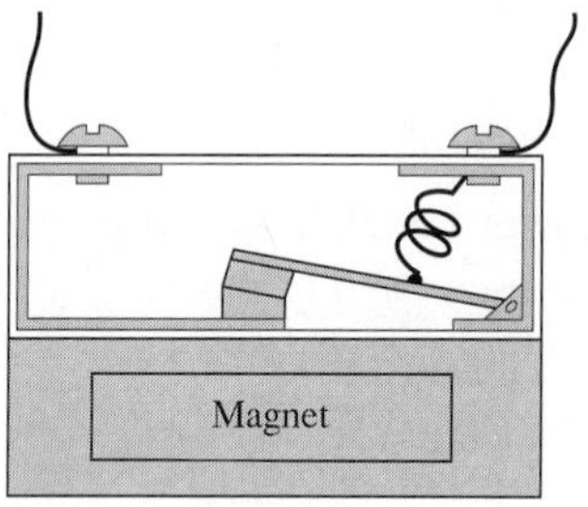

(a) Contact is closed.

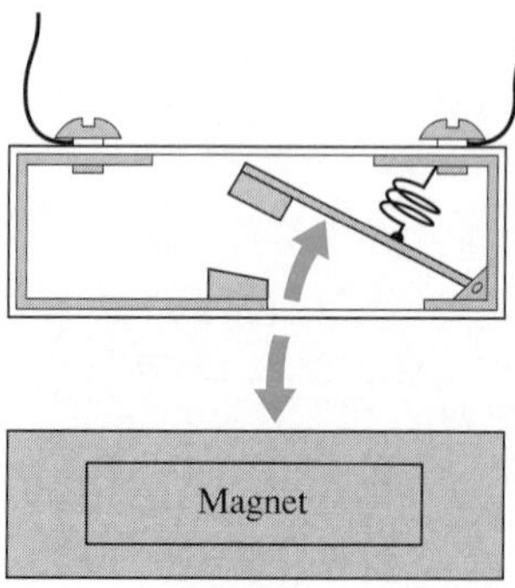

(b) Contact opens.

FIGURE 7–6
Operation of a magnetic switch.

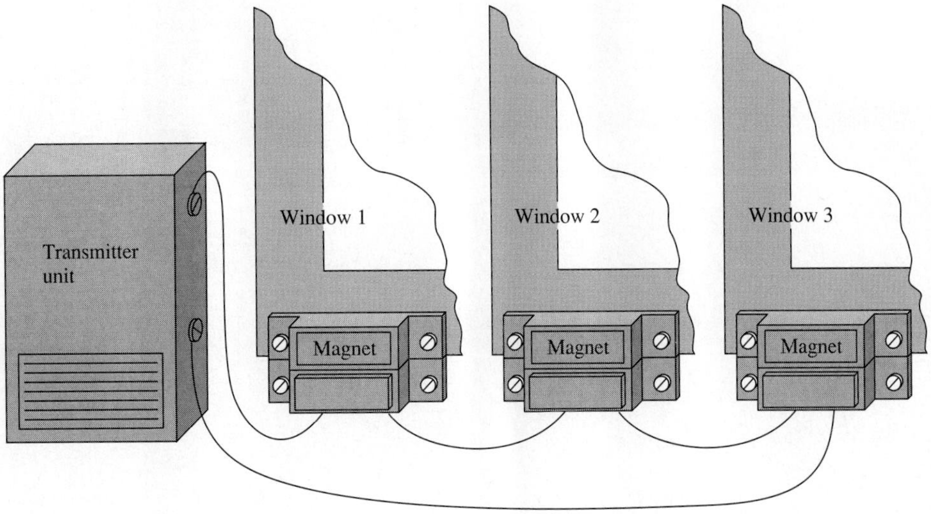

FIGURE 7–7
Connection of a typical perimeter alarm system.

Switches of this type are commonly used in perimeter alarm systems to detect entry into a building through windows or doors. As Figure 7–7 shows, several openings can be protected by magnetic switches wired to a common transmitter. When any one of the switches opens, the transmitter is activated and sends a signal to a central receiver and alarm unit.

SECTION 7–1 REVIEW

1. When the north poles of two magnets are placed close together, do they repel or attract each other?
2. What is magnetic flux?
3. What is the flux density when $\phi = 4.5\ \mu\text{Wb}$ and $A = 5 \times 10^{-3}\ \text{m}^2$?

7–2 ■ ELECTROMAGNETISM

Electromagnetism is the production of a magnetic field by current in a conductor. Many types of useful devices such as tape recorders, electric motors, speakers, solenoids, and relays are based on electromagnetism.

After completing this section, you should be able to

- **Explain the principles of electromagnetism**
  - □ Determine the direction of the magnetic lines of force
  - □ Define *permeability*
  - □ Define *reluctance*
  - □ Define *magnetomotive force*
  - □ Describe a basic electromagnet

Current produces a magnetic field, called an **electromagnetic field,** around a conductor, as illustrated in Figure 7–8. The invisible lines of force of the magnetic field form a concentric circular pattern around the conductor and are continuous along its length.

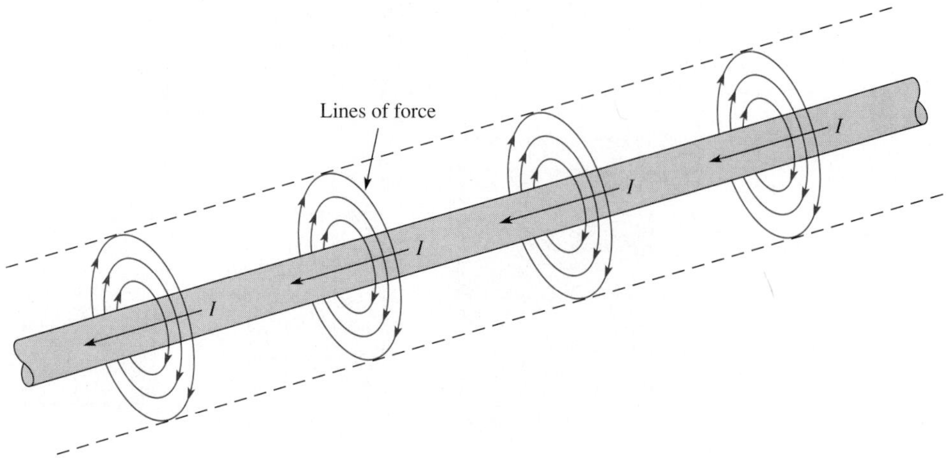

FIGURE 7–8
Magnetic field around a current-carrying conductor.

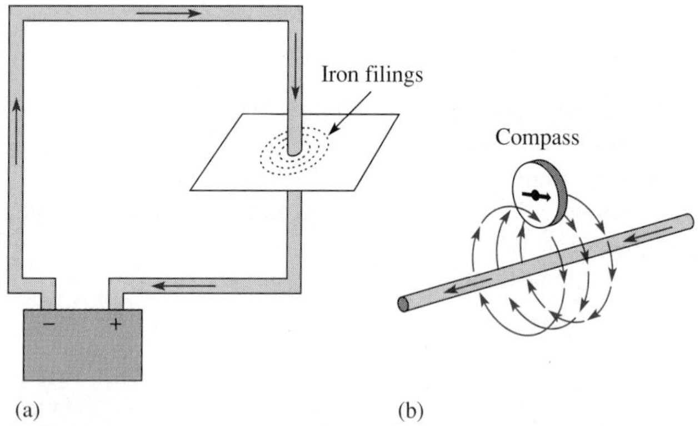

FIGURE 7–9
Visible effects of an electromagnetic field.

Although the magnetic field cannot be seen, it is capable of producing visible effects. For example, if a current-carrying wire is inserted through a sheet of paper in a perpendicular direction, iron filings placed on the surface of the paper arrange themselves along the magnetic lines of force in concentric rings, as illustrated in Figure 7–9(a). Part (b) of the figure illustrates that the north pole of a compass placed in the electromagnetic field will point in the direction of the lines of force. The field is stronger closer to the conductor and becomes weaker with increasing distance from the conductor.

Direction of the Lines of Force

The direction of the lines of force surrounding the conductor are indicated in Figure 7–10. When the direction of current is right to left, as in part (a), the lines are in a clock-

FIGURE 7–10
Magnetic lines of force around a current-carrying conductor.

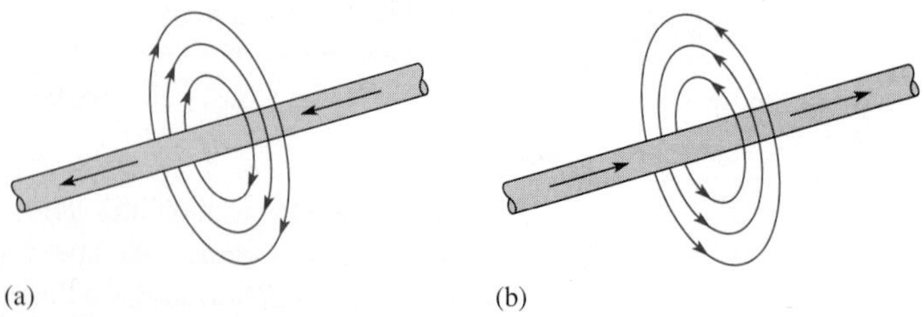

wise direction. When current is left to right, as in part (b), the lines are in a counterclockwise direction.

Left-Hand Rule

An aid to remembering the direction of the lines of force is illustrated in Figure 7–11. Imagine that you are grasping the conductor with your left hand, with your thumb pointing in the direction of current. Your fingers indicate the direction of the magnetic lines of force.

FIGURE 7–11
Illustration of left-hand rule.

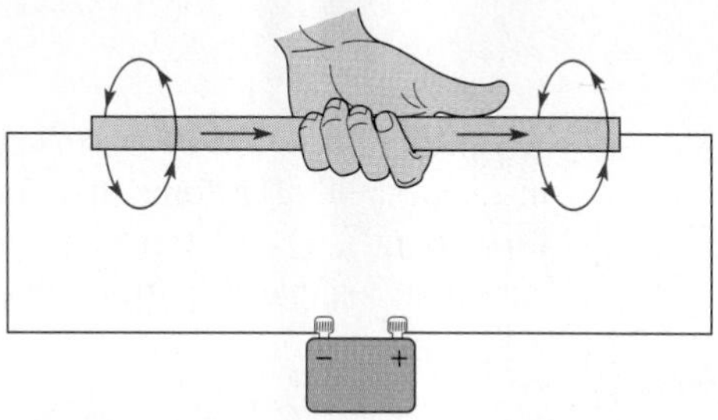

Electromagnetic Properties

Several important properties relating to electromagnetic fields are now discussed.

Permeability ***(μ)*** The ease with which a magnetic field can be established in a given material is measured by the **permeability** of that material. The higher the permeability, the more easily a magnetic field can be established.

The symbol of permeability is μ, and its value varies depending on the type of material. The permeability of a vacuum (μ_0) is $4\pi \times 10^{-7}$ Wb/At·m (weber/ampere-turn·meter) and is used as a reference. Ferromagnetic materials typically have permeabilities hundreds of times larger than that of a vacuum, indicating that a magnetic field can be set up with relative ease in these materials. Ferromagnetic materials include iron, steel, nickel, cobalt, and their alloys.

The relative permeability (μ_r) of a material is the ratio of its absolute permeability (μ) to the permeability of a vacuum (μ_0). Since μ_r is a ratio, it has no units.

$$\mu_r = \frac{\mu}{\mu_0} \qquad (7\text{–}2)$$

Reluctance ***($\mathcal{R}$)*** The opposition to the establishment of a magnetic field in a material is called **reluctance** ($\mathcal{R}$). The value of reluctance is directly proportional to the length (l) of the magnetic path, and inversely proportional to the permeability (μ) and to the cross-sectional area (A) of the material as expressed by the following equation:

$$\mathcal{R} = \frac{l}{\mu A} \qquad (7\text{–}3)$$

Reluctance in magnetic circuits in analogous to resistance in electric circuits. The unit of reluctance can be derived using l in meters. A (area) in square meters, and μ in Wb/At·m as follows:

$$\mathcal{R} = \frac{l}{\mu A} = \frac{\text{m}}{(\text{Wb/At·m})(\text{m}^2)} = \frac{\text{At}}{\text{Wb}}$$

At/Wb is ampere-turns/weber.

EXAMPLE 7–2 What is the reluctance of a material that has a length of 0.05 m, a cross-sectional area of 0.012 m^2, and a permeability of 3500×10^{-6} Wb/At·m?

Solution

$$\mathcal{R} = \frac{l}{\mu A} = \frac{0.05\text{ m}}{(3500 \times 10^{-6}\text{ Wb/At·m})(0.012\text{ m}^2)} = 1190\text{ At/Wb}$$

Related Exercise What happens to the reluctance if l is doubled and A is halved?

Magnetomotive Force (mmf) As you have learned, current in a conductor produces a magnetic field. The force that produces the magnetic field is called the **magnetomotive force** (mmf). The unit of mmf, the **ampere-turn** (At), is established on the basis of the current in a single loop (turn) of wire. The formula for mmf is as follows:

$$F_m = NI \tag{7–4}$$

where F_m is the magnetomotive force, N is the number of turns of wire, and I is the current in amperes.

Figure 7–12 illustrates that a number of turns of wire carrying a current around a magnetic material create a force that sets up flux lines through the magnetic path. The amount of flux depends on the magnitude of the mmf and on the reluctance of the material, as expressed by the following equation:

$$\phi = \frac{F_m}{\mathcal{R}} \tag{7–5}$$

This is known as the *Ohm's law for magnetic circuits* because the flux (ϕ) is analogous to current, the mmf (F_m) is analogous to voltage, and the reluctance ($\mathcal{R}$) is analogous to resistance.

FIGURE 7–12
A basic magnetic circuit.

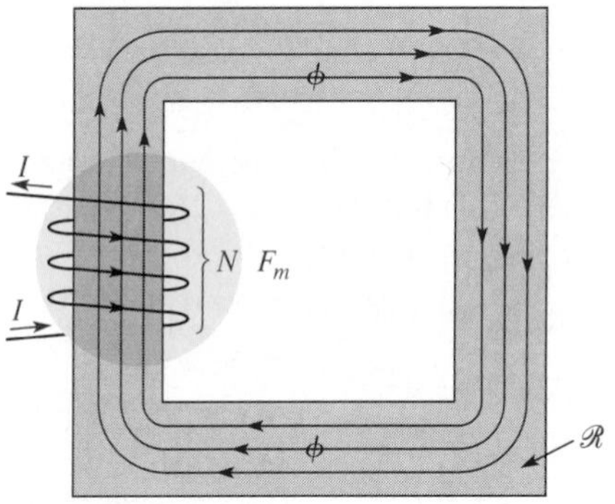

EXAMPLE 7–3 How much flux is established in the magnetic path of Figure 7–13 if the reluctance of the material is 0.28×10^5 At/Wb?

FIGURE 7–13

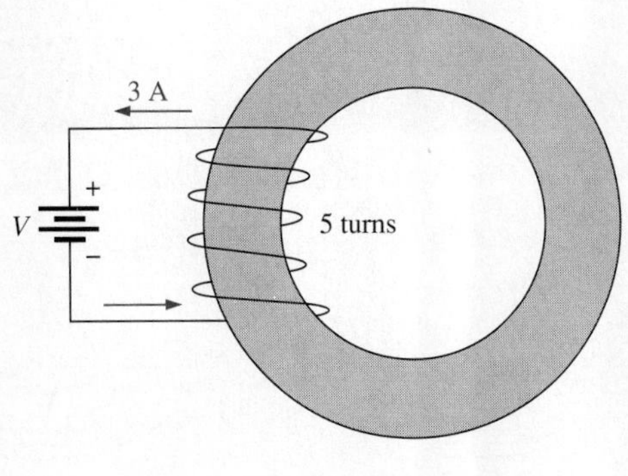

Solution

$$\phi = \frac{F_m}{\mathcal{R}} = \frac{NI}{\mathcal{R}} = \frac{(5\text{ t})(3\text{ A})}{0.28 \times 10^5\text{ At/Wb}} = 5.36 \times 10^{-4}\text{ Wb} = 536\ \mu\text{Wb}$$

Related Exercise How much flux is established in the magnetic path of Figure 7–13 if the reluctance is 7.5×10^3 At/Wb, the number of turns is 30, and the current is 1.8 A?

EXAMPLE 7–4

There are two amperes of current through a wire with 5 turns.
(a) What is the mmf?
(b) What is the reluctance of the circuit if the flux is 250 μWb?

Solution

(a) $N = 5$ and $I = 2$ A

$F_m = NI = (5\text{ t})(2\text{ A}) = 10$ At

(b) $\mathcal{R} = \frac{F_m}{\phi} = \frac{10\text{ At}}{250\ \mu\text{Wb}} = 0.04 \times 10^6\text{ At/Wb} = 4.0 \times 10^4\text{ At/Wb}$

Related Exercise Rework the example for $I = 850$ mA, $N = 50$, and $\phi = 500\ \mu$Wb.

The Electromagnet

An electromagnet is based on the properties that you have just learned. A basic electromagnet is simply a coil of wire wound around a core material that can be easily magnetized.

The shape of the electromagnet can be designed for various applications. For example, Figure 7–14 shows a U-shaped magnetic core. When the coil of wire is connected to a battery and current flows, as shown in part (a), a magnetic field is established as indicated. If the current is reversed, as shown in part (b), the direction of the magnetic field is also reversed. The closer the north and south poles are brought together, the smaller the air gap between them becomes, and the easier it becomes to establish a magnetic field, because the reluctance is lessened.

A good example of one application of an electromagnet is the process of recording on magnetic tape. In this situation, the recording head is an electromagnet with a narrow air gap, as shown in Figure 7–15. Current sets up a magnetic field across the air gap, and

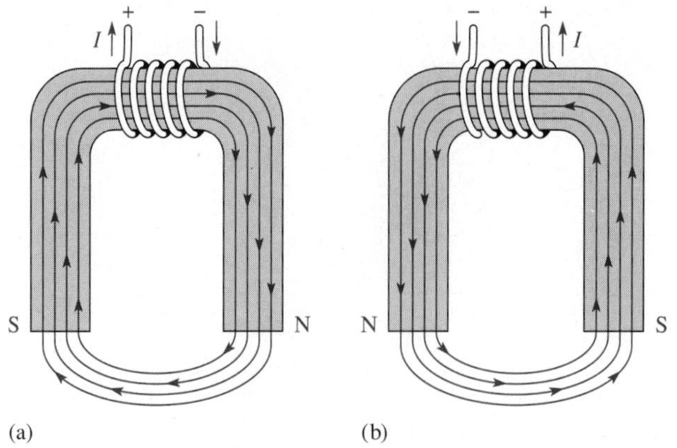

FIGURE 7–14
Reversing the current in the coil causes the electromagnetic field to reverse.

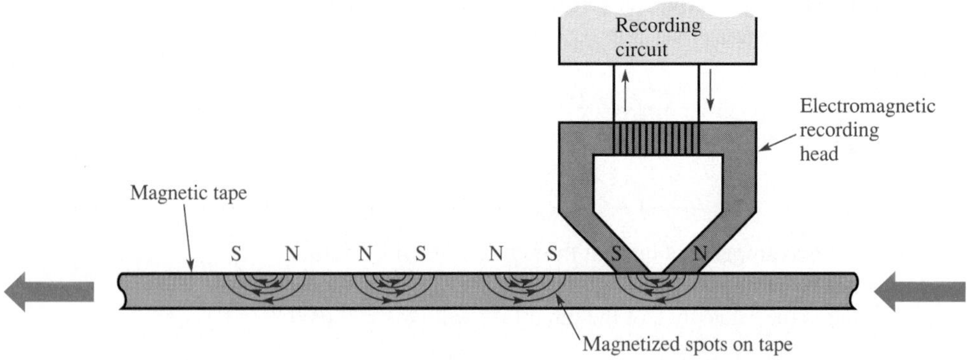

FIGURE 7–15
An electromagnetic recording head records information on magnetic tape by magnetizing the tape as it passes by.

as the recording head passes over the magnetic tape, the tape is permanently magnetized. In digital recording, for example, the tape is magnetized in one direction for a binary 1 and in the other direction for a binary 0, as illustrated in the figure. This magnetization in different directions is accomplished by reversing the coil current in the recording head.

SECTION 7–2 REVIEW

1. Explain the difference between magnetism and electromagnetism.
2. What happens to the magnetic field in an electromagnet when the current through the coil is reversed?

7–3 ■ ELECTROMAGNETIC DEVICES

In the last section, you learned that the recording head is a type of electromagnetic device. Now, several other common devices are introduced.

After completing this section, you should be able to

- **Describe the principle of operation for several types of electromagnetic devices**
  - ☐ Discuss how a solenoid works
  - ☐ Discuss how a relay works
  - ☐ Discuss how a speaker works
  - ☐ Discuss the basic analog meter movement

The recording head is one example of an electromagnetic device. Several other examples are now presented.

The Solenoid

The **solenoid** is a type of electromagnetic device that has a movable iron core called a plunger. The movement of this iron core depends on both an electromagnetic field and a mechanical spring force. The basic structure of a solenoid is shown in Figure 7–16. It consists of a cylindrical coil of wire wound around a nonmagnetic hollow form. A stationary iron core is fixed in position at the end of the shaft and a sliding iron core is attached to the stationary core with a spring.

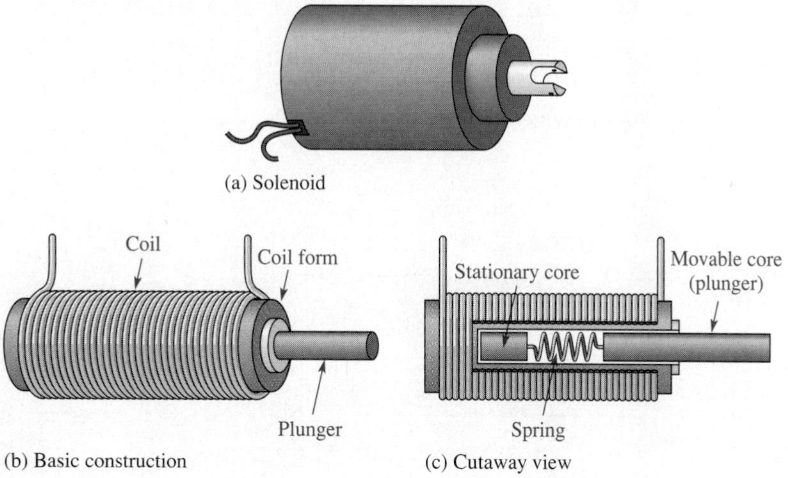

FIGURE 7–16
Basic solenoid structure.

In the at-rest, or unenergized, state, the plunger is extended as shown in Figure 7–17(a). The solenoid is energized by current through the coil, as shown in part (b). The current sets up an electromagnetic field that magnetizes both iron cores as indicated. The south pole of the stationary core attracts the north pole of the movable core causing it to slide inward, thus retracting the plunger and compressing the spring. As long as the coil current flows, the plunger remains retracted by the attractive force of the magnetic fields. When the current is cut off, the magnetic fields collapse and the force of the compressed spring pushes the plunger back out. The solenoid is used for many applications, such as opening and closing valves, automobile door locks, and so on.

The Relay

Relays differ from solenoids in that the electromagnetic action is used to open or close electrical contacts rather than to provide mechanical movement. Figure 7–18 shows the basic operation of a relay with one normally open (NO) contact and one normally closed (NC) contact (single pole–double throw). When there is no coil current, the armature is

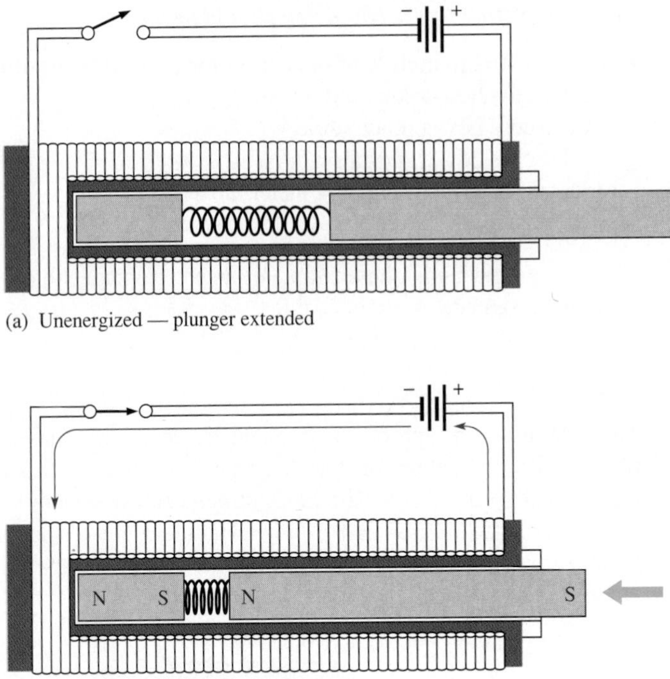

(a) Unenergized — plunger extended

(b) Energized — plunger retracted

FIGURE 7–17
Basic solenoid operation.

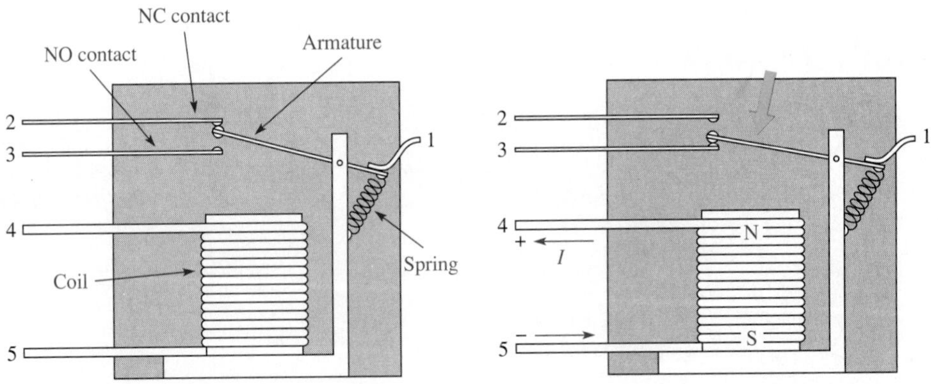

(a) Unenergized: continuity from 1 to 2

(b) Energized: continuity from 1 to 3

FIGURE 7–18
Basic structure of a single-pole–double-throw relay.

held against the upper contact by the spring, thus providing continuity from terminal 1 to terminal 2, as shown in Figure 7–18(a). When energized with coil current, the armature is pulled down by the attractive force of the electromagnetic field and makes connection with the lower contact to provide continuity from terminal 1 to terminal 3, as shown in Figure 7–18(b).

A typical relay and its schematic symbol are shown in Figure 7–19.

The Speaker

Permanent-magnet speakers are commonly used in stereos, radios, and TVs, and their operation is based on the principle of electromagnetism. A typical speaker is constructed

FIGURE 7–19
A typical relay.

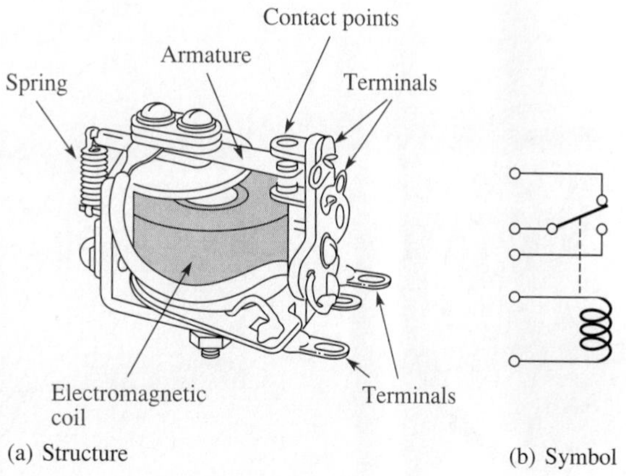

with a permanent magnet and an electromagnet, as shown in Figure 7–20(a). The cone of the speaker consists of a paper-like diaphragm to which is attached a hollow cylinder with a coil around it, forming an electromagnet. One of the poles of the permanent magnet is positioned within the cylindrical coil. When current flows through the coil in one direction, the interaction of the permanent magnetic field with the electromagnetic field causes the cylinder to move to the right, as indicated in Figure 7–20(b). Current through the coil in the other direction causes the cylinder to move to the left, as shown in Figure 7–20(c).

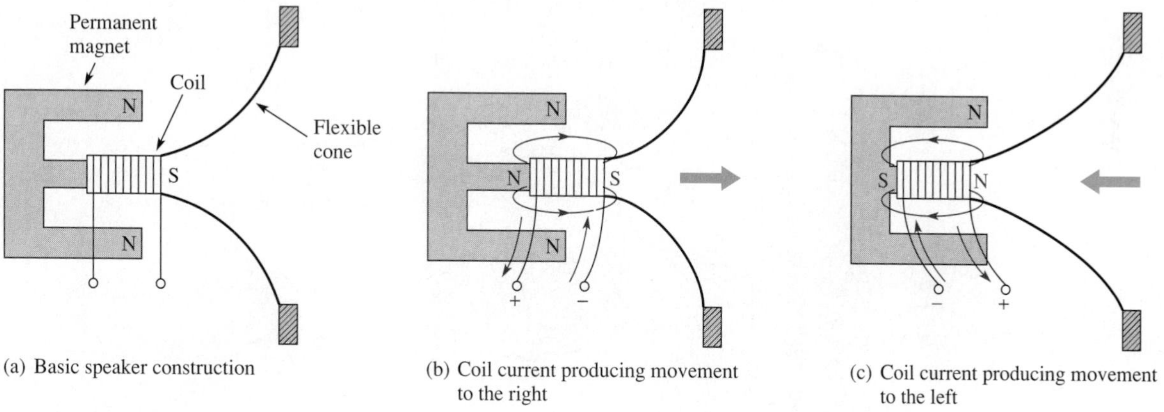

FIGURE 7–20
Basic speaker operation.

The movement of the coil cylinder causes the flexible diaphragm also to move in or out, depending on the direction of the coil current. The amount of coil current determines the intensity of the magnetic field, which controls the amount that the diaphragm moves.

As shown in Figure 7–21, when an audio signal voltage (voice or music) is applied to the speaker coil, the current varies proportionally in both direction and amount. In response, the diaphragm will vibrate in and out by varying amounts and at varying rates. Vibration in the diaphragm causes the air that is in contact with it to vibrate in the same manner. These air vibrations move through the air as sound waves.

Analog Meter Movement

The d'Arsonval movement is the most common type used in analog multimeters. In this type of meter movement, the pointer is deflected in proportion to the amount of current

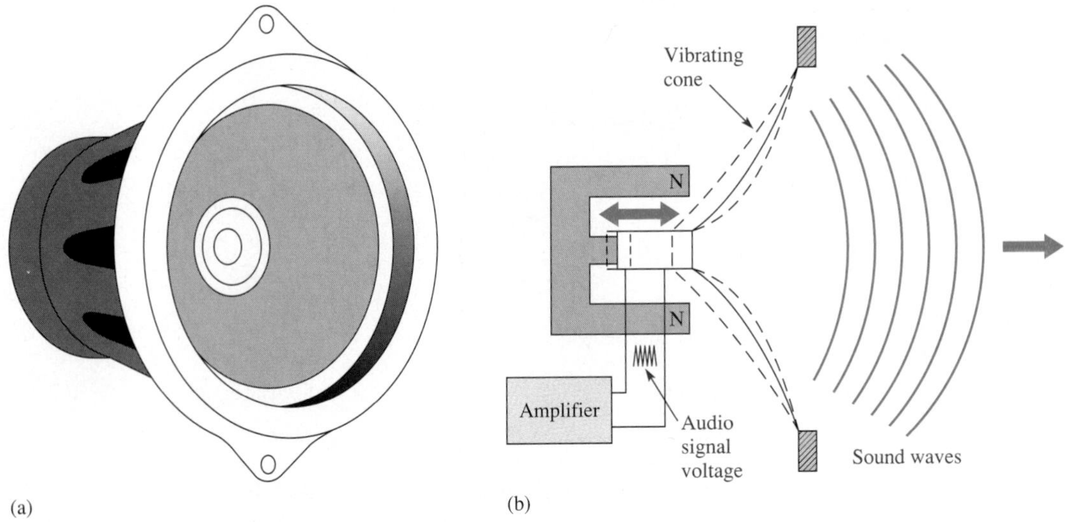

FIGURE 7–21
The speaker converts audio signal voltages into sound waves.

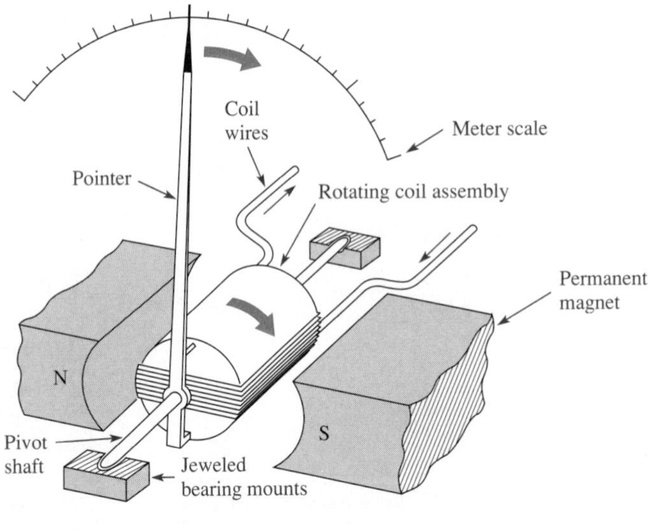

FIGURE 7–22
The basic d'Arsonval meter movement.

through a coil. Figure 7–22 shows a basic d'Arsonval meter movement. It consists of a coil of wire wound on a bearing-mounted assembly that is placed between the poles of a permanent magnet. A pointer is attached to the moving assembly. With no current through the coil, a spring mechanism keeps the pointer at its left-most (zero) position. When current flows through the coil, electromagnetic forces act on the coil, causing a rotation to the right. The amount of rotation depends on the amount of current.

Figure 7–23 illustrates how the interaction of magnetic fields produces rotation of the coil assembly. The current flows inward at the "cross" and outward at the "dot" in the single winding shown. The inward current produces a counterclockwise electromagnetic field that reinforces the permanent magnetic field below it. The result is an upward force on the left side of the coil as shown. A downward force is developed on the right side of the coil, where the current is outward. These forces produce a clockwise rotation of the coil assembly and are opposed by a spring mechanism. The indicated forces and the spring force are balanced at the value of the current. When current is removed, the spring force returns the pointer to its zero position.

FIGURE 7–23
When the electromagnetic field interacts with the permanent magnetic field, forces are exerted on the rotating coil assembly, causing it to move clockwise and thus deflecting the pointer.

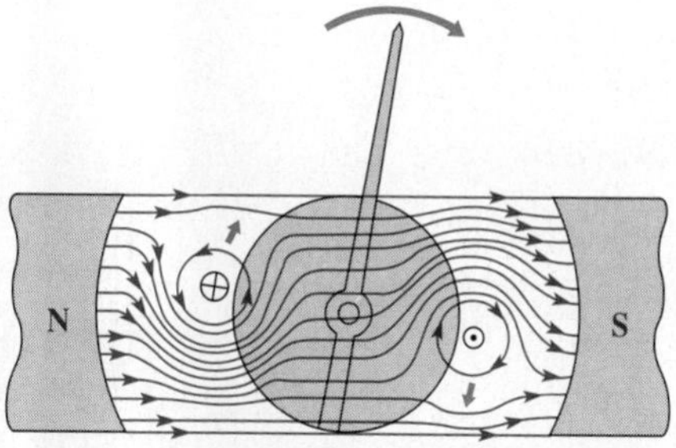

SECTION 7–3 REVIEW

1. Explain the difference between a solenoid and a relay.
2. What is the movable part of a solenoid called?
3. What is the movable part of a relay called?
4. Upon what basic principle is the d'Arsonval meter movement based?

7–4 ▪ MAGNETIC HYSTERESIS

When a magnetizing force is applied to a material, the magnetic flux density in the material changes in a certain way, which we will now examine.

After completing this section, you should be able to

- **Explain magnetic hysteresis**
  - ☐ State the formula for magnetizing force
  - ☐ Discuss a hysteresis curve
  - ☐ Define *retentivity*

Magnetizing Force (H)

The **magnetizing force** in a material is defined to be the magnetomotive force (F_m) per unit length (l) of the material, as expressed by the following equation. The unit of magnetizing force (H) is ampere-turns per meter (At/m).

$$H = \frac{F_m}{l} \qquad (7\text{–}6)$$

where $F_m = NI$. Note that the magnetizing force depends on the number of turns (N) of the coil of wire, the current (I) through the coil, and the length (l) of the material. It does not depend on the type of material.

Since $\phi = F_m/\mathcal{R}$, as F_m increases, the flux increases. Also, the magnetizing force (H) increases. Recall that the flux density (B) is the flux per unit cross-sectional area ($B = \phi/A$), so B is also proportional to H. The curve showing how these two quantities (B and H) are related is called the B-H curve, or the hysteresis curve. The parameters that influence both B and H are illustrated in Figure 7–24.

The Hysteresis Curve and Retentivity

Hysteresis is a characteristic of a magnetic material whereby a change in magnetization lags the application of a magnetizing force. The magnetizing force (H) can be readily in-

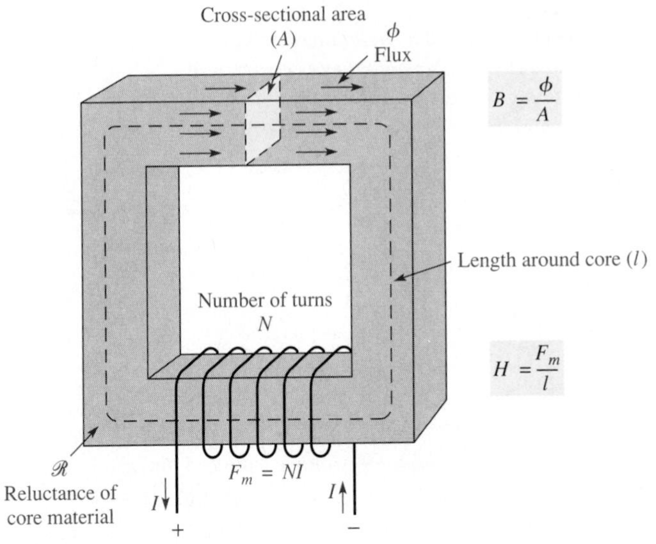

FIGURE 7–24
Parameters that determine the magnetizing force (H) and the flux density (B).

creased or decreased by varying the current through the coil of wire, and it can be reversed by reversing the voltage polarity across the coil.

Figure 7–25 illustrates the development of the hysteresis curve. We start by assuming a magnetic core is unmagnetized so that $B = 0$. As the magnetizing force (H) is increased from zero, the flux density (B) increases proportionally as indicated by the curve in Figure 7–25(a). When H reaches a certain value, the value of B begins to level off. As H continues to increase, B reaches a saturation value (B_{sat}) when H reaches a value (H_{sat})

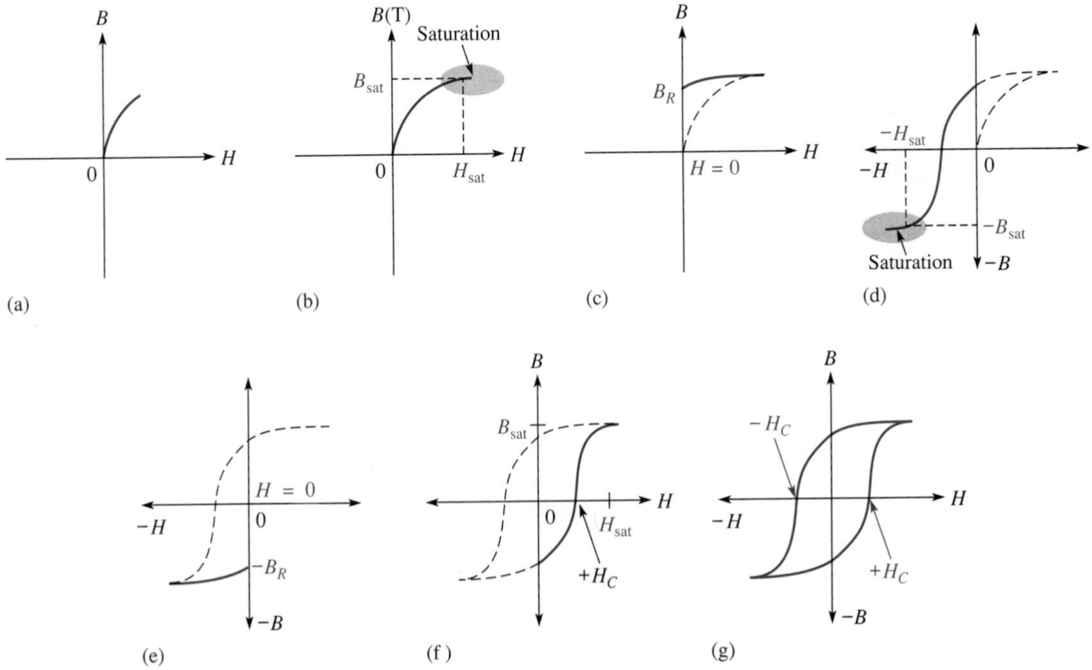

FIGURE 7–25
Development of a magnetic hysteresis curve.

as illustrated in Figure 7–25(b). Once saturation is reached, a further increase in H will not increase B.

Now, if H is decreased to zero, B will fall back along a different path to a residual value (B_R), as shown in Figure 7–25(c). This indicates that the material continues to be magnetized even with the magnetizing force removed ($H = 0$). The ability of a material, once magnetized, to maintain a magnetized state without the presence of a magnetizing force is called **retentivity**. The retentivity of a material is indicated by the ratio of B_R to B_{sat}.

Reversal of the magnetizing force is represented by negative values of H on the curve and is achieved by reversing the current in the coil of wire. An increase in H in the negative direction causes saturation to occur at a value ($-H_{sat}$) where the flux density is at its maximum negative value as indicated in Figure 7–25(d).

When the magnetizing force is removed ($H = 0$), the flux density goes to its negative residual value ($-B_R$) as shown in Figure 7–25(e). From the $-B_R$ value, the flux density follows the curve indicated in part (f) back to its maximum positive value when the magnetizing force equals H_{sat} in the positive direction.

The complete *B-H* curve is shown in Figure 7–25(g) and is called the *hysteresis curve*. The magnetizing force required to make the flux density zero is called the coercive force, H_C.

Materials with a low retentivity do not retain a magnetic field very well while those with high retentivities exhibit values of B_R very close to the saturation value of B. Depending on the application, retentivity in a magnetic material can be an advantage or a disadvantage. In permanent magnets and memory cores, for example, high retentivity is required. In ac motors, retentivity is undesirable because the residual magnetic field must be overcome each time the current reverses, thus wasting energy.

SECTION 7–4 REVIEW

1. For a given wirewound core, how does an increase in current through the coil affect the flux density?
2. Define *retentivity*.

7–5 ▪ ELECTROMAGNETIC INDUCTION

When a conductor is moved through a magnetic field, a voltage is produced across the conductor. This principle is known as electromagnetic induction and the resulting voltage is an induced voltage. Electromagnetic induction is what makes transformers, electrical generators, and many other devices possible.

After completing this section, you should be able to

- **Discuss the principle of electromagnetic induction**
  - ☐ Explain how voltage is induced in a conductor in a magnetic field
  - ☐ Determine polarity of an induced voltage
  - ☐ Discuss forces on a conductor in a magnetic field
  - ☐ State Faraday's law
  - ☐ State Lenz's law

Relative Motion

When a wire is moved across a magnetic field, there is a relative motion between the wire and the magnetic field. Likewise, when a magnetic field is moved past a stationary wire, there is also relative motion. In either case, there is an induced voltage (v_{ind}) in the wire

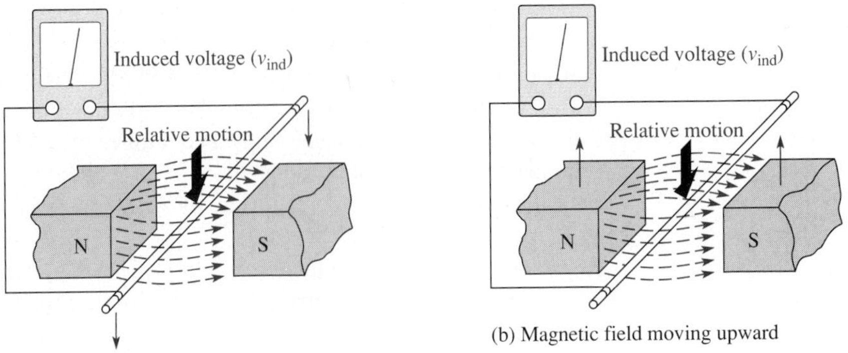

(a) Wire moving downward

(b) Magnetic field moving upward

FIGURE 7–26
Relative motion between a wire and a magnetic field.

there is also relative motion. In either case, there is an induced voltage (v_{ind}) in the wire as a result of this motion, as Figure 7–26 indicates. The lowercase v stands for instantaneous voltage.

The amount of the induced voltage depends on the rate at which the wire and the magnetic field move with respect to each other: The faster the relative speed, the greater the induced voltage.

Polarity of the Induced Voltage

If the conductor in Figure 7–26 is moved first one way and then another in the magnetic field, a reversal of the polarity of the induced voltage will be observed. As the wire is moved downward, a voltage is induced with the polarity indicated in Figure 7–27(a). As the wire is moved upward, the polarity is as indicated in part (b) of the figure.

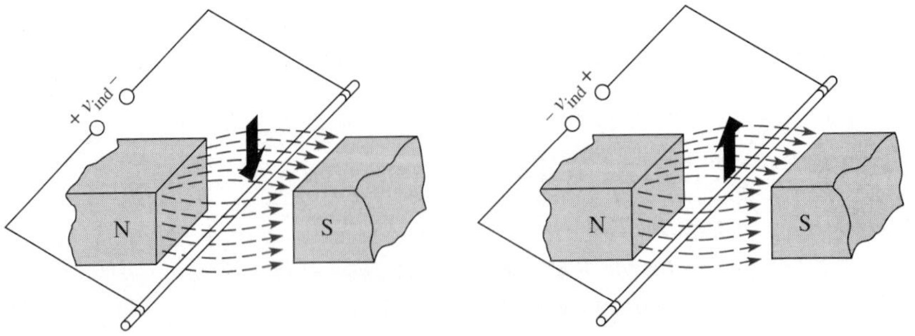

(a) (b)

FIGURE 7–27
Polarity of induced voltage depends on direction of motion.

Induced Current

When a load resistor is connected to the wire in Figure 7–27, the voltage induced by the relative motion in the magnetic field will cause a current in the load, as shown in Figure 7–28. This current is called the induced current (i_{ind}). The lowercase i stands for instantaneous current.

The action of producing a voltage and a current in a load by moving a conductor across a magnetic field is the basis for electrical generators. Also, the existence of a conductor in a moving magnetic field is fundamental to the concept of inductance in an electrical circuit.

FIGURE 7–28
Induced current in a load as the wire moves through the magnetic field.

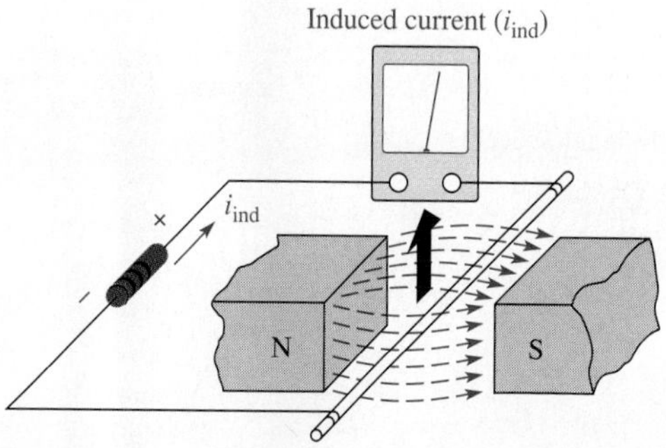

Forces on a Current-Carrying Conductor in a Magnetic Field (Motor Action)

Figure 7–29(a) shows current inward through a wire in a magnetic field. The electromagnetic field set up by the current interacts with the permanent magnetic field; as a result, the permanent lines of force above the wire tend to be deflected down under the wire, because they are opposite in direction to the electromagnetic lines of force. Therefore, the flux density above is reduced, and the magnetic field is weakened. The flux density below the conductor is increased, and the magnetic field is strengthened. An upward force on the conductor results, and the conductor tends to move toward the weaker magnetic field.

Figure 7–29(b) shows the current outward, resulting in a force on the conductor in the downward direction. This force is the basis for electrical motors.

FIGURE 7–29
Forces on a current-carrying conductor in a magnetic field.

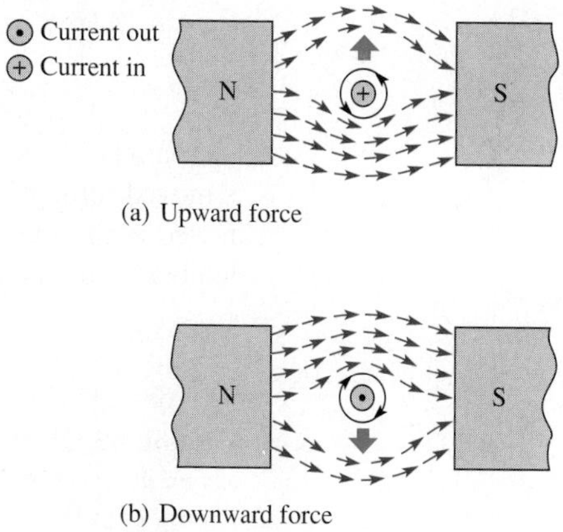

Faraday's Law

Michael Faraday discovered the principle of **electromagnetic induction** in 1831. He found that moving a magnet through a coil of wire induced a voltage across the coil, and that when a complete path was provided, the induced voltage caused an induced current, as you have seen. Faraday's two observations are as follows:

1. The amount of voltage induced in a coil is directly proportional to the rate of change of the magnetic field with respect to the coil.
2. The amount of voltage induced in a coil is directly proportional to the number of turns of wire in the coil.

Faraday's first observation is demonstrated in Figure 7–30, where a bar magnet is moved through a coil, thus creating a changing magnetic field. In part (a) of the figure, the magnet is moved at a certain rate, and a certain induced voltage is produced as indicated. In part (b), the magnet is moved at a faster rate through the coil, creating a greater induced voltage.

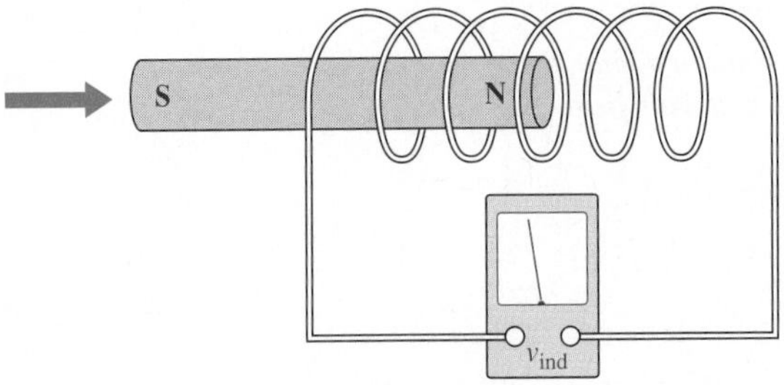

(a) As magnet moves slowly to the right, magnetic field is changing with respect to coil, and a voltage is induced.

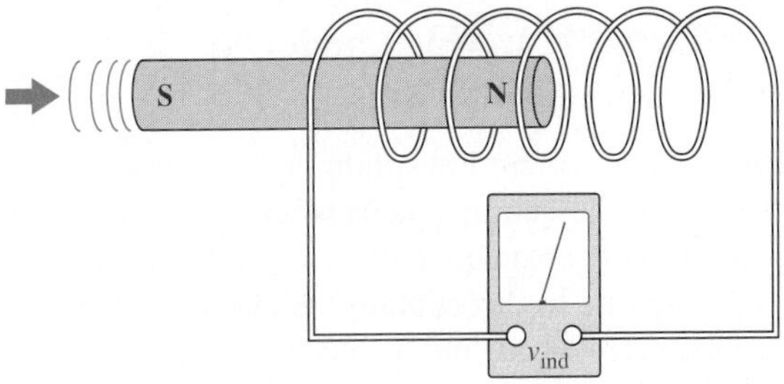

(b) As magnet moves more rapidly to the right, magnetic field is changing more rapidly with respect to coil, and a greater voltage is induced.

FIGURE 7–30

A demonstration of Faraday's first observation: The amount of induced voltage is directly proportional to the rate of change of the magnetic field with respect to the coil.

Faraday's second observation is demonstrated in Figure 7–31. In part (a), the magnet is moved through the coil and a voltage is induced as shown. In part (b), the magnet is moved at the same speed through a coil that has a greater number of turns. The greater number of turns creates a greater induced voltage.

FIGURE 7–31

A demonstration of Faraday's second observation: The amount of induced voltage is directly proportional to the number of turns in the coil.

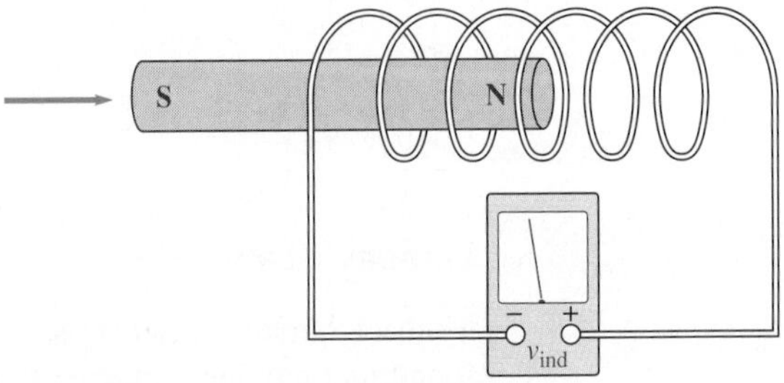

(a) Magnet moves through a coil and induces a voltage.

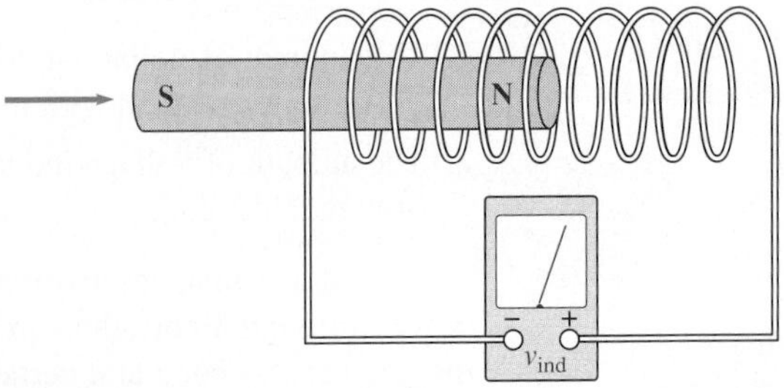

(b) Magnet moves at same rate through a coil with more turns (loops) and induces a greater voltage.

Faraday's law is stated as follows:

The voltage induced across a coil of wire equals the number of turns in the coil times the rate of change of the magnetic flux.

Lenz's Law

You have learned that a changing magnetic field induces a voltage in a coil that is directly proportional to the rate of change of the magnetic field and the number of turns in the coil. **Lenz's law** defines the polarity or direction of the induced voltage.

When the current through a coil changes, an induced voltage is created as a result of the changing magnetic field, and the direction of the induced voltage is such that it always opposes the change in current.

SECTION 7–5 REVIEW

1. What is the induced voltage across a stationary conductor in a stationary magnetic field?
2. When the speed at which a conductor is moved through a magnetic field is increased, does the induced voltage increase, decrease, or remain the same?
3. When there is current through a conductor in a magnetic field, what happens?

7–6 ■ APPLICATIONS OF ELECTROMAGNETIC INDUCTION

In this section, two interesting applications of electromagnetic induction are discussed—an automotive crankshaft position sensor and a dc generator. Although there are many varied applications, these two are representative.

After completing this section, you should be able to

- **Describe some applications of electromagnetic induction**
  - ☐ Explain how a crankshaft position sensor works
  - ☐ Explain how a dc generator works

Automotive Crankshaft Position Sensor

An interesting automotive application is a type of engine sensor that detects the crankshaft position directly using electromagnetic induction. The electronic engine controller in many automobiles uses the position of the crankshaft to set ignition timing and, sometimes, to adjust the fuel control system. Figure 7–32 shows the basic concept. A steel disk is attached to the engine's crankshaft by an extension rod; the protruding tabs on the disk represent specific crankshaft positions.

As illustrated in Figure 7–32, as the disk rotates with the crankshaft, the tabs periodically pass through the air gap of the permanent magnet. Since steel has a much lower reluctance than does air (a magnetic field can be established in steel much more easily than in air), the magnetic flux suddenly increases as a tab comes into the air gap, causing a voltage to be induced across the coil. This process is illustrated in Figure 7–33. The electronic engine control circuit uses the induced voltage as an indicator of the crankshaft position.

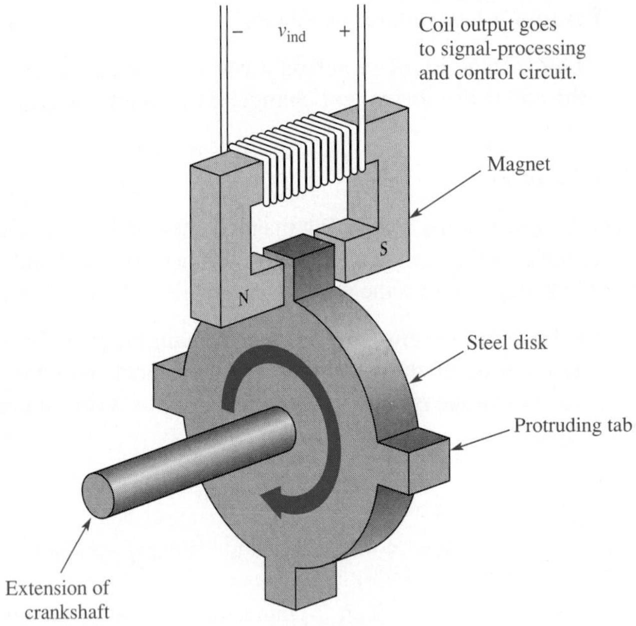

FIGURE 7–32

A crankshaft position sensor that produces a voltage when a tab passes through the air gap of the magnet.

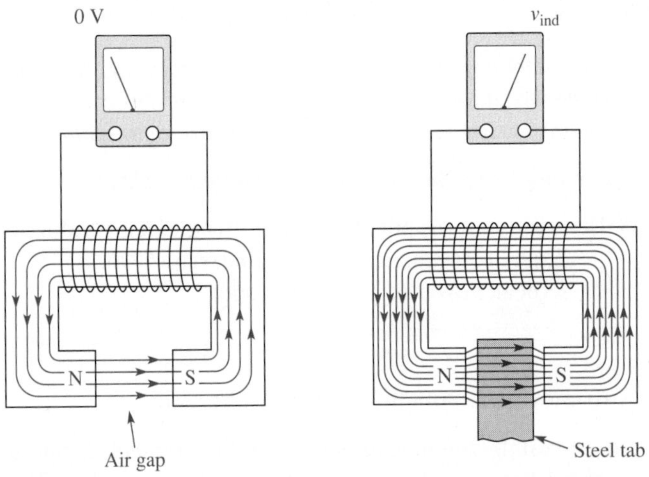

(a) There is no changing magnetic field, so there is no induced voltage.

(b) Insertion of the steel tab reduces the reluctance of the air gap, causing the magnetic flux to increase and thus inducing a voltage.

FIGURE 7–33

As the tab passes through the air gap of the magnet, the coil senses a change in the magnetic field, and a voltage is induced.

A DC Generator

Figure 7–34 shows a simplified dc generator consisting of a single loop of wire in a permanent magnetic field. Notice that each end of the loop is connected to a split-ring arrangement. This conductive metal ring is called a *commutator*. As the loop is rotated in the magnetic field, the split commutator ring also rotates. Each half of the split ring rubs against the fixed contacts, called *brushes*, and connects the loop to an external circuit.

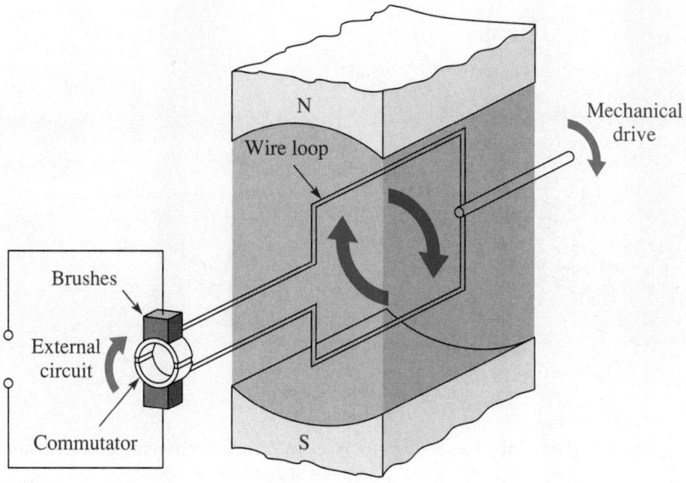

FIGURE 7–34
A basic dc generator.

As the loop rotates through the magnetic field, it "cuts" through the flux lines at varying angles, as illustrated in Figure 7–35. At position *A* in its rotation, the loop of wire is effectively moving parallel with the magnetic field. Therefore, at this instant, the rate at which it is cutting through the magnetic flux lines is zero.

FIGURE 7–35
End view of loop cutting through the magnetic field.

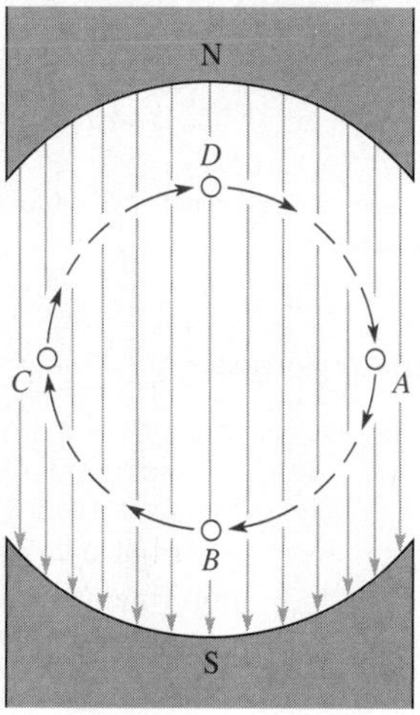

As the loop moves from position *A* to position *B*, it cuts through the flux lines at an increasing rate. At position *B*, it is moving effectively perpendicular to the magnetic field and thus is cutting through a maximum number of lines. As the loop rotates from position *B* to position *C*, the rate at which it cuts the flux lines decreases to minimum (zero) at *C*. From position *C* to position *D*, the rate at which the loop cuts the flux lines increases to a maximum at *D* and then back to a minimum again at *A*.

As you previously learned, when a wire moves through a magnetic field, a voltage is induced, and by Faraday's law, the amount of induced voltage is proportional to the number of loops (turns) in the wire and the rate at which it is moving with respect to the magnetic field. Now you know that the angle at which the wire moves with respect to the magnetic flux lines determines the amount of induced voltage, because the rate at which the wire cuts through the flux lines depends on the angle of motion.

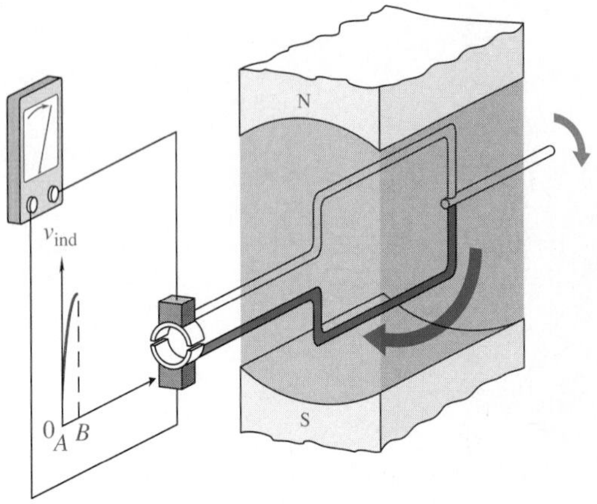

(a) Position B: Loop is moving perpendicular to flux lines, and voltage is maximum.

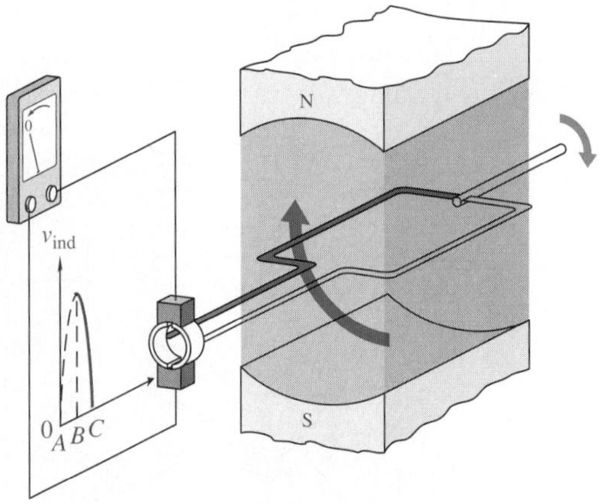

(b) Position C: Loop is moving parallel with flux lines, and voltage is zero.

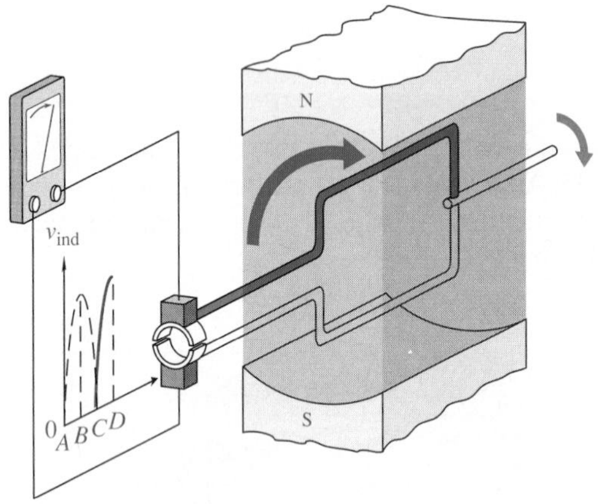

(c) Position D: Loop is moving perpendicular to flux lines, and voltage is maximum.

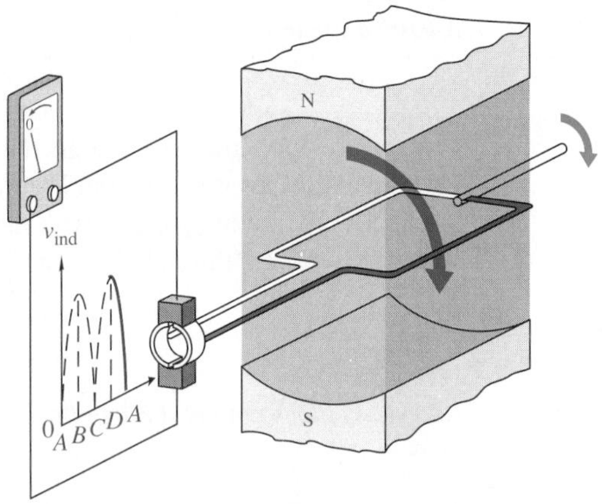

(d) Position A: Loop is moving parallel with flux lines, and voltage is zero.

FIGURE 7–36
Operation of a basic dc generator.

Figure 7–36 illustrates how a voltage is induced in the external circuit as the single loop rotates in the magnetic field. Assume that the loop is in its instantaneous horizontal position, so the induced voltage is zero. As the loop continues in its rotation, the induced voltage builds up to a maximum at position B, as shown in part (a) of the figure. Then, as the loop continues from B to C, the voltage decreases to zero at C, as shown in part (b).

During the second half of the revolution, shown in Figure 7–36(c) and (d), the brushes switch to opposite commutator sections, so the polarity of the voltage remains the same across the output. Thus, as the loop rotates from position C to position D and then back to A, the voltage increases from zero at C to a maximum at D and back to zero at A.

Figure 7–37 shows how the induced voltage varies as the loop goes through several rotations (three in this case). This voltage is a dc voltage because its polarity does not change. However, the voltage is pulsating between zero and its maximum value.

When more loops are added, the voltages induced across each loop are combined across the output. Since the voltages are offset from each other, they do not reach their maximum or zero values at the same time. A smoother dc voltage results, as shown in Figure 7–38 for two loops. The variations can be further smoothed out by filters to achieve a nearly constant dc voltage. (Filters are discussed later in the book.)

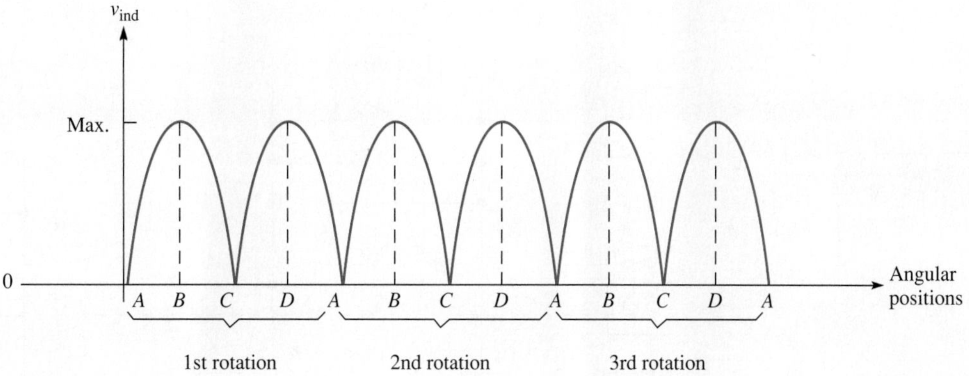

FIGURE 7–37
Induced voltage over three rotations of the loop.

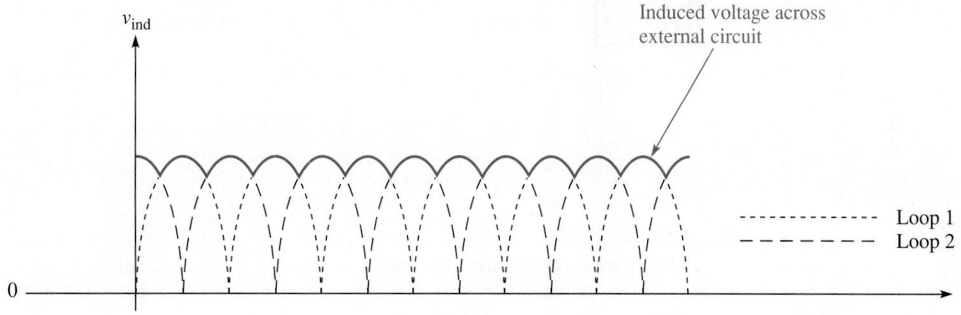

FIGURE 7–38
The induced voltage for a two-loop generator. There is much less variation in the induced voltage.

SECTION 7–6 REVIEW

1. If the steel disk in the crankshaft position sensor has stopped with a tab in the magnet's air gap, what is the induced voltage?
2. What happens to the induced voltage if the loop in the basic dc generator suddenly begins rotating at a faster speed?

7–7 ■ APPLICATION ASSIGNMENT

The relay is a common type of electromagnetic device that is used in many types of control applications. With a relay, a lower voltage, such as from a battery, can be used to switch a much higher voltage, such as the 110 V from an ac outlet. In this section, you will see how a relay can be used in a basic burglar alarm system.

The schematic in Figure 7–39 shows a basic simplified intrusion alarm system that uses a relay to turn on an audible alarm (siren) and lights. The system operates from a 9 V battery so that even if power to the house is off, the audible alarm will still work.

The magnetic detection switches are normally open (NO) magnetic switches that are parallel connected and located in the windows and doors. The relay is a three-pole, double-throw device that operates with a coil voltage of 9 V dc and draws approximately 50 mA. When an intrusion occurs, one of the switches closes and allows current to flow from the battery through the relay coil, which energizes the relay and causes the three sets of normally open contacts to close. Closure of contact *A* turns on the alarm, which draws 2 A from the battery. Closure of contact *C* turns on a light circuit in the house. Closure of contact *B* latches the relay and keeps it energized even if the intruder closes the door or window through which entry was made. If not for contact *B* in parallel with the detection switches, the alarm and lights would go off as soon as the window or door was shut behind the intruder.

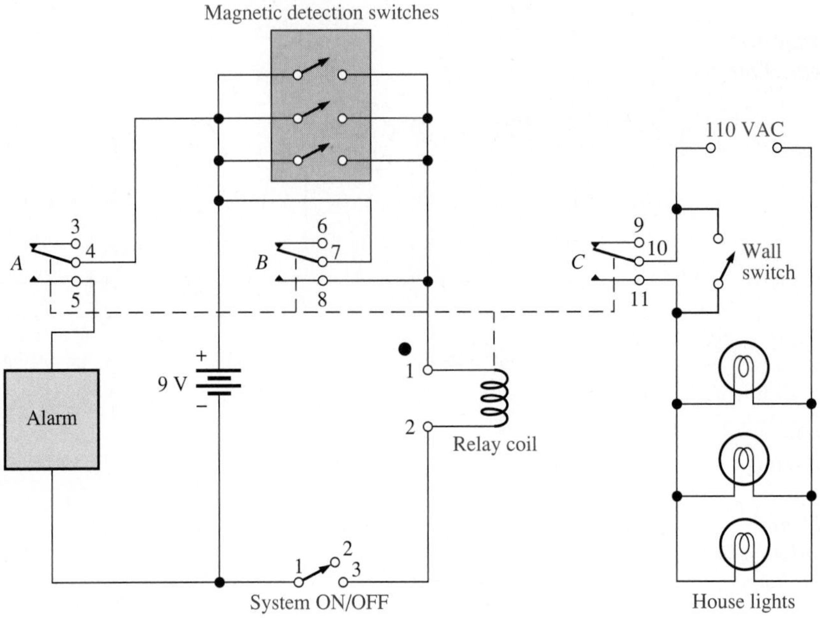

FIGURE 7–39
Simplified burglar alarm system.

The relay contacts are not physically remote in relation to the coil as the schematic indicates. The schematic is drawn this way for functional clarity. The entire relay is housed in the package shown in Figure 7–40. Also shown are the pin diagram and internal schematic for the relay.

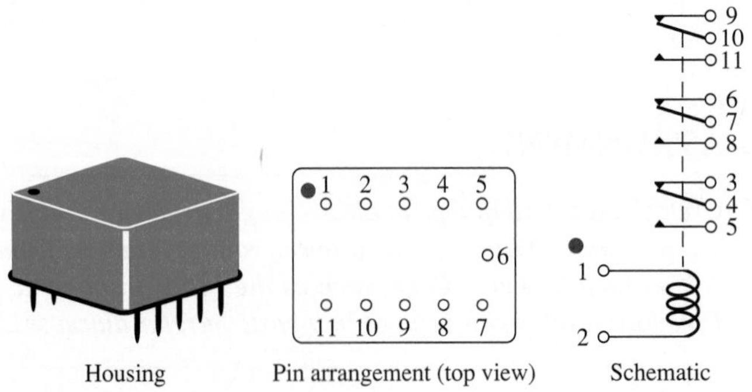

FIGURE 7–40
Triple-pole–double-throw relay.

Step 1: Connect the System

Develop a connection diagram and point-to-point wire list for interconnecting the components in Figure 7–41 to create the alarm system shown in the schematic of Figure 7–39. The connection points on the components are indicated by letters.

Step 2: Test the System

Develop a step-by-step procedure to check out the completely wired burglar alarm system.

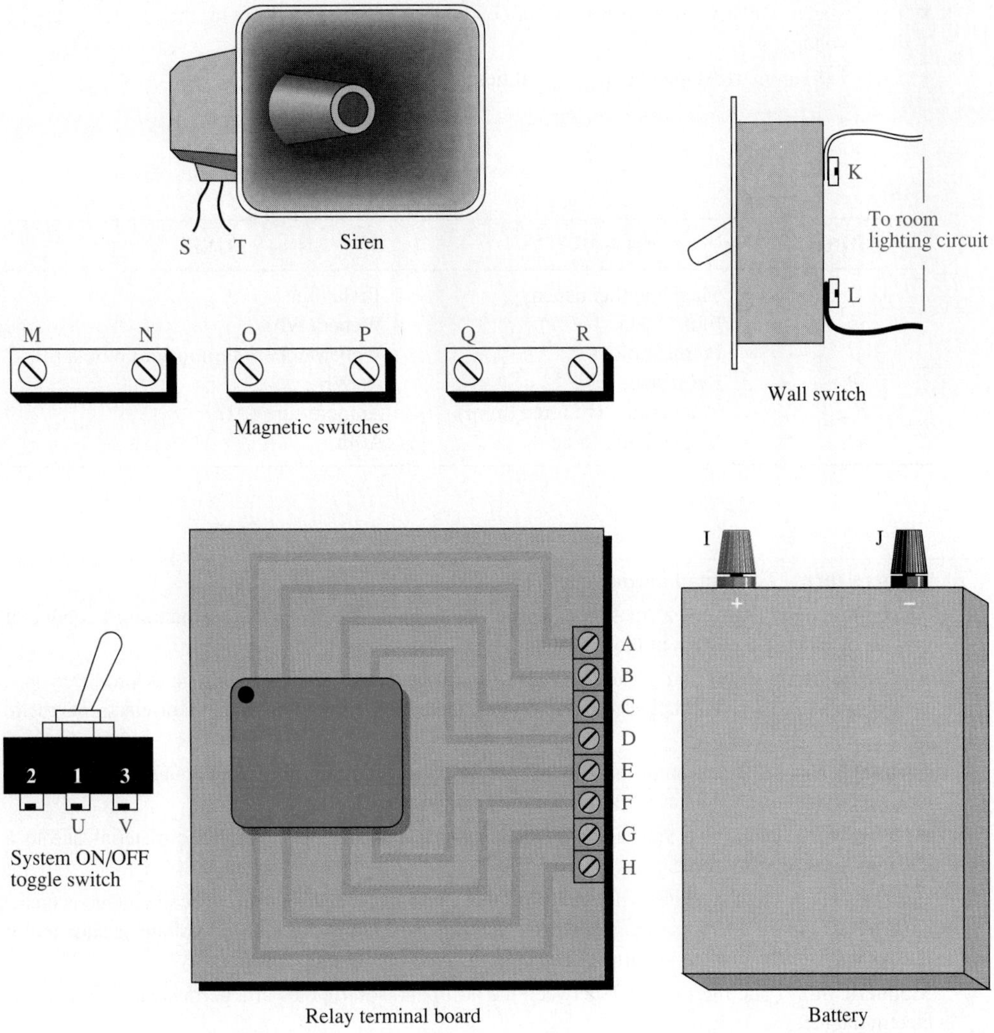

FIGURE 7–41
Array of burglar alarm components.

SECTION 7–7 REVIEW

1. How many magnetic detection switches in Figure 7–39 must be closed to activate the system?
2. What is the purpose of each of the three relay contacts?

■ SUMMARY

- Unlike magnetic poles attract each other, and like poles repel each other.
- Materials that can be magnetized are called *ferromagnetic*.
- When current flows through a conductor, it produces an electromagnetic field around the conductor.
- You can use the left-hand rule to establish the direction of the electromagnetic lines of force around a conductor.
- An electromagnet is basically a coil of wire around a magnetic core.
- When a conductor moves within a magnetic field, or when a magnetic field moves relative to a conductor, a voltage is induced across the conductor.

- The faster the relative motion between a conductor and a magnetic field, the greater is the induced voltage.
- Table 7–1 summarizes the magnetic quantities.

TABLE 7–1

SYMBOL	QUANTITY	UNIT
B	Magnetic flux density	Tesla (T)
ϕ	Flux	Weber (Wb)
μ	Permeability	Weber/ampere-turn·meter (Wb/At·m)
$\mathcal{R}$	Reluctance	At/Wb
F_m	Magnetomotive force (mmf)	Ampere-turn (At)
H	Magnetizing force	At/m

▪ GLOSSARY

Ampere-turn The unit of magnetomotive force (mmf).

Electromagnetic field A formation of a group of magnetic lines of force surrounding a conductor created by electrical current in the conductor.

Electromagnetic induction The phenomenon or process by which a voltage is produced in a conductor when there is relative motion between the conductor and a magnetic or electromagnetic field.

Faraday's law A law stating that the voltage induced across a coil of wire equals the number of turns in the coil times the rate of change of the magnetic flux.

Hysteresis A lagging in the values of resulting magnetization in a magnetic material due to a changing magnetizing force.

Lenz's law A law that states when the current through a coil changes, an induced voltage is created as a result of the changing magnetic field, and the direction of the induced voltage is such that it always opposes the change in current. The current cannot change instantaneously.

Magnetic flux The lines of force between the north and south poles of a permanent magnet or an electromagnet.

Magnetic flux density The number of lines of force per unit area in a magnetic field.

Magnetizing force The amount of mmf per unit length of magnetic material.

Magnetomotive force The force that produces a magnetic field.

Permeability The measure of ease with which a magnetic field can be established in a material.

Relay An electromagnetically controlled mechanical device in which electrical contacts are opened or closed by a magnetizing current.

Reluctance The opposition to the establishment of a magnetic field.

Retentivity The ability of a material, once magnetized, to maintain a magnetized state without the presence of a magnetizing force.

Solenoid An electromagnetically controlled device in which the mechanical movement of a shaft or plunger is activated by a magnetizing current.

Tesla The unit of flux density.

Weber The unit of magnetic flux.

▪ FORMULAS

(7–1) $$B = \frac{\phi}{A}$$ Magnetic flux density

(7–2) $$\mu_r = \frac{\mu}{\mu_0}$$ Relative permeability

(7–3) $$\mathcal{R} = \frac{l}{\mu A}$$ Reluctance

(7–4) $$F_m = NI$$ Magnetomotive force

(7–5) $\phi = \frac{F_m}{\mathcal{R}}$ Magnetic flux

(7–6) $H = \frac{F_m}{l}$ Magnetic flux intensity

■ SELF-TEST

1. When the south poles of two bar magnets are brought close together, there will be
(a) a force of attraction **(b)** a force of repulsion
(c) an upward force **(d)** no force

2. A magnetic field is made up of
(a) positive and negative charges **(b)** magnetic domains
(c) flux lines **(d)** magnetic poles

3. The direction of a magnetic field is from
(a) north pole to south pole **(b)** south pole to north pole
(c) inside to outside the magnet **(d)** from to back

4. Reluctance in a magnetic circuit is analogous to
(a) voltage in an electric circuit **(b)** current in an electric circuit
(c) power in an electric circuit **(d)** resistance in an electric circuit

5. The unit of magnetic flux is the
(a) tesla **(b)** weber **(c)** electron-volt **(d)** ampere-turn

6. The unit of magnetomotive force is the
(a) tesla **(b)** weber **(c)** ampere-turn **(d)** electron-volt

7. The unit of flux density is the
(a) tesla **(b)** weber **(c)** ampere-turn **(d)** electron-volt

8. The electromagnetic activation of a movable shaft is the basis for
(a) relays **(b)** circuit breakers **(c)** magnetic switches **(d)** solenoids

9. When current flows through a wire placed in a magnetic field,
(a) the wire will overheat **(b)** the wire will become magnetized
(c) a force is exerted on the wire **(d)** the magnetic field will be cancelled

10. A coil of wire is placed in a changing magnetic field. If the number of turns in the coil is increased, the voltage induced across the coil will
(a) remained unchanged **(b)** decrease
(c) increase **(d)** be excessive

11. If a conductor is moved back and forth at a constant rate in a constant magnetic field, the voltage induced in the conductor will
(a) remain constant **(b)** reverse polarity
(c) be reduced **(d)** be increased

12. In the crankshaft position sensor in Figure 7–32, the induced voltage across the coil is caused by
(a) current in the coil
(b) rotation of the disk
(c) a tab passing through the magnetic field
(d) acceleration of the disk's rotational speed

■ BASIC PROBLEMS

SECTION 7–1 The Magnetic Field

1. The cross-sectional area of a magnetic field is increased, but the flux remains the same. Does the flux density increase or decrease?

2. In a certain magnetic field the cross-sectional area is 0.5 m^2 and the flux is 1500 μWb. What is the flux density?

3. What is the flux in a magnetic material when the flux density is 2500×10^{-6} T and the cross-sectional area is 150 cm^2?

SECTION 7–2 Electromagnetism

4. What happens to the compass needle in Figure 7–9 when the current through the conductor is reversed?

5. What is the relative permeability of a ferromagnetic material whose absolute permeability is 750×10^{-6} Wb/At·m?

6. Determine the reluctance of a material with a length of 0.28 m and a cross-sectional area of 0.08 m^2 if the absolute permeability is 150×10^{-7} Wb/At·m.
7. What is the magnetomotive force in a 50 turn coil of wire with 3 A through it?

SECTION 7–3 Electromagnetic Devices

8. Typically, when a solenoid is activated, is the plunger extended or retracted?
9. **(a)** What force moves the plunger when a solenoid is activated?
   (b) What force causes the plunger to return to its at-rest position?
10. Explain the sequence of events in the circuit of Figure 7–42 starting when switch 1 (SW_1) is closed.
11. What causes the pointer in a d'Arsonval movement to deflect when there is current through the coil?

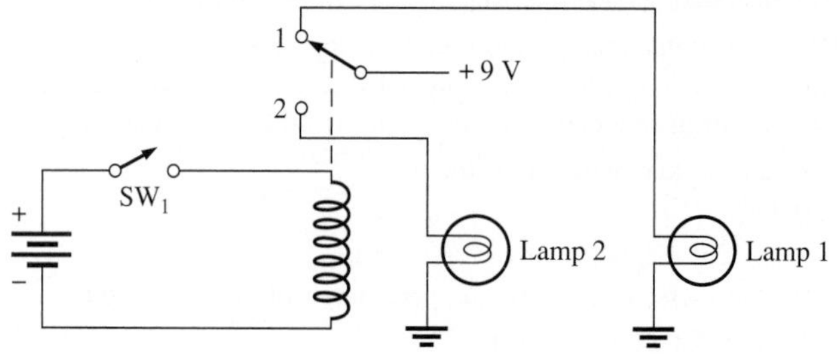

FIGURE 7–42

SECTION 7–4 Magnetic Hysteresis

12. What is the magnetizing force in Problem 7 if the length of the core is 0.2 m?
13. How can the flux density in Figure 7–43 be changed without altering the physical characteristics of the core?
14. In Figure 7–43; determine the following:
    (a) H **(b)** ϕ **(c)** B
15. Determine from the hysteresis curves in Figure 7–44 which material has the most retentivity.

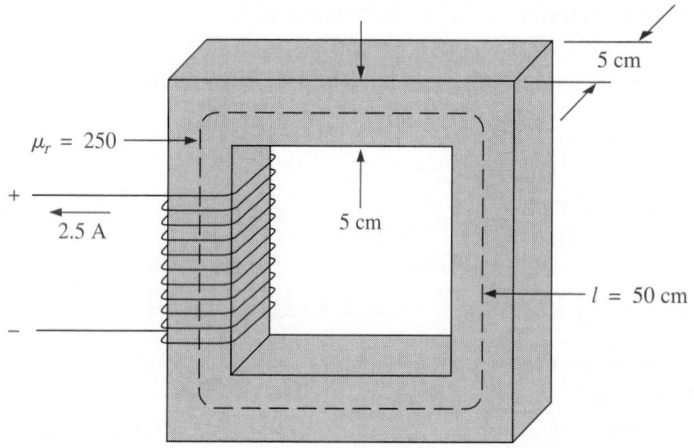

FIGURE 7–43

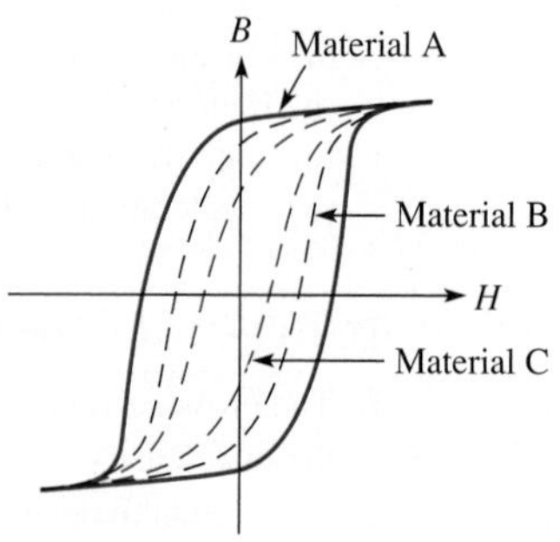

FIGURE 7–44

SECTION 7–5 Electromagnetic Induction

16. According to Faraday's law, what happens to the induced voltage across a given coil if the rate of change of magnetic flux doubles?
17. The voltage induced across a certain coil is 100 mV. A 100 Ω resistor is connected to the coil terminals. What is the induced current?

SECTION 7–6 Applications of Electromagnetic Induction

18. In Figure 7–32, why is there no induced voltage when the steel disk is not rotating?

19. Explain the purpose of the commutator and brushes in Figure 7–34.

■ ADVANCED PROBLEMS

20. A basic one-loop dc generator is rotated at 60 rps. How many times each second does the dc output voltage peak (reach a maximum)?

21. Assume that another loop, 90° from the first loop, is added to the dc generator in Problem 20. Make a graph of voltage versus time to show how the output voltage appears. Let the maximum voltage be 10 V.

■ ANSWERS TO SECTION REVIEWS

Section 7–1

1. The north poles repel.

2. Magnetic flux is the group of lines of force that make up a magnetic field.

3. $B = \phi/A = 900 \times 10^{-6}$ T

Section 7–2

1. Electromagnetism is produced by current through a conductor. An electromagnetic field exists only when current flows. A magnetic field exists independently of current.

2. The direction of the magnetic field also reverses when the current is reversed.

Section 7–3

1. A solenoid provides mechanical movement of a shaft. A relay provides contact closure.

2. The movable part of a solenoid is the plunger.

3. The movable part of a relay is the armature.

4. The d'Arsonval meter movement is based on the interaction of magnetic fields.

Section 7–4

1. In a wirewound core, an increase in current increases the flux density.

2. Retentivity is the ability of a material to remain magnetized after removal of the magnetizing force.

Section 7–5

1. The induced voltage is zero.

2. The induced voltage increases.

3. A force is exerted on the current-carrying conductor in a magnetic field.

Section 7–6

1. The induced voltage is zero.

2. The induced voltage increases.

Section 7–7

1. One or more magnetic detection switches must be closed.

2. Contact A activates alarm, contact B latches the relay, contact C turns on house lights.

■ ANSWERS TO RELATED EXERCISES FOR EXAMPLES

7–1 $B = 12000 \times 10^{-6}$ T

7–2 Reluctance quadruples.

7–3 7.2 mWb

7–4 8.5×10^{4} At/Wb

PART TWO

AC CIRCUITS

8

INTRODUCTION TO ALTERNATING CURRENT AND VOLTAGE

This chapter provides an introduction to alternating current (ac) circuits. Alternating voltages and currents fluctuate with time and periodically change polarity and direction according to certain patterns called **waveforms.** Particular emphasis is given to the sine wave because of its basic importance in ac circuits. Other types of waveforms are also introduced, including pulse, triangular, and sawtooth. The use of the oscilloscope for displaying and measuring waveforms is discussed.

■ CHAPTER OBJECTIVES

- □ Identify a sinusoidal waveform and measure its characteristics
- □ Describe how sine waves are generated
- □ Determine the voltage and current values of sine waves
- □ Describe angular relationships of sine waves
- □ Mathematically analyze a sinusoidal waveform
- □ Apply the basic circuit laws to ac resistive circuits
- □ Determine total voltages which are composed of both ac and dc components
- □ Use the oscilloscope to measure waveforms
- □ Identify the characteristics of basic nonsinusoidal waveforms

■ APPLICATION ASSIGNMENT

As an electronic technician working for a company that designs and manufactures laboratory instruments, you have been assigned to the testing of a new function generator that produces various types of time-varying voltages with adjustable parameters. The project supervisor has assigned you to measure and record the limits of operation of a prototype model. After studying this chapter, you should be able to complete the application assignment in Section 8–10.

8–1 ■ THE SINE WAVE

The sine wave is a common type of alternating current (ac) and alternating voltage. It is also referred to as a sinusoidal waveform, or, simply, sinusoid. The electrical service provided by the power companies is in the form of sinusoidal voltage and current. In addition, other types of waveforms are composites of many individual sine waves called harmonics.

After completing this section, you should be able to

- **Identify a sinusoidal waveform and measure its characteristics**
  - □ Define and determine the period
  - □ Define and determine the frequency
  - □ Relate the period and the frequency

Sine waves are produced by two types of sources: rotating electrical machines (ac generators) or electronic oscillator circuits, which are in instruments known as electronic signal generators. Figure 8–1 shows the symbol used to represent either source of sine wave voltage.

FIGURE 8–1
Symbol for a sine wave voltage source.

Figure 8–2 is a graph showing the general shape of a sine wave, which can be either an **alternating current** or voltage. Voltage (or current) is displayed on the vertical axis and time (t) is displayed on the horizontal axis. Notice how the voltage (or current) varies with time. Starting at zero, the voltage (or current) increases to a positive maximum (peak), returns to zero, and then increases to a negative maximum (peak) before returning again to zero, thus completing one full cycle.

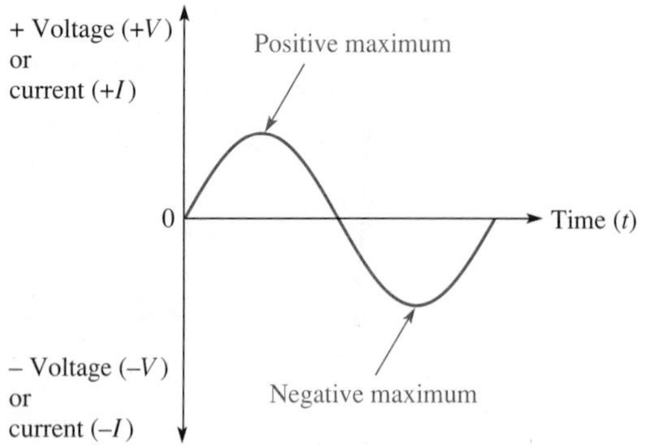

FIGURE 8–2
Graph of one cycle of sine wave.

Polarity of a Sine Wave

As you have seen, a sine wave changes polarity at its zero value; that is, it alternates between positive and negative values. When a sine wave voltage is applied to a resistive circuit, as in Figure 8–3, an alternating sine wave current results. When the voltage changes polarity, the current correspondingly changes direction as indicated.

During the positive alternation of the applied voltage V_s, the current is in the direction shown in Figure 8–3(a). During a negative alternation of the applied voltage, the current is in the opposite direction, as shown in Figure 8–3(b). The combined positive and negative alternations make up one **cycle** of a sine wave.

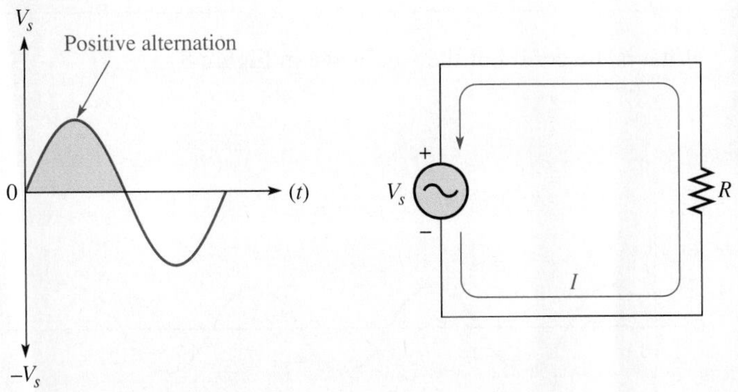

(a) Positive voltage: current direction as shown

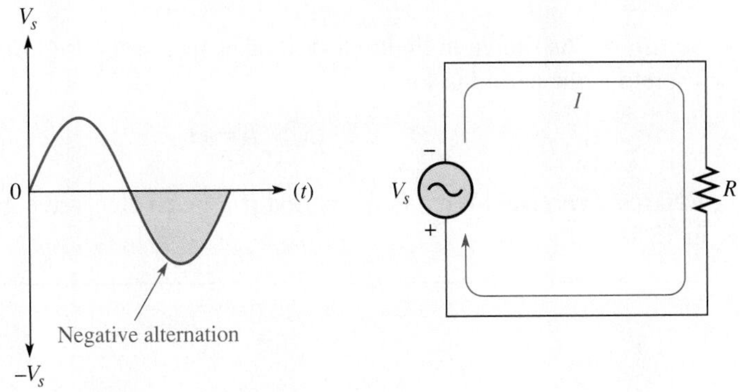

(b) Negative voltage: current reverses direction

FIGURE 8–3
Alternating current and voltage.

Period of a Sine Wave

The time required for a given sine wave to complete one full cycle is called the period (T).

Figure 8–4(a) illustrates the **period** of a sine wave. Typically, a sine wave continues to repeat itself in identical cycles, as shown in Figure 8–4(b). Since all cycles of a repetitive sine wave are the same, the period is always a fixed value for a given sine wave. The period of a sine wave can be measured from any point in a given cycle to the corresponding point in the next cycle.

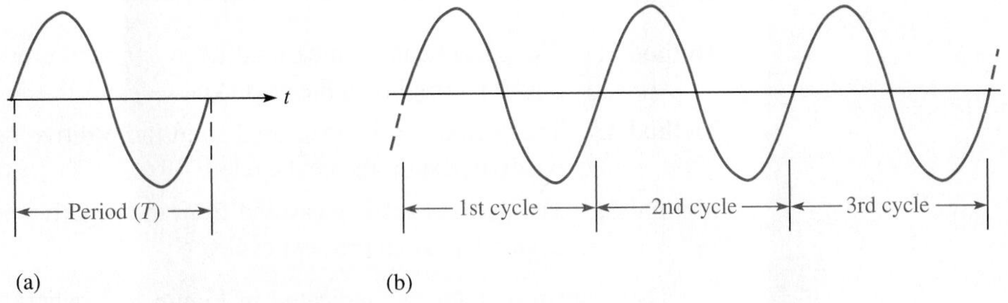

FIGURE 8–4
The period of a given sine wave is the same for each cycle.

EXAMPLE 8–1

What is the period of the sine wave in Figure 8–5?

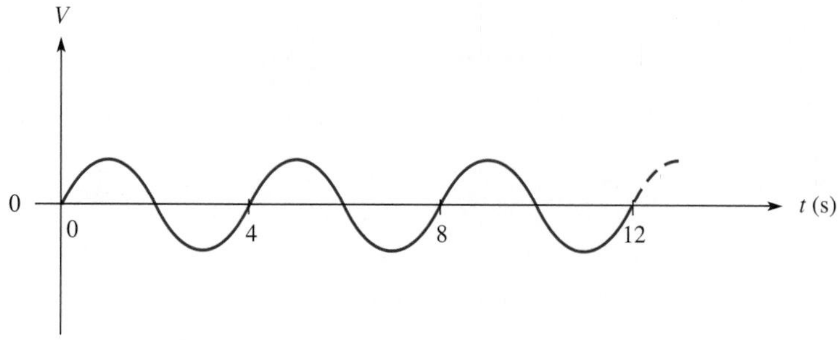

FIGURE 8–5

Solution As shown in Figure 8–5, it takes four seconds (4 s) to complete each cycle. Therefore, the period is 4 s.

$$T = 4 \text{ s}$$

Related Exercise What is the period if a given sine wave goes through five cycles in 12 s?

EXAMPLE 8–2

Show three possible ways to measure the period of the sine wave in Figure 8–6. How many cycles are shown?

FIGURE 8–6

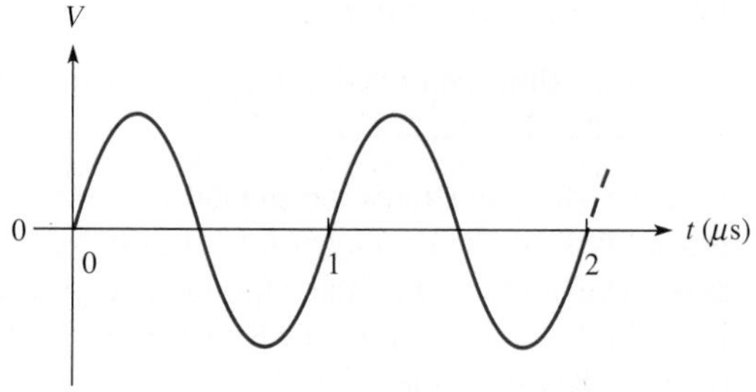

Solution

Method 1: The period can be measured from one zero crossing to the corresponding zero crossing in the next cycle.

Method 2: The period can be measured from the positive peak in one cycle to the positive peak in the next cycle.

Method 3: The period can be measured from the negative peak in one cycle to the negative peak in the next cycle.

These measurements are indicated in Figure 8–7, where two cycles of the sine wave are shown. Keep in mind that you obtain the same value for the period no matter which of these points on the waveform you use.

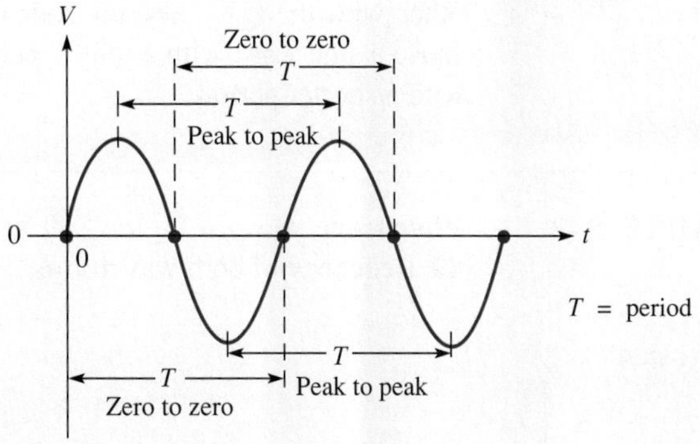

FIGURE 8–7
Measurement of the period of a sine wave.

Related Exercise If a positive peak occurs at 1 ms and the next positive peak occurs at 2.5 ms, what is the period?

Frequency of a Sine Wave

Frequency (f) is the number of cycles that a sine wave completes in 1 s.

The more cycles completed in one second, the higher the frequency. Figure 8–8 shows two sine waves. The sine wave in part (a) completes two full cycles in one second. The one in part (b) completes four cycles in one second. Therefore, the sine wave in part (b) has twice the frequency of the one in part (a).

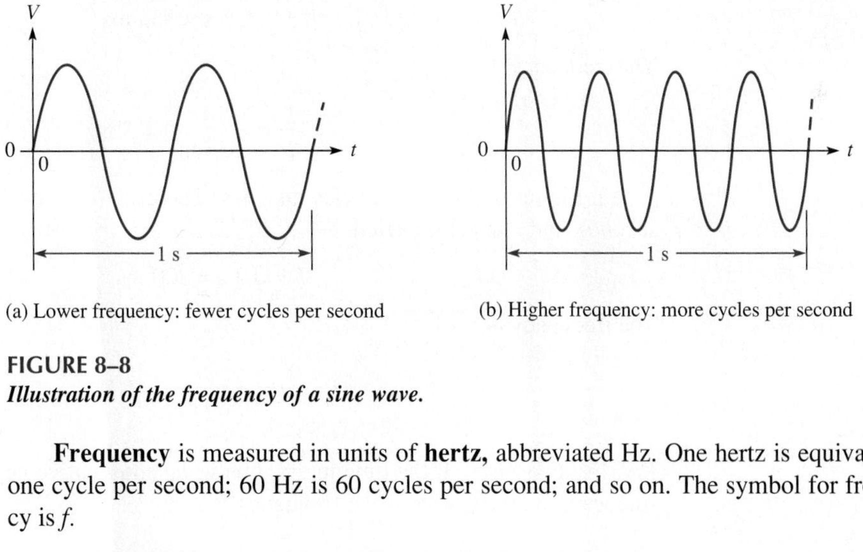

(a) Lower frequency: fewer cycles per second

(b) Higher frequency: more cycles per second

FIGURE 8–8
Illustration of the frequency of a sine wave.

Frequency is measured in units of **hertz,** abbreviated Hz. One hertz is equivalent to one cycle per second; 60 Hz is 60 cycles per second; and so on. The symbol for frequency is f.

Relationship of Frequency and Period

The relationship between frequency and period is very important. The formulas for this relationship are as follows:

$$f = \frac{1}{T} \qquad \textbf{(8–1)}$$

$$T = \frac{1}{f} \qquad \textbf{(8–2)}$$

There is a reciprocal relationship between f and T. Knowing one, you can calculate the other with the [1/x] key on your calculator. This inverse relationship makes sense because a sine wave with a longer period goes through fewer cycles in one second than one with a shorter period.

EXAMPLE 8–3

Which sine wave in Figure 8–9 has the higher frequency? Determine the period and the frequency of both waveforms.

FIGURE 8–9

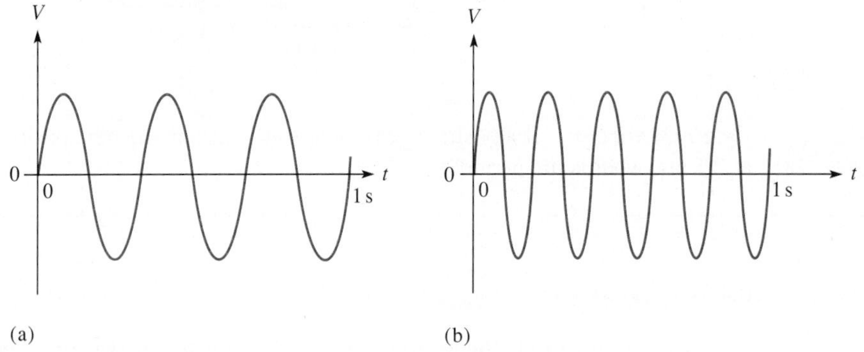

Solution The sine wave in Figure 8–9(b) has the higher frequency because it completes more cycles in 1 s than does the one in part (a).

In Figure 8–9(a), three cycles take 1 s. Therefore, one cycle takes 0.333 s (one-third second), and this is the period.

$$T = 0.333 \text{ s} = 333 \text{ ms}$$

The frequency is

$$f = \frac{1}{T} = \frac{1}{333 \text{ ms}} = 3 \text{ Hz}$$

In Figure 8–9(b), five cycles take 1 s. Therefore, one cycle takes 0.2 s (one-fifth second), and this is the period.

$$T = 0.2 \text{ s} = 200 \text{ ms}$$

The frequency is

$$f = \frac{1}{T} = \frac{1}{200 \text{ ms}} = 5 \text{ Hz}$$

Related Exercise If the time interval between consecutive negative peaks of a given sine wave is 50 μs, what is the frequency?

EXAMPLE 8–4

The period of a certain sine wave is 10 ms. What is the frequency?

Solution Use Equation (8–1).

$$f = \frac{1}{T} = \frac{1}{10 \text{ ms}} = \frac{1}{10 \times 10^{-3} \text{ s}} = 100 \text{ Hz}$$

Related Exercise A certain sine wave goes through four cycles in 20 ms. What is the frequency?

EXAMPLE 8–5

The frequency of a sine wave is 60 Hz. What is the period?

Solution Use Equation (8–2).

$$T = \frac{1}{f} = \frac{1}{60\text{ Hz}} = 16.7\text{ ms}$$

Related Exercise If $f = 1$ kHz, what is T?

SECTION 8–1 REVIEW

1. Describe one cycle of a sine wave.
2. At what point does a sine wave change polarity?
3. How many maximum points does a sine wave have during one cycle?
4. How is the period of a sine wave measured?
5. Define *frequency,* and state its unit.
6. Determine f when $T = 5\ \mu s$.
7. Determine T when $f = 120$ Hz.

8–2 ■ SINE WAVE VOLTAGE SOURCES

As mentioned in last section, two basic methods of generating sine wave voltages are electromagnetic and electronic. Sine waves are produced electromagnetically by ac generators and electronically by oscillator circuits.

After completing this section, you should be able to

- **Describe how sine waves are generated**
  - □ Discuss the basic operation of an ac generator
  - □ Discuss factors that affect frequency in ac generators
  - □ Discuss factors that affect voltage in ac generators

The AC Generator

Figure 8–10 is a cutaway view of one type of ac generator. To illustrate the basic operation, Figure 8–11 shows a greatly simplified ac **generator** consisting of a single loop of

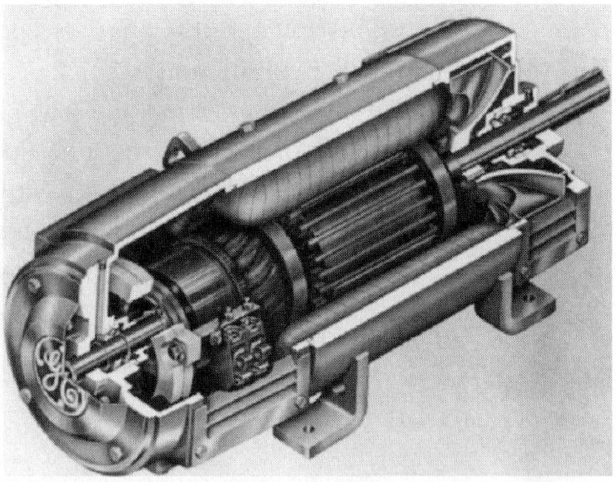

FIGURE 8–10
The ac generator, cutaway view (courtesy of General Electric).

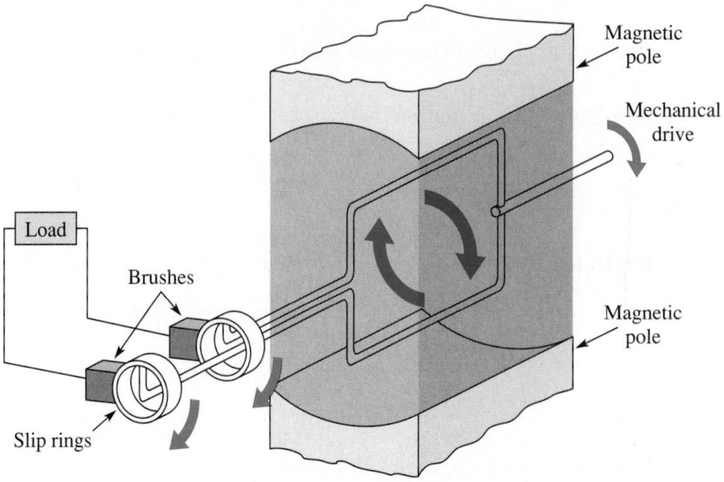

FIGURE 8–11
A basic ac generator.

wire in a permanent magnetic field. Notice that each end of the loop is connected to a separate solid conductive ring called a *slip ring*. As the loop rotates in the magnetic field, the slip rings also rotate and rub against the brushes which connect the loop to an external load. Compare this generator to the basic dc generator in Figure 7–34, and note the difference in the ring and brush arrangements.

As you learned in Chapter 7, when a conductor moves through a magnetic field, a voltage is induced. Figure 8–12 illustrates how a sine wave voltage is produced by the basic ac generator as the loop rotates. An oscilloscope is used to display the voltage waveform.

To begin, Figure 8–12(a) shows the loop rotating through the first quarter of a revolution. It goes from an instantaneous horizontal position, where the induced voltage is zero, to an instantaneous vertical position, where the induced voltage is maximum. As shown on the display, this part of the rotation produces the first quarter of the sine wave cycle as the voltage builds up from zero to its positive maximum. Figure 8–12(b) shows the loop completing the first half of a revolution. During this part of the rotation, the voltage decreases from its positive maximum back to zero.

During the second half of the revolution, illustrated in Figures 8–12(c) and 8–12(d), the loop is cutting through the magnetic field in the opposite direction, so the voltage produced has a polarity opposite to that produced during the first half of the revolution. After one complete revolution of the loop, one full cycle of the sine wave voltage has been produced. As the loop continues to rotate, repetitive cycles of the sine wave are generated.

Frequency You have seen that one revolution of the conductor through the magnetic field in the basic ac generator (also called an *alternator*) produces one cycle of induced sinusoidal voltage. It is obvious that the rate at which the conductor is rotated determines the time for completion of one cycle. For example, if the conductor completes 60 revolutions in one second (rps), the period of the resulting sine wave is 1/60 s, corresponding to a frequency of 60 Hz. Thus, the faster the conductor rotates, the higher the resulting frequency of the induced voltage, as illustrated in Figure 8–13.

Another way of achieving a higher frequency is to increase the number of magnetic poles. In the previous discussion, two magnetic poles were used to illustrate the ac generator principle. During one revolution, the conductor passes under a north pole and a south

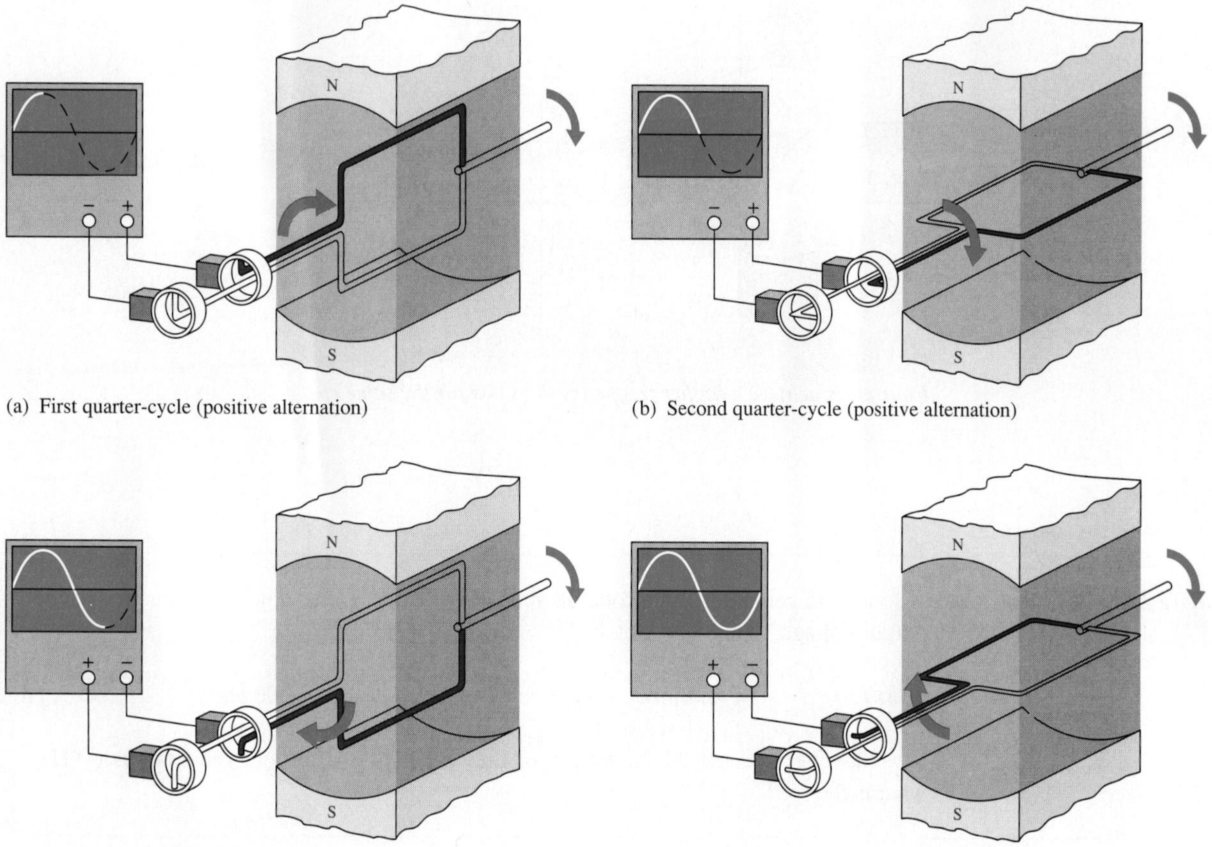

(a) First quarter-cycle (positive alternation)

(b) Second quarter-cycle (positive alternation)

(c) Third quarter-cycle (negative alternation)

(d) Fourth quarter-cycle (negative alternation)

FIGURE 8–12
One revolution of the loop generates one cycle of the sine wave voltage.

FIGURE 8–13
Frequency is directly proportional to the rate of rotation of the generator loop.

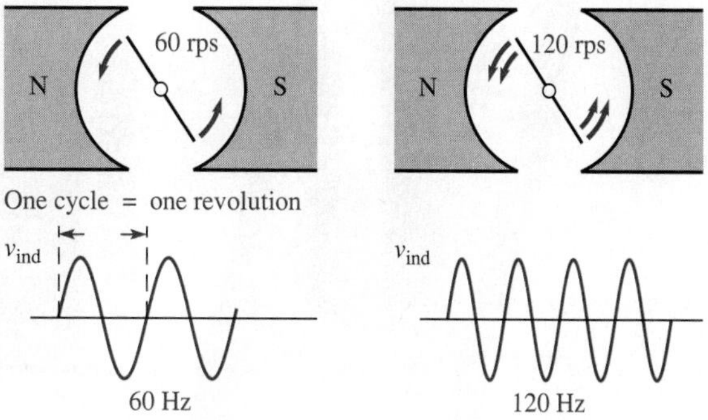

pole, thus producing one cycle of a sine wave. When four magnetic poles are used instead of two, as shown in Figure 8–14, one cycle is generated during one-half a revolution. This doubles the frequency for the same rate of rotation.

An expression for frequency in terms of the number of poles and the number of revolutions per second (rps) is as follows:

$$f = \text{number of pole pairs} \times (\text{rps}) \quad \textbf{(8–3)}$$

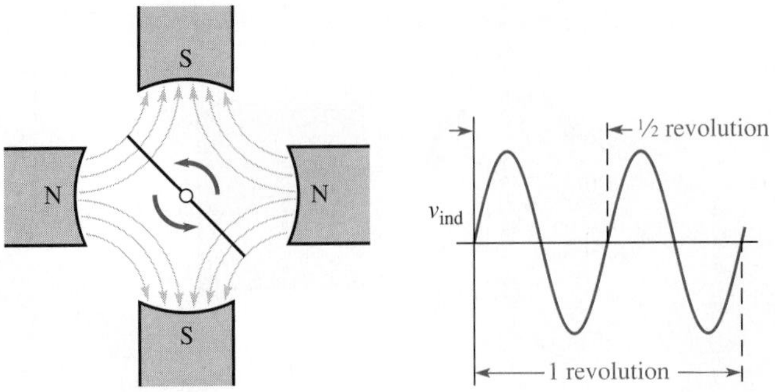

FIGURE 8–14
Four poles achieve a higher frequency than two for the same rps.

EXAMPLE 8–6

A four-pole generator has a rotation speed of 100 rps. Determine the frequency of the output voltage.

Solution f = number of pole pairs × (rps) = 2(100) = 200 Hz

Related Exercise If the frequency of the output of a four-pole generator is 60 Hz, what is the rps?

FIGURE 8–15
Typical signal generators. Parts (a) and (b) courtesy of B & K Precision. Part (c) courtesy of Hewlett-Packard.

Voltage Magnitude Recall from Chapter 7 that the amount of voltage induced in a conductor depends on the number of turns (N) and the rate of change with respect to the magnetic field. Therefore, when the speed of rotation of the conductor is increased, not only does the frequency of the induced voltage increase—so also does the magnitude. Since the frequency value normally is fixed, the most practical method of increasing the amount or magnitude of induced voltage is to increase the number of loops.

Signal Generators

The signal generator is an instrument that electronically produces sine waves for use in testing or controlling electronic circuits and systems. There are a variety of signal generators, ranging from special-purpose instruments that produce only one type of waveform in a limited frequency range, to programmable instruments that produce a wide range of frequencies and a variety of waveforms. All signal generators consist basically of an **oscillator,** which is an electronic circuit that produces sine wave voltages whose amplitude and frequency can be adjusted. Typical signal generators are shown in Figure 8–15.

SECTION 8–2 REVIEW

1. List two types of sine wave sources.
2. How are sine waves generated electromagnetically?
3. Upon what is the frequency of a generator's output voltage dependent?
4. Upon what is the magnitude of a generator's output voltage dependent?

8–3 ▪ VOLTAGE AND CURRENT VALUES OF SINE WAVES

There are several ways to express and measure the value of a sine wave in terms of its voltage or its current magnitude. These are instantaneous, peak, peak-to-peak, rms, and average values.

After completing this section, you should be able to

- **Determine the voltage and current values of sine waves**
  - □ Find the instantaneous value at any point
  - □ Find the peak value
  - □ Find the peak-to-peak value
  - □ Define *rms* and find the rms value
  - □ Explain why the average value of an alternating sine wave is always zero over a complete cycle
  - □ Find the half-cycle average value

Instantaneous Value

Figure 8–16 illustrates that at any point in time on a sine wave, the voltage (or current) has an **instantaneous value.** This instantaneous value is different at different points along the curve. Instantaneous values are positive during the positive alternation and negative during the negative alternation. Instantaneous values of voltage and current are symbolized by lowercase v and i, respectively, as shown in part (a). The curve is shown for voltage only, but applies equally for current when v's are replaced with i's. An example is shown in part (b), where the instantaneous voltage is 3.1 V at 1 μs, 7.07 V at 2.5 μs, 10 V at 5 μs, 0 V at 10 μs, −3.1 V at 11 μs, and so on.

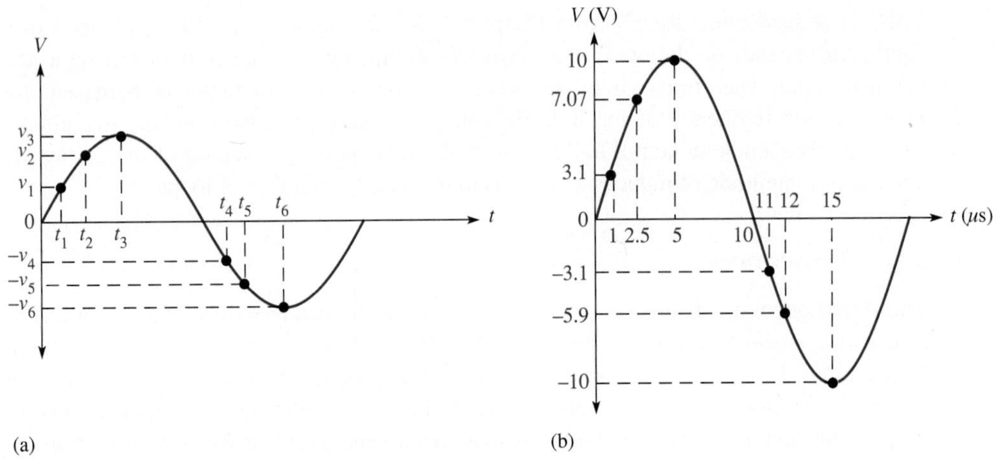

FIGURE 8–16
Example of instantaneous values of a sine wave voltage.

Peak Value

The **peak value** of a sine wave is the value of voltage (or current) at the positive or the negative maximum (peaks) with respect to zero. Since the peaks are equal in magnitude, a sine wave is characterized by a single peak value, as is illustrated in Figure 8–17. For a given sine wave, the peak value is constant and is represented by V_p or I_p. In this case, the peak value is 8 V.

Peak-to-Peak Value

The **peak-to-peak value** of a sine wave, as illustrated in Figure 8–18, is the voltage (or current) from the positive peak to the negative peak. It is always twice the peak value as expressed in the following equations:

$$V_{pp} = 2V_p \tag{8–4}$$

$$I_{pp} = 2I_p \tag{8–5}$$

Peak-to-peak values are represented by the symbols V_{pp} or I_{pp}. In Figure 8–18, the peak-to-peak value is 16 V.

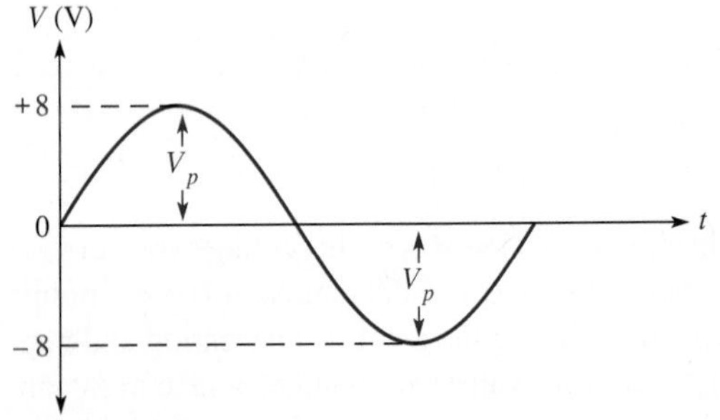

FIGURE 8–17
Example of the peak value of a sine wave voltage.

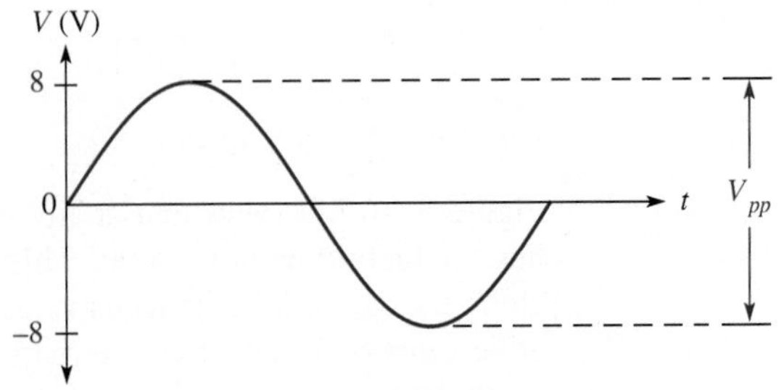

FIGURE 8–18
Example of the peak-to-peak value of a sine wave voltage.

RMS Value

The term *rms* stands for **root mean square.** It refers to the mathematical process by which this value is derived. The rms value is also referred to as the **effective value.** Most ac voltmeters display the rms value of a voltage. For example, the 110 V at your wall outlet is an rms value.

The rms value of a sine wave is actually a measure of the heating effect of the sine wave. For example, when a resistor is connected across an ac (sine wave) voltage source, as shown in Figure 8–19(a), a certain amount of heat is generated by the power in the resistor. Figure 8–19(b) shows the same resistor connected across a dc voltage source. In part (a), the value of the ac voltage can be adjusted so that the resistor gives off the same amount of heat as it does when connected to the dc source.

The rms value of a sine wave is equal to the dc voltage that produces the same amount of heat in a resistance as does the sinusoidal voltage.

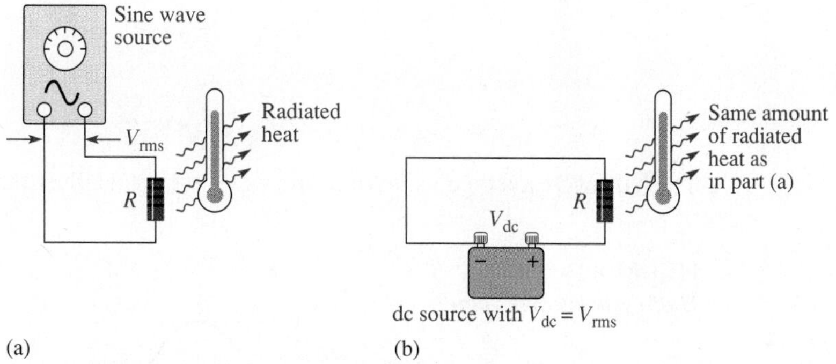

FIGURE 8–19
When the same amount of heat is being produced in both cases, the sine wave has an rms value equal to the dc voltage.

The peak value of a sine wave can be converted to the corresponding rms value using the following relationships for either voltage or current:

$$V_{rms} = 0.707V_p \quad \textbf{(8–6)}$$

$$I_{rms} = 0.707I_p \quad \textbf{(8–7)}$$

Using these formulas, you can also determine the peak value knowing the rms value as follows:

$$V_p = \frac{V_{rms}}{0.707}$$

$$V_p = 1.414V_{rms} \quad \textbf{(8–8)}$$

Similarly,

$$I_p = 1.414I_{rms} \quad \textbf{(8–9)}$$

To find the peak-to-peak value, simply double the peak value.

$$V_{pp} = 2.828V_{\text{rms}} \tag{8–10}$$

and

$$I_{pp} = 2.828I_{\text{rms}} \tag{8–11}$$

Average Value

The **average value** of a sine wave when taken over one complete cycle is always zero, because the positive values (above the zero crossing) offset the negative values (below the zero crossing).

To be useful for comparison purposes and in determining the average value of a rectified voltage such as found in power supplies, the **average value** of a sine wave is defined over a half-cycle rather than over a full cycle. The average value is expressed in terms of the peak value as follows for both voltage and current sine waves:

$$V_{\text{avg}} = 0.637V_p \tag{8–12}$$

$$I_{\text{avg}} = 0.637\,I_p \tag{8–13}$$

The half-cycle average value of a sine wave voltage is illustrated in Figure 8–20.

FIGURE 8–20
Half-cycle average value.

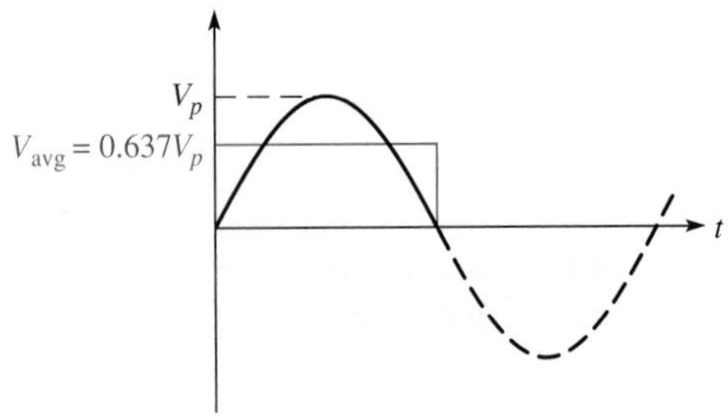

EXAMPLE 8–7 Determine V_p, V_{pp}, V_{rms}, and the half-cycle V_{avg} for the sine wave in Figure 8–21.

FIGURE 8–21

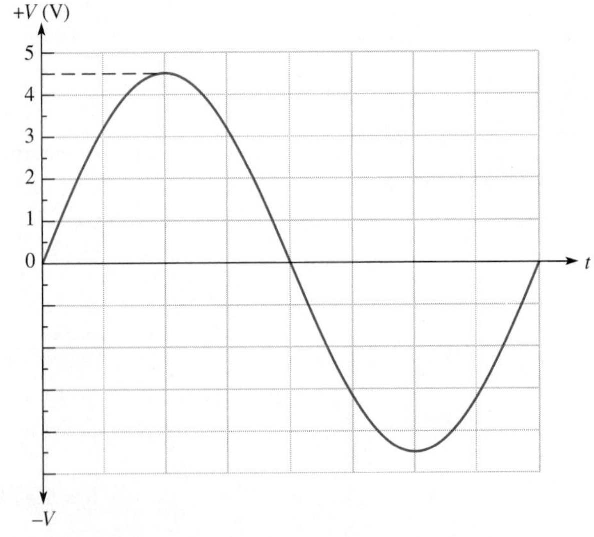

Solution As taken directly from the graph, $V_p = 4.5$ V. From this value calculate the other values:

$$V_{pp} = 2V_p = 2(4.5 \text{ V}) = 9 \text{ V}$$
$$V_{rms} = 0.707V_p = 0.707(4.5 \text{ V}) = 3.18 \text{ V}$$
$$V_{avg} = 0.637V_p = 0.637(4.5 \text{ V}) = 2.87 \text{ V}$$

Related Exercise If $V_p = 25$ V, determine V_{pp}, V_{rms}, and V_{avg} for a sine wave.

SECTION 8–3 REVIEW

1. Determine V_{pp} in each case when
   (a) $V_p = 1$ V **(b)** $V_{rms} = 1.414$ V **(c)** $V_{avg} = 3$ V
2. Determine V_{rms} in each case when
   (a) $V_p = 2.5$ V **(b)** $V_{pp} = 10$ V **(c)** $V_{avg} = 1.5$ V
3. Determine the half-cycle V_{avg} in each case when
   (a) $V_p = 10$ V **(b)** $V_{rms} = 2.3$ V **(c)** $V_{pp} = 60$ V

8–4 ■ ANGULAR MEASUREMENT OF A SINE WAVE

As you have seen, sine waves can be measured along the horizontal axis on a time basis; however, since the time for completion of one full cycle or any portion of a cycle is frequency-dependent, it is often useful to specify points on the sine wave in terms of an angular measurement expressed in degrees or radians. Angular measurement is independent of frequency.

After completing this section, you should be able to

- **Describe angular relationships of sine waves**
  - ☐ Show how to measure a sine wave in terms of angles
  - ☐ Define *radian*
  - ☐ Convert radians to degrees
  - ☐ Determine the phase of a sine wave

As previously discussed, sine wave voltage can be produced electromagnetically by rotating electromechanical machines. As the rotor of the ac generator goes through a full 360° of rotation, the resulting output is one full cycle of a sine wave voltage. Thus, the angular measurement of a sine wave can be related to the angular rotation of a generator, as shown in Figure 8–22.

Angular Measurement

A **degree** is an angular measurement corresponding to 1/360 of a circle or a complete revolution. A **radian** (rad) is the angular distance along the circumference of a circle equal to the radius of the circle. One radian is equivalent to 57.3°, as illustrated in Figure 8–23. In a 360° revolution, there are 2π radians.

The Greek letter π (pi) represents the ratio of the circumference of any circle to its diameter and has a constant value of approximately 3.1416.

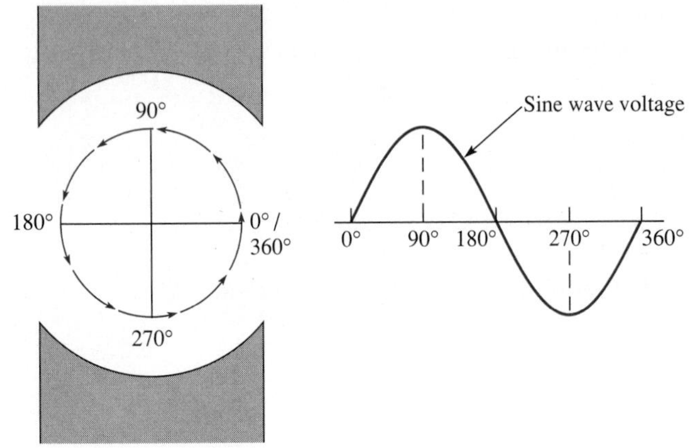

FIGURE 8–22

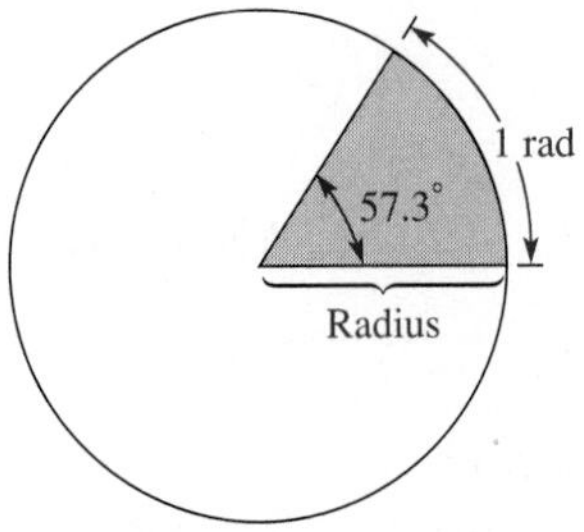

FIGURE 8–23
Angular measurement showing relationship of the radian to degrees.

Scientific calculators have a π key so that the actual numerical value does not have to be entered.

Table 8–1 lists several values of degrees and the corresponding radian values. These angular measurements are illustrated in Figure 8–24.

TABLE 8–1

DEGREES (°)	RADIANS (rad)
0	0
45	$\pi/4$
90	$\pi/2$
135	$3\pi/4$
180	π
225	$5\pi/4$
270	$3\pi/2$
315	$7\pi/4$
360	2π

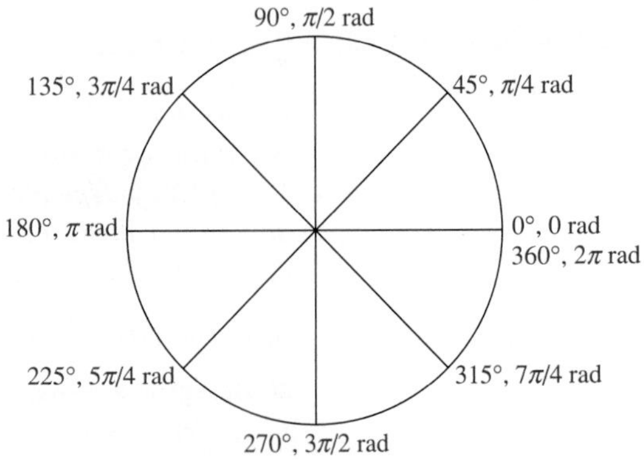

FIGURE 8–24
Angular measurements starting at 0° and going counterclockwise.

Radian/Degree Conversion

Radians can be converted to degrees using Equation (8–14).

$$\text{rad} = \left(\frac{\pi\ \text{rad}}{180^\circ}\right) \times \text{degrees} \qquad \textbf{(8–14)}$$

Similarly, degrees can be converted to radians with Equation (8–15).

$$\text{degrees} = \left(\frac{180^\circ}{\pi\ \text{rad}}\right) \times \text{rad} \qquad \textbf{(8–15)}$$

EXAMPLE 8–8 (a) Convert 60° to radians. (b) Convert $\pi/6$ rad to degrees.

Solution

(a) $\text{Rad} = \left(\dfrac{\pi \text{ rad}}{180°}\right)60° = \dfrac{\pi}{3} \text{ rad} = 1.047 \text{ rad}$

(b) $\text{Degrees} = \left(\dfrac{180°}{\pi \text{ rad}}\right)\left(\dfrac{\pi}{6} \text{ rad}\right) = 30°$

Related Exercise (a) Convert 15° to radians. (b) Convert 2π radians to degrees.

Sine Wave Angles

The angular measurement of a sine wave is based on 360° or 2π radians for a complete cycle. A half-cycle is 180° or π radians; a quarter-cycle is 90° or $\pi/2$ radians; and so on. Figure 8–25(a) shows angles in degrees over a full cycle of a sine wave; part (b) shows the same points in radians.

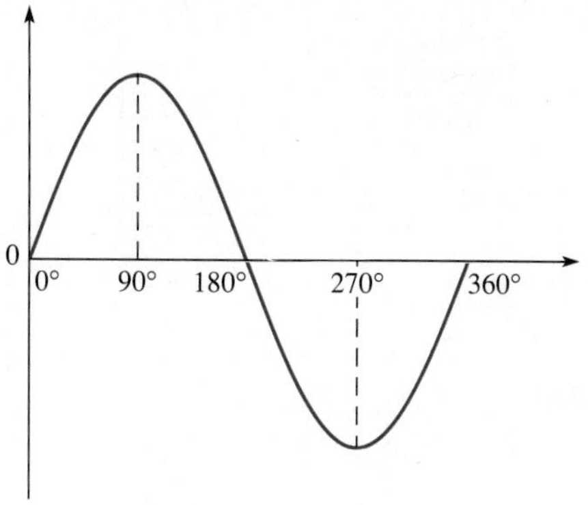

(a) Degrees

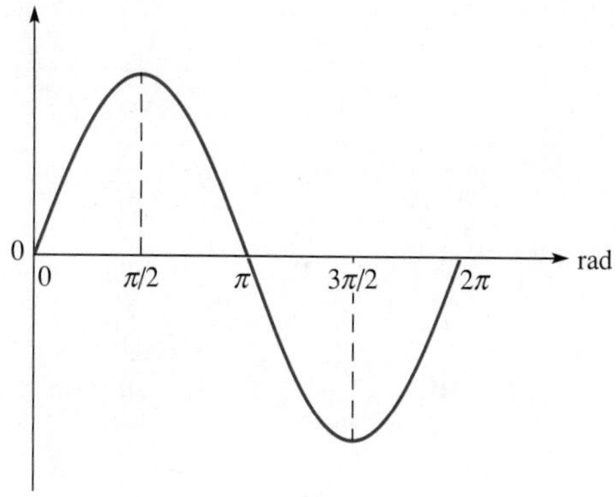

(b) Radians

FIGURE 8–25
Sine wave angles.

Phase of a Sine Wave

The **phase** of a sine wave is an angular measurement that specifies the position of that sine wave relative to a reference. Figure 8–26 shows one cycle of a sine wave to be used

FIGURE 8–26
Phase reference.

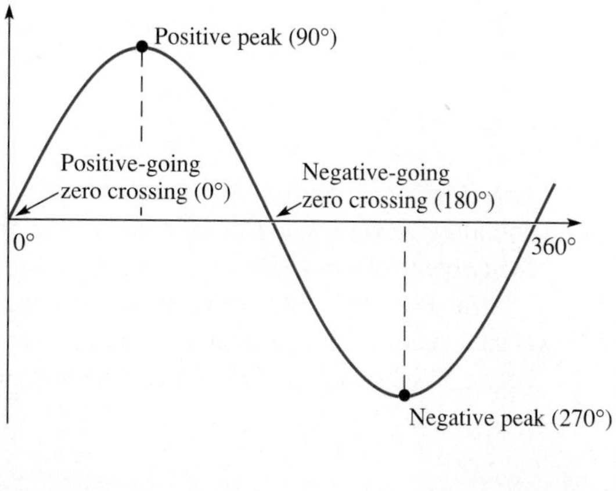

as the reference. Note that the first positive-going crossing of the horizontal axis (zero crossing) is at 0° (0 rad), and the positive peak is at 90° (π/2 rad). The negative-going zero crossing is at 180° (π rad), and the negative peak is at 270° (3π/2 rad). The cycle is completed at 360° (2π rad). When the sine wave is shifted left or right with respect to this reference, there is a phase shift.

Figure 8–27 illustrates phase shifts of a sine wave. In part (a), sine wave B is shifted to the right by 90° (π/2 rad). Thus, there is a phase angle of 90° between sine wave A and sine wave B. In terms of time, the positive peak of sine wave B occurs later than the positive peak of sine wave A, because time increases to the right along the horizontal axis. In this case, sine wave B is said to **lag** sine wave A by 90° or π/2 radians. In other words, sine wave A leads sine wave B by 90°.

In Figure 8–27(b), sine wave B is shown shifted left by 90°. Thus, again there is a phase angle of 90° between sine wave A and sine wave B. In this case, the positive peak of sine wave B occurs earlier in time than that of sine wave A; therefore, sine wave B is said to **lead** by 90°. In both cases there is a 90° phase angle between the two waveforms.

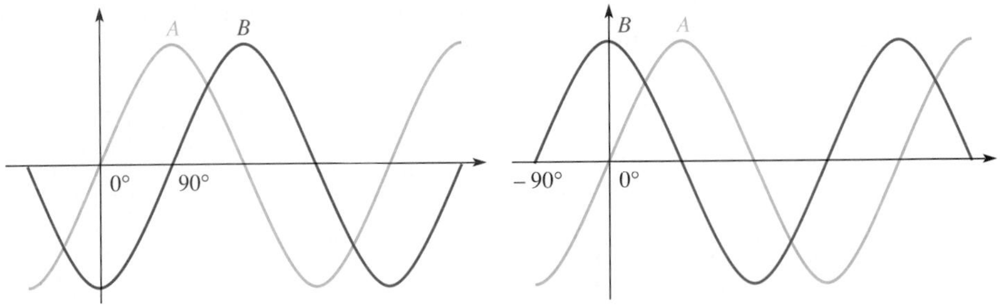

(a) A leads B by 90°, or B lags A by 90°.

(b) B leads A by 90°, or A lags B by 90°.

FIGURE 8–27
Illustration of a phase shift.

EXAMPLE 8–9

What are the phase angles between the two sine waves in Figure 8–28(a) and 8–28(b)?

FIGURE 8–28

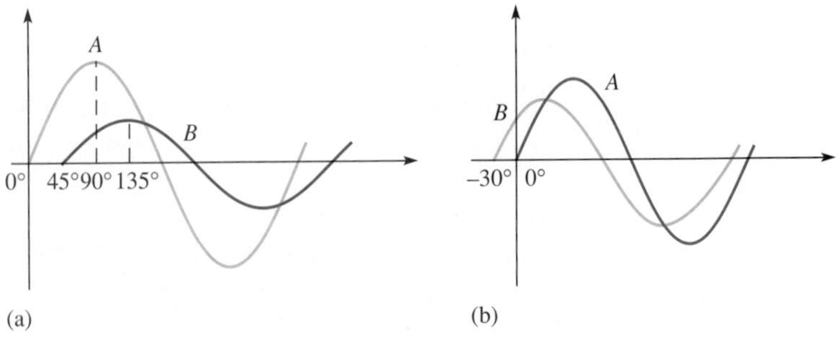

Solution In Figure 8–28(a) the zero crossing of sine wave A is at 0°, and the corresponding zero crossing of sine wave B is at 45°. There is a 45° phase angle between the two waveforms with sine wave A leading.

In Figure 8–28(b) the zero crossing of sine wave B is at –30°, and the corresponding zero crossing of sine wave A is at 0°. There is a 30° phase angle between the two waveforms with sine wave B leading.

Related Exercise If the positive-going zero crossing of one sine wave is at 15° relative to the 0° reference and that of the second sine wave is at 23° relative to the 0° reference, what is the phase angle between the sine waves?

SECTION 8–4 REVIEW

1. When the positive-going zero crossing of a sine wave occurs at 0°, at what angle does each of the following points occur?
   (a) positive peak **(b)** negative-going zero crossing
   (c) negative peak **(d)** end of first complete cycle
2. A half cycle is completed in ______ degrees, or ______ radians.
3. A full cycle is completed in ______ degrees, or ______ radians.
4. Determine the phase angle between the two sine waves in Figure 8–29.

FIGURE 8–29

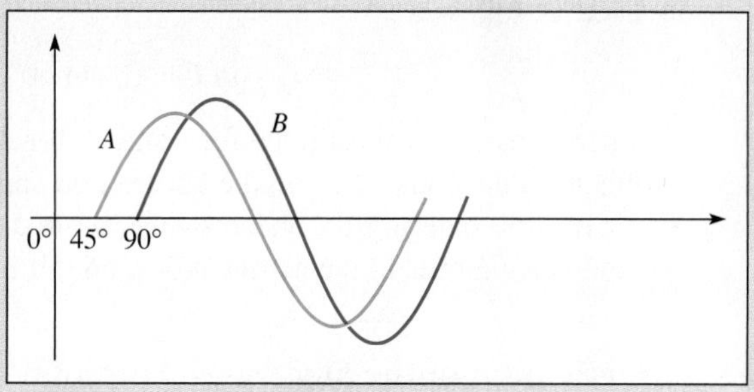

8–5 ■ THE SINE WAVE FORMULA

A sine wave can be graphically represented by voltage or current values on the vertical axis and by angular measurement (degrees or radians) along the horizontal axis. This graph can be expressed mathematically, as you will see.

After completing this section, you should be able to

- **Mathematically analyze a sinusoidal waveform**
  - ☐ State the sine wave formula
  - ☐ Find instantaneous values using the formula

A generalized graph of one cycle of a sine wave is shown in Figure 8–30. The **amplitude,** A, is the maximum value of the voltage or current on the vertical axis, and angular values run along the horizontal axis. The variable y is an instantaneous value representing either voltage or current at a given angle, θ.

All sine wave curves follow a specific mathematical formula. The general expression for the sine wave curve in Figure 8–30 is

$$y = A \sin \theta \qquad \textbf{(8–16)}$$

This general expression states that any point on the sine wave, represented by an instantaneous value y, is equal to the maximum value A, times the sine (sin) of the angle θ at that point. For example, if a certain sine wave voltage has a peak value of

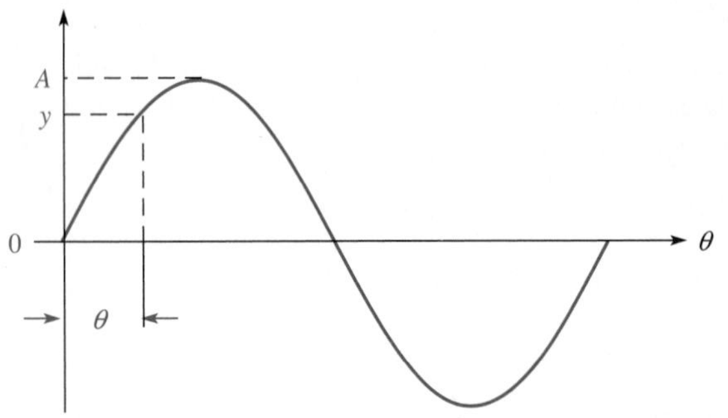

FIGURE 8–30
One cycle of a generalized sine wave showing amplitude and phase.

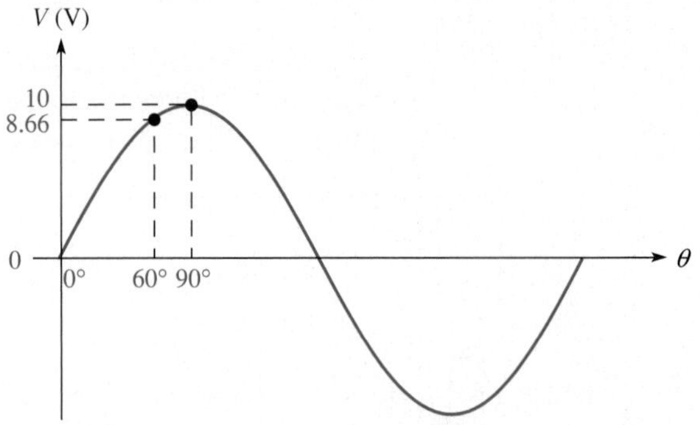

FIGURE 8–31
Illustration of the instantaneous value of a sine wave voltage at θ = 60°.

10 V, the instantaneous voltage at a point 60° along the horizontal axis can be calculated as follows:

$$v = V_p \sin \theta = 10 \sin 60° = 10(0.866) = 8.66 \text{ V}$$

where v has been substituted for y and V_p for A in the general formula. Figure 8–31 shows this instantaneous value on the curve. You can find the sine of any angle on your calculator by first entering the value of the angle and then pressing the [sin] key. Verify that your calculator is in the degree mode, and, if not, use the [DRG] key.

Derivation of the Sine Wave Formula

A rotating arrow called a **phasor** can be used to represent time-varying quantities such as sine waves in terms of their magnitude and angular measurement. One full cycle of a sine wave can be represented by rotation of a phasor through 360°.

Figure 8–32 shows how the phasor "traces out" the sine wave as it goes counterclockwise from 0° to 360°. You can relate this concept to the rotation in an ac generator as mentioned earlier.

Notice in Figure 8–32 that the length of the phasor arrow is equal to the peak value of the sine wave (observe the 90° and the 270° points). The angle of the phasor measured from 0° is the corresponding angular point on the sine wave.

Let's examine a phasor representation at one specific angle. Figure 8–33 shows the voltage phasor at an angular position of 45° and the corresponding point on the sine wave. The instantaneous value, v, of the sine wave at this point is related to both the position and the length of the phasor. The vertical distance from the phasor tip down to the horizontal axis represents the instantaneous value of the sine wave at that point.

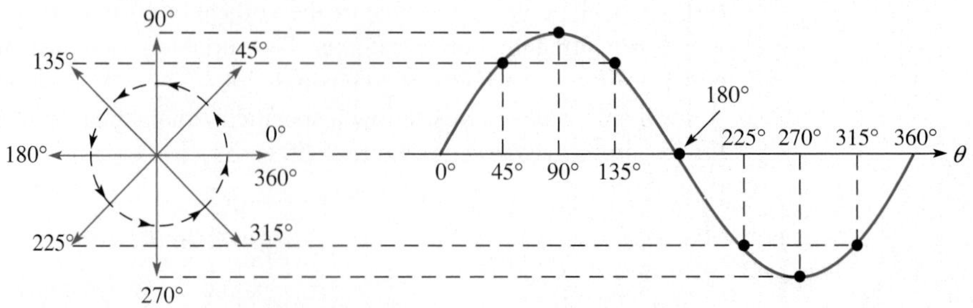

FIGURE 8–32
Sine wave represented by a rotating phasor arrow.

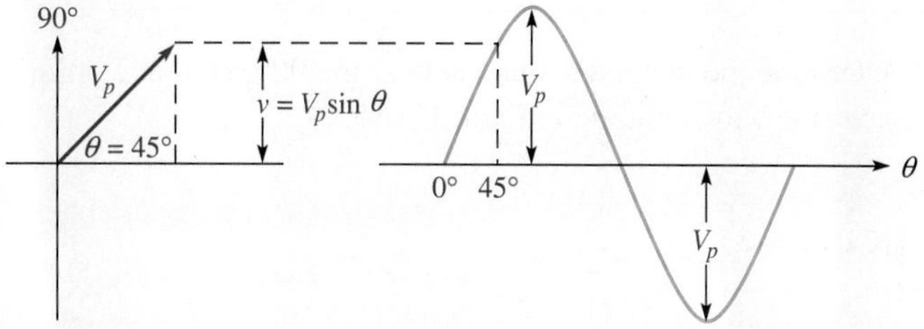

FIGURE 8–33
Right triangle derivation of sine wave formula, $v = V_p \sin \theta$.

Notice that when a vertical line is drawn from the phasor tip down to the horizontal axis, a **right triangle** is formed, as shown Figure 8–33. The length of the phasor is the **hypotenuse** of the triangle, and the vertical projection is the opposite side. From trigonometry, *the opposite side of a right triangle is equal to the hypotenuse times the sine of the angle* θ. In this case, the length of the phasor is the peak value of the sine wave voltage, V_p. Thus, the opposite side of the triangle, which is the instantaneous value, can be expressed as

$$v = V_p \sin \theta \qquad \textbf{(8–17)}$$

This formula also applies to a current sine wave as follows:

$$i = I_p \sin \theta \qquad \textbf{(8–18)}$$

Expressions for Phase-Shifted Sine Waves

When a sine wave is shifted to the right of the reference (lagging) by a certain angle, ϕ, as illustrated in Figure 8–34(a), the general expression is

$$y = A \sin(\theta - \phi) \qquad \textbf{(8–19)}$$

where y represents instantaneous voltage or current, and A represents the peak value (amplitude). When a sine wave is shifted to the left of the reference (leading) by a certain angle, ϕ, as shown in Figure 8–34(b), the general expression is

$$y = A \sin(\theta + \phi) \qquad \textbf{(8–20)}$$

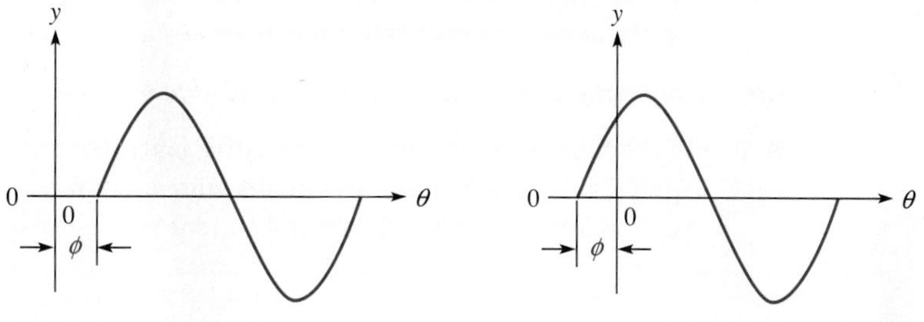

(a) $y = A \sin(\theta - \phi)$ (b) $y = A \sin(\theta + \phi)$

FIGURE 8–34
Shifted sine waves.

EXAMPLE 8–10 Determine the instantaneous value at the 90° reference point on the horizontal axis for each sine wave voltage in Figure 8–35.

FIGURE 8–35

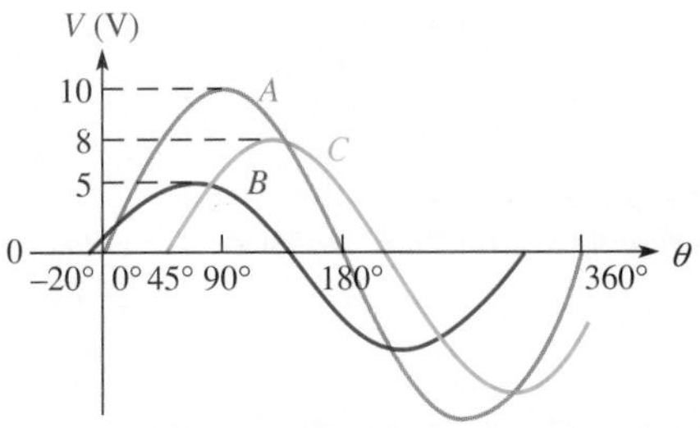

Solution Sine wave A is the reference. Sine wave B is shifted left 20° with respect to A, so it leads. Sine wave C is shifted right 45° with respect to A, so it lags.

$$v_A = V_p \sin\theta = 10 \sin 90° = 10(1) = 10 \text{ V}$$

$$v_B = V_p \sin(\theta + \phi_B)$$
$$= 5 \sin(90° + 20°) = 5 \sin 110° = 5(0.9397) = 4.7 \text{ V}$$

$$v_C = V_p \sin(\theta - \phi_C)$$
$$= 8 \sin(90° - 45°) = 8 \sin 45° = 8(0.707) = 5.66 \text{ V}$$

Related Exercise A sine wave has a peak value of 20 V. What is the instantaneous value at + 65° from the 0° reference?

SECTION 8–5 REVIEW

1. Determine the sine of the following angles:
   (a) 30° **(b)** 60° **(c)** 90°
2. Calculate the instantaneous value at 120° for the sine wave in Figure 8–31.
3. Determine the instantaneous value at the 45° point on the reference axis of a sine wave shifted 10° to the left of the zero reference (V_p = 10 V).

8–6 ■ OHM'S LAW AND KIRCHHOFF'S LAWS IN AC CIRCUITS

When time-varying ac voltages such as the sine wave are applied to a circuit, the circuit laws that you studied earlier still apply. Ohm's law and Kirchhoff's laws apply to ac circuits in the same way that they apply to dc circuits.

After completing this section, you should be able to

- **Apply the basic circuit laws to ac resistive circuits**
  - □ Apply Ohm's law to resistive circuits with ac sources
  - □ Apply Kirchhoff's voltage law and current law to resistive circuits with ac sources

If a sine wave voltage is applied across a resistor as shown in Figure 8–36, a sine wave current flows. The current is zero when the voltage is zero and is maximum when the voltage is maximum. When the voltage changes polarity, the current reverses direction. As a result, the voltage and current are said to be in phase with each other.

FIGURE 8–36
A sine wave voltage produces a sine wave current.

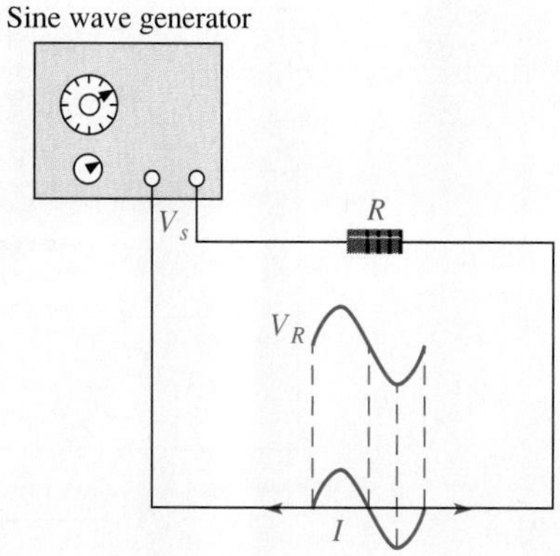

When using Ohm's law in ac circuits, remember that both the voltage and the current must be expressed consistently, that is, both as peak values, both as rms values, both as average values, and so on.

EXAMPLE 8–11 Determine the rms voltage across each resistor and the rms current in Figure 8–37. The source voltage is given as an rms value.

FIGURE 8–37

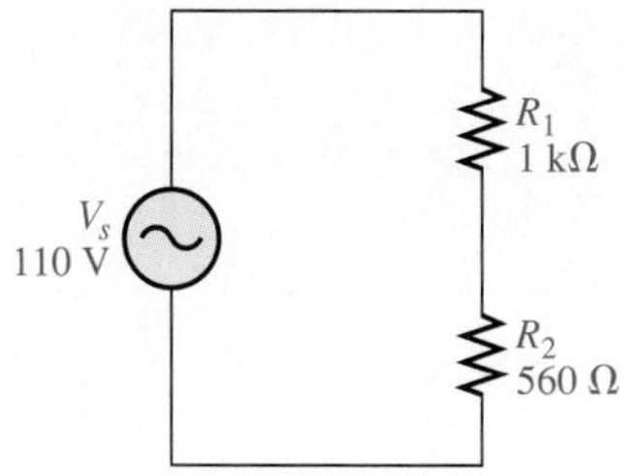

Solution The total resistance of the circuit is

$$R_T = R_1 + R_2 = 1\text{ k}\Omega + 560\ \Omega = 1.56\text{ k}\Omega$$

Use Ohm's law to find the rms current.

$$I_{\text{rms}} = \frac{V_{s(\text{rms})}}{R_T} = \frac{110\text{ V}}{1.56\text{ k}\Omega} = 70.5\text{ mA}$$

The rms voltage drop across each resistor is

$$V_{1(\text{rms})} = I_{\text{rms}}R_1 = (70.5\text{ mA})(1\text{ k}\Omega) = 70.5\text{ V}$$
$$V_{2(\text{rms})} = I_{\text{rms}}R_2 = (70.5\text{ mA})(560\ \Omega) = 39.5\text{ V}$$

Related Exercise Repeat this example for a source voltage of 10 V peak.

Kirchhoff's voltage and current laws apply to ac circuits as well as to dc circuits. Figure 8–38 illustrates Kirchhoff's voltage law in a resistive circuit that has a sine wave voltage source. As you can see, the source voltage is the sum of all the voltage drops across the resistors, just as in a dc circuit.

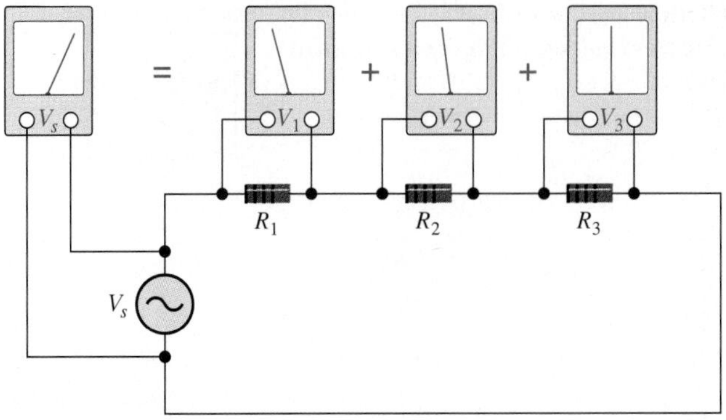

FIGURE 8–38
Illustration of Kirchhoff's voltage law in an ac circuit.

EXAMPLE 8–12

All values in Figure 8–39 are given in rms.

(a) Find the unknown peak voltage drop in Figure 8–39(a).

(b) Find the total rms current in Figure 8–39(b).

FIGURE 8–39

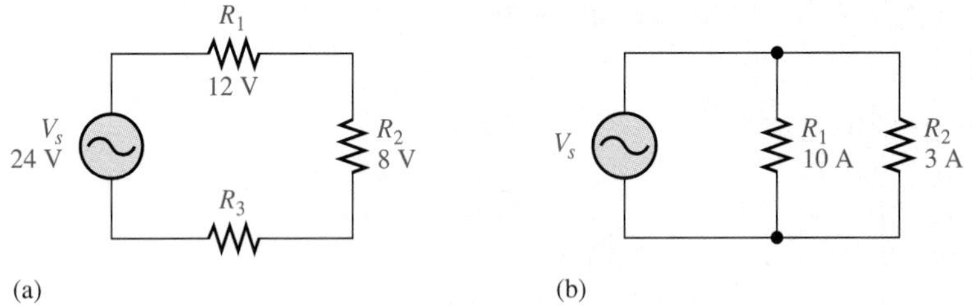

Solution

(a) Use Kirchhoff's voltage law to find V_3.

$$V_s = V_1 + V_2 + V_3$$

$$V_{3(\text{rms})} = V_{s(\text{rms})} - V_{1(\text{rms})} - V_{2(\text{rms})} = 24\text{ V} - 12\text{ V} - 8\text{ V} = 4\text{ V}$$

Convert rms to peak:

$$V_{3(p)} = 1.414V_{3(\text{rms})} = 1.414(4\text{ V}) = 5.66\text{ V}$$

(b) Use Kirchhoff's current law to find I_T.

$$I_{T(\text{rms})} = I_{1(\text{rms})} + I_{2(\text{rms})} = 10\text{ A} + 3\text{ A} = 13\text{ A}$$

Related Exercise A series circuit has the following voltage drops: $V_{1(\text{rms})} = 3.50$ V, $V_{2(p)} = 4.25$ V, $V_{3(\text{avg})} = 1.70$ V. Determine the peak-to-peak source voltage.

SECTION 8–6 REVIEW

1. A sine wave voltage with a half-cycle average value of 12.5 V is applied to a circuit with a resistance of 330 Ω. What is the peak current in the circuit?
2. The peak voltage drops in a series resistive circuit are 6.2 V, 11.3 V, and 7.8 V. What is the rms value of the source voltage?

8–7 ■ SUPERIMPOSED DC AND AC VOLTAGES

In many practical circuits, you will find both dc and ac voltages combined. An example of this is in amplifier circuits where ac signal voltages are superimposed on dc operating voltages.

After completing this section, you should be able to

- **Determine total voltages which are composed of both ac and dc components**

Figure 8–40 shows a dc source and an ac source in series. These two voltages will add algebraically to produce an ac voltage "riding" on a dc level, as measured across the resistor.

If V_{dc} is greater than the peak value of the sine wave, the combined voltage is a sine wave that never reverses polarity and is therefore nonalternating. That is, we have a sine wave riding on a dc level, as shown in Figure 8–41(a). If V_{dc} is less than the peak value of the sine wave, the sine wave will be negative during a portion of its lower half-cycle, as illustrated in Figure 8–41(b), and is therefore alternating. In either case the sine wave will reach a maximum voltage equal to $V_{dc} + V_p$, and it will reach a minimum voltage equal to $V_{dc} - V_p$.

FIGURE 8–40
Superimposed dc and ac voltages.

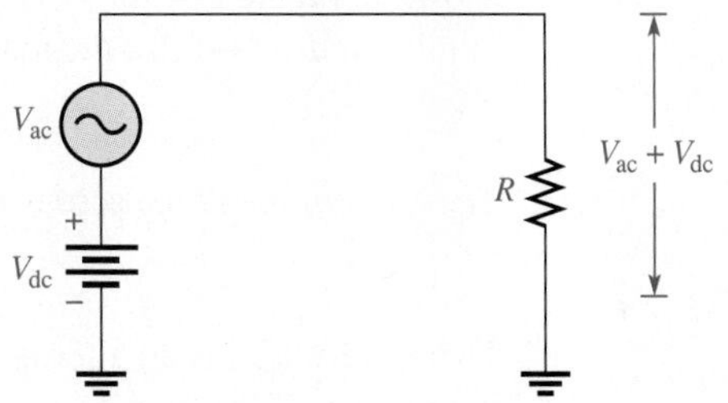

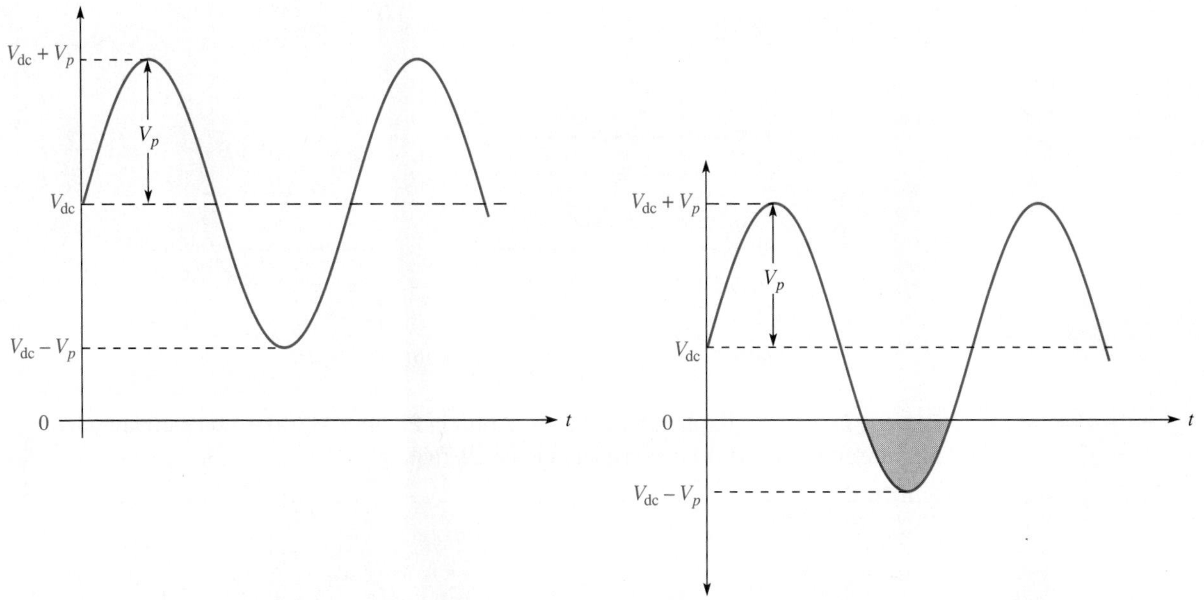

(a) $V_{dc} > V_p$. The sine wave never goes negative.

(b) $V_{dc} < V_p$. The sine wave reverses polarity during a portion of its cycle.

FIGURE 8–41
Sine waves with dc levels.

EXAMPLE 8–13 Determine the total voltage across the resistor in each circuit of Figure 8–42.

FIGURE 8–42

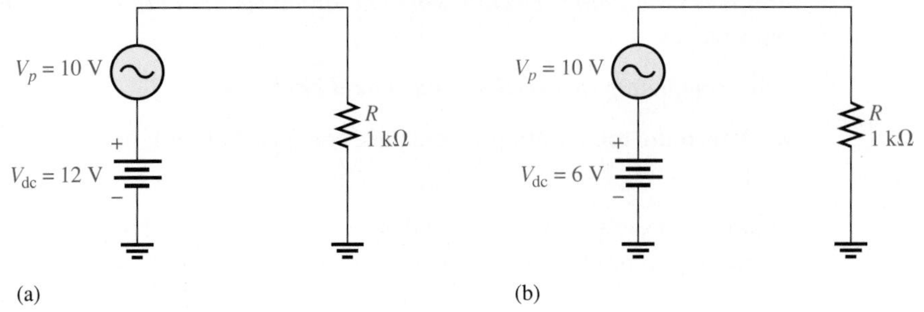

Solution In Figure 8–42(a), the maximum voltage across R is

$$V_{max} = V_{dc} + V_p = 12 \text{ V} + 10 \text{ V} = 22 \text{ V}$$

The minimum voltage across R is

$$V_{min} = V_{dc} - V_p = 12 \text{ V} - 10 \text{ V} = 2 \text{ V}$$

Therefore, $V_{R(tot)}$ is a nonalternating sine wave that varies from +22 V to +2 V, as shown in Figure 8–43(a).

In Figure 8–42(b), the maximum voltage across R is

$$V_{max} = V_{dc} + V_p = 6 \text{ V} + 10 \text{ V} = 16 \text{ V}$$

The minimum voltage across R is

$$V_{min} = V_{dc} - V_p = -4 \text{ V}$$

Therefore, $V_{R(tot)}$ is an alternating sine wave that varies from +16 V to −4 V, as shown in Figure 8–43(b).

FIGURE 8–43

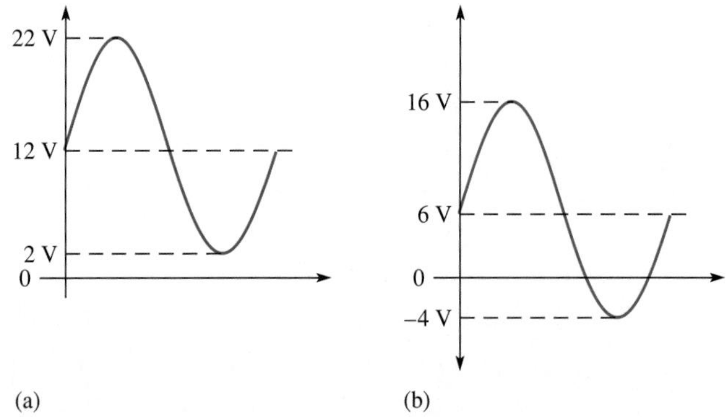

Related Exercise Explain why the waveform in Figure 8–43(a) is nonalternating but the waveform in part (b) is considered to be alternating.

SECTION 8–7 REVIEW

1. What is the maximum positive value of the resulting total voltage when a sine wave with $V_p = 5$ V is added to a dc voltage of +2.5 V?
2. Will the resulting voltage in Question 1 alternate polarity?
3. If the dc voltage in Question 1 is −2.5 V, what is the maximum positive value of the resulting total voltage?

8–8 ▪ THE OSCILLOSCOPE

The oscilloscope, or scope for short, is one of the most widely used and versatile test instruments. It displays on a screen the actual shape of voltages that are changing with time so that waveform measurements can be made and one waveform can be compared to another.

After completing this section, you should be able to

- **Use the oscilloscope to measure waveforms**
  - ☐ Describe a basic CRT
  - ☐ Identify basic oscilloscope controls
  - ☐ Explain how to measure amplitude
  - ☐ Explain how to measure period and frequency

Figure 8–44 shows a typical dual-channel oscilloscope. You can think of the **oscilloscope** as essentially a "graphing machine" that graphs out a voltage and shows you how it varies with time. The shape of the sine wave that you have seen throughout this chapter is simply a graph of voltage versus time. Of course, oscilloscopes can display any type of waveform, but we will use the sine wave to illustrate some basic concepts.

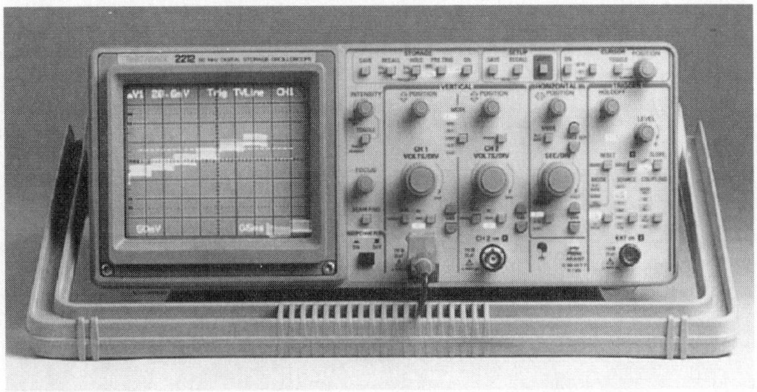

FIGURE 8–44
A typical oscilloscope (courtesy of Tektronix, Inc.).

The Oscilloscope Screen

Generally, the screen of an oscilloscope is divided into ten horizontal divisions and eight vertical divisions, as shown in Figure 8–45. Each major division is divided into five small or minor divisions on both the horizontal and the vertical axes. The vertical axis on the scope screen is the voltage scale and the horizontal axis is the time scale. Values of voltage are read vertically and values of time are read horizontally.

Cathode-Ray Tube (CRT)

The oscilloscope is built around the **cathode-ray tube (CRT),** which is the device that displays the waveforms. The screen of the scope is the front of the CRT.

The CRT is a vacuum tube device containing an electron gun that emits a narrow, focused beam of electrons. A phosphorescent coating on the face of the tube forms the screen. The beam is electronically focused and accelerated so that it strikes the screen, causing light to be emitted at the point of impact.

Figure 8–46 shows the basic construction of a CRT. The electron gun assembly contains a heater, a cathode, a control grid, and accelerating and focusing grids. The heater

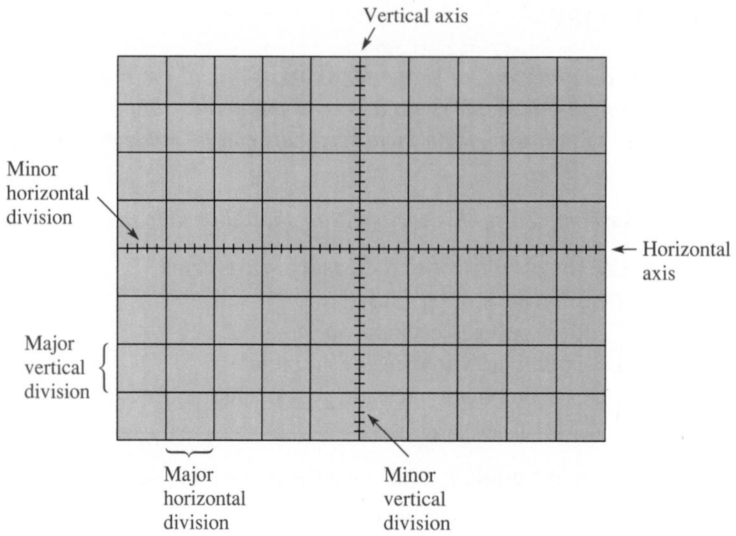

FIGURE 8–45
The oscilloscope screen.

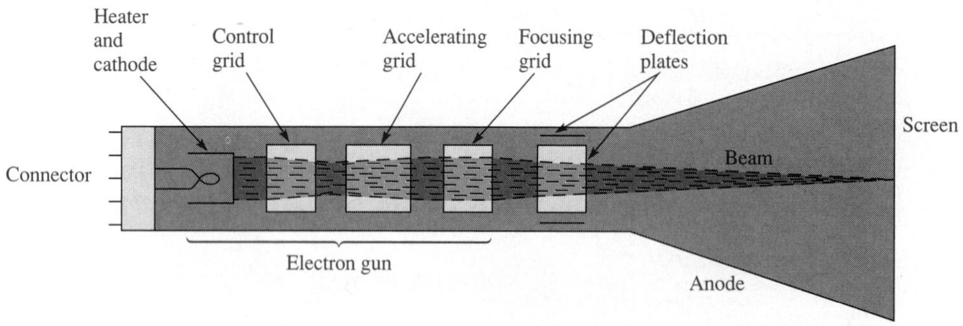

FIGURE 8–46
Basic construction of a CRT.

carries current that indirectly heats the cathode. The heated cathode emits electrons. The amount of voltage on the control grid determines the flow of electrons and thus the intensity of the beam. The electrons are accelerated by the accelerating grid and are focused by the focusing grid into a narrow beam that converges at the screen. The beam is further accelerated to a high speed after it leaves the electron gun by a high voltage on the anode surfaces of the CRT.

The purpose of the deflection plates in the CRT is to produce a "bending," or deflection, of the electron beam. This deflection allows the position of the point of impact on the screen to be varied. There are two sets of deflection plates: one set for vertical deflection, and the other set for horizontal deflection.

Figure 8–47 illustrates how a waveform (a sine wave in this case) is "plotted out" on the screen. A sawtooth voltage called the *horizontal sweep signal* is internally generated and applied to the horizontal deflection plates. Each cycle of the sawtooth causes the electron beam to sweep across the screen from left to right and then quickly return for another sweep. The external signal that is to be displayed on the screen is applied to the vertical deflection plates. This causes the electron beam to move up or down vertically as it is swept across the screen, tracing out a pattern that follows the variation of the external signal voltage over an interval of time.

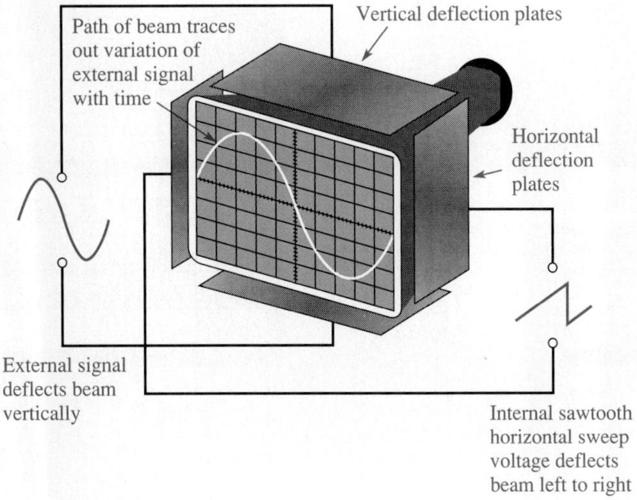

FIGURE 8–47
How a waveform is traced out on the scope screen.

Operation of the Oscilloscope

Figure 8–48 shows a typical oscilloscope like the one used in the Application Assignment sections. This is a dual-channel instrument, as most scopes are. *Dual-channel* means that you can display two waveforms from two different inputs on the screen at the same time.

The Volts/Division Control Notice that in Figure 8–48 there are two VOLTS/DIV switches, one for channel 1 (CH1) and one for channel 2 (CH2). The volts/division selector switch sets the number of volts to be represented by each major division on the vertical scale. For example, Figure 8–49(a) shows one cycle of a sine wave on the scope screen with the VOLTS/DIV switch set at 1 V. This means that each of the major vertical divisions is 1 V. The peak of the sine wave is three divisions high, and, since each division is 1 V, assuming a × 1 scope probe is used, the peak value of the sine wave is 3 V (3 divisions × 1 V/division = 3 V).

For the same sine wave, if the VOLTS/DIV setting is changed to 2 V, the sine wave will appear as shown in Figure 8–49(b). Notice that now the peak is 1½ major divisions high. Even though each division is now 2 V, the peak value is still 3 V because 1.5 divisions × 2 V/division = 3 V.

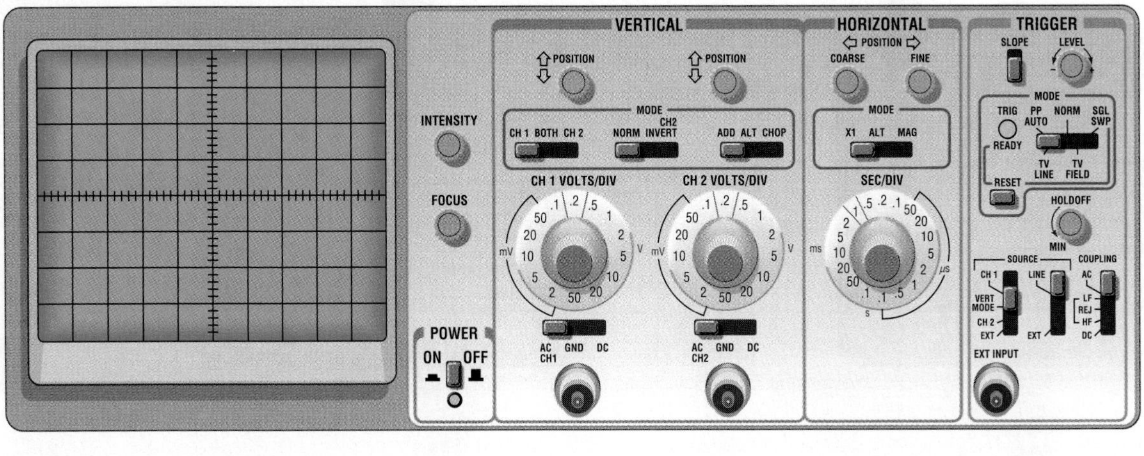

FIGURE 8–48
A typical dual-channel oscilloscope.

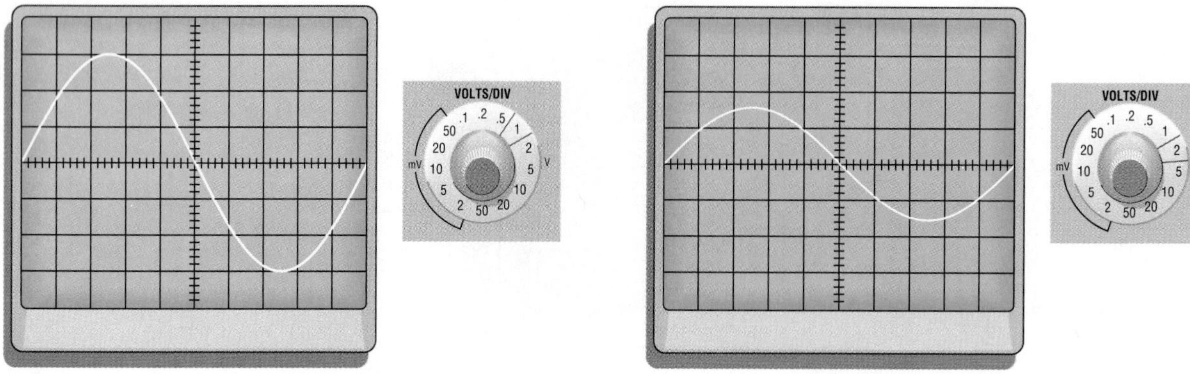

(a) (b)

FIGURE 8–49
Using the VOLTS/DIV control.

So, as you can see, it is relatively easy to measure the peak value by counting the number of major vertical divisions from the zero crossing to the peak of the waveform and multiplying by the VOLTS/DIV setting.

The Seconds/Division Control The SEC/DIV selector switch sets the number of seconds, milliseconds, or microseconds to be represented by each major division on the horizontal scale. It actually controls how fast the electron beam sweeps horizontally across the screen.

For example, Figure 8–50(a) shows one cycle of a sine wave displayed on the screen. In this case, the entire cycle covers ten horizontal divisions. The SEC/DIV switch is set at 10 μs. This means that each major horizontal division is 10 μs. Since a full cycle of the sine wave covers 10 divisions, the period of the sine wave is 100 μs (10 div × 10 μs/div = 100 μs). From this, $f = 1/100\ \mu s = 10$ kHz.

For the same sine wave, if the SEC/DIV setting is changed to 20 μs, the sine wave will appear as shown in Figure 8–50(b). Notice that now a full cycle covers only five horizontal divisions and there are two cycles on the screen. Even though each division is now 20 μs, the period is still 100 μs because 5 divisions × 20 μs/division = 100 μs.

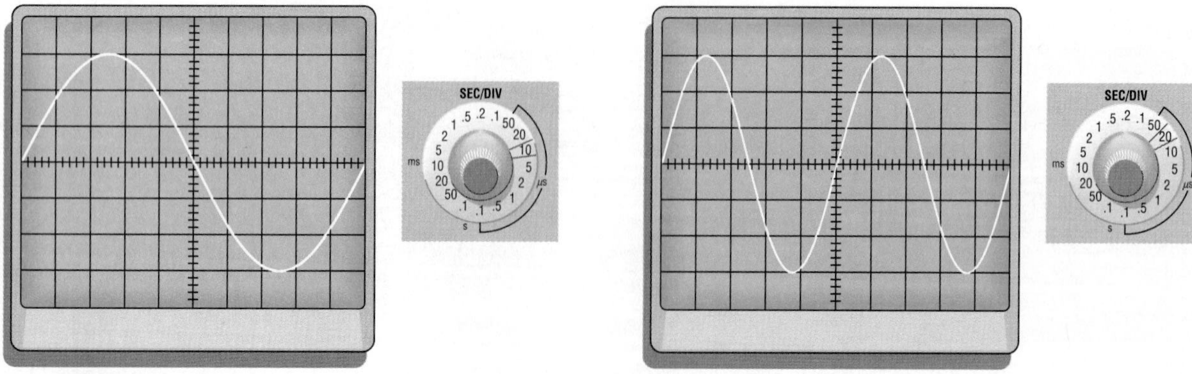

(a) (b)

FIGURE 8–50
Using the SEC/DIV control.

So, as you can see, it is easy to measure the period and then calculate the frequency by counting the number of major divisions covered by one cycle and then multiplying by the SEC/DIV setting. Some scopes automatically measure and display the frequency and other parameters.

Other Oscilloscope Controls

The following descriptions refer to the oscilloscope in Figure 8–48 but apply also to most general-purpose scopes.

Power Switch The power switch turns the power to the scope on and off. A light indicates when the power is on.

Intensity The intensity control knob varies the brightness of the trace on the screen. Caution should be used so that the intensity is not left too high for an extended period of time, especially when the beam forms a motionless dot on the screen. Damage to the screen can result from excessive intensity.

Focus This control focuses the beam so that it converges to a tiny point at the screen. An out-of-focus condition results in a fuzzy trace.

Horizontal Position These control knobs (coarse and fine) adjust the neutral horizontal position of the beam. They are used to reposition horizontally a waveform display for more convenient viewing or measurement.

Vertical Position The two vertical position controls move each trace up or down for easier measurement or observation.

AC-GND-DC Switch This switch, located below the VOLTS/DIV control, allows the input signal to be ac coupled, dc coupled, or grounded. The ac coupling eliminates any dc component on the input signal. The dc coupling permits dc values to be displayed. The ground position allows a 0 V reference to be established on the screen.

Signal Inputs The signals to be displayed are connected into the channel 1 (CH1) and/or channel 2 (CH2) input connectors. These connections are normally done with a special probe that minimizes the loading effect of the scope's input resistance on the circuit being measured. Oscilloscope voltage probes are generally either ×1 (nonattenuating) or ×10 (attenuates by 10). When a ×10 probe is used, the VOLTS/DIV setting must be multiplied by 10. To keep it simple, all applications in this book assume ×1 voltage probes.

Mode Switches These switches provide for displaying either or both channel inputs, inverting channel 2 signal, adding two waveforms, and selecting between alternate and chopped mode of sweep.

Trigger Control The **trigger** controls allow the beam to be triggered from various selected sources. The triggering of the beam causes it to begin its sweep across the screen. It can be triggered from an internally generated signal derived from an input signal, or from the line voltage, or from an externally applied trigger signal. The modes of triggering are auto, normal, single-sweep, and TV. In the auto mode, sweep occurs in the absence of an adequate trigger signal. In the normal mode, a trigger signal must be present for the sweep to occur. The TV mode provides triggering on the TV field or TV line signals. The slope switch allows the triggering to occur on either the positive-going slope or the negative-going slope of the trigger waveform. The level control selects the voltage level on the trigger signal at which the triggering occurs.

Basically, the trigger controls provide for synchronization of the horizontal sweep waveform and the input signal waveform. As a result, the display of the input signal is stable on the screen, rather than appearing to drift across the screen.

EXAMPLE 8–14 Determine the peak value and period of each sine wave in Figure 8–51 from the scope displays and the indicated settings for VOLTS/DIV and SEC/DIV.

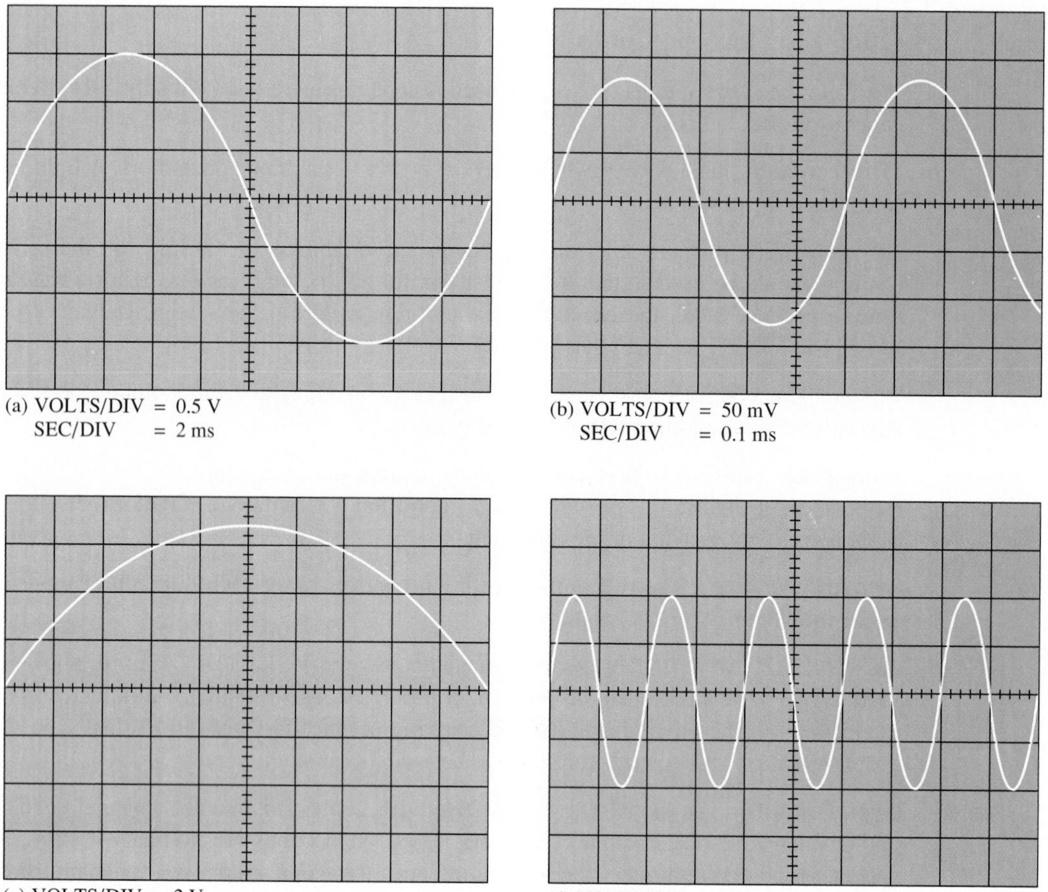

(a) VOLTS/DIV = 0.5 V
SEC/DIV = 2 ms

(b) VOLTS/DIV = 50 mV
SEC/DIV = 0.1 ms

(c) VOLTS/DIV = 2 V
SEC/DIV = 10 μs
Sine wave is centered vertically on the screen.

(d) VOLTS/DIV = 5 V
SEC/DIV = 2 μs

FIGURE 8–51

Solution Looking at the vertical scale in Figure 8–51(a),

$$V_p = 3 \text{ divisions} \times 0.5 \text{ V/division} = 1.5 \text{ V}$$

From the horizontal scale (one cycle covers ten divisions),

$$T = 10 \text{ divisions} \times 2 \text{ ms/division} = 20 \text{ ms}$$

Looking at the vertical scale in Figure 8–51(b),

$$V_p = 2.5 \text{ divisions} \times 50 \text{ mV/division} = 125 \text{ mV}$$

From the horizontal scale (one cycle covers six divisions),

$$T = 6 \text{ divisions} \times 0.1 \text{ ms/division} = 0.6 \text{ ms} = 600\ \mu\text{s}$$

Looking at the vertical scale in Figure 8–51(c),

$$V_p = 3.4 \text{ divisions} \times 2 \text{ V/division} = 6.8 \text{ V}$$

From the horizontal scale (one-half cycle covers ten divisions),

$$T = 20 \text{ divisions} \times 10\ \mu\text{s/division} = 200\ \mu\text{s}$$

Looking at the vertical scale in Figure 8–51(d),

$$V_p = 2 \text{ divisions} \times 5 \text{ V/division} = 10 \text{ V}$$

From the horizontal scale (one cycle covers two divisions),

$$T = 2 \text{ divisions} \times 2\ \mu\text{s/division} = 4\ \mu\text{s}$$

Related Exercise Determine the rms value and the frequency for each waveform displayed in Figure 8–51.

SECTION 8–8 REVIEW

1. What does *CRT* stand for?
2. On an oscilloscope, voltage is measured (horizontally, vertically) on the screen and time is measured (horizontally, vertically).
3. What can an oscilloscope do that a multimeter cannot?

8–9 ■ NONSINUSOIDAL WAVEFORMS

Sine waves are important in electronics, but they are by no means the only type of ac or time-varying waveform. Two other major types of waveforms, the pulse waveform and the triangular waveform, are discussed next.

After completing this section, you should be able to

- **Identify the characteristics of basic nonsinusoidal waveforms**
  - □ Discuss the properties of a pulse waveform
  - □ Define *duty cycle*
  - □ Discuss the properties of triangular and sawtooth waveforms
  - □ Discuss the harmonic content of a waveform

Pulse Waveforms

Basically, a **pulse** can be described as a very rapid transition (**leading edge**) from one voltage or current level (**baseline**) to another, and then, after an interval of time, a very rapid transition (**trailing edge**) back to the original baseline level. The transitions in level are called *steps.* An ideal pulse consists of two opposite-going steps of equal amplitude. When the leading or trailing edge is positive-going, it is called a **rising edge.** When the leading or trailing edge is negative-going, it is called a **falling edge.**

Figure 8–52(a) shows an ideal positive-going pulse consisting of two equal but opposite instantaneous steps separated by an interval of time called the **pulse width.** Figure 8–52(b) shows an ideal negative-going pulse. The height of the pulse measured from the baseline is its amplitude.

In many applications, analysis is simplified by treating all pulses as ideal (composed of instantaneous steps and perfectly rectangular in shape). Actual pulses, however, are never ideal. All pulses possess certain characteristics that cause them to be different from the ideal.

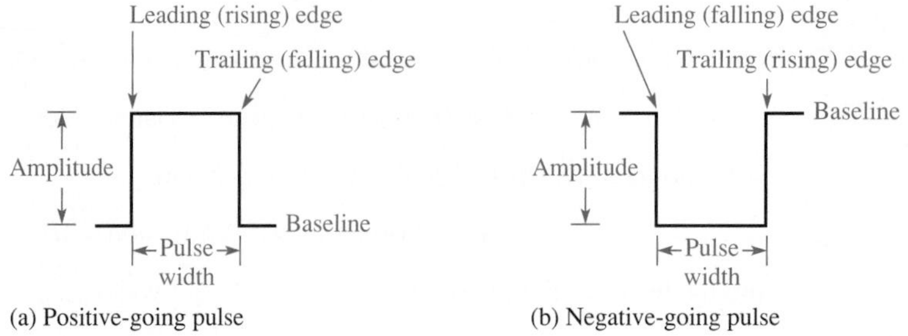

FIGURE 8–52
Ideal pulses.

In practice, pulses cannot change from one level to another instantaneously. Time is always required for a transition (step), as illustrated in Figure 8–53(a). As you can see, there is an interval of time during which the pulse is rising from its lower value to its higher value. This interval is called the **rise time, t_r.**

Rise time is the time required for the pulse to go from 10% of its full amplitude to 90% of its full amplitude.

The interval of time during which the pulse is falling from its higher value to its lower value is called the **fall time, t_f.**

Fall time is the time required for the pulse to go from 90% of its full amplitude to 10% of its full amplitude.

Pulse width (t_W) also requires a precise definition for the nonideal pulse, because the rising and falling edges are not vertical.

Pulse width is the time between the point on the rising edge, where the value is 50% of full amplitude, to the point on the falling edge, where the value is 50% of full amplitude.

Pulse width is shown in Figure 8–53(b).

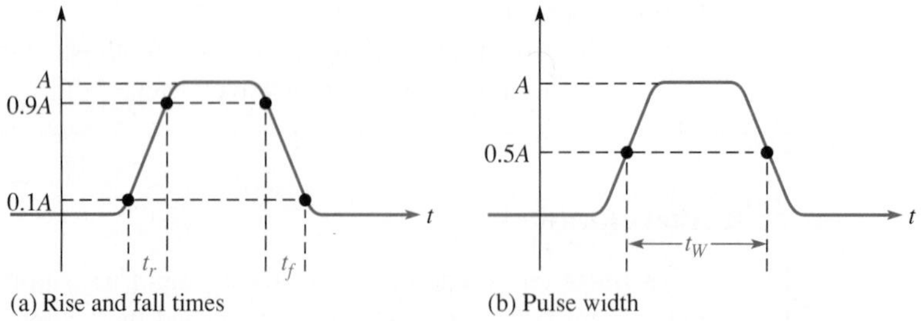

FIGURE 8–53
Nonideal pulse.

Repetitive Pulses

Any waveform that repeats itself at fixed intervals is **periodic.** Figure 8–54 shows some examples of periodic pulse waveforms. Notice that in each case, the pulses repeat at regular intervals. The rate at which the pulses repeat is the pulse repetition rate, which is the fundamental frequency of the waveform. The frequency can be expressed in hertz or in pulses per second. The time from one pulse to the corresponding point on the next pulse is the **period (T).**

FIGURE 8–54
Repetitive pulse waveforms.

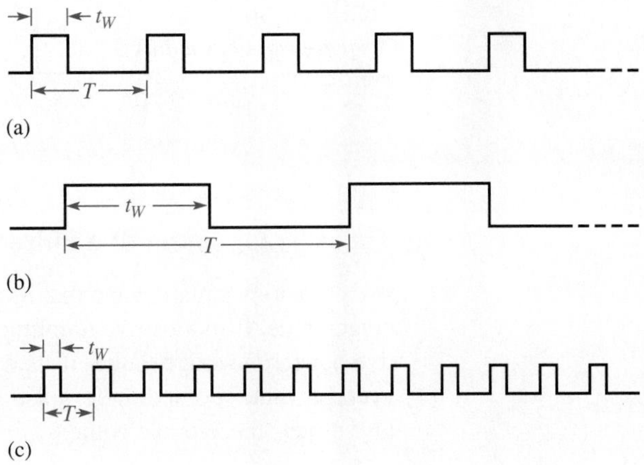

An important characteristic of periodic pulse waveforms is the **duty cycle.**

The duty cycle is the ratio of the pulse width (t_W) to the period (T) and is usually expressed as a percentage.

$$\text{Percent duty cycle} = \left(\frac{t_W}{T}\right)100\% \qquad \textbf{(8–21)}$$

EXAMPLE 8–15 Determine the period, frequency, and duty cycle for the pulse waveform in Figure 8–55.

FIGURE 8–55

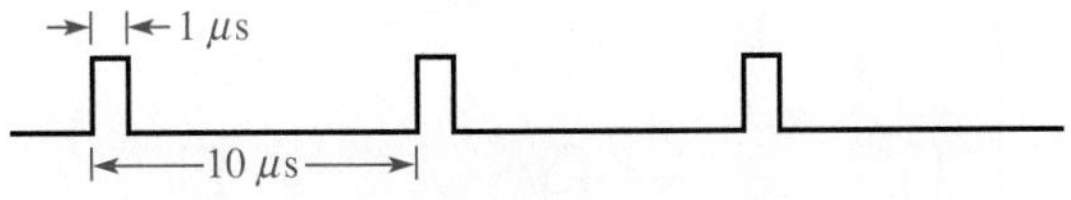

Solution As indicated in Figure 8–55, the period is

$$T = 10\ \mu\text{s}$$

Use Equations (8–1) and (8–21) to determine the frequency and duty cycle.

$$f = \frac{1}{T} = \frac{1}{10\ \mu\text{s}} = 0.1\ \text{MHz}$$

$$\text{Percent duty cycle} = \left(\frac{t_W}{T}\right)100\% = \left(\frac{1\ \mu\text{s}}{10\ \mu\text{s}}\right)100\% = 10\%$$

Related Exercise A certain pulse waveform has a frequency of 200 kHz and a pulse width of 0.25 μs. Determine the duty cycle.

Square Waves

A square wave is a pulse waveform with a duty cycle of 50%. Thus, the pulse width is equal to one-half of the period. A square wave is shown in Figure 8–56.

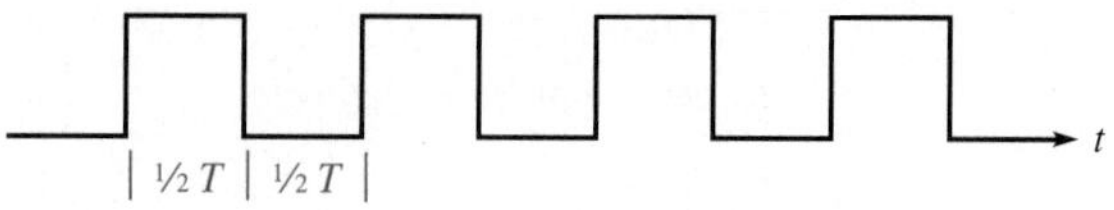

FIGURE 8–56
A square wave has a duty cycle of 50%.

The Average Value of a Pulse Waveform

An alternating square wave that has equal positive and negative amplitudes has a zero average value. If its positive amplitude is greater than its negative amplitude, it has a positive nonzero average value. If its negative amplitude is greater, it has a negative nonzero average value. For periodic pulse waveforms that are not square waves, the duty cycle also affects the average value.

The average value (V_{avg}) of a pulse waveform is equal to its baseline value plus the product of its duty cycle and its amplitude.

The lower level of the waveform is taken as the baseline. The formula is as follows:

$$V_{avg} = \text{baseline} + (\text{duty cycle})(\text{amplitude}) \quad \textbf{(8–22)}$$

The following example illustrates the calculation of the average value of pulse waveforms.

EXAMPLE 8–16 Determine the average value of each of the waveforms in Figure 8–57.

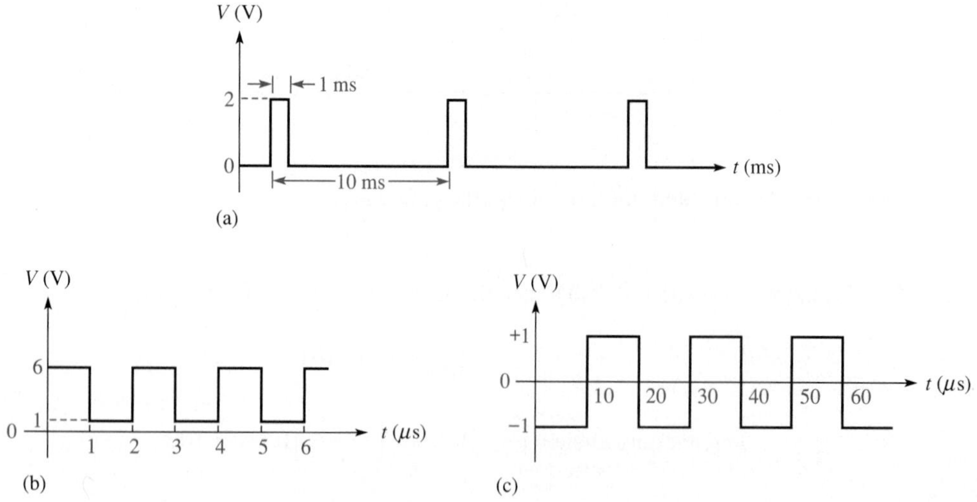

FIGURE 8–57

Solution In Figure 8–57(a), the baseline is at 0 V, the amplitude is 2 V, and the duty cycle is 10%. The average value is

$$V_{avg} = \text{baseline} + (\text{duty cycle})(\text{amplitude})$$
$$= 0\text{ V} + (0.1)(2\text{ V}) = 0.2\text{ V}$$

The waveform in Figure 8–57(b) has a baseline of +1V, an amplitude of 5 V, and a duty cycle of 50%. The average value is

$$V_{avg} = \text{baseline} + (\text{duty cycle})(\text{amplitude})$$
$$= 1\text{ V} + (0.5)(5\text{ V}) = 1\text{ V} + 2.5\text{ V} = 3.5\text{ V}$$

The waveform in Figure 8–57(c) is a square wave with a baseline of −1 V and an amplitude of 2 V. The average value is

$$V_{avg} = \text{baseline} + (\text{duty cycle})(\text{amplitude})$$
$$= -1\text{ V} + (0.5)(2\text{ V}) = -1\text{ V} + 1\text{ V} = 0\text{ V}$$

This is an alternating square wave, and, like an alternating sine wave, it has an average value of zero over a full cycle.

Related Exercise If the baseline of the waveform in Figure 8–57(a) is shifted to +1 V, what is the average value?

Triangular and Sawtooth Waveforms

Triangular and sawtooth waveforms are formed by voltage or current ramps. A **ramp** is a linear increase or decrease in the voltage or current. Figure 8–58 shows both positive- and negative-going ramps. In part (a) the ramp has a positive slope; in part (b), the ramp has a negative slope. The slope of a voltage ramp is $\pm V/t$ and is expressed in units of V/s. The slope of a current ramp is $\pm I/t$ and is expressed in units of A/s.

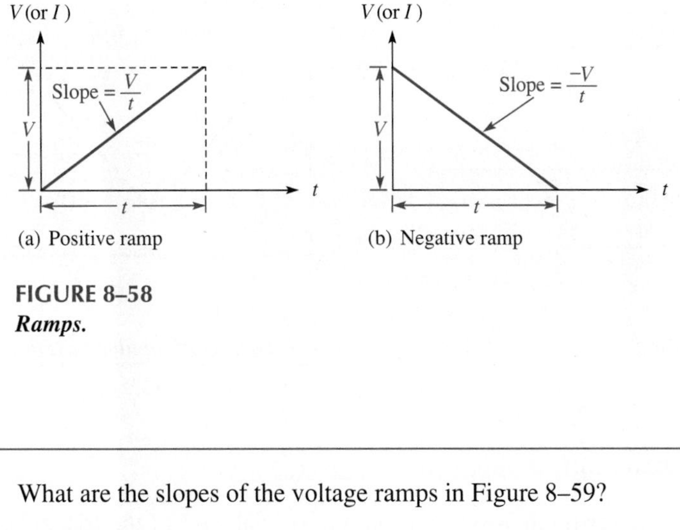

FIGURE 8–58
Ramps.

EXAMPLE 8–17 What are the slopes of the voltage ramps in Figure 8–59?

FIGURE 8–59

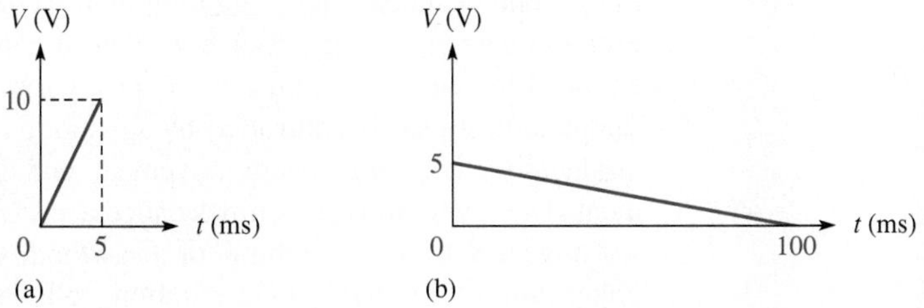

Solution In Figure 8–59(a), the voltage increases from 0 V to +10 V in 5 ms. Thus, $V = 10$ V and $t = 5$ ms. The slope is

$$\frac{V}{t} = \frac{10\text{ V}}{5\text{ ms}} = 2\text{ V/ms} = 2\text{ kV/s}$$

In Figure 8–59(b), the voltage decreases from +5 V to 0 V in 100 ms. Thus, $V =$ -5 V and $t = 100$ ms. The slope is

$$\frac{V}{t} = \frac{-5\text{ V}}{100\text{ ms}} = -0.05\text{ V/ms} = -50\text{ V/s}$$

Related Exercise A certain voltage ramp has a slope of +12 V/μs. If the ramp starts at zero, what is the voltage at 0.01 ms?

Triangular Waveforms

Figure 8–60 shows that a **triangular waveform** is composed of positive- and negative-going ramps having equal slopes. The period of this waveform is measured from one peak to the next corresponding peak, as illustrated. This particular triangular waveform is alternating and has an average value of zero.

Figure 8–61 depicts a triangular waveform with a nonzero average value. The frequency for triangular waves is determined in the same way as for sine waves, that is, $f = 1/T$.

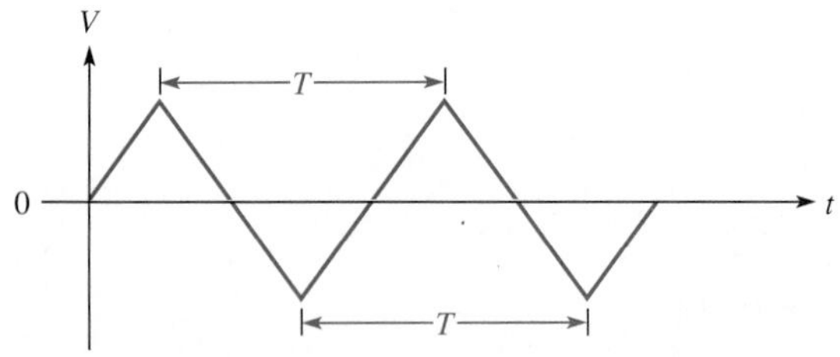

FIGURE 8–60
Alternating triangular waveform.

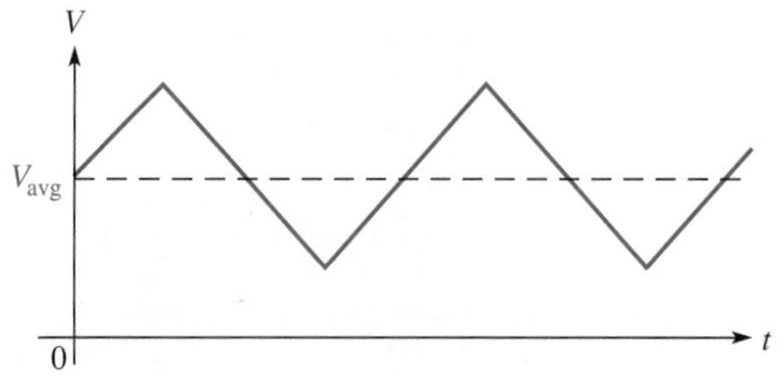

FIGURE 8–61
Nonalternating triangular waveform.

Sawtooth Waveforms

The **sawtooth wave** is actually a special case of the triangular wave consisting of two ramps, one of much longer duration than the other. Sawtooth waveforms are commonly used in many electronic systems, as you saw in the case of the oscilloscope. Another example is the electron beam that sweeps across the screen of your TV receiver, creating the picture; it also is controlled by sawtooth voltages and currents. One sawtooth wave produces the horizontal beam movement, and the other produces the vertical beam movement. The sawtooth is sometimes called a *sweep*.

Figure 8–62 is an example of a sawtooth wave. Notice that it consists of a positive-going ramp of relatively long duration, followed by a negative-going ramp of relatively short duration.

FIGURE 8–62
Alternating sawtooth waveform.

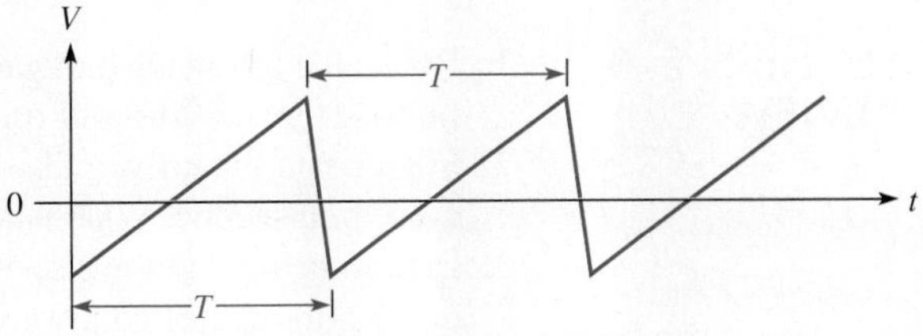

Harmonics

A repetitive nonsinusoidal waveform is composed of a fundamental frequency and harmonic frequencies. The **fundamental frequency** is the repetition rate of the waveform, and the **harmonics** are higher-frequency sine waves that are multiples of the fundamental.

Odd Harmonics *Odd harmonics* are frequencies that are odd multiples of the fundamental frequency of a waveform. For example, a 1 kHz square wave consists of a fundamental of 1 kHz and odd harmonics of 3 kHz, 5 kHz, 7 kHz, and so on. The 3 kHz frequency in this case is the third harmonic; the 5 kHz frequency is the fifth harmonic; and so on.

Even Harmonics *Even harmonics* are frequencies that are even multiples of the fundamental frequency. For example, if a certain wave has a fundamental of 200 Hz, the second harmonic is 400 Hz, the fourth harmonic is 800 Hz, the sixth harmonic is 1200 Hz, and so on. These are even harmonics.

Composite Waveform Any variation from a pure sine wave produces harmonics. A nonsinusoidal wave is a composite of the fundamental and the harmonics. Some types of waveforms have only odd harmonics, some have only even harmonics, and some contain both. The shape of the wave is determined by its harmonic content. Generally, only the fundamental and the first few harmonics are of significant importance in determining the waveshape.

A square wave is an example of a waveform that consists of a fundamental and only odd harmonics. When the instantaneous values of the fundamental and each odd harmonic are added algebraically at each point, the resulting curve will have the shape of a square wave, as illustrated in Figure 8–63. In part (a), the fundamental and the third harmonic produce a waveshape that begins to resemble a square wave. In part (b), the fundamental, third, and fifth harmonics produce a closer resemblance. When the seventh harmonic is included, as in part (c), the resulting waveshape becomes even more like a square wave. As more harmonics are included, a square wave is approached.

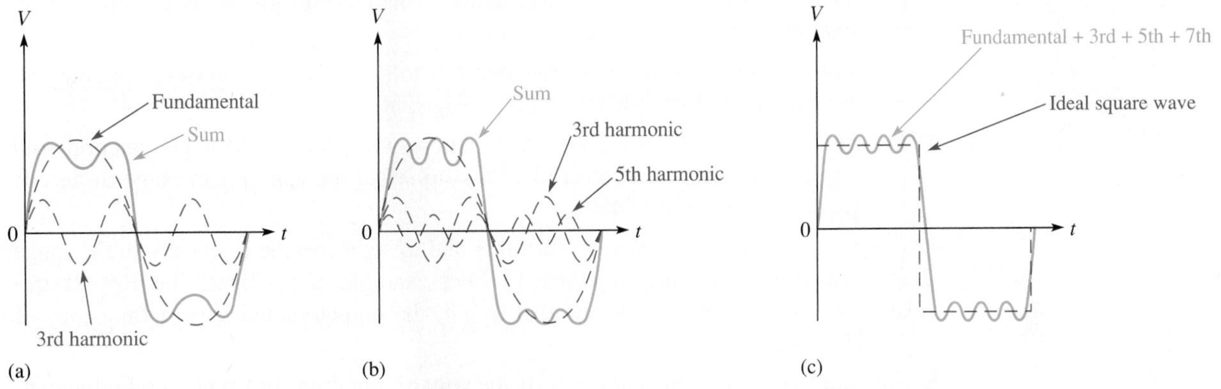

FIGURE 8–63
Odd harmonics combine to produce a square wave.

SECTION 8–9 REVIEW

1. Define the following parameters:
   (a) rise time (b) fall time (c) pulse width
2. In a certain repetitive pulse waveform, the pulses are 200 μs wide and occur once every millisecond. What is the frequency of this waveform?
3. Determine the duty cycle, amplitude, and average value of the waveform in Question 2 if the pulses go from −2 V to +6 V.
4. What is the second harmonic of a fundamental frequency of 1 kHz?

8–10 ■ APPLICATION ASSIGNMENT

You have been assigned to check out a new function generator that produces sine wave, triangular wave, and pulse outputs. For each type of output, you will use the oscilloscope to measure the minimum and maximum frequencies, the minimum and maximum amplitudes, the maximum positive and negative dc offsets, and the minimum and maximum duty cycles for the pulse waveform. You will record your measurements in a logical format.

Step 1: Familiarization with the Function Generator

The function generator is shown in Figure 8–64. Each control is marked by a circled number and a description follows.

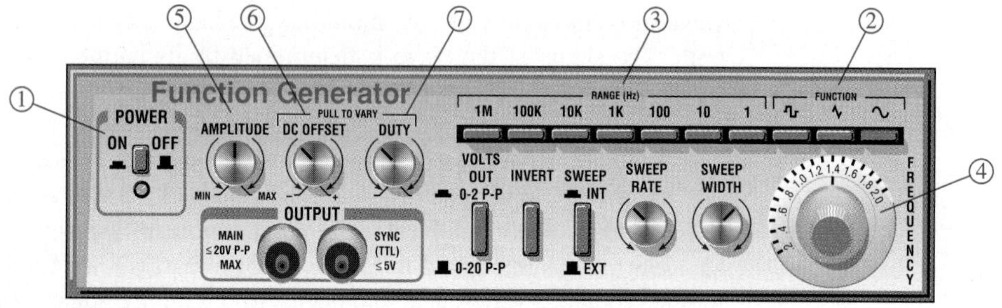

FIGURE 8–64
Function generator.

1. *Power on/off switch* Press this push-button switch to turn the power on. Press again to release the switch and turn the power off.
2. *Function switches* Press one of the switches to select either a sinusoidal, triangular, or pulse output. Only one switch can be engaged at a time.
3. *Frequency range switches* These push-button switches are used in conjunction with the frequency adjustment control (4). Press to select the appropriate range in decade increments from 1 Hz to 1 MHz.
4. *Frequency adjustment control* Turn this dial to set a specific frequency in the range selected by the push button switch (3). For example, if you press the 100 Hz frequency range switch and set the dial to 1.4, the output should have a frequency of 140 Hz (1.4 × 100).
5. *Amplitude control* This knob adjusts the voltage amplitude of the output. Fully counterclockwise is the minimum and fully clockwise is the maximum. It can be adjusted for any value of amplitude between minimum and maximum.

6. *DC offset control* This knob adjusts the dc level of the ac output. You can add a positive or a negative dc level to a waveform.
7. *Duty control* This knob adjusts the duty cycle of the pulse waveform output. The sine wave or triangular wave outputs are not affected by this control.

Step 2: Review the Operation and Controls of the Oscilloscope

Review Section 8–8 on the oscilloscope with particular emphasis on the controls for the scope in Figure 8–48. Note that when the channel 1 coupling switch is set to GND, the screen displays a horizontal trace at the 0 V level. This trace can be set on any of the vertical divisions as a zero-volt reference using the vertical position control. On each screen in the following figures, the zero-volt reference is indicated by "0 V".

Step 3: Measure the Sinusoidal Output

The output of the function generator is connected to the channel 1 (CH1) input of the scope and the sinusoidal output function is selected. The coupling switch on the scope is set to DC.

In Figure 8–65(a), the amplitude and frequency of the function generator are set to the minimum values. Measure and record these values. Express amplitude in both peak and rms. How does the measured frequency compare to the frequency setting on the function generator?

In Figure 8–65(b), the amplitude and frequency are set to the maximum values. Measure and record these values. Express amplitude in both peak and rms. How does the measured frequency compare to the frequency setting on the function generator?

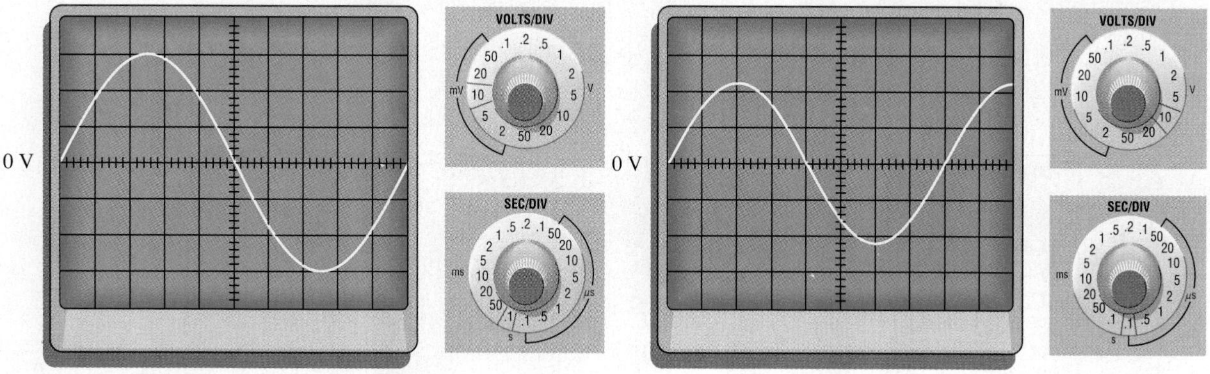

(a) (b)

FIGURE 8–65

Step 4: Measure the DC Offset

The sinusoidal amplitude and frequency of the function generator is set to an arbitrary value for dc offset measurement. The coupling switch on the scope is set to DC.

In Figure 8–66(a), the dc offset of the function generator is adjusted to the maximum positive value. Measure and record this value.

In Figure 8–66(b), the dc offset is adjusted to the maximum negative value. Measure and record this value.

Step 5: Measure the Triangular Output

The triangular output of the function generator is selected. The coupling switch on the scope is set to AC.

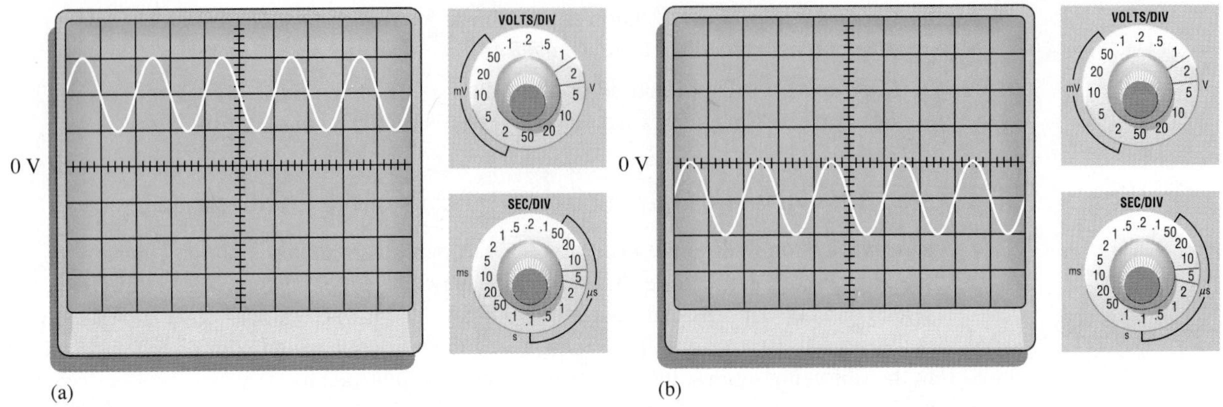

FIGURE 8–66

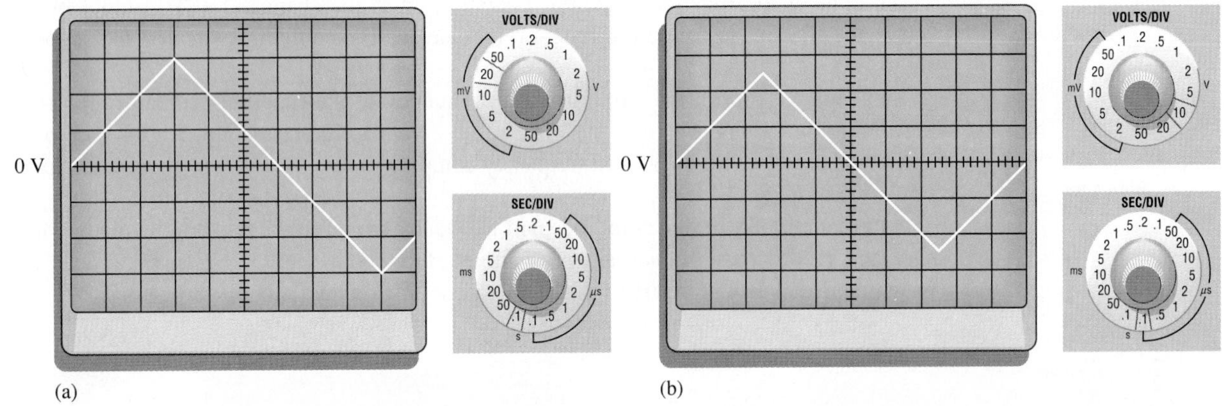

FIGURE 8–67

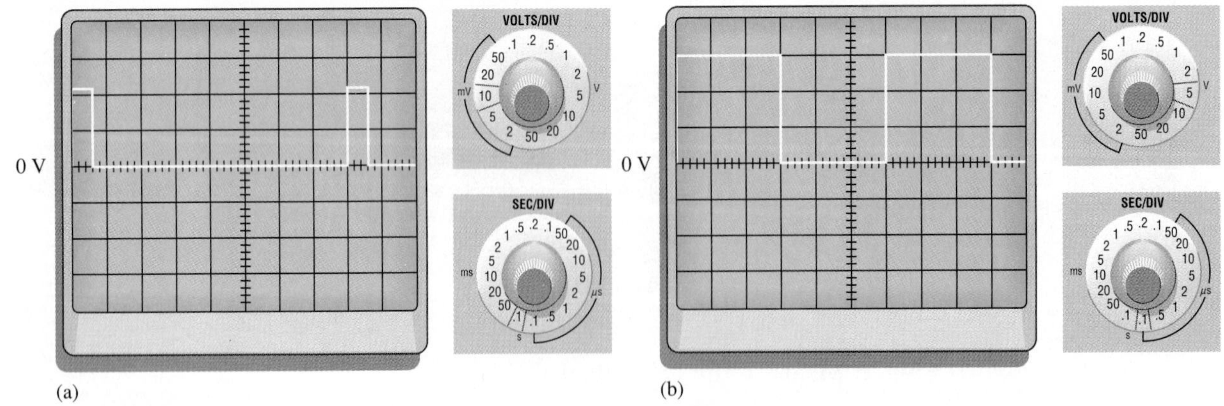

FIGURE 8–68

In Figure 8–67(a), the amplitude and frequency of the function generator are set to the minimum values. Measure and record these values.

In Figure 8–67(b), the amplitude and frequency are set to the maximum values. Measure and record these values.

Step 6: Measure the Pulse Output

The pulse output of the function generator is selected. The coupling switch on the scope is set to DC.

In Figure 8–68(a), the amplitude from the base line and frequency of the function generator are set to the minimum values. The duty cycle is adjusted to minimum. Measure and record these values. Express duty cycle as a percentage.

In Figure 8–68(b), the amplitude and frequency are set to the maximum values. The duty cycle is adjusted to maximum. Measure and record these values. Express duty cycle as a percentage.

SECTION 8–10 REVIEW

1. Explain the purpose of each position of the AC-GND-DC switch on each channel of the scope.
2. Generally, what setting of the SEC/DIV control on the scope should be used to obtain the most accurate measurement of frequency?
3. Generally, what setting of the VOLTS/DIV control on the scope should be used to obtain the most accurate measurement of amplitude?

■ SUMMARY

- The sine wave is a time-varying, periodic waveform.
- The sine wave is a common form of ac (alternating current).
- Alternating current changes direction in response to changes in the polarity of the source voltage.
- One cycle of an alternating sine wave consists of a positive alternation and a negative alternation.
- Two common sources of sine waves are the electromagnetic ac generator and the electronic oscillator circuit.
- A full cycle of a sine wave is 360°, or 2π radians. A half-cycle is 180°, or π radian. A quarter-cycle is 90°, or $\pi/2$ radians.
- A sine wave voltage can be generated by a conductor rotating in a magnetic field.
- Phase angle is the difference in degrees (or radians) between two sine waves.
- The angular position of a phasor represents the angle of a sine wave, and the length of the phasor represents the amplitude.
- When applying Ohm's or Kirchhoff's laws in ac circuits, voltages and currents must all be expressed with consistent values.
- Average power in a resistive ac circuit is determined using rms voltage and/or current values.
- A pulse consists of a transition from a baseline level to an amplitude level, followed by a transition back to the baseline level.
- A triangular or sawtooth wave consists of positive-going and negative-going ramps.
- Harmonic frequencies are odd or even multiples of the repetition rate of a nonsinusoidal waveform.
- Conversions of sine wave values are summarized in Table 8–2.

TABLE 8–2

TO CHANGE FROM	TO	MULTIPLY BY
Peak	rms	0.707
Peak	Peak-to-peak	2
Peak	Average	0.637
rms	Peak	1.414
Peak-to-peak	Peak	0.5
Average	Peak	1.57

▪ GLOSSARY

Alternating current (ac) Current that reverses direction in response to a change in source voltage polarity.

Amplitude The maximum value of a voltage or current.

Average value The average of a sine wave over one half-cycle. It is 0.637 times the peak value.

Baseline The normal level of a pulse waveform; the voltage level in the absence of a pulse.

Cathode-ray tube (CRT) A vacuum tube device containing an electron gun that emits a narrow focused beam of electrons onto a phosphor-coated screen.

Cycle One repetition of a periodic waveform.

Degree The unit of angular measure corresponding to 1/360 of a complete revolution.

Duty cycle A characteristic of a pulse waveform that indicates the percentage of time that a pulse is present during a cycle; the ratio of pulse width to period.

Effective value A measure of the heating effect of a sine wave; also known as the *rms* (root mean square) *value*.

Falling edge The negative-going transition of a pulse.

Fall time (t_f) The time interval required for a pulse to change from 90% to 10% of its full amplitude.

Frequency A measure of the rate of change of a periodic function; the number of cycles completed in 1 s. The unit of frequency is the hertz.

Fundamental frequency The repetition rate of a waveform.

Generator An energy source that produces electrical signals.

Harmonics The frequencies contained in a composite waveform, which are integer multiples of the repetition frequency (fundamental).

Hertz (Hz) The unit of frequency. One hertz equals one cycle per second.

Hypotenuse The longest side of a right triangle.

Instantaneous value The voltage or current value of a waveform at a given instant in time.

Lag To be behind; describes a condition of the phase or time relationship of waveforms in which one waveform is behind the other in phase or time.

Lead To be ahead; describes a condition of the phase or time relationship of waveforms in which one waveform is ahead of the other in phase or time; also, a wire or cable connection to a device or instrument.

Leading edge The first step or transition of a pulse.

Magnitude The value of a quantity, such as the number of volts of voltage or the number of amperes of current.

Oscillator An electronic circuit that produces a time-varying signal without an external input signal using positive feedback.

Oscilloscope A measurement instrument that displays signal waveforms on a screen.

Peak value The voltage or current value of a waveform at its maximum positive or negative points.

Peak-to-peak value The voltage or current value of a waveform measured from its minimum to its maximum points.

Period The time interval of one complete cycle of a given sine wave or any periodic waveform.

Periodic Characterized by a repetition at fixed time intervals.

Phase The relative displacement of a time-varying waveform in terms of its occurrence with respect to a reference.

Phasor A representation of a sine wave in terms of both magnitude and phase angle.

Pulse A type of waveform that consists of two equal and opposite steps in voltage or current separated by a time interval.

Pulse width The time interval between the opposite steps of an ideal pulse. For a nonideal pulse, the time between the 50% points on the leading and trailing edges.

Radian A unit of angular measurement. There are 2π rads in one complete revolution. One radian equals 57.3°.

Ramp A type of waveform characterized by a linear increase or decrease in voltage or current.

Right triangle A triangle with one 90° angle.

Rise time (t_r) The time interval required for a pulse to change from 10% to 90% of its amplitude.

Rising edge The positive-going transition of a pulse.

Root mean square (rms) The value of a sine wave that indicates its heating effect; also known as the *effective value*. It is equal to 0.707 times the peak value.

Sawtooth wave A type of electrical waveform composed of ramps; a special case of a triangular wave.

Trailing edge The second step or transition of a pulse.

Triangular wave A type of electrical waveform that consists of two ramps.

Trigger The activating mechanism of some electronic devices or instruments.

Waveform The pattern of variations of a voltage or current showing how the quantity changes with time.

■ FORMULAS

(8–1)	$f = \frac{1}{T}$	Frequency
(8–2)	$T = \frac{1}{f}$	Period
(8–3)	$f = \text{number of pole pairs} \times (\text{rps})$	Output frequency of a generator
(8–4)	$V_{pp} = 2V_p$	Peak-to-peak voltage (sine wave)
(8–5)	$I_{pp} = 2I_p$	Peak to peak current (sine wave)
(8–6)	$V_{rms} = 0.707V_p$	Root-mean-square voltage (sine wave)
(8–7)	$I_{rms} = 0.707I_p$	Root-mean-square current (sine wave)
(8–8)	$V_p = 1.414V_{rms}$	Peak voltage (sine wave)
(8–9)	$I_p = 1.414I_{rms}$	Peak current (sine wave)
(8–10)	$V_{pp} = 2.828V_{rms}$	Peak-to-peak voltage (sine wave)
(8–11)	$I_{pp} = 2.828I_{rms}$	Peak-to-peak current (sine wave)
(8–12)	$V_{avg} = 0.637V_p$	Half-cycle average voltage (sine wave)
(8–13)	$I_{avg} = 0.637I_p$	Half-cycle average current (sine wave)
(8–14)	$\text{rad} = \left(\frac{\pi\ \text{rad}}{180°}\right) \times \text{degrees}$	
(8–15)	$\text{degrees} = \left(\frac{180°}{\pi\ \text{rad}}\right) \times \text{rad}$	
(8–16)	$y = A \sin \theta$	General formula for a sine wave
(8–17)	$v = V_p \sin \theta$	Sine wave voltage
(8–18)	$i = I_p \sin \theta$	Sine wave current
(8–19)	$y = A \sin(\theta - \phi)$	Lagging sine wave
(8–20)	$y = A \sin(\theta + \phi)$	Leading sine wave
(8–21)	$\text{Percent duty cycle} = \left(\frac{t_W}{T}\right)100\%$	
(8–22)	$V_{avg} = \text{baseline} + (\text{duty cycle})(\text{amplitude})$	Average value of pulse waveform

■ SELF-TEST

1. The difference between alternating current (ac) and direct current (dc) is
   (a) ac changes value and dc does not **(b)** ac changes direction and dc does not
   (c) both (a) and (b) **(d)** neither (a) nor (b)
2. During each cycle, a sine wave reaches a peak value
   (a) one time **(b)** two times
   (c) four times **(d)** a number of times depending on the frequency

3. A sine wave with a frequency of 12 kHz is changing at a faster rate than a sine wave with a frequency of
(a) 20 kHz (b) 15,000 Hz (c) 10,000 Hz (d) 1.25 MHz
4. A sine wave with a period of 2 ms is changing at a faster rate than a sine wave with a period of
(a) 1 ms (b) 0.0025 s (c) 1.5 ms (d) 1000 μs
5. When a sine wave has a frequency of 60 Hz, in 10 s it goes through
(a) 6 cycles (b) 10 cycles (c) 1/16 cycle (d) 600 cycles
6. If the peak value of a sine wave is 10 V, the peak-to-peak value is
(a) 20 V (b) 5 V (c) 100 V (d) none of these
7. If the peak value of a sine wave is 20 V, the rms value is
(a) 14.14 V (b) 6.37 V (c) 7.07 V (d) 0.707 V
8. The average value of a 10 V peak sine wave over one complete cycle is
(a) 0 V (b) 6.37 V (c) 7.07 V (d) 5 V
9. The average half-cycle value of a sine wave with a 20 V peak is
(a) 0 V (b) 6.37 V (c) 12.74 V (d) 14.14 V
10. One sine wave has a positive-going zero crossing at 10° and another sine wave has a positive-going zero crossing at 45°. The phase angle between the two waveforms is
(a) 55° (b) 35° (c) 0° (d) none of these
11. The instantaneous value of a 15 A peak sine wave at a point 32° from its positive-going zero crossing is
(a) 7.95 A (b) 7.5 A (c) 2.13 A (d) 7.95 V
12. If the rms current through a 10 kΩ resistor is 5 mA, the rms voltage drop across the resistor is
(a) 70.7 V (b) 7.07 V (c) 5 V (d) 50 V
13. Two series resistors are connected to an ac source. If there is 6.5 V rms across one resistor and 3.2 V rms across the other, the peak source voltage is
(a) 9.7 V (b) 9.19 V (c) 13.72 V (d) 4.53 V
14. A 10 kHz pulse waveform consists of pulses that are 10 μs wide. Its duty cycle is
(a) 100% (b) 10% (c) 1% (d) not determinable
15. The duty cycle of a square wave
(a) varies with the frequency (b) varies with the pulse width
(c) both (a) and (b) (d) is 50%

▪ BASIC PROBLEMS

SECTION 8–1 The Sine Wave

1. Calculate the frequency for each of the following periods:
(a) 1 s (b) 0.2 s (c) 50 ms (d) 1 ms (e) 500 μs (f) 10 μs
2. Calculate the period for each of the following frequencies:
(a) 1 Hz (b) 60 Hz (c) 500 Hz (d) 1 kHz (e) 200 kHz (f) 5 MHz
3. A sine wave goes through 5 cycles in 10 μs. What is its period?
4. A sine wave has a frequency of 50 kHz. How many cycles does it complete in 10 ms?

SECTION 8–2 Sine Wave Voltage Sources

5. The conductive loop on the rotor of a simple two-pole, single-phase generator rotates at a rate of 250 rps. What is the frequency of the induced output voltage?
6. A certain four-pole generator has a speed of rotation of 3600 rpm. What is the frequency of the voltage produced by this generator?
7. At what speed of rotation must a four-pole generator be operated to produce a 400 Hz sine wave voltage?

SECTION 8–3 Voltage and Current Values of Sine Waves

8. A sine wave has a peak value of 12 V. Determine the following voltage values:
(a) rms (b) peak-to-peak (c) half-cycle average
9. A sinusoidal current has an rms value of 5 mA. Determine the following current values:
(a) peak (b) half-cycle average (c) peak-to-peak
10. For the sine wave in Figure 8–69, determine the peak, peak-to-peak, rms, and average values.

FIGURE 8–69

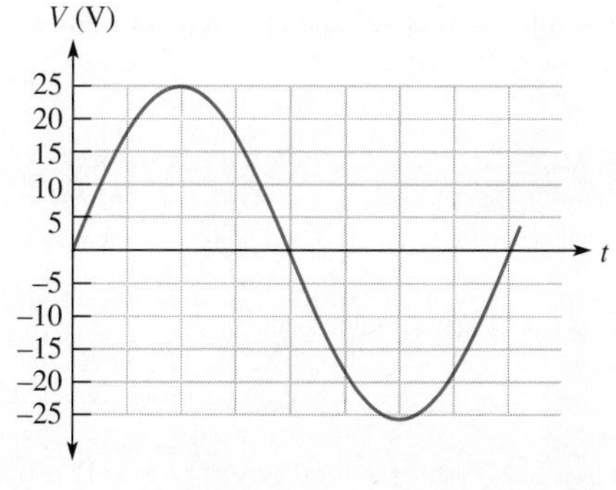

SECTION 8–4 Angular Measurement of a Sine Wave

11. Sine wave *A* has a positive-going zero crossing at 30°. Sine wave *B* has a positive-going zero crossing at 45°. Determine the phase angle between the two signals. Which signal leads?

12. One sine wave has a positive peak at 75°, and another has a positive peak at 100°. How much is each sine wave shifted in phase from the 0° reference? What is the phase angle between them?

13. Make a sketch of two sine waves as follows: Sine wave *A* is the reference, and sine wave *B* lags *A* by 90°. Both have equal amplitudes.

14. Convert the following angular values from degrees to radians:
(a) 30° **(b)** 45° **(c)** 78° **(d)** 135° **(e)** 200° **(f)** 300°

15. Convert the following angular values from radians to degrees:
(a) $\pi/8$ **(b)** $\pi/3$ **(c)** $\pi/2$ **(d)** $3\pi/5$ **(e)** $6\pi/5$ **(f)** 1.8π

SECTION 8–5 The Sine Wave Formula

16. A certain sine wave has a positive-going zero crossing at 0° and an rms value of 20 V. Calculate its instantaneous value at each of the following angles:
(a) 15° **(b)** 33° **(c)** 50° **(d)** 110° **(e)** 70° **(f)** 145° **(g)** 250° **(h)** 325°

17. For a particular 0° reference sinusoidal current, the peak value is 100 mA. Determine the instantaneous value at each of the following points:
(a) 35° **(b)** 95° **(c)** 190° **(d)** 215° **(e)** 275° **(f)** 360°

18. For a 0° reference sine wave with an rms value of 6.37 V, determine its instantaneous value at each of the following points:
(a) $\pi/8$ rad **(b)** $\pi/4$ rad **(c)** $\pi/2$ rad **(d)** $3\pi/4$ rad **(e)** π rad **(f)** $3\pi/2$ rad **(g)** 2π rad

19. Sine wave *A* lags sine wave *B* by 30°. Both have peak values of 15 V. Sine wave *A* is the reference with a positive-going crossing at 0°. Determine the instantaneous value of sine wave *B* at 30°, 45°, 90°, 180°, 200°, and 300°.

20. Repeat Problem 19 for the case when sine wave *A* leads sine wave *B* by 30°.

SECTION 8–6 Ohm's Law and Kirchhoff's Laws in AC Circuits

21. A sinusoidal voltage is applied to the resistive circuit in Figure 8–70. Determine the following:
(a) I_{rms} **(b)** I_{avg} **(c)** I_p **(d)** I_{pp} **(e)** i at the positive peak

22. Find the half-cycle average values of the voltages across R_1 and R_2 in Figure 8–71. All values shown are rms.

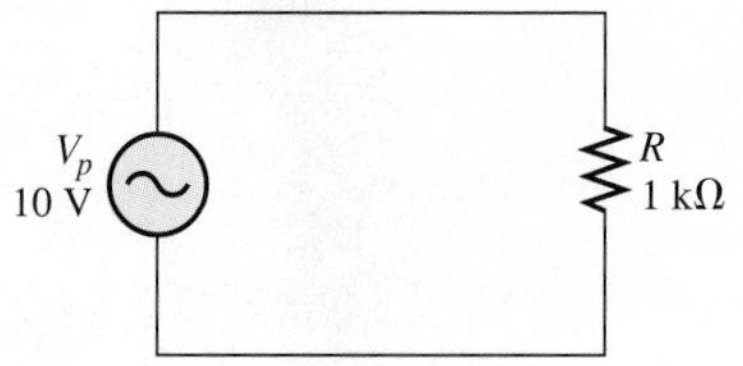

FIGURE 8–70

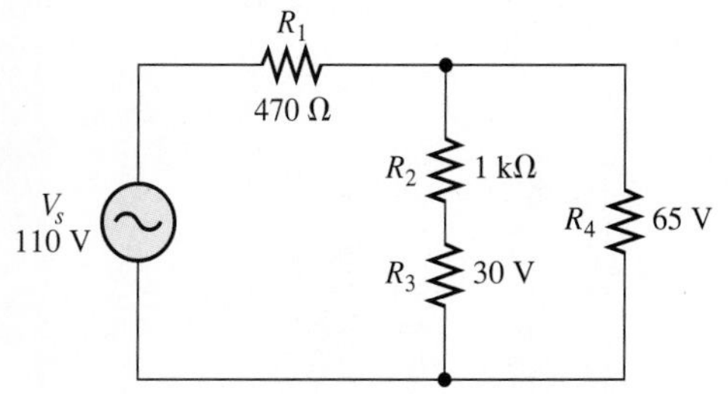

FIGURE 8–71

23. Determine the rms voltage across R_3 in Figure 8–72.

FIGURE 8–72

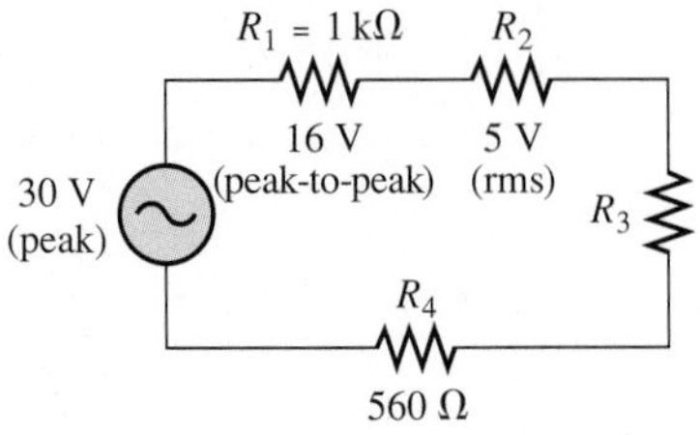

SECTION 8–7 Superimposed DC and AC Voltages

24. A sine wave with an rms value of 10.6 V is riding on a dc level of 24 V. What are the maximum and minimum values of the resulting waveform?

25. How much dc voltage must be added to a 3 V rms sine wave in order to make the resulting voltage nonalternating (no negative values)?

26. A 6 V peak sine wave is riding on a dc voltage of 8 V. If the dc voltage is lowered to 5 V, how far negative will the sine wave go?

SECTION 8–8 The Oscilloscope

27. Determine the peak value and the period of the sine wave displayed on the scope screen in Figure 8–73.

28. Determine the rms value and the frequency of the sine wave displayed on the scope screen in Figure 8–73.

29. Determine the rms value and the frequency of the sine wave displayed on the scope screen in Figure 8–74.

FIGURE 8–73

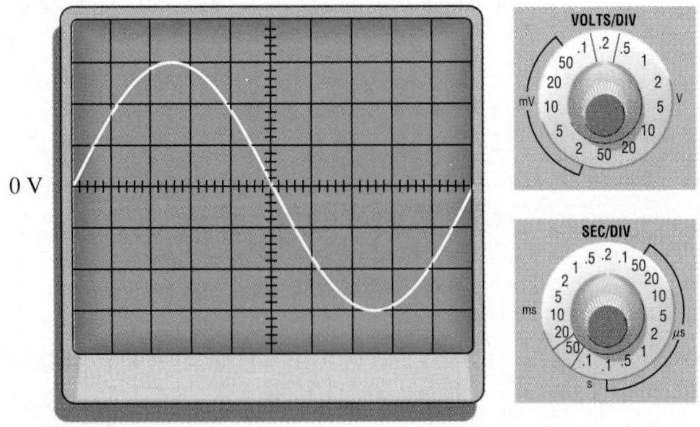

FIGURE 8–74

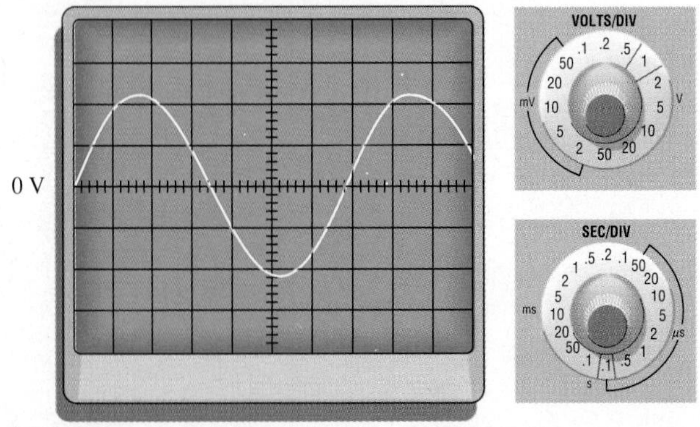

30. Find the amplitude, pulse width, and duty cycle for the pulse waveform displayed on the scope screen in Figure 8–75.

FIGURE 8–75

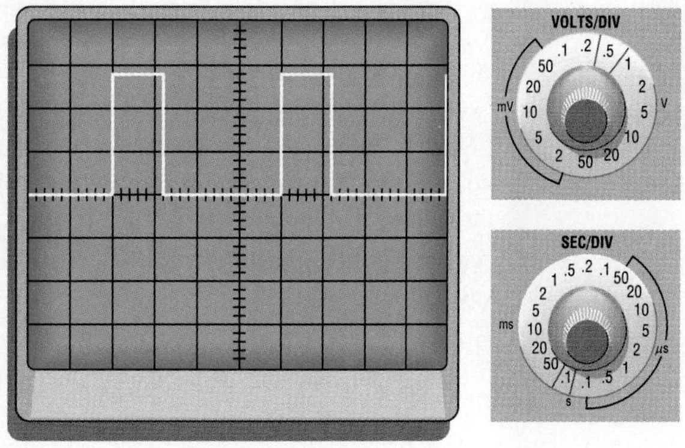

SECTION 8–9 Nonsinusoidal Waveforms

31. From the graph in Figure 8–76, determine the approximate values of t_r, t_f, t_W, and amplitude.

32. Determine the duty cycle for each pulse waveform in Figure 8–77.

33. Find the average value of each pulse waveform in Figure 8–77.

34. What is the frequency of each waveform in Figure 8–77?

35. What is the frequency of each sawtooth waveform in Figure 8–78?

FIGURE 8–76

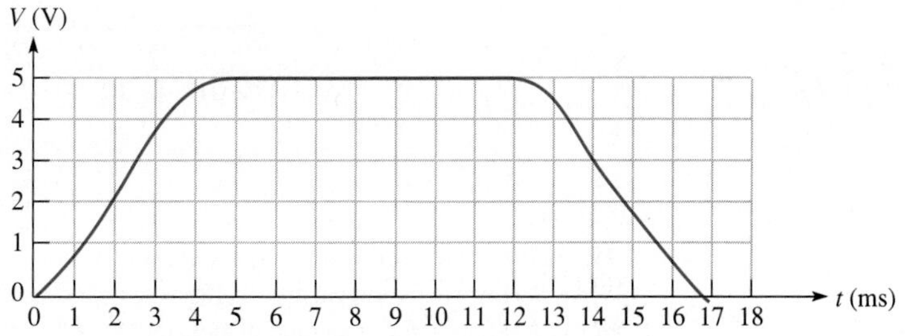

FIGURE 8–77

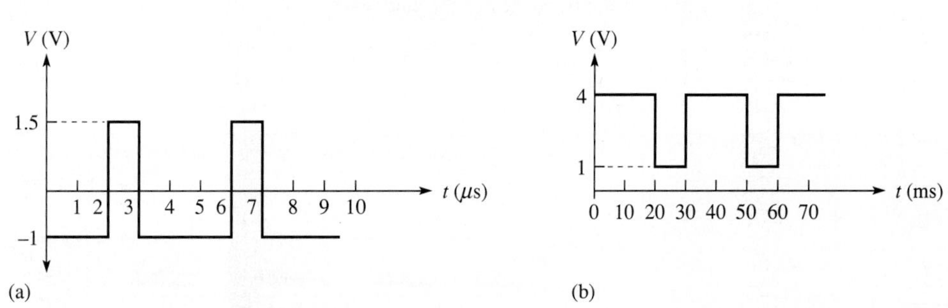

FIGURE 8–78

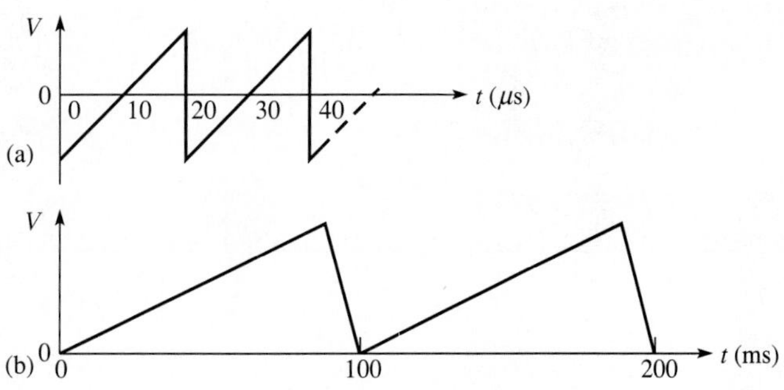

36. A square wave has a period of 40 μs. List the first six odd harmonics.
37. What is the fundamental frequency of the square wave mentioned in Problem 36?

▪ ADVANCED PROBLEMS

38. A certain sine wave has a frequency of 2.2 kHz and an rms value of 25 V. Assuming a given cycle begins (zero crossing) at $t = 0$ s, what is the change in voltage from 0.12 ms to 0.2 ms?
39. Figure 8–79 shows a sinusoidal voltage source in series with a dc source. Effectively, the two voltages are superimposed. Sketch the voltage across R_L. Determine the maximum current through R_L and the average voltage across R_L.
40. A nonsinusoidal waveform called a *stairstep* is shown in Figure 8–80. Determine its average value.

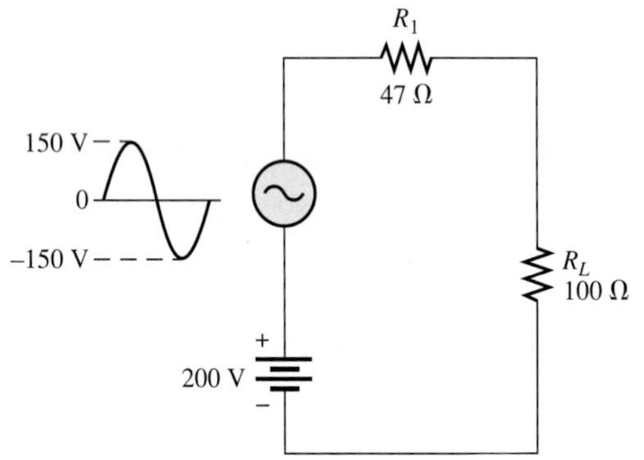

FIGURE 8–79

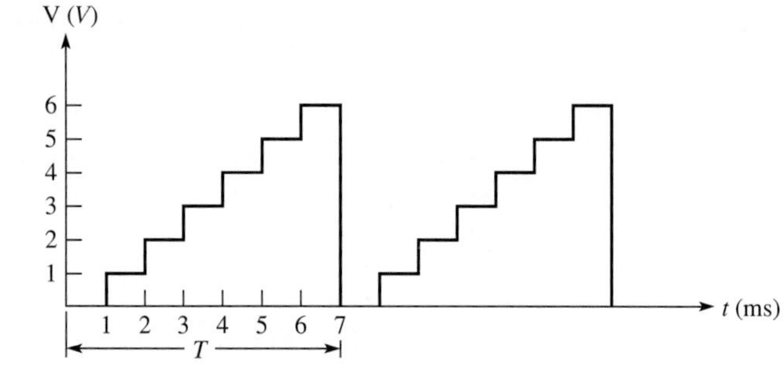

FIGURE 8–80

41. Refer to the oscilloscope in Figure 8–81.
(a) How many cycles are displayed?
(b) What is the rms value of the sine wave?
(c) What is the frequency of the sine wave?

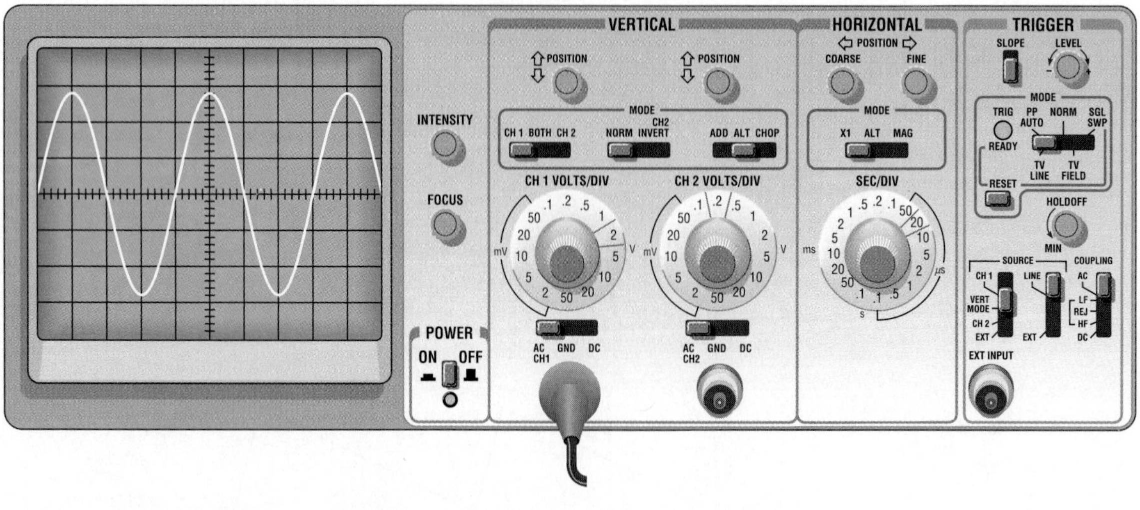

FIGURE 8–81

42. Accurately sketch on a grid representing the scope screen in Figure 8–81 how the sine wave will appear if the volts/division control is moved to the next higher setting.

43. Accurately sketch on a grid representing the scope screen in Figure 8–81 how the sine wave will appear if the seconds/division control is moved to the next lower setting.

44. Based on the instrument settings and an examination of the scope display and the circuit board in Figure 8–82 determine the frequency and peak value of the input signal and output signal. The waveform shown is channel 1. Sketch the channel 2 waveform as it would appear on the scope with the indicated settings.

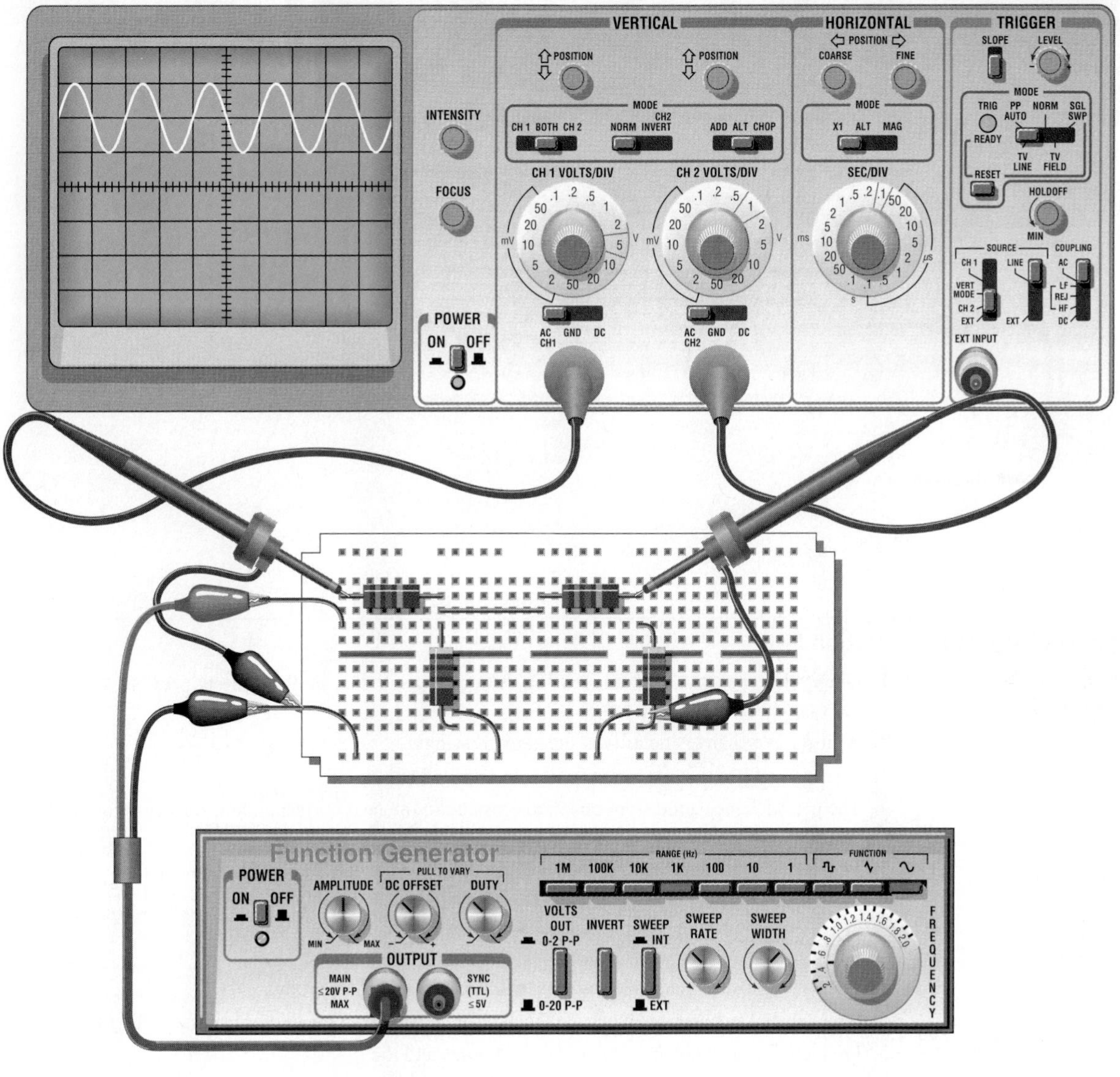

FIGURE 8–82

45. Examine the circuit board and the oscilloscope display in Figure 8–83 and determine the peak value and the frequency of the unknown input signal.

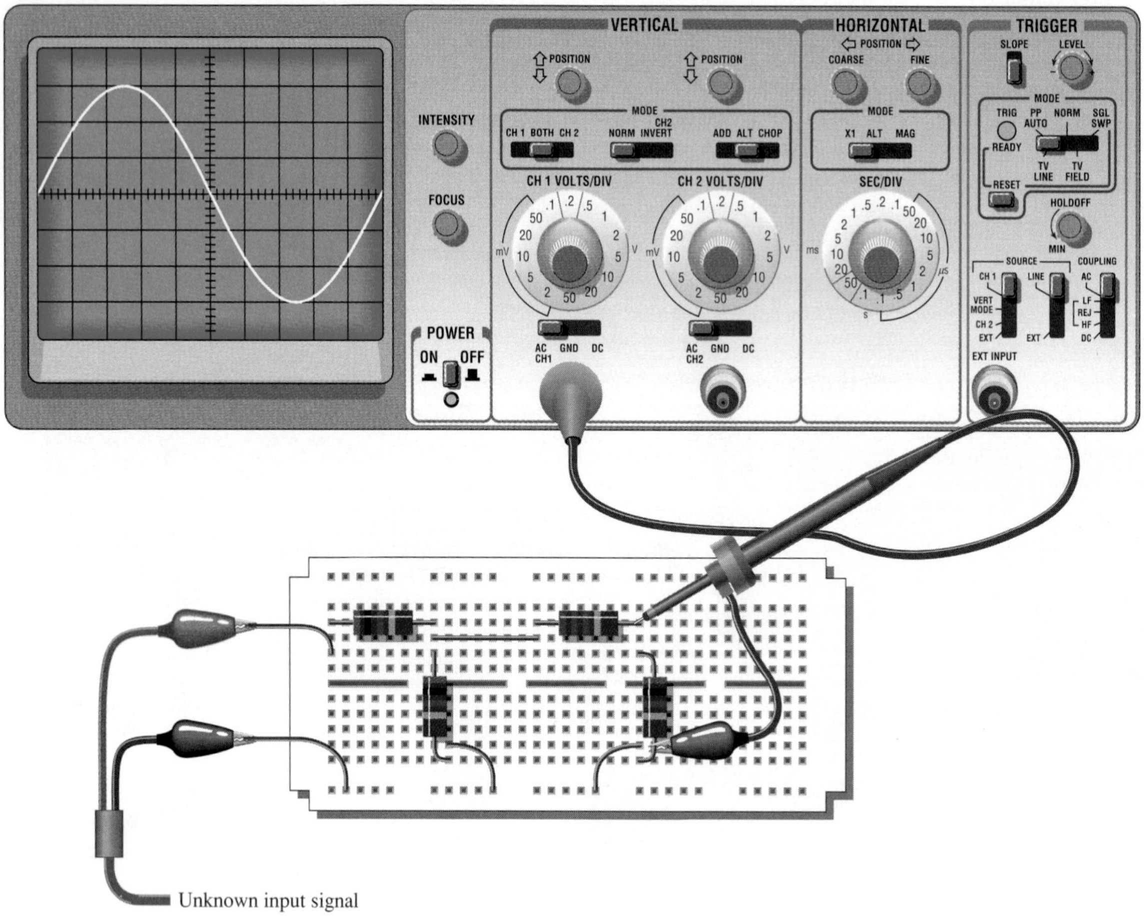

FIGURE 8–83

■ ANSWERS TO SECTION REVIEWS

Section 8–1

1. One cycle goes from the zero crossing through a positive peak, then through zero to a negative peak and back to the zero crossing.
2. A sine wave changes polarity at the zero crossings.
3. A sine wave has two maximum points in one cycle.
4. The period is measured from one zero crossing to the next corresponding zero crossing, or from one peak to the next corresponding peak.
5. Frequency is the number of cycles completed in one second and its unit is the hertz.
6. $f = 1/5\ \mu\text{s} = 200$ kHz **7.** $T = 1/120$ Hz $= 8.33$ ms

Section 8–2

1. Two types of sine wave sources are electromagnetic and electronic.
2. Sine waves are generated electromagnetically by rotating a conductor in a magnetic field.
3. Frequency is dependent on the number of poles and the rps.
4. Magnitude is dependent on the number of turns and the rps.

Section 8–3

1. **(a)** $V_{pp} = 2(1\text{ V}) = 2\text{ V}$ **(b)** $V_{pp} = 2(1.414)(1.414\text{ V}) = 4\text{ V}$ **(c)** $V_{pp} = 2(1.57)(3\text{ V}) = 9.42\text{ V}$
2. **(a)** $V_{\text{rms}} = (0.707)(2.5\text{ V}) = 1.77\text{ V}$ **(b)** $V_{\text{rms}} = (0.5)(0.707)(10\text{ V}) = 3.54\text{ V}$
   (c) $V_{\text{rms}} = (0.707)(1.57)(1.5\text{ V}) = 1.66\text{ V}$
3. **(a)** $V_{\text{avg}} = (0.637)(10\text{ V}) = 6.37\text{ V}$ **(b)** $V_{\text{avg}} = (0.637)(1.414)(2.3\text{ V}) = 2.07\text{ V}$
   (c) $V_{\text{avg}} = (0.637)(0.5)(60\text{ V}) = 19.1\text{ V}$

Section 8–4

1. **(a)** The positive peak is at 90°.
   (b) The negative-going zero crossing is at 180°.
   (c) The negative peak is at 270°.
   (d) The cycle ends at 360°.
2. A half-cycle is completed in 180° or π radians.
3. A full cycle is completed in 360° or 2π radians.
4. $\theta = 90° - 45° = 45°$

Section 8–5

1. **(a)** $\sin 30° = 0.5$ **(b)** $\sin 60° = 0.866$ **(c)** $\sin 90° = 1$
2. $v = 10 \sin 120° = 8.66$ V **3.** $v = 10 \sin(45° + 10°) = 8.19$ V

Section 8–6

1. $I_p = V_p/R = (1.57)(12.5\text{ V})/330\ \Omega = 59.5$ mA **2.** $V_{s(\text{rms})} = (0.707)(25.3\text{ V}) = 17.9$ V

Section 8–7

1. $+V_{\text{max}} = 5\text{ V} + 2.5\text{ V} = 7.5$ V **2.** Yes, it will alternate. **3.** $+V_{\text{max}} = 5\text{ V} - 2.5\text{ V} = 2.5$ V

Section 8–8

1. CRT stands for cathode-ray tube.
2. Voltage is measured vertically; time is measured horizontally.
3. An oscilloscope displays the waveform of a voltage. A multimeter provides rms value.

Section 8–9

1. **(a)** Rise time is the interval from 10% to 90% of the rising edge.
   (b) Fall time is the interval from 90% to 10% of the falling edge.
   (c) Pulse width is the interval from 50% of the leading edge to 50% of the trailing edge.
2. $f = 1/1\text{ ms} = 1$ kHz
3. D.C. = $(200\ \mu\text{s}/1\text{ ms})100\% = 20\%$; Amplitude = $6\text{ V} - (-2\text{ V}) = 8$ V;
   $V_{\text{avg}} = -2\text{ V} + (0.2)(8\text{ V}) = -0.4$ V
4. $2 \times 1\text{ kHz} = 2$ kHz

Section 8–10

1. AC: Couples ac voltage only; GND: Grounds input so trace is at 0 V; DC: Couples ac and dc voltages
2. Lowest SEC/DIV setting that allows frequency to be measured
3. Lowest VOLTS/DIV setting that allows voltage to be measured

■ ANSWERS TO RELATED EXERCISES FOR EXAMPLES

8–1	2.4 s	**8–2**	1.5 ms
8–3	20 kHz	**8–4**	200 Hz
8–5	1 ms	**8–6**	30 rps
8–7	$V_{pp} = 50$ V; $V_{\text{rms}} = 17.7$ V; $V_{\text{avg}} = 15.9$ V	**8–8**	**(a)** $\pi/12$ **(b)** 360°
8–9	8°	**8–10**	18.1 V
8–11	$I_{\text{rms}} = 4.53$ mA; $V_{1(\text{rms})} = 4.53$ V; $V_{2(\text{rms})} = 2.54$ V	**8–12**	23.7 V
8–13	The waveform in part (a) never goes negative. The waveform in part (b) goes negative for a portion of its cycle.	**8–14**	**(a)** $V_{\text{rms}} = 1.06$ V; $f = 50$ Hz **(b)** $V_{\text{rms}} = 88.4$ mV; $f = 1.67$ kHz **(c)** $V_{\text{rms}} = 4.81$ V; $f = 50$ kHz **(d)** $V_{\text{rms}} = 7.07$ V; $f = 250$ kHz
8–15	5%	**8–16**	1.2 V
8–17	120 V		

9

CAPACITORS

The capacitor is an electrical device that can store electrical charge, thereby creating an electric field which in turn stores energy; the measure of that energy-storing ability is called *capacitance*. In this chapter, the basic capacitor is introduced and its characteristics are studied. The physical construction and electrical properties of various types of capacitors are discussed.

The basic behavior of capacitors in both dc and ac circuits is studied, and series and parallel combinations are analyzed. Representative applications and methods of testing capacitors also are discussed.

■ CHAPTER OBJECTIVES

- ☐ Describe the basic structure and characteristics of a capacitor
- ☐ Discuss various types of capacitors
- ☐ Analyze series capacitors
- ☐ Analyze parallel capacitors
- ☐ Analyze capacitive dc switching circuits
- ☐ Analyze capacitive ac circuits
- ☐ Discuss some capacitor applications
- ☐ Test a capacitor

■ APPLICATION ASSIGNMENT

Capacitors are used in many applications. The application assignment in this chapter focuses on the use of capacitors in amplifier circuits to couple an ac voltage from one point to another while blocking dc voltage. You will determine if voltages measured on an amplifier circuit board with an oscilloscope are correct; and if they are not correct, you will specify the problem. After studying this chapter, you should be able to complete the application assignment in Section 9–9.

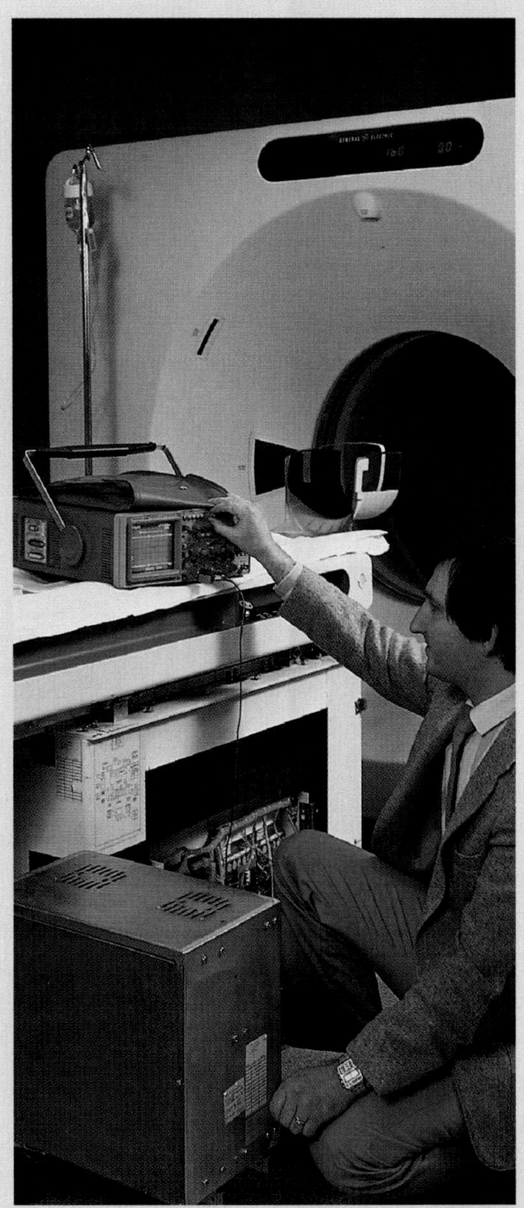

9–1 ■ THE BASIC CAPACITOR

In this section, the basic construction and characteristics of capacitors are examined.

After completing this section, you should be able to

- **Describe the basic structure and characteristics of a capacitor**
  - □ Explain how a capacitor stores charge
  - □ Define *capacitance* and state its unit
  - □ State Coulomb's law
  - □ Explain how a capacitor stores energy
  - □ Discuss voltage rating and temperature coefficient
  - □ Explain capacitor leakage
  - □ Specify how the physical characteristics affect the capacitance

Basic Construction

In its simplest form, a **capacitor** is an electrical device constructed of two parallel conductive plates separated by an insulating material called the **dielectric.** Connecting leads are attached to the parallel plates. A basic capacitor is shown in Figure 9–1(a), and the schematic symbol is shown in part (b).

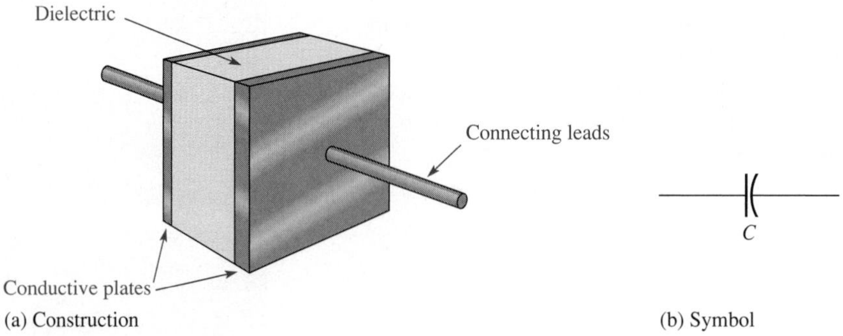

FIGURE 9–1
The basic capacitor.

How a Capacitor Stores Charge

In the neutral state, both plates of a capacitor have an equal number of free electrons, as indicated in Figure 9–2(a). When the capacitor is connected to a voltage source through a resistor, as shown in part (b), electrons (negative charge) are removed from plate A, and an equal number are deposited on plate B. As plate A loses electrons and plate B gains electrons, plate A becomes positive with respect to plate B. During this charging process, electrons flow only through the connecting leads and the source. No electrons flow through the dielectric of the capacitor because it is an insulator. The movement of electrons ceases when the voltage across the capacitor equals the source voltage, as indicated in Figure 9–2(c). If the capacitor is disconnected from the source, it retains the stored charge for a long period of time (the length of time depends on the type of capacitor) and still has the voltage across it, as shown in Figure 9–2(d). Actually, the charged capacitor can be considered as a temporary battery.

Capacitance

The amount of charge per unit of voltage that a capacitor can store is its **capacitance,** designated C. That is, capacitance is a measure of a capacitor's ability to store charge.

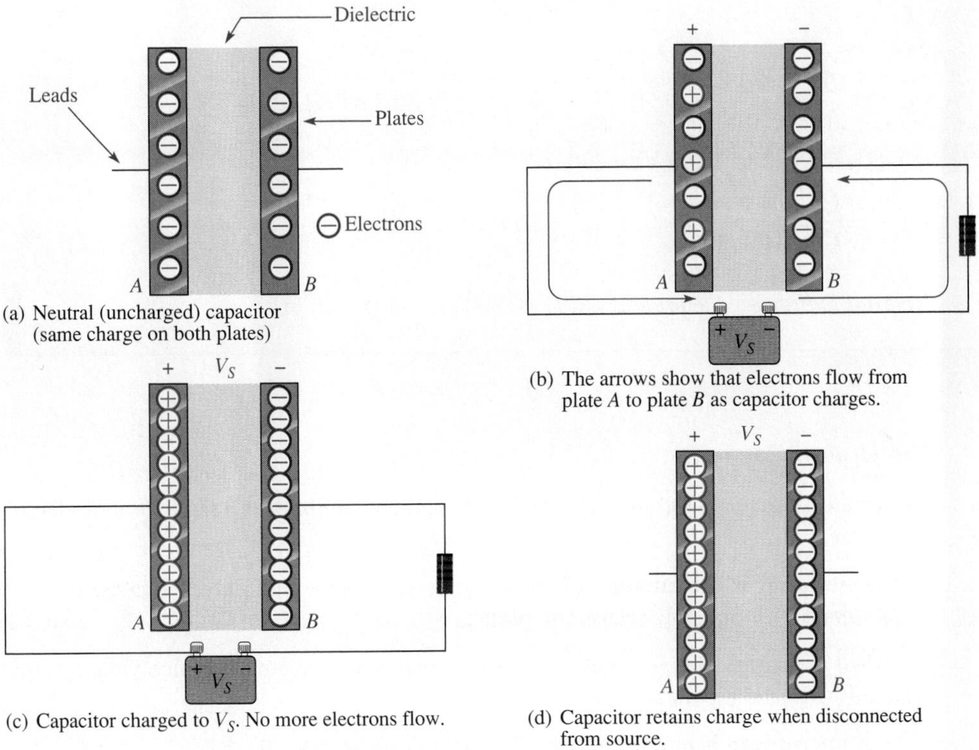

FIGURE 9–2
Illustration of a capacitor storing charge.

The more charge per unit of voltage that a capacitor can store, the greater its capacitance, as expressed by the following formula:

$$C = \frac{Q}{V} \tag{9–1}$$

where C is capacitance, Q is charge, and V is voltage.

By rearranging Equation (9–1), you can obtain two other forms for calculating Q and V.

$$Q = CV \tag{9–2}$$

$$V = \frac{Q}{C} \tag{9–3}$$

EXAMPLE 9–1

(a) A certain capacitor stores 50 microcoulombs (50 μC) when 10 V are applied across its plates. What is its capacitance?

(b) A 2 microfarad (2 μF) capacitor has 100 V across its plates. How much charge does it store?

(c) Determine the voltage across a 100 picofarad (100 pF) capacitor that is storing 2 microcoulombs (2 μC) of charge.

Solution

(a) $C = \frac{Q}{V} = \frac{50\ \mu C}{10\ V} = 5\ \mu F$ 5 0 EXP +/− 6 ÷ 1 0 =

(b) $Q = CV = (2\ \mu F)(100\ V) = 200\ \mu C$

(c) $V = \frac{Q}{C} = \frac{2\ \mu C}{100\ pF} = 0.02 \times 10^6 = 20 \times 10^3\ V = 20\ kV$

Related Exercise Determine V if $C = 1000$ pF and $Q = 10\ \mu$C.

The Unit of Capacitance

The **farad (F)** is the basic unit of capacitance, and the **coulomb (C)** is the unit of electrical charge.

One farad is the amount of capacitance when one coulomb of charge is stored with one volt across the plates.

Most capacitors that you will use in electronics work have capacitance values in microfarads (μF) and picofarads (pF).

A microfarad is one-millionth of a farad ($1\ \mu F = 1 \times 10^{-6}$ F).

A picofarad is one-trillionth of a farad ($1\ pF = 1 \times 10^{-12}$ F).

Conversions for farads, microfarads, and picofarads are given in Table 9–1.

TABLE 9–1
Conversions for farads, microfarads, and picofarads

TO CONVERT FROM	TO	MOVE THE DECIMAL POINT
Farads	Microfarads	6 places to right ($\times 10^6$)
Farads	Picofarads	12 places to right ($\times 10^{12}$)
Microfarads	Farads	6 places to left ($\times 10^{-6}$)
Microfarads	Picofarads	6 places to right ($\times 10^6$)
Picofarads	Farads	12 places to left ($\times 10^{-12}$)
Picofarads	Microfarads	6 places to left ($\times 10^{-6}$)

EXAMPLE 9–2

Convert the following values to microfarads:
(a) 0.00001 F (b) 0.005 F (c) 1000 pF (d) 200 pF

Solution

(a) $0.00001\ F \times 10^6 = 10\ \mu F$
(b) $0.005\ F \times 10^6 = 5000\ \mu F$
(c) $1000\ pF \times 10^{-6} = 0.001\ \mu F$
(d) $200\ pF \times 10^{-6} = 0.0002\ \mu F$

Related Exercise Convert 50,000 pF to microfarads.

EXAMPLE 9–3 Convert the following values to picofarads:

(a) 0.1×10^{-8} F **(b)** 0.000025 F **(c)** 0.01 μF **(d)** 0.005 μF

Solution

(a) $0.1 \times 10^{-8}\ \text{F} \times 10^{12} = 1000\ \text{pF}$ **(b)** $0.000025\ \text{F} \times 10^{12} = 25 \times 10^{6}\ \text{pF}$

(c) $0.01\ \mu\text{F} \times 10^{6} = 10{,}000\ \text{pF}$ **(d)** $0.005\ \mu\text{F} \times 10^{6} = 5000\ \text{pF}$

Related Exercise Convert 100 μF to picofarads.

How a Capacitor Stores Energy

A capacitor stores energy in the form of an electric field that is established by the opposite charges on the two plates. The electric field is represented by lines of force between the positive and negative charges and concentrated within the dielectric, as shown in Figure 9–3.

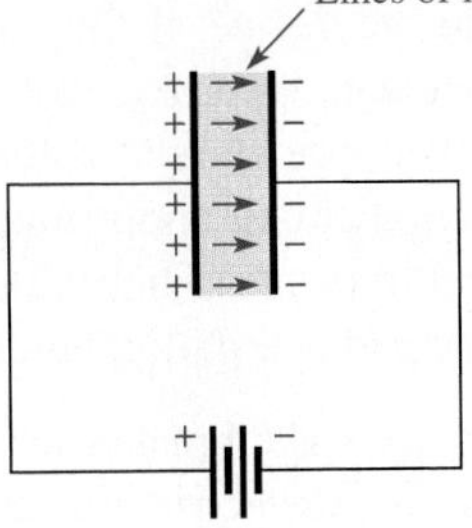

FIGURE 9–3
The electric field stores energy in a capacitor.

Coulomb's law states

> **A force exists between two charged bodies that is directly proportional to the product of the two charges and inversely proportional to the square of the distance between the bodies.**

This relationship is expressed as

$$F = \frac{kQ_1Q_2}{d^2} \qquad \textbf{(9–4)}$$

where F is the force in newtons, Q_1 and Q_2 are the charges in coulombs, d is the distance between the charges in meters, and k is a proportionality constant equal to 9×10^9.

Figure 9–4(a) illustrates the line of force between a positive and a negative charge. Figure 9–4(b) shows that many opposite charges on the plates of a capacitor create many

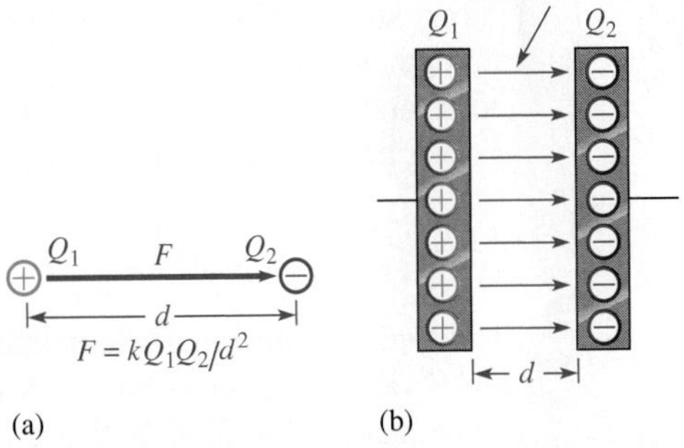

FIGURE 9–4
Lines of force are created by opposite charges.

lines of force, which form an electric field between the plates that stores energy within the dielectric.

The greater the forces between the charges on the plates of a capacitor, the more energy is stored. The amount of energy stored therefore is directly proportional to the capacitance because, from Coulomb's law, the more charge stored, the greater the force.

Also, from the equation $Q = CV$, the amount of charge stored is directly related to the voltage as well as the capacitance. The amount of energy stored is dependent on the square of the voltage across the plates of the capacitor. The formula for the energy stored by a capacitor is as follows:

$$W = \frac{1}{2}CV^2 \qquad \textbf{(9–5)}$$

where the energy (W) is in joules when C is in farads and V is in volts.

Voltage Rating

Every capacitor has a limit on the amount of voltage that it can withstand across its plates. The voltage rating specifies the maximum dc voltage that can be applied without risk of damage to the device. If this maximum voltage, commonly called the *breakdown voltage* or *working voltage,* is exceeded, permanent damage to the capacitor can result.

Both the capacitance and the voltage rating must be taken into consideration before a capacitor is used in a circuit application. The choice of capacitance value is based on particular circuit requirements. The voltage rating should always be well above the maximum voltage expected in a particular application.

Dielectric Strength The breakdown voltage of a capacitor is determined by the **dielectric strength** of the dielectric material used. The dielectric strength is expressed in volts/mil (1 mil = 0.001 in.). Table 9–2 lists typical values for several materials. Exact values vary depending on the specific composition of the material.

TABLE 9–2
Some common dielectric materials and their typical dielectric strengths

MATERIAL	DIELECTRIC STRENGTH (VOLTS/MIL)
Air	80
Oil	375
Ceramic	1000
Paper (paraffined)	1200
Teflon®	1500
Mica	1500
Glass	2000

The dielectric strength can best be explained by an example. Assume that a certain capacitor has a plate separation of 1 mil and that the dielectric material is ceramic. This particular capacitor can withstand a maximum voltage of 1000 V because its dielectric strength is 1000 V/mil. If the maximum voltage is exceeded, the dielectric may break down and conduct current, causing permanent damage to the capacitor. If the ceramic capacitor has a plate separation of 2 mils, its breakdown voltage is 2000 V.

Temperature Coefficient

The **temperature coefficient** indicates the amount and direction of a change in capacitance value with temperature. A positive temperature coefficient means that the capacitance increases with an increase in temperature or decreases with a decrease in temperature. A negative coefficient means that the capacitance decreases with an increase in temperature or increases with a decrease in temperature.

Temperature coefficients typically are specified in parts per million per degree Celsius (ppm/°C). For example, a negative temperature coefficient of 150 ppm/°C for a 1 μF capacitor means that for every degree rise in temperature, the capacitance decreases by 150 pF (there are one million picofarads in one microfarad).

Leakage

No insulating material is perfect. The dielectric of any capacitor will conduct some very small amount of current. Thus, the charge on a capacitor will eventually leak off. Some types of capacitors have higher leakages than others. An equivalent circuit for a nonideal capacitor is shown in Figure 9–5. The parallel resistor represents the extremely high resistance of the dielectric material through which leakage current flows.

FIGURE 9–5
Equivalent circuit for a nonideal capacitor.

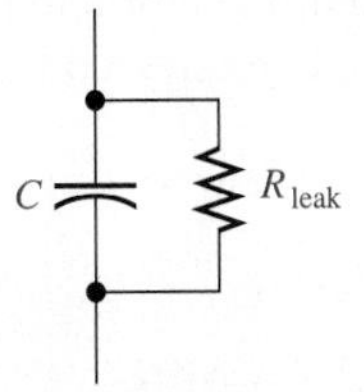

Physical Characteristics of Capacitors

The following parameters are important in establishing the capacitance and the voltage rating of a capacitor: plate area, plate separation, and dielectric constant.

Plate Area *Capacitance is directly proportional to the physical size of the plates as determined by the plate area, A.* A larger plate area produces a larger capacitance, and vice versa. Figure 9–6(a) shows that the plate area of a parallel plate capacitor is the area of one of the plates. If the plates are moved in relation to each other, as shown in Figure 9–6(b), the overlapping area determines the effective plate area. This variation in effective plate area is the basis for a certain type of variable capacitor.

Plate Separation *Capacitance is inversely proportional to the distance between the plates.* The plate separation is designated *d,* as shown in Figure 9–7. A greater separation of the plates produces a smaller capacitance, as illustrated in the figure. The breakdown voltage is directly proportional to the plate separation. The further the plates are separated, the greater the breakdown voltage.

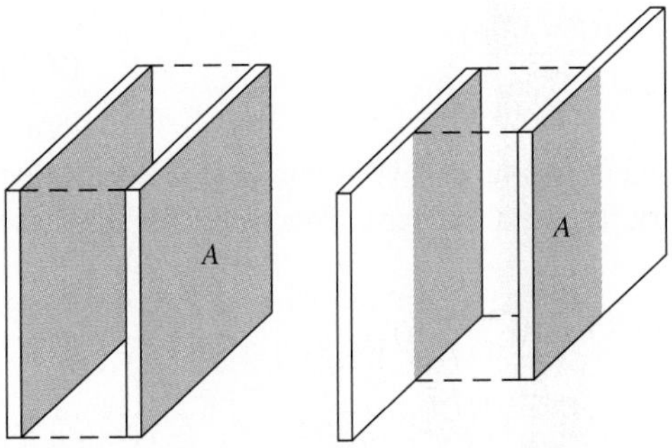

(a) Full plate area: more capacitance

(b) Reduced plate area: less capacitance

FIGURE 9–6
Capacitance is directly proportional to plate area (A).

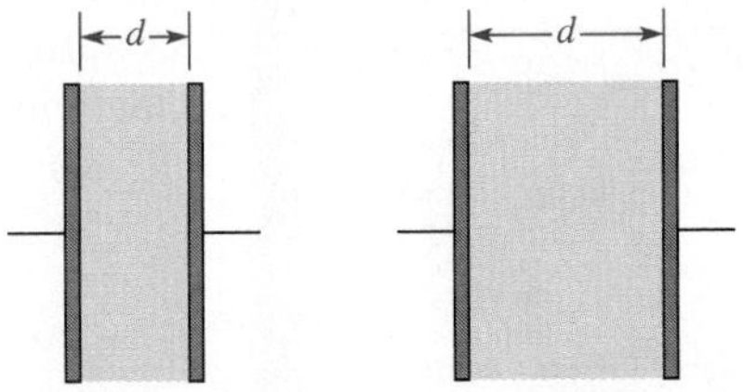

(a) More capacitance

(b) Less capacitance

FIGURE 9–7
Capacitance is inversely proportional to the distance d between the plates.

Dielectric Constant As you know, the insulating material between the plates of a capacitor is called the *dielectric*. Every dielectric material has the ability to concentrate the lines of force of the electric field existing between the oppositely charged plates of a capacitor and thus increase the capacity for energy storage. The measure of a material's ability to establish an electric field is called the **dielectric constant** or *relative permittivity,* symbolized by ϵ_r (the Greek letter epsilon).

Capacitance is directly proportional to the dielectric constant. The dielectric constant of a vacuum is defined as 1, and that of air is very close to 1. These values are used as a reference, and all other materials have values of ϵ_r specified with respect to that of a vacuum or air. For example, a material with $\epsilon_r = 8$ can have a capacitance eight times greater than that of air, with all other factors being equal.

Table 9–3 lists several common dielectric materials and typical dielectric constants for each. Values can vary because they depend on the specific composition of the material.

The dielectric constant (relative permittivity) is dimensionless, because it is a relative measure and is a ratio of the absolute permittivity, ϵ, of a material to the absolute permittivity, ϵ_0, of a vacuum, as expressed by the following formula:

$$\epsilon_r = \frac{\epsilon}{\epsilon_0} \qquad \text{(9–6)}$$

The value of ϵ_0 is 8.85×10^{-12} F/m (farads per meter).

TABLE 9–3
Some common dielectric materials and their typical dielectric constants

MATERIAL	TYPICAL ϵ_r VALUES
Air (vacuum)	1.0
Teflon®	2.0
Paper (paraffined)	2.5
Oil	4.0
Mica	5.0
Glass	7.5
Ceramic	1200

Formula for Capacitance in Terms of Physical Parameters

You have seen how capacitance is directly related to plate area A and the dielectric constant ϵ_r, and inversely related to plate separation d. An exact formula for calculating the capacitance in terms of these three quantities is as follows:

$$C = \frac{A\epsilon_r(8.85 \times 10^{-12}\text{ F/m})}{d} \qquad \text{(9–7)}$$

where A is in square meters (m^2), d is in meters (m), and C is in farads (F). Recall that 8.85×10^{-12} F/m is the absolute permittivity, ϵ_0, of a vacuum. Therefore, the absolute permittivity of a dielectric is

$$\epsilon = \epsilon_r \times \epsilon_0 = \epsilon_r \times (8.85 \times 10^{-12}\text{ F/m})$$

EXAMPLE 9–4

Determine the capacitance of a parallel plate capacitor having a plate area of 0.01 m^2 and a plate separation of 0.02 m. The dielectric is mica, which has a dielectric constant of 5.0.

Solution Use Equation (9–7).

$$C = \frac{A\epsilon_r(8.85 \times 10^{-12}\text{ F/m})}{d} = \frac{(0.01\text{ m}^2)(5.0)(8.85 \times 10^{-12}\text{ F/m})}{0.02\text{ m}} = 22.1\text{ pF}$$

The calculator sequence is

[•] [0] [1] [×] [5] [×] [8] [•] [8] [5] [EXP] [+/−] [1] [2] [÷] [•] [0] [2] [=]

Related Exercise Determine C where $A = 0.005\text{ m}^2$, $d = 0.008$ m, and ceramic is the dielectric.

SECTION 9–1 REVIEW

1. Define *capacitance.*
2. **(a)** How many microfarads in one farad?
   (b) How many picofarads in one farad?
   (c) How many picofarads in one microfarad?
3. Convert 0.0015 μF to picofarads. To farads.
4. How much energy in joules is stored by a 0.01 μF capacitor with 15 V across its plates?
5. **(a)** When the plate area of a capacitor is increased, does the capacitance increase or decrease?
   (b) When the distance between the plates is increased, does the capacitance increase or decrease?
6. The plates of a ceramic capacitor are separated by 10 mils. What is the typical breakdown voltage?
7. A ceramic capacitor has a plate area of 0.2 m^2. The thickness of the dielectric is 0.005 m. What is the capacitance?

9–2 ■ TYPES OF CAPACITORS

Capacitors normally are classified according to the type of dielectric material. The most common types of dielectric materials are mica, ceramic, plastic-film, and electrolytic (aluminum oxide and tantalum oxide). In this section, the characteristics and construction of each of these types of capacitors and the variable capacitors are examined.

After completing this section, you should be able to

- **Discuss various types of capacitors**
  - ☐ Describe the characteristics of mica, ceramic, plastic-film, and electrolytic capacitors
  - ☐ Describe types of variable capacitors
  - ☐ Identify capacitor labeling

Fixed Capacitors

Mica Capacitors Two types of mica capacitors are stacked-foil and silver-mica. The basic construction of the stacked-foil type is shown in Figure 9–8. It consists of alternate layers of metal foil and thin sheets of mica. The metal foil forms the plate, with alternate foil sheets connected together to increase the plate area. More layers are used to increase the plate area, thus increasing the capacitance. The mica/foil stack is encapsulated in an

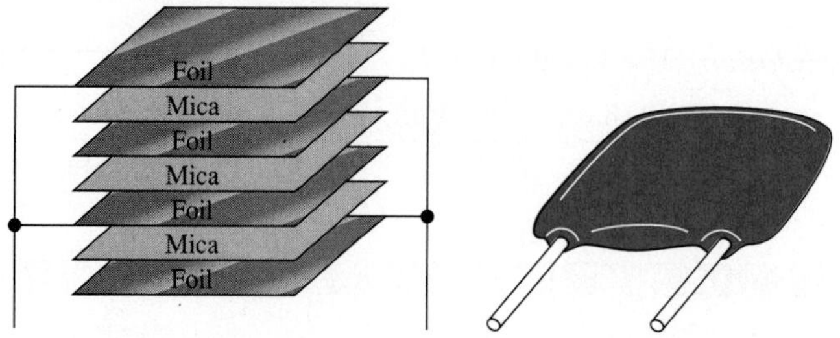

(a) Stacked layer arrangement

(b) Layers are pressed together and encapsulated

FIGURE 9–8
Construction of a typical radial-lead mica capacitor.

insulating material such as Bakelite®, as shown in Figure 9–8(b). The silver-mica capacitor is formed in a similar way by stacking mica sheets with silver electrode material screened on them.

Mica capacitors are available with capacitance values ranging from 1 pF to 0.1 μF and voltage ratings from 100 to 2500 V dc and higher. Temperature coefficients from −20 ppm/°C to +100 ppm/°C are common. Mica has a typical dielectric constant of 5.

Ceramic Capacitors Ceramic dielectrics provide very high dielectric constants (1200 is typical). As a result, comparatively high capacitance values can be achieved in a small physical size. Ceramic capacitors are commonly available in a ceramic disk form, as shown in Figure 9–9; in a multilayer radial-lead configuration, as shown in Figure 9–10; or in a leadless ceramic chip, as shown in Figure 9–11, for surface mounting on PC boards.

Ceramic capacitors typically are available in capacitance values ranging from 1 pF to 2.2 μF with voltage ratings up to 6 kV. A typical temperature coefficient for ceramic capacitors is 200,000 ppm/°C.

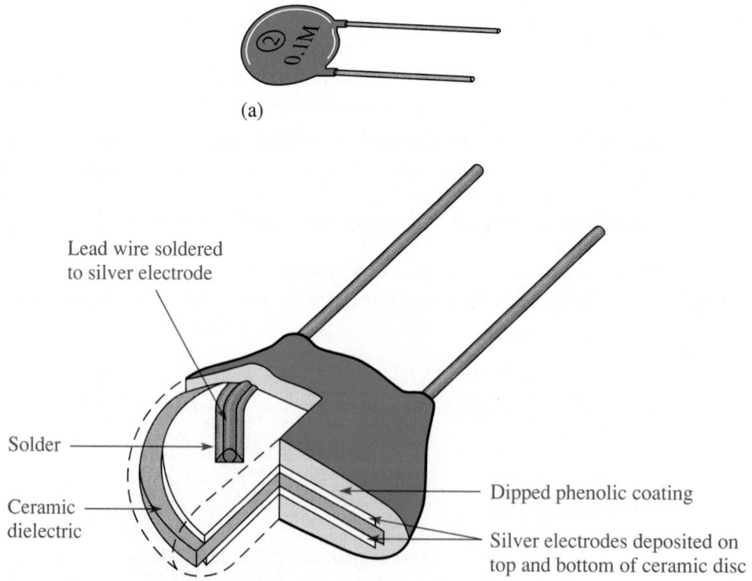

FIGURE 9–9
A ceramic disk capacitor and its basic construction.

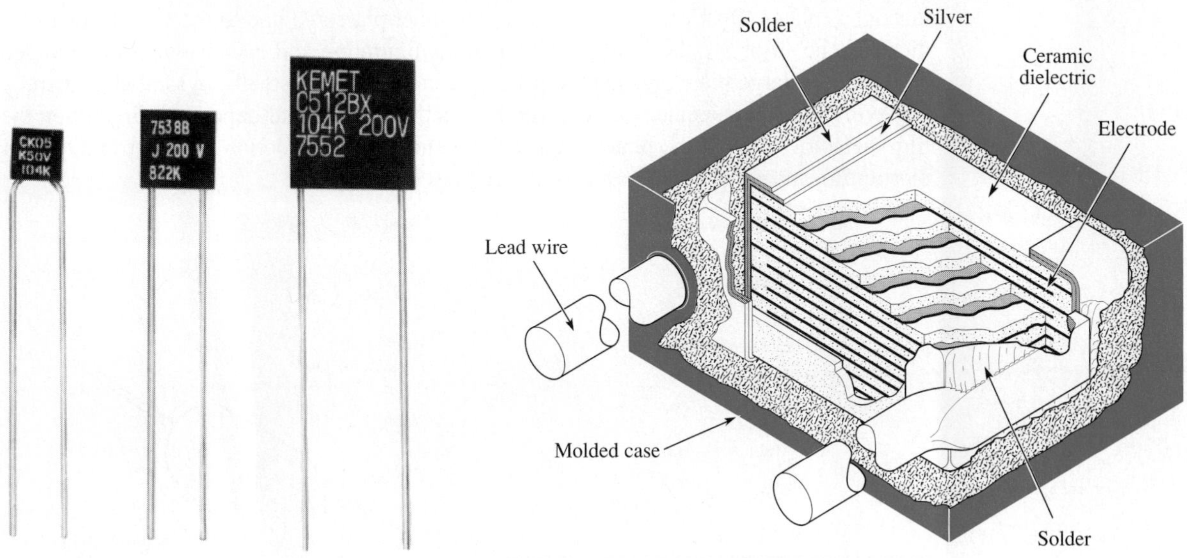

(a) (b)

FIGURE 9–10
Ceramic capacitors. (a) Typical capacitors with radial leads. (b) Construction view (courtesy of Union Carbide Electronics Corporation).

FIGURE 9–11
Construction view of a typical ceramic chip capacitor used for surface mounting on PC boards.

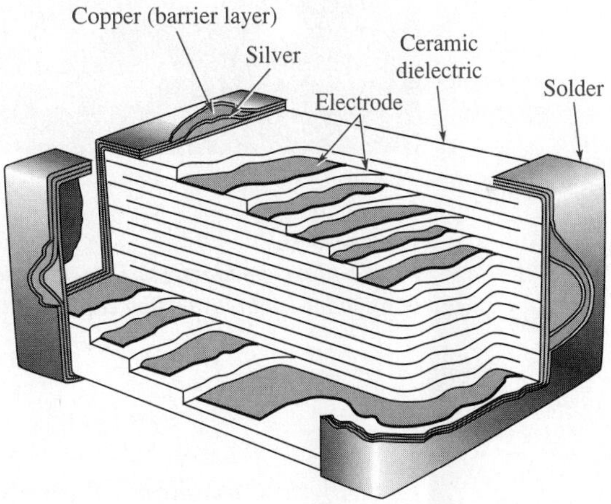

Plastic-Film Capacitors There are several types of plastic-film capacitors. Polycarbonate, propylene, polyester, polystyrene, polypropylene, and polycarbonate are some of the more common dielectric materials used. Some of these types have capacitance values up to 100 μF.

Figure 9–12 shows a common basic construction used in many plastic-film capacitors. A thin strip of plastic-film dielectric is sandwiched between two thin metal strips

FIGURE 9–12
Basic construction of axial-lead tubular plastic-film dielectric capacitors.

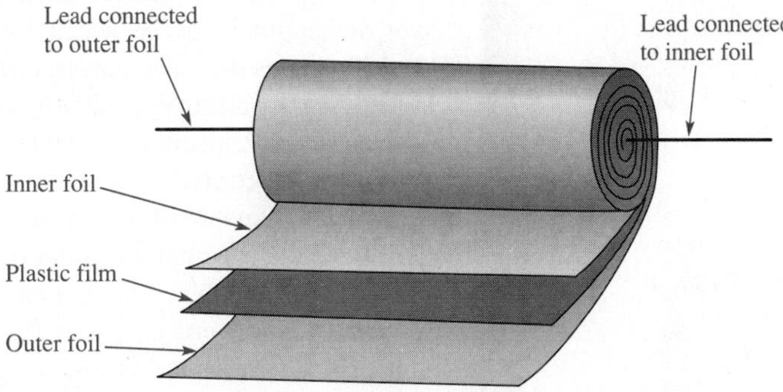

that act as plates. One lead is connected to the inner plate and one to the outer plate as indicated. The strips are then rolled in a spiral configuration and encapsulated in a molded case. Thus, a large plate area can be packaged in a relatively small physical size, thereby achieving large capacitance values. Another method uses metal deposited directly on the film dielectric to form the plates. Figure 9–13 shows a construction view for one type of plastic-film capacitor.

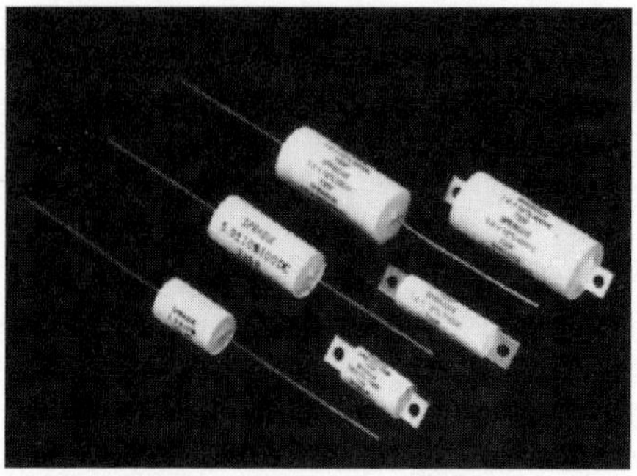

(a)

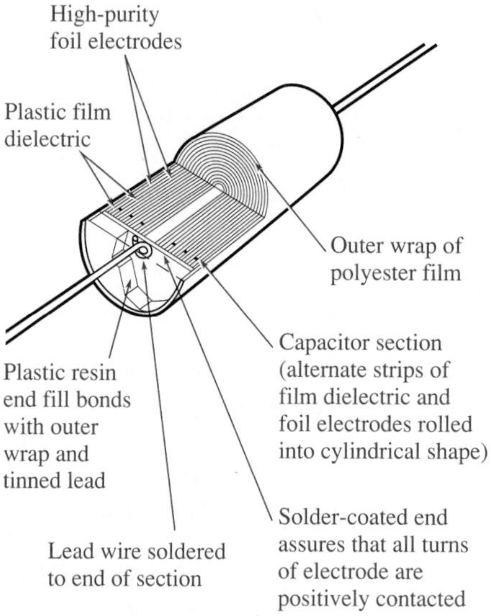

(b)

FIGURE 9–13

Plastic-film capacitors. (a) Typical example (courtesy of Siemens Corp.). (b) Construction view of an axial-lead plastic-film capacitor (courtesy of Union Carbide Electronics Corporation).

Electrolytic Capacitors Electrolytic capacitors are polarized so that one plate is positive and the other negative. These capacitors are used for high capacitance values up to over 200,000 μF, but they have relatively low breakdown voltages (350 V is a typical maximum) and high amounts of leakage.

Electrolytic capacitors offer much higher capacitance values than mica or ceramic capacitors, but their voltage ratings are typically lower. Aluminum electrolytics are probably the most commonly used type. While other capacitors use two similar plates, the electrolytic consists of one plate of aluminum foil and another plate made of a conducting electrolyte applied to a material such as plastic film. These two "plates" are separated by a layer of aluminum oxide which forms on the surface of the aluminum plate. Figure 9–14(a) illustrates the basic construction of a typical aluminum electrolytic capacitor with axial leads. Other electrolytics with radial leads are shown in Figure 9–14(b); the symbol for an electrolytic capacitor is shown in part (c).

Tantalum electrolytics can be in either a tubular configuration similar to Figure 9–14 or "tear drop" shape as shown in Figure 9–15 (page 352). In the tear drop configuration, the positive plate is actually a pellet of tantalum powder rather than a sheet of foil. Tantalum oxide forms the dielectric and manganese dioxide forms the negative plate.

Because of the process used for the insulating oxide dielectric, the metallic (aluminum or tantalum) plate is always positive with respect to the electrolyte plate, and,

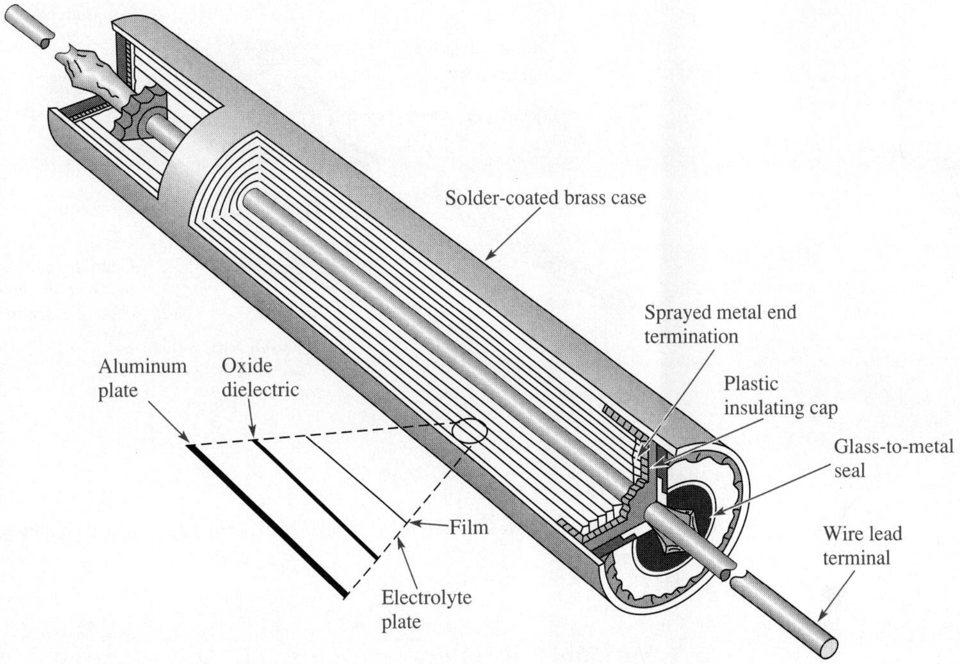

(a) Construction view of an axial-lead electrolytic capacitor

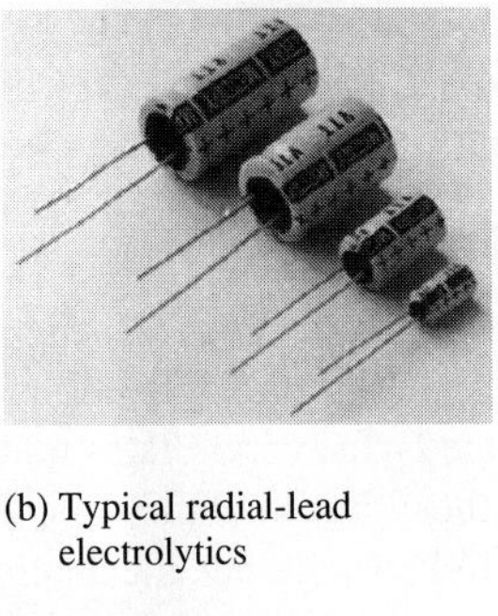

(b) Typical radial-lead electrolytics

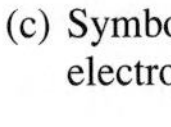

(c) Symbol for an electrolytic capacitor

FIGURE 9–14
Electrolytic capacitors.

thus all electrolytic capacitors are polarized. The metal plate (positive lead) is usually indicated by a plus sign or some other obvious marking and must always be connected in a dc circuit where the voltage across the capacitor does not change polarity. Reversal of the polarity of the voltage will usually result in complete destruction of the capacitor.

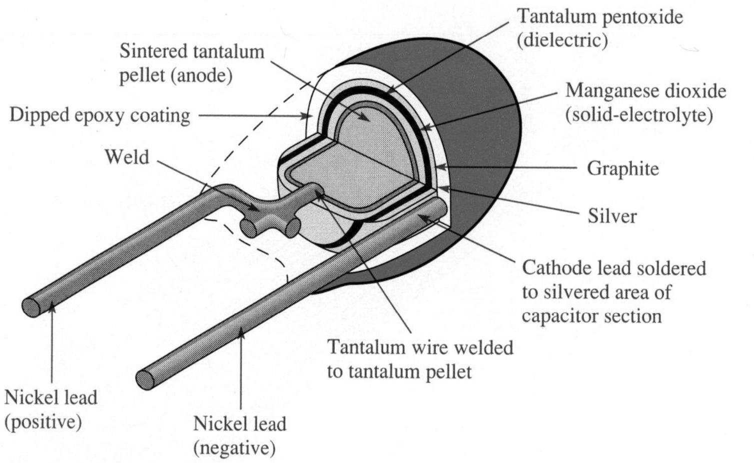

FIGURE 9–15
Construction view of a typical "tear drop" shaped tantalum electrolytic capacitor.

Variable Capacitors

Variable capacitors are used in a circuit when there is a need to adjust the capacitance value either manually or automatically, for example, in radio or TV tuners. The major types of variable or adjustable capacitors are now introduced. The schematic symbol for a variable capacitor is shown in Figure 9–16.

FIGURE 9–16
Schematic symbol for a variable capacitor.

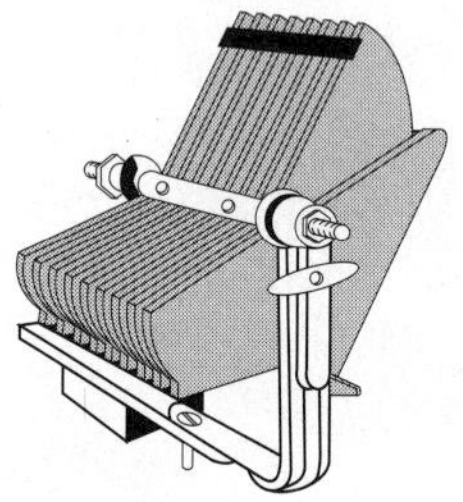

FIGURE 9–17
A typical variable air capacitor.

Air Capacitor Variable capacitors with air dielectrics, such as the one shown in Figure 9–17, are sometimes used as tuning capacitors in applications requiring frequency selection. This type of capacitor is constructed of several plates that mesh together. One set of plates can be moved relative to the other, thus changing the effective plate area and the capacitance. The movable plates are linked together mechanically so that they all move when a shaft is rotated.

Trimmers Adjustable capacitors that normally have screwdriver adjustments and are used for very fine adjustments in a circuit are called **trimmers.** Ceramic or mica is a common dielectric in these types of capacitors, and the capacitance usually is changed by adjusting the plate separation. Figure 9–18 shows some typical devices.

Varactors The varactor is a semiconductor device that exhibits a capacitance characteristic that is varied by changing the voltage across its terminals. This device usually is covered in detail in a course on electronic devices.

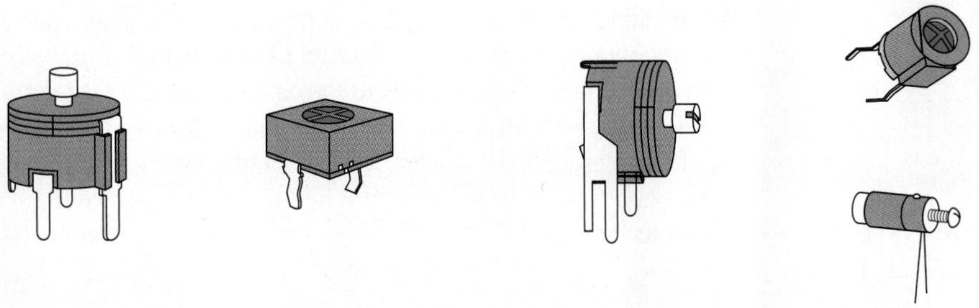

FIGURE 9–18
Trimmer capacitors.

Capacitor Labeling

Capacitor values are indicated on the body of the capacitor either by typographical labels or by color codes. Typographical labels consist of letters and numbers that indicate various parameters such as capacitance, voltage rating, and tolerance.

Some capacitors carry no unit designation for capacitance. In these cases, the units are implied by the value indicated and are recognized by experience. For example, a ceramic capacitor marked .001 or .01 has units of microfarads because picofarad values that small are not available. As another example, a ceramic capacitor labeled 50 or 330 has units of picofarads because microfarad units that large normally are not available in this type. In some cases a 3-digit designation is used. The first two digits are the first two digits of the capacitance value. The third digit is the number of zeros after the second digit. For example, 103 means 10,000 pF. In some instances, the units are labeled as pF or μF; sometimes the microfarad unit is labeled as MF or MFD.

Voltage rating appears on some types of capacitors with WV or WVDC and is omitted on others. When it is omitted, the voltage rating can be determined from information supplied by the manufacturer. The tolerance of the capacitor is usually labeled as a percentage, such as ±10%. The temperature coefficient is indicated by a *parts per million* marking. This type of label consists of a P or N followed by a number. For example, N750 means a negative temperature coefficient of 750 ppm/°C, and P30 means a positive temperature coefficient of 30 ppm/°C. An NP0 designation means that the positive and negative coefficients are zero; thus the capacitance does not change with temperature. Certain types of capacitors are color coded. Refer to Appendix D, for color code information.

SECTION 9–2 REVIEW

1. How are capacitors commonly classified?
2. What is the difference between a fixed and a variable capacitor?
3. What type of capacitor normally is polarized?
4. What precautions must be taken when a polarized capacitor is installed in a circuit?

9–3 ▪ SERIES CAPACITORS

In this section, you will see why the total capacitance of a series connection of capacitors is less than any of the individual capacitances.

After completing this section, you should be able to

- **Analyze series capacitors**
  - ☐ Determine total capacitance
  - ☐ Determine capacitor voltages

When capacitors are connected in series, the total capacitance is less than the smallest capacitance value because the effective plate separation increases. The calculation of total series capacitance is analogous to the calculation of total resistance of parallel resistors.

To start, we will use two capacitors in series to show how the total capacitance is determined. Figure 9–19 shows two capacitors, which initially are uncharged, connected in series with a dc voltage source. When the switch is closed, as shown in part (a), current begins to flow.

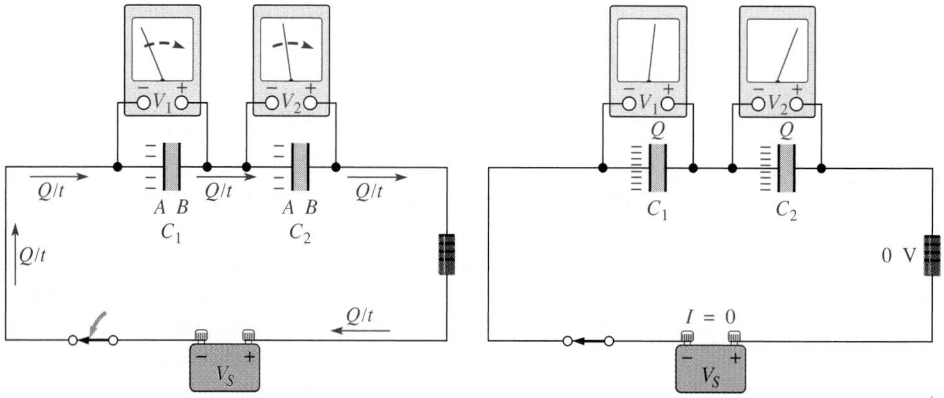

(a) While charging, $I = Q/t$ is the same at all points. (b) Both capacitors store the same amount of charge ($Q_T = Q_1 = Q_2$).

FIGURE 9–19
Capacitors in series produce a total capacitance that is less than the smallest value.

Recall that current is the same at all points in a series circuit and that current is defined as the rate of flow of charge ($I = Q/t$). In a certain period of time, a certain amount of charge moves through the circuit. Since current is the same everywhere in the circuit of Figure 9–19(a), the same amount of charge is moved from the negative side of the source to plate A of C_1, and from plate B of C_1 to plate A of C_2, and from plate B of C_2 to the positive side of the source. As a result, of course, the same amount of charge is deposited on the plates of both capacitors in a given period of time, and the total charge (Q_T) moved through the circuit in that period of time equals the charge stored by C_1 and also equals the charge stored by C_2:

$$Q_T = Q_1 = Q_2$$

As the capacitors charge, the voltage across each one increases as indicated.

Figure 9–19(b) shows the capacitors after they have been completely charged and the current has ceased. Both capacitors store an equal amount of charge (Q), and the voltage across each one depends on its capacitance value ($V = Q/C$). By Kirchhoff's voltage law, which applies to capacitive circuits as well as to resistive circuits, the sum of the capacitor voltages equals the source voltage:

$$V_S = V_1 + V_2$$

Using the formula $V = Q/C$, you can substitute into the formula for Kirchhoff's law and get the following relationship (where $Q = Q_T = Q_1 = Q_2$):

$$\frac{Q}{C_T} = \frac{Q}{C_1} + \frac{Q}{C_2}$$

The Q can be factored out of the right side of the equation and canceled with the Q on the left side as follows:

$$\frac{\cancel{Q}}{C_T} = \cancel{Q}\left(\frac{1}{C_1} + \frac{1}{C_2}\right)$$

Thus you can have the following relationship for two capacitors in series:

$$\frac{1}{C_T} = \frac{1}{C_1} + \frac{1}{C_2} \quad \textbf{(9–8)}$$

Taking the reciprocal of both sides of Equation (9–8) gives the formula for the total capacitance:

$$C_T = \frac{1}{\left(\frac{1}{C_1}\right) + \left(\frac{1}{C_2}\right)} \quad \textbf{(9–9)}$$

EXAMPLE 9–5 Find C_T in Figure 9–20.

FIGURE 9–20

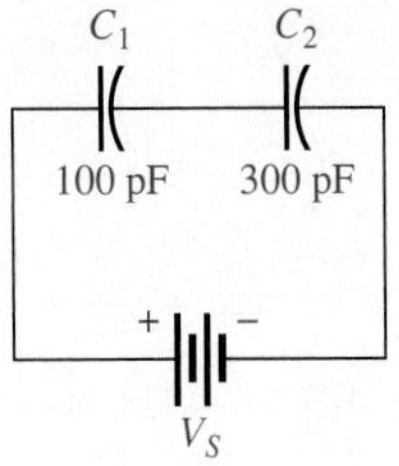

Solution

$$C_T = \frac{1}{\left(\frac{1}{C_1}\right) + \left(\frac{1}{C_2}\right)} = \frac{1}{\left(\frac{1}{100\text{ pF}}\right) + \left(\frac{1}{300\text{ pF}}\right)} = 75\text{ pF}$$

The calculator sequence is

[1] [0] [0] [EXP] [+/−] [1] [2] [1/x] [+] [3] [0] [0] [EXP] [+/−] [1] [2] [1/x] [=] [1/x]

Related Exercise Determine C_T if $C_1 = 470$ pF and $C_2 = 680$ pF in Figure 9–20.

General Formula for Total Series Capacitance

Equations (9–8) and (9–9) can be extended to any number of capacitors in series, as shown in Figure 9–21.

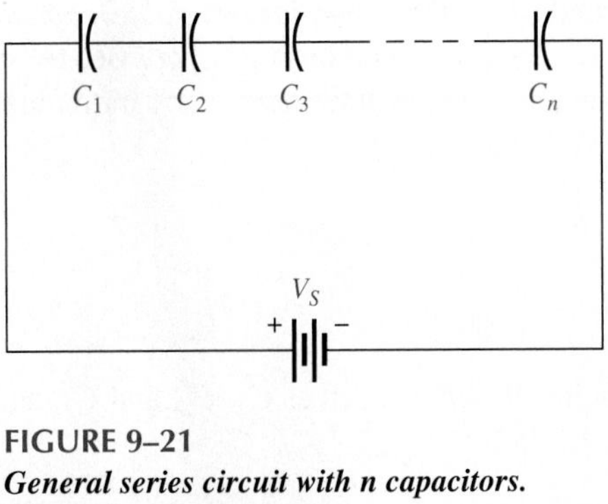

FIGURE 9–21
General series circuit with n capacitors.

The expanded formulas for total series capacitance are as follows; the subscript n can be any number.

$$\frac{1}{C_T} = \frac{1}{C_1} + \frac{1}{C_2} + \frac{1}{C_3} + \cdots + \frac{1}{C_n} \qquad \textbf{(9–10)}$$

$$C_T = \frac{1}{\left(\frac{1}{C_1}\right) + \left(\frac{1}{C_2}\right) + \left(\frac{1}{C_3}\right) + \cdots + \left(\frac{1}{C_n}\right)} \qquad \textbf{(9–11)}$$

Remember,

The total series capacitance is always less than the smallest capacitance.

EXAMPLE 9–6 Determine the total capacitance in Figure 9–22.

FIGURE 9–22

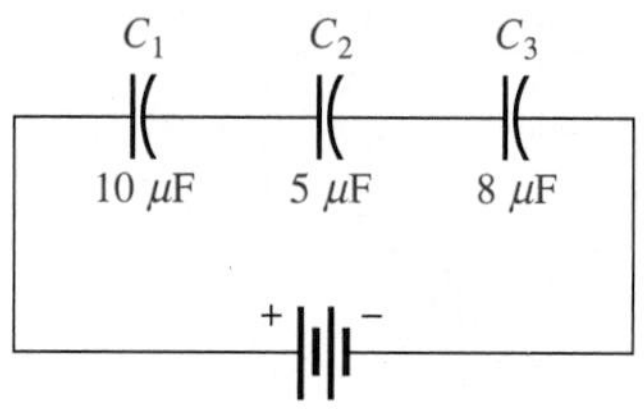

Solution

$$\frac{1}{C_T} = \frac{1}{C_1} + \frac{1}{C_2} + \frac{1}{C_3} = \frac{1}{10\ \mu\text{F}} + \frac{1}{5\ \mu\text{F}} + \frac{1}{8\ \mu\text{F}}$$

Taking the reciprocal of both sides yields the following:

$$C_T = \frac{1}{\left(\frac{1}{10\ \mu\text{F}}\right) + \left(\frac{1}{5\ \mu\text{F}}\right) + \left(\frac{1}{8\ \mu\text{F}}\right)} = \frac{1}{0.425}\ \mu\text{F} = 2.35\ \mu\text{F}$$

Related Exercise If a 4.7 μF capacitor is connected in series with the three existing capacitors in Figure 9–22, what is the value of C_T?

Capacitor Voltages

The voltage across each capacitor in a series connection depends on its capacitance value according to the formula $V = Q/C$. *The largest-value capacitor will have the smallest voltage* because of the reciprocal relationship. Likewise, *the smallest capacitance value will have the largest voltage.* The voltage across any individual capacitor in series can be determined using the following formula:

$$V_x = \left(\frac{C_T}{C_x}\right)V_S \qquad \textbf{(9–12)}$$

where C_x is any capacitor in series, such as C_1, C_2, and C_3, and so on, and V_x is the voltage across C_x.

EXAMPLE 9–7 Find the voltage across each capacitor in Figure 9–23.

FIGURE 9–23

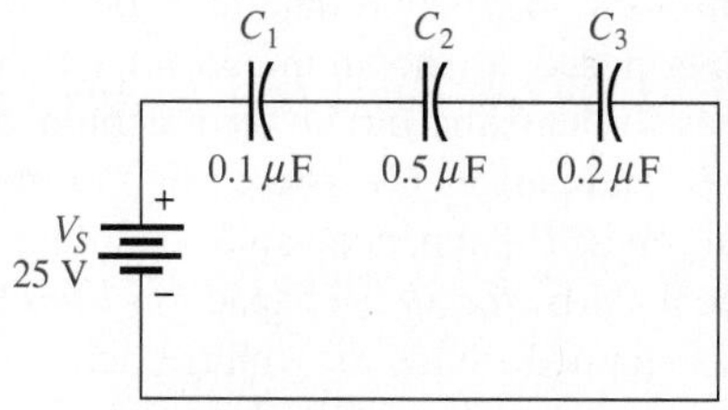

Solution Calculate the total capacitance.

$$\frac{1}{C_T} = \frac{1}{C_1} + \frac{1}{C_2} + \frac{1}{C_3} = \frac{1}{0.1\ \mu\text{F}} + \frac{1}{0.5\ \mu\text{F}} + \frac{1}{0.2\ \mu\text{F}}$$

$$C_T = \frac{1}{17}\ \mu\text{F} = 0.0588\ \mu\text{F}$$

The voltages are as follows:

$$V_1 = \left(\frac{C_T}{C_1}\right)V_S = \left(\frac{0.0588\ \mu\text{F}}{0.1\ \mu\text{F}}\right)25\ \text{V} = 14.7\ \text{V}$$

$$V_2 = \left(\frac{C_T}{C_2}\right)V_S = \left(\frac{0.0588\ \mu\text{F}}{0.5\ \mu\text{F}}\right)25\ \text{V} = 2.94\ \text{V}$$

$$V_3 = \left(\frac{C_T}{C_3}\right)V_S = \left(\frac{0.0588\ \mu\text{F}}{0.2\ \mu\text{F}}\right)25\ \text{V} = 7.35\ \text{V}$$

The calculator sequence for V_1 is

[•] [0] [5] [8] [8] [EXP] [+/–] [6] [÷] [•] [1] [EXP] [+/–] [6] [×] [2] [5] [=]

Related Exercise A 0.47 μF capacitor is connected in series with the existing capacitors in Figure 9–23. Determine the voltage across the new capacitor.

SECTION 9–3 REVIEW

1. Is the total capacitance of series capacitors less than or greater than the value of the smallest capacitor?
2. The following capacitors are in series: 100 pF, 250 pF, and 500 pF. What is the total capacitance?
3. A 0.01 μF and a 0.015 μF capacitor are in series. Determine the total capacitance.
4. Determine the voltage across the 0.01 μF capacitor in Question 3 if 10 V are connected across the two series capacitors.

9–4 ■ PARALLEL CAPACITORS

In this section, you will see why capacitances add when they are connected in parallel.

After completing this section, you should be able to

- **Analyze parallel capacitors**
  - □ Determine total capacitance

When capacitors are connected in parallel, the total capacitance is the sum of the individual capacitances because the effective plate area increases. The calculation of total parallel capacitance is analogous to the calculation of total series resistance.

Figure 9–24 shows two parallel capacitors connected to a dc voltage source. When the switch is closed, as shown in part (a), current begins to flow. A total amount of charge (Q_T) moves through the circuit in a certain period of time. Part of the total charge is stored by C_1 and part by C_2. The portion of the total charge that is stored by each capacitor depends on its capacitance value according to the relationship $Q = CV$.

Figure 9–24(b) shows the capacitors after they have been completely charged and the current has stopped. Since the voltage across both capacitors is the same, the larger capacitor stores more charge. If the capacitors are equal in value, they store an equal amount of charge. The charge stored by both of the capacitors together equals the total charge that was delivered from the source:

$$Q_T = Q_1 + Q_2$$

Since $Q = CV$, Equation (9–2), you can use substitution to get the following relationship:

$$C_T V_S = C_1 V_S + C_2 V_S$$

Because all the V_S terms are equal, they can be canceled. Therefore, the total capacitance for two capacitors in parallel is

$$C_T = C_1 + C_2 \qquad \textbf{(9–13)}$$

FIGURE 9–24
Capacitors in parallel produce a total capacitance that is the sum of the individual capacitances.

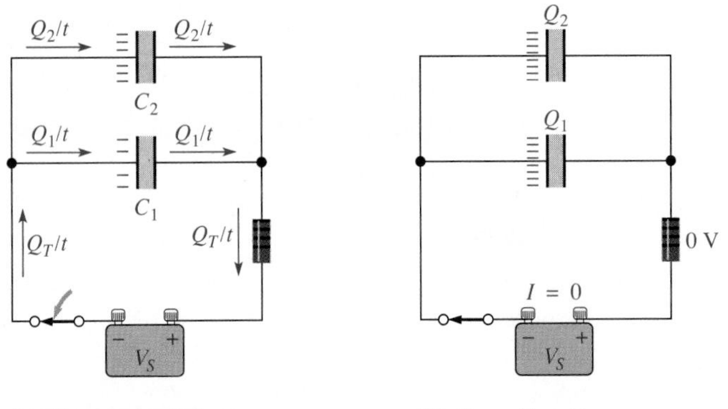

(a) The amount of charge on each capacitor is directly proportional to its capacitance value.

(b) $Q_T = Q_1 + Q_2$

EXAMPLE 9–8 What is the total capacitance in Figure 9–25? What is the voltage across each capacitor?

FIGURE 9–25

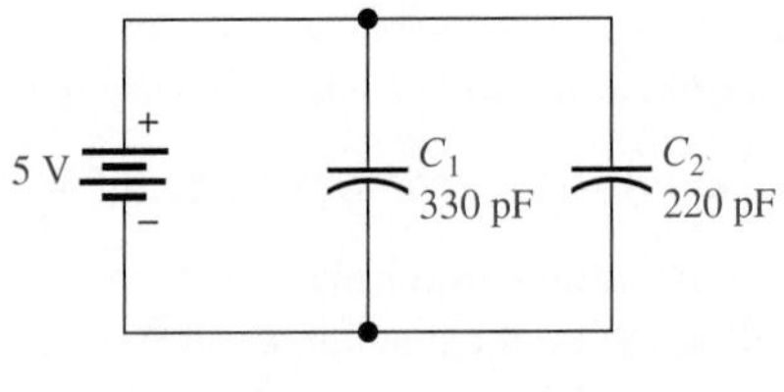

Solution The total capacitance is

$$C_T = C_1 + C_2 = 330 \text{ pF} + 220 \text{ pF} = 550 \text{ pF}$$

The voltage across each capacitor is $V_1 = V_2 = 5$ V.

Related Exercise What is C_T if a 100 pF capacitor is connected in parallel with C_1 and C_2 in Figure 9–25?

General Formula for Total Parallel Capacitance

Equation (9–13) can be extended to any number of capacitors in parallel, as shown in Figure 9–26. The expanded formula is as follows; the subscript n can be any number.

$$C_T = C_1 + C_2 + C_3 + \cdots + C_n \quad \textbf{(9–14)}$$

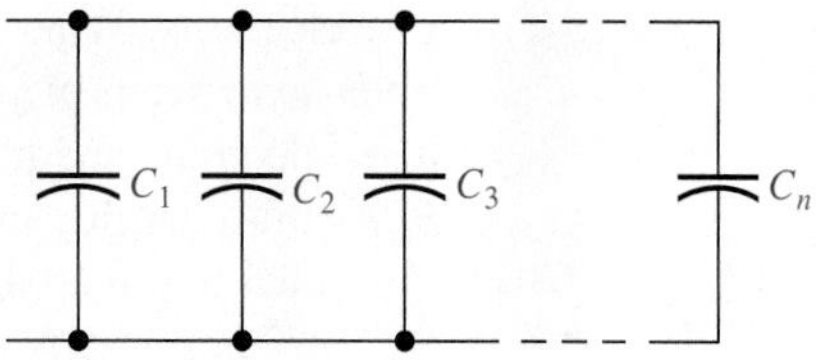

FIGURE 9–26
General parallel circuit with n capacitors.

EXAMPLE 9–9 Determine C_T in Figure 9–27.

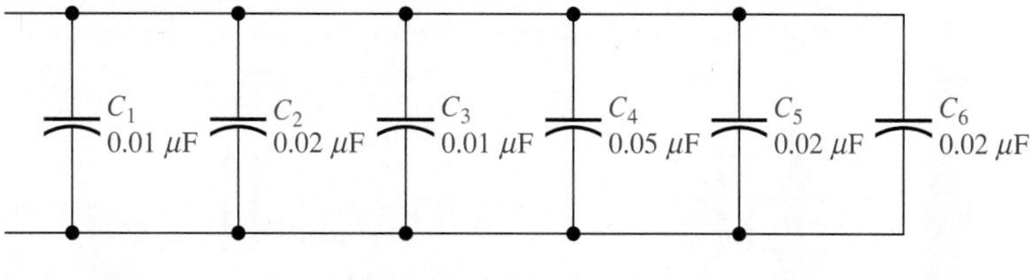

FIGURE 9–27

Solution

$$C_T = C_1 + C_2 + C_3 + C_4 + C_5 + C_6$$
$$= 0.01\ \mu\text{F} + 0.02\ \mu\text{F} + 0.01\ \mu\text{F} + 0.05\ \mu\text{F} + 0.02\ \mu\text{F} + 0.02\ \mu\text{F} = 0.13\ \mu\text{F}$$

Related Exercise If three more 0.01 μF capacitors are connected in parallel in Figure 9–27, what is C_T?

SECTION 9–4 REVIEW

1. How is total parallel capacitance determined?
2. In a certain application, you need 0.05 μF. The only values available are 0.01 μF, which are available in large quantities. How can you get the total capacitance that you need?
3. The following capacitors are in parallel: 10 pF, 5 pF, 33 pF, and 50 pF. What is C_T?

9–5 ■ CAPACITORS IN DC CIRCUITS

A capacitor will charge up when it is connected to a dc voltage source. The buildup of charge across the plates occurs in a predictable manner which is dependent on the capacitance and the resistance in a circuit.

After completing this section, you should be able to

- **Analyze capacitive dc switching circuits**
  - ☐ Describe the charging and discharging of a capacitor
  - ☐ Define *time constant*
  - ☐ Relate the time constant to charging and discharging
  - ☐ Write equations for the charging and discharging curves
  - ☐ Explain why a capacitor blocks constant dc

Charging a Capacitor

A capacitor charges when it is connected to a dc voltage source, as shown in Figure 9–28. The capacitor in part (a) is uncharged; that is, plate *A* and plate *B* have equal numbers of free electrons. When the switch is closed, as shown in part (b), the source moves electrons away from plate *A* through the circuit to plate *B* as the arrows indicate. As plate *A* loses electrons and plate *B* gains electrons, plate *A* becomes positive with respect to plate *B*. As this charging process continues, the voltage across the plates builds up rapidly until it is equal to the applied voltage, V_S, but opposite in polarity, as shown in part (c). When the capacitor is fully charged, there is no current.

A capacitor blocks constant dc.

When the charged capacitor is disconnected from the source, as shown in Figure 9–28(d), it remains charged for long periods of time, depending on its leakage resistance, and can cause severe electrical shock. The charge on an electrolytic capacitor generally leaks off more rapidly than in other types of capacitors.

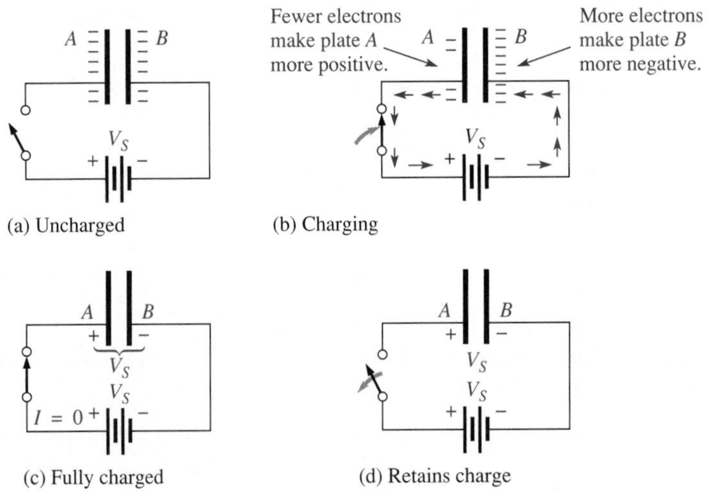

FIGURE 9–28
Charging a capacitor.

Discharging a Capacitor

When a wire is connected across a charged capacitor, as shown in Figure 9–29, the capacitor will discharge. In this particular case, a very low resistance path (the wire) is

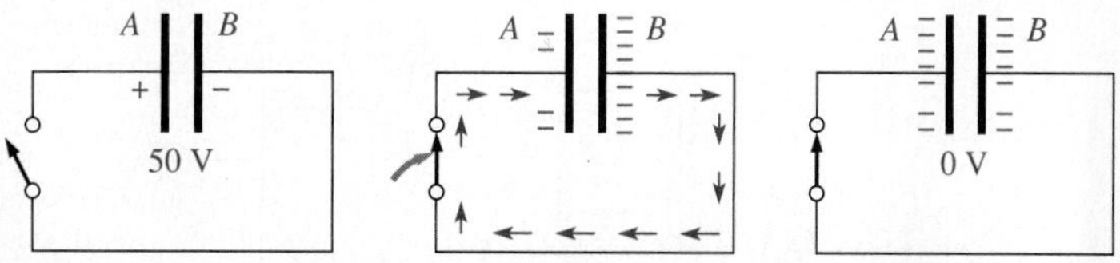

(a) Retains charge (b) Discharging (c) Uncharged

FIGURE 9–29
Discharging a capacitor.

connected across the capacitor with a switch. Before the switch is closed, the capacitor is charged to 50 V, as indicated in part (a). When the switch is closed, as shown in part (b), the excess electrons on plate B move through the circuit to plate A (indicated by the arrows); as a result of the current through the low resistance of the wire, the energy stored by the capacitor is dissipated in the wire. The charge is neutralized when the numbers of free electrons on both plates are again equal. At this time, the voltage across the capacitor is zero, and the capacitor is completely discharged, as shown in part (c).

Current and Voltage During Charging and Discharging

Notice in Figures 9–28 and 9–29 that the direction of the current during discharge is opposite to that of the charging current. It is important to understand that *there is no current through the dielectric of the capacitor during charging or discharging because the dielectric is an insulating material.* Current flows from one plate to the other only through the external circuit.

Figure 9–30(a) shows a capacitor connected in series with a resistor and a switch to a dc voltage source. Initially, the switch is open and the capacitor is uncharged with zero volts across its plates. At the instant the switch is closed, the current jumps to its maximum value and the capacitor begins to charge. The current is maximum initially because the capacitor has zero volts across it and, therefore, effectively acts as a short; thus, the current is limited only by the resistance. As time passes and the capacitor charges, the current decreases and the voltage across the capacitor (V_C) increases. The resistor voltage is proportional to the current during this charging period.

After a certain period of time, the capacitor reaches full charge. At this point, the current is zero and the capacitor voltage is equal to the dc source voltage, as shown in Figure 9–30(b). If the switch were opened now, the capacitor would retain its full charge (neglecting any leakage).

In Figure 9–30(c), the voltage source has been removed. When the switch is closed, the capacitor begins to discharge. Initially, the current jumps to a maximum but in a direction opposite to its direction during charging. As time passes, the current and capacitor voltage decrease. The resistor voltage is always proportional to the current. When the capacitor has fully discharged, the current and the capacitor voltage are zero.

Remember the following rules about capacitors in dc circuits:

1. Voltage across a capacitor *cannot* change instantaneously.

2. Current in a capacitive circuit *can* ideally change instantaneously.

3. A fully charged capacitor appears as an *open* to nonchanging current.

4. An uncharged capacitor appears as a *short* to an instantaneous change in current.

Now we will examine in more detail how the voltage and current change with time in a capacitive circuit.

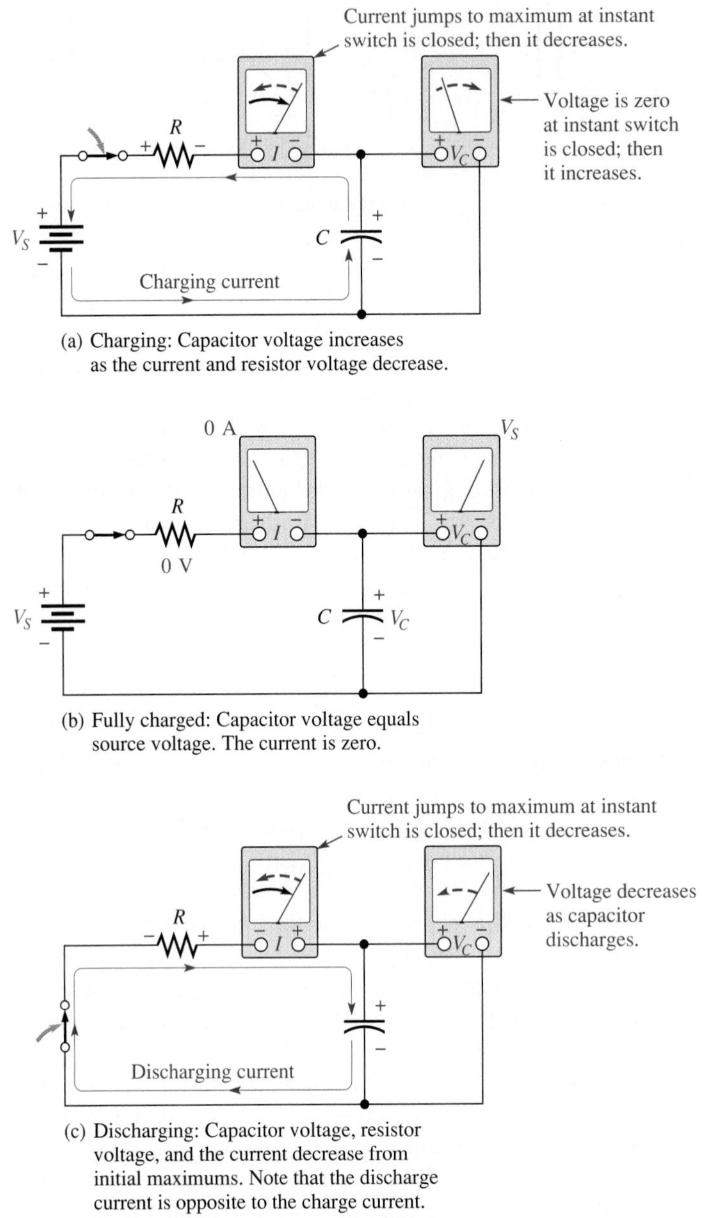

(a) Charging: Capacitor voltage increases as the current and resistor voltage decrease.

(b) Fully charged: Capacitor voltage equals source voltage. The current is zero.

(c) Discharging: Capacitor voltage, resistor voltage, and the current decrease from initial maximums. Note that the discharge current is opposite to the charge current.

FIGURE 9–30

Current and voltage in a charging and discharging capacitor.

The *RC* Time Constant

In a practical situation, there cannot be capacitance without some resistance in a circuit. It may simply be the small resistance of a wire, or it may be a designed-in resistance. Because of this, the charging and discharging characteristics of a capacitor must always be considered in light of the associated series resistance. The resistance introduces the element of *time* in the charging and discharging of a capacitor.

When a capacitor charges or discharges through a resistance, a certain time is required for the capacitor to charge fully or discharge fully. The voltage across a capacitor cannot change instantaneously because a finite time is required to move charge from one point to another. The rate at which the capacitor charges or discharges is determined by the **time constant** of the circuit.

The time constant of a series *RC* circuit is a time interval that equals the product of the resistance and the capacitance and is expressed in units of seconds when resistance is in ohms and capacitance is in farads.

The time constant is symbolized by τ (Greek letter tau), and the formula is as follows:

$$\tau = RC \qquad (9\text{–}15)$$

Recall that $I = Q/t$. The current is the amount of charge moved in a given time. When the resistance is increased, the charging current is reduced, thus increasing the charging time of the capacitor. When the capacitance is increased, the amount of charge increases; thus, for the same current, more time is required to charge the capacitor.

EXAMPLE 9–10 A series *RC* circuit has a resistance of 1 MΩ and a capacitance of 5 μF. What is the time constant?

Solution

$$\tau = RC = (1 \times 10^{6}\ \Omega)(5 \times 10^{-6}\ \text{F}) = 5\ \text{s}$$

Related Exercise A series *RC* circuit has a 270 kΩ resistor and a 3300 pF capacitor. What is the time constant?

During one time-constant interval, the charge on a capacitor changes approximately 63%.

An uncharged capacitor charges to 63% of its fully charged voltage in one time constant. When a capacitor is discharging, its voltage drops to approximately 37% (100% – 63%) of its initial value in one time constant. This change also corresponds to a 63% change.

The Charging and Discharging Curves

A capacitor charges and discharges following a nonlinear curve, as shown in Figure 9–31. In these graphs, the percentage of full charge is shown at each time-constant interval. This type of curve follows a precise mathematical formula and is called an *exponential curve*. The charging curve is an increasing exponential, and the discharging curve is a decreasing exponential. It takes five time constants to approximately reach the final value.

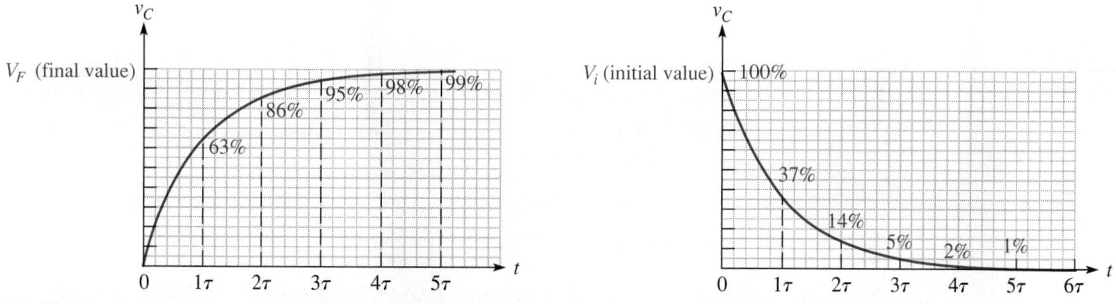

(a) Charging curve with percentages of final value

(b) Discharging curve with percentages of initial value

FIGURE 9–31

Charging and discharging exponential curves for the capacitor voltage in an RC circuit.

A five time-constant interval is accepted as the time to fully charge or discharge a capacitor and is called the *transient time*.

General Formula The general expressions for either increasing or decreasing exponential curves are given in the following equations for both voltage and current:

$$v = V_F + (V_i - V_F)e^{-t/\tau} \quad \textbf{(9–16)}$$

$$i = I_F + (I_i - I_F)e^{-t/\tau} \quad \textbf{(9–17)}$$

where V_F and I_F are the final values and V_i and I_i are the initial values. The lowercase letters v and i are the instantaneous values of the capacitor voltage and current at time t, and e is the base of natural logarithms with a value of 2.718. The e^x key or the INV and ln x keys on your calculator make it easy to work with these formulas.

Charging from Zero The formula for the special case in which an increasing exponential voltage curve begins at zero ($V_i = 0$), as shown in Figure 9–31(a), is given in Equation (9–18). It is developed as follows, starting with the general formula in Equation (9–16).

$$\begin{aligned} v &= V_F + (V_i - V_F)e^{-t/\tau} \\ &= V_F + (0 - V_F)e^{-t/RC} \\ &= V_F - V_F e^{-t/RC} \end{aligned}$$

$$v = V_F(1 - e^{-t/RC}) \quad \textbf{(9–18)}$$

Using Equation (9–18), you can calculate the value of the charging voltage of a capacitor at any instant of time. The same is true for an increasing current.

EXAMPLE 9–11 In Figure 9–32, determine the capacitor voltage 50 μs after the switch is closed if the capacitor initially is uncharged. Sketch the charging curve.

FIGURE 9–32

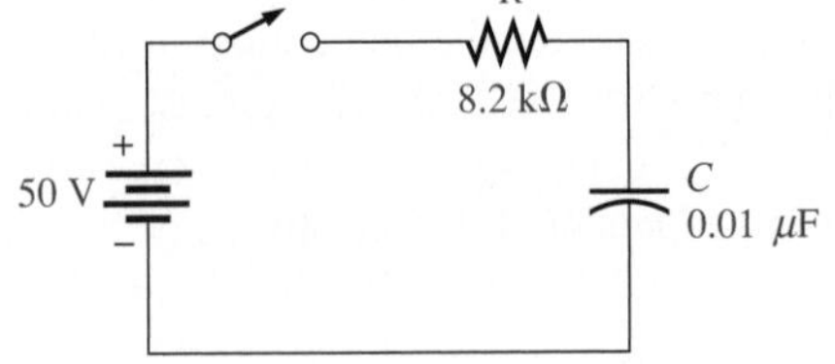

Solution The time constant is $RC = (8.2\ \text{k}\Omega)(0.01\ \mu\text{F}) = 82\ \mu\text{s}$. The voltage to which the capacitor will fully charge is 50 V (this is V_F). The initial voltage is zero. Notice that 50 μs is less than one time constant; so the capacitor will charge less than 63% of the full voltage in that time.

$$\begin{aligned} v_C &= V_F(1 - e^{-t/RC}) = 50\ \text{V}(1 - e^{-50\mu\text{s}/82\mu\text{s}}) \\ &= 50\ \text{V}(1 - e^{-0.61}) = 50\ \text{V}(1 - 0.543) = 22.8\ \text{V} \end{aligned}$$

Determine the value of $e^{-0.61}$ on the calculator by entering −0.61 and then pressing e^x or INV ln x .

The charging curve for the capacitor is shown in Figure 9–33.

FIGURE 9–33

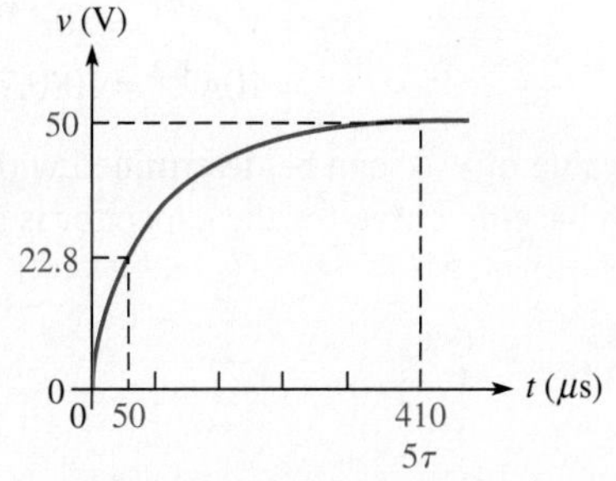

The calculator sequence is

5 0 ÷ 8 2 = +/− e^x +/− + 1 = × 5 0 =

Related Exercise Determine the capacitor voltage 15 μs after the switch is closed in Figure 9–32.

Discharging to Zero The formula for the special case in which a decreasing exponential voltage curve ends at zero, as shown in Figure 9–31(b), is derived from the general formula as follows:

$$v = V_F + (V_i - V_F)e^{-t/\tau}$$
$$= 0 + (V_i - 0)e^{-t/RC}$$

$$v = V_i e^{-t/RC} \qquad \textbf{(9–19)}$$

where V_i is the voltage at the beginning of the discharge. You can use this formula to calculate the discharging voltage at any instant, as Example 9–12 illustrates.

EXAMPLE 9–12

Determine the capacitor voltage in Figure 9–34 at a point in time 6 ms after the switch is closed. Sketch the discharging curve.

FIGURE 9–34

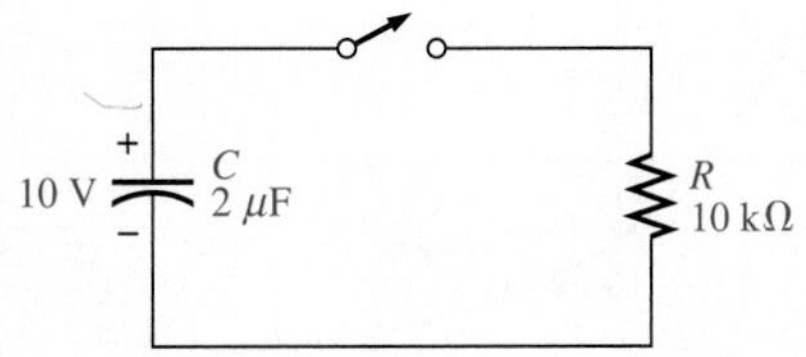

Solution The discharge time constant is

$$RC = (10\text{ k}\Omega)(2\ \mu\text{F}) = 20\text{ ms}$$

The initial capacitor voltage is 10 V. Notice that 6 ms is less than one time constant, so the capacitor will discharge less than 63%. Therefore, it will have a voltage greater than 37% of the initial voltage at 6 ms.

$$v_C = V_i e^{-t/RC} = 10e^{-6\text{ms}/20\text{ms}}$$
$$= 10e^{-0.3} = 10(0.741) = 7.41 \text{ V}$$

Again, the value of $e^{-0.3}$ can be determined with a calculator.

The discharging curve for the capacitor is shown in Figure 9–35.

FIGURE 9–35

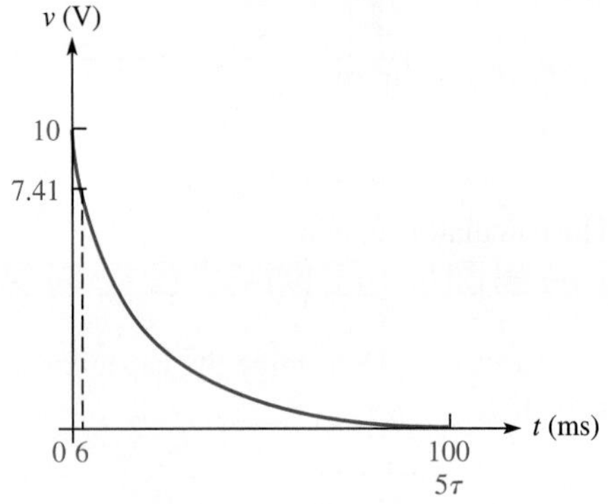

Related Exercise In Figure 9–34, change R to 2.2 kΩ and determine the capacitor voltage 1 ms after the switch is closed.

Graphical Method Using Universal Exponential Curves The universal curves in Figure 9–36 provide a graphic solution of the charge and discharge of capacitors. Example 9–13 illustrates this graphical method.

FIGURE 9–36
Normalized universal exponential curves.

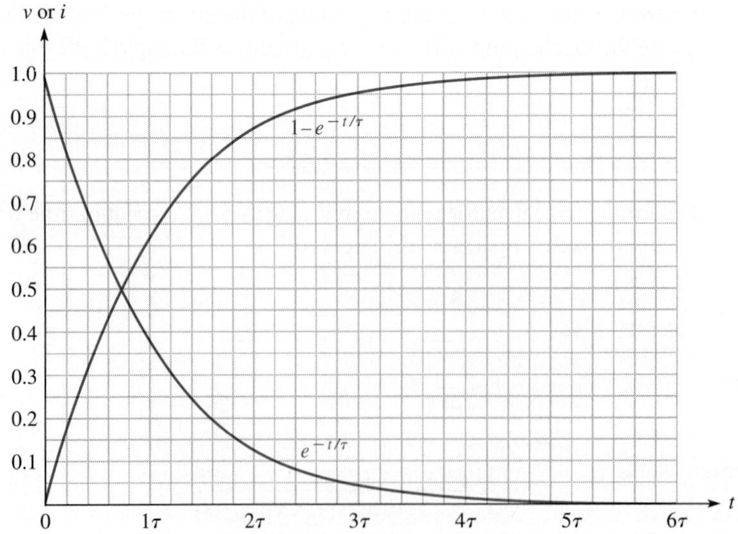

EXAMPLE 9–13 How long will it take the capacitor in Figure 9–37 to charge to 75 V? What is the capacitor voltage 2 ms after the switch is closed? Use the normalized universal curves in Figure 9–36 to determine the answers.

FIGURE 9–37

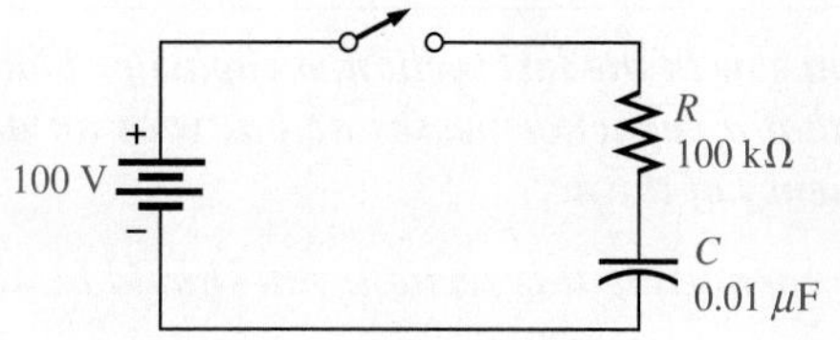

Solution The full charge voltage is 100 V, which is the 100% level (1.0) on the graph. 75 V is 75% of the maximum, or 0.75 on the graph. You can see that this value occurs at 1.4 time constants. One time constant is 1 ms. Therefore, the capacitor voltage reaches 75 V at 1.4 ms after the switch is closed.

The capacitor is at approximately 86 V in 2 ms. These graphic solutions are shown in Figure 9–38.

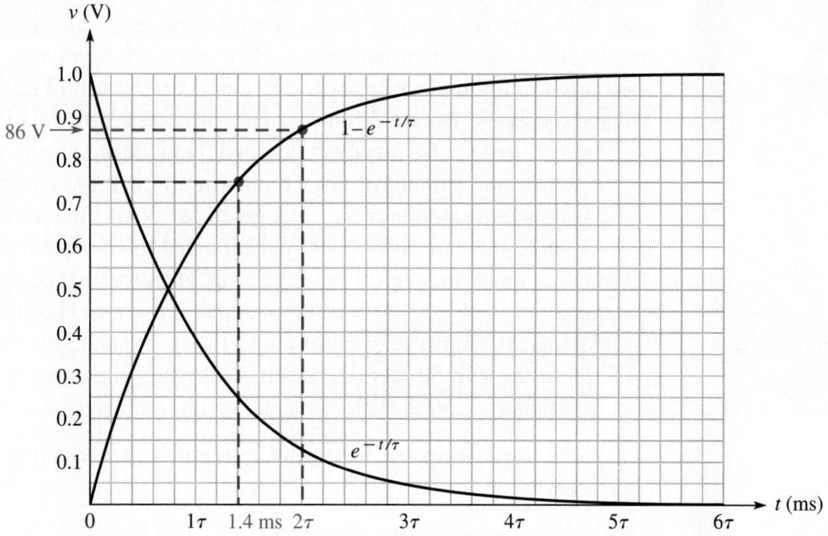

FIGURE 9–38

Related Exercise Using the normalized universal exponential curves, determine how long it will take the capacitor in Figure 9–37 to charge to 50 V from zero. What is v_C 3 ms after switch closure?

SECTION 9–5 REVIEW

1. Determine the time constant when $R = 1.2\ \text{k}\Omega$ and $C = 1000$ pF.
2. If the circuit in Question 1 is charged with a 5 V source, how long will it take the initially uncharged capacitor to reach full charge? At full charge, what is the capacitor voltage?
3. A certain circuit has a time constant of 1 ms. If it is charged with a 10 V battery, what will the capacitor voltage be at each of the following intervals: 2 ms, 3 ms, 4 ms, and 5 ms? The capacitor is initially uncharged.
4. A capacitor is charged to 100 V. If it is discharged through a resistor, what is the capacitor voltage at one time constant?

9–6 ▪ CAPACITORS IN AC CIRCUITS

As you saw in the last section, a capacitor blocks constant dc. You will learn in this section that a capacitor passes ac but with an amount of opposition that depends on the frequency of the ac.

After completing this section, you should be able to

- **Analyze capacitive ac circuits**
  - ☐ Define *capacitive reactance*
  - ☐ Determine the value of capacitive reactance in a given circuit
  - ☐ Explain why a capacitor causes a phase shift between voltage and current
  - ☐ Discuss instantaneous, true, and reactive power in a capacitor

Capacitive Reactance, X_C

In Figure 9–39, a capacitor is shown connected to a sinusoidal voltage source. When the source voltage is held at a constant amplitude value and its frequency is increased, the amplitude of the current increases. Also, when the frequency of the source is decreased, the current amplitude decreases. The reason is as follows: When the frequency of the voltage increases, its rate of change also increases. This relationship is illustrated in Figure 9–40, where the frequency is doubled. Now, if the rate at which the voltage is changing increases, the amount of charge moving through the circuit in a given period of time must also increase. More charge in a given period of time means more current. For example, a tenfold increase in frequency means that the capacitor is charging and discharging 10 times as much in a given time interval. Therefore, since the rate of charge movement has increased 10 times, the current must increase by 10 because $I = Q/t$.

An increase in the amount of current for a fixed amount of voltage indicates that opposition to the current has decreased. Therefore, the capacitor offers opposition to current, and that opposition varies *inversely* with frequency.

The opposition to sinusoidal current in a capacitor is called capacitive reactance.

The symbol for **capacitive reactance** is X_C, and its unit is the ohm (Ω).

You have just seen how frequency affects the opposition to current (capacitive reactance) in a capacitor. Now let's see how the capacitance itself affects the reactance. Figure 9–41(a) shows that when a sinusoidal voltage with a fixed amplitude and fixed fre-

Fixed amplitude ac voltage source.

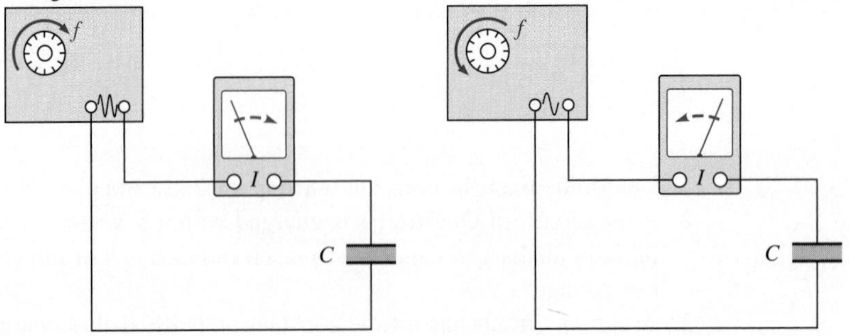

(a) Current increases when the frequency increases.

(b) Current decreases when the frequency decreases.

FIGURE 9–39

The current in a capacitive circuit varies directly with the frequency of the applied voltage.

FIGURE 9–40
Rate of change of a sine wave increases when frequency increases.

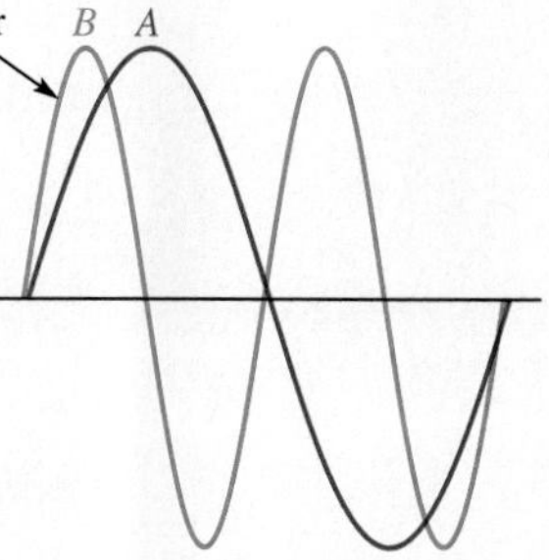

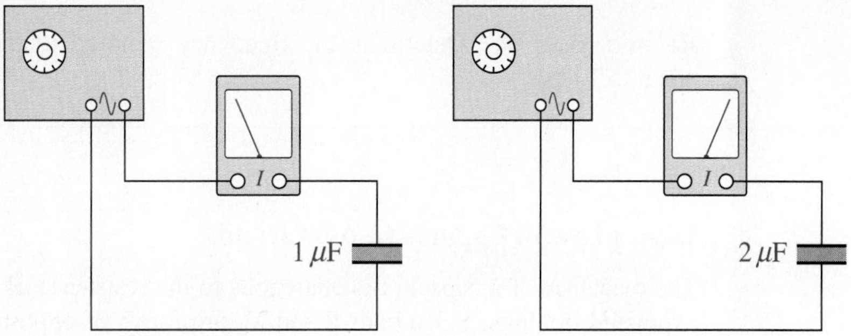

(a) Less capacitance, less current

(b) More capacitance, more current

FIGURE 9–41
For a fixed voltage and fixed frequency, the current varies directly with the capacitance value.

quency is applied to a 1 μF capacitor, a certain amount of alternating current flows. When the capacitance value is increased to 2 μF, the current increases, as shown in Figure 9–41(b). Thus, when the capacitance is increased, the opposition to current (capacitive reactance) decreases. Therefore, not only is the capacitive reactance *inversely* proportional to frequency, but it is also *inversely* proportional to capacitance. This relationship can be stated as follows:

X_C is proportional to $1/fC$.

It can be proven that the constant of proportionality that relates X_C to $1/fC$ is $1/2\pi$. Therefore, the formula for capacitive reactance (X_C) is

$$X_C = \frac{1}{2\pi fC} \tag{9–20}$$

The 2π term comes from the fact that, as you learned in Chapter 8, a sine wave can be described in terms of rotational motion, and one revolution contains 2π radians.

EXAMPLE 9–14 A sinusoidal voltage is applied to a capacitor, as shown in Figure 9–42. The frequency of the sine wave is 1 kHz. Determine the capacitive reactance.

FIGURE 9–42

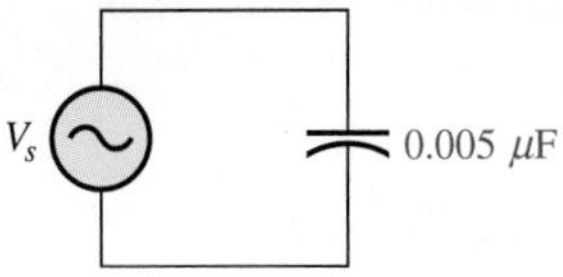

Solution

$$X_C = \frac{1}{2\pi fC} = \frac{1}{2\pi(1 \times 10^3\ \text{Hz})(0.005 \times 10^{-6}\ \text{F})} = 31.8\ \text{k}\Omega$$

The calculator sequence is

[2] [×] [2nd] [π] [×] [1] [EXP] [3] [×] [•] [0] [0] [5] [EXP] [+/−] [6] [=] [1/x]

Related Exercise Determine the frequency required to make X_C in Figure 9–42 equal to 10 kΩ.

Ohm's Law in Capacitive AC Circuits

The reactance of a capacitor is analogous to the resistance of a resistor. In fact, both are expressed in ohms. Since both R and X_C are forms of opposition to current, Ohm's law applies to capacitive circuits as well as to resistive circuits and is stated as follows for capacitive reactance:

$$V = IX_C \tag{9–21}$$

When applying Ohm's law in ac circuits, you must express both the current and the voltage in the same way, that is, both in rms, both in peak, and so on.

EXAMPLE 9–15 Determine the rms current in Figure 9–43.

FIGURE 9–43

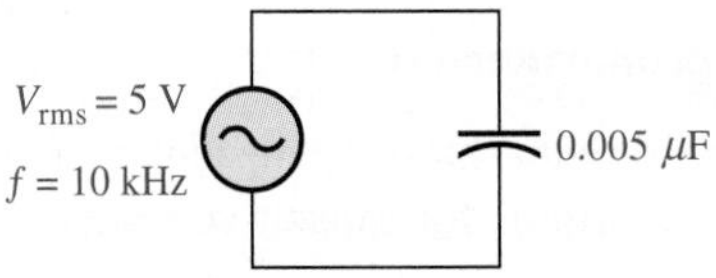

Solution First, find X_C.

$$X_C = \frac{1}{2\pi fC} = \frac{1}{2\pi(10 \times 10^3\ \text{Hz})(0.005 \times 10^{-6}\ \text{F})} = 3.18\ \text{k}\Omega$$

Then, apply Ohm's law.

$$V_{rms} = I_{rms} X_C$$

$$I_{rms} = \frac{V_{rms}}{X_C} = \frac{5\ \text{V}}{3.18\ \text{k}\Omega} = 1.57\ \text{mA}$$

Related Exercise Change the frequency in Figure 9–43 to 25 kHz and determine the rms value of the current.

Current Leads Capacitor Voltage by 90°

A sinusoidal voltage is shown in Figure 9–44. Notice that the rate at which the voltage is changing varies along the sine wave curve, as indicated by the "steepness" of the curve. At the zero crossings, the curve is changing at a faster rate than anywhere else along the curve. At the peaks, the curve has a zero rate of change because it has just reached its maximum and is at the point of changing direction.

The amount of charge stored by a capacitor determines the voltage across it. Therefore, the rate at which the charge is moved ($Q/t = I$) from one plate to the other determines the rate at which the voltage changes. When the current is changing at its maximum rate (at the zero crossings), the voltage is at its maximum value (peak). When the current is changing at its minimum rate (zero at the peaks), the voltage is at its minimum value (zero). This **phase** relationship is illustrated in Figure 9–45. As you can see, the current peaks occur a quarter of a cycle before the voltage peaks. Thus, the current leads the voltage by 90°.

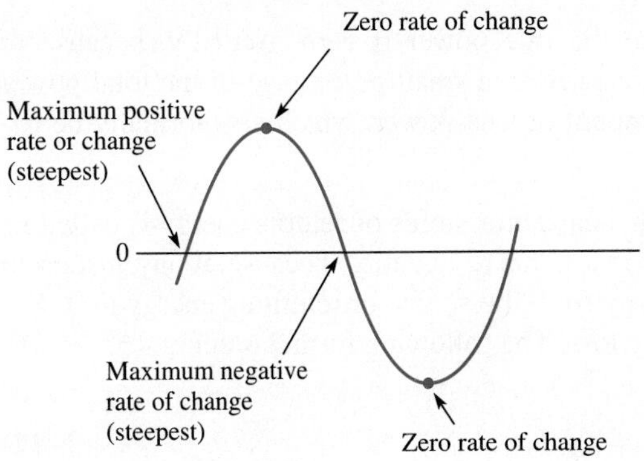

FIGURE 9–44
The rates of change of a sine wave.

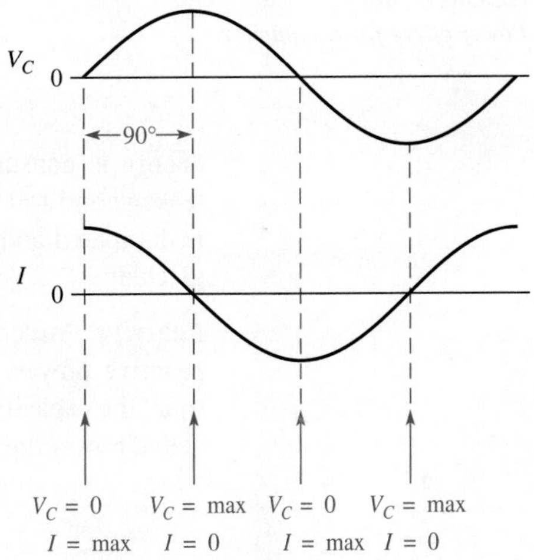

FIGURE 9–45
Current is always leading the capacitor voltage by 90°.

Power in a Capacitor

As discussed earlier in this chapter, a charged capacitor stores energy in the electric field within the dielectric. An ideal capacitor does not dissipate energy; it only stores it. When an ac voltage is applied to a capacitor, energy is stored by the capacitor during a portion of the voltage cycle. Then the stored energy is returned to the source during another portion of the cycle. There is no net energy loss. Figure 9–46 shows the power curve that results from one cycle of capacitor voltage and current.

Instantaneous Power (p) The product of instantaneous voltage, v, and instantaneous current, i, gives instantaneous power, p. At points where v or i is zero, p is also zero. When both v and i are positive, p is also positive. When either v or i is positive and the other negative, p is negative. When both v and i are negative, p is positive. As you can see, the power follows a sinusoidal-type curve. Positive values of power indicate that energy is stored by the capacitor. Negative values of power indicate that energy is returned from the capacitor to the source. Note that the power fluctuates at a frequency twice that of the voltage or current, as energy is alternately stored and returned to the source.

True Power (P_{true}) Ideally, all of the energy stored by a capacitor during the positive portion of the power cycle is returned to the source during the negative portion. No net

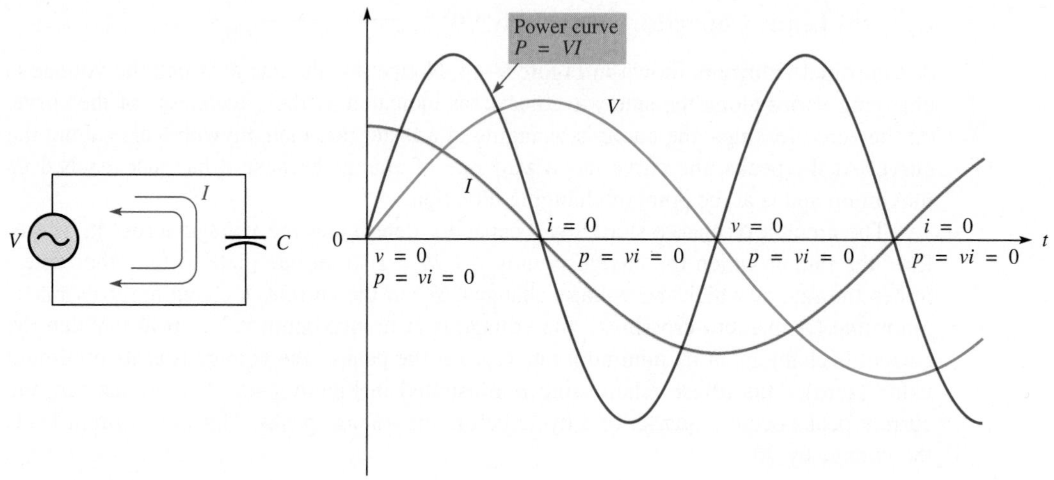

FIGURE 9–46
Power curve for a capacitor.

energy is consumed in the capacitor, so the true power is zero. Actually, because of leakage and foil resistance in a practical capacitor, a small percentage of the total power is dissipated and there is a very small amount of true power, which can normally be neglected.

Reactive Power (P_r) The rate at which a capacitor stores or returns energy is called its **reactive power, P_r.** The reactive power is a nonzero quantity, because at any instant in time, the capacitor is actually taking energy from the source or returning energy to it. Reactive power does not represent an energy loss. The following formulas apply:

$$P_r = V_{rms}I_{rms} \quad \textbf{(9–22)}$$

$$P_r = \frac{V_{rms}^2}{X_C} \quad \textbf{(9–23)}$$

$$P_r = I_{rms}^2 X_C \quad \textbf{(9–24)}$$

Notice that these equations are of the same form as those introduced in Chapter 3 for true power in a resistor. The voltage and current are expressed as rms values. The unit of reactive power is **volt-ampere reactive** (VAR).

EXAMPLE 9–16 Determine the true power and the reactive power in Figure 9–47.

FIGURE 9–47

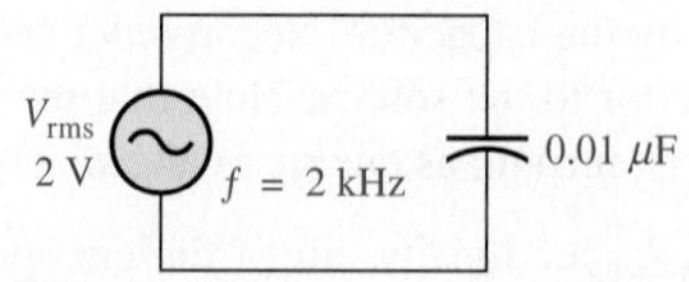

Solution The true power P_{true} is always zero for an ideal capacitor. The reactive power is determined by first finding the value for the capacitive reactance.

$$X_C = \frac{1}{2\pi fC} = \frac{1}{2\pi(2 \times 10^3\text{ Hz})(0.01 \times 10^{-6}\text{ F})} = 7.96\text{ k}\Omega$$

$$P_r = \frac{V_{rms}^2}{X_C} = \frac{(2\text{ V})^2}{7.96\text{ k}\Omega} = 503 \times 10^{-6}\text{ VAR} = 503\ \mu\text{VAR}$$

Related Exercise If the frequency is doubled in Figure 9–47, what are the true power and the reactive power?

SECTION 9–6 REVIEW

1. Calculate X_C for $f = 5$ kHz and $C = 50$ pF.
2. At what frequency is the reactance of a 0.1 μF capacitor equal to 2 kΩ?
3. Calculate the rms current in Figure 9–48.

FIGURE 9–48

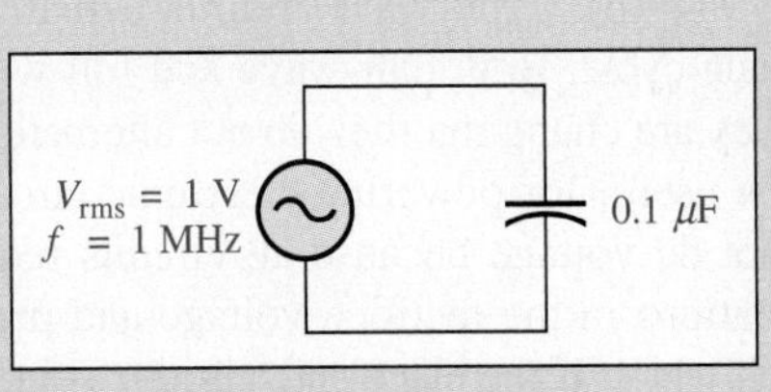

4. State the phase relationship between current and voltage in a capacitor.
5. A 1 μF capacitor is connected to an ac voltage source of 12 V rms. What is the true power?
6. In Question 5, determine reactive power at a frequency of 500 Hz.

9–7 ■ CAPACITOR APPLICATIONS

Capacitors are widely used in electrical and electronic applications. A few typical applications are discussed in this section to illustrate the usefulness of this component.

After completing this section, you should be able to

- **Discuss some capacitor applications**
  - ☐ Describe a power supply filter
  - ☐ Explain the purposes of coupling and by-pass capacitors
  - ☐ Discuss the basics of capacitors applied to tuned circuits, timing circuits, and computer memories

If you pick up any circuit board, open any power supply, or look inside any piece of electronic equipment, chances are you will find capacitors of one type or another. These components are used for a variety of reasons in both dc and ac applications.

Electrical Storage

One of the most basic applications of a capacitor is as a backup voltage source for low-power circuits such as certain types of semiconductor memories in computers. This particular application requires a very high capacitance value and negligible leakage.

The storage capacitor is connected between the dc power supply input to the circuit and ground. When the circuit is operating from its normal power supply, the capacitor remains fully charged to the dc power supply voltage. If the normal power source is disrupted, effectively removing the power supply from the circuit, the storage capacitor temporarily becomes the power source for the circuit.

The capacitor provides voltage and current to the circuit as long as its charge remains sufficient. As current is drawn by the circuit, charge is removed from the capacitor and the voltage decreases. For this reason, the storage capacitor can only be used as a temporary power source. The length of time that the capacitor can provide sufficient power to the circuit depends on the capacitance and the amount of current drawn by the circuit. The smaller the current and the higher the capacitance, the longer the time.

Power Supply Filtering

A basic dc power supply consists of a circuit known as a **rectifier** followed by a **filter.** The rectifier converts the 110 V, 60 Hz sinusoidal voltage available at a standard outlet to a pulsating dc voltage that can be either a half-wave rectified voltage or a full-wave rectified voltage, depending on the type of rectifier circuit. As shown in Figure 9–49(a), a half-wave rectifier removes each negative half-cycle of the sinusoidal voltage. As shown in Figure 9–49(b), a full-wave rectifier actually reverses the polarity of the negative portion of each cycle. Both half-wave and full-wave rectified voltages are dc because, even though they are changing, they do not alternate polarity.

To be useful for powering electronic circuits, the rectified voltage must be changed to constant dc voltage because all circuits require constant power. The filter eliminates the fluctuations in the rectified voltage and provides a smooth constant-value dc voltage to the load which is the electronic circuit, as indicated in Figure 9–50.

The Capacitor As a Power Supply Filter Capacitors are used as filters in dc power supplies because of their ability to store electrical charge. Figure 9–51(a) shows a dc power supply with a full-wave rectifier and a capacitor filter. The operation can be described from a charging and discharging point-of-view as follows. Assume the capacitor is initially uncharged. When the power supply is first turned on and the first cycle of the rectified voltage occurs, the capacitor will quickly charge through the low resistance of

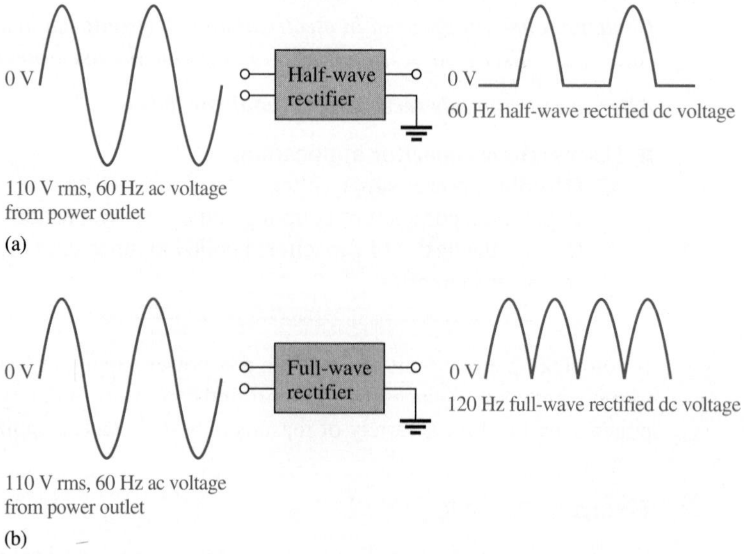

FIGURE 9–49
Half-wave and full-wave rectifier operation.

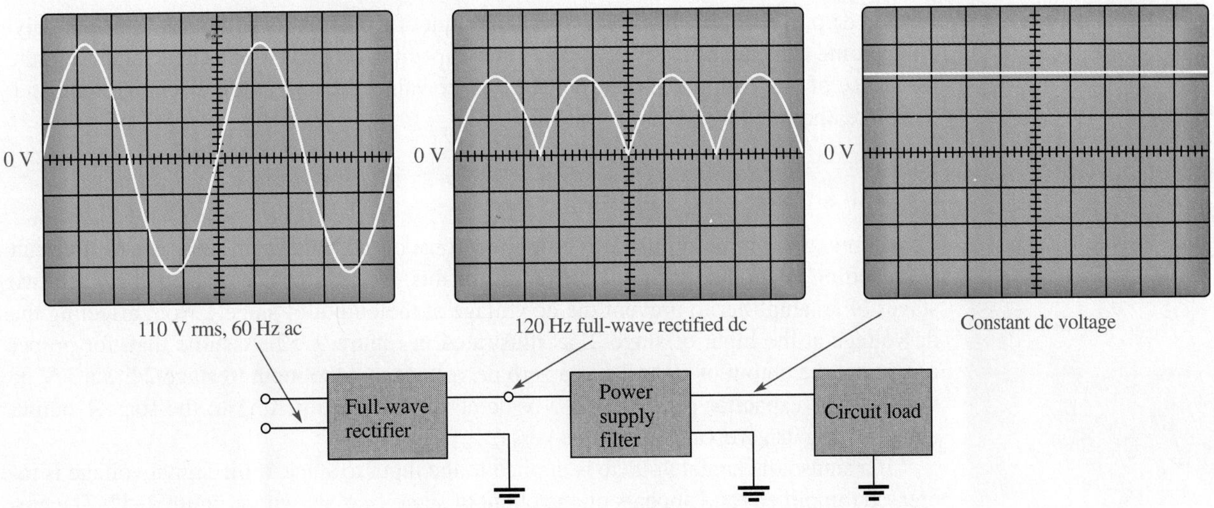

FIGURE 9–50
Basic block diagram and operation of a dc power supply.

the rectifier. The capacitor voltage will follow the rectified voltage curve up to the peak of the rectified voltage. As the rectified voltage passes the peak and begins to decrease, the capacitor will begin to discharge very slowly through the high resistance of the load circuit, as indicated in Figure 9–51(b). The amount of discharge is typically very small and is exaggerated in the figure for purposes of illustration. The next cycle of the rectified voltage will recharge the capacitor back to the peak value by replenishing the small amount of charge lost since the previous peak. This pattern of a small amount of charging and discharging continues as long as the power is on.

A rectifier is designed so that it allows current only in the direction to charge the capacitor. The capacitor will not discharge back through the rectifier but will only discharge a small amount through the relatively high resistance of the load. The small fluctuation in voltage due to the charging and discharging of the capacitor is called the **ripple voltage.**

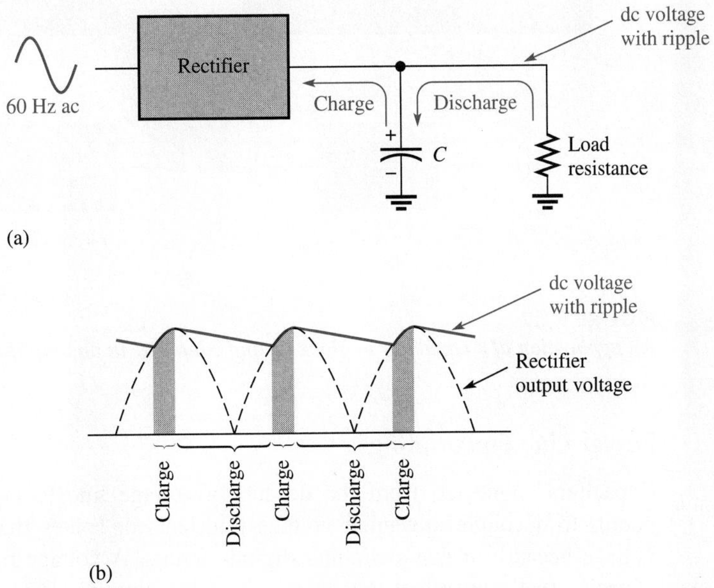

FIGURE 9–51
Basic operation of a power supply filter capacitor.

A good dc power supply has a very small amount of ripple on its dc output. Since the discharge time constant of a power supply filter capacitor depends on its capacitance and the resistance of the load, the higher the capacitance value, the longer the discharge time and, therefore, the smaller the ripple voltage.

DC Blocking and AC Coupling

Capacitors are commonly used to block the constant dc voltage in one part of a circuit from getting to another part. As an example of this, a capacitor is connected between two stages of an amplifier to prevent the dc voltage at the output of stage 1 from affecting the dc voltage at the input of stage 2, as illustrated in Figure 9–52. Assume that, for proper operation, the output of stage 1 has a zero dc voltage and the input to stage 2 has a 3 V dc voltage. The capacitor prevents the 3 V dc at stage 2 from getting to the stage 1 output and affecting its zero value, and vice versa.

If a sinusoidal signal voltage is applied to the input to stage 1, the signal voltage is increased (amplified) and appears on the ouput of stage 1, as shown in Figure 9–52. The amplified signal voltage is then coupled through the capacitor to the input of stage 2 where it is superimposed on the 3 V dc level and then again amplified by stage 2. In order for the signal voltage to be passed through the capacitor without being reduced, the capacitor must be large enough so that its reactance at the frequency of the signal voltage is negligible. In this type of application, the capacitor is known as a *coupling capacitor* which ideally appears as an open to dc and as a short to ac. As the signal frequency is reduced, the capacitive reactance increases and, at some point, the capacitive reactance becomes large enough to cause a significant reduction in ac voltage between stage 1 and stage 2.

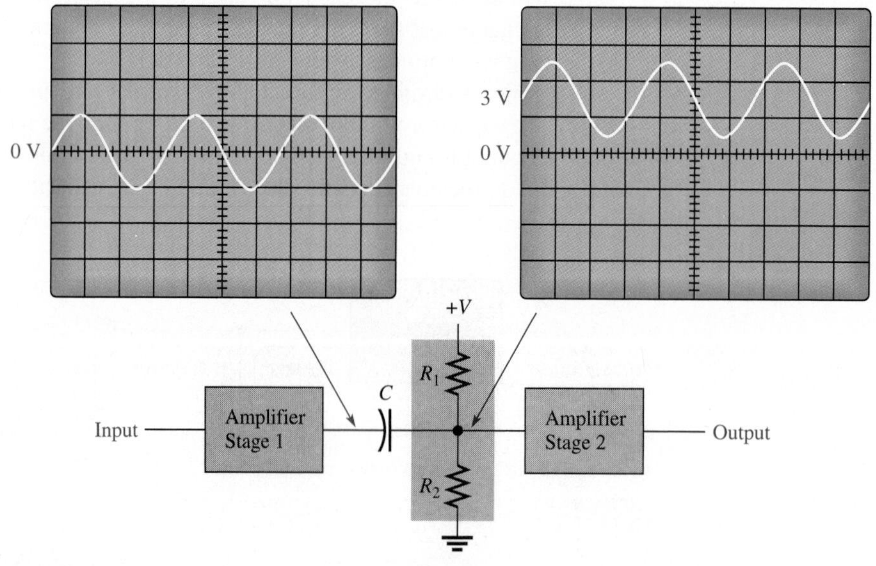

FIGURE 9–52
An application of a capacitor to block dc and couple ac in an amplifier.

Power Line Decoupling

Capacitors connected from the dc supply voltage line to ground are used on circuit boards to decouple unwanted voltage transients or spikes that occur on the dc supply voltage because of fast switching digital circuits. A voltage transient contains high frequencies that may affect the operation of the circuits. These transients are shorted to ground through the very low reactance of the decoupling capacitors. Several decoupling capacitors are often used at various points along the supply voltage line on a circuit board.

Bypassing

Another capacitor application is in bypassing an ac voltage around a resistor in a circuit without affecting the dc voltage across the resistor. In amplifier circuits, for example, dc voltages called *bias voltages* are required at various points. For the amplifier to operate properly, certain bias voltages must remain constant and, therefore, any ac voltages must be removed. A sufficiently large capacitor connected from a bias point to ground provides a low reactance path to ground for ac voltages, leaving the constant dc bias voltage at the given point. This bypass application is illustrated in Figure 9–53. At lower frequencies, the bypass capacitor becomes less effective because of its increased reactance.

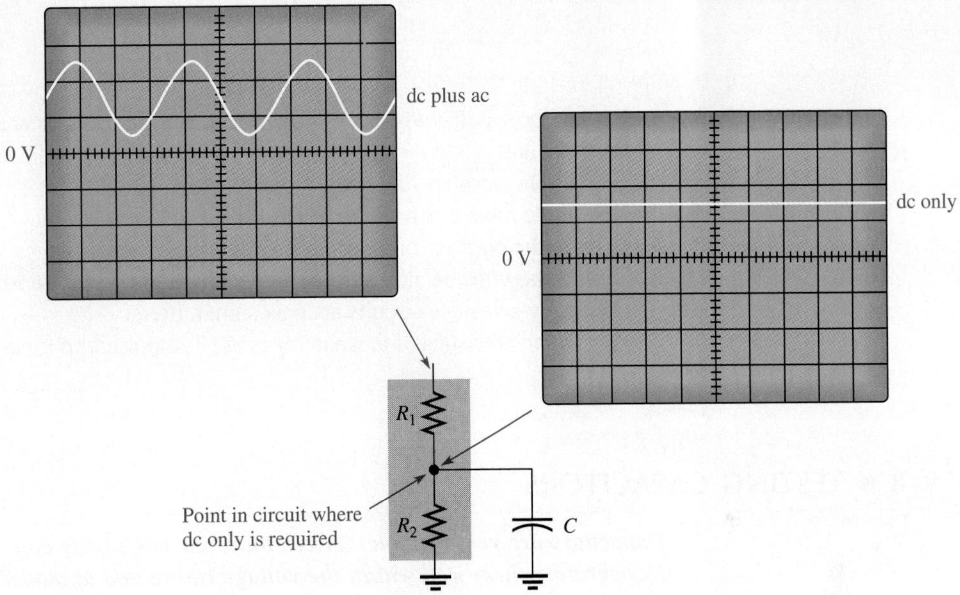

FIGURE 9–53
Example of the operation of a bypass capacitor.

Signal Filters

Capacitors are essential to the operation of a class of circuits called *filters* that are used for selecting one ac signal with a certain specified frequency from a wide range of signals with many different frequencies or for selecting a certain band of frequencies and eliminating all others. A common example of this application is in radio and television receivers where it necessary to select the signal transmitted from a given station and eliminate or filter out the signals transmitted from all the other stations in the area.

When you tune your radio or TV, you are actually changing the capacitance in the tuner circuit (which is a type of filter) so that only the signal from the station or channel you want passes through to the receiver circuitry. Capacitors are used in conjunction with resistors, inductors (covered in the next chapter), and other components in these types of filters. The topic of filters will be covered in detail in later chapters.

The main characteristic of a filter is its frequency selectivity, which is based on the fact that the reactance of a capacitor depends on frequency ($X_C = 1/2\pi fC$).

Timing Circuits

Another important area in which capacitors are used is in timing circuits that generate specified time delays or produce waveforms with specific characteristics. Recall that the time constant of a circuit with resistance and capacitance can be controlled by selecting appropriate values for R and C. The charging time of a capacitor can be used as a basic time delay in various types of circuits. An example is the circuit that controls the turn indicators on your car where the light flashes on and off at regular intervals.

Computer Memories

Dynamic computer memories use capacitors as the basic storage element for binary information, which consists of two digits, 1 and 0. A charged capacitor can represent a stored 1 and a discharged capacitor can represent a stored 0. Patterns of 1s and 0s that make up binary data are stored in a memory that consists of an array of capacitors with associated circuitry. You will study this topic in a computer or digital fundamentals course.

SECTION 9–7 REVIEW

1. Explain how half-wave or full-wave rectified dc voltages are smoothed out by a filter capacitor.
2. Explain the purpose of a coupling capacitor.
3. How large must a coupling capacitor be?
4. Explain the purpose of a decoupling capacitor.
5. Discuss how the relationship of frequency and capacitive reactance is important in frequency-selective circuits such as signal filters.
6. What characteristics of a capacitor is most important in time-delay applications.

9–8 ■ TESTING CAPACITORS

Capacitors are very reliable devices but their useful life can be extended significantly by operating them well within the voltage rating and at moderate temperatures. In this section, the basic types of failures are discussed and methods for checking for them are introduced.

After completing this section, you should be able to

- **Test a capacitor**
  - ☐ Perform an ohmmeter check
  - ☐ Explain what an LC meter is

Capacitor failures can be categorized into two areas: catastrophic and degradation. The catastrophic failures are usually a short circuit caused by dielectric breakdown or an open circuit caused by connection failure. Degradation usually results in a gradual decrease in leakage resistance, hence an increase in leakage current or an increase in equivalent series resistance or dielectric absorption.

Ohmmeter Check

When there is a suspected problem, the capacitor can be removed from the circuit and checked with an ohmmeter. First, to be sure that the capacitor is discharged, short its leads, as indicated in Figure 9–54(a). Connect the meter—set on a high ohms range such as R × 1M—to the capacitor, as shown in Figure 9–54(b), and observe the needle. It should initially indicate near zero ohms. Then it should begin to move toward the high-resistance end of the scale as the capacitor charges from the ohmmeter's battery, as shown in Figure 9–54(c). When the capacitor is fully charged, the meter will indicate an extremely high resistance, as shown in Figure 9–54(d).

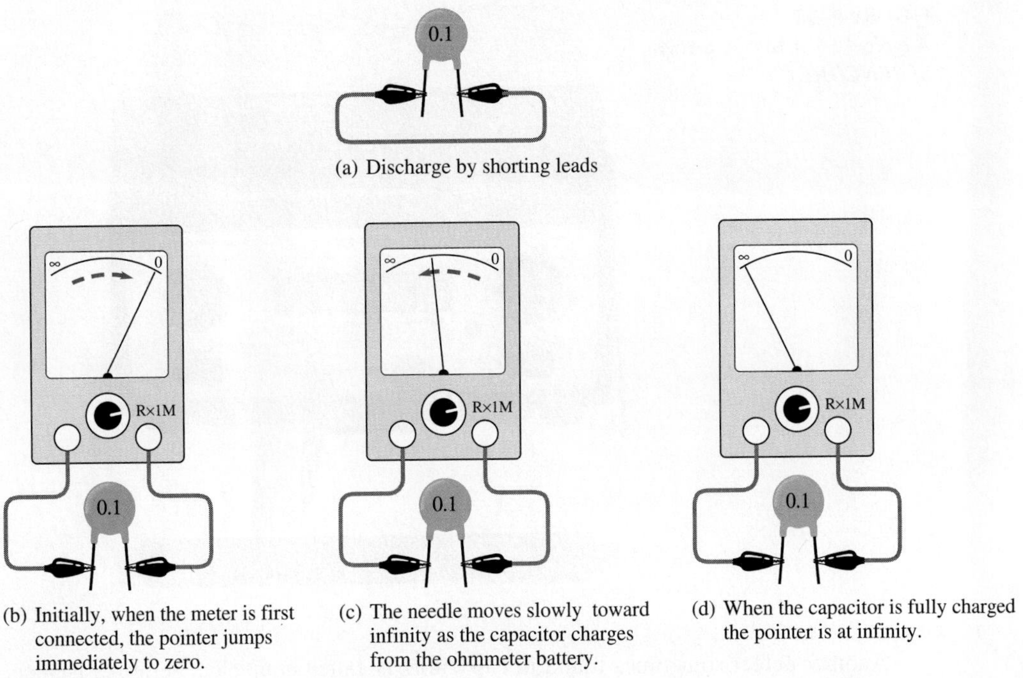

FIGURE 9–54
Checking a capacitor with an analog ohmmeter. This check shows a good capacitor.

As mentioned, the capacitor charges from the internal battery of the ohmmeter, and the meter responds to the charging current. The larger the capacitance value, the more slowly the capacitor will charge, as indicated by the needle movement. For small pF values, the meter response may be insufficient to indicate the fast charging action.

If the capacitor is internally shorted, the meter will go to zero and stay there. If it is leaky, the final meter reading will be much less than normal. Most capacitors have a resistance of several hundred megohms. The exception is the electrolytic, which may normally have less than one megohm of leakage resistance. If the capacitor is open, no charging action will be observed, and the meter will indicate an infinite resistance.

Testing for Capacitance Value and Other Parameters with an *LC* Meter

An *LC* meter such as the one shown in Figure 9–55 can be used to check the value of a capacitor. All capacitors change value over a period of time, some more than others. Ceramic capacitors, for example, often exhibit a 10% to 15% change in value during the first year. Electrolytic capacitors are particularly subject to value change due to drying of the electrolyte. In other cases, capacitors may be labeled incorrectly or the wrong value was installed in the circuit. Although a value change represents less than 25% of defective capacitors, a value check should be made to quickly eliminate this as a source of trouble when troubleshooting a circuit.

Typically, values from 1 pF to 200,000 μF can be measured by simply connecting the capacitor, pushing the appropriate button, and reading the value on the display.

Many *LC* meters can also be used to check for leakage current in capacitors. In order to check for leakage, a sufficient voltage must be applied across the capacitor to simulate operating conditions. This is automatically done by the test instrument. Over 40% of all defective capacitors have excessive leakage current and electrolytics are particularly susceptible to this problem.

The problem of dielectric absorption occurs mostly in electrolytic capacitors when they do not completely discharge during use and retain a residual charge. Approximately 25% of defective capacitors have exhibited this condition.

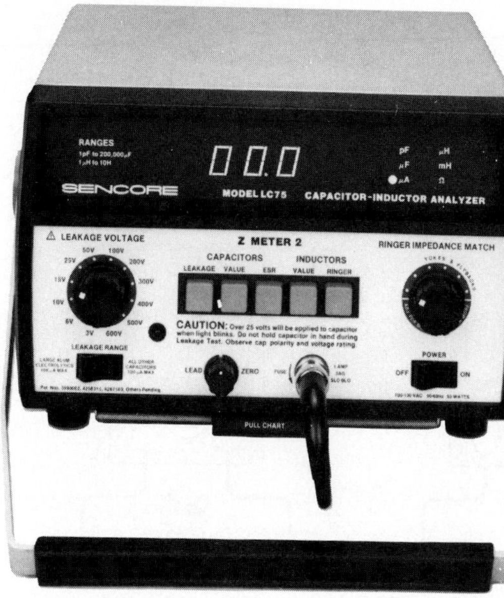

FIGURE 9–55
A typical LC meter (courtesy of SENCORE).

Another defect sometimes found in capacitors is called equivalent series resistance. This problem may be caused by a defective lead to plate contacts, resistive leads, or resistive plates and shows up only under ac conditions. This is the least common capacitor defect and occurs in less than 10% of all defects.

SECTION 9–8 REVIEW

1. How can a capacitor be discharged after removal from the circuit?
2. Describe how the needle of an ohmmeter responds when a good capacitor is checked.
3. List four common capacitor defects.

9–9 ■ APPLICATION ASSIGNMENT

As you learned in this chapter, capacitors are used in certain types of amplifiers for coupling ac signals and blocking dc voltages. In this assignment, an amplifier circuit board contains two coupling capacitors. Your assignment is to check certain voltages on three identical amplifier circuit boards to determine if the capacitors are working properly. A knowledge of amplifier circuits is not necessary for this assignment.

All amplifier circuits contain transistors that require dc voltages to establish proper operating conditions for amplifying ac signals. These dc voltages are referred to as bias voltages. As indicated in Figure 9–56(a), a common type of dc bias circuit used in amplifiers is the voltage divider formed by R_1 and R_2, which sets up the proper dc voltage at the input to the amplifier.

When an ac signal voltage is applied to the amplifier, the input coupling capacitor, C_1, prevents the internal resistance of the ac source from changing the dc bias voltage. Without the capacitor, the internal source resistance would appear in parallel with R_2 and drastically change the value of the dc voltage.

The coupling capacitance is chosen so that its reactance (X_C) at the frequency of the ac signal is very small compared to the bias resistor values. The coupling capacitance

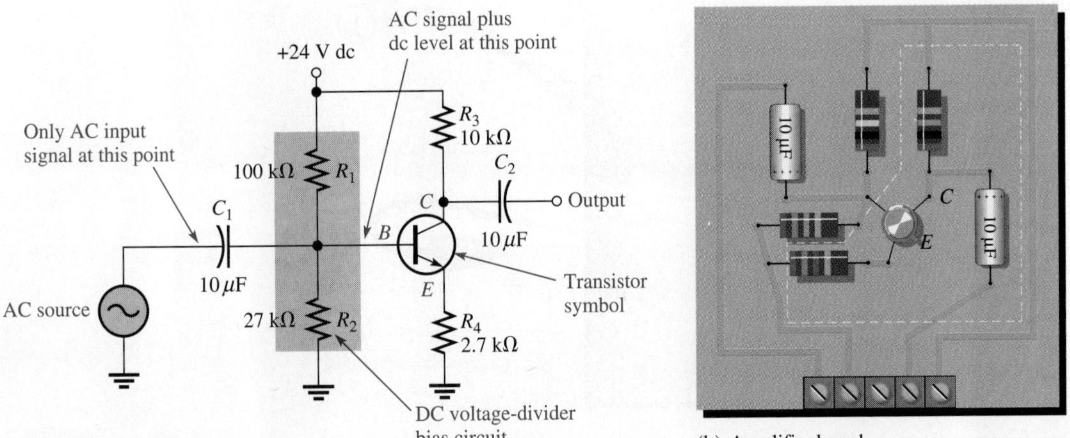

(a) Amplifier schematic (b) Amplifier board

FIGURE 9–56
Capacitively coupled amplifier.

therefore efficiently couples the ac signal from the source to the input of the amplifier. On the source side of the input coupling capacitor there is only ac but on the amplifier side there is ac plus dc (the signal voltage is riding on the dc bias voltage set by the voltage divider), as indicated in Figure 9–56(a). Capacitor C_2 is the output coupling capacitor which couples the amplified ac signal to another amplifier stage that would be connected to the output.

You will check three amplifier boards like the one in Figure 9–56(b) for the proper input voltages using an oscilloscope. If the voltages are incorrect, you will determine the most likely fault. For all measurements, assume the amplifier has no dc loading effect on the voltage-divider bias circuit.

Step 1: Compare the Printed Circuit Board with the Schematic

Check the printed circuit board in Figure 9–56(b) to make sure it agrees with the amplifier schematic in part (a).

Step 2: Test the Input to Board 1

The oscilloscope probe is connected from channel 1 to the board as shown in Figure 9–57.

The input signal from a sinusoidal voltage source is connected to the board and set to a frequency of 5 kHz with an amplitude of 1 V rms. Determine if the voltage and frequency displayed on the scope are correct. If the scope measurement is incorrect, specify the most likely fault in the circuit.

Step 3: Test the Input to Board 2

The oscilloscope probe is connected from channel 1 to board 2 the same as was shown in Figure 9–57 for board 1.

The input signal from the sinusoidal voltage source is the same as Step 2. Determine if the scope display in Figure 9–58 is correct. If the scope measurement is incorrect, specify the most likely fault in the circuit.

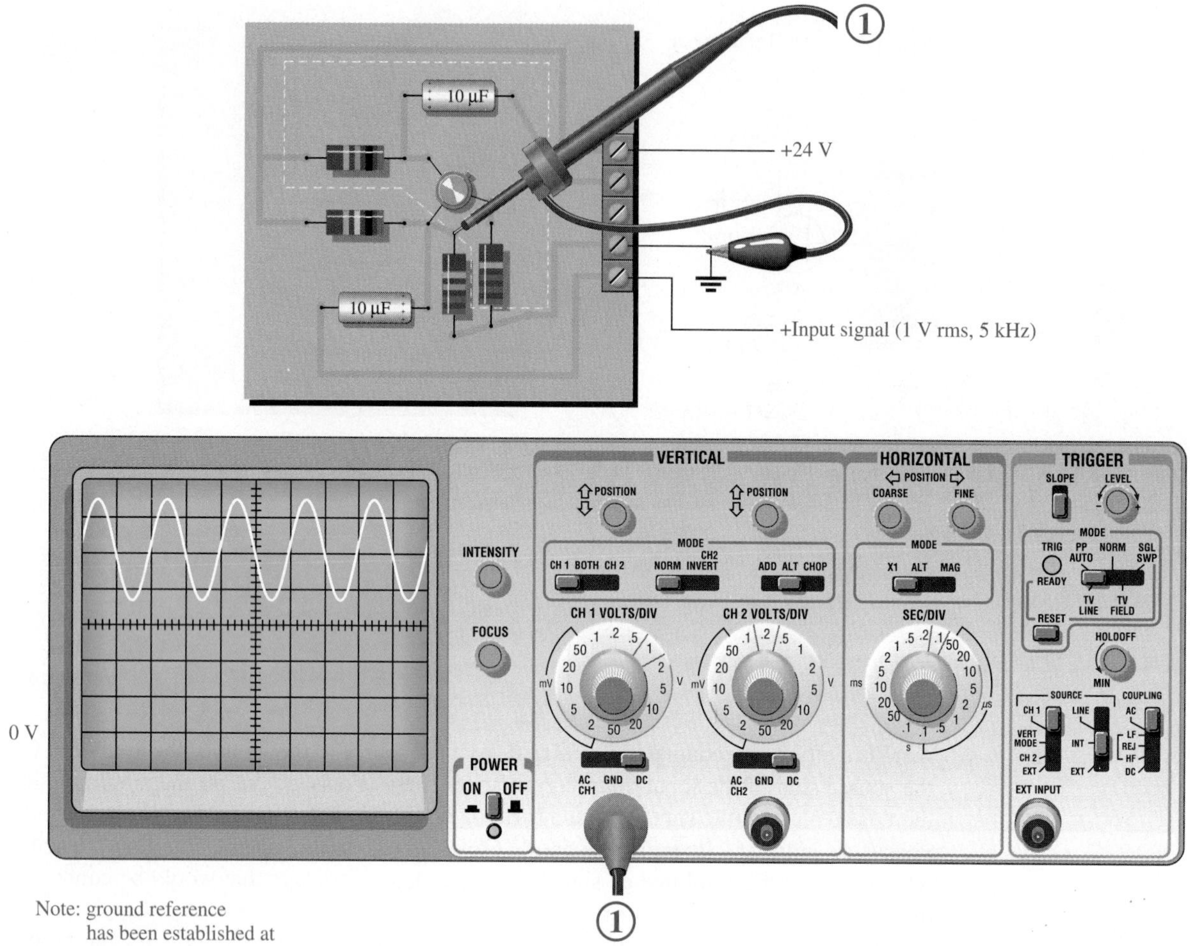

FIGURE 9–57
Testing board 1.

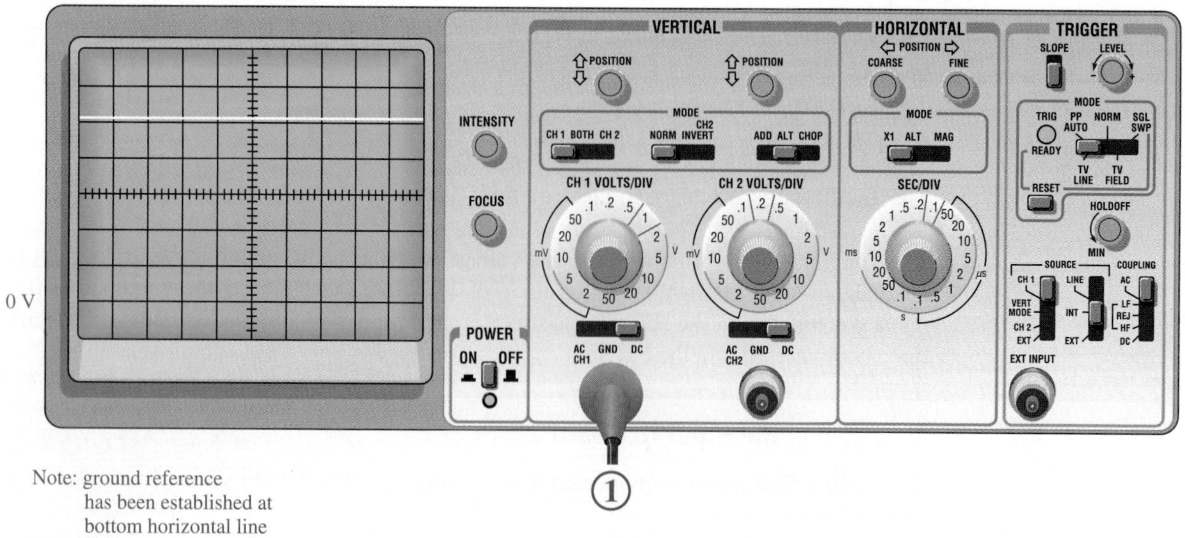

FIGURE 9–58
Testing board 2.

Step 4: Test the Input to Board 3

The oscilloscope probe is connected from channel 1 to board 3 the same as was shown in Figure 9–57 for board 1.

The input signal from the sinusoidal voltage source is the same as Step 3. Determine if the scope display in Figure 5–59 is correct. If the scope measurement is incorrect, specify the most likely fault in the circuit.

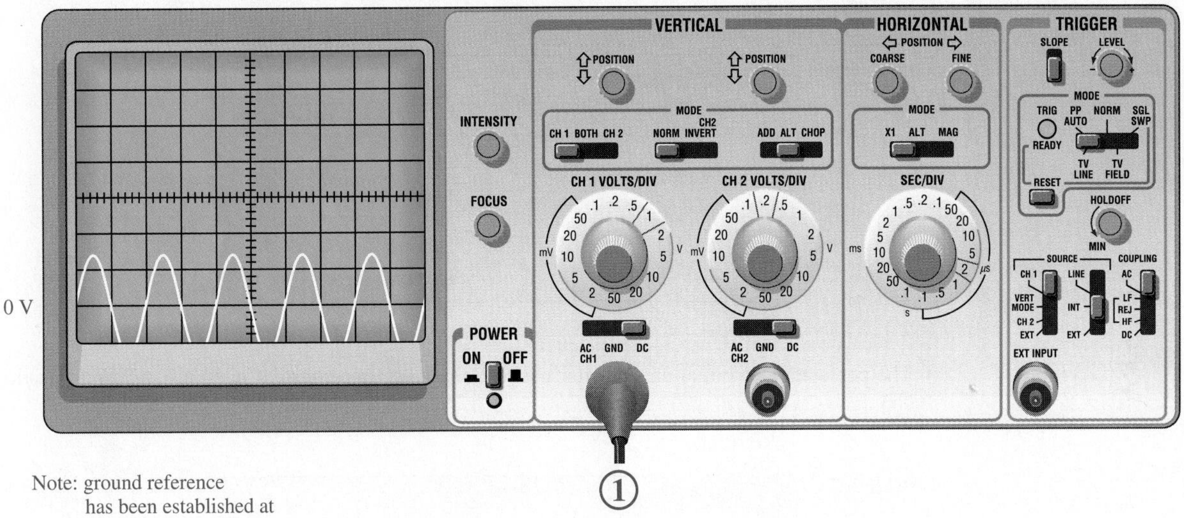

Note: ground reference has been established at bottom horizontal line

FIGURE 9–59
Testing board 3.

SECTION 9–9 REVIEW

1. Explain why the input coupling capacitor is necessary when connecting an ac source to the amplifier.
2. Capacitor C_2 in Figure 9–56 is an output coupling capacitor. Generally, what would you expect to measure at point C and at the output of the circuit when an input signal is applied to the amplifier?

▪ SUMMARY

- The opposition of a capacitor to ac is inversely proportional to frequency and capacitance.
- A capacitor is composed of two parallel conductive plates separated by an insulating material called the *dielectric.*
- A capacitor stores electrical charge on its plates.
- Energy is stored in a capacitor by the electric field created between the charged plates in the dielectric.
- Capacitance is measured in units of farads (F).
- Capacitance is directly proportional to the plate area and the dielectric constant and inversely proportional to the distance between the plates (the dielectric thickness.)
- The dielectric constant is an indication of the ability of a material to establish an electric field.
- The dielectric strength is one factor that determines the breakdown voltage of a capacitor.
- Capacitors are commonly classified according to the dielectric material. Typical materials are mica, ceramic, plastic, and electrolytic (aluminum oxide and tantalum oxide).
- The total capacitance of series capacitors is less than the smallest capacitance.
- The total capacitance of parallel capacitors is the sum of all the capacitances.

- A capacitor blocks constant dc.
- The time constant determines the charging and discharging time of a capacitor with resistance in series.
- In a circuit with capacitance and resistance, the voltage and current during charging and discharging make a 63% change during each time-constant interval.
- Five time constants are required for a capacitor to fully charge or discharge. This is called the *transient time.*
- The percentage of final charge after each charging time-constant interval is given in Table 9–4.

TABLE 9–4

NUMBER OF TIME CONSTANTS	% OF FINAL CHARGE
1	63
2	86
3	95
4	98
5	99 (considered 100%)

- The percentage of initial charge after each discharging time-constant interval is given in Table 9–5.

TABLE 9–5

NUMBER OF TIME CONSTANTS	% OF INITIAL CHARGE
1	37
2	14
3	5
4	2
5	1 (considered 0)

- Alternating current in a capacitor leads the voltage by 90°.
- A capacitor passes ac to an extent that depends on its reactance.
- Capacitive reactance (X_C) is inversely proportional to the frequency and to the capacitance value.
- Ideally, there is no energy loss in a capacitor and, thus, the true power (watts) is zero. However, in most capacitors there is some small energy loss due to leakage resistance.

■ GLOSSARY

Capacitance The ability of a capacitor to store electrical charge.

Capacitive reactance The opposition of a capacitor to sinusoidal current. The unit is the ohm.

Capacitor An electrical device consisting of two conductive plates separated by an insulating material and possessing the property of capacitance.

Coulomb (C) The unit of electrical charge.

Coulomb's law A physical law that states a force exists between two charged bodies that is directly proportional to the product of the two charges and inversely proportional to the square of the distance between them.

Dielectric The insulating material between the plates of a capacitor.

Dielectric constant A measure of the ability of a dielectric material to establish an electric field.

Dielectric strength A measure of the ability of a dielectric material to withstand voltage.

Farad (F) The unit of capacitance.

Filter A type of circuit that passes or blocks certain frequencies to the exclusion of all others.

Phase The relative displacement of a time-varying quantity with respect to a given reference.

Reactive power (P_r) The rate at which energy is stored and alternately returned to the source by a capacitor. The unit is the VAR.

Rectifier An electronic circuit that converts ac into pulsing dc; one part of a power supply.

Ripple voltage The small fluctuation in voltage due to the charging and discharging of a capacitor.

Temperature coefficient A constant specifying the amount of change in the value of a quantity for a given change in temperature.

Time constant A fixed interval, equal to the product of the *R* and *C* values, that determines the time response of a circuit.

Trimmer A small variable capacitor.

Volt-ampere reactive (VAR) The unit of reactive power.

■ FORMULAS

(9–1)	$C = \frac{Q}{V}$	Capacitance in terms of charge and voltage
(9–2)	$Q = CV$	Charge in terms of capacitance and voltage
(9–3)	$V = \frac{Q}{C}$	Voltage in terms of charge and capacitance
(9–4)	$F = \frac{kQ_1Q_2}{d^2}$	Coulomb's law (force between two charged bodies)
(9–5)	$W = \frac{1}{2}CV^2$	Energy stored by a capacitor
(9–6)	$\epsilon_r = \frac{\epsilon}{\epsilon_0}$	Dielectric constant (relative permittivity)
(9–7)	$C = \frac{A\epsilon_r(8.85 \times 10^{-12}\text{ F/m})}{d}$	Capacitance in terms of physical parameters
(9–8)	$\frac{1}{C_T} = \frac{1}{C_1} + \frac{1}{C_2}$	Reciprocal of total series capacitance (two capacitors)
(9–9)	$C_T = \frac{1}{\left(\frac{1}{C_1}\right) + \left(\frac{1}{C_2}\right)}$	Total series capacitance (two capacitors)
(9–10)	$\frac{1}{C_T} = \frac{1}{C_1} + \frac{1}{C_2} + \frac{1}{C_3} + \cdots + \frac{1}{C_n}$	Reciprocal of total series capacitance (general)
(9–11)	$C_T = \frac{1}{\left(\frac{1}{C_1}\right) + \left(\frac{1}{C_2}\right) + \left(\frac{1}{C_3}\right) + \cdots + \left(\frac{1}{C_n}\right)}$	Total series capacitance (general)
(9–12)	$V_x = \left(\frac{C_T}{C_x}\right)V_S$	Voltage across series capacitor
(9–13)	$C_T = C_1 + C_2$	Two capacitors in parallel
(9–14)	$C_T = C_1 + C_2 + \cdots + C_n$	*n* capacitors in parallel
(9–15)	$\tau = RC$	Time constant
(9–16)	$v = V_F + (V_i - V_F)e^{-t/\tau}$	Exponential voltage (general)
(9–17)	$i = I_F + (I_i - I_F)e^{-t/\tau}$	Exponential current (general)
(9–18)	$v = V_F(1 - e^{-t/RC})$	Increasing exponential voltage beginning at zero
(9–19)	$v = V_ie^{-t/RC}$	Decreasing exponential voltage ending at zero
(9–20)	$X_C = \frac{1}{2\pi fC}$	Capacitive reactance

(9–21)	$V = IX_C$	Ohm's law for a capacitor
(9–22)	$P_r = V_{rms}I_{rms}$	Reactive power in a capacitor
(9–23)	$P_r = \dfrac{V_{rms}^2}{X_C}$	Reactive power in a capacitor
(9–24)	$P_r = I_{rms}^2 X_C$	Reactive power in a capacitor

■ SELF-TEST

1. Which of the the following statement(s) accurately describe(s) a capacitor?
(a) The plates are conductive.
(b) The dielectric is an insulator between the plates.
(c) Constant dc flows through a fully charged capacitor.
(d) A practical capacitor stores charge indefinitely when disconnected from the source.
(e) none of the above
(f) all the above
(g) only (a) and (b)

2. Which one of the following statements is true?
(a) There is current through the dielectric of a charging capacitor.
(b) When a capacitor is connected to a dc voltage source, it will charge to the value of the source.
(c) An ideal capacitor can be discharged by disconnecting it from the voltage source.

3. A capacitance of 0.01 μF is larger than
(a) 0.00001 F **(b)** 100,000 pF **(c)** 1000 pF **(d)** all of the above

4. A capacitance of 1000 pF is smaller than
(a) 0.01 μF **(b)** 0.001 μF **(c)** 0.00000001 F **(d)** both (a) and (c)

5. When the voltage across a capacitor is increased, the stored charge
(a) increases **(b)** decreases **(c)** remains constant **(d)** fluctuates

6. When the voltage across a capacitor is doubled, the stored charge
(a) stays the same **(b)** is halved **(c)** increases by four **(d)** doubles

7. The voltage rating of a capacitor is increased by
(a) increasing the plate separation **(b)** decreasing the plate separation
(c) increasing the plate area **(d)** both (b) and (c)

8. The capacitance value is increased by
(a) decreasing plate area **(b)** increasing plate separation
(c) decreasing plate separation **(d)** increasing plate area
(e) both (a) and (b) **(f)** both (c) and (d)

9. A 1 μF, a 2.2 μF, and a 0.05 μF capacitor are connected in series. The total capacitance is less than
(a) 1 μF **(b)** 2.2 μF **(c)** 0.05 μF **(d)** 0.001 μF

10. Four 0.02 μF capacitors are in parallel. The total capacitance is
(a) 0.02 μF **(b)** 0.08 μF **(c)** 0.05 μF **(d)** 0.04 μF

11. An uncharged capacitor and a resistor are connected in series with a switch and a 12 V battery. At the instant the switch is closed, the voltage across the capacitor is
(a) 12 V **(b)** 6 V **(c)** 24 V **(d)** 0 V

12. In Question 11, the voltage across the capacitor when it is fully charged is
(a) 12 V **(b)** 6 V **(c)** 24 V **(d)** −6 V

13. In Question 11, the capacitor will reach full charge in a time equal to approximately
(a) RC **(b)** $5RC$ **(c)** $12RC$ **(d)** cannot be predicted

14. A sine wave voltage is applied across a capacitor. When the frequency of the voltage is increased, the current
(a) increases **(b)** decreases **(c)** remains constant **(d)** ceases

15. A capacitor and a resistor are connected in series to a sine wave generator. The frequency is set so that the capacitive reactance is equal to the resistance and, thus, an equal amount of voltage appears across each component. If the frequency is decreased,
(a) $V_R > V_C$ **(b)** $V_C > V_R$ **(c)** $V_R = V_C$ **(d)** $V_C < V_R$

16. An ohmmeter is connected across a discharged capacitor and the needle stabilizes at approximately 50 kΩ. The capacitor is
(a) good **(b)** charged **(c)** too large **(d)** leaky

BASIC PROBLEMS

SECTION 9–1 The Basic Capacitor

1. **(a)** Find the capacitance when $Q = 50\ \mu C$ and $V = 10$ V.
   (b) Find the charge when $C = 0.001\ \mu F$ and $V = 1$ kV.
   (c) Find the voltage when $Q = 2$ mC and $C = 200\ \mu F$.
2. Convert the following values from microfarads to picofarads:
   (a) 0.1 μF **(b)** 0.0025 μF **(c)** 5 μF
3. Convert the following values from picofarads to microfarads:
   (a) 1000 pF **(b)** 3500 pF **(c)** 250 pF
4. Convert the following values from farads to microfarads:
   (a) 0.0000001 F **(b)** 0.0022 F **(c)** 0.0000000015 F
5. What size capacitor is capable of storing 10 mJ of energy with 100 V across its plates?
6. A mica capacitor has a plate area of 0.04 m^2 and a dielectric thickness of 0.008 m. What is its capacitance?
7. An air capacitor has plates with an area of 0.1 m^2. The plates are separated by 0.01 m. Calculate the capacitance.
8. At ambient temperature (25°C), a certain capacitor is specified to be 1000 pF. It has a negative temperature coefficient of 200 ppm/°C. What is its capacitance at 75°C?
9. A 0.001 μF capacitor has a positive temperature coefficient of 500 ppm/°C. How much change in capacitance will a 25°C increase in temperature cause?

SECTION 9–2 Types of Capacitors

10. In the construction of a stacked-foil mica capacitor, how is the plate area increased?
11. What type of capacitor has the higher dielectric constant, mica or ceramic?
12. Show how to connect an electrolytic capacitor across points *A* and *B* in Figure 9–60.

FIGURE 9–60

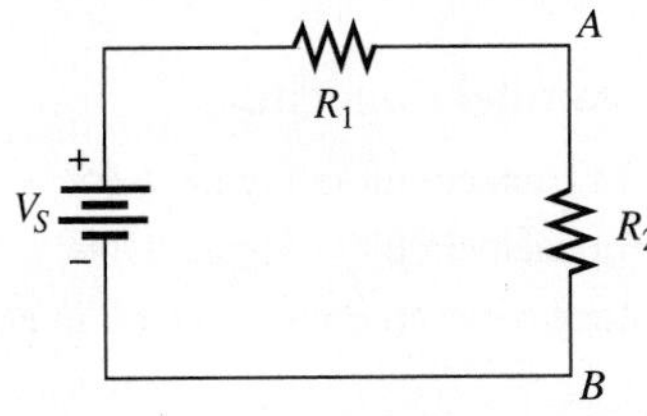

13. Determine the value of the typographically labeled ceramic disk capacitors in Figure 9–61.
14. Name two types of electrolytic capacitors. How do electrolytics differ from other capacitors?
15. Identify the parts of the ceramic disk capacitor shown in the cutaway view of Figure 9–62 by referring to Figure 9–9(b).

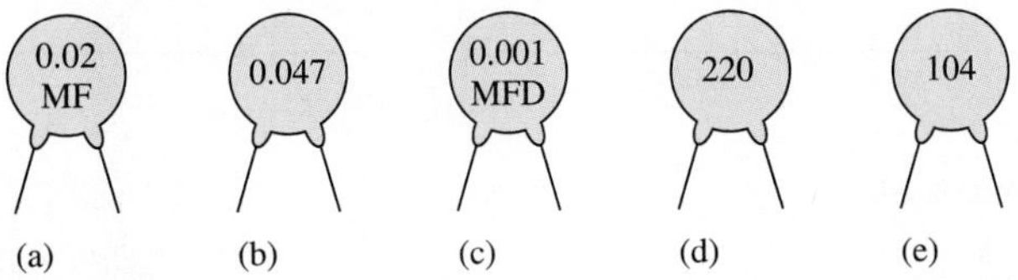

FIGURE 9–61

FIGURE 9–62

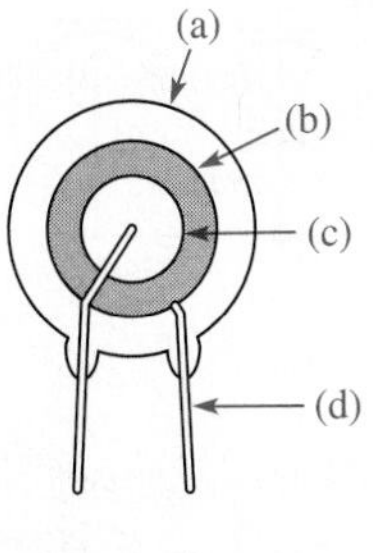

SECTION 9–3 Series Capacitors

16. Five 1000 pF capacitors are in series. What is the total capacitance?

17. Find the total capacitance for each circuit in Figure 9–63.

18. For each circuit in Figure 9–63, determine the voltage across each capacitor.

19. The total charge stored by the series capacitors in Figure 9–64 is 10 μC. Determine the voltage across each of the capacitors.

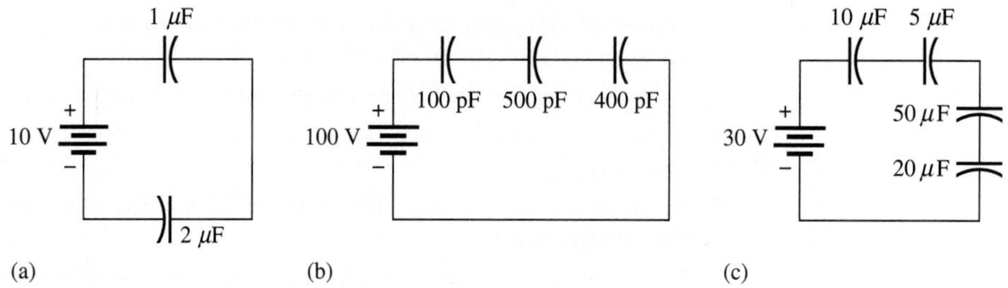

FIGURE 9–63

FIGURE 9–64

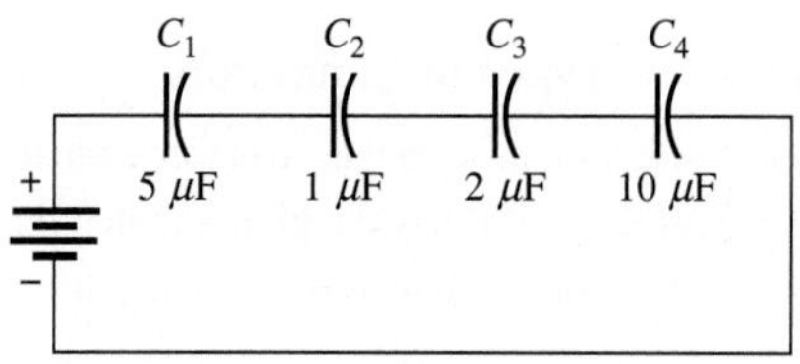

SECTION 9–4 Parallel Capacitors

20. Determine C_T for each circuit in Figure 9–65.

21. Determine C_T for each circuit in Figure 9–66.

22. What is the voltage between points *A* and *B* in each circuit in Figure 9–66?

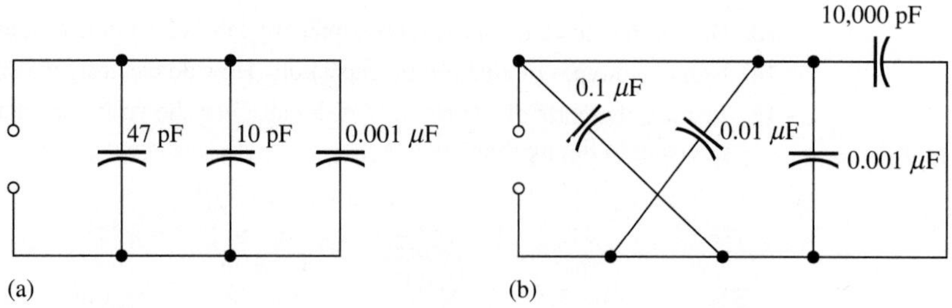

FIGURE 9–65

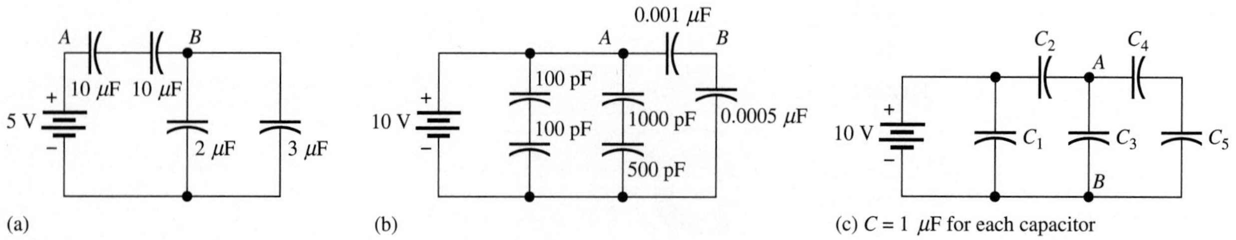

FIGURE 9–66

SECTION 9–5 Capacitors in DC Circuits

23. Determine the time constant for each of the following series RC combinations:
(a) $R = 100\ \Omega$, $C = 1\ \mu F$ **(b)** $R = 10\ M\Omega$, $C = 50\ pF$
(c) $R = 4.7\ k\Omega$, $C = 0.005\ \mu F$ **(d)** $R = 1.5\ M\Omega$, $C = 0.01\ \mu F$

24. Determine how long it takes the capacitor to reach full charge for each of the following combinations:
(a) $R = 50\ \Omega$, $C = 50\ \mu F$ **(b)** $R = 3300\ \Omega$, $C = 0.015\ \mu F$
(c) $R = 22\ k\Omega$, $C = 100\ pF$ **(d)** $R = 5\ M\Omega$, $C = 10\ pF$

25. In the circuit of Figure 9–67, the capacitor initially is uncharged. Determine the capacitor voltage at the following times after the switch is closed:
(a) 10 μs **(b)** 20 μs **(c)** 30 μs **(d)** 40 μs **(e)** 50 μs

26. In Figure 9–68, the capacitor is charged to 25 V. Find the capacitor voltage at the following times when the switch is closed:
(a) 1.5 ms **(b)** 4.5 ms **(c)** 6 ms **(d)** 7.5 ms

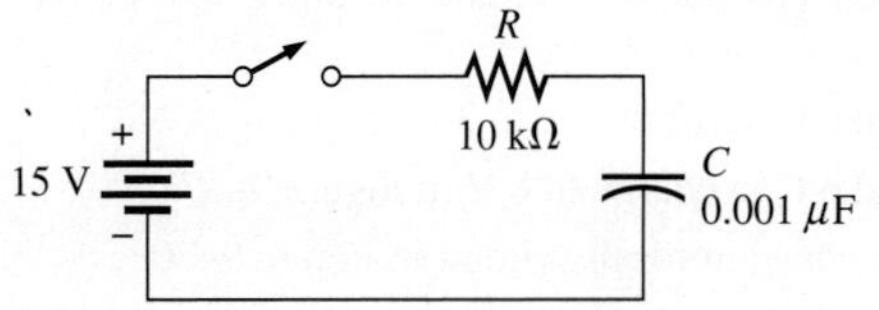

FIGURE 9–67

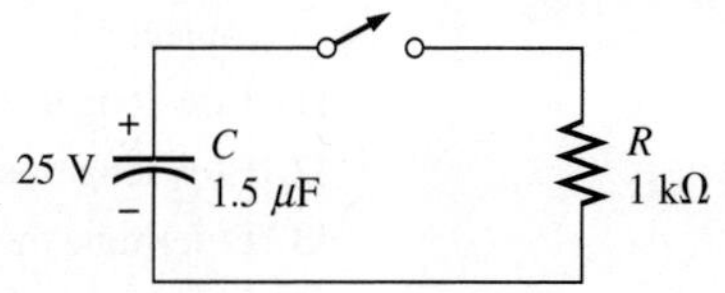

FIGURE 9–68

27. Repeat Problem 25 for the following time intervals:
(a) 2 μs **(b)** 5 μs **(c)** 15 μs

28. Repeat Problem 26 for the following time intervals:
(a) 0.5 ms **(b)** 1 ms **(c)** 2 ms

SECTION 9–6 Capacitors in AC Circuits

29. Determine X_C for a 0.047 μF capacitor at each of the following frequencies:
(a) 10 Hz **(b)** 250 Hz **(c)** 5 kHz **(d)** 100 kHz

30. What is the value of the total capacitive reactance in each circuit in Figure 9–69?

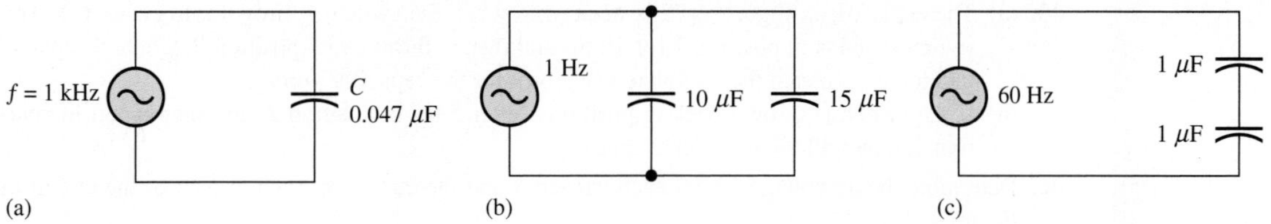

FIGURE 9–69

31. In Figure 9–66, each dc voltage source is replaced by a 10 V rms, 2 kHz ac source. Determine the reactance in each case.

32. In each circuit of Figure 9–69, what frequency is required to produce an X_C of 100 Ω? An X_C of 1 kΩ?

33. A sinusoidal voltage of 20 V rms produces an rms current of 100 mA when connected to a certain capacitor. What is the reactance?

34. A 10 kHz voltage is applied to a 0.0047 μF capacitor, and 1 mA of rms current is measured. What is the rms value of the voltage?

35. Determine the true power and the reactive power in Problem 34.

SECTION 9–7 Capacitor Applications

36. If another capacitor is connected in parallel with the existing capacitor in the power supply filter of Figure 9–51, how is the ripple voltage affected?

37. Ideally, what should the reactance of a bypass capacitor be in order to eliminate a 10 kHz ac voltage at a given point in an amplifier circuit?

SECTION 9–8 Testing Capacitors

38. Assume that you are checking a capacitor with an ohmmeter, and when you connect the leads across the capacitor, the pointer does not move from its left-end scale position. What is the problem?

39. In checking a capacitor with the ohmmeter, you find that the pointer goes all the way to the right end of the scale and stays there. What is the problem?

■ ADVANCED PROBLEMS

40. Two series capacitors (one 1 μF, the other of unknown value) are charged from a 12 V source. The 1 μF capacitor is charged to 8 V, and the other to 4 V. What is the value of the unknown capacitor?

41. How long does it take C to discharge to 3 V in Figure 9–68?

42. How long does it take C to charge to 8 V in Figure 9–67?

43. Determine the time constant for the circuit in Figure 9–70.

44. In Figure 9–71, the capacitor initially is uncharged. At $t = 10\ \mu$s after the switch is closed, the instantaneous capacitor voltage is 7.2 V. Determine the value of R.

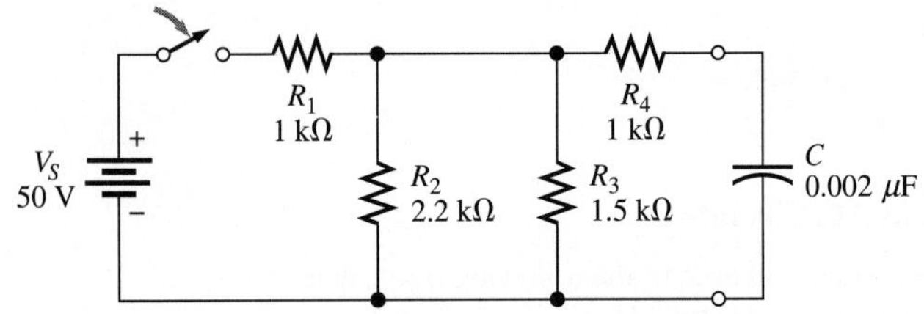

FIGURE 9–70

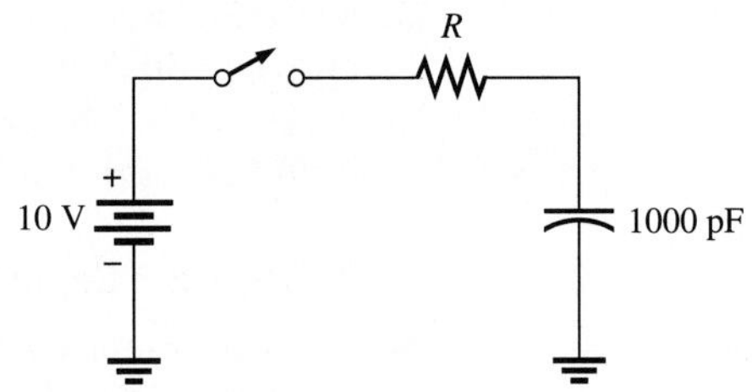

FIGURE 9–71

45. **(a)** The capacitor in Figure 9–72 is uncharged when the switch is thrown into position 1. The switch remains in position 1 for 10 ms and then is thrown into position 2, where it remains indefinitely. Sketch the complete waveform for the capacitor voltage.

(b) If the switch is thrown back to position 1 after 5 ms in position 2, and then is left in position 1, how will the waveform appear?

46. Determine the ac voltage across each capacitor and the current in each branch of the circuit in Figure 9–73.

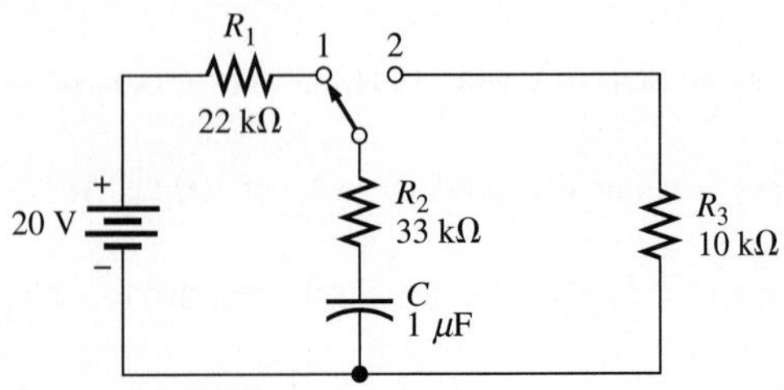

FIGURE 9–72

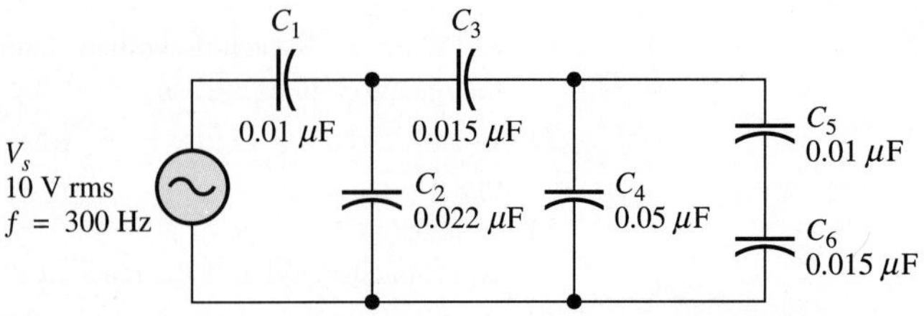

FIGURE 9–73

47. Find the value of C_1 in Figure 9–74.

48. How much does the voltage across C_5 and C_6 change when the ganged switch is thrown from position 1 to position 2 in Figure 9–75?

FIGURE 9–74

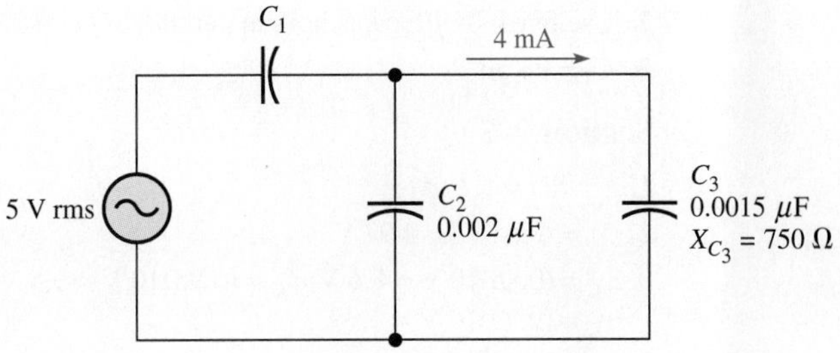

FIGURE 9–75

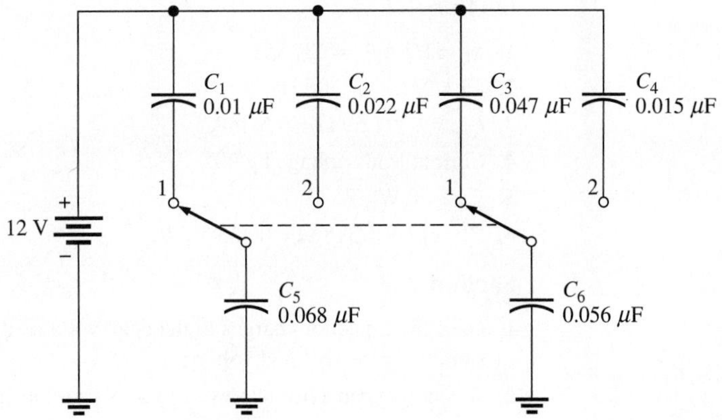

49. If C_4 in Figure 9–73 is open, determine the voltages that would be measured across the other capacitors.

▪ ANSWERS TO SECTION REVIEWS

Section 9–1

1. Capacitance is the ability (capacity) to store electrical charge.

2. **(a)** There are 1,000,000 microfarads in one farad.
(b) There are 1×10^{12} picofarads in one farad.
(c) There are 1,000,000 picofarads in one microfarad.

3. $0.0015\ \mu\text{F} \times 10^6 = 1500$ pF; $0.0015\ \mu\text{F} \times 10^{-6} = 0.0000000015$ F

4. $W = 1/2\ CV^2 = 1/2\ (0.01\ \mu\text{F})(15\ \text{V})^2 = 1.125\ \mu\text{J}$

5. **(a)** Capacitance increases when plate area is increased.
(b) Capacitance decreases when plate separation is increased.

6. (1000 V/mil)(10 mils) = 10 kV

7. $C = (0.2\ \text{m}^2)(1200)(8.85 \times 10^{-12}\ \text{F/m})/0.005\ \text{m} = 0.425\ \mu\text{F}$

Section 9–2

1. Capacitors are commonly classified by the dielectric material.

2. A fixed capacitance cannot be changed; a variable can.

3. Electrolytics are polarized.

4. Be sure that the voltage rating is sufficient and connect the positive end to the positive side of the circuit when installing a polarized capacitor.

Section 9–3

1. C_T of series capacitors is less than the smallest value.

2. $C_T = 62.5$ pF

3. $C_T = 0.006\ \mu F$
4. $V = (0.006\ \mu F/0.01\ \mu F)10\ V = 6\ V$

Section 9–4

1. The individual parallel capacitors are added to get C_T.
2. Use five 0.01 μF capacitors in parallel to get 0.05 μF.
3. $C_T = 98\ pF$

Section 9–5

1. $\tau = RC = 1.2\ \mu s$
2. $5\tau = 6\ \mu s$; $V_C = 4.97\ V$
3. $v_C = (0.86)10\ V = 8.6\ V$; $v_C = (0.95)10\ V = 9.5\ V$; $v_C = (0.98)10\ V = 9.8\ V$; $v_C = (0.99)10\ V = 9.9\ V$
4. $v_C = (0.37)(100\ V) = 37\ V$

Section 9–6

1. $X_C = 1/2\pi fC = 637\ k\Omega$
2. $f = 1/2\pi CX_C = 796\ Hz$
3. $I_{rms} = 1\ V/1.59\ \Omega = 629\ mA$
4. Current leads voltage by 90°.
5. $P_{true} = 0\ W$
6. $P_r = (12\ V)^2/318\ \Omega = 0.453\ VAR$

Section 9–7

1. Once the capacitor charges to the peak voltage, it discharges very little before the next peak thus smoothing the rectified voltage.
2. A coupling capacitor allows ac to pass from one point to another, but blocks constant dc.
3. A coupling capacitor must be large enough to have a negligible reactance at the frequency that is to be passed without opposition.
4. A decoupling capacitor shorts power line voltage transients to ground.
5. X_C is inversely proportional to frequency and so is the filter's ability to pass ac signals.
6. The time constant is used in delay applications.

Section 9–8

1. Short its leads to discharge.
2. Initially, the needle jumps to zero; then it moves to the high-resistance end of the scale when the capacitor is good.
3. Shorted, open, leakage, and dielectric absorption are possible capacitor faults.

Section 9–9

1. The coupling capacitor prevents the source from affecting the dc voltage but passes the ac input signal.
2. An ac voltage riding on a dc voltage is at point *C*.

■ ANSWERS TO RELATED EXERCISES FOR EXAMPLES

9–1	10 kV
9–2	0.05 μF
9–3	100,000,000 pF
9–4	6,638 pF
9–5	278 pF
9–6	1.57 μF
9–7	2.78 V
9–8	650 pF
9–9	0.16 μF
9–10	891 μs

9–11 8.36 V
9–12 7.97 V
9–13 0.7 ms; 95 V
9–14 3.18 kHz
9–15 3.93 mA
9–16 0 W; 1.01 mVAR

10

INDUCTORS

Inductance is the property of a coil of wire that opposes a change in current. The basis for inductance is the electromagnetic field that surrounds any conductor when there is current through it. The electrical component designed to have the property of inductance is called an *inductor, coil,* or *choke.* All of these terms refer to the same type of device.

In this chapter, the basic inductor is introduced and its characteristics are studied. Various types of inductors are covered in terms of their physical construction and their electrical properties. The basic behavior of inductors in both dc and ac circuits is studied, and series and parallel combinations are analyzed. A method of testing inductors is discussed.

■ CHAPTER OBJECTIVES

- □ Describe the basic structure and characteristics of an inductor
- □ Discuss various types of inductors
- □ Analyze series inductors
- □ Analyze parallel inductors
- □ Analyze inductive dc switching circuits
- □ Analyze inductive ac circuits
- □ Discuss some inductor applications
- □ Test an inductor

■ APPLICATION ASSIGNMENT

You are working on a piece of defective communications equipment and your supervisor asks you to check out the unmarked coils that have been removed from the system and determine their approximate inductance value. After studying this chapter, you should be able to complete the application assignment in Section 10–9.

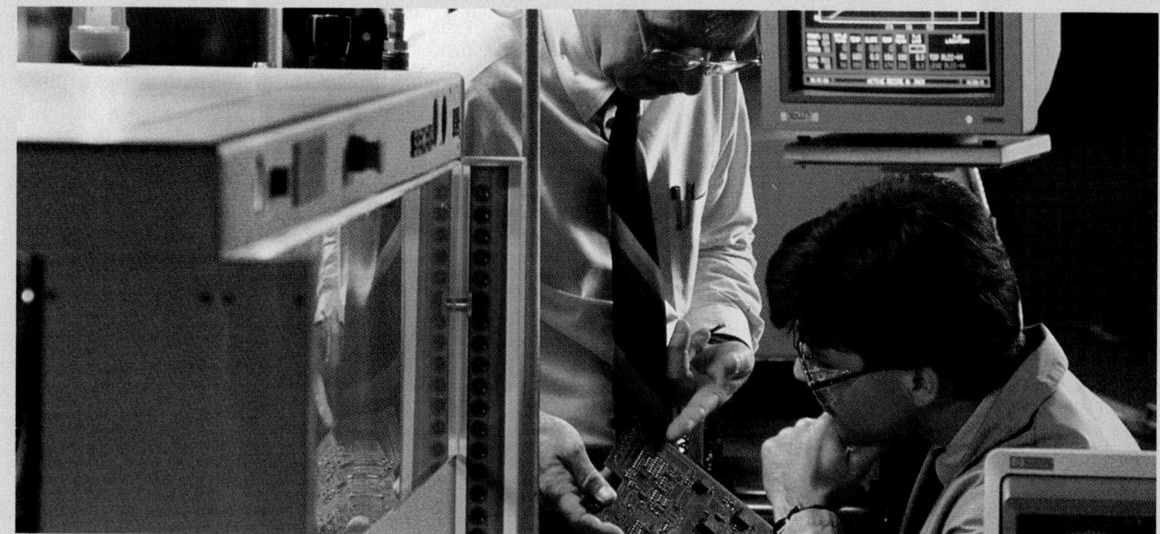

10–1 ■ THE BASIC INDUCTOR

In this section, the basic construction and characteristics of inductors are examined.

After completing this section, you should be able to

- **Describe the basic structure and characteristics of an inductor**
  - ☐ Explain how an inductor stores energy
  - ☐ Define *inductance* and state its unit
  - ☐ State Faraday's law
  - ☐ State Lenz's law
  - ☐ Discuss induced voltage
  - ☐ Specify how the physical characteristics affect inductance
  - ☐ Discuss winding resistance and winding capacitance

When a length of wire is formed into a coil, as shown in Figure 10–1, it becomes a basic **inductor.** Current through the coil produces an electromagnetic field. The magnetic lines of force around each loop (turn) in the **winding** of the coil effectively add to the lines of force around the adjoining loops, forming a strong magnetic field within and around the coil, as shown. The net direction of the total magnetic field creates a north and a south pole.

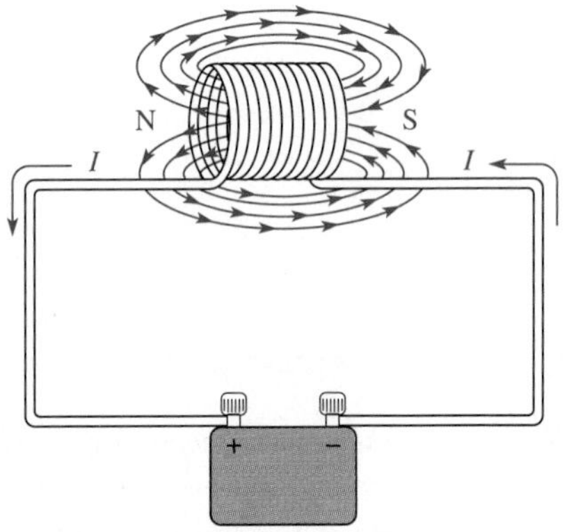

FIGURE 10–1
A coil of wire forms an inductor. When current flows through it, a three-dimensional electromagnetic field is created, surrounding the coil in all directions.

Self-Inductance

When there is current through an inductor, an electromagnetic field is established. When the current changes, the electromagnetic field also changes. An increase in current expands the field, and a decrease in current reduces it. Therefore, a changing current produces a changing electromagnetic field around the inductor (also known as **coil** or **choke**). In turn, the changing electromagnetic field causes an **induced voltage** across the coil in a direction to oppose the change in current. This property is called *self-inductance,* but is usually referred to as simply **inductance.** Inductance is symbolized by *L.*

> **Inductance is a measure of a coil's ability to establish an induced voltage as a result of a change in its current, and that induced voltage is in a direction to oppose that change in current.**

The Unit Inductance

The **henry,** symbolized by H, is the basic unit of inductance. By definition, the inductance is one henry when current through the coil, changing at the rate of one ampere per second, induces one volt across the coil. In many practical applications, millihenries (mH) and microhenries (μH) are the more common units. A schematic symbol for the inductor is shown in Figure 10–2.

FIGURE 10–2
A symbol for the inductor.

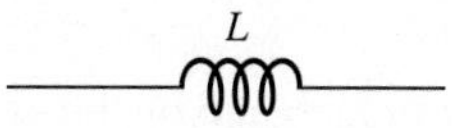

Energy Storage

An inductor stores energy in the magnetic field created by the current. The energy stored is expressed as follows:

$$W = \frac{1}{2}LI^2 \qquad \textbf{(10–1)}$$

As you can see, the energy stored is proportional to the inductance and the square of the current. When current (I) is in amperes and inductance (L) is in henries, the energy (W) is in joules.

Physical Characteristics of Inductors

The following characteristics are important in establishing the inductance of a coil: the permeability of the core material, the number of turns of wire, the length, and the cross-sectional area.

Core Material As discussed earlier, an inductor is basically a coil of wire. The material around which the coil is formed is called the **core.** Coils are wound on either nonmagnetic or magnetic materials. Examples of nonmagnetic materials are air, wood, copper, plastic, and glass. The permeabilities of these materials are the same as for a vacuum. Examples of magnetic materials are iron, nickel, steel, cobalt, or alloys. These materials have permeabilities that are hundreds or thousands of times greater than that of a vacuum and are classified as *ferromagnetic.* A ferromagnetic core provides a better path for the magnetic lines of force and thus permits a stronger magnetic field.

As you learned in Chapter 7, the permeability (μ) of the core material determines how easily a magnetic field can be established. *The inductance is directly proportional to the permeability of the core material.*

Physical Parameters As indicated in Figure 10–3, the number of turns of wire, the length, and the cross-sectional area of the core are factors in setting the value of inductance. The inductance is inversely proportional to the length of the core and directly proportional to the cross-sectional area. Also, the inductance is directly related to the number of turns squared. This relationship is as follows:

$$L = \frac{N^2 \mu A}{l} \qquad \textbf{(10–2)}$$

where L is the inductance in henries, N is the number of turns, μ is the permeability, A is the cross-sectional area in meters squared, and l is the core length in meters.

FIGURE 10–3
Factors that determine the inductance of a coil.

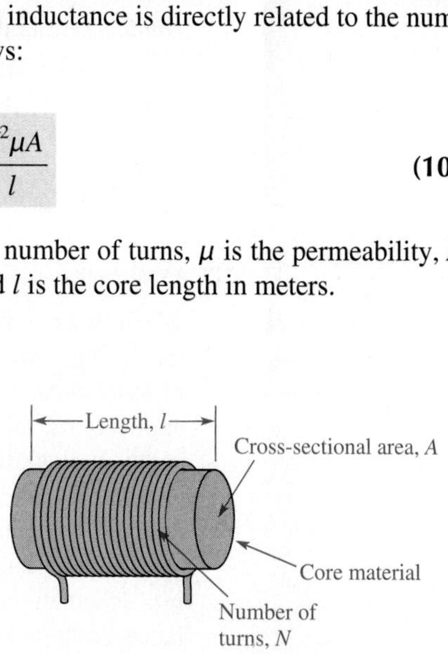

EXAMPLE 10–1 Determine the inductance of the coil in Figure 10–4. The permeability of the core is 0.25×10^{-3}.

FIGURE 10–4

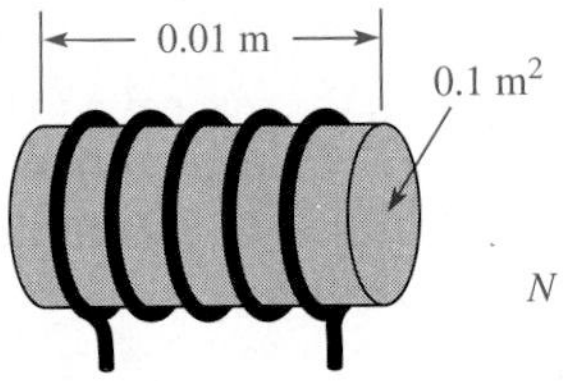

Solution

$$L = \frac{N^2\mu A}{l} = \frac{(5)^2(0.25 \times 10^{-3})(0.1)}{0.01} = 62.5 \text{ mH}$$

The calculator sequence is

5 x² × • 2 5 EXP +/− 3 × • 1 ÷ • 0 1 =

Related Exercise Determine the inductance of a coil with 12 turns around a core that is 0.05 m long and has a cross-sectional area of 0.15 m². The permeability is 0.25×10^{-3}.

Winding Resistance When a coil is made of a certain material, for example, insulated copper wire, that wire has a certain resistance per unit of length. When many turns of wire are used to construct a coil, the total resistance may be significant. This inherent resistance is called the *dc resistance* or the *winding resistance* (R_W). Although this resistance is distributed along the length of the wire, as shown in Figure 10–5(a), it is sometimes indicated in a schematic as resistance appearing in series with the inductance of the coil, as shown in Figure 10–5(b). In many applications, the winding resistance can be ignored and the coil considered as an ideal inductor. In other cases, the resistance must be considered.

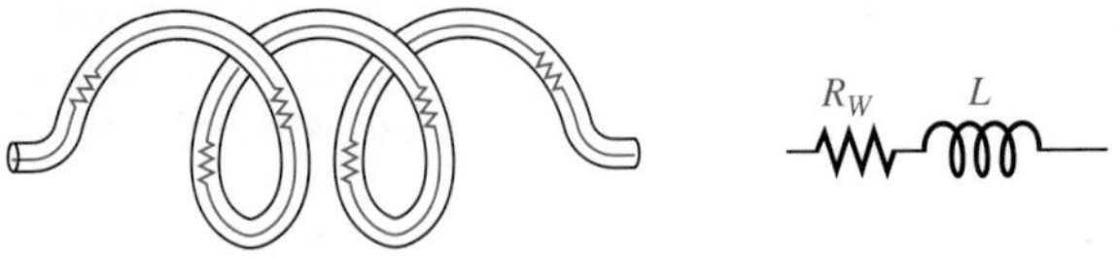

FIGURE 10–5
Winding resistance of a coil.

(a) The wire has resistance distributed along its length. (b) Equivalent circuit

Winding Capacitance

When two conductors are placed side by side, there is always some capacitance between them. Thus, when many turns of wire are placed close together in a coil, a certain amount of stray capacitance, called *winding capacitance* (C_W), is a natural side effect. In many applications, this winding capacitance is very small and has no significant effect. In other cases, particularly at high frequencies, it may become quite important.

The equivalent circuit for an inductor with both its winding resistance (R_W) and its winding capacitance (C_W) is shown in Figure 10–6. The capacitance effectively acts in parallel. The total of the stray capacitances between each loop of the winding is indicated in a schematic as a capacitance appearing in parallel with the coil and its winding resistance, as shown in Figure 10–6(b).

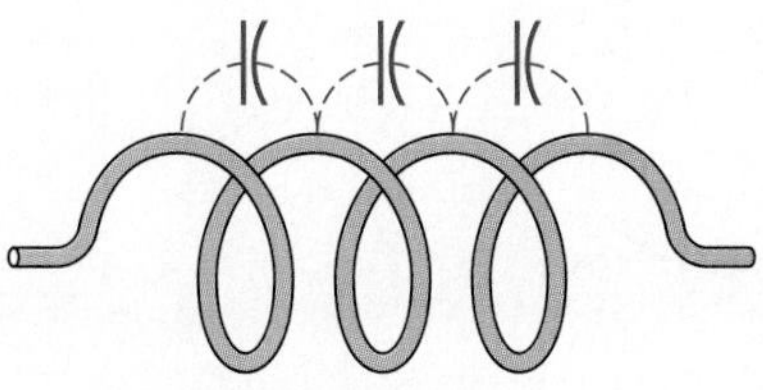

(a) Stray capacitance between each loop appears as a total parallel capacitance.

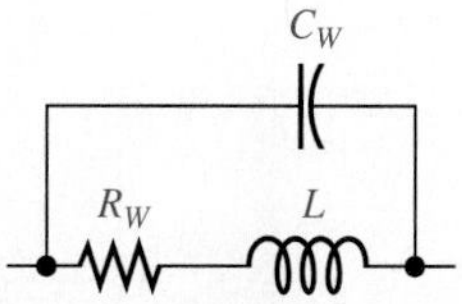

(b) Equivalent circuit

FIGURE 10–6
Winding capacitance of a coil.

Review of Faraday's law

The principle of Faraday's law was introduced in Chapter 7 and is reviewed here because of its importance in the study of inductors. Faraday found that by moving a magnet through a coil of wire, a voltage was induced across the coil, and that when a complete path was provided, the induced voltage caused an induced current.

The amount of induced voltage is directly proportional to the rate of change of the magnetic field with respect to the coil.

This principle is illustrated in Figure 10–7, where a bar magnet is moved through a coil of wire. An induced voltage is indicated by the voltmeter connected across the coil. The faster the magnet is moved, the greater is the induced voltage.

When a wire is formed into a certain number of loops or turns and is exposed to a changing magnetic field, a voltage is induced across the coil. The induced voltage is proportional to the number of turns of wire in the coil, *N,* and to the rate at which the magnetic field changes.

FIGURE 10–7
Induced voltage is created by a changing magnetic field.

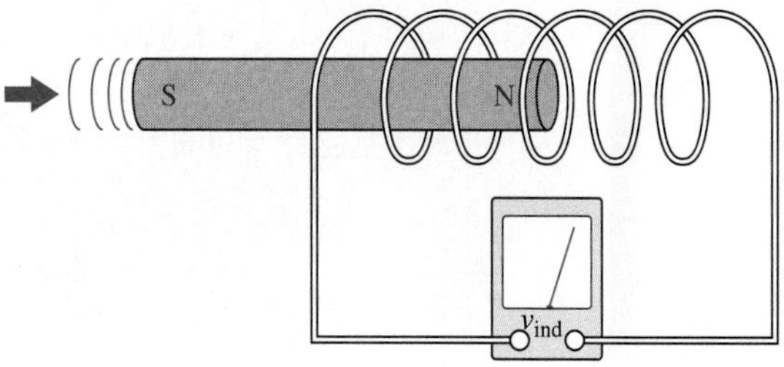

Lenz's Law

Lenz's law adds to Faraday's law by defining the direction of induced voltage as follows:

When the current through a coil changes and an induced voltage is created as a result of the changing magnetic field, the direction of the induced voltage is such that it always opposes the change in current.

Figure 10–8 illustrates Lenz's law. In part (a), the current is constant and is limited by R_1. There is no induced voltage because the magnetic field is unchanging. In part (b), the switch suddenly is closed, placing R_2 in parallel with R_1 and thus reducing the resistance. Naturally, the current tries to increase and the magnetic field begins to expand, but the induced voltage opposes this attempted increase in current for an instant.

In Figure 10–8(c), the induced voltage gradually decreases, allowing the current to increase. In part (d), the current has reached a constant value as determined by the parallel resistors, and the induced voltage is zero. In part (e), the switch has been sud-

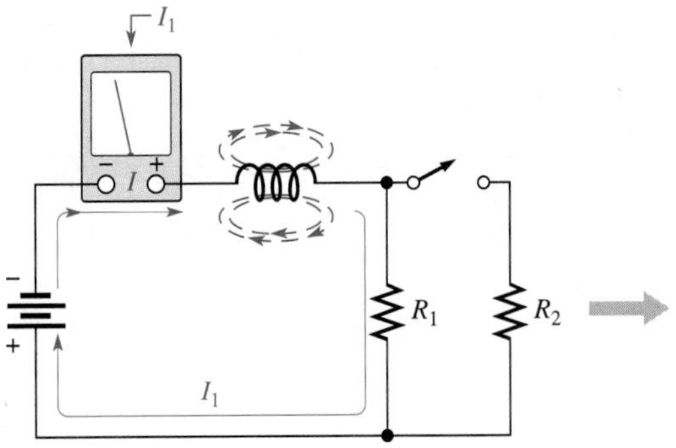

(a) Switch open: Constant current and constant magnetic field; no induced voltage.

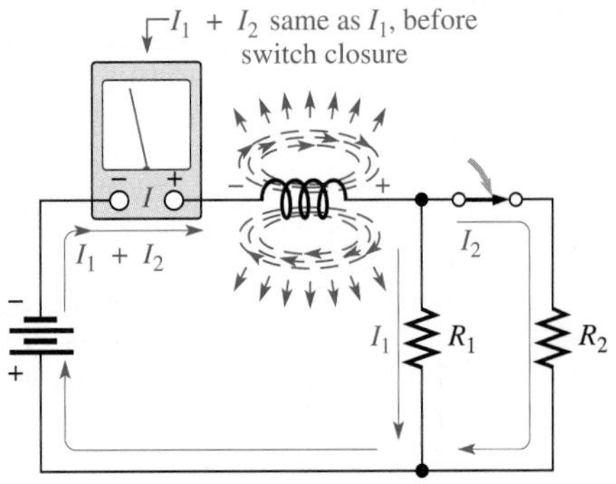

(b) At instant of switch closure: Expanding magnetic field induces voltage, which opposes increase in total current.

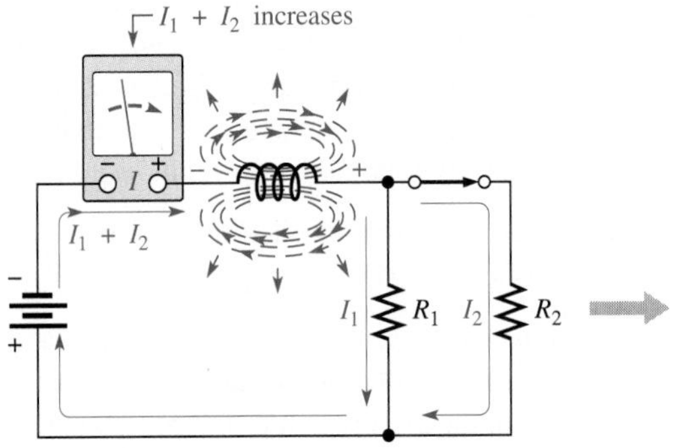

(c) Right after switch closure: The rate of expansion of the magnetic field decreases, allowing the current to increase exponentially as induced voltage decreases.

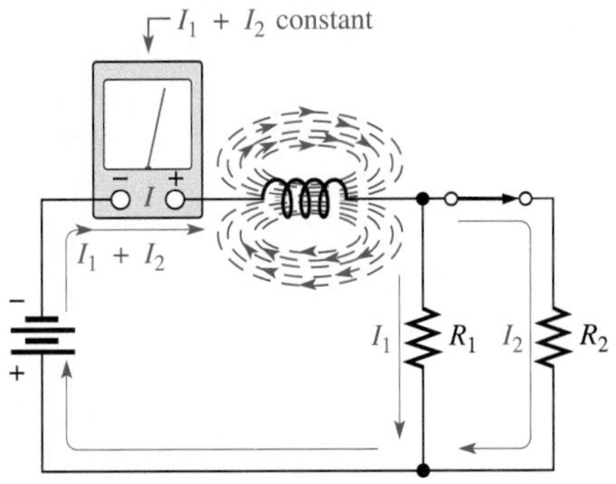

(d) Switch remains closed: Current and magnetic field reach constant value.

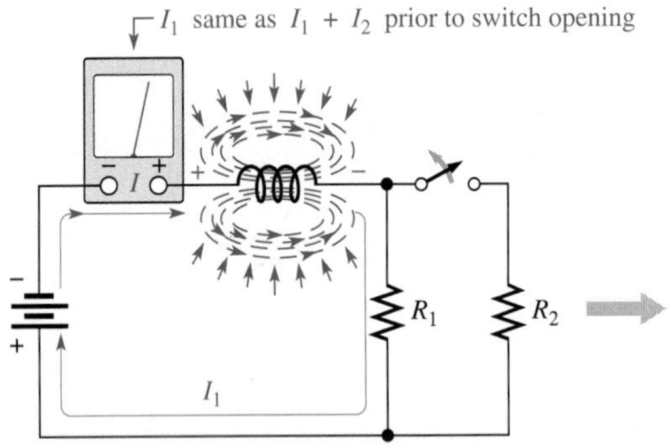

(e) At instant of switch opening: Magnetic field begins to collapse, creating an induced voltage, which opposes decrease in current.

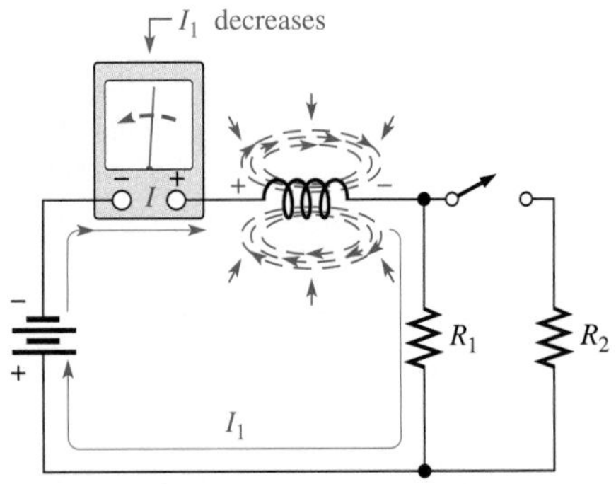

(f) After switch opening: Rate of collapse of magnetic field decreases, allowing current to decrease exponentially back to original value.

FIGURE 10–8
Demonstration of Lenz's law: When the current tries to change suddenly, the electromagnetic field changes and induces a voltage in a direction that opposes that change in current.

denly opened, and, for an instant, the induced voltage prevents any decrease in current. In part (f), the induced voltage gradually decreases, allowing the current to decrease back to a value determined by R_1. Notice that the induced voltage has a polarity that opposes any current change. The polarity of the induced voltage is opposite that of the battery voltage for an increase in current and aids the battery voltage for a decrease in current.

SECTION 10–1 REVIEW

1. List the parameters that contribute to the inductance of a coil.
2. Describe what happens to L when
   (a) N is increased
   (b) the core length is increased
   (c) the cross-sectional area of the core is decreased
   (d) a ferromagnetic core is replaced by an air core
3. Explain why inductors inevitably have some winding resistance.
4. Explain why inductors inevitably have some winding capacitance.

10–2 ■ TYPES OF INDUCTORS

Inductors normally are classified according to the type of core material. In this section, the basic types of inductors are examined.

After completing this section, you should be able to

- **Discuss various types of inductors**
  - ☐ Describe the basic types of fixed inductors
  - ☐ Distinguish between fixed and variable inductors

Inductors are made in a variety of shapes and sizes. Basically, they fall into two general categories: fixed and variable. The standard schematic symbols are shown in Figure 10–9.

Both fixed and variable inductors can be classified according to the type of core material. Three common types are the air core, the iron core, and the ferrite core. Each has a unique symbol, as shown in Figure 10–10.

Adjustable (variable) inductors usually have a screw-type adjustment that moves a sliding core in and out, thus changing the inductance. A wide variety of inductors exists, some of which are shown in Figure 10–11.

(a) Fixed (b) Variable

FIGURE 10–9
Symbols for (a) fixed and (b) variable inductors.

(a) Air core (b) Iron core (c) Ferrite core

FIGURE 10–10
Inductor symbols.

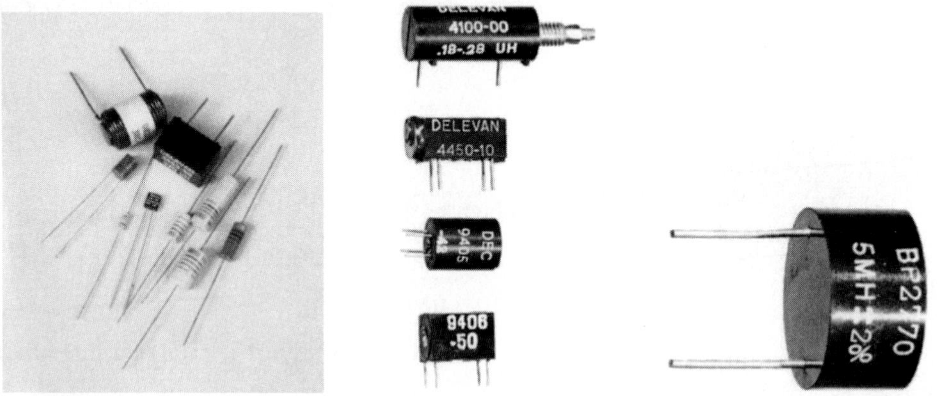

(a) (b) (c)

FIGURE 10–11
Typical inductors. (a) Fixed molded inductors; (b) variable coils; (c) toroid inductor. Parts (a) and (b) courtesy of Delevan/American Precision. Part (c) courtesy of Dale Electronics.

SECTION 10–2 REVIEW

1. Name two general categories of inductors.
2. Identify the inductor symbols in Figure 10–12.

FIGURE 10–12

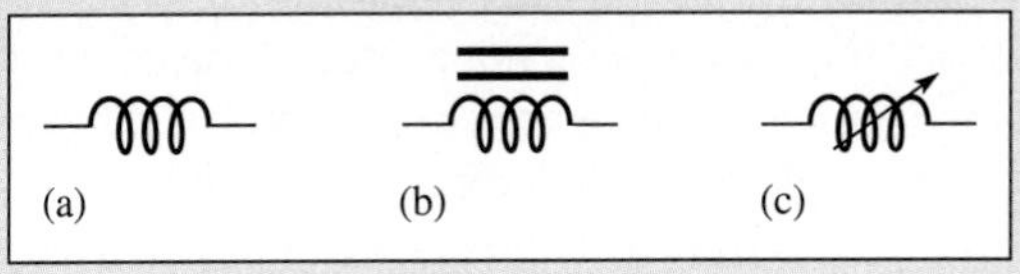

10–3 ■ SERIES INDUCTORS

In this section, you will see that when inductors are connected in series, the total inductance increases.

After completing this section, you should be able to

- **Analyze series inductors**
  - ☐ Determine total inductance

When inductors are connected in series, as in Figure 10–13, the total inductance, L_T, is the sum of the individual inductances. The formula for L_T is expressed in the following equation for the general case of n inductors in series:

$$L_T = L_1 + L_2 + L_3 + \cdots + L_n \tag{10–3}$$

Notice that the formula for inductance in series is similar to the formula for resistance in series (Equation (4–1)).

FIGURE 10–13
Inductors in series.

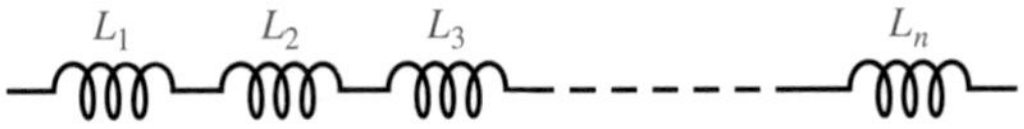

EXAMPLE 10–2 Determine the total inductance for each of the series connections in Figure 10–14.

FIGURE 10–14

Solution In Figure 10–14(a),

$$L_T = 1\text{ H} + 2\text{ H} + 1.5\text{ H} + 5\text{ H} = 9.5\text{ H}$$

In Figure 10–14(b),

$$L_T = 5\text{ mH} + 2\text{ mH} + 10\text{ mH} + 1\text{ mH} = 18\text{ mH}$$

Note: $1000\ \mu\text{H} = 1\text{ mH}$

Related Exercise What is the total inductance of ten 50 μH inductors in series?

SECTION 10–3 REVIEW

1. State the rule for combining inductors in series.
2. What is L_T for a series connection of 100 μH, 500 μH, and 2 mH?
3. Five 100 mH coils are connected in series. What is L_T?

10–4 ■ PARALLEL INDUCTORS

In this section, you will see that when inductors are connected in parallel, total inductance is reduced.

After completing this section, you should be able to

- **Analyze parallel inductors**
  - □ Determine total inductance

When inductors are connected in parallel, as in Figure 10–15, the total inductance is less than the smallest inductance. The general formula states that the reciprocal of the total inductance is equal to the sum of the reciprocals of the individual inductances:

$$\frac{1}{L_T} = \frac{1}{L_1} + \frac{1}{L_2} + \frac{1}{L_3} + \cdots + \frac{1}{L_n} \qquad \textbf{(10–4)}$$

Total inductance, L_T, can be found by taking the reciprocals of both sides of Equation (10–4):

$$L_T = \frac{1}{\left(\frac{1}{L_1}\right) + \left(\frac{1}{L_2}\right) + \left(\frac{1}{L_3}\right) + \cdots + \left(\frac{1}{L_n}\right)} \qquad \textbf{(10–5)}$$

This formula for total inductance in parallel is similar to that for total parallel resistance or total series capacitance.

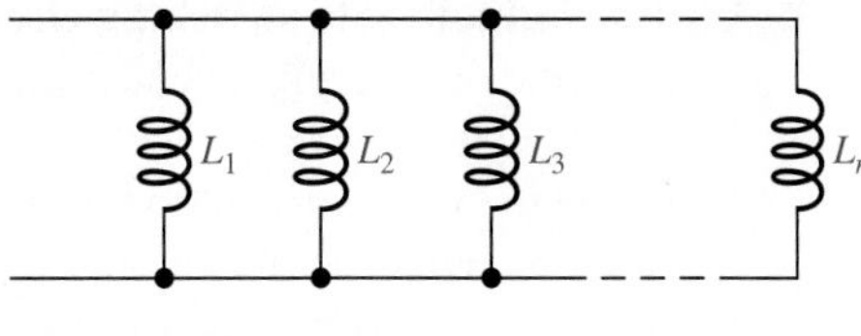

FIGURE 10–15
Inductors in parallel.

EXAMPLE 10–3 Determine L_T in Figure 10–16.

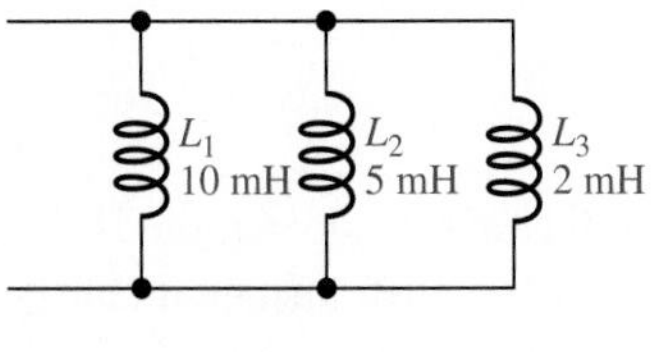

FIGURE 10–16

Solution

$$L_T = \cfrac{1}{\cfrac{1}{L_1} + \cfrac{1}{L_2} + \cfrac{1}{L_3}}$$

$$= \cfrac{1}{\left(\cfrac{1}{10\text{ mH}}\right) + \left(\cfrac{1}{5\text{ mH}}\right) + \left(\cfrac{1}{2\text{ mH}}\right)} = \frac{1}{0.8}\text{ mH} = 1.25\text{ mH}$$

The calculator sequence is

1 0 EXP +/− 3 1/x + 5 EXP +/− 3 1/x + 2 EXP +/− 3 1/x = 1/x

Related Exercise Determine L_T for the following inductors in parallel: 50 μH, 80 μH, 100 μH, and 150 μH.

SECTION 10–4 REVIEW

1. Compare the total inductance in parallel with the smallest-value individual inductor.
2. The calculation of total parallel inductance is similar to that for parallel resistance (T or F).
3. Determine L_T for each parallel combination:
   - **(a)** 100 mH, 50 mH, and 10 mH
   - **(b)** 40 μH and 60 μH
   - **(c)** Ten 1 H coils

10–5 ■ INDUCTORS IN DC CIRCUITS

An inductor will energize when it is connected to a dc voltage source. The buildup of current through the inductor occurs in a predictable manner, which is dependent on both the inductance and the resistance in a circuit.

After completing this section, you should be able to

- **Analyze inductive dc switching circuits**
  - □ Describe the energizing and deenergizing of an inductor
  - □ Define *time constant*
  - □ Relate the time constant to the energizing and deenergizing of an inductor
  - □ Describe induced voltage
  - □ Write the exponential equations for current in an inductor

When there is constant direct current in an inductor, there is no induced voltage. There is, however, a voltage drop due to the winding resistance of the coil. The inductance itself appears as a short to dc. Energy is stored in the magnetic field according to the formula $W = \frac{1}{2}LI^2$. The only energy loss occurs in the winding resistance ($P = I^2R_W$). This condition is illustrated in Figure 10–17.

The *RL* Time Constant

Because the inductor's basic action is to oppose a change in its current, it follows that current cannot change instantaneously in an inductor. A certain time is required

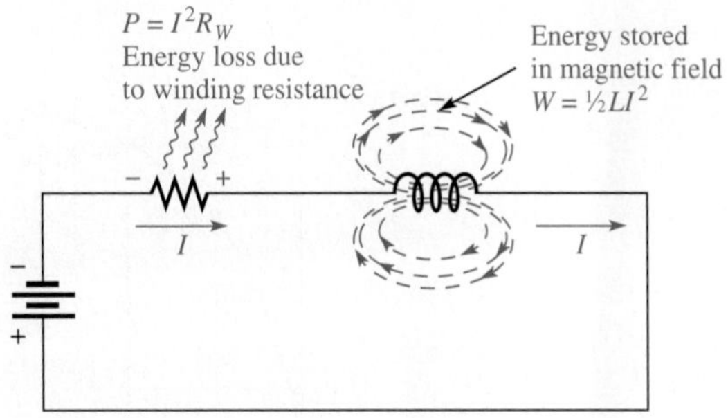

FIGURE 10–17
Energy storage and loss in an inductor. The only dc voltage drop across the coil is due to the winding resistance.

for the current to make a change from one value to another. The rate at which the current changes is determined by the **time constant.** The time constant for a series *RL* circuit is

$$\tau = \frac{L}{R} \qquad \textbf{(10–6)}$$

where τ is in seconds when L is in henries and R is in ohms.

EXAMPLE 10–4 A series *RL* circuit has a resistance of 1 kΩ and an inductance of 1 mH. What is the time constant?

Solution

$$\tau = \frac{L}{R} = \frac{1\text{ mH}}{1\text{ k}\Omega} = \frac{1 \times 10^{-3}\text{ H}}{1 \times 10^{-3}\ \Omega} = 1 \times 10^{-6}\text{ s} = 1\ \mu\text{s}$$

Related Exercise Find the time constant for $R = 2.2$ kΩ and $L = 500\ \mu$H.

Energizing Current in an Inductor

In a series *RL* circuit, the current will increase to 63% of its full value in one time-constant interval after the switch is closed. This buildup of current is analogous to the buildup of capacitor voltage during the charging in an *RC* circuit; they both follow an exponential curve and reach the approximate percentages of final value as indicated in Table 10–1 and as illustrated in Figure 10–18.

TABLE 10–1
Percentage of final current after each time-constant interval during current buildup

NUMBER OF TIME CONSTANTS	% OF FINAL VALUE
1	63
2	86
3	95
4	98
5	99 (considered 100%)

The change in current over five time-constant intervals is illustrated in Figure 10–19. When the current reaches its final value at approximately 5τ, it ceases to change. At this

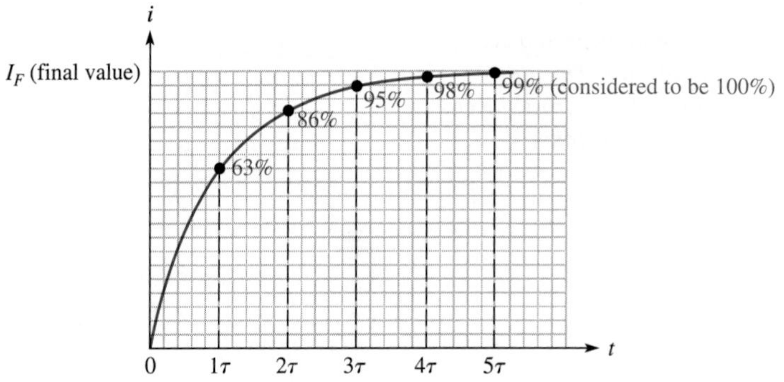

FIGURE 10–18
Energizing current in an inductor.

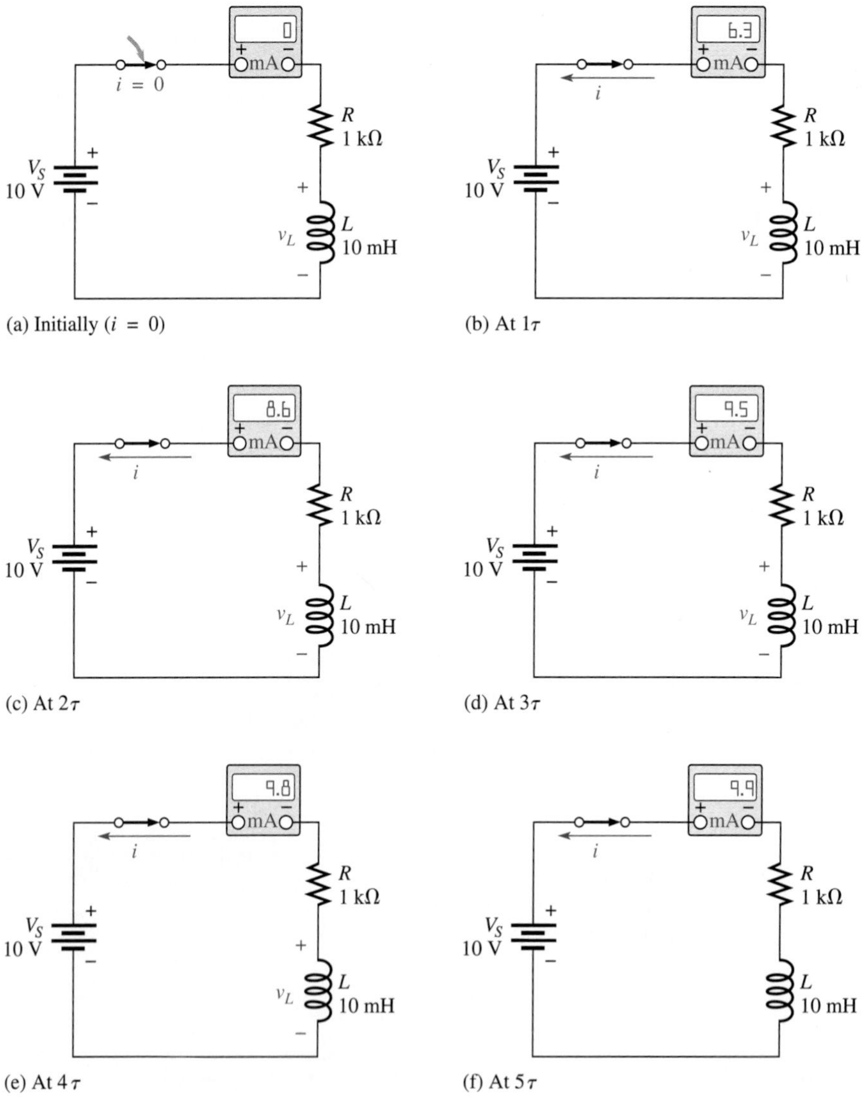

FIGURE 10–19
Illustration of the exponential buildup of current in an inductor. The current increases another 63% during each time-constant interval. A voltage (v_L) is induced in the coil that tends to oppose the increase in current.

time, the inductor acts as a short (except for winding resistance) to the constant current. The final value of the current is

$$I_F = \frac{V_S}{R_W} = \frac{10\text{ V}}{1\text{ k}\Omega} = 10\text{ mA}$$

EXAMPLE 10–5 Calculate the time constant for Figure 10–20. Then determine the current and the time at each time-constant interval, measured from the instant the switch is closed.

FIGURE 10–20

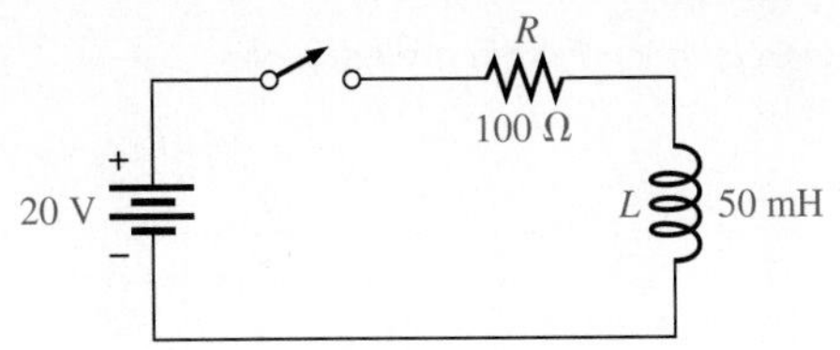

Solution The time constant is

$$\tau = \frac{L}{R} = \frac{50\text{ mH}}{100\ \Omega} = 500\ \mu\text{s}$$

The final current is

$$I_F = \frac{V_S}{R} = \frac{20\text{ V}}{100\ \Omega} = 200\text{ mA}$$

Using the time-constant percentage values from Table 10–1,

At $1\tau = 500\ \mu$s: $i = 0.63(200\text{ mA}) = 126\text{ mA}$
At $2\tau = 1$ ms: $i = 0.86(200\text{ mA}) = 172\text{ mA}$
At $3\tau = 1.5$ ms: $i = 0.95(200\text{ mA}) = 190\text{ mA}$
At $4\tau = 2$ ms: $i = 0.98(200\text{ mA}) = 196\text{ mA}$
At $5\tau = 2.5$ ms: $i = 0.99(200\text{ mA}) = 198\text{ mA} \cong 200\text{ mA}$

Related Exercise Repeat the calculations if $R = 680\ \Omega$ and $L = 100\ \mu$H.

Deenergizing Current in an Inductor

Current in an inductor decreases exponentially according to the approximate percentage values shown in Table 10–2 and in Figure 10–21.

Figure 10–22(a) shows a constant current of 1 A (1000 mA) through an inductor. Switch 1 (SW_1) is opened and switch 2 (SW_2) is closed simultaneously, and for an instant

TABLE 10–2
Percentage of initial current after each time-constant interval while current is decreasing

NUMBER OF TIME CONSTANTS	% OF INITIAL VALUE
1	37
2	14
3	5
4	2
5	1 (considered 0)

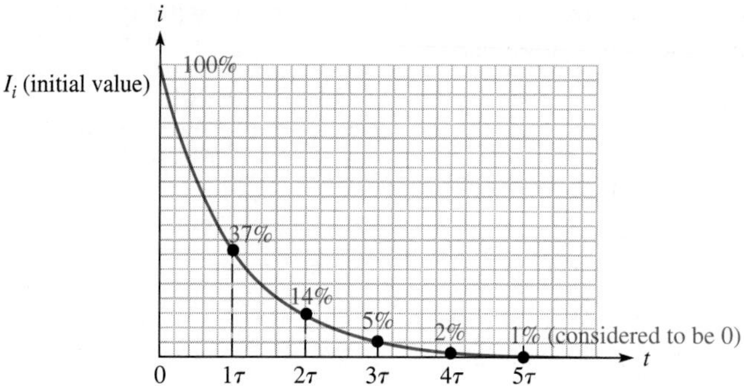

FIGURE 10–21
Deenergizing current in an inductor.

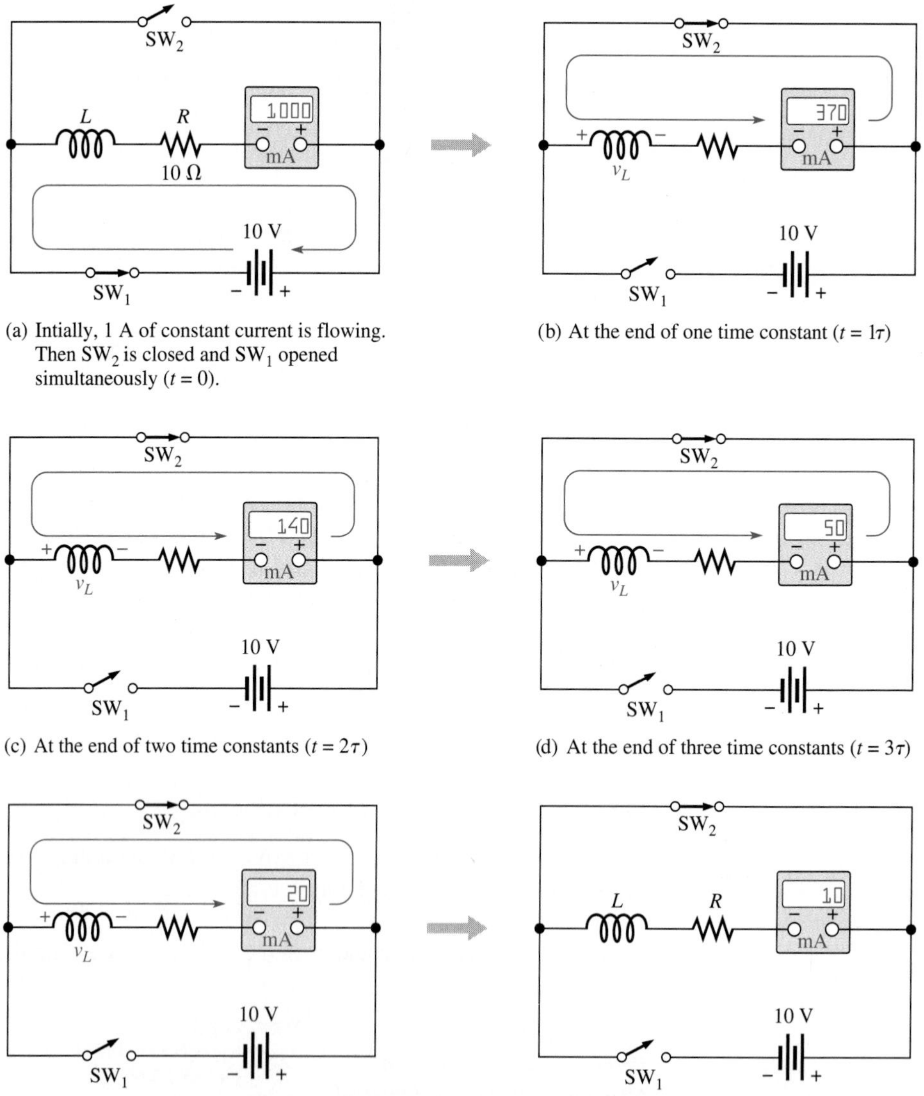

(a) Intially, 1 A of constant current is flowing. Then SW_2 is closed and SW_1 opened simultaneously ($t = 0$).

(b) At the end of one time constant ($t = 1\tau$)

(c) At the end of two time constants ($t = 2\tau$)

(d) At the end of three time constants ($t = 3\tau$)

(e) At the end of four time constants ($t = 4\tau$)

(f) At the end of five time constants ($t = 5\tau$). Since only 1% of the current is left, this value is taken as the final zero value.

FIGURE 10–22
Illustration of the exponential decrease of current in an inductor. The current decreases another 63% during each time constant interval.

the induced voltage keeps the 1 A flowing. During the first time-constant interval, the current decreases by 63% down to 370 mA (37% of its initial value), as indicated in Figure 10–22(b). During the second time-constant interval, the current decreases by another 63% to 140 mA (14% of its initial value), as shown in Figure 10–22(c). The continued decrease in the current is illustrated by the remaining parts of Figure 10–22. Part (f) shows that only 1% of the initial current is left at the end of five time constants. Traditionally, this value is accepted as the final value and is approximated as zero current. Notice that until after the five time constants have elapsed, there is an induced voltage across the coil which is trying to keep the current flowing. This voltage follows a decreasing exponential curve, as will be discussed later.

EXAMPLE 10–6 In Figure 10–23, switch 1 (SW_1) is opened at the instant that switch 2 (SW_2) is closed.

(a) What is the time constant?
(b) What is the initial coil current at the instant of switching?
(c) What is the coil current at 1τ?

Assume steady-state current through the coil prior to switch change.

FIGURE 10–23

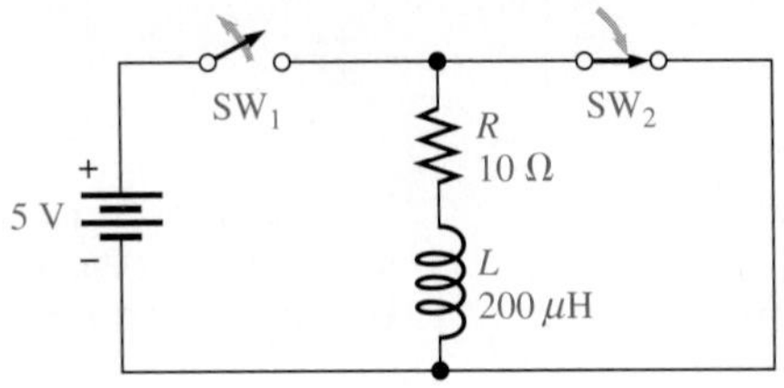

Solution

(a) $\tau = \dfrac{L}{R} = \dfrac{200\ \mu\text{H}}{10\ \Omega} = 20\ \mu\text{s}$

(b) Current cannot change instantaneously in an inductor. Therefore, the current at the instant of the switch change is the same as the steady-state current:

$$I = \frac{5\text{ V}}{10\ \Omega} = 500\text{ mA}$$

(c) At 1τ, the current has decreased to 37% of its initial value:

$$i = 0.37(500\text{ mA}) = 185\text{ mA}$$

Related Exercise Change R to 47 Ω and L to 1 mH in Figure 10–23 and repeat each calculation.

Induced Voltage in the Series *RL* Circuit

As you know, when current changes in an inductor, a voltage is induced. We now examine what happens to the voltages across the resistor and the coil in a series circuit when a change in current occurs.

Look at the circuit in Figure 10–24(a). When the switch is open, there is no current, and the resistor voltage and the coil voltage are both zero. At the instant the switch is closed, as indicated in part (b), V_R is zero and V_L is 10 V. The reason for this change is

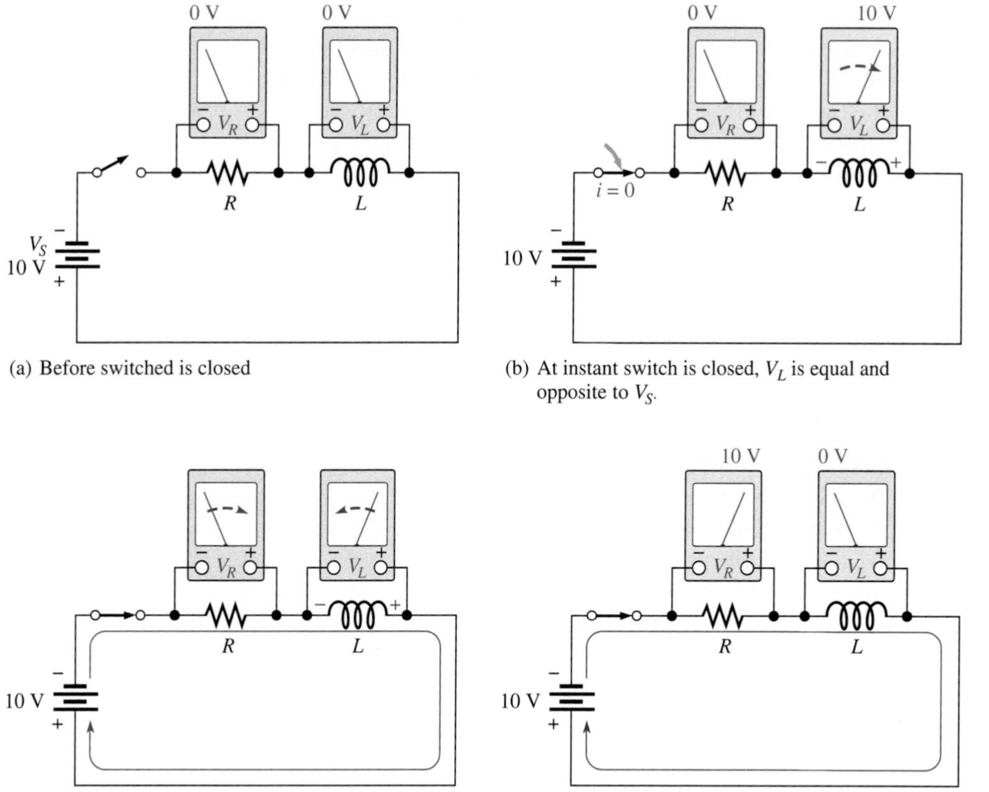

(a) Before switched is closed

(b) At instant switch is closed, V_L is equal and opposite to V_S.

(c) During the first five time constants, V_R increases exponentially with current, and V_L decreases exponentially.

(d) After the first five time constants, $V_R = 10$ V and $V_R \approx 0$ V. The current is at a constant maximum value.

FIGURE 10–24
Illustration of how the coil voltage and the resistor voltage change in response to an increase in current.

that the induced voltage across the coil is equal and opposite to the applied voltage, preventing the current from changing instantaneously. Therefore, *at the instant of switch closure, L effectively acts as an open with all the applied voltage across it.*

During the first five time constants, the current is building up exponentially, and the induced coil voltage is decreasing. The resistor voltage increases with the current, as Figure 10–24(c) illustrates.

After five time constants have elapsed, the current has reached its final value, V_S/R. At this time, all of the applied voltage is dropped across the resistor and none across the coil. Thus, L effectively acts as a short to nonchanging current, as Figure 10–24(d) illustrates. Keep in mind that the inductor always reacts to a change in current by creating an induced voltage in order to counteract that change.

Now let's examine the case illustrated in Figure 10–25, where the current is switched out, and the inductor discharges through another path. Part (a) shows the steady-state condition, and part (b) illustrates the instant at which the source is removed by opening SW_1 and the discharge path is connected by the closure of SW_2. There was 1 A through L prior to this. Notice that 10 V are induced in L in the direction to aid the 1 A in an effort to keep it from changing. Then, as shown in Figure 10–25(c), the current decays exponentially, and so do V_R and V_L. After 5τ, as shown in Figure 10–25(d), all the energy stored in the magnetic field of L is dissipated, and all values are zero.

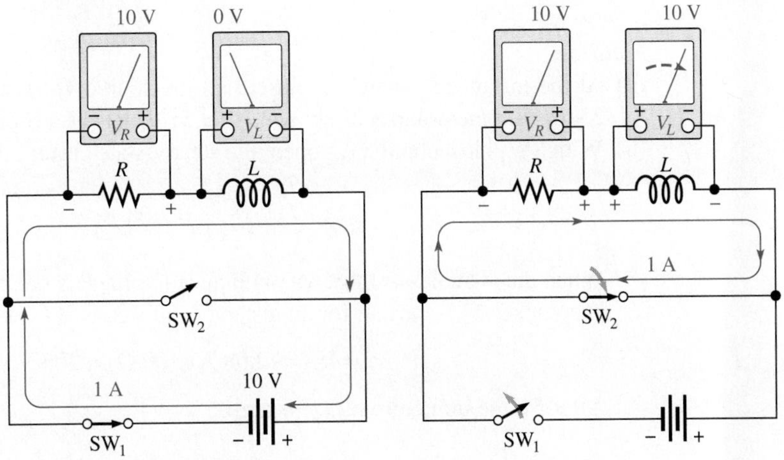

(a) Initially, a constant 1 A flows and the voltage is dropped across R.

(b) At the instant that SW_1 is opened and SW_2 closed, 10 V is induced across L.

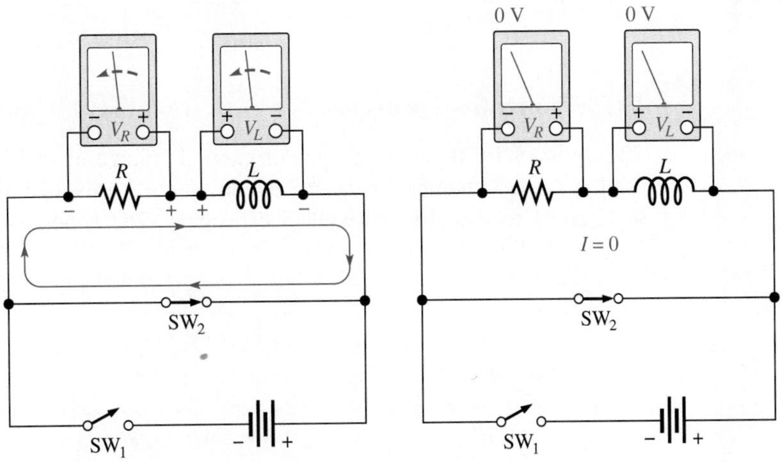

(c) During the five-time-constant interval, V_R and V_L decrease exponentially with the current.

(d) After the five time constants, V_R, V_L, and I are all zero.

FIGURE 10–25
Illustration of how the coil voltage and the resistor voltage change in response to a decrease in current.

EXAMPLE 10–7

(a) In Figure 10–26(a), what is v_L at the instant the switch (SW) is closed? What is v_L after 5τ?

(b) In Figure 10–26(b), what is v_L at the instant SW_1 opens and SW_2 closes? What is v_L after 5τ?

FIGURE 10–26

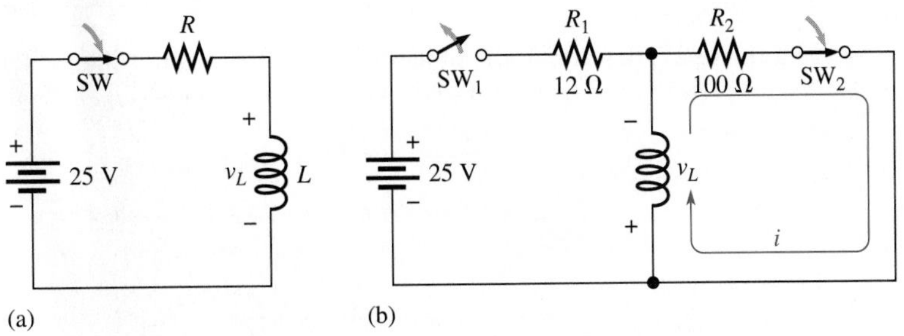

Solution

(a) At the instant the switch is closed, all the source voltage is across L. Thus, $v_L =$ 25 V, with the polarity as shown. After 5τ, L acts as a short, so $v_L = 0$ V.

(b) With SW_1 closed and SW_2 open, the steady-state current is

$$I = \frac{25\text{ V}}{12\ \Omega} = 2.08\text{ A}$$

When the switches are thrown, an induced voltage is created across L sufficient to keep this 2.08 A current for an instant. In this case, it takes an induced voltage

$$v_L = IR_2 = (2.08\text{ A})(100\ \Omega) = 208\text{ V}$$

After 5τ, the inductor voltage is zero.

Related Exercise Repeat part (a) if the voltage source is 9 V. Repeat part (b) if R_1 is 27 Ω with the source voltage at 25 V.

The Exponential Formulas

The formulas for the exponential current and voltage in an *RL* circuit are similar to those used in the last chapter for the *RC* circuit, and the universal exponential curves in Figure 9–36 apply to inductors as well as capacitors. The general formulas for *RL* circuits are stated as follows:

$$v = V_F + (V_i - V_F)e^{-Rt/L} \qquad (10\text{–}7)$$

$$i = I_F + (I_i - I_F)e^{-Rt/L} \qquad (10\text{–}8)$$

where V_F and I_F are the final values, V_i and I_i are the initial values, and v and i are the instantaneous values of the inductor voltage or current at time t.

Increasing Current The formula for the special case in which an increasing exponential current curve begins at zero ($I_i = 0$) is

$$i = I_F(1 - e^{-Rt/L}) \qquad (10\text{–}9)$$

Using Equation (10–9), you can calculate the value of the increasing inductor current at any instant of time. The same is true for voltage.

EXAMPLE 10–8 In Figure 10–27, determine the inductor current 30 μs after the switch is closed.

FIGURE 10–27

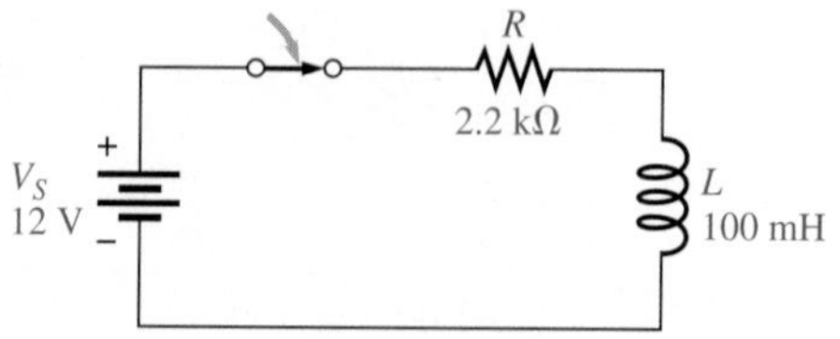

Solution The time constant is

$$\tau = \frac{L}{R} = \frac{100\text{ mH}}{2.2\text{ k}\Omega} = 45.5\ \mu\text{s}$$

The final current is

$$I_F = \frac{V_S}{R} = \frac{12\text{ V}}{2.2\text{ k}\Omega} = 5.45\text{ mA}$$

The initial current is zero. Notice that 30 μs is less than one time constant, so the current will reach less than 63% of its final value in that time.

$$i_L = I_F(1 - e^{-Rt/L}) = 5.45\text{ mA}(1 - e^{-0.66}) = 5.45\text{ mA}(1 - 0.517) = 2.63\text{ mA}$$

The calculator sequence is

3 0 EXP +/− 6 ÷ 4 5 • 5 EXP +/− 6 = +/− 2nd ln x +/−
+ 1 = × 5 • 4 5 EXP +/− 3 =

Related Exercise In Figure 10–27, determine the inductor current 55 μs after the switch is closed.

Decreasing Current The formula for the special case in which a decreasing exponential current has a final value of zero is as follows:

$$i = I_i e^{-Rt/L} \tag{10–10}$$

This formula can be used to calculate the deenergizing current at any instant, as the following example shows.

EXAMPLE 10–9 Determine the inductor current in Figure 10–28 at a point in time 2 ms after the switches are thrown (SW_1 opened and SW_2 closed).

FIGURE 10–28

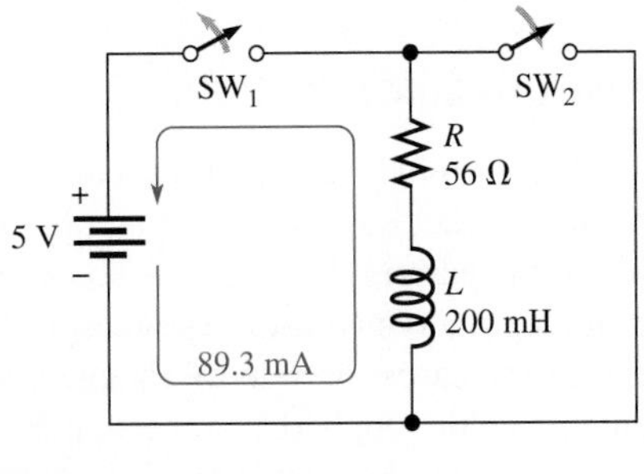

Solution The deenergizing time constant is

$$\tau = \frac{L}{R} = \frac{200\text{ mH}}{56\ \Omega} = 3.57\text{ ms}$$

The initial current in the inductor is 89.3 mA. Notice that 2 ms is less than one time constant, so the current will show a decrease less than 63%. Therefore, the current will be greater than 37% of its initial value at 2 ms after the switches are thrown.

$$i_L = I_i e^{-Rt/L} = (89.3\text{ mA})e^{-0.56} = 51.0\text{ mA}$$

The calculator sequence is

[2] [EXP] [+/−] [3] [÷] [3] [•] [5] [7] [EXP] [+/−] [3] [=] [+/−] [2nd] [ln x]
[×] [8] [9] [•] [3] [EXP] [+/−] [3] [=]

Related Exercise Determine the inductor current in Figure 10–28 at a point in time 6 ms after SW_1 is opened and SW_2 is closed if the source voltage is changed to 10 V.

SECTION 10–5 REVIEW

1. A 15 mH inductor with a winding resistance of 10 Ω has a constant direct current of 10 mA through it. What is the voltage drop across the inductor?
2. A 20 V dc source is connected to a series *RL* circuit with a switch. At the instant of switch closure, what are the values of v_R and v_L?
3. In the same circuit, after a time interval equal to 5τ from switch closure, what are v_R and v_L?
4. In a series *RL* circuit where $R = 1$ kΩ and $L = 500$ μH, what is the time constant? Determine the current 0.25 μs after a switch connects 10 V across the circuit.

10–6 ■ INDUCTORS IN AC CIRCUITS

You will learn in this section that an inductor passes ac but with an amount of opposition that depends on the frequency of the ac.

After completing this section, you should be able to

- **Analyze inductive ac circuits**
  - □ Define *inductive reactance*
  - □ Determine the value of inductive reactance in a given circuit
  - □ Discuss instantaneous, true, and reactive power in an inductor

Inductive Reactance, X_L

In Figure 10–29, an inductor is shown connected to a sinusoidal voltage source. When the source voltage is held at a constant amplitude value and its frequency is increased, the amplitude of the current decreases. Also, when the frequency of the source is decreased, the current amplitude increases. The reason is as follows: When the frequency of the applied voltage increases, its rate of change also increases, as you already know. Now, if the frequency of the applied voltage is increased, the frequency of the current also increases. According to Faraday's and Lenz's laws, this increase in frequency induces more voltage across the inductor in a direction to oppose the current and cause it to decrease in amplitude. Similarly, a decrease in frequency will cause an increase in current.

A decrease in the amount of current for an increase in frequency with a fixed amount of voltage indicates that opposition to the current has increased. Thus, the inductor offers opposition to current which varies directly with frequency.

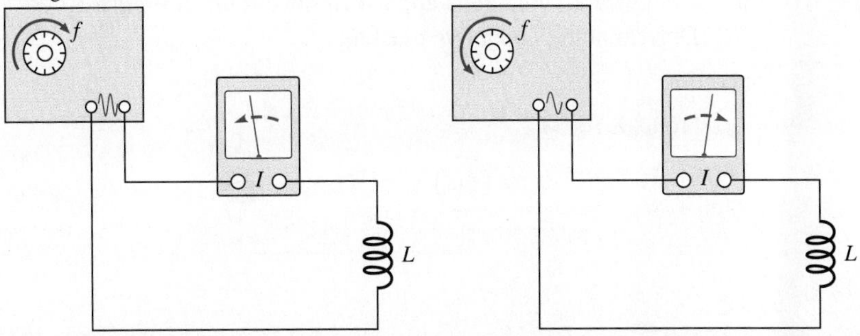

(a) Current decreases when the frequency increases.

(b) Current increases when the frequency decreases.

FIGURE 10–29
The current in an inductive circuit varies inversely with the frequency of the applied voltage.

The opposition to sinusoidal current in an inductor is called inductive reactance.

The symbol for **inductive reactance** is X_L, and its unit is the ohm (Ω).

You have just seen how frequency affects the opposition to current (inductive reactance) in an inductor. Now let's see how the inductance, L, affects the reactance. Figure 10–30(a) shows that when a sinusoidal voltage with a fixed amplitude and fixed frequency is applied to a 1 mH inductor, a certain amount of alternating current flows. When the inductance value is increased to 2 mH, the current decreases, as shown in part (b). Thus, when the inductance is increased, the opposition to current (inductive reactance) increases. So not only is the inductive reactance *directly* proportional to frequency, but it is also *directly* proportional to inductance. This relationship can be stated as follows:

X_L is proportional to fL.

It can be proven that the constant of proportionality is 2π, so the formula for inductive reactance (X_L) is

$$X_L = 2\pi f L \quad \textbf{(10–11)}$$

As with capacitive reactance, the 2π term comes from the relationship of the sine wave to rotational motion.

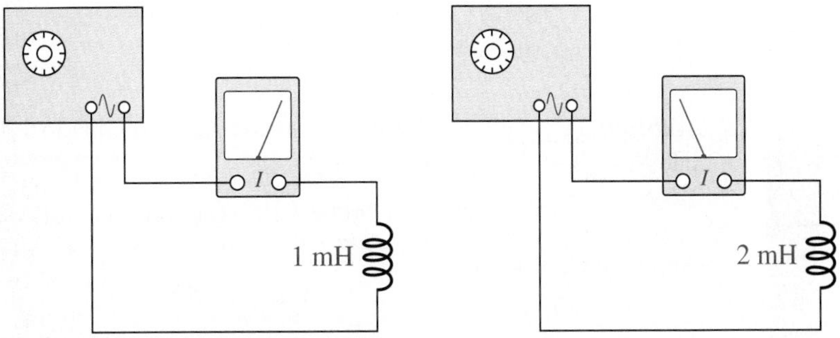

(a) Less inductance, more current

(b) More inductance, less current.

FIGURE 10–30
For a fixed voltage and fixed frequency, the current varies inversely with the inductance value.

EXAMPLE 10–10 A sinusoidal voltage is applied to the circuit in Figure 10–31. The frequency is 1 kHz. Determine the inductive reactance.

FIGURE 10–31

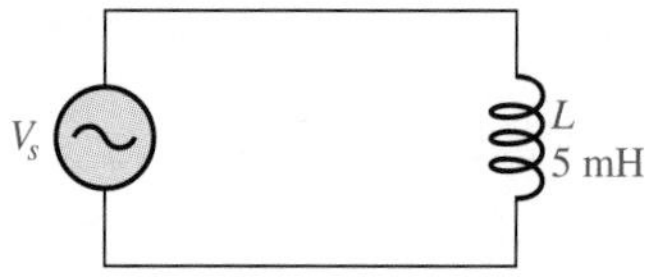

Solution Convert 1 kHz to 1×10^3 Hz and 5 mH to 5×10^{-3} H. Then, the inductive reactance is

$$X_L = 2\pi fL = 2\pi(1 \times 10^3\text{ Hz})(5 \times 10^{-3}\text{ H}) = 31.4\ \Omega$$

Related Exercise What is X_L in Figure 10–31 if the frequency is increased to 3.5 kHz?

Ohm's Law in Inductive AC Circuits

The reactance of an inductor is analogous to the resistance of a resistor. In fact, X_L, just like X_C and R, is expressed in ohms. Since inductive reactance is a form of opposition to current. Ohm's law applies to inductive circuits as well as to resistive circuits and capacitive circuits, and it is stated as follows:

$$V = IX_L \tag{10–12}$$

When applying Ohm's law in ac circuits, you must express both the current and the voltage in the same way, that is, both in rms, both in peak, and so on.

EXAMPLE 10–11 Determine the rms current in Figure 10–32.

FIGURE 10–32

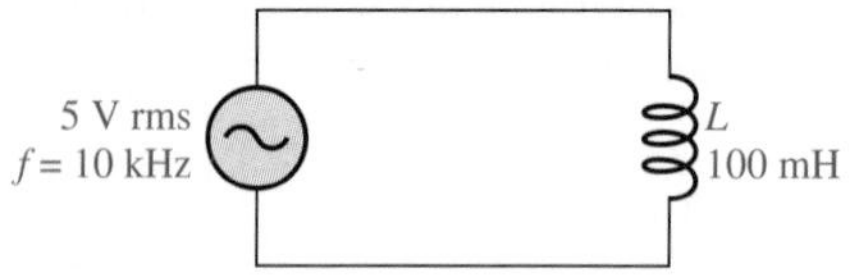

Solution Convert 10 kHz to 10×10^3 Hz and 100 mH to 100×10^{-3} H. Then calculate X_L:

$$X_L = 2\pi fL = 2\pi(10 \times 10^3\text{ Hz})(100 \times 10^{-3}\text{ H}) = 6283\ \Omega$$

Applying Ohm's law,

$$I_{rms} = \frac{V_{rms}}{X_L} = \frac{5\text{ V}}{6283\ \Omega} = 796\ \mu\text{A}$$

Related Exercise Determine the rms current in Figure 10–32 for the following values: $V_{rms} = 12$ V, $f = 4.9$ kHz, and $L = 680\ \mu$H.

Current Lags Inductor Voltage by 90°

As you know, a sinusoidal voltage has a maximum rate of change at its zero crossings and a zero rate of change at the peaks. From Faraday's law (Chapter 7) you know that the amount of voltage induced across a coil is directly proportional to the rate at which the current is changing. Therefore, the coil voltage is maximum at the zero crossings of the current where the rate of change of the current is the greatest. Also, the amount of voltage is zero at the peaks of the current where the rate of change is zero. This phase relationship is illustrated in Figure 10–33. As you can see, the current peaks occur a quarter cycle after the voltage peaks. Thus, the current lags the voltage by 90°.

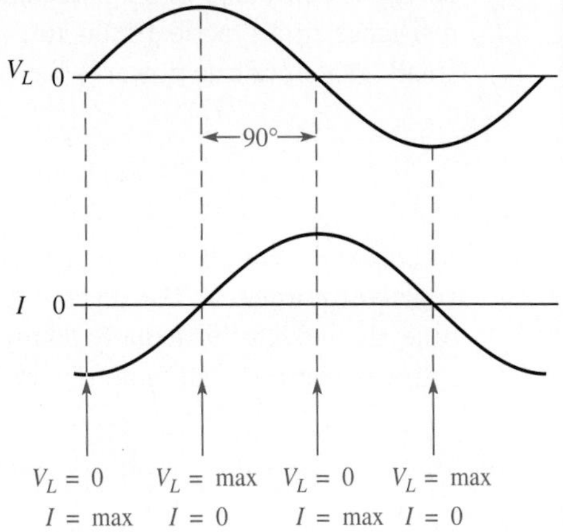

FIGURE 10–33
Current is always lagging the inductor voltage by 90°.

Power in an Inductor

As discussed earlier, an inductor stores energy in its magnetic field when there is current through it. An ideal inductor (assuming no winding resistance) does not dissipate energy; it only stores it. When an ac voltage is applied to an inductor, energy is stored by the inductor during a portion of the cycle; then the stored energy is returned to the source during another portion of the cycle. There is no net energy loss. Figure 10–34 shows the power curve that results from one cycle of inductor current and voltage.

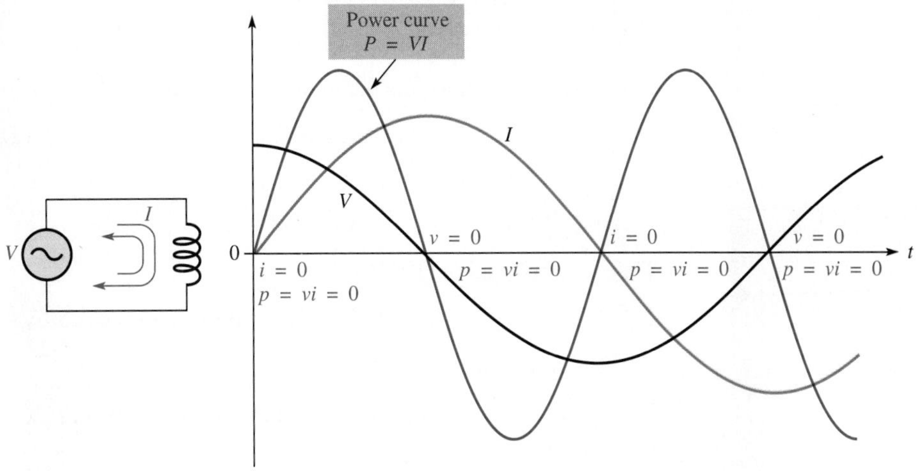

FIGURE 10–34
Power curve for an inductor.

Instantaneous Power **(p)** The product of instantaneous voltage, v, and instantaneous current, i, gives instantaneous power, p. At points where v or i is zero, p is also zero. When both v and i are positive, p is also positive. When either v or i is positive and the other negative, p is negative. When both v and i are negative, p is positive. As you can see in Figure 10–34, the power follows a sinusoidal-type curve. Positive values of power indicate that energy is stored by the inductor. Negative values of power indicate that energy is returned from the inductor to the source. Note that the power fluctuates at a frequency twice that of the voltage or current, as energy is alternately stored and returned to the source.

True Power **(P_{true})** Ideally, all of the energy stored by an inductor during the positive portion of the power cycle is returned to the source during the negative portion. No net energy is consumed in the inductance, so the power is zero. Actually, because of winding resistance in a practical inductor, some power is always dissipated, and there is a very small amount of true power, which can normally be neglected.

$$P_{true} = I^2_{rms}R_W \tag{10–13}$$

Reactive Power **(P_r)** The rate at which an inductor stores or returns energy is called its **reactive power, P_r**. The reactive power is a nonzero quantity, because at any instant in time, the inductor is actually taking energy from the source or returning energy to it. Reactive power does not represent an energy loss. The following formulas apply:

$$P_r = V_{rms}I_{rms} \tag{10–14}$$

$$P_r = \frac{V^2_{rms}}{X_L} \tag{10–15}$$

$$P_r = I^2_{rms}X_L \tag{10–16}$$

The unit of reactive power is volt-ampere reactive (VAR).

EXAMPLE 10–12 A 10 V rms signal with a frequency of 1 kHz is applied to a 10 mH coil with a winding resistance of 5 Ω. Determine the reactive power (P_r) and the true power (P_{true}).

Solution First, calculate the inductive reactance and current values:

$$X_L = 2\pi fL = 2\pi(1\text{ kHz})(10\text{ mH}) = 62.8\ \Omega$$

$$I = \frac{V_S}{X_L} = \frac{10\text{ V}}{62.8\ \Omega} = 159\text{ mA}$$

Then using Equation (10–16),

$$P_r = I^2X_L = (159\text{ mA})^2(62.8\ \Omega) = 1.59\text{ VAR}$$

The true power is

$$P_{true} = I^2R_W = (159\text{ mA})^2\,(5\ \Omega) = 126\text{ mW}$$

Related Exercise What happens to the reactive power if the frequency increases?

The Quality Factor (Q) of a Coil

The **quality factor (Q)** is the ratio of the reactive power in the inductor to the true power in the winding resistance of the coil or the resistance in series with the coil. It is a ratio of the power in L to the power in R_W. The quality factor is very important in resonant circuits, which are studied in Chapter 14. A formula for Q is developed as follows:

$$Q = \frac{\text{reactive power}}{\text{true power}} = \frac{P_r}{P_{\text{true}}} = \frac{I^2 X_L}{I^2 R_W}$$

In a series circuit, I is the same in L and R; thus, the I^2 terms cancel, leaving

$$Q = \frac{X_L}{R_W} \qquad \textbf{(10–17)}$$

When the resistance is just the winding resistance of the coil, the circuit Q and the coil Q are the same. Note that Q is a ratio of like units and, therefore, has no unit itself.

SECTION 10–6 REVIEW

1. State the phase relationship between current and voltage in an inductor.
2. Calculate X_L for $f = 5$ kHz and $L = 100$ mH.
3. At what frequency is the reactance of a 50 μH inductor equal to 800 Ω?
4. Calculate the rms current in Figure 10–35.

FIGURE 10–35

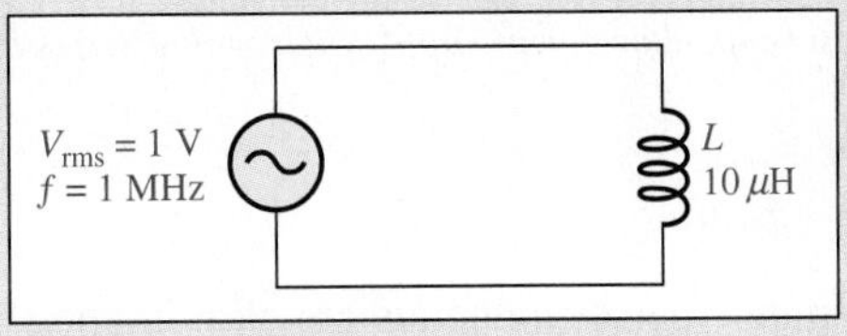

5. A 50 mH inductor is connected to a 12 V rms source. What is the true power? What is the reactive power at a frequency of 1 kHz?

10–7 ■ INDUCTOR APPLICATIONS

Inductors are not as versatile as capacitors and tend to be more limited in their applications due, in part, to size and cost factors. However, there are many practical uses for inductors (coils) such as those discussed in Chapter 7. Recall that relay and solenoid coils, recording and pick-up heads, and sensing elements were introduced as electromagnetic applications of coils. In this section, some additional uses of inductors are presented.

After completing this section, you should be able to

- **Discuss some inductor applications**
  - ☐ Describe a power supply filter
  - ☐ Explain the purpose of an rf choke
  - ☐ Discuss the basics of tuned circuits

Power Supply Filter

In Chapter 9, you saw that a capacitor is used to filter the pulsating dc on the output of the rectifier in a power supply. The final output voltage was a dc voltage with a small amount of ripple. In many cases, an inductor is used in the filter, as shown in Figure 10–36(a), to smooth out the ripple voltage. The inductor, placed in series with the load as shown, tends to oppose the current fluctuations caused by the ripple voltage, and thus the voltage developed across the load is more constant, as shown in Figure 10–36(b).

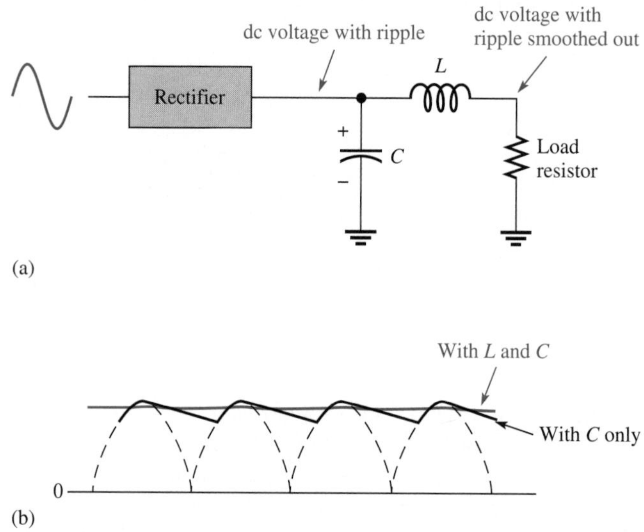

FIGURE 10–36
Basic capacitor power supply filter with a series inductor.

RF Choke

Certain types of inductors are used in applications where radio frequencies (rf) must be prevented from getting into parts of a system, such as the power supply or the audio section of a receiver. In these situations, an inductor is used as a series filter and "chokes" off any unwanted rf signals that may be picked up on a line. This filtering action is based on the fact that the reactance of a coil increases with frequency. When the frequency of the current is sufficiently high, the reactance of the coil becomes extremely large and essentially blocks the current. A basic illustration of an inductor used as an rf choke is shown in Figure 10–37.

Tuned Circuits

As mentioned in Chapter 9, inductors are used in conjunction with capacitors to provide frequency selection in communications systems. These tuned circuits allow a narrow

FIGURE 10–37
An inductor used as an rf choke to minimize interfering signals on the power supply line.

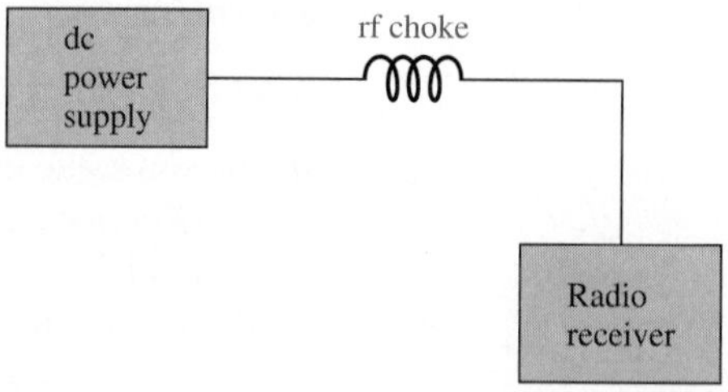

band of frequencies to be selected while all other frequencies are rejected. The tuners in your TV and radio receivers are based on this principle and permit you to select one channel or station out of the many that are available.

Frequency selectivity is based on the fact that the reactances of both capacitors and inductors depend on the frequency and on the interaction of these two components when connected in series or parallel. Since the capacitor and the inductor produce opposite phase shifts, their combined opposition to current can be used to obtain a desired response at a selected frequency. Tuned (resonant) *RLC* circuits are covered in detail in Chapter 14.

SECTION 10–7 REVIEW

1. Explain how the ripple voltage from a power supply filter can be reduced through use of an inductor.
2. How does an inductor connected in series act as an rf choke?

10–8 ■ TESTING INDUCTORS

In this section, two basic types of failures in inductors are discussed and methods for testing inductors are introduced.

After completing this section, you should be able to

- **Test an inductor**
  - ☐ Perform an ohmmeter check for an open winding
  - ☐ Perform an ohmmeter check for shorted windings

The most common failure in an inductor is an open coil. To check for an open, remove the coil from the circuit. If there is an open, an ohmmeter check will indicate infinite resistance, as shown in Figure 10–38(a). If the coil is good, the ohmmeter will show the winding resistance. The value of the winding resistance depends on the wire size and length of the coil. It can be anywhere from one ohm to several hundred ohms. Figure 10–38(b) shows a good reading.

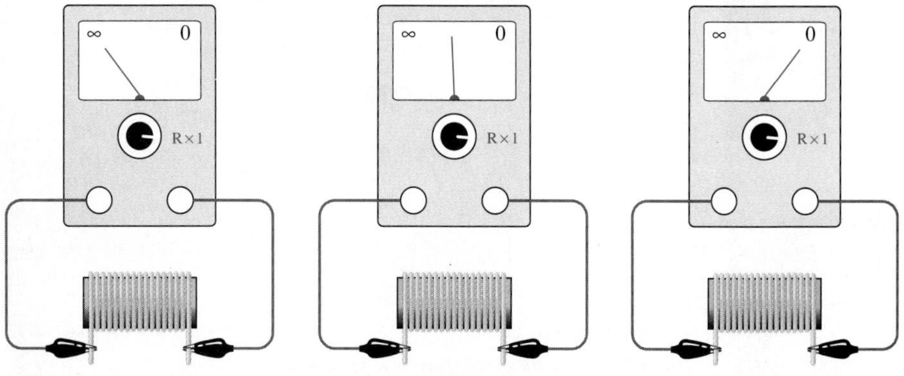

(a) Coil open: Meter reads infinity (b) Coil good: Meter reads winding resistance (c) Shorted winding: Meter reads zero or a lower than normal winding resistance depending on the number of turns that are shorted

FIGURE 10–38
Checking a coil by measuring the resistance.

Occasionally, when an inductor is overheated with excessive current, the wire insulation will melt, and some coils will short together. This produces a reduction in the inductance by reducing the effective number of turns and a corresponding reduction in winding resistance, as shown in Figure 10–38(c).

SECTION 10–8 REVIEW

1. When a coil is checked, a reading of infinity on the ohmmeter indicates a partial short (T or F).
2. An ohmmeter check of a good coil will indicate the value of the inductance (T or F).

10–9 ▪ APPLICATION ASSIGNMENT

Your supervisor has handed you two unmarked coils and wants to know their inductance values. You cannot find an inductance bridge, which is an instrument for measuring inductance directly. After scratching your head for awhile, you decide to use the time-constant characteristic of inductive circuits to determine the unknown inductances. A test setup consisting of a square wave generator and an oscilloscope is used to make the measurements.

The method is to place the coil in series with a resistor of known value and measure the time constant by applying a square wave to the circuit and observing the resulting voltage across the resistor with an oscilloscope. Knowing the time constant and the resistance value, you can calculate the inductance L.

Each time the square wave input voltage goes high, the inductor is energized; and each time the square wave goes back to zero, the inductor is deenergized. The time it takes for the exponential resistor voltage to increase to approximately its final value equals to five time constants. This operation is illustrated in Figure 10–39. To make sure that the winding resistance of the coil can be neglected, it must be measured; and the value of the resistor used in the circuit must be selected to be considerably larger than the winding resistance.

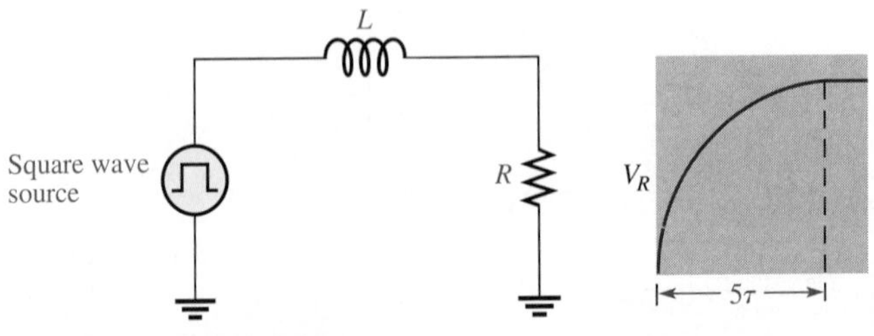

FIGURE 10–39
Circuit for time-constant measurement.

Step 1: Measure the Coil Resistance and Select a Series Resistor

Assume that the winding resistance has been measured with an ohmmeter and found to be 85 Ω. To make the winding resistance negligible, a 10 kΩ series resistor is used in the circuit.

Step 2: Determine the Inductance of Coil 1

Refer to Figure 10–40. To determine the inductance, a 10 V square wave is applied to the breadboarded circuit. The frequency of the square wave is adjusted so that the inductor has time to fully energize during each square wave pulse; the scope is set to view a complete energizing curve as shown. Determine the approximate circuit time constant from the scope display and calculate the inductance of coil 1.

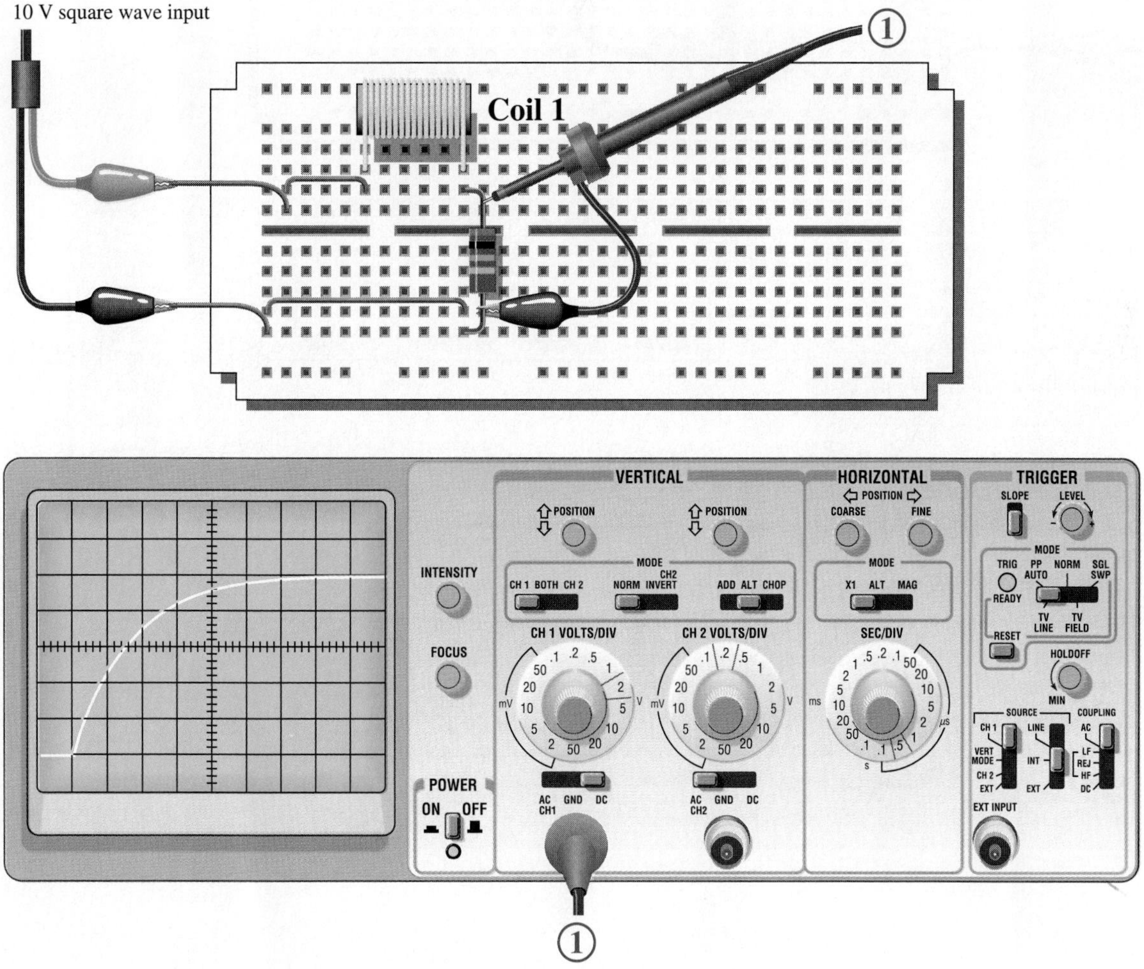

FIGURE 10–40
Testing coil 1.

Step 3: Determine the Inductance of Coil 2

Refer to Figure 10–41. To determine the inductance, a 10 V square wave is applied to the breadboarded circuit. The frequency of the square wave is adjusted so that the inductor has time to fully energize during each square wave pulse; the scope is set to view a complete energizing curve as shown. Determine the approximate circuit time constant from the scope display and calculate the inductance of coil 2. Discuss any difficulty you find with this method.

Step 4: Determine Another Way to Find Unknown Inductance

Determination of the time constant is not the only way that you can find an unknown inductance. Specify a method using a sinusoidal input voltage instead of the square wave.

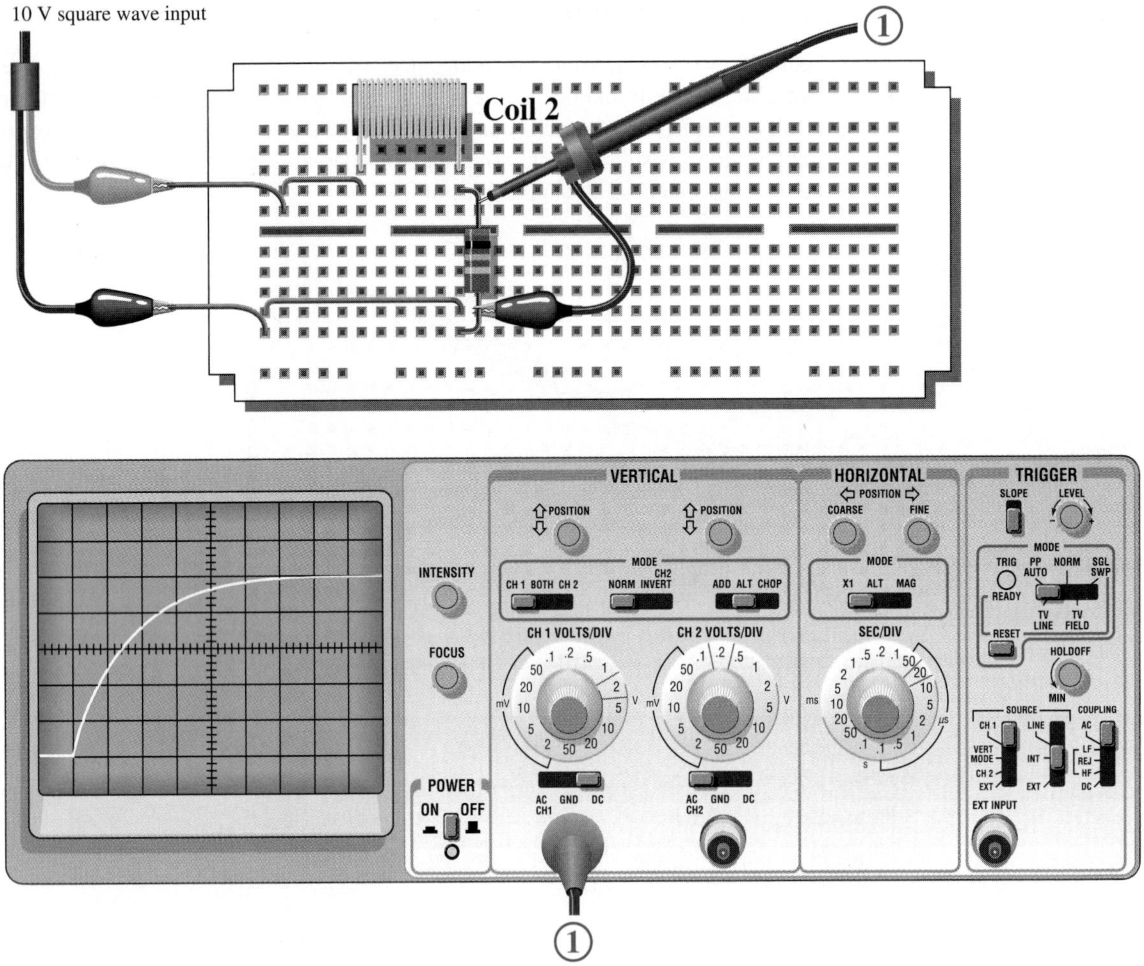

FIGURE 10–41
Testing coil 2.

SECTION 10–9 REVIEW

1. What is the maximum square wave frequency that can be used in Figure 10–40?
2. What is the maximum square wave frequency that can be used in Figure 10–41?
3. What happens if the frequency exceeds the maximum you determined in Questions 1 and 2? Explain how your measurements would be affected.

▪ SUMMARY

- Self-inductance is a measure of a coil's ability to establish an induced voltage as a result of a change in its current.
- An inductor opposes a change in its own current.
- Faraday's law states that relative motion between a magnetic field and a coil induces a voltage across the coil.
- Lenz's law states that the polarity of induced voltage is such that the resulting induced current is in a direction that opposes the change in the magnetic field that produced it.
- Energy is stored by an inductor in its magnetic field.
- One henry is the amount of inductance when current, changing at the rate of one ampere per second, induces one volt across the inductor.

- Inductance is directly proportional to the square of the turns, the permeability, and the cross-sectional area of the core. It is inversely proportional to the length of the core.
- The permeability of a core material is an indication of the ability of the material to establish a magnetic field.
- Inductors add in series.
- Total parallel inductance is less than that of the smallest inductor in parallel.
- The time constant for a series *RL* circuit is the inductance divided by the resistance.
- In an *RL* circuit, the voltage and current in an energizing or deenergizing inductor make a 63% change during each time-constant interval.
- Energizing and deenergizing current and voltage follow exponential curves.
- Voltage leads current by 90° in an inductor.
- Inductive reactance (X_L) is directly proportional to frequency and inductance.
- The true power in an inductor is zero; that is, there is no energy loss in an ideal inductor, only in its winding resistance.

GLOSSARY

Choke An inductor. The term is more commonly used in connection with inductors used to block or choke off high frequencies.

Coil A common term for an inductor.

Core The structure around which the winding of an inductor is formed. The core material influences the electromagnetic characteristics of the inductor.

Henry (H) The unit of inductance.

Induced voltage Voltage produced as a result of a changing magnetic field.

Inductance (L) The property of an inductor that produces an opposition to any change in current.

Inductive reactance (X_L) The opposition of an inductor to sinusoidal current. The unit is the ohm.

Inductor An electrical device formed by a wire wound around a core having the property of inductance; also known as a coil or a choke.

Lenz's law A physical law that states when the current through a coil changes, an induced voltage is created in a direction to oppose the change in current. The current cannot change instantaneously.

Quality factor (Q) The ratio of reactive power to true power.

Reactive power (P_r) The rate at which energy is alternately stored and returned to the source by an inductor.

Time constant A fixed time interval, set by the *L* and *R* values, that determines the time response of a circuit.

Winding The loops of wire in an inductor.

FORMULAS

(10–1) $$W = \frac{1}{2}LI^2$$ Energy stored by an inductor

(10–2) $$L = \frac{N^2\mu A}{l}$$ Inductance in terms of physical parameters

(10–3) $$L_T = L_1 + L_2 + L_3 + \cdots + L_n$$ Series inductance

(10–4) $$\frac{1}{L_T} = \frac{1}{L_1} + \frac{1}{L_2} + \frac{1}{L_3} + \cdots + \frac{1}{L_n}$$ Reciprocal of total parallel inductance

(10–5) $$L_T = \frac{1}{\left(\frac{1}{L_1}\right) + \left(\frac{1}{L_2}\right) + \left(\frac{1}{L_3}\right) + \cdots + \left(\frac{1}{L_n}\right)}$$ Total parallel inductance

(10–6) $$\tau = \frac{L}{R}$$ Time constant

(10–7) $$v = V_F + (V_i - V_F)e^{-Rt/L}$$ Exponential voltage (general)

(10–8) $$i = I_F + (I_i - I_F)e^{-Rt/L}$$ Exponential current (general)

(10–9)	$i = I_F(1 - e^{-Rt/L})$	Increasing exponential current beginning at zero
(10–10)	$i = I_i e^{-Rt/L}$	Decreasing exponential current ending at zero
(10–11)	$X_L = 2\pi fL$	Inductive reactance
(10–12)	$V = IX_L$	Ohm's law
(10–13)	$P_{true} = I^2_{rms}R_W$	True power
(10–14)	$P_r = V_{rms}I_{rms}$	Reactive power
(10–15)	$P_r = \dfrac{V^2_{rms}}{X_L}$	Reactive power
(10–16)	$P_r = I^2_{rms}X_L$	Reactive power
(10–17)	$Q = \dfrac{X_L}{R_W}$	Quality factor

▪ SELF-TEST

1. An inductance of 0.05 μH is larger than
   (a) 0.0000005 H **(b)** 0.000005 H **(c)** 0.000000008 H **(d)** 0.00005 mH
2. An inductance of 0.33 mH is smaller than
   (a) 33 μH **(b)** 330 μH **(c)** 0.05 mH **(d)** 0.0005 H
3. When the current through an inductor increases, the amount of energy stored in the electromagnetic field
   (a) decreases **(b)** remains constant **(c)** increases **(d)** doubles
4. When the current through an inductor doubles, the stored energy
   (a) doubles **(b)** quadruples **(c)** is halved **(d)** does not change
5. The winding resistance of a coil can be decreased by
   (a) reducing the number of turns **(b)** using a larger wire
   (c) changing the core material **(d)** either (a) or (b)
6. The inductance of an iron-core coil increases if
   (a) the number of turns is increased **(b)** the iron core is removed
   (c) the length of the core is increased **(d)** larger wire is used
7. Four 10 mH inductors are in series. The total inductance is
   (a) 40 mH **(b)** 2.5 mH **(c)** 40,000 μH **(d)** both (a) and (c)
8. A 1 mH, a 3.3 mH, and a 0.1 mH inductor are connected in parallel. The total inductance is
   (a) 4.4 mH **(b)** greater than 3.3 mH **(c)** less than 0.1 mH **(d)** both (a) and (b)
9. An inductor, a resistor, and a switch are connected in series to a 12 V battery. At the instant the switch is closed, the inductor voltage is
   (a) 0 V **(b)** 12 V **(c)** 6 V **(d)** 4 V
10. A sine wave voltage is applied across an inductor. When the frequency of the voltage is increased, the current
    (a) decreases **(b)** increases **(c)** does not change **(d)** momentarily goes to zero
11. An inductor and a resistor are in series with a sinusoidal voltage source. The frequency is set so that the inductive reactance is equal to the resistance. If the frequency is increased, then
    (a) $V_R > V_L$ **(b)** $V_L < V_R$ **(c)** $V_L = V_R$ **(d)** $V_L > V_R$
12. An ohmmeter is connected across an inductor and the pointer indicates an infinite value. The inductor is
    (a) good **(b)** open **(c)** shorted **(d)** resistive

▪ BASIC PROBLEMS

SECTION 10–1 The Basic Inductor

1. Convert the following to millihenries:
   (a) 1 H **(b)** 250 μH **(c)** 10 μH **(d)** 0.0005 H
2. Convert the following to microhenries:
   (a) 300 mH **(b)** 0.08 H **(c)** 5 mH **(d)** 0.00045 mH
3. How many turns are required to produce 30 mH with a coil wound on a cylindrical core having a cross-sectional area of 10×10^{-5} m² and a length of 0.05 m? The core has a permeability of 1.2×10^{-6}.

4. A 12 V battery is connected across a coil with a winding resistance of 12 Ω. How much current is there in the coil?
5. How much energy is stored by a 100 mH inductor with a current of 1 A?
6. The current through a 100 mH coil is changing at a rate of 200 mA/s. How much voltage is induced across the coil?

SECTION 10–3 Series Inductors

7. Five inductors are connected in series. The lowest value is 5 μH. If the value of each inductor is twice that of the preceding one, and if the inductors are connected in order of ascending values, what is the total inductance?
8. Suppose that you require a total inductance of 50 mH. You have available a 10 mH coil and a 22 mH coil. How much additional inductance do you need?

SECTION 10–4 Parallel Inductors

9. Determine the total parallel inductance for the following coils in parallel: 75 μH, 50 μH, 25 μH, and 15 μH.
10. You have a 12 mH inductor, and it is your smallest value. You need an inductance of 8 mH. What value can you use in parallel with the 12 mH to obtain 8 mH?
11. Determine the total inductance of each circuit in Figure 10–42.
12. Determine the total inductance of each circuit in Figure 10–43.

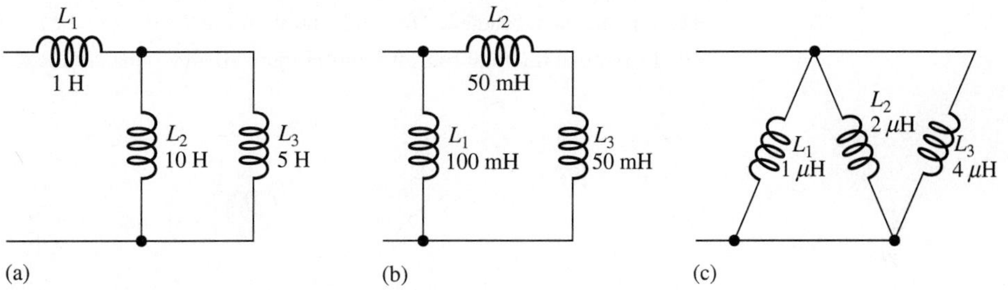

FIGURE 10–42

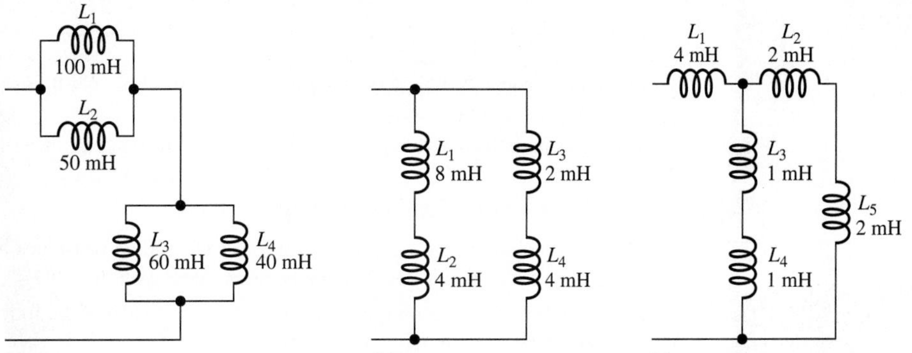

FIGURE 10–43

SECTION 10–5 Inductors in DC Circuits

13. Determine the time constant for each of the following series *RL* combinations:
    (a) $R = 100\ \Omega, L = 100\ \mu\text{H}$ **(b)** $R = 4.7\ \text{k}\Omega, L = 10\ \text{mH}$ **(c)** $R = 1.5\ \text{M}\Omega, L = 3\ \text{H}$
14. In a series *RL* circuit, determine how long it takes the current to build up to its full value for each of the following:
    (a) $R = 56\ \Omega, L = 50\ \mu\text{H}$ **(b)** $R = 3300\ \Omega, L = 15\ \text{mH}$ **(c)** $R = 22\ \text{k}\Omega, L = 100\ \text{mH}$

15. In the circuit of Figure 10–44, there is initially no current. Determine the inductor voltage at the following times after the switch is closed:
(a) 10 μs (b) 20 μs (c) 30 μs (d) 40 μs (e) 50 μs

16. In Figure 10–45, there are 100 mA through the coil. When SW_1 is opened and SW_2 simultaneously closed, find the inductor voltage at the following times:
(a) initially (b) 1.5 ms (c) 4.5 ms (d) 6 ms

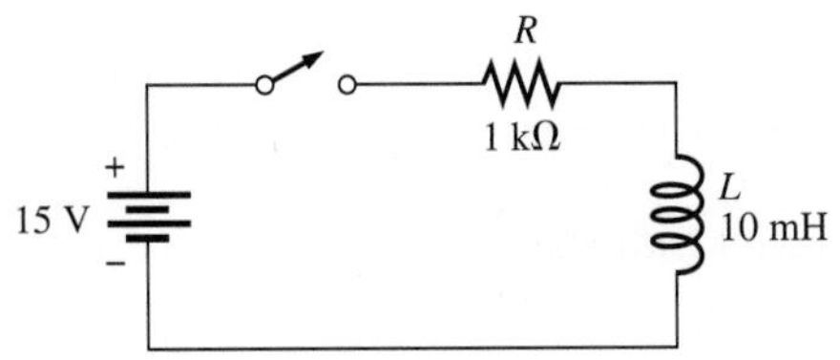

FIGURE 10–44

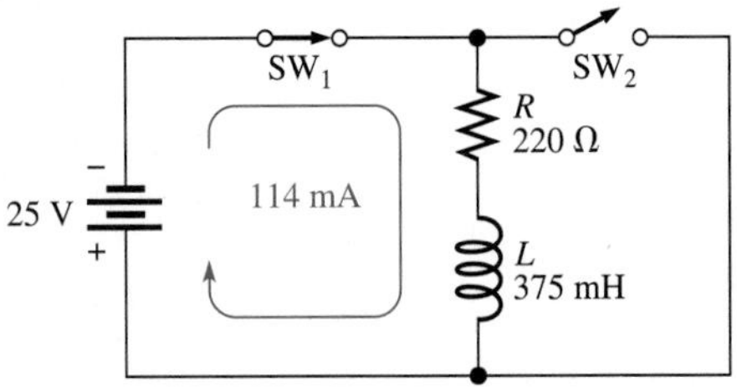

FIGURE 10–45

SECTION 10–6 Inductors in AC Circuits

17. Find the total reactance for each circuit in Figure 10–42 when a voltage with a frequency of 5 kHz is applied across the terminals.

18. Find the total reactance for each circuit in Figure 10–43 when a 400 Hz voltage is applied.

19. Determine the total rms current in Figure 10–46. What are the currents through L_2 and L_3?

FIGURE 10–46

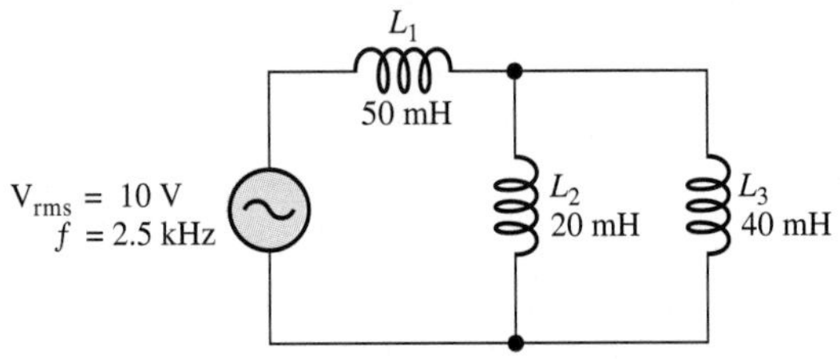

20. What frequency will produce a total rms current of 500 mA in each circuit of Figure 10–43 with an rms input voltage of 10 V?

21. Determine the reactive power in Figure 10–46 neglecting the winding resistance.

SECTION 10–8 Testing Inductors

22. A certain coil that is supposed to have a 5 Ω winding resistance is measured with an ohmmeter. The meter indicates 2.8 Ω. What is the problem with the coil?

23. Determine the meter indication corresponding to each of the following failures in a coil:
(a) open (b) shorted (c) partially shorted

■ ADVANCED PROBLEMS

24. Determine the time constant for the circuit in Figure 10–47. See Appendix E.

25. Find the inductor current 10 μs after the switch is thrown from position 1 to position 2 in Figure 10–48. For simplicity, assume that the switch makes contact with position 2 at the same instant it breaks contact with position 1.

26. In Figure 10–49, SW_1 is opened and SW_2 is closed at the same instant (t_0). What is the instantaneous voltage across R_2 at t_0?

27. Determine the value of I_{L2} in Figure 10–50.

28. What is the total inductance between points A and B for each switch position in Figure 10–51?

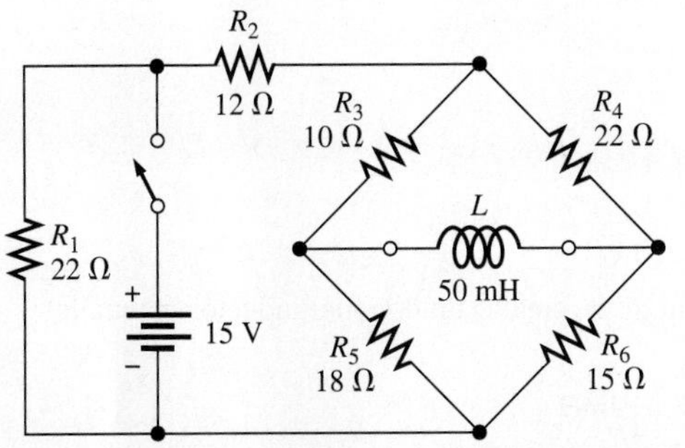

FIGURE 10–47

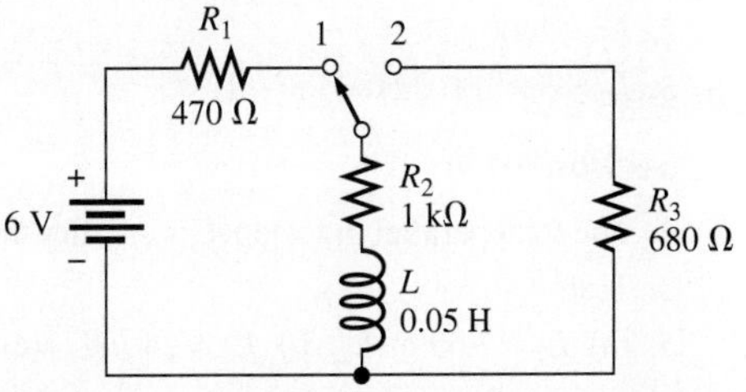

FIGURE 10–48

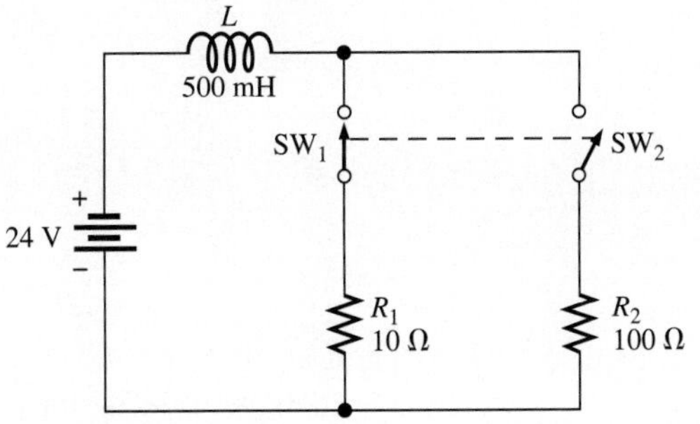

FIGURE 10–49

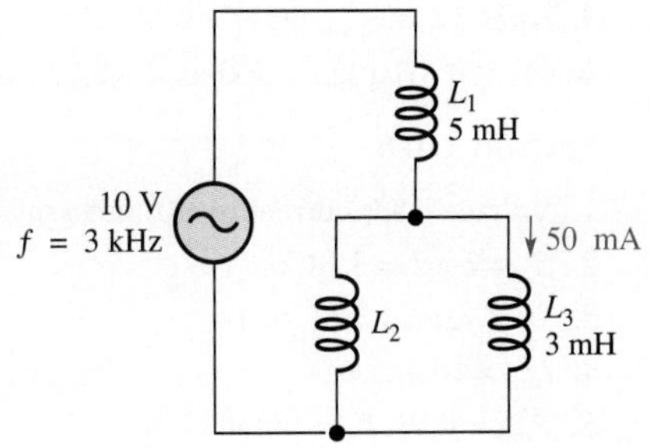

FIGURE 10–50

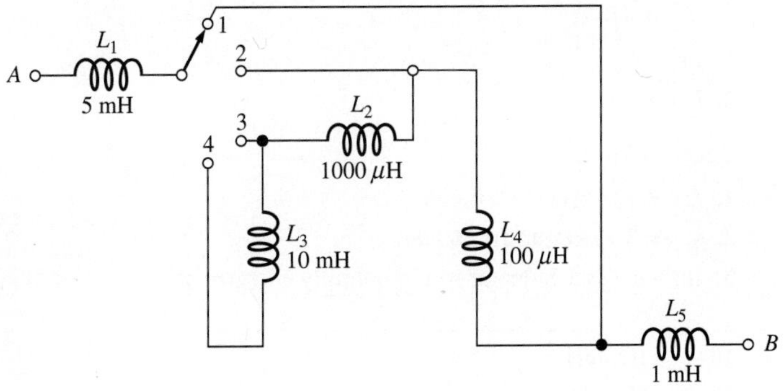

FIGURE 10–51

■ ANSWERS TO SECTION REVIEWS

Section 10–1

1. Parameters that determine inductance are turns, permeability, cross-sectional area, and length.
2. **(a)** When N increases, L increases. **(b)** When core length increases, L decreases. **(c)** When area decreases, L decreases. **(d)** For an air core, L decreases.
3. All wire has some resistance, and since inductors are made from turns of wire, there is always resistance in the winding.
4. Adjacent turns act as plates and produce a small capacitance.

Section 10–2

1. Two categories of inductors are fixed and variable.
2. **(a)** air core **(b)** iron core **(c)** variable

Section 10–3

1. Inductances are added in series.
2. $L_T = 2600\ \mu H$
3. $L_T = 5 \times 100\ mH = 500\ mH$

Section 10–4

1. The total parallel inductance is smaller than that of the smallest individual inductor in parallel.
2. True
3. **(a)** $L_T = 7.69\ mH$ **(b)** $L_T = 24\ \mu H$ **(c)** $L_T = 100\ mH$

Section 10–5

1. $V_L = (10\ mA)(10\ \Omega) = 100\ mV$
2. Initially, $v_R = 0\ V$, $v_L = 20\ V$
3. After 5τ, $v_R = 20\ V$, $v_L = 0\ V$
4. $\tau = 500\ \mu H/1\ k\Omega = 500\ ns$, $i = 3.93\ mA$

Section 10–6

1. Voltage leads current by 90° in an inductor.
2. $X_L = 2\pi fL = 3.14\ k\Omega$
3. $f = X_L/2\pi L = 2.55\ MHz$
4. $I_{rms} = 15.9\ mA$
5. $P_{true} = 0\ W$, $P_r = 458\ mVAR$

Section 10–7

1. The inductor tends to level out the ripple because of its opposition to changes in current.
2. The inductive reactance is very high at radio frequencies and blocks these frequencies.

Section 10–8

1. False
2. False

Section 10–9

1. $f_{max} = 125\ kHz$ ($5\tau = 4\ ms$)
2. $f_{max} = 3.13\ kHz$ ($5\tau = 160\ \mu s$)
3. If $f > f_{max}$, the inductor would not fully energize because $T/2 < 5\tau$.

▪ ANSWERS TO RELATED EXERCISES FOR EXAMPLES

10–1 108 mH

10–2 500 μH

10–3 20.3 μH

10–4 227 ns

10–5 $I_F = 29.4\ mA$, $\tau = 147\ ns$
At 1τ, $i = 18.5\ mA$
At 2τ, $i = 25.3\ mA$
At 3τ, $i = 27.9\ mA$
At 4τ, $i = 28.8\ mA$
At 5τ, $i = 29.1\ mA$

10–6 **(a)** 21.3 μs **(b)** 106 mA **(c)** 39.2 mA

10–7 **(a)** 9 V; 0 V **(b)** 92.6 V; 0 V

10–8 3.82 mA

10–9 33.3 mA

10–10 110 Ω

10–11 574 mA

10–12 P_r decreases

11

TRANSFORMERS

In the last chapter, self-inductance was studied. In this chapter, another inductive effect, called *mutual inductance,* is introduced. Mutual inductance occurs when two or more coils are very close together. It is based on the principle of magnetic coupling. A basic transformer is formed by two coils that are magnetically coupled but have no electrical connection between them. The transfer of energy from one coil to the other occurs entirely through a changing magnetic field. There are many applications for transformers, and a basic understanding of their operation is essential.

■ CHAPTER OBJECTIVES

- ☐ Explain mutual inductance
- ☐ Describe how a transformer is constructed and how it works
- ☐ Explain how a step-up transformer works
- ☐ Explain how a step-down transformer works
- ☐ Discuss the effect of a resistive load across the secondary winding
- ☐ Discuss the concept of a reflected load in a transformer
- ☐ Discuss impedance matching with transformers
- ☐ Explain how the transformer acts as an isolation device
- ☐ Describe a practical transformer
- ☐ Describe several types of transformers
- ☐ Troubleshoot transformers

■ APPLICATION ASSIGNMENT

Your assignment is to troubleshoot a type of dc power supply that uses a transformer to couple the ac voltage from a standard outlet. By making voltage measurements at various points, you will determine if there is a fault and be able to specify the part of the power supply that is faulty. After studying this chapter, you should be able to complete the application assignment in Section 11–12.

11–1 ■ MUTUAL INDUCTANCE

When two coils are placed close to each other, a changing electromagnetic field produced by the current in one coil will cause an induced voltage in the second coil because of the mutual inductance.

After completing this section, you should be able to

- **Explain mutual inductance**
  - □ Discuss magnetic coupling
  - □ Define *electrical isolation*
  - □ Define *coefficient of coupling*
  - □ Identify the factors that affect mutual inductance and state the formula

Recall that the electromagnetic field surrounding a coil expands, collapses, and reverses as the current increases, decreases, and reverses.

When a second coil is placed very close to the first coil so that the changing magnetic flux lines cut through the second coil, the coils are magnetically coupled and a voltage is induced, as indicated in Figure 11–1. When two coils are magnetically coupled, they provide **electrical isolation** because there is no electrical connection between them, only a magnetic link. If the current in the first coil is a sine wave, the voltage induced in the second coil is also a sine wave. The amount of voltage induced in the second coil as a result of the current in the first coil is dependent on the **mutual inductance, L_M,** between the two coils. The mutual inductance is established by the inductance of each coil (L_1 and L_2) and by the amount of coupling (k) between the two coils.

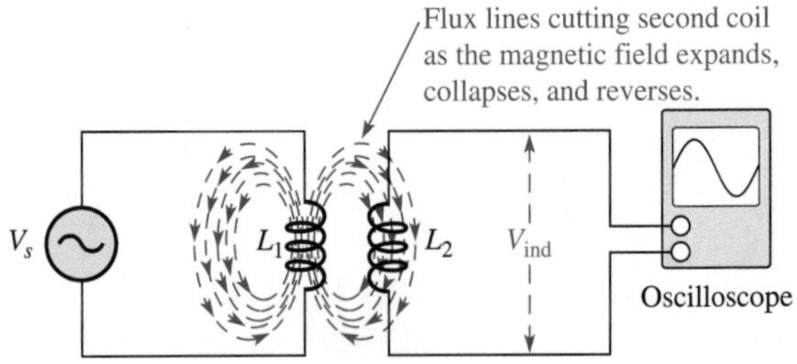

FIGURE 11–1
A voltage is induced in the second coil as a result of the changing current in the first coil producing a changing magnetic field that links the second coil.

Coefficient of Coupling

The **coefficient of coupling, *k*,** between two coils is the ratio of the flux lines produced by coil 1 that link coil 2 (ϕ_{12}) to the total flux produced by coil 1 (ϕ_1).

$$k = \frac{\phi_{12}}{\phi_1} \tag{11–1}$$

For example, if half of the total flux produced by coil 1 links coil 2, then $k = 0.5$. A greater value of k means that more voltage is induced in coil 2 for a certain rate of change of current in coil 1. Note that k has no units. Recall that the unit of magnetic lines of force (flux) is the weber, abbreviated Wb.

The coefficient k depends on the physical closeness of the coils and the type of core material on which they are wound. Also, the construction and shape of the cores are factors.

Formula for Mutual Inductance

The three factors that influence mutual inductance, L_M, (k, L_1, and L_2) are shown in Figure 11–2. The formula for mutual inductance is

$$L_M = k\sqrt{L_1 L_2} \tag{11–2}$$

FIGURE 11–2
The mutual inductance of two coils.

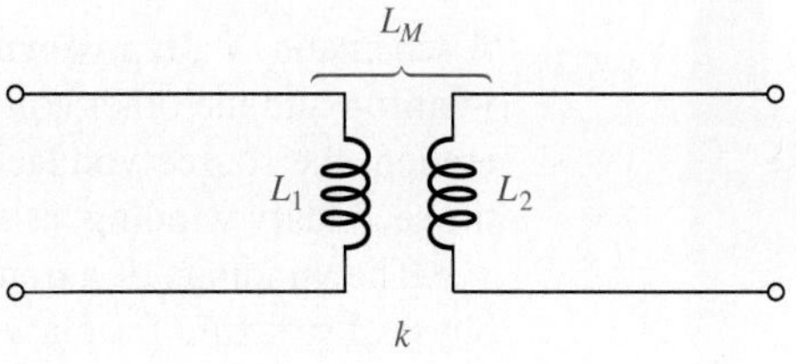

EXAMPLE 11–1 One coil produces a total magnetic flux of 50 μWb, and 20 μWb link coil 2. What is k?

Solution

$$k = \frac{\phi_{12}}{\phi_1} = \frac{20\ \mu\text{Wb}}{50\ \mu\text{Wb}} = 0.4$$

Related Exercise Determine k when $\phi_1 = 500\ \mu$Wb and $\phi_{12} = 375\ \mu$Wb.

EXAMPLE 11–2 Two coils are wound on a single core, and the coefficient of coupling is 0.3. The inductance of coil 1 is 10 μH, and the inductance of coil 2 is 15 μH. What is L_M?

Solution

$$L_M = k\sqrt{L_1 L_2} = 0.3\sqrt{(10\ \mu\text{H})(15\ \mu\text{H})} = 3.67\ \mu\text{H}$$

Related Exercise Determine the mutual inductance when $k = 0.5$, $L_1 = 1$ mH, and $L_2 = 600\ \mu$H.

SECTION 11–1 REVIEW

1. Define *mutual inductance.*
2. Two 50 mH coils have $k = 0.9$. What is L_M?
3. If k is increased, what happens to the voltage induced in one coil as a result of a current change in the other coil?

11–2 ■ THE BASIC TRANSFORMER

A basic transformer is an electrical device constructed of two coils placed in close proximity to each other so that there is a mutual inductance between the two coils.

After completing this section, you should be able to

- **Describe how a transformer is constructed and how it works**
  - ☐ Identify the parts of a basic transformer
  - ☐ Discuss the importance of the core material
  - ☐ Define *primary winding* and *secondary winding*
  - ☐ Define *turns ratio*
  - ☐ Discuss how the direction of windings affect voltage polarities

---

A schematic of a **transformer** is shown in Figure 11–3(a). One coil is called the **primary winding** and the other coil is called the **secondary winding** as indicated. In standard operation, the source voltage is applied to the primary winding, and a load is connected to the secondary winding, as shown in Figure 11–3(b).

The windings of a transformer are formed around the **core.** The core provides both a physical structure for placement of the windings and a magnetic path so that the magnetic flux lines are concentrated close to the coils. There are three general categories of core material: air, ferrite, and iron. The schematic symbol for each type is shown in Figure 11–4.

Air-core and ferrite-core transformers generally are used for high-frequency applications and consist of windings on an insulating shell which is hollow (air) or constructed of ferrite, such as depicted in Figure 11–5. The wire is typically covered by a varnish-type coating to prevent the windings from shorting together. The amount of **magnetic coupling** between the primary winding and the secondary winding is set by the type of core material and by the relative positions of the windings. In Figure 11–5(a), the windings are loosely coupled because they are separated, and in part (b) they are tightly coupled because they are overlapping. The tighter the coupling, the greater the induced voltage in the secondary for a given current in the primary.

Iron-core transformers generally are used for audio frequency (af) and power applications. These transformers consist of windings on a core constructed from laminated sheets of ferromagnetic material insulated from each other, as shown in Figure 11–6. This construction provides an easy path for the magnetic flux and increases the amount of coupling between the windings. Figures 11–6(a) and 11–6(b) show the basic construction of two major configurations of iron-core transformers. In the core-type construction, shown in part (a), the windings are on separate legs of the laminated core. In the shell-type construction, shown in part (b), both windings are on the same leg. A variety of transformers are shown in Figure 11–7.

FIGURE 11–3
The basic transformer.

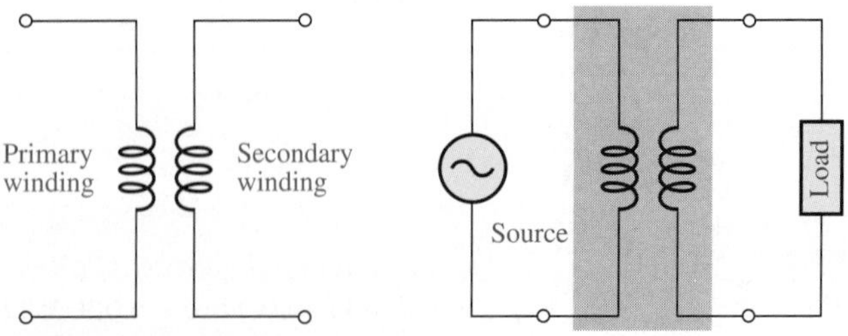

(a) Schematic symbol (b) Source/load connections

FIGURE 11–4
Schematic symbols based on type of core.

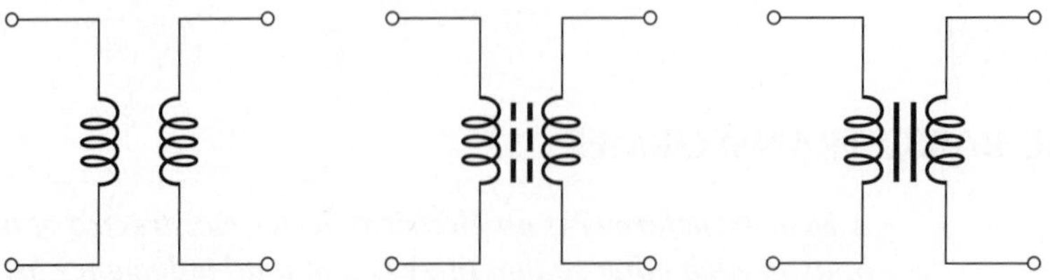

(a) Air core (b) Ferrite core (c) Iron core

FIGURE 11–5
Transformers with cylindrical-shaped cores.

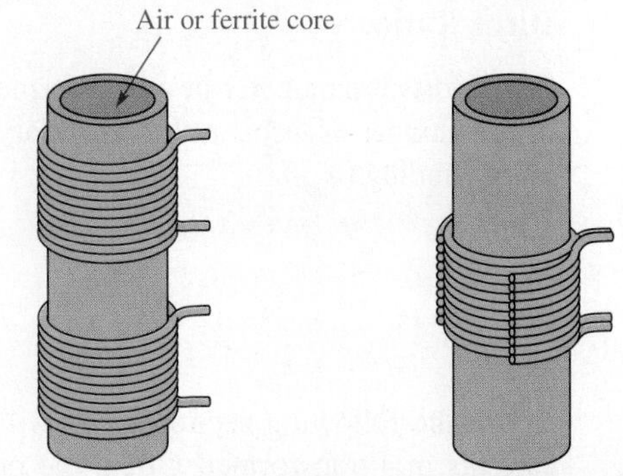

(a) Loosely coupled windings

(b) Tightly coupled windings. Cutaway view shows both windings.

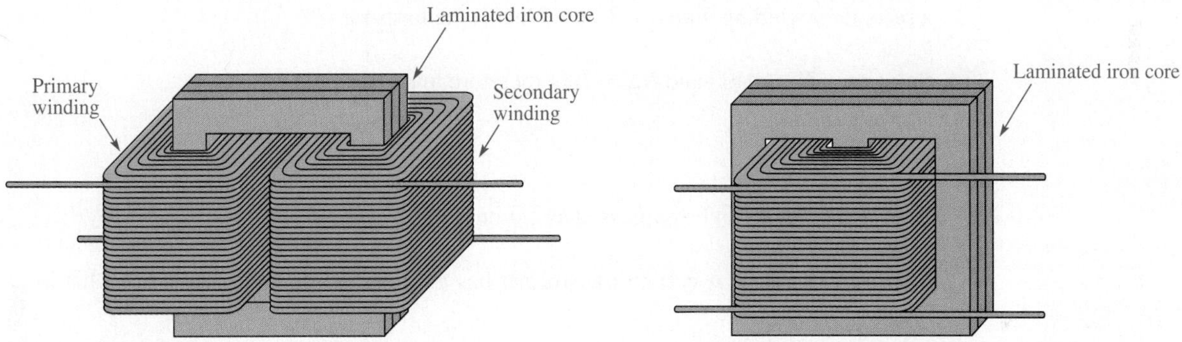

(a) Core-type has each winding on a separate leg.

(b) Shell-type has both windings on the same leg.

FIGURE 11–6
Iron-core transformer construction with multilayer windings.

FIGURE 11–7
Some common types of transformers (courtesy of Litton Triad-Utrad).

Turns Ratio

An important parameter of a transformer is its **turns ratio.** The turns ratio (n) is the ratio of the number of turns in the secondary winding (N_{sec}) to the number of turns in the primary winding (N_{pri}).

$$n = \frac{N_{sec}}{N_{pri}} \qquad \textbf{(11–3)}$$

In the following sections, you will see how the turns ratio affects the voltages and currents in a transformer. Often, the primary and secondary voltages are specified on a transformer, while the turns ratio is not.

EXAMPLE 11–3

A certain transformer used in a radar system has a primary winding with 100 turns and a secondary winding with 400 turns. What is the turns ratio?

Solution $N_{sec} = 400$ and $N_{pri} = 100$; therefore, the turns ratio is

$$n = \frac{N_{sec}}{N_{pri}} = \frac{400}{100} = 4$$

A turns ratio of 4 can be expressed as 1:4 on a schematic.

Related Exercise A certain transformer has a turns ratio of 10. If $N_{pri} = 50$, what is N_{sec}?

Direction of Windings

Another important transformer parameter is the direction in which the windings are placed around the core. As illustrated in Figure 11–8, the direction of the windings determines the polarity of the secondary voltage with respect to the primary voltage. Phase dots can be used to indicate polarities, as shown in Figure 11–9. If the primary is positive at the dot, the secondary is positive at the dot. If the primary is negative at the dot, the secondary is negative at the dot.

FIGURE 11–8
Relative polarities of the voltages are determined by the direction of the windings.

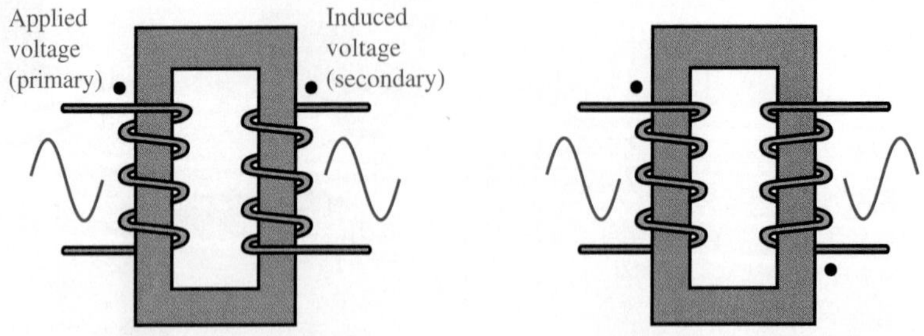

(a) The primary and secondary voltages are in phase when the windings are in the same effective direction around the magnetic path.

(b) The primary and secondary voltages are 180° out of phase when the windings are in the opposite direction.

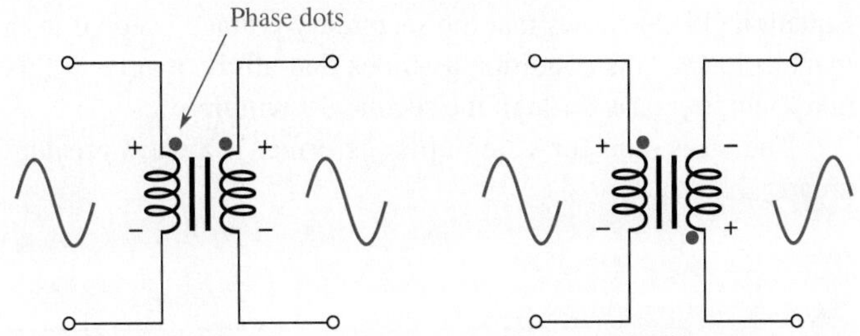

(a) Voltages are in phase. (b) Voltages are out of phase.

FIGURE 11–9
Phase dots indicate corresponding polarities of primary and secondary voltages.

SECTION 11–2 REVIEW

1. Upon what principle is the operation of a transformer based?
2. Define *turns ratio*.
3. Why are the directions of the windings of a transformer important?
4. A certain transformer has a primary winding with 500 turns and a secondary winding with 250 turns. What is the turns ratio?

11–3 ■ THE STEP-UP TRANSFORMER

A step-up transformer has more turns in its secondary winding than in its primary winding and is used to increase ac voltage.

After completing this section, you should be able to

- **Explain how a step-up transformer works**
  - □ State the relationship between primary and secondary voltages and the number of turns in the primary and secondary windings
  - □ Identify a step-up transformer by its turns ratio

A transformer in which the secondary voltage is greater than the primary voltage is called a **step-up transformer.** The amount that the voltage is stepped up depends on the turns ratio. For any transformer,

> **The ratio of secondary voltage (V_{sec}) to primary voltage (V_{pri}) is equal to the ratio of the number of secondary turns (N_{sec}) to the number of primary turns (N_{pri}).**

$$\frac{V_{sec}}{V_{pri}} = \frac{N_{sec}}{N_{pri}} \tag{11–4}$$

From this relationship,

$$V_{sec} = \left(\frac{N_{sec}}{N_{pri}}\right)V_{pri} \tag{11–5}$$

Equation (11–5) shows that the secondary voltage is equal to the turns ratio times the primary voltage. This condition assumes that all the magnetic flux lines produced in the primary winding pass through the secondary winding.

The turns ratio for a step-up transformer is always greater than 1 since N_{sec} is always greater than N_{pri}.

EXAMPLE 11–4 The transformer in Figure 11–10 has a 200 turn primary winding and a 600 turn secondary winding. What is the voltage across the secondary?

FIGURE 11–10

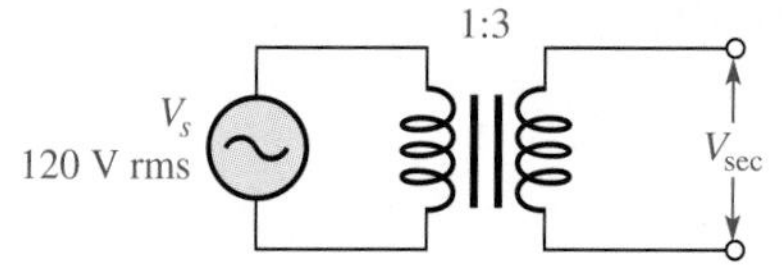

Solution The secondary voltage is

$$V_{sec} = \left(\frac{N_{sec}}{N_{pri}}\right)V_{pri} = \left(\frac{600}{200}\right)120\text{ V} = 3(120\text{ V}) = 360\text{ V}$$

Note that the turns ratio of 3 is indicated on the schematic as 1:3, meaning that there are three secondary turns for each primary turn.

The calculator sequence for V_{sec} is

[6] [0] [0] [÷] [2] [0] [0] [×] [1] [2] [0] [=]

Related Exercise The transformer in Figure 11–10 is changed to one with $N_{pri} = 75$ and $N_{sec} = 300$. Determine V_{sec}.

SECTION 11–3 REVIEW

1. What does a step-up transformer do?
2. If the turns ratio is 5, how much greater is the secondary voltage than the primary voltage?
3. When 240 V ac are applied to a transformer with a turns ratio of 10, what is the secondary voltage?

11–4 ■ THE STEP-DOWN TRANSFORMER

A step-down transformer has more turns in its primary winding than in its secondary winding and is used to decrease ac voltage.

After completing this section, you should be able to

- **Explain how a step-down transformer works**
  - ☐ Identify a step-down transformer by its turns ratio

A transformer in which the secondary voltage is less than the primary voltage is called a **step-down transformer.** The amount by which the voltage is stepped down depends on the turns ratio. Equation (11–5) also applies to a step-down transformer; the turns ratio, however, is always less than 1 since N_{sec} is always less than N_{pri}.

EXAMPLE 11–5 The transformer in Figure 11–11 is part of a laboratory power supply and has 50 turns in the primary and 10 turns in the secondary. What is the secondary voltage?

FIGURE 11–11

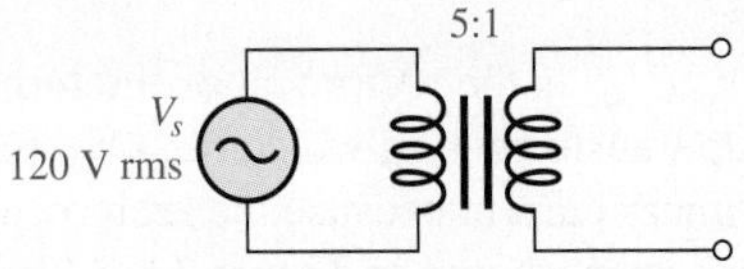

Solution The secondary voltage is

$$V_{sec} = \left(\frac{N_{sec}}{N_{pri}}\right)V_{pri} = \left(\frac{10}{50}\right)120\text{ V} = 0.2(120\text{ V}) = 24\text{ V}$$

Related Exercise The transformer in Figure 11–11 is changed to one with $N_{pri} = 250$ and $N_{sec} = 120$. Determine the secondary voltage.

SECTION 11–4 REVIEW

1. What does a step-down transformer do?
2. One hundred twenty volts ac are applied to the primary winding of a transformer with a turns ratio of 0.5. What is the secondary voltage?
3. A primary voltage of 120 V ac is reduced to 12 V ac. What is the turns ratio?

11–5 ■ LOADING THE SECONDARY WINDING

When a resistive load is connected to the secondary winding of a transformer, the load current depends on both the primary current and the turns ratio.

After completing this section, you should be able to

- **Discuss the effect of a resistive load across the secondary winding**
  - ☐ Determine the secondary current when a step-up transformer is loaded
  - ☐ Determine the secondary current when a step-down transformer is loaded
  - ☐ Discuss power in a transformer

When a load resistor (R_L) is connected to the secondary winding, as shown in Figure 11–12, secondary current will flow because of the voltage induced in the secondary coil. It can be shown that the ratio of the primary current (I_{pri}) to the secondary current (I_{sec}) is equal to the turns ratio, as expressed in the following equation:

$$\frac{I_{pri}}{I_{sec}} = \frac{N_{sec}}{N_{pri}} \quad \textbf{(11–6)}$$

Inverting both sides of the equation gives

$$\frac{I_{sec}}{I_{pri}} = \frac{N_{pri}}{N_{sec}}$$

Solving for I_{sec},

$$I_{sec} = \left(\frac{N_{pri}}{N_{sec}}\right)I_{pri} \qquad \textbf{(11–7)}$$

Note that N_{pri}/N_{sec} is the reciprocal of the turns ratio. Thus, as shown in Figure 11–12(a), for a step-up transformer, in which N_{sec}/N_{pri} is greater than 1, the secondary current is less than the primary current because the reciprocal of the turns ratio is less than 1. For a step-down transformer, shown in Figure 11–12(b), N_{sec}/N_{pri} is less than 1, and I_{sec} is greater than I_{pri} because the reciprocal of the turns ratio is greater than 1.

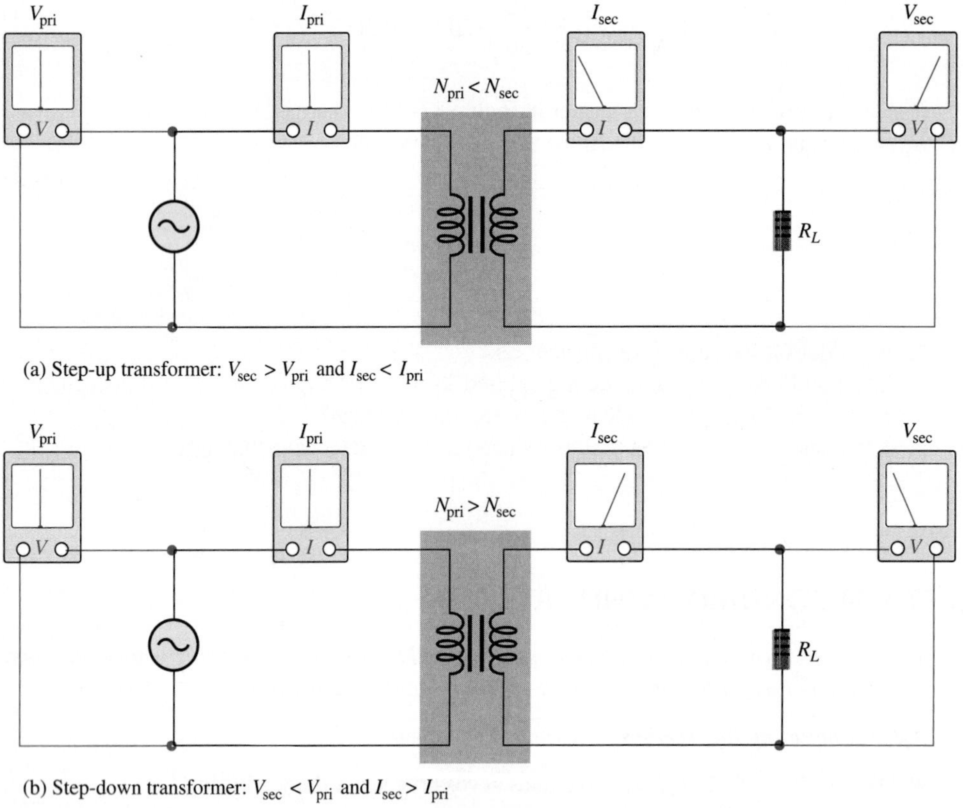

(a) Step-up transformer: $V_{sec} > V_{pri}$ and $I_{sec} < I_{pri}$

(b) Step-down transformer: $V_{sec} < V_{pri}$ and $I_{sec} > I_{pri}$

FIGURE 11–12
Illustration of voltages and currents in a transformer with a loaded secondary winding.

EXAMPLE 11–6 The transformers in Figures 11–13(a) and 11–13(b) have loaded secondaries. If the primary current is 100 mA in each case, how much current flows through the load?

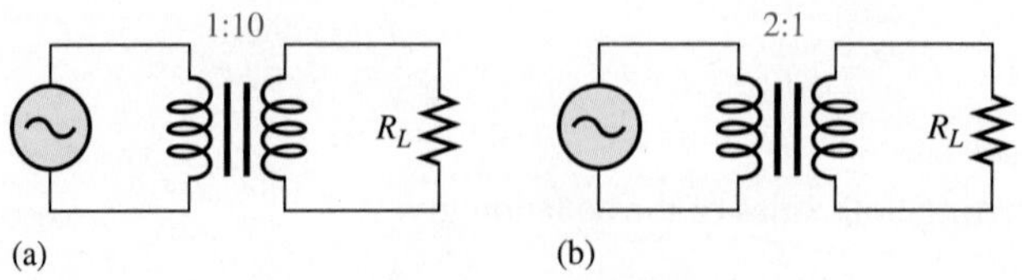

FIGURE 11–13

Solution In Figure 11–13(a), the secondary load current is

$$I_L = I_{sec} = \left(\frac{N_{pri}}{N_{sec}}\right)I_{pri} = 0.1(100\text{ mA}) = 10\text{ mA}$$

In Figure 11–13(b), the secondary load current is

$$I_L = I_{sec} = \left(\frac{N_{pri}}{N_{sec}}\right)I_{pri} = 2(100\text{ mA}) = 200\text{ mA}$$

Related Exercise What is the secondary current in Figure 11–13(a) if the turns ratio is doubled? What is the secondary current in Figure 11–13(b) if the turns ratio is halved? Assume I_{pri} remains at 100 mA in both cases.

Primary Power Equals Secondary Power

When a load is connected to the secondary winding of a transformer, the power transferred to the load can never be greater than the power in the primary winding. For an ideal transformer, the power in the secondary winding (P_{sec}) equals the power in the primary winding (P_{pri}). When losses are considered, the secondary power is always less.

Power is dependent on voltage and current, and there can be no increase in power in a transformer. Therefore, if the voltage is stepped up, the current is stepped down and vice versa. In an ideal transformer, the secondary power is the same as the primary power regardless of the turns ratio, as shown in the following steps:

The primary power is

$$P_{pri} = V_{pri}I_{pri}$$

The secondary power is

$$P_{sec} = V_{sec}I_{sec}$$

From Equations (11–5) and (11–7),

$$I_{sec} = \left(\frac{N_{pri}}{N_{sec}}\right)I_{pri} \quad \text{and} \quad V_{sec} = \left(\frac{N_{sec}}{N_{pri}}\right)V_{pri}$$

By substitution,

$$P_{sec} = \left(\frac{\cancel{N_{sec}}}{\cancel{N_{pri}}}\right)\left(\frac{\cancel{N_{pri}}}{\cancel{N_{sec}}}\right)V_{pri}I_{pri}$$

Canceling the terms,

$$P_{sec} = V_{pri}I_{pri} = P_{pri}$$

This result is closely approached in practice because of the very high efficiencies of transformers.

SECTION 11–5 REVIEW

1. If the turns ratio of a transformer is 2, is the secondary current greater than or less than the primary current? By how much?
2. A transformer has 100 primary turns and 25 secondary turns, and I_{pri} is 0.5 A. What is the value of I_{sec}?
3. In Question 2, how much primary current is necessary to produce a secondary load current of 10 A?

11–6 ■ REFLECTED LOAD

From the viewpoint of the primary circuit, a load connected across the secondary winding of a transformer appears to have a resistance that is not necessarily equal to the actual resistance of the load. The actual load is essentially "reflected" into the primary circuit altered by the turns ratio. This reflected load is what the primary source effectively sees and it determines the amount of primary current.

After completing this section, you should be able to

- **Discuss the concept of a reflected load in a transformer**
  - ☐ Define *reflected resistance*
  - ☐ Explain how the turns ratio affects the reflected resistance
  - ☐ Calculate reflected resistance

The concept of the **reflected load** is illustrated in Figure 11–14. The load (R_L) in the secondary circuit of a transformer is reflected into the primary circuit by transformer action. The load appears to the source in the primary circuit to be a resistance (R_{pri}) with a value determined by the turns ratio and the actual value of the load resistance. The resistance R_{pri} is called the **reflected resistance.**

The resistance in the primary circuit of Figure 11–14 is $R_{pri} = V_{pri}/I_{pri}$. The resistance in the secondary circuit is $R_L = V_{sec}/I_{sec}$. From Equations (11–4) and (11–6), you know that $V_{sec}/V_{pri} = N_{sec}/N_{pri}$ and $I_{pri}/I_{sec} = N_{sec}/N_{pri}$. Using these relationships, a formula for R_{pri} in terms of R_L is determined as follows:

$$\frac{R_{pri}}{R_L} = \frac{V_{pri}/I_{pri}}{V_{sec}/I_{sec}} = \left(\frac{V_{pri}}{V_{sec}}\right)\left(\frac{I_{sec}}{I_{pri}}\right) = \left(\frac{N_{pri}}{N_{sec}}\right)\left(\frac{N_{pri}}{N_{sec}}\right) = \frac{N_{pri}^2}{N_{sec}^2} = \left(\frac{N_{pri}}{N_{sec}}\right)^2$$

Solving for R_{pri},

$$R_{pri} = \left(\frac{N_{pri}}{N_{sec}}\right)^2 R_L \qquad \textbf{(11–8)}$$

Equation (11–8) shows that the resistance reflected into the primary circuit is the square of the reciprocal of the turns ratio times the load resistance.

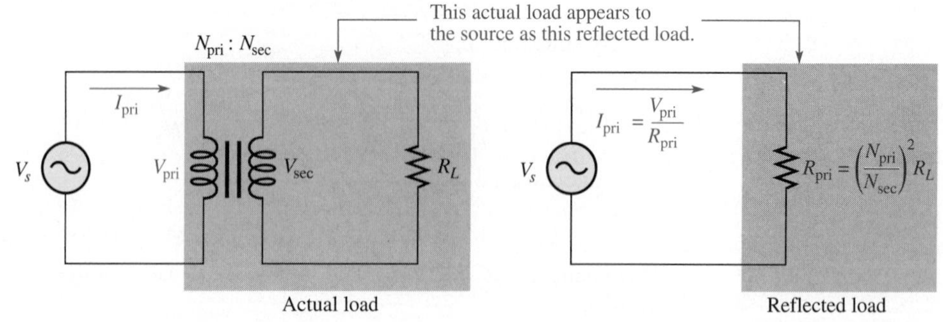

FIGURE 11–14
Reflected load in a transformer circuit.

EXAMPLE 11–7 Figure 11–15 shows a source that is transformer-coupled to a load resistor of 100 Ω. The transformer has 100 primary turns and 400 secondary turns for a turns ratio of 4. What is the reflected resistance seen by the source?

FIGURE 11–15

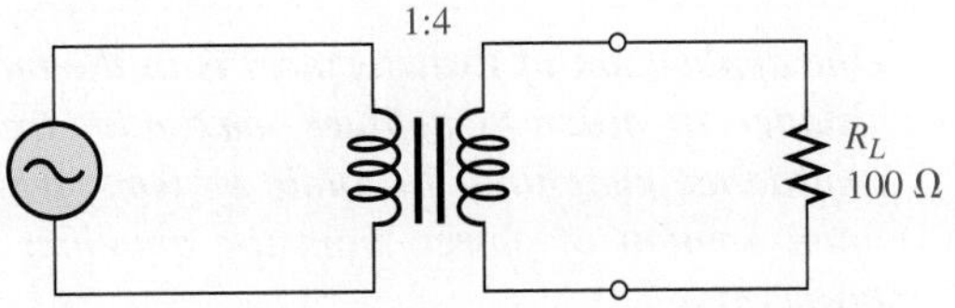

Solution The reflected resistance is determined as follows:

$$R_{pri} = \left(\frac{N_{pri}}{N_{sec}}\right)^2 R_L = \left(\frac{100}{400}\right)^2 100\ \Omega = \left(\frac{1}{4}\right)^2 100\ \Omega = \left(\frac{1}{16}\right) 100\ \Omega = 6.25\ \Omega$$

The source sees a resistance of 6.25 Ω just as if it were connected directly, as shown in the equivalent circuit of Figure 11–16.

The calculator sequence is

[1] [0] [0] [÷] [4] [0] [0] [=] [x^2] [×] [1] [0] [0] [=]

FIGURE 11–16

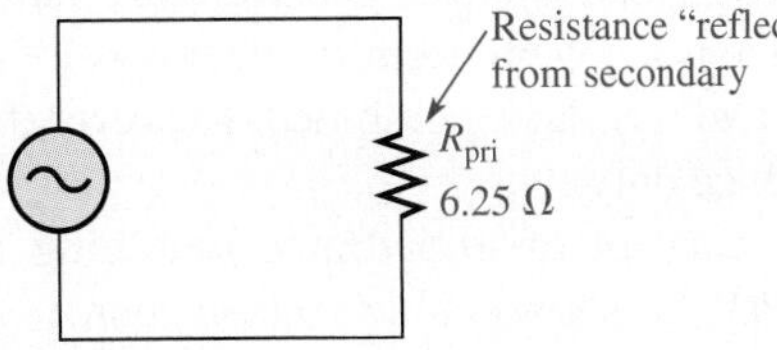

Related Exercise If the turns ratio in Figure 11–15 is 10 and R_L is 600 Ω, what is the reflected resistance?

EXAMPLE 11–8 In Figure 11–15, if a transformer is used having 40 primary turns and 10 secondary turns, what is the reflected resistance? Note that the turns ratio (n) is 0.25.

Solution The reflected resistance is

$$R_{pri} = \left(\frac{N_{pri}}{N_{sec}}\right)^2 R_L = \left(\frac{40}{10}\right)^2 100\ \Omega = (4)^2 100\ \Omega = 1600\ \Omega$$

Related Exercise To achieve a reflected resistance of 800 Ω, what turns ratio is required in Figure 11–15?

Examples 11–7 and 11–8 illustrate that in a step-up transformer ($n > 1$), the reflected resistance is less than the actual load resistance; and in a step-down transformer ($n < 1$), the reflected resistance is greater than the load resistance.

SECTION 11–6 REVIEW

1. Define *reflected resistance.*
2. What transformer characteristic determines the reflected resistance?
3. A given transformer has a turns ratio of 10, and the load is 50 Ω. How much resistance is reflected into the primary?
4. What is the turns ratio required to reflect a 4 Ω load resistance into the primary as 400 Ω?

11–7 ■ MATCHING THE LOAD AND SOURCE RESISTANCES

One application of transformers is in the matching of a load resistance to a source resistance in order to achieve maximum transfer of power. This technique is called impedance matching. In audio systems, transformers are often used to get the maximum amount of power from the amplifier to the speaker by proper selection of the turns ratio.

After completing this section, you should be able to

- **Discuss impedance matching with transformers**
  - □ Give a general definition of *impedance*
  - □ Discuss the maximum power transfer theorem
  - □ Define *impedance matching*
  - □ Explain the purpose of impedance matching
  - □ Describe a practical application

The term *impedance* will become very familiar to you in the next chapter. Basically, impedance is a general term for the opposition to current, including the effects of both resistance and reactance combined. However, for the time being, we will confine our usage to resistance only.

The concept of **impedance matching** is illustrated in the basic circuit of Figure 11–17. Part (a) shows an ac voltage source with a series resistance representing its internal resistance. Some internal resistance is inherent in all sources due to their internal circuitry or physical makeup. When the source is connected directly to a load, as shown in part (b), often the objective is to transfer as much of the power produced by the source to the load as possible. However, a certain amount of the power produced by the source is lost in its internal resistance, and the remaining power goes to the load.

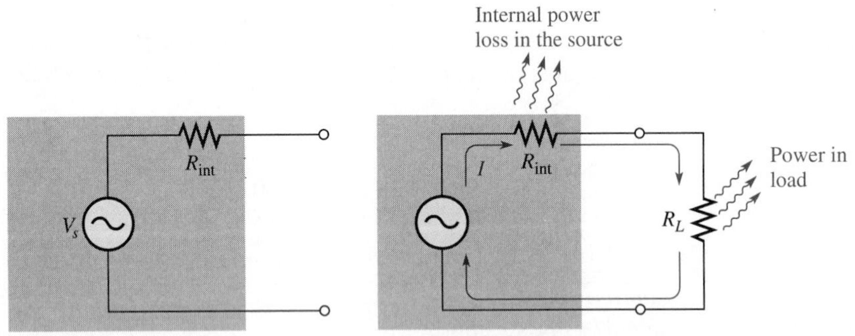

(a) Voltage source with internal resistence, R_{int}

(b) A portion of the total power is lost in R_{int}

FIGURE 11–17
Power transfer from a nonideal voltage source to a load.

Maximum Power Transfer Theorem

The **maximum power transfer theorem** is important when you need to know the value of the load at which the most power is delivered from the source. The theorem can be stated as follows:

> **When a source is connected to a load, maximum power is delivered to the load when the load resistance is equal to the internal source resistance.**

We will demonstrate this theorem by finding the load power for various values of load resistance and a specific internal source resistance in Example 11–9.

EXAMPLE 11–9 The source in Figure 11–18 has an internal source resistance of 75 Ω. Determine the load power for each of the following values of load resistance:

(a) 25 Ω **(b)** 50 Ω **(c)** 75 Ω **(d)** 100 Ω **(e)** 125 Ω

Draw a graph showing the load power versus the load resistance.

FIGURE 11–18

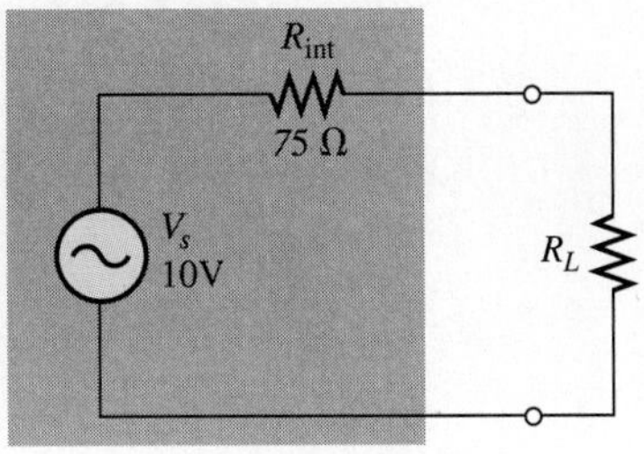

Solution Use Ohm's law ($I = V/R$) and the power formula ($P = I^2R$) to find the load power for each value of load resistance.

(a) For $R_L = 25\ \Omega$,

$$I = \frac{V_s}{R_{int} + R_L} = \frac{10\ \text{V}}{100\ \Omega} = 100\ \text{mA}$$

$$P_L = I^2R_L = (100\ \text{mA})^2(25\ \Omega) = 250\ \text{mW}$$

(b) For $R_L = 50\ \Omega$,

$$I = \frac{V_s}{R_{int} + R_L} = \frac{10\ \text{V}}{125\ \Omega} = 80\ \text{mA}$$

$$P_L = I^2R_L = (80\ \text{mA})^2(50\ \Omega) = 320\ \text{mW}$$

(c) For $R_L = 75\ \Omega$,

$$I = \frac{V_s}{R_{int} + R_L} = \frac{10\ \text{V}}{150\ \Omega} = 66.7\ \text{mA}$$

$$P_L = I^2R_L = (66.7\ \text{mA})^2(75\ \Omega) = 334\ \text{mW}$$

(d) For $R_L = 100\ \Omega$,

$$I = \frac{V_s}{R_{int} + R_L} = \frac{10\ \text{V}}{175\ \Omega} = 57.1\ \text{mA}$$

$$P_L = I^2R_L = (57.1\ \text{mA})^2(100\ \Omega) = 326\ \text{mW}$$

(e) For $R_L = 125\ \Omega$,

$$I = \frac{V_s}{R_{int} + R_L} = \frac{10\ \text{V}}{200\ \Omega} = 50\ \text{mA}$$

$$P_L = I^2R_L = (50\ \text{mA})^2(125\ \Omega) = 313\ \text{mW}$$

Notice that the load power is greatest when $R_L = 75\ \Omega$, which is the same as the internal source resistance. When the load resistance is less than or greater than this value, the load power drops off, as the curve in Figure 11–19 graphically illustrates.

The calculator sequence for $R_L = 25\ \Omega$ is

[1] [0] [÷] [(] [7] [5] [+] [2] [5] [)] [=] [x^2] [×] [2] [5] [=]

FIGURE 11–19
Curve showing that the load power is maximum when $R_L = R_{int}$.

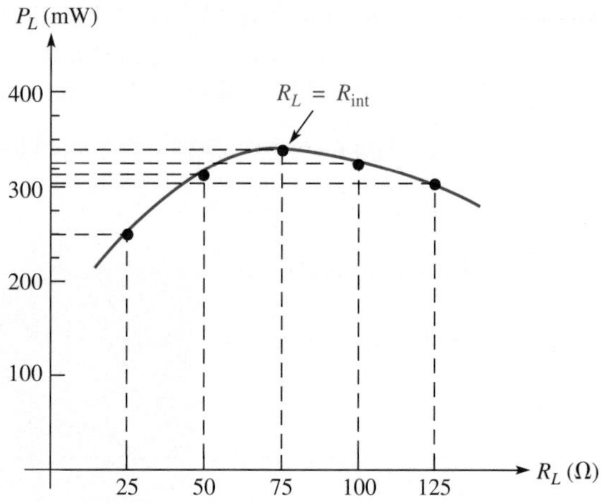

Related Exercise What is the maximum power that can be transfered from a 10 V source with an internal source resistance of 30 Ω?

A Practical Application

In most practical situations, the internal source resistance of various types of sources is fixed. Also, in many cases, the resistance of a device that acts as a load is fixed and cannot be altered. If you need to connect a given source to a given load, remember that only by chance will their resistances match. In this situation a transformer comes in handy. You can use the reflected-resistance characteristic of a transformer to make the load resistance appear to have the same value as the source resistance, thereby "fooling" the source into "thinking" that there is a match.

Let's take a practical, everyday situation to illustrate. The typical resistance of the input to a TV receiver is 300 Ω. An antenna must be connected to this input by a lead-in cable in order to receive TV signals. In this situation, the antenna and the lead-in act as the source, and the input resistance of the TV receiver is the load, as illustrated in Figure 11–20.

FIGURE 11–20
An antenna directly coupled to a TV receiver.

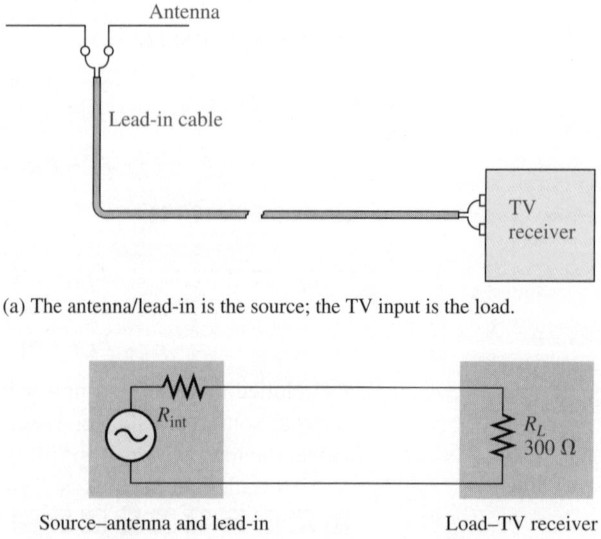

(a) The antenna/lead-in is the source; the TV input is the load.

(b) Circuit equivalent of antenna and TV receiver system

It is common for an antenna system to have a characteristic resistance of 75 Ω. Thus, if the 75 Ω source (antenna and lead-in) is connected directly to the 300 Ω TV input, maximum power will not be delivered to the input to the TV, and you will have poor signal reception. The solution is to use a matching transformer connected as indicated in Figure 11–21, in order to match the 300 Ω load resistance to the 75 Ω source resistance.

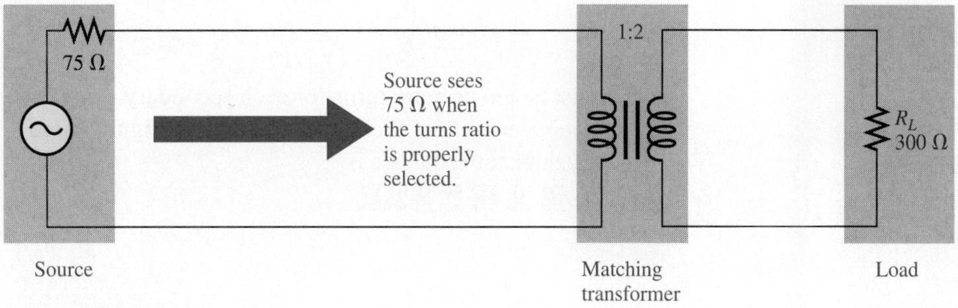

FIGURE 11–21
Example of a load matched to a source by transformer coupling for maximum power transfer.

To match the resistances, you must select a proper value of turns ratio. You want the 300 Ω load resistance to look like 75 Ω to the source. Use Equation (11–8) to obtain a formula to determine the turns ratio when you know the values for R_L and R_{pri}:

$$R_{pri} = \left(\frac{N_{pri}}{N_{sec}}\right)^2 R_L$$

Transpose terms and divide both sides by R_L:

$$\left(\frac{N_{pri}}{N_{sec}}\right)^2 = \frac{R_{pri}}{R_L}$$

Then take the square root of both sides:

$$\frac{N_{pri}}{N_{sec}} = \sqrt{\frac{R_{pri}}{R_L}}$$

Invert both sides to get the following formula:

$$\frac{N_{sec}}{N_{pri}} = \sqrt{\frac{R_L}{R_{pri}}} \qquad \textbf{(11–9)}$$

Finally, solve for this particular turns ratio:

$$n = \frac{N_{sec}}{N_{pri}} = \sqrt{\frac{300\ \Omega}{75\ \Omega}} = \sqrt{4} = 2$$

Therefore, a matching transformer with a turns ratio of 2 must be used in this application.

EXAMPLE 11–10 An amplifier has an 800 Ω internal resistance. In order to provide maximum power to an 8 Ω speaker, what turns ratio must be used in the coupling transformer?

Solution The reflected resistance must equal 800 Ω. Thus, from Equation (11–9), the turns ratio can be determined.

$$n = \frac{N_{sec}}{N_{pri}} = \sqrt{\frac{R_L}{R_{pri}}} = \sqrt{\frac{8\ \Omega}{800\ \Omega}} = \sqrt{0.01} = 0.1$$

There must be ten primary turns for each secondary turn because $n = 0.1$. The diagram and its equivalent reflected circuit are shown in Figure 11–22.

The calculator sequence is

[8] [÷] [8] [0] [0] [=] [2nd] [√x]

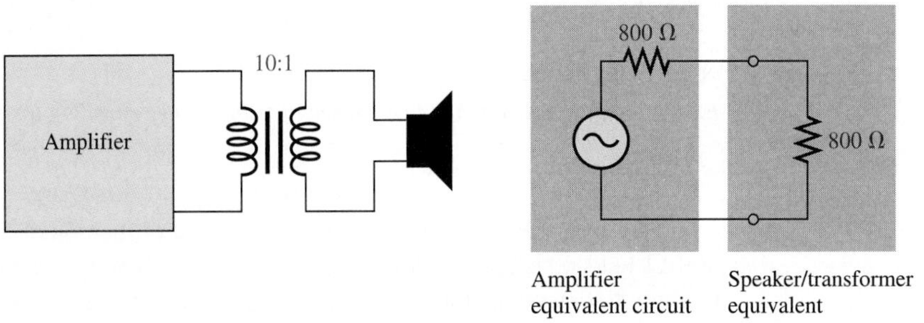

FIGURE 11–22

Related Exercise What must be the turns ratio in Figure 11–22 to provide maximum power to two 8 Ω speakers in parallel?

SECTION 11–7 REVIEW

1. What does impedance matching mean?
2. What is the advantage of matching the load resistance to the resistance of a source?
3. A transformer has 100 primary turns and 50 secondary turns. What is the reflected resistance with 100 Ω across the secondary?

11–8 ■ THE TRANSFORMER AS AN ISOLATION DEVICE

Transformers are useful in providing electrical isolation between the primary circuit and the secondary circuit because there is no electrical connection between the two windings. In a transformer, energy is transferred entirely by magnetic coupling.

After completing this section, you should be able to

- **Explain how the transformer acts as an isolation device**
  - □ Discuss dc isolation
  - □ Discuss power line isolation

DC Isolation

If a nonchanging direct current flows through the primary circuit of a transformer, nothing happens in the secondary circuit, as indicated in Figure 11–23(a). The reason is that the primary current must be changing in order to create a changing magnetic field. This will cause voltage to be induced in the secondary circuit, as indicated in Figure 11–23(b). Therefore, the transformer isolates the secondary circuit from any dc voltage in the primary circuit.

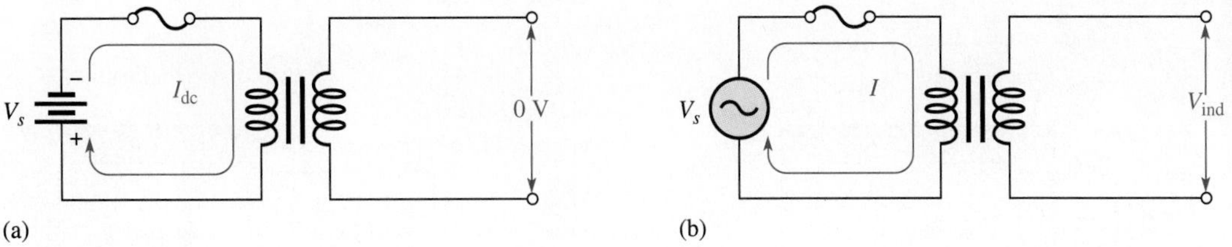

FIGURE 11–23
DC isolation and ac coupling.

For example, in some amplifiers, transformers are used to prevent dc voltages at certain points in the circuit, while allowing the signal (ac) voltage to get through. Figure 11–24 is a simplified illustration of transformer coupling in a two-stage audio amplifier.

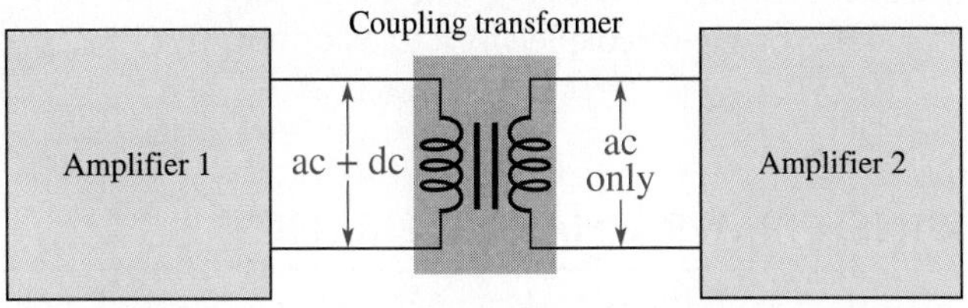

FIGURE 11–24
Audio amplifier stages with transformer coupling for dc isolation.

Power Line Isolation

Transformers are often used to electrically isolate electronic equipment from the 60 Hz, 110 V ac power line. The reason for using an isolation transformer to couple the 60 Hz ac to an instrument is to prevent a possible shock hazard if the 110 V line is connected to the metal chassis of the equipment. This condition is possible if the line cord plug can be inserted into an outlet either way. Incidentally, to prevent this situation, many plugs have keyed prongs so that they can be plugged in only one way.

Figure 11–25 illustrates how a transformer can prevent the metal chassis from being connected to the 110 V line rather than to neutral (ground), no matter how the cord is plugged into the outlet. When an isolation transformer is used, the secondary circuit is said to be "floating" because it is not referenced to the power line ground. Should a person come in contact with the secondary voltage, there is no complete current path back to ground, and therefore there is no shock hazard. Current must flow through your body in order for you to receive an electrical shock.

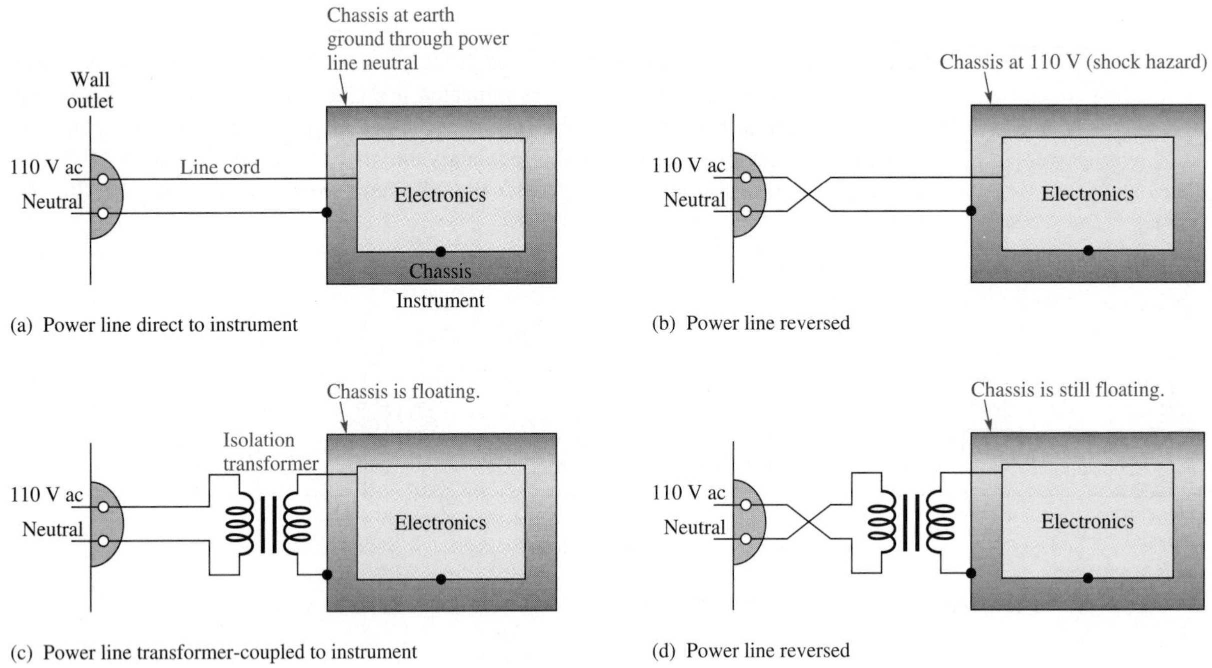

FIGURE 11–25
The use of an isolation transformer to prevent a shock hazard.

SECTION 11–8 REVIEW

1. What does the term *electrical isolation* mean?
2. Can a dc voltage be coupled by a transformer?

11–9 ■ NONIDEAL TRANSFORMER CHARACTERISTICS

Up to this point, transformer operation has been discussed from an ideal point of view, and this approach is valid when you are learning new concepts. However, you should be aware of the nonideal characteristics of practical transformers and how they affect performance.

After completing this section, you should be able to

- **Describe a practical transformer**
  - ☐ List and describe the nonideal characteristics
  - ☐ Explain power rating of a transformer
  - ☐ Define *efficiency* of a transformer

So far, the transformer has been considered as an ideal device. That is, the winding resistance, the winding capacitance, and nonideal core characteristics were all neglected and the transformer was treated as if it had an efficiency of 100%. For studying the basic concepts and in many applications, the ideal model is valid. However, the practical transformer has several nonideal characteristics of which you should be aware.

Winding Resistance

Both the primary and the secondary windings of a practical transformer have winding resistance. (You learned about the winding resistance of inductors in Chapter 10.) The

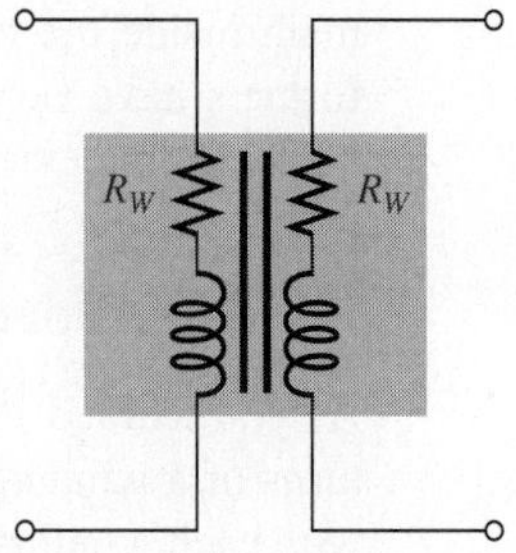

FIGURE 11–26
Winding resistance in a practical transformer.

winding resistances of a practical transformer are represented as resistors in series with the windings as shown in Figure 11–26.

Winding resistance in a practical transformer results in less voltage across a secondary load. Voltage drops due to the winding resistance effectively subtract from the primary and secondary voltages and result in a load voltage that is less than that predicted by the relationship $V_{sec} = (N_{sec}/N_{pri})V_{pri}$. In most cases the effect is relatively small and can be neglected.

Losses in the Core

There is always some energy loss in the core material of a practical transformer. This loss is seen as a heating of ferrite and iron cores, but it does not occur in air cores. Part of this energy is consumed in the continuous reversal of the magnetic field due to the changing direction of the primary current; this energy loss is called *hysteresis loss.* The rest of the energy loss is caused by eddy currents induced in the core material by the changing magnetic flux. The eddy currents flow in circular patterns in the core resistance, thus causing the energy loss. This loss is greatly reduced by the use of laminated construction of iron cores. The thin layers of ferromagnetic material are insulated from each other to minimize the buildup of eddy currents by confining them to a small area and to keep core losses to a minimum.

Magnetic Flux Leakage

In an ideal transformer, all the magnetic flux produced by the primary current is assumed to pass through the core to the secondary winding and vice versa. In a practical transformer, some of the magnetic flux lines break out of the core and pass through the surrounding air back to the other end of the winding, as illustrated in Figure 11–27 for the magnetic field produced by the primary current. Magnetic flux leakage results in a reduced secondary voltage.

The percentage of magnetic flux that actually reaches the secondary determines the coefficient of coupling of the transformer. For example, if nine out of ten flux lines re-

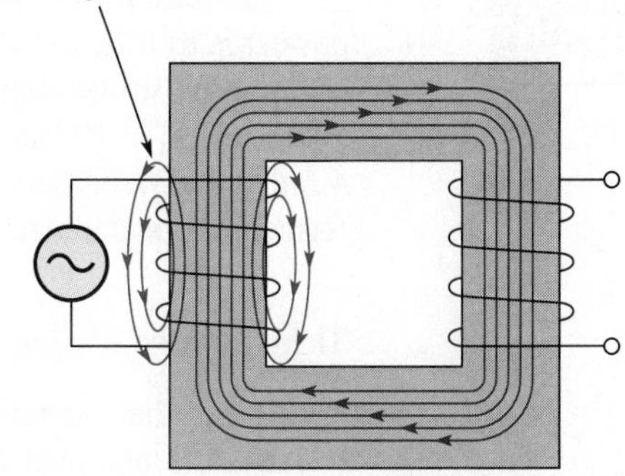

FIGURE 11–27
Flux leakage in a practical transformer.

main inside the core, the coefficient of coupling is 0.90 or 90%. Most iron-core transformers have very high coefficients of coupling (greater than 0.99), whereas ferrite-core and air-core devices have lower values.

Winding Capacitance

As you learned in Chapter 10, there is always some stray capacitance between adjacent turns of a winding. These stray capacitances result in an effective capacitance in parallel with each winding of a transformer, as indicated in Figure 11–28.

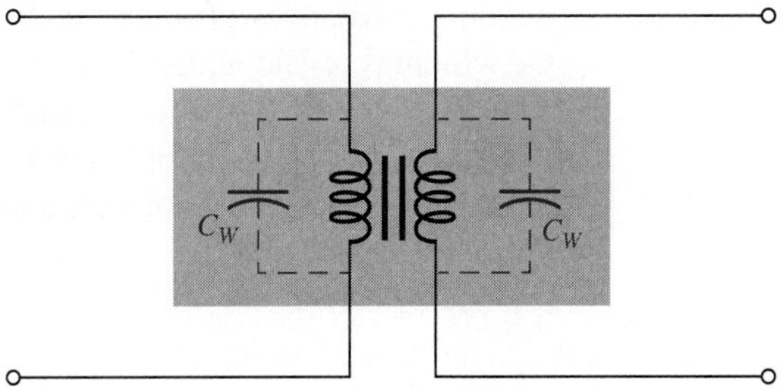

FIGURE 11–28
Winding capacitance in a practical transformer.

These stray capacitances have very little effect on the transformer's operation at low frequencies, because the reactances (X_C) are very high. However, at higher frequencies, the reactances decrease and begin to produce a bypassing effect across the primary winding and across the secondary load. As a result, less of the total primary current flows through the primary winding, and less of the total secondary current flows through the load. This effect reduces the load voltage as the frequency goes up.

Transformer Power Rating

A transformer is typically rated in volt-amperes (VA), primary/secondary voltage, and operating frequency. For example, a given transformer rating may be specified as 2 kVA, 500/50, 60 Hz. The 2 kVA value is the **apparent power rating.** The 500 and the 50 can be either secondary or primary voltages.

Let's assume, for example, that 50 V is the secondary voltage. In this case the load current is

$$I_L = \frac{P_{sec}}{V_{sec}} = \frac{2\text{ kVA}}{50\text{ V}} = 40\text{ A}$$

On the other hand, if 500 V is the secondary voltage, then

$$I_L = \frac{P_{sec}}{V_{sec}} = \frac{2\text{ kVA}}{500\text{ V}} = 4\text{ A}$$

These are the maximum currents that the secondary can handle in either case.

The reason that the power rating is in volt-amperes (VA) rather than in watts (true power) is as follows: If the transformer load is purely capacitive or purely inductive, the true power (watts) delivered to the load is ideally zero. However, the current for V_{sec} = 500 V and $X_C = 100\ \Omega$ at 60 Hz, for example, is 5 A. This current exceeds the maximum of 4 A that the 2 kVA secondary can handle, and even though the true power is zero, the transformer may be damaged. So it is meaningless to specify power in watts for transformers.

Transformer Efficiency

Recall that the secondary power is equal to the primary power in an ideal transformer. Because the nonideal characteristics just discussed result in a power loss in the transformer, the secondary (output) power is always less than the primary (input) power. The **efficiency** (η) of a transformer is a measure of the percentage of the input power that is delivered to the output:

$$\eta = \left(\frac{P_{out}}{P_{in}}\right)100\% \qquad \textbf{(11–10)}$$

Most power transformers have efficiencies in excess of 95%.

EXAMPLE 11–11 A certain type of transformer has a primary current of 5 A and a primary voltage of 4800 V. The secondary current is 90 A and the secondary voltage is 240 V. Determine the efficiency of this transformer.

Solution The input power is

$$P_{in} = V_{pri}I_{pri} = (4800\text{ V})(5\text{ A}) = 24\text{ kVA}$$

The output power is

$$P_{out} = V_{sec}I_{sec} = (240\text{ V})(90\text{ A}) = 21.6\text{ kVA}$$

The efficiency is

$$\eta = \left(\frac{P_{out}}{P_{in}}\right)100\% = \left(\frac{21.6\text{ kVA}}{24\text{ kVA}}\right)100\% = 90\%$$

Related Exercise A transformer has a primary current of 8 A with a primary voltage of 440 V. The secondary current is 30 A and the secondary voltage is 100 V. What is the efficiency?

SECTION 11–9 REVIEW

1. Explain how a practical transformer differs from the ideal model.
2. The coefficient of coupling of a certain transformer is 0.85. What does this mean?
3. A certain transformer has a rating of 10 kVA. If the secondary voltage is 250 V, how much load current can the transformer handle?

11–10 ▪ OTHER TYPES OF TRANSFORMERS

There are several important variations of the basic transformer. They include tapped transformers, multiple-winding transformers, and autotransformers.

After completing this section, you should be able to

- **Describe several types of transformers**
  - ☐ Describe center-tapped transformers
  - ☐ Describe multiple-winding transformers
  - ☐ Describe autotransformers

Tapped Transformers

A schematic of a transformer with a center-tapped secondary winding is shown in Figure 11–29(a). The **center tap (CT)** is equivalent to two secondary windings with half the total voltage across each.

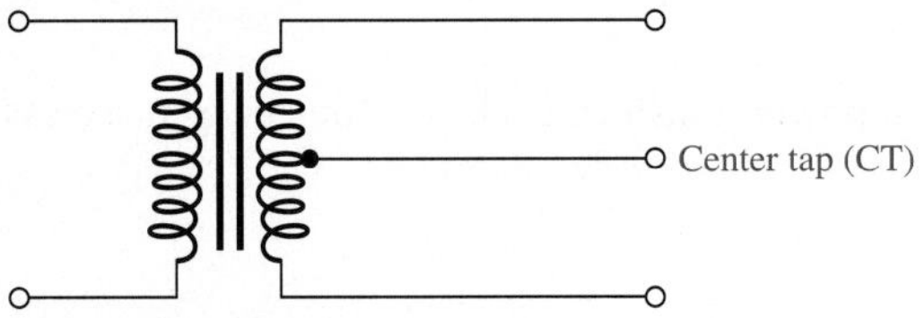

(a) Center-tapped transformer

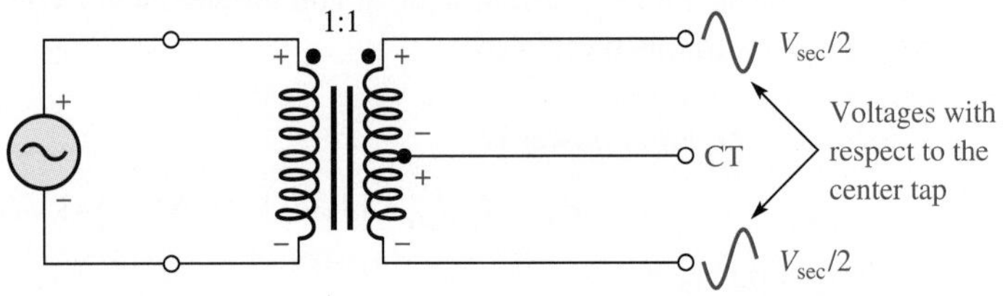

(b) Output voltages with respect to the center tap are 180° out of phase with each other and are one-half the magnitude of the secondary voltage.

FIGURE 11–29
Operation of a center-tapped transformer.

The voltages between either end of the secondary winding and the center tap are, at any instant, equal in magnitude but opposite in polarity, as illustrated in Figure 11–29(b). Here, for example, at some instant on the sine wave voltage, the polarity across the entire secondary winding is as shown (top end +, bottom −). At the center tap, the voltage is less positive than the top end but more positive than the bottom end of the secondary winding. Therefore, measured with respect to the center tap, the top end of the secondary winding is positive, and the bottom end is negative. This center-tapped feature is used in some power supply rectifiers in which the ac voltage is converted in dc, as illustrated in Figure 11–30.

Some tapped transformers have taps on the secondary winding at points other than the electrical center. Also, multiple primary and secondary taps are sometimes used in certain applications. Examples of these types of transformers are shown in Figure 11–31.

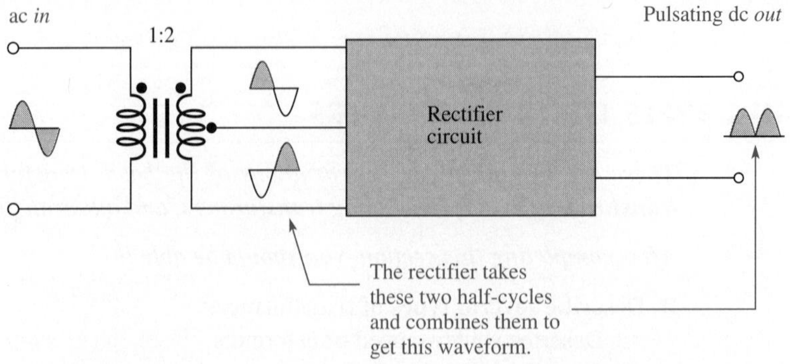

FIGURE 11–30
Application of a center-tapped transformer in ac-to-dc conversion.

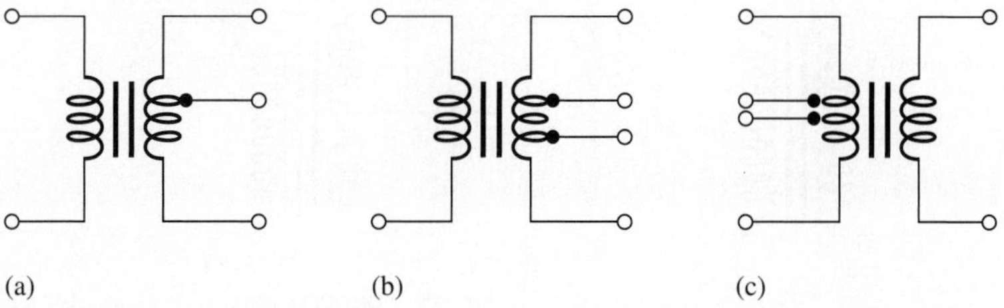

FIGURE 11–31
Tapped transformers.

One example of a transformer with a multiple-tap primary winding and a center-tapped secondary winding is the utility-pole transformer used by power companies to step down the high voltage from the power line to 110 V/220 V service for residential and commercial customers, as shown in Figure 11–32. The multiple taps on the primary winding are used for minor adjustments in the turns ratio in order to overcome line voltages that are slightly too high or too low.

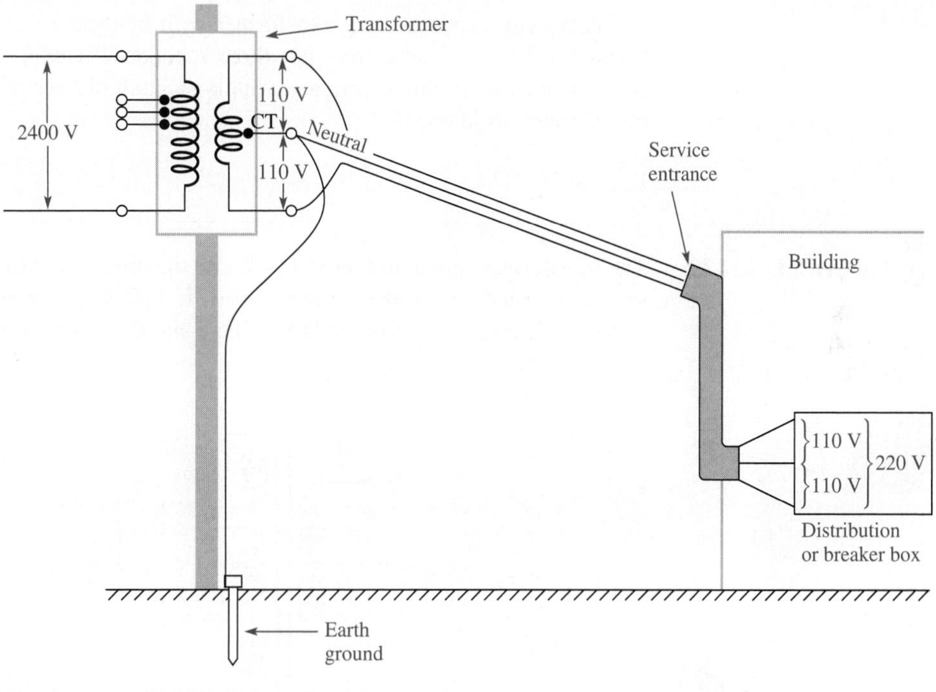

FIGURE 11–32
Utility-pole transformer in a typical power distribution system.

Multiple-Winding Transformers

Some transformers are designed to operate from either 110 V ac or 220 V ac lines. These transformers usually have two primary windings, each of which is designed for 110 V ac. When the two are connected in series, the transformer can be used for 220 V ac operation, as illustrated in Figure 11–33.

More than one secondary winding can be wound on a common core. Transformers with several secondary windings are often used to achieve several voltages by either stepping up or stepping down the primary voltage. These types are commonly used in power supply applications in which several voltage levels are required for the operation of an electronic instrument.

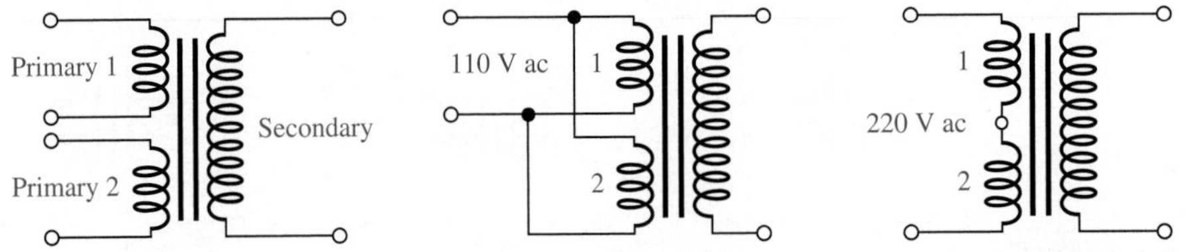

(a) Two primary windings (b) Primary windings in parallel for 110 V ac operation (c) Primary windings in series for 220 V ac operation

FIGURE 11–33
Multiple-primary transformer.

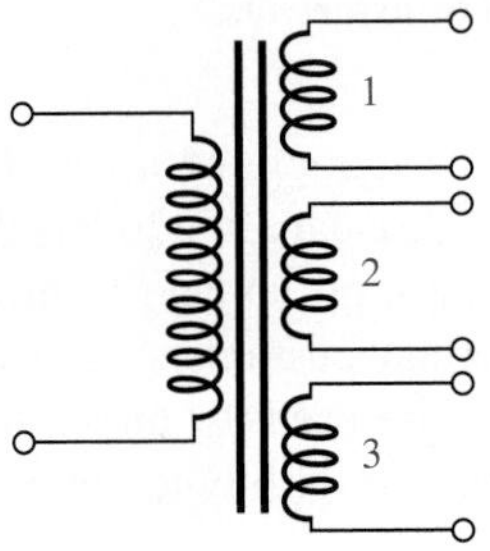

FIGURE 11–34
Multiple-secondary transformer.

A typical schematic of a transformer with multiple secondary windings is shown in Figure 11–34; this transformer has three secondary windings. Sometimes you will find combinations of multiple primary windings, multiple secondary windings, and tapped transformers all in one unit.

EXAMPLE 11–12

The transformer shown in Figure 11–35 has the numbers of turns indicated. One of the secondary windings is also center tapped. If 110 V ac are connected to the primary winding, determine each secondary voltage and the voltages with respect to CT on the middle secondary winding.

FIGURE 11–35

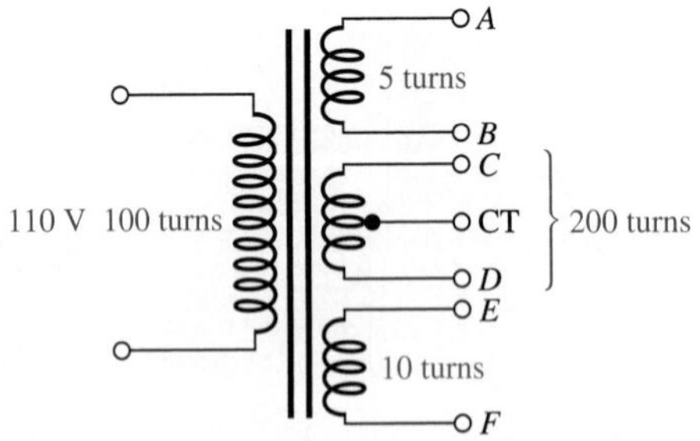

Solution

$$V_{AB} = \left(\frac{5}{100}\right)110\text{ V} = 5.5\text{ V}$$

$$V_{CD} = \left(\frac{200}{100}\right)110\text{ V} = 220\text{ V}$$

$$V_{(\text{CT})C} = V_{(\text{CT})D} = \frac{220\text{ V}}{2} = 110\text{ V}$$

$$V_{EF} = \left(\frac{10}{100}\right)110\text{ V} = 11\text{ V}$$

Related Exercise Repeat the calculations for a primary winding with 50 turns.

Autotransformers

In an **autotransformer,** one winding serves as both the primary and the secondary windings. The winding is tapped at the proper points to achieve the desired turns ratio for stepping up or stepping down the voltage.

Autotransformers differ from conventional transformers in that there is no electrical isolation between the primary and the secondary circuits because both are on one winding. Autotransformers normally are smaller and lighter than equivalent conventional transformers because they require a much lower kVA rating for a given load. Many autotransformers provide an adjustable tap using a sliding contact mechanism so that the output voltage can be varied (these are often called *variacs*). Figure 11–36 shows a typical autotransformer and several schematic symbols.

The following example illustrates why an autotransformer has a kVA requirement that is less than the input or output kVA.

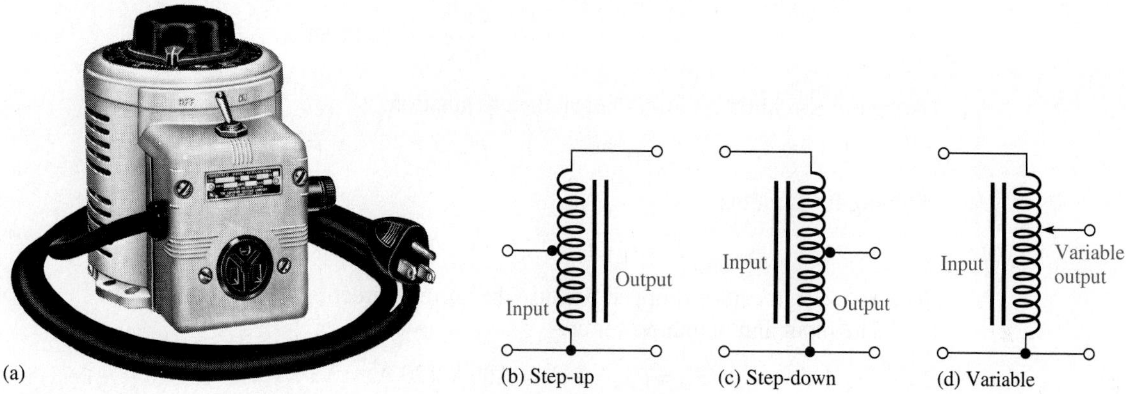

FIGURE 11–36
The autotransformer. Part (a) courtesy of Superior Electric Co.

EXAMPLE 11–13 A certain autotransformer is used to change a source voltage of 220 V to a load voltage of 160 V across an 8 Ω load resistance. Determine the input and output power in kVA, and show that the actual kVA requirement is less than this value. Assume that this transformer is ideal.

Solution The circuit is shown in Figure 11–37 with the voltages and currents indicated. The current directions have been assigned arbitrarily for convenience.

FIGURE 11–37

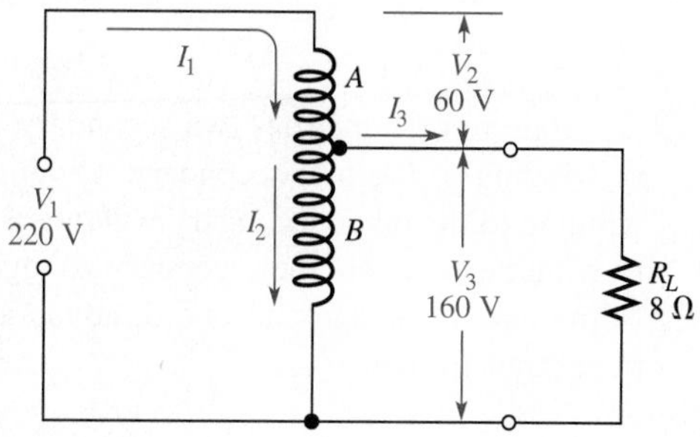

The load current I_3 is determined as follows:

$$I_3 = \frac{V_3}{R_L} = \frac{160\text{ V}}{8\ \Omega} = 20\text{ A}$$

The input power is the total source voltage (V_1) times the total current from the source (I_1):

$$P_{\text{in}} = V_1 I_1$$

The output power is the load voltage V_3 times the load current I_3:

$$P_{\text{out}} = V_3 I_3$$

For an ideal transformer, $P_{\text{in}} = P_{\text{out}}$; thus,

$$V_1 I_1 = V_3 I_3$$

Solving for I_1 yields

$$I_1 = \frac{V_3 I_3}{V_1} = \frac{(160\text{ V})(20\text{ A})}{220\text{ V}} = 14.55\text{ A}$$

Applying Kirchhoff's current law at the tap junction,

$$I_1 = I_2 + I_3$$

Solving for I_2 yields

$$I_2 = I_1 - I_3 = 14.55\text{ A} - 20\text{ A} = -5.45\text{ A}$$

The minus sign can be dropped because the current directions are arbitrary.

The input and output power are

$$P_{\text{in}} = P_{\text{out}} = V_3 I_3 = (160\text{ V})(20\text{ A}) = 3.2\text{ kVA}$$

The power in winding A is

$$V_2 I_1 = (60\text{ V})(14.55\text{ A}) = 873\text{ VA}$$

The power in winding B is

$$V_3 I_2 = (160\text{ V})(5.45\text{ A}) = 872\text{ VA}$$

Thus, the power rating required for each winding is less than the power that is delivered to the load. The slight difference in the calculated powers in windings A and B is due to rounding.

Related Exercise What happens to the kVA requirement if the load is changed to 4 Ω?

SECTION 11–10 REVIEW

1. A certain transformer has two secondary windings. The turns ratio from the primary winding to the first secondary winding is 10. The turns ratio from the primary winding to the other secondary winding is 0.2. If 220 V ac are applied to the primary winding, what are the secondary voltages?
2. Name one advantage and one disadvantage of an autotransformer over a conventional transformer.

11–11 ■ TROUBLESHOOTING

The common failures in transformers are opens, shorts, or partial shorts in either the primary or the secondary windings. One cause of such failures is the operation of the device under conditions that exceed its ratings. A few transformer failures and the associated symptoms are covered in this section.

After completing this section, you should be able to

- **Troubleshoot transformers**
  - ☐ Find an open primary or secondary winding
  - ☐ Find a shorted or partially shorted primary or secondary winding

Open Primary Winding

When there is an open primary winding, there is no primary current and, therefore, no induced voltage or current in the secondary. This condition is illustrated in Figure 11–38(a), and the method of checking with an ohmmeter is shown in part (b).

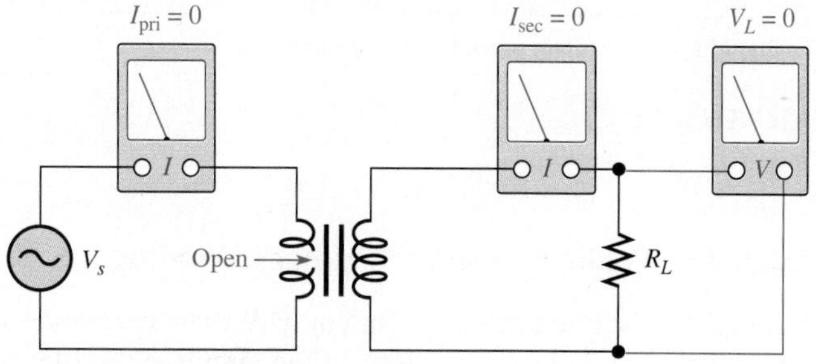

(a) Conditions when the primary winding is open

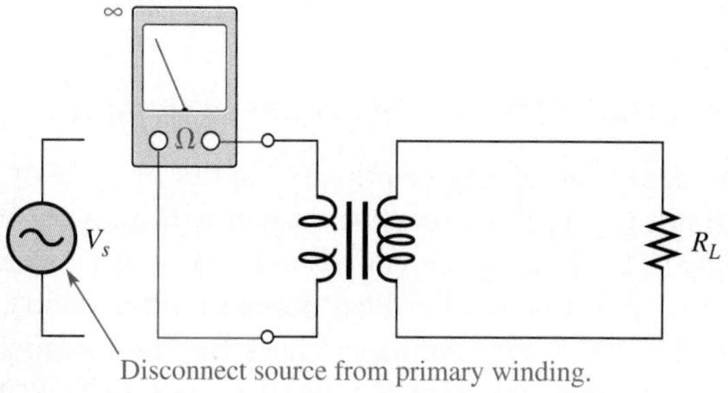

(b) Checking the primary winding with an ohmmeter

FIGURE 11–38
Open primary winding.

Open Secondary Winding

When there is an open secondary winding, there is no current in the secondary circuit and, as a result, no voltage across the load. Also, an open secondary winding causes the primary current to be very small (only a small magnetizing current flows). In fact, the primary current may be practically zero. This condition is illustrated in Figure 11–39(a), and the ohmmeter check is shown in part (b).

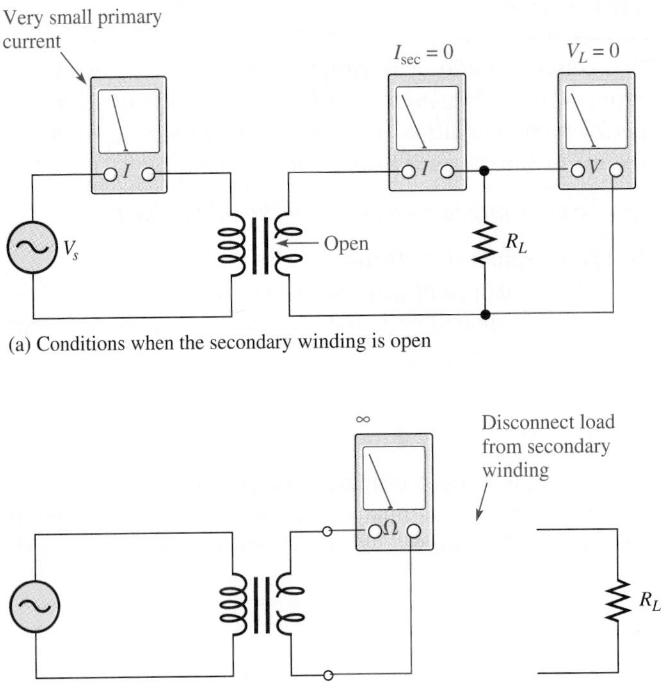

(a) Conditions when the secondary winding is open

(b) Checking the secondary winding with the ohmmeter

FIGURE 11–39
Open secondary winding.

Shorted or Partially Shorted Primary Winding

A completely shorted primary winding will draw excessive current from the source and, unless there is a breaker or a fuse in the circuit, either the source or the transformer or both will burn out. A partial short in the primary winding can cause higher than normal or even excessive primary current.

Shorted or Partially Shorted Secondary Winding

In this case, there is an excessive primary current because of the low reflected resistance due to the short. Often, this excessive current will burn out the primary winding and result in an open. The short-circuit current in the secondary winding causes the load current to be zero (full short) or smaller than normal (partial short), as demonstrated in Figure 11–40(a) and 11–40(b). The ohmmeter check for this condition is shown in part (c).

Normally, when a transformer fails, it is very difficult to repair, and therefore the simplest procedure is to replace it.

FIGURE 11–40
Shorted secondary winding.

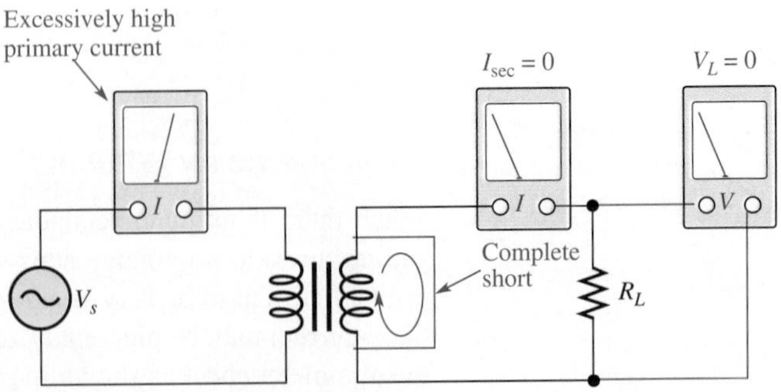

(a) Secondary winding completely shorted

FIGURE 11–40
Continued

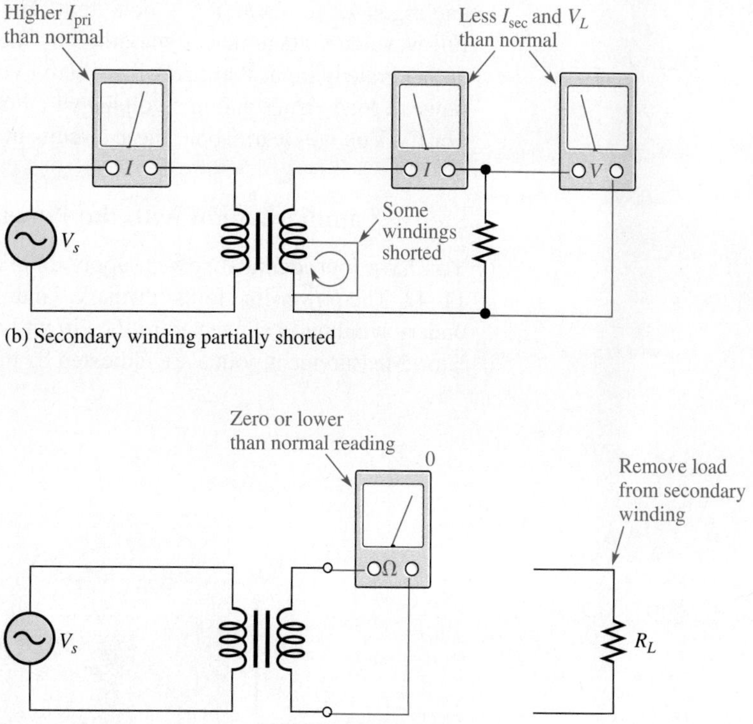

(b) Secondary winding partially shorted

(c) Checking the secondary winding with the ohmmeter

SECTION 11–11 REVIEW

1. Name two possible failures in a transformer.
2. What is often the cause of transformer failure?

11–12 ■ APPLICATION ASSIGNMENT

A common application of the transformer is in dc power supplies. The transformer is used to couple the ac line voltage into the power supply circuitry where it is converted to a dc voltage. Your assignment is to troubleshoot four identical transformer-coupled dc power supplies and, based on a series of measurements, determine the fault, if any, in each.

The transformer in the power supply schematic of Figure 11–41 steps the 110 V rms at the ac outlet down to a level that can be converted by the diode bridge rectifier, filtered,

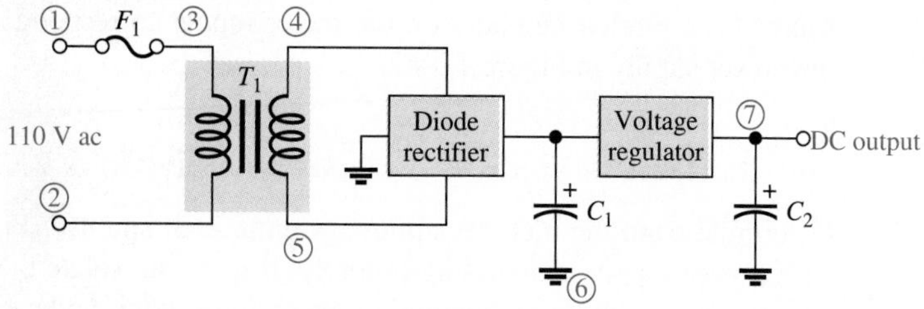

FIGURE 11–41
Basic transformer-coupled dc power supply.

and regulated to obtain a 6 V dc output. The diode rectifier changes the ac to a pulsating full-wave dc voltage that is smoothed by the capacitor filter C_1. The voltage regulator is an integrated circuit that takes the filtered voltage and provides a constant 6 V dc over a range of load values and line voltage variations. Additional filtering is provided by capacitor C_2. You will learn about these circuits in a later course.

Step 1: Familiarization with the Power Supply

You have four identical power supply units to troubleshoot like the one shown in Figure 11–42. The power line to the primary winding of transformer T_1 is fused by F_1. The secondary winding is connected to the circuit board containing the rectifier, filter, and regulator. Measurement points are indicated by the circled numbers.

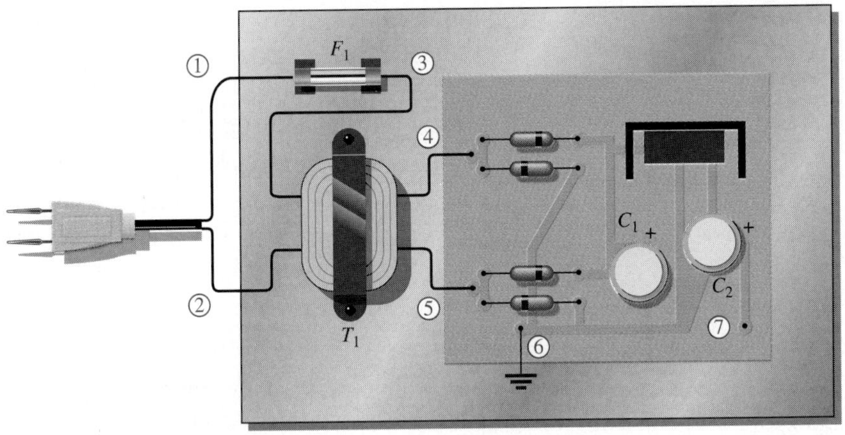

FIGURE 11–42
Power supply unit (top view).

Step 2: Measure Voltages on Power Supply Unit 1

After plugging the power supply into a standard wall outlet, an autoranging portable multimeter is used to measure the voltages. In an autoranging meter, the appropriate measurement range is automatically selected instead of being manually selected as in a standard multimeter.

Determine from the meter readings in Figure 11–43 whether or not the power supply is operating properly. If it is not, isolate the problem to one of the following: The circuit board containing the rectifier, filter, and regulator, the transformer, the fuse, or the power source. The circled numbers on the meter inputs correspond to the numbered points on the power supply in Figure 11–42.

Step 3: Measure Voltages on Power Supply Units 2, 3, and 4

Determine from the meter readings for units 2, 3, and 4 in Figure 11–44 whether or not each power supply is operating properly. If it is not, isolate the problem to one of the following: the circuit board containing the rectifier, filter, and regulator, the transformer, the fuse, or the power source. Only the meter displays and corresponding measurement points are shown.

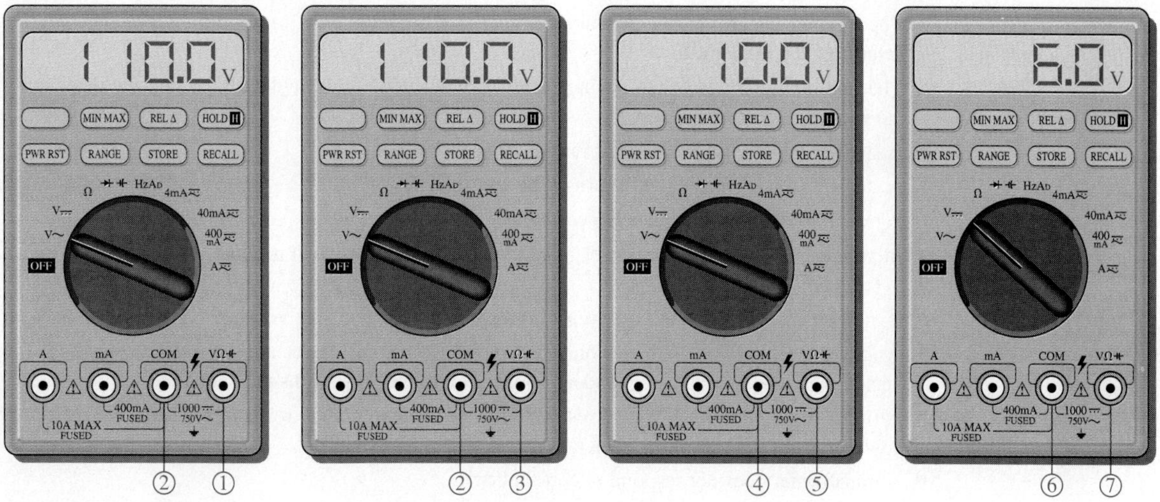

FIGURE 11–43
Voltage measurements on power supply unit 1.

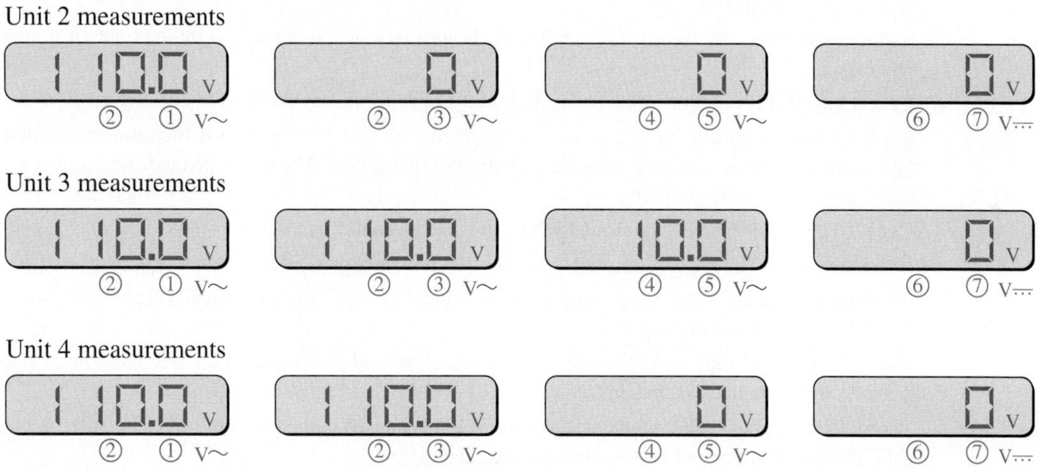

FIGURE 11–44
Measurements for power supply units 2, 3, and 4.

SECTION 11–12 REVIEW

1. In the case where the transformer was found to be faulty, how can you determine the specific fault (open windings or shorted windings)?
2. What type of fault in the transformer could cause the fuse to blow?

■ SUMMARY

- A transformer consists of two or more coils that are magnetically coupled on a common core.
- There is mutual inductance between two magnetically coupled coils.
- When current in one coil changes, voltage is induced in the other coil.
- The primary is the winding connected to the source, and the secondary is the winding connected to the load.

- The number of turns in the primary winding and the number of turns in the secondary winding determine the turns ratio.
- The relative polarities of the primary and secondary voltages are determined by the direction of the windings around the core.
- A step-up transformer has a turns ratio greater than 1.
- A step-down transformer has a turns ratio less than 1.
- A transformer cannot increase power.
- In an ideal transformer, the power from the source (input power) is equal to the power delivered to the load (output power).
- If the voltage is stepped up, the current is stepped down, and vice versa.
- A load connected across the secondary winding of a transformer appears to the source as a reflected load having a value dependent on the reciprocal of the turns ratio squared.
- A transformer can match a load resistance to an internal source resistance to achieve maximum power transfer to the load by selection of the proper turns ratio.
- A transformer does not respond to constant dc.
- Energy losses in a practical transformer result from winding resistances, hysteresis loss in the core, eddy currents in the core, and flux leakage.

GLOSSARY

Apparent power rating The method of rating transformers in which the power capability is expressed in volt-amperes (VA).

Autotransformer A transformer in which the primary and secondary windings are in a single winding.

Center tap (CT) A connection at the midpoint of the secondary winding of a transformer.

Coefficient of coupling (k) A constant associated with transformers that is the ratio of secondary flux to primary flux. The ideal value of 1 indicates that all the flux in the primary winding is coupled into the secondary winding.

Core The physical structure around which the windings are placed.

Efficiency (η) The ratio of output power to input power expressed as a percentage.

Electrical isolation The condition that exists when two coils are magnetically linked but have no electrical connection between them.

Impedance matching A technique used to match a load resistance to an internal source resistance in order to achieve a maximum transfer of power.

Magnetic coupling The magnetic connection between two coils as a result of the changing magnetic flux lines of one coil cutting through the second coil.

Maximum power transfer theorem A theorem that states the maximum power is transferred from a source to a load when the load resistance equals the internal source resistance.

Mutual inductance (L_M) The inductance between two separate coils, such as a transformer.

Primary winding The input winding of a transformer; also called *primary.*

Reflected load The load as it appears to the source in the primary circuit.

Reflected resistance The resistance in the secondary circuit reflected into the primary circuit.

Secondary winding The output winding of a transformer; also called *secondary.*

Step-down transformer A transformer in which the secondary voltage is less than the primary voltage.

Step-up transformer A transformer in which the secondary voltage is greater than the primary voltage.

Transformer A device formed by two or more windings that are magnetically coupled to each other and providing a transfer of power electromagnetically from one winding to another.

Turns ratio The ratio of turns in the secondary winding to turns in the primary winding.

FORMULAS

(11–1)	$k = \dfrac{\phi_{12}}{\phi_1}$	Coefficient of coupling
(11–2)	$L_M = k\sqrt{L_1 L_2}$	Mutual inductance

(11–3) $n = \dfrac{N_{sec}}{N_{pri}}$ Turns ratio

(11–4) $\dfrac{V_{sec}}{V_{pri}} = \dfrac{N_{sec}}{N_{pri}}$ Voltage ratio

(11–5) $V_{sec} = \left(\dfrac{N_{sec}}{N_{pri}}\right)V_{pri}$ Secondary voltage

(11–6) $\dfrac{I_{pri}}{I_{sec}} = \dfrac{N_{sec}}{N_{pri}}$ Current ratio

(11–7) $I_{sec} = \left(\dfrac{N_{pri}}{N_{sec}}\right)I_{pri}$ Secondary current

(11–8) $R_{pri} = \left(\dfrac{N_{pri}}{N_{sec}}\right)^2 R_L$ Reflected resistance

(11–9) $\dfrac{N_{sec}}{N_{pri}} = \sqrt{\dfrac{R_L}{R_{pri}}}$ Turns ratio for impedance matching

(11–10) $\eta = \left(\dfrac{P_{out}}{P_{in}}\right)100\%$ Transformer efficiency

■ SELF-TEST

1. A transformer is used for
(a) dc voltages **(b)** ac voltages **(c)** both dc and ac

2. The turns ratio of a transformer affects the following:
(a) primary voltage **(b)** dc voltage **(c)** secondary voltage **(d)** none of these

3. If the windings of a certain transformer with a turns ratio of 1 are in opposite directions around the core, the secondary voltage is
(a) in phase with the primary voltage **(b)** less than the primary voltage
(c) greater than the primary voltage **(d)** out of phase with the primary voltage

4. When the turns ratio of a transformer is 10 and the primary ac voltage is 6 V, the secondary voltage is
(a) 60 V **(b)** 0.6 V **(c)** 6 V **(d)** 36 V

5. When the turns ratio of a transformer is 0.5 and the primary ac voltage is 100 V, the secondary voltage is
(a) 200 V **(b)** 50 V **(c)** 10 V **(d)** 100 V

6. A certain transformer has 500 turns in the primary winding and 2500 turns in the secondary winding. The turns ratio is
(a) 0.2 **(b)** 2.5 **(c)** 5 **(d)** 0.5

7. If 10 W of power are applied to the primary winding of an ideal transformer with a turns ratio of 5, the power delivered to the secondary load is
(a) 50 W **(b)** 0.5 W **(c)** 0 W **(d)** 10 W

8. In a certain loaded transformer, the secondary voltage is one-third the primary voltage. The secondary current is
(a) one-third the primary current **(b)** three times the primary current
(c) equal to the primary current **(d)** both (b) and (d)

9. When a 1 kΩ load resistor is connected across the secondary winding of a transformer with a turns ratio of 2, the source "sees" a reflected load of
(a) 250 Ω **(b)** 2 kΩ **(c)** 4 kΩ **(d)** 1 kΩ

10. In Question 9, if the turns ratio is 0.5, the source "sees" a reflected load of
(a) 1 kΩ **(b)** 2 kΩ **(c)** 4 kΩ **(d)** 500 Ω

11. The turns ratio required to match a 50 Ω source to a 200 Ω load is
(a) 0.25 **(b)** 0.5 **(c)** 4 **(d)** 2

12. Maximum power is transferred from a source to a load when
(a) $R_L > R_s$ **(b)** $R_L < R_s$ **(c)** $R_L = R_s$ **(d)** $R_L = nR_s$

13. When a 12 V battery is connected across the primary winding of a transformer with a turns ratio of 4, the secondary voltage is
(a) 0 V **(b)** 12 V **(c)** 48 V **(d)** 3 V

14. A certain transformer has a turns ratio of 1 and a 0.95 coefficient of coupling. When 1 V ac is applied to the primary, the secondary voltage is
(a) 1 V (b) 1.95 V (c) 0.95 V

▪ BASIC PROBLEMS

SECTION 11–1 Mutual Inductance

1. What is the mutual inductance when $k = 0.75$, $L_1 = 1\ \mu\text{H}$, and $L_2 = 4\ \mu\text{H}$?
2. Determine the coefficient of coupling when $L_M = 1\ \mu\text{H}$, $L_1 = 8\ \mu\text{H}$, and $L_2 = 2\ \mu\text{H}$.

SECTION 11–2 The Basic Transformer

3. What is the turns ratio of a transformer having 120 primary turns and 360 secondary turns?
4. (a) What is the turns ratio of a transformer having 250 primary turns and 1000 secondary turns?
   (b) What is the turns ratio when the primary winding has 400 turns and the secondary winding has 100 turns?
5. Determine the phase of the secondary voltage with respect to the primary voltage for each transformer in Figure 11–45.

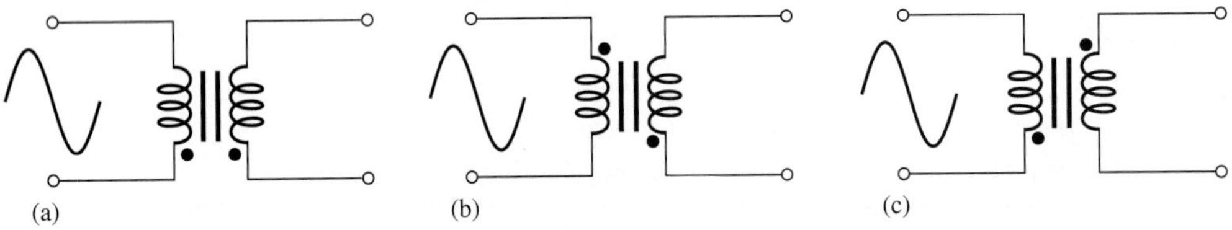

FIGURE 11–45

SECTION 11–3 The Step-Up Transformer

6. If 120 V ac are connected to the primary of a transformer with 100 primary turns and 150 secondary turns, what is the secondary voltage?
7. A certain transformer has 250 turns in its primary winding. In order to double the voltage, how many turns must be in the secondary winding?
8. How many primary volts must be applied to a transformer with a turns ratio of 10 to obtain a secondary voltage of 60 V ac?
9. For each transformer in Figure 11–46, sketch the secondary voltage showing its relationship to the primary voltage. Also indicate the amplitude.

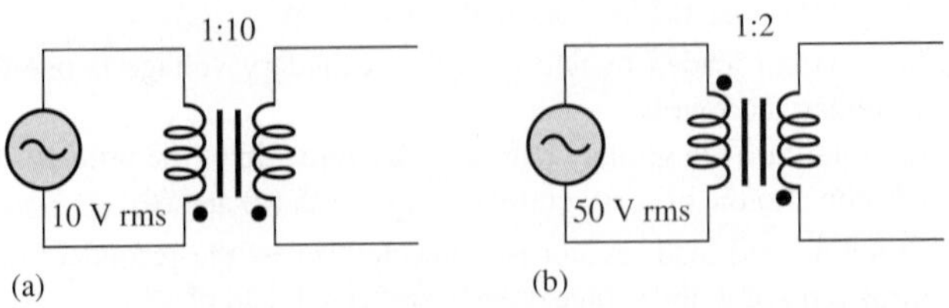

FIGURE 11–46

SECTION 11–4 The Step-Down Transformer

10. To step 120 V down to 30 V, what must the turns ratio be?
11. The primary winding of a transformer has 1200 V across it. What is the secondary voltage if the turns ratio is 0.2?
12. How many primary volts must be applied to a transformer with a turns ratio of 0.1 to obtain a secondary voltage of 6 V ac?

SECTION 11–5 Loading the Secondary Winding

13. Determine I_s in Figure 11–47.

14. Determine the following quantities in Figure 11–48:
(a) secondary voltage **(b)** secondary current
(c) primary current **(d)** power in the load

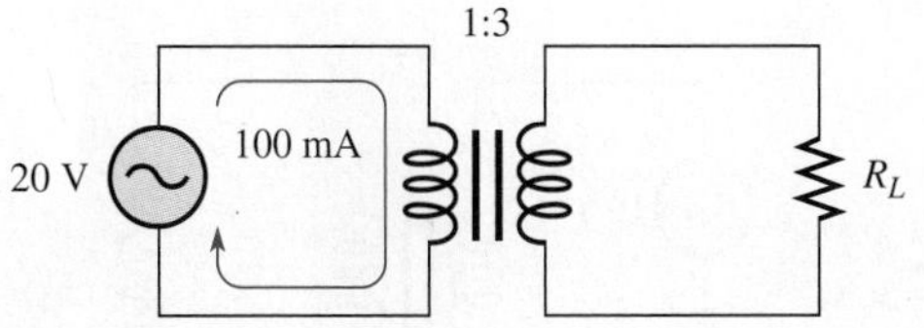

FIGURE 11–47

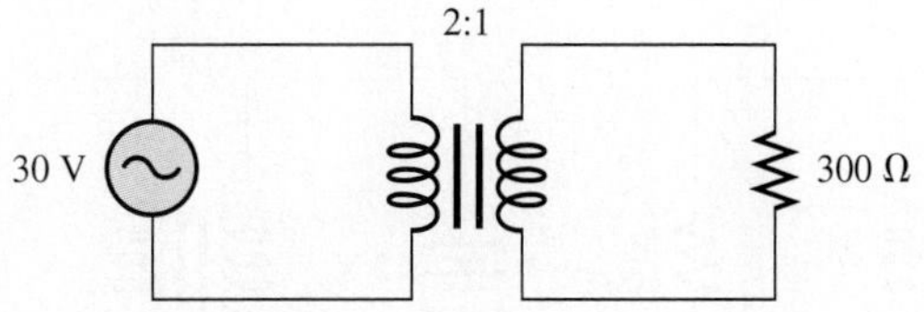

FIGURE 11–48

SECTION 11–6 Reflected Load

15. What is the load resistance as seen by the source in Figure 11–49?

16. What is the resistance reflected into the primary circuit in Figure 11–50?

17. What is the primary current (rms) in Figure 11–50 if the rms source voltage is 115 V?

18. What must the turns ratio be in Figure 11–51 in order to reflect 300 Ω into the primary circuit?

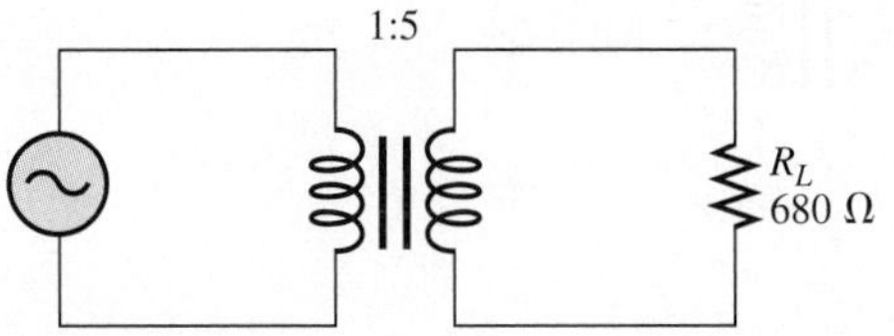

FIGURE 11–49

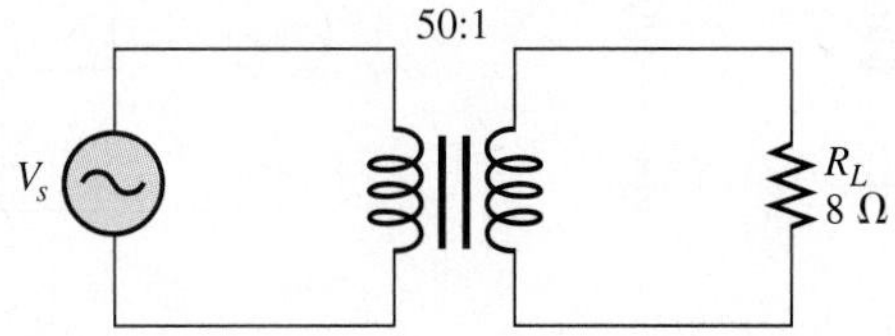

FIGURE 11–50

FIGURE 11–51

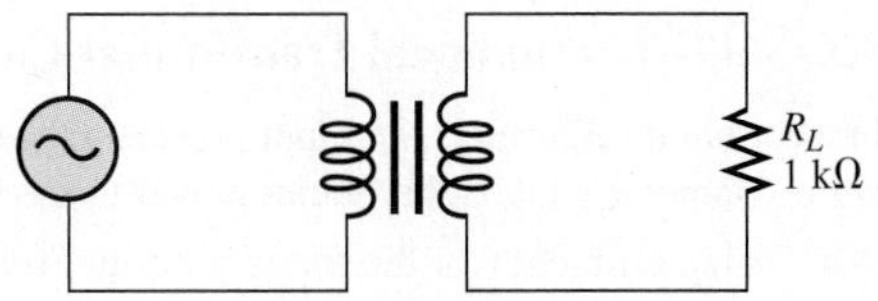

SECTION 11–7 Matching the Load and Source Resistances

19. For the circuit in Figure 11–52, find the turns ratio required to deliver maximum power to the 4 Ω speaker.

20. In Figure 11–52, what is the maximum power in watts delivered to the speaker?

21. Determine the value to which R_L must be adjusted in Figure 11–53 for maximum power transfer. The internal source resistance is 50 Ω.

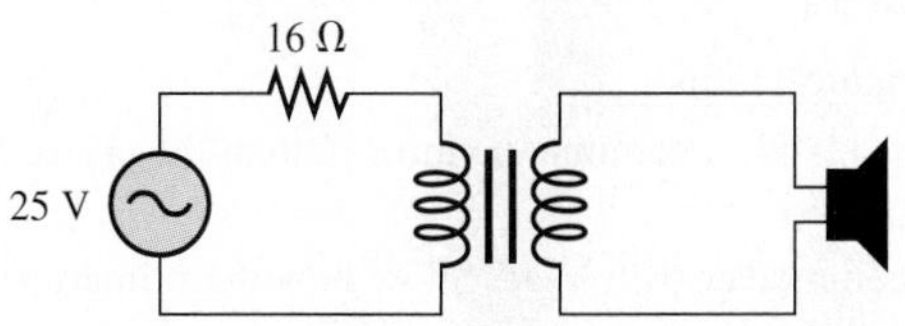

FIGURE 11–52

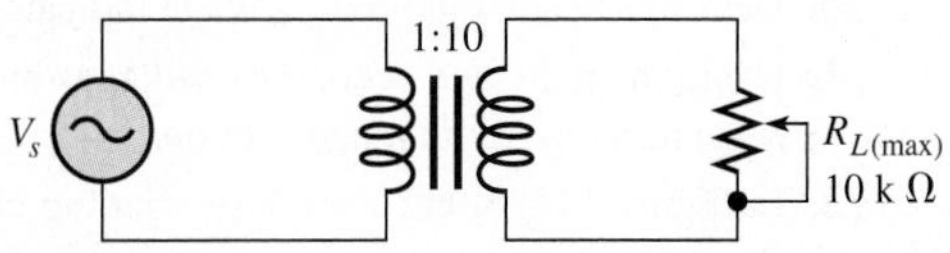

FIGURE 11–53

SECTION 11–8 The Transformer As an Isolation Device

22. What is the voltage across the load in each circuit of Figure 11–54?

23. If the bottom of each secondary winding in Figure 11–54 were grounded, would the values of the load voltages be changed?

24. Determine the unspecified meter readings in Figure 11–55.

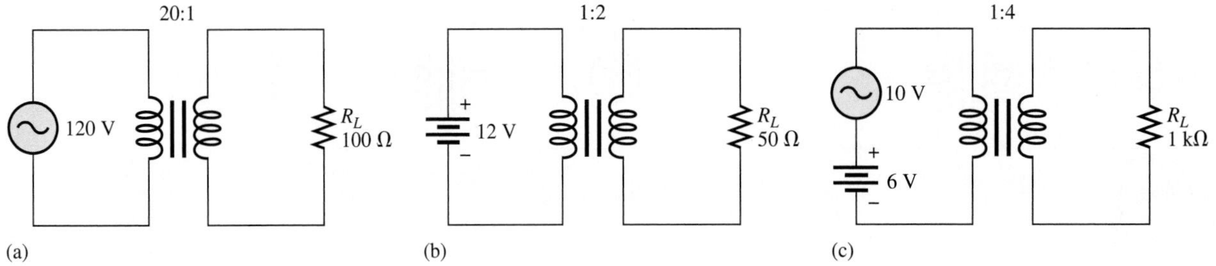

FIGURE 11–54

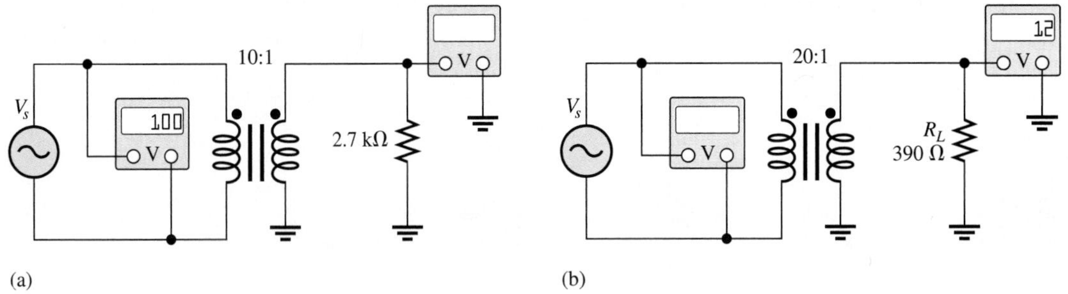

FIGURE 11–55

SECTION 11–9 Nonideal Transformer Characteristics

25. In a certain transformer, the input power to the primary is 100 W. If 5.5 W are lost in the winding resistances, what is the output power to the load, neglecting any other losses?

26. What is the efficiency of the transformer in Problem 25?

27. Determine the coefficient of coupling for a transformer in which 2% of the total flux generated in the primary does not pass through the secondary.

28. A certain transformer is rated at 1 kVA. It operates on 60 Hz, 120 V ac. The secondary voltage is 600 V.

(a) What is the maximum load current?

(b) What is the smallest R_L that you can drive?

(c) What is the largest capacitor that can be connected as a load?

29. What kVA rating is required for a transformer that must handle a maximum load current of 10 A with a secondary voltage of 2.5 kV?

SECTION 11–10 Other Types of Transformers

30. Determine each unknown voltage indicated in Figure 11–56.

31. Using the indicated secondary voltages in Figure 11–57, determine the turns ratio of the primary winding to each tapped section.

32. In Figure 11–58, each primary winding can accommodate 120 V ac. Show how the primaries should be connected for 240 V ac operation. Determine each secondary voltage.

33. Find the secondary voltage for each autotransformer in Figure 11–59.

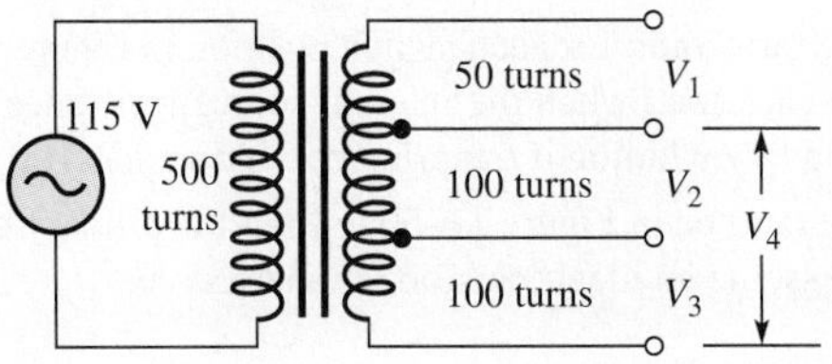

FIGURE 11–56

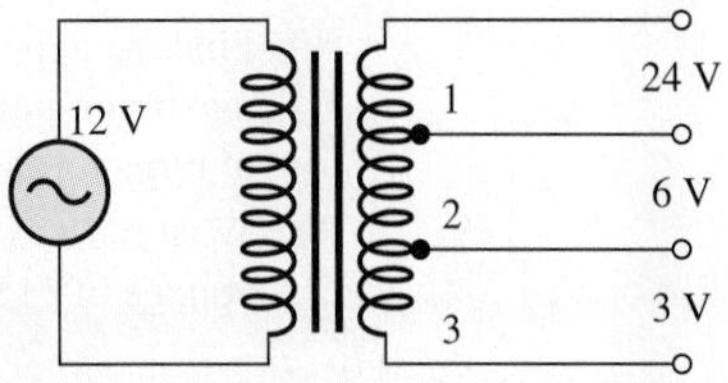

FIGURE 11–57

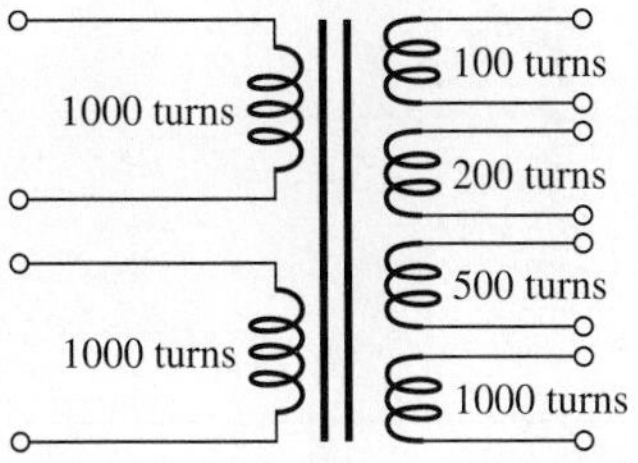

FIGURE 11–58

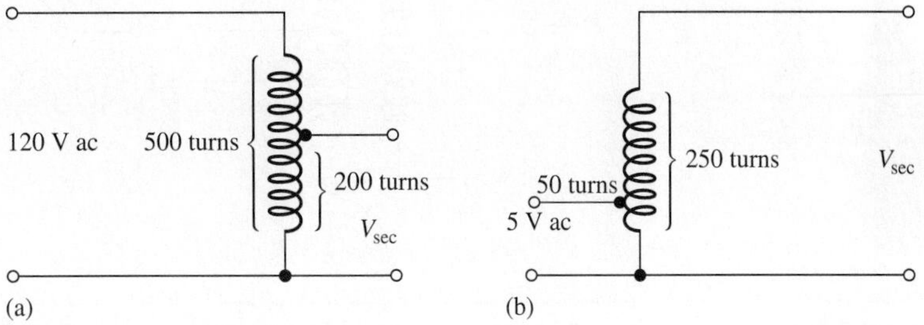

FIGURE 11–59

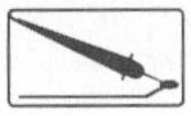

SECTION 11–11 Troubleshooting

34. When you apply 120 V ac across the primary winding of a transformer and check the voltage across the secondary winding, you get 0 V. Further investigation shows no primary or secondary current. List the possible faults. What is your next step in isolating the problem?

35. What is likely to happen if the primary winding of a transformer shorts?

36. In checking out a transformer circuit you find that the secondary voltage is less than it should be, although it is not zero. What is the most likely fault?

■ ADVANCED PROBLEMS

37. For the loaded, tapped-secondary transformer in Figure 11–60, determine the following:
 (a) all load voltages and currents
 (b) the resistance looking into the primary

38. A certain transformer is rated at 5 kVA, 2400/120 V, at 60 Hz.
 (a) What is the turns ratio if the 120 V is the secondary voltage?
 (b) What is the current rating of the secondary if 2400 V is the primary voltage?
 (c) What is the current rating of the primary if 2400 V is the primary voltage?

39. Determine the voltage measured by each voltmeter in Figure 11–61. The bench-type meters have one terminal that connects to ground as indicated.

FIGURE 11–60

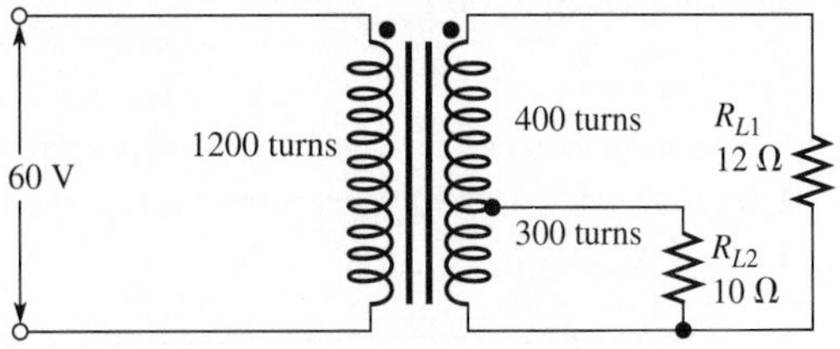

FIGURE 11–61

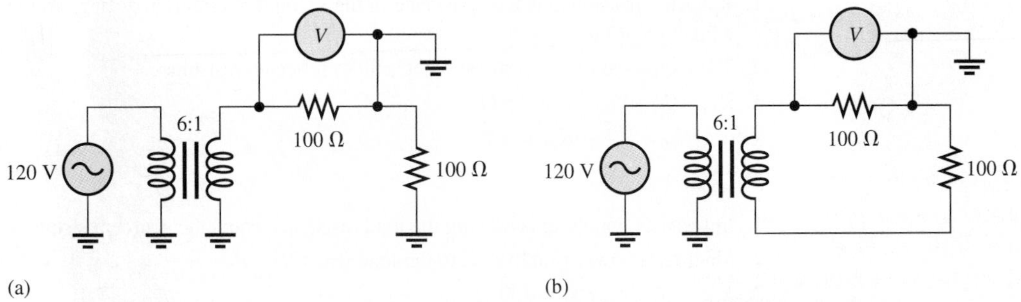

40. Find the appropriate turns ratio for each switch position in Figure 11–62 in order to transfer the maximum power to each load when the internal source resistance is 10 Ω. Specify the number of turns for the secondary winding if the primary winding has 100 turns.

41. What must the turns ratio be in Figure 11–50 to limit the primary current to 3 mA with a source voltage of 115 V? Assume an ideal transformer and source.

FIGURE 11–62

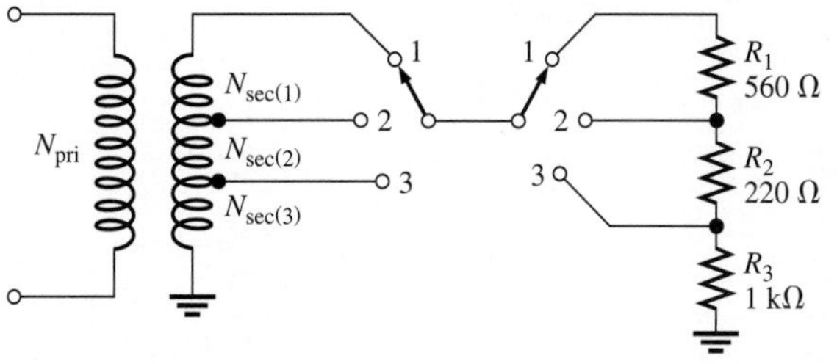

▪ ANSWERS TO SECTION REVIEWS

Section 11–1

1. Mutual inductance is the inductance between two coils, established by the amount of coupling between the coils.

2. $L_M = k\sqrt{L_1L_2} = 45$ mH

3. Induced voltage increases when k is increased.

Section 11–2

1. The operation of transformers is based on the principle of mutual inductance.

2. Turns ratio is the ratio of secondary turns to primary turns.

3. Directions of the windings determine voltage polarities.

4. $n = N_{sec}/N_{pri} = 0.5$

Section 11–3

1. A step-up transformer increases voltage.

2. The secondary voltage is five times greater.

3. $V_{sec} = (N_{sec}/N_{pri})V_{pri} = 2400$ V

Section 11–4

1. A step-down transformer decreases voltage.

2. $V_{sec} = (N_{sec}/N_{pri})V_{pri} = 60$ V

3. $n = 12\text{ V}/120\text{ V} = 0.1$

Section 11–5

1. The secondary current is half the primary current.

2. $I_{sec} = (N_{pri}/N_{sec})I_{pri} = 2$ A

3. $I_{pri} = (N_{sec}/N_{pri})I_{sec} = 2.5$ A

Section 11–6

1. Reflected resistance is the resistance in the secondary circuit as seen from the primary circuit as a function of the turns ratio.

2. The reciprocal of the turns ratio determines reflected resistance.

3. $R_{pri} = (N_{pri}/N_{sec})^2R_L = 0.5\ \Omega$

4. $N_{sec}/N_{pri} = \sqrt{R_L/R_{pri}} = 0.1$

Section 11–7

1. Impedance matching is making the load resistance equal the source resistance.

2. Maximum power is delivered to the load when $R_L = R_S$.

3. $R_{pri} = (N_{pri}/V_{sec})^2R_L = 400\ \Omega$

Section 11–8

1. Electrical isolation means there is no electrical connection between the primary and secondary circuits.
2. Transformers do not couple dc.

Section 11–9

1. In a practical transformer, energy loss reduces the efficiency. An ideal transformer has an efficiency of 100%.
2. 85% of the magnetic flux generated in the primary winding passes through the secondary winding when the coefficient of coupling is 0.85.
3. $I_L = 10\text{ kVA}/250\text{ V} = 40\text{ A}$

Section 11–10

1. $V_{sec} = 10(220\text{ V}) = 2200\text{ V}$; $V_{sec} = 0.2(220\text{ V}) = 44\text{ V}$
2. Autotransformers are smaller and lighter for same rating. Autotransformers provide no electrical isolation.

Section 11–11

1. Possible failures are open windings and shorted windings.
2. Operating above rated values causes transformer failure.

Section 11–12

1. Use an ohmmeter to check for open windings. Shorted windings are indicated by an incorrect secondary voltage.
2. A short will cause the fuse to blow.

■ ANSWERS TO RELATED EXERCISES FOR EXAMPLES

11–1 0.75

11–2 387 μH

11–3 500 turns

11–4 480 V

11–5 57.6 V

11–6 5 mA; 400 mA

11–7 6 Ω

11–8 0.354

11–9 833 mW

11–10 0.0707 or 14.14:1

11–11 85.2%

11–12 $V_{AB} = 12\text{ V}$; $V_{CD} = 480\text{ V}$; $V_{(CT)C} = V_{(CT)D} = 240\text{ V}$; $V_{EF} = 24\text{ V}$

11–13 Increases to 2.13 kVA

12

RC CIRCUITS

An *RC* circuit contains both resistance and capacitance. It is one of the basic types of reactive circuits that you will study. In this chapter, basic series and parallel *RC* circuits and their responses to sine wave voltages are covered. Series-parallel combinations are also examined. Power considerations in *RC* circuits are introduced, and practical aspects of power ratings are discussed. Several basic areas of application are presented to give you an idea of how simple combinations of resistors and capacitors can be applied.

■ CHAPTER OBJECTIVES

- ☐ Describe the relationship between current and voltage in an *RC* circuit
- ☐ Determine impedance and phase angle in a series *RC* circuit
- ☐ Analyze a series *RC* circuit
- ☐ Determine impedance and phase angle in a parallel *RC* circuit
- ☐ Analyze a parallel *RC* circuit
- ☐ Analyze series-parallel *RC* circuits
- ☐ Determine power in *RC* circuits
- ☐ Discuss some basic *RC* applications
- ☐ Troubleshoot *RC* circuits

■ APPLICATION ASSIGNMENT

Your assignment is to run frequency response measurements on a capacitively coupled amplifier used in a certain communications system. You will be concentrating on the *RC* input circuit of the amplifier and how it responds to different frequencies. The results of your measurements will be graphically documented with a frequency response curve. After studying this chapter, you should be able to complete the application assignment in Section 12–10.

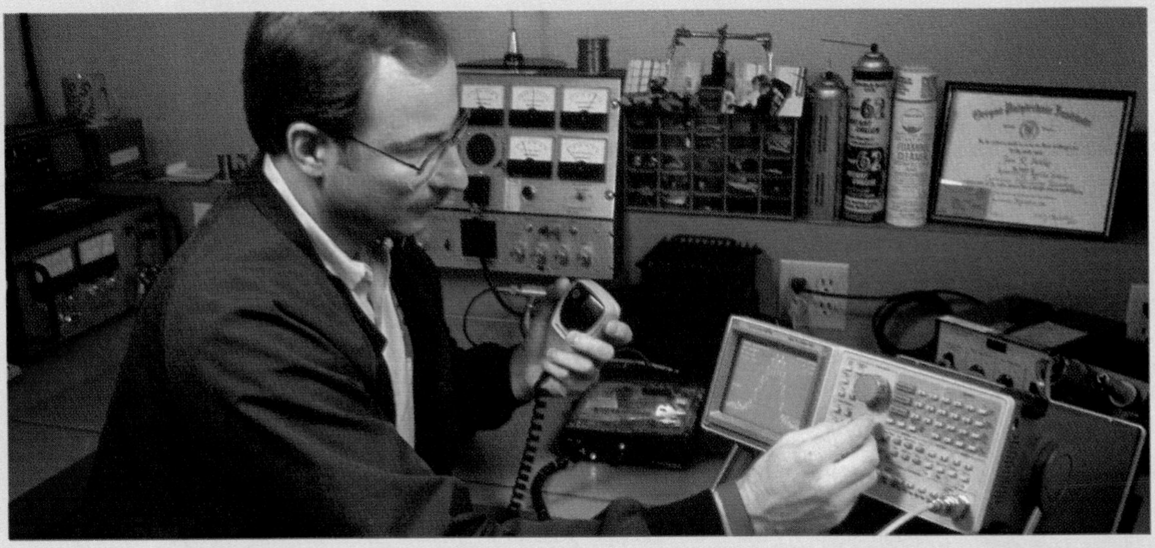

12–1 ▪ SINUSOIDAL RESPONSE OF *RC* CIRCUITS

When a sinusoidal voltage is applied to an RC circuit, each resulting voltage drop and the current in the circuit are also sinusoidal and have the same frequency as the applied voltage. The capacitance causes a phase shift between the voltage and current that depends on the relative values of the resistance and the capacitive reactance.

After completing this section, you should be able to

- **Describe the relationship between current and voltage in an *RC* circuit**
  - ☐ Discuss voltage and current waveforms
  - ☐ Discuss phase shift
  - ☐ Describe types of signal generators

As shown in Figure 12–1, the resistor voltage, the capacitor voltage, and the current are all sine waves with the frequency of the source.

Phase shifts are introduced because of the capacitance. As you will learn, the resistor voltage and current are in phase with each other and **lead** the source voltage in phase. The capacitor voltage **lags** the source voltage. The phase angle between the current and the capacitor voltage is always 90°. These generalized phase relationships are indicated in Figure 12–1.

FIGURE 12–1
Illustration of sinusoidal response with general phase relationships of V_R, V_C, and I relative to the source voltage. V_R and I are in phase; V_R leads V_s; V_C lags V_s; and V_R and V_C are 90° out of phase.

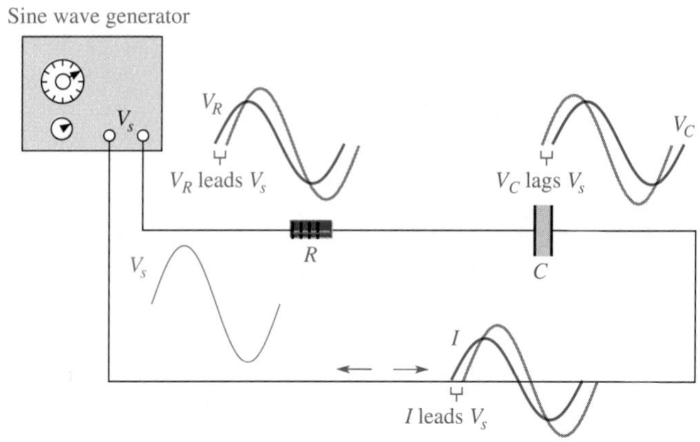

The amplitudes and the phase relationships of the voltages and current depend on the values of the resistance and the capacitive reactance (defined in Section 9–6). When a circuit is purely resistive, the phase angle between the applied (source) voltage and the total current is zero. When a circuit is purely capacitive, the phase angle between the applied voltage and the total current is 90°, with the current leading the voltage. When there is a combination of both resistance and capacitive reactance in a circuit, the phase angle between the applied voltage and the total current is somewhere between zero and 90°, depending on the relative values of the resistance and the reactance.

Signal Generators

When a circuit is hooked up for a laboratory experiment or for troubleshooting, a signal generator similar to those shown in Figure 12–2 is used to provide the source voltage. These instruments, depending on their capability, are classified as sine wave generators which produce only sine waves; sine/square generators, which produce both sine waves and square waves; or function generators, which produce sine waves, pulse waveforms, and triangular (ramp) waveforms.

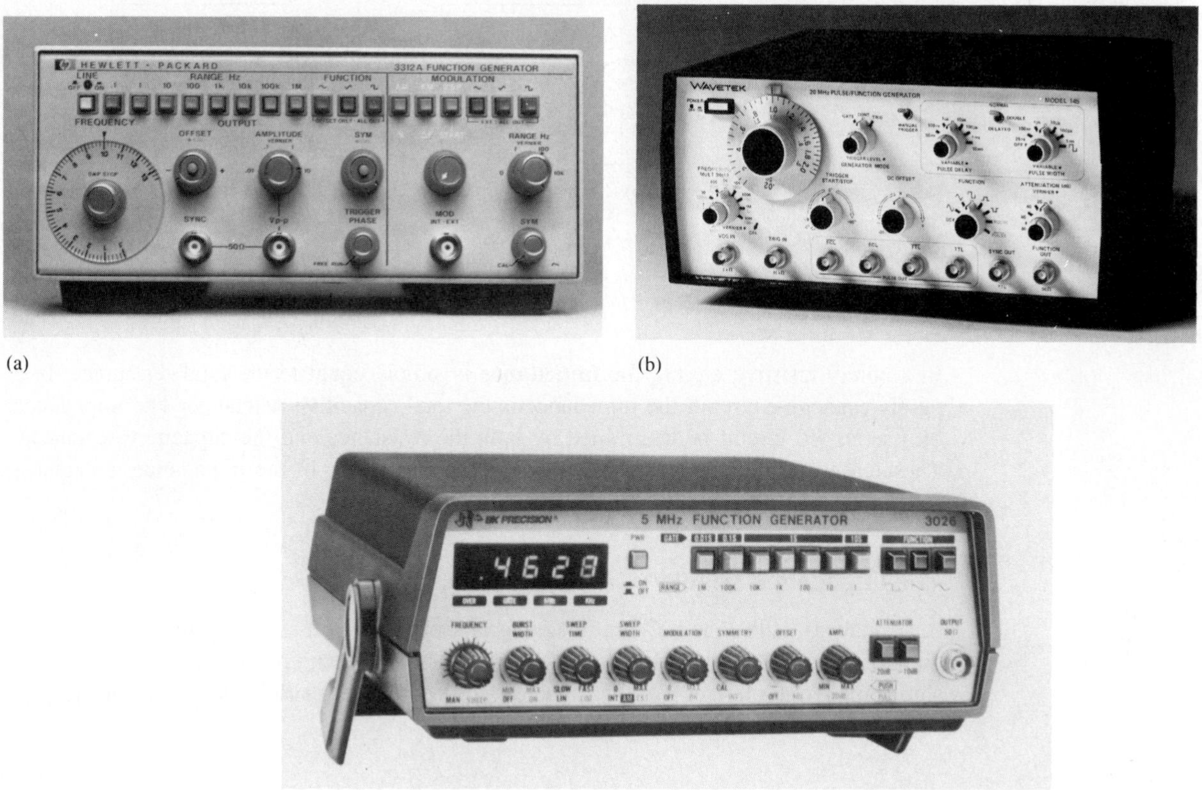

(a) (b) (c)

FIGURE 12–2

Typical signal (function) generators used in circuit testing and troubleshooting. Part (a) courtesy of Hewlett-Packard Company. Part (b) courtesy of Wavetek. Part (c) courtesy of B & K Precision.

SECTION 12–1 REVIEW

1. A 60 Hz sinusoidal voltage is applied to an *RC* circuit. What is the frequency of the capacitor voltage? What is the current?
2. What causes the phase shift between V_s and I in a series *RC* circuit?
3. When the resistance in an *RC* circuit is greater than the capacitive reactance, is the phase angle between the applied voltage and the total current closer to 0° or to 90°?

12–2 ■ IMPEDANCE AND PHASE ANGLE OF A SERIES *RC* CIRCUIT

The impedance of an RC circuit is the total opposition to sinusoidal current and its unit is the ohm. The phase angle is the phase difference between the total current and the source voltage.

After completing this section, you should be able to

- **Determine impedance and phase angle in a series *RC* circuit**
  - □ Define *impedance*
  - □ Draw an impedance triangle
  - □ Calculate total impedance magnitude
  - □ Calculate the phase angle

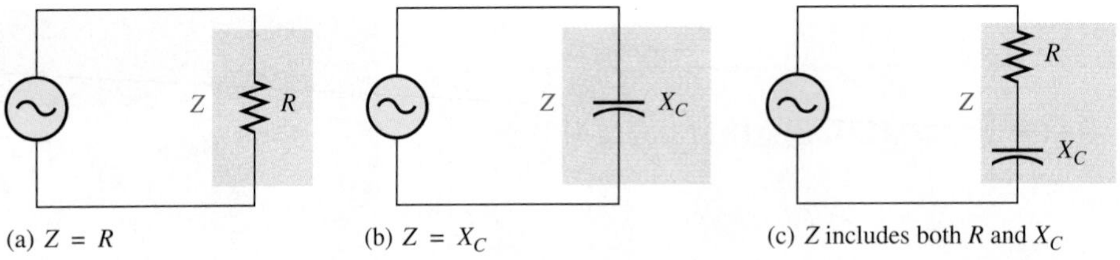

FIGURE 12–3
Three cases of impedance.

In a purely resistive circuit, the **impedance** is simply equal to the total resistance. In a purely capacitive circuit, the impedance is the total capacitive reactance. The impedance of a series *RC* circuit is determined by both the resistance and the capacitive reactance. These cases are illustrated in Figure 12–3. The magnitude of the impedance is symbolized by *Z*.

The Impedance Triangle

In ac analysis, both R and X_C are treated as phasor quantities, as shown in the phasor diagram of Figure 12–4(a), and X_C appearing at a –90° angle with respect to R. This relationship comes from the fact that the capacitor voltage in a series *RC* circuit lags the current, and thus the resistor voltage, by 90°.

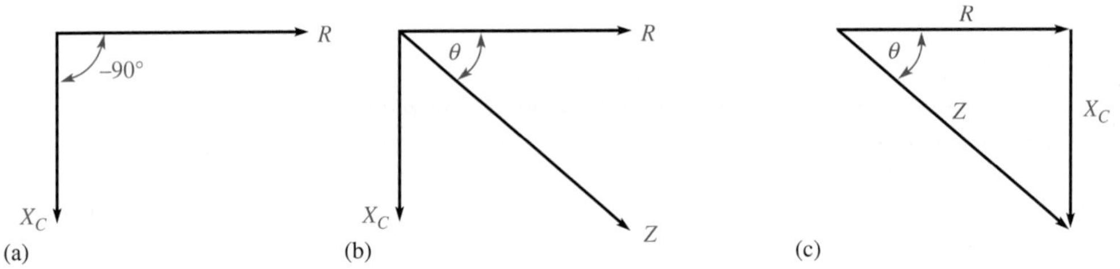

FIGURE 12–4
Development of the impedance triangle for a series RC circuit.

Since Z is the phasor sum of R and X_C, its phasor representation is as shown in Figure 12–4(b). A repositioning of the phasors, as shown in Figure 12–4(c), forms a right triangle, called the *impedance triangle.* The length of each phasor represents the magnitude in ohms, and the angle θ (the Greek letter theta) is the phase angle of the *RC* circuit and represents the phase difference between the applied (source) voltage and the current.

From right angle trigonometry (Pythagorean theorem), the magnitude of the impedance can be expressed in terms of the resistance and reactance as

$$Z = \sqrt{R^2 + X_C^2} \tag{12–1}$$

This is the magnitude of Z, as shown in Figure 12–5, and is expressed in ohms.

The phase angle, θ, is expressed as

$$\theta = \tan^{-1}\left(\frac{X_C}{R}\right) \tag{12–2}$$

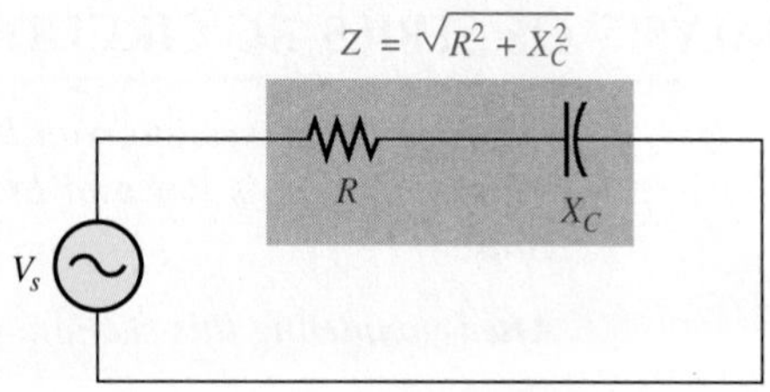

FIGURE 12–5
Impedance of a series RC circuit.

The symbol $\tan^{-1}$ stands for inverse tangent and can be found on some calculators by pressing [2nd] [tan] and on others by pressing [INV] [tan]. We will use the [2nd] [tan] sequence in this book. Another term for inverse tangent is *arctangent* (arctan).

EXAMPLE 12–1

Determine the impedance and the phase angle of the circuit in Figure 12–6.

FIGURE 12–6

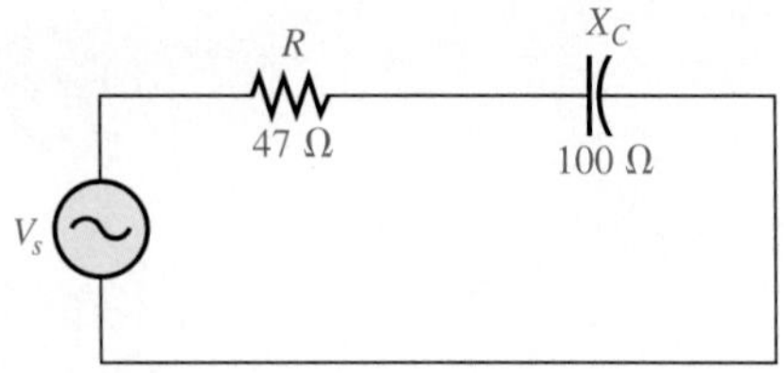

Solution The impedance is

$$Z = \sqrt{R^2 + X_C^2} = \sqrt{(47\ \Omega)^2 + (100\ \Omega)^2} = 110\ \Omega$$

The phase angle is

$$\theta = \tan^{-1}\left(\frac{X_C}{R}\right) = \tan^{-1}\left(\frac{100\ \Omega}{47\ \Omega}\right) = \tan^{-1}(2) = 64.8°$$

The applied voltage lags the current by 64.8°.

The calculator sequences are as follows:

Z: [4] [7] [x^2] [+] [1] [0] [0] [x^2] [=] [2nd] [$\sqrt{x}$]

θ: [1] [0] [0] [÷] [4] [7] [=] [2nd] [tan]

Related Exercise Find Z and θ for $R = 1\ \text{k}\Omega$ and $X_C = 2.2\ \text{k}\Omega$.

SECTION 12–2 REVIEW

1. Define *impedance.*
2. Does the source voltage lead or lag the current in a series *RC* circuit?
3. What causes the phase angle in the *RC* circuit?
4. A series *RC* circuit has a resistance of 33 kΩ and a capacitive reactance of 50 kΩ. What is the value of the impedance? What is the phase angle?

12–3 ■ ANALYSIS OF SERIES *RC* CIRCUITS

In the previous section, you learned how to express the impedance of a series RC circuit. Now, Ohm's law and Kirchhoff's voltage law are used in the analysis of RC circuits.

After completing this section, you should be able to

- **Analyze a series *RC* circuit**
  - ☐ Apply Ohm's law and Kirchhoff's voltage law to series *RC* circuits
  - ☐ Determine the phase relationships of the voltages and current
  - ☐ Show how impedance and phase angle vary with frequency

Ohm's Law

The application of Ohm's law to series *RC* circuits involves the use of the quantities of *Z*, *V*, and *I*. The three equivalent forms of Ohm's law are as follows:

$$V = IZ \tag{12–3}$$

$$I = \frac{V}{Z} \tag{12–4}$$

$$Z = \frac{V}{I} \tag{12–5}$$

The following two examples illustrate the use of Ohm's law.

EXAMPLE 12–2 If the current in Figure 12–7 is 0.2 mA, determine the source voltage and the phase angle.

FIGURE 12–7

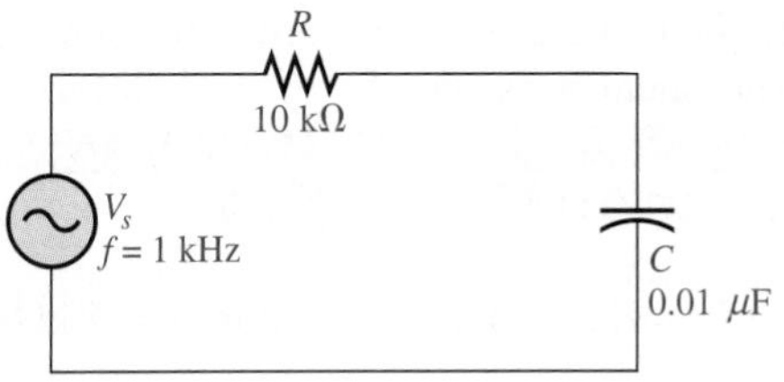

Solution The capacitive reactance is

$$X_C = \frac{1}{2\pi fC} = \frac{1}{2\pi(1000\text{ Hz})(0.01\ \mu\text{F})} = 15.9\text{ k}\Omega$$

The impedance is

$$Z = \sqrt{R^2 + X_C^2} = \sqrt{(10\text{ k}\Omega)^2 + (15.9\text{ k}\Omega)^2} = 18.8\text{ k}\Omega$$

Applying Ohm's law,

$$V_s = IZ = (0.2\text{ mA})(18.8\text{ k}\Omega) = 3.76\text{ V}$$

The phase angle is

$$\theta = \tan^{-1}\left(\frac{X_C}{R}\right) = \tan^{-1}\left(\frac{15.9\ \text{k}\Omega}{10\ \text{k}\Omega}\right) = 57.8^\circ$$

The source voltage has a magnitude of 3.76 V and lags the current by 57.8°.

The calculator sequence for V_s is

[2] [×] [2nd] [π] [×] [1] [0] [0] [0] [×] [•] [0] [1] [EXP] [+/−] [6] [=] [1/x]
[x²] [+] [1] [0] [EXP] [3] [x²] [=] [2nd] [√x] [×] [•] [2] [EXP] [+/−] [3] [=]

Related Exercise Determine V_s in Figure 12–7 if $f = 2$ kHz and $I = 200\ \mu$A.

EXAMPLE 12–3 Determine the current in the circuit of Figure 12–8.

FIGURE 12–8

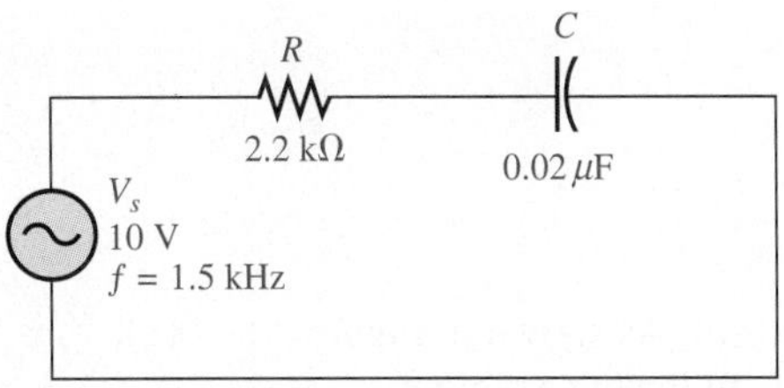

Solution The capacitive reactance is

$$X_C = \frac{1}{2\pi fC} = \frac{1}{2\pi(1.5\ \text{kHz})(0.02\ \mu\text{F})} = 5.3\ \text{k}\Omega$$

The impedance is

$$Z = \sqrt{R^2 + X_C^2} = \sqrt{(2.2\ \text{k}\Omega)^2 + (5.3\ \text{k}\Omega)^2} = 5.74\ \text{k}\Omega$$

Applying Ohm's law,

$$I = \frac{V}{Z} = \frac{10\ \text{V}}{5.74\ \text{k}\Omega} = 1.74\ \text{mA}$$

Related Exercise Determine the phase angle between V_s and I in Figure 12–8.

Relationships of the Current and Voltages in a Series *RC* Circuit

In a series circuit, the current is the same through both the resistor and the capacitor. Thus, the resistor voltage is in phase with the current, and the capacitor voltage lags the current by 90°. Therefore, there is a phase difference of 90° between the resistor voltage, V_R, and the capacitor voltage, V_C, as shown in the waveform diagram of Figure 12–9.

You know from Kirchhoff's voltage law that the sum of the voltage drops must equal the applied voltage. However, since V_R and V_C are not in phase with each other, they must be added as phasor quantities, with V_C lagging V_R by 90°, as shown in Figure

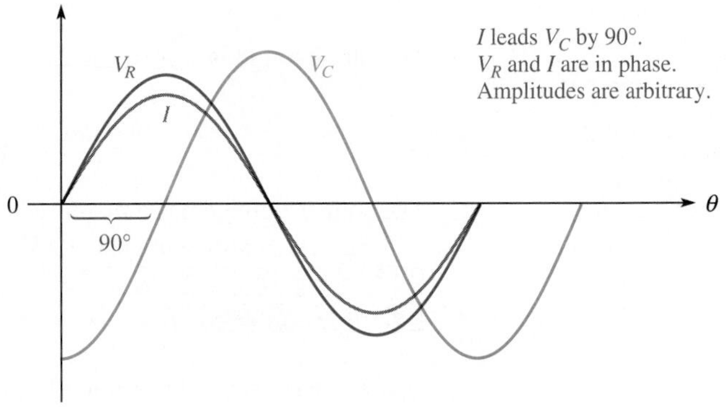

FIGURE 12–9
Phase relation of voltages and current in a series RC circuit.

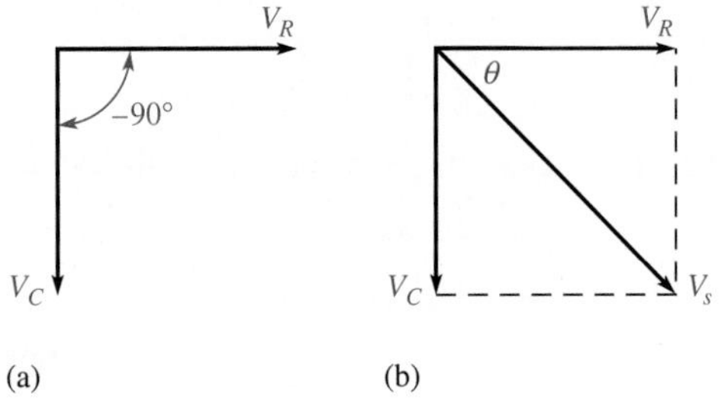

FIGURE 12–10
Voltage phasor diagram for a series RC circuit.

12–10(a). As shown in Figure 12–10(b), V_s is the phasor sum of V_R and V_C, as expressed in the following equation:

$$V_s = \sqrt{V_R^2 + V_C^2} \qquad \textbf{(12–6)}$$

The phase angle between the resistor voltage and the source voltage can be expressed as

$$\theta = \tan^{-1}\left(\frac{V_C}{V_R}\right) \qquad \textbf{(12–7)}$$

Since the resistor voltage and the current are in phase, θ in Equation (12–7) also represents the phase angle between the source voltage and the current and is equivalent to $\tan^{-1}(X_C/R)$.

Figure 12–11 shows a voltage and current phasor diagram representing the waveform diagram of Figure 12–9.

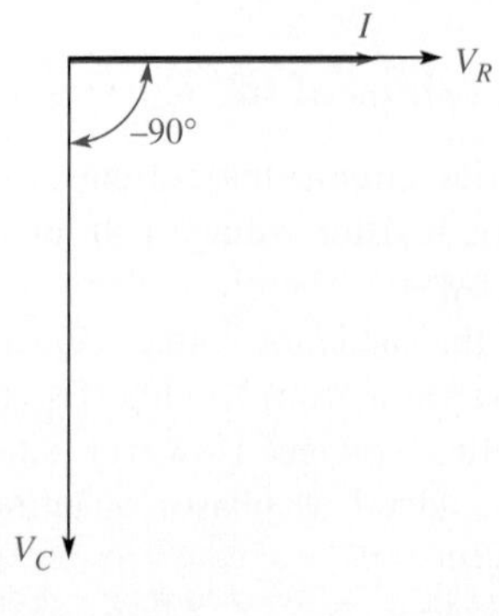

FIGURE 12–11
Voltage and current phasor diagram for the waveforms in Figure 12–9.

EXAMPLE 12–4 Determine the source voltage and the phase angle in Figure 12–12.

FIGURE 12–12

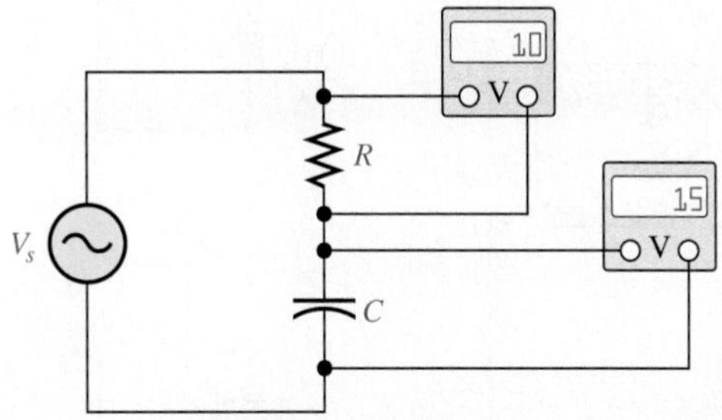

Solution Since V_R and V_C are 90° out of phase, they cannot be added directly. The source voltage is the phasor sum of V_R and V_C:

$$V_s = \sqrt{V_R^2 + V_C^2} = \sqrt{(10\text{ V})^2 + (15\text{ V})^2} = 18\text{ V}$$

The phase angle between the current and source voltage is

$$\theta = \tan^{-1}\left(\frac{V_C}{V_R}\right) = \tan^{-1}\left(\frac{15\text{ V}}{10\text{ V}}\right) = 56.3°$$

The calculator sequences are as follows:

V_s: 1 0 x^2 + 1 5 x^2 = 2nd $\sqrt{x}$

θ: 1 5 ÷ 1 0 = 2nd tan

Related Exercise In a certain series *RC* circuit, $V_s = 10$ V and $V_R = 7$ V. Find V_C.

Variation of Impedance with Frequency

As you know, capacitive reactance varies inversely with frequency. Since $Z = \sqrt{R^2 + X_C^2}$, you can see that when X_C increases, the entire term under the square root sign increases and thus the total impedance also increases; and when X_C decreases, the total impedance also decreases. Therefore, *in an RC circuit, Z is inversely dependent on frequency.*

Figure 12–13 illustrates how the voltages and current in a series *RC* circuit vary as the frequency increases or decreases, with the source voltage held at a constant value. In part (a), as the frequency is increased, X_C decreases; so less voltage is dropped across the capacitor. Also, Z decreases as X_C decreases, causing the current to increase. An increase in the current causes more voltage across R.

In Figure 12–13(b), as the frequency is decreased, X_C increases; so more voltage is dropped across the capacitor. Also, Z increases as X_C increases, causing the current to decrease. A decrease in the current causes less voltage across R.

Changes in Z and X_C can be observed as shown in Figure 12–14. As the frequency increases, the voltage across Z remains constant because V_s is constant. Also, the voltage across C decreases. The increasing current indicates that Z is decreasing. It does so because of the inverse relationship stated in Ohm's law ($Z = V_Z/I$). The increasing current also indicates that X_C is decreasing ($X_C = V_C/I$). The decrease in V_C corresponds to the decrease in X_C.

Variation of the Phase Angle with Frequency

Since X_C is the factor that introduces the phase angle in a series *RC* circuit, a change in X_C produces a change in the phase angle. As the frequency is increased, X_C be-

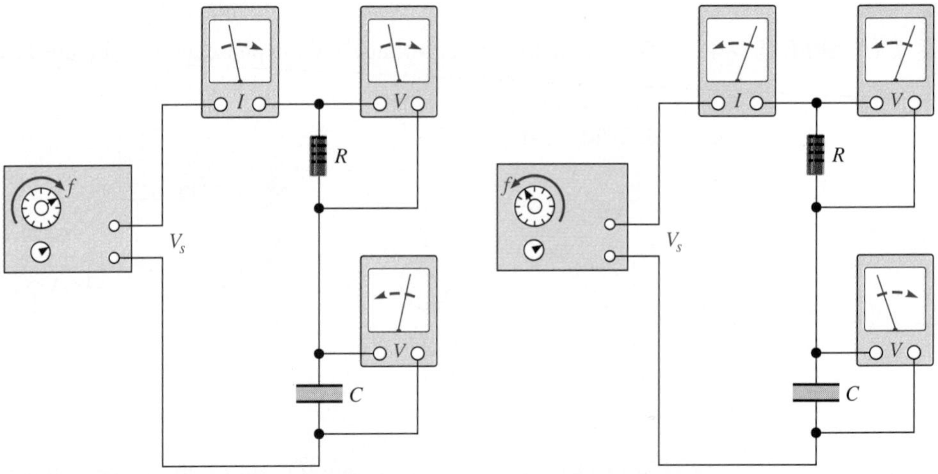

(a) As frequency is increased, I and V_R increase and V_C decreases.

(b) As frequency is decreased, I and V_R decrease and V_C increases.

FIGURE 12–13
An illustration of how the variation of impedance affects the voltages and current as the source frequency is varied. The source voltage is held at a constant amplitude.

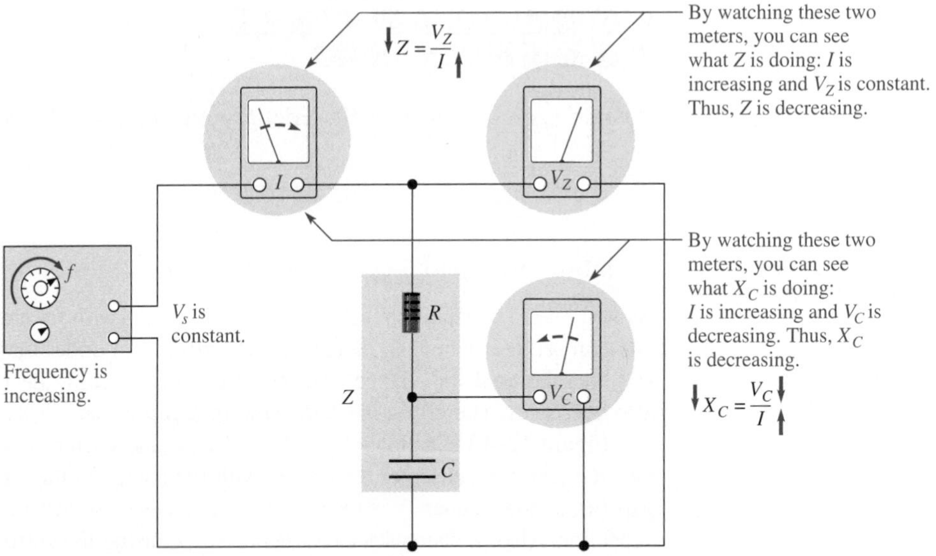

FIGURE 12–14
An illustration of how Z and X_C change with frequency.

comes smaller, and thus the phase angle decreases. As the frequency is decreased, X_C becomes larger, and thus the phase angle increases. This variation is illustrated in Figure 12–15 where a phase meter is connected across the capacitor to measure the angle between V_s and V_R. This angle is the phase angle of the circuit because I is in phase with V_R. By measuring the phase of V_R, you are effectively measuring the phase of I. You can use the phase-meter portion of a vector voltmeter such as the one shown in Figure 12–15(c) to measure the phase of one voltage relative to another. You can also see the variation of phase angle with frequency by using the impedance triangle as shown in Figure 12–16.

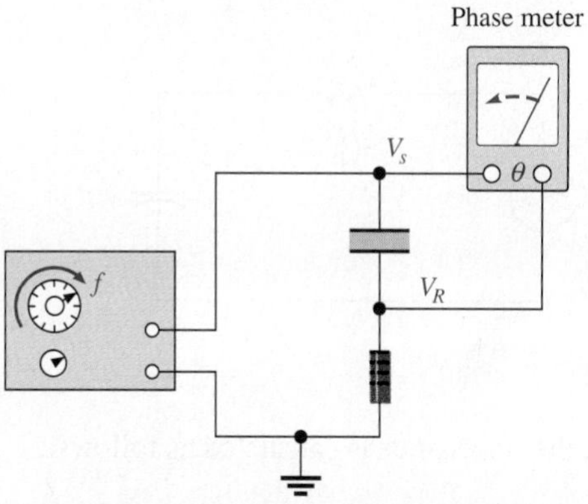

(a) As frequency is increased, the phase angle decreases.

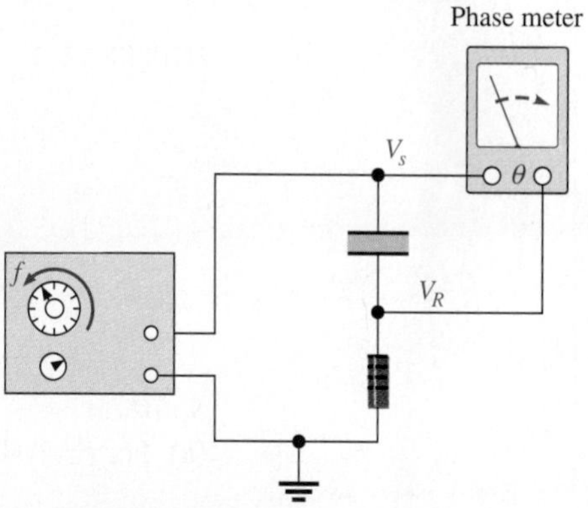

(b) As frequency is decreased, the phase angle increases.

(c)

FIGURE 12–15

The phase meter on a vector voltmeter shows the effect of frequency on the phase angle of a circuit. The phase meter indicates the phase angle between two voltages, V_s and V_R. Since V_R and I are in phase, this angle is the same as the angle between V_s and I. The vector voltmeter also indicates voltage magnitudes (photo courtesy of Hewlett-Packard Co.).

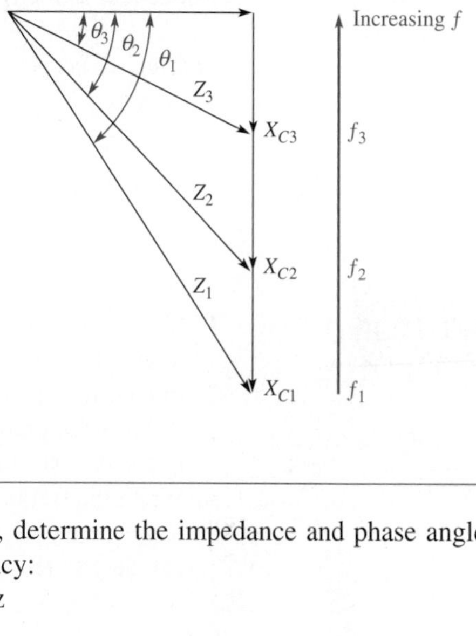

FIGURE 12–16

As the frequency increases, X_C decreases, Z decreases, and θ decreases. Each value of frequency can be visualized as forming a different impedance triangle.

EXAMPLE 12–5 For the series *RC* circuit in Figure 12–17, determine the impedance and phase angle for each of the following values of frequency:

(a) 10 kHz **(b)** 20 kHz **(c)** 30 kHz

FIGURE 12–17

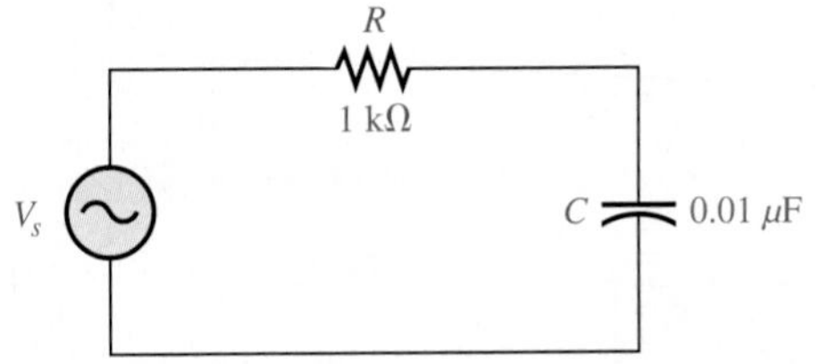

Solution

(a) For $f = 10$ kHz, the impedance is calculated as follows:

$$X_C = \frac{1}{2\pi fC} = \frac{1}{2\pi(10\text{ kHz})(0.01\ \mu\text{F})} = 1.59\text{ k}\Omega$$

$$Z = \sqrt{R^2 + X_C^2} = \sqrt{(1\text{ k}\Omega)^2 + (1.59\text{ k}\Omega)^2} = 1.88\text{ k}\Omega$$

The phase angle is

$$\theta = \tan^{-1}\left(\frac{X_C}{R}\right) = \tan^{-1}\left(\frac{1.59\text{ k}\Omega}{1\text{ k}\Omega}\right) = 57.8°$$

(b) For $f = 20$ kHz,

$$X_C = \frac{1}{2\pi(20\text{ kHz})(0.01\ \mu\text{F})} = 796\ \Omega$$

$$Z = \sqrt{(1\text{ k}\Omega)^2 + (796\ \Omega)^2} = 1.28\text{ k}\Omega$$

$$\theta = \tan^{-1}\left(\frac{796\ \Omega}{1\text{ k}\Omega}\right) = 38.5°$$

(c) For $f = 30$ kHz,

$$X_C = \frac{1}{2\pi(30\text{ kHz})(0.01\ \mu\text{F})} = 531\ \Omega$$

$$Z = \sqrt{(1\text{ k}\Omega)^2 + (531\ \Omega)^2} = 1.13\text{ k}\Omega$$

$$\theta = \tan^{-1}\left(\frac{531\ \Omega}{1\text{ k}\Omega}\right) = 28.0°$$

Notice that as the frequency increases, X_C, Z, and θ decrease.

Related Exercise Find the total impedance and phase angle in Figure 12–17 for $f = 1$ kHz.

SECTION 12–3 REVIEW

1. In a certain series *RC* circuit, $V_R = 4$ V and $V_C = 6$ V. What is the magnitude of the total voltage?
2. In Question 1, what is the phase angle?
3. What is the phase difference between the capacitor voltage and the resistor voltage in a series *RC* circuit?
4. When the frequency of the applied voltage in a series *RC* circuit is increased, what happens to each of the following?
(a) the capacitive reactance **(b)** the impedance **(c)** the phase angle

12–4 ■ IMPEDANCE AND PHASE ANGLE OF A PARALLEL *RC* CIRCUIT

In this section, you will learn how to determine the impedance and phase angle of a parallel RC circuit. Also conductance (G), capacitive susceptance (B_C), and total admittance (Y_T) are discussed because of their usefulness in parallel circuit analysis.

After completing this section, you should be able to

- **Determine impedance and phase angle in a parallel *RC* circuit**
  - □ Express total impedance in a product over sum form
  - □ Express the phase angle in terms of *R* and X_C
  - □ Determine the conductance, susceptance, and admittance in a parallel *RC* circuit
  - □ Convert admittance to impedance

Figure 12–18 shows a basic parallel *RC* circuit.

FIGURE 12–18
Parallel RC circuit.

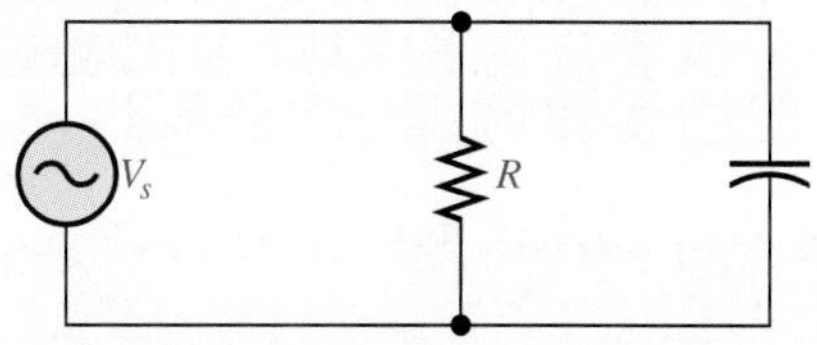

The expression for the impedance in Equation (12–8) is given in a product over sum form similar to the way two resistors in parallel can be expressed. In this case, the denominator is the phasor sum of *R* and X_C.

$$Z = \frac{RX_C}{\sqrt{R^2 + X_C^2}} \tag{12–8}$$

The phase angle between the applied voltage and the total current can be expressed in terms of *R* and X_C as shown in Equation (12–9).

$$\theta = \tan^{-1}\left(\frac{R}{X_C}\right) \tag{12–9}$$

This formula is derived from an equivalent formula using branch currents, which is introduced in the next section.

EXAMPLE 12–6 For each circuit in Figure 12–19, determine the impedance and the phase angle.

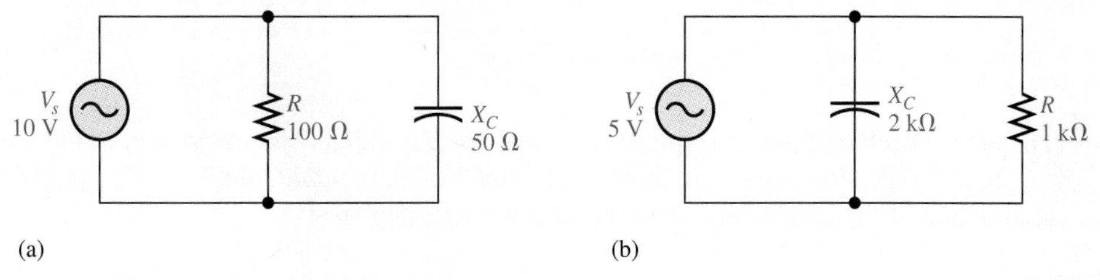

FIGURE 12–19

Solution For the circuit in Figure 12–19(a), the impedance and phase angle are

$$Z = \frac{RX_C}{\sqrt{R^2 + X_C^2}} = \frac{(100\ \Omega)(50\ \Omega)}{\sqrt{(100\ \Omega)^2 + (50\ \Omega)^2}} = 44.7\ \Omega$$

$$\theta = \tan^{-1}\left(\frac{R}{X_C}\right) = \tan^{-1}\left(\frac{100\ \Omega}{50\ \Omega}\right) = 63.4°$$

For the circuit in Figure 12–19(b),

$$Z = \frac{(1\ \text{k}\Omega)(2\ \text{k}\Omega)}{\sqrt{(1\ \text{k}\Omega)^2 + (2\ \text{k}\Omega)^2}} = 894\ \Omega$$

$$\theta = \tan^{-1}\left(\frac{1\ \text{k}\Omega}{2\ \text{k}\Omega}\right) = 26.6°$$

The calculator sequences for the circuit in Figure 12–19(a) are as follows:

Z: [1] [0] [0] [×] [5] [0] [÷] [(] [(] [1] [0] [0] [x²] [+] [5] [0] [x²] [)] [2nd] [√x] [)] [=]

θ: [1] [0] [0] [÷] [5] [0] [=] [2nd] [tan]

Related Exercise Determine Z in Figure 12–19(a) if the frequency is doubled.

Conductance, Susceptance, and Admittance

In working with parallel circuits, you will often find it easier to use conductance (G), capacitive susceptance (B_C), and admittance (Y) rather than resistance (R), capacitive reactance (X_C) and impedance (Z). **Conductance (G)** is the reciprocal of resistance and is expressed as

$$G = \frac{1}{R} \qquad \textbf{(12–10)}$$

Susceptance is the reciprocal of reactance; therefore, **capacitive susceptance (B_C)** is the reciprocal of capacitive reactance and is expressed as

$$B_C = \frac{1}{X_C} \qquad \textbf{(12–11)}$$

Admittance (Y) is the reciprocal of impedance and is expressed as

$$Y = \frac{1}{Z} \qquad \textbf{(12–12)}$$

The unit of each of these quantities is the siemens (S), which is the reciprocal of the ohm.

In a parallel RC circuit, as shown in Figure 12–20, the total admittance is simply the phasor sum of the conductance and the capacitive susceptance:

$$Y_T = \sqrt{G^2 + B_C^2} \qquad \textbf{(12–13)}$$

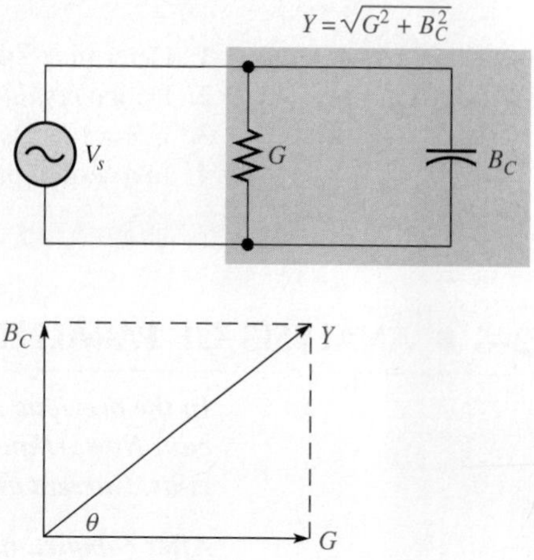

FIGURE 12–20
Admittance is the reciprocal of impedance in a parallel RC circuit.

EXAMPLE 12–7

Determine the admittance in Figure 12–21, and then convert it to impedance.

FIGURE 12–21

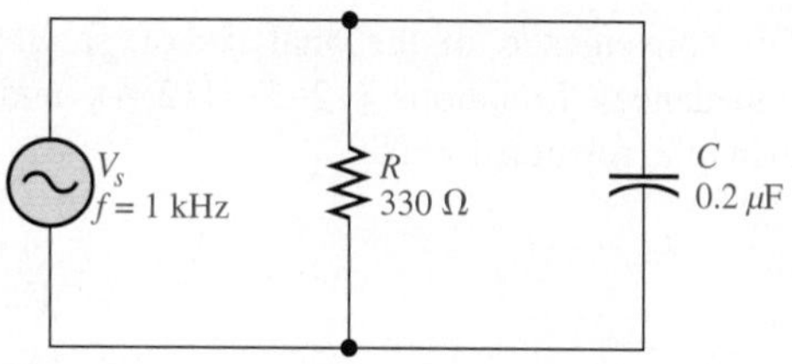

Solution To determine Y, first calculate the values for G and B_C. Since $R = 330\ \Omega$, then

$$G = \frac{1}{R} = \frac{1}{330\ \Omega} = 3.03\text{ mS}$$

The capacitive reactance is

$$X_C = \frac{1}{2\pi fC} = \frac{1}{2\pi(1000\text{ Hz})(0.2\ \mu\text{F})} = 796\ \Omega$$

The capacitive susceptance is

$$B_C = \frac{1}{X_C} = \frac{1}{796\ \Omega} = 1.26\text{ mS}$$

Therefore, the total admittance is

$$Y_T = \sqrt{G^2 + B_C^2} = \sqrt{(3.03\text{ mS})^2 + (1.26\text{ mS})^2} = 3.28\text{ mS}$$

Admittance is converted to impedance as follows:

$$Z = \frac{1}{Y_T} = \frac{1}{3.28\text{ mS}} = 305\ \Omega$$

The calculator sequence for Y is

3 • 0 3 EXP +/− 3 x^2 + 1 • 2 6 EXP +/− 3 x^2 = 2nd $\sqrt{x}$

Related Exercise Calculate the admittance in Figure 12–21 if f is increased to 2.5 kHz.

SECTION 12–4 REVIEW

1. Determine Z if a 1 kΩ resistance is in parallel with a 650 Ω capacitive reactance.
2. Define *conductance, capacitive susceptance,* and *admittance.*
3. If $Z = 100\ \Omega$, what is the value of Y?
4. In a certain parallel RC circuit, $R = 50\ \Omega$ and $X_C = 75\ \Omega$. Determine Y.

12–5 ■ ANALYSIS OF PARALLEL *RC* CIRCUITS

In the previous section, you learned how to express the impedance of a parallel RC circuit. Now, Ohm's law and Kirchhoff's current law are used in the analysis of RC circuits. Current and voltage relationships in a parallel RC circuit are examined.

After completing this section, you should be able to

- ***Analyze a parallel RC circuit***
  - □ Apply Ohm's law and Kirchhoff's current law to parallel *RC* circuits
  - □ Show how impedance and phase angle vary with frequency
  - □ Convert from a parallel circuit to an equivalent series circuit

For convenience in the analysis of parallel circuits, the Ohm's law formulas using impedance—Equations (12–3), (12–4), and (12–5)—can be rewritten for admittance using the relation $Y = 1/Z$:

$$V = \frac{I}{Y} \qquad \textbf{(12–14)}$$

$$I = VY \qquad \textbf{(12–15)}$$

$$Y = \frac{I}{V} \qquad \textbf{(12–16)}$$

EXAMPLE 12–8 Determine the total current and the phase angle in Figure 12–22.

FIGURE 12–22

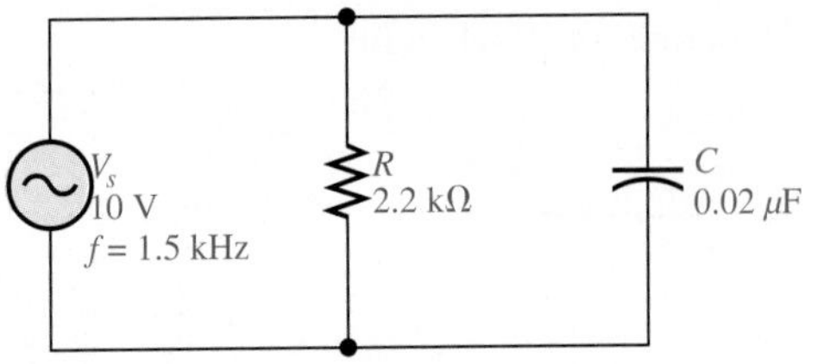

Solution First determine the total admittance. The capacitive reactance is

$$X_C = \frac{1}{2\pi fC} = \frac{1}{2\pi(1.5\text{ kHz})(0.02\ \mu\text{F})} = 5.31\text{ k}\Omega$$

The capacitive susceptance is

$$B_C = \frac{1}{X_C} = \frac{1}{5.31\ \text{k}\Omega} = 188\ \mu\text{S}$$

The conductance is

$$G = \frac{1}{R} = \frac{1}{2.2\ \text{k}\Omega} = 455\ \mu\text{S}$$

Therefore, the total admittance is

$$Y_T = \sqrt{G^2 + B_C^2} = \sqrt{(455\ \mu\text{S})^2 + (188\ \mu\text{S})^2} = 492\ \mu\text{S}$$

Applying Ohm's law, calculate the total current:

$$I_T = VY_T = (10\ \text{V})(492\ \mu\text{S}) = 4.92\ \text{mA}$$

The phase angle is

$$\theta = \tan^{-1}\left(\frac{R}{X_C}\right) = \tan^{-1}\left(\frac{2.2\ \text{k}\Omega}{5.31\ \text{k}\Omega}\right) = 22.5°$$

The total current is 4.92 mA, and it leads the applied voltage by 22.5°.

Related Exercise What is the total current if the frequency is doubled?

Relationships of the Currents and Voltages in a Parallel *RC* Circuit

Figure 12–23(a) shows all the currents and voltages in a basic parallel *RC* circuit. As you can see, the applied voltage, V_s, appears across both the resistive and the capacitive branches, so V_s, V_R, and V_C are all in phase and of the same magnitude. The total current, I_T, divides at the junction into the two branch currents, I_R and I_C.

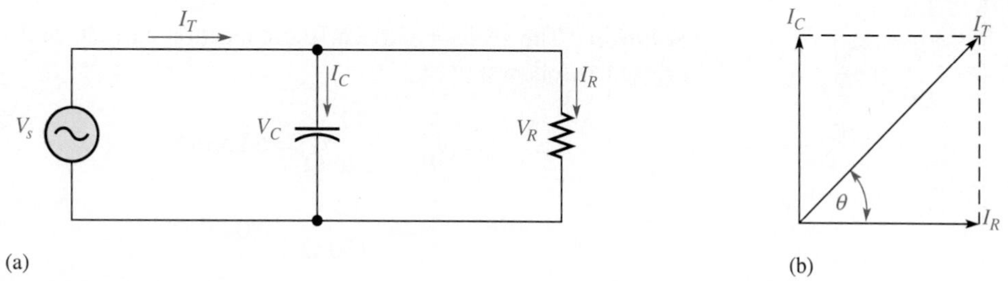

FIGURE 12–23
Currents and voltages in a parallel RC circuit. The current directions shown in (a) are instantaneous and, of course, reverse when the source voltage reverses.

The current through the resistor is in phase with the voltage. The current through the capacitor leads the voltage, and thus the resistive current, by 90°. By Kirchhoff's current law, the total current is the phasor sum of the two branch currents, as shown by the phasor diagram in Figure 12–23(b). The total current is expressed as

$$I_T = \sqrt{I_R^2 + I_C^2} \qquad \textbf{(12–17)}$$

The phase angle can be found from the current values as follows:

$$\theta = \tan^{-1}\left(\frac{I_C}{I_R}\right) \tag{12–18}$$

Equation (12–18) is equivalent to the previous expression, $\theta = \tan^{-1}(R/X_C)$ in Equation (12–9).

Figure 12–24 shows a complete current and voltage phasor diagram. Notice that I_C leads I_R by 90° and that I_R is in phase with the voltage ($V_s = V_R = V_C$).

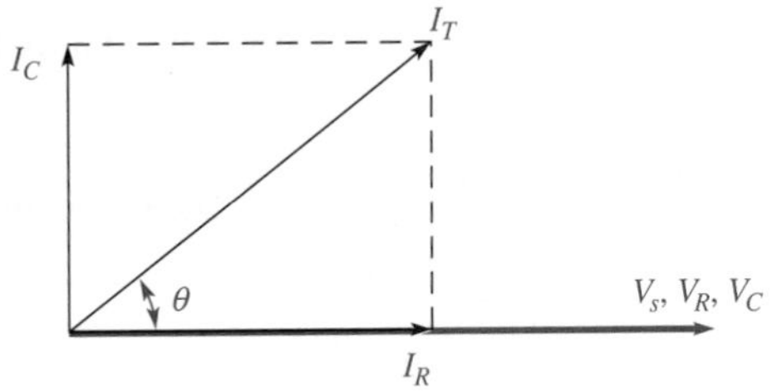

FIGURE 12–24
Current and voltage phasor diagram for a parallel RC circuit (amplitudes are arbitrary).

EXAMPLE 12–9

Determine the value of each current in Figure 12–25, and describe the phase relationship of each with the applied voltage.

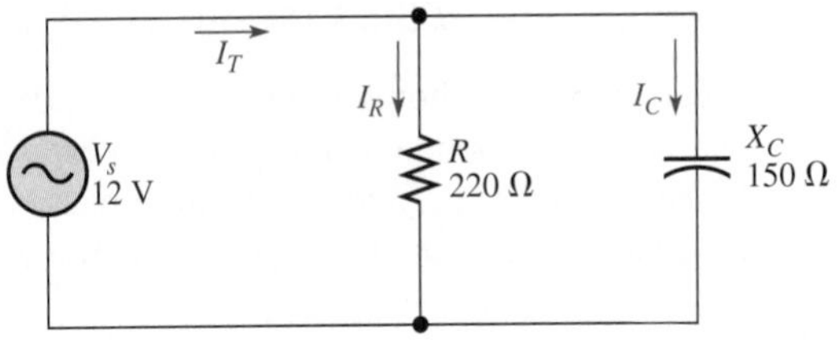

FIGURE 12–25

Solution The resistor current, the capacitor current, and the total current are expressed as follows:

$$I_R = \frac{V_s}{R} = \frac{12\text{ V}}{220\ \Omega} = 54.5\text{ mA}$$

$$I_C = \frac{V_s}{X_C} = \frac{12\text{ V}}{150\ \Omega} = 80\text{ mA}$$

$$I_T = \sqrt{I_R^2 + I_C^2} = \sqrt{(54.5\text{ mA})^2 + (80\text{ mA})^2} = 96.8\text{ mA}$$

The phase angle is

$$\theta = \tan^{-1}\left(\frac{I_C}{I_R}\right) = \tan^{-1}\left(\frac{80\text{ mA}}{54.5\text{ mA}}\right) = 55.7°$$

I_R is in phase with the voltage, I_C leads the voltage by 90°, and I_T leads the voltage by 55.7°.

Related Exercise In a certain parallel circuit, I_R = 100 mA and I_C = 60 mA. Determine the total current and the phase angle.

Conversion from Parallel to Series Form

For every parallel *RC* circuit, there is an equivalent series *RC* circuit for any given frequency. Two circuits are equivalent when they both present an equal impedance and phase angle.

To obtain the equivalent series circuit for a given parallel *RC* circuit, first find the impedance and phase angle of the parallel circuit. Then use the values of Z and θ to construct an impedance triangle shown in Figure 12–26. The vertical and horizontal sides of the triangle represent the equivalent series resistance and capacitive reactance as indicated. These values can be found using the following trigonometric relationships:

$$R_{eq} = Z \cos \theta \tag{12–19}$$

$$X_{C(eq)} = Z \sin \theta \tag{12–20}$$

The cosine (cos) and the sine (sin) functions are available on your calculator.

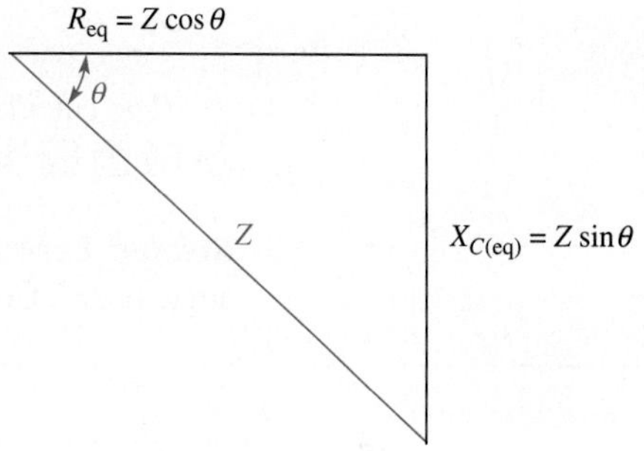

FIGURE 12–26
Impedance triangle for the series equivalent of a parallel RC circuit. Z and θ are the known values for the parallel circuit. R_{eq} and $X_{C(eq)}$ are the series equivalent values.

EXAMPLE 12–10 Convert the parallel circuit in Figure 12–27 to an equivalent series form.

FIGURE 12–27

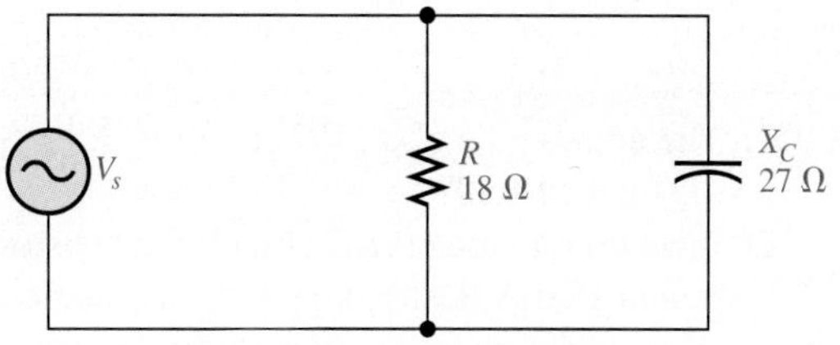

Solution First, find the total admittance of the parallel circuit as follows:

$$G = \frac{1}{R} = \frac{1}{18\ \Omega} = 55.6\text{ mS}$$

$$B_C = \frac{1}{X_C} = \frac{1}{27\ \Omega} = 37.0\text{ mS}$$

$$Y_T = \sqrt{G^2 + B_C^2} = \sqrt{(55.6\text{ mS})^2 + (37.0\text{ mS})^2} = 66.8\text{ mS}$$

Then, the total impedance is

$$Z_T = \frac{1}{Y_T} = \frac{1}{66.8\text{ mS}} = 15.0\ \Omega$$

The phase angle is

$$\theta = \tan^{-1}\left(\frac{R}{X_C}\right) = \tan^{-1}\left(\frac{18\ \Omega}{27\ \Omega}\right) = 33.6°$$

The equivalent series values are

$$R_{eq} = Z\cos\theta = 15.0\cos(33.6°) = 12.5\ \Omega$$
$$X_{C(eq)} = Z\sin\theta = 15.0\sin(33.6°) = 8.3\ \Omega$$

The series equivalent circuit is shown in Figure 12–28. The value of C can only be determined if the frequency is known.

FIGURE 12–28

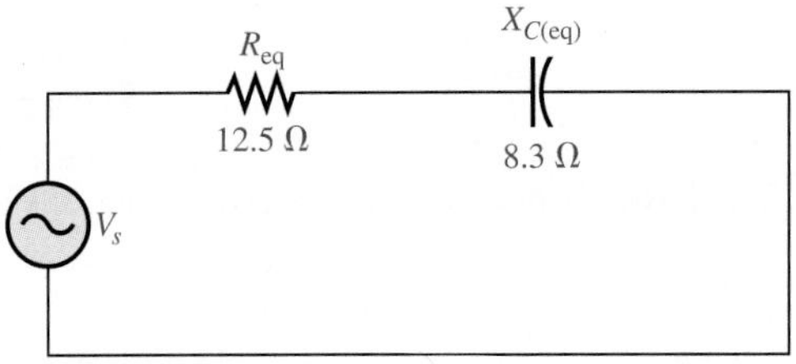

The calculator sequence for R_{eq} is

[3] [3] [•] [6] [cos] [×] [1] [5] [=]

Related Exercise The impedance of a parallel *RC* circuit is 10 kΩ and the phase angle is 26°. Convert to an equivalent series circuit.

Note that a parallel *RC* circuit becomes less reactive when X_C is increased. That is, the circuit phase angle becomes smaller. The reason for this effect is that when X_C is increased relative to *R*, less current flows through the capacitive branch, and although the in-phase or resistive current does not increase, it becomes a greater percentage of the total current.

SECTION 12–5 REVIEW

1. The admittance of an *RC* circuit is 3.5 mS, and the applied voltage is 6 V. What is the total current?
2. In a certain parallel *RC* circuit, the resistor current is 10 mA, and the capacitor current is 15 mA. Determine the phase angle and the total current.
3. What is the phase angle between the capacitor current and the applied voltage in a parallel *RC* circuit?

12–6 ■ SERIES-PARALLEL ANALYSIS

In this section, the concepts studied in the previous sections are used to analyze circuits with combinations of both series and parallel R and C elements.

After completing this section, you should be able to

- **Analyze series-parallel *RC* circuits**
  - □ Determine total impedance
  - □ Calculate currents and voltages
  - □ Measure impedance and phase angle

The following example demonstrates the procedures used to analyze a series-parallel circuit.

EXAMPLE 12–11 In the circuit of Figure 12–29, determine the following:
(a) total impedance **(b)** total current **(c)** phase angle by which I_T leads V_s

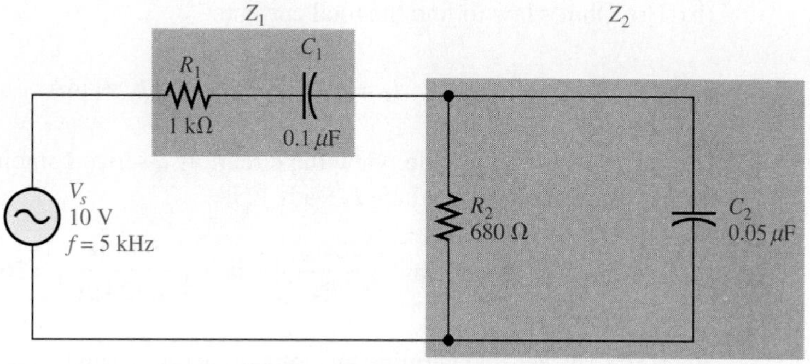

FIGURE 12–29

Solution

(a) First, calculate the magnitudes of capacitive reactances:

$$X_{C1} = \frac{1}{2\pi(5\text{ kHz})(0.1\ \mu\text{F})} = 318\ \Omega$$

$$X_{C2} = \frac{1}{2\pi(5\text{ kHz})(0.05\ \mu\text{F})} = 637\ \Omega$$

One approach is to find the series equivalent resistance and capacitive reactance for the parallel portion of the circuit, and then add the resistances to get total resistance and add the reactances to get total reactance. From this, the total impedance can be determined.

Find the impedance of the parallel portion (Z_2) by first finding the admittance:

$$G_2 = \frac{1}{R_2} = \frac{1}{680\ \Omega} = 1.47\text{ mS}$$

$$B_{C2} = \frac{1}{X_{C2}} = \frac{1}{637\ \Omega} = 1.57\text{ mS}$$

$$Y_2 = \sqrt{G_2^2 + B_{C2}^2} = \sqrt{(1.47\text{ mS})^2 + (1.57\text{ mS})^2} = 2.15\text{ mS}$$

$$Z_2 = \frac{1}{Y_2} = \frac{1}{2.15\text{ mS}} = 465\ \Omega$$

The phase angle associated with the parallel portion of the circuit is

$$\theta_p = \tan^{-1}\left(\frac{R_2}{X_{C2}}\right) = \tan^{-1}\left(\frac{680\ \Omega}{637\ \Omega}\right) = 46.9^\circ$$

The series equivalent values for the parallel portion are

$$R_{\text{eq}} = Z_2 \cos\theta_p = 465\cos(46.9^\circ) = 318\ \Omega$$

$$X_{C(\text{eq})} = Z_2 \sin\theta_p = 465\sin(46.9^\circ) = 340\ \Omega$$

The total resistance is

$$R_T = R_1 + R_{eq} = 1000\ \Omega + 318\ \Omega = 1.32\ \text{k}\Omega$$

The total reactance is

$$X_{C(tot)} = X_{C1} + X_{C(eq)} = 318\ \Omega + 340\ \Omega = 658\ \Omega$$

The total impedance is

$$Z_T = \sqrt{R_T^2 + X_{C(tot)}^2} = \sqrt{(1.32\ \text{k}\Omega)^2 + (658\ \Omega)^2} = 1.47\ \text{k}\Omega$$

(b) Use Ohm's law to find the total current:

$$I_T = \frac{V_S}{Z_T} = \frac{10\ \text{V}}{1.47\ \text{k}\Omega} = 6.80\ \text{mA}$$

(c) To get the phase angle, view the circuit as a series combination of R_T and $X_{C(tot)}$. The phase angle by which I_T leads V_s is

$$\theta = \tan^{-1}\left(\frac{X_{C(tot)}}{R_T}\right) = \tan^{-1}\left(\frac{658\ \Omega}{1.32\ \text{k}\Omega}\right) = 26.5°$$

Related Exercise Determine the voltages across Z_1 and Z_2 in Figure 12–29.

Measurement of Z_T and θ

Now, let's see how the values of Z_T and θ for the circuit in Example 12–11 can be determined by measurement. First, the total impedance is measured as outlined in the following steps and as illustrated in Figure 12–30 (other ways are also possible):

1. Using a sine wave generator, set the source voltage to a known value (10 V) and the frequency to 5 kHz. It is advisable to check the voltage with an ac voltmeter and the frequency with a frequency counter rather than relying on the marked values on the generator controls.
2. Connect an ac ammeter as shown in Figure 12–30, and measure the total current.
3. Calculate the total impedance by using Ohm's law.

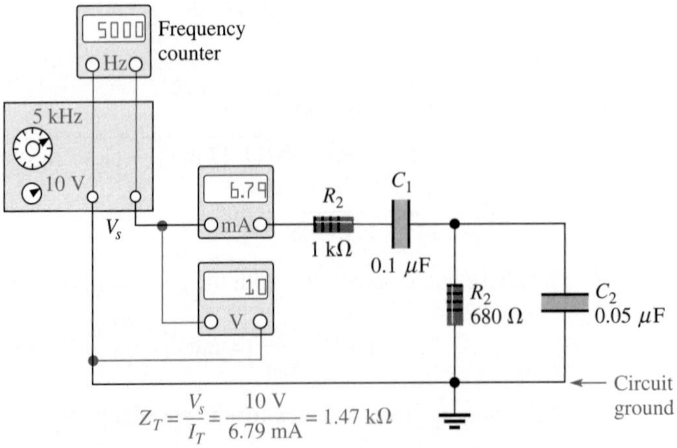

FIGURE 12–30
Determining Z_T by measurement of V_s and I_T.

Although you could use a phase meter to measure the phase angle, we will use an oscilloscope in this illustration because it is more commonly available. To measure the phase angle, the source voltage and the total current must be displayed on the screen in the proper time relationship. Two basic types of scope probes are available to measure the quantities with an oscilloscope: the voltage probe and the current probe. Although the current probe is a convenient device, it is often not as readily available as a voltage probe. For this reason, we will confine our phase measurement technique to the use of voltage probes in conjunction with the oscilloscope. A typical oscilloscope voltage probe has two points that are connected to the circuit: the probe tip and the ground lead. Thus, all voltage measurements must be referenced to ground.

Since only voltage probes are to be used, the total current cannot be measured directly. However, for phase measurement, the voltage across R_1 is in phase with the total current and can be used to establish the phase angle of the current.

Before proceeding with the actual phase measurement, there is a problem with displaying V_{R1}. If the scope probe is connected across the resistor, as indicated in Figure 12–31(a), the ground lead of the scope will short point B to ground, thus bypassing the rest of the components and effectively removing them from the circuit electrically, as illustrated in Figure 12–31(b) (assuming that the scope is not isolated from power line ground).

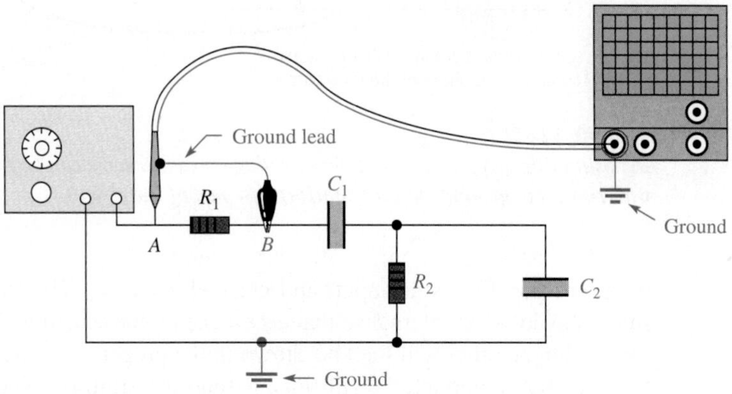

(a) Ground lead on scope probe grounds point B.

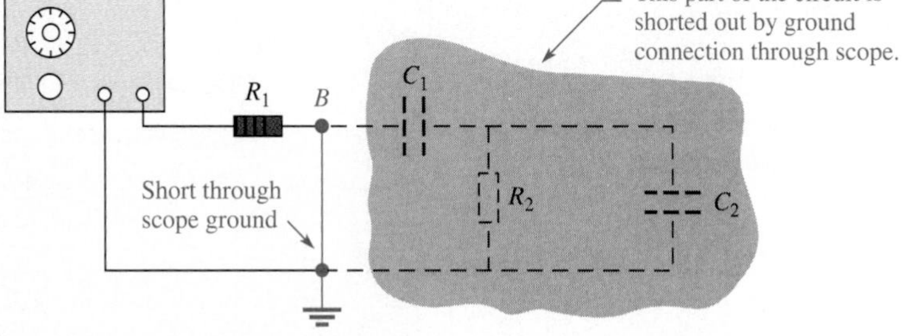

(b) The effect of grounding point B is to short out the rest of the circuit.

FIGURE 12–31

Effects of measuring directly across a component when the instrument and the circuit are grounded.

To avoid this problem, you can switch the generator output terminals so that one end of R_1 is connected to the ground terminal, as shown in Figure 12–32(a). Now the scope can be connected across it to display V_{R1}, as indicated in Figure 12–32(b). The other probe is connected across the voltage source to display V_s as indicated. Now channel 1 of

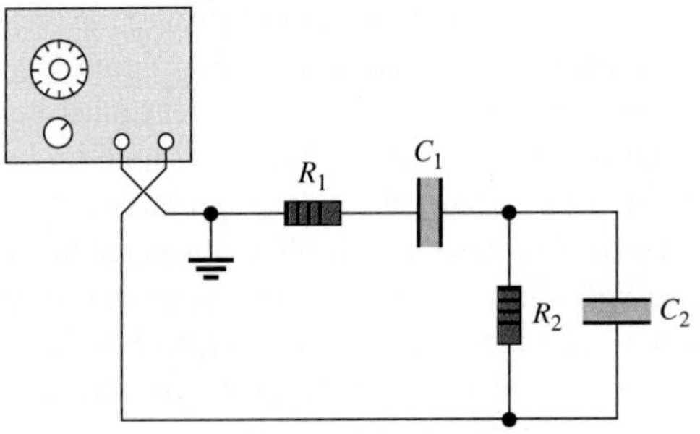

(a) Ground repositioned so that one end of R_1 is grounded.

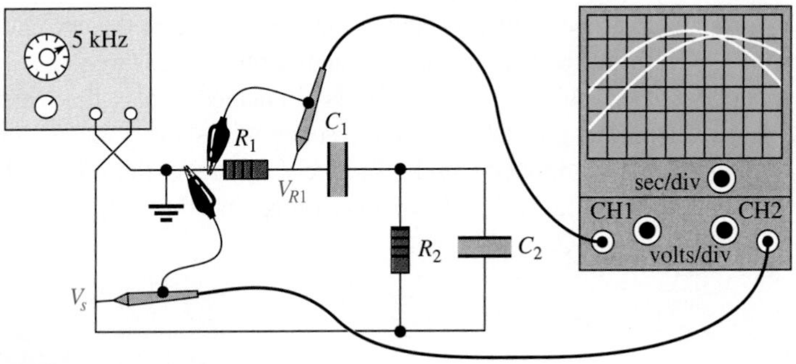

(b) The scope displays a half-cycle of V_{R1} and V_s. V_{R1} represents the phase of the total current.

FIGURE 12–32
Repositioning ground so that a direct voltage measurement can be made with respect to ground without shorting out part of the circuit.

the scope has V_{R1} as an input, and channel 2 has V_s. The trigger source switch on the scope should be on internal so that each trace on the screen will be triggered by one of the inputs and the other will then be shown in the proper time relationship to it. Since amplitudes are not important, the volts/div settings are arbitrary. The sec/div settings should be adjusted so that one half-cycle of the waveforms appears on the screen.

Before connecting the probes to the circuit, you must align the two horizontal lines (traces) so that they appear as a single line across the center of the screen. To do so,

FIGURE 12–33
Measurement of the phase angle on the oscilloscope.

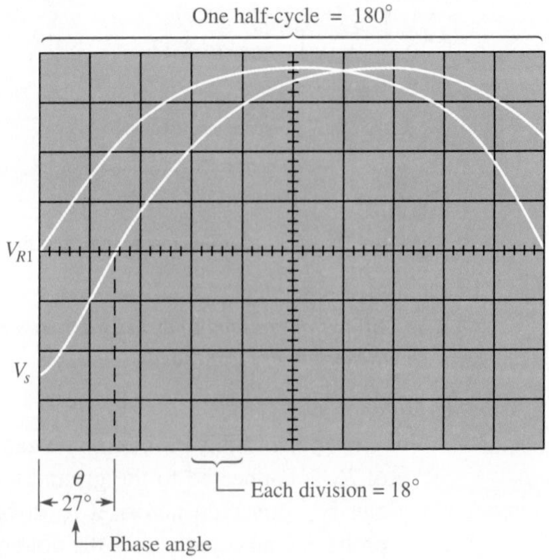

ground the probe tips and adjust the vertical position knobs to move the traces toward the center line of the screen until they are superimposed. This procedure ensures that both waveforms have the same zero crossing so that an accurate phase measurement can be made.

The resulting oscilloscope display is shown in Figure 12–33. Since there is 180° in one half-cycle, each of the ten horizontal divisions across the screen represents 18°. Thus, the horizontal distance between the corresponding points of the two waveforms is the phase angle in degrees as indicated.

SECTION 12–6 REVIEW

1. Determine the series equivalent circuit for the series-parallel circuit in Figure 12–29.
2. What is the voltage across R_1 in Figure 12–29?

12–7 ■ POWER IN *RC* CIRCUITS

In a purely resistive ac circuit, all of the energy delivered by the source is dissipated in the form of heat by the resistance. In a purely capacitive ac circuit, all of the energy delivered by the source is stored by the capacitor during a portion of the voltage cycle and then returned to the source during another portion of the cycle so that there is no net energy loss. When there is both resistance and capacitance, some of the energy is alternately stored and returned by the capacitance and some is dissipated by the resistance. The amount of energy loss is determined by the relative values of the resistance and the capacitive reactance.

After completing this section, you should be able to

- **Determine power in *RC* circuits**
  - □ Explain true and reactive power
  - □ Draw the power triangle
  - □ Define *power factor*
  - □ Explain apparent power
  - □ Calculate power in an *RC* circuit

It is reasonable to assume that when the resistance is greater than the reactance, more of the total energy delivered by the source is dissipated by the resistance than is stored by the capacitance. Likewise, when the reactance is greater than the resistance, more of the total energy is stored and returned than is dissipated.

The formulas for power in a resistor, sometimes called *true power* (P_{true}), and the power in a capacitor, called *reactive power* (P_r), are restated here. The unit of true power is the watt (W), and the unit of reactive power is the volt-ampere reactive (VAR).

$$P_{true} = I_T^2 R \quad \textbf{(12–21)}$$

$$P_r = I_T^2 X_C \quad \textbf{(12–22)}$$

The Power Triangle

The generalized impedance phasor diagram is shown in Figure 12–34(a). A phasor relationship for power can also be represented by a similar diagram because the respective

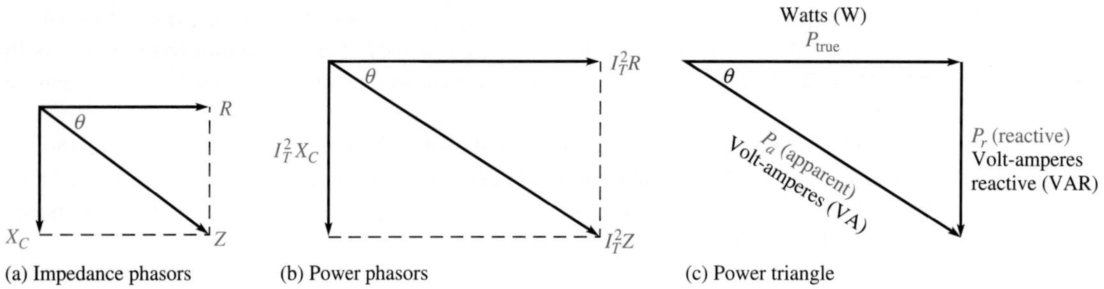

FIGURE 12–34
Development of the power triangle for an RC circuit.

magnitudes of the powers, P_{true} and P_r, differ from R and X_C by a factor of I_T^2, as shown in Figure 12–34(b).

The resultant power phasor, I_T^2Z, represents the **apparent power, P_a**. At any instant in time, P_a is the total power that appears to be transferred between the source and the *RC* circuit. Part of the apparent power is true power, and part of it is reactive power. The unit of apparent power is the volt-ampere (VA). The expression for apparent power is

$$P_a = I_T^2Z \qquad \textbf{(12–23)}$$

The power phasor diagram in Figure 12–34(b) can be rearranged in the form of a right triangle as shown in Figure 12–34(c), which is called the *power triangle.* Using the rules of trigonometry, P_{true} can be expressed as

$$P_{true} = P_a\cos\theta$$

Since P_a equals I_T^2Z or V_sI_T, the equation for the true power can be written as

$$P_{true} = V_sI_T\cos\theta \qquad \textbf{(12–24)}$$

where V_s is the applied voltage and I_T is the total current.

For the case of a purely resistive circuit, $\theta = 0°$ and $\cos 0° = 1$, so P_{true} equals V_sI_T. For the case of a purely capacitive circuit, $\theta = 90°$ and $\cos 90° = 0$, so P_{true} is zero. As you already know, there is no power dissipated in an ideal capacitor.

Power Factor

The term $\cos\theta$ is called the **power factor** and is stated as follows:

$$PF = \cos\theta \qquad \textbf{(12–25)}$$

As the phase angle between applied voltage and total current increases, the power factor decreases, indicating an increasingly reactive circuit. The smaller the power factor, the smaller the power dissipation.

The power factor can vary from 0 for a purely reactive circuit to 1 for a purely resistive circuit. In an *RC* circuit, the power factor is referred to as a leading power factor because the current leads the voltage.

EXAMPLE 12–12 Determine the power factor and the true power in the circuit of Figure 12–35.

FIGURE 12–35

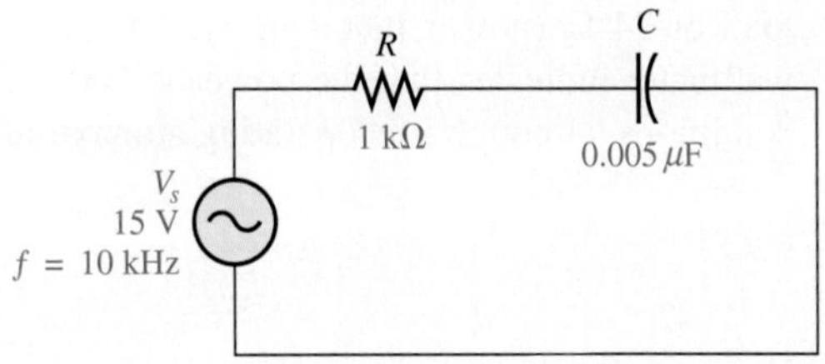

Solution Calculate the capacitive reactance and phase angle as follows:

$$X_C = \frac{1}{2\pi fC} = \frac{1}{2\pi(10\text{ kHz})(0.005\ \mu\text{F})} = 3.18\text{ k}\Omega$$

$$\theta = \tan^{-1}\left(\frac{X_C}{R}\right) = \tan^{-1}\left(\frac{3.18\text{ k}\Omega}{1\text{ k}\Omega}\right) = 72.5°$$

The power factor is

$$PF = \cos\theta = \cos(72.5°) = 0.301$$

The impedance is

$$Z = \sqrt{R^2 + X_C^2} = \sqrt{(1\text{ k}\Omega)^2 + (3.18\text{ k}\Omega)^2} = 3.33\text{ k}\Omega$$

Therefore, the current is

$$I = \frac{V_s}{Z} = \frac{15\text{ V}}{3.33\text{ k}\Omega} = 4.50\text{ mA}$$

The true power is

$$P_{\text{true}} = V_sI\cos\theta = (15\text{ V})(4.50\text{ mA})(0.301) = 20.3\text{ mW}$$

Related Exercise What is the power factor if *f* is reduced by half in Figure 12–35?

The Significance of Apparent Power

As mentioned, apparent power is the power that appears to be transferred between the source and the load, and it consists of two components: a true power component and a reactive power component.

In all electrical and electronic systems, it is the true power that does the work. The reactive power is simply shuttled back and forth between the source and load. Ideally, in terms of performing useful work, all of the power transferred to the load should be true power and none of it reactive power. However, in most practical situations the load has some reactance associated with it, and therefore you must deal with both power components.

In Chapter 11 we discussed the use of apparent power in relation to transformers. For any reactive load, there are two components of the total current: the resistive component and the reactive component. If you consider only the true power (watts) in a load, you are dealing with only a portion of the total current that the load demands from a source. In order to have a realistic picture of the actual current that a load will draw, you must consider apparent power (in VA).

A source such as an ac generator can provide current to a load up to some maximum value. *If the load draws more than this maximum value, the source can be damaged.* Figure 12–36(a) shows a 120 V generator that can deliver a maximum current of 5 A to a load. Assume that the generator is rated at 600 W and is connected to a purely resistive load of 24 Ω (power factor of 1). The ammeter shows that the current is 5 A, and the wattmeter indicates that the power is 600 W. The generator has no problem under these conditions, although it is operating at maximum current and power.

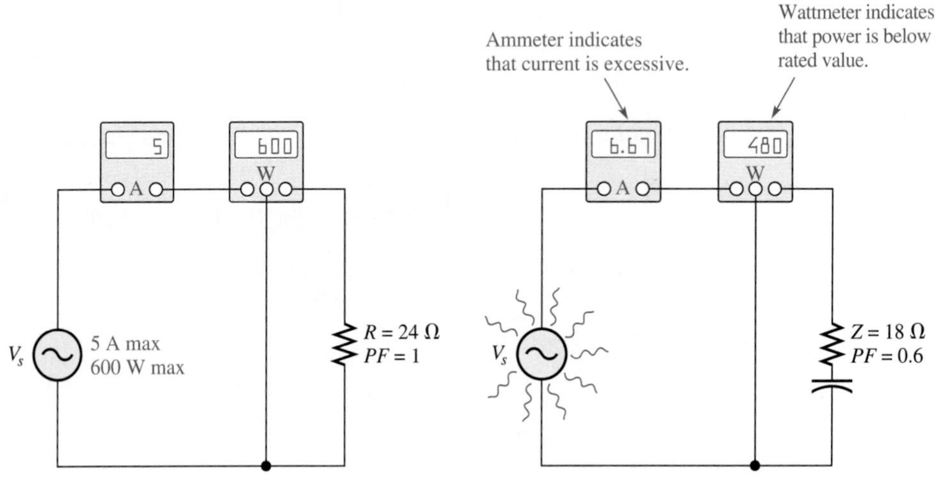

(a) Generator operating at its limits with a resistive load.

(b) Generator is in danger of internal damage due to excess current, even though the wattmeter indicates that the power is below the maximum wattage rating.

FIGURE 12–36

The wattage rating of a source is inappropriate when the load is reactive. The rating should be in VA rather than in watts.

Now, consider what happens if the load is changed to a reactive one with an impedance of 18 Ω and a power factor of 0.6, as indicated in Figure 12–36(b). The current is 120 V/18 Ω = 6.67 A, which exceeds the maximum. Even though the wattmeter reads 480 W, which is less than the power rating of the generator, the excessive current probably will cause damage. This example shows that a true power rating can be deceiving and is inappropriate for ac sources. This particular ac generator should be rated at 600 VA, a rating the manufacturer would use, rather than 600 W.

EXAMPLE 12–13

For the circuit in Figure 12–37, find the true power, the reactive power, and the apparent power. X_C has been determined to be 2 kΩ.

FIGURE 12–37

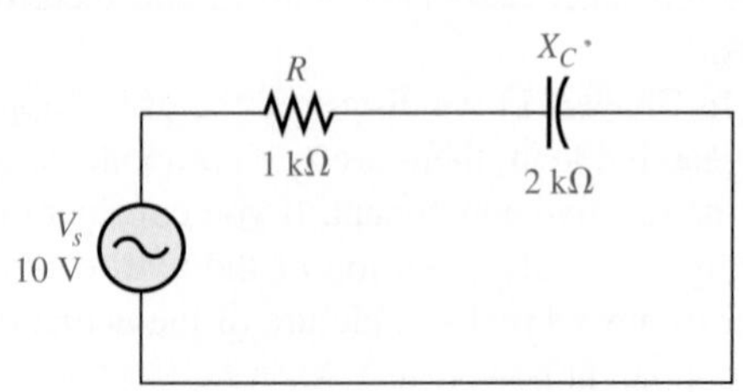

Solution First find the total impedance so that the current can be calculated.

$$Z_T = \sqrt{R^2 + X_C^2} = \sqrt{(1\text{ k}\Omega)^2 + (2\text{ k}\Omega)^2} = 2.24\text{ k}\Omega$$

$$I = \frac{V_s}{Z} = \frac{10\text{ V}}{2.24\text{ k}\Omega} = 4.46\text{ mA}$$

The phase angle, θ, is

$$\theta = \tan^{-1}\left(\frac{X_C}{R}\right) = \tan^{-1}\left(\frac{2\text{ k}\Omega}{1\text{ k}\Omega}\right) = 63.4°$$

The true power is

$$P_{\text{true}} = V_s I \cos\theta = (10\text{ V})(4.46\text{ mA})\cos(63.4°) = 20\text{ mW}$$

Note that the same result is realized using the formula $P_{\text{true}} = I^2R$.

The reactive power is

$$P_r = I^2X_C = (4.46\text{ mA})^2\,(2\text{ k}\Omega) = 39.8\text{ mVAR}$$

The apparent power is

$$P_a = I^2Z = (4.46\text{ mA})^2\,(2.24\text{ k}\Omega) = 44.6\text{ mVA}$$

The apparent power is also the phasor sum of P_{true} and P_r:

$$P_a = \sqrt{P_{\text{true}}^2 + P_r^2} = 44.6\text{ mVA}$$

Related Exercise What is the true power in Figure 12–37 if $X_C = 10\text{ k}\Omega$?

SECTION 12–7 REVIEW

1. To which component in an *RC* circuit is the power dissipation due?
2. If the phase angle, θ, is 45°, what is the power factor?
3. A certain series *RC* circuit has the following parameter values: $R = 330\ \Omega$, $X_C = 460\ \Omega$, and $I = 2$ A. Determine the true power, the reactive power, and the apparent power.

12–8 ■ BASIC APPLICATIONS

RC circuits are found in a variety of applications, often as part of a more complex circuit. Two major applications, phase shift networks and frequency-selective networks (filters), are covered in this section.

After completing this section, you should be able to

- **Discuss some basic *RC* applications**
  - □ Discuss and analyze the *RC* lag network
  - □ Discuss and analyze the *RC* lead network
  - □ Discuss how the *RC* circuit operates as a filter

The *RC* Lag Network

The *RC* lag network is a phase shift circuit in which the output voltage lags the input voltage by a specified angle, ϕ. Phase shift circuits are commonly used in electronic communication systems and in other application areas.

A basic *RC* lag network is shown in Figure 12–38(a). Notice that the output voltage is taken across the capacitor, and the input voltage is the total voltage applied across the circuit. The relationship of the voltages is shown in the phasor diagram in Figure 12–38(b). Keep in mind that the circuit phase angle is the angle between V_{in} and the current. Notice in Figure 12–38(b) that the output voltage, V_C, lags V_{in} by an angle, designated ϕ, that is the difference between 90° and the circuit phase angle θ. The angle ϕ is called the phase lag when the output voltage lags the input voltage and the phase lead when the output voltage leads the input.

Since $\theta = \tan^{-1}(X_C/R)$, the phase lag, ϕ, can be expressed as

$$\phi = 90° - \tan^{-1}\left(\frac{X_C}{R}\right) \tag{12–26}$$

When the input and output waveforms are displayed on an oscilloscope, a relationship similar to that in Figure 12–39 is observed. The exact amount of phase lag between the input and the output depends on the values of the resistance and the capacitive reactance. The magnitude of the output voltage depends on these values also.

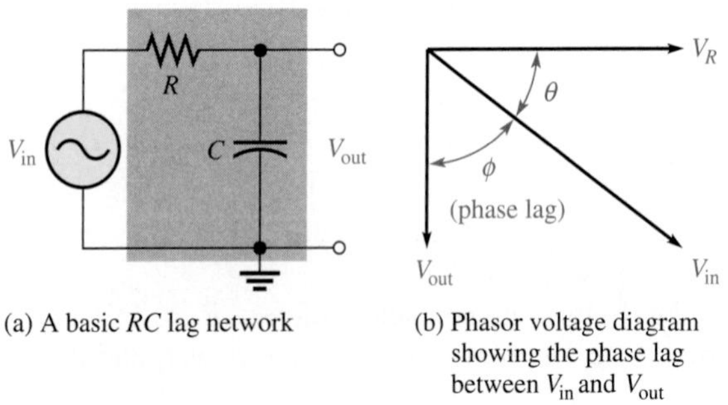

FIGURE 12–38
The RC lag network ($V_{out} = V_C$).

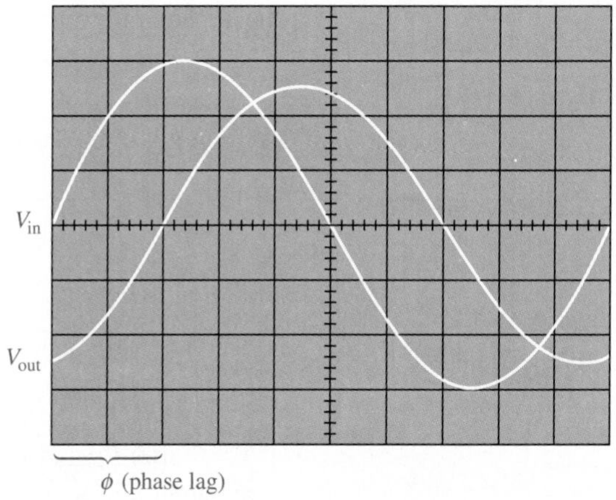

FIGURE 12–39
General oscilloscope display of the input and output waveforms of a lag network (V_{out} lags V_{in}).

EXAMPLE 12–14 Determine the amount of phase lag from input to output in the lag network in Figure 12–40.

FIGURE 12–40

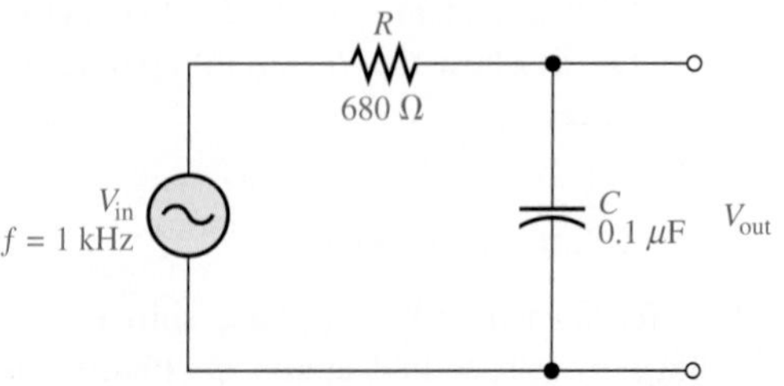

Solution First determine the capacitive reactance.

$$X_C = \frac{1}{2\pi fC} = \frac{1}{2\pi(1\ \text{kHz})(0.1\ \mu\text{F})} = 1.59\ \text{k}\Omega$$

The phase lag between the output and the input is

$$\phi = 90° - \tan^{-1}\left(\frac{X_C}{R}\right) = 90° - \tan^{-1}\left(\frac{1.59\ \text{k}\Omega}{680\ \Omega}\right) = 23.2°$$

The output lags the input by 23.2°.

Related Exercise In a lag network, what happens to the phase lag if the frequency increases?

The phase-lag network can be viewed as a voltage divider with a portion of the input voltage dropped across R and a portion across C. The output voltage can be determined with the following formula:

$$V_{out} = \left(\frac{X_C}{\sqrt{R^2 + X_C^2}}\right)V_{in} \qquad (12\text{–}27)$$

EXAMPLE 12–15 For the lag network in Figure 12–40 of Example 12–14, determine the output voltage in phasor form when the input voltage has an rms value of 10 V. Sketch the input and output waveforms showing the proper relationships. The values for X_C (1.59 kΩ) and ϕ (23.2°) were found in Example 12–14.

Solution Use Equation (12–27) to determine the output voltage for the lag network in Figure 12–40.

$$V_{out} = \left(\frac{X_C}{\sqrt{R^2 + X_C^2}}\right)V_{in} = \left(\frac{1.59\ \text{k}\Omega}{\sqrt{(680\ \Omega)^2 + (1.59\ \text{k}\Omega)^2}}\right)10\ \text{V} = 9.2\ \text{V}$$

The waveforms are shown in Figure 12–41.

FIGURE 12–41

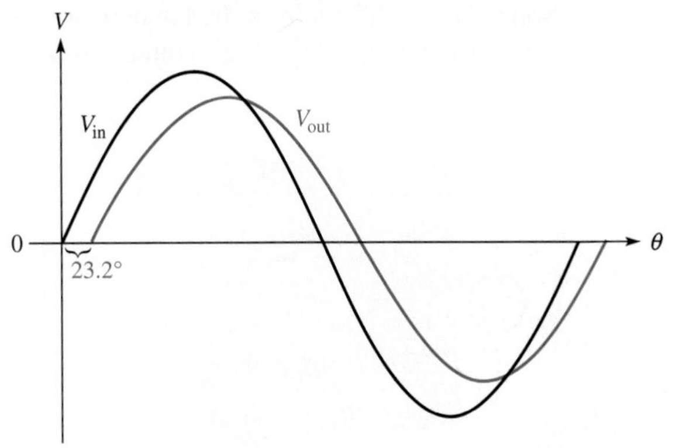

Related Exercise In a lag network, what happens to the output voltage if the frequency increases?

Effects of Frequency on the Lag Network Since the circuit phase angle, θ, decreases as frequency increases, the phase lag between the input and the output voltages increases. You can see this relationship by examining Equation (12–26). Also, the magnitude of V_{out} decreases as the frequency increases, because X_C becomes smaller and less of the total input voltage is dropped across the capacitor. Figure 12–42 illustrates the changes in phase lag and output voltage magnitude with frequency.

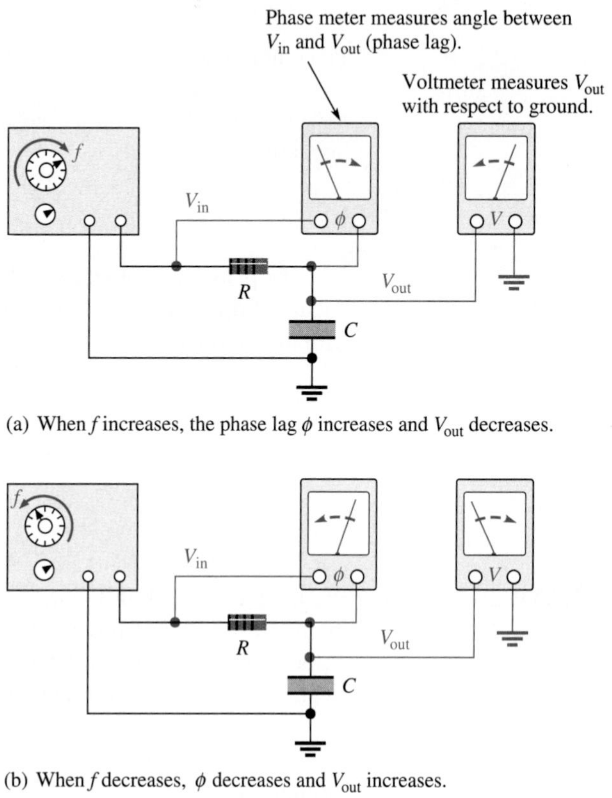

FIGURE 12–42
Illustration of how the frequency affects the phase lag and the output voltage in an RC lag network, with the amplitude of V_{in} held constant.

The *RC* Lead Network

The *RC* lead network is a phase shift circuit in which the output voltage leads the input voltage by a specified angle, ϕ. A basic *RC* lead network is shown in Figure 12–43(a). Notice how it differs from the lag network. Here the output voltage is taken across the resistor. The relationship of the voltages is given in the phasor diagram in Figure 12–43(b).

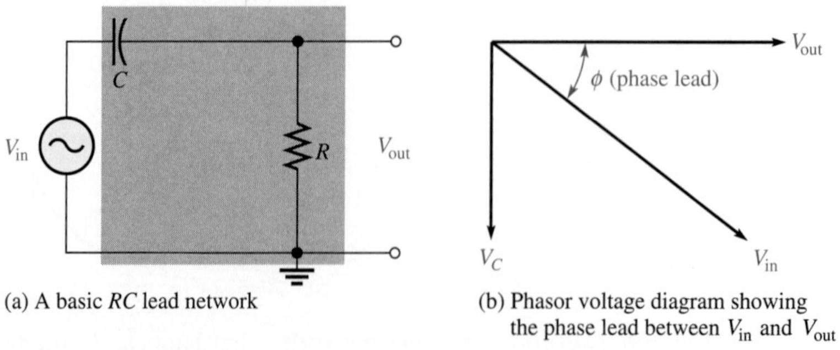

FIGURE 12–43
The RC lead network ($V_{out} = V_R$).

The output voltage, V_R, leads V_{in} by an angle that is the same as the circuit phase angle, because V_R and I are in phase with each other.

When the input and output waveforms are displayed on an oscilloscope, a relationship similar to that in Figure 12–44 is observed. Of course, the exact amount of phase lead and the output voltage magnitude depend on the values of R and X_C. The phase lead, ϕ, is expressed as follows:

$$\phi = \tan^{-1}\left(\frac{X_C}{R}\right) \quad \textbf{(12–28)}$$

The output voltage is expressed as

$$V_{out} = \left(\frac{R}{\sqrt{R^2 + X_C^2}}\right)V_{in} \quad \textbf{(12–29)}$$

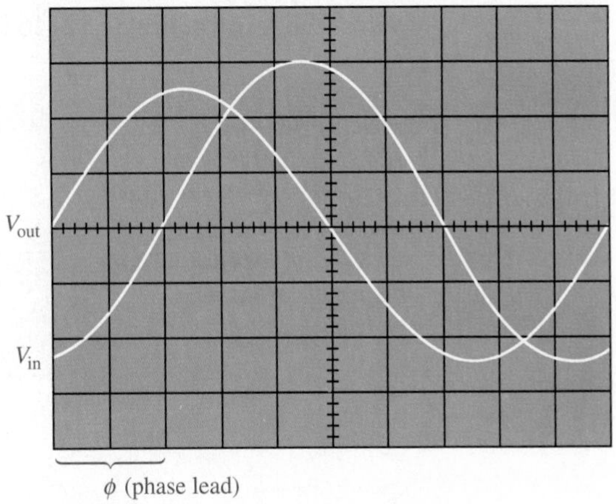

FIGURE 12–44
General oscilloscope display of the input and output waveforms of a lead network (V_{out} leads V_{in}).

EXAMPLE 12–16 Calculate the phase lead and the output voltage for the circuit in Figure 12–45.

FIGURE 12–45

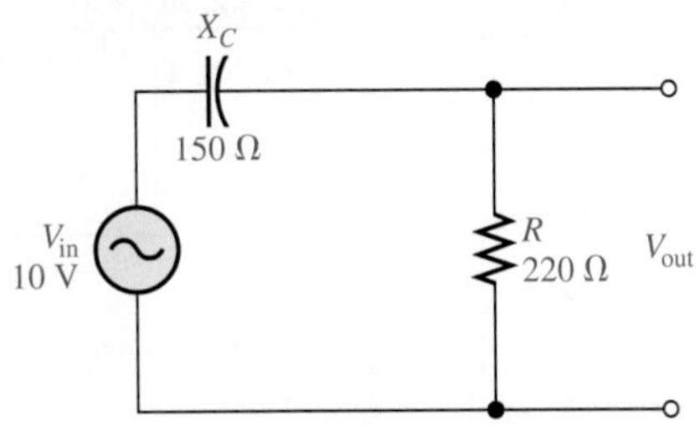

Solution The phase lead is

$$\phi = \tan^{-1}\left(\frac{X_C}{R}\right) = \tan^{-1}\left(\frac{150\ \Omega}{220\ \Omega}\right) = 34.3°$$

The output leads the input by 34.3°.

Use Equation (12–29) to determine the output voltage:

$$V_{out} = \left(\frac{R}{\sqrt{R^2 + X_C^2}}\right)V_{in} = \left(\frac{220\ \Omega}{\sqrt{(220\ \Omega)^2 + (150\ \Omega)^2}}\right)10\ \text{V} = 8.26\ \text{V}$$

The calculator sequence for V_{out} is

[2] [2] [0] [÷] [(] [(] [2] [2] [0] [x^2] [+] [1] [5] [0] [x^2] [)] [2nd] [$\sqrt{x}$] [)] [=] [×] [1] [0] [=]

Related Exercise How does an increase in R affect the phase lead and the output voltage in Figure 12–45?

Effects of Frequency on the Lead Network Since the phase lead is the same as the circuit phase angle θ, it decreases as frequency increases. The output voltage increases with frequency because as X_C becomes smaller, more of the input voltage is dropped across the resistor. Figure 12–46 illustrates this relationship.

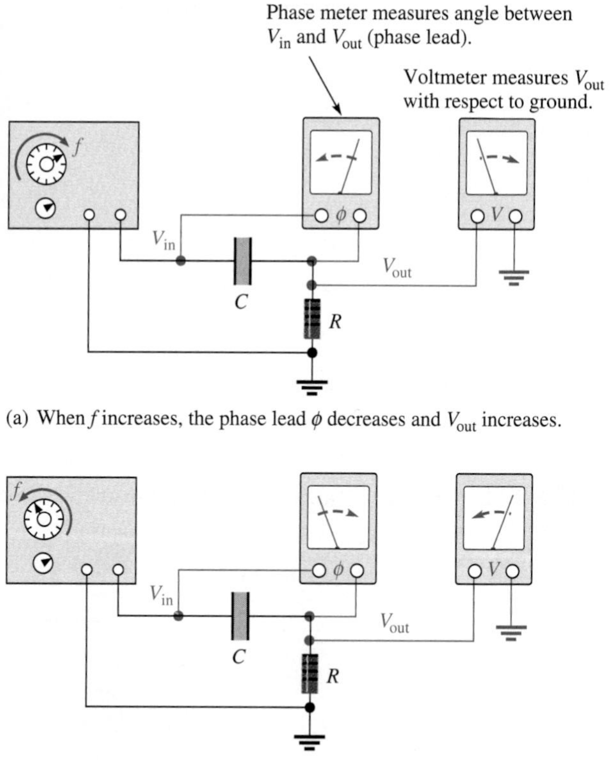

(a) When f increases, the phase lead ϕ decreases and V_{out} increases.

(b) When f decreases, ϕ increases and V_{out} decreases.

FIGURE 12–46
Illustration of how the frequency affects the phase lead and output voltage in an RC lead network with the amplitude of V_{in} held constant.

The *RC* Circuit As a Filter

Filters are frequency-selective circuits that permit signals of certain frequencies to pass from the input to the output while blocking all others. That is, all frequencies but the selected ones are filtered out.

Series *RC* circuits exhibit a frequency-selective characteristic and therefore act as simple passive filters. There are two types. The first one that we examine, called a **low-pass filter,** is realized by taking the output across the capacitor, just as in a lag network.

The second type, called a **high-pass filter,** is implemented by taking the output across the resistor, as in a lead network.

Low-Pass Filter You have seen what happens to the phase and the output voltage in the lag network. In terms of its filtering action, we are interested primarily in how the magnitude of the output voltage varies with frequency.

To illustrate low-pass filter action, Figure 12–47 shows a specific series of measurements in which the frequency starts at zero (dc) and is increased in increments up to 20 kHz. At each value of frequency, the output voltage is measured. As you can see, the capacitive reactance decreases as frequency goes up, thus dropping less voltage across the capacitor while the input voltage is held at a constant 10 V throughout each step. Table 12–1 summarizes the variation of the circuit parameters with frequency.

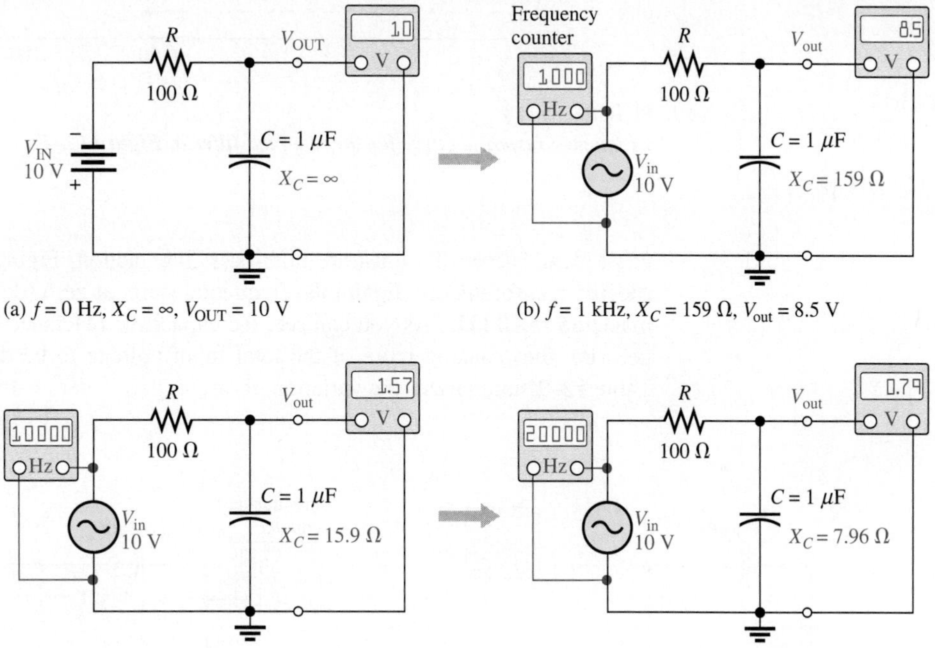

(a) $f = 0$ Hz, $X_C = \infty$, $V_{OUT} = 10$ V

(b) $f = 1$ kHz, $X_C = 159\ \Omega$, $V_{out} = 8.5$ V

(c) $f = 10$ kHz, $X_C = 15.9\ \Omega$, $V_{out} = 1.57$ V

(d) $f = 20$ kHz, $X_C = 7.96\ \Omega$, $V_{out} = 0.79$ V

FIGURE 12–47

Example of low-pass filter action. As frequency increases, V_{out} decreases.

TABLE 12–1

f	X_C (Ω)	Z_T (Ω)	I (mA)	V_{out} (V)
0	∞	∞	0	10
1 kHz	159	187.8	53.25	8.5
10 kHz	15.9	101.3	98.7	1.57
20 kHz	7.96	100.3	99.7	0.79

The **frequency response** of the low-pass filter in Figure 12–47 is shown in Figure 12–48, where the measured values are plotted on a graph of V_{out} versus frequency, and a smooth curve is drawn connecting the points. This graph, called a *response curve,* shows that the output voltage is greater at the lower frequencies and decreases as the frequency increases.

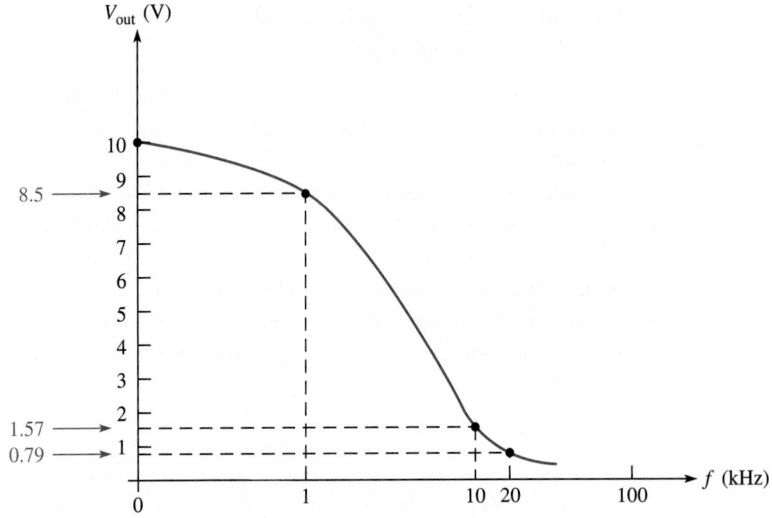

FIGURE 12–48
Frequency response curve for the low-pass filter in Figure 12–47.

High-Pass Filter To illustrate high-pass filter action, Figure 12–49 shows a series of specific measurements. Again, the frequency starts at zero (dc) and is increased in increments up to 10 kHz. As you can see, the capacitive reactance decreases as the frequency goes up, thus causing more of the total input voltage to be dropped across the resistor. Table 12–2 summarizes the variation of circuit parameters with frequency.

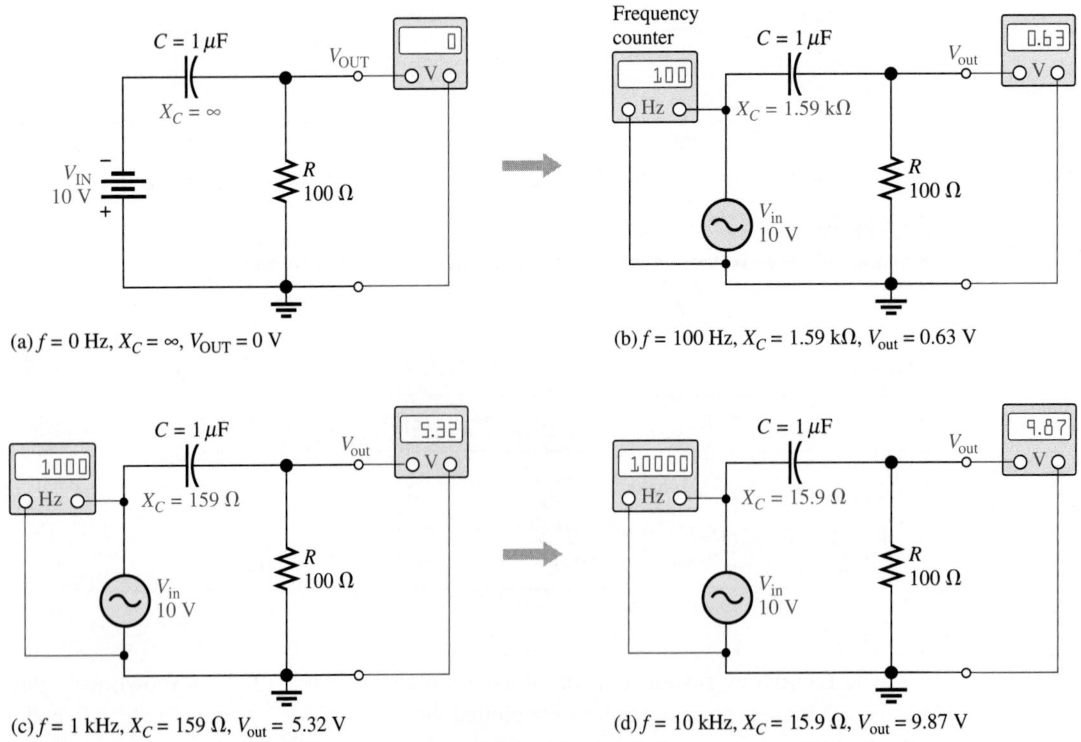

FIGURE 12–49
Example of high-pass filter action. As frequency increases, V_{out} increases.

TABLE 12–2

f	X_C (Ω)	Z_T (Ω)	I (mA)	V_{out} (V)
0	∞	∞	0	0
100 Hz	1590	1593	6.3	0.63
1 kHz	159	187.8	53.2	5.32
10 kHz	15.9	101.3	98.7	9.87

In Figure 12–50, the measured values for the high-pass filter shown in Figure 12–49 have been plotted to produce a response curve for this circuit. As you can see, the output voltage is greater at the higher frequencies and decreases as the frequency is reduced.

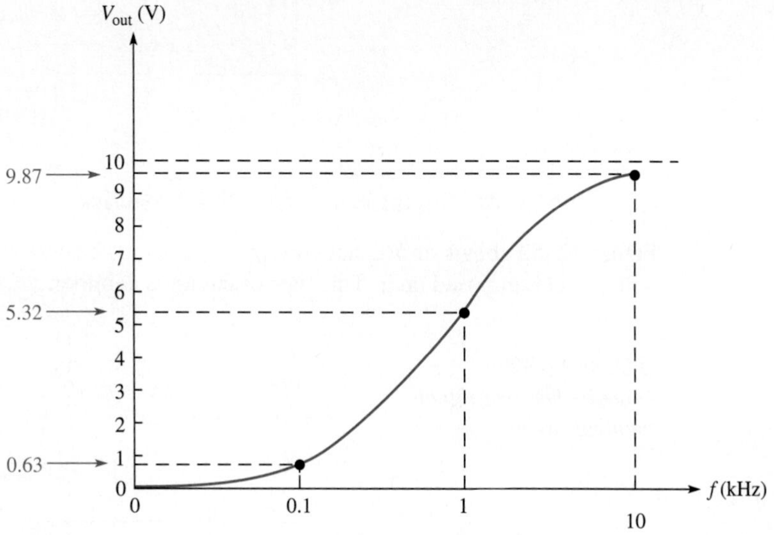

FIGURE 12–50
Frequency response curve for the high-pass filter in Figure 12–49.

The Cutoff Frequency and the Bandwidth of a Filter The frequency at which the capacitive reactance equals the resistance in a low-pass or high-pass RC filter is called the **cutoff frequency** and is designated f_c. This condition is expressed as $1/(2\pi f_c C) = R$. Solving for f_c, we get the following formula:

$$f_c = \frac{1}{2\pi RC} \qquad \textbf{(12–30)}$$

At f_c, the output voltage of the filter is 70.7% of its maximum value. It is standard practice to consider the cutoff frequency as the limit of a filter's performance in terms of passing or rejecting frequencies. For example, in a high-pass filter, all frequencies above f_c are considered to be passed by the filter, and all those below f_c are considered to be rejected. The reverse is true for a low-pass filter.

The range of frequencies that are considered to be passed by a filter is called the **bandwidth.** Figure 12–51 illustrates the bandwidth and the cutoff frequency for a low-pass filter.

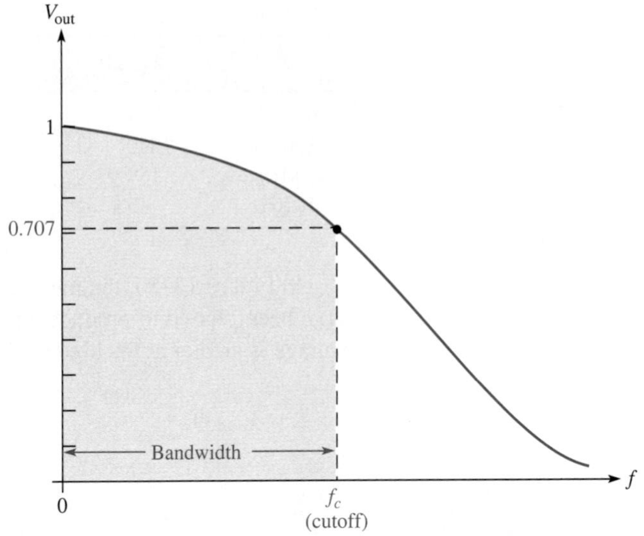

FIGURE 12–51
Normalized general response curve of a low-pass filter showing the cutoff frequency and the bandwidth.

Coupling an AC Signal into a DC Bias Network

Figure 12–52 shows an *RC* network that is used to create a dc voltage level with an ac voltage superimposed on it. This type of circuit is commonly found in amplifiers in which

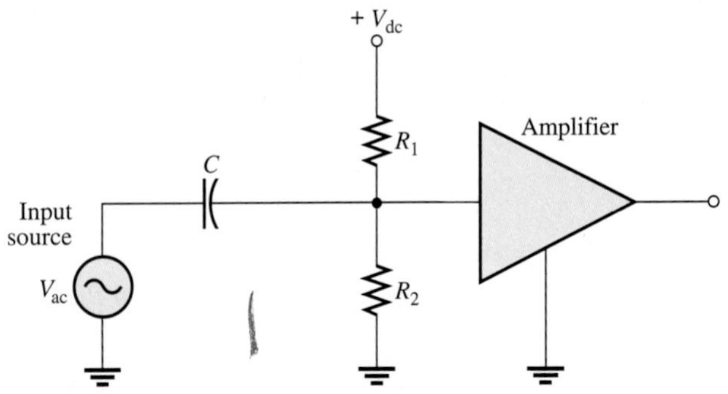

FIGURE 12–52
Amplifier bias and signal-coupling circuit.

the dc voltage is required to **bias** the amplifier to the proper operating point and the signal voltage to be amplified is coupled through a capacitor and superimposed on the dc level. The capacitor prevents the low internal resistance of the signal source from affecting the dc bias voltage.

In this type of application, a relatively high value of capacitance is selected so that for the frequencies to be amplified, the reactance is very small compared to the resistance of the bias network. When the reactance is very small (ideally zero), there is practically no phase shift or signal voltage dropped across the capacitor. Therefore, all of the signal voltage passes from the source to the input to the amplifier.

Figure 12–53 illustrates the application of the superposition principle to the circuit in Figure 12–52. In part (a), the ac source has been effectively removed from the circuit and replaced with a short to represent its ideal internal resistance. Since *C* is open to dc, the voltage at point *A* is determined by the voltage-divider action of R_1 and R_2 and the dc voltage source.

In Figure 12–53(b), the dc source has been effectively removed from the circuit and replaced with a short to represent its ideal internal resistance. Since *C* appears as a short at the frequency of the ac, the signal voltage is coupled directly to point *A* and appears across the parallel combination of R_1 and R_2. Figure 12–53(c) illustrates that the combined effect of the superposition of the dc and the ac voltages results in the signal voltage “riding” on the dc level.

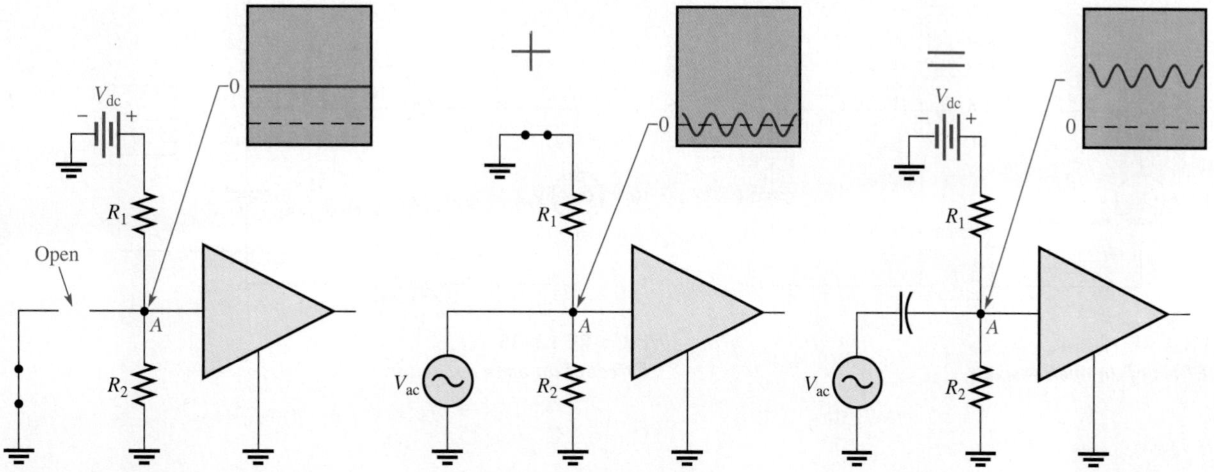

(a) dc equivalent: ac source replaced by short. C is open to dc. R_1 and R_2 act as dc voltage divider.

(b) ac equivalent: dc source is replaced by short. C is short to ac. All of V_{ac} is coupled to point A.

(c) dc + ac: Voltages are superimposed at point A.

FIGURE 12–53
The superposition of dc and ac voltages in an RC bias and coupling circuit.

SECTION 12–8 REVIEW

1. A certain *RC* lag network consists of a 4.7 kΩ resistor and a 0.022 μF capacitor. Determine the phase lag between input and output at a frequency of 3 kHz.
2. An *RC* lead network has the same component values as the lag network in Question 1. What is the magnitude of the output voltage at 3 kHz when the input is 10 V rms?
3. When an *RC* circuit is used as a low-pass filter, across which component is the output taken?

12–9 ▪ TROUBLESHOOTING

In this section, the effects that typical component failures or degradation have on the response of basic RC circuits are considered.

After completing this section, you should be able to

- **Troubleshoot *RC* circuits**
  - □ Find an open resistor
  - □ Find an open capacitor
  - □ Find a shorted capacitor
  - □ Find a leaky capacitor

Effect of an Open Resistor

It is very easy to see how an open resistor affects the operation of a basic series *RC* circuit, as shown in Figure 12–54. Obviously, there is no path for current, so the capacitor voltage remains at zero; thus, the total voltage, V_s, appears across the open resistor.

Effect of an Open Capacitor

When the capacitor is open, there is no current; thus, the resistor voltage remains at zero. The total source voltage is across the open capacitor, as shown in Figure 12–55.

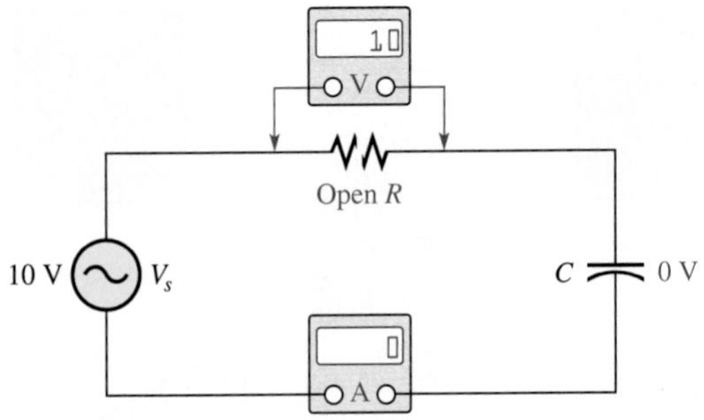

FIGURE 12–54
Effect of an open resistor.

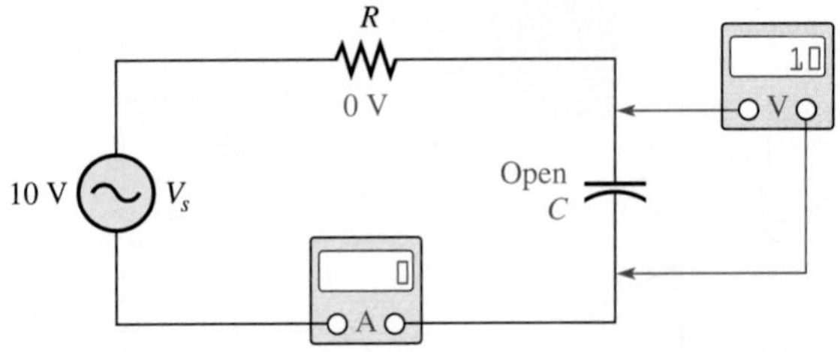

FIGURE 12–55
Effect of an open capacitor.

Effect of a Shorted Capacitor

When a capacitor shorts out, the voltage across it is zero, the current equals V_s/R, and the total voltage appears across the resistor, as shown in Figure 12–56.

FIGURE 12–56
Effect of a shorted capacitor.

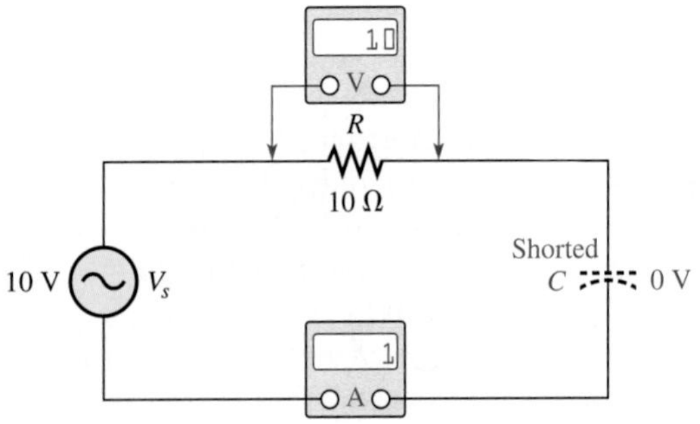

Effect of a Leaky Capacitor

When a capacitor exhibits a high leakage current, the leakage resistance effectively appears in parallel with the capacitor, as shown in Figure 12–57(a). When the leakage resistance is comparable in value to the circuit resistance, *R,* the circuit response is drastically affected. The circuit, looking from the capacitor toward the source, can be Thevenized, as shown in Figure 12–57(b). The Thevenin equivalent resistance is *R* in parallel with R_{leak} (the source appears as a short), and the Thevenin equivalent voltage is determined by the voltage-divider action of *R* and R_{leak}.

$$R_{\text{th}} = R \parallel R_{\text{leak}}$$

$$R_{\text{th}} = \frac{RR_{\text{leak}}}{R + R_{\text{leak}}} \qquad \textbf{(12–31)}$$

$$V_{\text{th}} = \frac{R_{\text{leak}}V_{\text{in}}}{R + R_{\text{leak}}} \qquad \textbf{(12–32)}$$

As you can see, the voltage to which the capacitor will charge is reduced since $V_{\text{th}} < V_s$. Also, the circuit time constant is reduced, and the current is increased. The Thevenin equivalent circuit is shown in Figure 12–57(c).

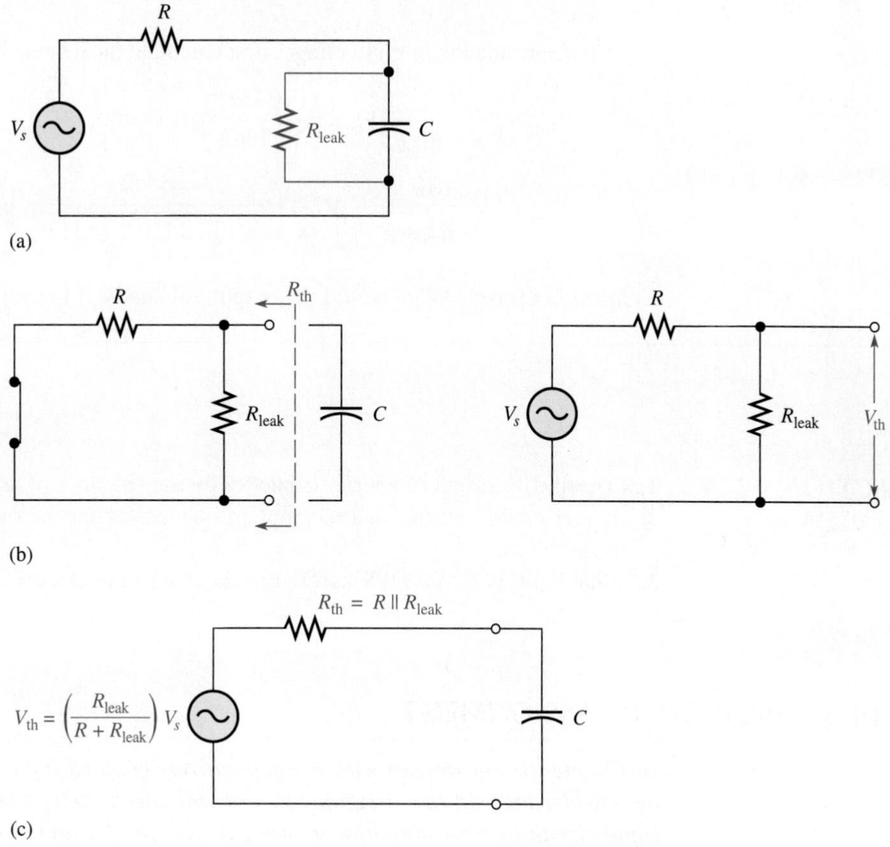

FIGURE 12–57
Effect of a leaky capacitor.

EXAMPLE 12–17 Assume that the capacitor in Figure 12–58 is degraded to a point where its leakage resistance is 10 kΩ. Determine the phase shift from input to output and the output voltage under the degraded condition.

FIGURE 12–58

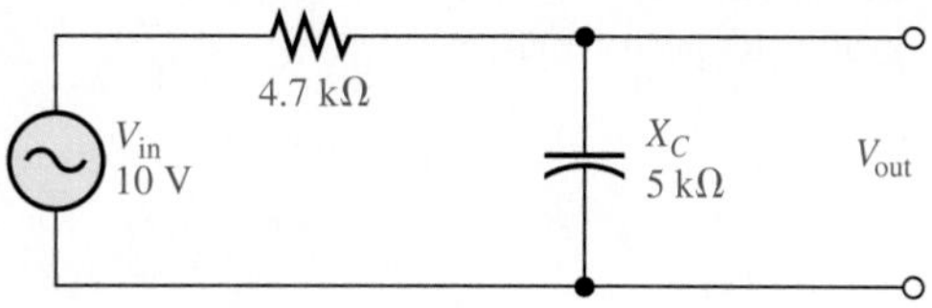

Solution The effective circuit resistance is

$$R_{th} = \frac{RR_{leak}}{R + R_{leak}} = \frac{(4.7\ \text{k}\Omega)(10\ \text{k}\Omega)}{14.7\ \text{k}\Omega} = 3.2\ \text{k}\Omega$$

The phase lag is

$$\phi = 90° - \tan^{-1}\left(\frac{X_C}{R_{th}}\right) = 90° - \tan^{-1}\left(\frac{5\ \text{k}\Omega}{3.2\ \text{k}\Omega}\right) = 32.6°$$

To determine the output voltage, first calculate the Thevenin equivalent voltage.

$$V_{th} = \left(\frac{R_{leak}}{R + R_{leak}}\right)V_{in} = \left(\frac{10\text{ k}\Omega}{14.7\text{ k}\Omega}\right)10\text{ V} = 6.80\text{ V}$$

$$V_{out} = \left(\frac{X_C}{\sqrt{R_{th}^2 + X_C^2}}\right)V_{th} = \left(\frac{5\text{ k}\Omega}{\sqrt{(3.2\text{ k}\Omega)^2 + (5\text{ k}\Omega)^2}}\right)6.8\text{ V} = 5.73\text{ V}$$

Related Exercise What would the output voltage be if the capacitor were not leaky?

SECTION 12–9 REVIEW

1. Describe the effect of a leaky capacitor on the response of an *RC* circuit.
2. In a series *RC* circuit, if all the applied voltage appears across the capacitor, what is the problem?
3. What faults can cause 0 V across the capacitor in a series *RC* circuit?

12–10 ■ APPLICATION ASSIGNMENT

In Chapter 9, you worked with the capacitively coupled input to an amplifier with voltage-divider bias. In this assignment, you will check the voltage of a similar amplifier's input circuit to determine how it changes with the frequency of the input signal. If too much voltage is dropped across the capacitor, the overall performance of the amplifier is adversely affected. To do this assignment, you do not have to be familiar with the details of the amplifier circuit but you should review the application assignment in Chapter 9 before proceeding.

As you learned in Chapter 9, the coupling capacitor (C_1) in Figure 12–59 passes the input signal voltage to the input of the amplifier (point *A* to point *B*) without affecting the dc level at point *B* produced by the resistive voltage divider (R_1 and R_2). If the input frequency is high enough so that the reactance of the coupling capacitor is negligibly small, essentially no ac signal voltage is dropped across the capacitor. As the signal frequency is reduced, the capacitive reactance increases and more of the signal voltage is dropped across the capacitor. This lowers the overall voltage gain of the amplifier and thus degrades its performance.

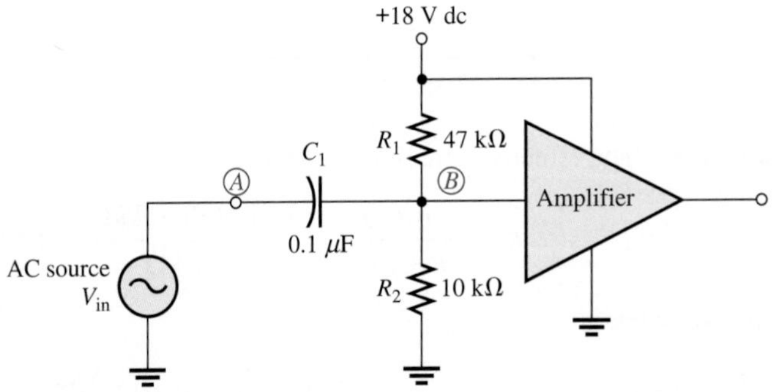

FIGURE 12–59
A capacitively coupled amplifier.

The amount of signal voltage that is coupled from the input source (point A) to the amplifier input (point B) is determined by the values of the capacitor and the dc bias resistors (assuming the amplifier has no loading effect) in Figure 12–59. These components actually form a high-pass RC filter, as shown in Figure 12–60. The voltage-divider bias resistors are effectively in parallel with each other as far as the ac source is concerned. The lower end of R_2 goes to ground and the upper end of R_1 goes to the dc supply voltage as shown in Figure 12–60(a). Since there is no ac voltage at the +18 V dc terminal, the upper end of R_1 is at 0 V ac which is referred to as *ac ground.* The development of the circuit into an effective high-pass RC filter is shown in parts (b) and (c).

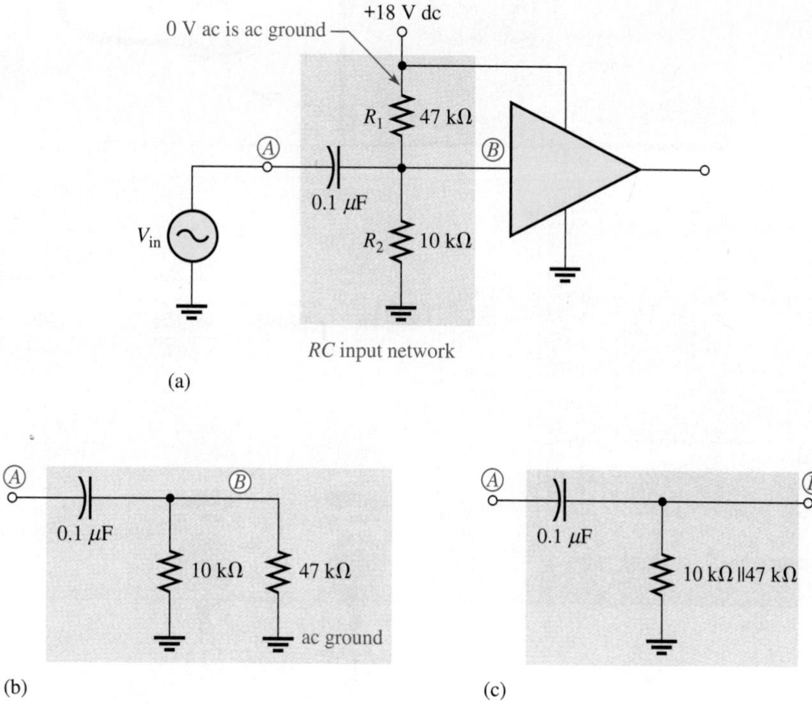

FIGURE 12–60
The RC input circuit acts effectively like a high-pass RC filter.

Step 1: Evaluate the Amplifier Input Circuit

Determine the value of the equivalent resistance of the input circuit. Assume the amplifier has no loading effect on the input circuit.

Step 2: Measure the Response at Frequency f_1

Refer to Figure 12–61. The input signal voltage is applied to the amplifier circuit board and displayed on channel 1 of the oscilloscope and channel 2 is connected to a point on the circuit board. Determine to what point on the circuit the channel 2 probe is connected, the frequency, and the voltage that should be displayed.

Step 3: Measure the Response at Frequency f_2

Refer to Figure 12–62 and the circuit board in Figure 12–61. The input signal voltage displayed on channel 1 of the oscilloscope is applied to the amplifier circuit board. Determine the frequency and the voltage that should be displayed on channel 2.

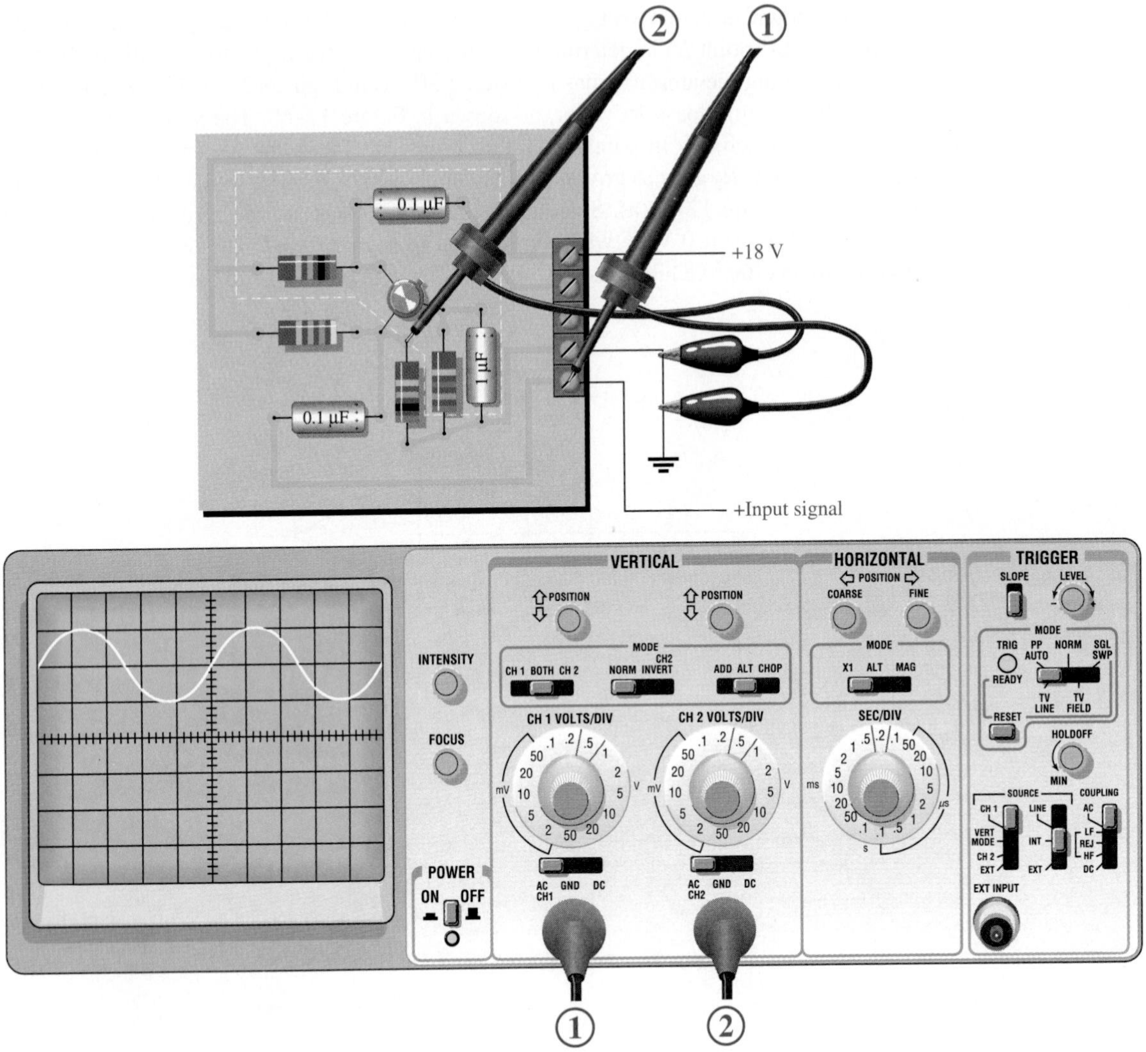

FIGURE 12–61
Measuring the input circuit response at frequency f_1. Circled numbers relate scope inputs to the probes. The channel 1 waveform is shown.

State the difference between the channel 2 waveforms determined in Step 2 and Step 3. Explain the reason for the difference.

Step 4: Measure the Response at Frequency f_3

Refer to Figure 12–63 and the circuit board in Figure 12–61. The input signal voltage displayed on channel 1 of the oscilloscope is applied to the amplifier circuit board. Determine the frequency and the voltage that should be displayed on channel 2.

State the difference between the channel 2 waveforms determined in Step 2 and Step 4. Explain the reason for the difference.

Step 5: Plot a Response Curve for the Amplifier Input Circuit

Determine the frequency at which the signal voltage at point *B* in Figure 12–59 is 70.7% of its maximum value. Plot the response curve using this voltage value and the values at frequencies f_1, f_2, and f_3. How does this curve show that the input circuit acts as a high-pass filter? What can you do to the circuit to lower the frequency at which the voltage is 70.7% of maximum without affecting the dc bias voltage?

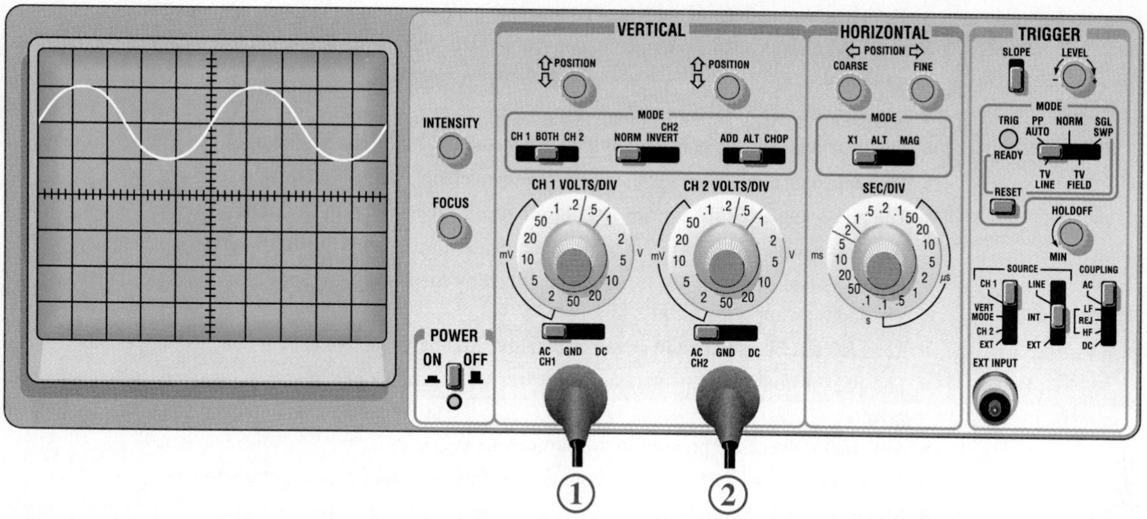

FIGURE 12–62
Measuring the input circuit response at frequency f_2. The channel 1 waveform is shown.

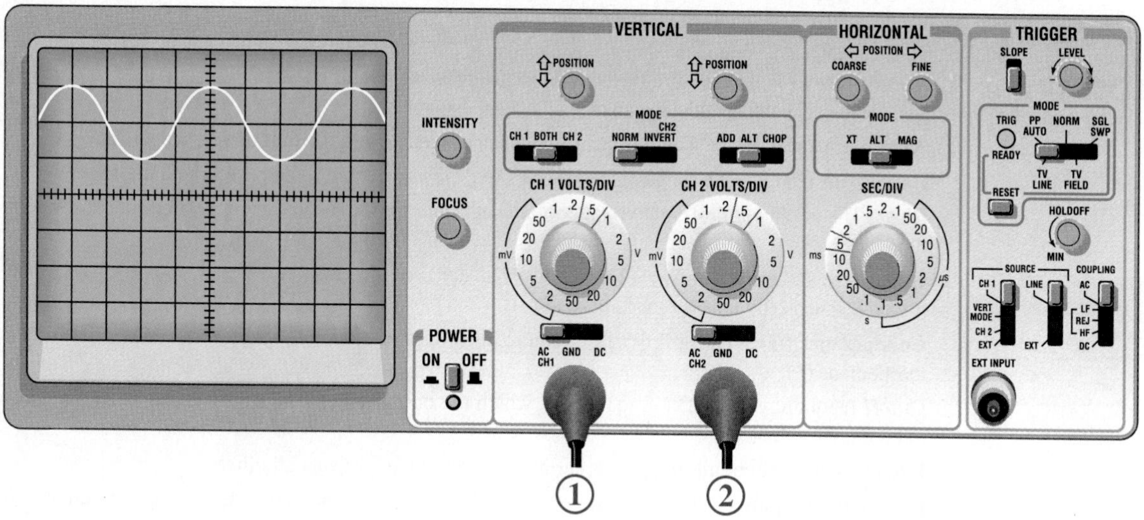

FIGURE 12–63
Measuring the input circuit response at frequency f_3. The channel 1 waveform is shown.

SECTION 12–10 REVIEW

1. Explain the effect on the response of the amplifier input circuit of reducing the value of the coupling capacitor.
2. What is the voltage at point *B* in Figure 12–59 if the coupling capacitor opens when the ac input signal is 10 mV rms?
3. What is the voltage at point *B* in Figure 12–59 if resistor R_1 is open when the ac input signal is 10 mV rms?

■ SUMMARY

- When a sine wave voltage is applied to an *RC* circuit, the current and all the voltage drops are also sine waves.
- Total current in an *RC* circuit always leads the source voltage.
- The resistor voltage is always in phase with the current.
- The capacitor voltage always lags the current by 90°.
- In an *RC* circuit, the impedance is determined by both the resistance and the capacitive reactance combined.
- Impedance is expressed in units of ohms.
- The circuit phase angle is the angle between the total current and the applied (source) voltage.
- The impedance of a series *RC* circuit varies inversely with frequency.
- The phase angle (θ) of a series *RC* circuit varies inversely with frequency.
- For each parallel *RC* circuit, there is an equivalent series circuit for any given frequency.
- The impedance of a circuit can be determined by measuring the applied voltage and the total current and then applying Ohm's law.
- In an *RC* circuit, part of the power is resistive and part reactive.
- The phasor combination of resistive power (true power) and reactive power is called *apparent power.*
- Apparent power is expressed in volt-amperes (VA).
- The power factor (*PF*) indicates how much of the apparent power is true power.
- A power factor of 1 indicates a purely resistive circuit, and a power factor of 0 indicates a purely reactive circuit.
- In a lag network, the output voltage lags the input voltage in phase.
- In a lead network, the output voltage leads the input voltage.
- A filter passes certain frequencies and rejects others.

■ GLOSSARY

Admittance (Y) A measure of the ability of a reactive circuit to permit current; the reciprocal of impedance. The unit is the siemens (S).

Apparent power (P_a) The product of volts and amperes with the unit of volt-amperes (VA).

Bandwidth (BW) The range of frequencies that are considered to be passed by a filter.

Bias The application of a dc voltage to an electronic device to produce a desired mode of operation.

Capacitive susceptance (B_C) The ability of a capacitor to permit current; the reciprocal of capacitive reactance. The unit is the siemens (S).

Conductance (G) The ability of a circuit to allow current; the reciprocal of resistance. The unit is the siemens (S).

Cutoff frequency (f_c) The frequency at which the output voltage of a filter is 70.7% of the maximum output voltage.

Filter A type of circuit that passes certain frequencies and rejects all others.

Frequency response The way the output voltage of a circuit varies as the input frequency is changed.

High-pass filter A certain type of filter in which higher frequencies are passed and lower frequencies are rejected.

Impedance (Z) The total opposition to sinusoidal current expressed in ohms.

Lag To be behind; describes a condition of the phase or time relationship of waveforms in which one waveform is behind the other in phase or time.

Lead To be in front of; describes a condition of the phase or time relationship of waveforms in which one waveform is ahead of the other in phase or time.

Low-pass filter A certain type of filter in which lower frequencies are passed and higher frequencies are rejected.

Power factor The relationship between volt-amperes and true power or watts. Volt-amperes multiplied by the power factor equals true power.

▪ FORMULAS

(12–1)	$Z = \sqrt{R^2 + X_C^2}$	Series *RC* impedance
(12–2)	$\theta = \tan^{-1}\left(\frac{X_C}{R}\right)$	Series *RC* phase angle
(12–3)	$V = IZ$	Ohm's law
(12–4)	$I = \frac{V}{Z}$	Ohm's law
(12–5)	$Z = \frac{V}{I}$	Ohm's law
(12–6)	$V_s = \sqrt{V_R^2 + V_C^2}$	Total voltage in series *RC*
(12–7)	$\theta = \tan^{-1}\left(\frac{V_C}{V_R}\right)$	Series *RC* phase angle
(12–8)	$Z = \frac{RX_C}{\sqrt{R^2 + X_C^2}}$	Parallel *RC* impedance
(12–9)	$\theta = \tan^{-1}\left(\frac{R}{X_C}\right)$	Parallel *RC* phase angle
(12–10)	$G = \frac{1}{R}$	Conductance
(12–11)	$B_C = \frac{1}{X_C}$	Capacitive susceptance
(12–12)	$Y = \frac{1}{Z}$	Admittance
(12–13)	$Y_T = \sqrt{G^2 + B_C^2}$	Total admittance
(12–14)	$V = \frac{I}{Y}$	Ohm's law
(12–15)	$I = VY$	Ohm's law
(12–16)	$Y = \frac{I}{V}$	Ohm's law
(12–17)	$I_T = \sqrt{I_R^2 + I_C^2}$	Total current in parallel *RC* circuits
(12–18)	$\theta = \tan^{-1}\left(\frac{I_C}{I_R}\right)$	Parallel *RC* phase angle
(12–19)	$R_{eq} = Z \cos \theta$	Equivalent series resistance
(12–20)	$X_{C(eq)} = Z \sin \theta$	Equivalent series reactance
(12–21)	$P_{true} = I_T^2 R$	True power (W)
(12–22)	$P_r = I_T^2 X_C$	Reactive power (VAR)
(12–23)	$P_a = I_T^2 Z$	Apparent power (VA)
(12–24)	$P_{true} = V_s I_T \cos \theta$	True Power
(12–25)	$PF = \cos \theta$	Power factor
(12–26)	$\phi = 90° - \tan^{-1}\left(\frac{X_C}{R}\right)$	Phase lag
(12–27)	$V_{out} = \left(\frac{X_C}{\sqrt{R^2 + X_C^2}}\right)V_{in}$	Output of lag network
(12–28)	$\phi = \tan^{-1}\left(\frac{X_C}{R}\right)$	Phase lead
(12–29)	$V_{out} = \left(\frac{R}{\sqrt{R^2 + X_C^2}}\right)V_{in}$	Output of lead network

(12–30) $f_c = \dfrac{1}{2\pi RC}$ Cutoff frequency of an *RC* filter

(12–31) $R_{th} = \dfrac{RR_{leak}}{R + R_{leak}}$ Thevenin equivalent resistance

(12–32) $V_{th} = \dfrac{R_{leak}V_{in}}{R + R_{leak}}$ Thevenin equivalent voltage

▪ SELF-TEST

1. In a series *RC* circuit, the voltage across the resistance is
(a) in phase with the source voltage **(b)** lagging the source voltage by 90°
(c) in phase with the current **(d)** lagging the current by 90°

2. In a series *RC* circuit, the voltage across the capacitor is
(a) in phase with the source voltage **(b)** lagging the resistor voltage by 90°
(c) in phase with the current **(d)** lagging the source voltage by 90°

3. When the frequency of the voltage applied to a series *RC* circuit is increased, the impedance
(a) increases **(b)** decreases **(c)** remains the same **(d)** doubles

4. When the frequency of the voltage applied to a series *RC* circuit is decreased, the phase angle
(a) increases **(b)** decreases **(c)** remains the same **(d)** becomes erratic

5. In a series *RC* circuit when the frequency and the resistance are doubled, the impedance
(a) doubles **(b)** is halved **(c)** is quadrupled **(d)** cannot be determined without values

6. In a series *RC* circuit, 10 V rms is measured across the resistor and 10 V rms is also measured across the capacitor. The rms source voltage is
(a) 20 V **(b)** 14.14 V **(c)** 28.28 V **(d)** 10 V

7. The voltages in Problem 6 are measured at a certain frequency. To make the resistor voltage greater than the capacitor voltage, the frequency
(a) must be increased **(b)** must be decreased **(c)** is held constant **(d)** has no effect

8. When $R = X_C$, the phase angle is
(a) 0° **(b)** +90° **(c)** –90° **(d)** 45°

9. To decrease the phase angle below 45°, the following condition must exist:
(a) $R = X_C$ **(b)** $R < X_C$ **(c)** $R > X_C$ **(d)** $R = 10X_C$

10. When the frequency of the source voltage is increased, the impedance of a parallel *RC* circuit
(a) increases **(b)** decreases **(c)** does not change

11. In a parallel *RC* circuit, there is 1 A rms through the resistive branch and 1 A rms through the capacitive branch. The total rms current is
(a) 1 A **(b)** 2 A **(c)** 2.28 A **(d)** 1.414 A

12. A power factor of 1 indicates that the circuit phase angle is
(a) 90° **(b)** 45° **(c)** 180° **(d)** 0°

13. For a certain load, the true power is 100 W and the reactive power is 100 VAR. The apparent power is
(a) 200 VA **(b)** 100 VA **(c)** 141.4 VA **(d)** 141.4 W

14. Energy sources are normally rated in
(a) watts **(b)** volt-amperes **(c)** volt-amperes reactive **(d)** none of these

15. If the bandwidth of a certain low-pass filter is 1 kHz, the cutoff frequency is
(a) 0 Hz **(b)** 500 Hz **(c)** 2 kHz **(d)** 1000 Hz

▪ BASIC PROBLEMS

SECTION 12–1 Sinusoidal Response of *RC* Circuits

1. A 8 kHz sinusoidal voltage is applied to a series *RC* circuit. What is the frequency of the voltage across the resistor? The capacitor?

2. What is the waveshape of the current in the circuit of Problem 1?

SECTION 12–2 Impedance and Phase Angle of a Series *RC* Circuit

3. Find the impedance of each circuit in Figure 12–64.

4. Determine the impedance and the phase angle in each circuit in Figure 12–65.

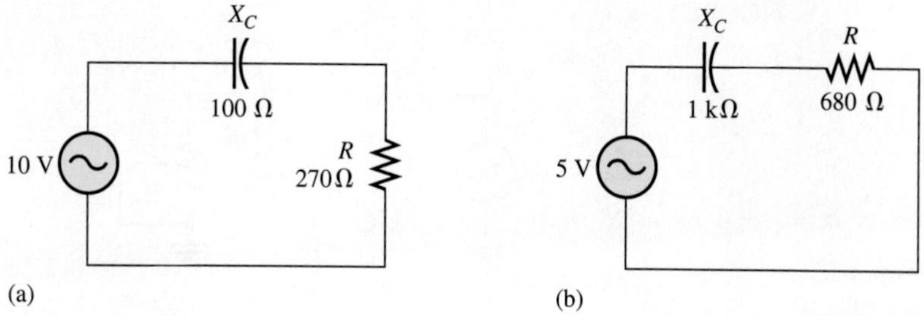

FIGURE 12–64

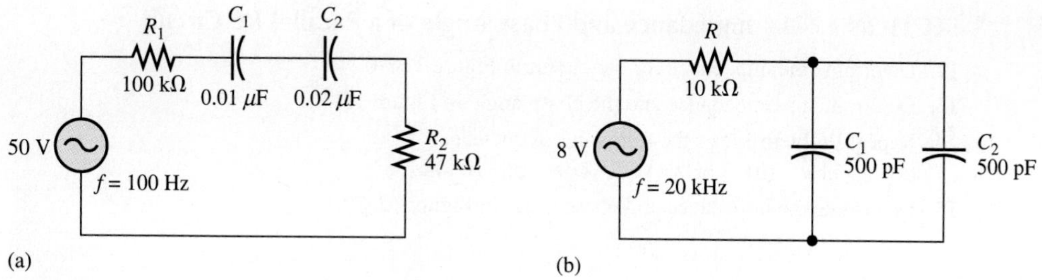

FIGURE 12–65

5. For the circuit of Figure 12–66, determine the impedance for each of the following frequencies: **(a)** 100 Hz **(b)** 500 Hz **(c)** 1 kHz **(d)** 2.5 kHz

6. Repeat Problem 5 for $C = 0.005\ \mu F$.

FIGURE 12–66

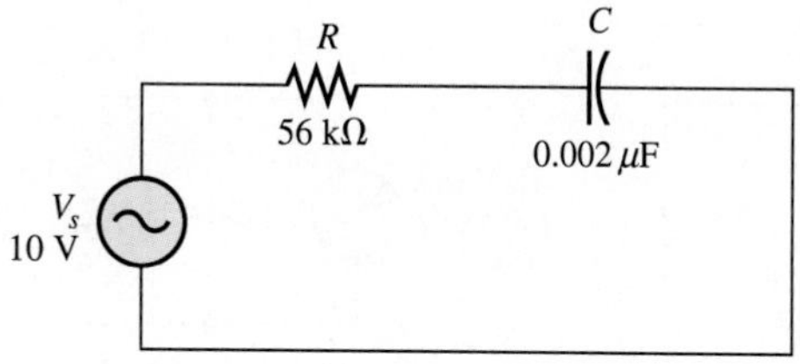

SECTION 12–3 Analysis of Series *RC* Circuits

7. Calculate the total current in each circuit of Figure 12–64.

8. Repeat Problem 7 for the circuits in Figure 12–65.

9. For the circuit in Figure 12–67, draw the phasor diagram showing all voltages and the total current. Indicate the phase angles.

10. For the circuit in Figure 12–68, determine the following:
(a) Z **(b)** I **(c)** V_R **(d)** V_C

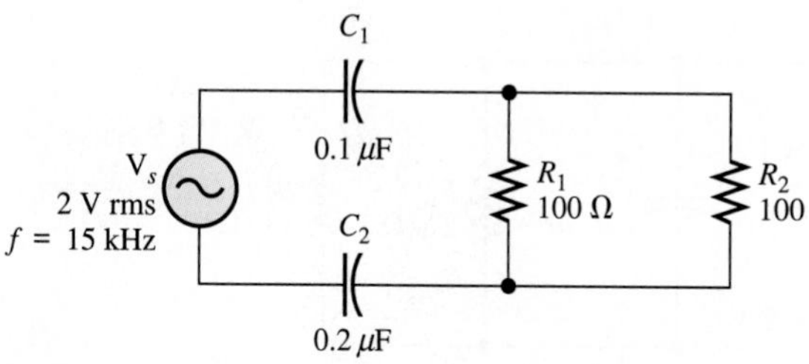

FIGURE 12–67

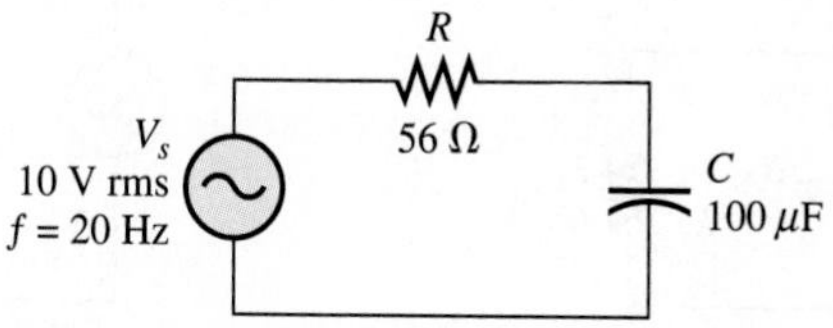

FIGURE 12–68

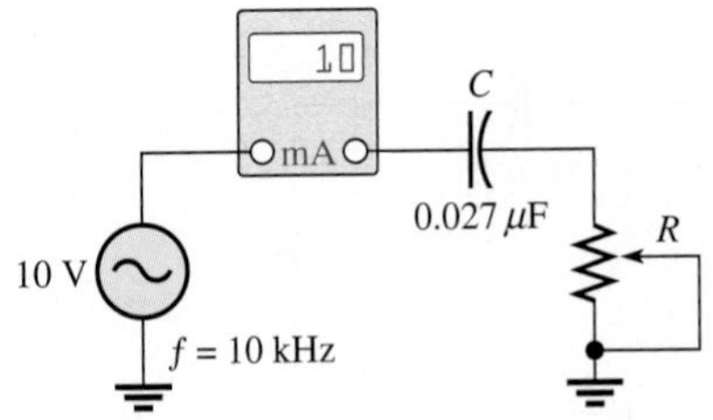

FIGURE 12–69

11. To what value must the rheostat be set in Figure 12–69 to make the total current 10 mA? What is the resulting phase angle?

SECTION 12–4 Impedance and Phase Angle of a Parallel *RC* Circuit

12. Determine the impedance for the circuit in Figure 12–70.
13. Determine the impedance and the phase angle in Figure 12–71.
14. Repeat Problem 13 for the following frequencies:
    (a) 1.5 kHz **(b)** 3 kHz **(c)** 5 kHz **(d)** 10 kHz
15. Determine the impedance and phase angle in Figure 12–72.

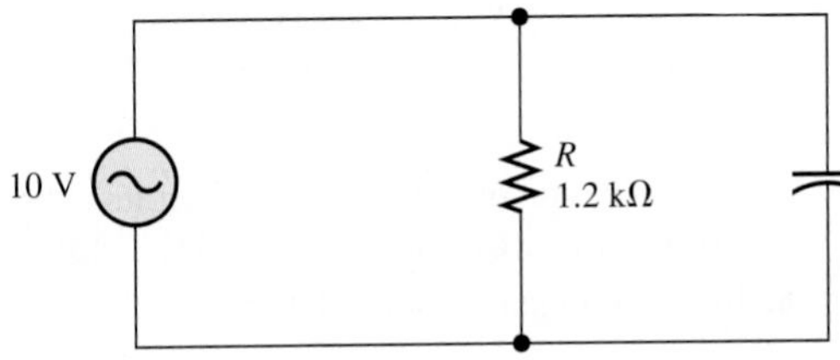

FIGURE 12–70

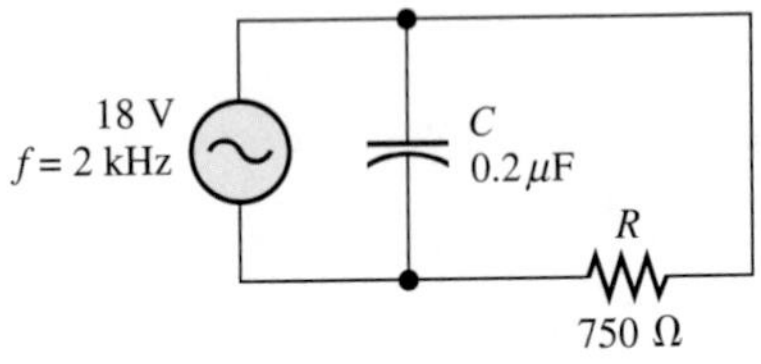

FIGURE 12–71

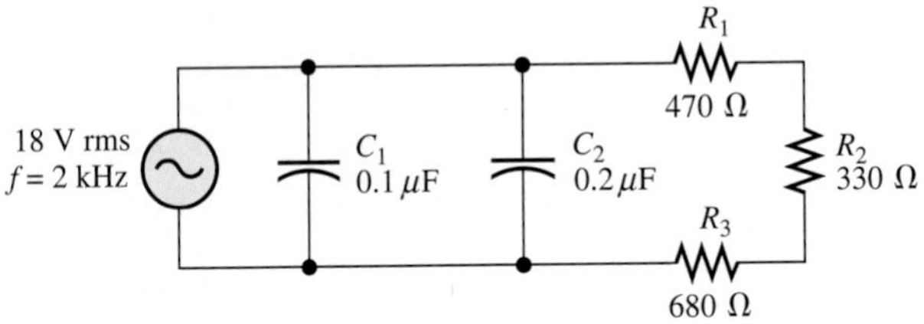

FIGURE 12–72

SECTION 12–5 Analysis of Parallel *RC* Circuits

16. For the circuit in Figure 12–73, find all the currents and voltages.
17. For the parallel circuit in Figure 12–74, find each branch current and the total current. What is the phase angle between the applied voltage and the total current?

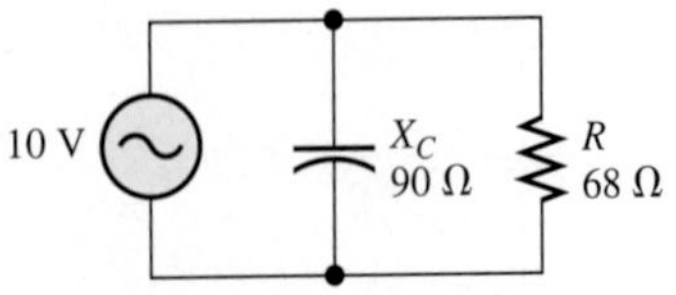

FIGURE 12–73

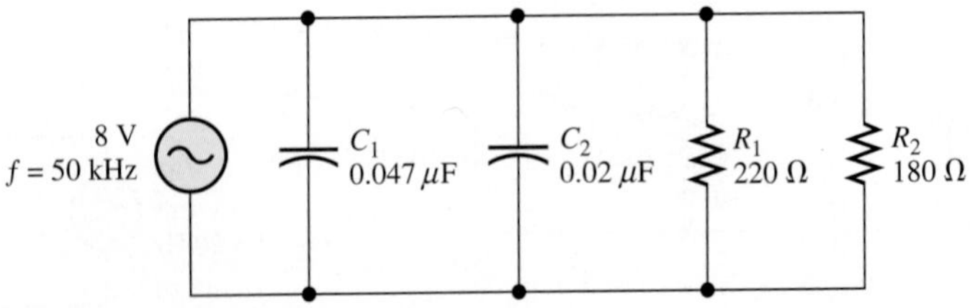

FIGURE 12–74

18. For the circuit in Figure 12–75, determine the following:
(a) Z **(b)** I_R **(c)** I_C **(d)** I_T **(e)** θ

19. Repeat Problem 18 for $R = 5$ kΩ, $C = 0.05$ μF, and $f = 500$ Hz.

20. Convert the circuit in Figure 12–76 to an equivalent series form.

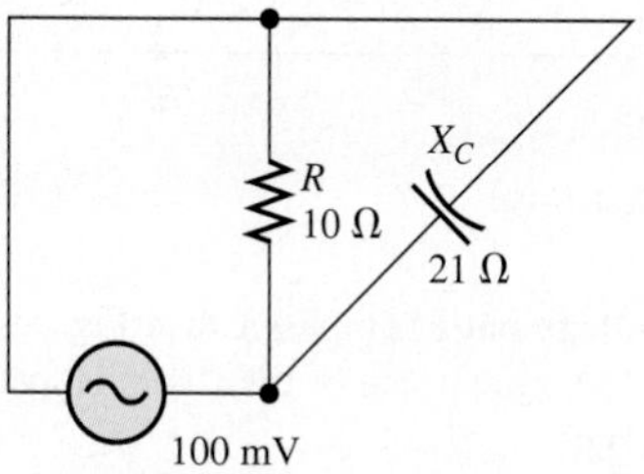

FIGURE 12–75

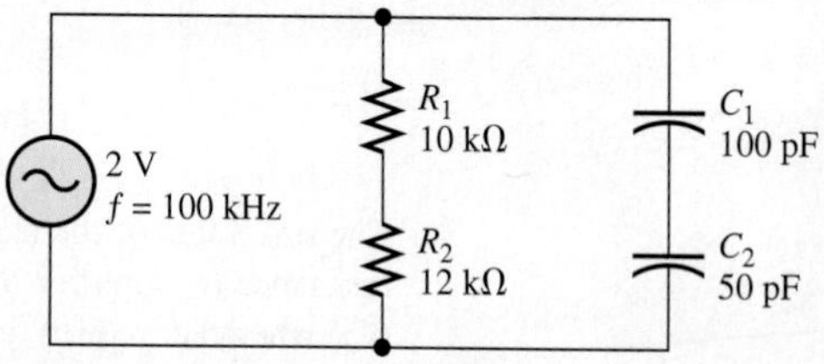

FIGURE 12–76

SECTION 12–6 Series-Parallel Analysis

21. Determine the voltages across each element in Figure 12–77. Sketch the voltage phasor diagram.

22. Is the circuit in Figure 12–77 predominantly resistive or predominantly capacitive?

23. Find the current through each branch and the total current in Figure 12–77. Sketch the current phasor diagram.

24. For the circuit in Figure 12–78, determine the following:
(a) I_T **(b)** θ **(c)** V_{R1} **(d)** V_{R2} **(e)** V_{R3} **(f)** V_C

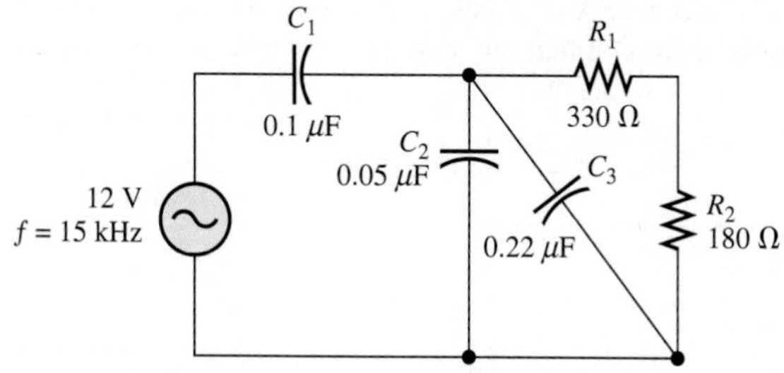

FIGURE 12–77

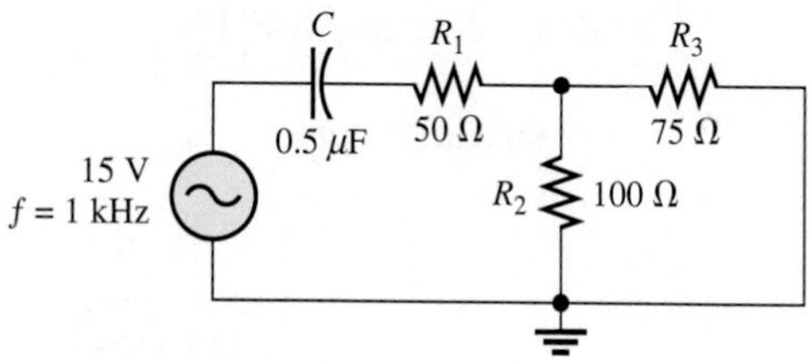

FIGURE 12–78

SECTION 12–7 Power in *RC* Circuits

25. In a certain series *RC* circuit, the true power is 2 W, and the reactive power is 3.5 VAR. Determine the apparent power.

26. In Figure 12–68, what is the true power and the reactive power?

27. What is the power factor for the circuit of Figure 12–76?

28. Determine P_{true}, P_r, P_a, and PF for the circuit in Figure 12–78. Sketch the power triangle.

SECTION 12–8 Basic Applications

29. For the lag network in Figure 12–79, determine the phase lag between the input voltage and the output voltage for each of the following frequencies:
(a) 1 Hz **(b)** 100 Hz **(c)** 1 kHz **(d)** 10 kHz

30. The lag network in Figure 12–79 also acts as a low-pass filter. Draw a response curve for this circuit by plotting the output voltage versus frequency for 0 Hz to 10 kHz in 1 kHz increments.

31. Repeat Problem 29 for the lead network in Figure 12–80.

32. Plot the frequency response curve for the lead network in Figure 12–80 for a frequency range of 0 Hz to 10 kHz in 1 kHz increments.

33. Draw the voltage phasor diagram for each circuit in Figures 12–79 and 12–80 for a frequency of 5 kHz with $V_{in} = 1$ V rms.

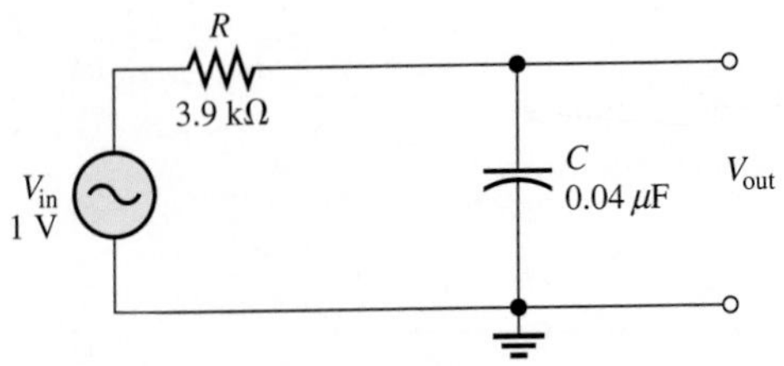

FIGURE 12–79

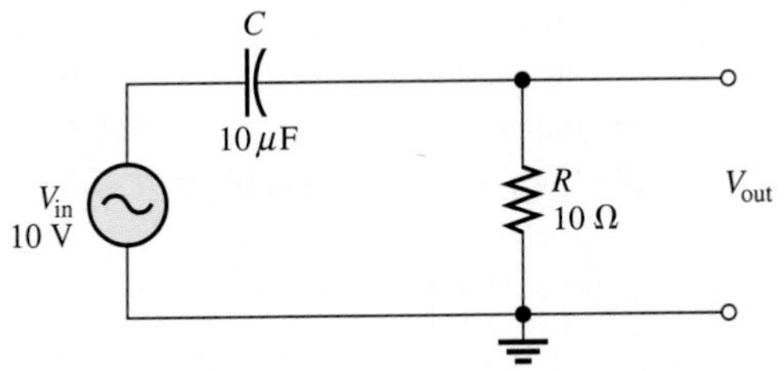

FIGURE 12–80

34. The rms value of the signal voltage out of amplifier A in Figures 12–81 is 50 mV. If the input resistance to amplifier B is 10 kΩ, how much of the signal is lost due to the coupling capacitor (C_c) when the frequency is 3 kHz?
35. Determine the cutoff frequency for each circuit in Figures 12–79 and 12–80.
36. Determine the bandwidth of the circuit in Figure 12–79.

FIGURE 12–81

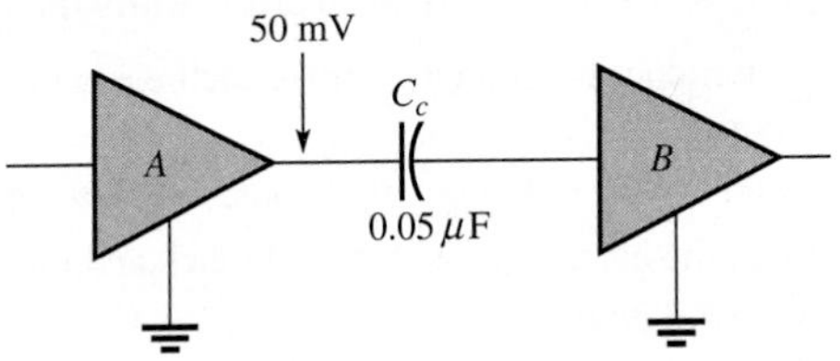

SECTION 12–9 Troubleshooting

37. Assume that the capacitor in Figure 12–82 is excessively leaky. Show how this degradation affects the output voltage and phase angle, assuming that the leakage resistance is 5 kΩ and the frequency is 10 Hz.

FIGURE 12–82

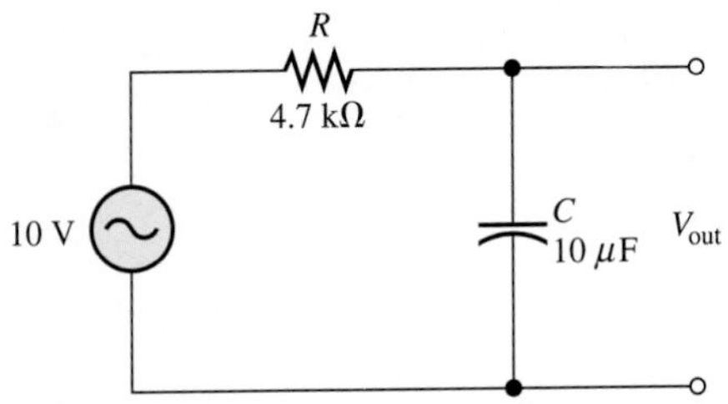

38. Each of the capacitors in Figure 12–83 has developed a leakage resistance of 2 kΩ. Determine the output voltages under this condition for each circuit.
39. Determine the output voltage for the circuit in Figure 12–83(a) for each of the following failure modes, and compare it to the correct output:
    (a) R_1 open **(b)** R_2 open **(c)** C open **(d)** C shorted
40. Determine the output voltage for the circuit in Figure 12–83(b) for each of the following failure modes, and compare it to the correct output.
    (a) C open **(b)** C shorted **(c)** R_1 open **(d)** R_2 open **(e)** R_3 open

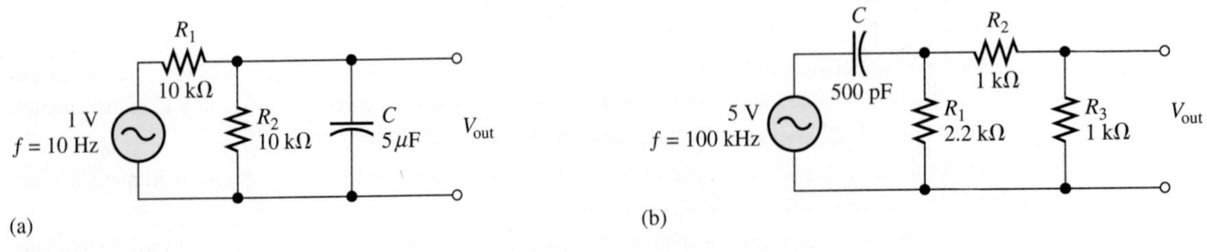

FIGURE 12–83

■ ADVANCED PROBLEMS

41. A single 220 V, 60 Hz source drives two loads. Load *A* has an impedance of 50 Ω and a power factor of 0.85. Load *B* has an impedance of 72 Ω and a power factor of 0.95.
(a) How much current does each load draw?
(b) What is the reactive power in each load?
(c) What is the true power in each load?
(d) What is the apparent power in each load?
(e) Which load has more voltage drop along the lines connecting it to the source?

42. What value of coupling capacitor is required in Figure 12–84 so that the signal voltage at the input of amplifier 2 is at least 70.7% of the signal voltage at the output of amplifier 1 when the frequency is 20 Hz? Disregard the input resistance of the amplifier.

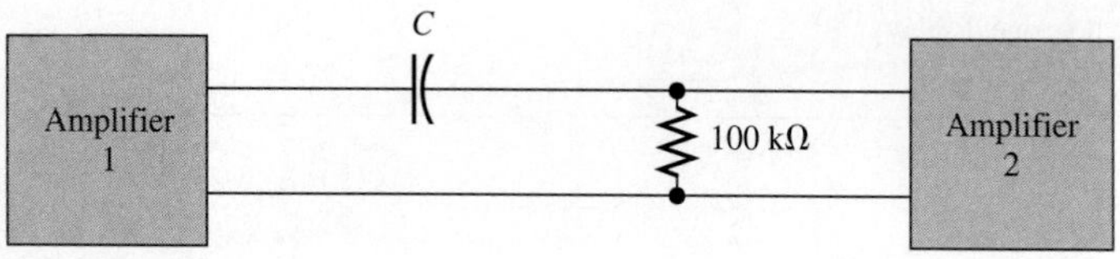

FIGURE 12–84

43. Determine the value to which R_1 must be adjusted to get a phase angle of 30° between the source voltage and the total current in Figure 12–85.

44. Sketch the voltage and current phasor diagram for Figure 12–86.

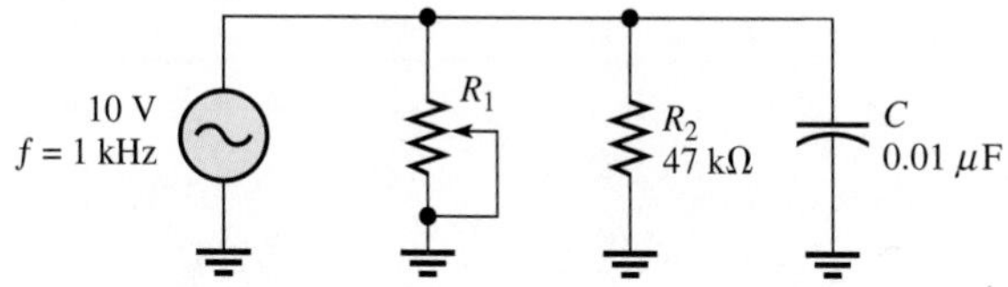

FIGURE 12–85

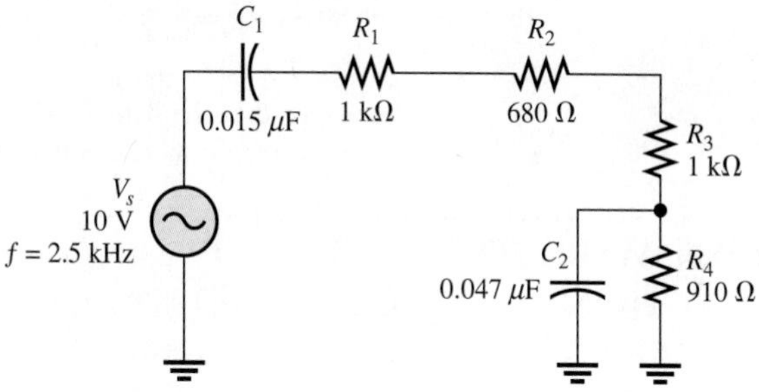

FIGURE 12–86

45. A certain load dissipates 1.5 kW of power with an impedance of 12 Ω and a power factor of 0.75. What is its reactive power? What is its apparent power?

46. Determine the series element or elements that are in the block of Figure 12–87 to meet the following overall circuit requirements:
(a) $P_{true} = 400$ W **(b)** leading power factor (I_T leads V_s)

47. Determine the value of C_2 in Figure 12–88 when $V_A = V_B$.

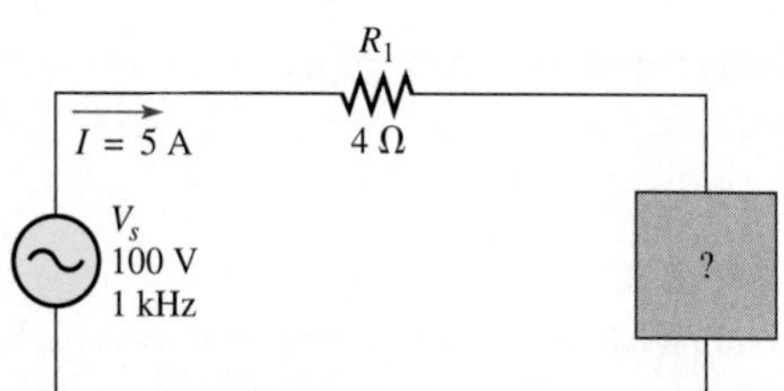

FIGURE 12–87

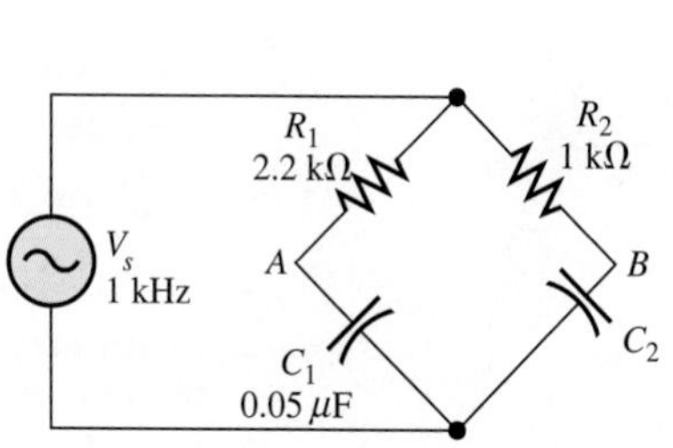

FIGURE 12–88

FIGURE 12–89

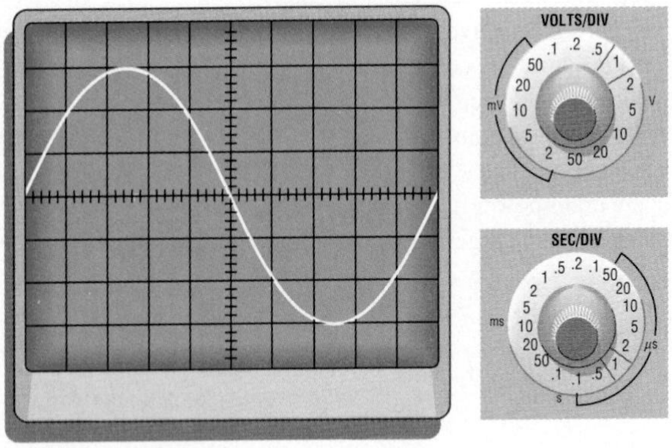

(a) Oscilloscope display

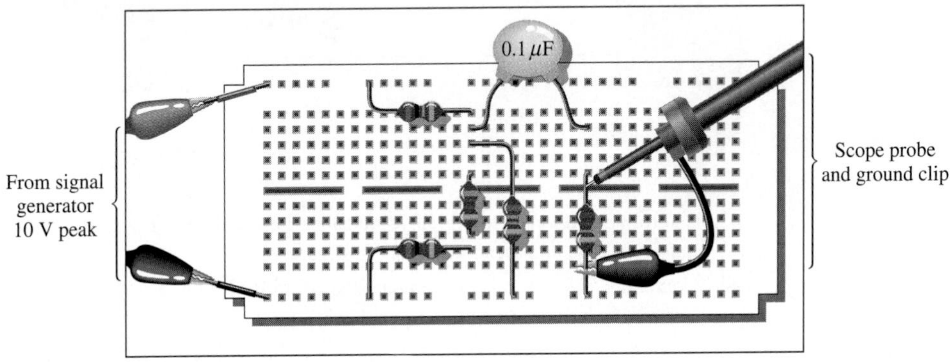

(b) Circuit with leads connected

48. Draw the schematic for the circuit in Figure 12–89 and determine if the waveform on the scope is correct. If there is a fault in the circuit, identify it.

■ ANSWERS TO SECTION REVIEWS

Section 12–1

1. V_C is 60 Hz, I is 60 Hz.

2. Capacitive reactance

3. θ is closer to 0° when $R > X_C$.

Section 12–2

1. Impedance is the opposition to sinusoidal current.

2. V_s lags I.

3. The capacitive reactance produces the phase angle.

4. $Z = \sqrt{R^2 + X_C^2} = 59.9\ \Omega$; $\theta = \tan^{-1}(X_C/R) = 56.6°$

Section 12–3

1. $V_T = \sqrt{V_R^2 + V_C^2} = 7.2$ V

2. $\theta = \tan^{-1}(V_C/V_R) = 56.3°$

3. $\theta = 90°$

4. **(a)** X_C decreases when f increases. **(b)** Z decreases when f increases.
(c) θ decreases when f increases.

Section 12–4

1. $Z = RX_C/\sqrt{R^2 + X_C^2} = 545\ \Omega$

2. *Conductance* is the reciprocal of resistance; *capacitive susceptance* is the reciprocal of capacitive reactance; and *admittance* is the reciprocal of impedance.

3. $Y = 1/Z = 10$ mS

4. $Y = \sqrt{G^2 + B_C^2} = 24$ mS

Section 12–5

1. $I_T = V_s Y = 21$ mA
2. $\theta = \tan^{-1}(I_C/I_R) = 56.3°$; $I_T = \sqrt{I_R^2 + I_C^2} = 18$ mA
3. $\theta = 90°$

Section 12–6

1. See Figure 12–90.
2. $V_1 = I_T R_1 = 6.80$ V

FIGURE 12–90

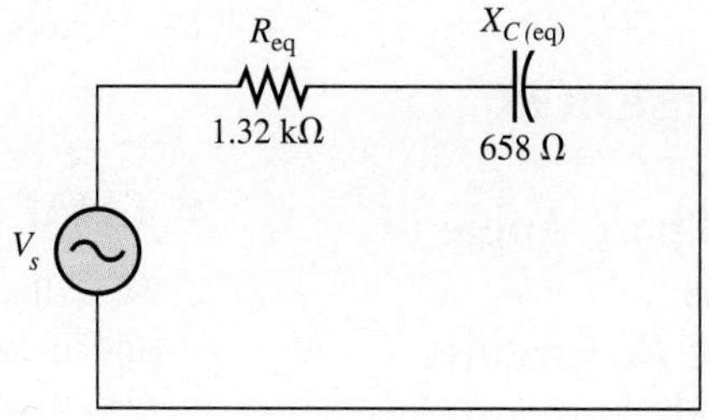

Section 12–7

1. Power is in the resistance.
2. $PF = \cos 45° = 0.707$
3. $P_{true} = I_T^2 R = 1.32$ kW; $P_r = I_T^2 X_C = 1.84$ kVAR; $P_a = I_T^2 Z = 2.26$ kVA

Section 12–8

1. $\phi = 90° - \tan^{-1}(X_C/R) = 62.8°$
2. $V_{out} = (X_C/\sqrt{R^2 + X_C^2})V_{in} = 8.9$ V rms
3. Output is across the capacitor.

Section 12–9

1. The leakage resistance acts in parallel with C, which alters the circuit time constant.
2. The capacitor is open.
3. A shorted capacitor or an open resistor can cause 0 V across the capacitor.

Section 12–10

1. A lower value coupling capacitor will increase the frequency at which a significant drop in voltage occurs.
2. $V_B = 3.16$ V dc
3. $V_B = 10$ mV rms

▪ ANSWERS TO RELATED EXERCISES FOR EXAMPLES

12–1 2.42 kΩ; 65.6°

12–2 2.56 V

12–3 67.5°

12–4 7.14 V

12–5 15.9 kΩ; 86.4°

12–6 24.3 Ω

12–7 4.36 mS

12–8 5.91 mA

12–9 117 mA; 31°

12–10 $R_{eq} = 8.99$ kΩ; $X_{C(eq)} = 4.38$ kΩ

12–11 $V_1 = 7.14$ V; $V_2 = 3.16$ V

12–12 0.155

12–13 990 μW

12–14 Phase lag ϕ increases.

12–15 Output voltage decreases.

12–16 Phase lead ϕ decreases; output voltage increases.

12–17 7.29 V

13

RL CIRCUITS

An *RL* circuit contains both resistance and inductance. In this chapter, basic series and parallel *RL* circuits and their responses to sine wave voltages are covered. In addition, series-parallel combinations are examined. Power considerations in *RL* circuits are introduced, and practical aspects of the power factor are discussed. A method of improving the power factor is presented, and basic areas of application are covered. As you study this chapter, note both the differences and the similarities in the response of *RL* circuits compared to *RC* circuits.

■ CHAPTER OBJECTIVES

- ☐ Describe the relationship between current and voltage in an *RL* circuit
- ☐ Determine impedance and phase angle in a series *RL* circuit
- ☐ Analyze a series *RL* circuit
- ☐ Determine impedance and phase angle in a parallel *RL* circuit
- ☐ Analyze a parallel *RL* circuit
- ☐ Analyze series-parallel *RL* circuits
- ☐ Determine power in *RL* circuits
- ☐ Discuss some basic *RL* applications
- ☐ Troubleshoot *RL* circuits

■ APPLICATION ASSIGNMENT

Your assignment is to identify the types of *RL* circuits contained in two sealed modules that have been removed from a communications system. You will use your knowledge of *RL* circuits and basic measurements to determine the circuit arrangement and component values. After studying this chapter, you should be able to complete the application assignment in Section 13–10.

13–1 ■ SINUSOIDAL RESPONSE OF *RL* CIRCUITS

As with the RC circuit, all currents and voltages in an RL circuit are sinusoidal when the input voltage is sinusoidal. The inductance causes a phase shift between the voltage and the current that depends on the relative values of the resistance and the inductive reactance.

After completing this section, you should be able to

- **Describe the relationship between current and voltage in an *RL* circuit**
  - ☐ Discuss voltage and current waveforms
  - ☐ Discuss phase shift

As you will learn, the resistor voltage and the current lag the source voltage. The inductor voltage leads the source voltage. The phase angle between the current and the inductor voltage is always 90°. These generalized phase relationships are indicated in Figure 13–1. Notice that they are opposite from those of the *RC* circuit, as discussed in Chapter 12.

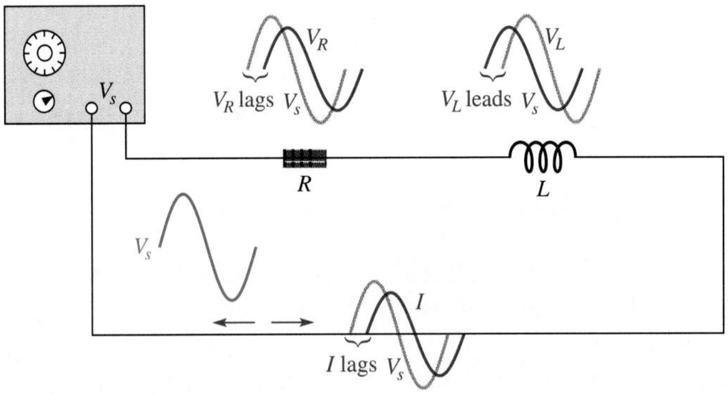

FIGURE 13–1
Illustration of sinusoidal response with general phase relationships of V_R, V_L, and I relative to V_s. V_R and I are in phase, V_R lags V_s, and V_L leads V_s. V_R and V_L are 90° out of phase with each other.

The amplitudes and the phase relationships of the voltages and current depend on the values of the resistance and the inductive reactance (defined in Section 10–6). When a circuit is purely inductive, the phase angle between the applied voltage and the total current is 90°, with the current lagging the voltage. When there is a combination of both resistance and inductive reactance in a circuit, the phase angle is somewhere between zero and 90°, depending on the relative values of *R* and X_L.

SECTION 13–1 REVIEW

1. A 1 kHz sinusoidal voltage is applied to an *RL* circuit. What is the frequency of the resulting current?
2. When the resistance in an *RL* circuit is greater than the inductive reactance, is the phase angle between the applied voltage and the total current closer to 0° or to 90°?

13–2 ■ IMPEDANCE AND PHASE ANGLE OF SERIES *RL* CIRCUITS

The impedance of an RL circuit is the total opposition to sinusoidal current and its unit is the ohm. The phase angle is the phase difference between the total current and the source voltage.

After completing this section, you should be able to

- **Determine impedance and phase angle in a series *RL* circuit**
  - ☐ Draw the impedance triangle
  - ☐ Calculate impedance magnitude
  - ☐ Calculate the phase angle

The impedance of a series *RL* circuit is determined by the resistance and the inductive reactance, as indicated in Figure 13–2.

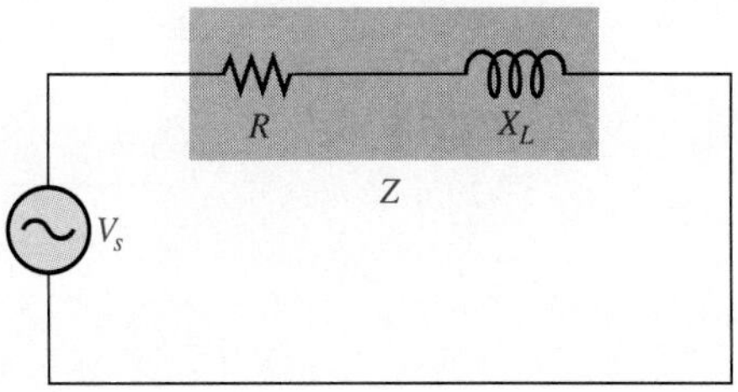

FIGURE 13–2
Impedance of a series RL circuit.

The Impedance Triangle

In ac analysis, both R and X_L are treated as phasor quantities, as shown in the phasor diagram of Figure 13–3(a), with X_L appearing at a +90° angle with respect to R. This relationship comes from the fact that the inductor voltage leads the current, and thus the resistor voltage, by 90°. Since Z is the phasor sum of R and X_L, its phasor representation is as shown in Figure 13–3(b). A repositioning of the phasors, as shown in part (c), forms a right triangle. This formation, as you have learned, is called the *impedance triangle.* The length of each phasor represents the magnitude of the quantity, and θ is the phase angle between the applied voltage and the current in the *RL* circuit.

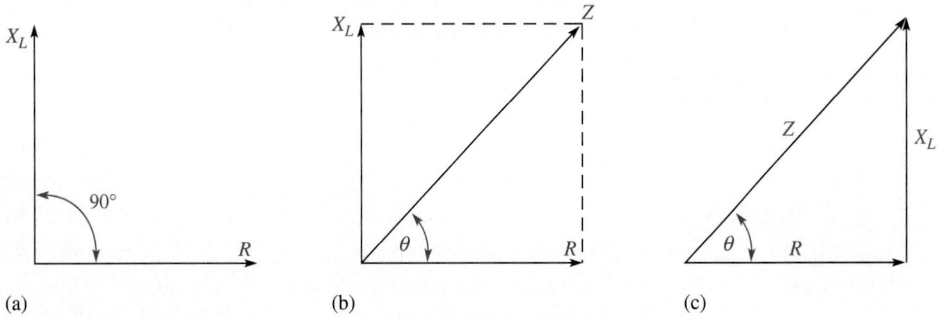

FIGURE 13–3
Development of the impedance triangle for a series RL circuit.

The impedance, Z, of the series RL circuit can be expressed in terms of the resistance and reactance as

$$Z = \sqrt{R^2 + X_L^2} \qquad \textbf{(13–1)}$$

where Z is expressed in ohms.

The phase angle, θ, is expressed as

$$\theta = \tan^{-1}\left(\frac{X_L}{R}\right) \qquad \textbf{(13–2)}$$

EXAMPLE 13–1

Determine the impedance and phase angle of the circuit in Figure 13–4.

FIGURE 13–4

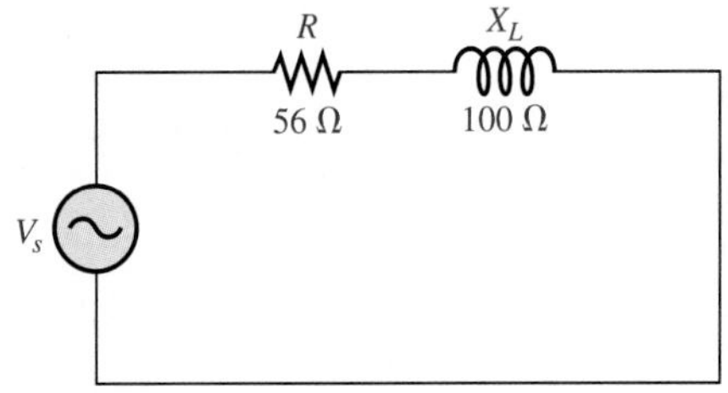

Solution The impedance is

$$Z = \sqrt{R^2 + X_L^2} = \sqrt{(56\ \Omega)^2 + (100\ \Omega)^2} = 115\ \Omega$$

The phase angle is

$$\theta = \tan^{-1}\left(\frac{X_L}{R}\right) = \tan^{-1}\left(\frac{100\ \Omega}{56\ \Omega}\right) = 60.8°$$

The applied voltage leads the current by 60°.

The calculator sequences are as follows:

Z: 5 6 x² + 1 0 0 x² = 2nd √x

θ: 1 0 0 ÷ 5 6 = 2nd tan

Related Exercise In a series RL circuit, R = 1.8 kΩ and X_L = 950 Ω. Determine the impedance and phase angle.

SECTION 13–2 REVIEW

1. Does the source voltage lead or lag the current in a series RL circuit?
2. What causes the phase angle in the RL circuit?
3. How does the response of an RL circuit differ from that of an RC circuit?
4. A series RL circuit has a resistance of 33 kΩ and an inductive reactance of 50 kΩ. Determine Z and θ.

13–3 ▪ ANALYSIS OF SERIES *RL* CIRCUITS

In the previous section, you learned how to express the impedance of a series RL circuit. Now, Ohm's law and Kirchhoff's voltage law are used in the analysis of RL circuits.

After completing this section, you should be able to

- **Analyze a series *RL* circuit**
  - ☐ Apply Ohm's law and Kirchhoff's voltage law to series *RL* circuits
  - ☐ Determine the phase relationships of the voltages and current
  - ☐ Show how impedance and phase angle vary with frequency

Ohm's Law

The application of Ohm's law to series *RL* circuits involves the use of the quantities of *Z*, *V*, and *I*. The three equivalent forms of Ohm's law were stated in Chapter 12 for *RC* circuits. They apply also to *RL* circuits and are restated here for convenience:

$$V = IZ \qquad I = \frac{V}{Z} \qquad Z = \frac{V}{I}$$

The following example illustrates the use of Ohm's law.

EXAMPLE 13–2 The current in Figure 13–5 is 200 μA. Determine the source voltage.

FIGURE 13–5

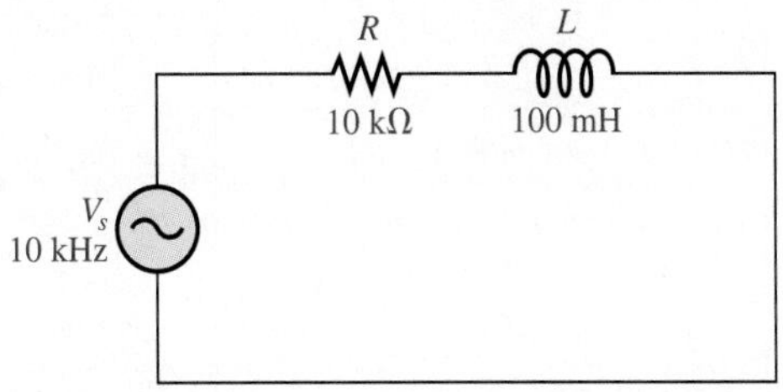

Solution From Equation (10–11), the inductive reactance is

$$X_L = 2\pi fL = 2\pi(10\text{ kHz})(100\text{ mH}) = 6.28\text{ k}\Omega$$

The impedance is

$$Z = \sqrt{R^2 + X_L^2} = \sqrt{(10\text{ k}\Omega)^2 + (6.28\text{ k}\Omega)^2} = 11.8\text{ k}\Omega$$

Applying Ohm's law,

$$V_s = IZ = (200\ \mu\text{A})(11.8\text{ k}\Omega) = 2.36\text{ V}$$

Related Exercise If the source voltage in Figure 13–5 were 5 V, what would be the current?

Relationships of the Current and Voltages in a Series *RL* Circuit

In a series *RL* circuit, the current is the same through both the resistor and the **inductor.** Thus, the resistor voltage is in phase with the current, and the inductor voltage

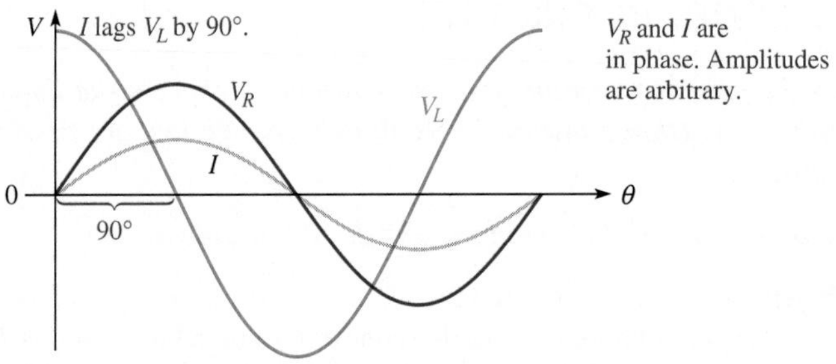

FIGURE 13–6
Phase relation of current and voltages in a series RL circuit.

leads the current by 90°. Therefore, there is a phase difference of 90° between the resistor voltage, V_R, and the inductor voltage, V_L, as shown in the waveform diagram of Figure 13–6.

From Kirchhoff's voltage law, the sum of the voltage drops must equal the applied voltage. However, since V_R and V_L are not in phase with each other, they must be added as phasor quantities with V_L leading V_R by 90°, as shown in Figure 13–7(a). As shown in part (b), V_s is the phasor sum of V_R and V_L. This equation can be expressed as

$$V_s = \sqrt{V_R^2 + V_L^2} \tag{13–3}$$

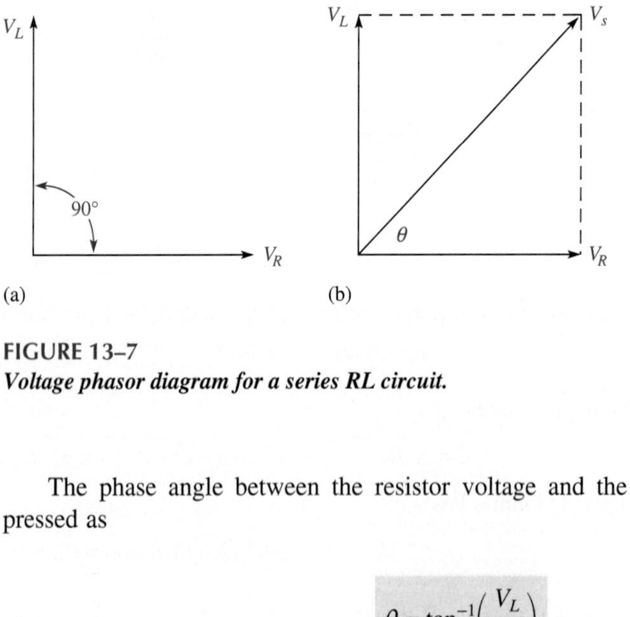

FIGURE 13–7
Voltage phasor diagram for a series RL circuit.

The phase angle between the resistor voltage and the source voltage can be expressed as

$$\theta = \tan^{-1}\left(\frac{V_L}{V_R}\right) \tag{13–4}$$

Since the resistor voltage and the current are in phase, θ in Equation (13–4) also represents the phase angle between the source voltage and the current and is equivalent to $\tan^{-1}(X_L/R)$.

Figure 13–8 shows a voltage and current phasor diagram that represents the waveform diagram of Figure 13–6.

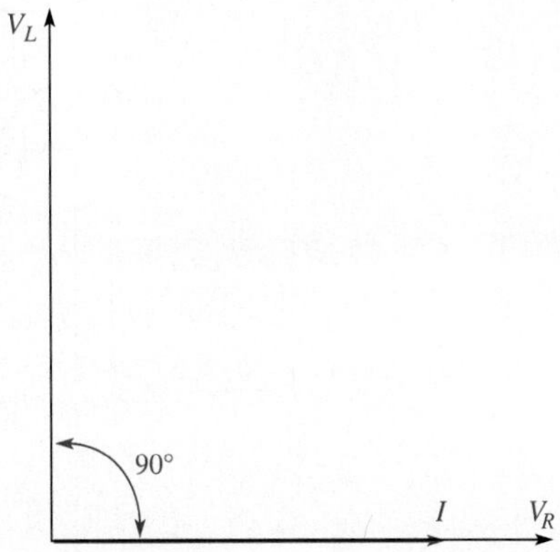

FIGURE 13–8
Voltage and current phasor diagram for the waveforms in Figure 13–6.

EXAMPLE 13–3 Determine the source voltage and the phase angle in Figure 13–9.

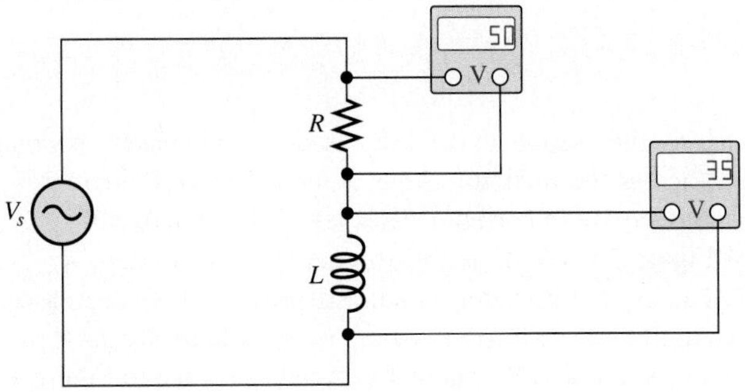

FIGURE 13–9

Solution Since V_R and V_L are 90° out of phase, they cannot be added directly and must be added as phasor quantities. The source voltage is

$$V_s = \sqrt{V_R^2 + V_L^2} = \sqrt{(50\text{ V})^2 + (35\text{ V})^2} = 61\text{ V}$$

The phase angle between the current and the source voltage is

$$\theta = \tan^{-1}\left(\frac{V_L}{V_R}\right) = \tan^{-1}\left(\frac{35\text{ V}}{50\text{ V}}\right) = 35°$$

Related Exercise With the information given, can you determine the current in Figure 13–9?

Variation of Impedance with Frequency

As you know, inductive reactance varies directly with frequency. When X_L increases, the total impedance also increases; and when X_L decreases, the total impedance decreases. Thus, *in an RL circuit, Z is directly dependent on frequency.*

Figure 13–10 illustrates how the voltages and current in a series *RL* circuit vary as the frequency increases or decreases, with the source voltage held at a constant value. In

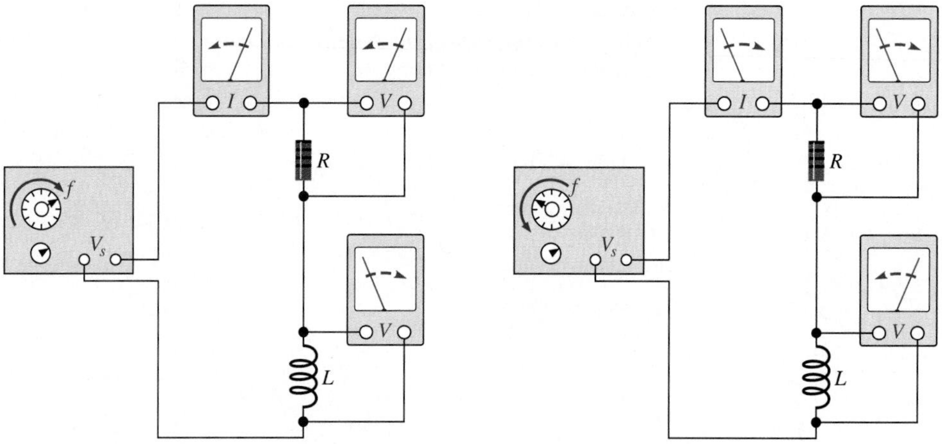

(a) As frequency is increased, I and V_R decrease and V_L increases.

(b) As frequency is decreased, I and V_R increase and V_L decreases.

FIGURE 13–10

An illustration of how the variation of impedance affects the voltages and current as the source frequency is varied. The source voltage is held at a constant amplitude.

part (a), as the frequency is increased, X_L increases; so more of the total voltage is dropped across the inductor. Also, Z increases as X_L increases, causing the current to decrease. A decrease in current causes less voltage across R.

In Figure 13–10(b), as the frequency is decreased, X_L decreases; so less voltage is dropped across the inductor. Also, Z decreases as X_L decreases, causing the current to increase. An increase in current causes more voltage across R.

Changes in Z and X_L can be observed as shown in Figure 13–11. As the frequency increases, the voltage across Z remains constant because V_s is constant, but the voltage across L increases. The decreasing current indicates that Z is increasing. It does so because of the inverse relationship stated in Ohm's law ($Z = V_Z/I$). The decreasing current also indicates that X_L is increasing. The increase in V_L corresponds to the increase in X_L.

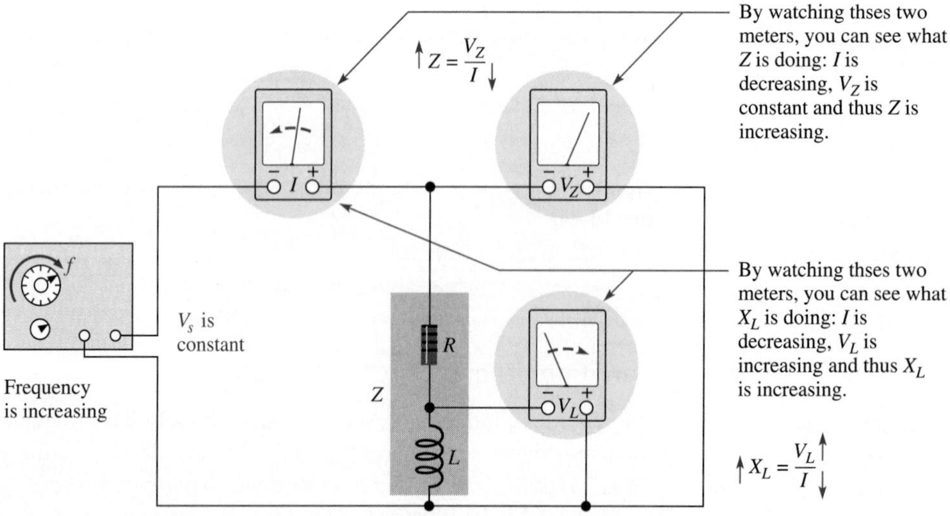

FIGURE 13–11

Observing changes in Z and X_L with frequency by watching the meters and recalling Ohm's law.

Variation of the Phase Angle with Frequency

Since X_L is the factor that introduces the phase angle in a series *RL* circuit, a change in X_L produces a change in the phase angle. As the frequency is increased, X_L becomes greater, and thus the phase angle increases. As the frequency is decreased, X_L becomes smaller, and thus the phase angle decreases. This relationship is illustrated in Figure 13–12 where a phase meter is connected to V_s and to V_R. This connection gives the angle between V_s and V_R, which is the phase angle of the circuit because I is in phase with V_R. By measuring the phase of V_R, you are effectively measuring the phase of I.

The variations of phase angle with frequency are illustrated with the impedance triangle as shown in Figure 13–13.

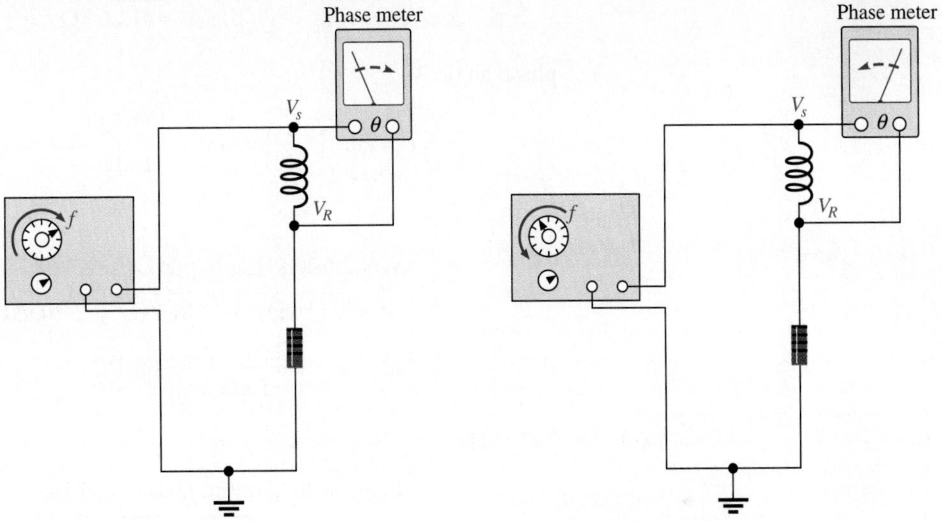

(a) As frequency is increased, the phase angle increases.

(b) As frequency is decreased, the phase angle decreases.

FIGURE 13–12
A phase meter indicates that the phase angle between V_s and V_R varies directly with frequency. V_R is the same phase as I.

FIGURE 13–13
As the frequency increases, the phase angle θ increases.

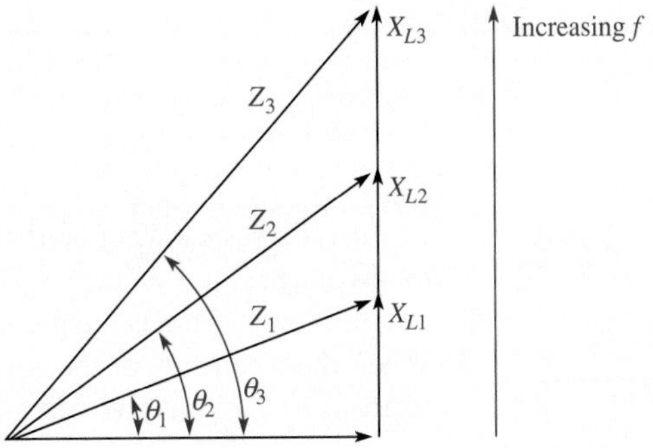

EXAMPLE 13–4

For the series *RL* circuit in Figure 13–14, determine the impedance and the phase angle for each of the following frequencies:

(a) 10 kHz **(b)** 20 kHz **(c)** 30 kHz

FIGURE 13–14

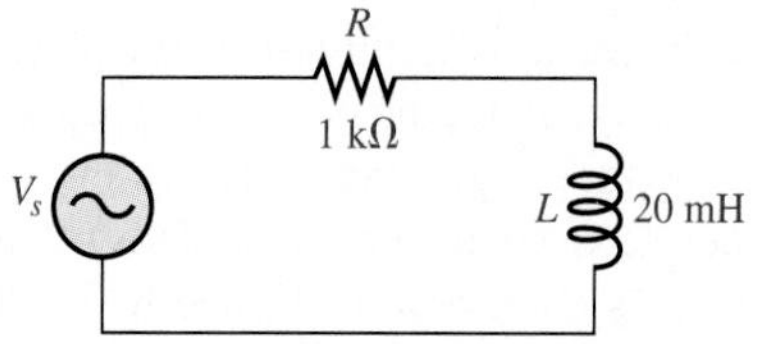

Solution

(a) For $f = 10$ kHz, the impedance is calculated as follows:

$$X_L = 2\pi fL = 2\pi(10 \text{ kHz})(20 \text{ mH}) = 1.26 \text{ k}\Omega$$

$$Z = \sqrt{R^2 + X_L^2} = \sqrt{(1 \text{ k}\Omega)^2 + (1.26 \text{ k}\Omega)^2} = 1.61 \text{ k}\Omega$$

The phase angle is

$$\theta = \tan^{-1}\left(\frac{X_L}{R}\right) = \tan^{-1}\left(\frac{1.26 \text{ k}\Omega}{1 \text{ k}\Omega}\right) = 51.6^\circ$$

(b) For $f = 20$ kHz,

$$X_L = 2\pi(20 \text{ kHz})(20 \text{ mH}) = 2.51 \text{ k}\Omega$$

$$Z = \sqrt{(1 \text{ k}\Omega)^2 + (2.51 \text{ k}\Omega)^2} = 2.70 \text{ k}\Omega$$

$$\theta = \tan^{-1}\left(\frac{2.51 \text{ k}\Omega}{1 \text{ k}\Omega}\right) = 68.3^\circ$$

(c) For $f = 30$ kHz,

$$X_L = 2\pi(30 \text{ kHz})(20 \text{ mH}) = 3.77 \text{ k}\Omega$$

$$Z = \sqrt{(1 \text{ k}\Omega)^2 + (3.77 \text{ k}\Omega)^2} = 3.90 \text{ k}\Omega$$

$$\theta = \tan^{-1}\left(\frac{3.77 \text{ k}\Omega}{1 \text{ k}\Omega}\right) = 75.1^\circ$$

Notice that as the frequency increases, X_L, Z, and θ also increase.

Related Exercise Determine Z and θ in Figure 13–14 if $f = 100$ kHz.

SECTION 13–3 REVIEW

1. In a certain series *RL* circuit, $V_R = 2$ V and $V_L = 3$ V. What is the magnitude of the total voltage?
2. In Question 1, what is the phase angle?
3. When the frequency of the applied voltage in a series *RL* circuit is increased, what happens to the inductive reactance? To the impedance? To the phase angle?

13–4 ■ IMPEDANCE AND PHASE ANGLE OF PARALLEL *RL* CIRCUITS

In this section, you will learn how to determine the impedance and phase angle of a parallel RL circuit. Also, inductive susceptance and admittance of a parallel RL circuit are introduced.

After completing this section, you should be able to

- **Determine impedance and phase angle in a parallel *RL* circuit**
  - ☐ Express total impedance in a product over sum form
  - ☐ Express the phase angle in terms of R and X_L
  - ☐ Determine susceptance and admittance in a parallel *RL* circuit
  - ☐ Convert admittance to impedance

A basic parallel *RL* circuit is shown in Figure 13–15. The expression for the impedance is

$$Z = \frac{RX_L}{\sqrt{R^2 + X_L^2}} \tag{13-5}$$

The phase angle between the applied voltage and the total current is

$$\theta = \tan^{-1}\left(\frac{R}{X_L}\right) \tag{13-6}$$

FIGURE 13–15
Parallel RL circuit.

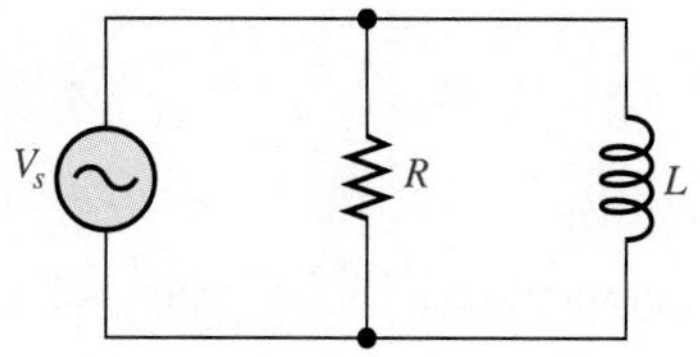

EXAMPLE 13–5 For each circuit in Figure 13–16, determine the impedance and the phase angle.

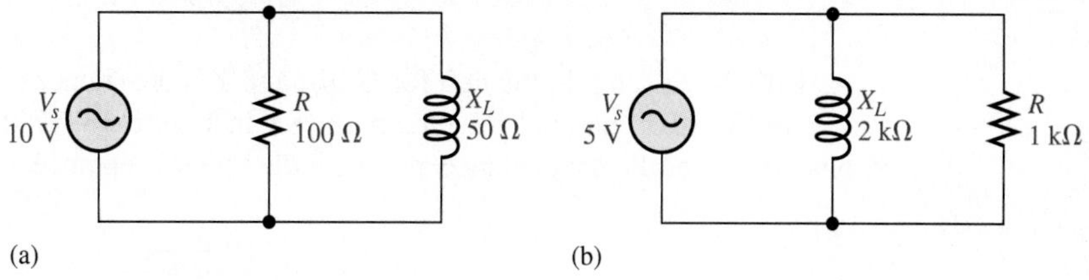

FIGURE 13–16

Solution For the circuit in Figure 13–16(a), the impedance and phase angle are

$$Z = \frac{RX_L}{\sqrt{R^2 + X_L^2}} = \frac{(100\ \Omega)(50\ \Omega)}{\sqrt{(100\ \Omega)^2 + (50\ \Omega)^2}} = 44.7\ \Omega$$

$$\theta = \tan^{-1}\left(\frac{R}{X_L}\right) = \tan^{-1}\left(\frac{100\ \Omega}{50\ \Omega}\right) = 63.4°$$

For the circuit in Figure 13–16(b),

$$Z = \frac{(1\ \text{k}\Omega)(2\ \text{k}\Omega)}{\sqrt{(1\ \text{k}\Omega)^2 + (2\ \text{k}\Omega)^2}} = 894\ \Omega$$

$$\theta = \tan^{-1}\left(\frac{1\ \text{k}\Omega}{2\ \text{k}\Omega}\right) = 26.6°$$

The voltage leads the current, as opposed to the parallel *RC* case where the voltage lags the current.

The calculator sequences for the circuit in Figure 13–16(a) are as follows:

Z: [1] [0] [0] [×] [5] [0] [÷] [(] [(] [1] [0] [0] [x²] [+] [5] [0] [x²] [)] [2nd] [√x] [)] [=]

θ: [1] [0] [0] [÷] [5] [0] [=] [2nd] [tan]

Related Exercise In a parallel circuit $R = 10\ \text{k}\Omega$ and $X_L = 14\ \text{k}\Omega$. Find Z and θ.

Conductance, Susceptance, and Admittance

As you know from Section 12–4, conductance (G) is the reciprocal of resistance, susceptance (B) is the reciprocal of reactance, and admittance (Y) is the reciprocal of impedance.

For parallel *RL* circuits, conductance (G) is expressed as

$$G = \frac{1}{R} \qquad \textbf{(13–7)}$$

Inductive susceptance (B_L) is expressed as

$$B_L = \frac{1}{X_L} \qquad \textbf{(13–8)}$$

Admittance (Y) is expressed as

$$Y = \frac{1}{Z} \qquad \textbf{(13–9)}$$

As with the *RC* circuit, the unit for G, B_L, and Y is the siemens (S).

In the basic parallel *RL* circuit shown in Figure 13–17, the total admittance is the phasor sum of the conductance and the inductive susceptance and is expressed as follows:

$$Y_T = \sqrt{G^2 + B_L^2} \qquad \textbf{(13–10)}$$

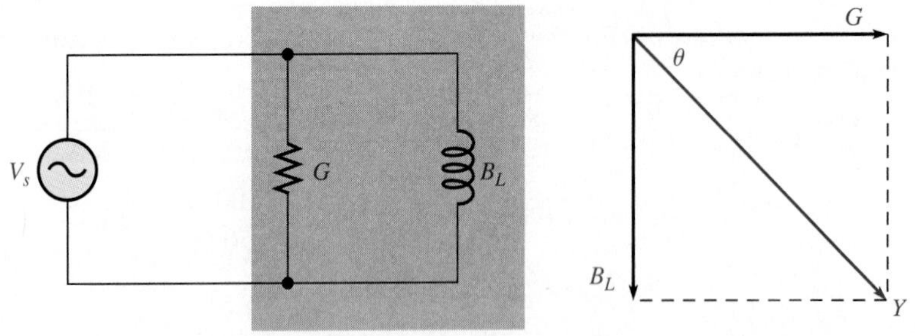

FIGURE 13–17
Admittance is the reciprocal of impedance in a parallel RL circuit.

EXAMPLE 13–6 Determine the total admittance in Figure 13–18; then convert it to impedance.

FIGURE 13–18

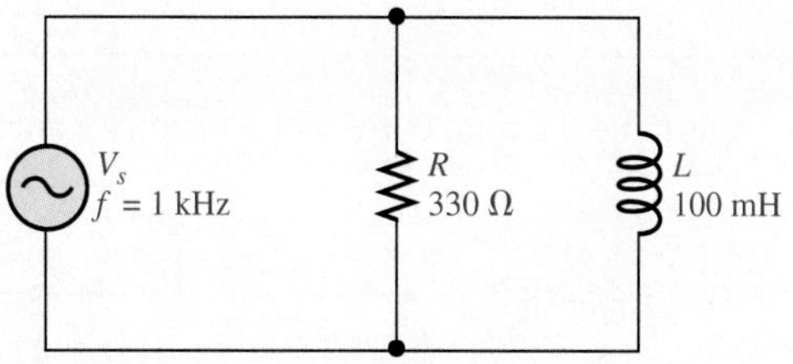

Solution To determine Y, first calculate the values for G and B_L. Since $R = 330\ \Omega$,

$$G = \frac{1}{R} = \frac{1}{330\ \Omega} = 3.03\text{ mS}$$

The inductive reactance is

$$X_L = 2\pi fL = 2\pi(1000\text{ Hz})(100\text{ mH}) = 628\ \Omega$$

The inductive susceptance is

$$B_L = \frac{1}{X_L} = 1.59\text{ mS}$$

Therefore, the total admittance is

$$Y_T = \sqrt{G^2 + B_L^2} = \sqrt{(3.03\text{ mS})^2 + (1.59\text{ mS})^2} = 3.42\text{ mS}$$

Converting to impedance,

$$Z = \frac{1}{Y_T} = \frac{1}{3.42\text{ mS}} = 292\ \Omega$$

Related Exercise What is the admittance of the circuit in Figure 13–18 if f is increased to 2 kHz?

SECTION 13–4 REVIEW

1. If $Y = 50$ mS, what is the value of Z?
2. In a certain parallel *RL* circuit, $R = 47\ \Omega$ and $X_L = 75\ \Omega$. Determine the admittance.
3. In the circuit of Question 2, does the total current lead or lag the applied voltage and by what phase angle?

13–5 ■ ANALYSIS OF PARALLEL *RL* CIRCUITS

In the previous section, you learned how to express the impedance of a parallel RL circuit. Now, Ohm's law and Kirchhoff's current law are used in the analysis of RL circuits. Current and voltage relationships in a parallel RL circuit are examined.

After completing this section, you should be able to

- **Analyze a parallel *RL* circuit**
  - ☐ Apply Ohm's law and Kirchhoff's current law to parallel *RL* circuits
  - ☐ Determine total current and phase angle

The following example applies Ohm's law to the analysis of a parallel *RL* circuit.

EXAMPLE 13–7 Determine the total current and the phase angle in the circuit of Figure 13–19.

FIGURE 13–19

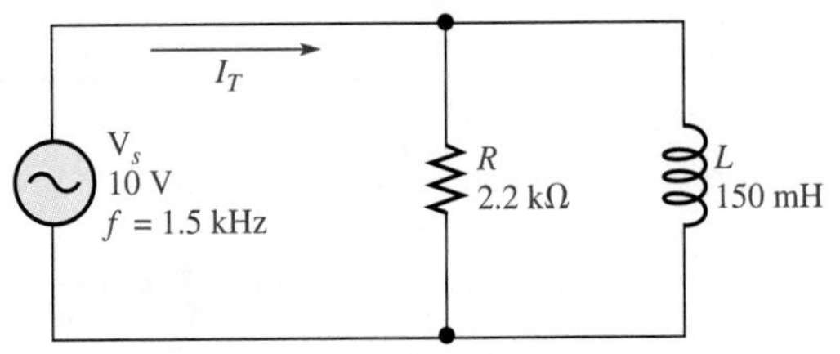

Solution First determine the total admittance. The inductive reactance is

$$X_L = 2\pi fL = 2\pi(1.5\text{ kHz})(150\text{ mH}) = 1.41\text{ k}\Omega$$

The inductive susceptance is

$$B_L = \frac{1}{X_L} = \frac{1}{1.41\text{ k}\Omega} = 709\ \mu\text{S}$$

The conductance is

$$G = \frac{1}{R} = \frac{1}{2.2\text{ k}\Omega} = 455\ \mu\text{S}$$

Therefore, the total admittance is

$$Y_T = \sqrt{G^2 + B_L^2} = \sqrt{(455\ \mu\text{S})^2 + (709\ \mu\text{S})^2} = 842\ \mu\text{S}$$

Applying Ohm's law to get the total current,

$$I_T = VY_T = (10\text{ V})(842\ \mu\text{S}) = 8.42\text{ mA}$$

The phase angle is

$$\theta = \tan^{-1}\left(\frac{R}{X_L}\right) = \tan^{-1}\left(\frac{2.2\text{ k}\Omega}{1.41\text{ k}\Omega}\right) = 57.3^\circ$$

The total current lags the source voltage by 57.3°.

Related Exercise Determine the total current and the phase angle if the frequency is reduced to 800 Hz in Figure 13–19.

Relationships of the Currents and Voltages in a Parallel *RL* Circuit

Figure 13–20(a) shows all the currents and voltages in a basic parallel *RL* circuit. As you can see, the applied voltage, V_s, appears across both the resistive and the inductive branches, so V_s, V_R, and V_L are all in phase and of the same magnitude. The total current, I_T, divides at the junction into the two branch currents, I_R and I_L.

The current through the resistor is in phase with the voltage. The current through the inductor lags the voltage and the resistor current by 90°. By Kirchhoff's current law, the

FIGURE 13–20
Currents and voltages in a parallel RL circuit. The current directions shown are instantaneous and, of course, reverse when the source voltage reverses.

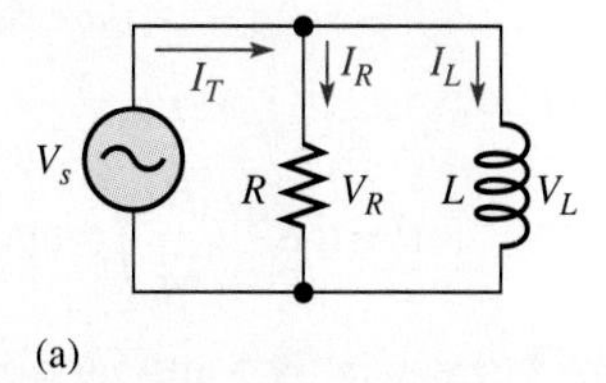

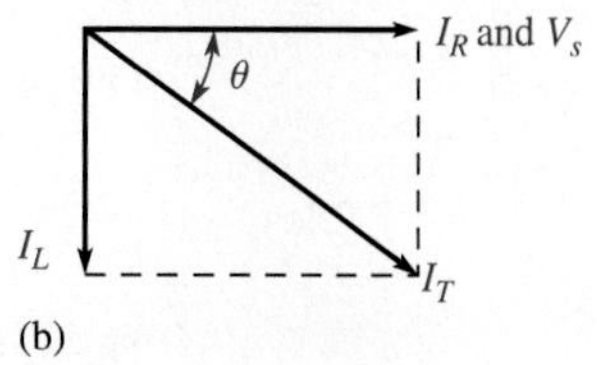

total current is the phasor sum of the two branch currents, as shown by the phasor diagram in Figure 13–20(b). The total current is expressed as

$$I_T = \sqrt{I_R^2 + I_L^2} \qquad \textbf{(13–11)}$$

The phase angle between the resistor current and the total current is

$$\theta = \tan^{-1}\left(\frac{I_L}{I_R}\right) \qquad \textbf{(13–12)}$$

Figure 13–21 shows a complete current and voltage phasor diagram.

FIGURE 13–21
Current and voltage phasor diagram for a parallel RL circuit (amplitudes are arbitrary).

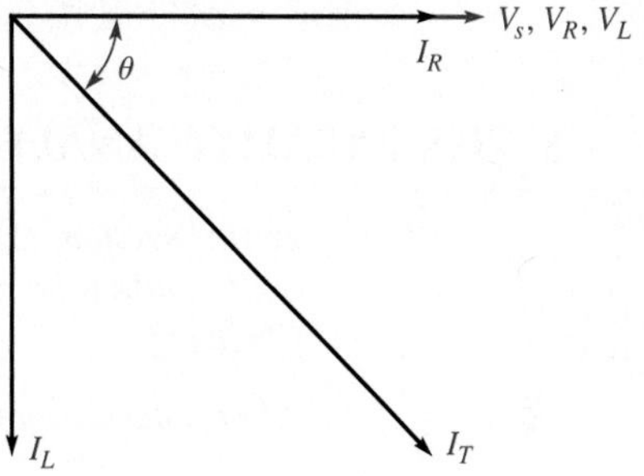

EXAMPLE 13–8

Determine the value of each current in Figure 13–22, and describe the phase relationship of each with the applied voltage.

FIGURE 13–22

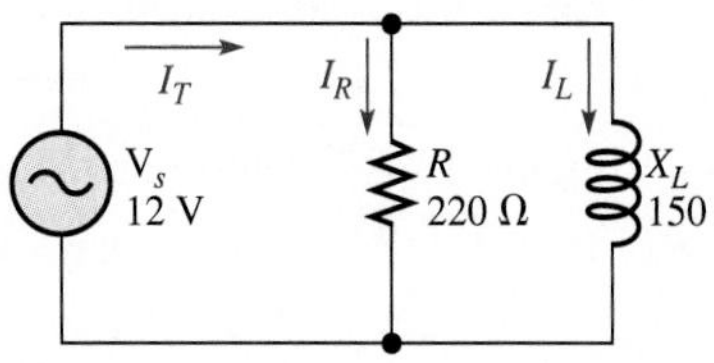

Solution The resistor current, the inductor current, and the total current are calculated as follows:

$$I_R = \frac{V_s}{R} = \frac{12\text{ V}}{220\ \Omega} = 54.5\text{ mA}$$

$$I_L = \frac{V_s}{X_L} = \frac{12\text{ V}}{150\ \Omega} = 80\text{ mA}$$

$$I_T = \sqrt{I_R^2 + I_L^2} = \sqrt{(54.5\text{ mA})^2 + (80\text{ mA})^2} = 96.8\text{ mA}$$

The phase angle is

$$\theta = \tan^{-1}\left(\frac{R}{X_L}\right) = \tan^{-1}\left(\frac{220\ \Omega}{150\ \Omega}\right) = 55.7°$$

The resistor current is 54.5 mA and is in phase with the applied voltage. The inductor current is 80 mA and lags the applied voltage by 90°. The total current is 96.8 mA and lags the voltage by 55.7°.

Related Exercise Find I_T and θ if $X_L = 300\ \Omega$ in Figure 13–22.

SECTION 13–5 REVIEW

1. The admittance of a parallel *RL* circuit is 4 mS, and the applied voltage is 8 V. What is the total current?
2. In a certain parallel *RL* circuit, the resistor current is 12 mA, and the inductor current is 20 mA. Determine the phase angle and the total current.
3. What is the phase angle between the inductor current and the applied voltage in a parallel *RL* circuit?

13–6 ▪ SERIES-PARALLEL ANALYSIS

In this section, the concepts studied in the previous sections are used to analyze circuits with combinations of both series and parallel R and L elements. Examples are used to illustrate.

After completing this section, you should be able to

- **Analyze series-parallel *RL* circuits**
  - ☐ Determine total impedance
  - ☐ Calculate currents and voltages

The following two examples illustrate the basic procedures used to analyze a series-parallel circuit.

EXAMPLE 13–9 In the circuit of Figure 13–23, determine the values of the following:
(a) Z_T **(b)** I_T **(c)** θ

FIGURE 13–23

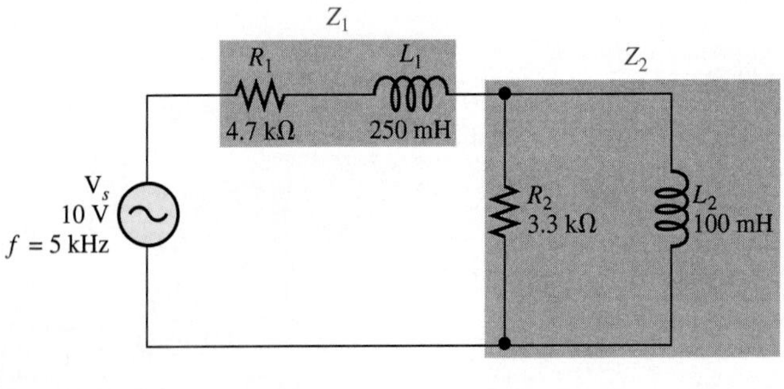

Solution

(a) First, calculate the inductive reactances:

$$X_{L1} = 2\pi f L_1 = 2\pi(5\text{ kHz})(250\text{ mH}) = 7.85\text{ k}\Omega$$
$$X_{L2} = 2\pi f L_2 = 2\pi(5\text{ kHz})(100\text{ mH}) = 3.14\text{ k}\Omega$$

One approach is to find the series equivalent resistance and inductive reactance for the parallel portion of the circuit. Then you can add the resistances to get the total resistance, add the reactances to get the total reactance, and then determine the total impedance.

Determine the impedance of the parallel portion (Z_2) as follows:

$$G_2 = \frac{1}{R_2} = \frac{1}{3.3\text{ k}\Omega} = 303\ \mu\text{S}$$

$$B_{L2} = \frac{1}{X_{L2}} = \frac{1}{3.14\text{ k}\Omega} = 318\ \mu\text{S}$$

$$Y_2 = \sqrt{G_2^2 + B_L^2} = \sqrt{(303\ \mu\text{S})^2 + (318\ \mu\text{S})^2} = 439\ \mu\text{S}$$

Then

$$Z_2 = \frac{1}{Y_2} = \frac{1}{439\ \mu\text{S}} = 2.28\text{ k}\Omega$$

The phase angle associated with the parallel portion of the circuit is

$$\theta_p = \tan^{-1}\left(\frac{R_2}{X_{L2}}\right) = \tan^{-1}\left(\frac{3.3\text{ k}\Omega}{3.14\text{ k}\Omega}\right) = 46.4°$$

The series equivalent values for the parallel portion are

$$R_{eq} = Z_2\cos\theta_p = (2.28\text{ k}\Omega)\cos(46.4°) = 1.57\text{ k}\Omega$$
$$X_{L(eq)} = Z_2\sin\theta_p = (2.28\text{ k}\Omega)\sin(46.4°) = 1.65\text{ k}\Omega$$

The total circuit resistance is

$$R_T = R_1 + R_{eq} = 4.7\text{ k}\Omega + 1.57\text{ k}\Omega = 6.27\text{ k}\Omega$$

The total circuit reactance is

$$X_{L(tot)} = X_{L1} + X_{L(eq)} = 7.85\text{ k}\Omega + 1.65\text{ k}\Omega = 9.50\text{ k}\Omega$$

The total circuit impedance is

$$Z_T = \sqrt{R_T^2 + X_{L(tot)}^2} = \sqrt{(6.27\text{ k}\Omega)^2 = (9.50\text{ k}\Omega)^2} = 11.4\text{ k}\Omega$$

(b) Use Ohm's law to find the total current:

$$I_T = \frac{V_s}{Z_T} = \frac{10\text{ V}}{11.4\text{ k}\Omega} = 877\ \mu\text{A}$$

(c) To find the phase angle, view the circuit as a series combination of R_T and $X_{L(tot)}$. The phase angle by which I_T lags V_s is

$$\theta = \tan^{-1}\left(\frac{X_{L(tot)}}{R_T}\right) = \tan^{-1}\left(\frac{9.5\text{ k}\Omega}{6.27\text{ k}\Omega}\right) = 56.6°$$

Related Exercise **(a)** Determine the voltage across the series part of the circuit in Figure 13–23. **(b)** Determine the voltage across the parallel part of the circuit.

EXAMPLE 13–10

Determine the voltage across each element in Figure 13–24. Sketch a voltage phasor diagram.

FIGURE 13–24

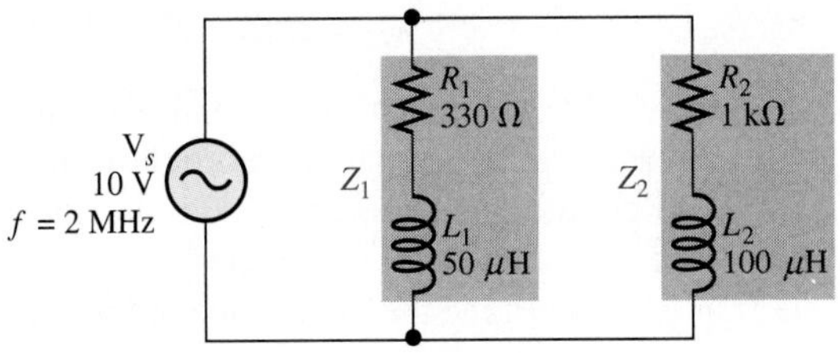

Solution First calculate X_{L1} and X_{L2}:

$$X_{L1} = 2\pi fL_1 = 2\pi(2\text{ MHz})(50\ \mu\text{H}) = 628\ \Omega$$
$$X_{L2} = 2\pi fL_2 = 2\pi(2\text{ MHz})(100\ \mu\text{H}) = 1.26\text{ k}\Omega$$

Now, determine the impedance of each branch:

$$Z_1 = \sqrt{R_1^2 + X_{L1}^2} = \sqrt{(330\ \Omega)^2 + (628\ \Omega)^2} = 709\ \Omega$$
$$Z_2 = \sqrt{R_2^2 + X_{L2}^2} = \sqrt{(1\text{ k}\Omega)^2 + (1.26\text{ k}\Omega)^2} = 1.61\text{ k}\Omega$$

Calculate each branch current:

$$I_1 = \frac{V_s}{Z_1} = \frac{10\text{ V}}{709\ \Omega} = 14.1\text{ mA}$$
$$I_2 = \frac{V_s}{Z_2} = \frac{10\text{ V}}{1.61\text{ k}\Omega} = 6.21\text{ mA}$$

Now, use Ohm's law to find the voltage across each element:

$$V_{R1} = I_1R_1 = (14.1\text{ mA})(330\ \Omega) = 4.65\text{ V}$$
$$V_{L1} = I_1X_{L1} = (14.1\text{ mA})(628\ \Omega) = 8.85\text{ V}$$
$$V_{R2} = I_2R_2 = (6.21\text{ mA})(1\text{ k}\Omega) = 6.21\text{ V}$$
$$V_{L2} = I_2X_{L2} = (6.21\text{ mA})(1.26\text{ k}\Omega) = 7.82\text{ V}$$

Now determine the angles associated with each parallel branch:

$$\theta_1 = \tan^{-1}\left(\frac{X_{L1}}{R_1}\right) = \tan^{-1}\left(\frac{628\ \Omega}{330\ \Omega}\right) = 62.3^\circ$$
$$\theta_2 = \tan^{-1}\left(\frac{X_{L2}}{R_2}\right) = \tan^{-1}\left(\frac{1.26\text{ k}\Omega}{1\text{ k}\Omega}\right) = 51.6^\circ$$

Thus, I_1 lags V_s by 62.3°, and I_2 lags V_s by 51.6°, as indicated in Figure 13–25(a), where the negative signs indicate lagging angles.

The phase relationships of the voltages are determined as follows:

- ☐ V_{R1} is in phase with I_1 and therefore lags V_s by 62.3°.
- ☐ V_{L1} leads I_1 by 90°, so its angle is 90° – 62.3° = 27.7°.
- ☐ V_{R2} is in phase with I_2 and therefore lags V_s by 51.6°.
- ☐ V_{L2} leads I_2 by 90°, so its angle is 90° – 51.6° = 38.4°.

These phase relationships are shown in Figure 13–25(b).

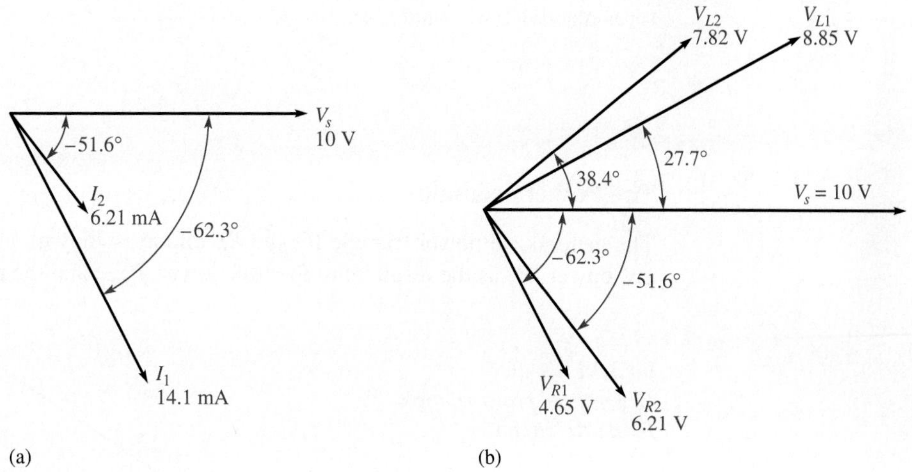

FIGURE 13–25

Related Exercise What effect does an increase in frequency have on the total current in Figure 13–24?

SECTION 13–6 REVIEW

1. Determine the total current for the circuit in Figure 13–24.
2. What is the total impedance of the circuit in Figure 13–24?

13–7 ■ POWER IN *RL* CIRCUITS

In a purely resistive ac circuit, all of the energy delivered by the source is dissipated in the form of heat by the resistance. In a purely inductive ac circuit, all of the energy delivered by the source is stored by the inductor in its magnetic field during a portion of the voltage cycle and then returned to the source during another portion of the cycle so that there is no net energy loss. When there is both resistance and inductance, some of the energy is alternately stored and returned by the inductance and some is dissipated by the resistance. The amount of energy loss is determined by the relative values of the resistance and the inductive reactance.

After completing this section, you should be able to

- **Determine power in *RL* circuits**
  - □ Explain true and reactive power
  - □ Draw the power triangle
  - □ Define *power factor*
  - □ Explain power factor correction

When the resistance is greater than the inductive reactance, more of the total energy delivered by the source is dissipated by the resistance than is stored and returned by the inductor. When the reactance is greater than the resistance, more of the total energy is stored and returned than is dissipated.

As you know, the power in a resistor is called the true power. The power in an inductor is reactive power and is expressed as

$$P_r = I^2X_L \tag{13–13}$$

The Power Triangle

The generalized power triangle for the *RL* circuit is shown in Figure 13–26. The **apparent power, P_a**, is the resultant of the true power, P_{true}, and the reactive power, P_r.

FIGURE 13–26
Generalized power triangle for an RL circuit.

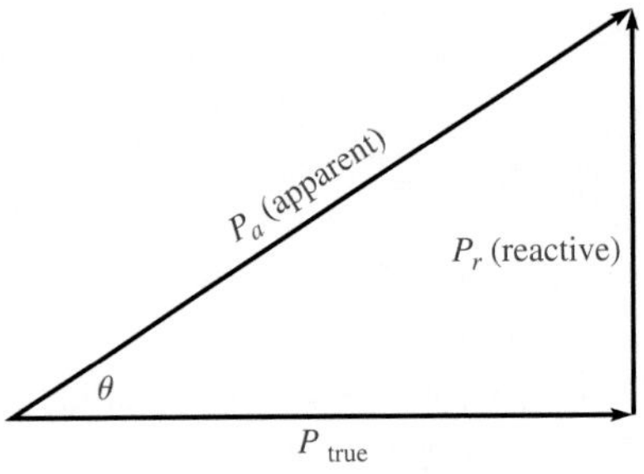

Recall that the power factor equals the cosine of θ ($PF = \cos\theta$). As the phase angle between the applied voltage and the total current increases, the power factor decreases, indicating an increasingly reactive circuit. The smaller the power factor, the smaller the true power is compared to the reactive power. The power factor of inductive loads is called a *lagging power factor* because the current lags the source voltage.

EXAMPLE 13–11 Determine the power factor, the true power, the reactive power, and the apparent power in Figure 13–27.

FIGURE 13–27

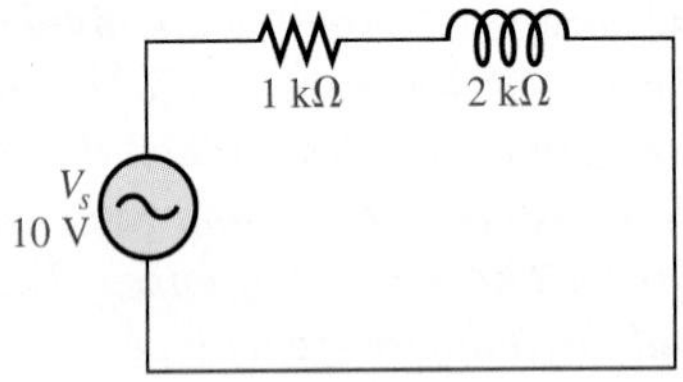

Solution The impedance of the circuit is

$$Z = \sqrt{R^2 + X_L^2} = \sqrt{(1\text{ k}\Omega)^2 + (2\text{ k}\Omega)^2} = 2.24\text{ k}\Omega$$

The current is

$$I = \frac{V_s}{Z} = \frac{10\text{ V}}{2.24\text{ k}\Omega} = 4.46\text{ mA}$$

The phase angle is

$$\theta = \tan^{-1}\left(\frac{X_L}{R}\right) = \tan^{-1}\left(\frac{2\text{ k}\Omega}{1\text{ k}\Omega}\right) = 63.4°$$

Therefore, the power factor is

$$PF = \cos\theta = \cos(63.4°) = 0.448$$

The true power is

$$P_{true} = V_sI\cos\theta = (10\text{ V})(4.46\text{ mA})(0.448) = 20\text{ mW}$$

The reactive power is

$$P_r = I^2X_L = (4.46\text{ mA})^2(2\text{ k}\Omega) = 39.8\text{ mVAR}$$

The apparent power is

$$P_a = I^2Z = (4.46\text{ mA})^2(2.24\text{ k}\Omega) = 44.6\text{ mVA}$$

Related Exercise If the frequency in Figure 13–27 is increased, what happens to P_{true}, P_r, and P_a?

Significance of the Power Factor

As you learned in Chapter 12, the **power factor** (*PF*) is very important in determining how much useful power (true power) is transferred to a load. The highest power factor is 1, which indicates that all of the current to a load is in phase with the voltage (resistive). When the power factor is 0, all of the current to a load is 90° out of phase with the voltage (reactive).

Generally, a power factor as close to 1 as possible is desirable, because then most of the power transferred from the source to the load is useful or true power. True power goes only one way—from source to load—and performs work on the load in terms of energy utilization. Reactive power simply goes back and forth between the source and the load with no net work being done. Energy must be used in order for work to be done.

Many practical loads have inductance as a result of their particular function, and it is essential for their proper operation. Examples are transformers, electric motors, and speakers, to name a few. Therefore, inductive (and capacitive) loads are a fact of life and we must live with them.

To see the effect of the power factor on system requirements, refer to Figure 13–28 which shows a representation of a typical inductive load consisting effectively of induc-

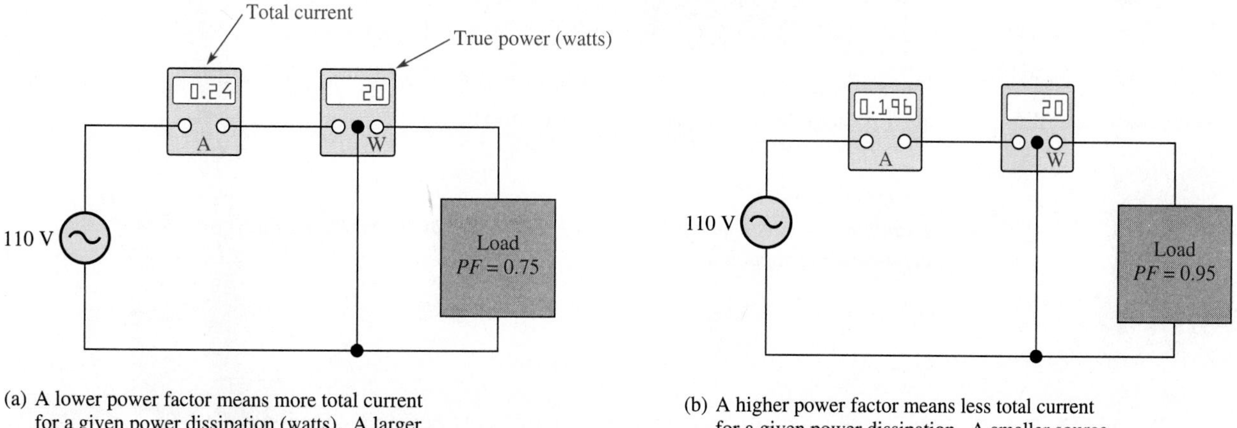

(a) A lower power factor means more total current for a given power dissipation (watts). A larger source is required to deliver the true power (watts).

(b) A higher power factor means less total current for a given power dissipation. A smaller source can deliver the same true power (watts).

FIGURE 13–28

Illustration of the effect of the power factor on system requirements such as source rating (VA) and conductor size.

tance and resistance in parallel. Part (a) shows a load with a relatively low power factor (0.75), and part (b) shows a load with a relatively high power factor (0.95). Both loads dissipate equal amounts of power as indicated by the wattmeters. Thus, an equal amount of work is done on both loads.

Although both loads are equivalent in terms of the amount of work done (true power), the low-power factor load in Figure 13–28(a) draws more current from the source than does the high-power factor load in Figure 13–28(b), as indicated by the ammeters. Therefore, the source in part (a) must have a higher VA rating than the one in part (b). Also, the lines connecting the source to the load must be a larger wire gage than those in part (b), a condition that becomes significant when very long transmission lines are required, such as in power distribution.

Figure 13–28 has demonstrated that a higher power factor is an advantage in delivering power more efficiently to a load.

Power Factor Correction

The power factor of an inductive load can be increased by the addition of a capacitor in parallel, as shown in Figure 13–29. The capacitor compensates for the phase lag of the total current by creating a capacitive component of current that is 180° out of phase with the inductive component. This has a canceling effect and reduces the phase angle, thus increasing the power factor as well as the total current, as illustrated in the figure.

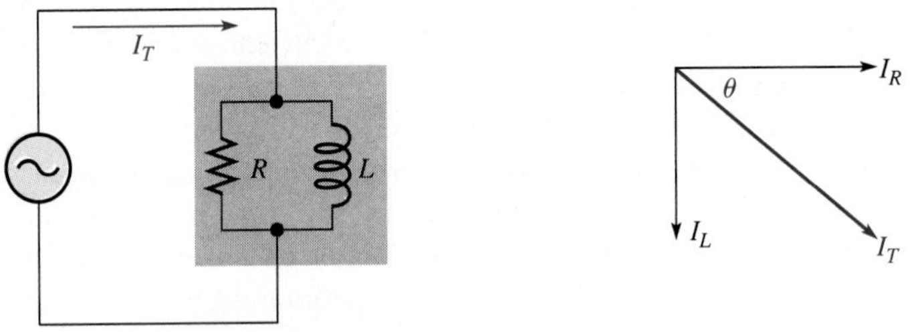

(a) Total current is the resultant of I_R and I_L.

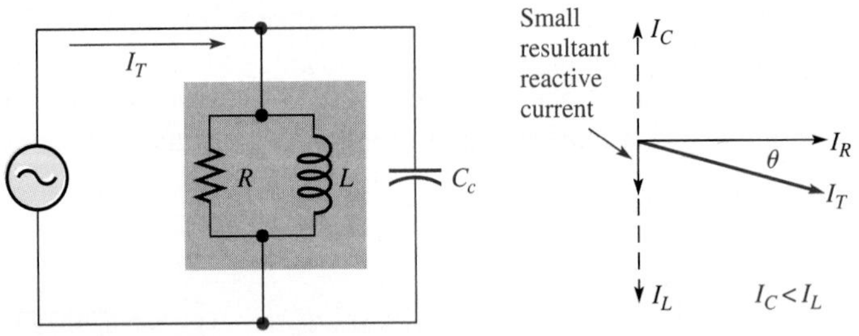

(b) I_C subtracts from I_L, leaving only a small reactive current, thus decreasing I_T and the phase angle.

FIGURE 13–29
Example of how the power factor can be increased by the addition of a compensating capacitor.

SECTION 13–7 REVIEW

1. To which component in an *RL* circuit is the power dissipation due?
2. Calculate the power factor when $\theta = 50°$.
3. A certain series *RL* circuit consists of a 470 Ω resistor and an inductive reactance of 620 Ω at the operating frequency. Determine P_{true}, P_r, and P_a when $I = 100$ mA.

13–8 ■ BASIC APPLICATIONS

Two basic applications, phase shift networks and frequency-selective networks (filters), are covered in this section.

After completing this section, you should be able to

- **Discuss some basic *RL* applications**
  - ☐ Discuss and analyze the *RL* lag network
  - ☐ Discuss and analyze the *RL* lead network
  - ☐ Describe how the *RL* circuit operates as a filter

The *RL* Lag Network

The *RL* lag network is a phase shift circuit in which the output voltage lags the input voltage by a specified angle, ϕ. A basic *RL* lag network is shown in Figure 13–30(a). Notice that the output voltage is taken across the resistor and the input voltage is the total voltage applied across the circuit. The relationship of the voltages is shown in the phasor diagram in Figure 13–30(b), and a waveform diagram is shown in Figure 13–30(c). Notice that the output voltage, V_R, lags V_{in} by an angle, designated ϕ, that is the same as the circuit phase angle. The angles are equal, of course, because V_R and I are in phase with each other.

The phase lag, ϕ, can be expressed as

$$\phi = \tan^{-1}\left(\frac{X_L}{R}\right) \qquad \textbf{(13–14)}$$

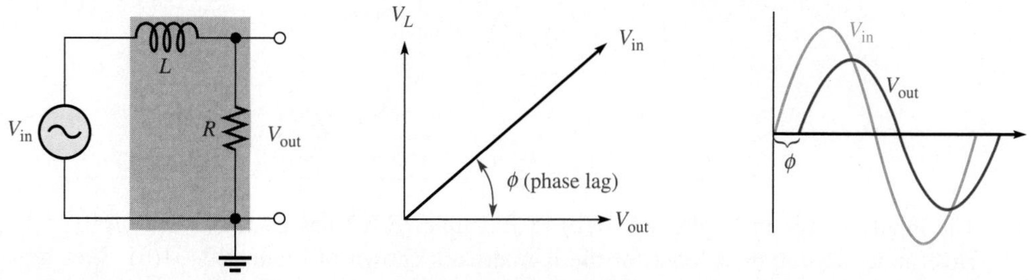

(a) A basic *RL* lag network

(b) Phasor voltage diagram showing phase lag between V_{in} and V_{out}

(c) Input and output waveforms

FIGURE 13–30
The RL lag network ($V_{out} = V_R$).

EXAMPLE 13–12 Calculate the phase lag for each circuit in Figure 13–31.

FIGURE 13–31

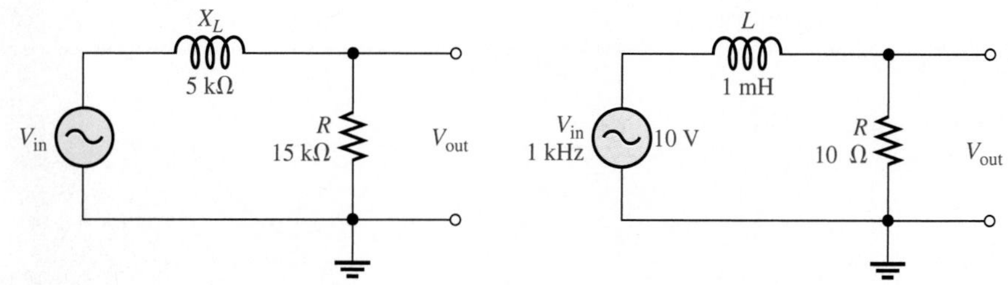

(a)

(b)

Solution For the circuit in Figure 13–31(a),

$$\phi = \tan^{-1}\left(\frac{X_L}{R}\right) = \tan^{-1}\left(\frac{5\text{ k}\Omega}{15\text{ k}\Omega}\right) = 18.4°$$

The output lags the input by 18.4°.

For the circuit in Figure 13–31(b), first determine the inductive reactance:

$$X_L = 2\pi fL = 2\pi(1\text{ kHz})(1\text{ mH}) = 6.28\ \Omega$$

The phase lag is

$$\phi = \tan^{-1}\left(\frac{X_L}{R}\right) = \tan^{-1}\left(\frac{6.28\ \Omega}{10\ \Omega}\right) = 32.1°$$

The output lags the input by 32.1°.

Related Exercise In a certain lag network, $R = 5.6\text{ k}\Omega$ and $X_L = 3.5\text{ k}\Omega$. Determine the phase lag between input and output.

The phase-lag network can be considered as a voltage divider with a portion of the input voltage dropped across L and a portion across R. The output voltage can be determined with the following formula:

$$V_{out} = \left(\frac{R}{\sqrt{R^2 + X_L^2}}\right)V_{in} \quad \textbf{(13–15)}$$

EXAMPLE 13–13

The input voltage in Figure 13–31(b) of Example 13–12 has an rms value of 10 V. Determine the output voltage for the lag network shown in Figure 13–31(b). Sketch the waveform relationships for the input and output voltages. The phase lag (32.1°) and X_L (6.28 Ω) were found in Example 13–12.

Solution Use Equation (13–15) to determine the output voltage for the lag network in Figure 13–31(b).

$$V_{out} = \left(\frac{R}{\sqrt{R^2 + X_L^2}}\right)V_{in} = \left(\frac{10\ \Omega}{11.8\ \Omega}\right)10\text{ V} = 8.47\text{ V rms}$$

The waveforms are shown in Figure 13–32.

FIGURE 13–32

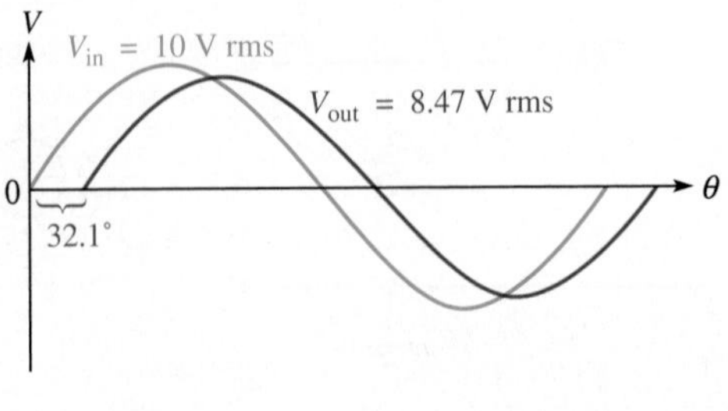

The calculator sequence is

[1] [0] [÷] [(] [(] [1] [0] [x^2] [+] [6] [•] [2] [8] [x^2] [)] [2nd] [$\sqrt{x}$] [)] [=] [×] [1] [0] [=]

Related Exercise In a lag network, $R = 4.7\text{ k}\Omega$ and $X_L = 6\text{ k}\Omega$. If the rms input voltage is 20 V, what is the output voltage?

Effects of Frequency on the Lag Network Since the circuit phase angle and the phase lag are the same, an increase in frequency causes an increase in phase lag. Also, an increase in frequency causes a decrease in the magnitude of the output voltage because X_L becomes greater and more of the total voltage is dropped across the inductor and less across the resistor. Figure 13–33 illustrates the changes in phase lag and output voltage magnitude with frequency.

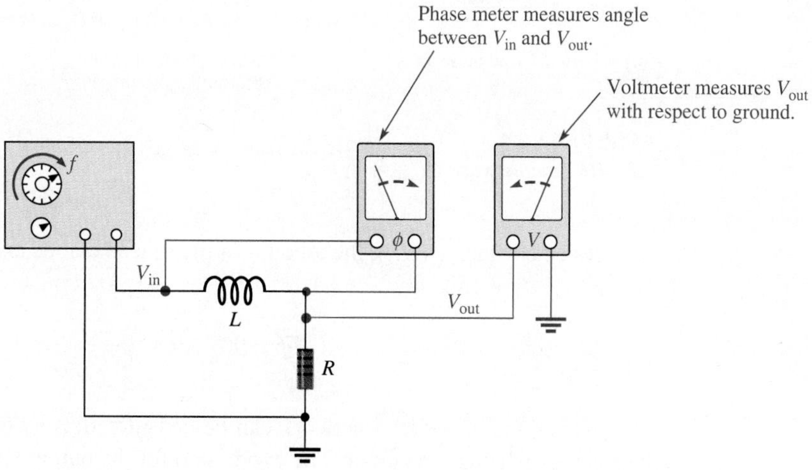

(a) When f increases, the phase lag ϕ increases and V_{out} decreases.

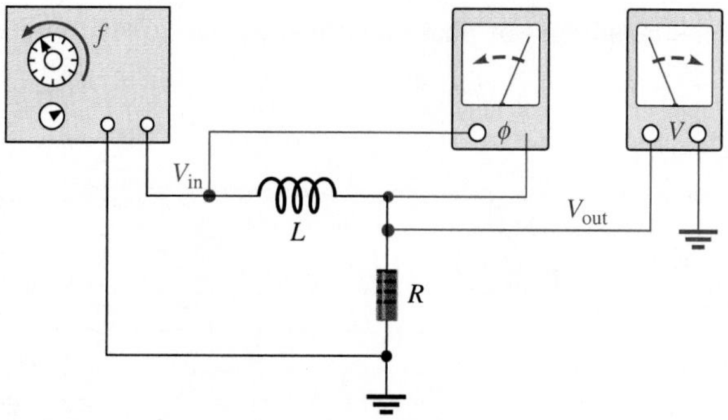

(b) When f decreases, ϕ decreases and V_{out} increases.

FIGURE 13–33

Illustration of how the frequency affects the phase lag and the output voltage in an RL lag network with the amplitude of V_{in} held constant.

The *RL* Lead Network

The *RL* lead network is a phase shift circuit in which the output voltage leads the input voltage by a specified angle, ϕ. A basic *RL* lead network is shown in Figure 13–34(a). Notice how this network differs from the lag network. Here the output voltage is taken across the inductor rather than across the resistor. The relationship of the voltages is shown in the phasor diagram of Figure 13–34(b) and in the waveform plot of Figure 13–34(c). Notice that the output voltage, V_L, leads V_{in} by an angle (phase lead) that is the difference between 90° and the circuit phase angle θ.

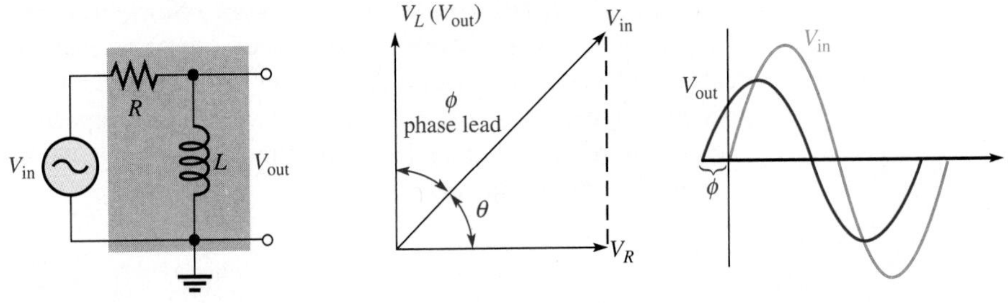

(a) A basic *RL* lead network

(b) Phasor voltage diagram showing V_{out} leading V_{in}

(c) Input and output waveforms

FIGURE 13–34
The RL lead network ($V_{out} = V_L$).

Since $\theta = \tan^{-1}(X_L/R)$, the amount of phase lead can be expressed as follows:

$$\phi = 90^\circ - \tan^{-1}\left(\frac{X_L}{R}\right) \tag{13–16}$$

Again, the phase-lead network can be considered as a voltage divider with the voltage across *L* being the output. The expression for the output voltage is

$$V_{out} = \left(\frac{X_L}{\sqrt{R^2 + X_L^2}}\right)V_{in} \tag{13–17}$$

EXAMPLE 13–14 Determine the amount of phase lead and output voltage in the lead network in Figure 13–35.

FIGURE 13–35

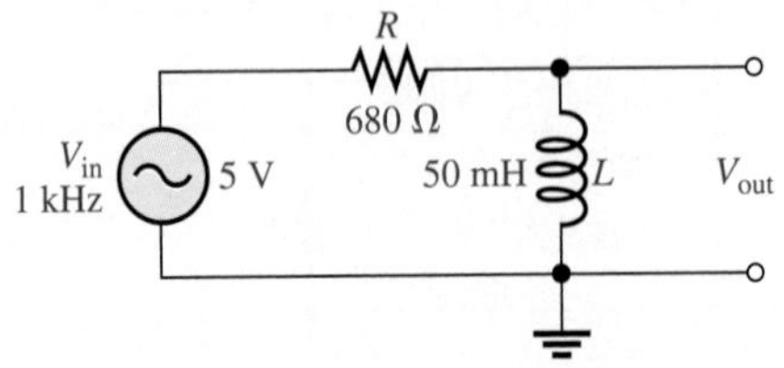

Solution First determine the inductive reactance:

$$X_L = 2\pi fL = 2\pi(1\text{ kHz})(50\text{ mH}) = 314\ \Omega$$

The phase lead is

$$\phi = 90° - \tan^{-1}\left(\frac{X_L}{R}\right) = 90° - \tan^{-1}\left(\frac{314\ \Omega}{680\ \Omega}\right) = 65.2°$$

The output leads the input by 65.2°.

The output voltage is

$$V_{out} = \left(\frac{X_L}{\sqrt{R^2 + X_L^2}}\right)V_{in} = \left(\frac{314\ \Omega}{\sqrt{(680\ \Omega)^2 + (314\ \Omega)^2}}\right)5\ \text{V} = 2.1\ \text{V}$$

Related Exercise In a certain lead network, $R = 2.2\ \text{k}\Omega$ and $X_L = 1\ \text{k}\Omega$. What is the phase lead?

Effects of Frequency on the Lead Network Since the circuit phase angle, θ, increases as frequency increases, the phase lead between the input and the output voltages decreases. You can see this relationship by examining Equation (13–16). Also, the amplitude of the output voltage increases as the frequency increases because X_L becomes greater and more of the total input voltage is dropped across the inductor. Figure 13–36 illustrates the changes in the phase lead and output voltage with frequency.

FIGURE 13–36

Illustration of how the frequency affects the phase lead and output voltage in an RL lead network with the amplitude of V_{in} held constant.

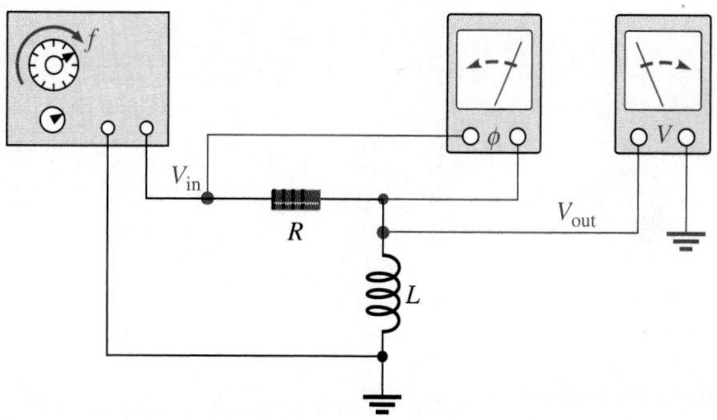

(a) When f increases, the phase lead ϕ decreases and V_{out} increases.

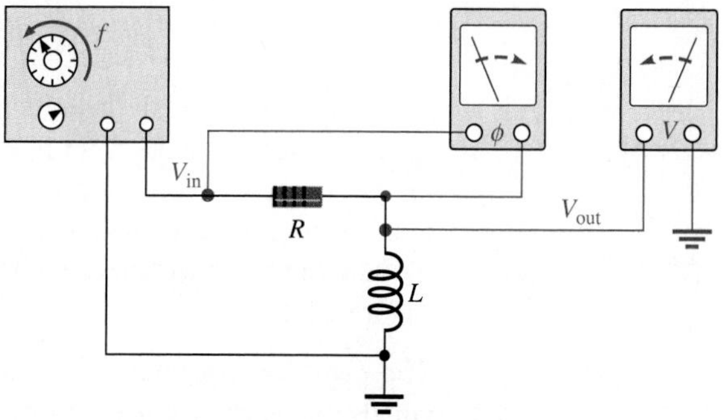

(b) When f decreases, ϕ increases and V_{out} decreases.

The *RL* Circuit As a Filter

Like *RC* circuits, series *RL* circuits also exhibit a frequency-selective characteristic and therefore act as basic filters.

Low-Pass Filter You have seen what happens to the phase and the output voltage in the lag network. In terms of its filtering action, the change in the amplitude of the output voltage with frequency is the primary consideration.

To illustrate low-pass filter action, Figure 13–37 shows a specific series of measurements in which the frequency starts at zero (dc) and is increased in increments up to 20 kHz. At each value of frequency, the output voltage is measured. As you can see, the inductive reactance increases as frequency goes up, thus causing less voltage to be dropped across the resistor while the input voltage is held at a constant 10 V throughout each step. The response curve for these particular values would appear the same as the response curve in Figure 12–48 for the *RC* low-pass filter.

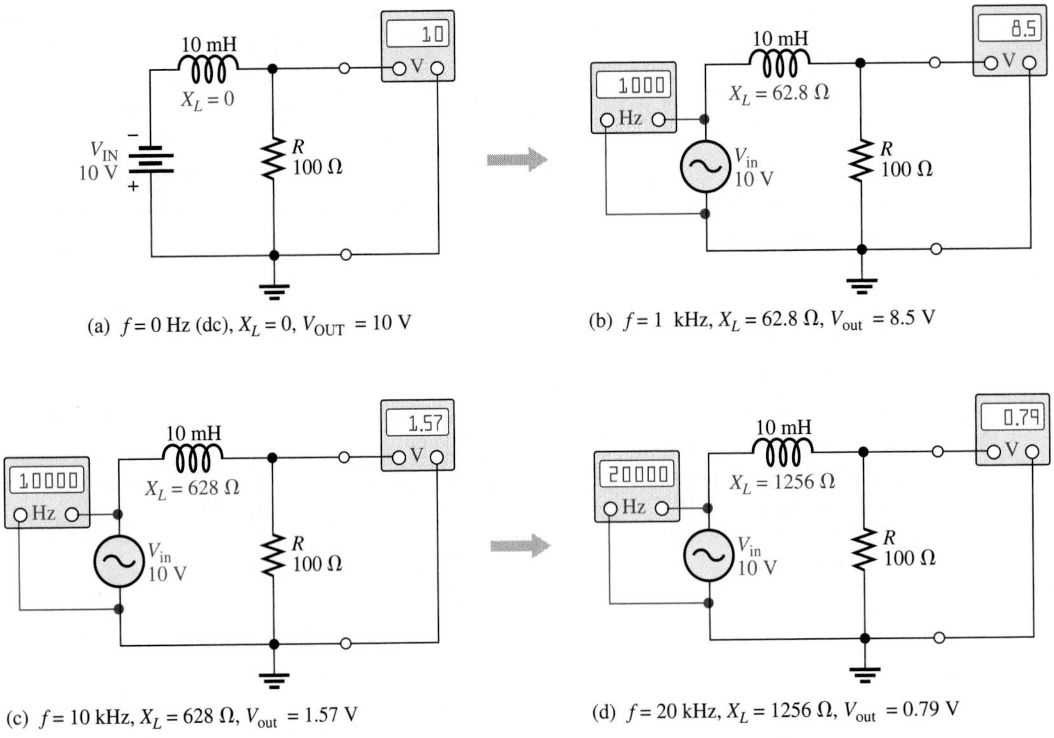

(a) $f = 0$ Hz (dc), $X_L = 0$, $V_{OUT} = 10$ V

(b) $f = 1$ kHz, $X_L = 62.8\ \Omega$, $V_{out} = 8.5$ V

(c) $f = 10$ kHz, $X_L = 628\ \Omega$, $V_{out} = 1.57$ V

(d) $f = 20$ kHz, $X_L = 1256\ \Omega$, $V_{out} = 0.79$ V

FIGURE 13–37
Example of low-pass filter action. Winding resistance has been neglected.

High-Pass Filter To illustrate *RL* high-pass filter action, Figure 13–38 shows a series of specific measurements. The frequency starts at zero (dc) and is increased in increments up to 10 kHz. As you can see, the inductive reactance increases as the frequency goes up, thus causing more voltage to be dropped across the inductor. Again, when the values are plotted, the response curve is the same as the one for the *RC* high-pass filter as shown in Figure 12–50.

The Cutoff Frequency of an RL Filter The frequency at which the inductive reactance equals the resistance in a low-pass or high-pass *RL* filter is called the *cutoff frequency* and is designated f_c. This condition is expressed as $2\pi f_c L = R$. Solving for f_c, you get the following formula:

$$f_c = \frac{R}{2\pi L} \tag{13–18}$$

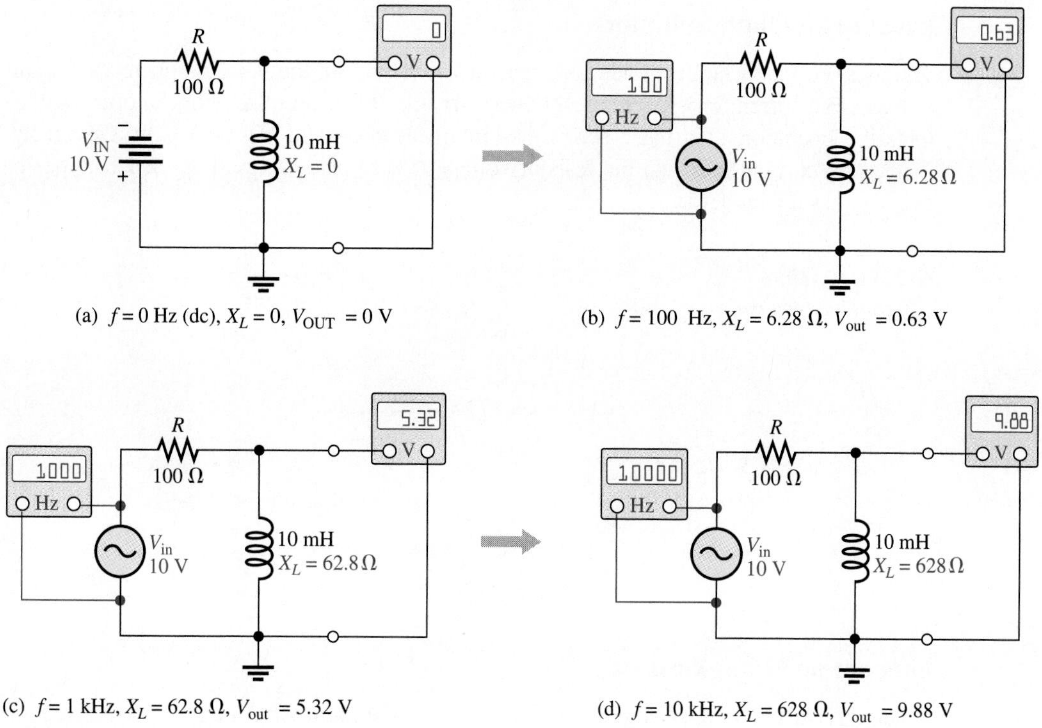

FIGURE 13–38
Example of high-pass filter action. Winding resistance has been neglected.

As with the *RC* filter, the output voltage is 70.7% of its maximum value at f_c. In a high-pass filter, all frequencies above f_c are considered to be passed by the filter, and all those below f_c are considered to be rejected. The reverse, of course, is true for a low-pass filter. Filter bandwidth, defined in Chapter 12, applies to both *RC* and *RL* filter circuits.

SECTION 13–8 REVIEW

1. A certain *RL* lead network consists of a 3.3 kΩ resistor and a 15 mH inductor. Determine the phase lead between input and output at a frequency of 5 kHz.
2. An *RL* lag network has the same component values as the lead network in Question 1. What is the magnitude of the output voltage at 5 kHz when the input is 10 V rms?
3. When an *RL* circuit is used as a low-pass filter, across which component is the output taken?

13–9 ■ TROUBLESHOOTING

In this section, the effects that typical component failures have on the response of basic RL circuits are considered.

After completing this section, you should be able to

- **Troubleshoot *RL* circuits**
  - □ Find an open inductor
  - □ Find an open resistor
  - □ Find an open in a parallel circuit
  - □ Find an inductor with shorted windings

Effect of an Open Inductor

The most common failure mode for inductors occurs when the winding opens as a result of excessive current or a mechanical contact failure. It is easy to see how an open coil affects the operation of a basic series *RL* circuit, as shown in Figure 13–39. Obviously, there is no current path—so the resistor voltage is zero, and the total applied voltage appears across the inductor.

FIGURE 13–39
Effect of an open coil.

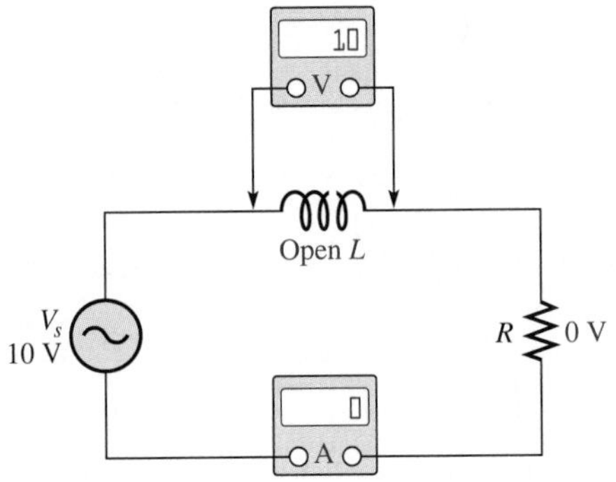

Effect of an Open Resistor

When the resistor is open, there is no current, and the inductor voltage is zero. The total input voltage is across the open resistor, as shown in Figure 13–40.

FIGURE 13–40
Effect of an open resistor.

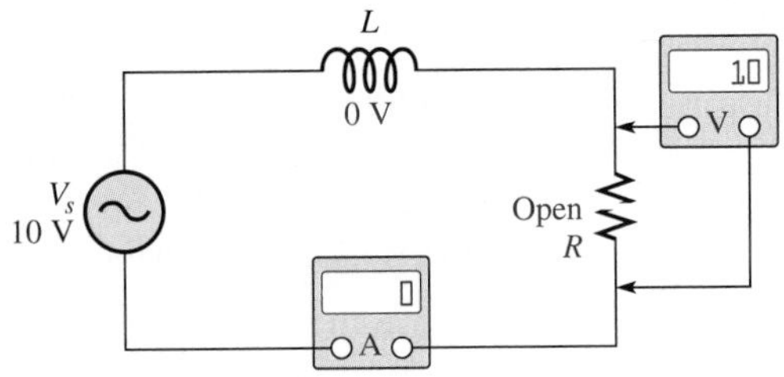

Open Components in Parallel Circuits

In a parallel *RL* circuit, an open resistor or inductor will cause the total current to decrease, because the total impedance will increase. Obviously, the branch with the open component will have zero current. Figure 13–41 illustrates these conditions.

FIGURE 13–41
Effect of an open component in a parallel circuit with V_s constant.

Effect of an Inductor with Shorted Windings

It is possible for some of the windings of coils to short together as a result of damaged insulation. This failure mode is much less likely than the open coil. Shorted windings result in a reduction in inductance because the inductance of a coil is proportional to the square of the number of turns.

SECTION 13–9 REVIEW

1. Describe the effect of an inductor with shorted windings on the response of a series *RL* circuit.
2. In the circuit of Figure 13–42, indicate whether I_T, V_{R1} and V_{R2} increase or decrease as a result of L opening.

FIGURE 13–42

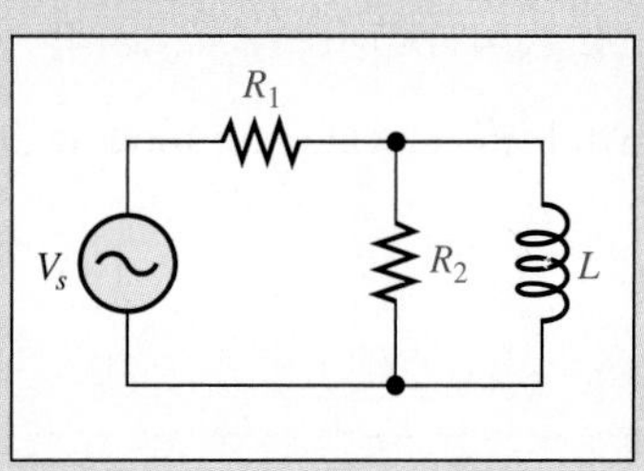

13–10 ■ APPLICATION ASSIGNMENT

You are given two sealed modules that have been removed from a communications system that is being modified. Each module has three terminals and are labeled as RL filters, but no specifications are given. Your assignment is to test the modules to determine the type of filters and the component values.

The sealed modules have three terminals labeled IN, GND, and OUT as shown in Figure 13–43. You will apply your knowledge of series *RL* circuits and some basic measurements to determine the internal circuit configuration and the component values.

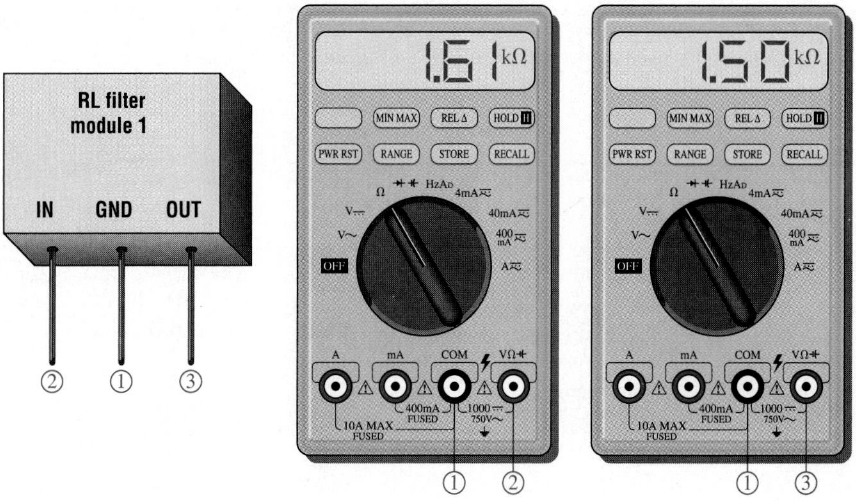

FIGURE 13–43
Ohmmeter measurements of module 1.

Step 1: Ohmmeter Measurements of Module 1

Determine the arrangement of the two components and the values of the resistor and winding resistance for module 1 indicated by the meter readings in Figure 13–43.

Step 2: AC Measurements of Module 1

Determine the inductance value for module 1 indicated by the test setup in Figure 13–44.

Step 3: Ohmmeter Measurements of Module 2

Determine the arrangement of the two components and the values of the resistor and the winding resistance for module 2 indicated by the meter readings in Figure 13–45.

Step 4: AC Measurements of Module 2

Determine the inductance value for module 2 indicated by the test setup in Figure 13–46.

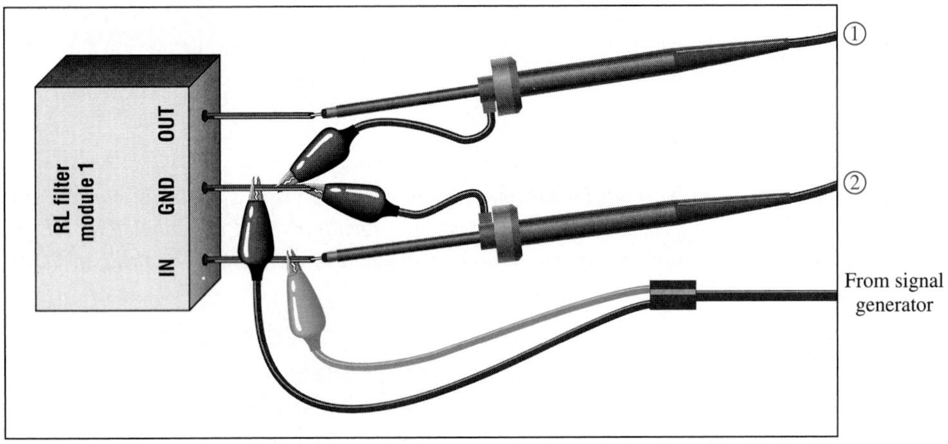

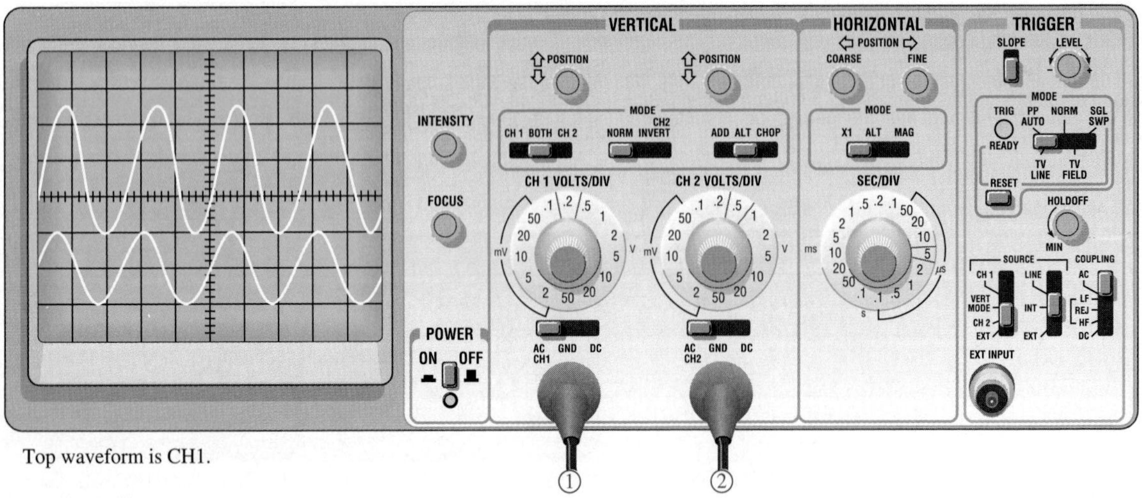

Top waveform is CH1.

FIGURE 13–44
AC measurements for module 1.

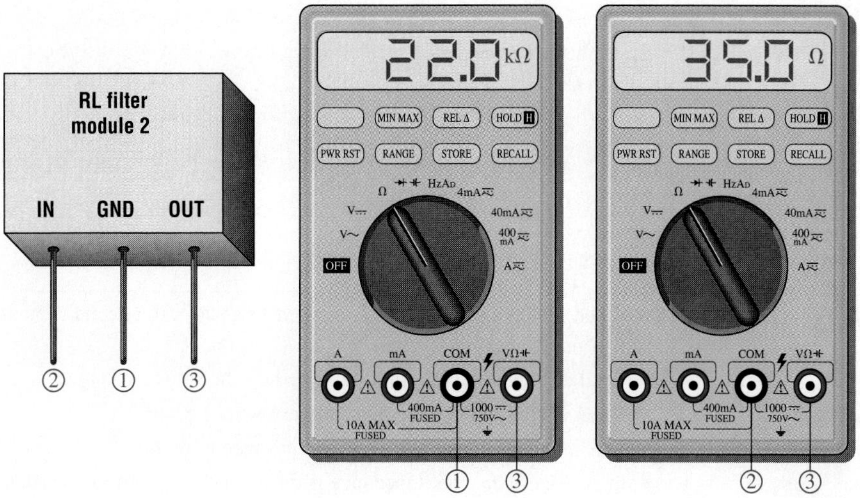

FIGURE 13–45
Ohmmeter measurements of module 2.

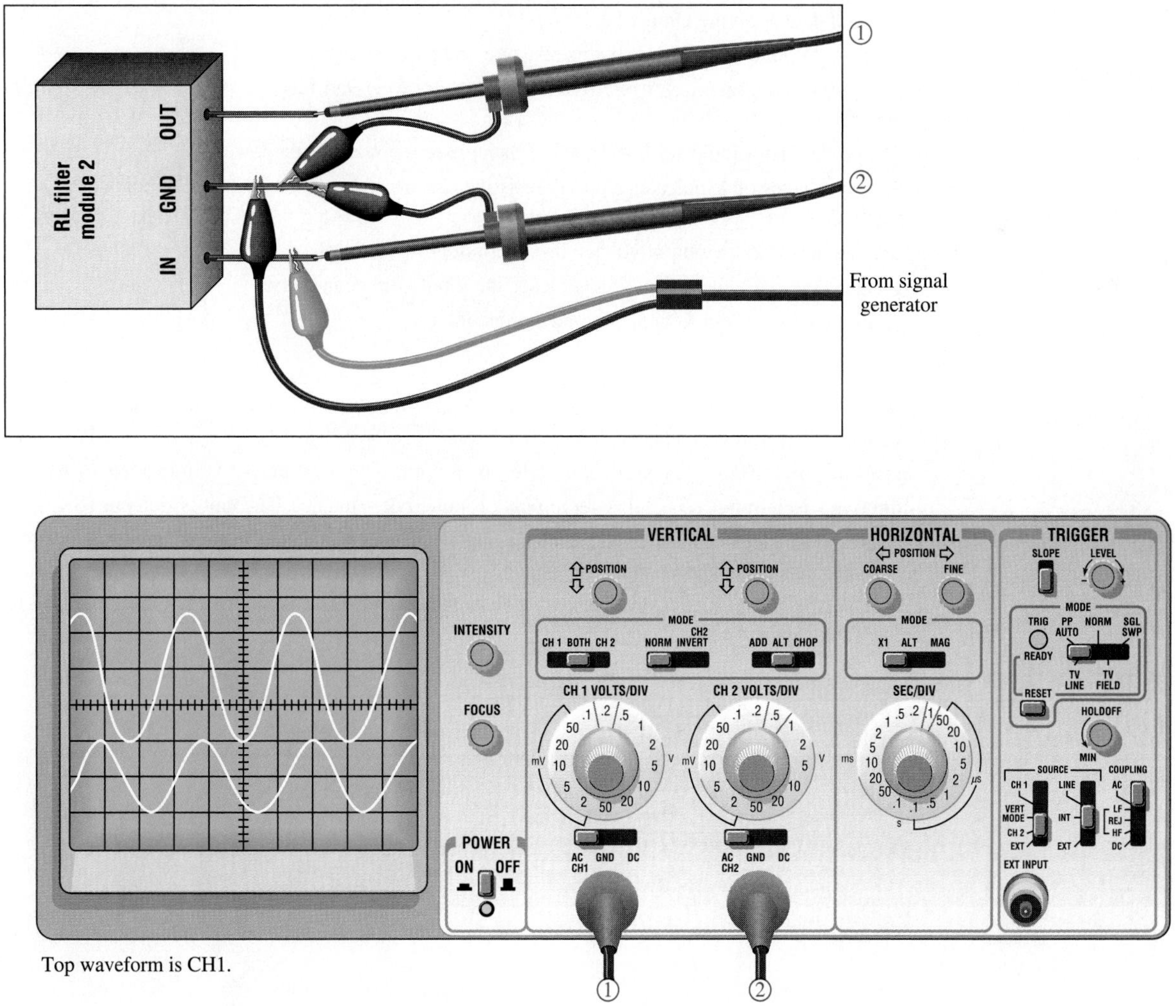

FIGURE 13–46
AC measurements for module 2.

SECTION 13–10 REVIEW

1. If the inductor in module 1 were open, what would you measure on the output with the test setup of Figure 13–44?
2. If the inductor in module 2 were open, what would you measure on the output with the test setup of Figure 13–46?

■ SUMMARY

- When a sine wave voltage is applied to an *RL* circuit, the current and all the voltage drops are also sine waves.
- Total current in an *RL* circuit always lags the source voltage.
- The resistor voltage is always in phase with the current.
- The inductor voltage always leads the current by 90°.
- In an *RL* circuit, the impedance is determined by both the resistance and the inductive reactance combined.
- Impedance is expressed in units of ohms.
- The impedance of an *RL* circuit varies directly with frequency.
- The phase angle (θ) of a series *RL* circuit varies directly with frequency.
- You can determine the impedance of a circuit by measuring the applied voltage and the total current and then applying Ohm's law.
- In an *RL* circuit, part of the power is resistive and part reactive.
- The phasor combination of resistive power (true power) and reactive power is called *apparent power.*
- The power factor indicates how much of the apparent power is true power.
- A power factor of 1 indicates a purely resistive circuit, and a power factor of 0 indicates a purely reactive circuit.
- In a lag network, the output voltage lags the input voltage in phase.
- In a lead network, the output voltage leads the input voltage in phase.
- A filter passes certain frequencies and rejects others.

■ GLOSSARY

Admittance (*Y*) A measure of the ability of a reactive circuit to permit current; the reciprocal of impedance. The unit is siemens (S).

Apparent power (P_a) The product of volts and amperes, with the unit of volt-amperes (VA).

Inductive susceptance (B_L) The reciprocal of inductive reactance. The unit is siemens (S).

Inductor An electrical device formed by a coil of wire around a core material having the property of inductance; also known as a choke or coil.

Power factor The relationship between volt-amperes and watts. Volt-amperes multiplied by the power factor equals watts.

■ FORMULAS

(13–1)	$Z = \sqrt{R^2 + X_L^2}$	Series *RL* impedance
(13–2)	$\theta = \tan^{-1}\left(\frac{X_L}{R}\right)$	Series *RL* phase angle
(13–3)	$V_s = \sqrt{V_R^2 + V_L^2}$	Total voltage in a series *RL* circuit
(13–4)	$\theta = \tan^{-1}\left(\frac{V_L}{V_R}\right)$	Series *RL* phase angle
(13–5)	$Z = \frac{RX_L}{\sqrt{R^2 + X_L^2}}$	Parallel *RL* impedance
(13–6)	$\theta = \tan^{-1}\left(\frac{R}{X_L}\right)$	Parallel *RL* phase angle

(13–7)	$G = \dfrac{1}{R}$	Conductance
(13–8)	$B_L = \dfrac{1}{X_L}$	Inductive susceptance
(13–9)	$Y = \dfrac{1}{Z}$	Admittance
(13–10)	$Y_T = \sqrt{G^2 + B_L^2}$	Total admittance
(13–11)	$I_T = \sqrt{I_R^2 + I_L^2}$	Total current in parallel *RL* circuit
(13–12)	$\theta = \tan^{-1}\left(\dfrac{I_L}{I_R}\right)$	Parallel *RL* phase angle
(13–13)	$P_r = I^2X_L$	Reactive power
(13–14)	$\phi = \tan^{-1}\left(\dfrac{X_L}{R}\right)$	Phase lag
(13–15)	$V_{out} = \left(\dfrac{R}{\sqrt{R^2 + X_L^2}}\right)V_{in}$	Output of lag network
(13–16)	$\phi = 90° - \tan^{-1}\left(\dfrac{X_L}{R}\right)$	Phase lead
(13–17)	$V_{out} = \left(\dfrac{X_L}{\sqrt{R^2 + X_L^2}}\right)V_{in}$	Output of lead network
(13–18)	$f_c = \dfrac{R}{2\pi L}$	Cutoff frequency of *RL* filter

▪ SELF-TEST

1. In a series *RL* circuit, the resistor voltage
   (a) leads the applied voltage **(b)** lags the applied voltage
   (c) is in phase with the applied voltage **(d)** is in phase with the current
   (e) both (a) and (d) **(f)** both (b) and (d)
2. When the frequency of the voltage applied to a series *RL* circuit is increased, the impedance
   (a) decreases **(b)** increases **(c)** does not change
3. When the frequency of the voltage applied to a series *RL* circuit is decreased, the phase angle
   (a) decreases **(b)** increases **(c)** does not change
4. If the frequency is doubled and the resistance is halved, the impedance of a series *RL* circuit
   (a) doubles **(b)** halves **(c)** remains constant **(d)** cannot be determined without values
5. To reduce the current in a series *RL* circuit, the frequency should be
   (a) increased **(b)** decreased **(c)** constant
6. In a series *RL* circuit, 10 V rms is measured across the resistor, and 10 V rms is measured across the inductor. The peak value of the source voltage is
   (a) 14.14 V **(b)** 28.28 V **(c)** 10 V **(d)** 20 V
7. The voltages in Question 6 are measured at a certain frequency. To make the resistor voltage greater than the inductor voltage, the frequency is
   (a) increased **(b)** decreased **(c)** doubled **(d)** not a factor
8. When the resistor voltage in a series *RL* circuit becomes greater than the inductor voltage, the phase angle
   (a) increases **(b)** decreases **(c)** is not affected
9. When the frequency is increased, the impedance of a parallel *RL* circuit
   (a) increases **(b)** decreases **(c)** remains constant
10. In a parallel *RL* circuit, there are 2 A rms in the resistive branch and 2 A rms in the inductive branch. The total rms current is
    (a) 4 A **(b)** 5.656 A **(c)** 2 A **(d)** 2.828 A
11. You are observing two voltage waveforms on an oscilloscope. The time base (sec/div) of the scope is adjusted so that one half-cycle of the waveforms covers the ten horizontal divisions. The positive-going zero crossing of one waveform is at the leftmost division, and the positive-

going zero crossing of the other is three divisions to the right. The phase angle between these two waveforms is
(a) 18° **(b)** 36° **(c)** 54° **(d)** 180°

12. Which of the following power factors results in less energy loss in an *RL* circuit?
(a) 1 **(b)** 0.9 **(c)** 0.5 **(d)** 0.1

13. If a load is purely inductive and the reactive power is 10 VAR, the apparent power is
(a) 0 VA **(b)** 10 VA **(c)** 14.14 VA **(d)** 3.16 VA

14. For a certain load, the true power is 10 W and the reactive power is 10 VAR. The apparent power is
(a) 5 VA **(b)** 20 VA **(c)** 14.14 VA **(d)** 100 VA

15. The cutoff frequency of a certain low-pass *RL* filter is 20 kHz. The filter's bandwidth is
(a) 20 kHz **(b)** 40 kHz **(c)** 0 kHz **(d)** unknown

▪ BASIC PROBLEMS

SECTION 13–1 Sinusoidal Response of *RL* Circuits

1. A 15 kHz sinusoidal voltage is applied to a series *RL* circuit. Determine the frequency of I, V_R, and V_L.

2. What are the waveshapes of I, V_R, and V_L in Problem 1?

SECTION 13–2 Impedance and Phase Angle of Series *RL* Circuits

3. Find the impedance of each circuit in Figure 13–47.

4. Determine the impedance and phase angle in each circuit in Figure 13–48.

5. In Figure 13–49, determine the impedance at each of the following frequencies:
(a) 100 Hz **(b)** 500 Hz **(c)** 1 kHz **(d)** 2 kHz

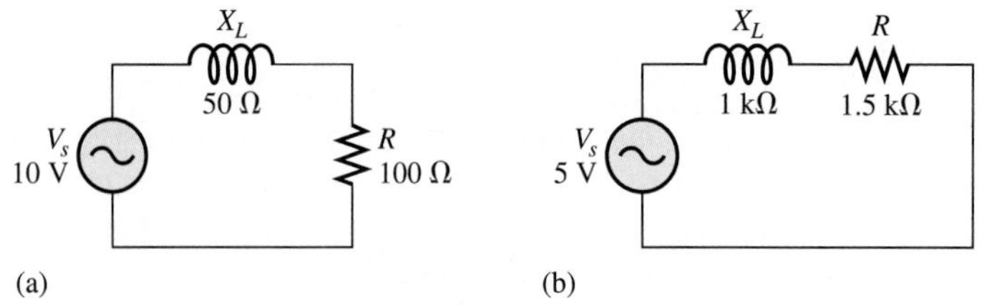

FIGURE 13–47

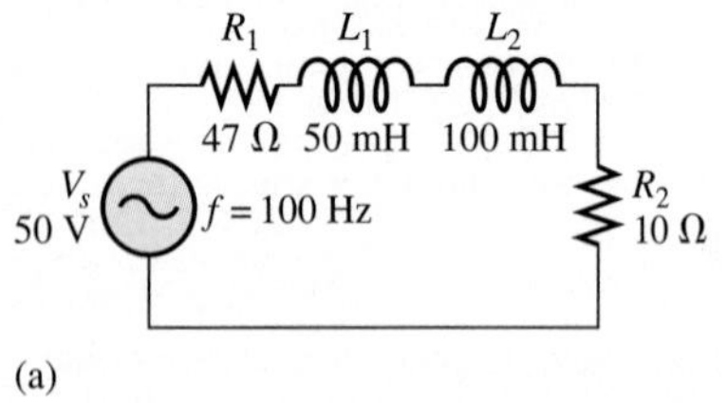

FIGURE 13–48

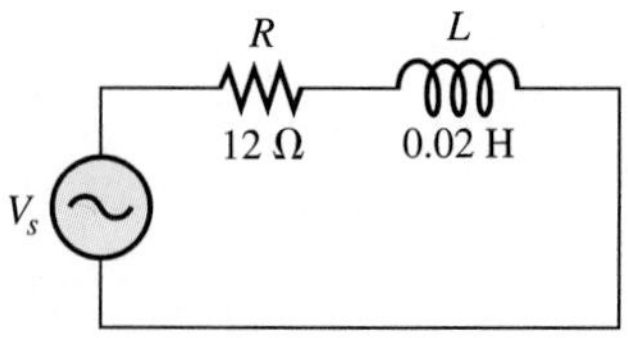

FIGURE 13–49

6. Determine the values of R and X_L in a series *RL* circuit for the following values of impedance and phase angle:
(a) $Z = 20\ \Omega$, $\theta = 45°$ **(b)** $Z = 500\ \Omega$, $\theta = 35°$
(c) $Z = 2.5\ \text{k}\Omega$, $\theta = 72.5°$ **(d)** $Z = 998\ \Omega$, $\theta = 45°$

SECTION 13–3 Analysis of Series *RL* Circuits

7. Find the current for each circuit of Figure 13–47.

8. Calculate the total current in each circuit of Figure 13–48.

9. Determine θ for the circuit in Figure 13–50.

10. If the inductance in Figure 13–50 is doubled, does θ increase or decrease, and by how many degrees?

11. Sketch the waveforms for V_s, V_R, and V_L in Figure 13–50. Show the proper phase relationships.

12. For the circuit in Figure 13–51, find V_R and V_L for each of the following frequencies:
(a) 60 Hz **(b)** 200 Hz **(c)** 500 Hz **(d)** 1 kHz

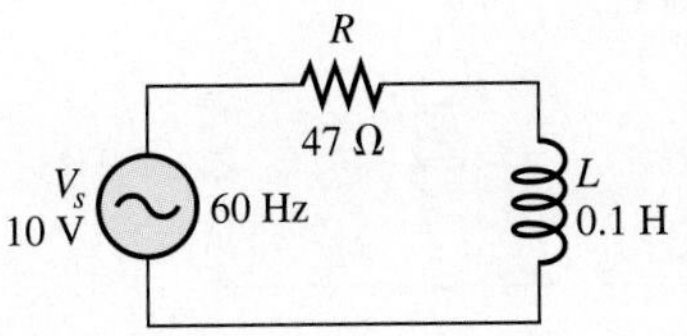

FIGURE 13–50

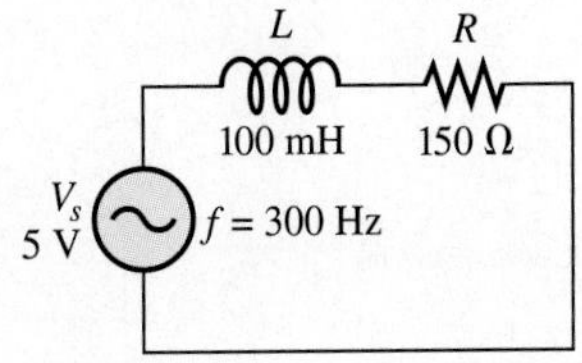

FIGURE 13–51

SECTION 13–4 Impedance and Phase Angle of Parallel *RL* Circuits

13. What is the impedance for the circuit in Figure 13–52?

14. Repeat Problem 13 for the following frequencies:
(a) 1.5 kHz **(b)** 3 kHz **(c)** 5 kHz **(d)** 10 kHz

15. At what frequency does X_L equal R in Figure 13–52?

FIGURE 13–52

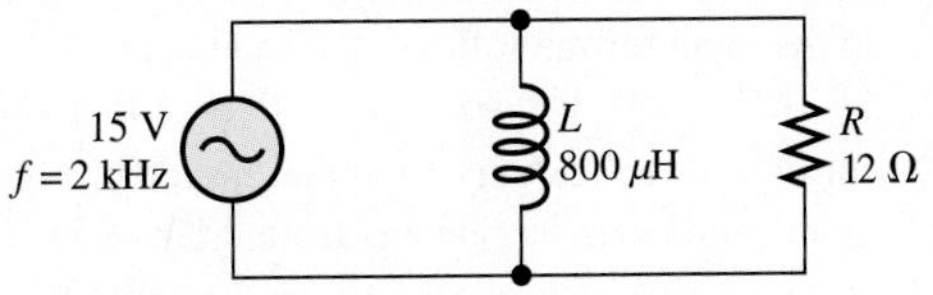

SECTION 13–5 Analysis of Parallel *RL* Circuits

16. Find the total current and each branch current in Figure 13–53.

17. Determine the following quantities in Figure 13–54:
(a) Z **(b)** I_R **(c)** I_L **(d)** I_T **(e)** θ

18. Convert the circuit in Figure 13–55 to an equivalent series form.

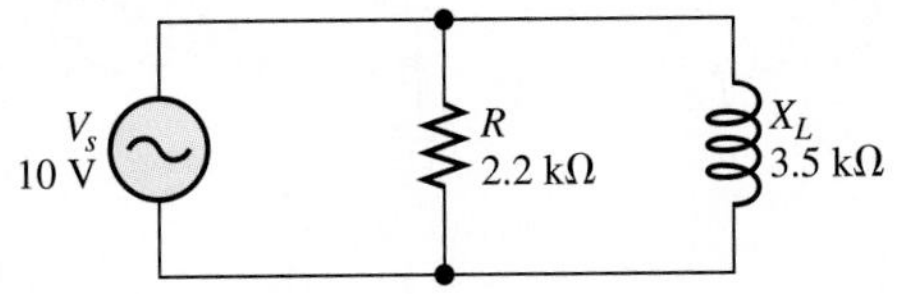

FIGURE 13–53

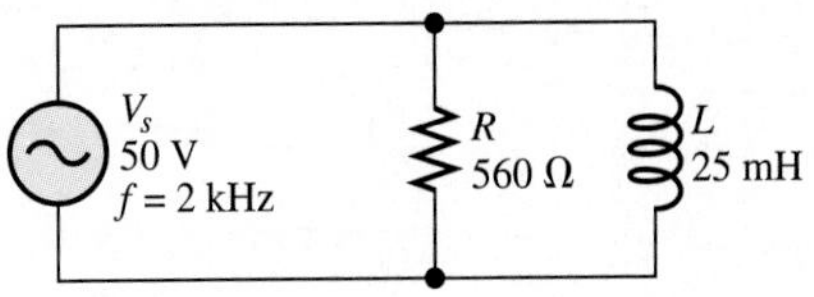

FIGURE 13–54

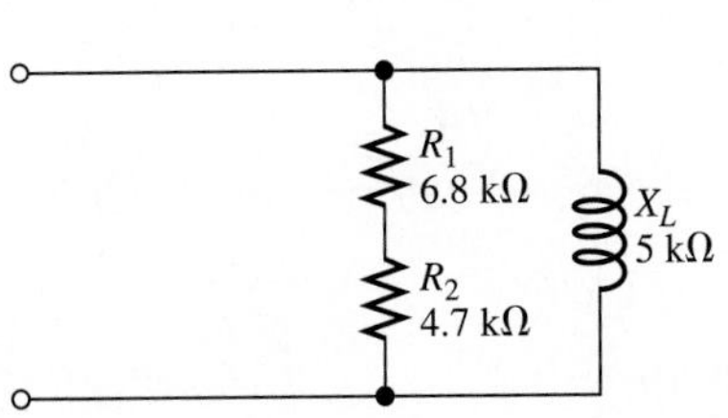

FIGURE 13–55

SECTION 13–6 Series-Parallel Analysis

19. Determine the voltage across each element in Figure 13–56.

20. Is the circuit in Figure 13–56 predominantly resistive or predominantly inductive?

21. Find the current in each branch and the total current in Figure 13–56.

FIGURE 13–56

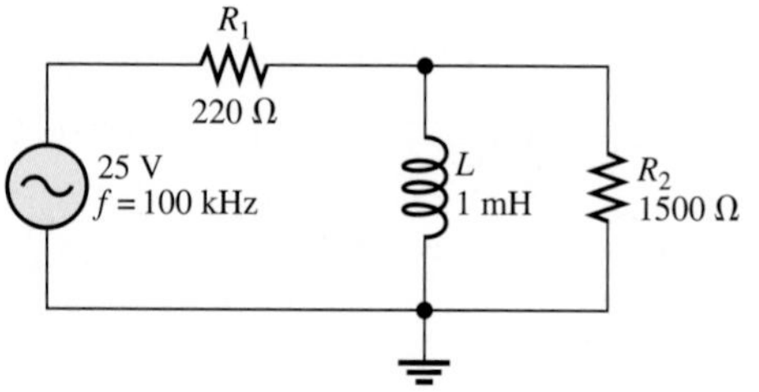

SECTION 13–7 Power in *RL* Circuits

22. In a certain *RL* circuit, the true power is 100 mW, and the reactive power is 340 mVAR. What is the apparent power?

23. Determine the true power and the reactive power in Figure 13–50.

24. What is the power factor in Figure 13–53?

25. Determine P_{true}, P_r, P_a, and *PF* for the circuit in Figure 13–56. Sketch the power triangle.

SECTION 13–8 Basic Applications

26. For the lag network in Figure 13–57, determine the phase lag of the output voltage with respect to the input for the following frequencies:
(a) 1 Hz **(b)** 100 Hz **(c)** 1 kHz **(d)** 10 kHz

27. Plot the response curve for the circuit in Figure 13–57. Show the output voltage versus frequency in 1 kHz increments from 0 Hz to 5 kHz.

28. Repeat Problem 26 for the lead network in Figure 13–58.

29. Using the same procedure as in Problem 27, plot the response curve for Figure 13–58.

30. Sketch the voltage phasor diagram for each circuit in Figures 13–57 and 13–58 for a frequency of 8 kHz.

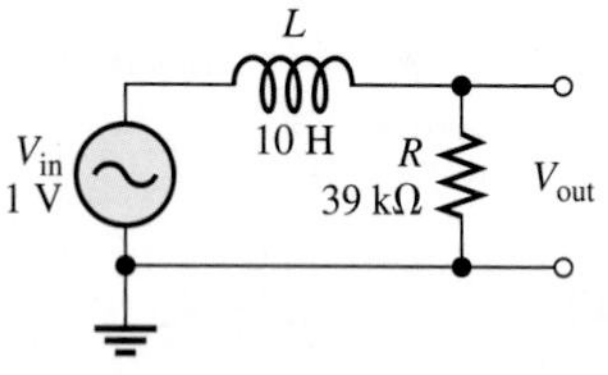

FIGURE 13–57

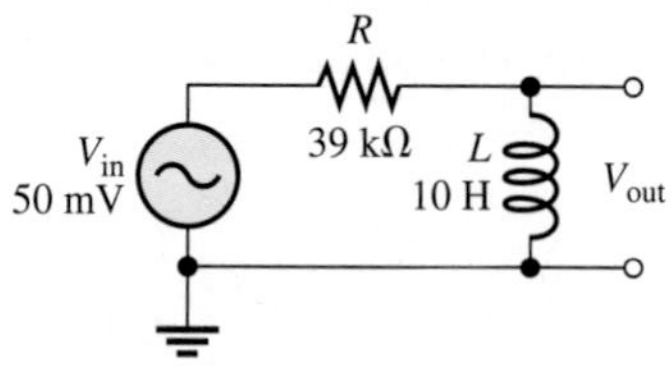

FIGURE 13–58

SECTION 13–9 Troubleshooting

31. Determine the voltage across each element in Figure 13–59 if L_1 were open.

FIGURE 13–59

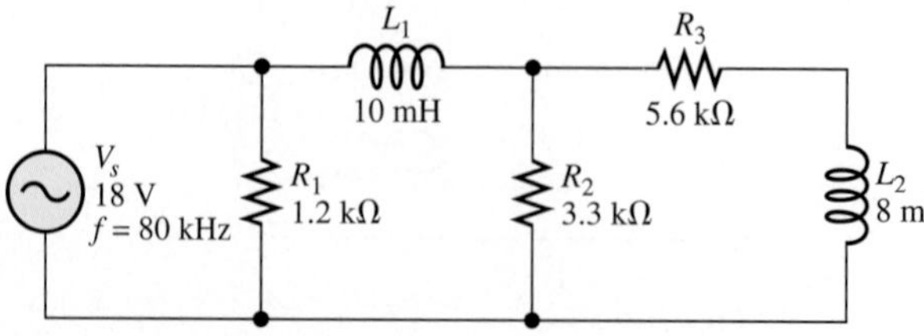

32. Determine the output voltage in Figure 13–60 for each of the following failure modes:
(a) L_1 open **(b)** L_2 open **(c)** R_1 open **(d)** a short across R_2

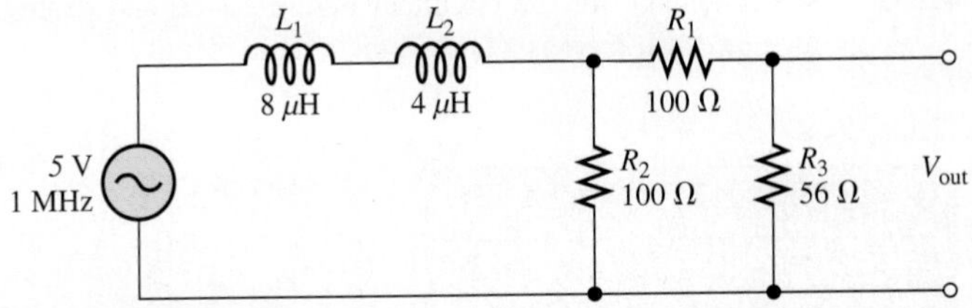

FIGURE 13–60

▪ ADVANCED PROBLEMS

33. Determine the voltage across the inductors in Figure 13–61.

34. Is the circuit in Figure 13–61 predominantly resistive or predominantly inductive?

35. Find the total current in Figure 13–61.

36. For the circuit in Figure 13–62, determine the following:
(a) Z_T **(b)** I_T **(c)** θ **(d)** V_L **(e)** V_{R3}

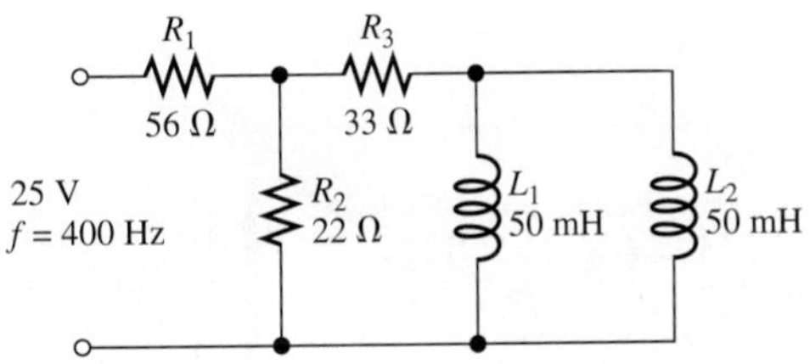

FIGURE 13–61

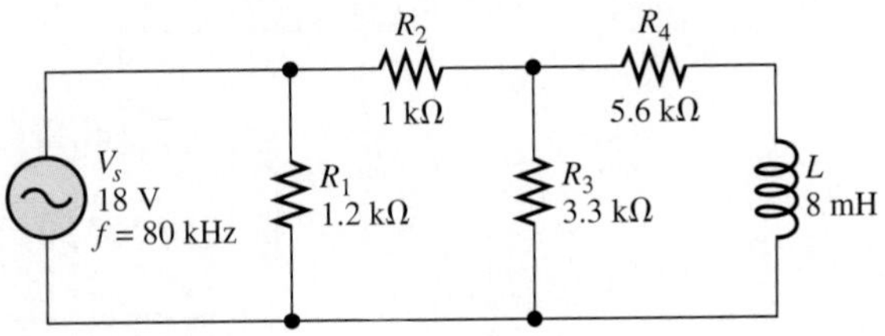

FIGURE 13–62

37. For the circuit in Figure 13–63, determine the following:
(a) I_{R1} **(b)** I_{L1} **(c)** I_{L2} **(d)** I_{R2}

38. Determine the phase shift and attenuation (ratio of V_{out} to V_{in}) from the input to the output for the network in Figure 13–64.

39. Determine the attenuation from the input to the output for the circuit in Figure 13–65.

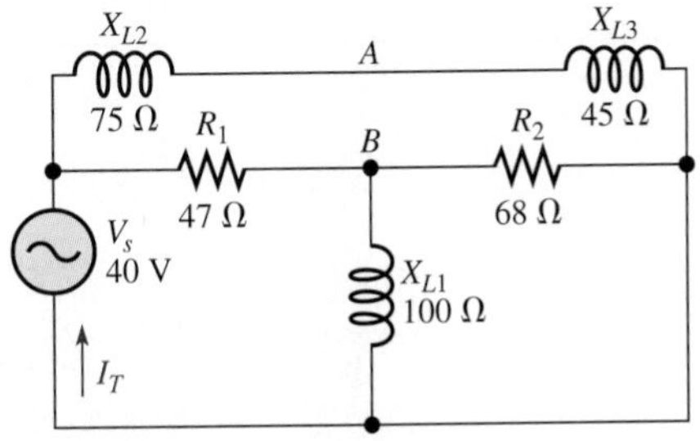

FIGURE 13–63

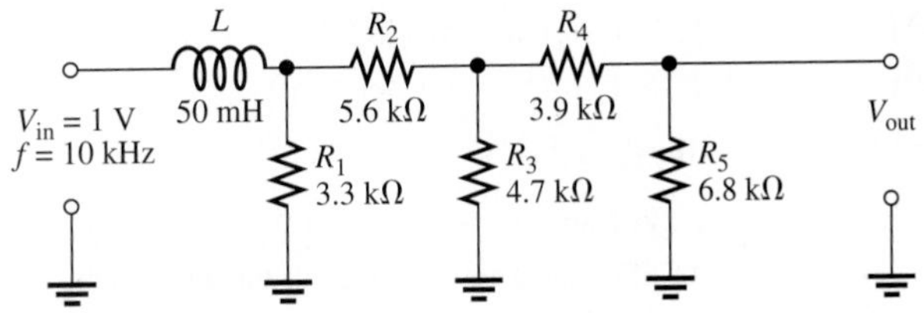

FIGURE 13–64

FIGURE 13–65

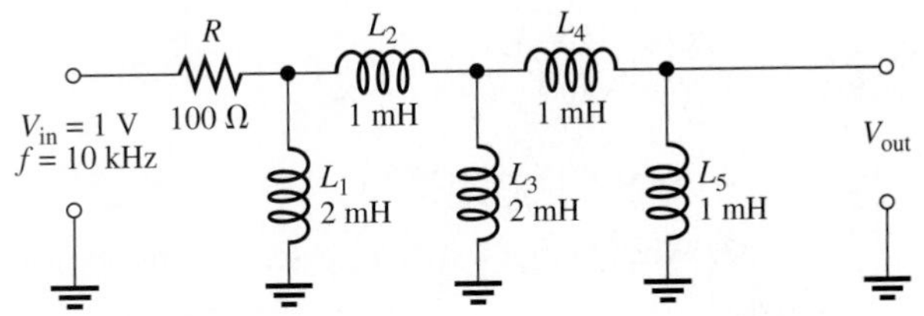

40. Design an ideal inductive switching circuit that will provide a momentary voltage of 2.5 kV from a 12 V dc source when a switch is thrown instantaneously from one position to another. The drain on the source must not exceed 1 A.

41. Draw the schematic for the circuit in Figure 13–66 and determine if the waveforms on the scope are correct. If there is a fault in the circuit, identify it.

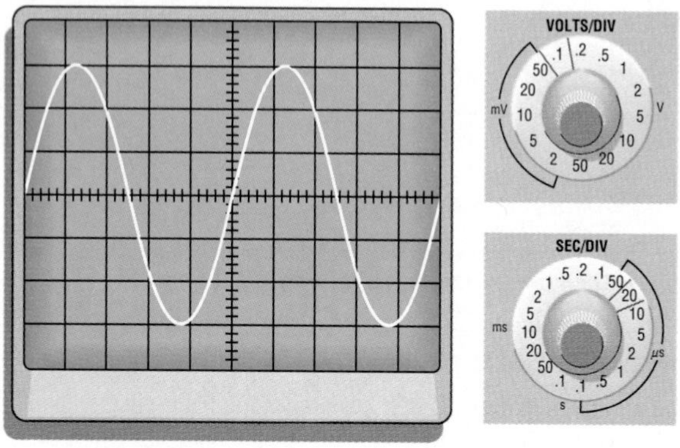

(a) Oscilloscope display

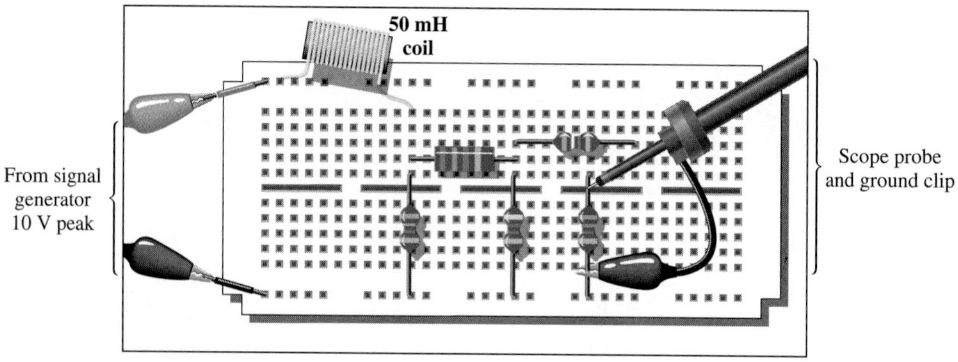

(b) Circuit with leads connected

FIGURE 13–66

▪ ANSWERS TO SECTION REVIEWS

Section 13–1

1. The current is at 1 kHz.
2. θ is closer to 0° when $R > X_L$.

Section 13–2

1. V_s leads I.
2. The inductive reactance produces the phase angle.
3. *RL* has an opposite phase angle compared to *RC*.
4. $Z = \sqrt{R^2 + X_L^2} = 59.9\ \text{k}\Omega$; $\theta = \tan^{-1}\left(\frac{X_L}{R}\right) = 56.6°$

Section 13–3

1. $V_T = \sqrt{V_R^2 + V_L^2} = 3.61\ \text{V}$
2. $\theta = \tan^{-1}\left(\frac{X_L}{R}\right) = 56.3°$
3. When f increases, X_L increases, Z increases, and θ increases.

Section 13–4

1. $Z = \dfrac{1}{Y} = 20\ \Omega$
2. $Y = \sqrt{G^2 + B_L^2} = 25$ mS
3. I_T lags V_s by 32.1°.

Section 13–5

1. $I_T = V_s Y = 32$ mA
2. $\theta = \tan^{-1}(I_L/I_R) = 59.0°$; $I_T = \sqrt{I_R^2 + I_L^2} = 23.3$ mA
3. $\theta = 90°$

Section 13–6

1. $I_T = \sqrt{I_1^2 + I_2^2} = 20.2$ mA
2. $Z = V_s/I_T = 495\ \Omega$

Section 13–7

1. Power is in the resistor.
2. $PF = \cos 50° = 0.643$
3. $P_{\text{true}} = I^2R = 4.7$ W; $P_r = I^2X_L = 6.2$ VAR; $P_a = \sqrt{P_{\text{true}}^2 + P_r^2} = 7.78$ VA

Section 13–8

1. $\phi = 90° - \tan^{-1}(X_L/R) = 81.9°$
2. $V_{\text{out}} = \left(R/\sqrt{R^2 + X_L^2}\right) V_{\text{in}} = 1.41$ V
3. Output is across the resistor.

Section 13–9

1. Shorted windings reduce L and thereby reduce X_L at any given frequency.
2. I_T decreases, V_{R1} decreases, and V_{R2} increases when L opens.

Section 13–10

1. 0 V on output
2. Output same as input

■ ANSWERS TO RELATED EXERCISES FOR EXAMPLES

13–1 2.04 kΩ; 27.8°
13–2 423 μA
13–3 No
13–4 12.6 kΩ; 85.5°
13–5 8.14 kΩ; 35.5°
13–6 3.13 mS
13–7 14.0 mA
13–8 67.6 mA; 36.3°
13–9 **(a)** 8.04 V **(b)** 2.00 V
13–10 Current decreases.
13–11 P_{true}, P_r, and P_a decrease.
13–12 32°
13–13 12.3 V rms
13–14 65.6°

14

RLC CIRCUITS AND RESONANCE

In this chapter, you will learn about frequency response of *RLC* circuits—that is, circuits with combinations of resistance, capacitance, and inductance. Both series and parallel *RLC* circuits, including the concepts of series and parallel resonance will be discussed.

Because it is the basis for frequency selectivity, resonance in electrical circuits is very important to the operation of many types of electronic systems, particularly in the area of communications. For example, the ability of a radio or television receiver to select a certain frequency transmitted by a certain station and to eliminate frequencies from other stations is based on the principle of resonance.

The operation of both band-pass and band-stop filters is based on resonance of circuits containing inductance and capacitance, and these filters are discussed in this chapter. Also, certain system applications are presented.

■ CHAPTER OBJECTIVES

- ☐ Determine the impedance and phase angle of a series *RLC* circuit
- ☐ Analyze series *RLC* circuits
- ☐ Analyze a circuit for series resonance
- ☐ Analyze series resonant filters
- ☐ Analyze parallel *RLC* circuits
- ☐ Analyze a circuit for parallel resonance
- ☐ Analyze the operation of parallel resonant filters
- ☐ Discuss some system applications of resonant circuits

■ APPLICATION ASSIGNMENT

Your application assignment in this chapter requires that you plot a frequency response curve for a resonant filter with unknown characteristics. From the frequency response measurements, you will identify the type of filter and determine the resonant frequency and the bandwidth. After studying this chapter, you should be able to complete the application assignment in Section 14–9.

14–1 ▪ IMPEDANCE AND PHASE ANGLE OF SERIES *RLC* CIRCUITS

A series RLC circuit contains both inductance and capacitance. Since inductive reactance and capacitive reactance have opposite effects on the circuit phase angle, the total reactance is less than either individual reactance.

After completing this section, you should be able to

- **Determine the impedance and phase angle of a series *RLC* circuit**
  - ☐ Calculate total reactance
  - ☐ Determine whether a circuit is predominately inductive or capacitive

---

A series *RLC* circuit is shown in Figure 14–1. It contains resistance, inductance, and capacitance.

FIGURE 14–1
Series RLC circuit.

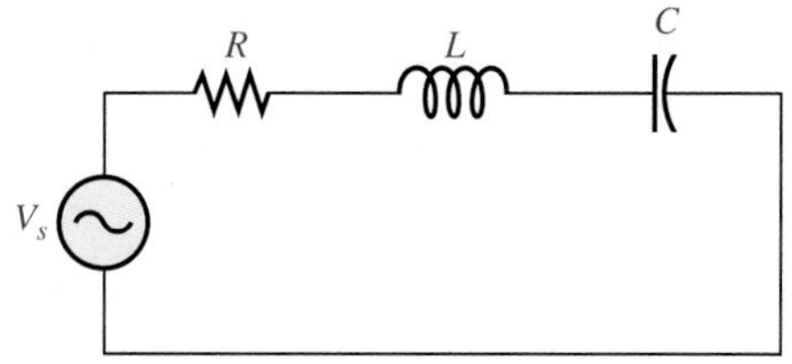

As you know, inductive reactance (X_L) causes the total current to lag the applied voltage. Capacitive reactance (X_C) has the opposite effect: It causes the current to lead the voltage. Thus, X_L and X_C tend to offset each other. When they are equal, they cancel, and the total reactance is zero. In any case, the total reactance in the series circuit is

$$X_T = |X_L - X_C| \qquad \textbf{(14–1)}$$

The term $|X_L - X_C|$ means the absolute value of the difference of the two reactances. That is, the sign of the result is considered positive no matter which reactance is greater. For example, $3 - 7 = -4$, but the absolute value is

$$|3 - 7| = 4$$

When $X_L > X_C$, the circuit is predominantly inductive, and when $X_C > X_L$, the circuit is predominantly capacitive.

The total impedance for the series *RLC* circuit is given by

$$Z_T = \sqrt{R^2 + (X_L - X_C)^2} \qquad \textbf{(14–2)}$$

and the phase angle between V_s and I is

$$\theta = \tan^{-1}\left(\frac{X_T}{R}\right) \qquad \textbf{(14–3)}$$

EXAMPLE 14–1 Determine the total impedance and the phase angle in Figure 14–2.

FIGURE 14–2

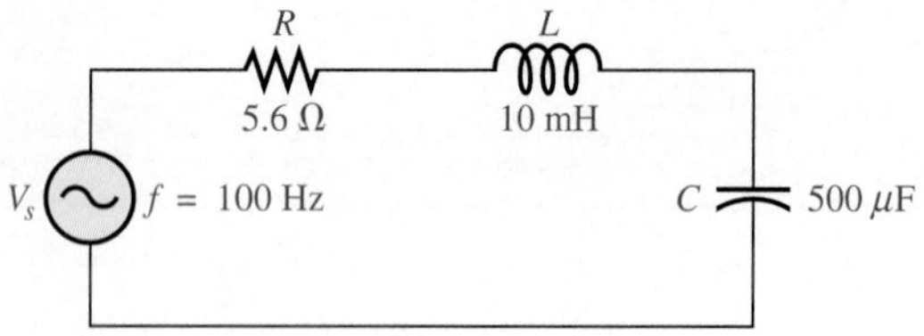

Solution First find X_C and X_L:

$$X_C = \frac{1}{2\pi fC} = \frac{1}{2\pi(100 \text{ Hz})(500\ \mu\text{F})} = 3.18\ \Omega$$

$$X_L = 2\pi fL = 2\pi(100 \text{ Hz})(10 \text{ mH}) = 6.28\ \Omega$$

In this case, X_L is greater than X_C, and thus the circuit is more inductive than capacitive. The magnitude of the total reactance is

$$X_T = |X_L - X_C| = |6.28\ \Omega - 3.18\ \Omega| = 3.10\ \Omega \qquad \text{(inductive)}$$

The total circuit impedance is

$$Z_T = \sqrt{R^2 + X_T^2} = \sqrt{(5.6\ \Omega)^2 + (3.10\ \Omega)^2} = 6.40\ \Omega$$

The phase angle (between I and V_s) is

$$\theta = \tan^{-1}\left(\frac{X_T}{R}\right) = \tan^{-1}\left(\frac{3.10\ \Omega}{5.6\ \Omega}\right) = 29.0° \qquad \text{(current lagging } V_s\text{)}$$

The calculator sequences are as follows:

Z: [2] [×] [2nd] [π] [×] [1] [0] [0] [×] [1] [0] [EXP] [+/−] [3] [=] [STO] [2] [×]
[2nd] [π] [×] [1] [0] [0] [×] [5] [0] [0] [EXP] [+/−] [6] [=] [1/x] [+/−] [SUM] [RCL]
[x^2] [+] [5] [•] [6] [x^2] [=] [2nd] [$\sqrt{x}$]

θ: [EXP] [÷] [5] [•] [6] [=] [2nd] [tan]

Related Exercise Increase the frequency in Figure 14–2 to 200 Hz and determine Z and θ.

As you have seen, when the inductive reactance is greater than the capacitive reactance, the circuit appears inductive, and the current lags the applied voltage. When the capacitive reactance is greater, the circuit appears capacitive, and the current leads the applied voltage.

SECTION 14–1 REVIEW

1. Explain how you can determine if a series *RLC* circuit is inductive or capacitive.
2. In a given series *RLC* circuit, X_C is 150 Ω and X_L is 80 Ω. What is the total reactance in ohms? Is it inductive or capacitive?
3. Determine the impedance for the circuit in Question 2 when $R = 45\ \Omega$. What is the phase angle? Is the current leading or lagging the applied voltage?

14–2 ▪ ANALYSIS OF SERIES *RLC* CIRCUITS

Recall that capacitive reactance varies inversely with frequency and that inductive reactance varies directly with frequency. In this section, the combined effects of the reactances as a function of frequency are examined.

After completing this section, you should be able to

- **Analyze series *RLC* circuits**
  - □ Determine current in a series *RLC* circuit
  - □ Determine the voltages in a series *RLC* circuit
  - □ Determine the phase angle

As shown in Figure 14–3, for a typical series *RLC* circuit the total reactance behaves as follows: Starting at a very low frequency, X_C is high, X_L is low, and the circuit is predominantly capacitive. As the frequency is increased, X_C decreases and X_L increases until a value is reached where $X_C = X_L$ and the two reactances cancel, making the circuit purely resistive. This condition is **series resonance** and will be studied in Section 14–3. As the frequency is increased further, X_L becomes greater than X_C, and the circuit is predominantly inductive. Example 14–2 illustrates how the impedance and phase angle change as the source frequency is varied.

Note that the graph of X_L is a straight line and the graph of X_C is curved in Figure 14–3. The reason is as follows: The general equation for a straight line is $y = mx + b$, where m is the slope of the line and b is the y-axis intercept point. The formula $X_L = 2\pi fL$ fits this general straight-line formula, where $y = X_L$ (a variable), $m = 2\pi L$ (a constant), $x = f$ (a variable), and $b = 0$ as follows: $X_L = 2\pi Lf + 0$.

The X_C curve is called a *hyperbola,* and the general equation of a hyperbola is $xy = k$. The equation for capacitive reactance, $X_C = 1/2\pi fC$, can be rearranged as $X_C f = 1/2\pi C$ where $x = X_C$ (a variable), $y = f$ (a variable), and $k = 1/2\pi C$ (a constant).

FIGURE 14–3
How X_C and X_L vary with frequency.

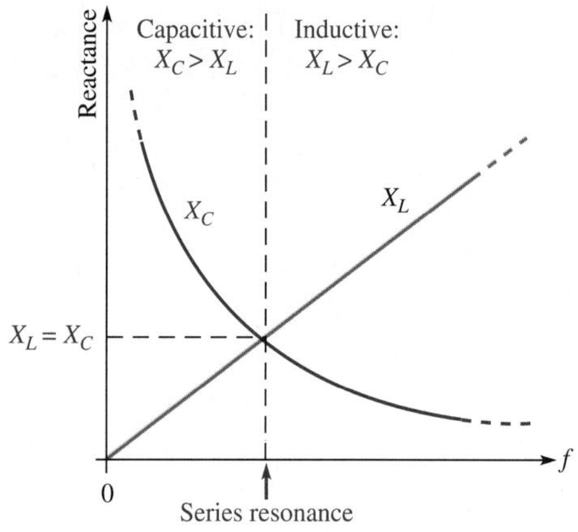

EXAMPLE 14–2

For each of the following frequencies of the source voltage, find the impedance and the phase angle for the circuit in Figure 14–4. Note how the impedance and the phase angle change with frequency.
(a) $f = 1$ kHz **(b)** $f = 3.5$ kHz **(c)** $f = 5$ kHz

FIGURE 14–4

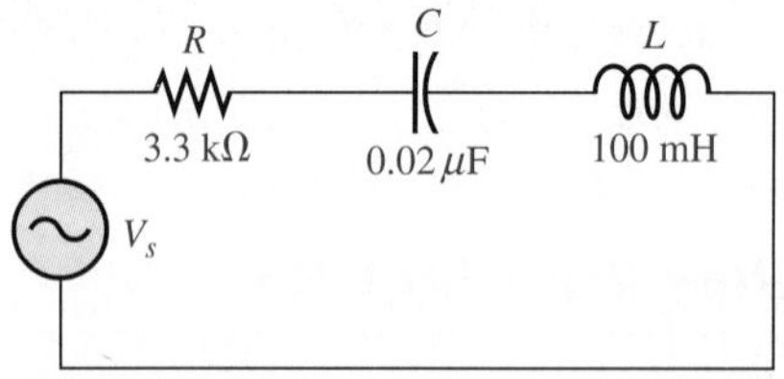

Solution
(a) At $f = 1$ kHz,

$$X_C = \frac{1}{2\pi fC} = \frac{1}{2\pi(1\text{ kHz})(0.02\ \mu\text{F})} = 7.96\text{ k}\Omega$$

$$X_L = 2\pi fL = 2\pi(1\text{ kHz})(100\text{ mH}) = 628\ \Omega$$

The circuit is highly capacitive because X_C is much larger than X_L. The impedance is

$$Z = \sqrt{R^2 + (X_L - X_C)^2} = \sqrt{(3.3\ \text{k}\Omega)^2 + (628\ \Omega - 7.96\ \text{k}\Omega)^2} = 8.04\ \text{k}\Omega$$

The phase angle is

$$\theta = \tan^{-1}\left(\frac{X_T}{R}\right) = \tan^{-1}\left(\frac{|X_L - X_C|}{R}\right) = \tan^{-1}\left(\frac{7.33\ \text{k}\Omega}{3.3\ \text{k}\Omega}\right) = 65.8°$$

I leads V_s by 65.8°.

(b) At $f = 3.5$ kHz,

$$X_C = \frac{1}{2\pi(3.5\ \text{kHz})(0.02\ \mu\text{F})} = 2.27\ \text{k}\Omega$$

$$X_L = 2\pi(3.5\ \text{kHz})(100\ \text{mH}) = 2.20\ \text{k}\Omega$$

The circuit is very close to being purely resistive but is still slightly capacitive because X_C is slightly larger than X_L. The impedance and phase angle are

$$Z = \sqrt{(3.3\ \text{k}\Omega)^2 + (2.20\ \text{k}\Omega - 2.27\ \text{k}\Omega)^2} = 3.30\ \text{k}\Omega$$

$$\theta = \tan^{-1}\left(\frac{X_T}{R}\right) = \tan^{-1}\left(\frac{70\ \Omega}{3.3\ \text{k}\Omega}\right) = 1.22°$$

I leads V_s by 1.22°.

(c) At $f = 5$ kHz,

$$X_C = \frac{1}{2\pi(5\ \text{kHz})(0.02\ \mu\text{F})} = 1.59\ \text{k}\Omega$$

$$X_L = 2\pi(5\ \text{kHz})(100\ \text{mH}) = 3.14\ \text{k}\Omega$$

The circuit is now predominantly inductive because $X_L > X_C$. The impedance and phase angle are

$$Z = \sqrt{(3.3\ \text{k}\Omega)^2 + (3.14\ \text{k}\Omega - 1.59\ \text{k}\Omega)^2} = 3.65\ \text{k}\Omega$$

$$\theta = \tan^{-1}\left(\frac{X_T}{R}\right) = \tan^{-1}\left(\frac{1.55\ \text{k}\Omega}{3.3\ \text{k}\Omega}\right) = 25.2°$$

I lags V_s by 25.2°.

Notice how the circuit changed from capacitive to inductive as the frequency increased. The phase condition changed from the leading current to the lagging current. It is interesting to note that both the impedance and the phase angle decreased to a minimum and then began increasing again as the frequency went up.

Related Exercise Determine Z for $f = 7$ kHz in Figure 14–4.

In a series *RLC* circuit, the capacitor voltage and the inductor voltage are always 180° out of phase with each other. For this reason, V_C and V_L subtract from each other, and thus the voltage across L and C combined is always less than the larger individual voltage across either element, as illustrated in Figure 14–5 and in the waveform diagram of Figure 14–6.

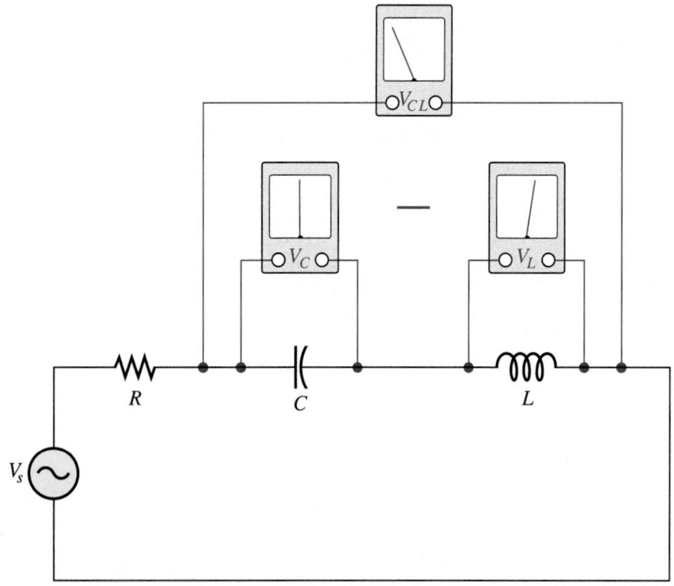

FIGURE 14–5
The voltage across the series combination of C and L is always less than the larger individual voltage.

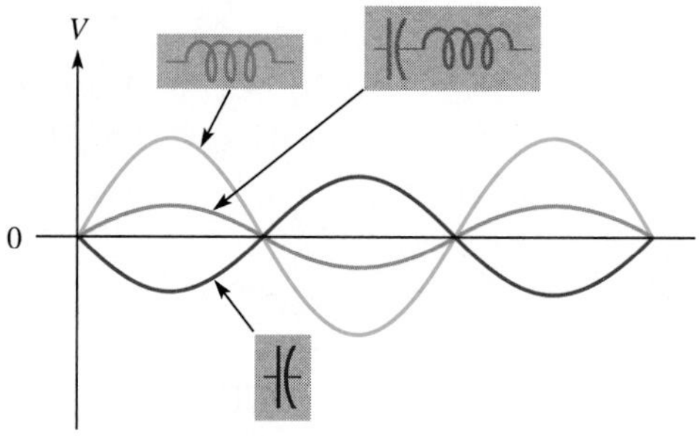

FIGURE 14–6
Inductor voltage and capacitor voltage effectively subtract.

In the next example, Ohm's law is used to find the voltages in a series *RLC* circuit.

EXAMPLE 14–3

Find the voltages across each element in Figure 14–7, and draw a complete voltage phasor diagram. Also find the voltage across *L* and *C* combined.

FIGURE 14–7

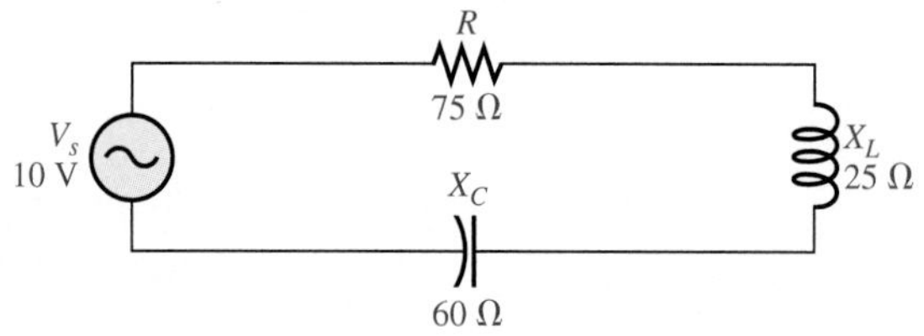

Solution First find the total impedance:

$$Z = \sqrt{R^2 + (X_L - X_C)^2} = \sqrt{(75\ \Omega)^2 + (25\ \Omega - 60\ \Omega)^2}$$
$$= \sqrt{(75\ \Omega)^2 + (35\ \Omega)^2} = 82.8\ \Omega$$

Apply Ohm's law to find the current:

$$I = \frac{V_s}{Z} = \frac{10\text{ V}}{82.8\ \Omega} = 121\text{ mA}$$

Now apply Ohm's law to find the voltages across *R, L,* and *C*:

$$V_R = IR = (121\text{ mA})(75\ \Omega) = 9.08\text{ V}$$
$$V_L = IX_L = (121\text{ mA})(25\ \Omega) = 3.03\text{ V}$$
$$V_C = IX_C = (121\text{ mA})(60\ \Omega) = 7.26\text{ V}$$

The voltage across L and C combined is

$$V_{LC} = V_C - V_L = 7.26\text{ V} - 3.03\text{ V} = 4.23\text{ V}$$

The circuit phase angle is

$$\theta = \tan^{-1}\left(\frac{X_T}{R}\right) = \tan^{-1}\left(\frac{|X_L - X_C|}{R}\right) = \tan^{-1}\left(\frac{35\ \Omega}{75\ \Omega}\right) = 25°$$

Since the circuit is capacitive ($X_C > X_L$), the current leads the source voltage by 25°.

The phasor diagram is shown in Figure 14–8. Notice that V_L is leading V_R by 90°, and V_C is lagging V_R by 90°. Also, there is a 180° phase difference between V_L and V_C. If the current phasor were shown, it would be at the same angle as V_R.

FIGURE 14–8

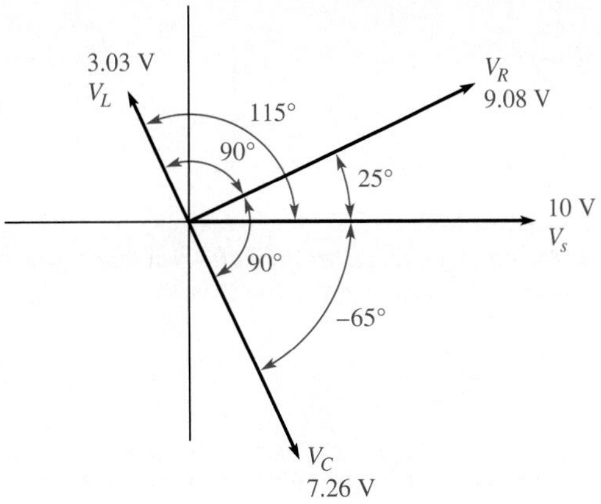

Related Exercise What will happen to the current if the frequency in Figure 14–7 is increased?

SECTION 14–2 REVIEW

1. The following voltages occur in a certain series *RLC* circuit: $V_R = 24$ V, $V_L = 15$ V, $V_C = 45$ V. Determine the source voltage.
2. When $R = 10\ \Omega$, $X_C = 18\ \Omega$, and $X_L = 12\ \Omega$, does the current lead or lag the applied voltage? Why?
3. Determine the total reactance in Question 2.

14–3 ■ SERIES RESONANCE

In a series RLC circuit, series resonance occurs when $X_L = X_C$. The frequency at which resonance occurs is called the resonant frequency and is designated f_r.

After completing this section, you should be able to

- **Analyze a circuit for series resonance**
  - □ Define *series resonance*
  - □ Determine the impedance at resonance
  - □ Explain why the reactances cancel at resonance
  - □ Determine the series resonant frequency
  - □ Calculate the current, voltages, and phase angle at resonance

Figure 14–9 illustrates the series resonant condition. Since $X_L = X_C$, the reactances effectively cancel, and the impedance is purely resistive. These resonant conditions are stated in the following equations:

$$X_L = X_C \quad (14\text{–}4)$$

$$Z_r = R \quad (14\text{–}5)$$

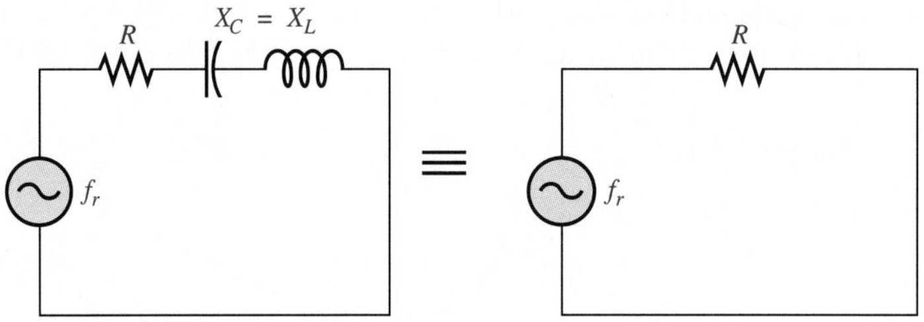

FIGURE 14–9
At the resonant frequency (f_r), the reactances are equal in magnitude and effectively cancel, leaving Z = R.

EXAMPLE 14–4 For the series *RLC* circuit in Figure 14–10, determine X_C and Z at resonance.

FIGURE 14–10

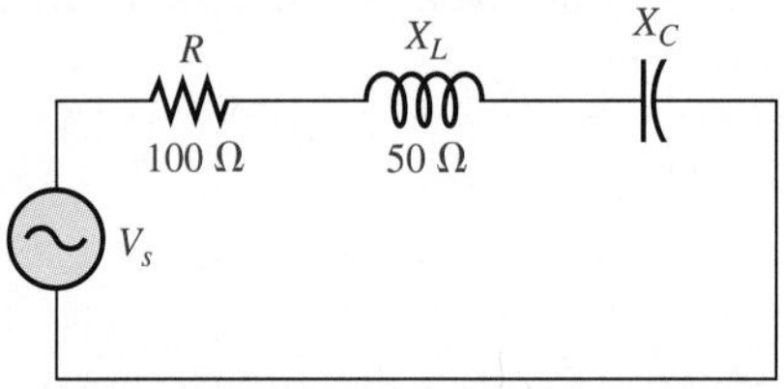

Solution X_L equals X_C at resonance; therefore,

$$X_L = X_C = 50\ \Omega$$

Since the reactances cancel each other, the impedance equals the resistance.

$$Z_r = R = 100\ \Omega$$

Related Exercise Just below the resonant frequency, is the circuit more inductive or more capacitive?

X_L and X_C Effectively Cancel at Resonance

At the series **resonant frequency,** the voltages across C and L are equal in magnitude because the reactances are equal and because the same current flows through both since they are in series ($IX_C = IX_L$). Also, V_L and V_C are always 180° out of phase with each other.

During any given cycle, the polarities of the voltages across C and L are opposite, as shown in Figures 14–11(a) and 14–11(b). The equal and opposite voltages across C and L cancel, leaving zero volts from point A to point B as shown in the figure. Since there is no voltage drop from A to B but there is still current, the total reactance must be zero, as indicated in Figure 14–11(c). Also, the voltage phasor diagram in part (d) shows that V_C and V_L are equal in magnitude and 180° out of phase with each other.

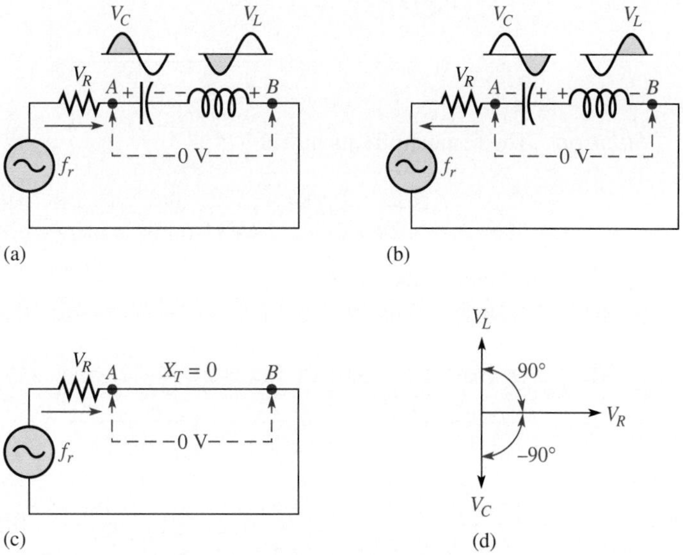

FIGURE 14–11
At the resonant frequency, f_r, the voltages across C and L are equal in magnitude. Since they are 180° out of phase with each other, they cancel, leaving 0 V across the LC combination. The section of the circuit from A to B effectively looks like a short at resonance (neglecting winding resistance).

The Series Resonant Frequency

For a given series RLC circuit, resonance occurs at only one specific frequency. A formula for this resonant frequency is developed as follows:

$$X_L = X_C$$

Substitute the reactance formulas:

$$2\pi f_r L = \frac{1}{2\pi f_r C}$$

Solve for the resonant frequency (f_r):

$$(2\pi f_r L)(2\pi f_r C) = 1$$
$$4\pi^2 f_r^2 LC = 1$$
$$f_r^2 = \frac{1}{4\pi^2 LC}$$

Take the square root of both sides. The formula for resonant frequency is

$$f_r = \frac{1}{2\pi\sqrt{LC}} \qquad \textbf{(14–6)}$$

EXAMPLE 14–5 Find the series resonant frequency for the circuit in Figure 14–12.

FIGURE 14–12

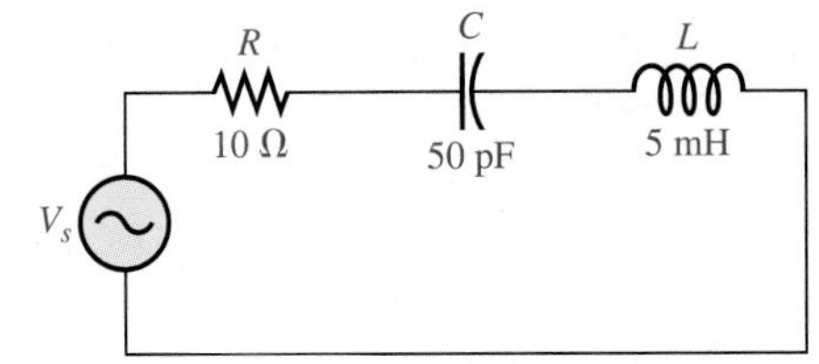

Solution The resonant frequency is

$$f_r = \frac{1}{2\pi\sqrt{LC}} = \frac{1}{2\pi\sqrt{(5\text{ mH})(50\text{ pF})}} = 318\text{ kHz}$$

The calculator sequence is

[5] [EXP] [+/−] [3] [×] [5] [0] [EXP] [+/−] [1] [2] [=] [2nd] [√x] [×] [2] [×] [2nd] [π] [=] [1/x]

Related Exercise If $C = 0.01\ \mu$F in Figure 14–12, what is the resonant frequency?

Voltages and Current in a Series *RLC* Circuit

It is interesting to see how the current and the voltages in a series *RLC* circuit vary as the frequency is increased from below the resonant frequency, through resonance, and then above resonance. Figure 14–13 illustrates the general response of a circuit in terms of changes in the current and in the voltage drops.

Below the Resonant Frequency Figure 14–13(a) indicates the response as the source frequency is increased from zero toward f_r. At $f = 0$ Hz (dc), the capacitor appears open and blocks the current. Thus, there is no voltage across *R* or *L*, and the entire source voltage appears across *C*. The impedance of the circuit is infinitely large at 0 Hz because X_C is infinite (*C* is open). As the frequency begins to increase, X_C decreases and X_L increases, causing the total reactance, $X_C - X_L$, to decrease. As a result, the impedance decreases and the current increases. As the current increases, V_R also increases, and both V_C and V_L increase. The voltage across *C* and *L* combined decreases from its maximum value of V_s because as V_C and V_L approach the same value, their difference becomes less.

At the Resonant Frequency When the frequency reaches its resonant value, f_r, as shown in Figure 14–13(b), V_C and V_L can each be much larger than the source voltage. V_C and V_L cancel leaving 0 V across the combination, because they are equal in magnitude but opposite in phase. At this point the total impedance is equal to *R* and is at its minimum value because the total reactance is zero. Thus, the current is at its maximum value, V_s/R, and V_R is at its maximum value which is equal to the source voltage.

Above the Resonant Frequency As the frequency is increased above resonance, as indicated in Figure 14–13(c), X_L continues to increase and X_C continues to decrease, causing the total reactance, $X_L - X_C$, to increase. As a result, there is an increase in impedance and a decrease in current. As the current decreases, V_R also decreases and V_C and V_L decrease. As V_C and V_L decrease, their difference becomes greater, so V_{CL} increases. As the frequency becomes very high, the current approaches zero, both V_R and V_C approach zero, and V_L approaches V_s.

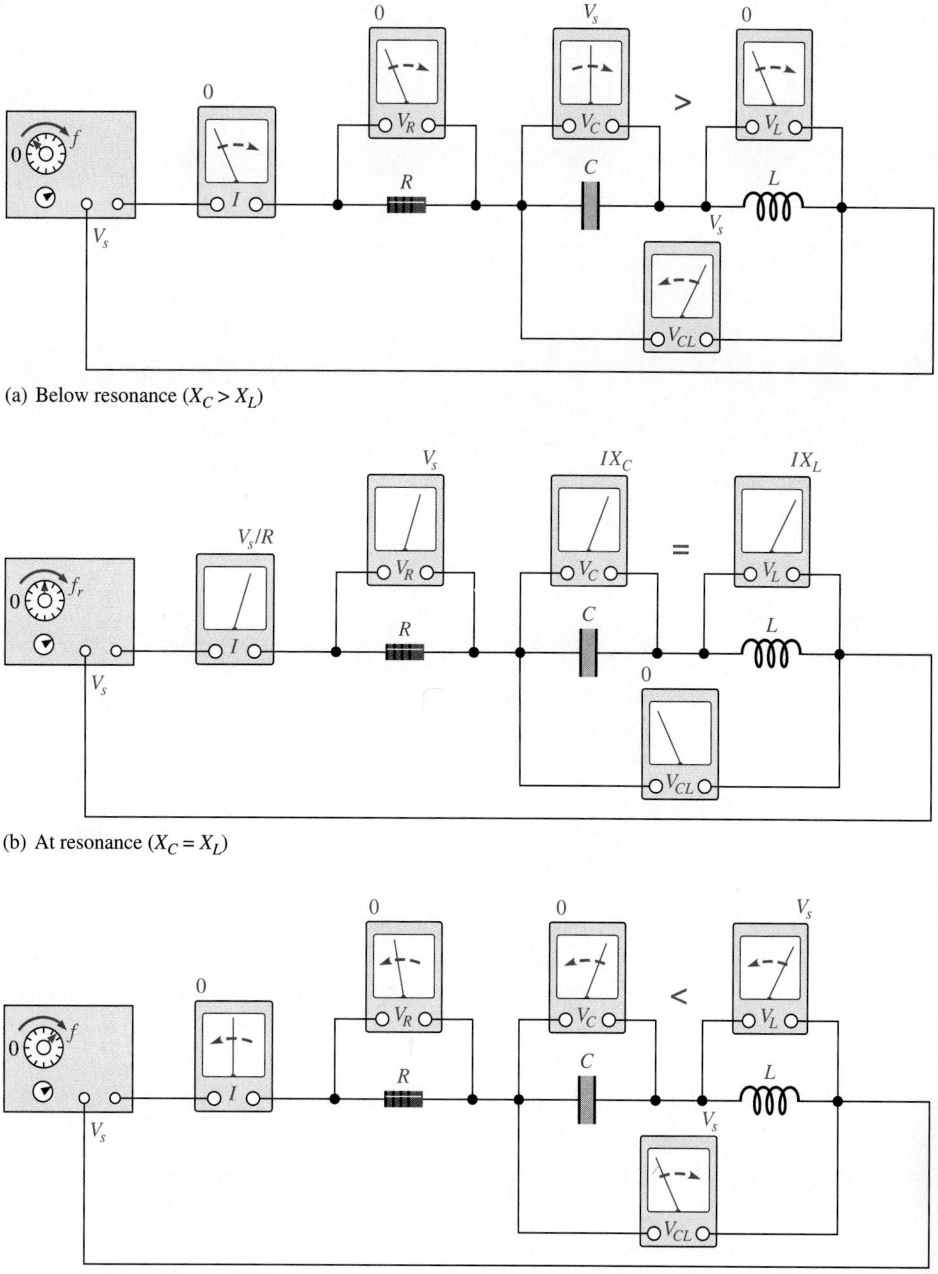

(a) Below resonance ($X_C > X_L$)

(b) At resonance ($X_C = X_L$)

(c) Above resonance ($X_L > X_C$)

FIGURE 14–13

An illustration of how the voltages and the current respond in a series RLC circuit as the frequency is increased from below to above its resonant value. The source voltage is held at a constant amplitude.

Figure 14–14 (a) and (b) illustrate the responses of voltage and current to increasing frequency. As frequency is increased, the current increases below resonance, reaches a peak at the resonant frequency, and then decreases above resonance. The resistor voltage responds in the same way as the current.

The general slopes of the V_C and V_L curves are indicated in Figure 14–14 (c) and (d). The voltages are maximum at resonance but drop off above and below f_r. The voltages

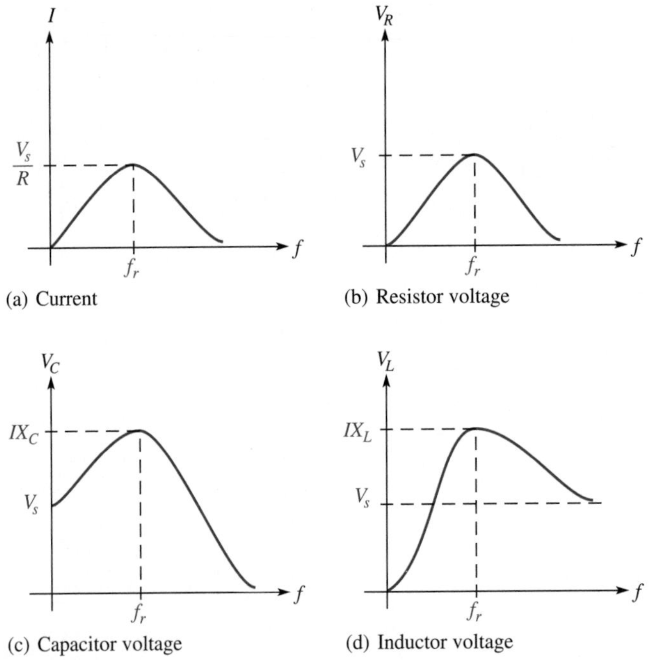

FIGURE 14–14
Current and voltages as functions of frequency in a series RLC circuit.

across L and C at resonance are exactly equal in magnitude but 180° out of phase, so they cancel. Thus, the total voltage across both L and C is zero, and $V_R = V_s$ at resonance. Individually, V_L and V_C can be much greater than the source voltage. Keep in mind that V_L and V_C are always opposite in polarity regardless of the frequency, but only at resonance are their magnitudes equal. The voltage across the C and L combination decreases as the frequency increases below resonance, reaching a minimum of zero at the resonant frequency; then it increases above resonance.

EXAMPLE 14–6

Find I, V_R, V_L, and V_C at resonance in Figure 14–15.

FIGURE 14–15

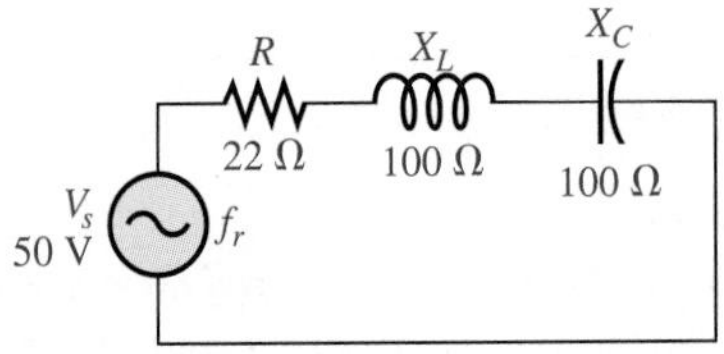

Solution At resonance, I is maximum and equal to V_s/R:

$$I = \frac{V_s}{R} = \frac{50\ \text{V}}{22\ \Omega} = 2.27\ \text{A}$$

Applying Ohm's law, obtain the following voltages:

$$V_R = IR = (2.27\ \text{A})(22\ \Omega) = 50\ \text{V}$$
$$V_L = IX_L = (2.27\ \text{A})(100\ \Omega) = 227\ \text{V}$$
$$V_C = IX_C = (2.27\ \text{A})(100\ \Omega) = 227\ \text{V}$$

Notice that all of the source voltage is dropped across the resistor. Also, of course, V_L and V_C are equal in magnitude but opposite in phase. This causes the voltage to cancel, making the total reactive voltage zero.

Related Exercise What is the current at resonance in Figure 14–15 if $X_L = X_C = 1\ \text{k}\Omega$?

The Impedance of a Series *RLC* Circuit

Figure 14–16 shows a general graph of impedance versus frequency superimposed on the curves for X_C and X_L. At zero frequency, both X_C and Z are infinitely large and X_L is zero because the capacitor looks like an open at 0 Hz and the inductor looks like a short. As the frequency increases, X_C decreases and X_L increases. Since X_C is larger than X_L at frequencies below f_r, Z decreases along with X_C. At f_r, $X_C = X_L$ and $Z = R$. At frequencies above f_r, X_L becomes increasingly larger than X_C, causing Z to increase.

FIGURE 14–16
Series RLC impedance as a function of frequency.

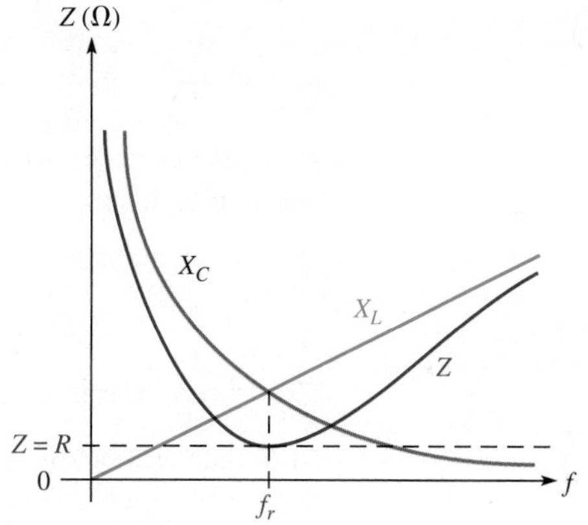

EXAMPLE 14–7 For the circuit in Figure 14–17, determine the impedance at the following frequencies:
(a) f_r **(b)** 1000 Hz below f_r **(c)** 1000 Hz above f_r

FIGURE 14–17

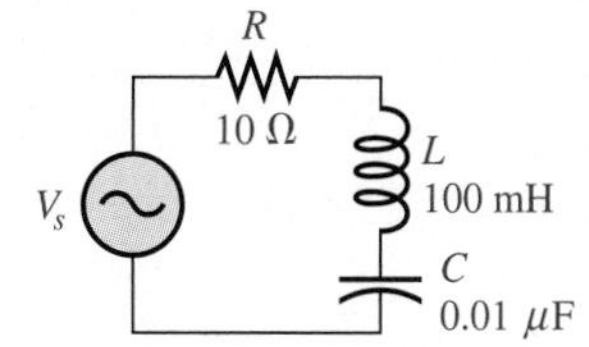

Solution

(a) At f_r, the impedance is equal to R:

$$Z = R = 10\ \Omega$$

To determine the impedance above and below f_r, first calculate the resonant frequency as follows:

$$f_r = \frac{1}{2\pi\sqrt{LC}} = \frac{1}{2\pi\sqrt{(100\text{ mH})(0.01\ \mu\text{F})}} = 5.03\text{ kHz}$$

(b) At 1000 Hz below f_r, the frequency and reactances are as follows:

$$f = f_r - 1 \text{ kHz} = 5.03 \text{ kHz} - 1 \text{ kHz} = 4.03 \text{ kHz}$$

$$X_C = \frac{1}{2\pi fC} = \frac{1}{2\pi(4.03 \text{ kHz})(0.01\ \mu\text{F})} = 3.95 \text{ k}\Omega$$

$$X_L = 2\pi fL = 2\pi(4.03 \text{ kHz})(100 \text{ mH}) = 2.53 \text{ k}\Omega$$

Therefore, the impedance at $f_r - 1$ kHz is

$$Z = \sqrt{R^2 + (X_L - X_C)^2} = \sqrt{(10\ \Omega)^2 + (2.53 \text{ k}\Omega - 3.95 \text{ k}\Omega)^2} = 1.42 \text{ k}\Omega$$

(c) At 1000 Hz above f_r,

$$f = 5.03 \text{ kHz} + 1 \text{ kHz} = 6.03 \text{ kHz}$$

$$X_C = \frac{1}{2\pi(6.03 \text{ kHz})(0.01\ \mu\text{F})} = 2.64 \text{ k}\Omega$$

$$X_L = 2\pi(6.03 \text{ kHz})(100 \text{ mH}) = 3.79 \text{ k}\Omega$$

Therefore, the impedance at $f_r + 1$ kHz is

$$Z = \sqrt{(10\ \Omega)^2 + (3.79 \text{ k}\Omega - 2.64 \text{ k}\Omega)^2} = 1.15 \text{ k}\Omega$$

X_L is greater than X_C, so Z is inductive.

Related Exercise What happens to the impedance if f is decreased below 4.03 kHz? Above 6.03 kHz?

The Phase Angle of a Series *RLC* Circuit

At frequencies below resonance, $X_C > X_L$, and the current leads the source voltage, as indicated in Figure 14–18(a). The phase angle decreases as the frequency approaches the resonant value and is 0° at resonance, as indicated in part (b). At frequencies above reso-

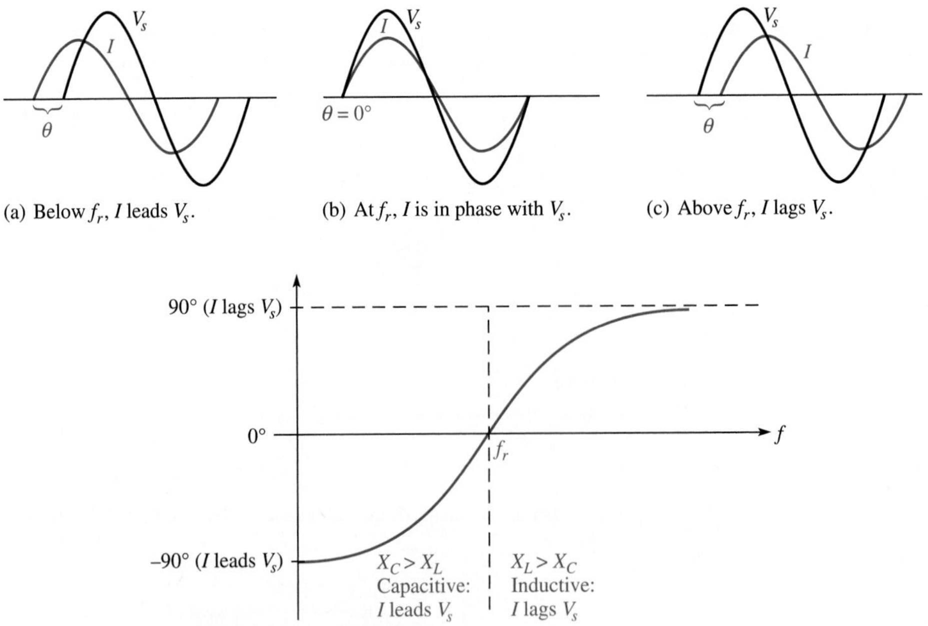

(a) Below f_r, I leads V_s.

(b) At f_r, I is in phase with V_s.

(c) Above f_r, I lags V_s.

(d) Phase angle versus frequency.

FIGURE 14–18
The phase angle as a function of frequency in a series RLC circuit.

nance, $X_L > X_C$, and the current lags the source voltage, as indicated in part (c). As the frequency goes higher, the phase angle approaches 90°. A plot of phase angle versus frequency is shown in Figure 14–18(d).

SECTION 14–3 REVIEW

1. What is the condition for series resonance?
2. Why is the current maximum at the resonant frequency?
3. Calculate the resonant frequency for $C = 1000$ pF and $L = 1000$ μH.
4. In Question 3, is the circuit inductive or capacitive at 50 kHz?

14–4 ■ SERIES RESONANT FILTERS

A common use of series RLC circuits is in filter applications. In this section, you will learn the basic configurations for band-pass and band-stop filters and several important filter characteristics.

After completing this section, you should be able to

- **Analyze series resonant filters**
  - ☐ Identify a basic series resonant band-pass filter
  - ☐ Define and determine *bandwidth*
  - ☐ Define the *half-power frequency*
  - ☐ Discuss dB measurement
  - ☐ Define *selectivity*
  - ☐ Discuss filter quality factor (Q)
  - ☐ Identify a series resonant band-stop filter

The Band-Pass Filter

A basic series resonant **band-pass filter** is shown in Figure 14–19. Notice that the series *LC* portion is placed between the input and the output and that the output is taken across the resistor.

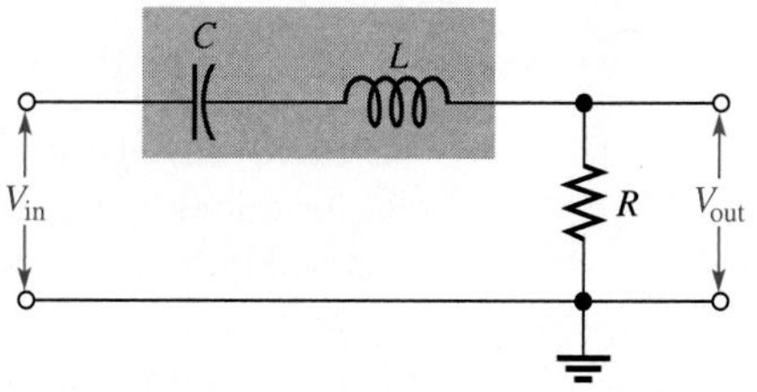

FIGURE 14–19
A basic series resonant band-pass filter.

A band-pass filter allows signals at the resonant frequency and at frequencies within a certain band (or range) extending below and above the resonant value to pass from input to output without a significant reduction in amplitude. Signals at frequencies lying outside this specified band (called the **pass band**) are reduced in amplitude to below a certain level and are considered to be rejected by the filter.

The filtering action is the result of the impedance characteristic of the filter. As you learned in Section 14–3, the impedance is minimum at resonance and has increasingly higher values below and above the resonant frequency. At very low frequencies, the impedance is very high and tends to block the current. As the frequency increases, the impedance drops, allowing more current and thus more voltage across the output resistor. At the resonant frequency, the impedance is very low and equal to the winding resistance of the coil. At this point a maximum current is flowing and the resulting output voltage is maximum. As the frequency goes above resonance, the impedance again increases, causing the current and the resulting output voltage to drop. Figure 14–20 illustrates the general frequency response of a series resonant band-pass filter.

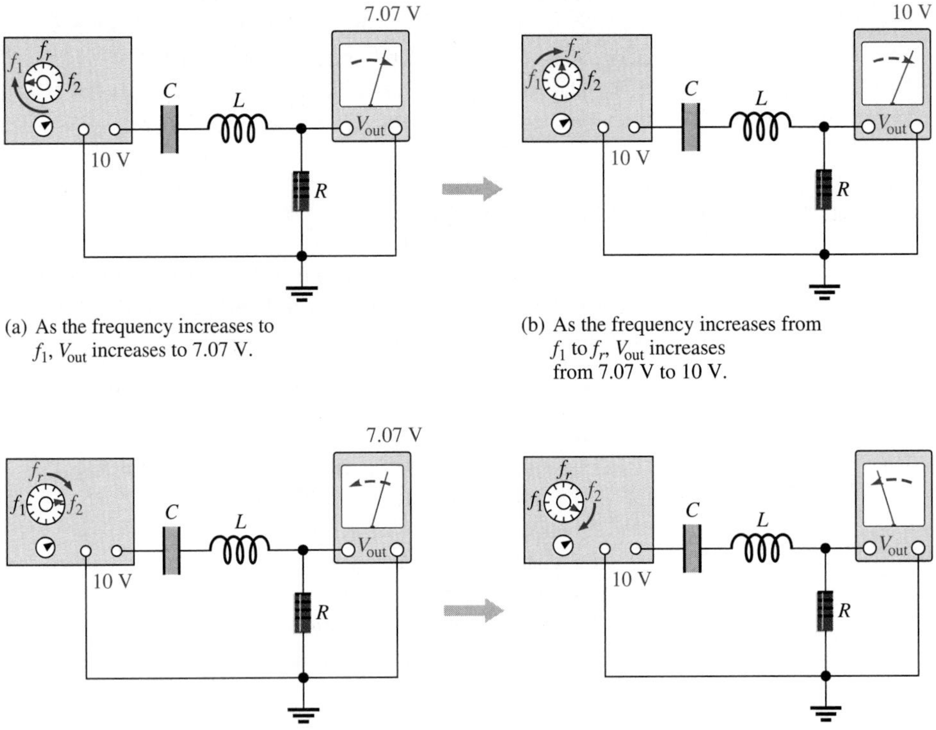

(a) As the frequency increases to f_1, V_{out} increases to 7.07 V.

(b) As the frequency increases from f_1 to f_r, V_{out} increases from 7.07 V to 10 V.

(c) As the frequency increases from f_r to f_2, V_{out} decreases from 10 V to 7.07 V.

(d) As the frequency increases above f_2, V_{out} decreases below 7.07 V.

FIGURE 14–20
Example of the frequeny response of a series resonant band-pass filter with the input voltage at a constant 10 V rms. The winding resistance of the coil is neglected.

Bandwidth of the Pass Band

The **bandwidth (BW)** of a band-pass filter is the range of frequencies for which the current (or output voltage) is equal to or greater than 70.7% of its value at the resonant frequency. Figure 14–21 shows the bandwidth on the response curve for the band-pass filter.

The frequencies at which the output of a filter is 70.7% of its maximum are the **cutoff frequencies.** Notice in Figure 14–21 that frequency f_1, which is below f_r, is the frequency at which I (or V_{out}) is 70.7% of the resonant value (I_{max}); f_1 is commonly called the *lower cutoff frequency.* At frequency f_2, above f_r, the current (or V_{out}) is again 70.7% of its maximum; f_2 is called the *upper cutoff frequency.* Other names for f_1 and f_2 are *–3 dB frequencies, band frequencies,* and *half-power frequencies.* (The term *decibel,* abbreviated dB, is defined on page 590).

The formula for calculating the bandwidth is as follows:

$$BW = f_2 - f_1 \quad (14\text{–}7)$$

The unit of bandwidth is the hertz (Hz), the same as for frequency.

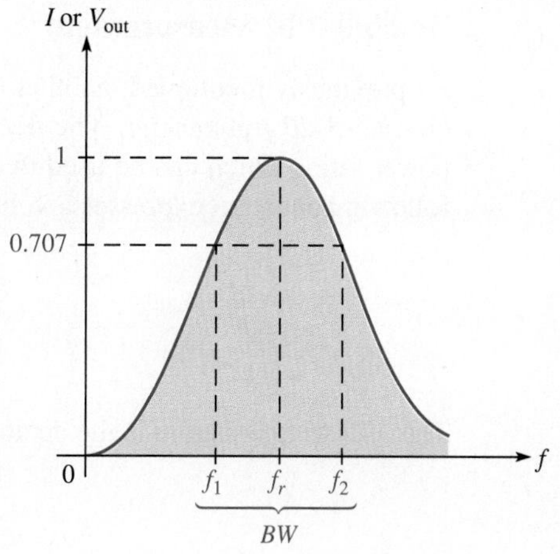

FIGURE 14–21
Generalized response curve of a series resonant band-pass filter.

EXAMPLE 14–8 A certain series resonant band-pass filter has a maximum current of 100 mA at the resonant frequency. What is the value of the current at the cutoff frequencies?

Solution Current at the cutoff frequencies is 70.7% of maximum:

$$I_{f1} = I_{f2} = 0.707 I_{max} = 0.707(100 \text{ mA}) = 70.7 \text{ mA}$$

Related Exercise Will a change in the cutoff frequencies affect the current at these frequencies if the maximum current remains 100 mA?

EXAMPLE 14–9 A resonant circuit has a lower cutoff frequency of 8 kHz and an upper cutoff frequency of 12 kHz. Determine the bandwidth.

Solution

$$BW = f_2 - f_1 = 12 \text{ kHz} - 8 \text{ kHz} = 4 \text{ kHz}$$

Related Exercise What is the bandwidth when $f_1 = 1$ MHz and $f_2 = 1.2$ MHz?

Half-Power Points of the Filter Response

As previously mentioned, the upper and lower cutoff frequencies are sometimes called the **half-power frequencies.** This term is derived from the fact that the true power from the source at these frequencies is one-half the power delivered at the resonant frequency. The following steps show that this relationship is true for a series resonant circuit.

At resonance,

$$P_{max} = I^2_{max} R$$

The power at f_1 or f_2 is

$$P_{f1} = I^2_{f1} R = (0.707 I_{max})^2 R = (0.707)^2 I^2_{max} R$$
$$= 0.5 I^2_{max} R = 0.5 P_{max}$$

Decibel (dB) Measurement

As previously mentioned, another common term for the upper and lower cutoff frequencies is *−3 dB frequencies.* The **decibel** (dB) is a logarithmic measurement of voltage or power ratios which can be used to express the input-to-output relationship of a filter. The following equation expresses a voltage ratio in decibels:

$$\text{dB} = 20\log\left(\frac{V_{out}}{V_{in}}\right) \tag{14–8}$$

The following equation is the decibel formula for a power ratio:

$$\text{dB} = 10\log\left(\frac{P_{out}}{P_{in}}\right) \tag{14–9}$$

EXAMPLE 14–10 At a certain frequency, the output voltage of a filter is 5 V and the input is 10 V. Express the voltage ratio in decibels.

Solution

$$20\log\left(\frac{V_{out}}{V_{in}}\right) = 20\log\left(\frac{5\text{ V}}{10\text{ V}}\right) = 20\log(0.5) = -6.02\text{ dB}$$

The calculator sequence is

5 ÷ 1 0 = log × 2 0 =

Related Exercise Express 0.2 in decibels.

The −3 dB Frequencies The output of a filter is said to be down 3 dB at the cutoff frequencies. As you know, this frequency is the point at which the output voltage is 70.7% of the maximum voltage at resonance. We can show that the 70.7% point is the same as 3 dB below maximum (or −3 dB) as follows. The maximum voltage is the 0 dB reference:

$$20\log\left(\frac{0.707V_{max}}{V_{max}}\right) = 20\log(0.707) = -3\text{ dB}$$

Selectivity of a Band-Pass Filter

The response curve in Figure 14–21 is also called a *selectivity curve.* **Selectivity** defines how well a resonant circuit responds to a certain frequency and discriminates against all others. *The narrower the bandwidth, the greater the selectivity.*

We normally assume that a resonant circuit accepts frequencies within its bandwidth and completely eliminates frequencies outside the bandwidth. Such is not actually the case, however, because signals with frequencies outside the bandwidth are not completely eliminated. Their magnitudes, however, are greatly reduced. The further the frequencies are from the cutoff frequencies, the greater is the reduction, as illustrated in Figure 14–22(a). An ideal selectivity curve is shown in Figure 14–22(b).

As you can see in Figure 14–23, another factor that influences selectivity is the steepness of the slopes of the curve. The faster the curve drops off at the cutoff frequencies, the more selective the circuit is, because the frequencies outside the pass band are more quickly reduced (attenuated).

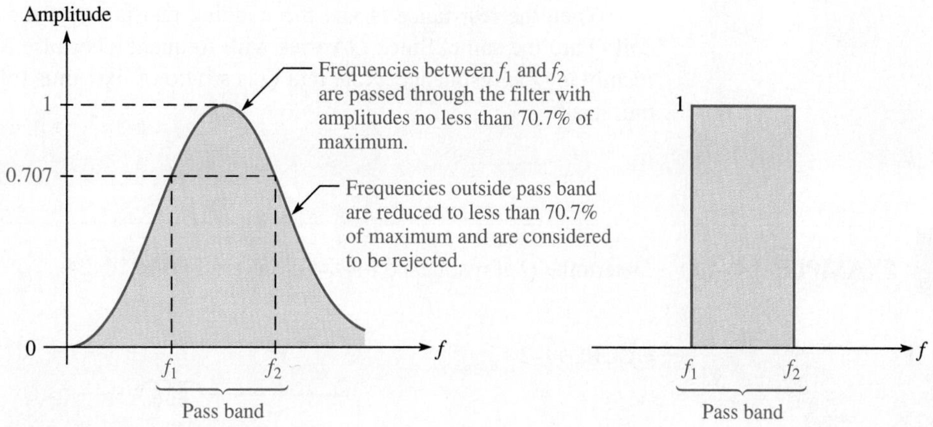

FIGURE 14–22
Generalized selectivity curve of a band-pass filter.

FIGURE 14–23
Comparative selectivity curves.

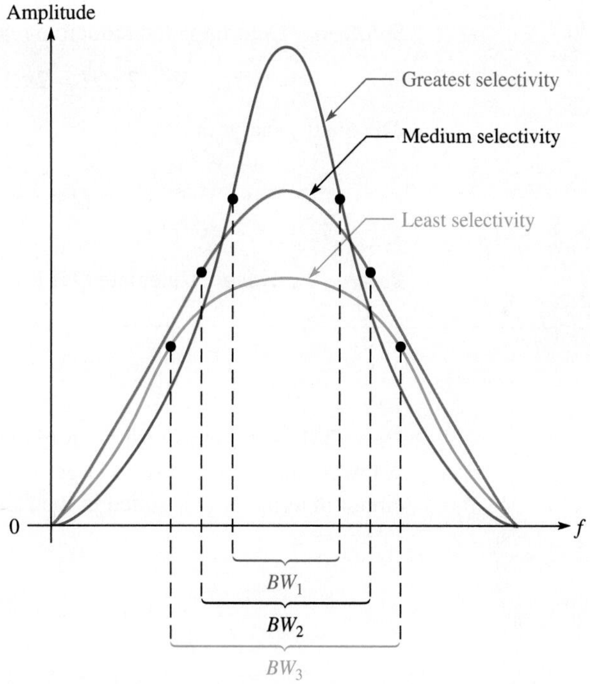

The Quality Factor (Q) of a Resonant Circuit and Its Effect on Bandwidth

The **quality factor (*Q*)** is the ratio of the reactive power in the inductor to the true power in the winding resistance of the coil or the resistance in series with the coil. It is a ratio of the power in L to the power in R. The quality factor is very important in resonant circuits. A formula for Q is developed as follows:

$$Q = \frac{\text{energy stored}}{\text{energy dissipated}} = \frac{\text{reactive power}}{\text{true power}} = \frac{I^2 X_L}{I^2 R}$$

In a series circuit, I is the same in L and R; thus, the I^2 terms cancel, leaving

$$Q = \frac{X_L}{R} \tag{14–10}$$

When the resistance is just the winding resistance of the coil, the circuit Q and the coil Q are the same. Since Q varies with frequency because X_L varies, we are interested mainly in Q at resonance. Note that Q is a ratio of like units (ohms) and, therefore, has no unit itself.

EXAMPLE 14–11

Determine Q at resonance for the circuit in Figure 14–24.

FIGURE 14–24

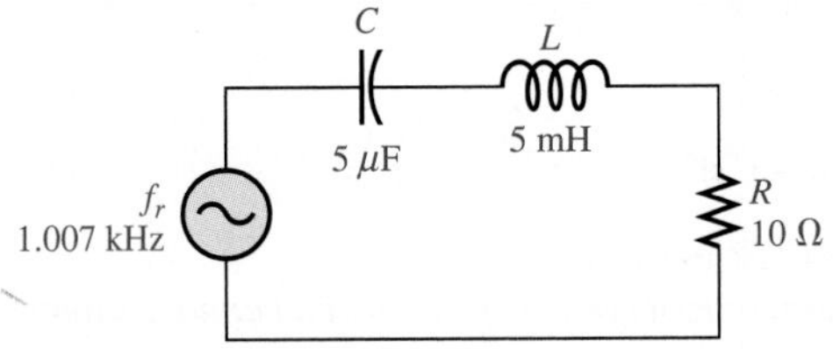

Solution Determine the inductive reactance:

$$X_L = 2\pi f_r L = 2\pi(1.007\text{ kHz})(5\text{ mH}) = 31.6\ \Omega$$

The quality factor is

$$Q = \frac{X_L}{R} = \frac{31.6\ \Omega}{10\ \Omega} = 3.16$$

Related Exercise Calculate Q if R in Figure 14–24 is decreased to 2.7 Ω.

How Q Affects Bandwidth A higher value of circuit Q results in a smaller bandwidth. A lower value of Q causes a larger bandwidth. A formula for the bandwidth of a resonant circuit in terms of Q is stated in the following equation:

$$BW = \frac{f_r}{Q} \tag{14–11}$$

EXAMPLE 14–12

What is the bandwidth of the filter in Figure 14–25?

FIGURE 14–25

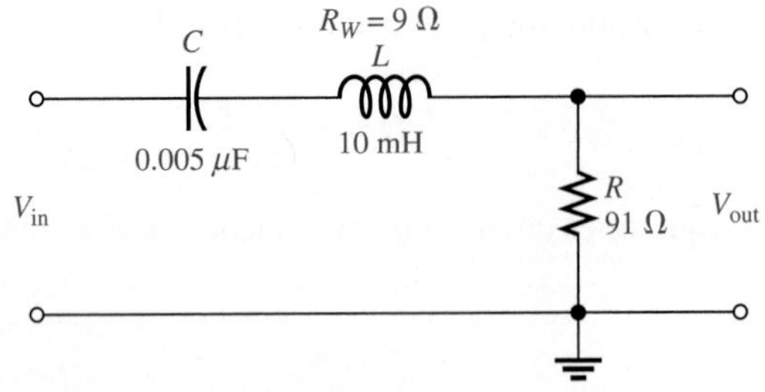

Solution The total resistance is

$$R_T = R + R_W = 91\ \Omega + 9\ \Omega = 100\ \Omega$$

The bandwidth is found as follows:

$$f_r = \frac{1}{2\pi\sqrt{LC}} = \frac{1}{2\pi\sqrt{(10\ \text{mH})(0.005\ \mu\text{F})}} = 22.5\ \text{kHz}$$

$$X_L = 2\pi f_r L = 2\pi(22.5\ \text{kHz})(10\ \text{mH}) = 1.41\ \text{k}\Omega$$

$$Q = \frac{X_L}{R_T} = \frac{1.41\ \text{k}\Omega}{100\ \Omega} = 14.1$$

$$BW = \frac{f_r}{Q} = \frac{22.5\ \text{kHz}}{14.1} = 1.6\ \text{kHz}$$

Related Exercise Determine the bandwidth if L is changed to 50 mH in Figure 14–25.

The Band-Stop Filter

The basic series resonant **band-stop filter** is shown in Figure 14–26. Notice that the output voltage is taken across the *LC* portion of the circuit. This filter is still a series *RLC* circuit, just as the band-pass filter is. The difference is that in this case, the output voltage is taken across the combination of *L* and *C* rather than across *R*.

The band-stop filter rejects signals with frequencies between the upper and lower cutoff frequencies and passes those signals with frequencies below and above the cutoff values, as shown in the response curve of Figure 14–27. The range of frequencies between the lower and upper cutoff points is called the **stop band.** This type of filter is also referred to as a *band-elimination filter, band-reject filter,* or a *notch filter.*

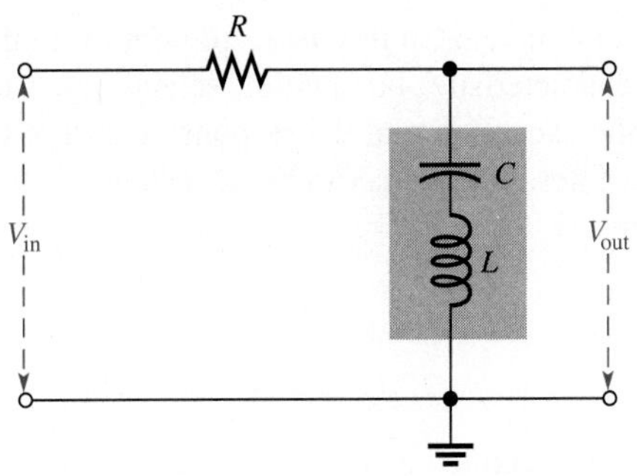

FIGURE 14–26
A basic series resonant band-stop filter.

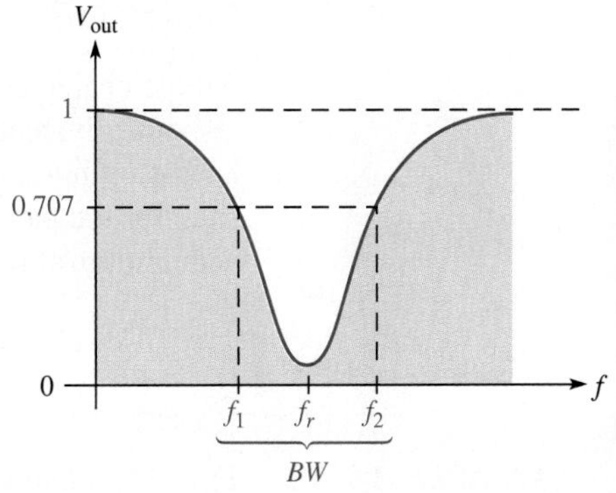

FIGURE 14–27
Generalized response curve for a band-stop filter.

At very low frequencies, the *LC* combination appears as a near open due to the high X_C, thus allowing most of the input voltage to pass through to the output. As the frequency increases, the impedance of the *LC* combination decreases until, at resonance, it is zero (ideally). Thus, the input signal is shorted to ground, and there is very little output voltage. As the frequency goes above its resonant value, the *LC* impedance increases, allowing an increasing amount of voltage to be dropped across it. The general frequency response of a series resonant band-stop filter is illustrated in Figure 14–28.

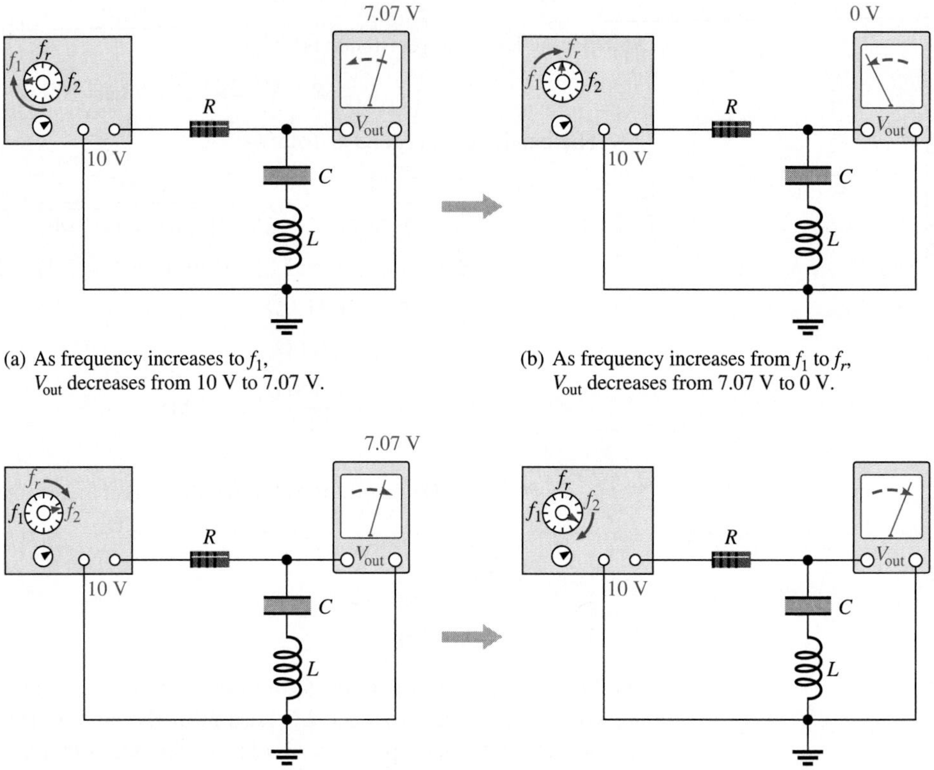

(a) As frequency increases to f_1, V_{out} decreases from 10 V to 7.07 V.

(b) As frequency increases from f_1 to f_r, V_{out} decreases from 7.07 V to 0 V.

(c) As frequency increases from f_r to f_2, V_{out} increases from 0 V to 7.07 V.

(d) As frequency increases above f_2, V_{out} increases toward 10 V.

FIGURE 14–28

Example of the frequency response of a series resonant band-stop filter with V_{in} at a constant 10 V rms. The winding resistance is neglected.

Characteristics of the Band-Stop Filter

All the characteristics that have been discussed in relation to the band-pass filter (current response, impedance characteristic, bandwidth, selectivity, and *Q*) apply equally to the band-stop filter, with the exception that the response curve of the output voltage is opposite. For the band-pass filter, V_{out} is maximum at resonance. For the band-stop filter, V_{out} is minimum at resonance.

EXAMPLE 14–13 Find the output voltage at f_r and the bandwidth in Figure 14–29.

FIGURE 14–29

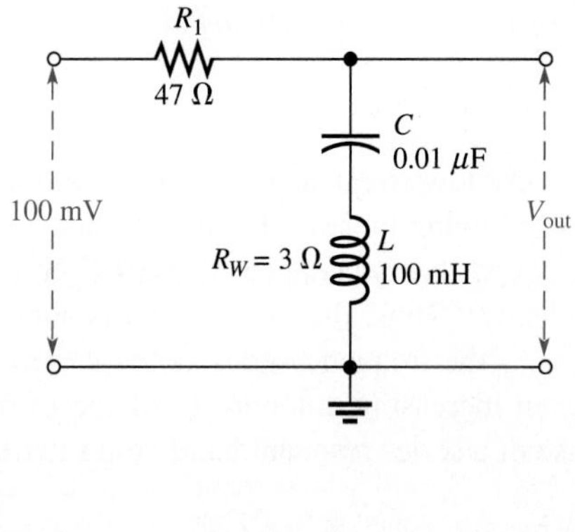

Solution Since $X_L = X_C$ at resonance, then

$$V_{out} = \left(\frac{R_W}{R_1 + R_W}\right)V_{in} = \left(\frac{3\ \Omega}{50\ \Omega}\right)100\text{ mV} = 6.0\text{ mV}$$

The bandwidth is found as follows:

$$f_r = \frac{1}{2\pi\sqrt{LC}} = \frac{1}{2\pi\sqrt{(100\text{ mH})(0.01\ \mu\text{F})}} = 5.03\text{ kHz}$$

$$X_L = 2\pi f_r L = 2\pi(5.03\text{ kHz})(100\text{ mH}) = 3160\ \Omega$$

$$Q = \frac{X_L}{R} = \frac{X_L}{R_1 + R_W} = \frac{3160\ \Omega}{50\ \Omega} = 63.2$$

$$BW = \frac{f_r}{Q} = \frac{5.03\text{ kHz}}{63.2} = 79.6\text{ Hz}$$

Related Exercise What happens to V_{out} if the frequency is increased above resonance? Below resonance?

SECTION 14–4 REVIEW

1. The output voltage of a certain band-pass filter is 15 V at the resonant frequency. What is its value at the cutoff frequencies?
2. For a certain band-pass filter, f_r = 120 kHz and Q = 12. What is the bandwidth of the filter?
3. In a band-stop filter, is the current minimum or maximum at resonance? Is the output voltage minimum or maximum at resonance?

14–5 ■ PARALLEL *RLC* CIRCUITS

In this section, you will learn how to determine the impedance and phase angle of a parallel RLC circuit. The current relationships are also covered.

After completing this section, you should be able

- **Analyze parallel *RLC* circuits**
  - ☐ Calculate the impedance
  - ☐ Calculate the phase angle
  - ☐ Determine all currents
  - ☐ Convert from a series-parallel *RLC* circuit to an equivalent parallel form

Impedance and Phase Angle

The circuit in Figure 14–30 consists of the parallel combination of *R*, *L*, and *C*. To find the admittance, add the conductance (*G*) and the total susceptance (B_T) as phasor quanti-

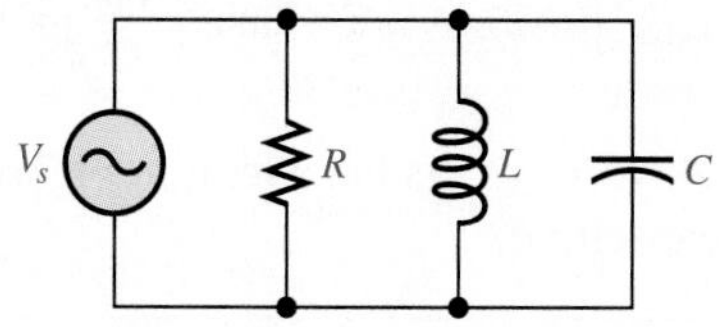

FIGURE 14–30
Parallel RLC circuit.

ties. B_T is the difference of the capacitive susceptance and the inductive susceptance. Thus,

$$Y = \sqrt{G^2 + (B_C - B_L)^2} \tag{14–12}$$

The total impedance is the reciprocal of the admittance:

$$Z_T = \frac{1}{Y}$$

The phase angle of the circuit is given by the following formula:

$$\theta = \tan^{-1}\left(\frac{B_T}{G}\right) \tag{14–13}$$

When the frequency is above its resonant value ($X_C < X_L$), the impedance of the circuit in Figure 14–30 is predominantly capacitive because the capacitive current is greater, and the total current leads the source voltage. When the frequency is below its resonant value ($X_L < X_C$), the impedance of the circuit is predominantly inductive, and the total current lags the source voltage.

EXAMPLE 14–14 Determine the total impedance and the phase angle in Figure 14–31.

FIGURE 14–31

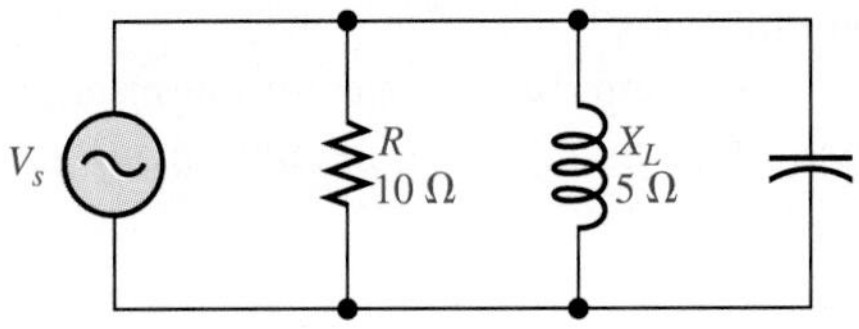

Solution First determine the admittance as follows:

$$G = \frac{1}{R} = \frac{1}{10\ \Omega} = 100\text{ mS}$$

$$B_C = \frac{1}{X_C} = \frac{1}{10\ \Omega} = 100\text{ mS}$$

$$B_L = \frac{1}{X_L} = \frac{1}{5\ \Omega} = 200\text{ mS}$$

$$Y = \sqrt{G^2 + (B_C - B_L)^2} = \sqrt{(100\text{ mS})^2 + (100\text{ mS} - 200\text{ mS})^2}$$
$$= \sqrt{(100\text{ mS})^2 + (-100\text{ mS})^2} = 141.4\text{ mS}$$

From Y, you can get Z_T:

$$Z_T = \frac{1}{Y} = \frac{1}{141.4\text{ mS}} = 7.07\ \Omega$$

The phase angle is

$$\theta = \tan^{-1}\left(\frac{B_T}{G}\right) = \tan^{-1}\left(\frac{100\text{ mS}}{100\text{ mS}}\right) = 45°$$

Related Exercise Is the circuit in Figure 14–31 predominately inductive or predominately capacitive?

Current Relationships

In a parallel *RLC* circuit, the current in the capacitive branch and the current in the inductive branch are always 180° out of phase with each other (neglecting any coil resistance). For this reason, I_C and I_L subtract from each other, and thus the total current into the parallel branches of L and C is always less than the largest individual branch current, as illustrated in Figure 14–32 and in the waveform diagram of Figure 14–33. Of course, the current in the resistive branch is always 90° out of phase with both reactive currents, as shown in the current phasor diagram of Figure 14–34. The total current can be expressed as

$$I_T = \sqrt{I_R^2 + (I_C - I_L)^2} \qquad \textbf{(14–14)}$$

The phase angle can also be expressed in terms of the branch currents as

$$\theta = \tan^{-1}\left(\frac{I_{CL}}{I_R}\right) \qquad \textbf{(14–15)}$$

where I_{CL} is $|I_C - I_L|$.

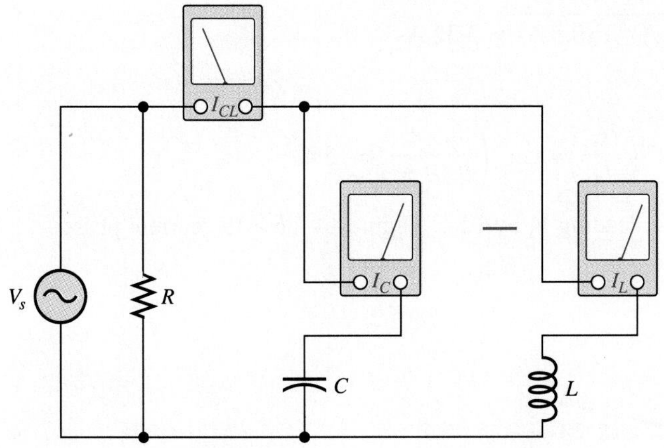

FIGURE 14–32
The total current into the parallel combination of C and L is the difference of the two branch currents ($I_{CL} = I_C - I_L$).

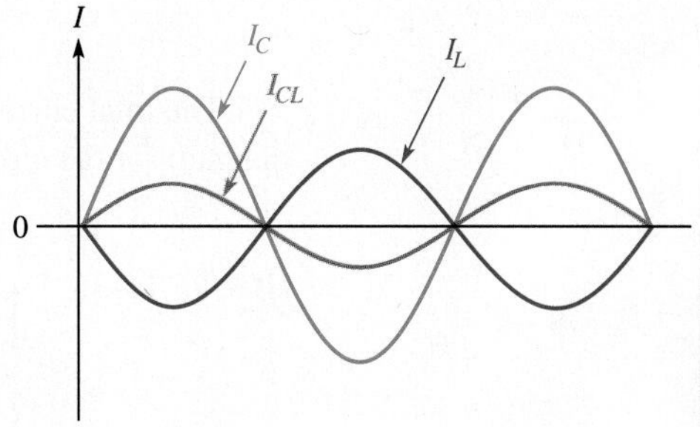

FIGURE 14–33
I_C and I_L effectively subtract.

FIGURE 14–34
Current phasor diagram for a parallel RLC circuit.

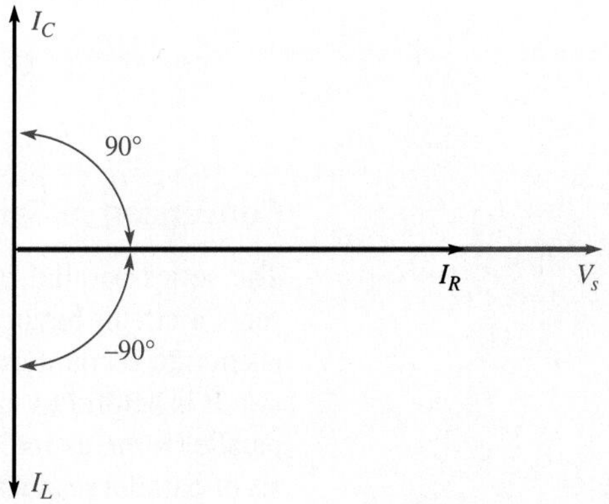

EXAMPLE 14–15 Find each branch current and the total current in Figure 14–35, and show their relationship.

FIGURE 14–35

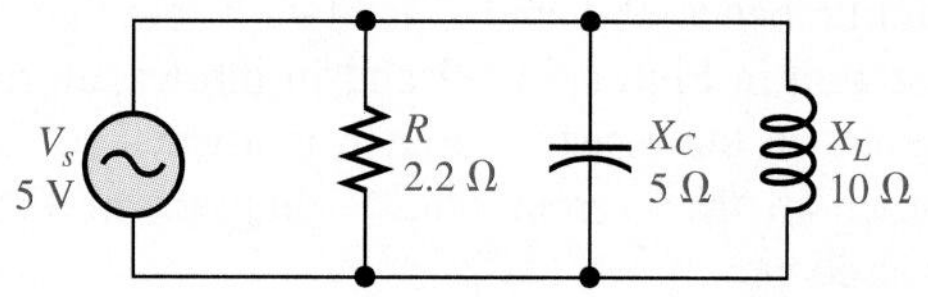

Solution Each branch current can be found using Ohm's law:

$$I_R = \frac{V_s}{R} = \frac{5\text{ V}}{2.2\ \Omega} = 2.27\text{ A}$$

$$I_C = \frac{V_s}{X_C} = \frac{5\text{ V}}{5\ \Omega} = 1\text{ A}$$

$$I_L = \frac{V_s}{X_L} = \frac{5\text{ V}}{10\ \Omega} = 0.5\text{ A}$$

The total current is the phasor sum of the branch currents:

$$I_T = \sqrt{I_R^2 + (I_C - I_L)^2} = \sqrt{(2.27\text{ A})^2 + (1\text{ A} - 0.5\text{ A})^2}$$
$$= \sqrt{(2.27\text{ A})^2 + (0.5\text{ A})^2} = 2.32\text{ A}$$

The phase angle is

$$\theta = \tan^{-1}\left(\frac{I_{CL}}{I_R}\right) = \tan^{-1}\left(\frac{0.5\text{ A}}{2.27\text{ A}}\right) = 12.4°$$

The total current is 2.32 A, leading V_s by 12.4°. Figure 14–36 is the current phasor diagram for the circuit.

FIGURE 14–36

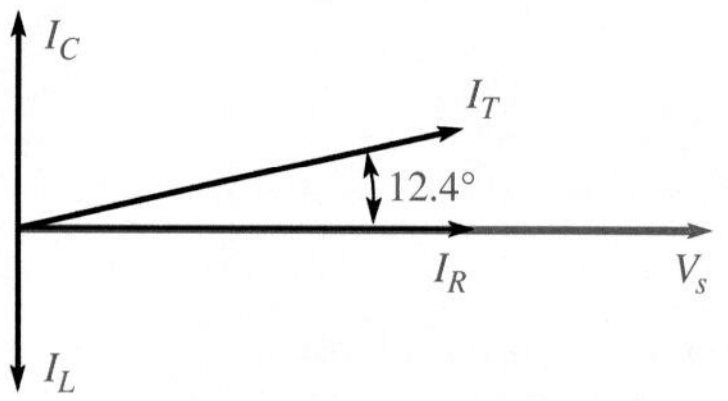

Related Exercise Will the total current increase or decrease if the frequency in Figure 14–35 is increased? Why?

Conversion of Series-Parallel to Parallel

The series-parallel configuration shown in Figure 14–37 is important because it represents a circuit having parallel L and C branches, with the winding resistance of the coil taken into account as a series resistance in the L branch.

It is helpful to view the series-parallel circuit in Figure 14–37 in an equivalent purely parallel form, as indicated in Figure 14–38. This equivalent form will simplify the analysis of parallel resonant characteristics as discussed in Section 14–6.

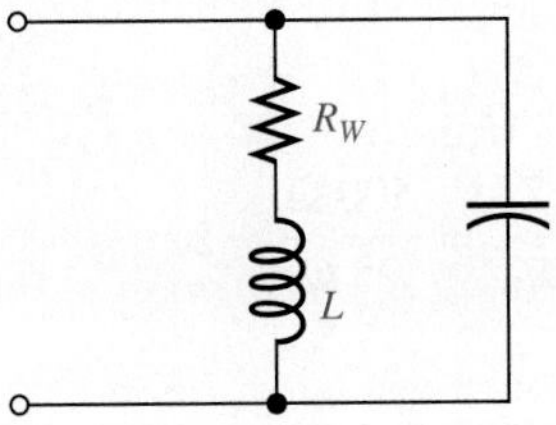

FIGURE 14–37
An important series-parallel RLC circuit ($Q = X_L/R_W$).

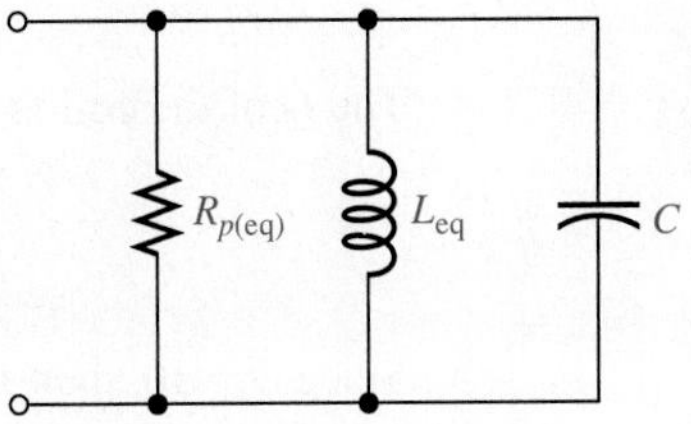

FIGURE 14–38
Parallel equivalent form of the circuit in Figure 14–37.

The equivalent inductance L_{eq} and the equivalent parallel resistance $R_{p(eq)}$ are given by the following formulas:

$$L_{eq} = L\left(\frac{Q^2 + 1}{Q^2}\right) \tag{14–16}$$

$$R_{p(eq)} = R_W(Q^2 + 1) \tag{14–17}$$

where Q is the quality factor of the coil, X_L/R_W. Derivations of these formulas are quite involved and thus are not given here. Notice in the equations that for a $Q \geq 10$, the value of L_{eq} is approximately the same as the original value of L. For example, if $L = 10$ mH, then

$$L_{eq} = 10\text{ mH}\left(\frac{10^2 + 1}{10^2}\right) = 10\text{ mH}(1.01) = 10.1\text{ mH}$$

The equivalency of the two circuits means that at a given frequency, when the same value of voltage is applied to both circuits, the same total current flows in both circuits and the phase angles are the same. Basically, an equivalent circuit simply makes circuit analysis more convenient.

EXAMPLE 14–16

Convert the series-parallel circuit in Figure 14–39 to an equivalent parallel form at the given frequency.

FIGURE 14–39

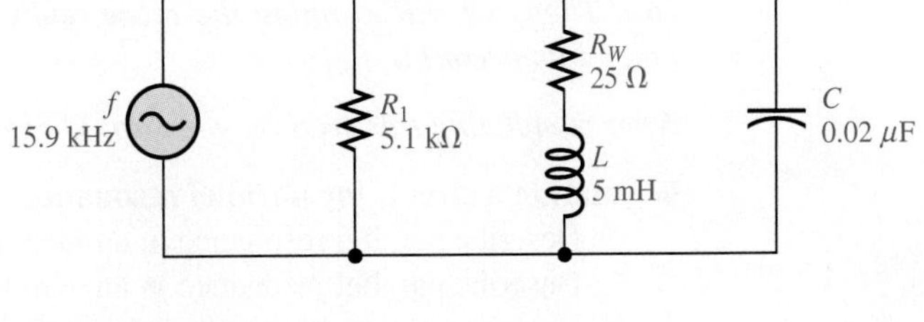

Solution Determine the inductive reactance:

$$X_L = 2\pi fL = 2\pi(15.9\text{ kHz})(5\text{ mH}) = 500\ \Omega$$

The Q of the coil is

$$Q = \frac{X_L}{R_W} = \frac{500\ \Omega}{25\ \Omega} = 20$$

Since $Q > 10$, then $L_{eq} \cong L = 5$ mH.

The equivalent parallel resistance is

$$R_{p(eq)} = R_W(Q^2 + 1) = (25\ \Omega)(20^2 + 1) = 10.0\ \text{k}\Omega$$

This equivalent resistance ($R_{p(eq)}$) appears in parallel with R_1 as shown in Figure 14–40(a). When combined, they give a total parallel resistance ($R_{p(tot)}$) of 3.38 kΩ, as indicated in Figure 14–40(b).

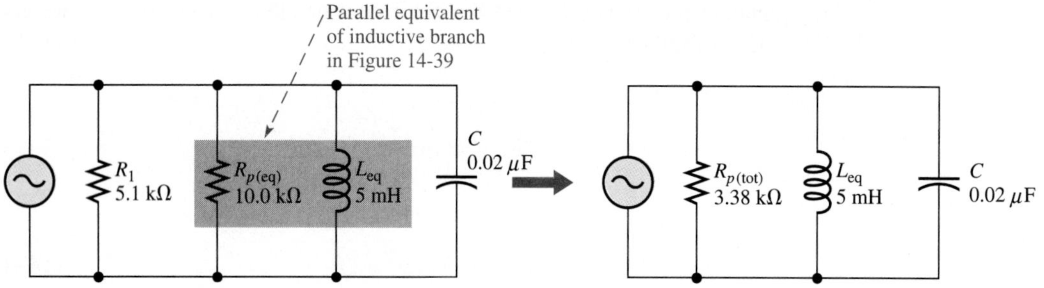

(a) Parallel equivalent of the circuit in Figure 14-39

(b) $R_{p(tot)} = R_1 \parallel R_{p(eq)} = 3.38$ kΩ

FIGURE 14–40

Related Exercise Find the equivalent parallel circuit if $R_W = 10\ \Omega$ in Figure 14–39.

SECTION 14–5 REVIEW

1. In a three-branch parallel circuit, $R = 150\ \Omega$, $X_C = 100\ \Omega$, and $X_L = 50\ \Omega$ at a certain frequency. Determine the current in each branch when $V_s = 12$ V.
2. Is the circuit in Question 1 capacitive or inductive? Why?
3. Find the equivalent parallel inductance and resistance for a 20 mH coil with a winding resistance of 10 Ω at a frequency of 1 kHz.

14–6 ▪ PARALLEL RESONANCE

In this section, we will first look at the resonant condition in an ideal parallel LC circuit. Then, we will examine the more realistic case where the resistance of the coil is taken into account.

After completing this section, you should be able to

- **Analyze a circuit for parallel resonance**
  - ☐ Describe parallel resonance in an ideal circuit
  - ☐ Describe parallel resonance in an nonideal circuit
  - ☐ Explain how impedance varies with frequency
  - ☐ Determine current and phase angle at resonance
  - ☐ Determine parallel resonant frequency
  - ☐ Discuss the effects of loading a parallel resonant circuit

Condition for Ideal Parallel Resonance

Ideally, **parallel resonance** occurs when $X_L = X_C$. The frequency at which resonance occurs is called the *resonant frequency,* just as in the series case. When $X_L = X_C$, the two branch currents, I_C and I_L, are equal in magnitude, and, of course, they are always 180° out of phase with each other. Thus, the two currents cancel and the total current is zero, as presented in Figure 14–41. In this ideal case, the winding resistance of the coil is assumed to be zero.

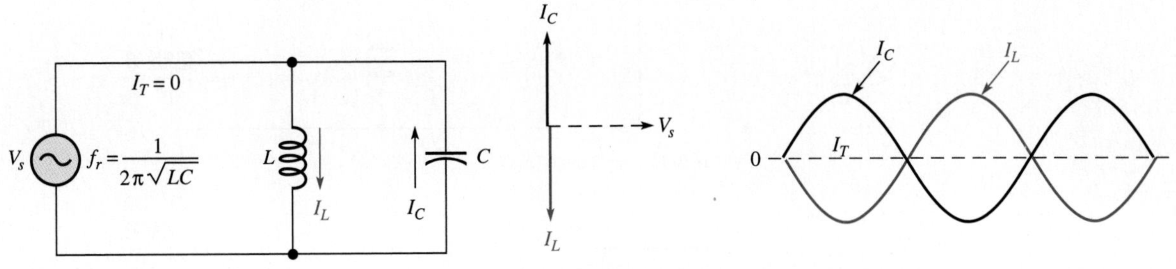

(a) Parallel circuit at resonance ($X_C = X_L$, $Z = \infty$)

(b) Current phasors

(c) Current waveforms

FIGURE 14–41
An ideal parallel LC circuit at resonance.

Since the total current is zero, the impedance of the parallel *LC* circuit is infinitely large (∞). These ideal resonant conditions are stated in the following equations:

$$X_L = X_C$$

$$Z_r = \infty$$

The Parallel Resonant Frequency

For an ideal parallel resonant circuit, the frequency at which resonance occurs is determined by the same formula as in series resonant circuits:

$$f_r = \frac{1}{2\pi\sqrt{LC}} \qquad \textbf{(14–18)}$$

Currents in a Parallel Resonant Circuit

It is interesting to see how the currents in a parallel *LC* circuit vary as the frequency is increased from below the resonant value, through resonance, and then above the resonant value. Figure 14–42 illustrates the general response of an ideal circuit in terms of changes in the currents. Part (a) indicates the response as the source frequency is increased from zero toward f_r. At very low frequencies, X_C is very high and X_L is very low, so most of the current is through *L*. As the frequency increases, the current through *L* decreases and the current through *C* increases, causing the total current to decrease. At all times, I_L and I_C are 180° out of phase with each other, and thus the total current is the difference of the two branch currents. During this time, the impedance is increasing, as indicated by the decrease in total current.

When the frequency reaches its resonant value, f_r, as shown in Figure 14–42(b), X_C and X_L are equal, so I_C and I_L cancel because they are equal in magnitude but opposite in phase. At this point the total current is at its minimum value of zero. Since I_T is zero, Z is infinite. Thus, the ideal parallel *LC* circuit appears as an open at the resonant frequency.

FIGURE 14–42

An illustration of how the currents respond in a parallel LC circuit as the frequency is varied from below resonance to above resonance. The source voltage amplitude is constant.

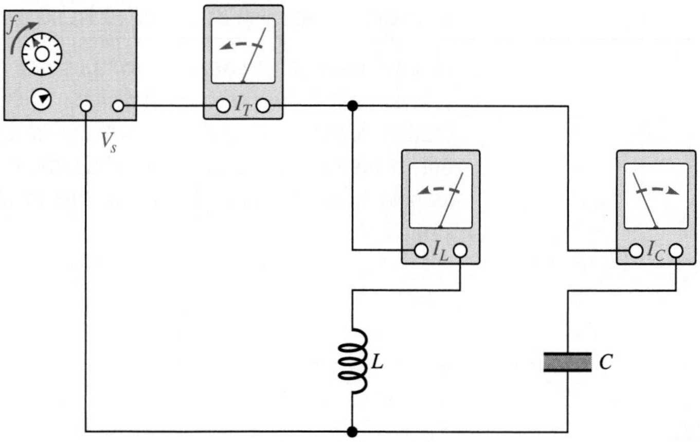

(a) Below resonance ($X_C > X_L$)

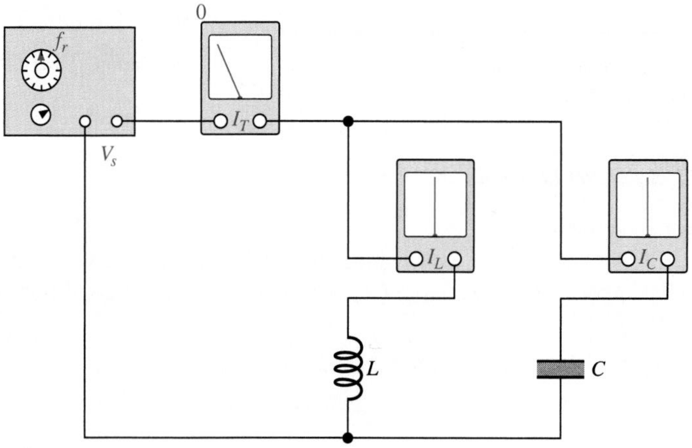

(b) At resonance ($X_C = X_L$)

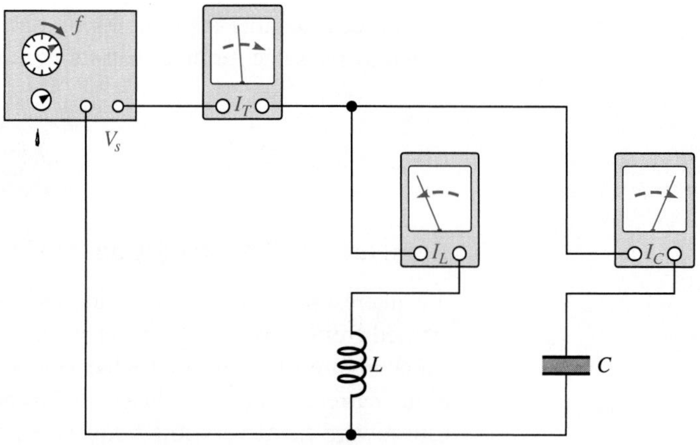

(c) Above resonance ($X_L > X_C$)

As the frequency is increased above resonance, as indicated in Figure 14–42(c), X_C continues to decrease and X_L continues to increase, causing the branch currents again to be unequal, with I_C being larger. As a result, total current increases and impedance decreases. As the frequency becomes very high, the impedance becomes very small due to the dominance of a very small X_C in parallel with a very large X_L.

In summary, the current dips to a minimum as the impedance peaks to a maximum at parallel resonance. The expression for the total current into the L and C branches is

$$I_T = |I_C - I_L| \quad \textbf{(14–19)}$$

EXAMPLE 14–17 Find the resonant frequency and the branch currents in the ideal (winding resistance neglected) parallel LC circuit of Figure 14–43.

FIGURE 14–43

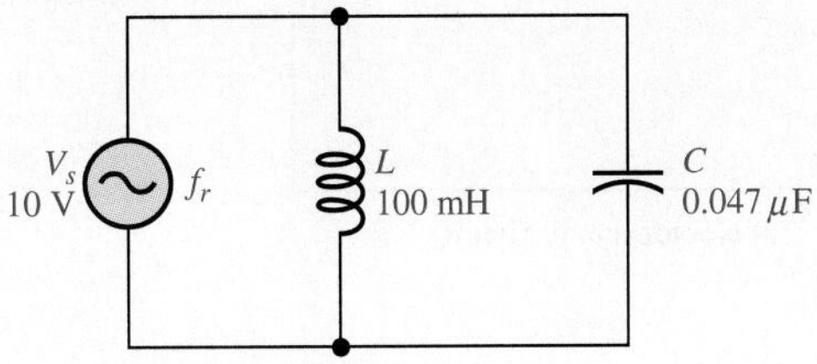

Solution The resonant frequency is

$$f_r = \frac{1}{2\pi\sqrt{LC}} = \frac{1}{2\pi\sqrt{(100\text{ mH})(0.047\ \mu\text{F})}} = 2.32\text{ kHz}$$

Determine the branch currents as follows:

$$X_L = 2\pi f_r L = 2\pi(2.32\text{ kHz})(100\text{ mH}) = 1.46\text{ k}\Omega$$

$$X_C = X_L = 1.46\text{ k}\Omega$$

$$I_L = \frac{V_s}{X_L} = \frac{10\text{ V}}{1.46\text{ k}\Omega} = 6.85\text{ mA}$$

$$I_C = I_L = 6.85\text{ mA}$$

The total current into the parallel LC circuit is

$$I_T = I_C - I_L = 0\text{ A}$$

Related Exercise Find f_r and all the currents at resonance in Figure 14–43 if C is changed to 0.022 μF.

The Parallel Resonant *LC* Circuit Is Often Called a Tank Circuit

The term **tank circuit** refers to the fact that the parallel resonant circuit stores energy in the magnetic field of the coil and in the electric field of the capacitor. The stored energy is transferred back and forth between the capacitor and the coil on alternate half-cycles as the current goes first one way and then the other when the indicator deenergizes and the capacitor charges, and vice versa. This concept is illustrated in Figure 14–44.

FIGURE 14–44
Energy storage in an ideal parallel resonant tank circuit.

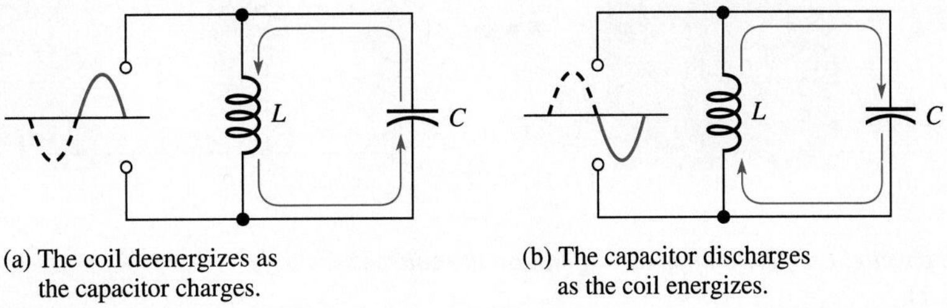

Parallel Resonant Conditions in a Nonideal Circuit

So far, the resonance of an ideal parallel LC circuit has been examined. Now we will consider resonance in a tank circuit with the resistance of the coil taken into account. Figure 14–45 shows a nonideal tank circuit and its parallel RLC equivalent.

FIGURE 14–45
A practical treatment of parallel resonant circuits includes the coil resistance.

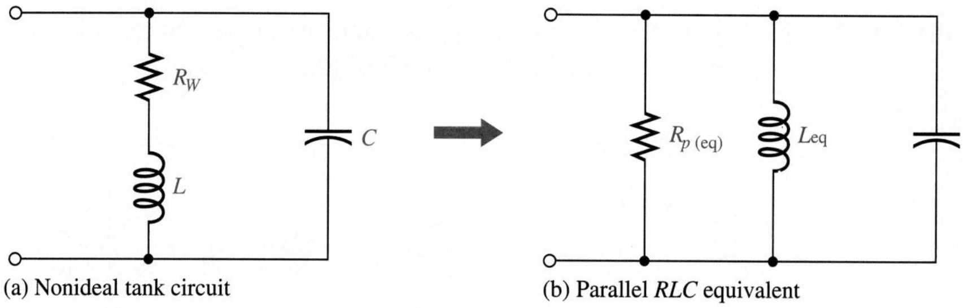

The Q of the circuit at resonance is simply the Q of the coil:

$$Q = \frac{X_L}{R_W}$$

The expressions for the equivalent parallel resistance and the equivalent inductance were given in Equations (14–17) and (14–16) as

$$R_{p(\text{eq})} = R_W(Q^2 + 1)$$

$$L_{\text{eq}} = L\left(\frac{Q^2 + 1}{Q^2}\right)$$

Recall that for $Q \geq 10$, $L_{\text{eq}} \cong L$.

At parallel resonance,

$$X_{L(\text{eq})} = X_C$$

In the parallel equivalent circuit, $R_{p(\text{eq})}$ is in parallel with an ideal coil and a capacitor, so the L and C branches act as an ideal tank circuit which has an infinite impedance at resonance as shown in Figure 14–46. Therefore, the total impedance of the nonideal tank circuit at resonance can be expressed as simply the equivalent parallel resistance:

$$Z_r = R_W(Q^2 + 1) \qquad \textbf{(14–20)}$$

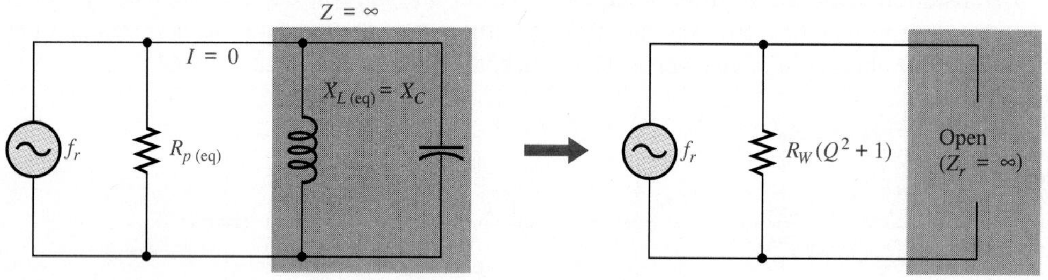

FIGURE 14–46
At resonance, the parallel LC portion appears open and the source sees only $R_{p(eq)} = R_W(Q^2 = 1)$.

EXAMPLE 14–18 Determine the impedance of the circuit in Figure 14–47 at the resonant frequency ($f_r \cong$ 17,794 Hz).

FIGURE 14–47

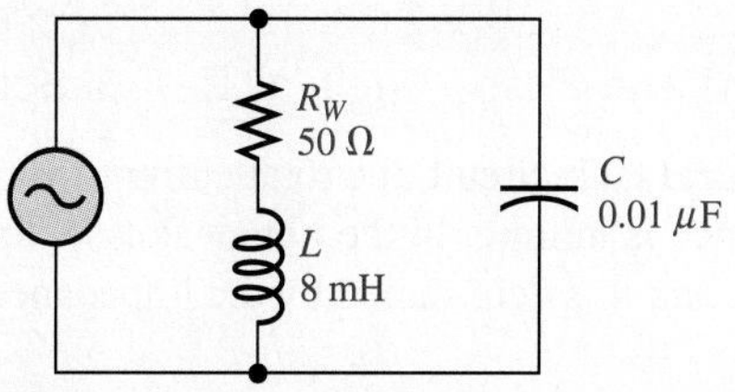

Solution First calculate the quality factor Q as follows:

$$X_L = 2\pi f_r L = 2\pi(17{,}794\text{ Hz})(8\text{ mH}) = 894\ \Omega$$

$$Q = \frac{X_L}{R_W} = \frac{894\ \Omega}{50\ \Omega} = 17.9$$

The impedance is

$$Z_r = R_W(Q^2 + 1) = 50\ \Omega(17.9^2 + 1) = 16.1\text{ k}\Omega$$

The calculator sequence is

2 × 2nd π × 1 7 7 9 4 × 8 EXP +/− 3 = ÷ 5 0
= x^2 + 1 = × 5 0 =

Related Exercise Determine Z_r in Figure 14–47 if the winding resistance is 10 Ω.

Variation of the Impedance with Frequency

The impedance of a parallel resonant circuit is maximum at the resonant frequency and decreases at lower and higher frequencies, as indicated by the curve in Figure 14–48.

At very low frequencies, X_L is very small and X_C is very high, so the total impedance is essentially equal to that of the inductive branch. As the frequency goes up, the

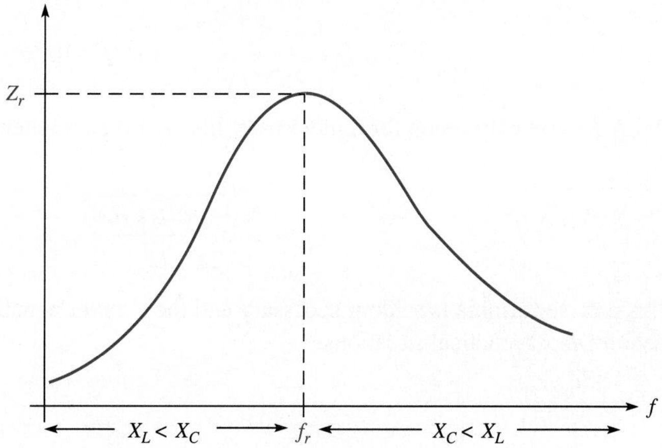

FIGURE 14–48
Generalized impedance curve for a parallel resonant circuit. The circuit is inductive below f_r, resistive at f_r, and capacitive above f_r.

impedance also increases, and the inductive reactance dominates (because it is less than X_C) until the resonant frequency is reached. At this point, of course, $X_L \cong X_C$ (for $Q > 10$) and the impedance is at its maximum. As the frequency goes above resonance, the capacitive reactance dominates (because it is less than X_L) and the impedance decreases.

Current and Phase Angle at Resonance

In the ideal tank circuit, the total current from the source at resonance is zero because the impedance is infinite. In the nonideal case, there is some total current at the resonant frequency, and it is determined by the impedance at resonance:

$$I_T = \frac{V_s}{Z_r} \tag{14–21}$$

The phase angle of the parallel resonant circuit is 0° because the impedance is purely resistive at the resonant frequency.

Parallel Resonant Frequency in a Nonideal Circuit

As you know, when the coil resistance is considered, the resonant condition is

$$X_{L(\text{eq})} = X_C$$

which can be expressed as

$$2\pi f_r L\left(\frac{Q^2+1}{Q^2}\right) = \frac{1}{2\pi f_r C}$$

Solving for f_r in terms of Q,

$$f_r = \frac{1}{2\pi\sqrt{LC}}\sqrt{\frac{Q^2}{Q^2+1}}$$

When $Q \geq 10$, the term with the Q factors is approximately 1:

$$\sqrt{\frac{Q^2}{Q^2+1}} = \sqrt{\frac{100}{101}} = 0.995 \cong 1$$

Therefore, the parallel resonant frequency is approximately the same as the series resonant frequency as long as Q is equal to or greater than 10:

$$f_r \cong \frac{1}{2\pi\sqrt{LC}} \qquad \text{for } Q \geq 10$$

A precise expression for f_r in terms of the circuit component values is

$$f_r = \frac{\sqrt{1-(R_W^2C/L)}}{2\pi\sqrt{LC}} \tag{14–22}$$

This precise formula is seldom necessary and the simpler equation $f_r = 1/2\pi\sqrt{LC}$ is sufficient for most practical situations.

EXAMPLE 14–19 Find the frequency, impedance, and total current at resonance for the circuit in Figure 14–49 using Equation (14–22).

FIGURE 14–49

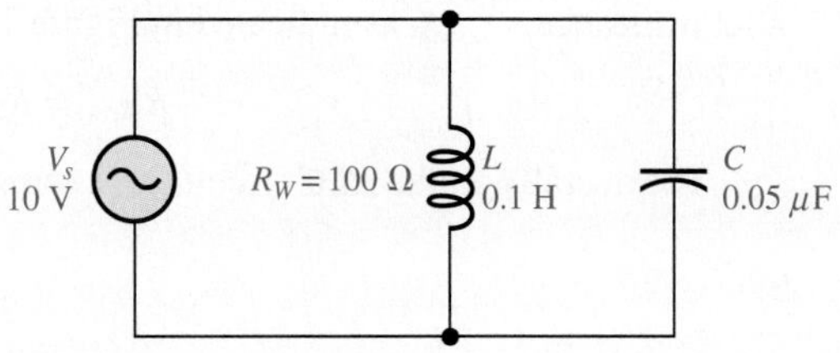

Solution The precise resonant frequency is

$$f_r = \frac{\sqrt{1 - (R_W^2 C/L)}}{2\pi\sqrt{LC}} = \frac{\sqrt{1 - [(100\ \Omega)^2(0.05\ \mu\text{F})/0.1\ \text{H}]}}{2\pi\sqrt{(0.05\ \mu\text{F})(0.1\ \text{H})}} = 2.25\ \text{kHz}$$

Calculate impedance as follows:

$$X_L = 2\pi f_r L = 2\pi(2.25\ \text{kHz})(0.1\ \text{H}) = 1.4\ \text{k}\Omega$$

$$Q = \frac{X_L}{R_W} = \frac{1.4\ \text{k}\Omega}{100\ \Omega} = 14$$

$$Z_r = R_W(Q^2 + 1) = 100\ \Omega(14^2 + 1) = 19.7\ \text{k}\Omega$$

The total current is

$$I_T = \frac{V_s}{Z_r} = \frac{10\ \text{V}}{19.7\ \text{k}\Omega} = 508\ \mu\text{A}$$

The calculator sequence for f_r is

[1] [−] [(] [1] [0] [0] [x²] [×] [•] [0] [5] [EXP] [+/−] [6] [÷] [•] [1]
[=] [)] [=] [2nd] [√x] [÷] [2] [÷] [2nd] [π] [÷] [(] [(] [•] [0] [5]
[EXP] [+/−] [6] [×] [•] [1] [)] [2nd] [√x] [=]

Related Exercise Repeat this example using the formula $f_r = 1/2\pi\sqrt{LC}$ and compare the results.

An External Parallel Load Resistance Affects a Tank Circuit

There are many practical situations in which an external load resistance appears in parallel with a tank circuit as shown in Figure 14–50(a). Obviously, the external resistor (R_L) will dissipate more of the energy delivered by the source and thus will lower the overall Q (Q_O) of the circuit. The external resistor effectively appears in parallel with the equiva-

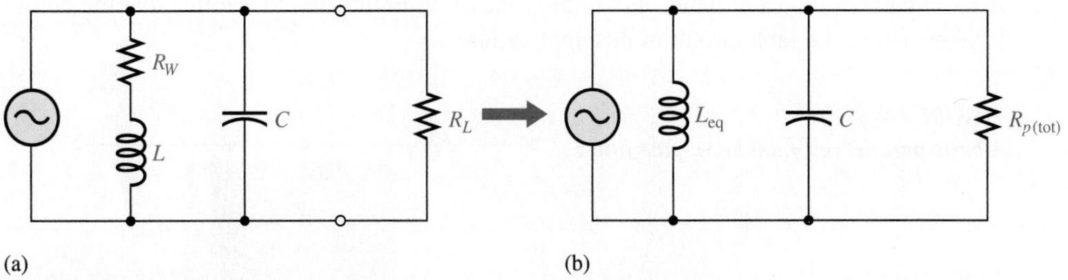

(a) (b)

FIGURE 14–50
Tank circuit with a parallel load resistor and its equivalent circuit.

lent parallel resistance of the coil, $R_{p(eq)}$, and both are combined to determine a total parallel resistance, $R_{p(tot)}$, as indicated in Figure 14–50(b):

$$R_{p(tot)} = R_L \parallel R_{p(eq)}$$

The overall parallel *RLC* circuit Q is expressed differently from the Q of a series circuit:

$$Q_O = \frac{R_{p(tot)}}{X_{L(eq)}} \quad \textbf{(14–23)}$$

As you can see, the effect of loading the tank circuit is to reduce its overall Q (which is equal to the coil Q when unloaded).

SECTION 14–6 REVIEW

1. Is the impedance minimum or maximum at parallel resonance?
2. Is the current minimum or maximum at parallel resonance?
3. At ideal parallel resonance, $X_L = 1.5$ kΩ. What is X_C?
4. A tank circuit has the following values: $R_W = 4$ Ω, $L = 50$ mH, and $C = 10$ pF. Calculate f_r and Z_r.
5. If $Q = 25$, $L = 50$ mH, and $C = 1000$ pF, what is f_r?
6. In Question 5, if $Q = 2.5$, what is f_r?
7. In a certain tank circuit, the coil resistance is 20 Ω. What is the total impedance at resonance if $Q = 20$?

14–7 ▪ PARALLEL RESONANT FILTERS

Parallel resonant circuits are commonly applied to band-pass and band-stop filters. In this section, we examine these applications.

After completing this section, you should be able to

- **Analyze the operation of parallel resonant filters**
  - ☐ Show how a band-pass filter is implemented
  - ☐ Define *bandwidth*
  - ☐ Explain how loading affects selectivity
  - ☐ Show how a band-stop filter is implemented
  - ☐ Determine the resonant frequency, bandwidth, and output voltage of band-pass and band-stop parallel resonant filters

The Band-Pass Filter

A basic parallel resonant band-pass filter is shown in Figure 14–51. Notice that the output is taken across the tank circuit in this application.

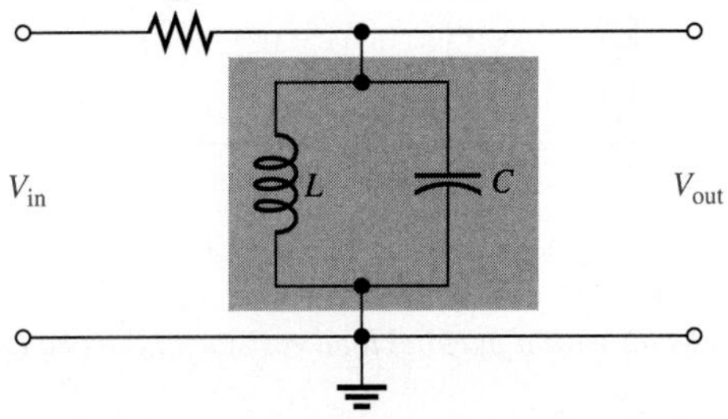

FIGURE 14–51
A basic parallel resonant band-pass filter.

The bandwidth and cutoff frequencies for a parallel resonant band-pass filter are defined in the same way as for the series resonant circuit, and the formulas given in Section 14–4 still apply. General band-pass frequency response curves showing both V_{out} and I_T versus frequency are given in Figures 14–52(a) and 14–52(b), respectively.

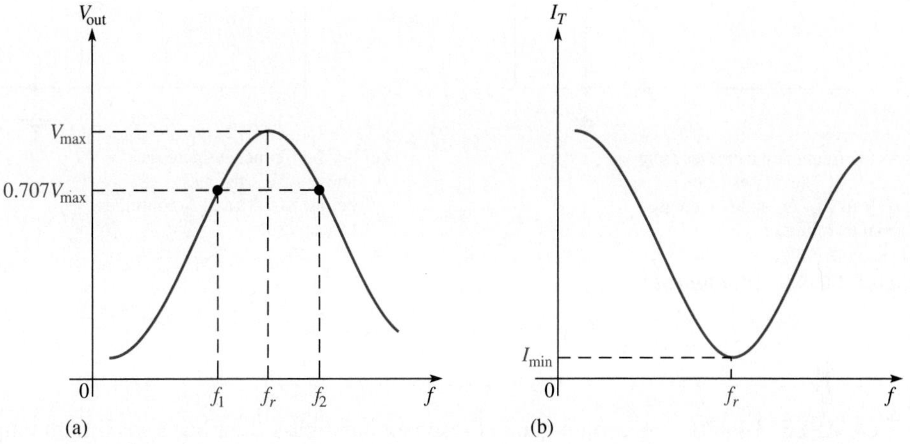

FIGURE 14–52
Generalized frequency response curves for a parallel resonant band-pass filter.

The filtering action is as follows: At very low frequencies, the impedance of the tank circuit is very low, and therefore only a small amount of voltage is dropped across it. As the frequency increases, the impedance of the tank circuit increases, and, as a result, the output voltage increases. When the frequency reaches its resonant value, the impedance is at its maximum and so is the output voltage. As the frequency goes above resonance, the impedance begins to decrease, causing the output voltage to decrease. The general response of a parallel resonant band-pass filter is illustrated in Figure 14–53. This illustration pictorially shows how the current and the output voltage change with frequency. (Parts (c) and (d) of Figure 14–53 are on p. 610.)

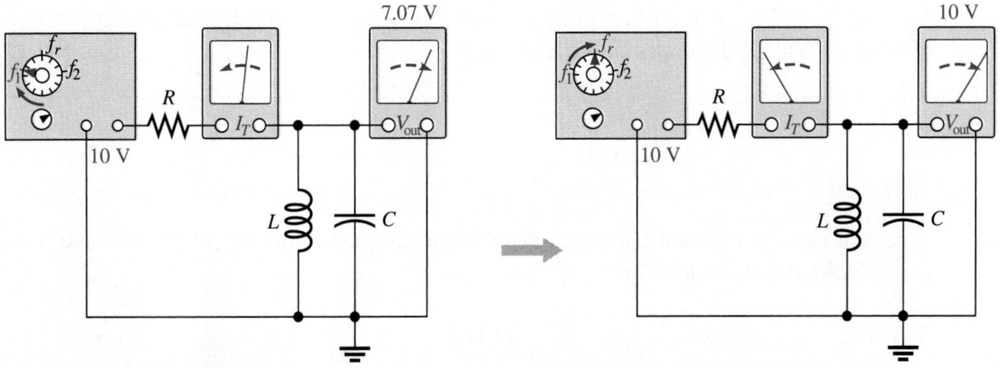

(a) As the frequency increases to f_1, V_{out} increases to 7.07 V, and I_T decreases.

(b) As the frequency increases from f_1 to f_r, V_{out} increases from 7.07 V to 10 V, and I_T decreases to its minimum value.

FIGURE 14–53
Example of the response of a parallel resonant band-pass filter with the input voltage at a constant 10 V rms.

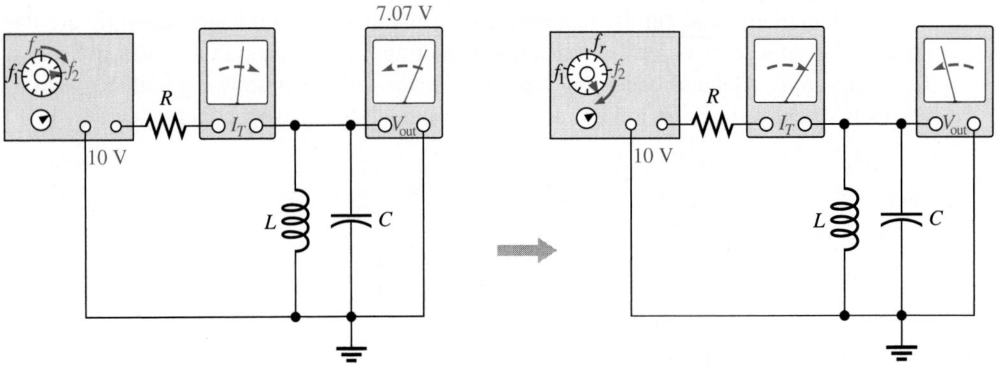

(c) As the frequency increases from f_r to f_2, V_{out} decreases from 10 V to 7.07 V, and I_T increases from its minimum.

(d) As the frequency increases above f_2, V_{out} decreases below 7.07 V, and I_T continues to increase.

FIGURE 14–53 ***(Continued)***

EXAMPLE 14–20

A certain parallel resonant band-pass filter has a maximum output voltage of 4 V at f_r. What is the value of V_{out} at the cutoff frequencies?

Solution V_{out} at the cutoff frequencies is 70.7% of maximum:

$$V_{out(1)} = V_{out(2)} = 0.707V_{out(max)} = 0.707(4\text{ V}) = 2.828\text{ V}$$

Related Exercise What is V_{out} at the cutoff frequencies if V_{out} at the resonant frequency is 10 V?

EXAMPLE 14–21

A parallel resonant circuit has a lower cutoff frequency of 3.5 kHz and an upper cutoff frequency of 6 kHz. What is the bandwidth?

Solution

$$BW = f_2 - f_1 = 6\text{ kHz} - 3.5\text{ kHz} = 2.5\text{ kHz}$$

Related Exercise A filter has a lower cutoff frequency of 520 kHz and a bandwidth of 10 kHz. What is the upper cutoff frequency?

EXAMPLE 14–22

A certain parallel resonant band-pass filter has a resonant frequency of 12 kHz and a Q of 10. What is its bandwidth?

Solution

$$BW = \frac{f_r}{Q} = \frac{12\text{ kHz}}{10} = 1.2\text{ kHz}$$

Related Exercise A certain parallel-resonant band-pass filter has a bandwidth of 4 MHz and a resonant frequency of 100 MHz. What is its Q?

Loading Affects the Selectivity of a Parallel Resonant Band-Pass Filter

When a resistive load is connected across the output of a filter as shown in Figure 14–54(a), the Q of the filter is reduced. Since $BW = f_r/Q$, the bandwidth is increased, thus reducing the selectivity. Also, the impedance of the filter at resonance is decreased because R_L effectively appears in parallel with $R_{p(eq)}$. Thus, the maximum output voltage is reduced by the voltage-divider effect of $R_{p(tot)}$ and the internal source resistance R_s, as illustrated in Figure 14–54(b). Figure 14–54(c) shows the general effect on the filter response curve.

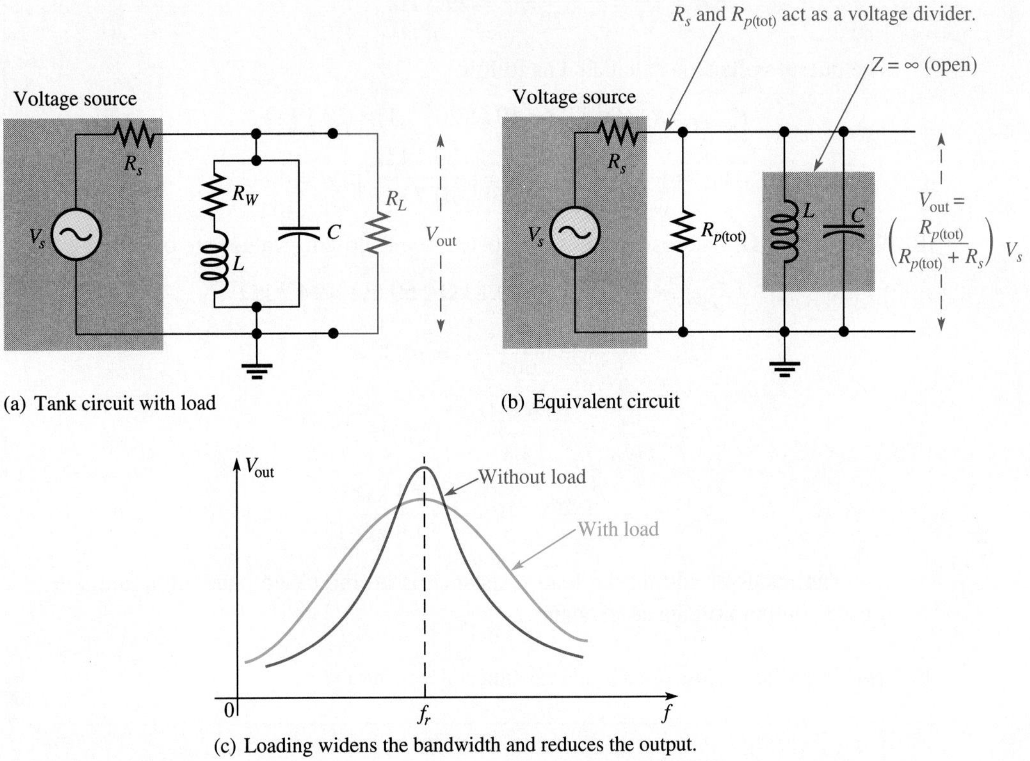

FIGURE 14–54
Effects of loading on a parallel resonant band-pass filter.

EXAMPLE 14–23 **(a)** Determine f_r, BW, and V_{out} at resonance for the unloaded filter in Figure 14–55. The source resistance is 600 Ω.
(b) Repeat part (a) when the filter is loaded with a 50 kΩ resistance, and compare the results.

FIGURE 14–55

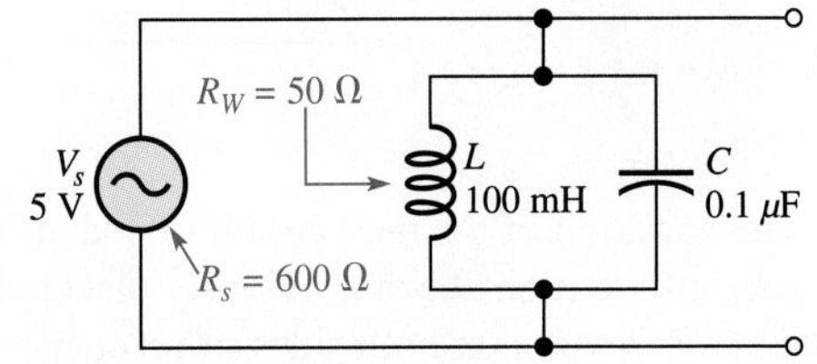

Solution

(a) $f_r \cong \dfrac{1}{2\pi\sqrt{LC}} = \dfrac{1}{2\pi\sqrt{(100\text{ mH})(0.1\ \mu\text{F})}} = 1.59\text{ kHz}$

At resonance, the bandwidth is calculated as follows:

$$X_L = 2\pi f_r L = 2\pi(1.59\text{ kHz})(100\text{ mH}) = 999\ \Omega$$

$$Q = \frac{X_L}{R_W} = \frac{999\ \Omega}{50\ \Omega} = 20$$

$$BW = \frac{f_r}{Q} = \frac{1.59\text{ kHz}}{20} = 79.5\text{ Hz}$$

The output voltage is calculated as follows:

$$R_{p(\text{eq})} = R_W(Q^2 + 1) = 50\ \Omega(20^2 + 1) = 20.1\text{ k}\Omega$$

$$V_{out} = \left(\frac{R_{p(\text{eq})}}{R_{p(\text{eq})} + R_s}\right)V_s = \left(\frac{20.1\text{ k}\Omega}{20.7\text{ k}\Omega}\right)5\text{ V} = 4.86\text{ V}$$

(b) When a 50 kΩ load resistance is connected, the following values are obtained:

$$R_{p(\text{tot})} = R_{p(\text{eq})} \parallel R_L = 20.1\text{ k}\Omega \parallel 50\text{ k}\Omega = 14.3\text{ k}\Omega$$

$$Q_O = \frac{R_{p(\text{tot})}}{X_L} = \frac{14.3\text{ k}\Omega}{999\ \Omega} = 14.3$$

$$BW = \frac{f_r}{Q_O} = \frac{1.59\text{ kHz}}{14.3} = 111\text{ Hz}$$

$$V_{out} = \left(\frac{R_{p(\text{tot})}}{R_{p(\text{tot})} + R_s}\right)V_s = \left(\frac{14.3\text{ k}\Omega}{14.9\text{ k}\Omega}\right)5\text{ V} = 4.80\text{ V}$$

The result of adding the load resistance is an increased bandwidth and a decreased output voltage at resonance.

Related Exercise How does load resistance affect the Q?

The Band-Stop Filter

A basic parallel resonant band-stop filter is shown in Figure 14–56. The output is taken across a load resistor in series with the tank circuit.

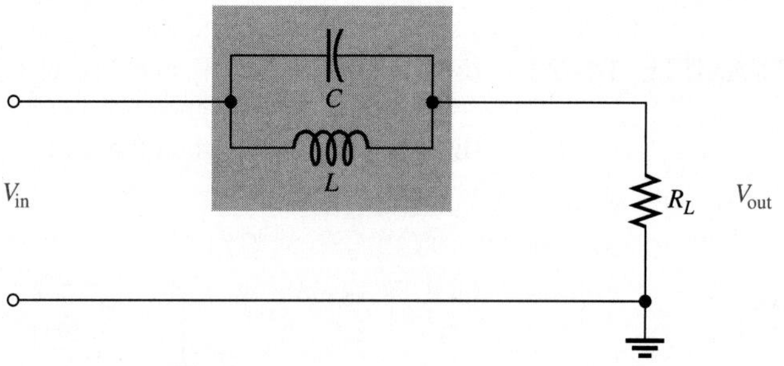

FIGURE 14–56
A basic parallel resonant band-stop filter.

The variation of the tank circuit impedance with frequency produces the familiar current response that has been previously discussed; that is, the current is minimum at resonance and increases on both sides of resonance. Since the output voltage is across the series load resistor, it follows the current, thus creating the band-stop response characteristic as indicated in Figure 14–57.

FIGURE 14–57
Band-stop filter response.

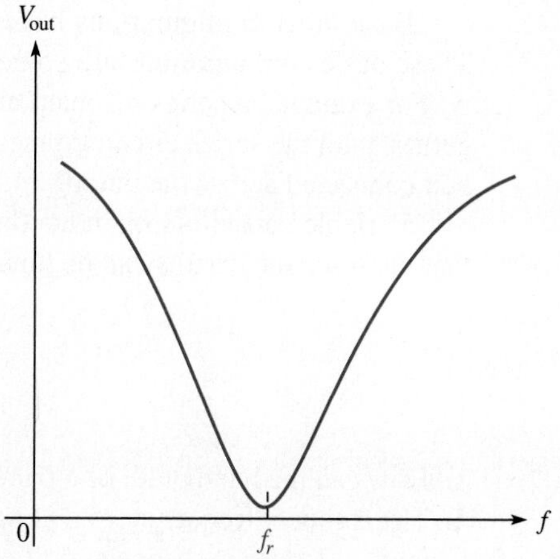

Actually, the band-stop filter in Figure 14–56 can be viewed as a voltage divider created by Z_r of the tank and the load resistance. Thus, the output voltage at f_r is

$$V_{out} = \left(\frac{R_L}{R_L + Z_r}\right)V_{in}$$

EXAMPLE 14–24 Find f_r and the output voltage at resonance for the band-stop filter in Figure 14–58.

FIGURE 14–58

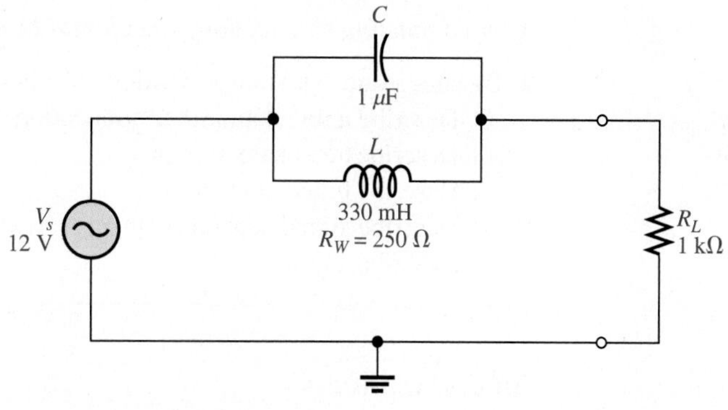

Solution The resonant frequency and impedance are calculated as follows:

$$f_r = \frac{\sqrt{1 - (R_W^2 C/L)}}{2\pi\sqrt{LC}} = 249\text{ Hz}$$

$$X_L = 2\pi f_r L = 516\ \Omega$$

$$Q = \frac{X_L}{R_W} = \frac{516\ \Omega}{250\ \Omega} = 2.06$$

$$Z_r = R_W(Q^2 + 1) = 1.31\text{ k}\Omega$$

At resonance,

$$V_{out} = \left(\frac{R_L}{R_L + Z_r}\right)V_s = \left(\frac{1\text{ k}\Omega}{2.31\text{ k}\Omega}\right)12\text{ V} = 5.19\text{ V}$$

Related Exercise Why wouldn't the formula for ideal resonant frequency ($f_r = 1/2\pi\sqrt{LC}$) work in this particular case?

Basic filter configurations have been introduced in this section and in Section 14–4. These basic configurations are sometimes used in combination to increase filter selectivity. For example, a series resonant and a parallel resonant band-pass filter can be used together, with the series circuit connected between the input and output and the parallel circuit connected across the output.

Variable capacitors or inductors are sometimes used in certain filter applications so that the resonant circuits can be tuned over a range of resonant frequencies.

SECTION 14–7 REVIEW

1. How can the bandwidth of a parallel resonant filter be increased?
2. The resonant frequency of a certain high-Q ($Q > 10$) filter is 5 kHz. If the Q is lowered to 2, does f_r change and, if so, to what value?
3. What is the impedance of a tank circuit at its resonant frequency if $R_W = 75\ \Omega$ and $Q = 25$?

14–8 ■ SYSTEM APPLICATIONS

Resonant circuits are used in a wide variety of applications, particularly in communication systems. In this section, we will look briefly at a few common communication systems applications. The purpose in this section is not to explain how the systems work, but to illustrate the importance of resonant circuits in electronic communication.

After completing this section, you should be able to

- **Discuss some system applications of resonant circuits**
  - ☐ Describe a tuned amplifier application
  - ☐ Describe antenna coupling
  - ☐ Describe tuned amplifiers
  - ☐ Describe signal separation in a receiver
  - ☐ Describe a radio receiver

Tuned Amplifiers

A tuned amplifier is a circuit that amplifies signals within a specified band. Typically, a parallel resonant circuit is used in conjunction with an amplifier to achieve the selectivity. In terms of the general operation, input signals with frequencies that range over a wide band are accepted on the amplifier's input and are amplified. The resonant circuit allows only a relatively narrow band of those frequencies to be passed on. The variable capacitor allows tuning over the range of input frequencies so that a desired frequency can be selected, as indicated in Figure 14–59.

Antenna Input to a Receiver

Radio signals are sent out from a transmitter via electromagnetic waves which propagate through the atmosphere. When the electromagnetic waves cut across the receiving antenna, small voltages are induced. Out of all the wide range of electromagnetic frequencies, only one frequency or a limited band of frequencies must be extracted. Figure 14–60 shows a typical arrangement of an antenna coupled to the receiver input by a transformer. A variable capacitor is connected across the transformer secondary to form a parallel resonant circuit.

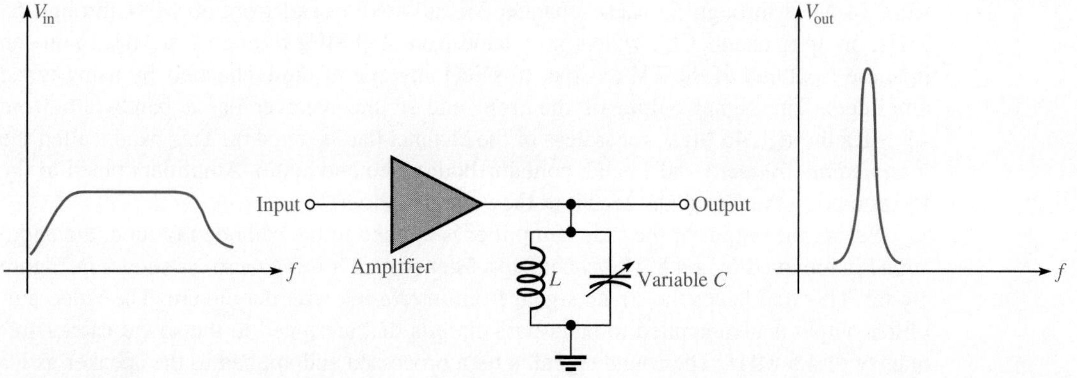

FIGURE 14–59
A basic tuned band-pass amplifier.

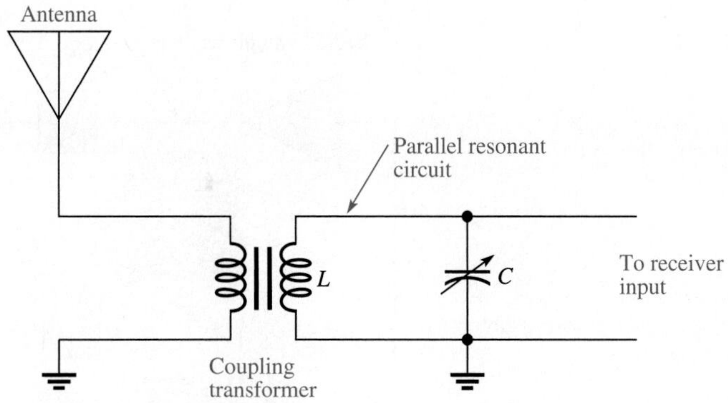

FIGURE 14–60
Resonant coupling from an antenna.

Double-Tuned Transformer Coupling in a Receiver

In some types of communication receivers, tuned amplifiers are transformer-coupled together to increase the amplification. Capacitors can be placed in parallel with the primary and secondary windings of the transformer, effectively creating two parallel resonant band-pass filters that are coupled together. This technique, illustrated in Figure 14–61, can result in a wider bandwidth and steeper slopes on the response curve, thus increasing the selectivity for a desired band of frequencies.

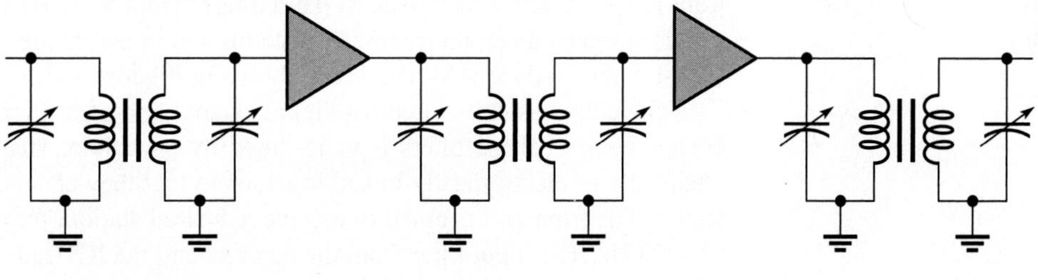

FIGURE 14–61
Double-tuned amplifiers.

Signal Reception and Separation in a TV Receiver

A television receiver must handle both video (picture) signals and audio (sound) signals. Each TV transmitting station is allotted a 6 MHz bandwidth. Channel 2 is allotted a band from 54 MHz through 59 MHz, channel 3 is allotted a band from 60 MHz through 65 MHz, on up to channel 13, which has a band from 210 MHz through 215 MHz. You can tune the front end of the TV receiver to select any one of these channels by using tuned amplifiers. The signal output of the front end of the receiver has a bandwidth from 41 MHz through 46 MHz, regardless of the channel that is tuned in. This band, called the *intermediate frequency* (IF) band, contains both video and audio. Amplifiers tuned to the IF band boost the signal and feed it to the video amplifier.

Before the output of the video amplifier is applied to the cathode-ray tube, the audio signal is removed by a 4.5 MHz band-stop filter (called a *wave trap*), as shown in Figure 14–62. This trap keeps the sound signal from interfering with the picture. The video amplifier output is also applied to band-pass circuits that are tuned to the sound carrier frequency of 4.5 MHz. The sound signal is then processed and applied to the speaker as indicated in Figure 14–62.

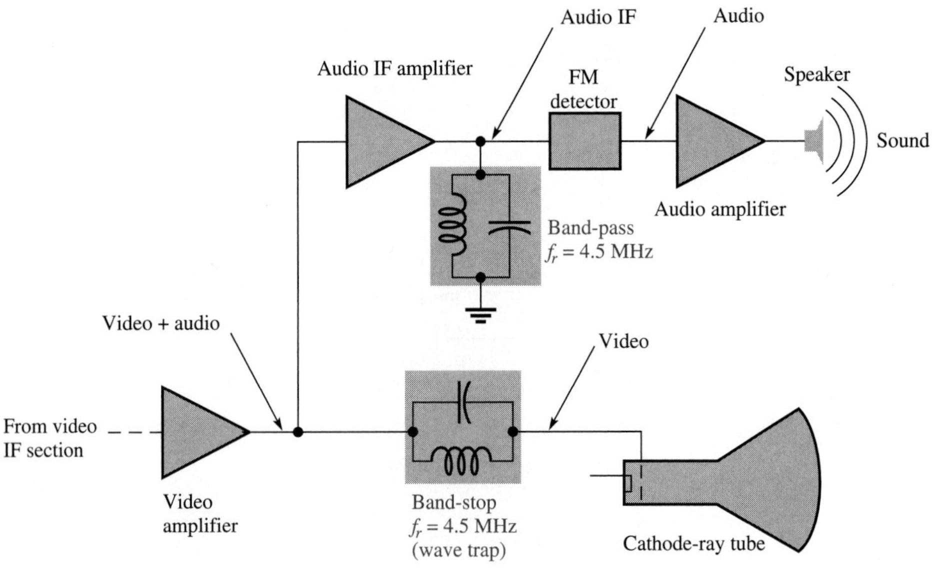

FIGURE 14–62
A simplified portion of a TV receiver showing filter usage.

Superheterodyne Receiver

Another good example of filter applications is in the common AM (amplitude modulation) receiver. The AM broadcast band ranges from 535 kHz to 1605 kHz. Each AM station is assigned a certain narrow bandwidth within that range. A simplified block diagram of a superheterodyne AM receiver is shown in Figure 14–63.

There are basically three parallel resonant band-pass filters in the front end of the receiver. Each of these filters is gang-tuned by capacitors; that is, the capacitors are mechanically or electronically linked together so that they change together as a station is selected. The front end is tuned to receive a desired station, for example, one that transmits at 600 kHz. The input filter from the antenna and the RF (radio frequency) amplifier filter select only a frequency of 600 kHz out of all the frequencies crossing the antenna. The actual audio (sound) signal is carried by the 600 kHz carrier frequency by modulating the amplitude of the carrier so that it follows the audio signal as indicated. The variation in the amplitude of the carrier corresponding to the audio signal is called the *envelope.* The 600 kHz is then applied to a circuit called the *mixer.* The local oscillator (LO) is tuned to a frequency that is 455 kHz above the selected frequency (1055 kHz, in this case). By a process called *heterodyning* or *beating,* the AM signal and the local oscillator signal

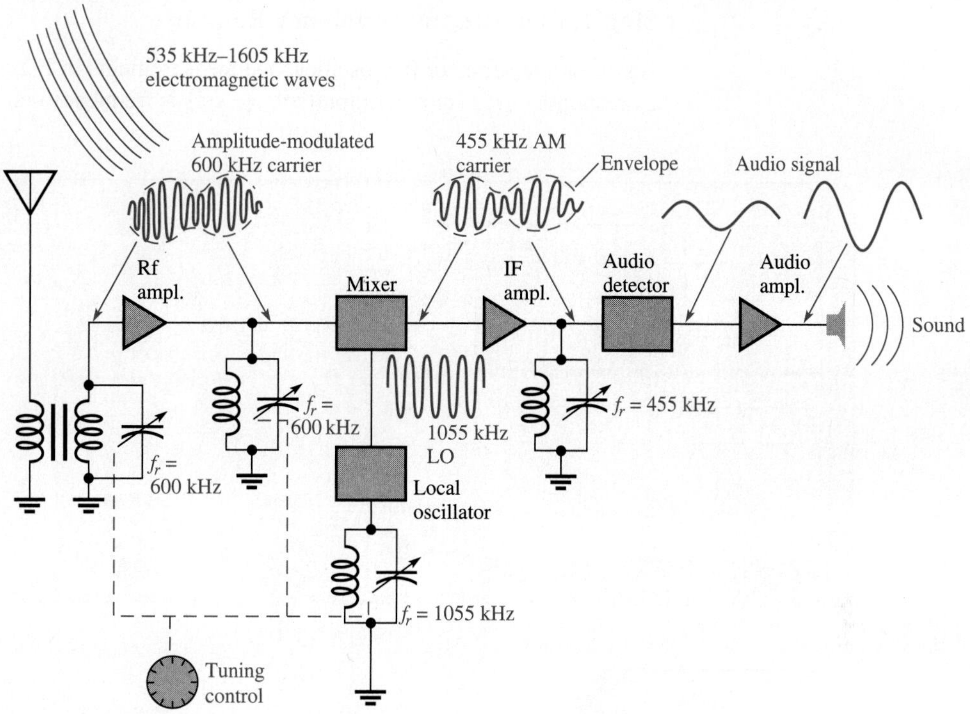

FIGURE 14–63
A simplified diagram of a superheterodyne AM radio broadcast receiver showing the application of tuned resonant circuits.

are mixed together, and the 600 kHz AM signal is converted to a 455 kHz AM signal (1055 kHz − 600 kHz = 455 kHz). The 455 kHz is the intermediate frequency (IF) for standard AM receivers. No matter which station within the broadcast band is selected, its frequency is always converted to the 455 kHz IF. The amplitude-modulated IF is applied to an audio detector which removes the IF, leaving only the envelope or audio signal. The audio signal is then amplified and applied to the speaker.

SECTION 14–8 REVIEW

1. Generally, why is a tuned filter necessary when a signal is coupled from an antenna to the input of a receiver?
2. What is a wave trap?
3. What is meant by *ganged tuning*?

14–9 ■ APPLICATION ASSIGNMENT

In the application assignment from the last chapter, you identified two types of filters in sealed housings. In this application assignment, you have another filter to evaluate which is some type of resonant filter with no specifications available. Your supervisor has requested that you come up with a frequency response curve and determine the resonant frequency and the bandwidth.

In this assignment you will use oscilloscope measurements to determine the frequency response characteristics of the filter. The internal circuitry and component values are of no concern in this case.

Step 1: Measure the Frequency Response

Based on the series of five oscilloscope measurements in Figure 14–64, create a frequency response curve (plot of output voltage versus frequency) for the filter.

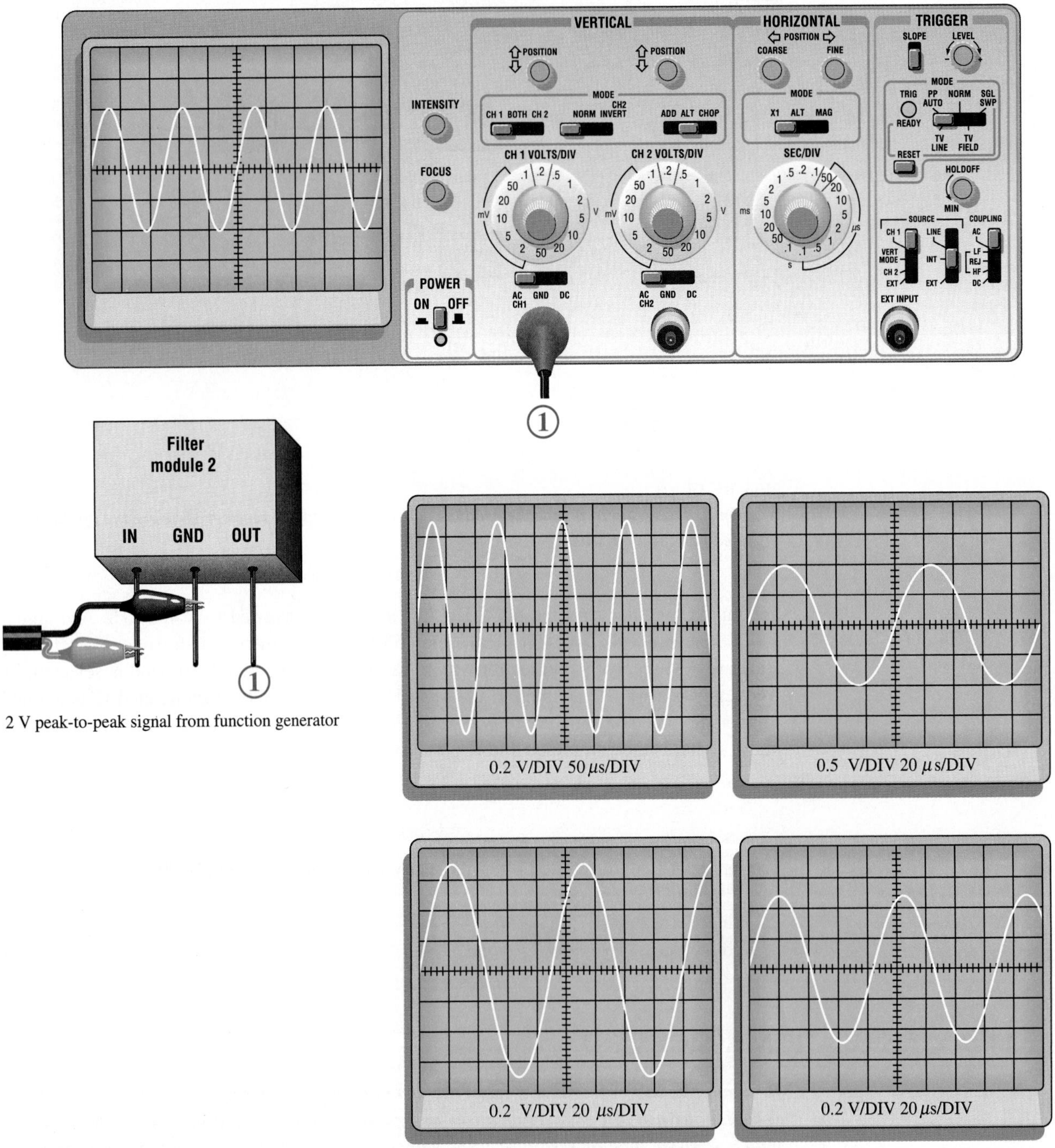

FIGURE 14–64
Frequency response measurements.

Step 2: Analyze the Response Curve

Specify the type of filter and determine the resonant frequency and the bandwidth.

SECTION 14–9 REVIEW

1. What are the half-power frequencies for the filter in this assignment?
2. From the measurements in Figure 14–64, can you determine the circuit arrangement or component values?

▪ SUMMARY

- X_L and X_C have opposing effects in an *RLC* circuit.
- In a series *RLC* circuit, the larger reactance determines the net reactance of the circuit.
- In a parallel *RLC* circuit, the smaller reactance determines the net reactance of the circuit.

For series resonance:

- The reactances are equal.
- The impedance is minimum and equal to the resistance.
- The current is maximum.
- The phase angle is zero.
- The voltages across *L* and *C* are equal in magnitude and, as always, 180° out of phase with each other and thus they cancel.

For parallel resonance:

- The reactances are approximately equal for $Q \geq 10$.
- The impedance is maximum.
- The current is minimum and, ideally, equal to zero.
- The phase angle is zero.
- The currents in the *L* and *C* branches are equal in magnitude and, as always, 180° out of phase with each other and thus they cancel.

▪ GLOSSARY

Band-pass filter A filter that passes frequencies within a certain range and blocks all others.

Band-stop filter A filter that blocks frequencies within a certain range.

Bandwidth (BW) The characteristic of certain electronic circuits that specifies the usable range of frequencies for which signals pass from input to output without significant reduction in amplitude.

Cutoff frequencies The frequencies at which the output of a filter is 70.7% of its maximum.

Decibel (dB) The unit of the logarithmic expression of a ratio, such as power or voltage ratios.

Half-power frequency The frequency at which the output of a filter is 70.7% of maximum.

Parallel resonance In a parallel *RLC* circuit, the condition where the impedance is maximum and the reactances are equal.

Pass band The range of frequencies passed by a filter.

Quality factor (*Q*) The ratio of reactive power to true power in a coil or a resonant circuit.

Resonant frequency The frequency at which a resonant condition occurs in a series or parallel *RLC* circuit.

Selectivity A measure of how well a filter passes certain frequencies and rejects others. The narrower the bandwidth, the greater the selectivity.

Series resonance In a series *RLC* circuit, the condition where the impedance is minimum and the reactances are equal.

Stop band The range of frequencies between the upper and lower cutoff points.

Tank circuit A parallel resonant circuit.

▪ FORMULAS

(14–1)	$X_T = \lvert X_L - X_C \rvert$	Total series reactance (absolute value)
(14–2)	$Z_T = \sqrt{R^2 + (X_L - X_C)^2}$	Total series *RLC* impedance
(14–3)	$\theta = \tan^{-1}\left(\frac{X_T}{R}\right)$	Series *RLC* phase angle

(14–4)	$X_L = X_C$	Condition for series resonance
(14–5)	$Z_r = R$	Series resonant impedance
(14–6)	$f_r = \dfrac{1}{2\pi\sqrt{LC}}$	Series resonant frequency
(14–7)	$BW = f_2 - f_1$	Bandwidth
(14–8)	$\text{dB} = 20 \log\left(\dfrac{V_{out}}{V_{in}}\right)$	Decibel formula for voltage ratio
(14–9)	$\text{dB} = 10 \log\left(\dfrac{P_{out}}{P_{in}}\right)$	Decibel formula for power ratio
(14–10)	$Q = \dfrac{X_L}{R}$	Series resonant Q
(14–11)	$BW = \dfrac{f_r}{Q}$	Bandwidth
(14–12)	$Y = \sqrt{G^2 + (B_C - B_L)^2}$	Parallel *RLC* admittance
(14–13)	$\theta = \tan^{-1}\left(\dfrac{B_T}{G}\right)$	Parallel *RLC* phase angle
(14–14)	$I_T = \sqrt{I_R^2 + (I_C - I_L)^2}$	Total parallel *RLC* current
(14–15)	$\theta = \tan^{-1}\left(\dfrac{I_{CL}}{I_R}\right)$	Parallel *RLC* phase angle
(14–16)	$L_{eq} = L\left(\dfrac{Q^2 + 1}{Q^2}\right)$	Equivalent parallel inductance
(14–17)	$R_{p(eq)} = R_W(Q^2 + 1)$	Equivalent parallel resistance
(14–18)	$f_r = \dfrac{1}{2\pi\sqrt{LC}}$	Ideal parallel resonant frequency
(14–19)	$I_T = \lvert I_C - I_L \rvert$	Total parallel *LC* current (absolute value)
(14–20)	$Z_r = R_W(Q^2 + 1)$	Impedance at parallel resonance
(14–21)	$I_T = \dfrac{V_s}{Z_r}$	Total current at parallel resonance
(14–22)	$f_r = \dfrac{\sqrt{1 - (R_W^2 C/L)}}{2\pi\sqrt{LC}}$	Parallel resonant frequency (exact)
(14–23)	$Q_O = \dfrac{R_{p(tot)}}{X_{L(eq)}}$	Parallel *RLC* Q

■ SELF-TEST

1. The total reactance of a series *RLC* circuit at resonance is
   (a) zero **(b)** equal to the resistance **(c)** infinity **(d)** capacitive
2. The phase angle of a series *RLC* circuit at resonance is
   (a) −90° **(b)** +90° **(c)** 0° **(d)** is dependent on the reactance
3. The impedance at the resonant frequency of a series *RLC* circuit with L = 15 mH, C = 0.015 μF, and R_W = 80 Ω is
   (a) 15 kΩ **(b)** 80 Ω **(c)** 30 Ω **(d)** 0 Ω
4. In a series *RLC* circuit that is operating below the resonant frequency, the current
   (a) is in phase with the applied voltage **(b)** lags the applied voltage
   (c) leads the applied voltage
5. If the value of C in a series *RLC* circuit is increased, the resonant frequency
   (a) is not affected **(b)** increases **(c)** remains the same **(d)** decreases
6. In a certain series resonant circuit, V_C = 150 V, V_L = 150 V, and V_R = 50 V. The value of the source voltage is
   (a) 150 V **(b)** 300 V **(c)** 50 V **(d)** 350 V

7. A certain series resonant band-pass filter has a bandwidth of 1 kHz. If the existing coil is replaced with one having a lower value of Q, the bandwidth will
   (a) increase (b) decrease (c) remain the same (d) be more selective
8. At frequencies below resonance in a parallel RLC circuit, the current
   (a) leads the source voltage (b) lags the source voltage
   (c) is in phase with the source voltage
9. The total current into the L and C branches of a parallel circuit at resonance is ideally
   (a) maximum (b) low (c) high (d) zero
10. To tune a parallel resonant circuit to a lower frequency, the capacitance should be
   (a) increased (b) decreased (c) left alone (d) replaced with inductance
11. The resonant frequency of a parallel circuit is the same as a series circuit when
   (a) the Q is very low (b) the Q is very high (c) there is no resistance
12. If the resistance in parallel with a parallel resonant filter is reduced, the bandwidth
   (a) disappears (b) decreases (c) becomes sharper (d) increases

■ BASIC PROBLEMS

SECTION 14–1 Impedance and Phase Angle of Series *RLC* Circuits

1. A certain series RLC circuit has the following values: $R = 10\ \Omega$, $C = 0.05\ \mu F$, and $L = 5$ mH. Determine the impedance and phase angle. What is the total reactance? $f = 5$ kHz.
2. Find the impedance in Figure 14–65.
3. If the frequency of the source voltage in Figure 14–65 is doubled from the value that produces the indicated reactances, how does the impedance change?

SECTION 14–2 Analysis of Series *RLC* Circuits

4. For the circuit in Figure 14–65, find I_T, V_R, V_L, and V_C.
5. Sketch the voltage phasor diagram for the circuit in Figure 14–65.
6. Analyze the circuit in Figure 14–66 for the following ($f = 25$ kHz):
   (a) I_T (b) P_{true} (c) P_r (d) P_a

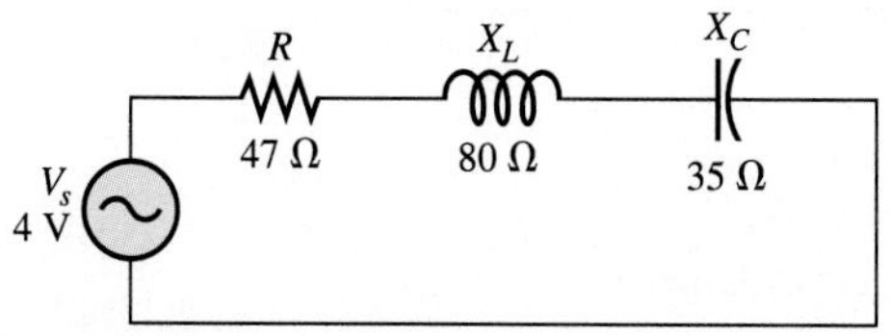

FIGURE 14–65

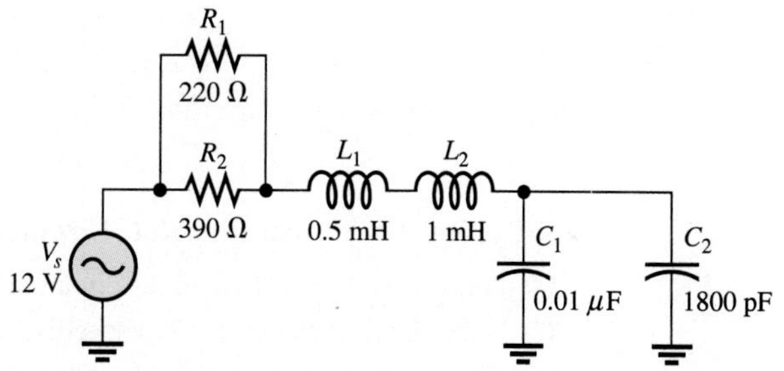

FIGURE 14–66

SECTION 14–3 Series Resonance

7. Find X_L, X_C, Z, and I at the resonant frequency in Figure 14–67.
8. A certain series resonant circuit has a maximum current of 50 mA and a V_L of 100 V. The applied voltage is 10 V. What is Z? What are X_L and X_C?
9. For the RLC circuit in Figure 14–68, determine the resonant frequency and the cutoff frequencies.
10. What is the value of the current at the half-power points in Figure 14–68?

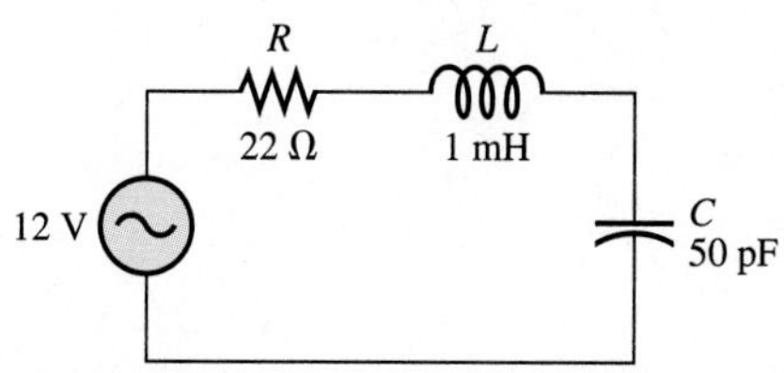

FIGURE 14–67

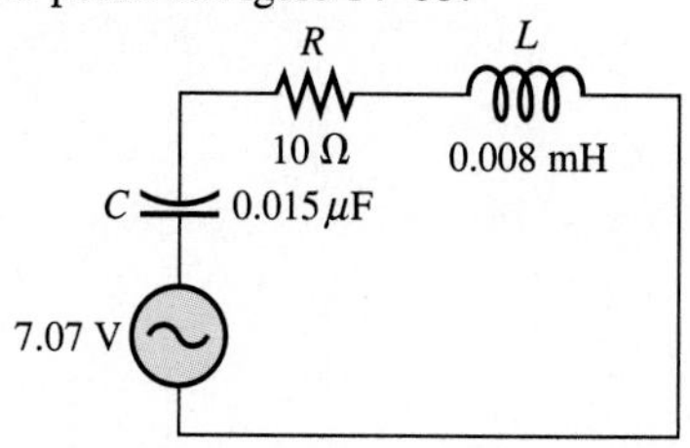

FIGURE 14–68

SECTION 14–4 Series Resonant Filters

11. Determine the resonant frequency for each filter in Figure 14–69. Are these filters band-pass or band-stop types?

12. Assuming that the coils in Figure 14–69 have a winding resistance of 10 Ω, find the bandwidth for each filter.

FIGURE 14–69

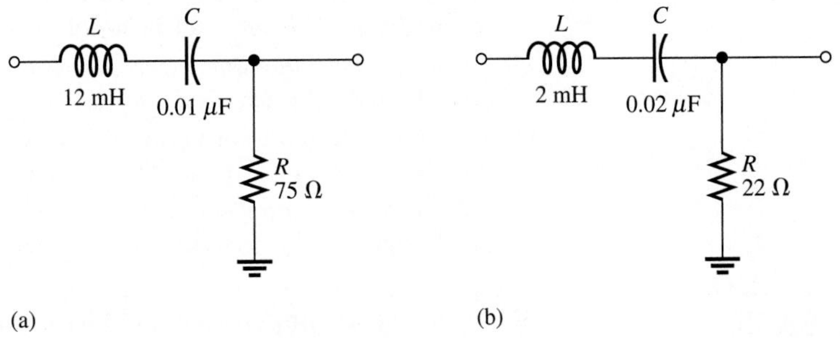

13. Determine the resonant frequency and bandwidth for each filter in Figure 14–70.

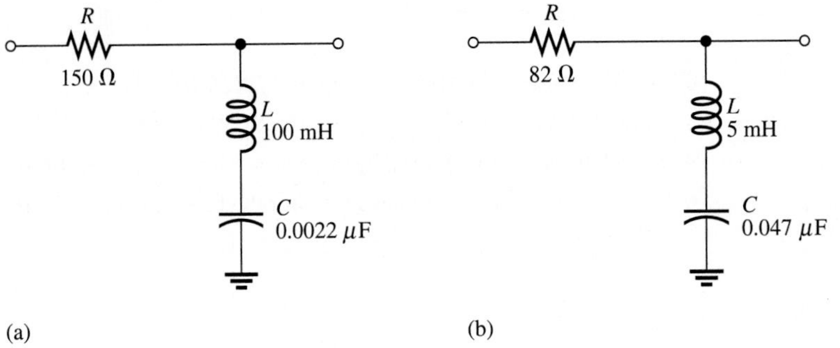

FIGURE 14–70

SECTION 14–5 Parallel *RLC* Circuits

14. Find the total impedance of the circuit in Figure 14–71.

15. Is the circuit in Figure 14–71 capacitive or inductive? Explain.

16. For the circuit in Figure 14–71, find all the currents and voltages.

FIGURE 14–71

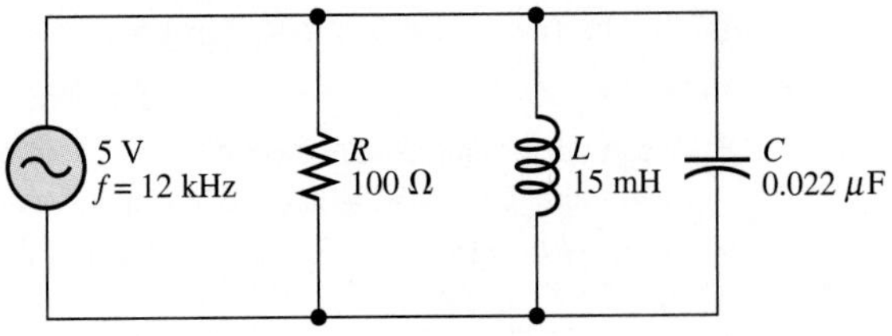

17. Find the total impedance for the circuit in Figure 14–72.

FIGURE 14–72

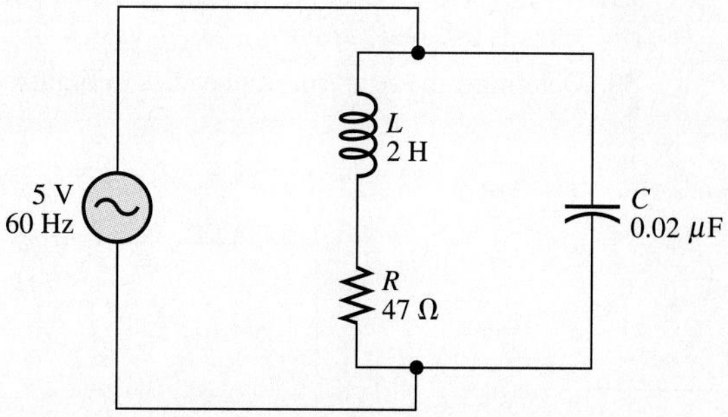

SECTION 14–6 Parallel Resonance

18. What is the impedance of an ideal parallel resonant circuit (no resistance in either branch)?

19. Find Z at resonance and f_r for the tank circuit in Figure 14–73.

20. How much current is drawn from the source in Figure 14–73 at resonance? What are the inductive current and the capacitive current at the resonant frequency?

FIGURE 14–73

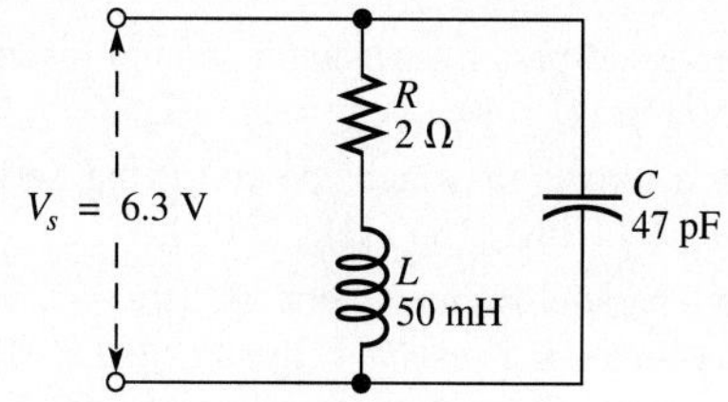

SECTION 14–7 Parallel Resonant Filters

21. At resonance, $X_L = 2\text{ k}\Omega$ and $R_W = 25\ \Omega$ in a parallel resonant band-pass filter. The resonant frequency is 5 kHz. Determine the bandwidth.

22. If the lower cutoff frequency is 2400 Hz and the upper cutoff frequency is 2800 Hz, what is the bandwidth?

23. In a certain resonant circuit, the power at resonance is 2.75 W. What is the power at the lower and upper cutoff frequencies?

24. What values of L and C should be used in a tank circuit to obtain a resonant frequency of 8 kHz? The bandwidth must be 800 Hz. The winding resistance of the coil is 10 Ω.

25. A parallel resonant circuit has a Q of 50 and a BW of 400 Hz. If Q is doubled, what is the bandwidth for the same f_r?

■ ADVANCED PROBLEMS

26. For each following case, express the voltage ratio in decibels:
(a) $V_{in} = 1$ V, $V_{out} = 1$ V **(b)** $V_{in} = 5$ V, $V_{out} = 3$ V
(c) $V_{in} = 10$ V, $V_{out} = 7.07$ V **(d)** $V_{in} = 25$ V, $V_{out} = 5$ V

27. Find the current through each component in Figure 14–74. Find the voltage across each component.

FIGURE 14–74

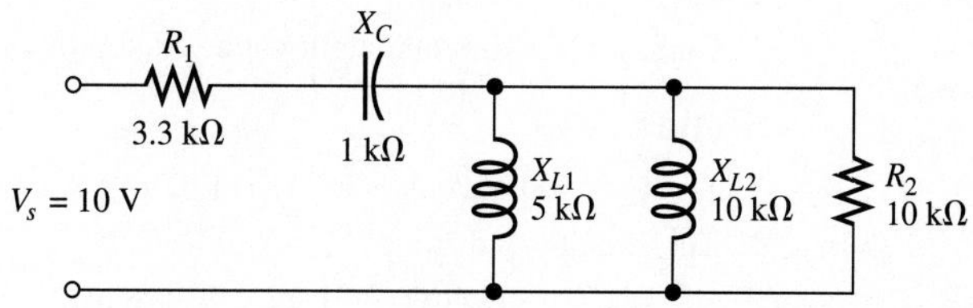

28. Determine whether there is a value of C that will make $V_{ab} = 0$ V in Figure 14–75. If not, explain.

29. If the value of C is 0.2 μF, how much current flows through each branch in Figure 14–75? What is the total current?

30. Determine the resonant frequencies in Figure 14–76 and find V_{out} at each resonant frequency.

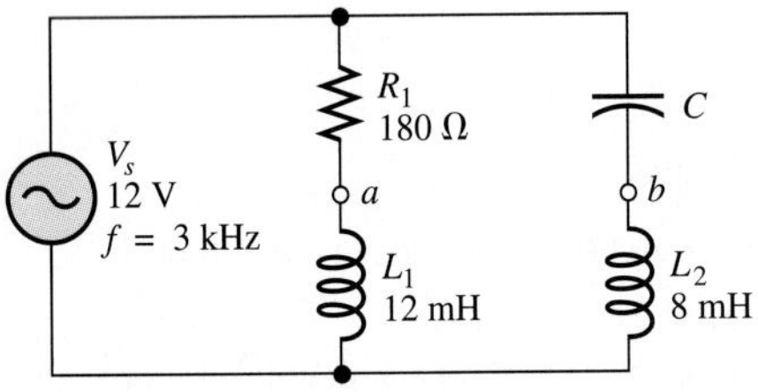

FIGURE 14–75

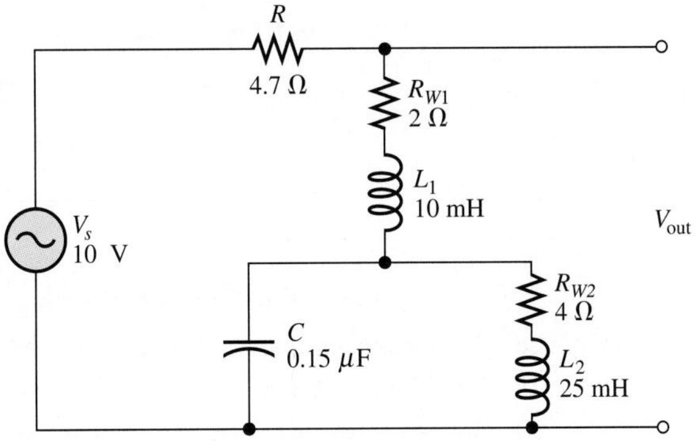

FIGURE 14–76

31. Design a band-pass filter using a parallel resonant circuit to meet the following specifications: $BW = 500$ Hz, $Q = 40$, $I_{C(max)} = 20$ mA, $V_{C(max)} = 2.5$ V.

32. Design a circuit in which the following series resonant frequencies are switch-selectable: 500 kHz, 1000 kHz, 1500 kHz, 2000 kHz.

33. Design a parallel-resonant network using a single coil and switch-selectable capacitors to produce the following resonant frequencies: 8 MHz, 9 MHz, 10 MHz, and 11 MHz. Assume a 10 μH coil with a winding resistance of 5 Ω.

▪ ANSWERS TO SECTION REVIEWS

Section 14–1

1. The circuit is inductive if $X_L > X_C$ and capacitive if $X_C > X_L$.

2. $X_T = X_C - X_L = 70\ \Omega$; The circuit is capacitive.

3. $Z = \sqrt{R^2 + X_T^2} = 83.2\ \Omega$; $\theta = \tan^{-1}(X_T/R) = 57.3°$; Current is leading V_s.

Section 14–2

1. $V_s = \sqrt{V_R^2 + (V_C - V_L)^2} = 38.4$ V

2. Current leads V_s because the circuit is capacitive.

3. $X_T = X_C - X_L = 6\ \Omega$

Section 14–3

1. Series resonance occurs when $X_L = X_C$.

2. The current is maximum because the impedance is minimum.

3. $f_r = 1/2\pi\sqrt{LC} = 159$ kHz

4. It is capacitive because $X_C > X_L$.

Section 14–4

1. $V_{out} = 0.707(15\text{ V}) = 10.6$ V

2. $BW = f_r/Q = 10$ kHz

3. Current is maximum; output voltage is minimum.

Section 14–5

1. $I_R = V_s/R = 80$ mA; $I_C = V_s/X_C = 120$ mA; $I_L = V_s/X_L = 240$ mA

2. The circuit is inductive ($X_L < X_C$).

3. $L_{eq} = L[(Q^2 + 1)/Q^2] = 20.1$ mH; $R_{eq} = R_W(Q^2 + 1) = 1598\ \Omega$

Section 14–6

1. The impedance is maximum.
2. The current is minimum.
3. At ideal parallel resonance, $X_C = X_L = 1.5\ \text{k}\Omega$.
4. $f_r = \sqrt{1 - R_W^2 C/L}/2\pi\sqrt{LC} = 225\ \text{kHz}$; $Z_r = R_W(Q^2 + 1) = 1249\ \text{M}\Omega$
5. $f_r = 1/2\pi\sqrt{LC} = 22.5\ \text{kHz}$
6. $f_r = \sqrt{Q^2/(Q^2 + 1)}/2\pi\sqrt{LC} = 20.9\ \text{kHz}$
7. $Z_r = R_W(Q^2 + 1) = 8.02\ \text{k}\Omega$

Section 14–7

1. The bandwidth can be increased by reducing the parallel resistance.
2. f_r changes to 4.47 kHz.
3. $Z_r = R_W(Q^2 + 1) = 47.0\ \text{k}\Omega$

Section 14–8

1. A tuned filter is used to select a narrow band of frequencies.
2. A wave trap is a band-stop filter.
3. Several variable capacitors (or inductors) whose values can be varied simultaneously with a common control is an example of ganged tuning.

Section 14–9

1. $f_{c(1)} = 8\ \text{kHz}$; $f_{c(2)} = 11.6\ \text{kHz}$
2. The circuit or component values cannot be determined from the data given.

▪ ANSWERS TO RELATED EXERCISES FOR EXAMPLES

14–1	12.3 Ω; 63.0°
14–2	4.64 kΩ
14–3	Current increases, then decreases.
14–4	More capacitive
14–5	22.5 kHz
14–6	2.27 A
14–7	Z increases; Z increases
14–8	No
14–9	200 kHz
14–10	−14 dB
14–11	11.7
14–12	319 Hz
14–13	V_{out} increases; V_{out} increases
14–14	Inductive
14–15	Increase
14–16	$R_{p(eq)} = 25\ \text{k}\Omega$; $L_{eq} = 5\ \text{mH}$; $C = 0.02\ \mu\text{F}$
14–17	$f_r = 3.39\ \text{kHz}$; $I_L = 6.85\ \text{mA}$; $I_C = 6.85\ \text{mA}$; $I_T = 0\ \text{A}$
14–18	79.9 kΩ
14–19	The differences are negligible.
14–20	7.07 V
14–21	530 kHz
14–22	25
14–23	Q decreases with less load resistance.
14–24	The Q is less than 10.

15

PULSE RESPONSE OF REACTIVE CIRCUITS

In Chapters 12 and 13, the frequency (sinusoidal) response of *RC* and *RL* circuits was covered. In this chapter, the response of *RC* and *RL* filters to pulse inputs is examined. With pulse inputs, the time response of the circuits is of primary importance. In the areas of pulse and digital circuits, the technician is often concerned with how a circuit responds over an interval of time to rapid changes in voltage or current. The relationship of the circuit time constant to the input pulse characteristics, such as pulse width and period, determines the wave shapes of the circuit voltages.

Before starting this chapter, you should review the material in Sections 9–5 and 10–5. Understanding exponential changes in voltages and currents in capacitors and inductors is crucial to the study of pulse response. Exponential formulas that were given in Chapter 9 are used throughout this chapter.

Integrator and *differentiator,* terms used throughout this coverage, refer to mathematical functions that are approximated by these circuits under certain conditions. Integration is an averaging process, and differentiation is a process for establishing a rate of change of a function. We will not deal with the purely mathematical aspects at this level of coverage.

■ CHAPTER OBJECTIVES

- □ Explain the operation of an *RC* integrator
- □ Analyze an *RC* integrator with a single input pulse
- □ Analyze an *RC* integrator with repetitive input pulses
- □ Analyze an *RC* differentiator with a single input pulse
- □ Analyze an *RC* differentiator with repetitive input pulses
- □ Analyze the operation of an *RL* integrator
- □ Analyze the operation of an *RL* differentiator
- □ Discuss some basic applications of integrators and differentiators
- □ Troubleshoot integrators and differentiators

■ APPLICATION ASSIGNMENT

For the application assignment in this chapter, you will breadboard a time-delay circuit and determine the component values required to meet certain specifications. You will also determine how to test the circuit for proper operation using a pulse generator and an oscilloscope. After studying this chapter, you should be able to complete the application assignment in Section 15–10.

15–1 ■ THE *RC* INTEGRATOR

A series RC circuit in which the output voltage is taken across the capacitor is known as an integrator in terms of pulse response. Recall that in terms of frequency response, it is a low-pass filter. The term, integrator, is derived from a mathematical function which this type of circuit approximates under certain conditions.

After completing this section, you should be able to

- **Explain the operation of an *RC* integrator**
    - □ Describe how the capacitor charges and discharges
    - □ Explain how a capacitor reacts to an instantaneous change in voltage or current
    - □ Describe the basic output voltage waveform

The Capacitor Charges and Discharges with a Pulse Input

When a pulse generator is connected to the input of an ***RC* integrator,** as shown in Figure 15–1, the capacitor will charge and discharge in response to the pulses. When the input goes from its low level to its high level, the capacitor charges toward the high level of the pulse through the resistor. This changing action is analogous to connecting a battery through a switch to the *RC* network, as illustrated in Figure 15–2(a). When the pulse goes from its high level back to its low level, the capacitor discharges back through the source. The resistance of the source is assumed to be negligible compared to *R*. This discharging action is analogous to replacing the source with a closed switch, as illustrated in Figure 15–2(b).

The capacitor will charge and discharge following an exponential curve. Its rate of charging and discharging, of course, depends on the *RC* **time constant** ($\tau = RC$).

FIGURE 15–1
An RC integrating circuit.

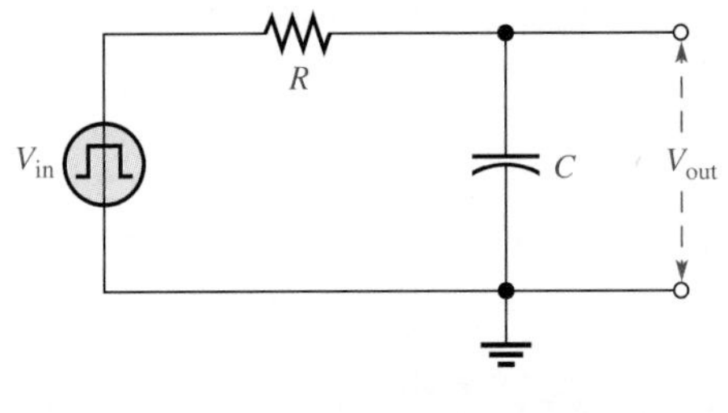

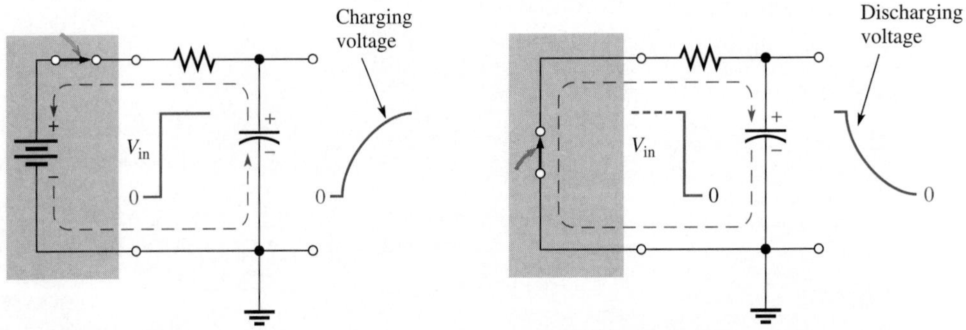

(a) When the input pulse goes high, the source effectively acts as a battery in series with a switch closure, thereby charging the capacitor.

(b) When the input pulse goes back low, the source effectively acts as a closed switch, providing a discharge path for the capacitor.

FIGURE 15–2
The equivalent action when a pulse source charges and discharges the capacitor.

For an ideal pulse, each edge is considered to occur instantaneously. Two basic rules of capacitor behavior help in understanding the **pulse response** of *RC* circuits:

1. The capacitor appears as a short to an instantaneous change in current and as an open to dc.
2. The voltage across the capacitor cannot change instantaneously—it can change only exponentially.

The Capacitor Voltage

In an *RC* integrator, the output is the capacitor voltage. The capacitor charges during the time that the input pulse is high. If the pulse is at its high level long enough, the capacitor will fully charge to the voltage amplitude of the pulse, as illustrated in Figure 15–3. The capacitor discharges during the time that the pulse is low. If the low time between pulses is long enough, the capacitor will fully discharge to zero, as shown in the figure. Then when the next pulse occurs, it will charge again.

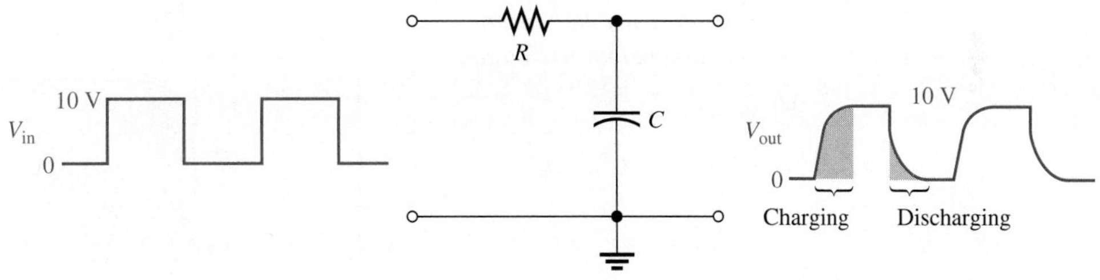

FIGURE 15–3
Illustration of a capacitor fully charging and discharging in response to a pulse input.

SECTION 15–1 REVIEW

1. Define the term *integrator* in relation to an *RC* circuit.
2. What causes a capacitor in an *RC* circuit to charge and discharge?

15–2 ■ SINGLE-PULSE RESPONSE OF *RC* INTEGRATORS

From the previous section, you have a general idea of how an RC integrator responds to a pulse input. In this section, the response to a single pulse is examined in detail.

After completing this section, you should be able to

- **Analyze an *RC* integrator with a single input pulse**
  - □ Discuss the importance of the circuit time constant
  - □ Define *transient time*
  - □ Determine the response when the pulse width is equal to or greater than five time constants
  - □ Determine the response when the pulse width is less than five time constants

Two conditions of pulse response must be considered:

1. When the input pulse width (t_W) is equal to or greater than five time constants ($t_W \geq 5\tau$)
2. When the input pulse width is less than five time constants ($t_W < 5\tau$)

Recall that five time constants is accepted as the time a capacitor needs to fully charge or fully discharge; this time is often called the **transient time.**

When the Pulse Width Is Equal to or Greater Than Five Time Constants

The capacitor will fully charge if the pulse width is equal to or greater than five time constants (5τ). This condition is expressed as $t_W \geq 5\tau$. At the end of the pulse, the capacitor fully discharges back through the source.

Figure 15–4 illustrates the output waveforms for various *RC* time constants and a fixed input pulse width. Notice that the shape of the output pulse approaches that of the input as the transient time is made small compared to the pulse width. In each case the output reaches the full amplitude of the input.

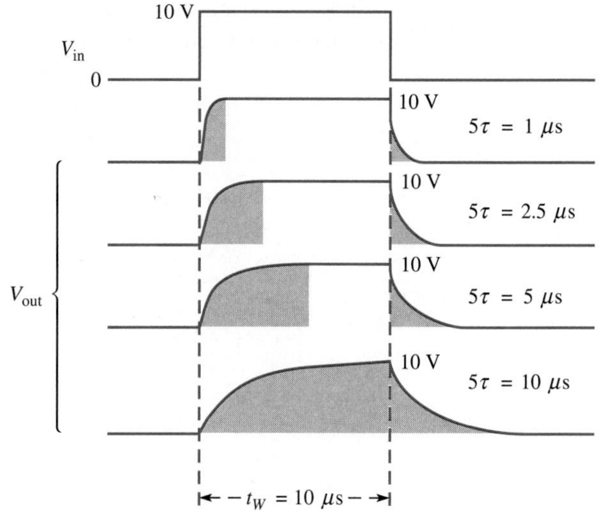

FIGURE 15–4
Variation of an integrator's output pulse shape with time constant. The shaded areas indicate when the capacitor is charging and discharging.

Figure 15–5 shows how a fixed time constant and a variable input pulse width affect the integrator output. Notice that as the pulse width is increased, the shape of the output pulse approaches that of the input. Again, this means that the transient time is short compared to the pulse width.

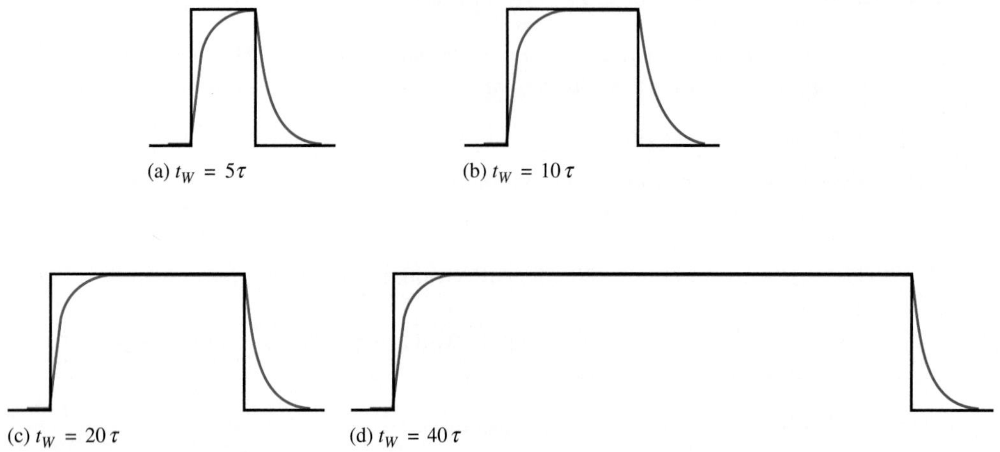

FIGURE 15–5
Variation of an integrator's output pulse shape with input pulse width (the time constant is fixed). Black is input and color is output.

When the Pulse Width Is Less Than Five Time Constants

Now let's examine the case in which the width of the input pulse is less than five time constants of the integrator. This condition is expressed as $t_W < 5\tau$.

As before, the capacitor charges for the duration of the pulse. The pulse width is the time the capacitor has for charging. However, because the pulse width is less than the time the capacitor needs to fully charge (5τ), the output voltage will *not* reach the full input voltage before the end of the pulse. The capacitor only partially charges, as illustrated in Figure 15–6 for several values of *RC* time constants. Notice that for longer time constants, the output reaches a lower voltage because the capacitor cannot charge as much during the pulse width. Of course, in our examples with a single-pulse input, the capacitor fully discharges after the pulse ends.

FIGURE 15–6
Capacitor voltage for various time constants which are longer than the input pulse width. Black is input and color is output.

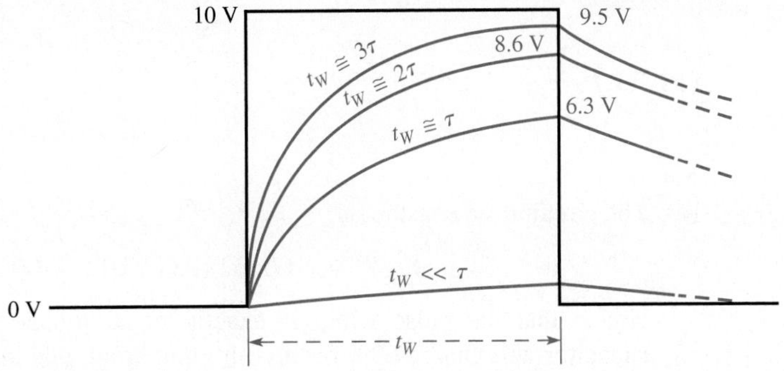

When the time constant is much greater than the input pulse width, the capacitor charges very little, and, as a result, the output voltage becomes almost negligible, as indicated in Figure 15–6.

Figure 15–7 illustrates the effect of reducing the input pulse width for a fixed time constant value. As the width is reduced, the output voltage becomes smaller because the capacitor has less time to charge. However, it takes the capacitor the same length of time (5τ) to discharge back to zero for each condition after the pulse is removed.

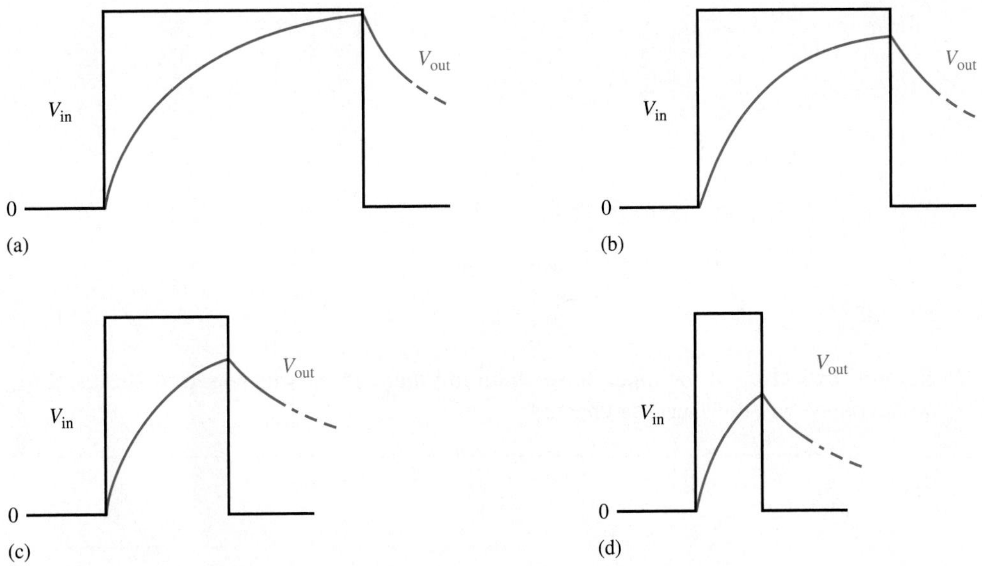

FIGURE 15–7
The capacitor charges less and less as the input pulse width is reduced. The time constant is fixed.

EXAMPLE 15–1

A single 10 V pulse with a width of 100 μs is applied to the integrator in Figure 15–8.

(a) To what voltage will the capacitor charge?

(b) How long will it take the capacitor to discharge?

(c) Sketch the output voltage.

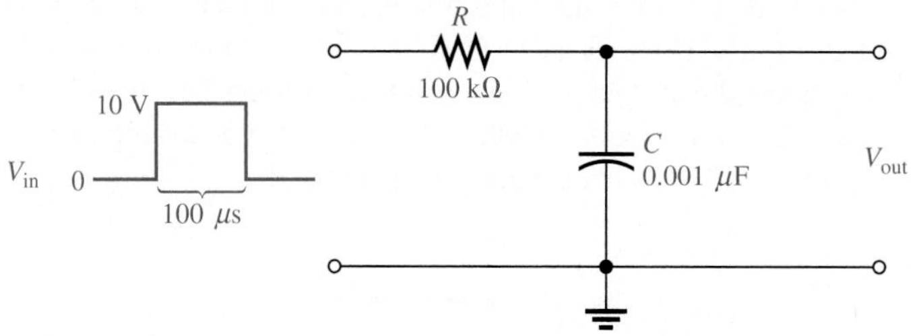

FIGURE 15–8

Solution

(a) The circuit time constant is

$$\tau = RC = (100\text{ k}\Omega)(0.001\ \mu\text{F}) = 100\ \mu\text{s}$$

Notice that the pulse width is exactly equal to one time constant. Thus, the capacitor will charge 63% of the full input amplitude in one time constant, so the output will reach a maximum voltage of 6.3 V.

(b) The capacitor discharges back through the source when the pulse ends. The total discharge time is

$$5\tau = 5(100\ \mu\text{s}) = 500\ \mu\text{s}$$

(c) The output charging and discharging curve is shown in Figure 15–9.

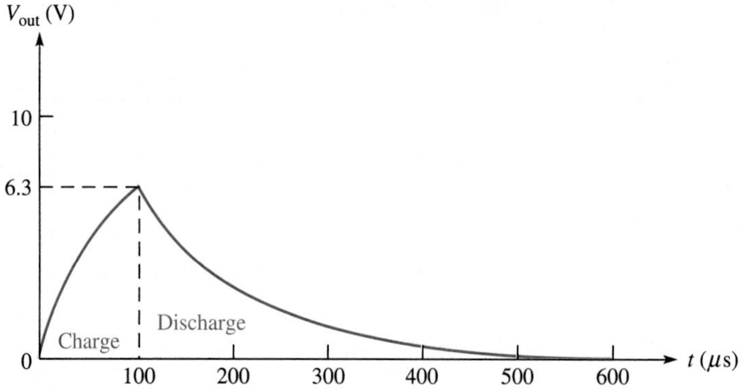

FIGURE 15–9

Related Exercise If the input pulse width in Figure 15–8 is increased to 200 μs, to what voltage will the capacitor charge?

EXAMPLE 15–2

Determine how much the capacitor in Figure 15–10 will charge when the single pulse is applied to the input.

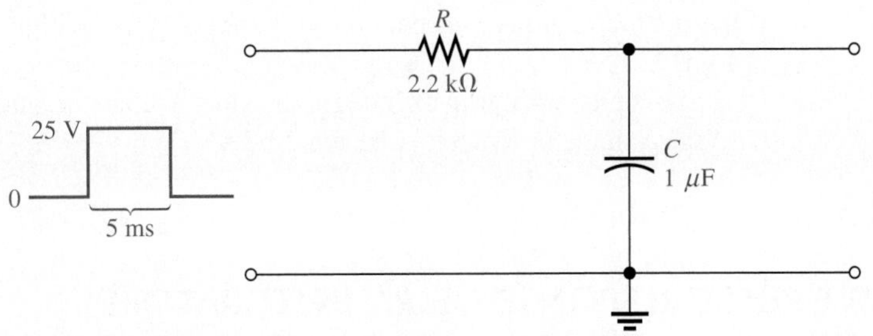

FIGURE 15–10

Solution Calculate the time constant:

$$\tau = RC = (2.2\ \text{k}\Omega)(1\ \mu\text{F}) = 2.2\ \text{ms}$$

Because the pulse width is 5 ms, the capacitor charges for approximately 2.27 time constants (5 ms/2.2 ms = 2.27). Use the exponential formula, Equation (9–18), to find the voltage to which the capacitor will charge. With $V_F = 25$ V and $t = 5$ ms, the calculation is done as follows:

$$v = V_F(1 - e^{-t/RC})$$
$$= (25\ \text{V})(1 - e^{-5/2.2}) = (25\ \text{V})(1 - e^{-2.27})$$
$$= (25\ \text{V})(1 - 0.103) = (25\ \text{V})(0.897) = 22.4\ \text{V}$$

These calculations show that the capacitor charges to 22.4 V during the 5 ms duration of the input pulse. It will discharge back to zero when the pulse goes back to zero.

The calculator sequence for v is

[5] [÷] [2] [•] [2] [=] [+/−] [2nd] [lnx] [+/−] [+] [1] [=] [×] [2] [5] [=]

Related Exercise Determine how much C will charge if the pulse width is increased to 10 ms.

SECTION 15–2 REVIEW

1. When an input pulse is applied to an *RC* integrator, what condition must exist in order for the output voltage to reach the full amplitude of the input?
2. For the circuit in Figure 15–11, which has a single input pulse, find the maximum output voltage and determine how long the capacitor will discharge.

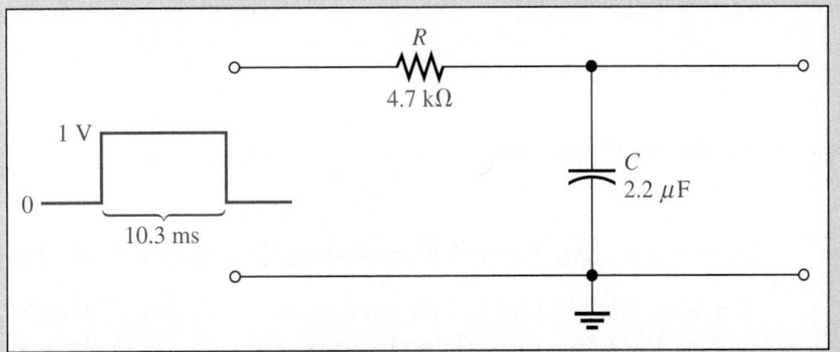

FIGURE 15–11

3. For Figure 15–11, sketch the approximate shape of the output voltage with respect to the input pulse.

4. If the integrator time constant equals the input pulse width, will the capacitor fully charge?
5. Describe the condition in an integrator under which the output voltage has the approximate shape of a rectangular input pulse.

15–3 ■ REPETITIVE-PULSE RESPONSE OF *RC* INTEGRATORS

In the last section, you learned how an RC integrator responds to a single-pulse input. These basic ideas are extended in this section to include the integrator response to repetitive pulses. In electronic systems, you will encounter repetitive-pulse waveforms much more often than single pulses. However, an understanding of the integrator's response to single pulses is necessary in order to understand how these circuits respond to repeated pulses.

After completing this section, you should be able to

- **Analyze an *RC* integrator with repetitive input pulses**
  - ☐ Determine the response when the capacitor does not fully charge or discharge
  - ☐ Define *steady state*
  - ☐ Describe the effect of a change in time constant on circuit response

If a **periodic** pulse waveform is applied to an *RC* integrator, as shown in Figure 15–12, *the output waveshape depends on the circuit time constant and the frequency and duty cycle of the input pulses.* The capacitor, of course, charges and discharges in response to a pulse input. The amount of charge and discharge of the capacitor depends both on the circuit time constant and on the input frequency and duty cycle, as mentioned.

If the pulse width and the time between pulses are each equal to or greater than five time constants, the capacitor will fully charge and fully discharge during each period of the input waveform. This case is shown in Figure 15–12.

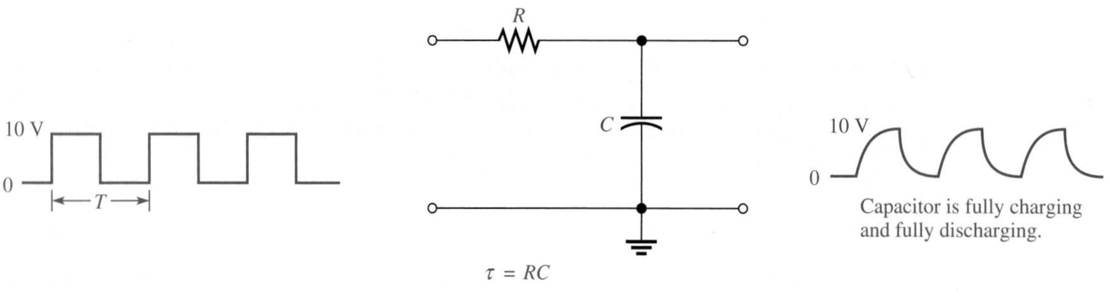

FIGURE 15–12
RC integrator with a repetitive-pulse waveform input.

When the Capacitor Does Not Fully Charge and Discharge

When the pulse width and the time between pulses are shorter than 5 time constants, as illustrated in Figure 15–13 for the case of a square wave, the capacitor will *not* completely charge or discharge. We will now examine the effects of this situation on the output voltage of the *RC* integrator.

For convenience, let's take an example of an *RC* integrator with a charging and discharging time constant equal to the pulse width of a 10 V square wave input, as in Figure 15–14. This choice will simplify the analysis and will demonstrate the basic action of the

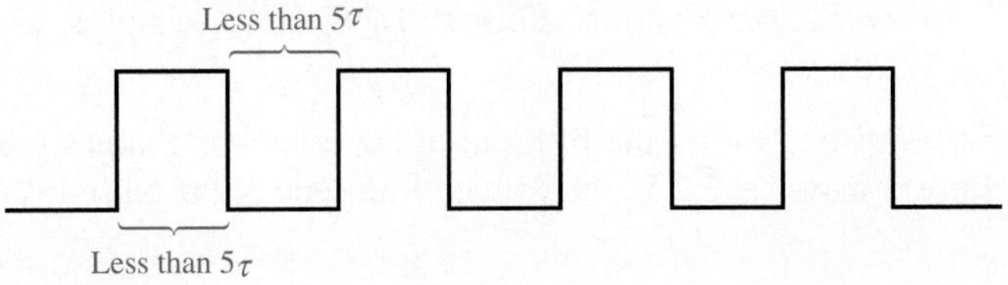

FIGURE 15–13
Waveform that does not allow full charge or discharge of the integrator capacitor.

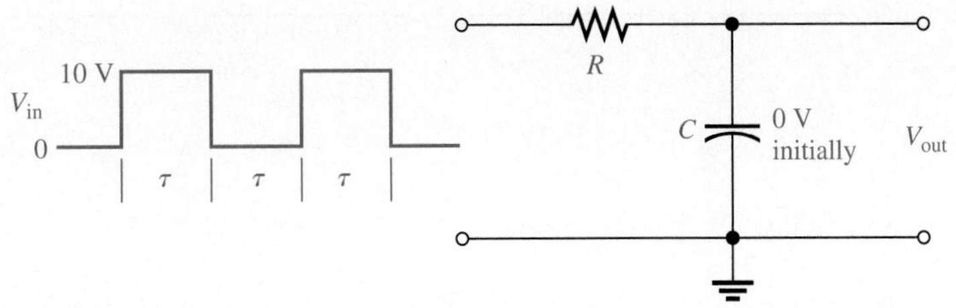

FIGURE 15–14
Integrator with a square wave input having a period equal to two time constants ($T = 2\tau$).

integrator under these conditions. At this point, we really do not care what the exact time constant value is because we know that an *RC* circuit charges 63% during one time constant interval.

We will assume that the capacitor in Figure 15–14 begins initially uncharged and examine the output voltage on a pulse-by-pulse basis. The results of this analysis are shown in Figure 15–15.

First Pulse During the first pulse, the capacitor charges. The output voltage reaches 6.3 V (63% of 10 V), as shown in Figure 15–15.

Between First and Second Pulses The capacitor discharges, and the voltage decreases to 37% of the voltage at the beginning of this interval: 0.37(6.3 V) = 2.33 V.

Second Pulse The capacitor voltage begins at 2.33 V and increases 63% of the way to 10 V. The total charging range is 10 V − 2.33 V = 7.67 V. The capacitor voltage will increase an additional 63% of 7.67 V, which is 4.83 V. Thus, at the end of the second pulse, the output voltage is 2.33 V + 4.83 V = 7.16 V, as indicated in Figure 15–15. Notice that the average is building up.

Between Second and Third Pulses The capacitor discharges during this time, and therefore the voltage decreases to 37% of the initial voltage by the end of the second pulse: 0.37(7.16 V) = 2.65 V.

Third Pulse At the start of the third pulse, the capacitor voltage is 2.65 V. The capacitor charges 63% of the way from 2.65 V to 10 V: 0.63(10 V − 2.65 V) = 4.63 V. Therefore, the voltage at the end of the third pulse is 2.65 V + 4.63 V = 7.28 V.

Between Third and Fourth Pulses The voltage during this interval decreases due to capacitor discharge. It will decrease to 37% of its value by the end of the third pulse. The final voltage in this interval is 0.37(7.28 V) = 2.69 V.

Fourth Pulse At the start of the fourth pulse, the capacitor voltage is 2.69 V. The voltage increases by 0.63(10 V − 2.69 V) = 4.605 V. Therefore, at the end of the fourth pulse, the capacitor voltage is 2.69 V + 4.605 V = 7.295 V. Notice that the values are leveling off as the pulses continue.

Between Fourth and Fifth Pulses Between these pulses, the capacitor voltage drops to 0.37(7.295 V) = 2.7 V.

Fifth Pulse During the fifth pulse, the capacitor charges 0.63(10 V − 2.7 V) = 4.6 V. Since it started at 2.7 V, the voltage at the end of the pulse is 2.7 V + 4.6 V = 7.3 V.

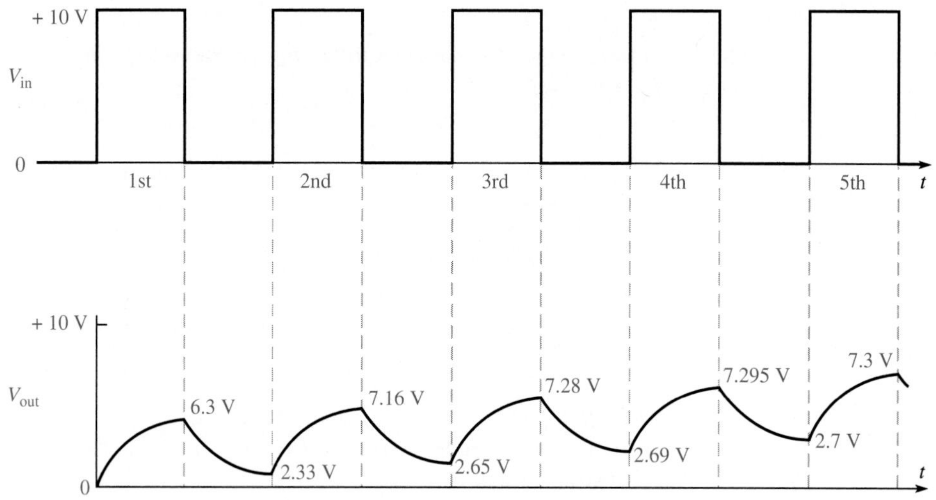

FIGURE 15–15
Input and output for the initially uncharged integrator in Figure 15–14.

Steady-State Response

In the preceding discussion, the output voltage gradually built up and then began leveling off. It takes approximately 5τ for the output voltage to build up to a constant average value, regardless of the number of pulses that may occur during that time interval. This interval is the transient time of the circuit. Once the output voltage reaches the average value of the input voltage, a **steady-state** condition is reached, which continues as long as the periodic input continues. This condition is illustrated in Figure 15–16 based on the values obtained in the preceding discussion.

The transient time for our example circuit is the time from the beginning of the first pulse to the end of the third pulse. The reason for this interval is that the capacitor voltage at the end of the third pulse is 7.28 V, which is about 99% of the final voltage.

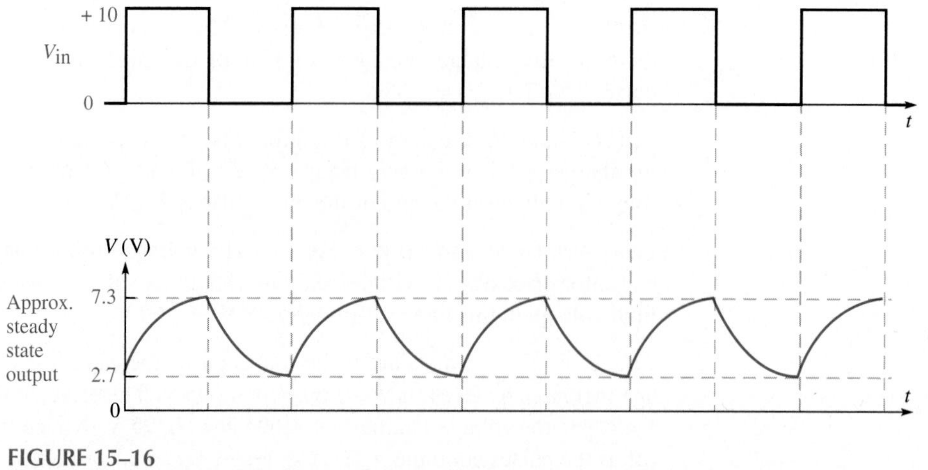

FIGURE 15–16
Output reaches steady state after 5τ.

The Effect of an Increase in Time Constant

What happens to the output voltage if the *RC* time constant of the integrator is increased with a variable resistor, as indicated in Figure 15–17? As the time constant is increased, the capacitor charges less during a pulse and discharges less between pulses. The result is a smaller fluctuation in the output voltage for increasing values of time constant, as shown in Figure 15–18.

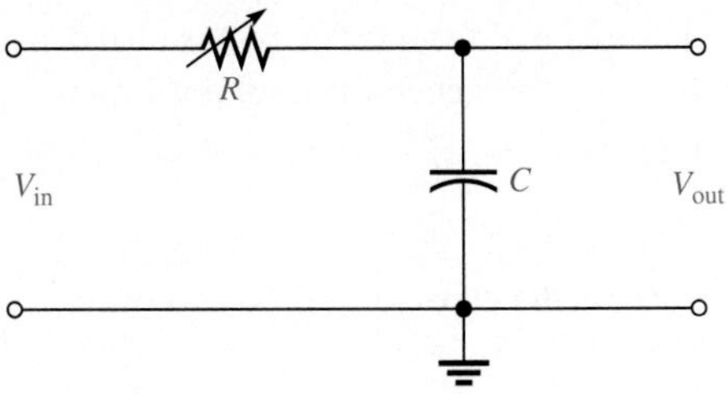

FIGURE 15–17
Integrator with a variable time constant.

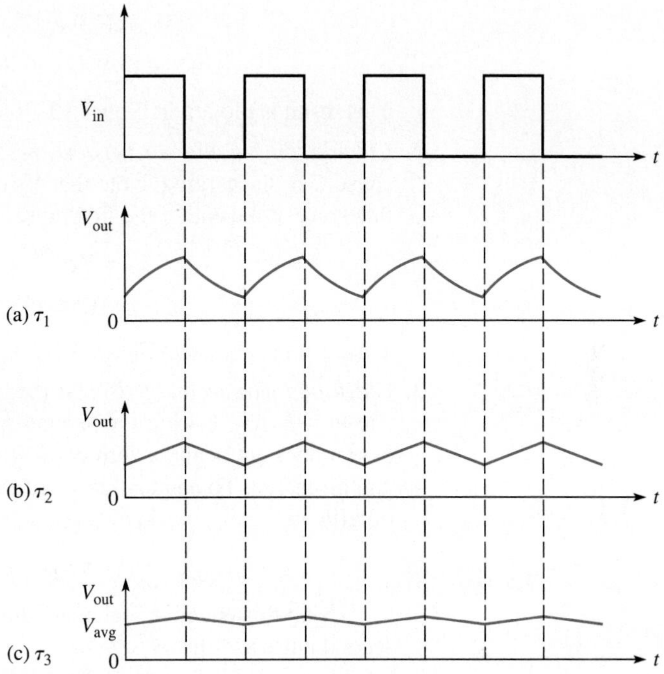

FIGURE 15–18
Effect of longer time constants on the output of an integrator ($\tau_3 > \tau_2 > \tau_1$).

As the time constant becomes extremely long compared to the pulse width, the output voltage approaches a constant dc voltage, as shown in Figure 15–18(c). This value is the average value of the input. For a square wave, it is one-half the amplitude.

EXAMPLE 15–3 Determine the output voltage waveform for the first two pulses applied to the integrator circuit in Figure 15–19. Assume that the capacitor is initially uncharged.

FIGURE 15–19

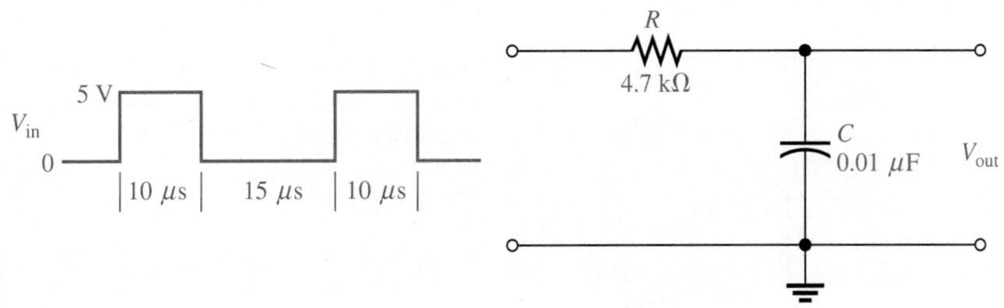

Solution First calculate the circuit time constant:

$$\tau = RC = (4.7\ \text{k}\Omega)(0.01\ \mu\text{F}) = 47\ \mu\text{s}$$

Obviously, the time constant is much longer than the input pulse width or the interval between pulses (notice that the input is not a square wave). Thus, in this case, the exponential formulas must be applied, and the analysis is relatively difficult. Follow the solution carefully.

1. *Calculation for first pulse:* Use Equation (9–18) because C is charging. Note that V_F is 5 V, and t equals the pulse width of 10 μs. Therefore,

$$v_C = V_F(1 - e^{-t/RC}) = (5\text{ V})(1 - e^{-10\mu s/47\mu s})$$
$$= (5\text{ V})(1 - 0.808) = 960\text{ mV}$$

   This result is plotted in Figure 15–20(a).

2. *Calculation for interval between first and second pulse:* Use Equation (9–19) because C is discharging. Note that V_i is 960 mV because C begins to discharge from this value at the end of the first pulse. The discharge time is 15 μs. Therefore,

$$v_C = V_i e^{-t/RC} = (960\text{ mV})^{-15\mu s/47\mu s}$$
$$= (960\text{ mV})(0.727) = 698\text{ mV}$$

   This result is shown in Figure 15–20(b).

3. *Calculation for second pulse:* At the beginning of the second pulse, the output voltage is 698 mV. During the second pulse, the capacitor will again charge. In this case it does not begin at zero volts. It already has 698 mV from the previous charge and discharge. To handle this situation, you must use the general formula of Equation (9–16):

$$v = V_F + (V_i - V_F)e^{-t/\tau}$$

   Using this equation, calculate the voltage across the capacitor at the end of the second pulse as follows:

$$v_C = V_F + (V_i - V_F)e^{-t/RC}$$
$$= 5\text{ V} + (698\text{ mV} - 5\text{ V})e^{-10\mu s/47\mu s}$$
$$= 5\text{ V} + (-4.30\text{ V})(0.808) = 5\text{ V} - 3.47\text{ V} = 1.53\text{ V}$$

   This result is shown in Figure 15–20(c).

FIGURE 15–20

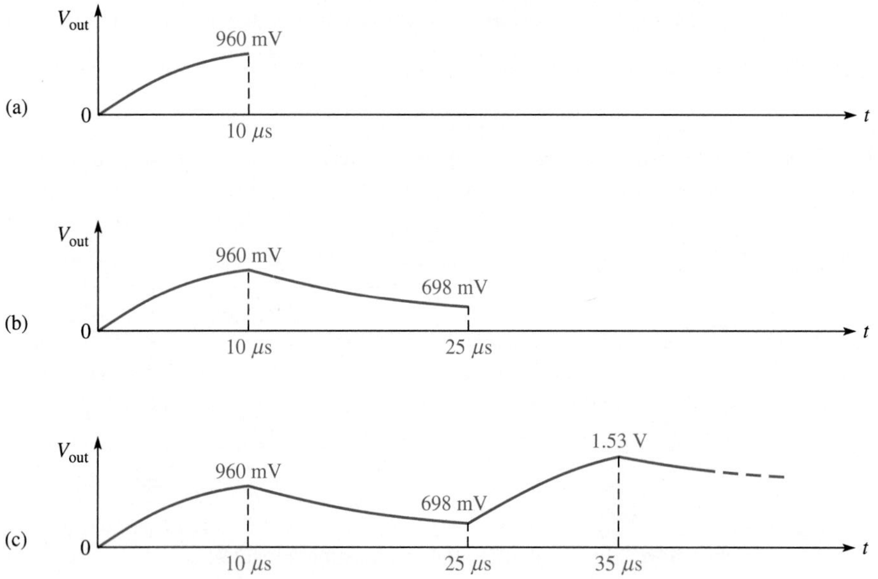

Notice that the output waveform builds up on successive input pulses. After approximately 5τ, it will reach its steady state and will fluctuate between a constant maximum and a constant minimum, with an average equal to the average value of the input. You can demonstrate this pattern by carrying the analysis in this example further.

Related Exercise Determine V_{out} at the beginning of the third pulse.

SECTION 15–3 REVIEW

1. What conditions allow an *RC* integrator capacitor to fully charge and discharge when a periodic pulse waveform is applied to the input?
2. What will the output waveform look like if the circuit time constant is extremely small compared to the pulse width of a square wave input?
3. When 5τ is greater than the pulse width of an input square wave, the time required for the output voltage to build up to a constant average value is called _______ .
4. Define *steady-state response.*
5. What does the average value of the output voltage of an integrator equal during steady state?

15–4 ■ SINGLE-PULSE RESPONSE OF *RC* DIFFERENTIATORS

A series RC circuit in which the output voltage is taken across the resistor is known as a differentiator in terms of pulse response. Recall that in terms of frequency response, a series RC circuit is a high-pass filter. The term, differentiator, is derived from a mathematical function which this type of circuit approximates under certain conditions.

After completing this section, you should be able to

- **Analyze an *RC* differentiator with a single input pulse**
  - □ Describe the response at the rising edge of the input pulse
  - □ Determine the response during and at the end of a pulse for various pulse width–time constant relationships

Figure 15–21 shows an *RC* **differentiator** with a pulse input. The same action occurs in the differentiator as in the integrator, except the output voltage is taken across the resistor rather than the capacitor. The capacitor charges exponentially at a rate depending on the *RC* time constant. The shape of the differentiator's resistor voltage is determined by the charging and discharging action of the capacitor.

FIGURE 15–21
An RC differentiating circuit.

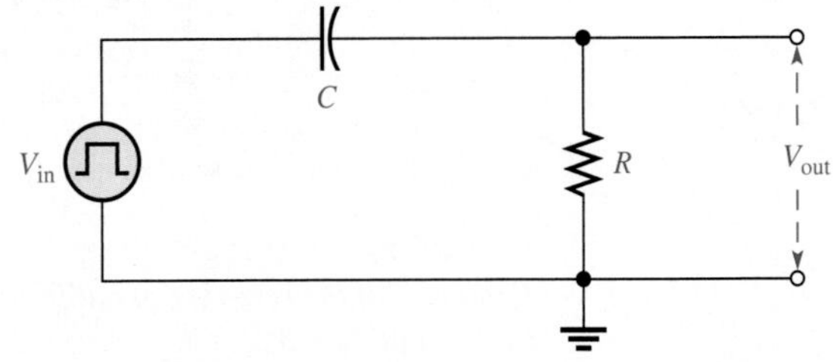

Pulse Response

To understand how the output voltage is shaped by a differentiator, you must consider the following:

1. The response to the rising pulse edge
2. The response between the rising and falling edges
3. The response to the falling pulse edge

Response to the Rising Edge of the Input Pulse Assume that the capacitor is initially uncharged prior to the rising pulse edge. Prior to the pulse, the input is zero volts. Thus, there are zero volts across the capacitor and also zero volts across the resistor, as indicated in Figure 15–22(a).

Now asssume that a 10 V pulse is applied to the input. When the rising edge occurs, point *A* goes to +10 V. Recall that the voltage across a capacitor cannot change instantaneously, and thus the capacitor appears instantaneously as a short. Therefore, if point *A*

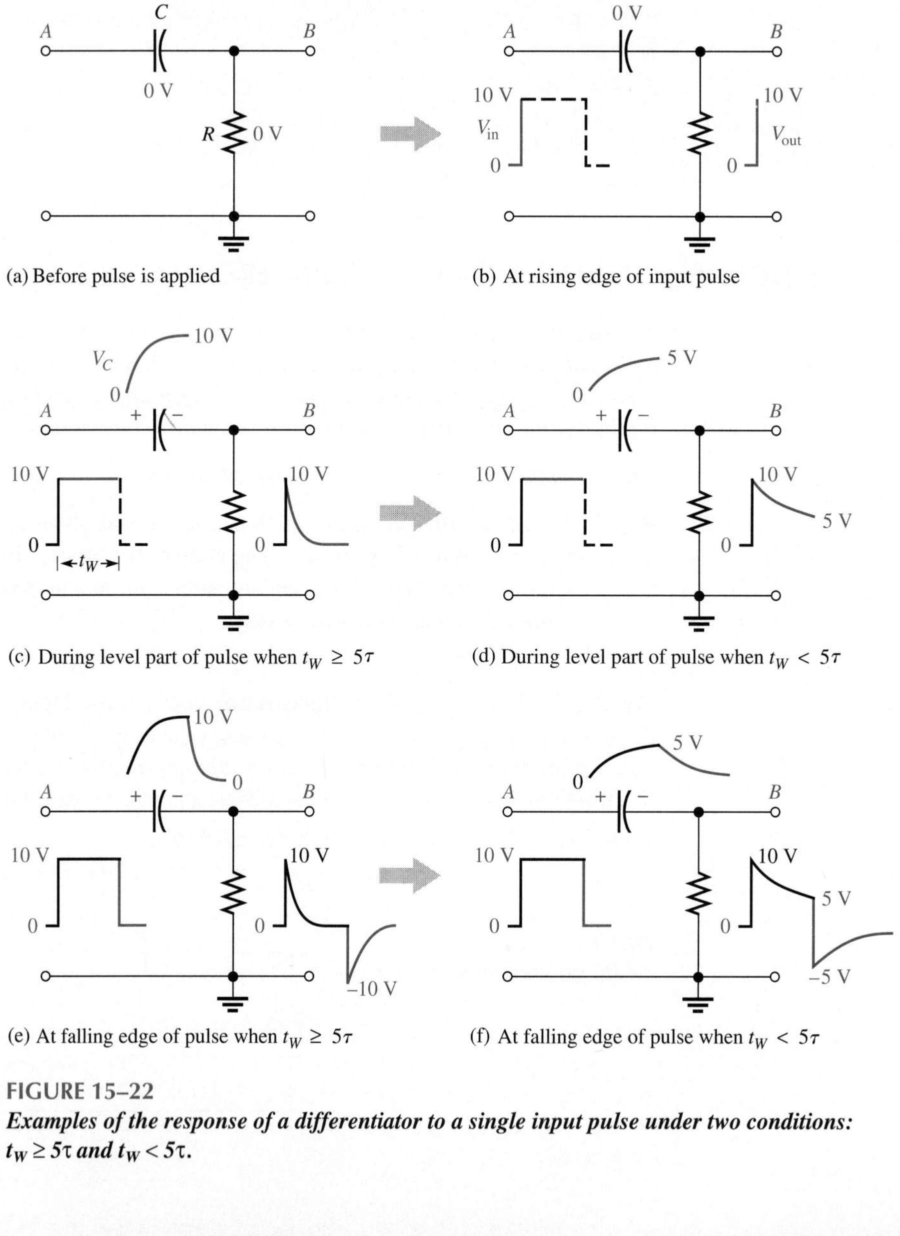

FIGURE 15–22
Examples of the response of a differentiator to a single input pulse under two conditions: $t_W \geq 5\tau$ and $t_W < 5\tau$.

goes instantly to +10 V, then point *B must* also go instantly to +10 V, keeping the capacitor voltage zero for the instant of the rising edge. The capacitor voltage is the voltage from point *A* to point *B*.

The voltage at point *B* with respect to ground is the voltage across the resistor (and the output voltage). Thus, the output voltage suddenly goes to +10 V in response to the rising pulse edge, as indicated in Figure 15–22(b).

Response During Pulse When $t_W \geq 5\tau$ While the pulse is at its high level between the rising edge and the falling edge, the capacitor is charging. When the pulse width is equal to or greater than five time constants ($t_W \geq 5\tau$), the capacitor has time to fully charge.

As the voltage across the capacitor builds up exponentially, the voltage across the resistor decreases exponentially until it reaches zero volts at the time the capacitor reaches full charge (+10 V in this case). This decrease in the resistor voltage occurs because the sum of the capacitor voltage and the resistor voltage at any instant must be equal to the applied voltage, in compliance with Kirchhoff's voltage law ($v_C + v_R = v_{in}$). This part of the response is illustrated in Figure 15–22(c).

Response During Pulse When $t_W < 5\tau$ When the pulse width is less than five time constants ($t_W < 5\tau$), the capacitor does not have time to fully charge. Its partial charge depends on the relation of the time constant and the pulse width.

Because the capacitor does not reach the full +10 V, the resistor voltage will not reach zero volts by the end of the pulse. For example, if the capacitor charges to +5 V during the pulse interval, the resistor voltage will decrease to +5 V, as illustrated in Figure 15–22(d).

Response to Falling Edge When $t_W \geq 5\tau$ Let's first examine the case in which the capacitor is fully charged at the end of the pulse ($t_W \geq 5\tau$). Refer to Figure 15–22(e). On the falling edge, the input pulse suddenly goes from +10 V back to zero. An instant before the falling edge, the capacitor is charged to 10 V, so point *A* is +10 V and point *B* is 0 V. Since the voltage across a capacitor cannot change instantaneously, when point *A* makes a transition from +10 V to zero on the falling edge, point *B must* also make a 10 V transition from zero to −10 V. This keeps the voltage across the capacitor at 10 V for the instant of the falling edge.

The capacitor now begins to discharge exponentially. As a result, the resistor voltage goes from −10 V to zero in an exponential curve, as indicated in Figure 15–22(e).

Response to Falling Edge When $t_W < 5\tau$ Next, let's examine the case in which the capacitor is only partially charged at the end of the pulse ($t_W < 5\tau$). For example, if the capacitor charges to +5 V, the resistor voltage at the instant before the falling edge is also +5 V, because the capacitor voltage plus the resistor voltage must add up to +10 V, as illustrated in Figure 15–22(d).

When the falling edge occurs, point *A* goes from +10 V to zero. As a result, point *B* goes from +5 V to −5 V, as illustrated in Figure 15–22(f). This decrease occurs, of course, because the capacitor voltage cannot change at the instant of the falling edge. Immediately after the falling edge, the capacitor begins to discharge to zero. As a result, the resistor voltage goes from −5 V to zero, as shown.

Summary of Differentiator Response to a Single Pulse

Perhaps a good way to summarize this section is to look at the general output waveforms of a differentiator as the time constant is varied from one extreme, when 5τ is much less than the pulse width, to the other extreme, when 5τ is much greater than the pulse width. These situations are illustrated in Figure 15–23. In part (a), the output consists of narrow positive and negative "spikes." In part (f), the output approaches the shape of the input. Various conditions between these extremes are illustrated in parts (b) through (e).

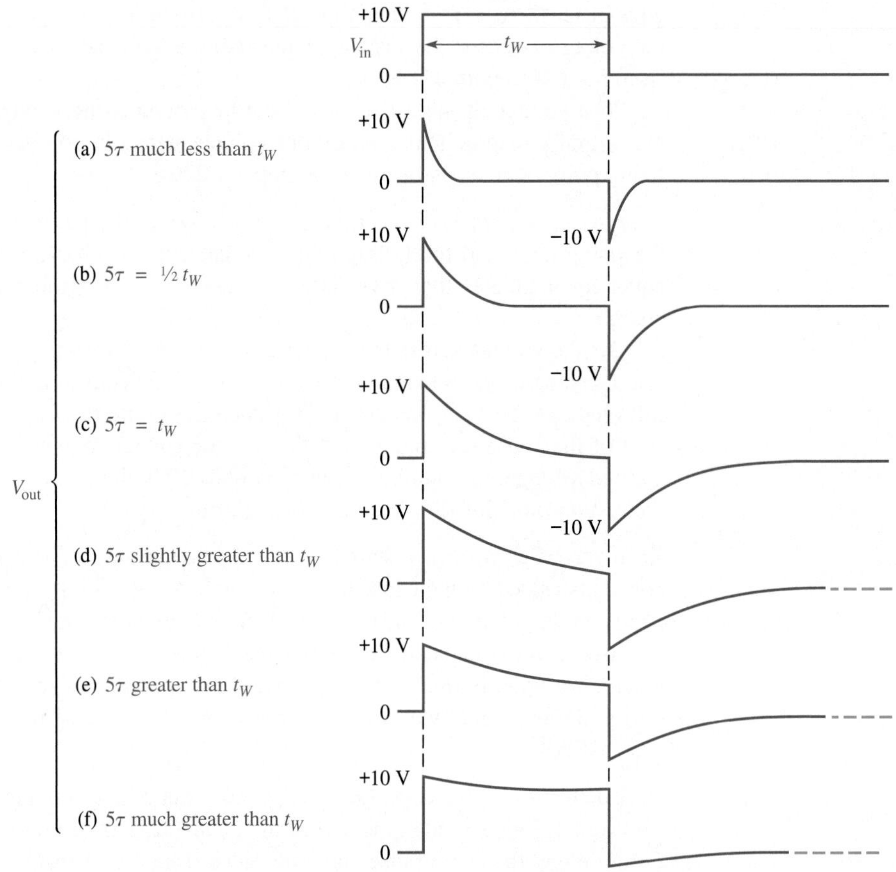

FIGURE 15–23
Effects of a change in time constant on the shape of the output voltage of a differentiator.

EXAMPLE 15–4

Sketch the output voltage for the circuit in Figure 15–24.

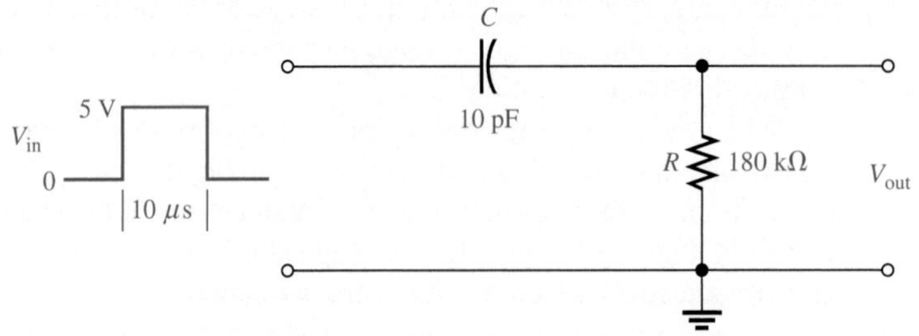

FIGURE 15–24

Solution First calculate the time constant:

$$\tau = RC = (180\ \text{k}\Omega)(10\ \text{pF}) = 1.8\ \mu\text{s}$$

In this case, $t_W > 5\tau$, so the capacitor reaches full charge in 9 μs (before the end of the pulse).

On the rising edge, the resistor voltage jumps to +5 V and then decreases exponentially to zero before the end of the pulse. On the falling edge, the resistor voltage jumps to −5 V and then goes back to zero exponentially. The resistor voltage is, of course, the output, and its shape is shown in Figure 15–25.

FIGURE 15–25

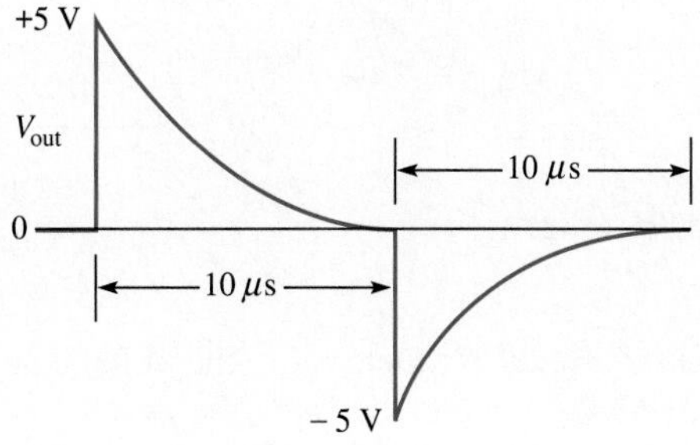

Related Exercise Sketch the output voltage if R = 18 kΩ and C = 47 pF in Figure 15–24.

EXAMPLE 15–5

Determine the output voltage waveform for the differentiator in Figure 15–26.

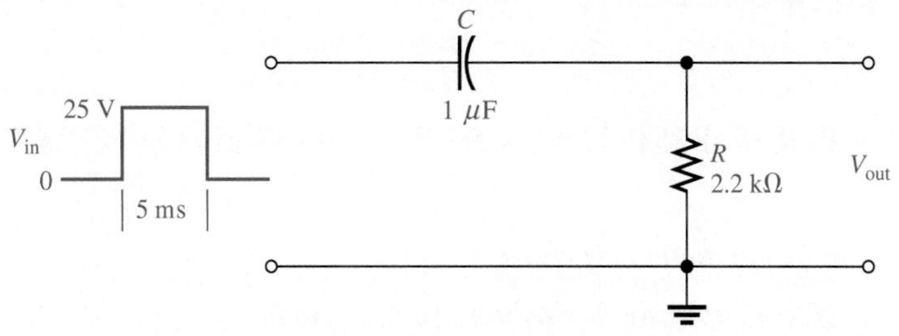

FIGURE 15–26

Solution First calculate the time constant:

$$\tau = (2.2\text{ k}\Omega)(1\ \mu\text{F}) = 2.2\text{ ms}$$

On the rising edge, the resistor voltage immediately jumps to +25 V. Because the pulse width is 5 ms, the capacitor charges for approximately 2.27 time constants during the pulse, and therefore does not reach full charge. Thus, you must use the formula for a decreasing exponential, Equation (9–19), in order to calculate to what voltage the output decreases by the end of the pulse:

$$v_{out} = V_i e^{-t/RC} = 25e^{-5\text{ms}/2.2\text{ms}} = 25(0.103) = 2.58\text{ V}$$

where V_i = 25 V and t = 5 ms. This calculation gives the resistor voltage at the end of the 5 ms pulse width interval.

On the falling edge, the resistor voltage immediately jumps from + 2.58 V down to −22.4 V (a 25 V transition). The resulting waveform of the output voltage is shown in Figure 15–27.

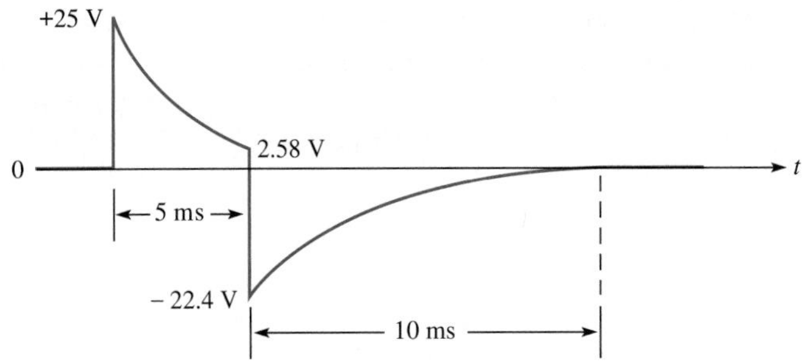

FIGURE 15–27

Related Exercise Determine the voltage at the end of the pulse in Figure 15–26 if R is 1.5 kΩ.

SECTION 15–4 REVIEW

1. Sketch the output of a differentiator for a 10 V input pulse when $5\tau = 0.5t_W$.
2. Under what condition does the output pulse shape most closely resemble the input pulse for a differentiator?
3. What does the differentiator output look like when 5τ is much less than the pulse width of the input?
4. If the resistor voltage in a differentiating circuit is down to +5 V at the end of a 15 V input pulse, to what negative value will the resistor voltage go in response to the falling edge of the input?

15–5 ■ REPETITIVE-PULSE RESPONSE OF *RC* DIFFERENTIATORS

The RC differentiator response to a single pulse, covered in the last section, is extended in this section to repetitive pulses.

After completing this section, you should be able to

- **Analyze an *RC* differentiator with repetitive input pulses**
  - □ Determine the response when the pulse width is less than five time constants

If a periodic pulse waveform is applied to an RC differentiating circuit, two conditions again are possible: $t_W \geq 5\tau$ or $t_W < 5\tau$. Figure 15–28 shows the output when $t_W = 5\tau$. As the time constant is reduced, both the positive and the negative portions of the output become narrower. Notice that the average value of the output is zero.

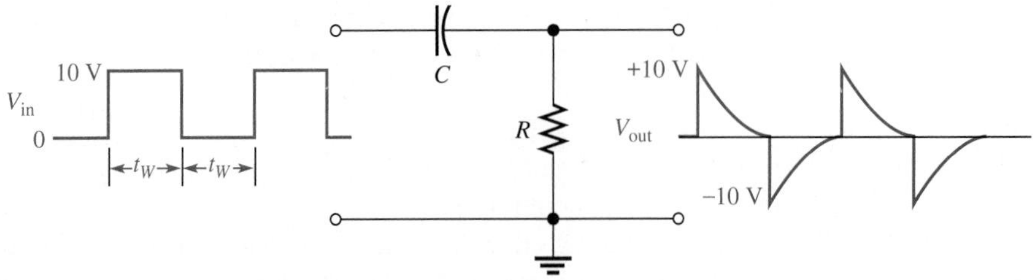

FIGURE 15–28
Example of differentiator response when $t_W = 5\tau$.

Figure 15–29 shows the steady-state output when $t_W < 5\tau$. As the time constant is increased, the positively and negatively sloping portions become flatter. For a very long time constant, the output approaches the shape of the input, but with an average value of zero. The average value of a waveform is its **dc component.** Because a capacitor blocks dc, the dc component of the input is prevented from passing through to the output.

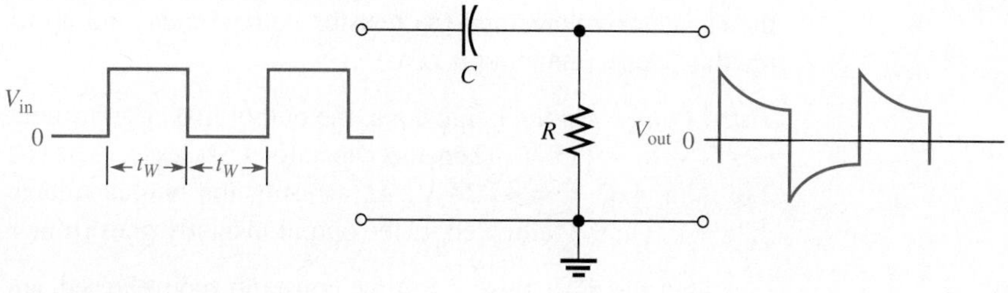

FIGURE 15–29
Example of differentiator response when $t_W < 5\tau$.

Like the integrator, the differentiator output takes time (5τ) to reach steady state. To illustrate the response, let's take an example in which the time constant equals the input pulse width.

Analysis of a Repetitive Waveform

At this point, we do not care what the circuit time constant is because we know that the resistor voltage will decrease to 37% of its maximum value during one pulse (1τ). Assume that the capacitor in Figure 15–30 begins initially uncharged and examine the output voltage on a pulse-by-pulse basis. The results of the analysis to follow are shown in Figure 15–31.

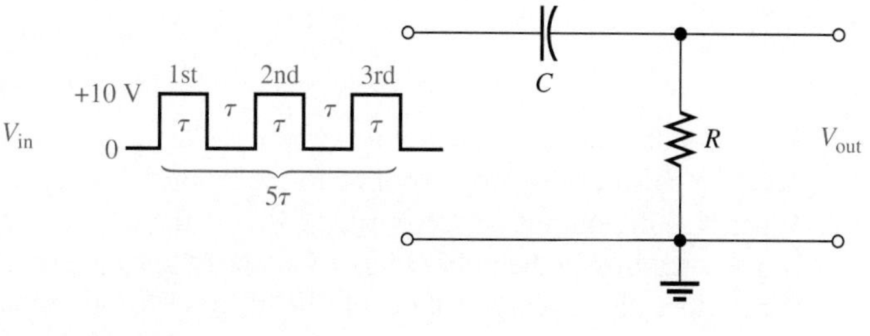

FIGURE 15–30
RC differentiator with $\tau = t_W$.

First Pulse On the rising edge, the output instantaneously jumps to +10 V. Then the capacitor partially charges to 63% of 10 V, which is 6.3 V. Thus, the output voltage must decrease to 3.7 V, as shown in Figure 15–31.

On the falling edge, the output instantaneously makes a negative-going 10 V transition to –6.3 V (–10 V + 3.7 V = –6.3 V).

Between First and Second Pulses The capacitor discharges to 37% of 6.3 V, which is 2.33 V. Thus, the resistor voltage, which starts at –6.3 V, must increase to –2.33 V. Why? Because at the instant prior to the next pulse, the input voltage is zero. Therefore, the sum of v_C and v_R must be zero (+2.33 V – 2.33 V = 0). Remember that $v_C + v_R = v_{in}$ at all times.

Second Pulse On the rising edge, the output makes an instantaneous, positive-going, 10 V transition from –2.33 V to 7.67 V. Then by the end of the pulse the capacitor charges 0.63 × (10 V − 2.33 V) = 4.83 V. Thus, the capacitor voltage increases from 2.33 V to 2.33 V + 4.83 V = 7.16 V. The output voltage drops to 0.37 × (7.67 V) = 2.84 V.

On the falling edge, the output instantaneously makes a negative-going transition from 2.84 V to −7.16 V, as shown in Figure 15–31.

Between Second and Third Pulses The capacitor discharges to 37% of 7.16 V, which is 2.65 V. Thus, the output voltage starts at −7.16 V and increases to −2.65 V, because the capacitor voltage and the resistor voltage must add up to zero at the instant prior to the third pulse (the input is zero).

Third Pulse On the rising edge, the output makes an instantaneous 10 V transition from −2.65 V to +7.35 V. Then the capacitor charges 0.63 × (10 V − 2.65 V) = 4.63 V to 2.65 V + 4.63 V = +7.28 V. As a result, the output voltage drops to 0.37 × 7.35 V = 2.72 V. On the falling edge, the output instantly goes from +2.72 V down to −7.28 V.

After the third pulse, five time constants have elapsed, and the output voltage is close to its steady state. Thus, it will continue to vary from a positive maximum of about +7.3 V to a negative maximum of about −7.3 V, with an average value of zero.

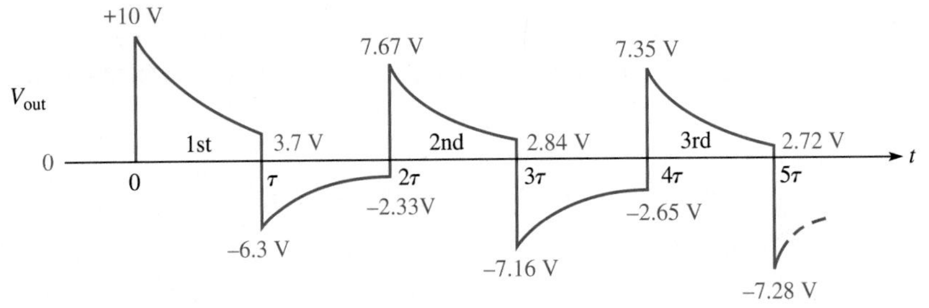

FIGURE 15–31
Differentiator output waveform during transient time for the circuit in Figure 15–30.

SECTION 15–5 REVIEW

1. What conditions allow an *RC* differentiator to fully charge and discharge when a periodic pulse waveform is applied to the input?
2. What will the output waveform look like if the circuit time constant is extremely small compared to the pulse width of a square wave input?
3. What does the average value of the differentiator output voltage equal during steady state?

15–6 ■ PULSE RESPONSE OF *RL* INTEGRATORS

A series RL circuit in which the output voltage is taken across the resistor is known as an integrator in terms of pulse response.

After completing this section, you should be able to

- **Analyze the operation of an *RL* integrator**
  - ☐ Determine the response to a single input pulse

Figure 15–32 shows an *RL* integrator. The output waveform is taken across the resistor and, under equivalent conditions, is the same shape as that for the *RC* integrator. Recall that in the *RC* case, the output was across the capacitor.

FIGURE 15–32
An RL integrating circuit.

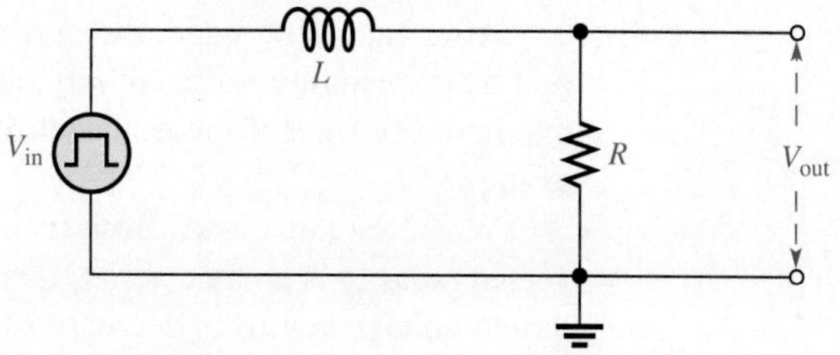

As you know, both edges of an ideal pulse are considered to occur instantaneously. Two basic rules for inductor behavior will aid in analyzing *RL* circuit responses to pulse inputs:

1. The inductor appears as an open to an instantaneous change in current and as a short (ideally) to dc.
2. The current in an inductor cannot change instantaneously—it can change only exponentially.

Response of the Integrator to a Single Pulse

When a pulse generator is connected to the input of the integrator and the voltage pulse goes from its low level to its high level, the inductor prevents a sudden change in current. As a result, the inductor acts as an open, and all the input voltage is across it at the instant of the rising pulse edge. This situation is indicated in Figure 15–33(a).

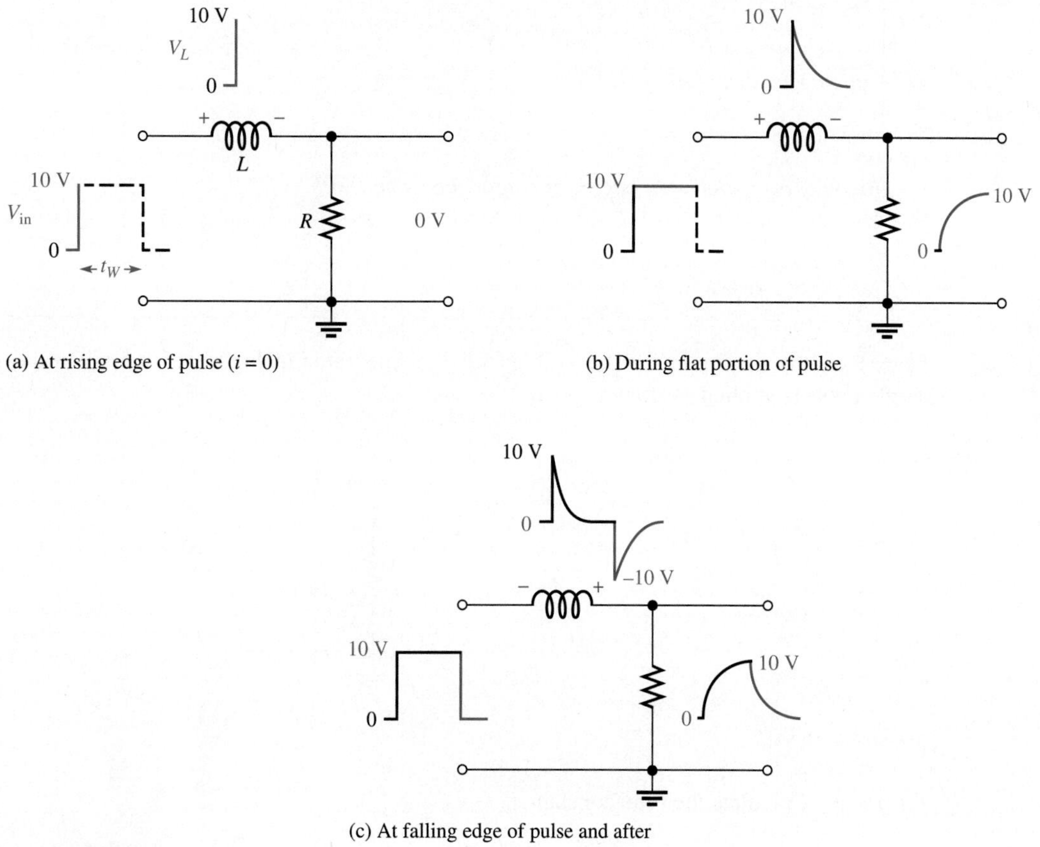

FIGURE 15–33
Illustration of the pulse response of an RL integrator ($t_W > 5\tau$).

After the rising edge, the current builds up, and the output voltage follows the current as it increases exponentially, as shown in Figure 15–33(b). The current can reach a maximum of V_p/R if the transient time is shorter than the pulse width ($V_p = 10$ V in this example).

When the pulse goes from its high level to its low level, an induced voltage with reversed polarity is created across the coil in an effort to keep the current equal to V_p/R. The output voltage begins to decrease exponentially, as shown in Figure 15–33(c).

The exact shape of the output depends on the L/R time constant as summarized in Figure 15–34 for various relationships between the time constant and the pulse width. You should note that the response of this *RL* circuit in terms of the shape of the output is identical to that of the *RC* integrator. The relationship of the L/R time constant to the input pulse width has the same effect as the *RC* time constant that we discussed earlier in this chapter. For example, when $t_W < 5\tau$, the output voltage will not reach its maximum possible value.

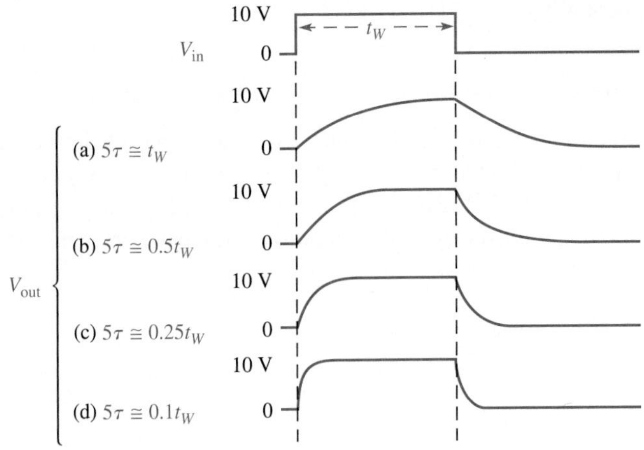

FIGURE 15–34
Illustration of the variation in integrator output pulse shape with time constant.

EXAMPLE 15–6

Determine the maximum output voltage for the integrator in Figure 15–35 when a single pulse is applied as shown.

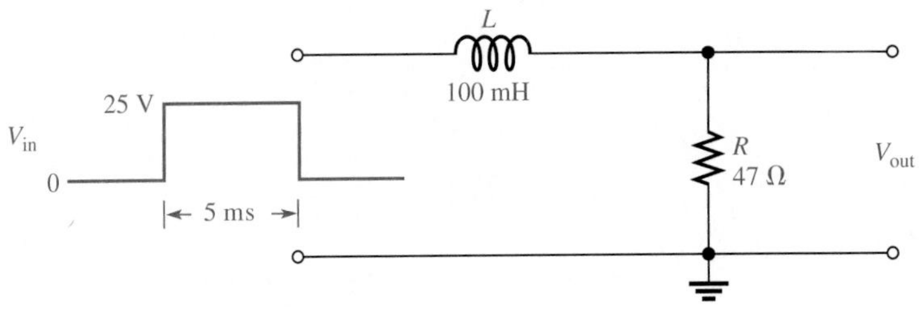

FIGURE 15–35

Solution Calculate the time constant:

$$\tau = \frac{L}{R} = \frac{100 \text{ mH}}{47\ \Omega} = 2.13 \text{ ms}$$

Because the pulse width is 5 ms, the inductor charges for approximately 2.35τ (5 ms/2.13 ms = 2.35). Use the exponential formula to calculate the voltage:

$$v_{out} = V_F(1 - e^{-t/\tau}) = 25(1 - e^{-5/2.13})$$
$$= 25(1 - e^{-2.35}) = 25(1 - 0.095) = 25(0.905) = 22.6 \text{ V}$$

Related Exercise What value must R be for the output voltage to reach 25 V by the end of the input pulse in Figure 15–35?

EXAMPLE 15–7 A pulse is applied to the *RL* integrator circuit in Figure 15–36. Determine the complete waveshapes and the values for I, V_R, and V_L.

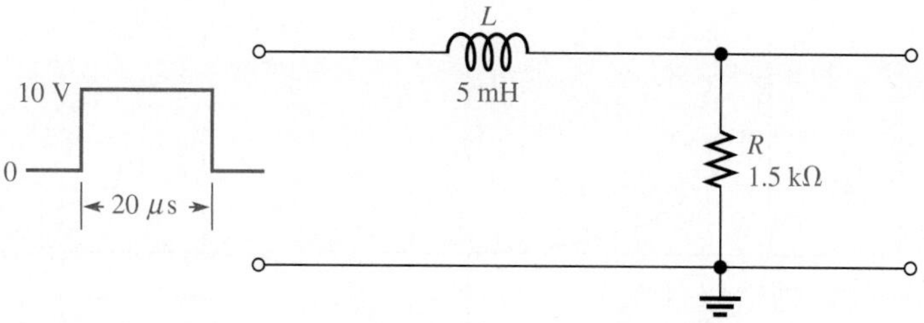

FIGURE 15–36

Solution The circuit time constant is

$$\tau = \frac{L}{R} = \frac{5 \text{ mH}}{1.5 \text{ k}\Omega} = 3.33\ \mu\text{s}$$

Since 5τ = 16.7 μs is less than t_W, the current will reach its maximum value and remain there until the end of the pulse.

At the rising edge of the pulse,

$$i = 0 \text{ A}$$
$$v_R = 0 \text{ V}$$
$$v_L = 10 \text{ V}$$

Since the inductor is initially an open, all of the input voltage appears across L.

During the pulse,

i increases exponentially to $\frac{V_p}{R} = \frac{10 \text{ V}}{1.5 \text{ k}\Omega} = 6.67$ mA in 16.7 μs

v_R increases exponentially to 10 V in 16.7 μs

v_L decreases exponentially to zero in 16.7 μs

At the falling edge of the pulse,

$$i = 6.67 \text{ mA}$$
$$v_R = 10 \text{ V}$$
$$v_L = -10 \text{ V}$$

After the pulse,

i decreases exponentially to zero in 16.7 μs
v_R decreases exponentially to zero in 16.7 μs
v_L decreases exponentially to zero in 16.7 μs

The waveforms are shown in Figure 15–37.

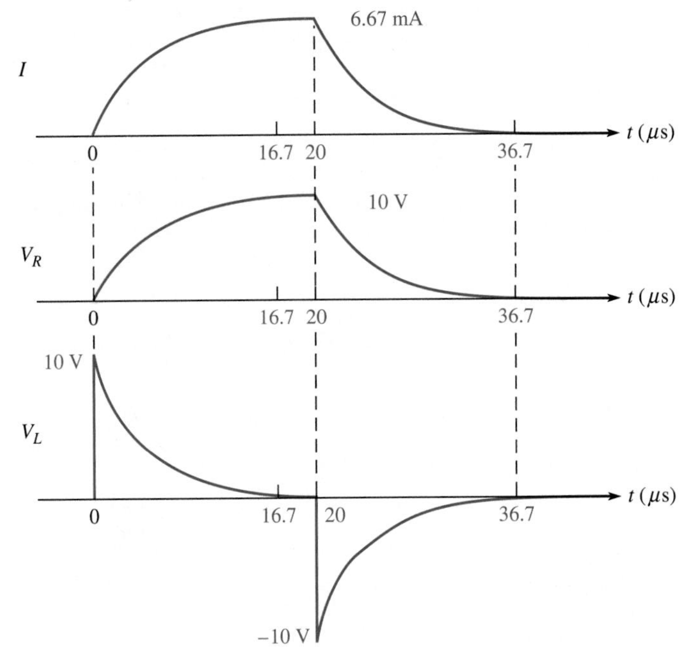

FIGURE 15–37

Related Exercise What will be the maximum output voltage if the amplitude of the input pulse is increased to 20 V in Figure 15–36?

EXAMPLE 15–8 A 10 V pulse with a width of 1 ms is applied to the integrator in Figure 15–38. Determine the voltage level that the output will reach during the pulse. If the source has an internal resistance of 30 Ω, how long will it take the output to decay to zero? Sketch the output voltage waveform.

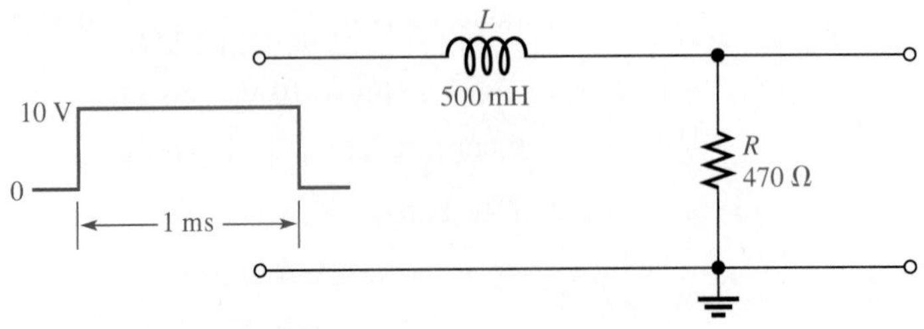

FIGURE 15–38

Solution The coil charges through the 30 Ω source resistance plus the 470 Ω external resistor. The time constant is

$$\tau = \frac{L}{R_T} = \frac{500 \text{ mH}}{470\ \Omega + 30\ \Omega} = \frac{500 \text{ mH}}{500\ \Omega} = 1 \text{ ms}$$

Notice that in this case the pulse width is exactly equal to τ. Thus, the output V_R will reach 63% of the full input amplitude in 1τ. Therefore, the output voltage gets to 6.3 V at the end of the pulse.

After the pulse is gone, the inductor discharges back through the 30 Ω source and the 470 Ω resistor. The source takes 5τ to completely discharge.

$$5\tau = 5(1 \text{ ms}) = 5 \text{ ms}$$

The output voltage is shown in Figure 15–39.

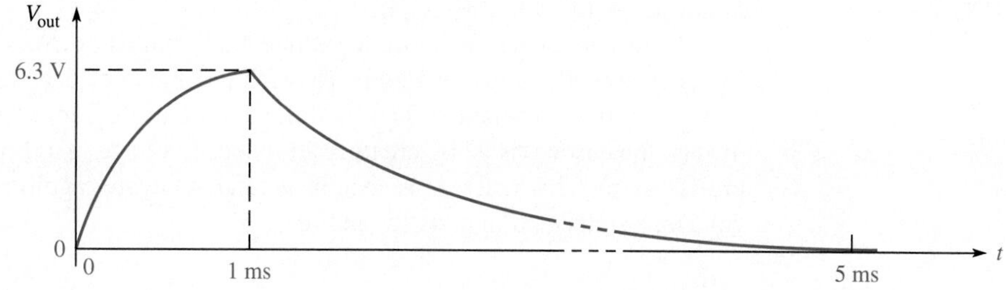

FIGURE 15–39

Related Exercise To what value must *R* be changed in Figure 15–38 to allow the output voltage to reach the input level during the pulse?

SECTION 15–6 REVIEW

1. In an *RL* integrator, across which component is the output voltage taken?
2. When a pulse is applied to an *RL* integrator, what condition must exist in order for the output voltage to reach the amplitude of the input?
3. Under what condition will the output voltage have the approximate shape of the input pulse?

15–7 ▪ PULSE RESPONSE OF *RL* DIFFERENTIATORS

A series RL circuit in which the output voltage is taken across the inductor is known as a differentiator.

After completing this section, you should be able to

- **Analyze the operation of an *RL* differentiator**
  - ☐ Determine the response to a single input pulse

Figure 15–40 shows an *RL* differentiator. Compare this circuit to the *RC* differentiator in which the output voltage is across the resistor.

FIGURE 15–40
An RL differentiating circuit.

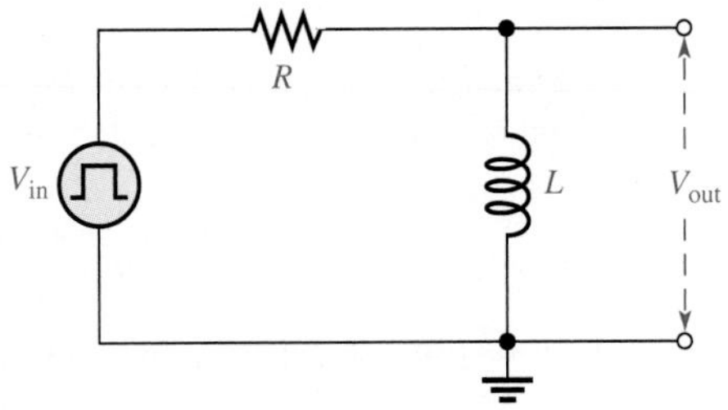

Response of the Differentiator to a Single Pulse

A pulse generator is connected to the input of the differentiator. Initially, before the pulse, there is no current in the circuit. When the input pulse goes from its low level to its high level, the inductor prevents a sudden change in current. It does this, as you know, with an induced voltage equal and opposite to the input. As a result, L looks like an open, and all of the input voltage appears across it at the instant of the rising edge, as shown in Figure 15–41(a) with a 10 V pulse.

During the pulse, the current exponentially builds up. As a result, the inductor voltage decreases, as shown in Figure 15–41(b). The rate of decrease, as you know, depends on the L/R time constant. When the falling edge of the input appears, the inductor reacts to keep the current as is, by creating an induced voltage in a direction as indicated in Figure 15–41(c). This reaction is seen as a sudden negative-going transition of the inductor voltage, as indicated in parts (c) and (d).

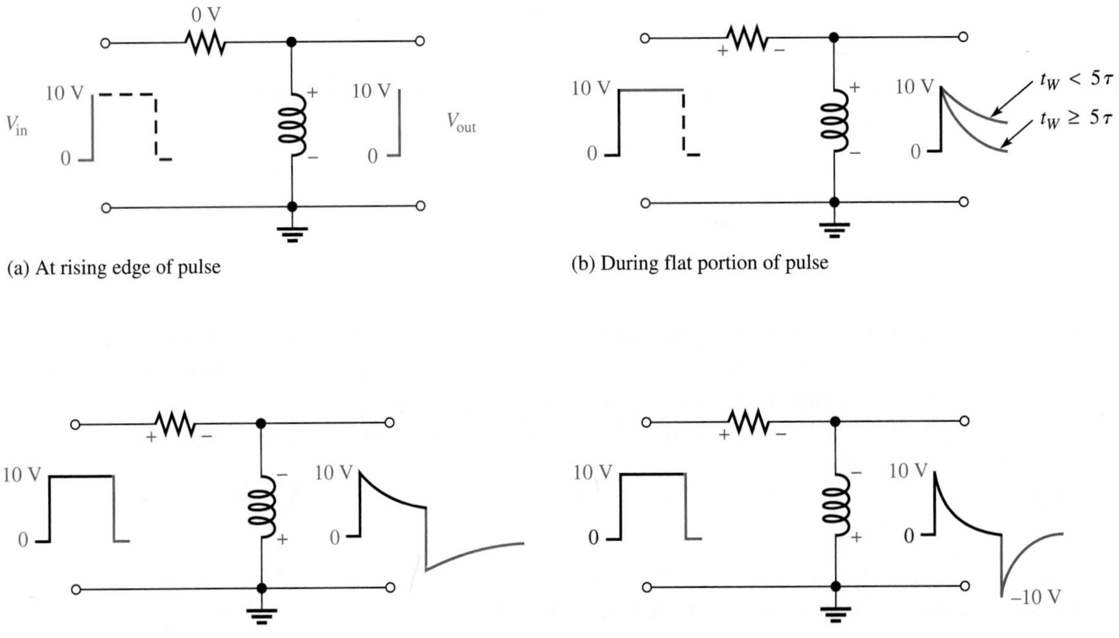

(a) At rising edge of pulse

(b) During flat portion of pulse

(c) At falling edge when $t_W < 5\tau$

(d) At falling edge when $t_W \geq 5\tau$

FIGURE 15–41
Illustration of the response of an RL differentiator for both time constant conditions.

Two conditions are possible, as indicated in Figure 15–41(c) and (d). In part (c), 5τ is greater than the input pulse width, and the output voltage does not have time to decay to zero. In part (d), 5τ is less than the pulse width, and so the output decays to zero before the end of the pulse. In this case a full, negative, 10 V transition occurs at the trailing edge.

Keep in mind that as far as the input and output waveforms are concerned, the RL integrator and differentiator perform the same as their RC counterparts.

A summary of the *RL* differentiator response for relationships of various time constants and pulse widths is shown in Figure 15–42.

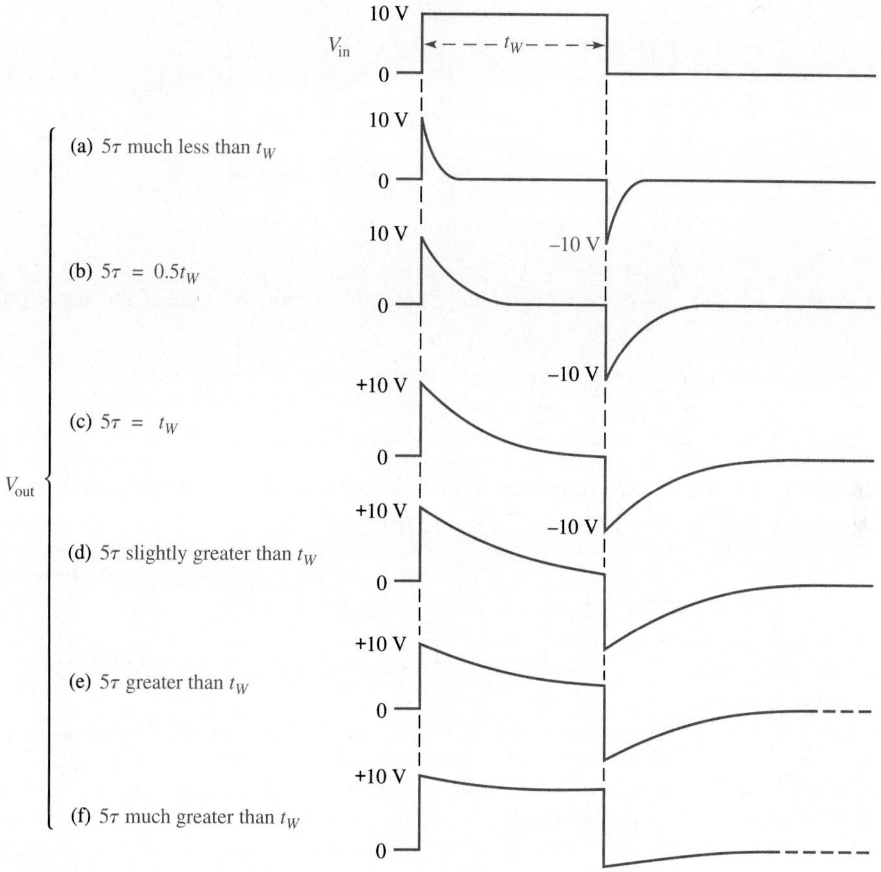

FIGURE 15–42
Illustration of the variation in output pulse shape with the time constant.

EXAMPLE 15–9 Sketch the output voltage for the circuit in Figure 15–43.

FIGURE 15–43

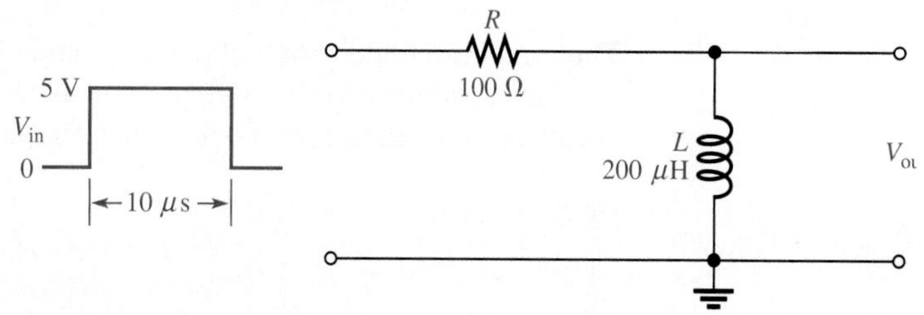

Solution First calculate the time constant:

$$\tau = \frac{L}{R} = \frac{200\ \mu\text{H}}{100\ \Omega} = 2\ \mu\text{s}$$

In this case, $t_W = 5\tau$, so the output will decay to zero at the end of the pulse.

On the rising edge, the inductor voltage jumps to +5 V and then decays exponentially to zero. It reaches approximately zero at the instant of the falling edge. On the falling edge of the input, the inductor voltage jumps to −5 V and then goes back to zero. The output waveform is shown in Figure 15–44.

FIGURE 15–44

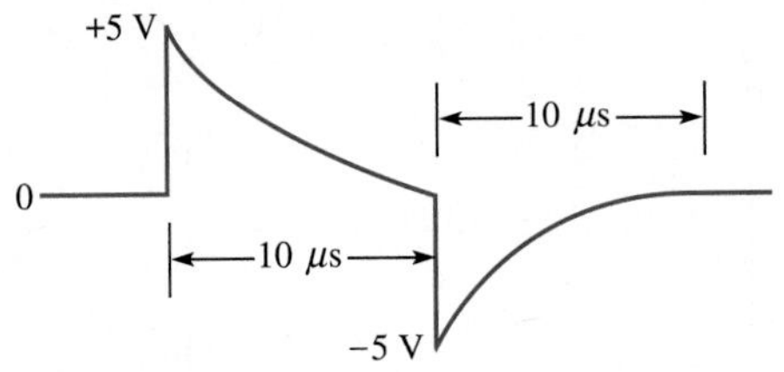

Related Exercise Sketch the output voltage if the input pulse width is reduced to 5 μs in Figure 15–43.

EXAMPLE 15–10 Determine the output voltage waveform for the differentiator in Figure 15–45.

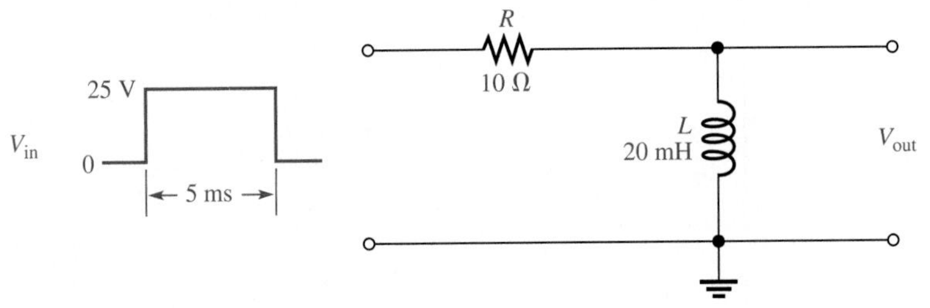

FIGURE 15–45

Solution First calculate the time constant:

$$\tau = \frac{L}{R} = \frac{20\text{ mH}}{10\ \Omega} = 2\text{ ms}$$

On the rising edge, the inductor voltage immediately jumps to +25 V. Because the pulse width is 5 ms, the inductor charges for only two and a half time constants (2.5τ), so you must use the formula for a decreasing exponential:

$$v_L = V_i e^{-t/\tau} = 25e^{-5\text{ms}/2\text{ms}} = 25e^{-2.5} = 25(0.082) = 2.05\text{ V}$$

This result is the inductor voltage at the end of the 5 ms input pulse.

On the falling edge, the output immediately drops from +2.05 V to −22.95 V (a 25 V negative-going transition). The complete output waveform is sketched in Figure 15–46.

FIGURE 15–46

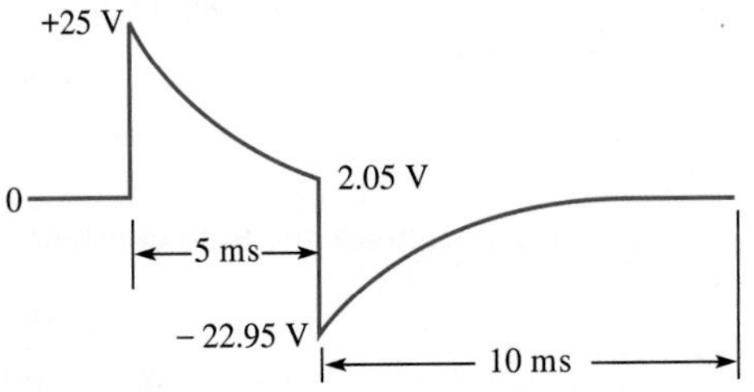

Related Exercise What must be the value of R for V_{out} to reach zero by the end of the input pulse in Figure 15–45?

SECTION 15–7 REVIEW

1. In an *RL* differentiator, across which component is the output taken?
2. Under what condition does the output pulse shape most closely resemble the input pulse?
3. If the inductor voltage in an *RL* differentiator is down to +2 V at the end of a +10 V input pulse, to what negative voltage will the output voltage go in response to the falling edge of the input?

15–8 ▪ APPLICATIONS

Although integrators and differentiators are found in many different kinds of applications, only three basic ones are discussed in this section to give you a general idea of how these circuits can be used.

After completing this section, you should be able to

- **Discuss some basic applications of integrators and differentiators**
  - ☐ Describe how *RC* integrators are used in timing circuits
  - ☐ Describe how an integrator can be used for converting pulse waveforms to dc
  - ☐ Describe how a differentiator can be used as a trigger pulse generator

Timing Circuits

RC integrators can be used in timing circuits to set a specified time interval for various purposes. By changing the time constant, the interval can be adjusted. For example, the *RC* integrator circuit in Figure 15–47(a) is used to provide a delay between the time a

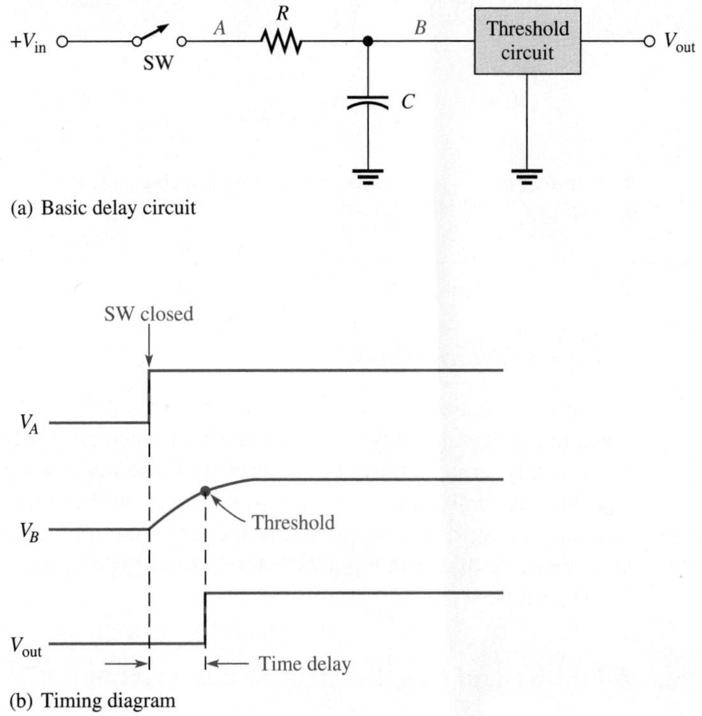

FIGURE 15–47
A basic time delay application of an RC integrator.

power-on switch is closed and the time a certain event is activated by a threshold circuit. The threshold circuit is designed to respond to the input voltage only when the input reaches a certain specified level. Don't worry how the threshold circuit works at this point; you will learn about this type of circuit in a later course.

When the switch is thrown, the capacitor begins to charge at a rate set by the RC time constant. After a certain time, the capacitor voltage reaches the threshold value and triggers (turns on) the circuit to activate a device such as a motor, relay, lamp, etc., depending on the particular application. This action is illustrated by the waveforms in Figure 15–47(b).

EXAMPLE 15–11 What is the time delay in the circuit of Figure 15–47(a) with $V_{in} = 9$ V, $R = 10$ MΩ, and $C = 0.47$ μF? Assume the threshold voltage is 5 V.

Solution Use the exponential formula, Equation (9–18), and solve for time as follows:

$$v = V_F(1 - e^{-t/RC})$$
$$v = V_F - V_F e^{-t/RC}$$
$$V_F - v = V_F e^{-t/RC}$$
$$e^{-t/RC} = \frac{V_F - v}{V_F}$$

Taking the natural log (ln) of both sides of the equation yields

$$-\frac{t}{RC} = \ln\left(\frac{V_F - v}{V_F}\right)$$
$$t = -RC\ln\left(\frac{V_F - v}{V_F}\right)$$

V_F is the final voltage to which C will charge and equals V_{in}. Substituting values and solving for t:

$$t = -(10\text{ M}\Omega)(0.47\ \mu\text{F})\ln\left(\frac{9\text{ V} - 5\text{ V}}{9\text{ V}}\right) = -(10\text{ M}\Omega)(0.47\ \mu\text{F})\ln\left(\frac{4\text{ V}}{9\text{ V}}\right) = 3.8\text{ s}$$

Related Exercise Find the time delay for the circuit in Figure 15–47 for $V_{in} = 24$ V, $R = 10$ kΩ, and $C = 1500$ pF.

A Pulse Waveform-to-DC Converter

An integrating circuit can be used to convert a pulse waveform to a constant dc value equal to the average of the waveform. This is accomplished by using a time constant that is extremely large compared to the period of the pulse waveform, as illustrated in Figure 15–48(a). Actually, there will be a slight ripple on the output as the capacitor charges and discharges a little. This ripple is less the larger the time constant and can approach a constant level, as shown in Figure 15–48(b). In terms of sinusoidal response, this circuit is a low-pass filter.

A Positive and Negative Trigger Pulse Generator

A differentiator circuit can be used to produce very short duration pulses (spikes) with both positive and negative polarities, as illustrated in Figure 15–49(a). When the time

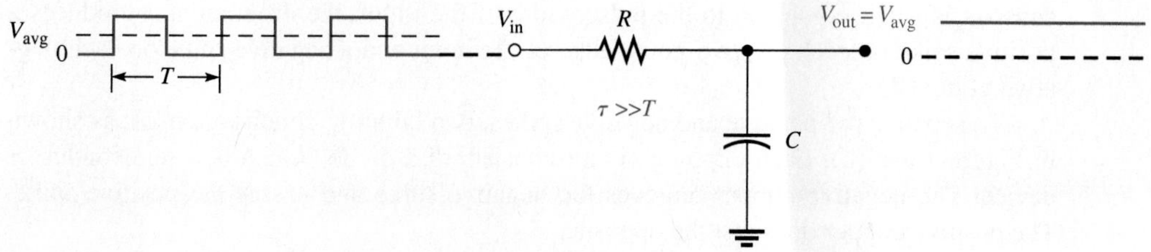

(a)

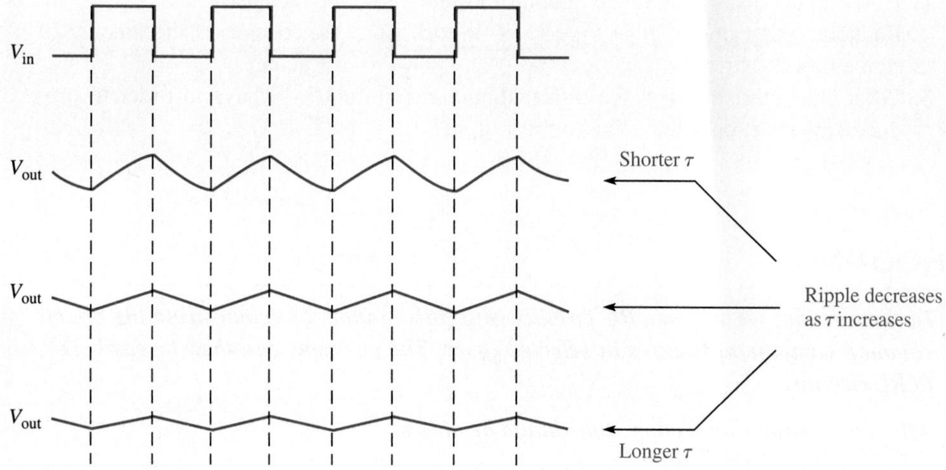

(b)

FIGURE 15–48
A long time constant integrator acts as a pulse-to-dc converter.

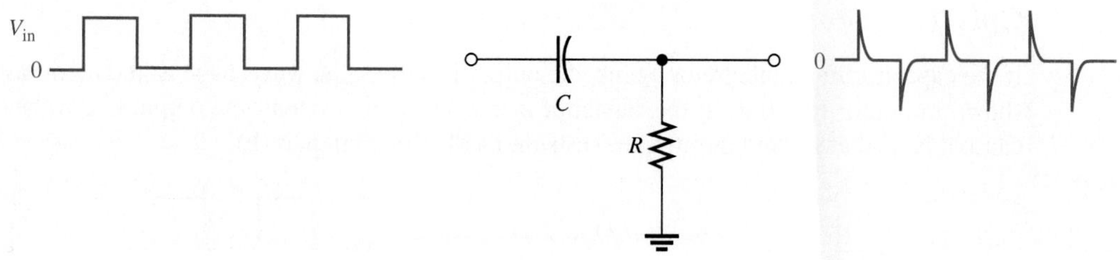

(a)

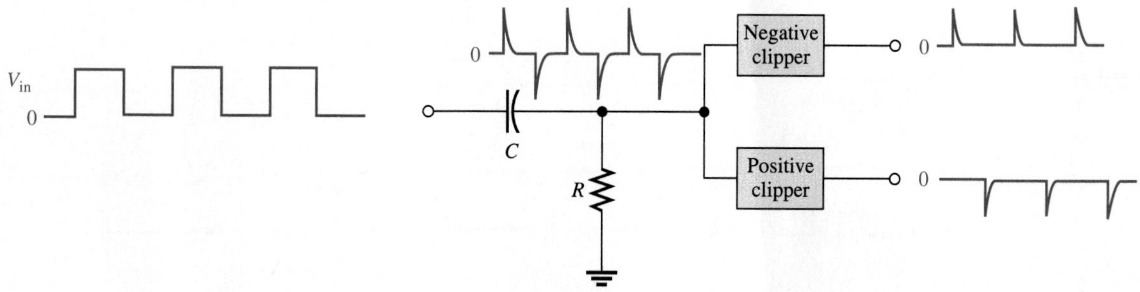

(b)

FIGURE 15–49
A very short time constant differentiator generates positive and negative spikes.

constant is small compared to the pulse width of the input, the differentiator produces a positive spike on each positive-going edge of the input and a negative spike on each negative-going edge.

To separate the positive and negative spikes, two clipping circuits are used, as shown in Figure 15–49(b). Each clipping circuit consists of a diode, which is a semiconductor device. The negative clipper removes the negative spike and passes the positive spike. The positive clipper does just the opposite.

SECTION 15–8 REVIEW

1. How can the delay time in the circuit of Figure 15–47 be increased?
2. Reducing the value of R in Figure 15–48 will (increase, decrease) the amount of ripple on the output.
3. What characteristic must the differentiator in Figure 15–49 have in order to produce very short duration spikes on its output?

15–9 ■ TROUBLESHOOTING

In this section, we will use RC circuits with pulse inputs to demonstrate the effects of common component failures in selected cases. The concepts can then be easily related to RL circuits.

After completing this section, you should be able to

- **Troubleshoot integrators and differentiators**
  - ☐ Recognize the effect of an open capacitor
  - ☐ Recognize the effect of a leaky capacitor
  - ☐ Recognize the effect of a shorted capacitor
  - ☐ Recognize the effect of an open resistor

Open Capacitor

If the capacitor in an integrator opens, the output has the same waveshape as the input, as shown in Figure 15–50(a). If the capacitor in a differentiator opens, the output is zero because it is held at ground through the resistor, as illustrated in part (b).

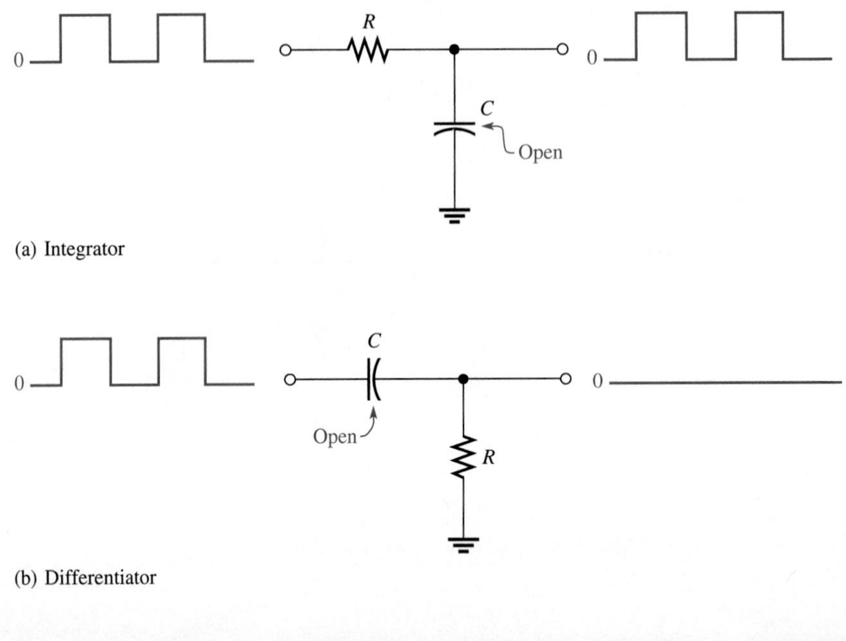

FIGURE 15–50
Examples of the effect of an open capacitor.

Leaky Capacitor

If the capacitor in an integrator becomes leaky, the time constant will be effectively reduced by the leakage resistance. When Thevenized, looking from C, R_{leak} appears in parallel with R through the zero resistance of the input source. The waveshape of the output voltage (across C) is altered from normal by a shorter charging time. Also, the amplitude of the output is reduced because R and R_{leak} effectively act as a voltage divider. These effects are illustrated in Figure 15–51(a).

If the capacitor in a differentiator becomes leaky, the time constant is reduced, just as in the integrator (they are both simply series RC circuits). When the capacitor reaches full charge, the output voltage (across R) is set by the effective voltage-divider action of R and R_{leak}, as shown in Figure 15–51(b).

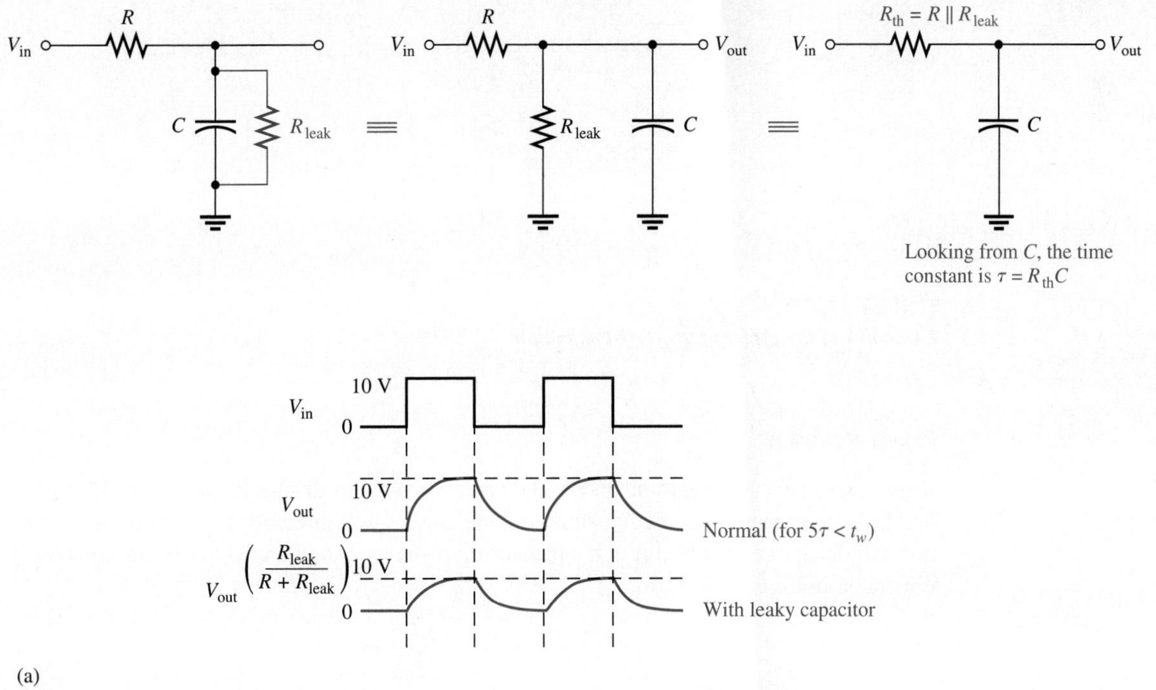

(a)

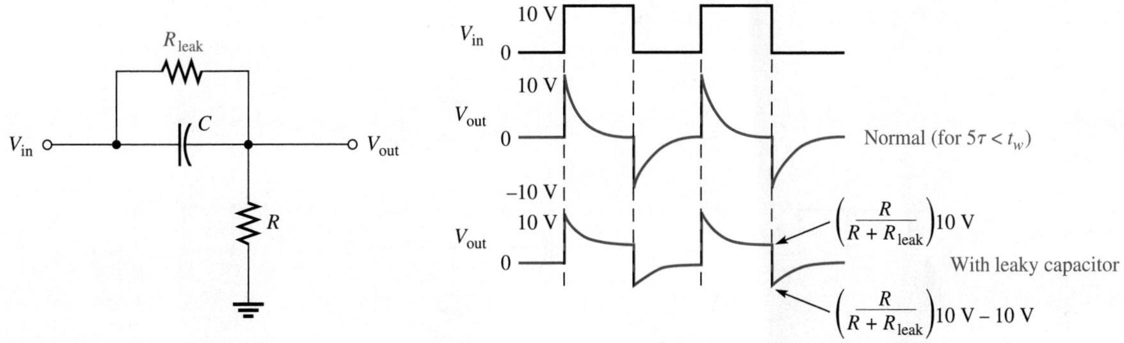

(b)

FIGURE 15–51
Examples of the effect of a leaky capacitor.

Shorted Capacitor

If the capacitor in an integrator shorts, the output is at ground as shown in Figure 15–52(a). If the capacitor in a differentiator shorts, the output is the same as the input, as shown in part (b).

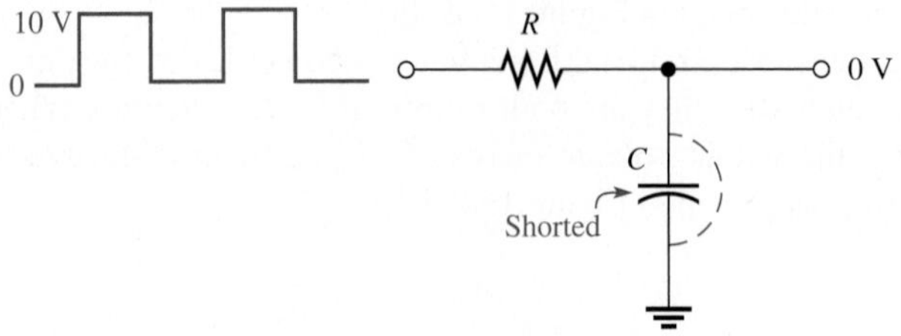

(a)

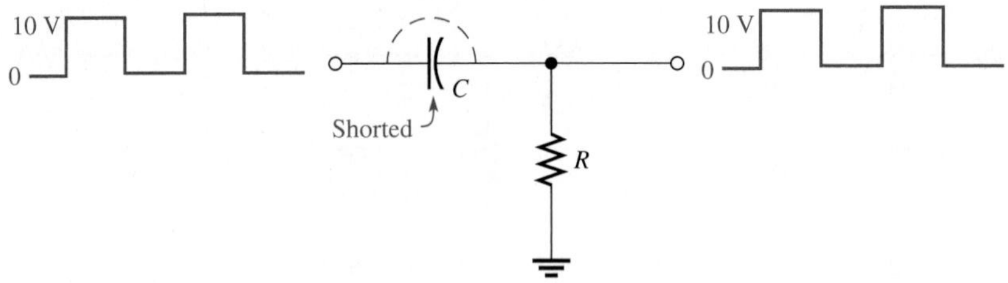

(b)

FIGURE 15–52
Examples of the effect of a shorted capacitor.

Open Resistor

If the resistor in an integrator opens, the capacitor has no discharge path, and, ideally, it will hold its charge. In an actual situation, the charge will gradually leak off or the capacitor will discharge slowly through a measuring instrument connected to the output. This is illustrated in Figure 15–53(a).

FIGURE 15–53
Examples of the effect of an open resistor.

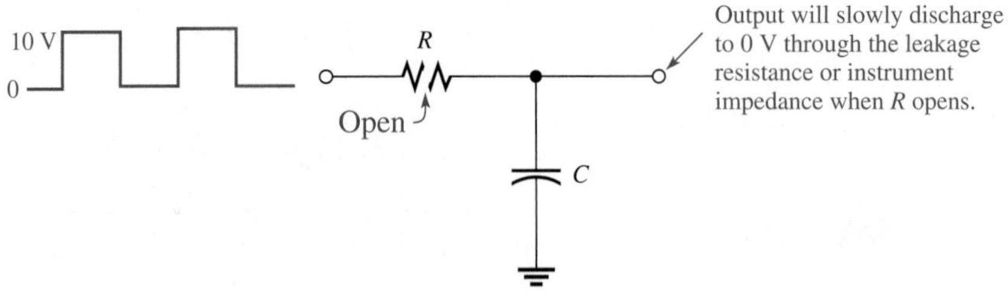

(a)

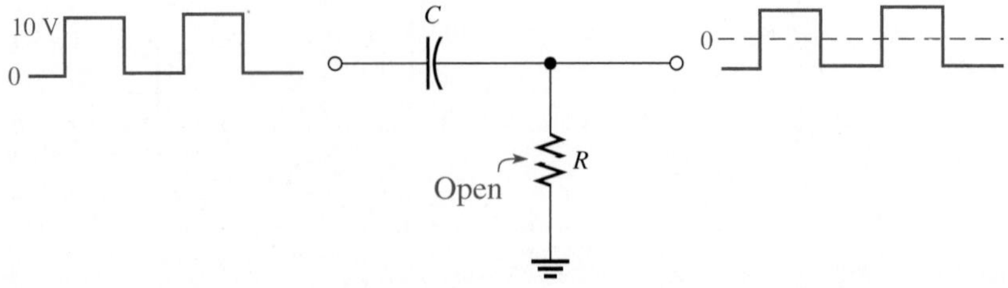

(b)

If the resistor in a differentiator opens, the output looks like the input except for the dc level because the capacitor now must charge and discharge through the extremely high resistance of the oscilloscope, as shown in Figure 15–53(b).

SECTION 15–9 REVIEW

1. An RC integrator has a zero output with a square wave input. What are the possible causes of this problem?
2. If the integrator capacitor is open, what is the output for a square wave input?
3. If the capacitor in a differentiator is shorted, what is the output for a square wave input?

15–10 ■ APPLICATION ASSIGNMENT

Your supervisor has given you the assignment of breadboarding and testing a time-delay circuit that will provide five switch-selectable delay times. An RC integrator is selected for this application. The input is a 5 V pulse of long duration which begins the time delay. The exponential output voltage goes to a threshold trigger circuit that is used to turn the power on to a portion of a system after any one of the five selectable time delays.

A schematic of the selectable time-delay integrator is shown in Figure 15–54. The RC integrator is driven by a positive pulse input and the output is the exponential voltage across the selected capacitor. The output voltage triggers a threshold circuit at the 3.5 V level, which then turns on the power to a portion of a system. The basic concept of operation is shown in Figure 15–55. In this application, the delay time of the integrator is spec-

FIGURE 15–54
Integrator delay circuit.

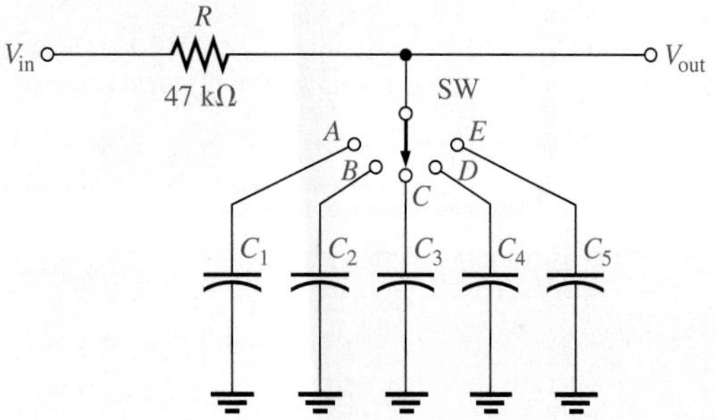

FIGURE 15–55
Illustration of the time-delay operation.

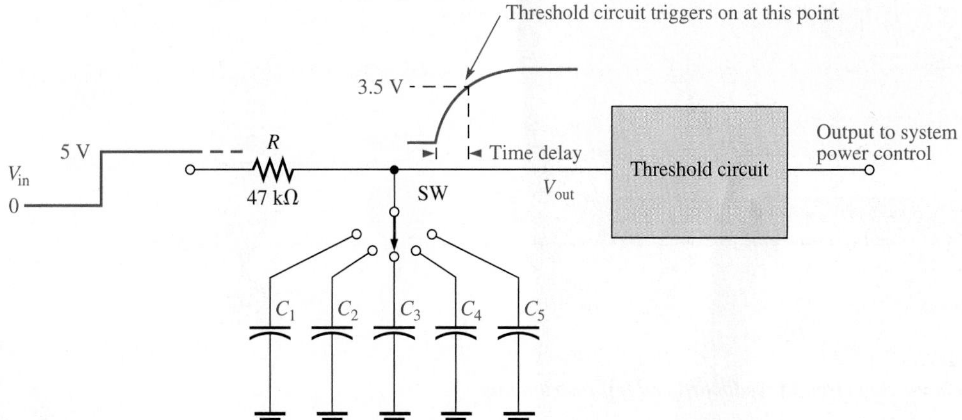

TABLE 15–1

SWITCH POSITION	DELAY TIME
A	10 ms
B	25 ms
C	40 ms
D	65 ms
E	85 ms

ified as the time from the rising edge of the input pulse to the point where the output voltage reaches 3.5 V. The specified time delays are listed in Table 15–1.

Step 1: Capacitor Values

Determine a value for each of the five capacitors in the time-delay circuit that will provide a delay time within 10% of the specified delay time. Select from the following list of standard values: 0.1 μF, 0.12 μF, 0.15 μF, 0.18 μF, 0.22 μF, 0.27 μF, 0.33 μF, 0.39 μF,

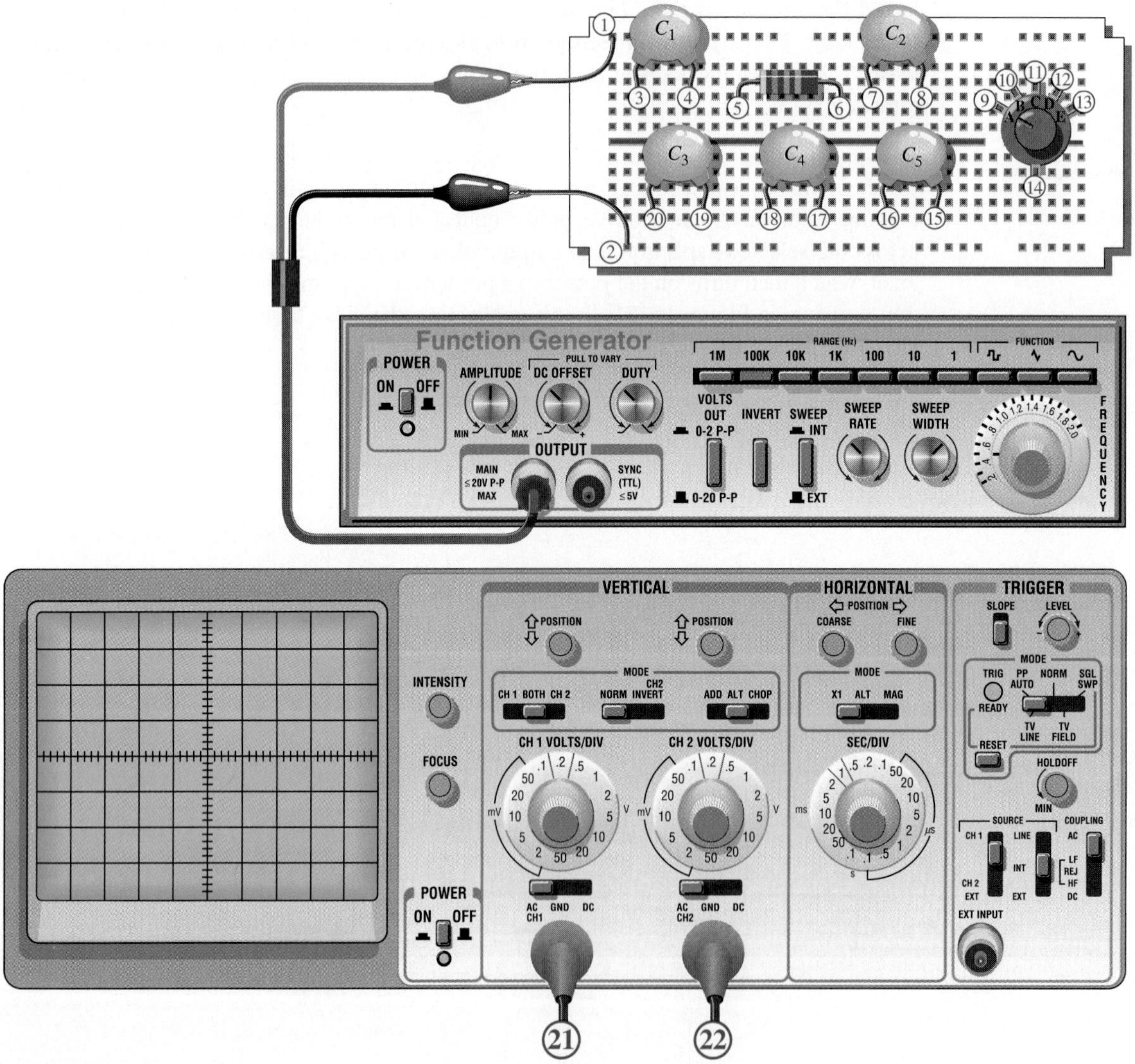

FIGURE 15–56
Time-delay circuit breadboard and test instruments.

0.47 μF, 0.56 μF, 0.68 μF, 0.82 μF, 1.0 μF, 1.2 μF, 1.5 μF, 1.8 μF, 2.2 μF, 2.7 μF, 3.3 μF, 3.9 μF, 4.7 μF, 5.6 μF, 6.8 μF, and 8.2 μF.

Step 2: Circuit Connections

Refer to Figure 15–56. The components for the time-delay circuit are assembled, but not connected, on the breadboard. Using the circled numbers, develop a point-to-point wiring list to show how the circuit and measurement instruments should be connected.

Step 3: Test Procedure and Instrument Settings

Develop a procedure for fully testing the time-delay circuit. Specify the amplitude, frequency, and duty cycle settings for the function generator in order to test each delay time. Specify the oscilloscope sec/div settings for measuring each of the five specified delay times.

Step 4: Measurement

Explain how you will verify that each switch setting produces the proper output delay time.

SECTION 15–10 REVIEW

1. Explain why the input pulse to the time-delay circuit must be of sufficiently long duration.
2. Indicate how you would modify the time-delay circuit so that each selected time delay had some adjustment range.

▪ SUMMARY

- In an *RC* integrating circuit, the output voltage is taken across the capacitor.
- In an *RC* differentiating circuit, the output voltage is taken across the resistor.
- In an *RL* integrating circuit, the output voltage is taken across the resistor.
- In an *RL* differentiating circuit, the output voltage is taken across the inductor.
- In an integrator, when the pulse width (t_W) of the input is much less than the transient time, the output voltage approaches a constant level equal to the average value of the input.
- In an integrator, when the pulse width of the input is much greater than the transient time, the output voltage approaches the shape of the input.
- In a differentiator, when the pulse width of the input is much less than the transient time, the output voltage approaches the shape of the input but with an average value of zero.
- In a differentiator, when the pulse width of the input is much greater than the transient time, the output voltage consists of narrow, positive-going and negative-going spikes occurring on the leading and trailing edges of the input pulses.

▪ GLOSSARY

DC component The average value of a pulse waveform.

Differentiator A circuit producing an output that approaches the mathematical derivative of the input.

Integrator A circuit producing an output that approaches the mathematical derivative of the input.

Periodic Characterized by a repetition at fixed intervals.

Pulse response The reaction of a circuit to a given input.

Steady state The equilibrium condition of a circuit that occurs after an initial transient time.

Time constant A fixed time interval, set by *R, C,* and *L* values, that determines the time response of a circuit.

Transient time An interval equal to approximately five time constants.

■ SELF-TEST

1. The output of an *RC* integrator is taken across the
   (a) resistor (b) capacitor (c) source (d) coil
2. When a 10 V input pulse with a width equal to one time constant is applied to an *RC* integrator, the capacitor charges to
   (a) 10 V (b) 5 V (c) 6.3 V (d) 3.7 V
3. When a 10 V pulse with a width equal to one time constant is applied to an *RC* differentiator, the capacitor charges to
   (a) 6.3 V (b) 10 V (c) 0 V (d) 3.7 V
4. In an *RC* integrator, the output pulse closely resembles the input pulse when
   (a) τ is much larger than the pulse width (b) τ is equal to the pulse width
   (c) τ is less than the pulse width (d) τ is much less than the pulse width
5. In an *RC* differentiator, the output pulse closely resembles the input pulse when
   (a) τ is much larger than the pulse width (b) τ is equal to the pulse width
   (c) τ is less than the pulse width (d) τ is much less than the pulse width
6. The positive and negative portions of a differentiator's output voltage are equal when
   (a) $5\tau < t_W$ (b) $5\tau > t_W$ (c) $5\tau = t_W$ (d) $5\tau > 0$
7. The output of an *RL* integrator is taken across the
   (a) resistor (b) coil (c) source (d) capacitor
8. The maximum current in an *RL* differentiator is
   (a) $I = \frac{V_p}{X_L}$ (b) $I = \frac{V_p}{Z}$ (c) $I = \frac{V_p}{R}$
9. The current in an *RL* differentiator reaches its maximum possible value when
   (a) $5\tau = t_W$ (b) $5\tau < t_W$ (c) $5\tau > t_W$ (d) $\tau = 0.5t_W$
10. If you have an *RC* and an *RL* differentiator with equal time constants sitting side-by-side and you apply the same input pulse to both,
    (a) the *RC* has the widest output pulse
    (b) the *RL* has the most narrow spikes on the output
    (c) the output of one is an increasing exponential and the output of the other is a decreasing exponential
    (d) you can't tell the difference by observing the output waveforms

■ BASIC PROBLEMS

SECTION 15–1 The *RC* Integrator

1. An integrating circuit has $R = 2.2\ \text{k}\Omega$ in series with $C = 0.05\ \mu\text{F}$. What is the time constant?
2. Determine how long it takes the capacitor in an integrating circuit to reach full charge for each of the following series *RC* combinations:
   (a) $R = 50\ \Omega$, $C = 50\ \mu\text{F}$ (b) $R = 3300\ \Omega$, $C = 0.015\ \mu\text{F}$
   (c) $R = 22\ \text{k}\Omega$, $C = 100$ pF (d) $R = 5\ \text{M}\Omega$, $C = 10$ pF

SECTION 15–2 Single-Pulse Response of *RC* Integrators

3. A 20 V pulse is applied to an *RC* integrator. The pulse width equals one time constant. To what voltage does the capacitor charge during the pulse? Assume that it is initially uncharged.
4. Repeat Problem 3 for the following values of t_W:
   (a) 2τ (b) 3τ (c) 4τ (d) 5τ
5. Sketch the approximate shape of an integrator output voltage where 5τ is much less than the pulse width of a 10 V square wave input. Repeat for the case in which 5τ is much larger than the pulse width.
6. Determine the output voltage for an integrator with a single input pulse, as shown in Figure 15–57. For repetitive pulses, how long will it take this circuit to reach steady state?

FIGURE 15–57

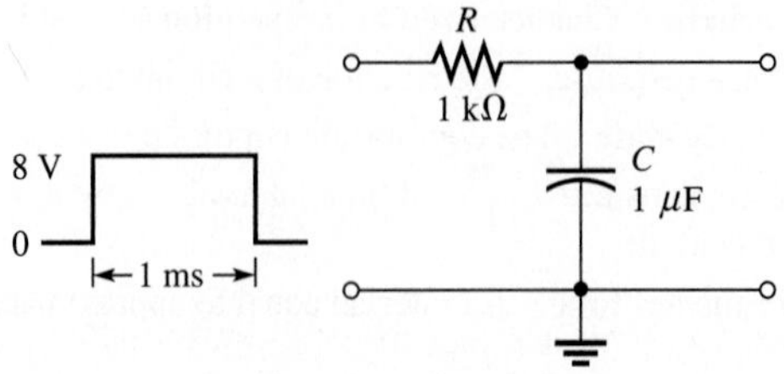

SECTION 15–3 Repetitive-Pulse Response of *RC* Integrators

7. Sketch the integrator output in Figure 15–58 showing maximum voltages.

8. A 1 V, 10 kHz pulse waveform with a duty cycle of 25% is applied to an integrator with $\tau = 25\ \mu s$. Graph the output voltage for three initial pulses. *C* is initially uncharged.

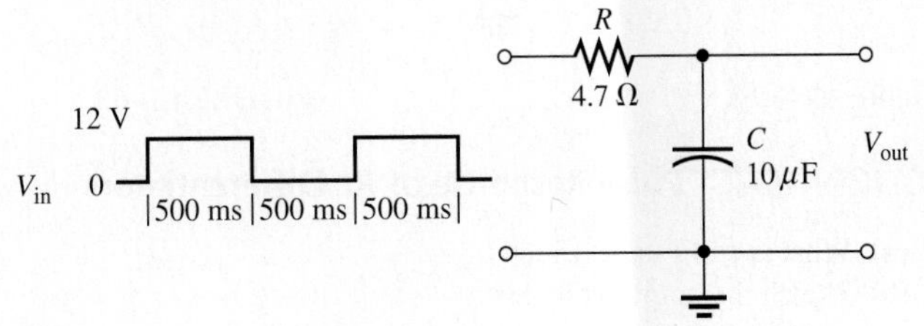

FIGURE 15–58

9. What is the steady-state output voltage of the integrator with a square wave input shown in Figure 15–59?

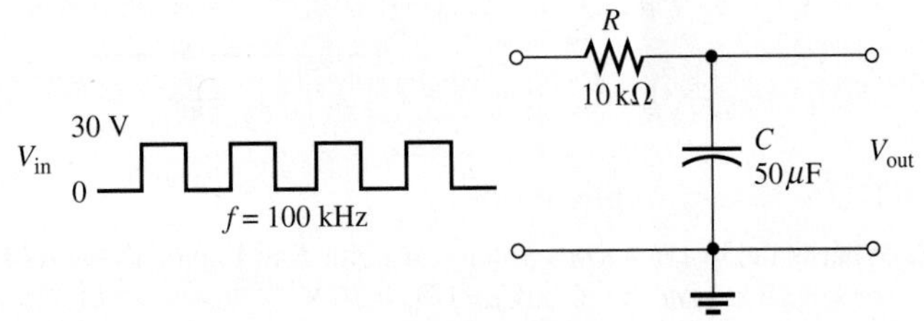

FIGURE 15–59

SECTION 15–4 Single-Pulse Response of *RC* Differentiators

10. Repeat Problem 5 for an *RC* differentiator.

11. Redraw the circuit in Figure 15–57 to make it a differentiator, and repeat Problem 6.

SECTION 15–5 Repetitive-Pulse Response of *RC* Differentiators

12. Sketch the differentiator output in Figure 15–60, showing maximum voltages.

13. What is the steady-state output voltage of the differentiator with the square wave input shown in Figure 15–61?

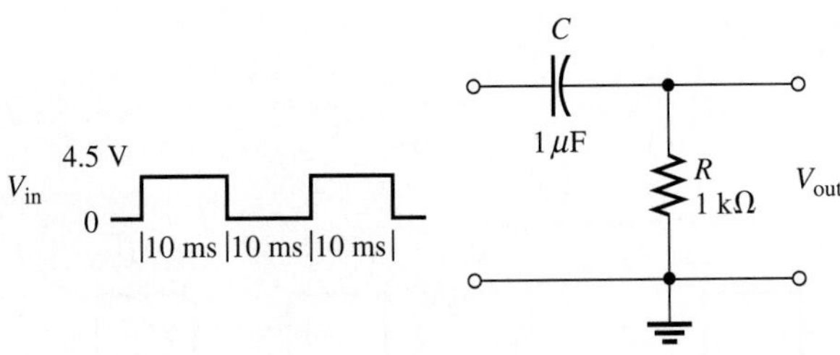

FIGURE 15–60

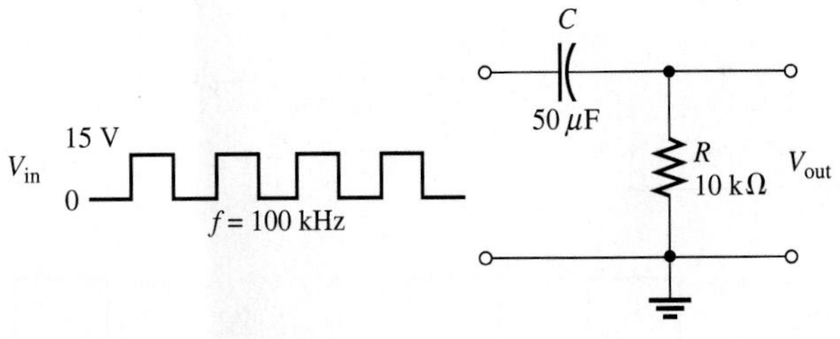

FIGURE 15–61

SECTION 15–6 Pulse Response of *RL* Integrators

14. Determine the output voltage for the circuit in Figure 15–62. A single input pulse is applied as shown.

15. Sketch the integrator output in Figure 15–63, showing maximum voltages.

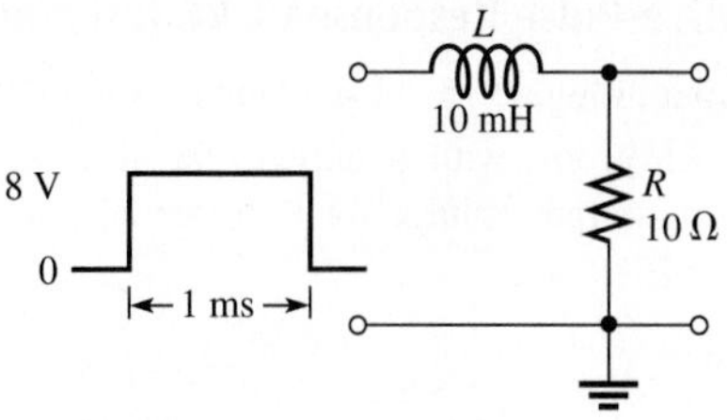

FIGURE 15–62

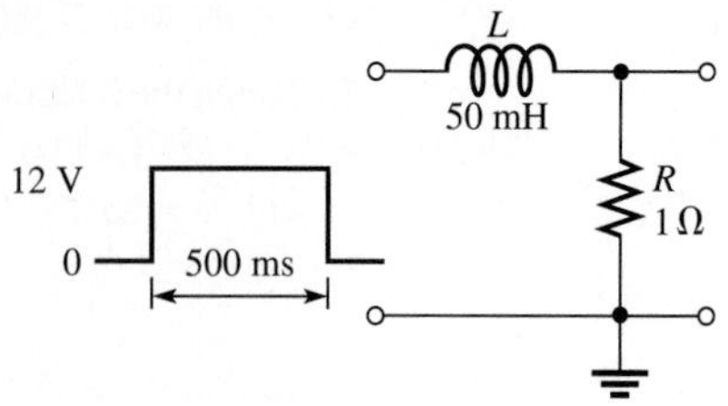

FIGURE 15–63

SECTION 15–7 Pulse Response of *RL* Differentiators

16. **(a)** What is τ in Figure 15–64?
(b) Sketch the output voltage.

17. Draw the output waveform if a periodic pulse waveform with $t_W = 25\ \mu s$ and $T = 60\ \mu s$ is applied to the circuit in Figure 15–64.

FIGURE 15–64

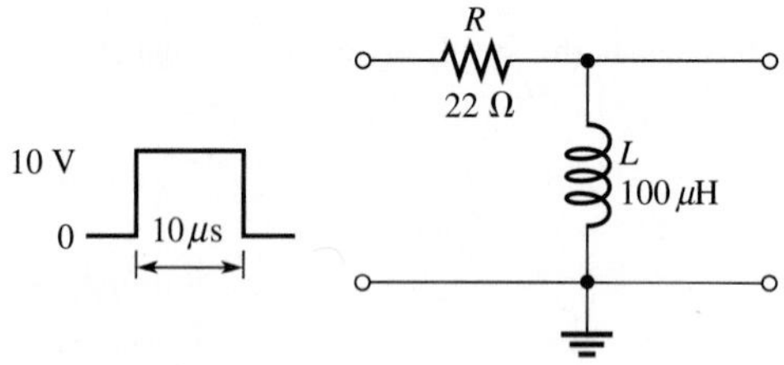

SECTION 15–8 Applications

18. What is the instantaneous voltage at point *B* in Figure 15–47 440 μs after the switch is closed for $R = 22\ k\Omega$ and $C = 0.001\ \mu F$? $V_{in} = 10$ V.

19. Ideally, what is the output of an *RC* integrator when the input is a square wave with an amplitude of 12 V? Assume that the time constant is much greater than the period of the input signal?

SECTION 15–9 Troubleshooting

20. Determine the most likely fault(s), in the circuit of Figure 15–65(a) for each set of waveforms in parts (b) through (d). V_{in} is a square wave with a period of 8 ms.

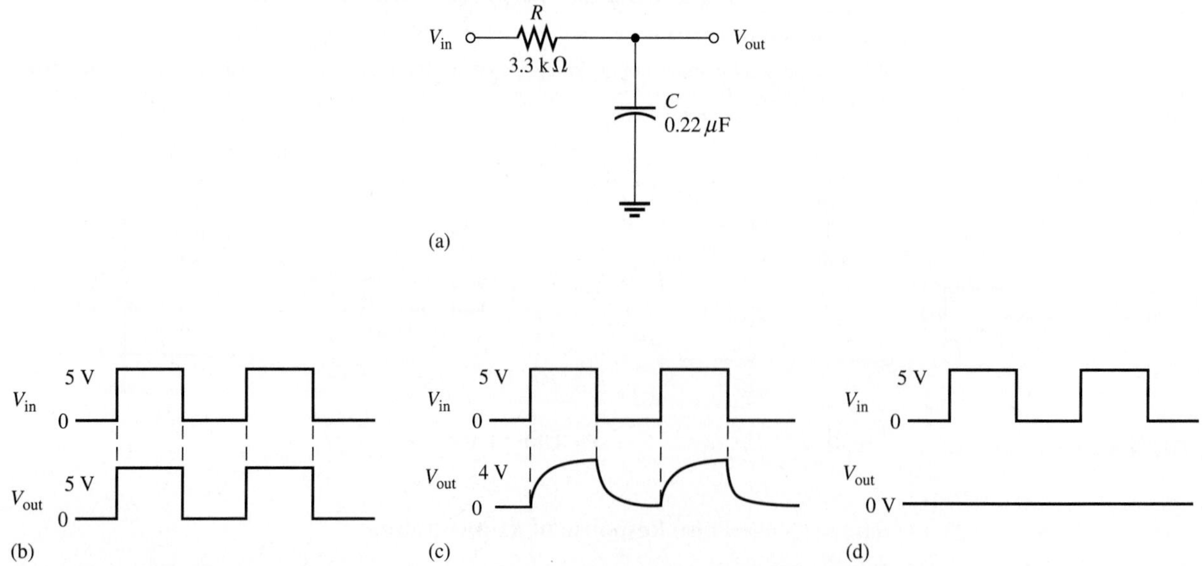

FIGURE 15–65

21. Determine the most likely fault(s), if any, in the circuit of Figure 15–66(a) for each set of waveforms in parts (b) through (d). V_{in} is a square wave with a period of 8 ms.

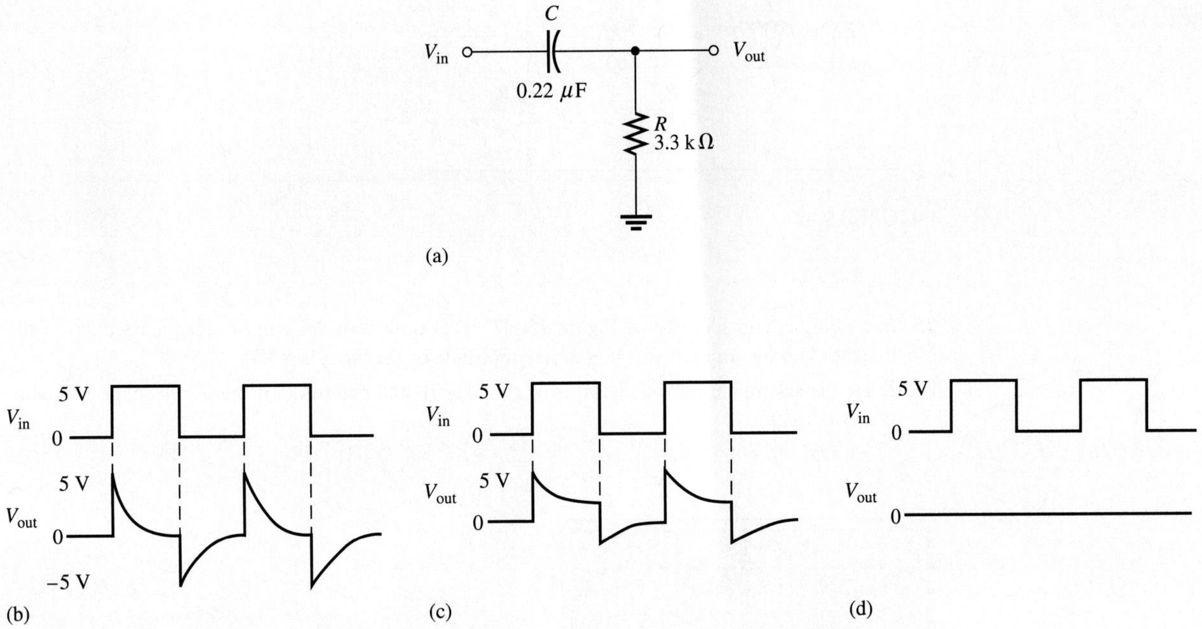

FIGURE 15–66

■ ADVANCED PROBLEMS

22. **(a)** What is τ in Figure 15–67?
(b) Sketch the output voltage.

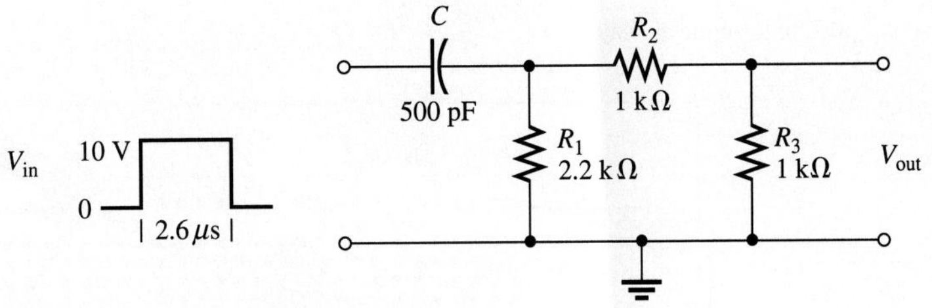

FIGURE 15–67

23. **(a)** What is τ in Figure 15–68?
(b) Sketch the output voltage.

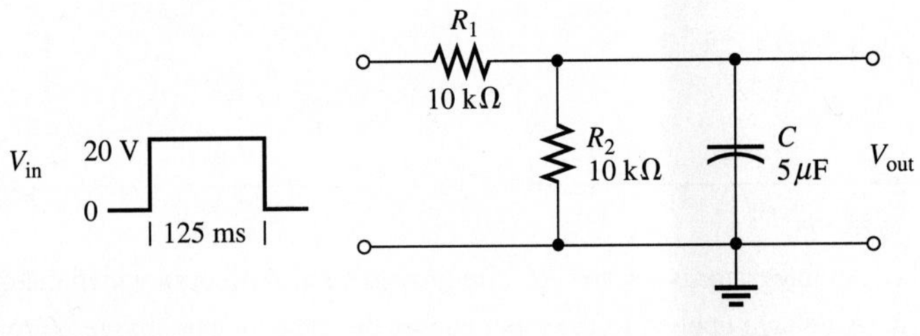

FIGURE 15–68

24. Determine the time constant in Figure 15–69. Is this circuit an integrator or a differentiator?

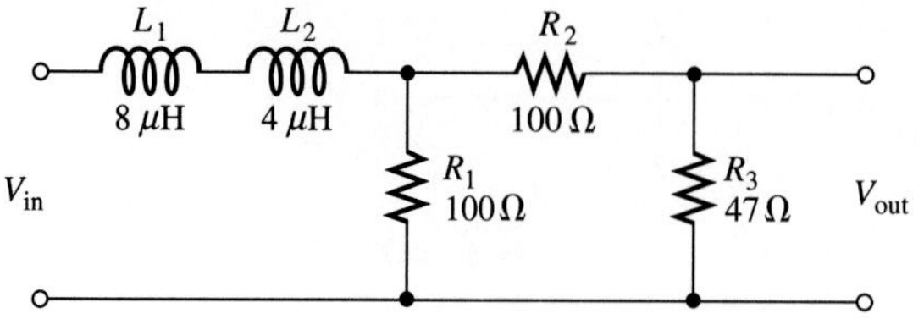

FIGURE 15–69

25. In a delay circuit like that of Figure 15–47, what time constant will produce a 1 s delay if the threshold of the circuit is 2.5 V and the amplitude of the input is 5 V?

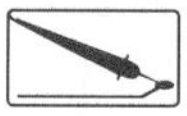

26. Draw the schematic for the circuit in Figure 15–70 and determine if the oscilloscope presentation is correct.

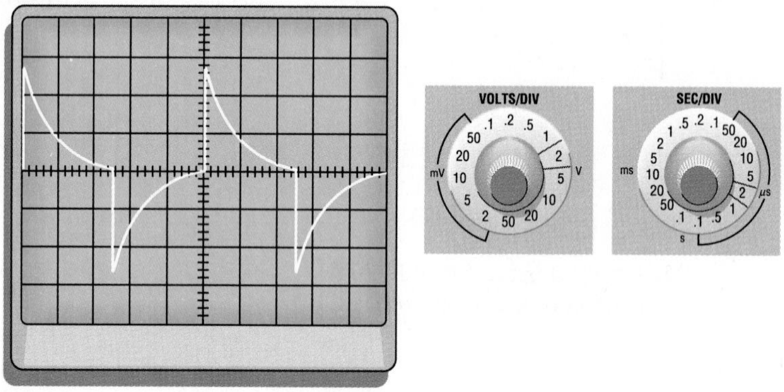

(a) Oscilloscope display

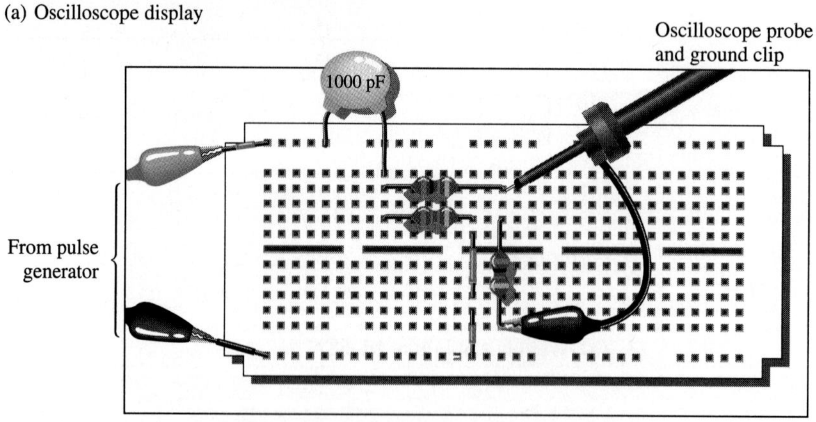

(b) Circuit board with instrument leads connected

FIGURE 15–70

▪ ANSWERS TO SECTION REVIEWS

Section 15–1

1. An integrator is a series *RC* circuit with a pulse input in which the output is across the capacitor.
2. A voltage applied to the input causes the capacitor to charge. Zero volts across the input causes the capacitor to discharge.

Section 15–2

1. The output of an integrator reaches full amplitude when $5\tau \leq t_W$.
2. $V_{out} = (0.63)1\text{ V} = 0.63\text{ V}$; $t_{disch} = 5\tau = 51.5\text{ ms}$.
3. See Figure 15–71.

FIGURE 15–71

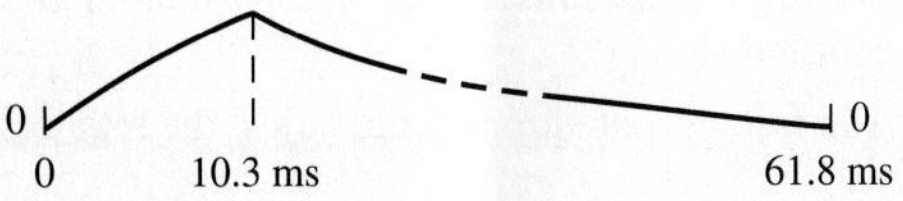

4. No, C will not fully charge.
5. The output is approximately the shape of the input when $5\tau << t_W$ (5τ much less than t_W).

Section 15–3

1. An integrator capacitor will fully charge and discharge when $5\tau \leq t_W$ and $5\tau \leq$ time between pulses.
2. The output looks like the input.
3. transient time
4. Steady state is the response after the transient time has passed.
5. The average value of the output voltage equals the average value of the input voltage.

Section 15–4

1. See Figure 15–72.

FIGURE 15–72

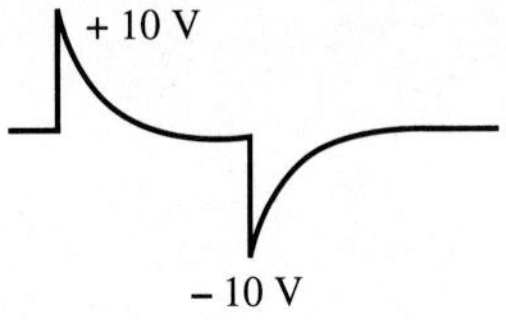

2. The output approximates the input when $5\tau >> t_W$.
3. The output consists of a positive and a negative spike.
4. $V_R = +5\text{ V} - 15\text{ V} = -10\text{ V}$

Section 15–5

1. C fully charges and discharges when $5\tau \leq t_W$ and $5\tau \leq$ time between pulses.
2. The output consists of positive and negative spikes.
3. $V_{out} = 0\text{ V}$

Section 15–6

1. The output is across the resistor.
2. The output reaches the input amplitude when $5\tau \leq t_W$.
3. The output approximates the input when $5\tau << t_W$.

Section 15–7

1. The output is across the inductor.
2. The output approximates the input when $5\tau >> t_W$.
3. $V_{out} = 2\text{ V} - 10\text{ V} = -8\text{ V}$

Section 15–8

1. Increase the time constant to increase delay time.
2. Reducing R will increase ripple.
3. A very short time constant produces very short duration spikes.

Section 15–9

1. Zero output of *RC* integrator indicates shorted capacitor or open resistor.
2. An open integrator capacitor results in an output exactly like the input.
3. A shorted differentiator capacitor results in an output exactly like the input.

Section 15–10

1. The input pulse must be sufficiently long to allow the output to reach 3.5 V for the longest time-constant setting.
2. Add a series potentiometer to adjust the total resistance.

▪ ANSWERS TO RELATED EXERCISES FOR EXAMPLES

15–1 8.65 V
15–2 24.7 V
15–3 1.11 V
15–4 See Figure 15–73.

FIGURE 15–73

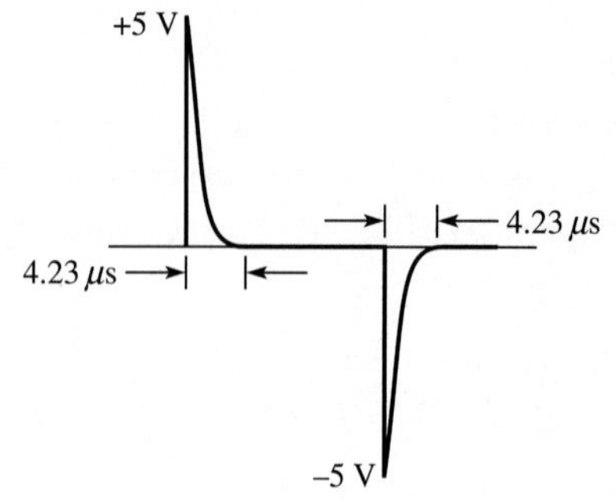

15–5 892 mV
15–6 100 Ω
15–7 20 V
15–8 2.5 kΩ
15–9 See Figure 15–74.

FIGURE 15–74

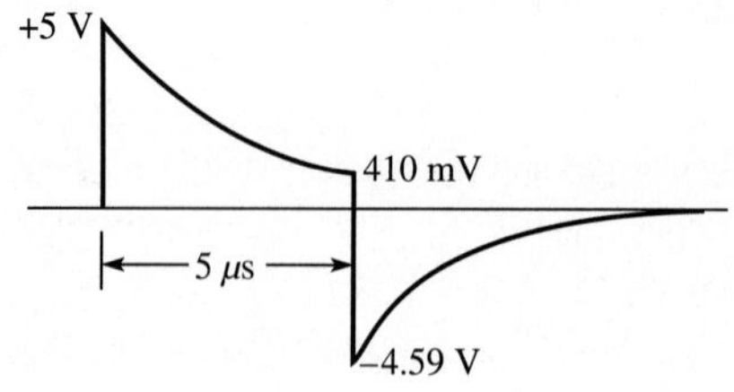

15–10 20 Ω
15–11 3.5 μs

PART THREE

DEVICES

16

INTRODUCTION TO SEMICONDUCTOR DEVICES

To acquire a basic understanding of semiconductors, you must have some knowledge of atomic theory and the structure of semiconductor materials. In this chapter, you will learn about atomic theory. We will discuss the basic materials used in manufacturing both **discrete devices,** such as diodes and transistors, and integrated circuits. We will introduce PN junctions, an important concept essential for the understanding of diode and transistor operation. Also, other diode characteristics are introduced, and you will learn how to properly use a diode in a circuit.

■ CHAPTER OBJECTIVES

- ☐ Discuss the basic structure of semiconductor materials
- ☐ Discuss covalent bonding in the silicon atom
- ☐ Explain how current occurs in a semiconductor
- ☐ Discuss the characteristics of doped semiconductors
- ☐ Describe the characteristics of a PN junction
- ☐ Explain how to bias a diode
- ☐ Describe the basic diode characteristics

■ APPLICATION ASSIGNMENT

Semiconductor diodes are widely used in many types of electronic applications. They are of great importance in power supplies, where they are used as rectifiers to convert ac to dc. Suppose your supervisor has given you four power supply units that have failed the final assembly test and has asked you to check the units with an ohmmeter to identify the faults. After studying this chapter, you should be able to complete the application assignment in Section 16–8.

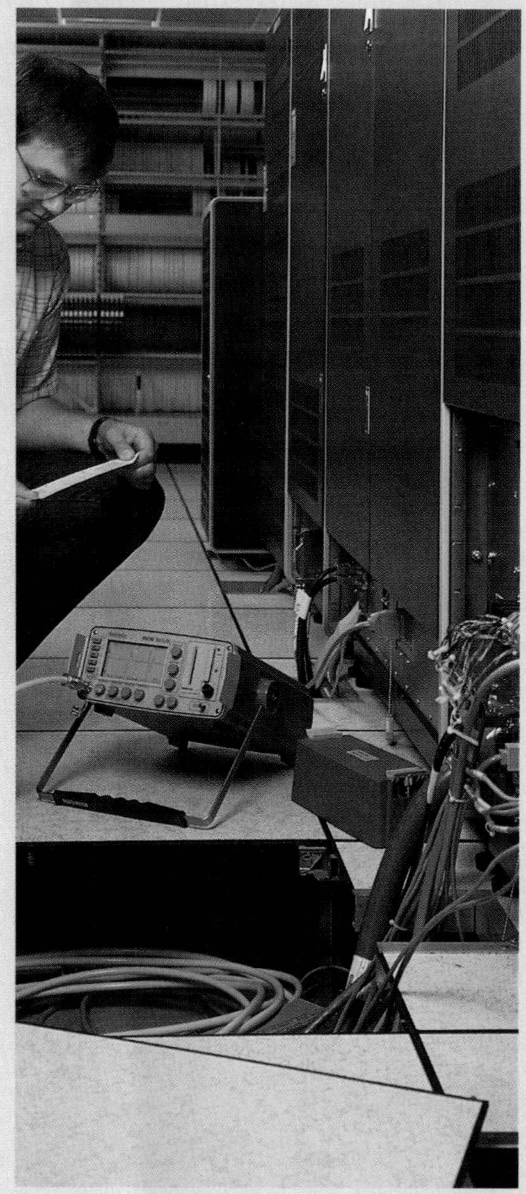

16–1 ■ THE ATOMIC STRUCTURE OF SEMICONDUCTORS

The basic concept of the structure of an atom was studied in Chapter 2. In this chapter, atomic theory is extended to semiconductor materials that are used in electronic devices such as diodes and transistors. This coverage lays the foundation for a good understanding of how semiconductor devices function.

After completing this section, you should be able to

- **Discuss the basic structure of semiconductor materials**
  - ☐ Discuss electron shells and orbits in an atom
  - ☐ Define a valence electron
  - ☐ Explain ionization
  - ☐ Describe the atomic structure of silicon and germanium

Electron Shells and Orbits

Electrons rotate around the nucleus of an **atom** at certain distances from the nucleus in paths called *orbits*. Electrons near the nucleus have less **energy** than those in more distant orbits. Only discrete (separate and distinct) values of electron energies exist within atomic structures. Therefore, electrons must orbit only at discrete distances from the nucleus.

Each discrete distance of an orbit from the nucleus corresponds to a certain energy level. In an atom, orbits are grouped into energy bands known as *shells*. A given atom has a fixed number of shells. Each shell has a fixed maximum number of electrons at permissible energy levels (orbits). The differences in energy levels within a shell are much smaller than the difference in energy between shells. The shells are designated *K, L, M, N,* and so on, with *K* being closest to the nucleus. This concept is illustrated in Figure 16–1.

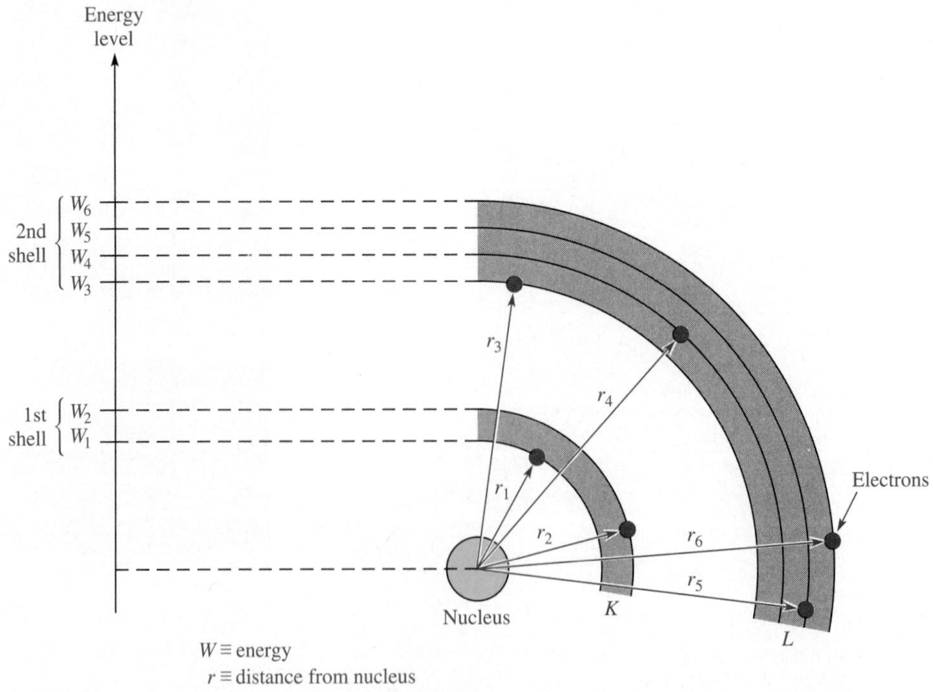

FIGURE 16–1
Energy levels increase as distance from the nucleus of the atom increases.

Valence Electrons

Electrons in orbits farther from the nucleus are less tightly bound to the atom than those closer to the nucleus. This is because the force of attraction between the positively charged nucleus and the negatively charged electron decreases with increasing distance.

Electrons with the highest energy levels exist in the outermost shell of an atom and are relatively loosely bound to the atom. These valence electrons contribute to chemical reactions and bonding within the structure of a material. The **valence** of an atom is the number of electrons in its outermost shell.

Ionization

When an atom absorbs energy from a heat source or from light, for example, the energy levels of the electrons are raised. When an electron gains energy, it moves to an orbit farther from the nucleus. Since the valence electrons possess more energy and are more loosely bound to the atom than inner electrons, they can jump to higher orbits more easily when external energy is absorbed.

If a valence electron acquires a sufficient amount of energy, it can be completely removed from the outer shell and the atom's influence. The departure of a valence electron leaves a previously neutral atom with an excess of positive charge (more protons than electrons). The process of losing a valence electron results in a positively charged atom called a positive **ion.** For example, the chemical symbol for hydrogen is H. When it loses its valence electron and becomes a positive ion, it is designated H^+. The escaped valence electron is called a **free electron.**

When a free electron falls into the outer shell of a neutral hydrogen atom, the atom becomes negatively charged (more electrons than protons) and is called a negative ion, designated H^-. The process whereby a neutral atom loses or gains an electron is called **ionization.**

Silicon and Germanium Atoms

Two types of semiconductor materials are **silicon** and **germanium.** Both the silicon and the germanium atoms have four valence electrons. They differ in that silicon has 14 protons in its nucleus and germanium has 32. Figure 16–2 shows a representation of the atomic structure for both materials. Silicon is much more widely used than germanium.

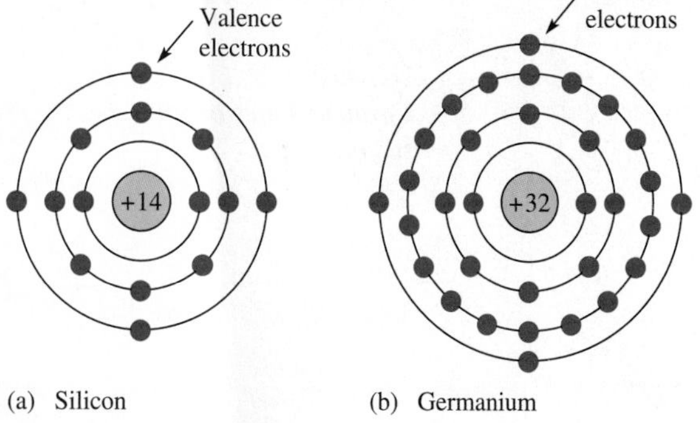

FIGURE 16–2
Silicon and germanium atoms.

SECTION 16–1 REVIEW

1. Define *orbit.*
2. Explain what an atomic shell is.
3. What is a valence electron?
4. How is a positive ion created?

16–2 ▪ ATOMIC BONDING

When silicon atoms combine into molecules to form a solid material, they arrange themselves in a fixed pattern called a crystal. The atoms within the crystal structure are held together by covalent bonds, which are created by interaction of the valence electrons of each atom. A solid chunk of silicon is a crystalline material.

After completing this section, you should be able to

- **Discuss covalent bonding in the silicon atom**
  - □ Explain how covalent bonds produce a crystal

Figure 16–3 shows how each silicon atom positions itself with four adjacent atoms. A silicon atom with its four valence electrons shares an electron with each of its four neighbors. This sharing of valence electrons produces the **covalent** bonds that hold the atoms together, because each shared electron is attracted equally by the two adjacent atoms which share it. Covalent bonding of a pure (intrinsic) silicon **crystal** is shown in Figure 16–4. Bonding for germanium is similar because it also has four valence electrons.

FIGURE 16–3
Covalent bonds in silicon.

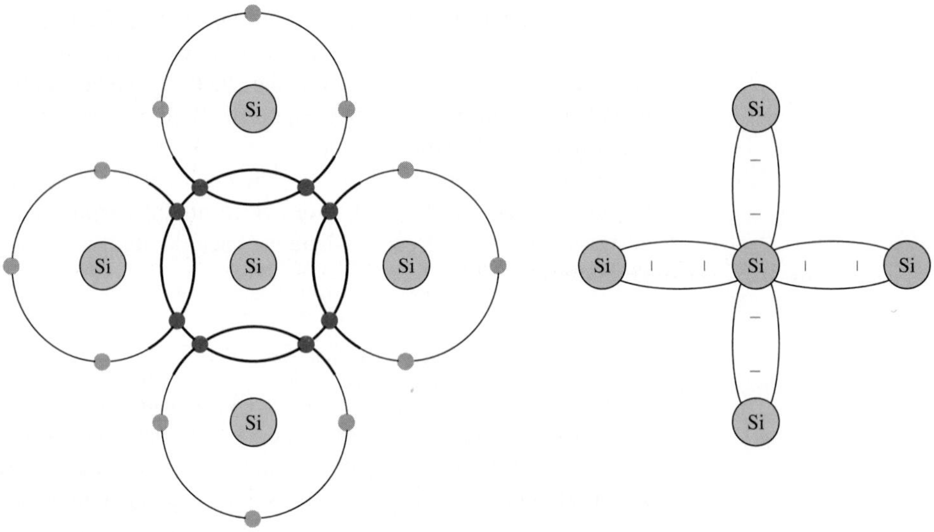

(a) Shared valence electrons form covalent bonds.

(b) Bonding diagram

FIGURE 16–4
Covalent bonds in a pure silicon crystal.

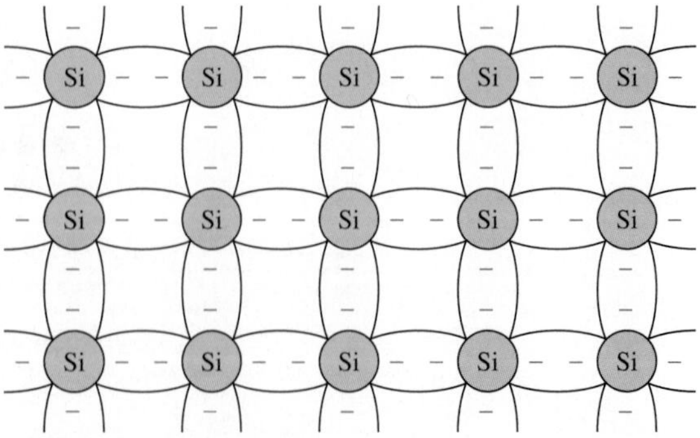

SECTION 16–2 REVIEW

1. Name two semiconductor materials.
2. What is a covalent bond?

16–3 ▪ CONDUCTION IN SEMICONDUCTOR MATERIALS

As you have seen, the electrons of an atom can exist only within prescribed energy bands. Each shell around the nucleus corresponds to a certain energy band and is separated from adjacent shells by energy gaps, in which no electrons can exist.

After completing this section, you should be able to

- **Explain how current occurs in a semiconductor**
  - ☐ Discuss the concept of the hole
  - ☐ Compare conduction properties of germanium and silicon
  - ☐ Explain how electrons produce current
  - ☐ Explain how holes are related to current
  - ☐ Compare conductors, semiconductors, and insulators

An energy band diagram is shown in Figure 16–5 for an unexcited silicon atom (no external energy). This condition occurs only at absolute zero temperature.

FIGURE 16–5
Energy band diagram for unexcited silicon atom.

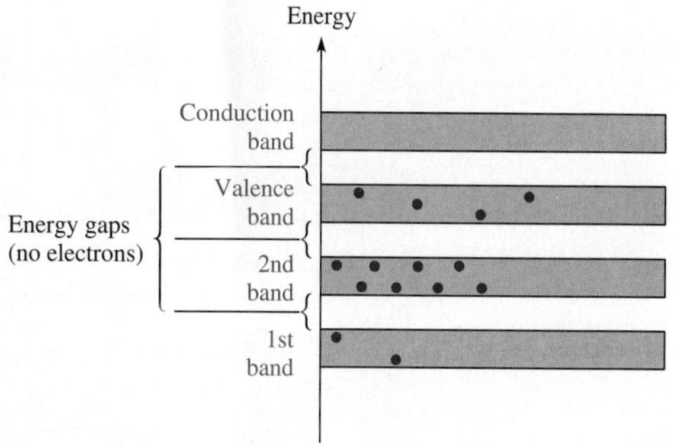

Conduction Electrons and Holes

A pure silicon crystal at room temperature derives heat (thermal) energy from the surrounding air, causing some valence electrons to gain sufficient energy to jump the gap from the valence band into the conduction band, becoming free electrons. This situation is illustrated in the energy diagram of Figure 16–6(a) and in the bonding diagram of Figure 16–6(b).

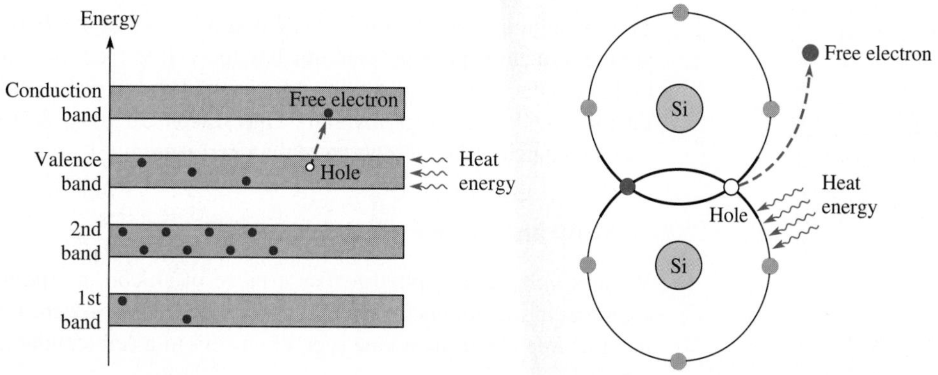

(a) Energy diagram (b) Bonding diagram

FIGURE 16–6
Creation of an electron-hole pair in an excited silicon atom.

When an electron jumps to the conduction band, a vacancy is left in the valence band. This vacancy is called a **hole.** For every electron raised to the conduction band by thermal or light energy, there is one hole left in the valence band, creating what is called an *electron-hole pair.* **Recombination** occurs when a conduction band electron loses energy and falls back into a hole in the valence band.

In summary, a piece of pure silicon at room temperature has, at any instant, a number of conduction band (free) electrons that are unattached to any atom and are essentially drifting randomly throughout the material. Also, an equal number of holes are created in the valence band when these electrons jump into the conduction band, as illustrated in Figure 16–7.

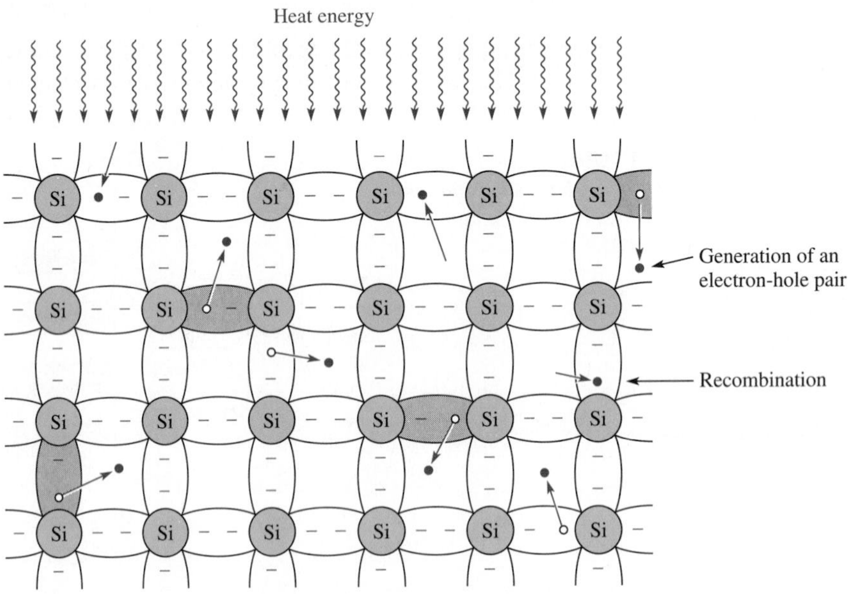

FIGURE 16–7
Electron-hole pairs in a silicon crystal.

Germanium Versus Silicon

The situation in a germanium crystal is similar to that in silicon except that, because of its atomic structure, pure germanium has more free electrons than silicon and therefore a higher conductivity. Silicon, however, is the favored **semiconductor** material and is used far more widely than germanium. One reason for its wide usage is that silicon can be used at a much higher temperature than germanium.

Electron and Hole Current

When a voltage is applied across a piece of silicon, as shown in Figure 16–8, the free electrons in the conduction band are easily attracted toward the positive end. This movement of free electrons is one type of current in a semiconductor material, called *electron current.*

Another current mechanism occurs at the valence level, where the holes created by the free electrons exist. Electrons remaining in the valence band are still attached to their

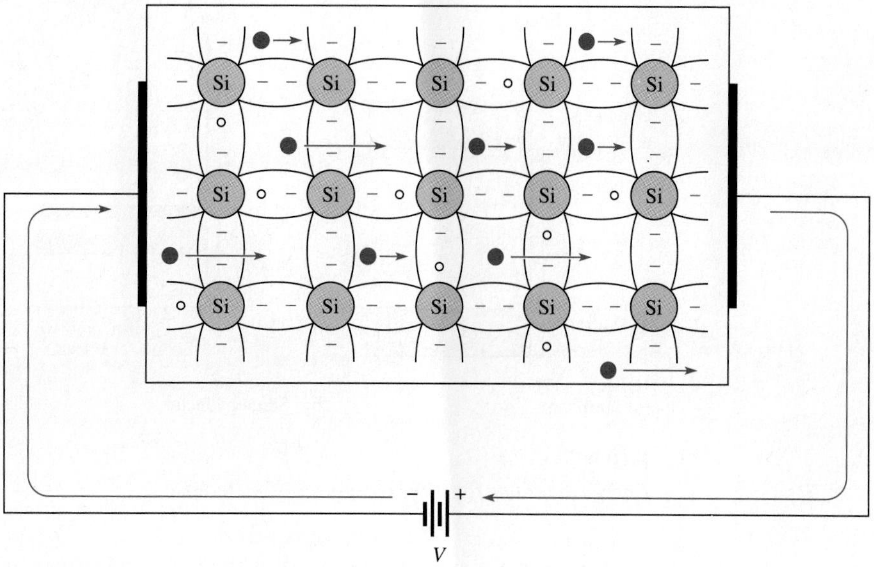

FIGURE 16–8
Free electron current in silicon.

atoms and are not free to move randomly in the crystal structure. However, a valence electron can "fall" into a nearby hole, with little change in its energy level, thus leaving another hole where it came from. The hole has effectively, although not physically, moved from one place to another in the crystal structure, as illustrated in Figure 16–9. This current is called *hole current.*

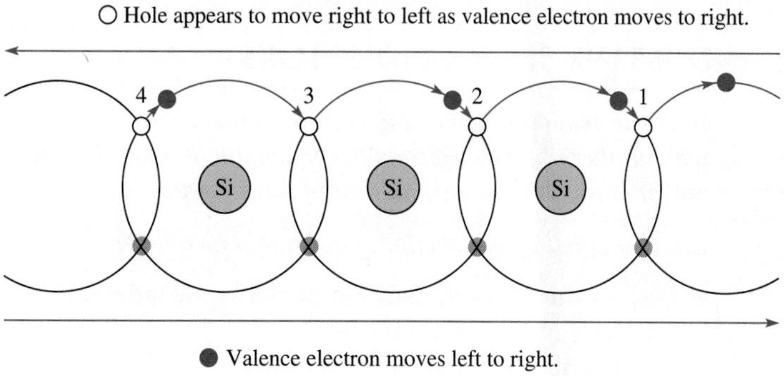

FIGURE 16–9
Hole current in silicon.

Semiconductors, Conductors, and Insulators

In an **intrinsic** (pure) **semiconductor,** there are relatively few free electrons, so neither silicon nor germanium is very useful in its intrinsic state. Pure semiconductors are neither insulators nor good conductors because current in a material depends directly on the number of free electrons.

A comparison of the energy bands in Figure 16–10 for the three types of materials shows the essential differences among them regarding conduction. The energy gap for an insulator is so wide that hardly any electrons acquire enough energy to jump into the conduction band. The valence band and the conduction band in a conductor (such as copper)

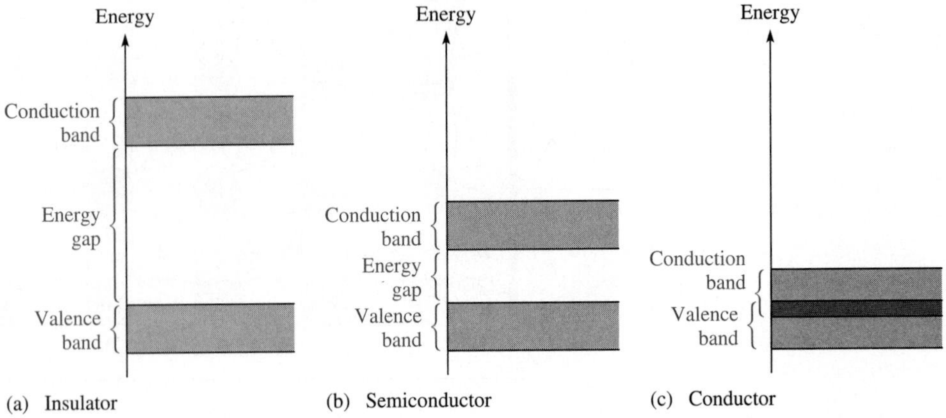

FIGURE 16–10
Energy diagrams for three categories of materials.

overlap so that there are always many conduction electrons, even without the application of external energy. A semiconductor, as Figure 16–10(b) shows, has an energy gap that is much narrower than that in an insulator.

SECTION 16–3 REVIEW

1. In the atomic structure of a semiconductor, within which energy band do free electrons exist? Valence electrons?
2. How are holes created in an intrinsic semiconductor?
3. Why is current established more easily in a semiconductor than in an insulator?

16–4 ■ N-TYPE AND P-TYPE SEMICONDUCTORS

Intrinsic semiconductor materials do not conduct current very well because of the limited number of free electrons in the conduction band. Thus, the resistivity of a semiconductor is much greater than that of a conductor.

After completing this section, you should be able to

- **Discuss the characteristics of doped semiconductors**
  - □ Describe the doping process
  - □ Describe an N-type semiconductor
  - □ Describe a P-type semiconductor

Doping

The resistivities of silicon and germanium can be drastically reduced and controlled by the addition of impurities to the pure semiconductor material. This process, called **doping,** increases the number of current carriers (electrons or holes), thus increasing the conductivity and decreasing the resistivity. The two categories of impurities are *N-type* and *P-type.*

N-Type Semiconductor

To increase the number of conduction band electrons in pure silicon, pentavalent impurity atoms are added. These are atoms with five valence electrons, such as arsenic, phosphorus, and antimony.

As illustrated in Figure 16–11, each pentavalent atom (antimony, in this case) forms covalent bonds with four adjacent silicon atoms. Four of the antimony atom's valence electrons are used to form the covalent bonds, leaving one extra electron. This extra electron becomes a conduction electron because it is not attached to any atom. The number of conduction electrons can be controlled by the amount of impurity added to the silicon.

FIGURE 16–11
Pentavalent impurity atom in a silicon crystal. An antimony (Sb) impurity atom is shown in the center.

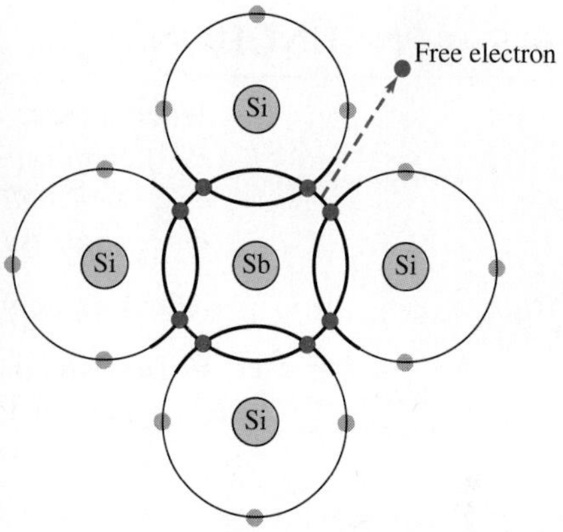

Since most of the current carriers are electrons, silicon (or germanium) doped in this way is an N-type semiconductor material where the *N* stands for the negative charge on an electron. The electrons are called the *majority carriers* in N-type material. Although the great majority of current carriers in N-type material are electrons, there are some holes. Holes in an N-type material are called *minority carriers.*

P-Type Semiconductor

To increase the number of holes in pure silicon, trivalent impurity atoms are added. These are atoms with three valence electrons, such as aluminum, boron, and gallium.

As illustrated in Figure 16–12, each trivalent atom (boron, in this case) forms covalent bonds with four adjacent silicon atoms. All three of the boron atom's valence electrons are used in the covalent bonds; and, since four electrons are required, a hole is formed with each trivalent atom. The number of holes can be controlled by the amount of trivalent impurity added to the silicon.

FIGURE 16–12
Trivalent impurity atom in a silicon crystal. A boron (B) impurity atom is shown in the center.

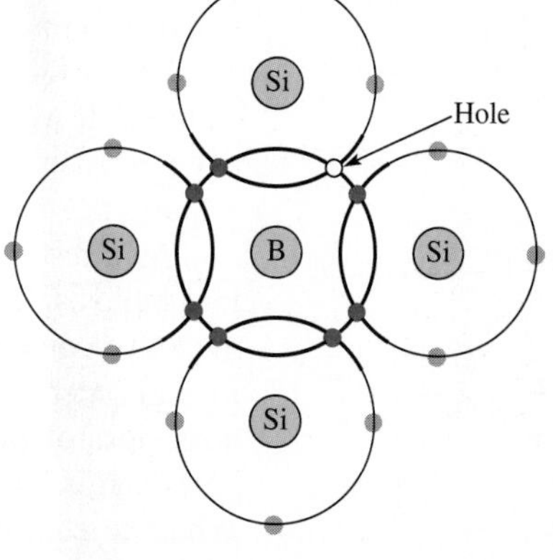

Since most of the current carriers are holes, silicon (or germanium) doped in this way is a P-type semiconductor material because holes can be thought of as positive charges. The holes are the majority carriers in P-type material. Although the great majority of current carriers in P-type material are holes, there are some electrons. Electrons in P-type material are the minority carriers.

SECTION 16–4 REVIEW

1. How is an N-type semiconductor formed?
2. How is a P-type semiconductor formed?
3. What is a majority carrier?

16–5 ■ PN JUNCTIONS

When a piece of silicon is doped so that half is N-type and the other half is P-type, a PN junction is formed between the two regions. A device formed in this way is known as a semiconductor diode. The PN junction is fundamental to the operation not only of diodes but also of transistors and other solid-state devices.

After completing this section, you should be able to

- **Describe the characteristics of a PN junction**
  - □ Define PN junction
  - □ Discuss the depletion layer in a PN junction
  - □ Explain the energy diagram of a PN junction

A **PN junction** in a **diode** is illustrated in Figure 16–13(a). The N region has many conduction electrons, and the P region has many holes, as shown in Figure 16–13(b).

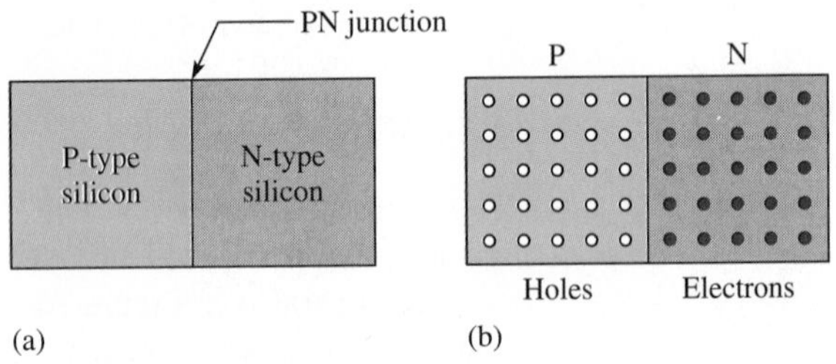

FIGURE 16–13
Basic PN structure in a diode at the instant of junction formation.

The Depletion Layer

With no external voltage, the conduction electrons in the N region are aimlessly drifting in all directions. At the instant of junction formation, some of the electrons near the junction drift across into the P region and recombine with holes near the junction.

For each electron that crosses the junction and recombines with a hole, a pentavalent atom is left with a net positive charge in the N region near the junction, making it a positive ion. Also, when the electron recombines with a hole in the P region, a trivalent atom acquires net negative charge, making it a negative ion.

As a result of this recombination process, a large number of positive and negative ions builds up near the PN junction. As this buildup occurs, the electrons in the N region must overcome both the attraction of the positive ions and the repulsion of the negative ions in order to migrate into the P region. Thus, as the ion layers build up, the area on both sides of the junction becomes essentially depleted of any conduction electrons or holes and is known as the *depletion layer.* This condition is illustrated in Figure 16–14. When an equilibrium condition is reached, the depletion layer has widened to a point where no more electrons can cross the PN junction.

The existence of the positive and negative ions on opposite sides of the junction creates a **barrier potential** (V_B) across the depletion layer, as indicated in Figure 16–14. At 25°C, the barrier potential is approximately 0.7 V for silicon and 0.3 V for germanium. As the junction temperature increases, the barrier potential decreases, and vice versa.

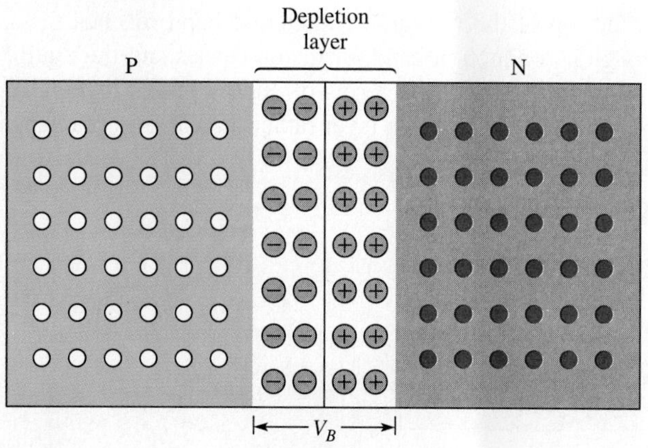

FIGURE 16–14
PN junction equilibrium condition in a diode.

Energy Diagram of the PN Junction

Now, we will look at the operation of the PN junction in terms of its energy level. First consider the PN junction at the instant of its formation. The energy bands of the trivalent impurity atoms in the P-type material are at a slightly higher level than those of the pentavalent impurity atom in the N-type material, as shown in the graph of Figure 16–15. They are higher because the core attraction for the valence electrons (+3) in the trivalent atom is less than the core attraction for the valence electrons (+5) in the pentavalent atom. Thus, the trivalent valence electrons are in a slightly higher obit and, thus, at a higher energy level.

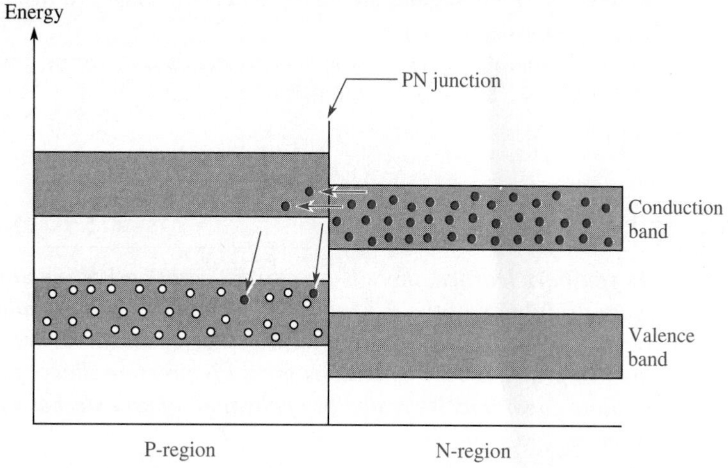

FIGURE 16–15
Graph of PN junction energy levels as diffusion begins.

Notice in Figure 16–15 that there is some overlap of the conduction bands in the P and N regions and also some overlap of the valence bands in the P and N regions. This overlap permits the electrons of higher energy near the top of the N-region conduction band to begin diffusing across the junction into the lower part of the P-region conduction band. As soon as an electron diffuses across the junction, it recombines with a hole in the valence band. As diffusion continues, the depletion layer begins to form. Also, the energy bands in the N region "shift" down as the electrons of higher energy are lost to diffusion.

When the top of the N-region conduction band reaches the same level as the bottom of the P-region conduction band, diffusion ceases and the equilibrium condition is reached. This condition is shown in terms of energy levels in Figure 16–16. There is an energy gradient across the depletion layer rather than an abrupt change in energy level.

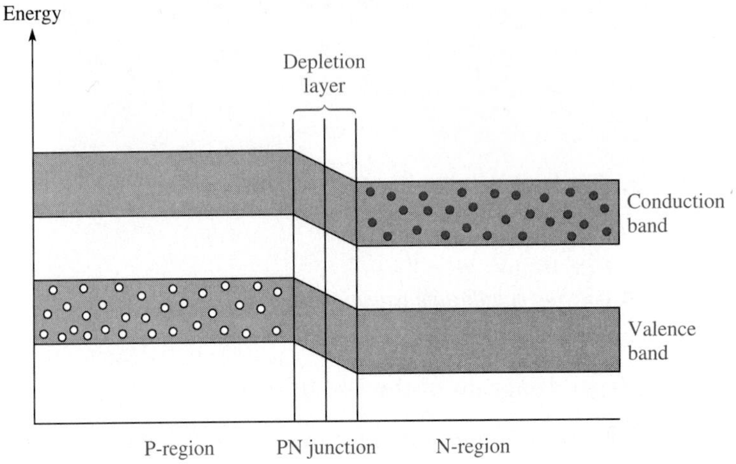

FIGURE 16–16
Graph of energy levels at equilibrium. The N-region bands are shifted down from their original positions when diffusion begins.

SECTION 16–5 REVIEW

1. What is a PN junction?
2. When P and N regions are joined, a depletion layer forms. Describe the characteristics of the depletion layer.
3. The barrier potential for silicon is greater than for germanium (T or F).
4. What is the barrier potential for silicon at 25°C?

16–6 ■ BIASING THE DIODE

As you have learned, there is no current across a PN junction at equilibrium. The primary usefulness of the PN junction diode is its ability to allow current in only one direction and to prevent current in the other direction as determined by the bias. There are two practical bias conditions for a PN junction diode: forward and reverse. Either of these conditions is created by application of an external voltage of the proper polarity and magnitude.

After completing this section, you should be able to

- **Explain how to bias a diode**
  - ☐ Discuss forward bias
  - ☐ Define barrier potential
  - ☐ Discuss reverse bias
  - ☐ Define *reverse breakdown*

Forward Bias

Forward bias is the condition that permits current across a PN junction. The term **bias** in electronics normally refers to a fixed dc voltage that sets the operating conditions for a

FIGURE 16–17
Forward-bias connection. The resistor limits the forward current in order to prevent damage to the diode.

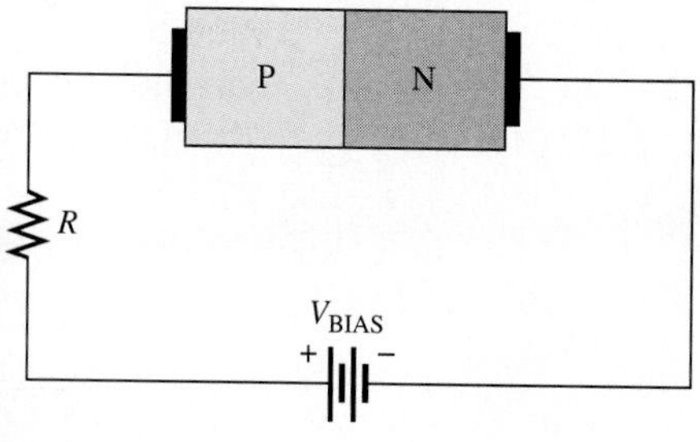

semiconductor device. Figure 16–17 shows a dc voltage connected in a direction to forward-bias the diode. Notice that the negative **terminal** of the battery is connected to the N region (called the **cathode**), and the positive terminal is connected to the P region (called the **anode**).

A discussion of the basic operation of forward bias follows: The negative terminal of the battery pushes the conduction band electrons in the N region toward the junction, while the positive terminal pushes the holes in the P region also toward the junction. Recall from Chapter 2 that like charges repel each other.

When it overcomes the barrier potential, the external voltage source provides the N-region electrons with enough energy to penetrate the depletion layer and cross the junction, where they combine with the P-region holes. As electrons leave the N region, more flow in from the negative terminal of the battery. Thus, current through the N region is formed by the movement of conduction electrons (majority carriers) toward the junction.

Once the conduction electrons enter the P region and combine with holes, they become valence electrons. Then they move as valence electrons from hole to hole toward the positive connection of the battery. The movement of these valence electrons is the same as the movement of holes in the opposite direction. Thus, current in the P region is formed by the movement of holes (majority carriers) toward the junction. Figure 16–18 illustrates current in a forward-biased diode.

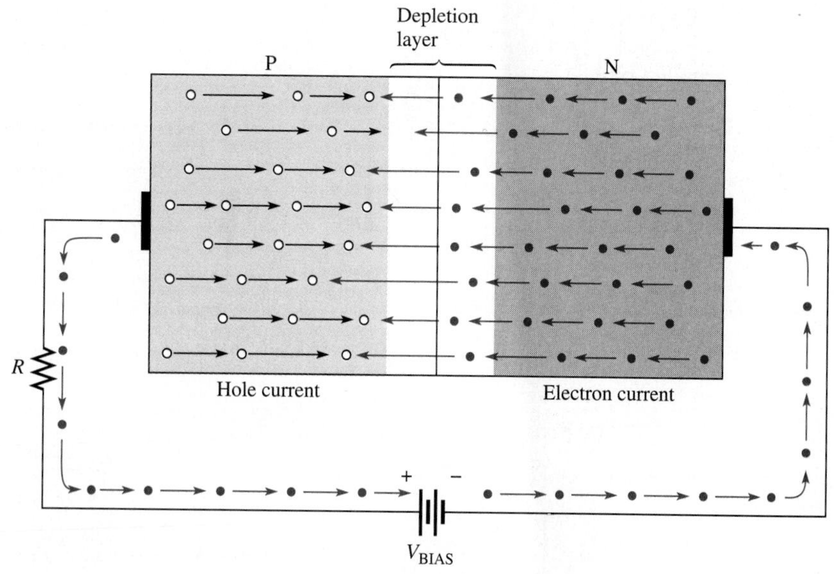

FIGURE 16–18
How current flows in a forward-biased diode.

Effect of Barrier Potential on Forward Bias The effect of the barrier potential in the depletion layer is to oppose forward bias. This is because the negative ions near the junction in the P region tend to prevent electrons from crossing the junction into the P region.

You can think of the barrier potential effect as simulating a small battery connected in a direction to oppose the forward-bias voltage, as shown in Figure 16–19. The resistances R_P and R_N represent the bulk resistances of the P and N materials.

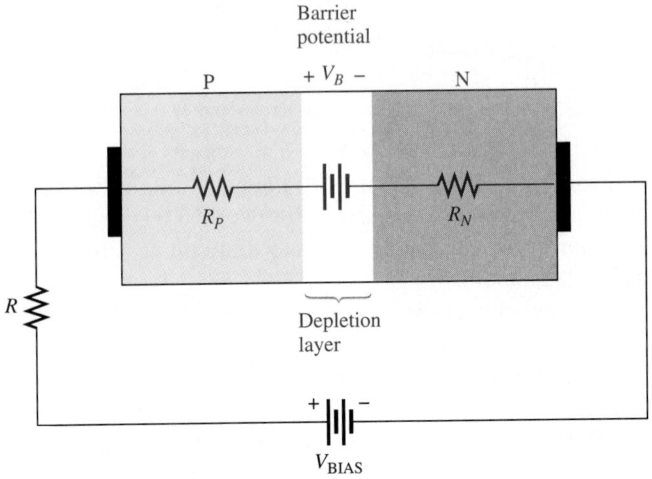

FIGURE 16–19
Barrier potential and bulk resistance equivalent for a PN junction diode.

The external bias voltage must overcome the effect of the barrier potential (V_B) before the diode conducts, as illustrated in Figure 16–20. Conduction occurs at approxi-

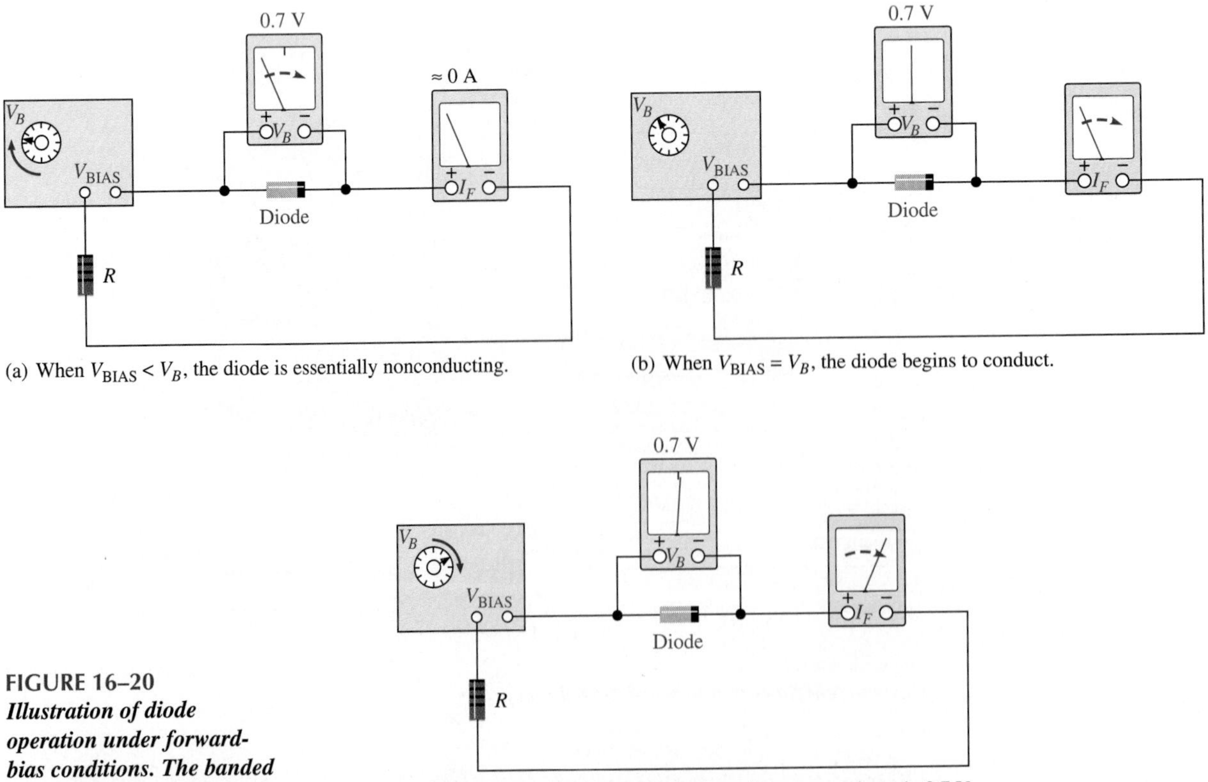

(a) When $V_{BIAS} < V_B$, the diode is essentially nonconducting.

(b) When $V_{BIAS} = V_B$, the diode begins to conduct.

(c) When $V_{BIAS} > V_B$, the diode drop remains at approximately 0.7 V (silicon), but the forward current (I_F) continues to increase.

FIGURE 16–20
Illustration of diode operation under forward-bias conditions. The banded end of the diode represents the N region (cathode).

mately 0.7 V for silicon and 0.3 V for germanium. Once the diode is conducting in the forward direction, the voltage drop across it remains at approximately the barrier potential and changes very little with changes in forward current (I_F) except for bulk resistance effects, as illustrated in Figure 16–20. The bulk resistances are usually only a few ohms and result in only a small voltage drop when the diode conducts. Often this drop can be neglected.

Energy Diagram for Forward Bias Forward bias raises the energy levels of the conduction electrons in the N region, allowing them to move into the P region and combine with holes in the valence band. This condition is shown in Figure 16–21.

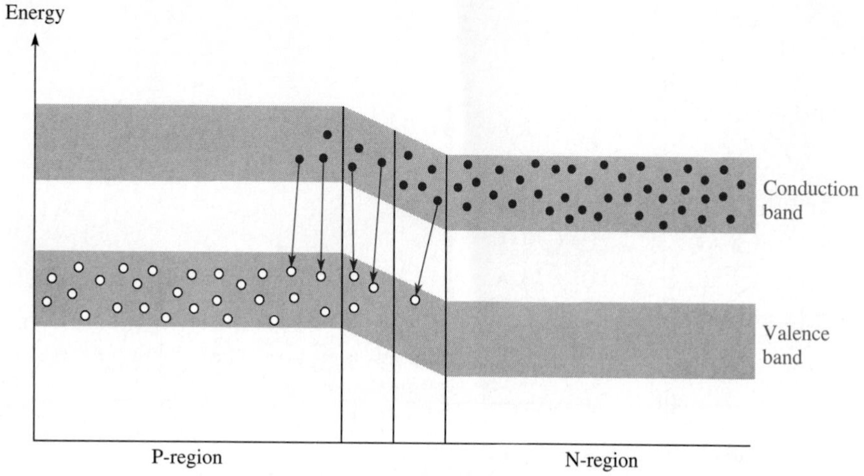

FIGURE 16–21
Energy diagram for forward bias, showing recombination in the depletion layer and in the P region as conduction electrons move across the junction.

Reverse Bias

Reverse bias is the condition that prevents current across the PN junction. Figure 16–22 shows a dc voltage source connected to reverse-bias the diode. Notice that the negative terminal of the battery is connected to the P region, and the positive terminal is connected to the N region.

A discussion of the basic operation for reverse bias follows: The negative terminal of the battery attracts holes in the P region away from the PN junction, while the positive terminal also attracts electrons away from the junction. As electrons and holes move

FIGURE 16–22
Reverse-bias connection.

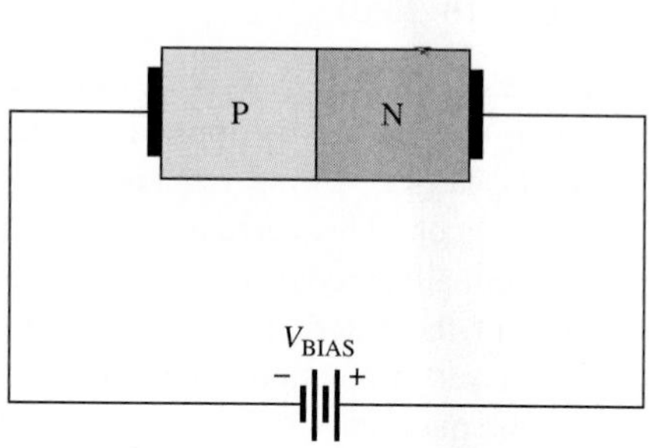

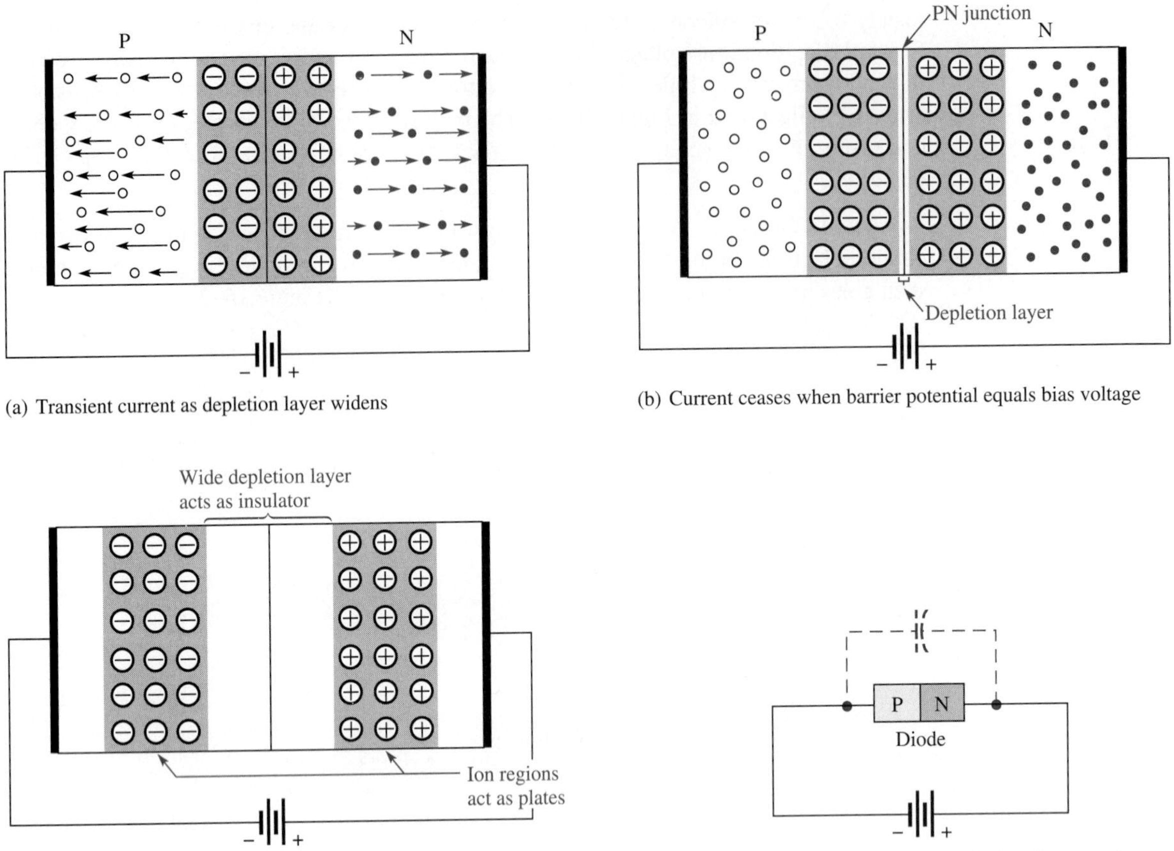

(a) Transient current as depletion layer widens

(b) Current ceases when barrier potential equals bias voltage

(c) Depletion layer widens as reverse bias increases

(d) Equivalent circuit with junction capacitance

FIGURE 16–23
Illustration of reverse bias.

away from the junction, the depletion layer widens; more positive ions are created in the N region, and more negative ions are created in the P region, as shown in Figure 16–23(a).

The depletion layer widens until the potential difference across it equals the external bias voltage. At this point, the holes and electrons stop moving away from the junction, and majority current ceases, as indicated in Figure 16–23(b).

The initial flow of majority carriers away from the junction is called *transient current* and lasts only for a very short time upon application of reverse bias.

When the diode is reverse-biased, the depletion layer effectively acts as an insulator between the layers of oppositely charged ions, forming an effective capacitance, as illustrated in Figure 16–23(c). Since the depletion layer widens with increased reverse-biased voltage, the capacitance decreases, and vice versa. This internal capacitance is called the *depletion-layer capacitance* and can be represented by an equivalent circuit as shown in Figure 16–23(d).

Reverse Current As you have learned, majority current very quickly becomes zero when reverse bias is applied. There is, however, a very small current produced by minority carriers during reverse bias. Germanium, as a rule, has a greater reverse current than silicon. This current is typically in the μA or nA range. A relatively small number of thermally produced electron-hole pairs exist in the depletion layer. Under the influence of the external voltage, some electrons manage to diffuse across the PN junction before recombination. This process establishes a small minority carrier current throughout the material.

The reverse current is dependent primarily on the junction temperature and not on the amount of reverse-biased voltage. A temperature increase causes an increase in reverse current.

Reverse Breakdown If the external reverse-bias voltage is increased to a large enough value, **reverse breakdown** occurs. The following describes what happens: Assume that one minority conduction band electron acquires enough energy from the external source to accelerate it toward the positive end of the diode. During its travel, it collides with an atom and imparts enough energy to knock a valence electron into the conduction band. There are now two conduction band electrons. Each will collide with an atom, knocking two more valance electrons into the conduction band. There are now four conduction band electrons which, in turn, knock four more into the conduction band. This rapid multiplication of conduction band electrons, known as an *avalanche effect,* results in a rapid buildup of reverse current.

Most diodes normally are not operated in reverse breakdown and can be damaged if they are. However, a particular type of diode known as a zener diode, (discussed in Chapter 17) is specially designed for reverse-breakdown operation.

SECTION 16–6 REVIEW

1. Name the two bias conditions.
2. Which bias condition produces majority carrier current?
3. Which bias condition produces a widening of the depletion region?
4. Minority carriers produce the current during reverse breakdown (T or F).

16–7 ▪ DIODE CHARACTERISTICS

As you learned in the last section, a diode conducts current when it is forward-biased if the bias voltage exceeds the barrier potential. The diode prevents current when it is reverse-biased at less than the breakdown voltage.

After completing this section, you should be able to

- **Describe the basic diode characteristics**
  - □ Explain the diode characteristic curve
  - □ Recognize the standard diode symbol
  - □ Test a diode with an ohmmeter
  - □ Understand diode approximations

Diode Characteristic Curve

Figure 16–24 is a graph of diode current versus voltage. The upper right quadrant of the graph represents the forward-biased condition. As you can see, there is very little forward current (I_F) for forward voltages (V_F) below the barrier potential. As the forward voltage approaches the value of the barrier potential (0.7 V for silicon and 0.3 V for germanium), the current begins to increase. Once the forward voltage reaches the barrier potential, the current increases drastically and must be limited by a series resistor. *The voltage across the forward-biased diode remains approximately equal to the barrier potential.*

The lower left quadrant of the graph represents the reverse-biased condition. As the reverse voltage (V_R) increases to the left, the current remains near zero until the breakdown voltage is reached. When breakdown occurs, there is a large reverse current which, if not limited, can destroy the diode. Typically, the breakdown voltage is greater than 50 V for most rectifier diodes. Remember that most diodes should not be operated in reverse breakdown.

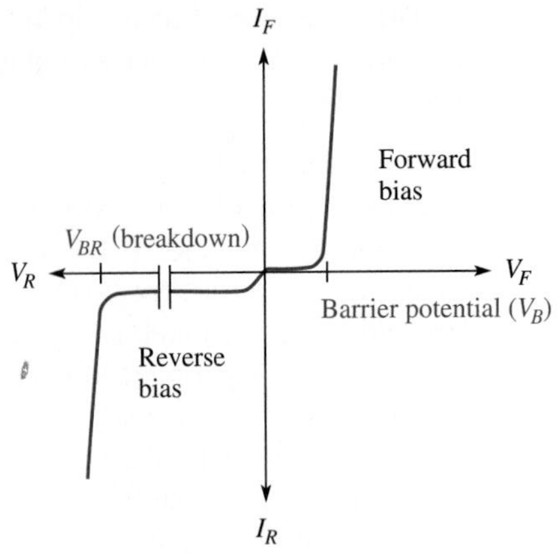

FIGURE 16–24
Diode characteristic curve.

Diode Symbol

Figure 16–25(a) is the standard schematic symbol for a general-purpose diode. The arrow points in the direction opposite the electron flow. The two terminals of the diode are the anode and cathode.

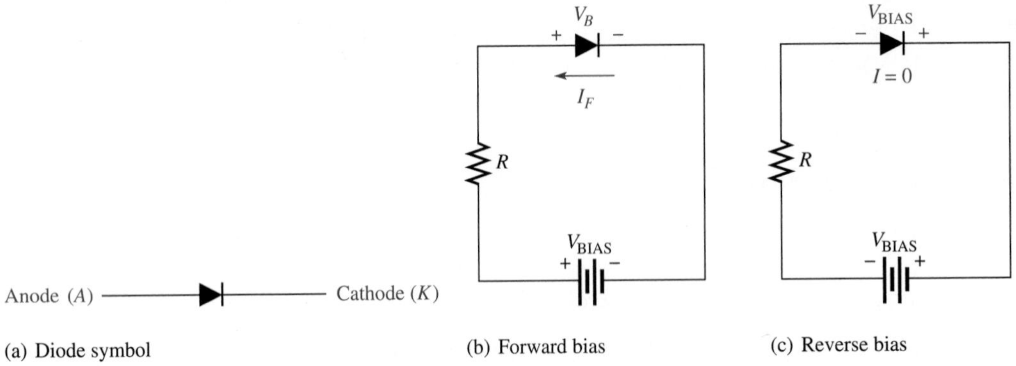

FIGURE 16–25
Diode schematic symbol and bias circuits. V_{BIAS} is the bias battery voltage, and V_B is the barrier potential.

When the anode is positive with respect to the cathode, the diode is forward-biased and current is from cathode to anode, as shown in Figure 16–25(b). Remember that when the diode is forward-biased, the barrier potential, V_B, always appears between anode and cathode, as indicated in the figure. When the anode is negative with respect to the cathode, the diode is reverse-biased, as shown in Figure 16–25(c) and there is no current.

Some typical diodes are shown in Figure 16–26(a) to illustrate the variety of physical structures. Figure 16–26(b) illustrates terminal identification.

Testing the Diode with an Ohmmeter

The internal battery in ohmmeters will forward-bias or reverse-bias a diode, permitting a quick and simple check for proper functioning based on relative readings. Many digital multimeters have a special diode test position.

To check the diode in the forward direction, the positive meter lead is connected to the anode and the negative lead to the cathode, as shown in Figure 16–27(a). When a diode is forward-biased, its internal resistance is low. When the meter leads are reversed, as shown in Figure 16–27(b), the internal ohmmeter battery reverse-biases the diode. The

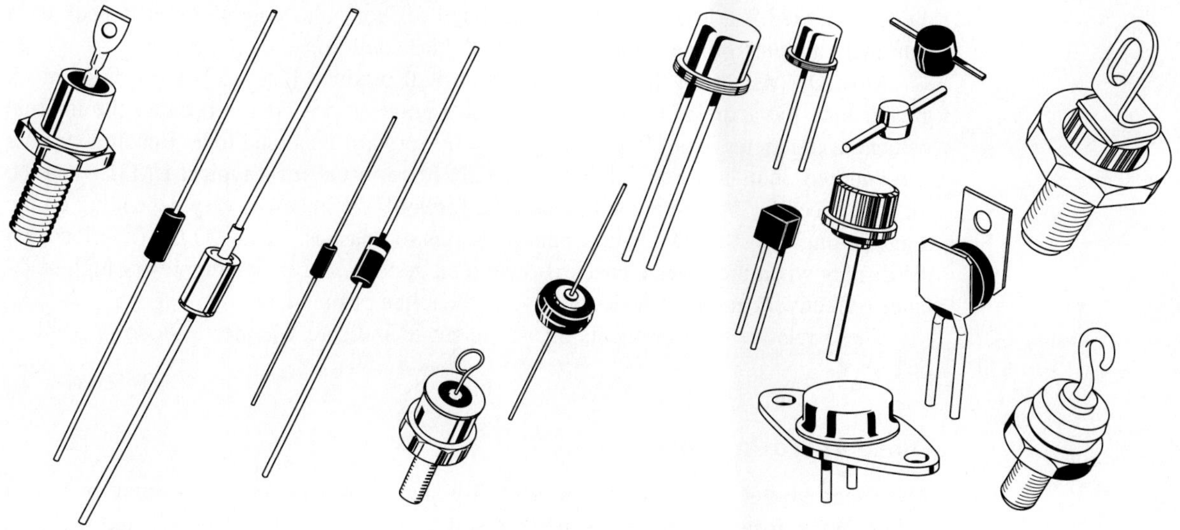

(a) A variety of package types

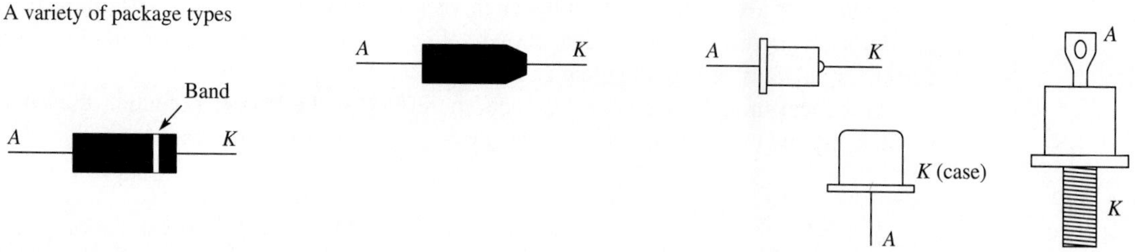

(b) Examples of terminal identification (*A* is anode; *K* is cathode}

FIGURE 16–26
Typical diodes.

FIGURE 16–27
Example of checking a diode with a DMM set on the "ohms" function.

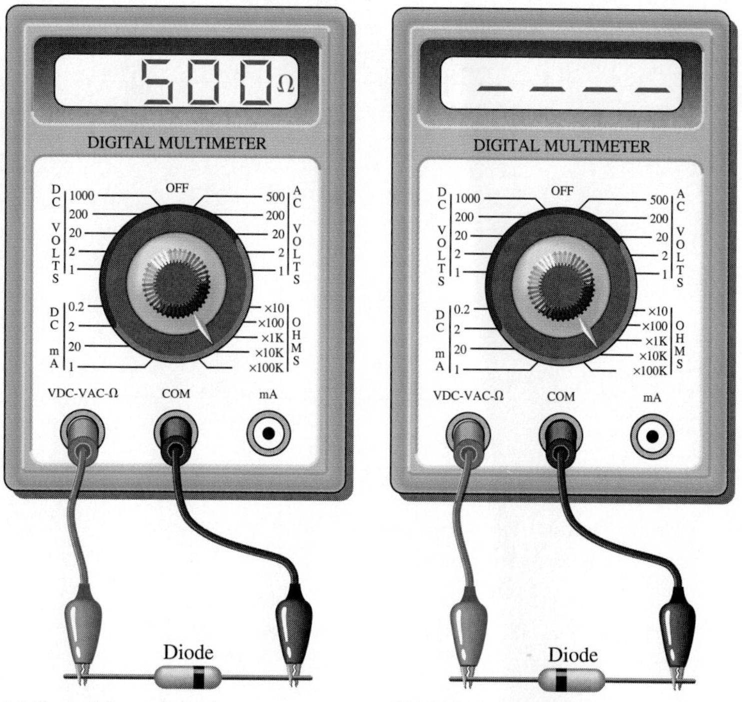

(a) Typical forward check (b) Typical reverse check

diode is shorted if a low resistance is indicated for both bias conditions; it is open if an extremely high resistance is indicated for both bias conditions.

Most DMMs set on the "ohms" function will produce forward-bias readings much greater than the actual forward resistance of a good diode. This is because the internal voltage of the meter is not large enough to fully forward-bias the diode. Readings from a few hundred ohms to a few kilohms range are not unusual for a typical DMM when the forward check is done on a good diode. The forward readings will vary depending on the range setting. Most DMMs will produce a series of "dashes" or an "O.L." (overload) on the display when the reverse check is done on a good diode, indicating a very high resistance (effectively an open) which is beyond the range of the meter.

These relative measurements are sufficient to indicate whether a diode is good or bad.

Diode Approximations

The Ideal Model The simplest way to visualize diode operation is to think of it as a switch. When forward-biased, the diode ideally acts as a closed (on) switch, and when reverse-biased, it acts as an open (off) switch, as shown in Figure 16–28. The characteristic curve for this approximation is also shown. Note that the forward voltage and the reverse current are always zero in the ideal case.

This ideal model, of course, neglects the effect of the barrier potential, the internal resistances, and other parameters. However, in many cases it is accurate enough.

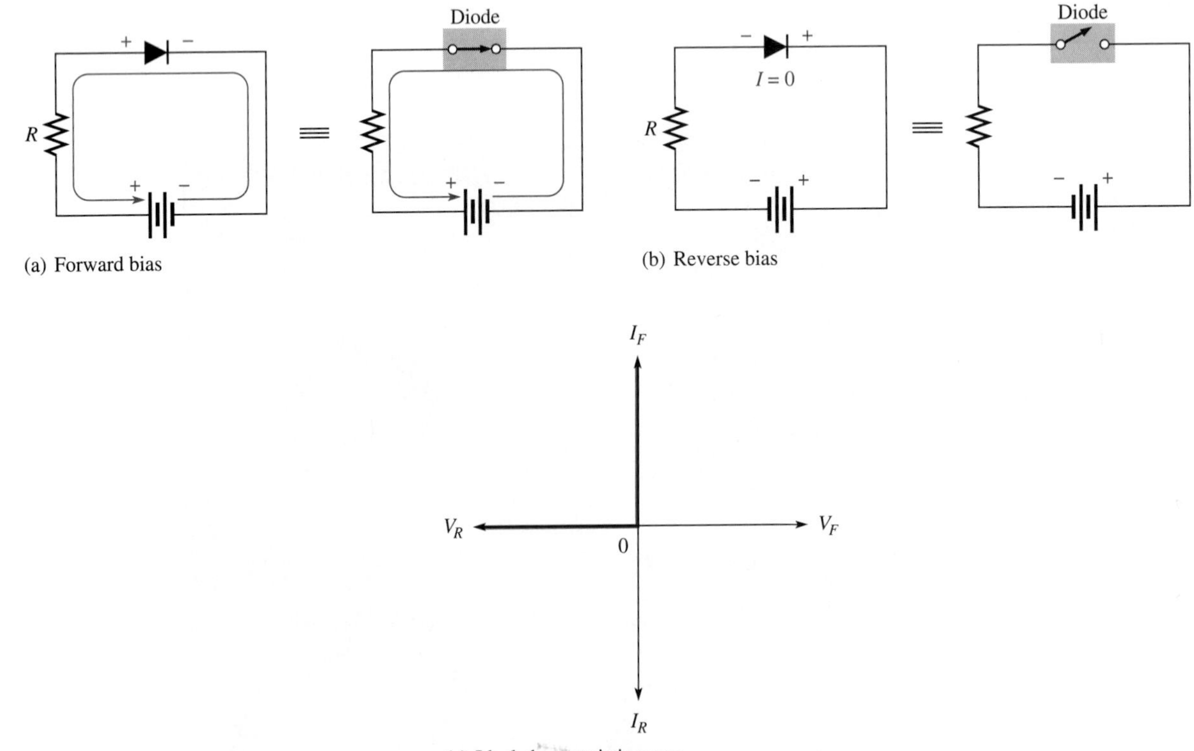

FIGURE 16–28
Ideal approximation of the diode as a switch.

The Barrier Potential Model The next higher level of accuracy is the barrier potential model. In this approximation, the forward-biased diode is represented as a closed switch in series with a small "battery" equal to the barrier potential V_B (0.7 V for Si and 0.3 V for Ge), as shown in Figure 16–29(a). The positive end of the battery is toward the anode. Keep in mind that the barrier potential is not a voltage source and cannot be measured with a voltmeter; rather it only has the effect of a battery when forward bias is applied because the forward bias voltage, V_{BIAS}, must overcome the barrier potential before the diode begins to conduct current. The reverse-biased diode is represented by an open switch, as in the ideal case, because the barrier potential does not affect reverse bias, as shown in Figure 16–29(b). The characteristic curve for this model is shown in Figure 16–29(c). In this textbook, the barrier potential is included in analysis unless otherwise stated.

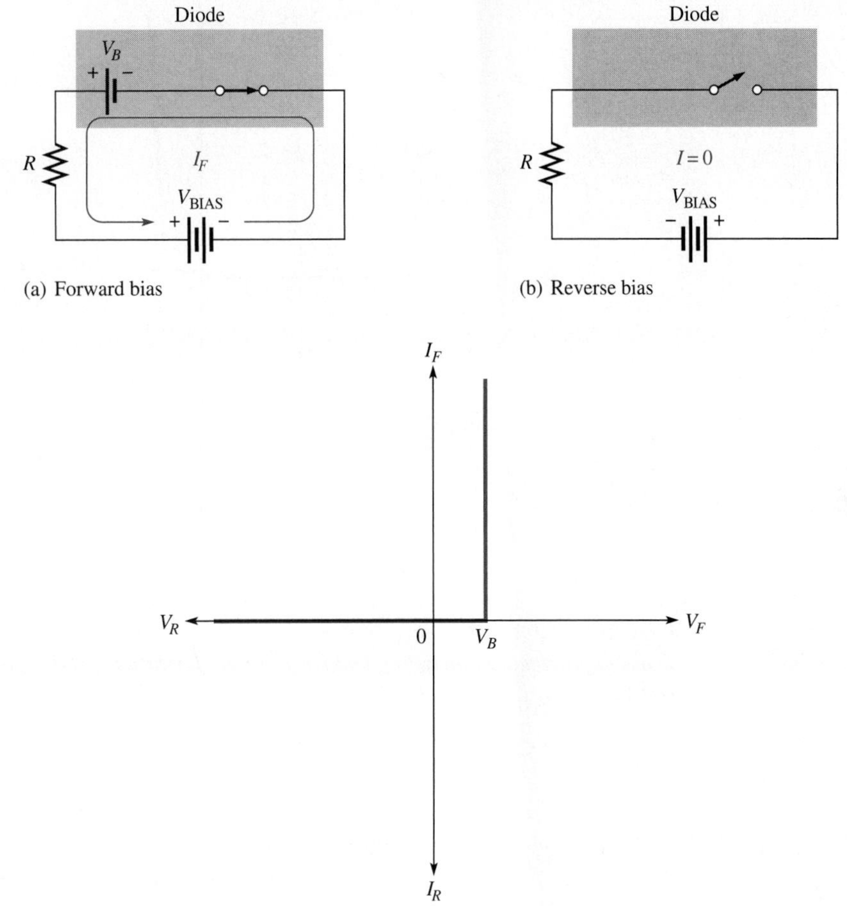

(a) Forward bias

(b) Reverse bias

(c) Characteristic curve

FIGURE 16–29
Diode approximation including barrier potential.

The Complete Model One more level of accuracy will be considered at this point. Figure 16–30(a) shows the forward-biased diode model with both the barrier potential and the low forward (bulk) resistance. Figure 16–30(b) shows how the high reverse resistance affects the reverse-biased model. The characteristic curve is shown in Figure 16–30(c).

Other parameters such as junction capacitance and breakdown voltage become important only under certain operating conditions and will be considered only where appropriate.

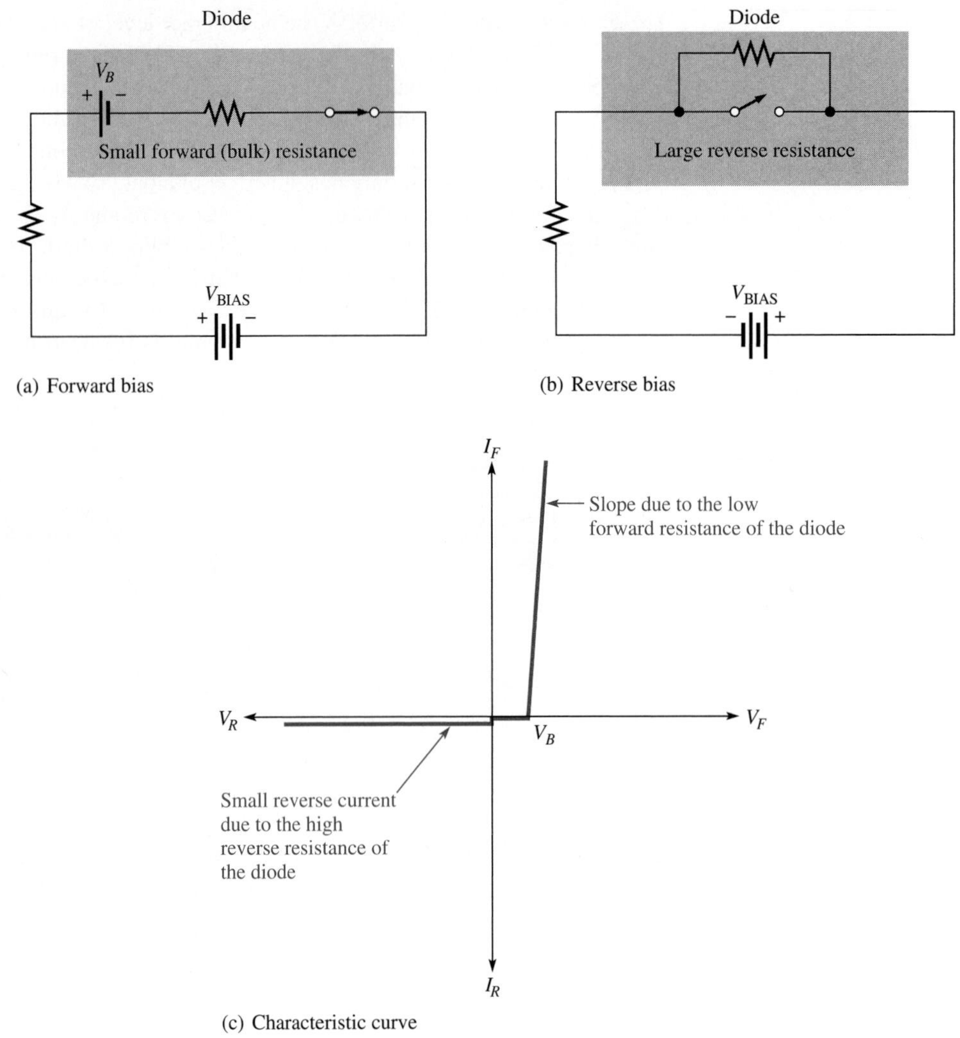

FIGURE 16–30
Diode approximation including barrier potential, forward resistance, and reverse leakage resistance.

SECTION 16–7 REVIEW

1. Sketch a rectifier diode symbol and label the terminals.
2. For a normal diode, the forward resistance is quite low and the reverse resistance is very high (T or F).
3. An open switch ideally represents a ______-biased diode. A closed switch ideally represents a ______-biased diode.

16–8 ▪ APPLICATION ASSIGNMENT

Your supervisor has given you the assignment of checking several production power supply units that have failed the final assembly test. You respond by saying that although you basically understand how transformers and individual diodes work, you know nothing about the overall operation of the power supply circuit or how diodes are applied. "No problem," says your supervisor. "Use a 3½-digit DMM set on the diode

test function to check for open diodes or transformer windings. You don't need to understand the power supply operation for this assignment. You will learn that later."

The dc power supply and its schematic are shown in Figure 16–31. The circuit consists of a power transformer, rectifier diodes, filter capacitors, and an integrated circuit voltage regulator. Isometric views of the capacitors and the voltage regulator attached to a finned heat sink are shown.

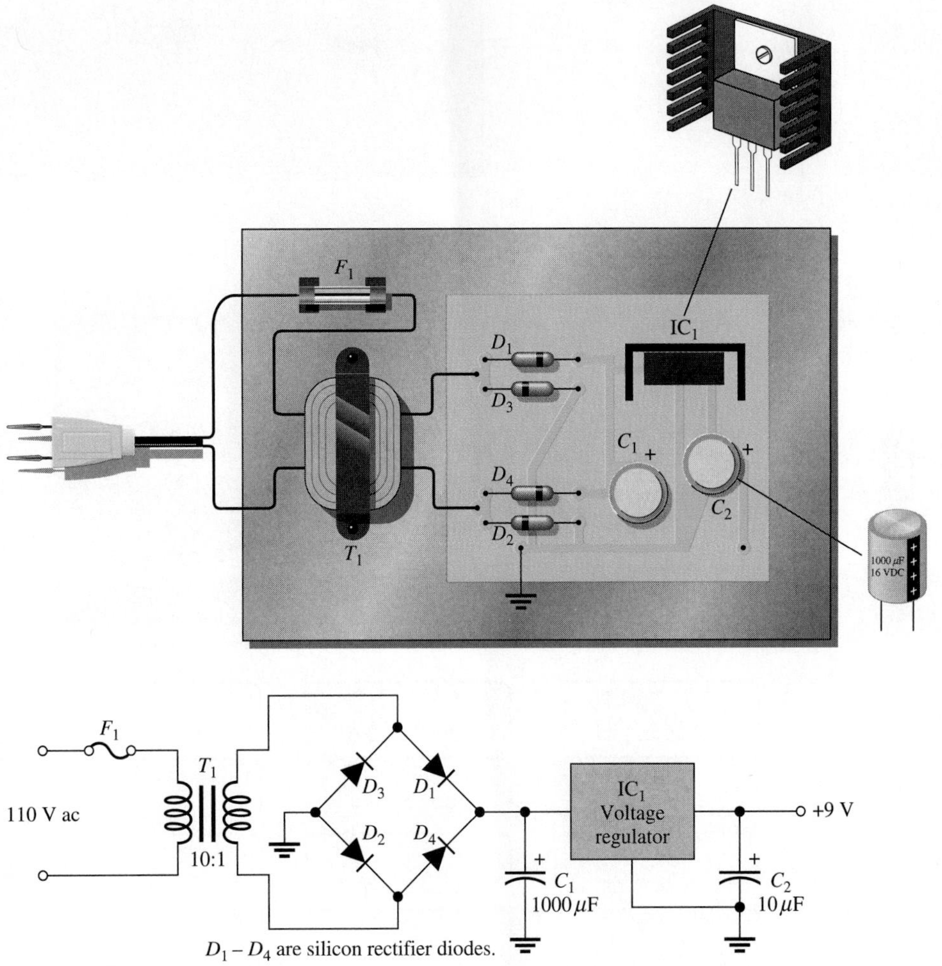

FIGURE 16–31
DC power supply and schematic.

Step 1: Identification of Components

Locate the fuse, the transformer, and the diodes in the power supply in Figure 16–31. Also identify the anode and the cathode of each diode.

Step 2: Basic Information

Before making DMM measurements, make sure the circuit is not connected to a power source (it must be unplugged). Assume that you have obtained the following specific information:

- ☐ The DMM will indicate a resistance of roughly 500 Ω when the diode is forward-biased. This reading will vary from one diode to the next.

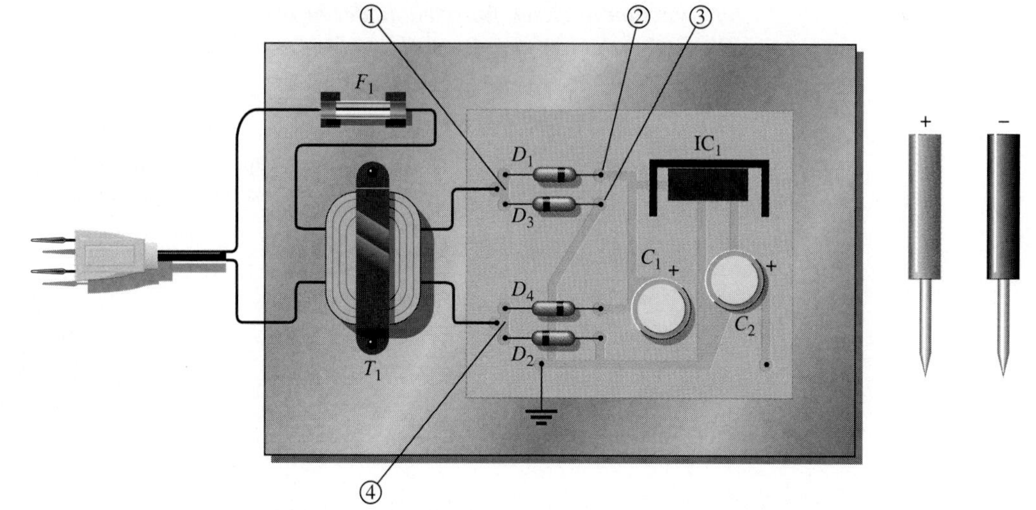

FIGURE 16–32
Power supply showing the numbered test points to which the DMM leads are to be connected.

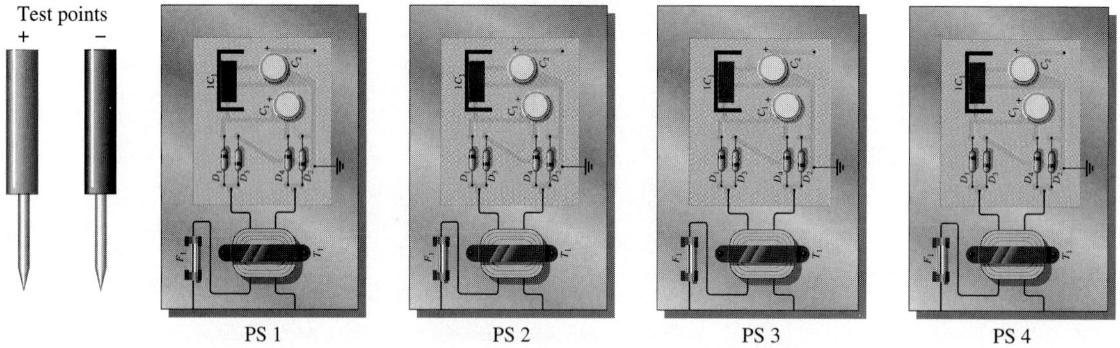

Test point +	Test point −	PS 1	PS 2	PS 3	PS 4
1	2	520 Ω	520 Ω	520 Ω	520 Ω
2	1	_ _ _ _	_ _ _ _	_ _ _ _	523 Ω
1	3	_ _ _ _	_ _ _ _	0 Ω	520 Ω
3	1	523 Ω	523 Ω	0 Ω	523 Ω
2	4	_ _ _ _	_ _ _ _	_ _ _ _	491 Ω
4	2	_ _ _ _	536 Ω	536 Ω	536 Ω
3	4	491 Ω	491 Ω	491 Ω	491 Ω
4	3	_ _ _ _	_ _ _ _	_ _ _ _	536 Ω
1	4	10 Ω	_ _ _ _	491 Ω	10 Ω

DMM readings

FIGURE 16–33
Results of the DMM check of each of the four power supply units.

- ☐ The DMM will indicate the very high reverse resistance of a diode by a series of dashes meaning that the measured value is out of the meter's range.
- ☐ The secondary winding resistance of the transformer is 10 Ω.

Step 3: Troubleshooting

Test points to which the leads are to be connected are numerically labeled in Figure 16–32. Determine the fault in each of the four power supply units based on the indicated DMM readings at the various combinations of test points shown in Figure 16–33. Each of the four diodes provides a slightly different forward reading.

SECTION 16–8 REVIEW

1. What faults could occur in the primary circuit of the power supply transformer and how would these faults be indicated by a DMM?
2. When measuring across a diode with a DMM, why is the polarity of the leads important?

▪ SUMMARY

- An atom is described as consisting of a nucleus containing protons and neutrons orbited by electrons.
- Protons are positive, neutrons are neutral, and electrons are negative.
- Atomic shells are energy bands.
- The outermost shell containing electrons is the valence shell.
- Silicon and germanium are semiconductor materials. Silicon is the most predominant.
- Atoms within a crystal structure are held together with covalent bonds.
- Electron-hole pairs are thermally produced.
- The process of adding impurities to an intrinsic (pure) semiconductor to increase and control conductivity is called *doping.*
- A P-type semiconductor is doped with trivalent impurity atoms.
- An N-type semiconductor is doped with pentavalent impurity atoms.
- The depletion layer is a region adjacent to the PN junction containing no majority carriers.
- Forward bias permits majority carrier current through the PN junction.
- Reverse bias prevents majority carrier current.
- A PN structure is called a *diode.*
- Reverse leakage current is due to thermally produced electron-hole pairs.
- Reverse breakdown occurs when the reverse-biased voltage exceeds a specified value.

▪ GLOSSARY

Anode The more positive terminal of a diode or other electronic device.

Atom The smallest particle of an element possessing the unique characteristics of that element.

Barrier potential The inherent voltage across the depletion layer of a PN junction.

Bias The application of dc voltage to a diode or other electronic device to produce a desired mode of operation.

Cathode The more negative terminal of a diode or other electronic device.

Covalent Related to the bonding of two or more atoms by the interaction of their valence electrons.

Crystal The pattern or arrangement of atoms forming a solid material.

Diode An electronic device that permits current in only one direction.

Discrete device An individual electrical or electronic component that must be used in combination with other components to form a complete functional circuit.

Doping The process of imparting impurities to an intrinsic semiconductor material in order to control its conduction characteristics.

Electron The basic particle of negative electrical charge in matter.

Energy The ability to do work.

Free electron A valence electron that has broken away from its parent atom and is free to move from atom to atom within the atomic structure of a material.

Germanium A semiconductor material.

Hole The absence of an electron.

Intrinsic semiconductor A pure material with relatively few free electrons.

Ion An atom that has gained or lost a valence electron resulting in a net positive or negative charge.

Ionization The removal or addition of an electron from or to an atom so that the resulting atom (ion) has a net positive or negative charge.

PN junction The boundary between N-type and P-type materials.

Recombination The process of a free electron falling into a hole in the valance band of an atom.

Reverse breakdown The condition of a diode in which excessive reverse-bias voltage causes a rapid buildup of reverse current.

Semiconductor A material that has a conductance value between that of a conductor and that of an insulator. Silicon and germanium are examples.

Silicon A semiconductor material used in diodes and transistors.

Terminal An external contact point on an electronic device.

Valence Related to the outer shell or orbit of an atom.

▪ SELF-TEST

1. The charge on electrons is
(a) positive **(b)** negative **(c)** neutral **(d)** variable

2. The number of valence electrons in both silicon and germanium is
(a) two **(b)** eight **(c)** four **(d)** eighteen

3. When an atom loses or gains a valence electron, the atom becomes
(a) covalent **(b)** neutral **(c)** a crystal **(d)** an ion

4. Atoms within a crystal are held together by
(a) atomic glue **(b)** subatomic particles **(c)** covalent bonds **(d)** the valence band

5. Free electrons exist in the
(a) valence band **(b)** conduction band **(c)** lowest band **(d)** recombination band

6. A hole is
(a) a vacancy in the valence band left by an electron
(b) a vacancy in the conduction band
(c) a positive electron
(d) a conduction band electron

7. The widest energy gap between the valence band and the conduction band occurs in
(a) semiconductors **(b)** insulators **(c)** conductors **(d)** a vacuum

8. The process of adding impurity atoms to a pure semiconductor material is called
(a) recombination **(b)** crystalization **(c)** bonding **(d)** doping

9. The two types of current in a semiconductor material are called
(a) positive current and negative current
(b) electron current and conventional current
(c) electron current and hole current
(d) forward current and reverse current

10. The majority carriers in an N-type semiconductor are
(a) electrons **(b)** holes **(c)** positive ions **(d)** negative ions

11. The PN junction is found in
(a) semiconductor diodes **(b)** transistors
(c) all semiconductor materials **(d)** both (a) and (b)

12. In a semiconductor diode, the region near the PN junction consisting of positive and negative ions is called the
(a) neutral zone **(b)** recombination region **(c)** depletion layer **(d)** diffusion area

13. A fixed dc voltage that sets the operating condition of a semiconductor device is called the
(a) bias (b) depletion voltage (c) battery (d) barrier potential
14. In a semiconductor diode, the two bias conditions are
(a) positive and negative (b) blocking and nonblocking
(c) open and closed (d) forward and reverse
15. When a diode is forward-biased, it is
(a) blocking current (b) conducting current
(c) similar to an open switch (d) similar to a closed switch
(e) both (a) and (c) (f) both (b) and (d)
16. The voltage across a forward-biased silicon diode is approximately
(a) 0.7 V (b) 0.3 V (c) 0 V (d) dependent on the bias voltage
17. In Figure 16–34, identify the forward-biased diode(s).
(a) D_1 (b) D_2 (c) D_3 (d) D_1 and D_3 (e) D_2 and D_3

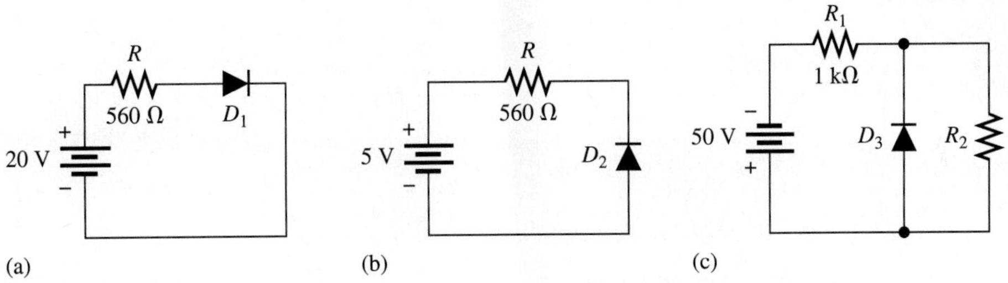

FIGURE 16–34

18. When the positive lead of an ohmmeter is connected to the cathode of a diode and the negative lead is connected to the anode, the meter reads
(a) a very low resistance
(b) an infinitely high resistance
(c) a high resistance initially, decreasing to about 100 Ω
(d) a gradually increasing resistance

▪ ANSWERS TO SECTION REVIEWS

Section 16–1

1. An orbit is the path of an electron around the nucleus of an atom.
2. A shell is an energy band into which orbits are grouped.
3. A valence electron is an electron in the outermost shell (valence band).
4. A positive ion is created when an atom loses a valence electron.

Section 16–2

1. Two semiconductors are silicon and germanium.
2. A covalent bond is the sharing of electrons with neighboring atoms.

Section 16–3

1. Free electrons exist in the conduction band; valence electrons exist in the valence band.
2. A hole is created when an electron is thermally raised to the conduction band, leaving a hole in the valence band.
3. The energy gap between the valence band and the conduction band is narrower for a semiconductor than for an insulator.

Section 16–4

1. An N-type semiconductor is formed by the addition of pentavalent atoms to the intrinsic semiconductor material.
2. A P-type semiconductor is formed by the addition of trivalent atoms to the intrinsic semiconductor material.
3. A majority carrier is the particle in greatest abundance: electrons in N-type material and holes in P-type material.

Section 16–5

1. A PN junction is the boundary between N-type and P-type materials.
2. The depletion layer is devoid of majority carriers and contains only positive and negative ions.
3. True
4. The barrier potential for silicon is 0.7 V.

Section 16–6

1. Two bias conditions are forward and reverse.
2. Forward bias produces majority carrier current.
3. Reverse bias causes the depletion area to widen.
4. True

Section 16–7

1. See Figure 16–35.

FIGURE 16–35

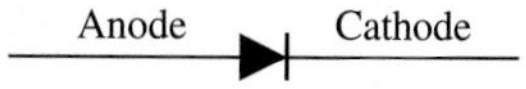

2. True
3. reverse; forward

Section 16–8

1. Open primary winding or open fuse; a typical DMM set on the ohms function will indicate an overload.
2. The polarity is important because the internal ohmmeter battery provides the bias voltage.

17

DIODES AND APPLICATIONS

In this chapter we discuss the applications of diodes in converting ac to dc by the process known as *rectification*. Half-wave and full-wave rectification are introduced, and you will study the basic circuits. The limitations of diodes used in rectifier applications are examined, and you will learn about diode clipping circuits and dc restoring (clamping) circuits.

In addition to rectifier diodes, zener diodes and their applications in voltage regulation are introduced. Varactor diodes, light-emitting diodes, and photodiodes and their applications also are discussed.

■ CHAPTER OBJECTIVES

- ☐ Analyze the operation of a half-wave rectifier
- ☐ Analyze the operation of a full-wave rectifier
- ☐ Describe the operation of rectifier filters
- ☐ Analyze the operation of diode clippers and clampers
- ☐ Explain the characteristics and applications of a zener diode
- ☐ Describe the basic operation of a varactor diode
- ☐ Discuss the operation and application of LEDs and photodiodes
- ☐ Interpret a typical data sheet
- ☐ Troubleshoot rectifier circuits

■ APPLICATION ASSIGNMENT

As a technician in an industrial manufacturing facility, you are responsible for maintaining and troubleshooting all of the automated production equipment. One particular system is used to count objects on a conveyor for control and inventory purposes. In order to check out and troubleshoot this system you must have a knowledge of power supply rectifiers, zener diodes, light-emitting diodes (LEDs), and photodiodes. After studying this chapter, you should be able to complete the application assignment in Section 17–10.

17–1 ■ HALF-WAVE RECTIFIERS

***Because of their ability to conduct current in one direction and block current in the other direction, diodes are used in circuits called* rectifiers *that convert ac voltage into dc voltage. Rectifiers are found in all dc power supplies that operate from an ac voltage source.* Power supplies *are an essential part of all electronic systems from the simplest to the most complex. In this section, you will study the most basic type of rectifier, the* half-wave rectifier.**

After completing this section, you should be able to

- **Analyze the operation of a half-wave rectifier**
  - ☐ Explain the process of half-wave rectification
  - ☐ Describe how the diode functions in a half-wave rectifier
  - ☐ Determine the average value of a half-wave voltage
  - ☐ Define *peak inverse voltage (PIV)*

Figure 17–1 illustrates the process called *half-wave rectification.* In part (a), an ac source is connected to a load resistor through a diode. Let's examine what happens during one cycle of the input voltage. When the sine wave input voltage goes positive, the diode is forward-biased and conducts current to the load resistor, as shown in part (b). The current produces a voltage across the load which has the same shape as the positive half-cycle of

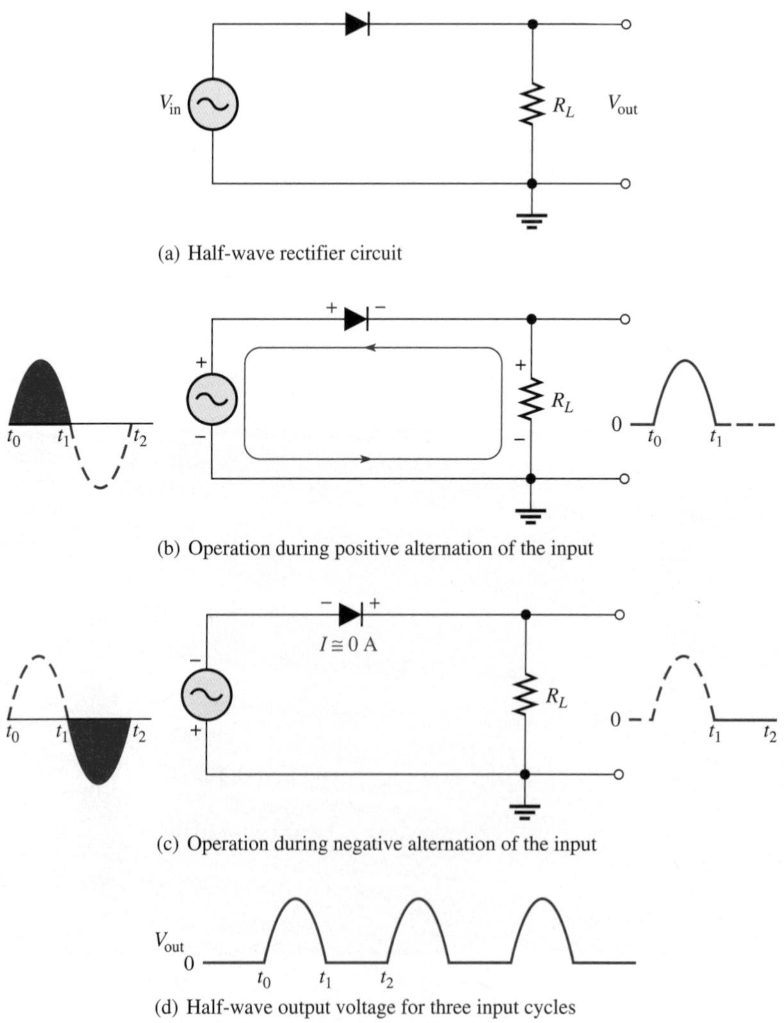

FIGURE 17–1
Operation of a half-wave rectifier.

the input voltage. When the input voltage goes negative during the second half of its cycle, the diode is reversed-biased. There is no current, so the voltage across the load resistor is zero, as shown in part (c). The net result is that only the positive half-cycles of the ac input voltage appear across the load, making the output a pulsating dc voltage, as shown in part (d).

Average Value of the Half-Wave Output Voltage

The average value of a half-wave output voltage is the value that would be indicated by a dc voltmeter. It can be calculated with the following equation where $V_{p(out)}$ is the peak value of the half-wave output voltage:

$$V_{AVG} = \frac{V_{p(out)}}{\pi} \tag{17–1}$$

Figure 17–2 shows the half-wave voltage with its average value indicated by the lower dashed line.

FIGURE 17–2
Average value of the half-wave rectified signal.

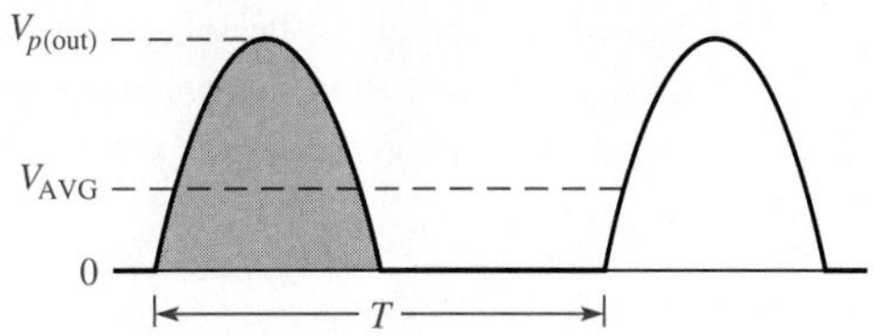

EXAMPLE 17–1 What is the average (dc) value of the half-wave rectified output voltage waveform in Figure 17–3?

FIGURE 17–3

Solution

$$V_{AVG} = \frac{V_{p(out)}}{\pi} = \frac{100\text{ V}}{\pi} = 31.8\text{ V}$$

Related Exercise Determine the average value of the half-wave output voltage if its peak amplitude is 12 V.

Effect of Barrier Potential on Half-Wave Rectifier Output Voltage

In the previous discussion, the diode was considered ideal. When the diode barrier potential is taken into account, here is what happens: During the positive half-cycle, the input voltage must overcome the barrier potential before the diode becomes forward-biased. For a silicon diode, this results in a half-wave output voltage with a peak value that is

0.7 V less than the peak value of the input voltage (0.3 V less for a germanium diode), as shown in Figure 17–4. The expression for peak output voltage is

$$V_{p(out)} = V_{p(in)} - 0.7\text{ V} \tag{17–2}$$

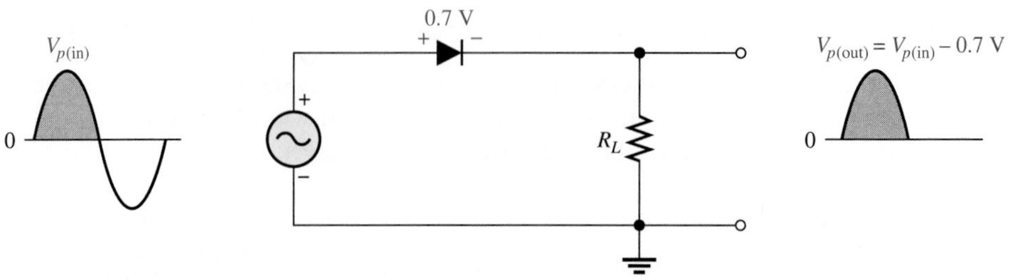

FIGURE 17–4
Effect of barrier potential on half-wave rectified output voltage (silicon diode shown).

In working with diode circuits, you will find that it is often practical to neglect the effect of barrier potential when the peak value of the applied voltage is much greater (at least ten times) than the barrier potential. As mentioned before, we will always use silicon diodes and consider the barrier potential unless stated otherwise.

EXAMPLE 17–2 Determine the peak output voltage of the silicon rectifier circuit in Figure 17–5 for the indicated input voltage.

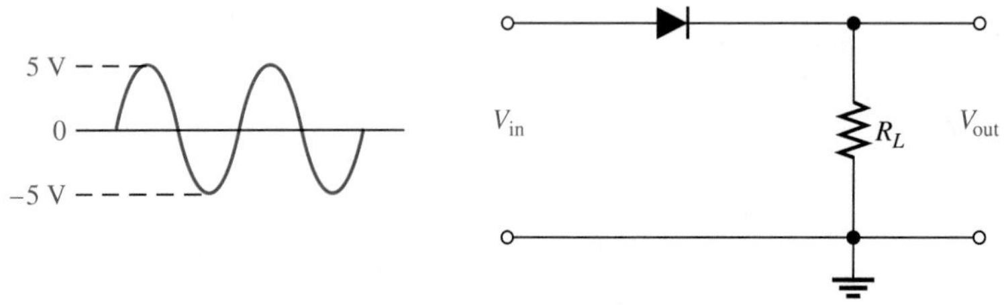

FIGURE 17–5

Solution The peak half-wave output voltage is

$$V_{p(out)} = V_{p(in)} - 0.7\text{ V} = 5\text{ V} - 0.7\text{ V} = 4.3\text{ V}$$

Related Exercise Determine the peak output voltage for the rectifier in Figure 17–5 if the peak input is 3 V.

Peak Inverse Voltage (PIV)

The maximum value of reverse voltage, sometimes designated as **peak inverse voltage (PIV),** occurs at the peak of the negative alternation of the input cycle when the diode is reverse-biased. This condition is illustrated in Figure 17–6. The PIV equals the peak value of the input voltage, and the diode must be capable of withstanding this amount of repetitive reverse voltage.

FIGURE 17–6
The PIV occurs at the peak of the half-cycle when the diode is reverse-biased. In this circuit, the PIV occurs at the time (t_p) of the peak of the negative half-cycle.

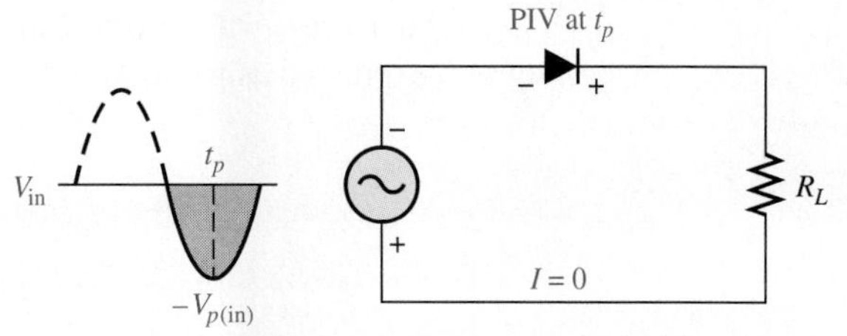

SECTION 17–1 REVIEW

1. At what point on the input cycle of a half-wave rectifier does the PIV occur?
2. For a half-wave rectifier, there is current through the load for approximately what percentage of the input cycle?
3. What is the average value of the output voltage shown in Figure 17–7?

FIGURE 17–7

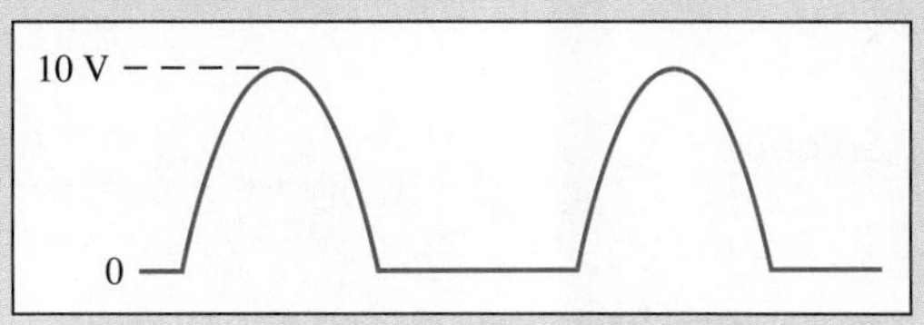

17–2 ■ FULL-WAVE RECTIFIERS

Although half-wave rectifiers have some applications, the full-wave rectifier is the most commonly used type in dc power supplies. In this section, you will use what you learned about half-wave rectification and expand it to full-wave rectifiers. You will learn about two types of full-wave rectifiers, center-tapped and bridge.

After completing this section, you should be able to

- **Analyze the operation of a full-wave rectifier**
  - □ Explain the process of full-wave rectification
  - □ Analyze the center-tapped full-wave rectifier
  - □ Analyze the full-wave bridge rectifier
  - □ Determine the PIV for each type of rectifier

The difference between full-wave and half-wave rectification is that a **full-wave rectifier** allows unidirectional current to the load during the entire input cycle, and the half-wave rectifier allows this only during one half of the cycle. The result of full-wave rectification is a dc output voltage that pulsates every half-cycle of the input, as shown in Figure 17–8.

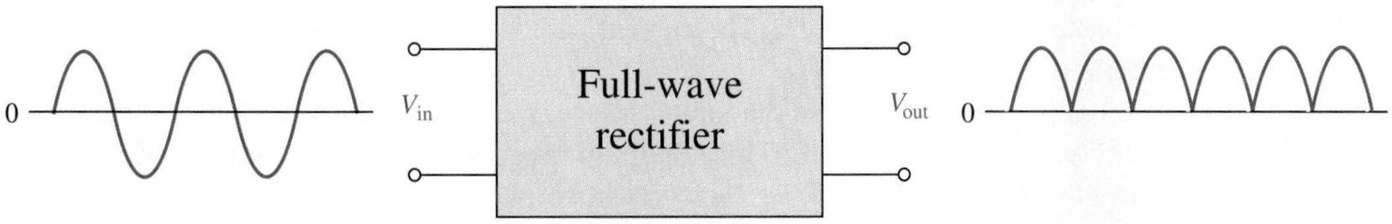

FIGURE 17–8
Full-wave rectification.

The average value for a full-wave rectified output voltage is twice that of the half-wave, expressed as follows:

$$V_{AVG} = \frac{2V_{p(out)}}{\pi} \tag{17–3}$$

EXAMPLE 17–3 Find the average value of the full-wave rectified output voltage in Figure 17–9.

FIGURE 17–9

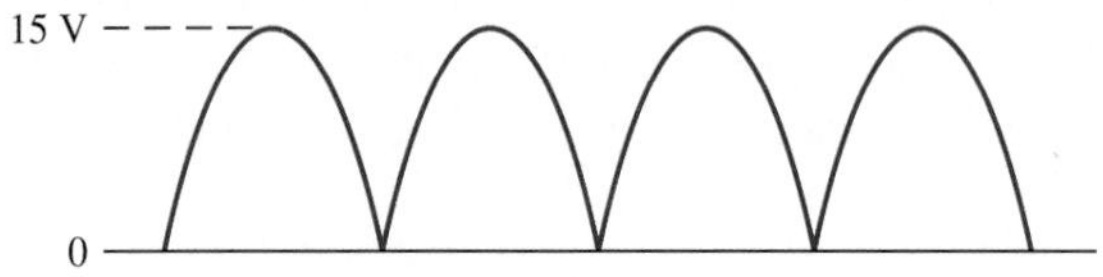

Solution

$$V_{AVG} = \frac{V_{p(out)}}{\pi} = \frac{2(15\text{ V})}{\pi} = 9.55\text{ V}$$

Related Exercise Find V_{AVG} for the full-wave rectified voltage if its peak is 155 V.

Center-Tapped Full-Wave Rectifier

The center-tapped (CT) full-wave rectifier circuit uses two diodes connected to the secondary of a center-tapped transformer, as shown in Figure 17–10. The input signal is coupled through the transformer to the secondary. Half of the secondary voltage appears between the **center tap** and each end of the secondary winding as shown.

FIGURE 17–10
A center-tapped full-wave rectifier.

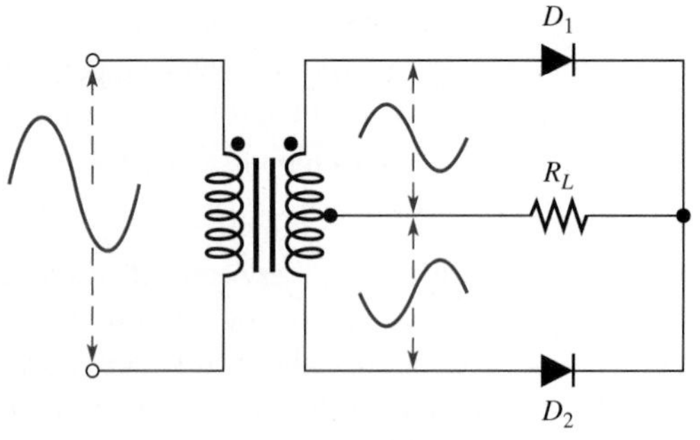

For a positive half-cycle of the input voltage, the polarities of the secondary voltages are as shown in Figure 17–11(a). This condition forward-biases the upper diode D_1 and reverse-biases the lower diode D_2. The current path is through D_1 and the load resistor, as indicated.

For a negative half-cycle of the input voltage, the voltage polarities on the secondary are as shown in Figure 17–11(b). This condition reverse-biases D_1 and forward-biases D_2. The current path is through D_2 and the load resistor, as indicated.

Because the current during both the positive and the negative portions of the input cycle is in the same direction through the load, the output voltage developed across the load is a full-wave rectified dc voltage.

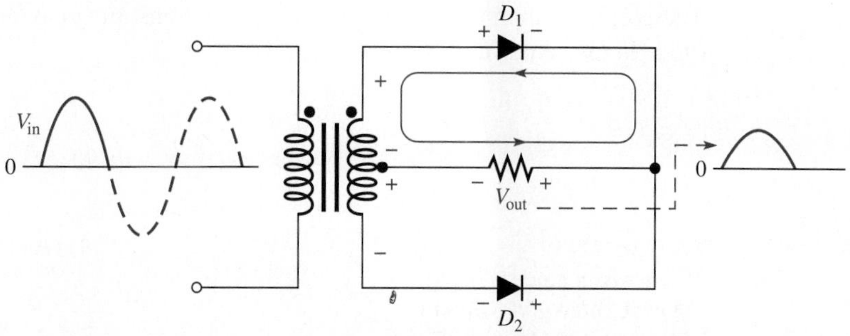

(a) During positive half-cycles, D_1 is forward-biased and D_2 is reverse-biased.

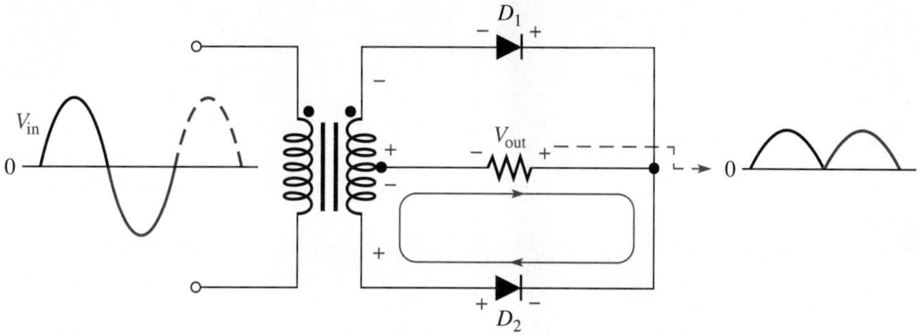

(b) During negative half-cycles, D_2 is forward-biased and D_1 is reverse-biased.

FIGURE 17–11

Basic operation of a center-tapped full-wave rectifier. Note that the current through the load resistor is in the same direction during the entire input cycle.

Effect of the Turns Ratio on Full-Wave Output Voltage If the turns ratio of the transformer is 1, the peak value of the rectified output voltage equals half the peak value of the primary input voltage less the barrier potential (diode drop). This value occurs because half of the input voltage appears across each half of the secondary winding.

In order to obtain an output voltage equal to the input (less the barrier potential), you must use a step-up transformer with a turns ratio of 1 to 2 (1:2). In this case, the total secondary voltage is twice the primary voltage, so the voltage across each half of the secondary is equal to the input.

Peak Inverse Voltage (PIV) Each diode in the full-wave rectifier is alternately forward-biased and then reverse-biased. The maximum reverse voltage that each diode must withstand is the peak value of the total secondary voltage (V_{sec}), as illustrated in Figure 17–12. When the total secondary voltage V_{sec} has the polarity shown, the anode of D_1 is $+V_{sec}/2$ and the anode of D_2 is $-V_{sec}/2$. Since D_1 is forward-biased, its cathode is 0.7 V less than its anode ($V_{sec}/2 - 0.7$ V); this is also the voltage on the cathode of D_2. The total reverse voltage across D_2 is

$$V_{D2} = \left(\frac{V_{sec}}{2} - 0.7\text{ V}\right) - \left(\frac{-V_{sec}}{2}\right) = V_{sec} - 0.7\text{ V}$$

Since $V_{p(out)} = V_{p(sec)}/2 - 0.7$ V, then

$$V_{p(sec)} = 2V_{p(out)} + 1.4\text{ V}$$

The peak inverse voltage in terms of the peak secondary voltage is

$$\text{PIV} = V_{p(sec)} - 0.7\text{ V}$$

Therefore, combining the two preceding equations, the peak inverse voltage across either diode in the center-tapped full-wave rectifier is

$$\text{PIV} = 2V_{p(out)} + 0.7\text{ V} \qquad \textbf{(17–4)}$$

FIGURE 17–12
PIV across a diode is twice the peak value of the output voltage plus the barrier potential.

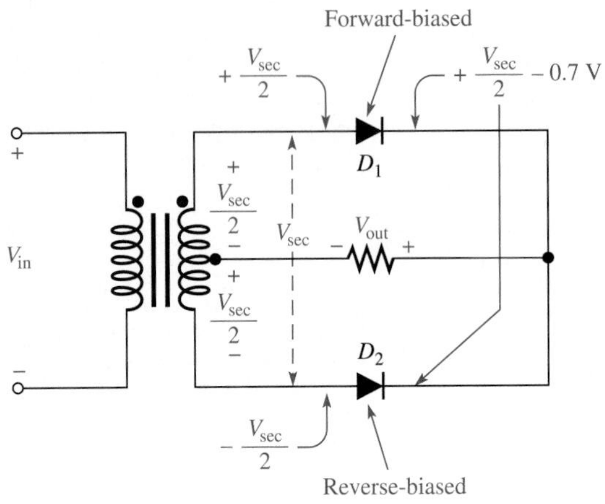

EXAMPLE 17–4

(a) Show the voltage waveforms across the secondary winding and across R_L when a 25 V peak sine wave is applied to the primary winding in Figure 17–13. This circuit is the same as the one in Figure 17–10, except the center tap and one end of R_L are connected to ground.

(b) What minimum PIV rating must the diodes have?

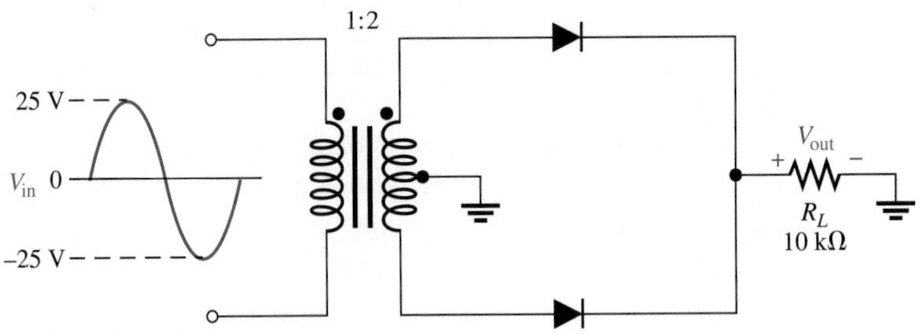

FIGURE 17–13

Solution

(a) The waveforms are shown in Figure 17–14.

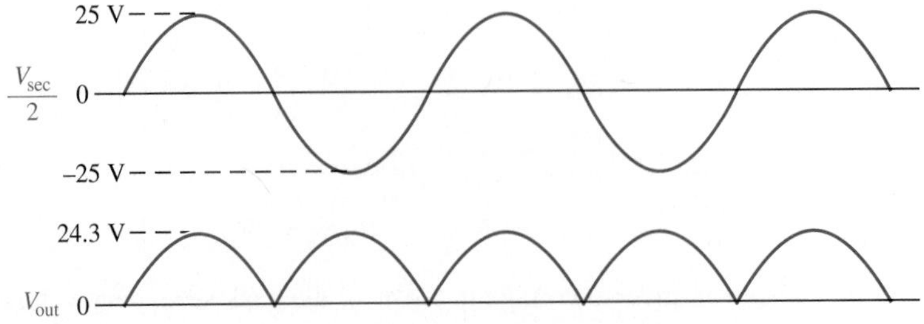

FIGURE 17–14

(b) The minimum PIV rating for each diode is calculated as follows:

$$V_{p(sec)} = \left(\frac{N_{sec}}{N_{pri}}\right)V_{p(in)} = (2)25\text{ V} = 50\text{ V}$$

$$V_{p(out)} = \frac{V_{p(sec)}}{2} - 0.7\text{ V} = \frac{50\text{ V}}{2} - 0.7\text{ V} = 24.3\text{ V}$$

$$\text{PIV} = 2V_{p(out)} + 0.7\text{ V} = 2(24.3\text{ V}) + 0.7\text{ V} = 49.3\text{ V}$$

Related Exercise What diode PIV rating is required to handle a peak input of 160 V in Figure 17–13?

Full-Wave Bridge Rectifier

The full-wave bridge rectifier uses four diodes, as shown in Figure 17–15. When the input cycle is positive as in part (a), diodes D_1 and D_2 are forward-biased and conduct current in the direction shown. A voltage is developed across R_L which looks like the positive half of the input cycle. During this time, diodes D_3 and D_4 are reverse-biased.

When the input cycle is negative, as in Figure 17–15(b), diodes D_3 and D_4 are forward-biased and conduct current in the same direction through R_L as during the positive half-cycle. During the negative half-cycle, D_1 and D_2 are reverse-biased. A full-wave rectified output voltage appears across R_L as a result of this action.

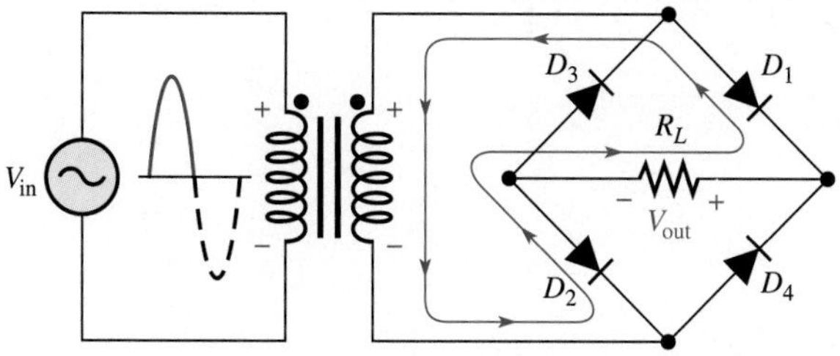

(a) During positive half-cycle of the input, D_1 and D_2 are forward-biased and conduct current. D_3 and D_4 are reverse-biased.

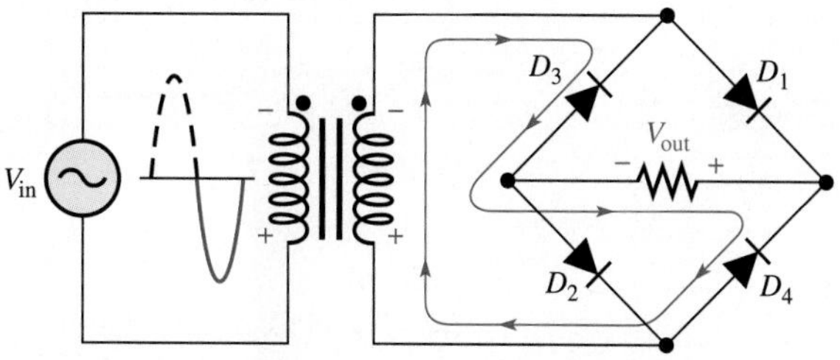

(b) During negative half-cycle, D_3 and D_4 are forward-biased and conduct current. D_1 and D_2 are reverse-biased.

FIGURE 17–15
Operation of full-wave bridge rectifier.

Bridge Output Voltage As you can see in Figure 17–15, two diodes are always in series with the load resistor during both the positive and the negative half-cycles. When these diode drops (barrier potentials) are taken into account, the output voltage is a full-wave rectified voltage with a peak value equal to the peak secondary voltage less the two diode drops.

$$V_{p(out)} = V_{p(sec)} - 1.4\text{ V} \quad \textbf{(17–5)}$$

Peak Inverse Voltage (PIV) Assuming that the input is in its positive half-cycle when D_1 and D_2 are forward-biased, let's examine the reverse voltage across D_3 and D_4. In Figure 17–16, you can see that D_3 and D_4 have a peak inverse voltage equal to the peak secondary voltage less one diode drop ($V_{p(sec)} - 0.7$ V). Since the peak secondary voltage is equal to the peak output voltage plus two diode drops ($V_{p(out)} + 1.4$ V), we have

$$\text{PIV} = V_{p(sec)} - 0.7\text{ V} = V_{p(out)} + 1.4\text{ V} - 0.7\text{ V}$$

$$\text{PIV} = V_{p(out)} + 0.7\text{ V} \quad \textbf{(17–6)}$$

The PIV rating of the bridge diodes is less than half that required for the center-tapped configuration.

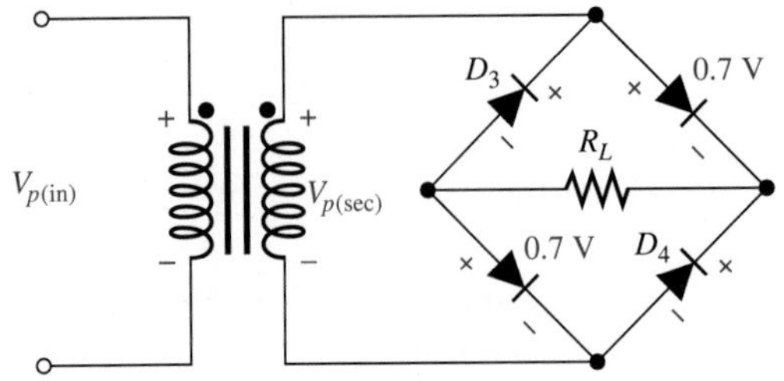

FIGURE 17–16
PIV in a bridge rectifier during the positive half-cycle of the input voltage.

EXAMPLE 17–5

(a) Determine the peak output voltage for the bridge rectifier in Figure 17–17.
(b) What minimum PIV rating is required for the silicon diodes?

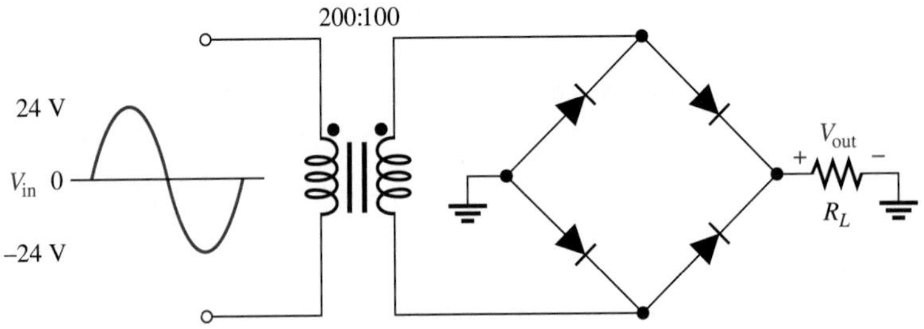

FIGURE 17–17

Solution
(a) The peak output voltage is

$$V_{p(\text{out})} = V_{p(\text{sec})} - 1.4\text{ V} = \left(\frac{N_{\text{sec}}}{N_{\text{pri}}}\right)V_{p(\text{in})} - 1.4\text{ V} = \left(\frac{100}{200}\right)24\text{ V} - 1.4\text{ V}$$
$$= 12\text{ V} - 1.4\text{ V} = 10.6\text{ V}$$

(b) The PIV for each diode is

$$\text{PIV} = V_{p(\text{out})} + 0.7\text{ V} = 10.6\text{ V} + 0.7\text{ V} = 11.3\text{ V}$$

Related Exercise Determine the peak output voltage for the bridge rectifier in Figure 17–17 if the peak primary voltage is 160 V. What is the PIV rating for the diodes?

SECTION 17–2 REVIEW

1. What is the average value of a full-wave rectified voltage with a peak value of 60 V?
2. Which type of full-wave rectifier has the greater output voltage for the same input voltage and transformer turns ratio?
3. For a given output voltage, the PIV for bridge rectifier diodes is less than for center-tapped rectifier diodes (T or F).

17–3 ■ RECTIFIER FILTERS

A power supply filter greatly reduces the fluctuations in the output voltage of a half-wave or full-wave rectifier and produces a nearly constant-level dc voltage. The reason for filtering is that electronic circuits require a constant source of dc voltage and current to provide power and biasing for proper operation. Filtering is done using capacitors, inductors, or combinations of both, as you will see in this section.

After completing this section, you should be able to

- **Describe the operation of rectifier filters**
  - ☐ Explain the operation of a capacitor-input filter
  - ☐ Determine the PIV for the diode in a filtered rectifier
  - ☐ Define ripple and discuss its cause
  - ☐ Calculate the ripple factor
  - ☐ Discuss surge current in a capacitor filter
  - ☐ Explain the operation of an inductor-input filter

In most power supply applications, the standard 60 Hz ac power line voltage must be converted to a sufficiently constant dc voltage. The 60 Hz pulsating dc output of a half-wave rectifier or the 120 Hz pulsating output of a full-wave rectifier must be filtered to virtually eliminate the large voltage variations. Figure 17–18 illustrates the filtering concept showing a smooth dc output voltage. The full-wave rectifier voltage is applied to the input of the **filter**, and, ideally, a constant dc level appears on the output.

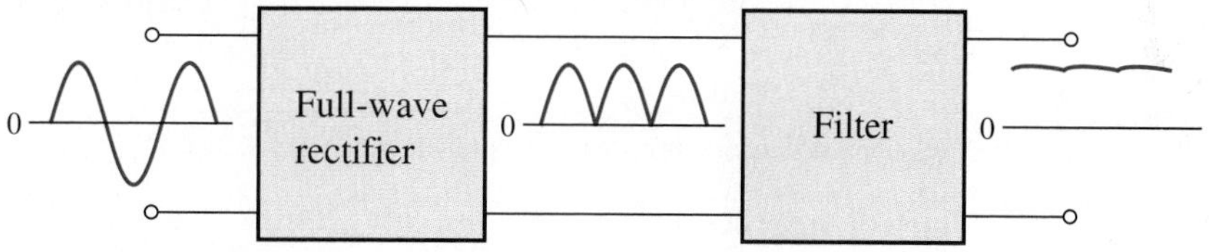

FIGURE 17–18

Basic block diagram of a power supply that converts 60 Hz ac to dc.

Capacitor-Input Filter

A half-wave rectifier with a capacitor-input filter is shown in Figure 17–19. We will use the half-wave rectifier to illustrate the filtering principle; then we will expand the concept to the full-wave rectifier.

During the positive first quarter-cycle of the input, the diode is forward-biased, allowing the capacitor to charge to within a diode drop of the input peak, as illustrated in Figure 17–19(a). When the input begins to decrease below its peak, as shown in part(b), the capacitor retains its charge and the diode becomes reverse-biased. During the remaining part of the cycle, the capacitor can discharge only through the load resistance at a rate determined by the R_LC time constant. The larger the time constant, the less the capacitor will discharge.

Because the capacitor charges to a peak value of $V_{p(\text{in})} - 0.7\text{ V}$, the peak inverse voltage of the diode in this application is

$$\text{PIV} = 2V_{p(\text{in})} - 0.7\text{ V} \qquad \textbf{(17–7)}$$

During the first quarter of the next cycle, as illustrated in Figure 17–19(c), the diode again will become forward-biased when the input voltage exceeds the capacitor voltage by approximately a diode drop.

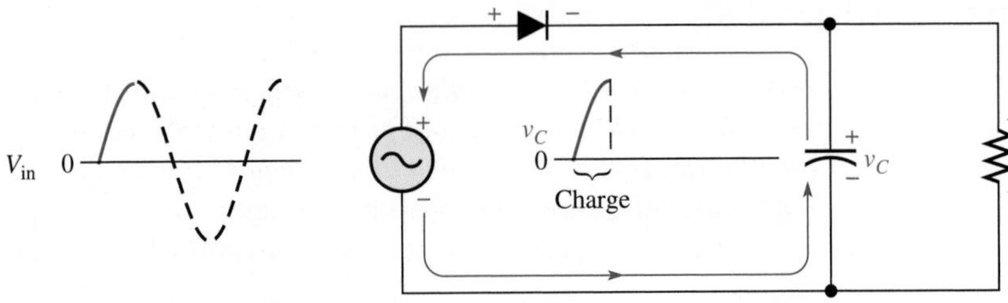

(a) Initial charging of capacitor (diode is forward-biased).

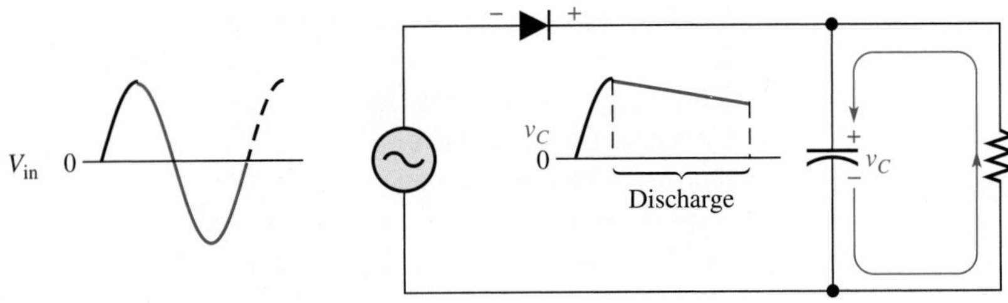

(b) Capacitor discharges through R_L after peak of the positive alternation (diode is reverse-biased).

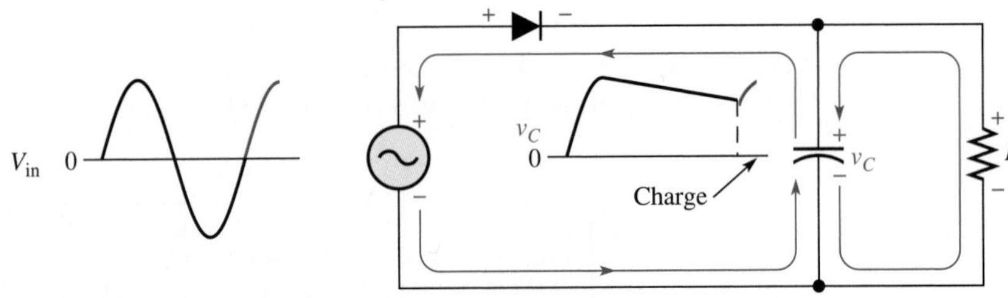

(c) Capacitor charges back to the peak (diode is forward-biased).

FIGURE 17–19
Operation of a half-wave rectifier with a capacitor-input filter.

Ripple Voltage As you have seen, the capacitor quickly charges at the beginning of a cycle and slowly discharges after the positive peak (when the diode is reverse-biased). The variation in the output voltage due to the charging and discharging is called the **ripple voltage.** The smaller the ripple, the better the filtering action, as illustrated in Figure 17–20.

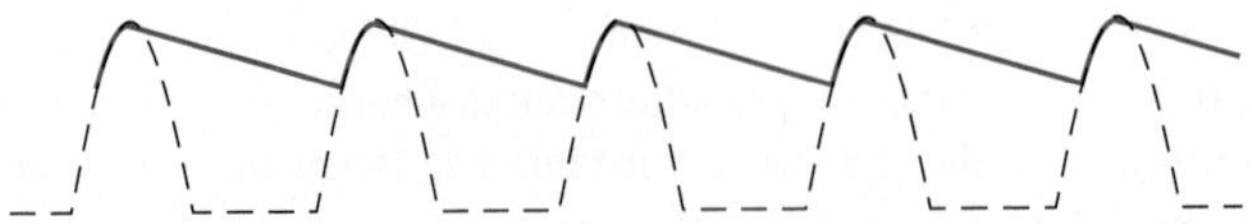

(a) Greater ripple

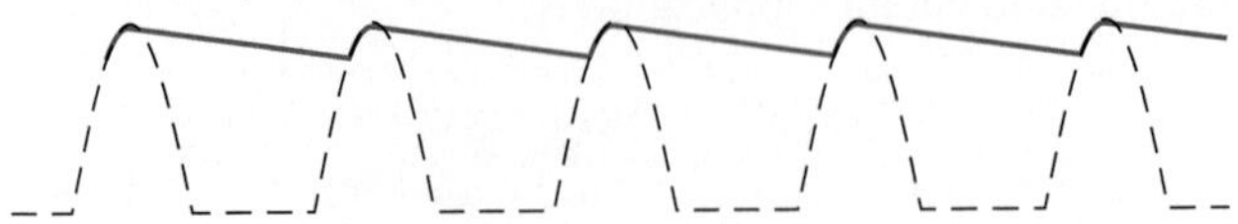

(b) Less ripple

FIGURE 17–20
Half-wave ripple voltage (solid line).

For a given input frequency, the output frequency of a full-wave rectifier is twice that of a half-wave rectifier. As a result, a full-wave rectifier is easier to filter. When filtered, the full-wave rectified voltage has less ripple than does a half-wave signal for the same load resistance and capacitor values. Less ripple occurs because the capacitor discharges less during the shorter interval between full-wave pulses, as shown in Figure 17–21. A good rule of thumb for effective filtering is to make $R_L C \geq 10T$, where T is the period of the rectified voltage.

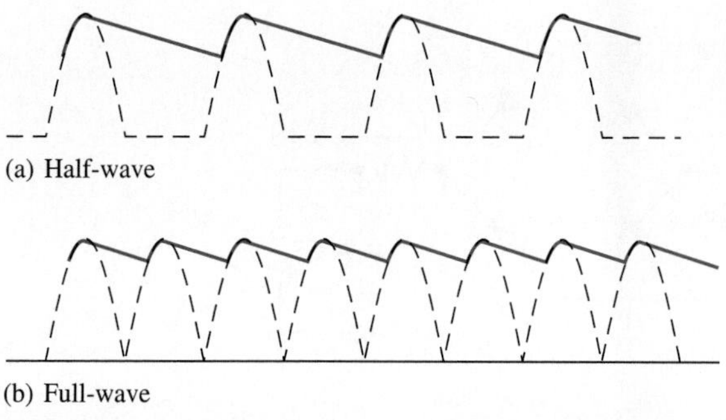

FIGURE 17–21
Comparison of ripple voltages for half-wave and full-wave signals with same filter and same input frequency.

The ripple factor (r) is an indication of the effectiveness of the filter and is defined as the ratio of the ripple voltage (V_r) to the dc (average) value of the filter output voltage (V_{dc}). These parameters are illustrated in Figure 17–22.

$$r = \left(\frac{V_r}{V_{dc}}\right)100\% \qquad \textbf{(17–8)}$$

The lower the ripple factor, the better the filter. The ripple factor can be decreased by increasing the value of the filter capacitor.

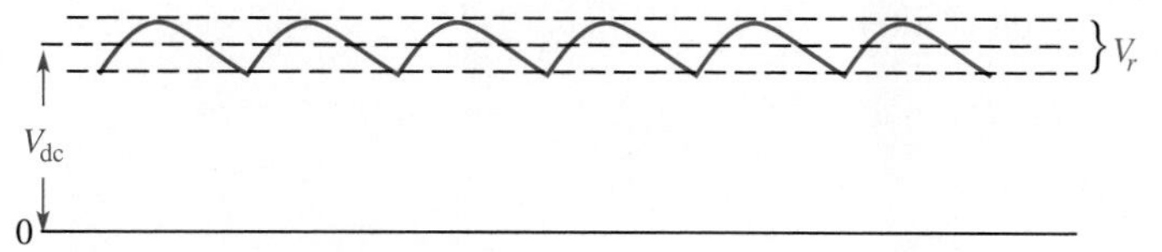

FIGURE 17–22
V_r and V_{dc} determine the ripple factor.

Surge Current in the Capacitor Filter Before the switch in Figure 17–23(a) is closed, the filter capacitor is uncharged. At the instant the switch is closed, voltage is connected to the bridge and the capacitor appears as a short, as shown. An initial surge of current is produced through the two forward-biased diodes. The worst-case situation occurs when the switch is closed at a peak of the secondary voltage and a maximum surge current $I_{S(max)}$ is produced, as illustrated in the figure.

It is possible that the surge current could destroy the diodes, and for this reason a surge-limiting resistor is sometimes connected, as shown in Figure 17–23(b). The value of this resistor must be small compared to R_L. Also, the diodes must have a forward current rating such that they can withstand the momentary surge of current.

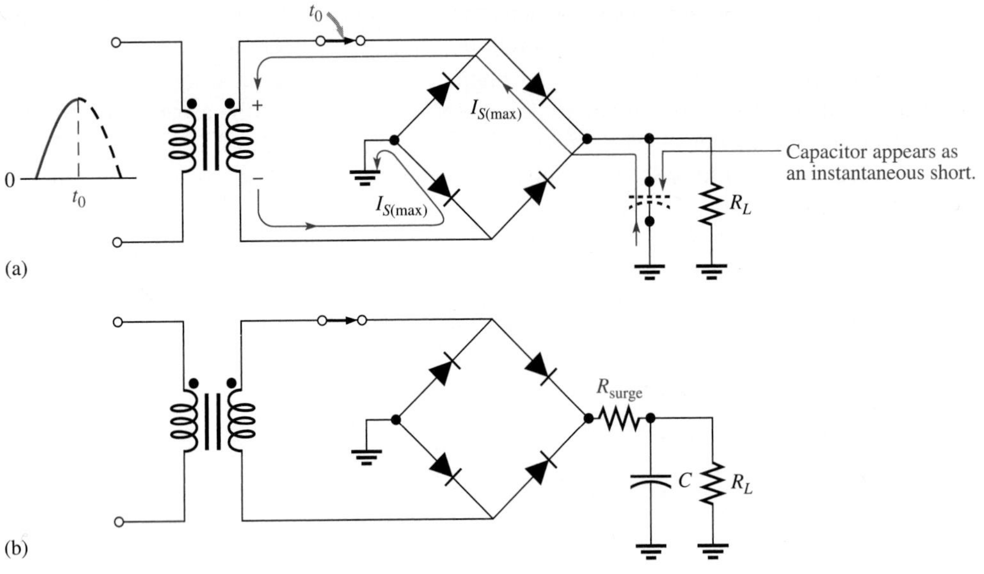

FIGURE 17–23
Surge current in a capacitor-input filter.

A general rule-of-thumb for estimating the value of the surge-limiting resistor is

$$R_{surge} > \frac{V_{p(sec)} - 1.4\text{ V}}{I_{FSM}} \tag{17–9}$$

where I_{FSM} is the specified nonrepetitive maximum forward surge current for the diode.

Inductor-Input Filter

When a choke is added to the filter input, as in Figure 17–24, a reduction in the ripple voltage is achieved. The choke has a high reactance at the ripple frequency, and the capacitive reactance is low compared to both X_L and R_L. The two reactances form an ac voltage divider that tends to significantly reduce the ripple from that of a straight capacitor-input filter.

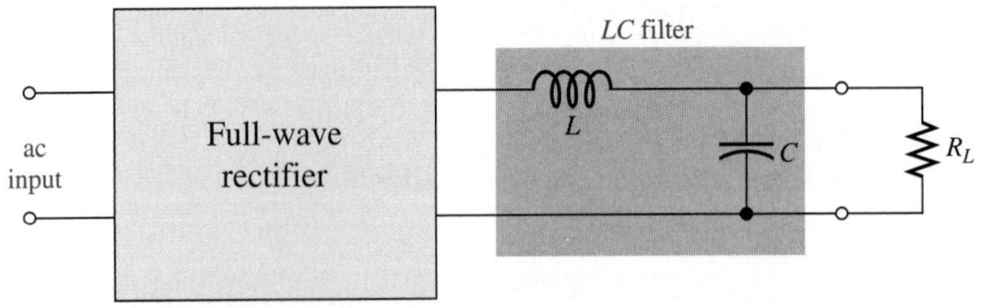

FIGURE 17–24
Rectifier with an LC filter.

It should be noted at this time that an inductor-input filter produces an output with a dc value approximately equal to the average value of the rectified input. The capacitor-input filter, however, produces an output with a dc value approximately equal to the peak value of the input.

Another point of comparison is that the amount of ripple voltage in the capacitor-input filter varies inversely with the load resistance. Ripple voltage in the *LC* filter is essentially independent of the load resistance and depends only on X_L and X_C, as long as X_C is sufficiently less than R_L.

π-Type and T-type Filters

A one-section π-type filter is shown in Figure 17–25(a). It can be thought of as a capacitor filter followed by an *LC* filter. The T-type filter in Figure 17–25(b) is basically an *LC* filter followed by an inductor filter. They combine the peak filtering action of the single-capacitor or inductor filter with the reduced ripple and load independence of the *LC* filter.

FIGURE 17–25
π-type and T-type LC filters.

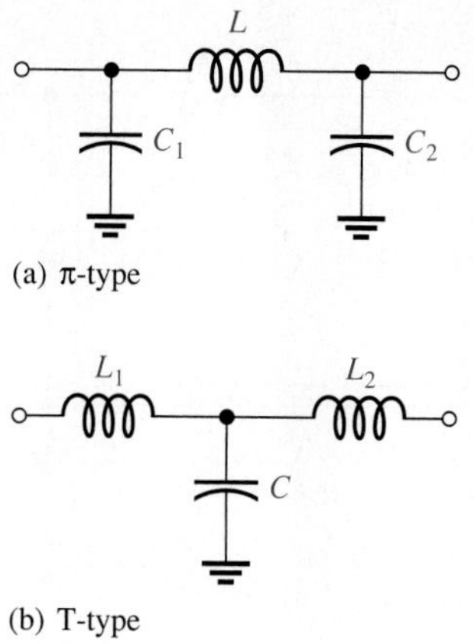

SECTION 17–3 REVIEW

1. A 60 Hz sine wave is applied to a half-wave rectifier. What is the output frequency? What is the output frequency for a full-wave rectifier?
2. What causes the ripple voltage on the output of a capacitor-input filter?
3. The load resistance of a capacitor-filtered full-wave rectifier is reduced. What effect does this reduction have on the ripple voltage?
4. Name one advantage of an *LC* filter over a capacitor filter. Name one disadvantage.

17–4 ■ DIODE CLIPPING AND CLAMPING CIRCUITS

Diode circuits are sometimes used to clip off portions of signal voltages above or below certain levels; these circuits are called clippers or limiters. Another type of diode circuit is used to restore a dc level to an electrical signal; these are called clampers. Both diode circuits will be examined in this section.

After completing this section, you should be able to

- **Analyze the operation of diode clippers and clampers**
  - □ Describe how both unbiased and biased clippers operate
  - □ Discuss a common clamper application

Diode Clippers

Diode **clippers** cut off voltage above or below specified levels. Figure 17–26(a) shows a diode circuit that clips off the positive part of the input signal. As the input signal goes positive, the diode becomes forward-biased. Since the cathode is at ground potential (0 V), the anode cannot exceed 0.7 V (assuming silicon). Thus, the output voltage across the diode is clipped at +0.7 V when the input exceeds this value.

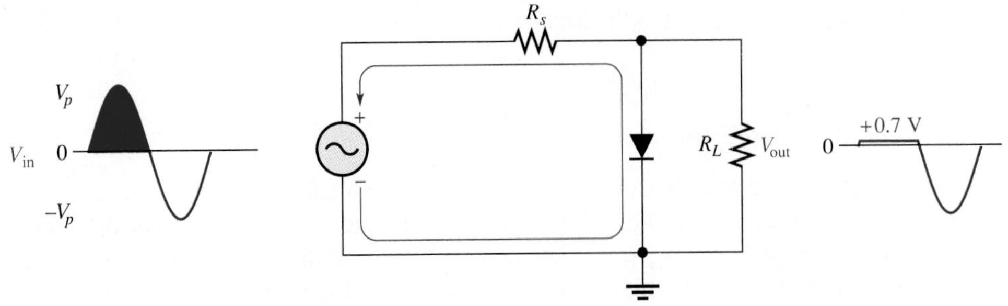

(a) Clipping of the positive alternation

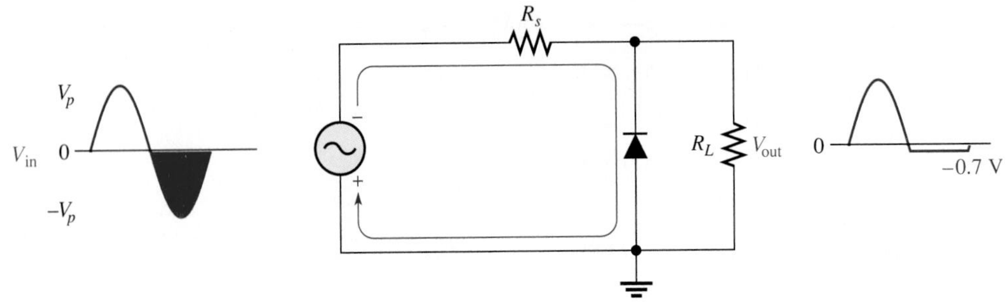

(b) Clipping of the negative alternation

FIGURE 17–26
Diode clipping (limiting) circuits.

When the input goes back below 0.7 V, the diode reverse-biases and appears as an open. The output voltage looks like the negative part of the input, but with a magnitude determined by the R_s and R_L voltage divider as follows:

$$V_{out} = \left(\frac{R_L}{R_s + R_L}\right)V_{in}$$

If R_s is small compared to R_L, then $V_{out} \cong V_{in}$. If the clipper circuit is unloaded, $R_L = \infty$ and $V_{out} = V_{in}$.

Turn the diode around, as in Figure 17–26(b), and the negative part of the input is clipped off. When the diode is forward-biased during the negative part of the input, the output voltage across the diode is held at −0.7 V by the diode drop. When the input goes above −0.7 V, the diode is no longer forward-biased and a voltage appears across R_L proportional to the input.

EXAMPLE 17–6

What would you expect to see displayed on an oscilloscope connected as shown in Figure 17–27? The time base on the scope is set to show one and one-half cycles.

FIGURE 17–27

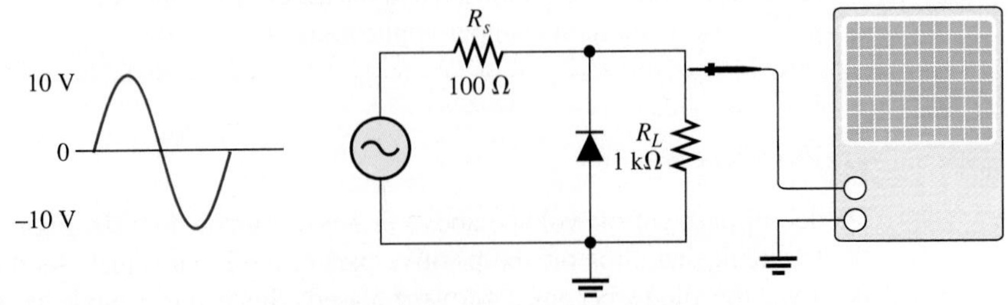

Solution The diode conducts when the input voltage goes below −0.77 V. Thus, you have a negative clipper with a peak output voltage determined by the following equation:

$$V_{p(\text{out})} = \left(\frac{R_L}{R_s + R_L}\right)V_{p(\text{in})} = \left(\frac{1\text{ k}\Omega}{1.1\text{ k}\Omega}\right)10\text{ V} = 9.09\text{ V}$$

The scope will display an output waveform as shown in Figure 17–28.

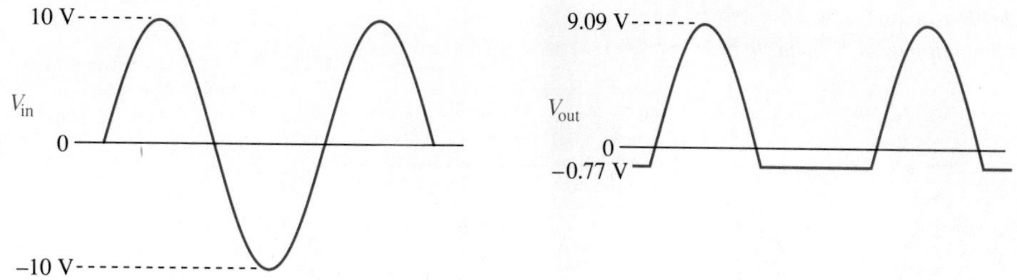

FIGURE 17–28
Waveforms for Figure 17–27.

Related Exercise Describe the output waveform for Figure 17–27 if the diode is germanium and R_L is changed to 680 Ω.

Adjustment of the Clipping Level To adjust the level at which a signal voltage is clipped, add a bias voltage in series with the diode, as shown in Figure 17–29. The voltage at point *A* must equal V_{BIAS} + 0.7 V before the diode will conduct. Once the diode begins to conduct, the voltage at point *A* with respect to ground is limited to V_{BIAS} + 0.7 V so that the output voltage above this level is clipped off, as shown in the figure.

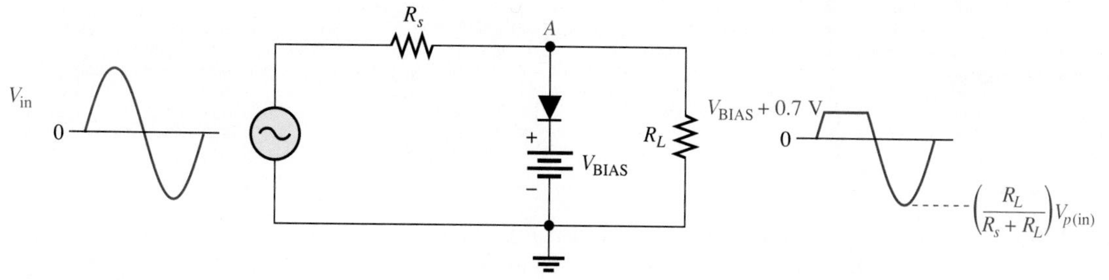

FIGURE 17–29
Positively biased diode clipper.

If the bias voltage is varied up or down, the clipping level changes correspondingly, as shown in Figure 17–30. If the polarity of the bias voltage is reversed, as in Figure 17–31, output voltages more positive than $-V_{\text{BIAS}}$ + 0.7 V are clipped, resulting in an output waveform as shown. The diode is reverse-biased only when the voltage at point *A* with respect to ground goes more negative than $-V_{\text{BIAS}}$ + 0.7 V.

If it is necessary to clip off voltage below a specified negative level, then the diode and bias battery must be connected as in Figure 17–32. In this case, the voltage at point *A* with respect to ground must go more negative than $-V_{\text{BIAS}}$ − 0.7 V to forward-bias the diode and initiate clipping action, as shown.

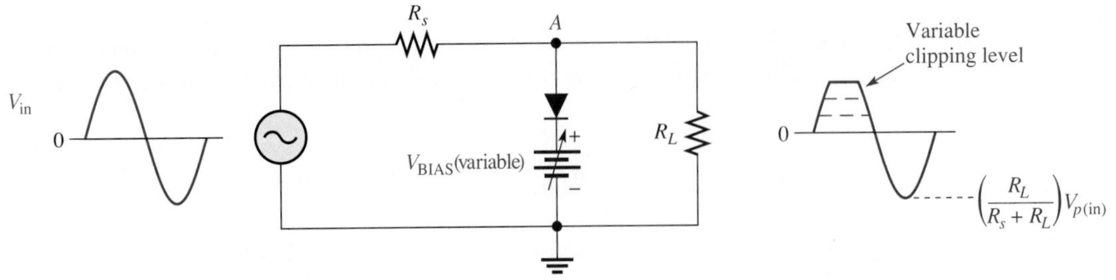

FIGURE 17–30
Positive clipper with variable bias.

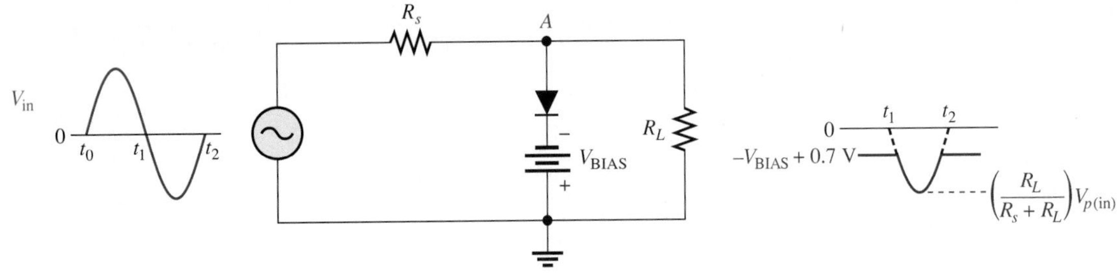

FIGURE 17–31
Positive clipper with bias voltage reversed.

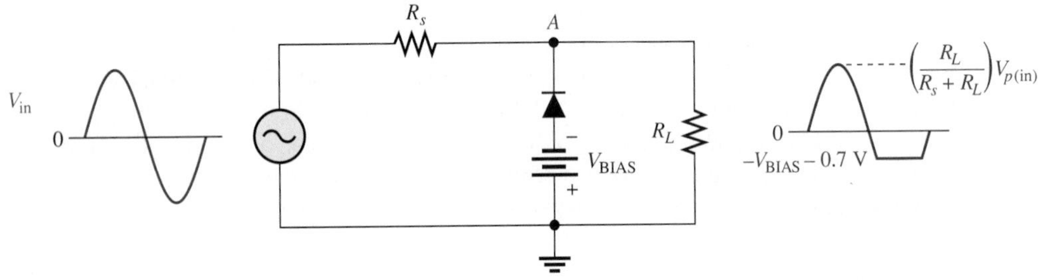

FIGURE 17–32
Negatively biased diode clipper.

EXAMPLE 17–7 Figure 17–33 shows an unloaded circuit combining a positive-biased clipper with a negative-biased clipper. Determine the output waveform.

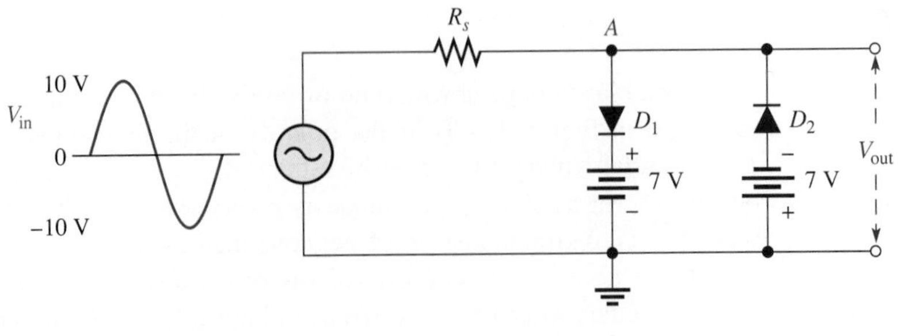

FIGURE 17–33

Solution When the voltage at point A with respect to ground reaches + 7.7 V, diode D_1 conducts and clips the waveform at +7.7 V. Diode D_2 does not conduct until the voltage reaches −7.7 V. Therefore, positive voltages above +7.7 V and negative voltages below −7.7 V are clipped off. The resulting output waveform is shown in Figure 17–34.

FIGURE 17–34
Output waveform for Figure 17–33.

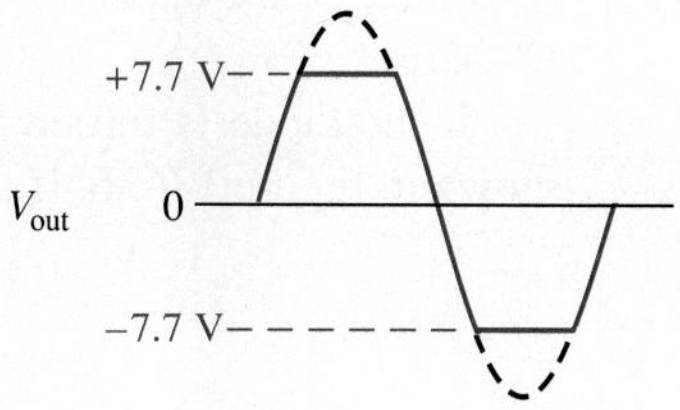

Related Exercise Determine the output voltage in Figure 17–33 if both dc bias sources are 10 V and the input has a peak value of 20 V.

Diode Clampers

A diode **clamper,** sometimes known as a *dc restorer,* adds a dc level to an ac signal. Figure 17–35 shows a diode clamper that inserts a positive dc level. To understand the operation of this circuit, start with the first negative half-cycle of the input voltage. When the input initially goes negative, the diode is forward-biased, allowing the capacitor to charge to near the peak of the input ($V_{p(in)}$ − 0.7 V), as shown in Figure 17–35(a). Just past the negative peak, the diode becomes reverse-biased because the cathode is held near $V_{p(in)}$ by the charge on the capacitor.

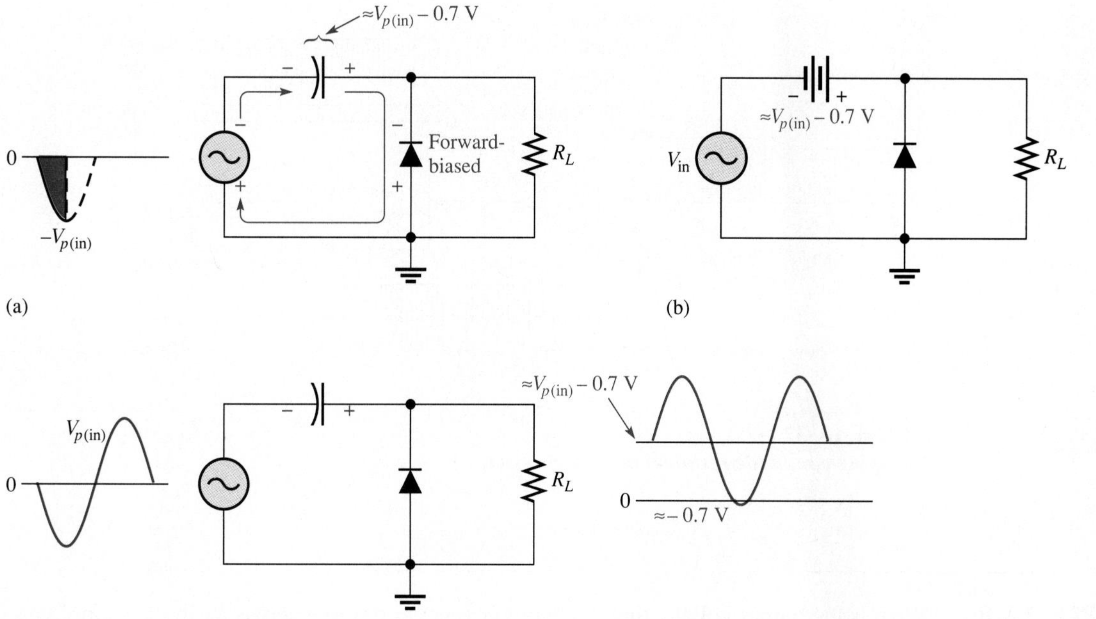

FIGURE 17–35
Positive clamping operation.

The capacitor can discharge only through the high resistance of R_L. Thus, from the peak of one negative half-cycle to the next, the capacitor discharges very little. The amount that is discharged, of course, depends on the value of R_L. For good clamping action, the RC time constant should be at least ten times the period of the input frequency.

The net effect of the clamping action is that the capacitor retains a charge approximately equal to the peak value of the input less the diode drop. The capacitor voltage acts essentially as a battery in series with the input signal, as shown in Figure 17–35(b). The dc voltage of the capacitor adds to the input voltage by superposition, as shown in Figure 17–35(c).

If the diode is turned around, a negative dc voltage is added to the input signal, as shown in Figure 17–36. If necessary, the diode can be biased to adjust the clamping level.

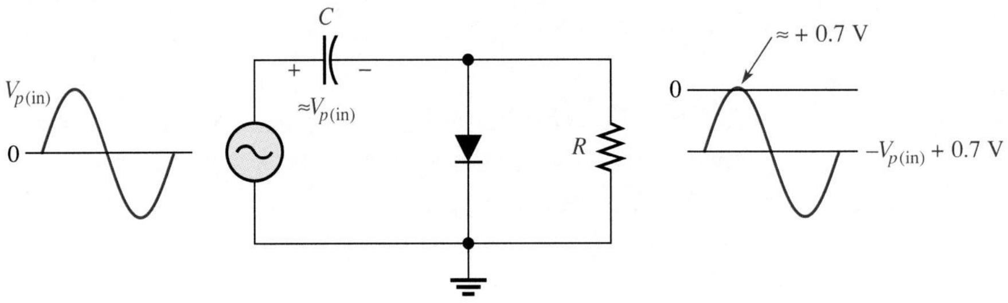

FIGURE 17–36
Negative clamping.

A Clamper Application A clamping circuit is often used in television receivers as a dc restorer. The incoming composite video signal is normally processed through capacitively coupled amplifiers that eliminate the dc component, thus losing the black and white reference levels and the blanking level. Before being applied to the picture tube, these reference levels must be restored. Figure 17–37 illustrates this process in a general way.

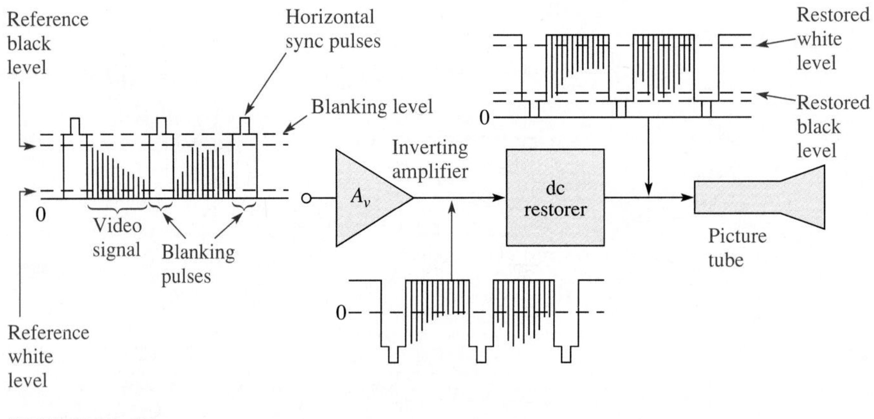

FIGURE 17–37
Clamping circuit (dc restorer) in a TV receiver.

EXAMPLE 17–8

What is the output voltage that you would expect to observe across R_L in the clamping circuit of Figure 17–38? Assume that RC is large enough to prevent significant capacitor discharge.

FIGURE 17–38

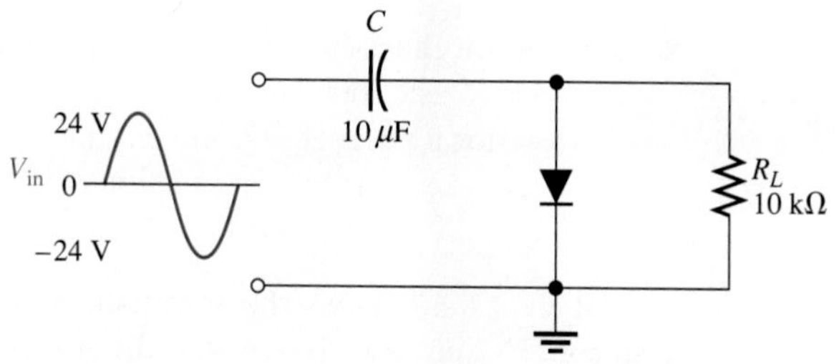

Solution Ideally, a negative dc value equal to the input peak less the diode drop is inserted by the clamping circuit:

$$V_{dc} = -(V_{p(in)} - 0.7\text{ V}) = -(24\text{ V} - 0.7\text{ V}) = -23.3\text{ V}$$

Actually, the capacitor will discharge slightly between peaks, and, as a result, the output voltage will have an average value of slightly less negative than that calculated above.

The output waveform goes to 0.7 V above ground, as shown in Figure 17–39.

FIGURE 17–39
Output waveform for Figure 17–38.

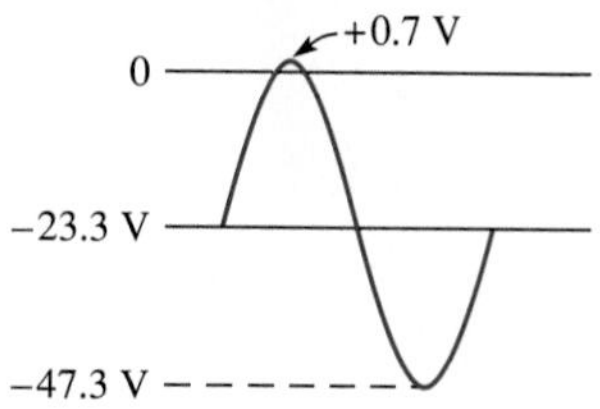

Related Exercise What is the output voltage that you would observe across R_L in Figure 17–38 if $C = 22\ \mu$F and $R_L = 18$ kΩ?

SECTION 17–4 REVIEW

1. Discuss how diode clippers and diode clampers differ in terms of their function.
2. What is the difference between a positive clipper and a negative clipper?
3. What is the maximum voltage across an unbiased positive silicon diode clipper during the positive alternation of the input voltage?
4. To limit the output of a positive clipper to 5 V when a 10 V peak input is applied, what value must the bias voltage be if silicon diodes are used in the circuit?
5. What component in a clamper circuit effectively acts as a battery?

17–5 ■ ZENER DIODES

A major application for zener diodes is voltage regulation in dc power supplies. In this section, you will see how the zener diode maintains a nearly constant dc voltage under the proper operating conditions. You will learn the conditions and limitations for properly using the zener diode and the factors that affect its performance.

After completing this section, you should be able to

- **Explain the characteristics and applications of a zener diode**
  - ☐ Discuss zener breakdown
  - ☐ Describe a zener equivalent circuit
  - ☐ Discuss zener voltage regulation

Figure 17–40 shows the schematic symbol for a zener diode. The **zener diode** is a silicon PN junction device that differs from the rectifier diode in that it is designed for operation in the reverse breakdown region. The breakdown voltage of a zener diode is set by carefully controlling the doping level during the manufacturing process. From the discussion of the diode characteristic curve in the last chapter, recall that when a diode reaches reverse breakdown, its voltage remains almost constant even though the current may change drastically. This volt-ampere characteristic is shown again in Figure 17–41.

FIGURE 17–40
Zener diode symbol.

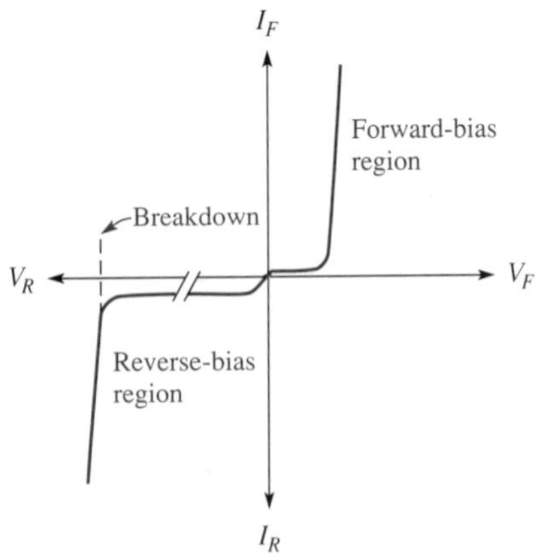

FIGURE 17–41
General diode characteristic.

Zener Breakdown

Two types of reverse breakdown in a zener diode are avalanche and zener. The avalanche breakdown also occurs in rectifier diodes at a sufficiently high reverse voltage. Zener breakdown occurs in a zener diode at low reverse voltages. A zener diode is heavily doped to reduce the breakdown voltage, causing a very narrow depletion layer. As a result, an intense electric field exists within the depletion layer. Near the breakdown voltage (V_Z), the field is intense enough to pull electrons from their valence bands and create current.

Zener diodes with breakdown voltages of less than approximately 5 V operate predominantly in zener breakdown. Those with breakdown voltages greater than approximately 5 V operate predominantly in avalanche breakdown. Both types, however, are called *zener diodes.* Zeners with breakdown voltages of 1.8 V to 200 V are commercially available.

Breakdown Characteristics Figure 17–42 shows the reverse portion of the characteristic curve of a zener diode. Notice that as the reverse voltage (V_R) is increased, the reverse current (I_R) remains extremely small up to the "knee" of the curve. At this point, the breakdown effect begins; the zener resistance (R_Z) begins to decrease as the current (I_Z) increases rapidly. From the bottom of the knee, the breakdown voltage (V_Z) remains es-

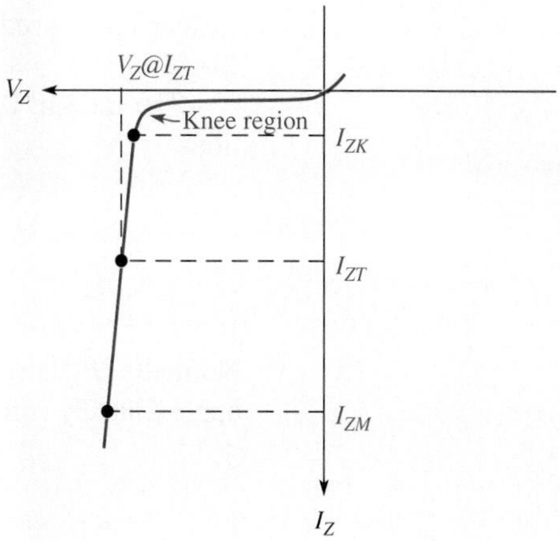

FIGURE 17–42
Reverse characteristic of a zener diode. V_Z is usually specified at the zener test current, I_{ZT}, and is designated V_{ZT}.

sentially constant. This regulating ability is the key feature of the zener diode: *A zener diode maintains an essentially constant voltage across its terminals over a specified range of reverse current values.*

A minimum value of reverse current, I_{ZK}, must be maintained in order to keep the diode in regulation. You can see on the curve that when the reverse current is reduced below the knee of the curve, the voltage changes drastically and regulation is lost. Also, there is a maximum current, I_{ZM}, above which the diode may be damaged.

Thus, basically, the zener diode maintains a nearly constant voltage across its terminals for values of reverse current ranging from I_{ZK} to I_{ZM}. A nominal zener test voltage, V_{ZT} is usually specified on a data sheet at a value of reverse current called the *zener test current,* I_{ZT}.

Zener Equivalent Circuit

Figure 17–43(a) shows the ideal approximation of a zener diode in reverse breakdown. It acts simply as a battery having a value equal to the zener voltage. Figure 17–43(b) represents the practical equivalent of a zener, where the zener resistance (R_Z) is included. The zener resistance is actually an ac resistance because it is dependent on the ratio of a change in voltage to a change in current and can be different for different portions of the characteristic curve. Since the voltage curve is not ideally vertical, a change in reverse

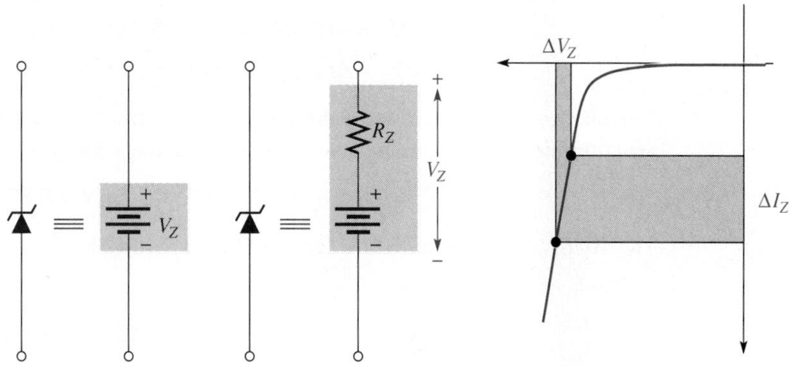

(a) Ideal (b) Practical (c) $R_Z = \dfrac{\Delta V_Z}{\Delta I_Z}$

FIGURE 17–43
Zener equivalent circuits.

current (ΔI_Z) produces a small change in zener voltage (ΔV_Z), as illustrated in Figure 17–43(c).

The ratio of ΔV_Z to ΔI_Z is the zener resistance, as expressed in the following equation:

$$R_Z = \frac{\Delta V_Z}{\Delta I_Z} \tag{17–10}$$

Normally, R_Z is specified at I_{ZT}, the zener test current. In most cases, this value of R_Z is approximately constant over the full range of reverse-current values.

EXAMPLE 17–9 A certain zener diode exhibits a 50 mV change in V_Z for a 2 mA change in I_Z. What is the zener resistance?

Solution

$$R_Z = \frac{\Delta V_Z}{\Delta I_Z} = \frac{50 \text{ mV}}{2 \text{ mA}} = 25\ \Omega$$

Related Exercise Calculate the zener resistance if the zener voltage changes 100 mV for a 20 mA change in zener current.

EXAMPLE 17–10 A certain zener diode has a resistance of 5 Ω. The data sheet gives $V_{ZT} = 6.8$ V at $I_{ZT} =$ 20 mA, $I_{ZK} = 1$ mA, and $I_{ZM} = 50$ mA. What is the voltage across its terminals when the current is 30 mA? What is the voltage when $I = 10$ mA?

Solution Figure 17–44 represents the diode. The 30 mA current is a 10 mA increase (ΔI_Z) above $I_{ZT} = 20$ mA.

$$\Delta I_Z = +10 \text{ mA}$$
$$\Delta V_Z = \Delta I_Z R_Z = (10 \text{ mA})(5\ \Omega) = +50 \text{ mV}$$

The change in voltage due to the increase in current above the I_{ZT} value causes the zener terminal voltage to increase. The zener voltage for $I_Z = 30$ mA is

$$V_Z = 6.8 \text{ V} + \Delta V_Z = 6.8 \text{ V} + 50 \text{ mV} = 6.85 \text{ V}$$

The 10 mA current is a 10 mA decrease below $I_{ZT} = 20$ mA.

$$\Delta I_Z = -10 \text{ mA}$$
$$\Delta V_Z = \Delta I_Z R_Z = (-10 \text{ mA})(5\ \Omega) = -50 \text{ mV}$$

The change in voltage due to the decrease in current below I_{ZT} causes the zener terminal voltage to decrease. The zener voltage for $I_Z = 10$ mA is

$$V_Z = 6.8 \text{ V} - \Delta V_Z = 6.8 \text{ V} - 50 \text{ mV} = 6.75 \text{ V}$$

FIGURE 17–44

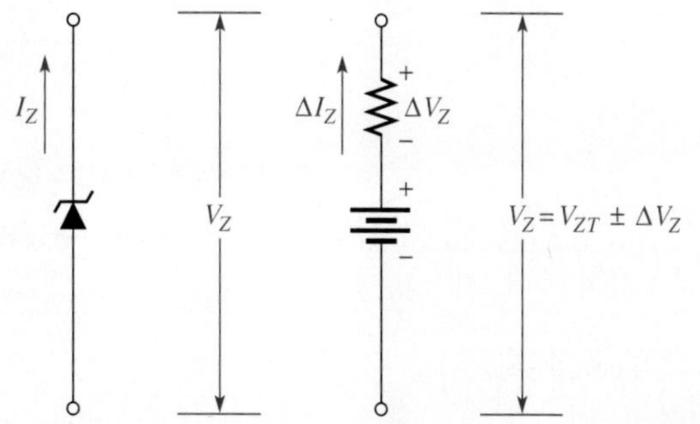

Related Exercise Repeat the analysis for a current of 20 mA and for a current of 80 mA using a diode with $V_{ZT} = 12$ V at $I_{ZT} = 50$ mA, $I_{ZK} = 0.5$ mA, $I_{ZM} = 100$ mA, and $R_Z = 20\ \Omega$.

Zener Voltage Regulation

Zener diodes are widely used for **voltage regulation.** Figure 17–45 illustrates how a zener diode can be used to regulate a varying dc voltage to keep it at a constant level. This process is called *input* or **line regulation.**

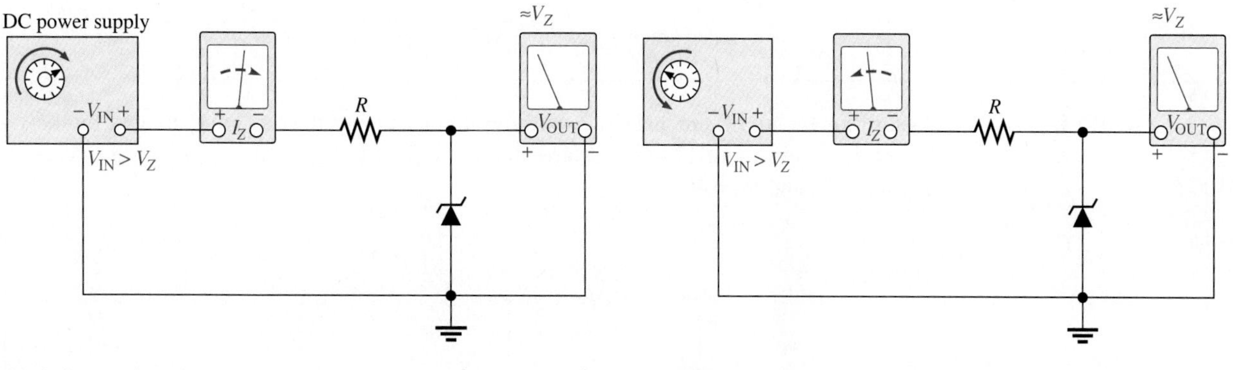

(a) As input voltage increases, V_{OUT} remains constant ($I_{ZK} < I_Z < I_{ZM}$). (b) As input voltage decreases, V_{OUT} remains constant ($I_{ZK} < I_Z < I_{ZM}$).

FIGURE 17–45
Zener regulation of a varying input voltage.

As the input voltage varies (within limits), the zener diode maintains an essentially constant voltage across the output terminals. However, as V_{IN} changes, I_Z will change proportionally, and therefore the limitations on the input variation are set by the minimum and maximum current values with which the zener can operate and on the condition that $V_{IN} > V_Z$. R is the series current-limiting resistor.

For example, suppose that the zener diode in Figure 17–46 can maintain regulation over a range of current values from 4 mA to 40 mA. For the minimum current, the voltage across the 1 kΩ resistor is

$$V_R = (4\text{ mA})(1\text{ k}\Omega) = 4\text{ V}$$

Since

$$V_R = V_{IN} - V_Z$$

then

$$V_{IN} = V_R + V_Z = 4\text{ V} + 10\text{ V} = 14\text{ V}$$

For the maximum current, the voltage across the 1 kΩ resistor is

$$V_R = (40\text{ mA})(1\text{ k}\Omega) = 40\text{ V}$$

Therefore,

$$V_{IN} = 40\text{ V} + 10\text{ V} = 50\text{ V}$$

As you can see, this zener diode can regulate an input voltage from 14 V to 50 V and maintain approximately a 10 V output. The output will vary slightly because of the zener resistance.

FIGURE 17–46

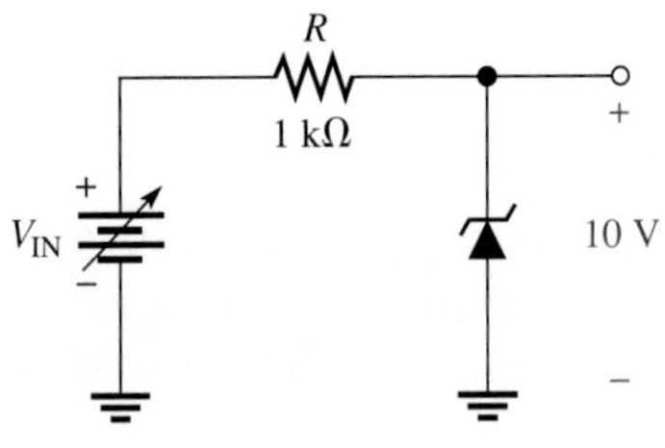

EXAMPLE 17–11 Determine the minimum and the maximum input voltages that can be regulated by the zener diode in Figure 17–47. Assume that $I_{ZK} = 1$ mA, $I_{ZM} = 15$ mA, $V_{ZT} = 5.1$ V at $I_{ZT} = 8$ mA, and $R_Z = 10\ \Omega$.

FIGURE 17–47

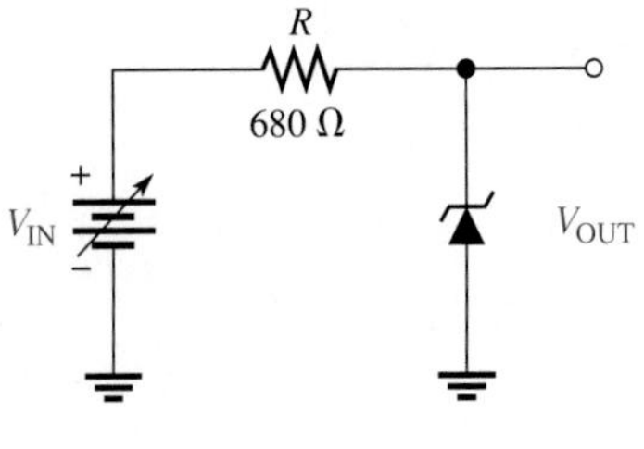

Solution The equivalent circuit is shown in Figure 17–48.

FIGURE 17–48
Equivalent of the circuit in Figure 17–47.

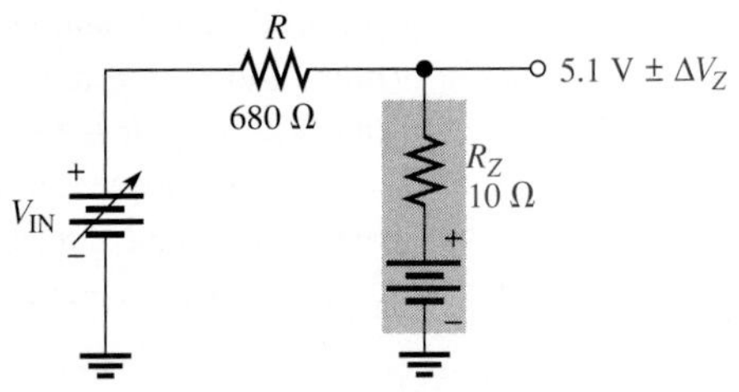

At $I_{ZK} = 1$ mA, the output voltage is

$$V_{OUT} = V_{ZT} - \Delta V_Z = 5.1\text{ V} - \Delta I_Z R_Z = 5.1\text{ V} - (I_{ZT} - I_{ZK})R_Z$$
$$= 5.1\text{ V} - (7\text{ mA})(10\ \Omega) = 5.1\text{ V} - 0.07\text{ V} = 5.03\text{ V}$$

Therefore,

$$V_{IN(min)} = I_{ZK}R + V_{OUT} = (1\text{ mA})(680\ \Omega) + 5.03\text{ V} = 5.71\text{ V}$$

At $I_{ZM} = 15$ mA, the output voltage is

$$V_{OUT} = 5.1\text{ V} + \Delta V_Z = 5.1\text{ V} + (7\text{ mA})(10\ \Omega) = 5.1\text{ V} + 0.07\text{ V} = 5.17\text{ V}$$

Therefore,

$$V_{IN(max)} = I_{ZM}R + V_{OUT} = (15\text{ mA})(680\ \Omega) + 5.17\text{ V} = 15.4\text{ V}$$

Related Exercise Determine the minimum and maximum input voltages that can be regulated by the zener diode in Figure 17–47. $I_{ZK} = 0.5$ mA, $I_{ZM} = 60$ mA, $V_{ZT} = 6.8$ V at $I_{ZT} = 30$ mA, and $R_Z = 18\ \Omega$.

Regulation with a Varying Load Figure 17–49 shows a zener regulator with a variable load resistor across the terminals. The zener diode maintains a constant voltage across R_L as long as the zener current is greater than I_{ZK} and less than I_{ZM}. This process is called **load regulation.**

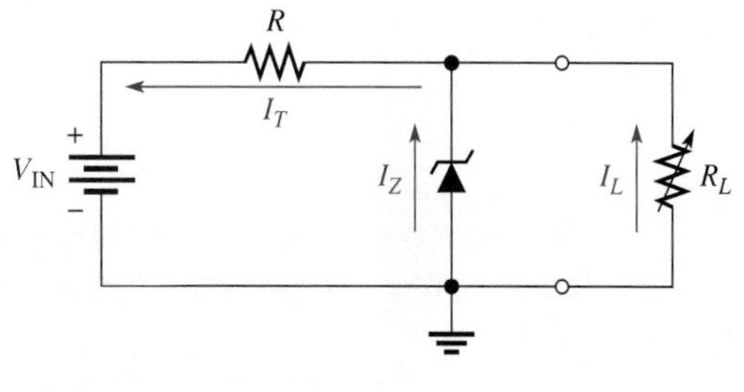

FIGURE 17–49
Zener regulation with a variable load.

No Load to Full Load When the output terminals are open ($R_L = \infty$), the load current is zero and all of the current is through the zener. When a load resistor is connected, part of the total current is through the zener and part through R_L.

As R_L is decreased, I_L goes up and I_Z goes down. The zener diode continues to regulate until I_Z reaches its minimum value, I_{ZK}. At this point the load current is maximum. The following example illustrates.

EXAMPLE 17–12 Determine the minimum and the maximum load currents for which the zener diode in Figure 17–50 will maintain regulation. What is the minimum R_L that can be used? $V_Z = 12$ V, $I_{ZK} = 3$ mA, and $I_{ZM} = 90$ mA. Assume that $R_Z = 0\ \Omega$ and V_Z remains a constant 12 V over the range of current values for simplicity.

FIGURE 17–50

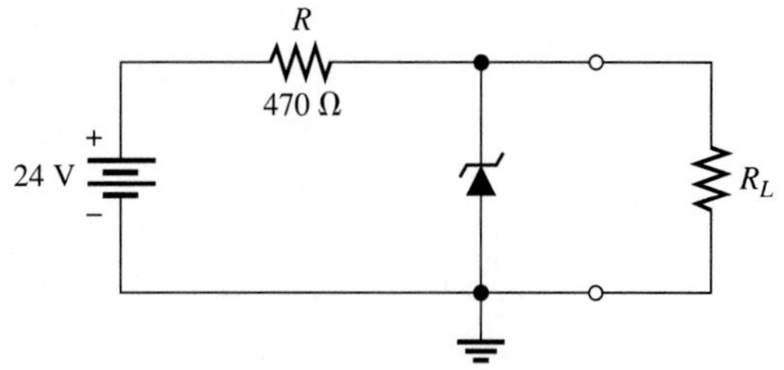

Solution When $I_L = 0$ A, I_Z is maximum and equal to the total circuit current I_T:

$$I_Z = \frac{V_{IN} - V_Z}{R} = \frac{24\text{ V} - 12\text{ V}}{470\ \Omega} = 25.5\text{ mA}$$

Since this is much less than I_{ZM}, 0 A is an acceptable minimum for I_L. That is,

$$I_{L(\text{min})} = 0\text{ A}$$

The maximum value of I_L occurs when I_Z is minimum (equal to I_{ZK}), so solve for $I_{L(\text{max})}$ as follows:

$$I_{L(\text{max})} = I_T - I_{Z(\text{min})} = 25.5\text{ mA} - 3\text{ mA} = 22.5\text{ mA}$$

The minimum value of R_L is

$$R_{L(\text{min})} = \frac{V_Z}{I_{L(\text{max})}} = \frac{12\text{ V}}{22.5\text{ mA}} = 533\ \Omega$$

Related Exercise Find the minimum and maximum load currents for which the circuit in Figure 17–50 will maintain regulation. Determine the minimum R_L that can be used. $V_Z = 3.3$ V (constant), $I_{ZK} = 2$ mA, $I_{ZM} = 75$ mA, and $R_Z = 0\ \Omega$.

Percent Regulation The percent regulation is a figure of merit used to specify the performance of a voltage **regulator.** It can be in terms of input (line) regulation or load regulation.

The *percent line regulation* specifies how much change occurs in the output voltage for a given change in input voltage. It is usually expressed as a percent change in V_{OUT} for a 1 V change in V_{IN} (%/V).

$$\text{Percent line regulation} = \frac{\Delta V_{OUT}}{\Delta V_{IN}} \times 100\% \qquad \textbf{(17–11)}$$

The *percent load regulation* specifies how much change occurs in the output voltage over a certain range of load current values, usually from minimum current (no load, NL) to maximum current (full load, FL). It is normally expressed as a percentage and can be calculated with the following formula:

$$\text{Percent load regulation} = \frac{V_{NL} - V_{FL}}{V_{FL}} \times 100\% \qquad \textbf{(17–12)}$$

where V_{NL} is the output voltage with no load, and V_{FL} is the output voltage with full (maximum) load.

EXAMPLE 17–13 A certain regulator has a no-load output voltage of 6 V and a full-load output voltage of 5.82 V. What is the percent load regulation?

Solution

$$\text{Percent load regulation} = \frac{V_{NL} - V_{FL}}{V_{FL}} \times 100\%$$

$$= \frac{6\text{ V} - 5.82\text{ V}}{5.82\text{ V}} \times 100\% = 3.09\%$$

Related Exercise

(a) If the no-load output voltage of a regulator is 24.8 V and the full-load output is 23.9 V, what is the percent load regulation?

(b) If the output voltage changes 0.035 V for a 2 V change in the input voltage, what is the line regulation expressed as %/V?

SECTION 17–5 REVIEW

1. Zener diodes are normally operated in the breakdown region (T or F).
2. A certain 10 V zener diode has a resistance of 8 Ω at 30 mA. What is the terminal voltage?
3. Explain the difference between line regulation and load regulation.
4. In a zener diode regulator, for what value of load resistance is the zener current a maximum?
5. A zener regulator has an output voltage of 12 V with no load and 11.9 V with full load. What is the percent load regulation?

17–6 ■ VARACTOR DIODES

Varactor diodes are also known as variable-capacitance diodes because the junction capacitance varies with the amount of reverse-bias voltage. Varactors are specifically designed to take advantage of this variable-capacitance characteristic. The capacitance can be changed by changing the reverse voltage. These devices are commonly used in electronic tuning circuits used in communications systems.

After completing this section, you should be able to

- **Describe the basic operation of a varactor diode**
  - □ Explain how a varactor produces a variable capacitance characteristic
  - □ Discuss some varactor applications

A **varactor** is basically a reverse-biased PN junction that utilizes the inherent capacitance of the depletion layer. The depletion layer, created by the reverse bias, acts as a capacitor dielectric because of its nonconductive characteristic. The P and N regions are conductive and act as the capacitor plates, as illustrated in Figure 17–51.

When the reverse-bias voltage increases, the depletion layer widens, effectively increasing the dielectric thickness (d) and thus decreasing the capacitance. When the reverse-bias voltage decreases, the depletion layer narrows, thus increasing the capacitance. This action is shown in Figures 17–52(a) and 17–52(b). A general curve of capacitance versus voltage is shown in Figure 17–52(c).

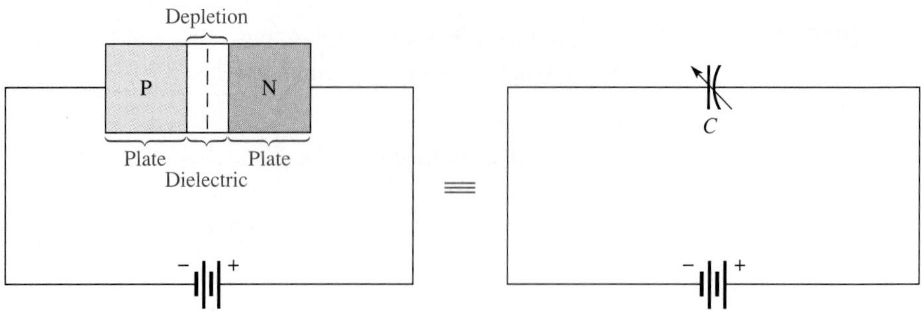

FIGURE 17–51
The reverse-biased varactor diode acts as a variable capacitor.

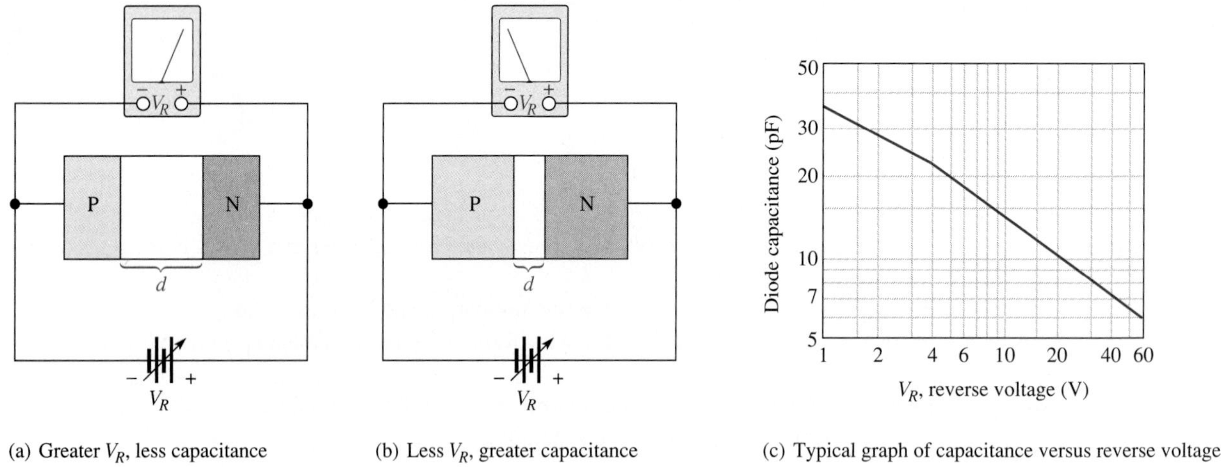

FIGURE 17–52
Varactor diode capacitance varies with reverse voltage.

Recall that capacitance is determined by the plate area (A), dielectric constant (ϵ), and dielectric thickness, (d) as expressed in the following formula:

$$C = \frac{A\epsilon}{d} \tag{17–13}$$

In a varactor diode, the capacitance parameters are controlled by the method of doping in the depletion layer and the size and geometry of the diode's construction. Varactor capacitances typically range from a few picofarads to a few hundred picofarads.

Figure 17–53(a) shows a common symbol for a varactor, and part (b) shows a simplified equivalent circuit. R_S is the reverse series resistance, and C_V is the variable capacitance.

FIGURE 17–53
Varactor diode.

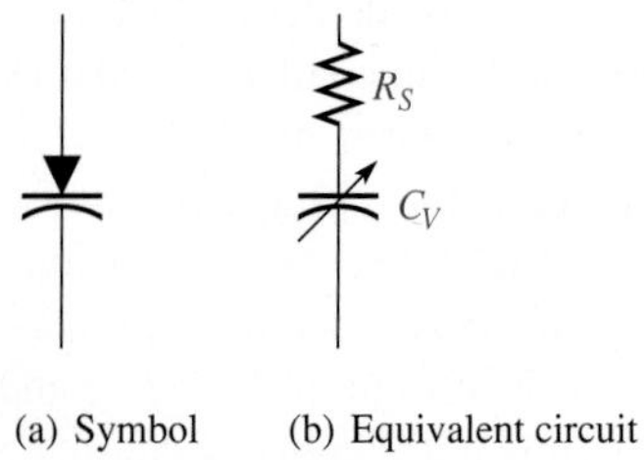

Applications

A major application of varactors is in tuning circuits. For example, electronic tuners in TV and other commercial receivers utilize varactors as one of their elements.

When used in a resonant circuit, the varactor acts as a variable capacitor, thus allowing the resonant frequency to be adjusted by a variable voltage level, as illustrated in Figure 17–54 where two varactor diodes provide the total variable capacitance in a parallel resonant (tank) circuit. A variable dc voltage (V_C) controls the reverse bias and therefore the capacitance of the diodes.

FIGURE 17–54
Varactors in a resonant circuit.

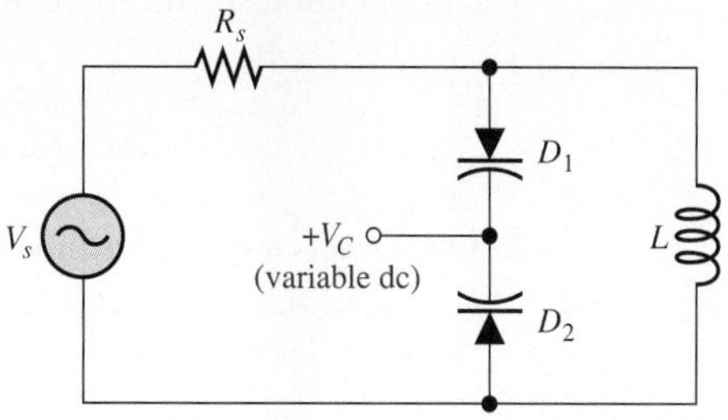

Recall that the resonant frequency of the tank circuit is

$$f_r = \frac{1}{2\pi\sqrt{LC}} \tag{17–14}$$

This approximation is valid for $Q > 10$.

EXAMPLE 17–14 The capacitance of a certain varactor can be varied from 5 pF to 50 pF. The diode is used in a tuned circuit similar to that shown in Figure 17–54. Determine the tuning range for the circuit if $L = 10$ mH.

Solution The equivalent circuit is shown in Figure 17–55. Notice that the varactor capacitances are in series.

FIGURE 17–55

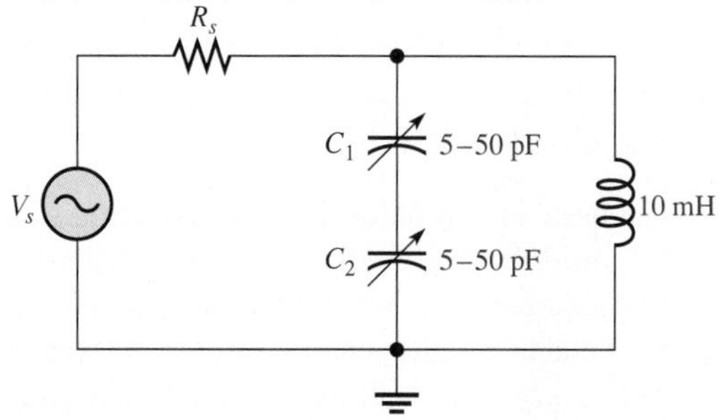

The minimum total capacitance is

$$C_{T(\min)} = \frac{C_{1(\min)}C_{2(\min)}}{C_{1(\min)} + C_{2(\min)}} = \frac{(5\text{ pF})(5\text{ pF})}{10\text{ pF}} = 2.5\text{ pF}$$

The maximum resonant frequency therefore is

$$f_{r(\max)} = \frac{1}{2\pi\sqrt{LC_{T(\min)}}} = \frac{1}{2\pi\sqrt{(10\text{ mH})(2.5\text{ pF})}} = 1\text{ MHz}$$

The maximum total capacitance is

$$C_{T(\max)} = \frac{C_{1(\max)}C_{2(\max)}}{C_{1(\max)} + C_{2(\max)}} = \frac{(50\text{ pF})(50\text{ pF})}{100\text{ pF}} = 25\text{ pF}$$

The minimum resonant frequency therefore is

$$f_{r(\min)} = \frac{1}{2\pi\sqrt{LC_{T(\max)}}} = \frac{1}{2\pi\sqrt{(10\text{ mH})(25\text{ pF})}} = 318\text{ kHz}$$

Related Exercise Determine the tuning range for Figure 17–55 if $L = 100$ mH.

SECTION 17–6 REVIEW

1. What is the purpose of a varactor diode?
2. Based on the general curve in Figure 17–52(c), what happens to the diode capacitance when the reverse voltage is increased?

17–7 ■ LEDs AND PHOTODIODES

In this section, two types of optoelectronic devices—the light-emitting diode (LED) and the photodiode—are introduced. As the name implies, the LED is a light emitter. The photodiode, on the other hand, is a light detector. We will examine the characteristics of both devices, and you will see an example of their use in a system application at the end of the chapter.

After completing this section, you should be able to

- **Discuss the operation and application of LEDs and photodiodes**
  - ☐ Describe electroluminescence
  - ☐ List the semiconductor materials used in LEDs
  - ☐ Explain how the photodiode can be used as a variable resistance device

The Light-Emitting Diode (LED)

The basic operation of the **light-emitting diode (LED)** is as follows: When the device is forward-biased, electrons cross the PN junction from the N-type material and recombine with holes in the P-type material. Recall that these free electrons are in the conduction band and at a higher energy level than the holes in the valence band. When recombination takes place, the recombining electrons release energy in the form of heat and light. A large exposed surface area on one layer of the semiconductor material permits the photons to be emitted as visible light. Figure 17–56 illustrates this process which is called *electroluminescence*.

The semiconductor materials used in LEDs are gallium arsenide (GaAs), gallium arsenide phosphide (GaAsP), and gallium phosphide (GaP). Silicon and germanium are not used because they are essentially heat-producing materials and are very poor at producing light. GaAs LEDs emit infrared (IR) radiation, GaAsP produces either red or yellow light, and GaP emits red or green light. The symbol for an LED is shown in Figure 17–57.

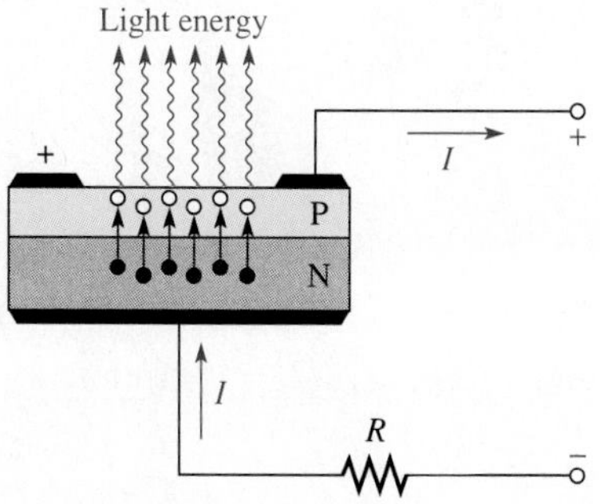

FIGURE 17–56
Electroluminescence in an LED.

FIGURE 17–57
Symbol for an LED.

The LED emits light in response to a sufficient forward current, as shown in Figure 17–58(a). The amount of light output is directly proportional to the forward current, as indicated in Figure 17–58(b). Typical LEDs are shown in part (c).

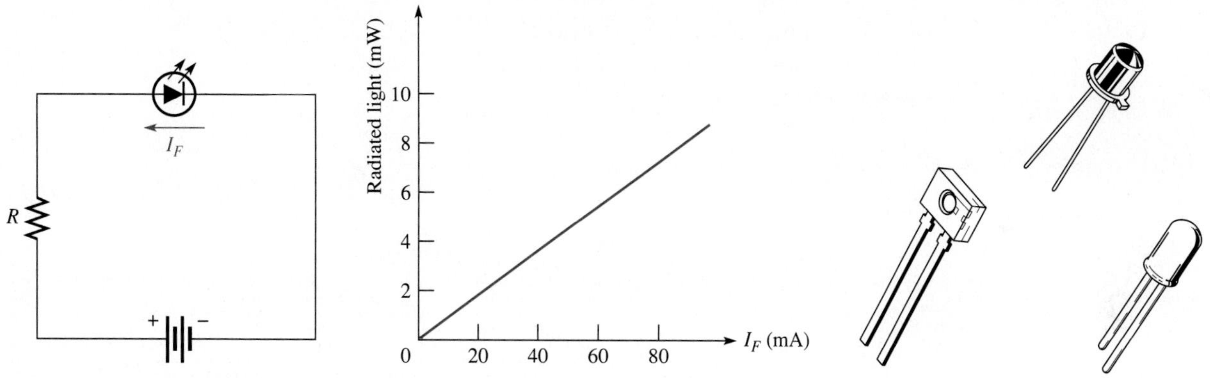

(a) Forward-biased operation

(b) Typical light output versus forward current

(c) Typical LEDs

FIGURE 17–58
Light-emitting diodes (LEDs).

Applications LEDs are commonly used for indicator lamps and readout displays on a wide variety of instruments, ranging from consumer appliances to scientific apparatus. A common type of display device using LEDs is the 7-segment display. Combinations of the segments form the ten decimal digits. Also, IR-emitting diodes are employed in optical coupling applications, often in conjunction with fiber optics.

The Photodiode

The **photodiode** is a PN junction device that operates in reverse bias, as shown in Figure 17–59(a). Note the schematic symbol for the photodiode. The photodiode has a small transparent window that allows light to strike the PN junction. Typical photodiodes are shown in Figure 17–59(b), and an alternate symbol is shown in part (c).

Recall that when reverse-biased, a rectifier diode has a very small reverse leakage current. The same is true for the photodiode. The reverse-biased current is produced by thermally generated electron hole pairs in the depletion layer, which are swept across the junction by the electric field created by the reverse voltage. In a rectifier diode, the reverse current increases with temperature due to an increase in the number of electron hole pairs.

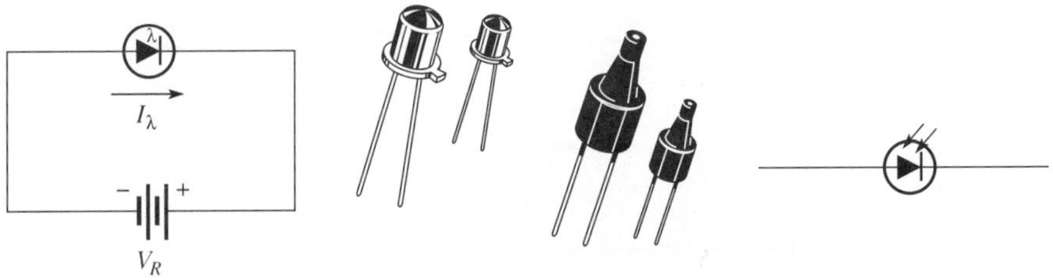

(a) Reverse-bias operation (b) Typical devices (c) Alternate symbol

FIGURE 17–59
Photodiode.

In a photodiode, the reverse current increases with the light intensity at the PN junction. When there is no incident light, the reverse current (I_λ) is almost negligible and is called the *dark current*. An increase in the amount of light energy (measured in lumens per square meter, lm/m^2) produces an increase in the reverse current, as shown by the graph in Figure 17–60(a). For a given value of reverse-bias voltage, Figure 17–60(b) shows a set of characteristic curves for a typical photodiode.

FIGURE 17–60
Typical photodiode characteristics.

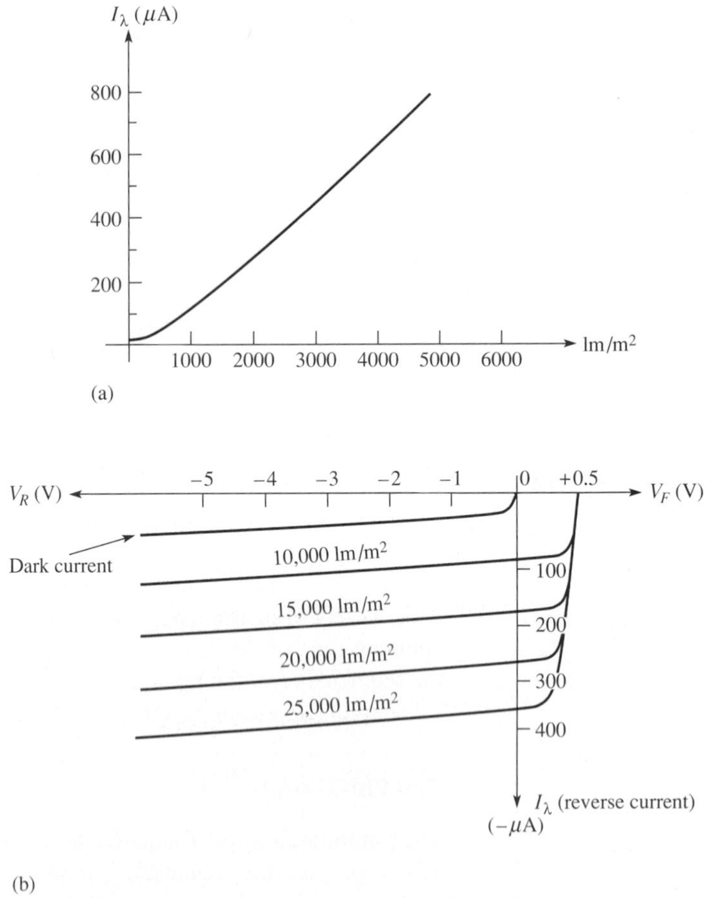

From the characteristic curve in Figure 17–60(b), the dark current for this particular device is approximately 25 μA at a reverse-bias voltage of 3 V. Therefore, the reverse resistance of the device with no incident light is

$$R_R = \frac{V_R}{I_\lambda} = \frac{3\text{ V}}{25\ \mu\text{A}} = 120\text{ k}\Omega$$

At 25,000 lm/m^2, the current is approximately 375 μA at −3 V. The reverse resistance under this condition is

$$R_R = \frac{V_R}{I_\lambda} = \frac{3\text{ V}}{375\ \mu\text{A}} = 8\text{ k}\Omega$$

These calculations show that the photodiode can be used as a variable-resistance device controlled by light intensity.

Figure 17–61 illustrates that the photodiode allows essentially no reverse current (except for a very small dark current) when there is no incident light. When a light beam strikes the photodiode, it conducts an amount of reverse current that is proportional to the light intensity.

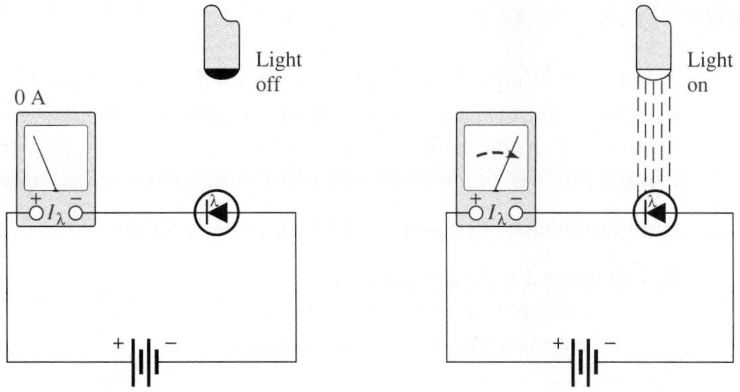

(a) No light, no current

(b) Where there is incident light, reverse current flows.

FIGURE 17–61
Operation of a photodiode.

A simple photodiode application is depicted in Figure 17–62. Here a beam of light continuously passes across a conveyor belt and into a transparent window behind which is a photodiode circuit. When the light beam is interrupted by an object passing by on the conveyor belt, the sudden reduction in diode current activates a control circuit that ad-

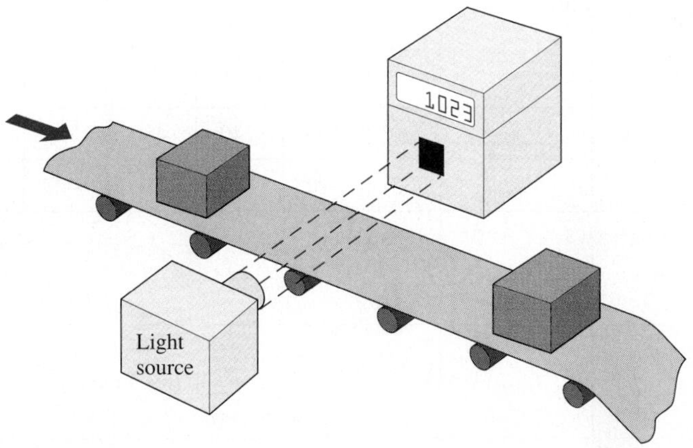

FIGURE 17–62
A photodiode circuit used in a system that counts objects as they pass on a conveyor belt.

vances a counter by one. The total count of objects that have passed that point is displayed by the counter. This basic concept can be extended and used for production control, shipping, and monitoring of activity on production lines.

SECTION 17–7 REVIEW

1. How does an LED differ from a photodiode?
2. List the semiconductor materials used in LEDs.
3. There is a very small reverse current in a photodiode under no-light conditions. What is this current called?

17–8 ▪ THE DIODE DATA SHEET

A manufacturer's data sheet gives detailed information on a device so that it can be used properly in a given application. A typical data sheet provides maximum ratings, electrical characteristics, mechanical data, and graphs of various parameters. We use a specific rectifier diode as an example to illustrate a typical data sheet.

After completing this section, you should be able to

- **Interpret a typical data sheet**
  - □ Determine maximum ratings
  - □ Determine electrical characteristics

Table 17–1 shows the maximum ratings for a certain series of rectifier diodes (1N4001 through 1N4007). These are the absolute maximum values under which the diode can be operated without damage to the device. For greatest reliability and longer life, the diode should always be operated well under these maximums. Generally, the maximum ratings are specified at 25°C and must be adjusted downward for greater temperatures.

TABLE 17–1
Maximum ratings

RATING	SYMBOL	1N4001	1N4002	1N4003	1N4004	1N4005	1N4006	1N4007	UNIT
Peak repetitive reverse voltage Working peak reverse voltage DC blocking voltage	V_{RRM} V_{RWM} V_R	50	100	200	400	600	800	1000	V
Nonrepetitive peak reverse voltage	V_{RSM}	60	120	240	480	720	1000	1200	V
rms reverse voltage	$V_{R(rms)}$	35	70	140	280	420	560	700	V
Average rectified forward current (single-phase, resistive load, 60 Hz, $T_A = 75°C$)	I_0	1.0							A
Nonrepetitive peak surge current (surge applied at rated load conditions)	I_{FSM}	30 (for 1 cycle)							A
Operating and storage junction temperature range	T_J, T_{stg}	−65 to +175							°C

An explanation of some of the parameters from Table 17–1 follows:

V_{RRM} The maximum reverse peak voltage that can be applied repetitively across the diode. Notice that in this case, it is 50 V for the 1N4001 and 1 kV for the 1N4007.

V_R The maximum reverse dc voltage that can be applied across the diode.

V_{RSM} The maximum reverse peak value of nonrepetitive (one-cycle) voltage that can be applied across the diode.

I_0 The maximum average value of a 60 Hz full-wave rectified forward current.

I_{FSM} The maximum peak value of nonrepetitive (one-cycle) forward current. The graph in Figure 17–63 expands on this parameter to show values for more than one cycle at temperatures of 25°C and 175°C. The dashed lines represent values where typical failures occur.

FIGURE 17–63
Nonrepetitive surge capability.

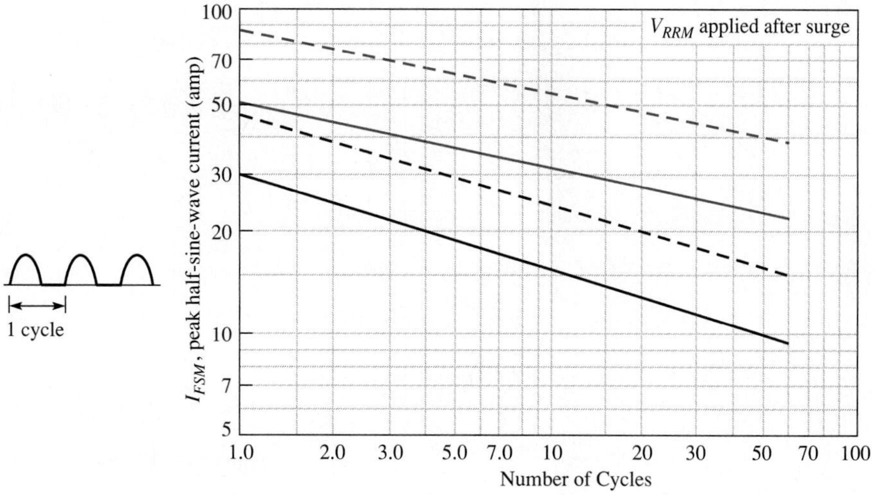

– – – Typical failures when surge applied at no-load conditions $T_J = 25°C$

—— Design limits when surge applied at no-load conditions $T_J = 25°C$

– – – Typical failures when surge applied at rated-load conditions $T_J = 175°C$

—— Design limits when surge applied at rated-load conditions $T_J = 175°C$

Notice what happens on the lower solid line when ten cycles of I_{FSM} are applied. The limit is 15 A rather than the one-cycle value of 30 A.

Table 17–2 lists typical and maximum values of certain electrical characteristics. These items differ from the maximum ratings in that they are not selected by design but are the result of operating the diode under specified conditions. A brief explanation of these parameters follows:

v_F The instantaneous voltage across the forward-biased diode when the forward current is 1 A at 25°C. Figure 17–64 shows how the forward voltages vary with forward current.

$V_{F(avg)}$ The maximum forward voltage drop averaged over a full cycle.

I_R The maximum current when the diode is reverse-biased with a dc voltage.

$I_{R(avg)}$ The maximum reverse current averaged over one cycle (when reverse-biased with an ac voltage).

The mechanical data for these particular diodes as they appear on a typical data sheet are shown in Figure 17–65.

TABLE 17–2
Electrical characteristics

CHARACTERISTICS AND CONDITIONS	SYMBOL	TYPICAL	MAXIMUM	UNIT
Maximum instantaneous forward voltage drop ($I_F = 1$ A, $T_J = 25°$C)	v_F	0.93	1.1	V
Maximum full-cycle average forward voltage drop ($I_0 = 1$ A, $T_L = 75°$C)	$V_{F(avg)}$	—	0.8	V
Maximum reverse current (rated dc voltage)	I_R			μA
$T_J = 25°$C		0.05	10.0	
$T_J = 100°$C		1.0	50.0	
Maximum full-cycle average reverse current ($I_0 = 1$ A, $T_L = 75°$C)	$I_{R(avg)}$	—	30.0	μA

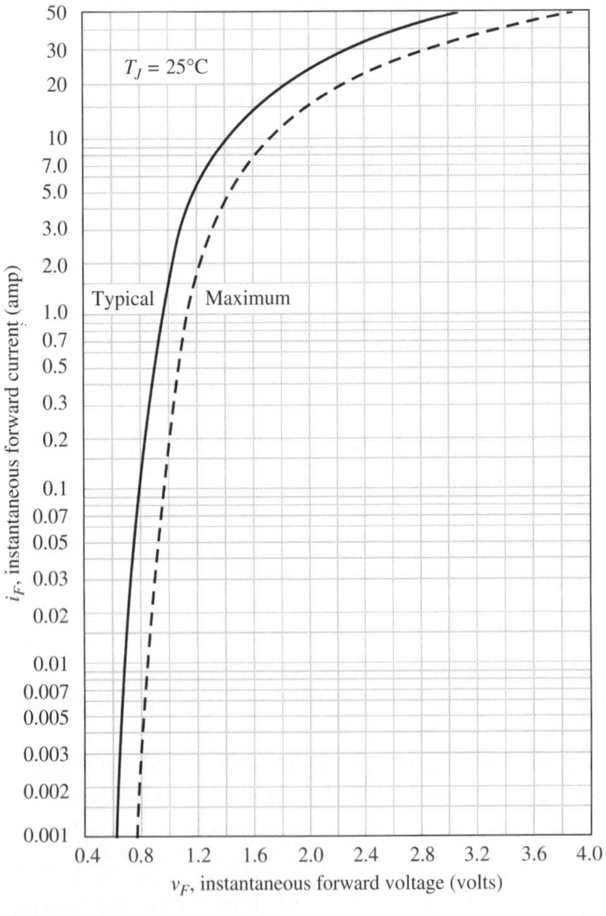

FIGURE 17–64
Forward voltage.

	Millimeters		Inches	
DIM	Min	Max	Min	Max
A	5.97	6.60	0.235	0.260
B	2.79	3.05	0.110	0.120
D	0.76	0.86	0.030	0.034
K	27.94	-	1.100	-

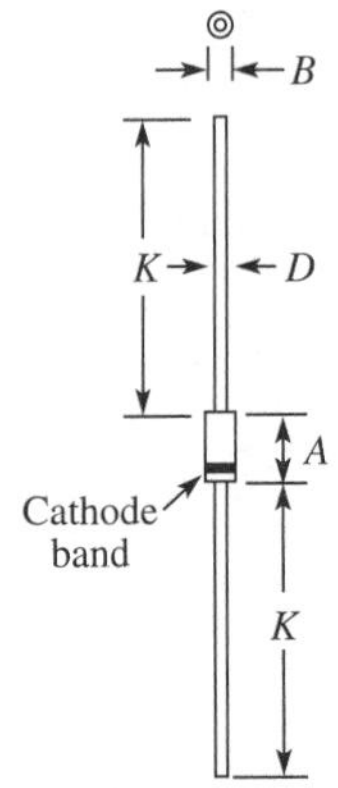

Mechanical characteristics

Case: Transfer molded plastic
Maximum lead temperature for soldering purposes: 350°C, 3/8″ from case for 10 seconds at 5 lbs. tension
Finish: all external surfaces are corrosion-resistant, leads are readily solderable
Polarity: Cathode indicated by band
Weight: 0.40 grams (approximately)

FIGURE 17–65
Mechanical data.

SECTION 17–8 REVIEW

1. List the three diode ratings categories.
2. Identify each of the following parameters: **(a)** V_F **(b)** I_R **(c)** I_0

17–9 ■ TROUBLESHOOTING

Several types of failures can occur in power supply rectifiers. In this section, we will examine some possible failures and the effects they would have on a circuit's operation.

After completing this section, you should be able to

- **Troubleshoot rectifier circuits**
  - □ Determine an open diode
  - □ Determine a shorted diode
  - □ Determine an open filter capacitor
  - □ Determine a shorted or leaky filter capacitor

Open Diode

A half-wave rectifier with a diode that has opened (a common failure mode) is shown in Figure 17–66. In this case, you would measure 0 V dc across the load resistor, as depicted.

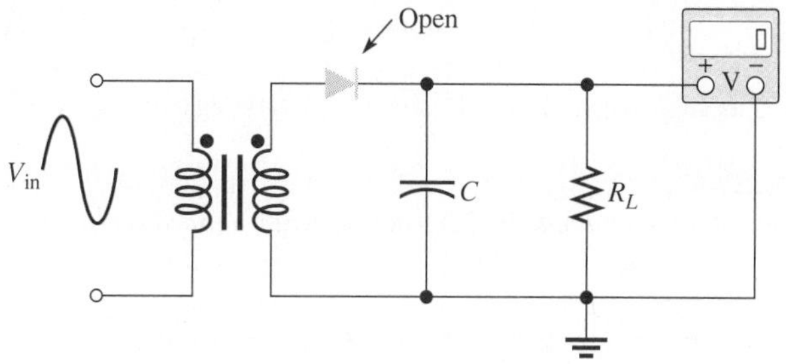

FIGURE 17–66
Test for an open diode in a half-wave rectifier with capacitor-input filter.

Now consider the full wave, center-tapped rectifier in Figure 17–67. Assume that diode D_1 has failed open. With an oscilloscope connected to the output, as shown in part (a), you would observe the following: You would see a larger-than-normal ripple voltage at a frequency of 60 Hz rather than 120 Hz. Disconnecting the filter capacitor, you would observe a half-wave rectified voltage, as in part (b). Now let's examine the reason for these observations. If diode D_1 is open, there will be current through R_L only during the negative half-cycle of the input signal. During the positive half-cycle, an open path prevents current through R_L. The result is a half-wave voltage, as illustrated.

With the filter capacitor in the circuit, the half-wave signal will allow it to discharge more than it would with a normal full-wave signal, resulting in a larger ripple voltage. Basically, the same observations would be made for an open failure of diode D_2.

An open diode in a bridge rectifier would create symptoms identical to those just discussed for the center-tapped rectifier. As illustrated in Figure 17–68, the open diode would prevent current through R_L during half of the input cycle (in this case, the negative half). As a result, there would be a half-wave output and an increased ripple voltage at 60 Hz, as discussed before.

Shorted Diode

A shorted diode is one that has failed such that it has a very low resistance in both directions. If a diode suddenly became shorted, it is likely that a sufficiently high current

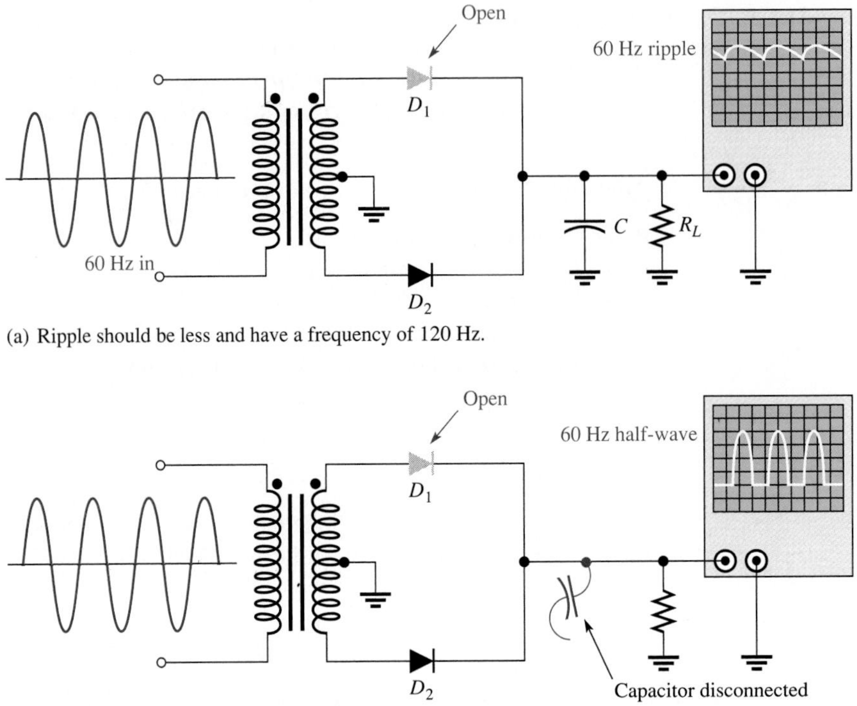

(a) Ripple should be less and have a frequency of 120 Hz.

(b) With C removed, output should be a full-wave 120 Hz signal.

FIGURE 17–67
Symptoms of an open diode in a full-wave, center-tapped rectifier.

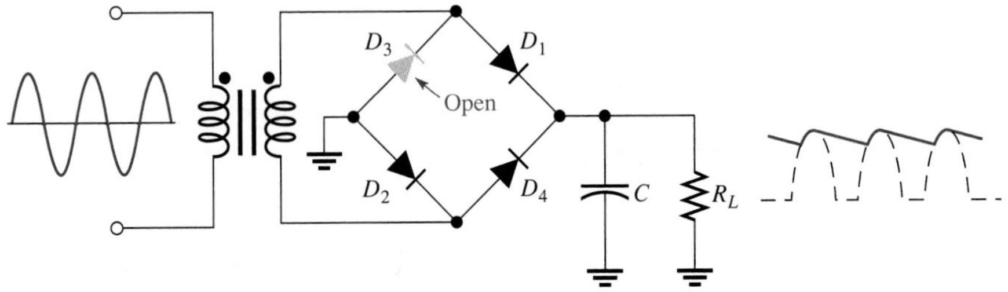

FIGURE 17–68
Effect of an open diode in a bridge rectifier.

would exist during one half of the input cycle such that the fuse in the primary circuit would open. An unfused primary could cause the shorted diode to burn open or the other diode in series with it to burn open. Also, the transformer could be damaged, as illustrated for the case of a bridge rectifier in Figure 17–69 with D_1 shorted.

In part (a) of Figure 17–69, current is supplied to the load through the shorted diode during the first positive half-cycle, just as though it were forward-biased. During the negative half-cycle, the current is shorted through D_1 and D_4, as shown in part (b). Again, damage to the transformer is possible. Also, it is likely that this excessive current would burn either or both of the diodes open. If only one of the diodes opened, you would still observe a half-wave voltage on the output. If both diodes opened, there would be no voltage developed across the load. These conditions are illustrated in Figure 17–70.

Open Filter Capacitor

An open filter capacitor would result in a full-wave rectified voltage on the output because the filtering action of the capacitor is lost.

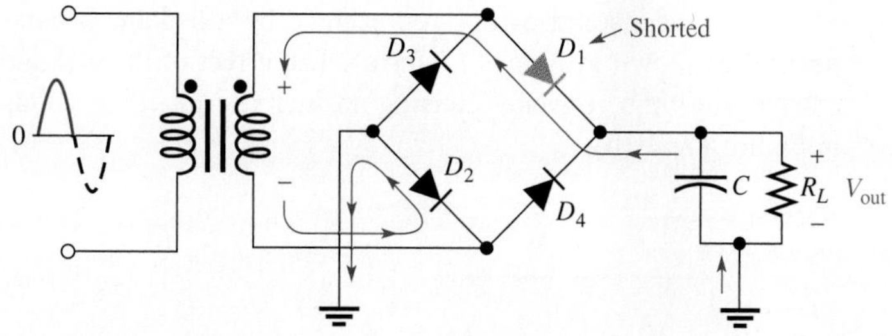

(a) Positive half-cycle: The shorted diode acts as a forward-biased diode, so the load current is normal.

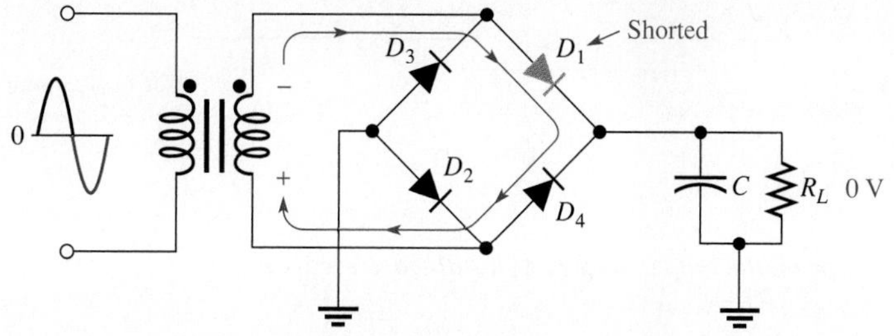

(b) Negative half-cycle: The shorted diode produces a short circuit across the source. As a result D_1, D_4, or the transformer secondary will probably burn open.

FIGURE 17–69
Effect of a shorted diode in a bridge rectifier.

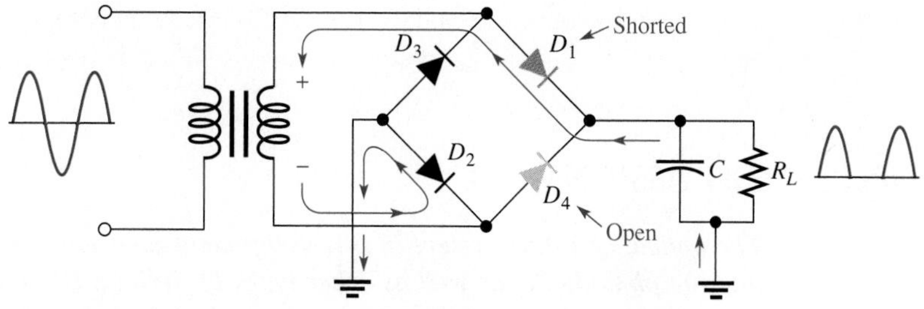

(a) One open diode produces a half-wave output.

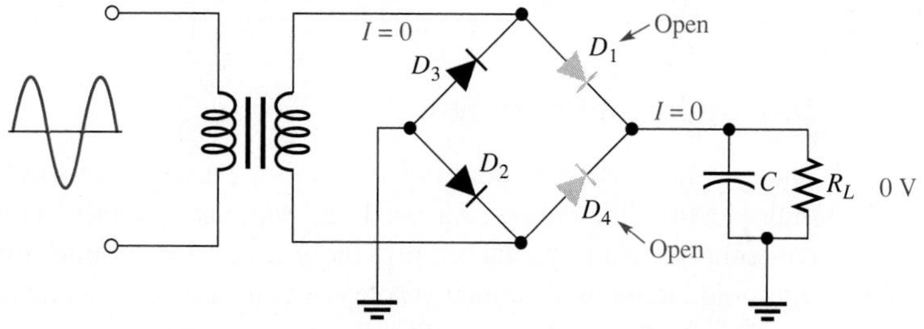

(b) Two open diodes produce a 0 V output.

FIGURE 17–70
Effect of open diodes.

Shorted or Leaky Filter Capacitor

A shorted capacitor would most likely cause some or all of the diodes in a full-wave rectifier to open due to excessive current. In any event, there would be no dc voltage on the output.

A leaky capacitor can be represented by a leakage resistance in parallel with the capacitor, as shown in Figure 17–71(a). The effect of the leakage resistance is to reduce the discharging time constant, causing an increase in ripple voltage on the output, as shown in Figure 17–71(b).

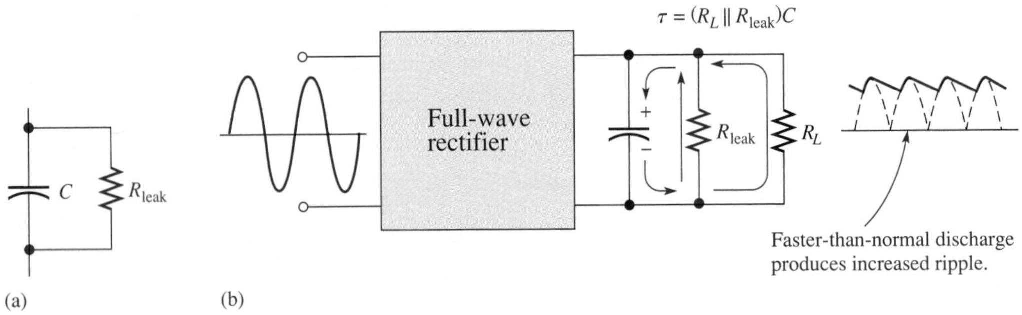

FIGURE 17–71
Effect of a leaky filter capacitor on the output voltage of a full-wave rectifier.

SECTION 17–9 REVIEW

1. What effect would an open D_2 produce in the rectifier of Figure 17–67?
2. You are checking a 60 Hz full-wave bridge rectifier and observe that the output has a 60 Hz ripple. What failure(s) do you suspect?
3. You observe that the output ripple voltage of a full-wave rectifier is much greater than normal but its frequency is still 120 Hz. What component do you suspect?

17–10 ▪ APPLICATION ASSIGNMENT

The optical counting system in this assignment incorporates the zener diode, the LED, and the photodiode, as well as other types of devices. Of course, at this point, you are interested in only how the zener and the optical diodes are used. The system power supply is similar to the one you worked with in the last chapter except that there is a zener regulator in this one rather than the integrated circuit regulator.

Description of the System

The system is used to count objects as they pass on a moving conveyor and can be used in production line monitoring, loading, shipping, inventory control, and the like. The system consists of the power supply, the infrared (IR) emitter circuit, the infrared detector, threshold circuit, and counter/display circuits as shown in Figure 17–72.

The power supply and infrared emitter are housed in a single unit and positioned on one side of the conveyor. The IR detector/counter unit is positioned directly opposite on the other side. The dc power supply provides power to both units. The emitter produces a continuous pattern of infrared light directed toward the detector. As long as the detector is receiving light, nothing happens. When an object passes between the two units and blocks the infrared light, the detector reacts. This reaction is sensed by the threshold circuit, which sends a signal to a digital counter, advancing its count by one. The total count can be displayed and/or sent to a central computer.

In this section, you are not concerned with the details of the threshold circuit or the digital counter and display circuit. In fact, they are beyond the scope of this book and per-

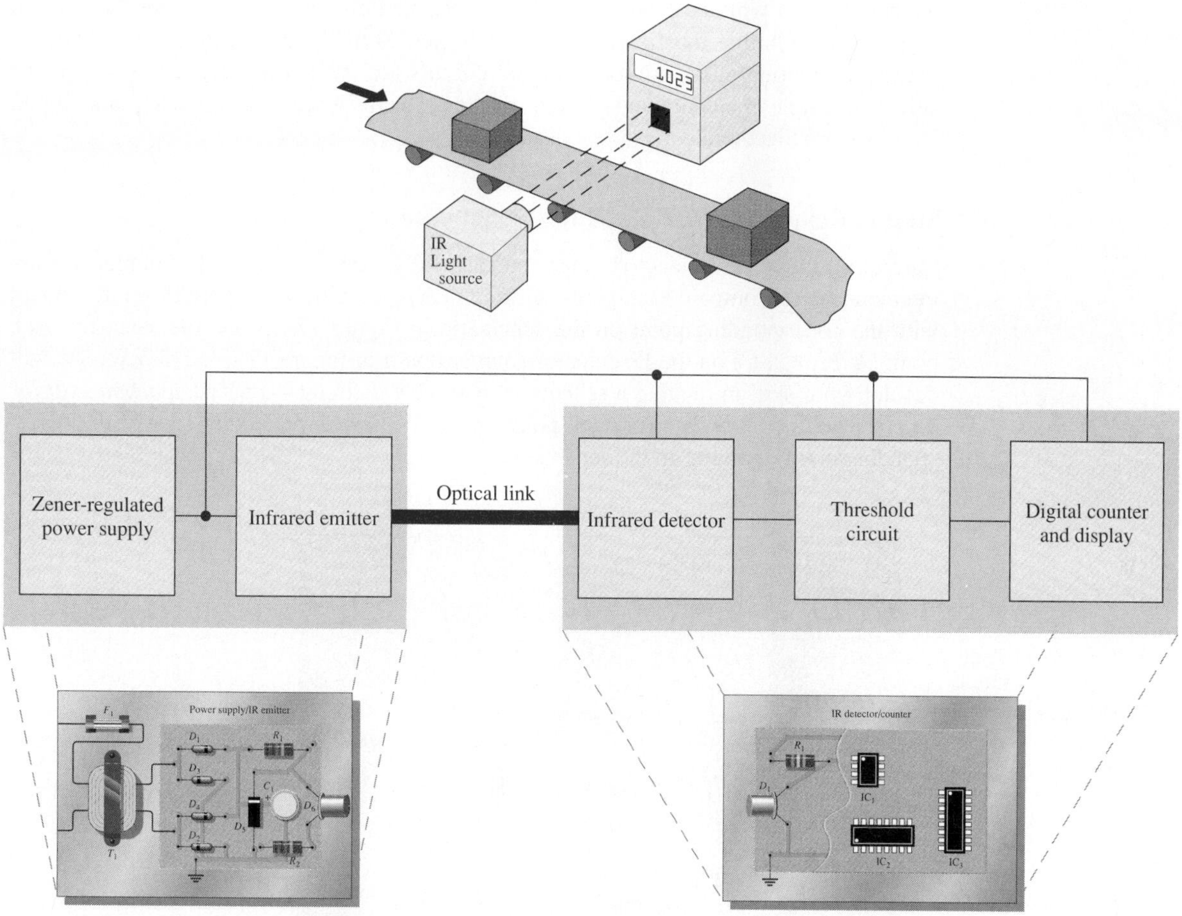

FIGURE 17–72
The optical counting system showing the system concept, block diagram, and circuit boards.

haps beyond your background at this time. They are included only to make a complete system and to show that the circuits you are concerned with must **interface** with other types of circuits in order to perform the required system function. At this point, you are strictly interested in how some of the devices covered in this chapter function in the system.

Physically, the power supply/IR emitter unit consists of one printed circuit board that contains the regulated dc power supply and the infrared emitter circuit. The IR detector/counter unit consists of one printed circuit board that contains the infrared detector circuit, a threshold circuit, and the digital counter and display circuits. All of the circuits on this board except the infrared detector are covered in another course, and you do not need to know the details of how they work for the purposes of this assignment.

Step 1: Identify the Components

As you can see in Figure 17–72, the power supply/IR emitter circuit board is very similar to the power supply board that you worked with in the last chapter except for the regulator and the infrared emitter circuit. The dc voltage regulator is made up of a resistor R1 and zener diode D5. The infrared emitter circuit consists of R2 and D6, which is an infrared LED. The infrared detector circuit on the IR detector/counter board is formed by R1 and infrared photodiode D1. As you can see, the LED and the photodiode are identical in appearance and can only be distinguished by a part number stamped on the case or on the package from which they are taken. The digital portion of this board is shown in the shaded area without detail. The digital integrated circuits, IC1, IC2, and IC3, are in SOIC (Small Outline Integrated Circuit) packages. SOIC is a surface-mount technology packaging configuration. Although digital circuits are studied in other courses, you see here an example of a common situation in many systems where linear (analog) and digital circuits work together.

Step 2: Relate the PC Boards to the Schematics

Carefully follow the conductive traces on each PC board to see how the components are interconnected. Compare each point (*A* through *H*) on the power supply/IR emitter board with the corresponding point on the schematic in Figure 17–73. Do the same for each point (*A* through *C*) on the IR detector/counter board in Figure 17–74. This exercise will develop your skill in relating a schematic to an actual circuit—*a very important skill for an electronics technician.* For each point on the PC board, place the letter of the corresponding point or points on the schematic.

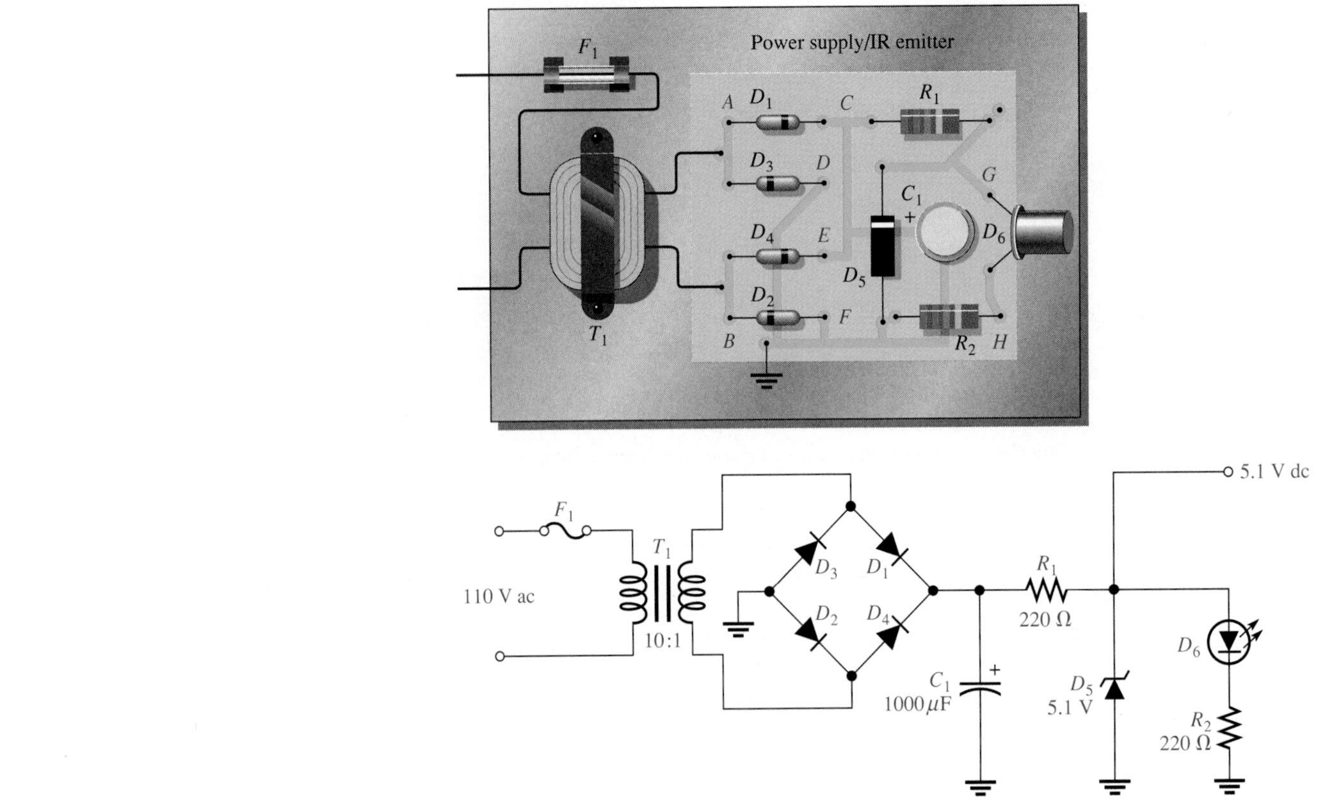

FIGURE 17–73

Step 3: Analyze the Power Supply and IR Emitter

With the power supply/IR emitter board plugged into an ac outlet, determine the dc voltages at points 1, 2, and 3 indicated in Figure 17–75. Assume that the circuit is operating properly.

Step 4: Analyze the IR Detector

With the dc supply voltage from the power supply connected to the IR detector/counter board and with light blocked from the photodiode lens, determine the voltage at point 4 in

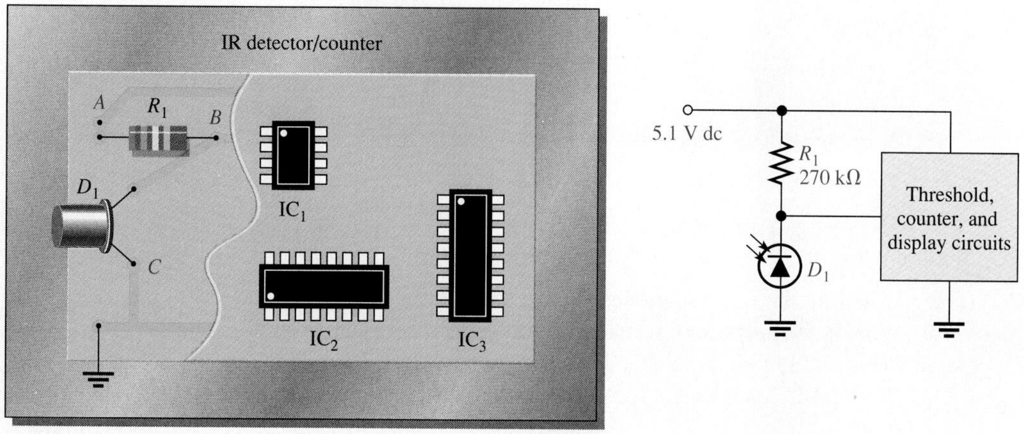

FIGURE 17–74

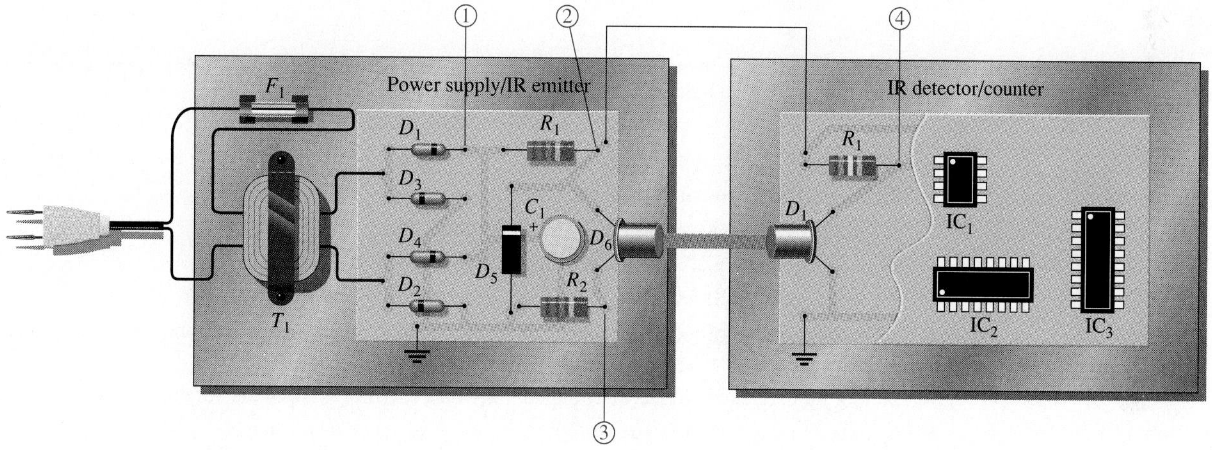

FIGURE 17–75

Figure 17–75. Assume that the dark current is negligible. Neglect any input current to the threshold circuit.

Step 5: Test the Total System

With the two units optically aligned and with the power on, determine the voltage at point 4 in the detector circuit. Assume that a reverse current of 10 μA flows through the photodiode with incident light. Neglect any input current drawn by the threshold circuit. What is the voltage at point 4 when the system detects a passing object?

Step 6: Troubleshoot the System

For each of the following measurements, determine the most probable cause if a faulty reading is indicated. Refer to Figure 17–75.

1. No voltage at point 1.
2. No voltage at point 2.
3. A voltage at point 2 that varies from 0 V to about 5 V with a frequency of 120 Hz.
4. A constant dc voltage of approximately 7.8 V at point 2.
5. A constant dc voltage of approximately 4.3 V at point 2.

6. A floating level at point 4.
7. Approximately 5.1 V at point 4 with no light obstruction between the emitter and the detector.
8. A voltage of 0 V at point 4.

SECTION 17–10 REVIEW

1. List all of the types of diodes used in this system.
2. What is the purpose of resistor R2 on the power supply/IR emitter board?
3. What would you do to change the dc voltage to 8.2 V in this system?
4. Refer to the data sheets in Appendix F to specify the appropriate device for Question 3.
5. Does the LED emit continuous IR or does it turn on and off as objects pass by?
6. What is the purpose of the photodiode in this system?

▪ SUMMARY

- The single diode in a half-wave rectifier conducts for half of the input cycle.
- The output frequency of a half-wave rectifier equals the input frequency.
- The average (dc) value of a half-wave rectified signal is 0.318 ($1/\pi$) times its peak value.
- The PIV is the maximum voltage appearing across the diode in reverse bias.
- Each diode in a full-wave rectifier conducts for half of the input cycle.
- The output frequency of a full-wave rectifier is twice the input frequency.
- The basic types of full-wave rectifier are center-tapped and bridge.
- The output voltage of a center-tapped, full-wave rectifier is approximately one-half of the total secondary voltage.
- The PIV for each diode in a center-tapped, full-wave rectifier is twice the output voltage.
- The output voltage of a bridge rectifier equals the total secondary voltage.
- The PIV for each diode in a bridge rectifier is half that required for the center-tapped configuration and is approximately equal to the peak output voltage.
- A capacitor-input filter provides a dc output approximately equal to the peak of the input.
- Ripple voltage is caused by the charging and discharging of the filter capacitor.
- The smaller the ripple, the better the filter.
- An *LC* filter provides improved ripple reduction over the capacitor-input filter.
- An inductor-input filter produces a dc output voltage approximately equal to the average value of the rectified input.
- Diode clippers cut off voltage above and below specified levels.
- Diode clampers add a dc level to an ac signal.
- The zener diode operates in reverse breakdown.
- There are two breakdown mechanisms in a zener diode; avalanche breakdown and zener breakdown.
- When $V_Z < 5$ V, zener breakdown is predominant.
- When $V_Z > 5$ V, avalanche breakdown is predominant.
- A zener diode maintains an essentially constant voltage across its terminals over a specified range of zener currents.
- Zener diodes are used as shunt voltage regulators.
- Regulation of output voltage over a range of input voltages is called *input* or *line regulation.*
- Regulation of output voltage over a range of load currents is called *load regulation.*
- The smaller the percent regulation, the better.
- A varactor diode acts as a variable capacitor under reverse-biased conditions.
- The capacitance of a varactor varies inversely with reverse-biased voltage.
- Diode symbols are shown in Figure 17–76.

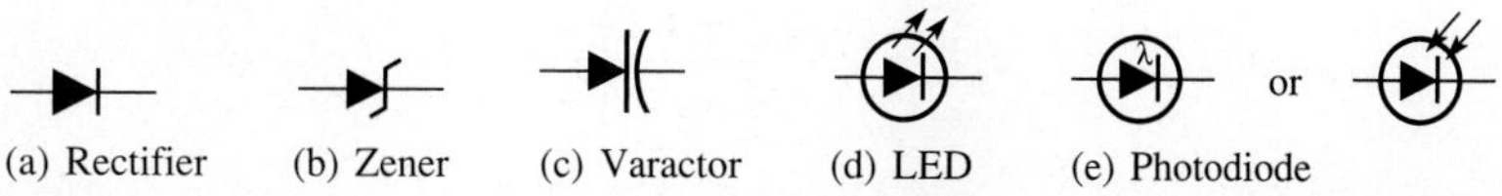

FIGURE 17–76
Diode symbols.

■ GLOSSARY

Center-tap (CT) A connection at the midpoint of the secondary of a transformer.

Clamper A circuit that adds a dc level to an ac signal; a dc restorer.

Clipper A circuit that removes part of a waveform above or below a specified level; a limiter.

Filter A type of electrical circuit that passes certain frequencies and rejects all others.

Full-wave rectifier A circuit that converts an alternating sine wave into a pulsating dc consisting of both halves of a sine wave for each input cycle.

Half-wave rectifier A circuit that converts an alternating sine wave into a pulsating dc consisting of one-half of a sine wave for each input cycle.

Interface To make the output of one type of circuit compatible with the input of another so that they can operate properly together.

Light-emitting diode (LED) A type of diode that emits light when forward current flows.

Line regulation The percent change in output voltage for a given change in line (input) voltage.

Load regulation The percent change in output voltage for a given change in load current.

Peak inverse voltage (PIV) The maximum value of reverse voltage which occurs at the peak of the input cycle when the diode is reverse-biased.

Photodiode A diode whose reverse resistance changes with incident light.

Power supply An electronic instrument that produces voltage, current, and power from the ac power line or batteries in a form suitable for use in various applications to power electronic equipment.

Rectifier An electronic circuit that converts ac into pulsating dc.

Regulator An electronic circuit that maintains an essentially constant output voltage with a changing input voltage or load.

Ripple voltage The variation in the dc voltage on the output of a filtered rectifier caused by the slight charging and discharging action of the filter capacitor.

Varactor A diode that is used as a voltage-variable capacitor.

Voltage regulation The process of maintaining an essentially constant output voltage over variations in input voltage or load.

Zener diode A type of diode that operates in reverse breakdown (called zener breakdown) to provide voltage regulation.

■ FORMULAS

(17–1) $$V_{AVG} = \frac{V_{p(out)}}{\pi}$$ Half-wave average value

(17–2) $$V_{p(out)} = V_{p(in)} - 0.7\text{ V}$$ Peak half-wave rectifier output voltage

(17–3) $$V_{AVG} = \frac{2V_{p(out)}}{\pi}$$ Full-wave average value

(17–4) $$\text{PIV} = 2V_{p(out)} + 0.7\text{ V}$$ Diode peak inverse voltage, center-tapped rectifier

(17–5) $$V_{p(out)} = V_{p(sec)} - 1.4\text{ V}$$ Bridge full-wave output (including diode drops)

(17–6) $$\text{PIV} = V_{p(out)} + 0.7\text{ V}$$ Diode peak inverse voltage, bridge rectifier

(17–7) $$\text{PIV} = 2V_{p(in)} - 0.7\text{ V}$$ PIV, half-wave rectifier with capacitor-input filter

(17–8) $$r = \left(\frac{V_r}{V_{dc}}\right)100\%$$ Ripple factor

(17–9)	$R_{surge} > \dfrac{V_{p(sec)} - 1.4\text{ V}}{I_{FSM}}$	Surge limiting resistance for capacitor filter
(17–10)	$R_Z = \dfrac{\Delta V_Z}{\Delta I_Z}$	Zener resistance
(17–11)	$\text{Percent line regulation} = \dfrac{\Delta V_{OUT}}{\Delta V_{IN}} \times 100\%$	Line regulation
(17–12)	$\text{Percent load regulation} = \dfrac{V_{NL} - V_{FL}}{V_{FL}} \times 100\%$	Load regulation
(17–13)	$C = \dfrac{A\epsilon}{d}$	Capacitance formula
(17–14)	$f_r = \dfrac{1}{2\pi\sqrt{LC}}$	Resonant frequency

■ SELF-TEST

1. The process of converting ac to pulsating dc is called
   (a) clipping (b) charging (c) rectification (d) filtering
2. The output frequency of a half-wave rectifier with a 60 Hz sinusoidal input is
   (a) 30 Hz (b) 60 Hz (c) 120 Hz (d) 0 Hz
3. The number of diodes used in a half-wave rectifier is
   (a) one (b) two (c) three (d) four
4. If a 75 V peak sine wave is applied to a half-wave rectifier, the peak inverse voltage across the diode is
   (a) 75 V (b) 150 V (c) 37.5 V (d) 0 V
5. The output frequency of a full-wave rectifier with a 60 Hz sinusoidal input is
   (a) 30 Hz (b) 60 Hz (c) 120 Hz (d) 0 Hz
6. Two types of full-wave rectifier are
   (a) single diode and dual diode (b) primary and secondary
   (c) forward-biased and reverse-biased (d) center-tapped and bridge
7. When a diode in a center-tapped rectifier opens, the output is
   (a) 0 V (b) half-wave rectified (c) reduced in amplitude (d) unaffected
8. During the positive half-cycle of the input voltage in a bridge rectifier,
   (a) one diode is forward-biased (b) all diodes are forward-biased
   (c) all diodes are reverse-biased (d) two diodes are forward-biased
9. The process of changing a half-wave or a full-wave rectified voltage to a constant dc voltage is called
   (a) filtering (b) ac to dc conversion (c) damping (d) ripple suppression
10. The small variation in the output voltage of a dc power supply is called
    (a) average voltage (b) surge voltage (c) residual voltage (d) ripple voltage
11. One measure of the effectiveness of a power supply filter is the
    (a) surge current (b) ripple factor
    (c) peak value of the output (d) average value of the output
12. A diode clipping circuit
    (a) removes part of a waveform
    (b) inserts a dc level
    (c) produces an output equal to the average value of the input
    (d) increases the peak value of the input
13. A clamping circuit is also known as a(an)
    (a) averaging circuit (b) inverter (c) dc restorer (d) ac restorer
14. The zener diode is designed for operation in
    (a) zener breakdown (b) forward bias (c) reverse bias (d) avalanche breakdown
15. Zener diodes are widely used as
    (a) current limiters (b) power distributors (c) voltage regulators (d) variable resistors
16. Varactor diodes are used as
    (a) variable resistors (b) variable current sources
    (c) variable inductors (d) variable capacitors

17. LEDs are based on the principle of
    (a) forward bias (b) electroluminescence
    (c) photon sensitivity (d) electron-hole recombination
18. In a photodiode, light produces
    (a) reverse current (b) forward current
    (c) electroluminescence (d) dark current

■ PROBLEMS

SECTION 17–1 Half-Wave Rectifiers

1. Calculate the average value of a half-wave rectified voltage with a peak value of 200 V.
2. Sketch the waveforms for the load current and voltage for Figure 17–77. Show the peak values. (The diode is silicon.)
3. Can a diode with a PIV rating of 5 V be used in the circuit of Figure 17–77?
4. Determine the peak power delivered to R_L in Figure 17–78.

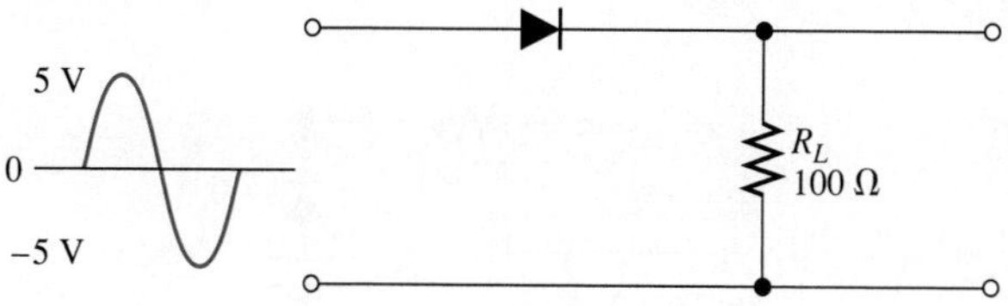

FIGURE 17–77

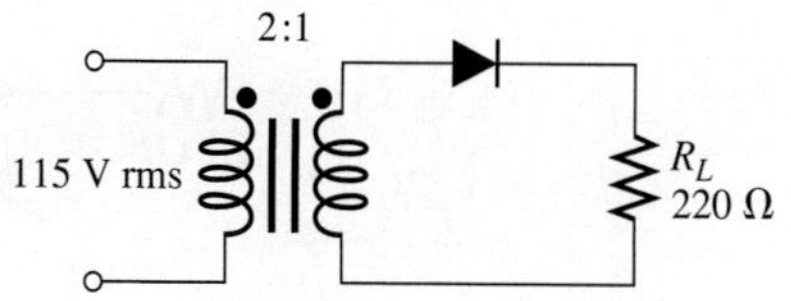

FIGURE 17–78

SECTION 17–2 Full-Wave Rectifiers

5. Calculate the average value of a full-wave rectified voltage with a peak value of 75 V.
6. Consider the circuit in Figure 17–79.
    (a) What type of circuit is this?
    (b) What is the total peak secondary voltage?
    (c) Find the peak voltage across each half of the secondary.
    (d) Sketch the voltage waveform across R_L.
    (e) What is the peak current through each diode?
    (f) What is the PIV for each diode?

FIGURE 17–79

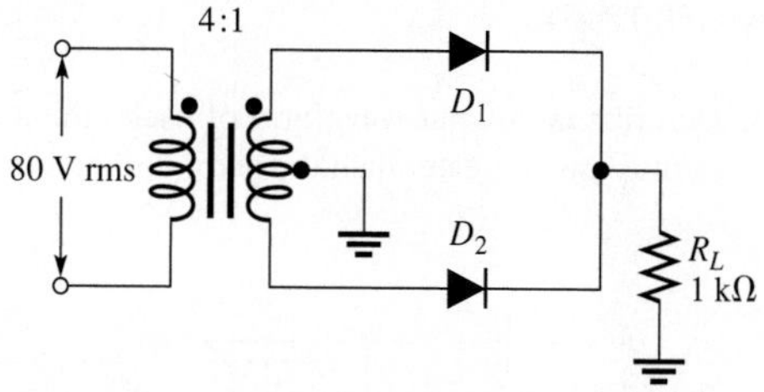

7. Calculate the peak voltage rating of each half of a center-tapped transformer used in a full-wave rectifier that has an average output voltage of 110 V.
8. Show how to connect the diodes in a center-tapped rectifier in order to produce a negative-going full-wave voltage across the load resistor.
9. What PIV rating is required for the diodes in a bridge rectifier that produces an average output voltage of 50 V?

SECTION 17–3 Rectifier Filters

10. The ideal dc output voltage of a capacitor-input filter is the (peak, average) value of the rectified input.

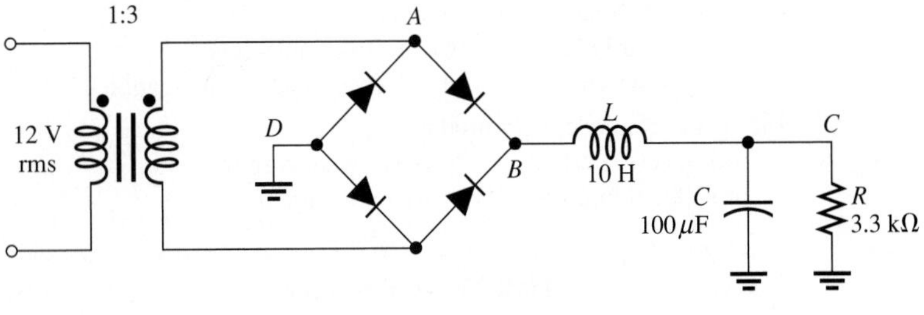

FIGURE 17–80

11. Refer to Figure 17–80 and sketch the following voltage waveforms in relationship to the input waveform: V_{AB}, V_{AD}, V_{BD}, and V_{CD}.

SECTION 17–4 Diode Clipping and Clamping Circuits

12. Sketch the output waveforms for each circuit in Figure 17–81.

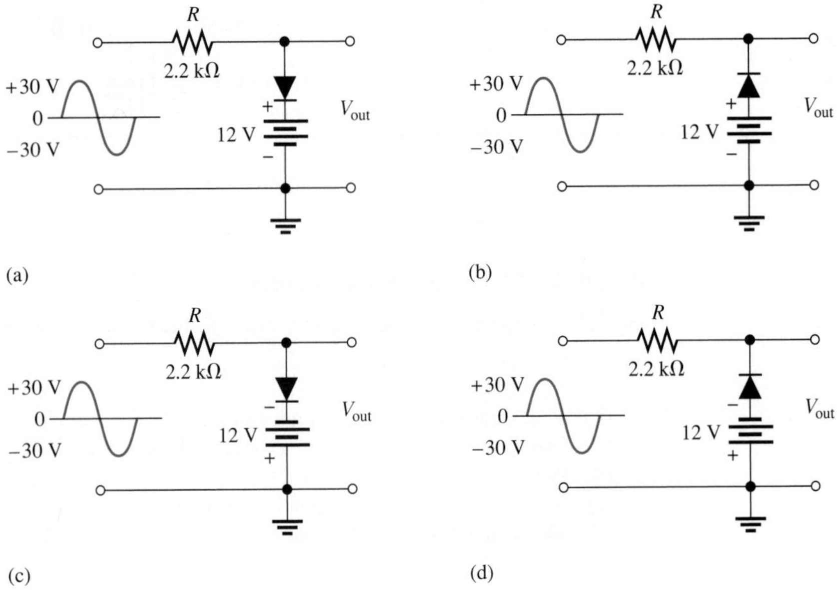

FIGURE 17–81

13. Describe the output waveform of each circuit in Figure 17–82. Assume that the *RC* time constant is much greater than the period of the input.

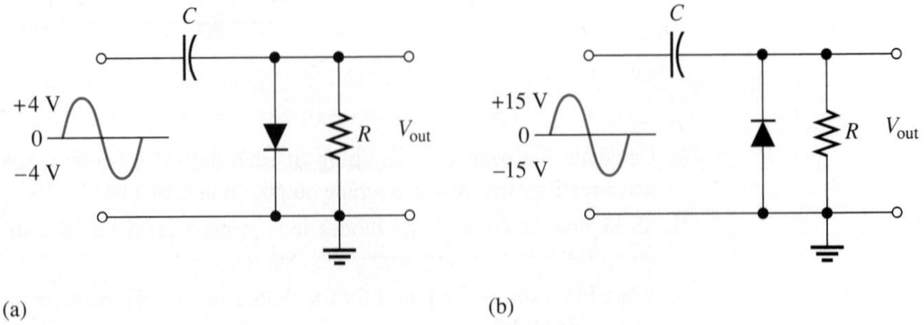

FIGURE 17–82

SECTION 17–5 Zener Diodes

14. A certain zener diode has a $V_Z = 7.5$ V and an $R_Z = 5\ \Omega$. Sketch the equivalent circuit.

FIGURE 17–83

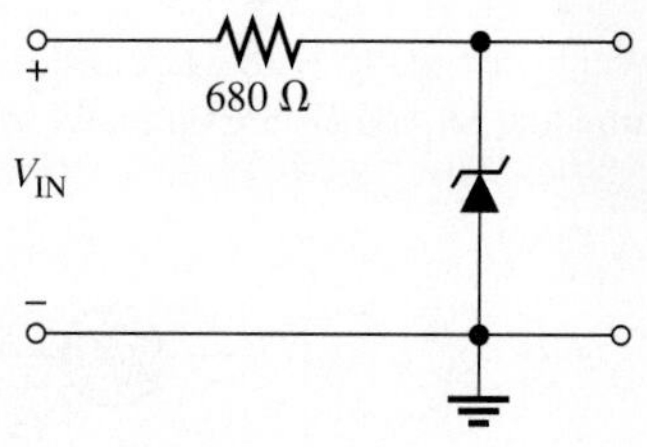

15. Determine the minimum input voltage required for regulation to be established in Figure 17–83. Assume an ideal zener diode with $I_{ZK} = 1.5$ mA and $V_Z = 14$ V.

16. To what value must R be adjusted in Figure 17–84 to make $I_Z = 40$ mA? Assume that $V_Z = 12$ V at 30 mA and $R_Z = 30\ \Omega$.

17. A loaded zener regulator is shown in Figure 17–85. For $V_Z = 5.1$ V at 35 mA, $I_{ZK} = 1$ mA, $I_{ZM} = 70$ mA, and $R_Z = 12\ \Omega$, determine the minimum and maximum permissible load currents.

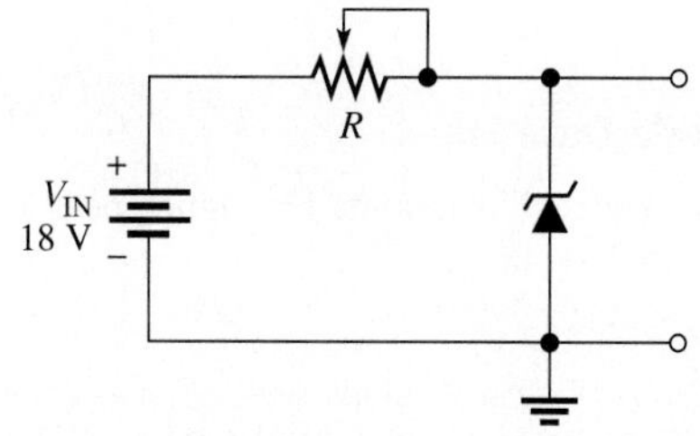

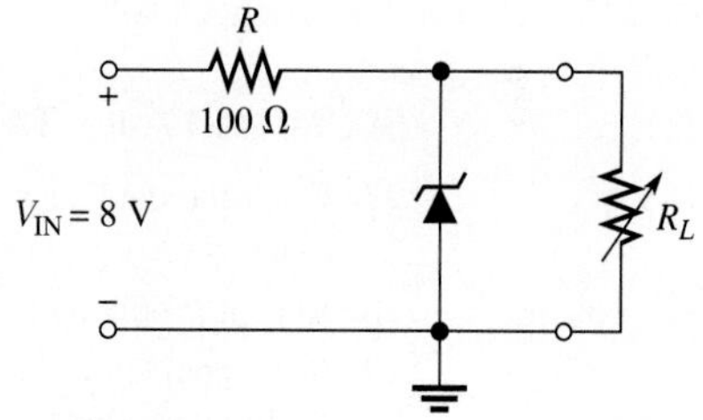

FIGURE 17–84

FIGURE 17–85

18. Find the percent load regulation in Problem 17.

19. For the circuit of Problem 17, assume that the input voltage is varied from 6 V to 12 V. Determine the percent input regulation with no load and with maximum load.

20. The no-load output voltage of a certain zener regulator is 8.23 V, and the full-load output is 7.98 V. Calculate the percent load regulation.

SECTION 17–6 Varactor Diodes

21. Figure 17–86 is a curve of reverse voltage versus capacitance for a certain varactor. Determine the change in capacitance if V_R varies from 5 V to 20 V.

22. Refer to Figure 17–86 and determine the value of V_R that produces 25 pF.

23. What capacitance value is required for each of the varactors in Figure 17–87 to produce a resonant frequency of 1 MHz?

24. At what value must the control voltage be set in Problem 23 if the varactors have the characteristic curve in Figure 17–86?

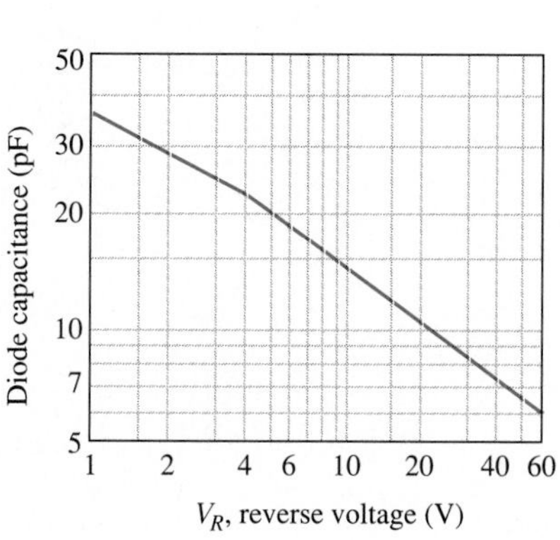

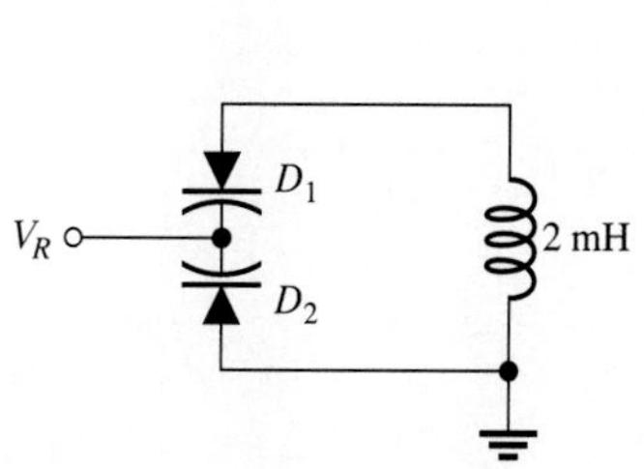

FIGURE 17–86

FIGURE 17–87

SECTION 17–7 LEDs and Photodiodes

25. When the switch in Figure 17–88 is closed, will the microammeter reading increase or decrease? Assume that D_1 and D_2 are optically coupled.

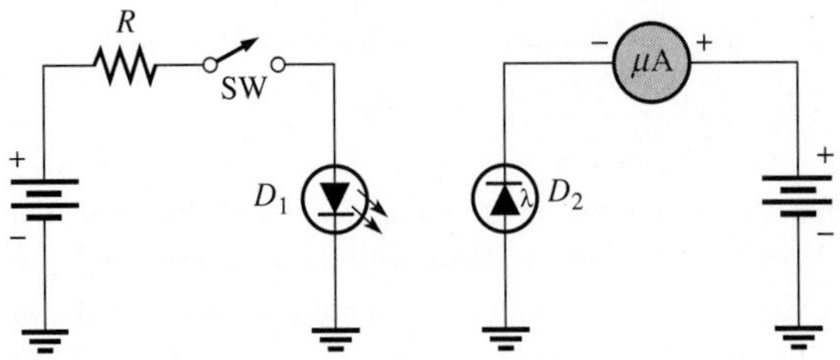

FIGURE 17–88

26. With no incident light, a certain amount of reverse current flows in a photodiode. What is this current called?

SECTION 17–8 The Diode Data Sheet

27. From the data sheet in Appendix F, determine how much peak inverse voltage that a 1N4001 diode can withstand.

28. Repeat Problem 27 for a 1N4004.

29. If the peak output voltage of a full-wave bridge rectifier is 50 V, determine the minimum value of the surge limiting resistor required when 1N4001 diodes are used.

SECTION 17–9 Troubleshooting

30. From the meter readings in Figure 17–89, determine if the rectifier circuit is functioning properly. If it is not, determine the most likely failure(s).

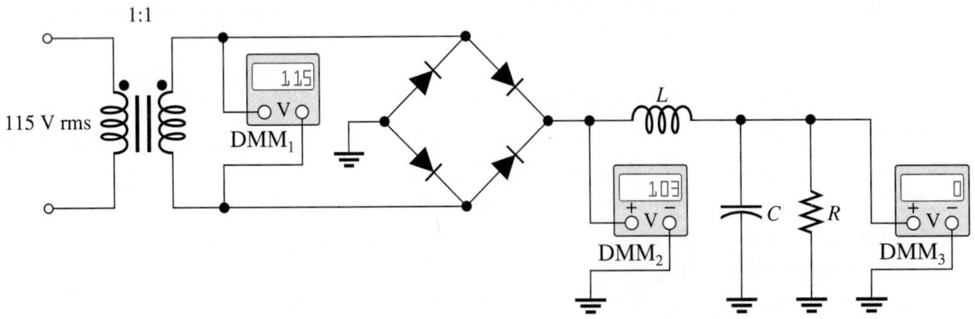

FIGURE 17–89

31. Each part of Figure 17–90 shows oscilloscope displays of rectifier output voltages. In each case, determine whether or not the rectifier is functioning properly and, if it is not, identify the most likely failure(s).

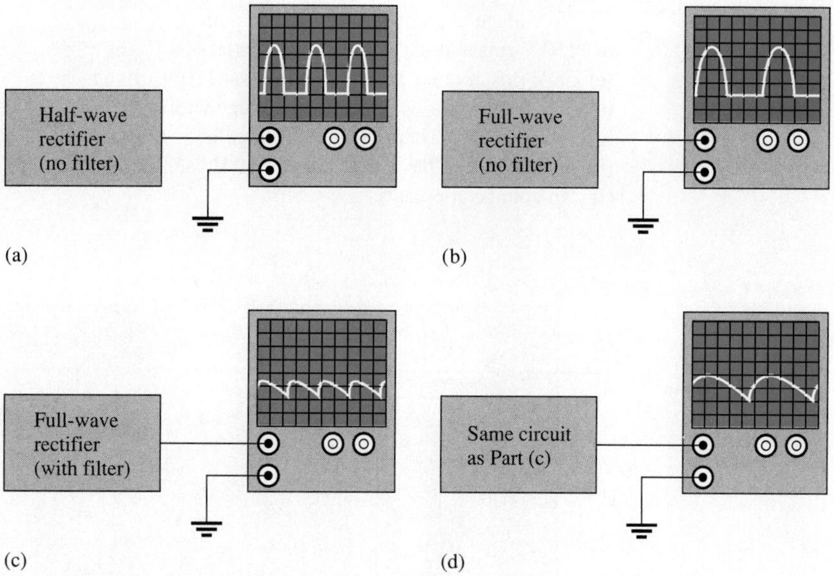

FIGURE 17–90

32. For each set of measured voltages at the points (1, 2, and 3) indicated in Figure 17–91, determine if they are correct and if not identify the most likely fault(s). State what you would do to correct the problem once it is isolated.

(a) $V_1 = 110$ V rms, $V_2 \cong 30$ V dc, $V_3 \cong 12$ V dc
(b) $V_1 = 110$ V rms, $V_2 \cong 30$ V dc, $V_3 \cong 30$ V dc
(c) $V_1 = 0$ V, $V_2 = 0$ V, $V_3 = 0$ V
(d) $V_1 = 110$ V rms, $V_2 \cong 30$ V peak full-wave 120 Hz voltage, $V_3 \cong 12$ V 120 Hz pulsating voltage
(e) $V_1 = 110$ V rms, $V_2 = 0$ V, $V_3 = 0$ V

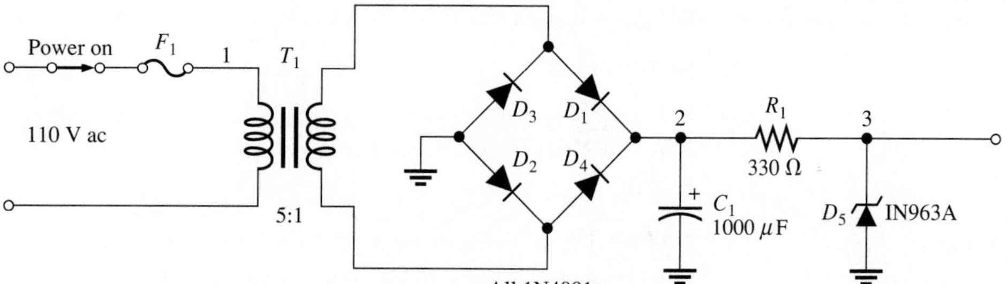

FIGURE 17–91

33. Determine the most likely failure in the circuit board of Figure 17–92 for each of the following symptoms. State the corrective action you would take in each case. The transformer has a turns ratio of 1.

(a) No voltage at point 1
(b) No voltage at point 2; 110 V rms at point 1
(c) No voltage at point 3; 110 V rms at point 1
(d) 150 V rms at point 2; input is correct at 110 V rms
(e) 68 V rms at point 3; input is correct at 110 V rms
(f) A pulsating full-wave rectified voltage with a peak of 155.5 V at point 4
(g) Excessive 120 Hz ripple voltage at point 5
(h) Ripple voltage has a frequency of 60 Hz at point 4
(i) No voltage at point 6

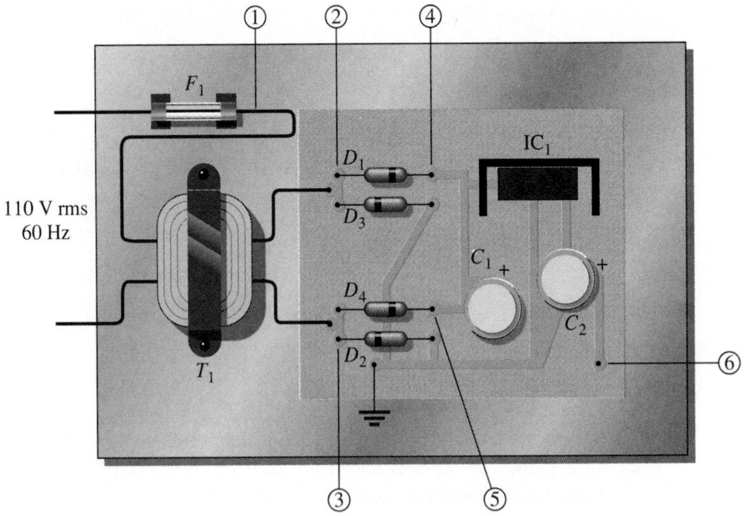

FIGURE 17–92

■ ANSWERS TO SECTION REVIEWS

Section 17–1

1. The PIV occurs at 270°.
2. There is load current for 50% of the input cycle.
3. $V_{AVG} = V_{p(out)}/\pi = 3.18$ V

Section 17–2

1. $V_{AVG} = 2V_{p(out)}/\pi = 38.2$ V
2. A bridge rectifier has the greater V_{out}.
3. True

Section 17–3

1. 60 Hz for half-wave; 120 Hz for full-wave.
2. Ripple voltage is caused by the capacitor charging and discharging slightly.
3. Reducing load resistance increases ripple voltage.
4. An *LC* filter reduces ripple voltage but has less dc voltage output than a capacitor-input filter.

Section 17–4

1. Clippers clip off or remove portions of a waveform. Clampers insert a dc level.
2. A positive clipper clips off positive voltages. A negative clipper clips off negative voltages.
3. 0.7 V appears across the diode in the clipper during the positive alternation.
4. The bias voltage must be 5 V – 0.7 V = 4.3 V.
5. The capacitor acts as a battery in a clamper.

Section 17–5

1. True
2. $V_Z = 10\text{ V} + (30\text{ mA})(8\ \Omega) = 10.2\text{ V}$
3. *Line regulation:* Constant output voltage for varying input voltage. *Load regulation:* Constant output voltage for varying load current.
4. I_Z is maximum for maximum R_L.
5. % load reg. = $[(V_{NL} - V_{FL})/V_{FL}] \times 100\% = 0.833\%$

Section 17–6

1. A varactor acts as a variable capacitor.
2. Diode capacitance decreases when reverse voltage is increased.

Section 17–7

1. LEDs give off light when forward-biased; photodiodes respond to light when reverse-biased.
2. Gallium arsenide, gallium arsenide phosphide, and gallium phosphide are LED materials.
3. Dark current is the small photodiode reverse current with no light.

Section 17–8

1. Three diode ratings categories are forward current and voltage, reverse current and voltage, and thermal.
2. **(a)** V_F: maximum dc forward voltage drop
   (b) I_R: maximum reverse direct current
   (c) I_0: average maximum forward current

Section 17–9

1. An open D_2 causes a half-wave output.
2. A 60 Hz ripple indicates an open diode.
3. An increase in ripple voltage indicates a leaky filter capacitor.

Section 17–10

1. Rectifier, zener, light-emitting, photo
2. R_2 limits the LED current.
3. Use an 8.2 V zener diode.
4. 1N756 or 1N959A
5. The LED emits a continuous IR beam.
6. The photodiode senses the absence of the IR beam.

■ ANSWERS TO RELATED EXERCISES FOR EXAMPLES

17–1 3.82 V
17–2 2.3 V
17–3 98.7 V
17–4 319 V
17–5 $V_{p(out)} = 78.6$ V; PIV = 79.3 V
17–6 A positive peak of 8.72 V and clipped at – 0.3 V
17–7 A sine wave clipped at +10.7 V and –10.7 V
17–8 Same as Figure 17–39
17–9 5 Ω
17–10 $V_Z = 11.4$ V at 20 mA; $V_Z = 12.6$ V at 80 mA
17–11 $V_{IN(min)} = 6.61$ V; $V_{IN(max)} = 48.1$ V
17–12 $I_{L(min)} = 0$ A; $I_{L(max)} = 42$ mA; $R_{L(min)} = 79\ \Omega$
17–13 **(a)** 3.77% **(b)** 1.75%/V
17–14 101 kHz to 318 kHz

18

TRANSISTORS AND THYRISTORS

Two basic types of transistors will be discussed in this chapter: the bipolar junction transistor (BJT) and the field-effect transistor (FET). The two major application areas of amplification and switching are introduced. Also, the unijunction transistor (UJT) will be introduced and common types of thyristors presented.

■ CHAPTER OBJECTIVES

- ☐ Describe the basic structure and operation of bipolar junction transistors
- ☐ Analyze a transistor circuit with voltage-divider bias
- ☐ Explain the operation of a bipolar transistor as an amplifier
- ☐ Analyze a transistor switching circuit
- ☐ Discuss important transistor parameters and maximum ratings
- ☐ Describe the basic structure and operation of a JFET
- ☐ Explain the operation of a JFET in terms of its characteristics
- ☐ Explain the basic structure and operation of MOSFETs
- ☐ Analyze several biasing arrangements for JFETs and MOSFETs
- ☐ Discuss the operation and application of the UJT
- ☐ Explain the basic structure and operation of the SCR, triac, and diac
- ☐ Recognize transistor packages and identify the terminals
- ☐ Troubleshoot transistors

■ APPLICATION ASSIGNMENT

You have been assigned to work on a proportional heat-control system that maintains a liquid in a tank at a precise temperature for a certain industrial production process. You are asked to analyze and troubleshoot a particular transistor circuit in the system that produces a dc output voltage proportional to the temperature of the substance in the tank. This circuit uses the variable resistance of a temperature sensor as part of its bias circuit. The output of this temperature detector circuit is converted to digital form for precise control of the burner that heats the tank. If the temperature of the substance tends to increase above a certain value, the circuitry causes less fuel to be fed to the burner and thus reduces the temperature. If the temperature tends to decrease, the circuitry causes more fuel to be fed to the burner and thus raises the temperature. This type of system maintains the substance at a nearly constant temperature and is typical of many industrial process-control systems. After studying this chapter, you should be able to complete the application assignment in Section 18–14.

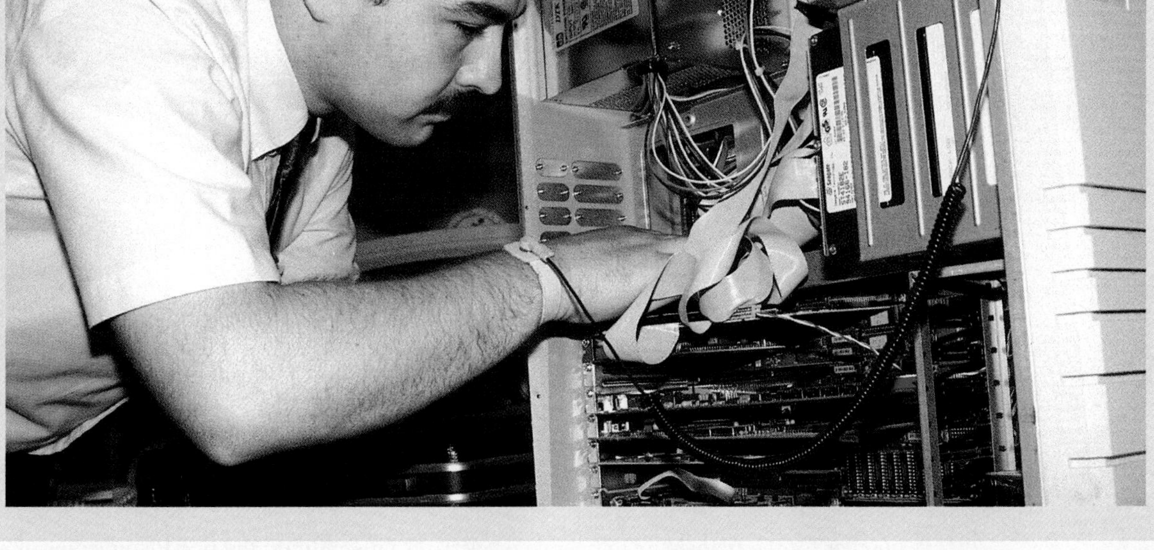

18–1 ■ BIPOLAR JUNCTION TRANSISTORS (BJTs)

The basic structure of the bipolar junction transistor, BJT, determines its operating characteristics. In this section, you will see how semiconductor materials are joined to form a transistor, and you will learn the standard transistor symbols. Also, you will see how important dc bias is to the operation of transistors in terms of setting up proper currents and voltages in a transistor circuit. Finally, two important parameters, α_{dc} and β_{dc}, are introduced.

After completing this section, you should be able to

- **Describe the basic structure and operation of bipolar junction transistors**
  - ☐ Explain the difference between NPN and PNP transistors
  - ☐ Explain transistor biasing
  - ☐ Discuss transistor currents and their relationships
  - ☐ Define β_{dc}
  - ☐ Define α_{dc}
  - ☐ Discuss the transistor voltages
  - ☐ Calculate currents and voltages

The **bipolar junction transistor** (BJT) is constructed with three doped semiconductor regions separated by two PN junctions. The three regions are called **emitter, base,** and **collector.** The two types of bipolar transistors are shown in Figure 18–1. One type consists of two N regions separated by a P region (NPN), and the other consists of two P regions separated by an N region (PNP).

The PN junction joining the base region and the emitter region is called the *base-emitter junction.* The junction joining the base region and the collector region is called the *base-collector* junction, as indicated in Figure 18–1(a). A wire lead connects to each of the three regions, as shown. These leads are labeled *E, B,* and *C* for emitter, base, and collector, respectively. The base material is lightly doped and very narrow compared to the heavily doped emitter and collector materials. The reason for this is discussed in the next section.

Figure 18–2 shows the schematic symbols for the NPN and PNP bipolar transistors. The term **bipolar** refers to the use of both holes and electrons as carriers in the transistor structure.

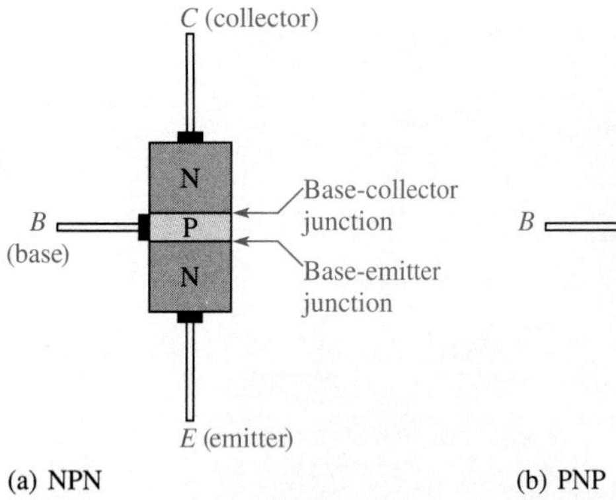

FIGURE 18–1
Basic construction of bipolar transistors.

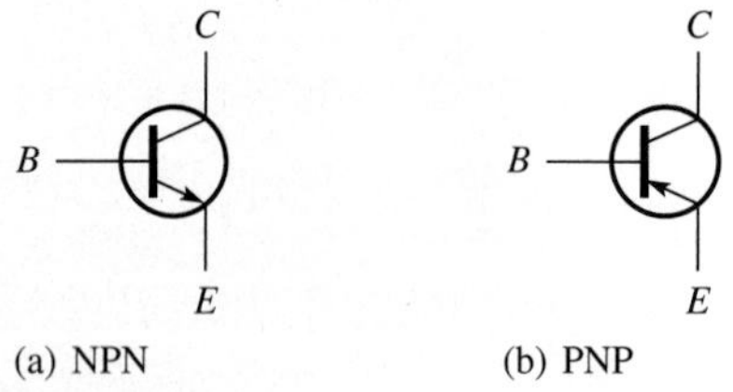

FIGURE 18–2
Transistor symbols.

Transistor Biasing

In order for the **transistor** to operate properly as an amplifier, the two PN junctions must be correctly biased with external dc voltages. We will use the NPN transistor to illustrate

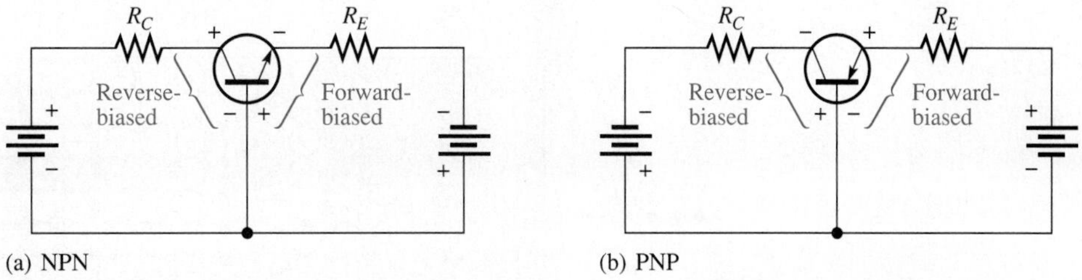

FIGURE 18–3
Forward-reverse bias of a bipolar transistor.

transistor biasing. The operation of the PNP is the same as for the NPN except that the roles of the electrons and holes, the bias voltage polarities, and the current directions are all reversed. Figure 18–3 shows the proper bias arrangement for both NPN and PNP transistors. Notice that in both cases the base-emitter (*BE*) junction is forward-biased and the base-collector (*BC*) junction is reverse-biased. This is called *forward-reverse bias.*

Now, let's examine what happens inside the transistor when it is forward-reverse biased. The forward bias from base to emitter narrows the *BE* depletion layer, and the reverse bias from base to collector widens the *BC* depletion layer, as depicted in Figure 18–4(a). The N-type emitter region is teeming with conduction-band (free) electrons that easily diffuse across the *BE* junction into the P-type base region, just as in a forward-biased diode.

The base region is lightly doped and very thin so that it has a very limited number of holes. Thus, only a small percentage of all the electrons flowing across the *BE* junction combine with the available holes. These relatively few recombined electrons flow out of the base lead as valence electrons, forming the small base current, I_B, as shown in Figure 18–4(b).

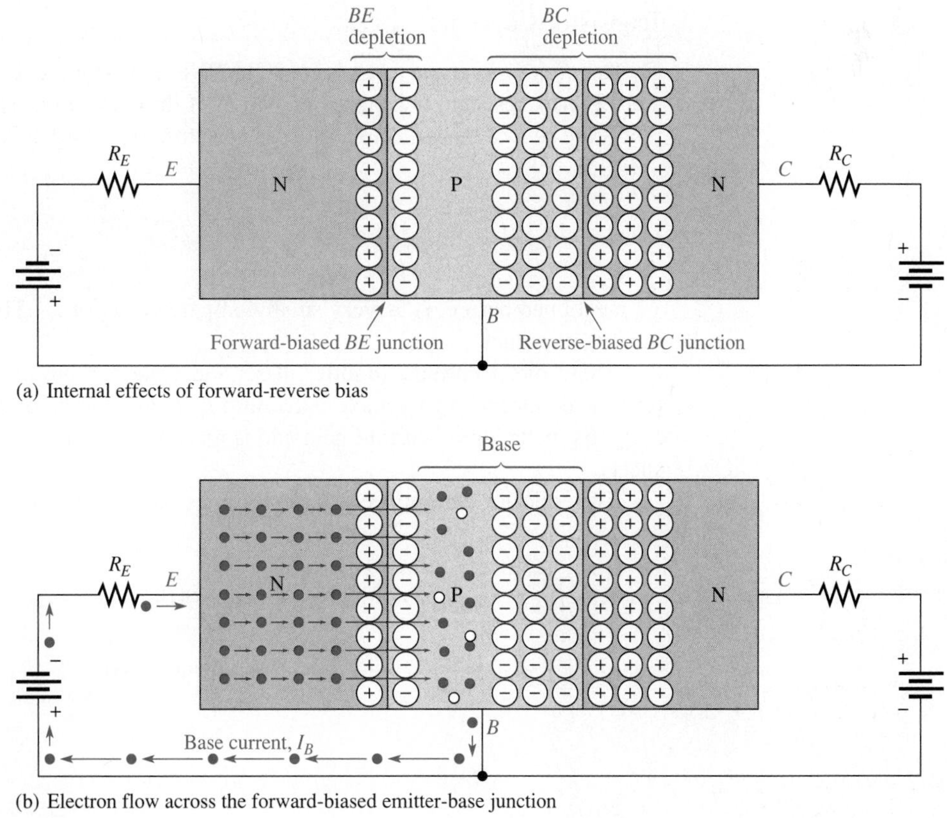

FIGURE 18–4
Illustration of BJT action. The base region is very narrow, but it is shown wider here for clarity.

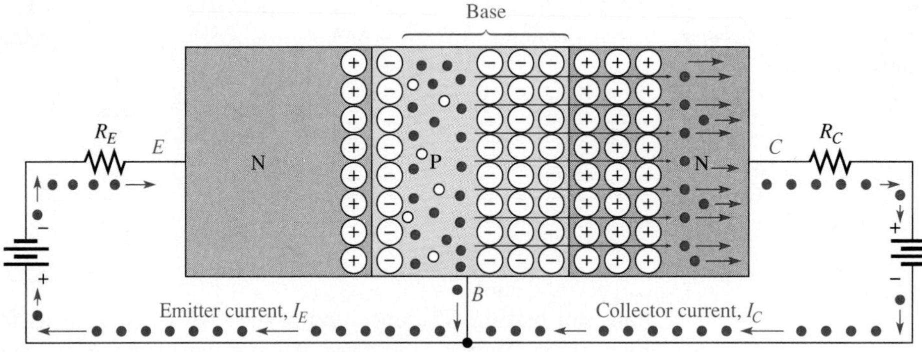

(c) Electron flow across the reverse-biased collector-base junction

FIGURE 18–4 ***Continued***

Most of the electrons flowing from the emitter into the base region diffuse into the *BC* depletion layer. Once in this layer, they are pulled across the *BC* junction by the depletion layer field set up by the force of attraction between the positive and negative ions. Actually, you can think of the electrons as being pulled across the reverse-biased *BC* junction by the attraction of the positive ions on the other side, as illustrated in Figure 18–4(c). The electrons now move through the collector region, out through the collector lead, and into the positive terminal of the external dc source, thereby forming the collector current, I_C, as shown. The amount of collector current depends directly on the amount of base current and is essentially independent of the dc collector voltage.

Transistor Currents

The directions of current in an NPN and a PNP transistor are as shown in Figures 18–5(a) and 18–5(b), respectively. An examination of these diagrams shows that the emitter current is the sum of the collector and base currents, expressed as follows:

$$I_E = I_C + I_B \tag{18–1}$$

As mentioned before, I_B is very small compared to I_E or I_C. The capital-letter subscripts indicate dc values.

These direct currents (emitter, base, and collector) are also related by two parameters: the dc **alpha (α_{dc}),** which is the ratio I_C/I_E and the dc **beta (β_{dc})**, which is the ratio I_C/I_B. β_{dc} is the direct current gain and is usually designated as h_{FE} on the transistor data sheets.

The collector current is equal to α_{dc} times the emitter current:

$$I_C = \alpha_{dc} I_E \tag{18–2}$$

where α_{dc} typically has a value between 0.95 and 0.99.

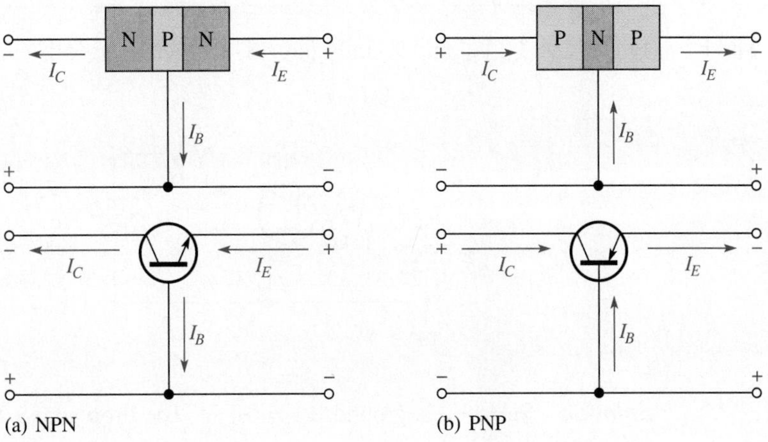

FIGURE 18–5
Current directions in a transistor.

The collector current is equal to the base current multiplied by β_{dc}:

$$I_C = \beta_{dc} I_B \tag{18–3}$$

where β_{dc} typically has a value between 20 and 200.

Transistor Voltages

The three dc voltages for the biased transistor in Figure 18–6 are the emitter voltage (V_E) the collector voltage (V_C), and the base voltage (V_B). These voltages are with respect to ground. The collector voltage is equal to the dc supply voltage, V_{CC}, less the drop across R_C:

$$V_C = V_{CC} - I_C R_C \tag{18–4}$$

The base voltage is equal to the emitter voltage plus the base-emitter junction barrier potential (V_{BE}), which is about 0.7 V for a silicon transistor:

$$V_B = V_E + V_{BE} \tag{18–5}$$

In the configuration of Figure 18–6, the emitter is the common terminal, so $V_E = 0$ V.

FIGURE 18–6
Bias voltages.

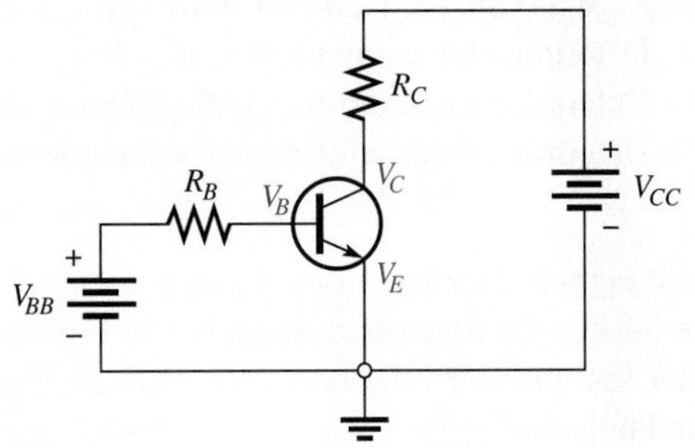

EXAMPLE 18–1 Find I_B, I_C, I_E, V_B, and V_C in Figure 18–7, where β_{dc} is 50.

FIGURE 18–7

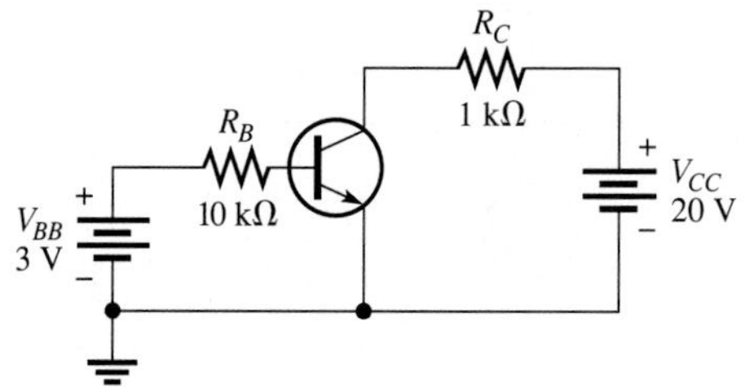

Solution Since V_E is ground, $V_B = 0.7$ V. The drop across R_B is $V_{BB} - V_B$, so I_B is calculated as follows:

$$I_B = \frac{V_{BB} - V_B}{R_B} = \frac{3\text{ V} - 0.7\text{ V}}{10\text{ k}\Omega} = 230\ \mu\text{A}$$

Now you can find I_C, I_E, and V_C.

$$I_C = \beta_{dc} I_B = 50(230\ \mu\text{A}) = 11.5\text{ mA}$$
$$I_E = I_C + I_B = 11.5\text{ mA} + 230\ \mu\text{A} = 11.73\text{ mA}$$
$$V_C = V_{CC} - I_C R_C = 20\text{ V} - (11.5\text{ mA})(1\text{ k}\Omega) = 8.5\text{ V}$$

Related Exercise Determine I_B, I_C, I_E, V_{CE}, and V_{CB} in Figure 18–7 for the following values: $R_B = 22$ kΩ, $R_C = 220$ Ω, $V_{BB} = 6$ V, $V_{CC} = 9$ V, and $\beta_{dc} = 90$.

SECTION 18–1 REVIEW

1. Name the three transistor terminals.
2. Define *forward-reverse bias.*
3. What is β_{dc}?
4. If I_B is 10 μA and β_{dc} is 100, what is the collector current?

18–2 ■ VOLTAGE-DIVIDER BIAS

Next, you will study a method of biasing a transistor for linear operation using a resistive voltage divider. This is the most widely used biasing method, for reasons that you will discover in this section.

After completing this section, you should be able to

- **Analyze a transistor circuit with voltage-divider bias**
    - ☐ Determine base input resistance
    - ☐ Determine base, emitter, and collector voltages
    - ☐ Determine base, emitter, and collector currents

The voltage-divider bias configuration uses only a single dc source to provide forward-reverse bias to the transistor, as shown in Figure 18–8. Resistors R_1 and R_2 form a voltage divider that provides the base bias voltage. Resistor R_E allows the emitter to rise above ground potential.

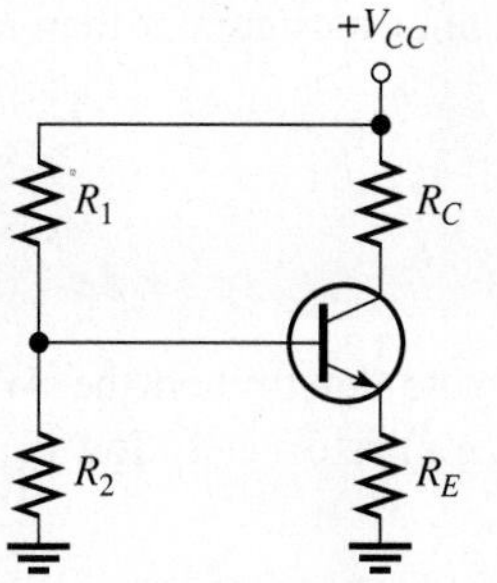

FIGURE 18–8
Voltage-divider bias.

The voltage divider is loaded by the resistance as viewed from the base of the transistor. In some cases, this loading effect is significant in determining the base bias voltage. We will now examine this arrangement in more detail.

Input Resistance at the Base

The approximate input resistance of the transistor, viewed from the base of the transistor in Figure 18–9, is derived as follows:

$$R_{IN} = \frac{V_B}{I_B}$$

Neglecting the V_{BE} of 0.7 V,

$$V_B \cong V_E = I_E R_E$$

Since $I_E \cong I_C$ when $\alpha_{dc} \cong 1$, then

$$I_E \cong \beta_{dc} I_B$$

Substituting yields

$$V_B \cong \beta_{dc} I_B R_E$$

$$R_{IN} \cong \frac{\beta_{dc} I_B R_E}{I_B}$$

The I_B terms cancel, leaving

$$R_{IN} \cong \beta_{dc} R_E \quad \textbf{(18–6)}$$

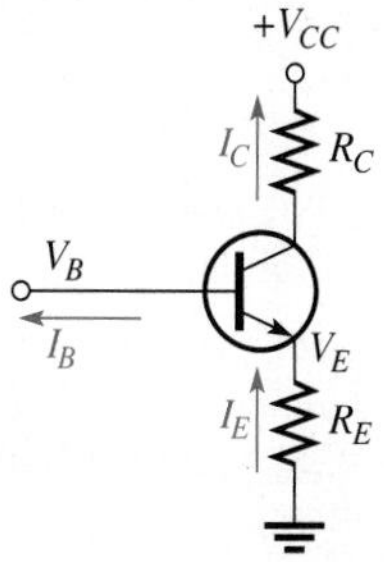

FIGURE 18–9
Circuit for deriving input resistance.

Base Voltage

Now, using the voltage-divider formula, the following equation gives the base voltage for the circuit in Figure 18–8:

$$V_B = \left(\frac{R_2 \parallel R_{IN}}{R_1 + R_2 \parallel R_{IN}} \right) V_{CC} \quad \textbf{(18–7)}$$

If R_{IN} is at least ten times greater than R_2, Equation (18–7) can be simplified as follows:

$$V_B \cong \left(\frac{R_2}{R_1 + R_2}\right) V_{CC} \qquad \textbf{(18–8)}$$

Once you have determined the base voltage, you can determine the emitter voltage V_E (for an NPN transistor) as follows:

$$V_E = V_B - 0.7\text{ V} \qquad \textbf{(18–9)}$$

EXAMPLE 18–2

Determine V_B, V_E, V_C, V_{CE}, I_B, I_E, and I_C in Figure 18–10.

FIGURE 18–10

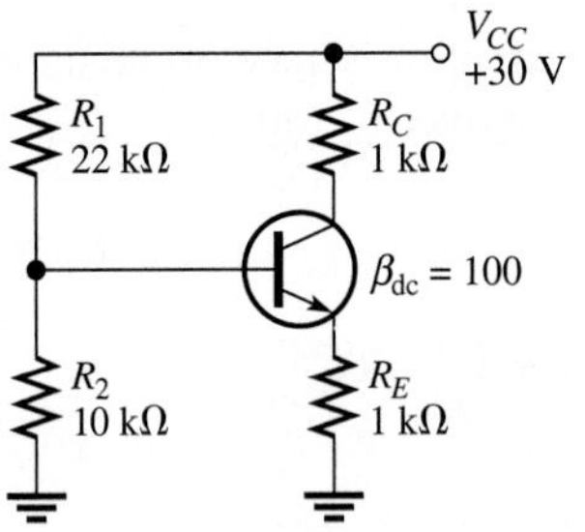

Solution The input resistance at the base is

$$R_{IN} \cong \beta_{dc}R_E = 100(1\text{ k}\Omega) = 100\text{ k}\Omega$$

Since R_{IN} is ten times greater than R_2, you can neglect it and the base voltage is approximately

$$V_B \cong \left(\frac{R_2}{R_1 + R_2}\right)V_{CC} = \left(\frac{10\text{ k}\Omega}{32\text{ k}\Omega}\right)30\text{ V} = 9.38\text{ V}$$

Therefore, $V_E = V_B - 0.7\text{ V} = 8.68\text{ V}$.

Now that you know V_E, you can find I_E by Ohm's law:

$$I_E = \frac{V_E}{R_E} = \frac{8.68\text{ V}}{1\text{ k}\Omega} = 8.68\text{ mA}$$

Since α_{dc} is so close to 1 for most transistors, it is a good approximation to assume that $I_C \cong I_E$. Thus,

$$I_C \cong 8.68\text{ mA}$$

Using $I_C = \beta_{dc}I_B$ and solving for I_B,

$$I_B = \frac{I_C}{\beta_{dc}} = \frac{8.68\text{ mA}}{100} = 86.8\ \mu\text{A}$$

Since you know I_C, you can find V_C:

$$V_C = V_{CC} - I_CR_C = 30\text{ V} - (8.68\text{ mA})(1\text{ k}\Omega) = 30\text{ V} - 8.68\text{ V} = 21.3\text{ V}$$

Since V_{CE} is the collector-to-emitter voltage, it is the difference of V_C and V_E:

$$V_{CE} = V_C - V_E = 21.3\text{ V} - 8.68\text{ V} = 12.6\text{ V}$$

Related Exercise Determine V_B, V_E, V_C, V_{CE}, I_B, I_E, and I_C taking into account R_{IN} at the base in Figure 18–10.

SECTION 18–2 REVIEW

1. How many dc voltage sources are required for voltage-divider bias?
2. In Figure 18–10, if R_2 is 4.7 kΩ, what is the value of V_B?
3. If V_{CC} is reduced to 20 V in Figure 18–10, what is I_C?

18–3 ■ THE BIPOLAR TRANSISTOR AS AN AMPLIFIER

The purpose of dc bias is to allow a transistor to operate as an amplifier. Thus, a transistor can be used to produce a larger signal using a smaller signal as a "pattern." In this section, we will discuss how a transistor acts as an amplifier. The subject of amplifiers is covered in more detail in the next chapter.

After completing this section, you should be able to

- **Explain the operation of a bipolar transistor as an amplifier**
  - ☐ Discuss the collector characteristic curves
  - ☐ Define *cutoff* and *saturation*
  - ☐ Analyze the load line operation of an amplifier
  - ☐ Discuss the significance of the Q point
  - ☐ Explain the signal operation of an amplifier
  - ☐ Calculate the voltage gain of an amplifier

First, the parameters and dc operating conditions important in the operation of a transistor **amplifier** are examined.

Collector Characteristic Curves

With a circuit such as that in Figure 18–11(a), a set of curves can be generated to show how I_C varies with V_{CE} for various values of I_B. These curves are the *collector characteristic curves.*

Notice that both V_{BB} and V_{CC} are adjustable. If V_{BB} is set to produce a specific value of I_B and V_{CC} is zero, then $I_C = 0$ and $V_{CE} = 0$.

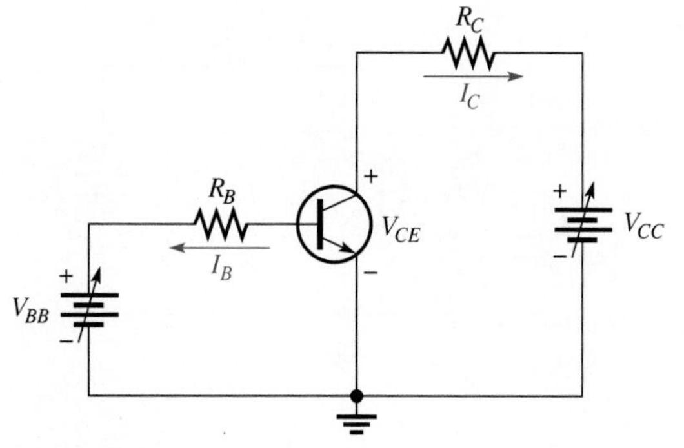

(a) Circuit

FIGURE 18–11
Collector characteristic curves.

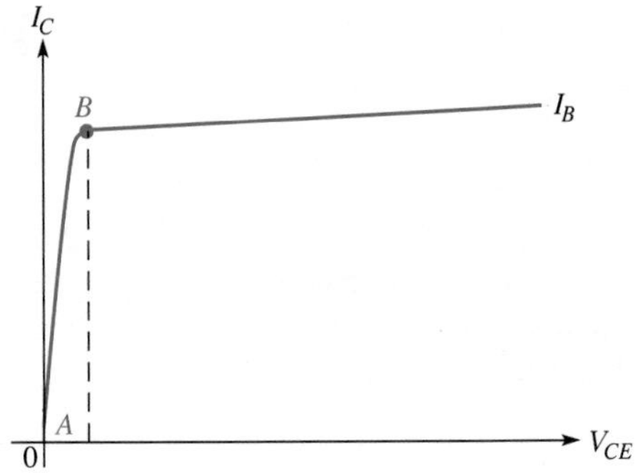

(b) I_C curve for one value of I_B with V_{CE} changing

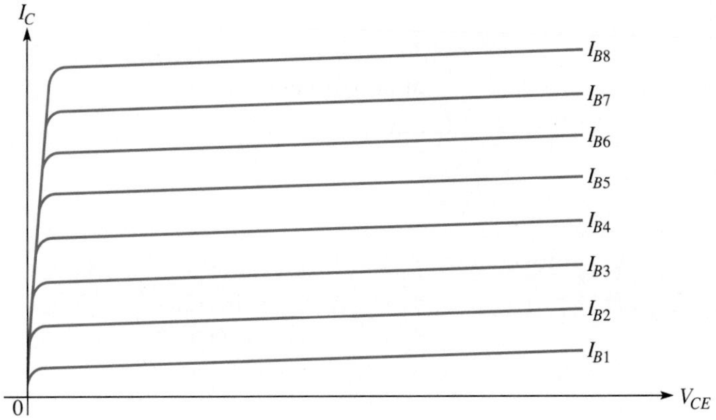

(c) Family of collector curves ($I_{B1} < I_{B2} < I_{B3}$, etc.)

FIGURE 18–11 ***Continued***

As V_{CC} is gradually increased, V_{CE} will increase and so will I_C, as indicated on the portion of the curve between points A and B in Figure 18–11(b).

When V_{CE} reaches approximately 0.7 V, the base-collector junction becomes reverse-biased and I_C reaches its full value determined by the relationship $I_C = \beta_{dc} I_B$. At this point, I_C levels off to an almost constant value as V_{CE} continues to increase. This action appears to the right of point B on the curve. Actually, I_C increases slightly as V_{CE} increases due to widening of the base-collector depletion layer which results in fewer holes for recombination in the base region.

By using other values of I_B, we can produce additional I_C versus V_{CE} curves, as shown in Figure 18–11(c). These curves constitute a family of collector curves for a given transistor.

EXAMPLE 18–3

Sketch the family of collector characteristic curves for the circuit in Figure 18–12 for $I_B = 5\ \mu A$ to $25\ \mu A$ in $5\ \mu A$ increments. Assume that $\beta_{dc} = 100$.

FIGURE 18–12

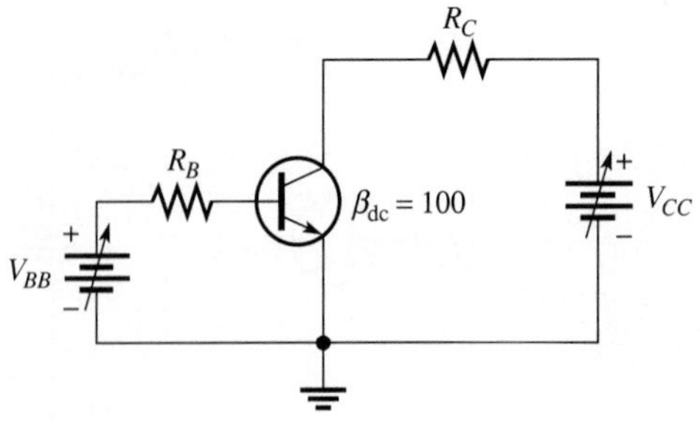

Solution Using the relationship $I_C = \beta_{dc} I_B$, values of I_C are calculated and tabulated in Table 18–1.

TABLE 18–1

I_B	I_C
5 μA	0.5 mA
10 μA	1.0 mA
15 μA	1.5 mA
20 μA	2.0 mA
25 μA	2.5 mA

The resulting curves are plotted in Figure 18–13.

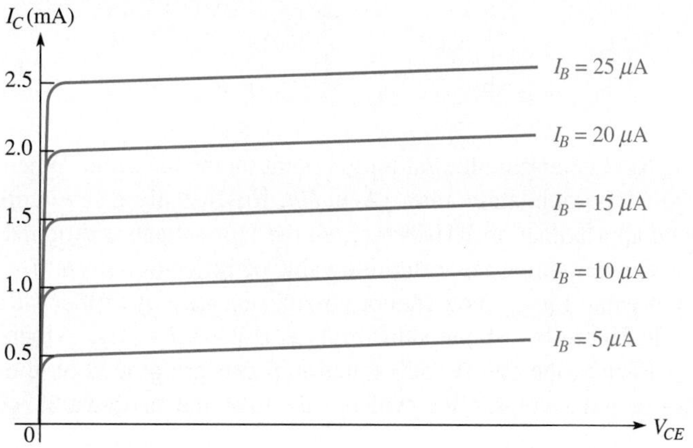

FIGURE 18–13

Related Exercise Where would the curve for $I_B = 0$ appear on the graph in Figure 18–13?

Cutoff and Saturation

When $I_B = 0$, the transistor is in **cutoff.** Under this condition, there is a very small amount of collector leakage current, I_{CEO}, due mainly to thermally produced carriers. In cutoff, both the base-emitter and the base-collector junctions are reverse-biased.

Now let's consider the condition known as **saturation.** When the base current is increased, the collector current also increases and V_{CE} decreases as a result of more drop across R_C. When V_{CE} reaches a value called $V_{CE(sat)}$, the base-collector junction becomes forward-biased and I_C can increase no further even with a continued increase in I_B. At the point of saturation, the relation $I_C = \beta_{dc} I_B$ is no longer valid.

For a transistor, $V_{CE(sat)}$ occurs somewhere below the knee of the collector curves and is usually only a few tenths of a volt for silicon transistors and is often assumed to be zero for analysis purposes.

Load Line Operation

A straight line drawn on the collector curves between the cutoff and saturation points of the transistor is called the *load line.* Once set up, the transistor always operates along this line. Thus, any value of I_C and the corresponding V_{CE} will fall on this line. Notice that the load line is determined by the collector circuit resistance and V_{CC} and not by the transistor itself.

Now we will set up a load line for the circuit in Figure 18–14. First determine the cutoff point on the load line. When the transistor is cut off, there is essentially no collector current. Thus, the collector-emitter voltage V_{CE} is equal to V_{CC}. In this case, $V_{CE} = 30$ V.

FIGURE 18–14

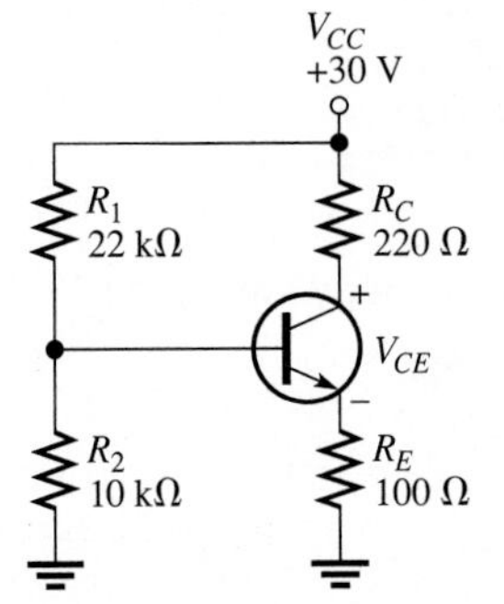

Next determine the saturation point on the load line. When the transistor is saturated, V_{CE} is approximately zero. (Actually, it is usually a few tenths of a volt, but zero is a good approximation.) Therefore, all the V_{CC} voltage is dropped across $R_C + R_E$. From this you can determine the saturation value of collector current, $I_{C(sat)}$. This value is the maximum value for I_C. You cannot possibly increase it further without changing V_{CC}, R_C, or R_E. In Figure 18–14, the value of $I_{C(sat)}$ is $V_{CC}/(R_C + R_E)$, which is 93.8 mA.

Finally, the cutoff and saturation points are plotted on the assumed curves in Figure 18–15, and a straight line, which is the load line, is drawn between them.

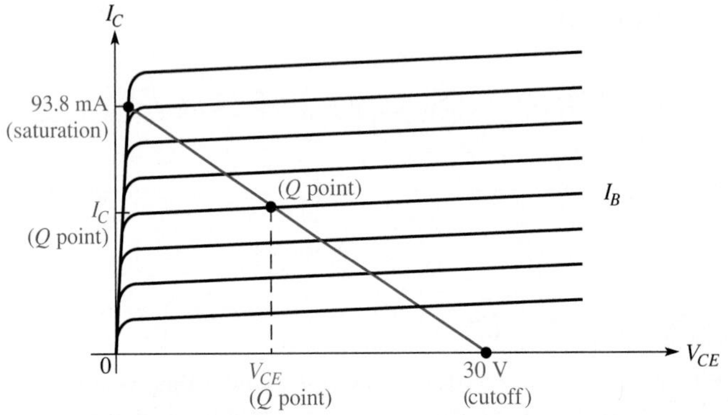

FIGURE 18–15
Load line for the circuit in Figure 18–14.

Q Point

The base current, I_B, is established by the base bias. The point at which the base current curve intersects the load line is the *quiescent* or ***Q* point** for the circuit. The coordinates of the *Q* point are the values for I_C and V_{CE} at that point, as illustrated in Figure 18–15.

Now we have completely described the dc operating conditions for the circuit. In the following paragraphs we discuss ac signal conditions.

Signal (ac) Operation of an Amplifier

The circuit in Figure 18–16 produces an output signal with the same waveform as the input signal but with a greater amplitude. This increase is called **amplification.** The fig-

FIGURE 18–16
An amplifier with voltage-divider bias with capacitively coupled input signal. V_{in} and V_{out} are with respect to ground.

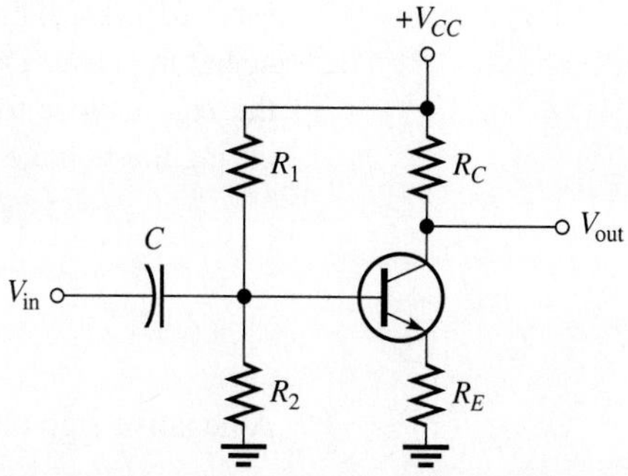

ure shows an input signal, V_{in}, capacitively coupled to the base. The collector voltage is the output signal, as indicated. The input signal voltage causes the base current to vary at the same frequency above and below its dc value. This variation in base current produces a corresponding variation in collector current. However, the variation in collector current is much larger than the variation in base current because of the current **gain** through the transistor. The ratio of the ac collector current (I_c) to the ac base current (I_b) is designated β_{ac} (the ac beta) or h_{fe}:

$$\beta_{ac} = \frac{I_c}{I_b} \qquad \textbf{(18–10)}$$

The value of β_{ac} normally differs slightly from that of β_{dc} for a given transistor. Remember that lowercase subscripts distinguish ac currents and voltages from dc currents and voltages.

Signal Voltage Gain of an Amplifier

Now let's take the amplifier in Figure 18–16 and examine its voltage gain with a signal input. The output voltage is the collector voltage. The variation in collector current produces a variation in the voltage across R_C and a resulting variation in the collector voltage, as shown in Figure 18–17.

FIGURE 18–17
Voltage amplification.

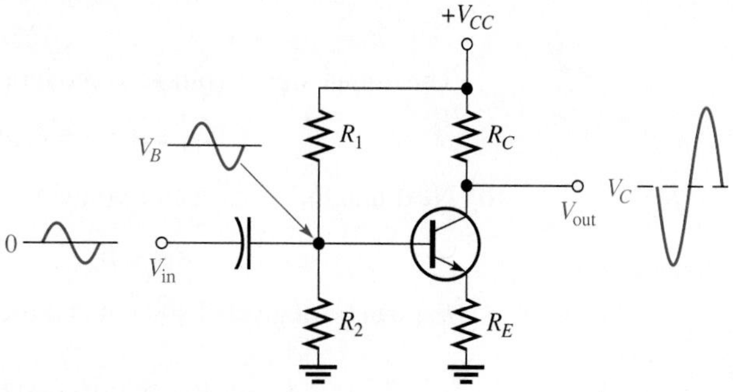

As the collector current increases, the I_cR_C drop increases. This increase produces a decrease in collector voltage because $V_c = V_{CC} - I_cR_C$. Likewise, as the collector current decreases, the I_cR_C drop decreases and produces an increase in collector voltage. Therefore, there is a 180° phase difference between the collector current and the collector voltage. The base voltage and collector voltage are also 180° out of phase, as indicated in Figure 18–17. This 180° phase difference between input and output is called an *inversion.*

The voltage gain A_v of the amplifier is V_{out}/V_{in}, where V_{out} is the signal voltage at the collector and V_{in} is the signal voltage at the base. Because the base-emitter junction is forward-biased, the signal voltage at the emitter is approximately equal to the signal voltage at the base. Thus, since $V_b \cong V_e$, the gain is approximately $V_c/V_e = I_cR_C/I_eR_E$. Because the α_{ac} is close to 1, I_c and I_e are very close to the same value. Therefore, they cancel, giving the voltage gain formula:

$$A_v \cong \frac{R_C}{R_E} \tag{18–11}$$

A negative sign on A_v is often used to indicate the inversion between input and output.

EXAMPLE 18–4

In Figure 18–18, a signal voltage of 50 mV rms is applied to the base.

(a) Determine the output voltage for the amplifier.
(b) Find the dc collector voltage on which the output signal voltage is riding.
(c) Sketch the output waveform.

FIGURE 18–18

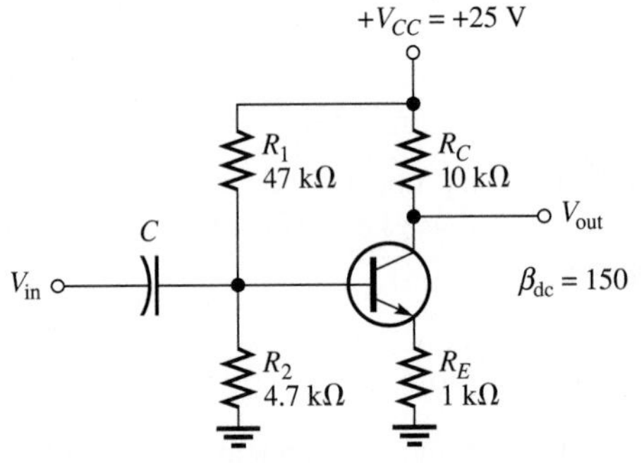

Solution

(a) The signal voltage gain is

$$A_v \cong \frac{R_C}{R_E} = \frac{10\ \text{k}\Omega}{1\ \text{k}\Omega} = 10$$

The output signal voltage is the input signal voltage times the voltage gain:

$$V_{out} = A_vV_{in} = (10)(50\ \text{mV}) = 500\ \text{mV rms}$$

(b) Next find the dc collector voltage:

$$R_{IN} \cong \beta_{dc}R_E = (150)(1\ \text{k}\Omega) = 150\ \text{k}\Omega$$

R_{IN} can be neglected since it is more than ten times R_2. Thus,

$$V_B \cong \left(\frac{R_2}{R_1 + R_2}\right)V_{CC} = \left(\frac{4.7\ \text{k}\Omega}{51.7\ \text{k}\Omega}\right)25\ \text{V} = 2.27\ \text{V}$$

$$I_C \cong I_E = \frac{V_E}{R_E} = \frac{V_B - 0.7\ \text{V}}{1\ \text{k}\Omega} = 1.57\ \text{mA}$$

$$V_C = V_{CC} - I_CR_C = 25\ \text{V} - (1.57\ \text{mA})(10\ \text{k}\Omega) = 9.3\ \text{V}$$

This value is the dc level of the output. The peak value of the output signal is

$$V_p = 1.414(500\ \text{mV}) = 707\ \text{mV}$$

(c) Figure 18–19 shows the output waveform with the signal voltage riding on the 9.3 V dc level.

FIGURE 18–19

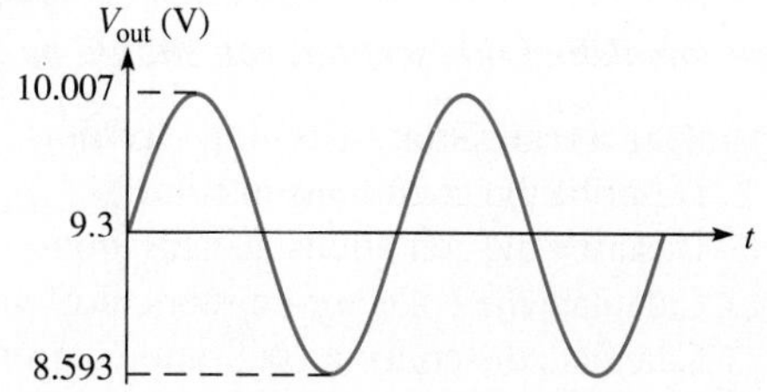

Related Exercise If R_C is changed to 12 kΩ in Figure 18–18, what is the output waveform?

Signal Operation on the Load Line

You can obtain a graphical picture of an amplifier's operation by showing an example of signal variations on a set of collector curves with a load line, as shown in Figure 18–20. Let's assume that the dc Q point values are as follows: $I_B = 40\ \mu A$, $I_C = 4$ mA, and $V_{CE} = 8$ V. The input signal varies the base current from a maximum of 50 μA to a minimum of 30 μA. The resulting variations in collector current and V_{CE} are shown on the graph. The operation is linear as long as the variations do not reach cutoff at the lower end of the load line or saturation at the upper end.

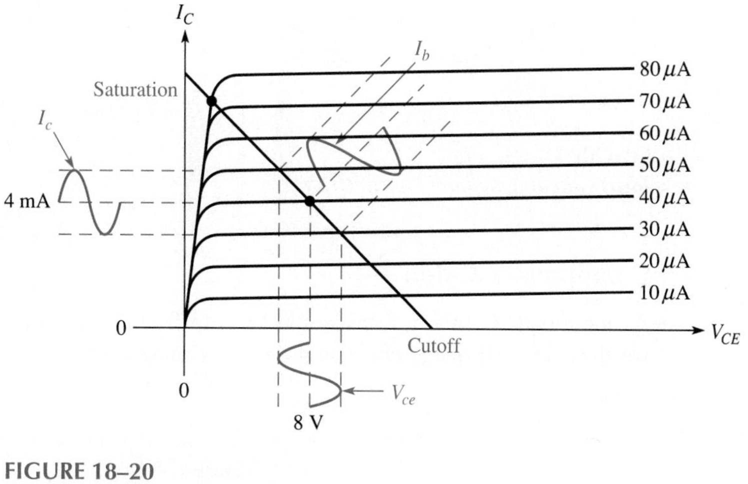

FIGURE 18–20
Signal operation on the load line.

SECTION 18–3 REVIEW

1. What does the term *amplification* mean?
2. What is the Q point?
3. A certain transistor circuit has $R_C = 47$ kΩ and $R_E = 2.2$ kΩ. What is the approximate voltage gain?

18–4 ■ THE BIPOLAR TRANSISTOR AS A SWITCH

In the previous section, we discussed the transistor as a linear amplifier. The second major application area is switching applications. When used as an electronic switch, a transistor normally is operated alternately in cutoff and saturation.

After completing this section, you should be able to

- **Analyze a transistor switching circuit**
  - ☐ Describe the conditions in cutoff
  - ☐ Describe the conditions in saturation
  - ☐ Calculate the collector-emitter cutoff voltage
  - ☐ Calculate the collector saturation current
  - ☐ Calculate the minimum base current to produce saturation

Figure 18–21 illustrates the basic operation of a transistor as a **switch.** In part (a), the transistor is cut off because the base-emitter junction is not forward-biased. In this condition, there is, ideally, an open between collector and emitter. In part (b), the transistor is saturated because the base-emitter junction is forward-biased, and the base current is large enough to cause the collector current to reach its saturated value. In this condition there is, ideally, a short between collector and emitter. Actually, a voltage drop of a few tenths of a volt normally occurs.

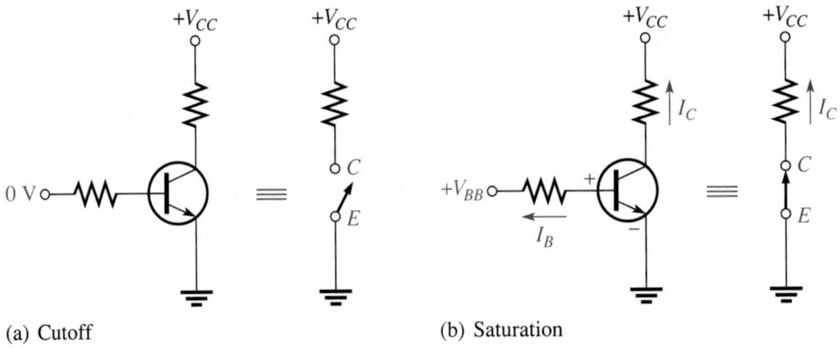

FIGURE 18–21
Ideal switching action of a transistor.

Conditions in Cutoff

As mentioned before, a transistor is in cutoff when the base-emitter junction is *not* forward-biased. All of the currents are approximately zero, and V_{CE} is approximately equal to V_{CC}:

$$V_{CE(\text{cutoff})} \cong V_{CC} \tag{18–12}$$

Conditions in Saturation

When the emitter junction is forward-biased and there is enough base current to produce a maximum collector current, the transistor is saturated. Since V_{CE} is approximately zero at saturation, the collector current is

$$I_{C(\text{sat})} \cong \frac{V_{CC}}{R_C} \tag{18–13}$$

The minimum value of base current needed to produce saturation is

$$I_{B(\text{min})} = \frac{I_{C(\text{sat})}}{\beta_{\text{dc}}} \qquad \textbf{(18–14)}$$

EXAMPLE 18–5

(a) For the transistor switching circuit in Figure 18–22, what is V_{CE} when $V_{\text{IN}} = 0$ V?
(b) What minimum value of I_B is required to saturate this transistor if the β_{dc} is 200?
(c) Calculate the maximum value of R_B when $V_{\text{IN}} = 5$ V.

FIGURE 18–22

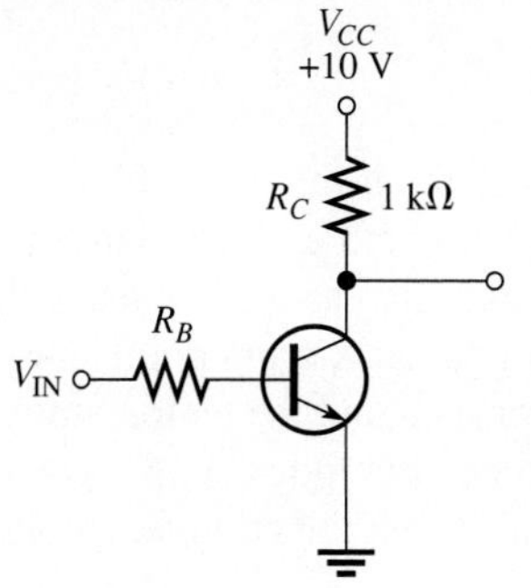

Solution

(a) When $V_{\text{IN}} = 0$ V, the transistor is off and $V_{CE} = V_{CC} = 10$ V.
(b) When the transistor is saturated, $V_{CE} \cong 0$ V, so

$$I_{C(\text{sat})} \cong \frac{V_{CC}}{R_C} = \frac{10\text{ V}}{1\text{ k}\Omega} = 10\text{ mA}$$

$$I_{B(\text{min})} = \frac{I_{C(\text{sat})}}{\beta_{\text{dc}}} = \frac{10\text{ mA}}{200} = 50\ \mu\text{A}$$

This is the value of I_B necessary to drive the transistor to the point of saturation. Any further increase in I_B will drive the transistor deeper into saturation but will not increase I_C.

(c) When the transistor is saturated, $V_{BE} = 0.7$ V. The voltage across R_B is

$$V_{\text{IN}} - 0.7\text{ V} = 4.3\text{ V}$$

The maximum value of R_B needed to allow a minimum I_B of 50 μA is calculated by Ohm's law as follows:

$$R_{B(\text{max})} = \frac{V_{\text{IN}} - 0.7\text{ V}}{I_B} = \frac{4.3\text{ V}}{50\ \mu\text{A}} = 86\text{ k}\Omega$$

Related Exercise Determine the minimum value of I_B required to saturate the transistor in Figure 18–22 if β_{dc} is 125 and $V_{CE(\text{sat})}$ is 0.2 V.

SECTION 18–4 REVIEW

1. When a transistor is used as a switching device, in what two states is it operated?
2. When does the collector current reach its maximum value?
3. When is the collector current approximately zero?
4. Name the two conditions that produce saturation.
5. When is V_{CE} equal to V_{CC}?

18–5 ■ BJT PARAMETERS AND RATINGS

Two parameters, β_{dc} and α_{dc}, were introduced in Section 18–1. Although these parameters are related, β_{dc} is the most useful, and we will discuss it further in this section. Maximum transistor ratings are also discussed because they are important in establishing the limits within which a transistor must operate.

After completing this section, you should be able to

- **Discuss important transistor parameters and maximum ratings**
  - □ Explain how β_{dc} varies with I_C and temperature
  - □ Calculate I_C and V_{CE} based on maximum power dissipation

More about β_{dc}

Because the β_{dc} is such an important bipolar transistor parameter, we need to examine it further. β_{dc} varies with both collector current and temperature. Keeping the junction temperature constant and increasing I_C causes β_{dc} to increase to a maximum. A further increase in I_C beyond this maximum point causes β_{dc} to decrease.

If I_C is held constant and the temperature is varied, β_{dc} changes directly with the temperature. If the temperature goes up, β_{dc} goes up, and vice versa. Figure 18–23 shows the variation of β_{dc} with I_C and junction temperature (T_J) for a typical transistor.

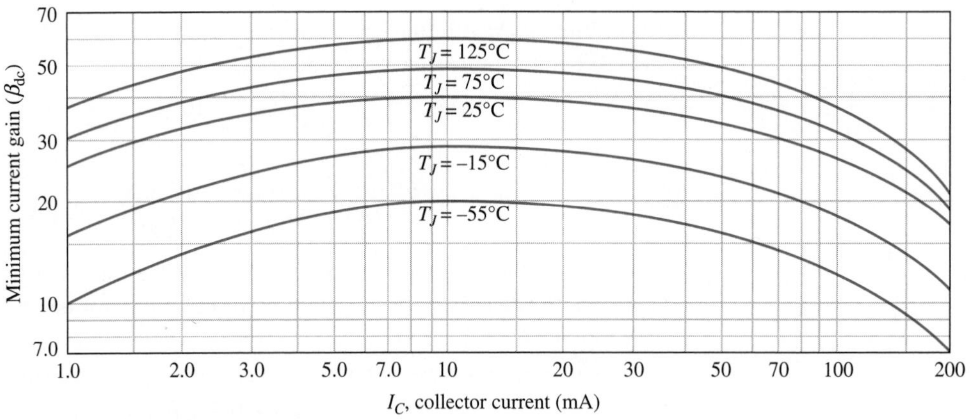

FIGURE 18–23
Variation of β_{dc} with I_C for several temperatures.

A transistor data sheet usually specifies β_{dc} (h_{FE}) at specific I_C values. Even at fixed values of I_C and temperature, β_{dc} varies from device to device for a given transistor. The β_{dc} specified at a certain value of I_C is usually the minimum value, $\beta_{dc(min)}$, although the maximum and typical values are also sometimes specified. A typical transistor data sheet is shown in Appendix F.

Maximum Ratings

Like any other electronic device, the transistor has limitations on its operation. These limitations are stated in the form of maximum ratings and are normally specified on the manufacturer's data sheet. Typically, maximum ratings are given for collector-to-base voltage (V_{CB}), collector-to-emitter voltage (V_{CE}), emitter-to-base voltage (V_{EB}), collector current (I_C), and power dissipation (P_D).

The product of V_{CE} and I_C must not exceed the maximum power dissipation. Both V_{CE} and I_C cannot be maximum at the same time. If V_{CE} is maximum, I_C can be calculated as

$$I_C = \frac{P_{D(max)}}{V_{CE}} \quad (18\text{–}15)$$

If I_C is maximum, we can calculate V_{CE} by rearranging Equation (18–15) as follows:

$$V_{CE} = \frac{P_{D(max)}}{I_C} \quad (18\text{–}16)$$

For a given transistor, a maximum power dissipation curve can be plotted on the collector curves, as shown in Figure 18–24(a). These values are tabulated in part (b). For this transistor, $P_{D(max)}$ is 0.5 W, $V_{CE(max)}$ is 20 V, and $I_{C(max)}$ is 50 mA. The curve shows that this particular transistor cannot be operated in the shaded portion of the graph. $I_{C(max)}$ is the limiting rating between points A and B, $P_{D(max)}$ is the limiting rating between points B and C, and $V_{CE(max)}$ is the limiting rating between points C and D.

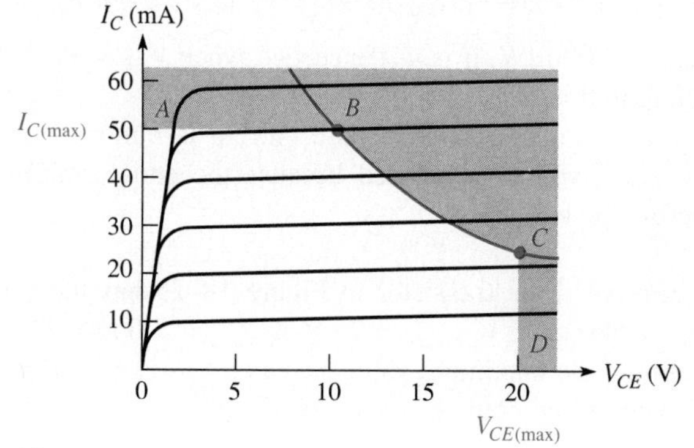

(a)

$P_{D(max)}$	V_{CE}	I_C
0.5 W	5 V	100 mA
0.5 W	10 V	50 mA
0.5 W	15 V	33 mA
0.5 W	20 V	25 mA

(b)

FIGURE 18–24
Maximum power dissipation curve.

EXAMPLE 18–6

The silicon transistor in Figure 18–25 has the following maximum ratings: $P_{D(max)}$ = 800 mW, $V_{CE(max)}$ = 15 V, $I_{C(max)}$ = 100 mA, $V_{CB(max)}$ = 20 V, and $V_{EB(max)}$ = 10 V. Determine the maximum value to which V_{CC} can be adjusted without exceeding a rating. Which rating would be exceeded first?

FIGURE 18–25

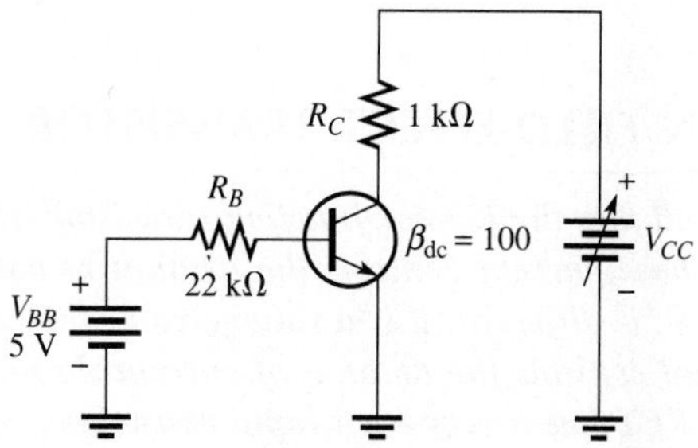

Solution First find I_B so that I_C can be determined.

$$I_B = \frac{V_{BB} - V_{BE}}{R_B} = \frac{5\text{ V} - 0.7\text{ V}}{22\text{ k}\Omega} = 195\ \mu\text{A}$$

$$I_C = \beta_{dc} I_B = (100)(195\ \mu\text{A}) = 19.5\text{ mA}$$

I_C is much less than $I_{C(max)}$ and will not change with V_{CC}. It is determined only by I_B and β_{dc}.

The voltage drop across R_C is

$$V_{R_C} = I_C R_C = (19.5\text{ mA})(1\text{ k}\Omega) = 19.5\text{ V}$$

Now you can determine the maximum value of V_{CC} when $V_{CE} = V_{CE(max)} = 15$ V:

$$V_{R_C} = V_{CC} - V_{CE}$$

Thus,

$$V_{CC(max)} = V_{CE(max)} + V_{R_C} = 15\text{ V} + 19.5\text{ V} = 34.5\text{ V}$$

V_{CC} can be increased to 34.5 V, under the existing conditions, before $V_{CE(max)}$ is exceeded. However, it is not known if $P_{D(max)}$ has been exceeded at this point. The power dissipation is

$$P_D = V_{CE(max)} I_C = (15\text{ V})(19.5\text{ mA}) = 293\text{ mW}$$

Since $P_{D(max)}$ is 800 mW, it is *not* exceeded when $V_{CE} = 34.5$ V. Thus, $V_{CE(max)}$ is the limiting rating in this case.

It should be noted that if the base current is removed, causing the transistor to turn off, $V_{CE(max)}$ will be exceeded because the entire supply voltage, V_{CC}, will be dropped across the transistor.

Related Exercise The transistor in Figure 18–25 has the following maximum ratings: $P_{D(max)} = 500$ mW, $V_{CE(max)} = 25$ V, $I_{C(max)} = 200$ mA, $V_{CB(max)} = 30$ V, $V_{EB(max)} = 15$ V. Determine the maximum value to which V_{CC} can be adjusted without exceeding a rating. Which rating would be exceeded first?

SECTION 18–5 REVIEW

1. Does the β_{dc} of a transistor increase or decrease with temperature?
2. Generally, what effect does an increase in I_C have on the β_{dc}?
3. What is the allowable collector current in a transistor with $P_{D(max)} = 32$ mW when $V_{CE} = 8$ V?

18–6 ■ THE JUNCTION FIELD-EFFECT TRANSISTOR (JFET)

Recall that the bipolar junction transistor (BJT) is a current-controlled device; that is, the base current controls the amount of collector current. The* field-effect transistor (FET) *is different; it is a voltage-controlled device in which the voltage at the gate terminal controls the amount of current through the device. Also, compared to the BJT, the FET has a very high input resistance, which makes it superior in certain applications.

After completing this section, you should be able to

- **Describe the basic structure and operation of a JFET**
  - ☐ Identify the standard JFET symbols
  - ☐ Explain the difference between N channel and P channel JFETs
  - ☐ Label the terminals of a JFET

The **junction field-effect transistor (JFET)** is a type of FET that operates with a reverse-biased junction to control current in the channel. Depending on their structure, JFETs fall into either of two categories: N-channel or P-channel. Figure 18–26(a) shows the basic structure of an N-channel JFET. Wire leads are connected to each end of the N-channel; the **drain** is at the upper end and the **source** is at the lower end. Two P-type regions are diffused in the N-type material to form a channel, and both P-type regions are connected to the **gate** lead. In the remaining structure diagrams, the interconnection of both P-type regions is omitted for simplicity, with a connection to only one shown. A P-channel JFET is shown in Figure 18–26(b).

FIGURE 18–26
Basic structure of the two types of JFET.

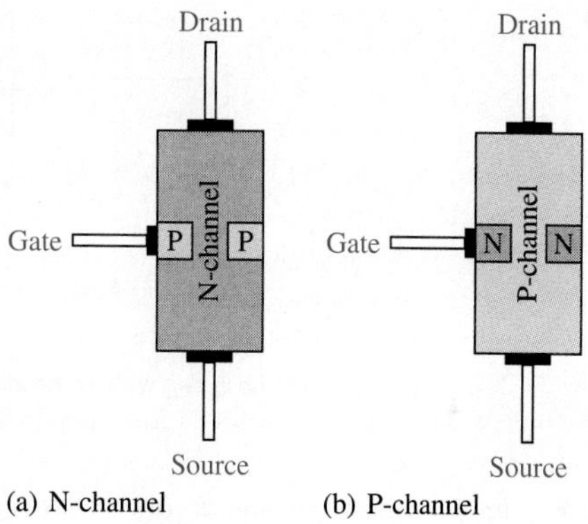

Basic Operation

To illustrate the operation of a JFET, bias voltages are shown applied to an N-channel device in Figure 18–27(a). V_{DD} provides a drain-to-source voltage and supplies current from drain to source. V_{GG} sets the reverse-biased voltage between the gate and the source, as shown.

The JFET is always operated with the gate-to-source PN junction reverse-biased. Reverse-biasing of the gate-source junction with a negative gate voltage produces a depletion region in the N channel and thus increases its resistance. The channel width can be controlled by varying the gate voltage, and thereby the amount of drain current, I_D, can also be controlled. This concept is illustrated in Figure 18–27(b) and (c).

The white areas surrounding the P material of the gate represent the depletion region created by the reverse bias. This depletion region is wider toward the drain end of the channel because the reverse-biased voltage between the gate and the drain is greater than that between the gate and the source.

JFET Symbols

The schematic symbols for both N-channel and P-channel JFETs are shown in Figure 18–28. Notice that the arrow on the gate points "in" for N-channel and "out" for P-channel.

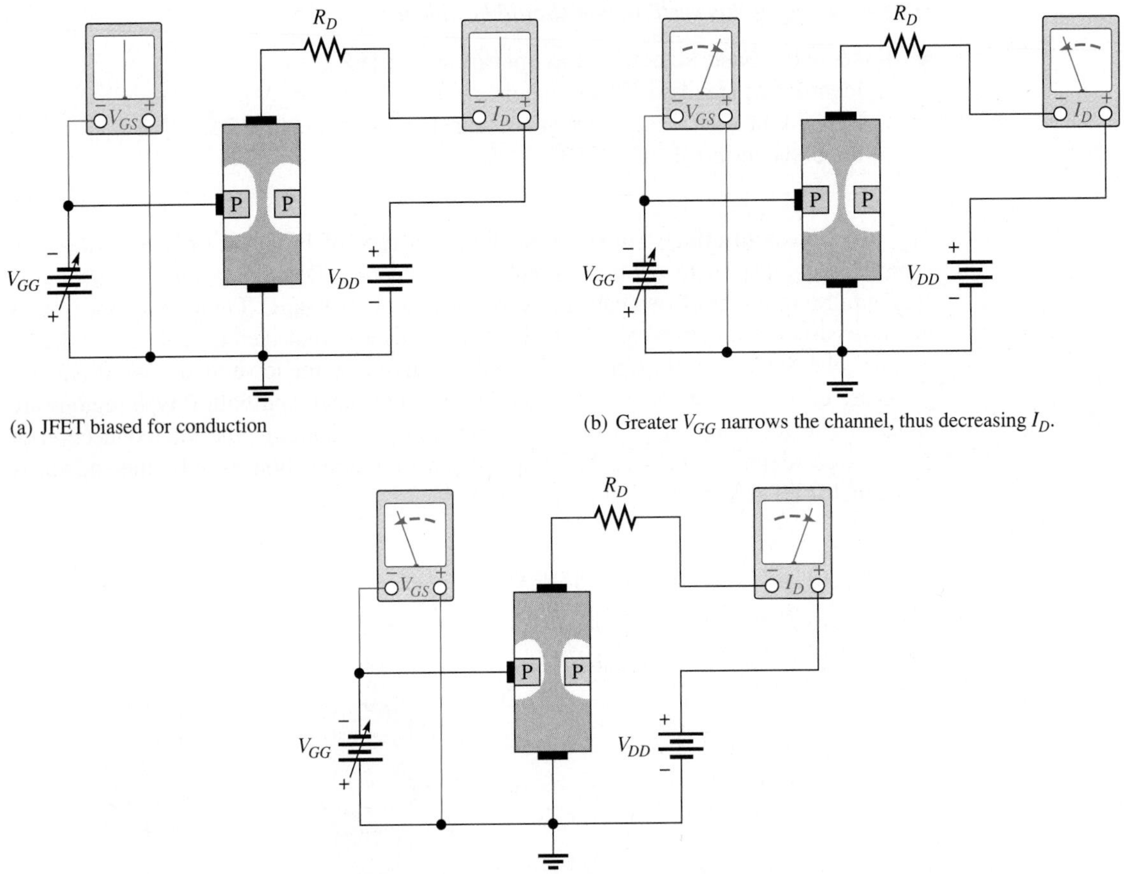

(a) JFET biased for conduction

(b) Greater V_{GG} narrows the channel, thus decreasing I_D.

(c) Less V_{GG} widens the channel and increases I_D.

FIGURE 18–27
Effects of V_{GG} on channel width and drain current ($V_{GG} = V_{GS}$).

FIGURE 18–28
JFET schematic symbols.

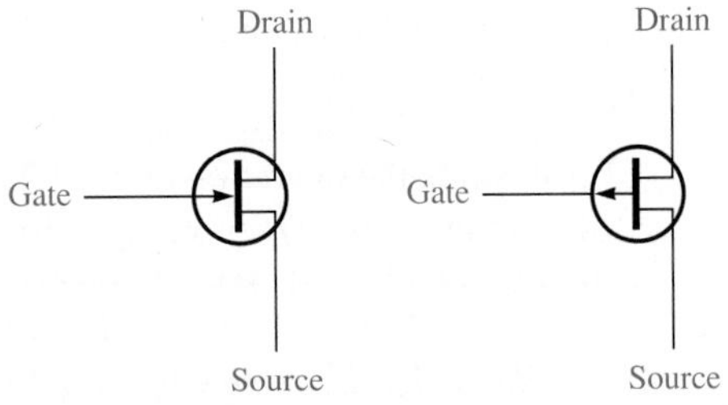

(a) N-channel

(b) P-channel

SECTION 18–6 REVIEW

1. Name the three terminals of a JFET.
2. An N-channel JFET requires a (positive, negative, or 0) V_{GS}.
3. How is the drain current controlled in a JFET?

18–7 ■ JFET CHARACTERISTICS

In this section, you will see how the JFET operates as a voltage-controlled, constant-current device. You will also learn about cutoff and pinch-off as well as JFET input resistance and capacitance.

After completing this section, you should be able to

- **Explain the operation of a JFET in terms of its characteristics**
  - □ Define *pinch-off*
  - □ Describe how V_{GS} controls I_D.
  - □ Discuss cutoff and how it differs from pinch-off

First let's consider the case where the gate-to-source voltage is 0 ($V_{GS} = 0$ V). This voltage is produced by shorting the gate to the source, as in Figure 18–29(a) where both are grounded. As V_{DD} (and thus V_{DS}) is increased from 0 V, I_D will increase proportionally, as shown in the graph of Figure 18–29(b) between points *A* and *B*. In this region, the channel resistance is essentially constant because the depletion region is not large enough to have significant effect. This region is called the *ohmic region* because V_{DS} and I_D are related by Ohm's law.

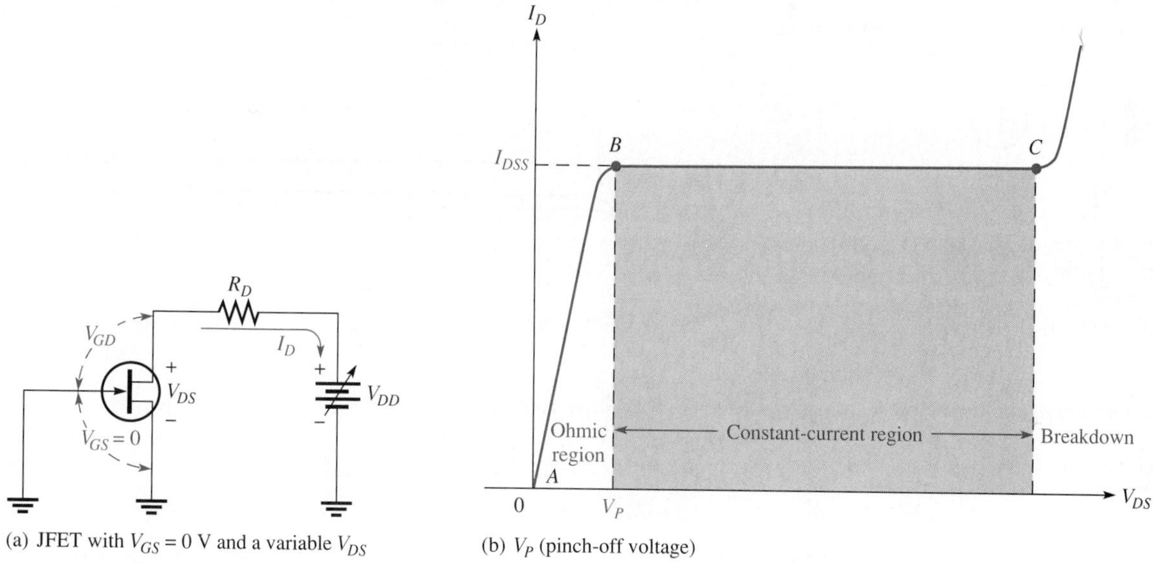

(a) JFET with $V_{GS} = 0$ V and a variable V_{DS}

(b) V_P (pinch-off voltage)

FIGURE 18–29
The drain characteristic curve of a JFET for $V_{GS} = 0$ V, showing pinch-off.

At point *B* in Figure 18–29(b), the curve levels off and I_D becomes essentially constant. As V_{DS} increases from point *B* to point *C*, the reverse-bias voltage from gate to drain (V_{GD}) produces a depletion region large enough to offset the increase in V_{DS}, thus keeping I_D relatively constant.

Pinch-Off

For $V_{GS} = 0$ V, the value of V_{DS} at which I_D becomes essentially constant (point *B* on the curve in Figure 18–29(b)) is the **pinch-off voltage,** V_P. For a given JFET, V_P has a fixed value. As you can see, a continued increase in V_{DS} above the pinch-off voltage produces an almost constant drain current. This value of drain current is I_{DSS} (Drain to Source current with gate Shorted) and is always specified on JFET data sheets. I_{DSS} is the *maximum* drain current that a specific JFET can produce regardless of the external circuit, and it is always specified for the condition, $V_{GS} = 0$ V.

Continuing along the graph in Figure 18–29(b), breakdown occurs at point *C* when I_D begins to increase very rapidly with any further increase in V_{DS}. Breakdown can result in irreversible damage to the device, so JFETs are always operated below breakdown and within the constant-current region (between points *B* and *C* on the graph).

V_{GS} Controls I_D

Let's connect a bias voltage, V_{GG}, from gate to source as shown in Figure 18–30(a). As V_{GS} is set to increasingly more negative values by adjusting V_{GG}, a family of drain characteristic curves is produced as shown in Figure 18–30(b). Notice that I_D decreases as the magnitude of V_{GS} is increased to larger negative values. Also notice that, for each increase in V_{GS}, the JFET reaches pinch-off (where constant current begins) at values of V_{DS} less then V_P. So, the amount of drain current is controlled by V_{GS}.

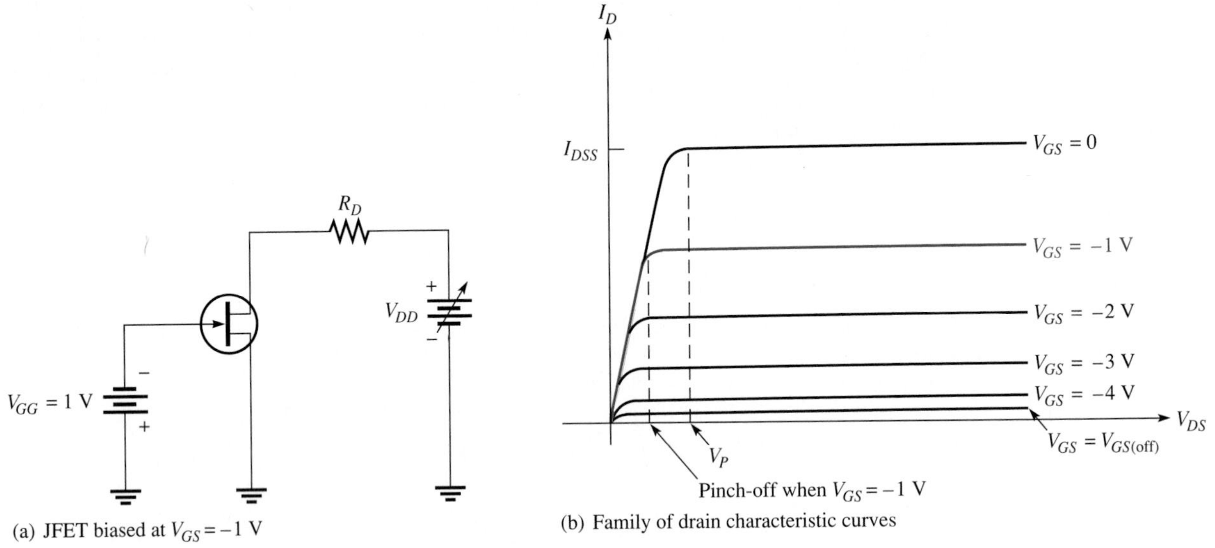

FIGURE 18–30
Pinch-off occurs at a lower V_{DS} as V_{GS} is increased to more negative values.

Cutoff

The value of V_{GS} that makes I_D approximately zero is the cutoff value, $V_{GS(off)}$. The JFET must be operated between $V_{GS} = 0$ V and $V_{GS(off)}$. For this range of gate-to-source voltages, I_D will vary from a maximum of I_{DSS} to a minimum of almost zero.

As you have seen, for an N-channel JFET, the more negative V_{GS} is, the smaller I_D becomes in the constant-current region. When V_{GS} has a sufficiently large negative value, I_D is reduced to zero. This cutoff effect is caused by the widening of the depletion region to a point where it completely closes the channel as shown in Figure 18–31.

FIGURE 18–31
JFET at cutoff.

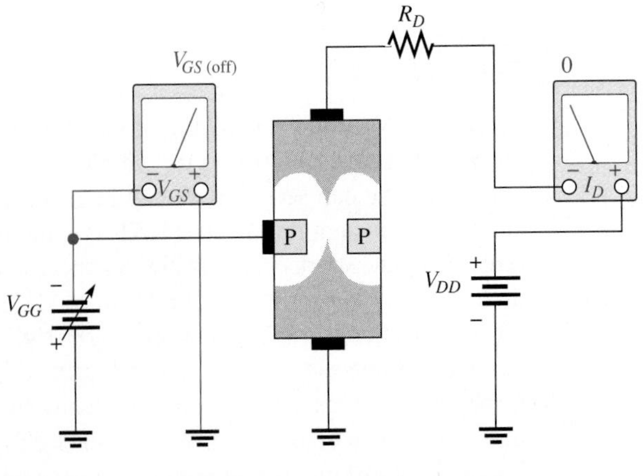

The basic operation of a P-channel JFET is the same as for an N-channel device except that it requires a negative V_{DD} and a positive V_{GS}.

Comparison of Pinch-Off and Cutoff

As you have seen, there is definitely a difference between pinch-off and cutoff. There is also a connection. V_P is the value of V_{DS} at which the drain current becomes constant and is always measured at $V_{GS} = 0$ V. However, pinch-off occurs for V_{DS} values less than V_P when V_{GS} is nonzero. So, although V_P is a constant, the minimum value of V_{DS} at which I_D becomes constant varies with V_{GS}.

$V_{GS(\text{off})}$ and V_P are always equal in magnitude but opposite in sign. A data sheet usually will give either $V_{GS(\text{off})}$ or V_P, but not both. However, when you know one, you have the other. For example, if $V_{GS(\text{off})} = -5$ V, then $V_P = +5$ V.

EXAMPLE 18–7 For the JFET in Figure 18–32, $V_{GS(\text{off})} = -3.5$ V and $I_{DSS} = 6$ mA. Determine the minimum value of V_{DD} required to put the device in the constant-current region of operation.

FIGURE 18–32

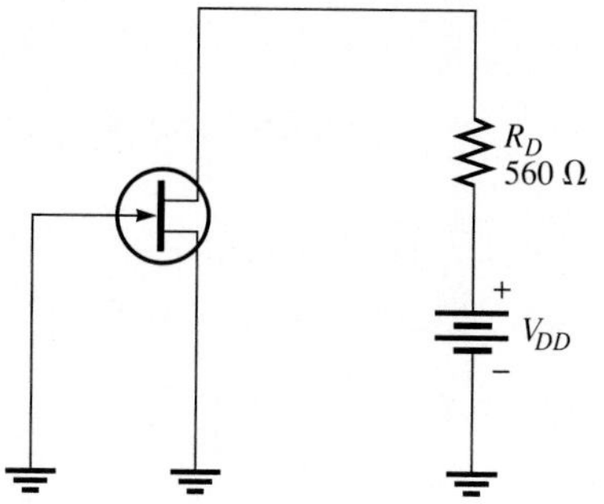

Solution Since $V_{GS(\text{off})} = -3.5$ V, $V_P = 3.5$ V. The minimum value of V_{DS} for the JFET to be in its constant-current region is

$$V_{DS} = V_P = 3.5 \text{ V}$$

In the constant-current region with $V_{GS} = 0$ V,

$$I_D = I_{DSS} = 6 \text{ mA}$$

The drop across the drain resistor is

$$V_{R_D} = (6 \text{ mA})(560 \ \Omega) = 3.36 \text{ V}$$

Applying Kirchhoff's law around the drain circuit gives

$$V_{DD} = V_{DS} + V_{R_D} = 3.5 \text{ V} + 3.36 \text{ V} = 6.86 \text{ V}$$

This is the value of V_{DD} to make $V_{DS} = V_P$ and put the device in the constant-current region.

Related Exercise If V_{DD} is increased to 15 V in Figure 18–32, what is the drain current?

JFET Input Resistance and Capacitance

A JFET operates with its gate-source junction reverse-biased. Therefore, the input resistance at the gate is very high. This high input resistance is one advantage of the JFET over the bipolar transistor. (Recall that a bipolar transistor operates with a forward-biased base-emitter junction.)

JFET data sheets often specify the input resistance by giving a value for the gate reverse current, I_{GSS}, at a certain gate-to-source voltage. The input resistance can then be determined using the following equation. The vertical lines indicate an absolute value (no sign).

$$R_{IN} = \left| \frac{V_{GS}}{I_{GSS}} \right| \tag{18–17}$$

For example, the 2N3970 data sheet lists a maximum I_{GSS} of 250 pA for $V_{GS} = -20$ V at 25°C. I_{GSS} increases with temperature, so the input resistance decreases.

The input capacitance C_{iss} of a JFET is considerably greater than that of a bipolar transistor because the JFET operates with a reverse-biased PN junction. Recall that a reverse-biased PN junction acts as a capacitor whose capacitance depends on the amount of reverse voltage. For example, the 2N3970 has a maximum C_{iss} of 25 pF for $V_{GS} = 0$ V.

EXAMPLE 18–8

A certain JFET has an I_{GSS} of 1 nA for $V_{GS} = -20$ V. Determine the input resistance.

Solution

$$R_{IN} = \left| \frac{V_{GS}}{I_{GSS}} \right| = \frac{20\text{ V}}{1\text{ nA}} = 20{,}000\text{ M}\Omega$$

Related Exercise Determine the minimum input resistance for the 2N3970 JFET.

SECTION 18–7 REVIEW

1. The drain-to-source voltage at the pinch-off point of a particular JFET is 7 V. If the gate-to-source voltage is 0 V, what is V_p?
2. The V_{GS} of a certain N-channel JFET is increased negatively. Does the drain current increase or decrease?
3. What value must V_{GS} have to produce cutoff in a P-channel JFET with a $V_p = -3$ V?

18–8 ▪ THE METAL-OXIDE SEMICONDUCTOR FET (MOSFET)

The metal-oxide semiconductor field-effect transistor (MOSFET) is the second category of field-effect transistor. The* MOSFET *differs from the JFET in that it has no PN junction structure; instead, the gate of the MOSFET is insulated from the channel by a silicon dioxide (SiO_2) layer. The two basic types of MOSFETs are depletion (D) and enhancement (E).

After completing this section, you should be able to

- **Explain the basic structure and operation of MOSFETs**
  - □ Discuss the D-MOSFET
  - □ Discuss the E-MOSFET
  - □ Recognize the symbols for both types of MOSFETs
  - □ Describe the proper handling of MOSFETs and discuss why it is necessary

Depletion (D-MOSFET)

Figure 18–33 illustrates the basic structure of D-MOSFETs. The drain and source are diffused into the substrate material and then connected by a narrow channel adjacent to the insulated gate. Both N-channel and P-channel devices are shown in the figure. We will use the N-channel device to describe the basic operation. The P-channel operation is the same, except the voltage polarities are opposite those of the N-channel.

FIGURE 18–33
Basic structure of D-MOSFETs.

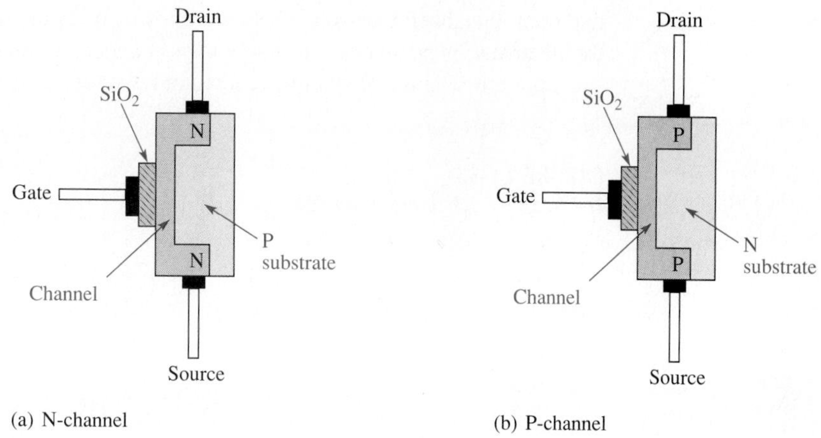

The D-MOSFET can be operated in either of two modes, the depletion mode or the enhancement mode, and is sometimes called a *depletion-enhancement MOSFET*. Since the gate is insulated from the channel, either a positive or a negative gate voltage can be applied. The MOSFET operates in the depletion mode when a negative gate-to-source voltage is applied and in the enhancement mode when a positive gate-to-source voltage is applied.

Depletion Mode Visualize the gate as one plate of a parallel plate capacitor and the channel as the other plate. The silicon dioxide insulating layer is the dielectric. With a negative gate voltage, the negative charges on the gate repel conduction electrons from the channel, leaving positive ions in their place. Thereby, the N-channel is depleted of some of its electrons, thus decreasing the channel conductivity. The greater the negative voltage on the gate, the greater the depletion of N-channel electrons. At a sufficiently negative gate-to-source voltage, $V_{GS(\text{off})}$, the channel is totally depleted and the drain current is zero. This depletion mode is illustrated in Figure 18–34(a).

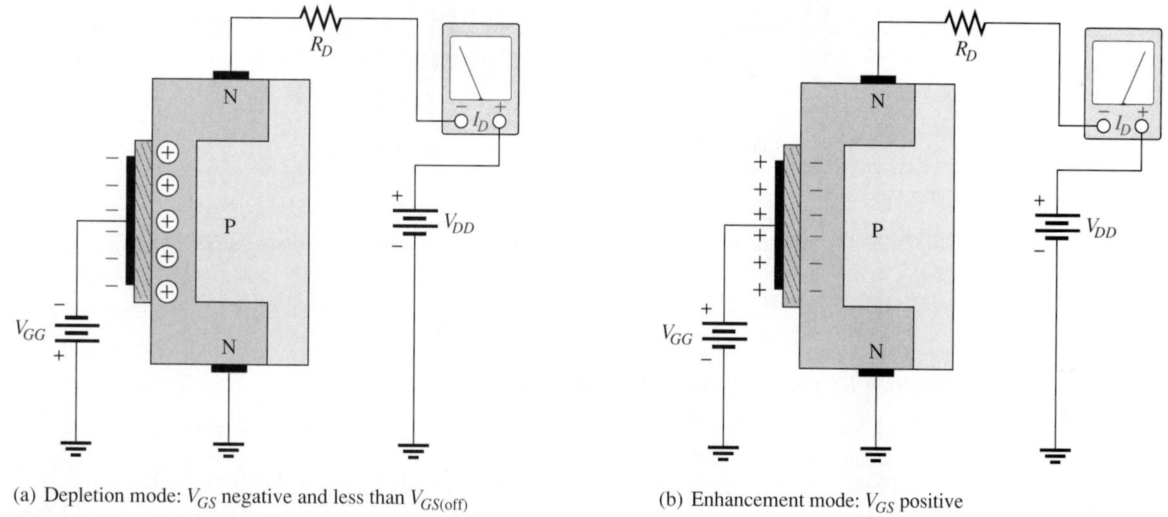

FIGURE 18–34
Operation of N-channel D-MOSFET.

Like the N-channel JFET, the N-channel D-MOSFET conducts drain current for gate-to-source voltages between $V_{GS(\text{off})}$ and 0 V. In addition, the D-MOSFET conducts for values of V_{GS} above 0 V.

Enhancement Mode With a positive gate voltage, more conduction electrons are attracted into the channel, thus increasing (enhancing) the channel conductivity, as illustrated in Figure 18–34(b).

D-MOSFET Symbols The schematic symbols for both the N-channel and the P-channel depletion-enhancement MOSFETs are shown in Figure 18–35. The substrate, indicated by the arrow, is normally (but not always) connected internally to the source. An inward substrate arrow is for N-channel, and an outward arrow is for P-channel.

FIGURE 18–35
D-MOSFET schematic symbols.

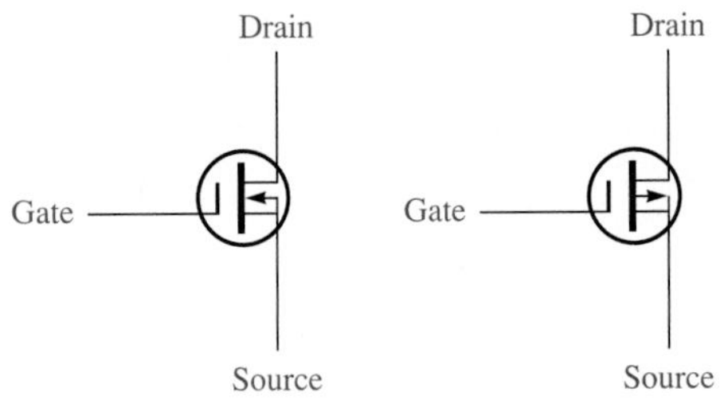

(a) N-channel (b) P-channel

Enhancement (E-MOSFET)

This type of MOSFET operates *only* in the enhancement mode and has no depletion mode. It differs in construction from the D-MOSFET in that it has no physical channel. Notice in Figure 18–36(a) that the substrate extends completely to the SiO_2 layer.

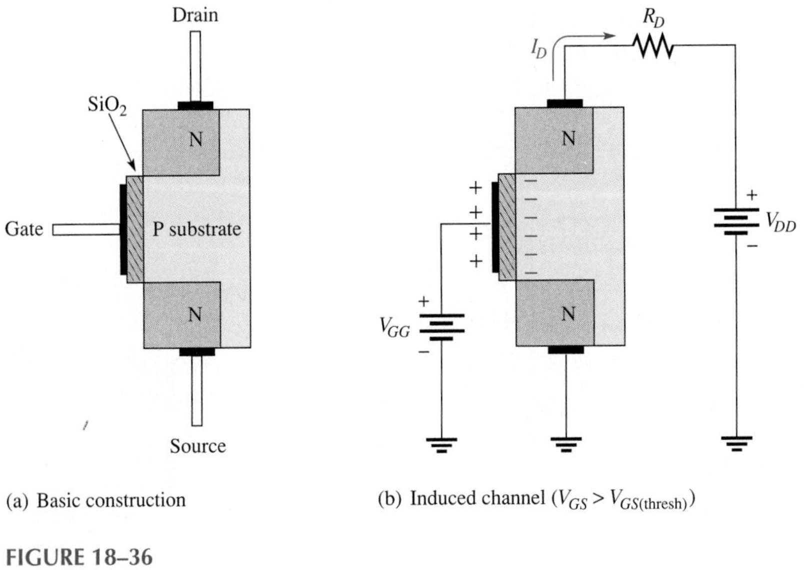

(a) Basic construction (b) Induced channel ($V_{GS} > V_{GS(\text{thresh})}$)

FIGURE 18–36
E-MOSFET construction and operation.

For an N-channel device, a positive gate voltage above a threshold value, $V_{GS(\text{thresh})}$, induces a channel by creating a thin layer of negative charges in the substrate region adjacent to the SiO_2 layer, as shown in Figure 18–36(b). The conductivity of the channel is enhanced by increasing the gate-to-source voltage, thus pulling more electrons into the channel. For any gate voltage below the threshold value, there is no channel.

The schematic symbols for the N-channel and P-channel E-MOSFETs are shown in Figure 18–37. The broken lines symbolize the absence of a physical channel.

FIGURE 18–37
E-MOSFET schematic symbols.

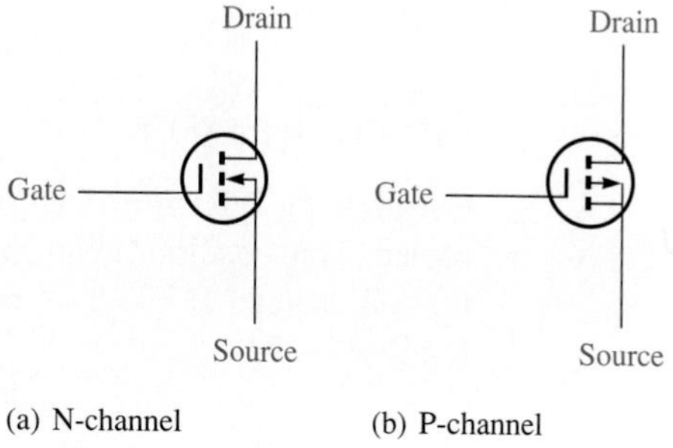

Handling Precautions

Because the gate of a MOSFET is insulated from the channel, the input resistance is extremely high (ideally infinite). The gate leakage current, I_{GSS}, for a typical MOSFET is in the pA range, whereas the gate reverse current for a typical JFET is in the nA range.

The input capacitance, of course, results from the insulated gate structure. Excess static charge can accumulate because the input capacitance combines with the very high input resistance and can result in damage to the device as a result of electrostatic discharge (ESD). To avoid ESD and possible damage, the following precautions should be taken:

1. MOS devices should be shipped and stored in conductive foam.
2. All instruments and metal benches used in assembly or test should be connected to earth ground (round prong of wall outlets).
3. The assembler's or handler's wrist should be connected to earth ground with a length of wire and a high-value series resistor.
4. Never remove a MOS device (or any other device, for that matter) from the circuit while the power is on.
5. Do not apply signals to a MOS device while the dc power supply is off.

SECTION 18–8 REVIEW

1. Name two types of MOSFETs, and describe the major difference in construction.
2. If the gate-to-source voltage in a depletion-enhancement MOSFET is 0 V, is there current from drain to source?
3. If the gate-to-source voltage in an E-MOSFET is 0 V, is there current from drain to source?

18–9 ▪ FET BIASING

Using some of the FET parameters discussed in the previous sections, we will now see how to dc bias FETs. The purpose of biasing is to select a proper dc gate-to-source voltage to establish a desired value of drain current.

After completing this section, you should be able to

- **Analyze several biasing arrangements for JFETs and MOSFETs**
  - □ Analyze a self-biased JFET circuit
  - □ Analyze a zero-biased D-MOSFET circuit
  - □ Analyze drain-feedback and voltage-divider biased E-MOSFET circuits

Self-Biasing a JFET

Recall that a JFET must be operated such that the gate-source junction is always reverse-biased. This condition requires a negative V_{GS} for an N-channel JFET and a positive V_{GS} for a P-channel JFET. This can be achieved using the self-bias arrangements shown in Figure 18–38.

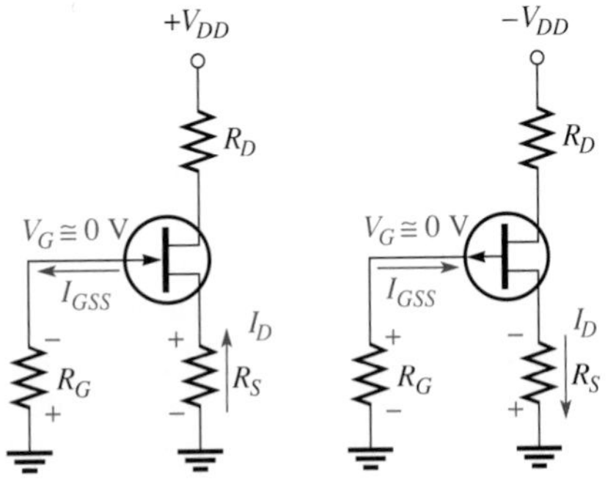

FIGURE 18–38
Self-biased JFETS ($I_S = I_D$ in all FETs).

Notice that the gate is biased at approximately 0 V by resistor R_G connected to ground. The reverse leakage current, I_{GSS}, does produce a very small voltage across R_G, but this can be neglected in most cases, and it can be assumed that R_G has no voltage drop across it as we will do.

For the N-channel JFET in Figure 18–38(a), I_D produces a voltage drop across R_S and makes the source positive with respect to ground. Since $V_G = 0$ and $V_S = I_D R_S$, the gate-to-source voltage is

$$V_{GS} = V_G - V_S = 0\text{ V} - I_D R_S$$

Thus,

$$V_{GS} = -I_D R_S \qquad \textbf{(18–18)}$$

For the P-channel JFET shown in Figure 18–38(b), the current through R_S produces a negative voltage at the source, making the gate positive with respect to the source. Therefore,

$$V_{GS} = +I_D R_S \qquad \textbf{(18–19)}$$

In the following analysis, the N-channel JFET is used for illustration. Keep in mind that analysis of the P-channel JFET is the same except for opposite polarity voltages.

The drain voltage with respect to ground is determined as follows:

$$V_D = V_{DD} - I_D R_D \qquad \textbf{(18–20)}$$

Since $V_S = I_D R_S$, the drain-to-source voltage is

$$V_{DS} = V_D - V_S$$

$$V_{DS} = V_{DD} - I_D(R_D + R_S) \quad \textbf{(18–21)}$$

EXAMPLE 18–9 Find V_{DS} and V_{GS} in Figure 18–39, given that $I_D = 5$ mA.

FIGURE 18–39

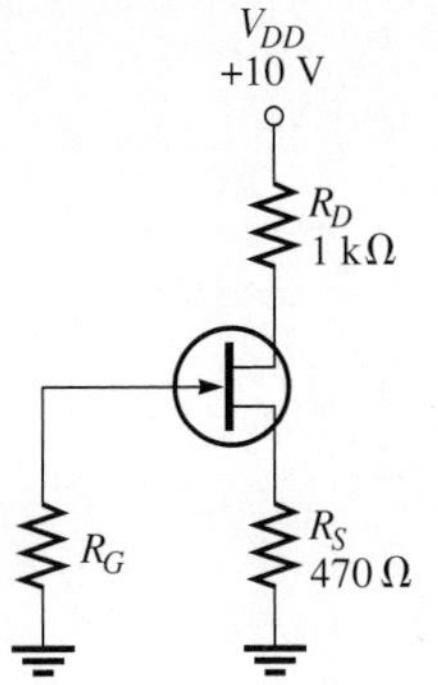

Solution $V_S = I_D R_S = (5 \text{ mA})(470\ \Omega) = 2.35$ V
$V_D = V_{DD} - I_D R_D = 10 \text{ V} - (5 \text{ mA})(1\text{ k}\Omega) = 10 \text{ V} - 5 \text{ V} = 5$ V

Therefore,

$$V_{DS} = V_D - V_S = 5 \text{ V} - 2.35 \text{ V} = 2.65 \text{ V}$$

Since $V_G = 0$ V,

$$V_{GS} = V_G - V_S = 0 \text{ V} - 2.35 \text{ V} = -2.35 \text{ V}$$

Related Exercise Determine V_{DS} and V_{GS} in Figure 18–39 when $I_D = 8$ mA. Assume that $R_D = 860\ \Omega$, $R_S = 390\ \Omega$, and $V_{DD} = 12$ V.

D-MOSFET Bias

Recall that depletion-enhancement MOSFETs can be operated with either positive or negative values of V_{GS}. A simple bias method is to set $V_{GS} = 0$ V so that an ac signal at the gate varies the gate-to-source voltage above and below this bias point. A MOSFET with zero bias is shown in Figure 18–40.

FIGURE 18–40
A zero-biased D-MOSFET.

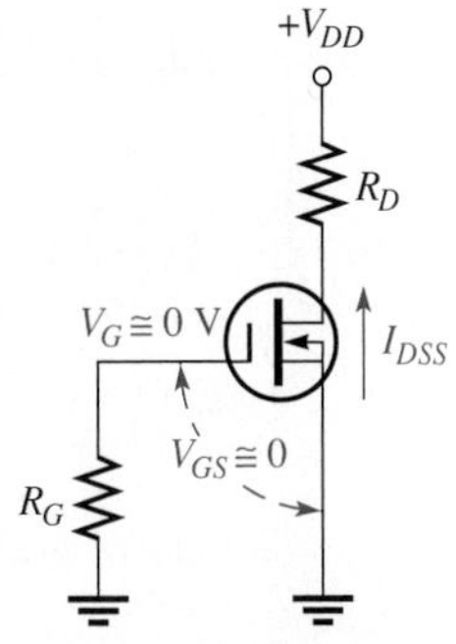

Since $V_{GS} = 0$ V, $I_D = I_{DSS}$ as indicated. The drain-to-source voltage is expressed as follows:

$$V_{DS} = V_{DD} - I_{DSS}R_D \tag{18–22}$$

EXAMPLE 18–10 Determine the drain current in the circuit of Figure 18–41. The MOSFET data sheet gives $V_{GS(\text{off})} = -8$ V and $I_{DSS} = 12$ mA.

FIGURE 18–41

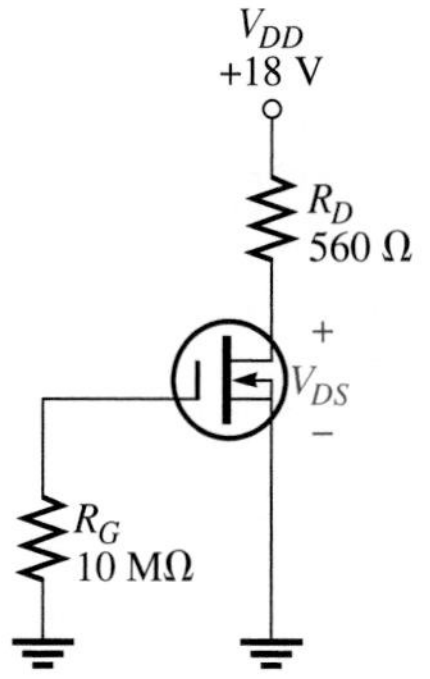

Solution Since $I_D = I_{DSS} = 12$ mA, the drain-to-source voltage is calculated as follows:

$$V_{DS} = V_{DD} - I_{DSS}R_D = 18\text{ V} - (12\text{ mA})(560\ \Omega) = 11.3\text{ V}$$

Related Exercise Find V_{DS} in Figure 18–41 when $V_{GS(\text{off})} = -10$ V and $I_{DSS} = 20$ mA.

E-MOSFET Bias

Recall that enhancement-only MOSFETs must have a V_{GS} greater than the threshold value, $V_{GS(\text{thresh})}$. Figure 18–42 shows two ways to bias an E-MOSFET where an N-channel device is used for illustration. In either bias arrangement, the purpose is to make the gate voltage more positive than the source by an amount exceeding $V_{GS(\text{thresh})}$.

FIGURE 18–42 ***E-MOSFET biasing arrangements.***

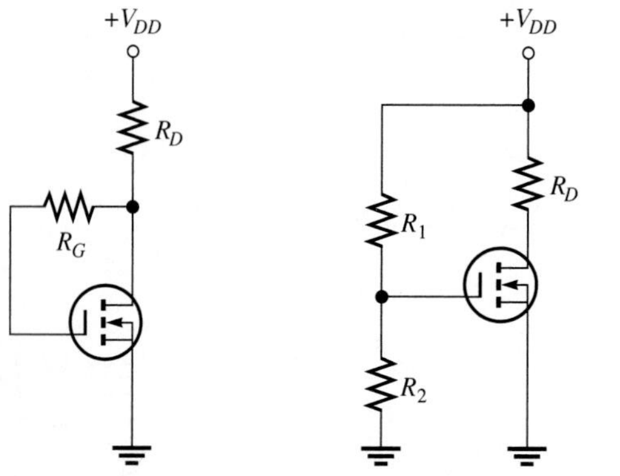

(a) Drain-feedback bias (b) Voltage-divider bias

In the drain feedback bias circuit in Figure 18–42(a), there is negligible gate current and, therefore, no voltage drop across R_G. As a result, $V_{GS} = V_{DS}$.

Equations for the voltage-divider bias in Figure 18–42(b) are as follows:

$$V_{GS} = \left(\frac{R_2}{R_1 + R_2}\right)V_{DD} \tag{18–23}$$

$$V_{DS} = V_{DD} - I_D R_D \tag{18–24}$$

EXAMPLE 18–11

Determine the amount of drain current in Figure 18–43. The MOSFET has a $V_{GS(\text{thresh})}$ of 3 V.

FIGURE 18–43

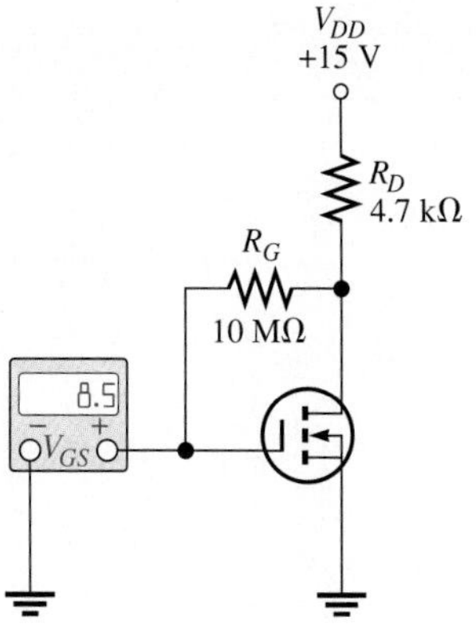

Solution The meter indicates that $V_{GS} = 8.5$ V. Since this is a drain feedback configuration, $V_{DS} = V_{GS} = 8.5$ V.

$$I_D = \frac{V_{DD} - V_{DS}}{R_D} = \frac{15\text{ V} - 8.5\text{ V}}{4.7\text{ k}\Omega} = 1.38\text{ mA}$$

Related Exercise Determine I_D if the meter in Figure 18–43 reads 5 V.

SECTION 18–9 REVIEW

1. A P-channel JFET must have a (positive, negative) V_{GS}.
2. In a certain self-biased N-channel JFET circuit, $I_D = 8$ mA and $R_S = 1$ kΩ. Determine V_{GS}.
3. For a D-MOSFET biased at $V_{GS} = 0$ V, is the drain current equal to 0 V, I_{GSS}, or I_{DSS}?
4. For an N-channel E-MOSFET with $V_{GS(\text{thresh})} = 2$ V, V_{GS} must be in excess of what value in order to conduct?

18–10 ■ UNIJUNCTION TRANSISTORS (UJTs)

The unijunction transistor (UJT) consists of an emitter and two bases. It has no collector. The UJT is used mainly in switching and timing applications.

After completing this section, you should be able to

- **Discuss the operation and application of the UJT**
  - ☐ Recognize the UJT symbol
  - ☐ Draw an equivalent circuit
  - ☐ Define standoff ratio
  - ☐ Explain how a UJT oscillator works

Figure 18–44(a) shows the construction of a **unijunction transistor (UJT).** The base contacts are made to the N-type bar. The emitter lead is connected to the P-region. The UJT schematic symbol is shown in Figure 18–44(b).

FIGURE 18–44
Unijunction transistor (UJT).

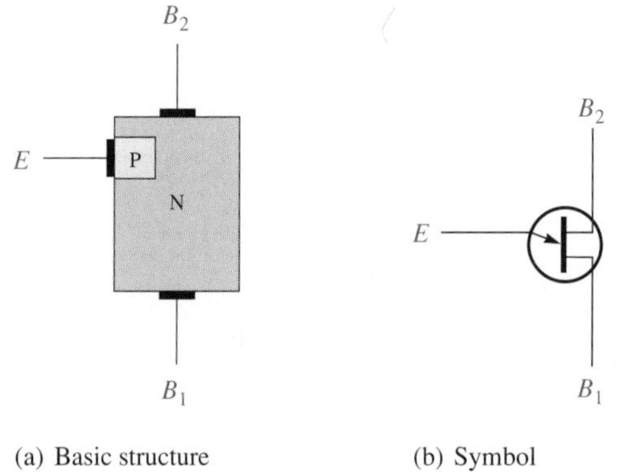

UJT Operation

In normal UJT operation, base 2 (B_2) and the emitter are biased positive with respect to base 1 (B_1). Figure 18–45 will illustrate the operation. This equivalent circuit represents the internal UJT characteristics. The total resistance between the two bases, R_{BB}, is the resistance of the N-type material. R_{B1} is the resistance between point k and base 1. R_{B2} is the resistance between point k and base 2. The sum of these two resistances makes up the total resistance, R_{BB}. The diode represents the PN junction between the emitter and the N-type material.

FIGURE 18–45
Equivalent circuit for a UJT.

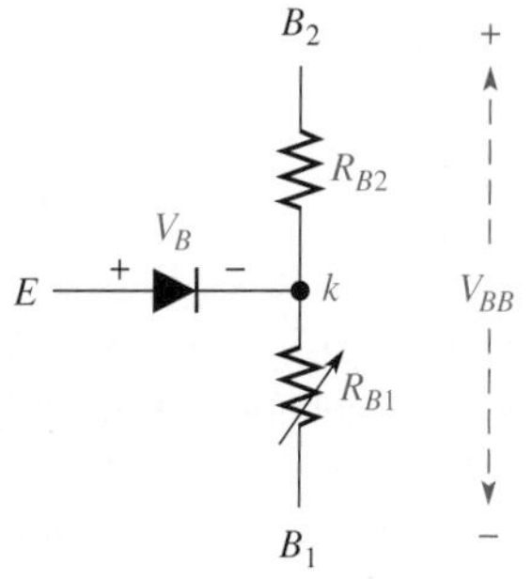

The ratio R_{B1}/R_{BB} is designated η (the Greek letter eta) and is defined as the *intrinsic standoff ratio.* It takes an emitter voltage of $V_B + \eta V_{BB}$ to turn the UJT on. This voltage is called the *peak voltage*. Once the device is on, resistance R_{B1} drops in value. Thus, as emitter current increases, emitter voltage decreases because of the decrease in R_{B1}. This characteristic is the negative resistance characteristic of the UJT.

As the emitter voltage decreases, it reaches a value called the *valley voltage*. At this point, the PN junction is no longer forward-biased, and the UJT turns off.

An Application

UJTs are commonly used in oscillator circuits. Figure 18–46(a) shows a typical circuit. Its operation is as follows: Initially the capacitor is uncharged and the UJT is off. When power is applied, the capacitor charges up exponentially. When it reaches the peak voltage, the UJT turns on and the capacitor begins to discharge as indicated. When the emitter reaches the valley voltage, the UJT turns off and the capacitor begins to charge again. The cycle repeats. Waveforms of the capacitor voltage and the R_1 voltage are shown in Figure 18–46(b).

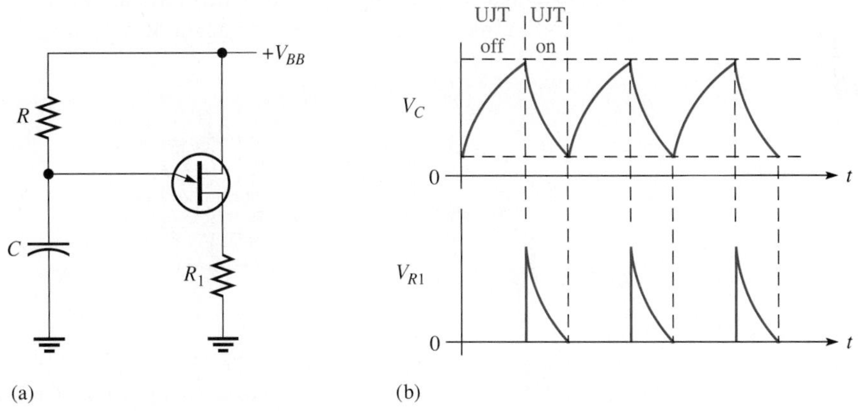

FIGURE 18–46
UJT oscillator circuit.

SECTION 18–10 REVIEW

1. What does "UJT" stand for?
2. What are the terminals of a UJT?
3. What is the intrinsic standoff ratio?

18–11 ■ THYRISTORS

***Thyristors** are devices constructed of four layers of semiconductor material. The types of thyristors covered in this section are the silicon-controlled rectifier (SCR), the diac, and the triac. These thyristor devices share certain characteristics. They act as open circuits capable of withstanding a certain rated voltage until triggered. When turned on (triggered), they become low-resistance current paths and remain so, even after the trigger is removed, until the current is reduced to a certain level or until they are turned off, depending on the type of device. Thyristors can be used to control the amount of ac power to a load and are used in lamp-dimming circuits, motor speed control, ignition systems, and charging circuits, to name a few. UJTs are often used as trigger devices for thyristors.*

After completing this section, you should be able to

- **Explain the basic structure and operation of the SCR, the triac, and the diac**
  - ☐ Recognize the SCR, triac, and diac symbols
  - ☐ Use a transistor analogy to describe SCR operation
  - ☐ Describe triac operation
  - ☐ Describe diac operation
  - ☐ Discuss typical thyristor applications

Silicon-Controlled Rectifiers (SCRs)

The **silicon-controlled rectifier (SCR)** has three terminals, as shown in the symbol in Figure 18–47. Like a rectifier diode, it is a unidirectional device, but the conduction of current is controlled by the gate, *G*. Current is from the cathode *K* to the anode *A* when the SCR is on. A positive voltage on the gate will turn the SCR on. The SCR will remain on as long as the current from the cathode to the anode is equal to or greater than a specified value called the *holding current*. When the current drops below the holding value, the SCR will turn off. It can be turned on again by a positive voltage on the gate.

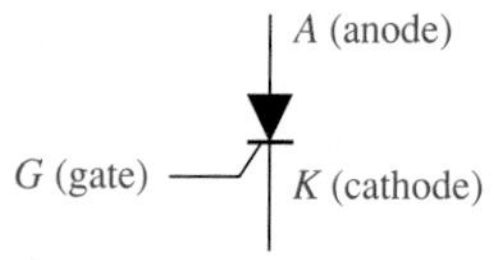

FIGURE 18–47
SCR symbol.

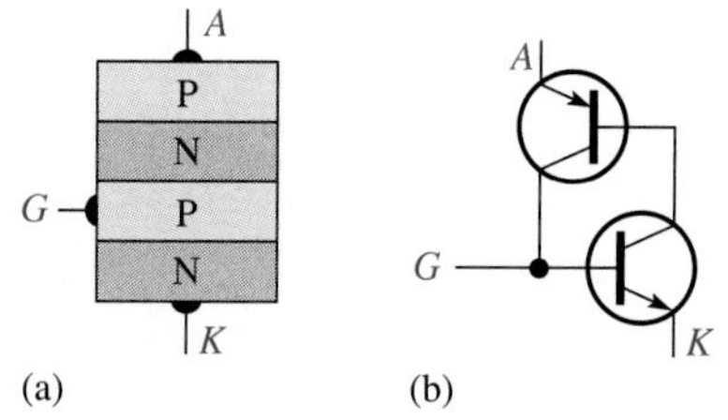

FIGURE 18–48
Construction of SCR and its transistor analogy.

SCR Construction The SCR is a four-layer device. It has two P-regions and two N-regions, as shown in Figure 18–48(a). This construction can be thought of as an NPN and a PNP transistor interconnected as in Figure 18–48(b). The upper N region is commonly shared by the base and the collector of the transistors.

An Application Figure 18–49 shows an SCR used to rectify and control the average power delivered to a load. The SCR rectifies the 60 Hz ac just as a conventional rectifier diode. It can conduct only during the positive half-cycle. The phase controller produces a trigger pulse at the SCR gate in order to turn it on during any portion of the positive half-cycle of the input. If the SCR is turned on earlier in the half-cycle, more average power is delivered to the load. If the SCR is fired later in the half-cycle, less average power is delivered.

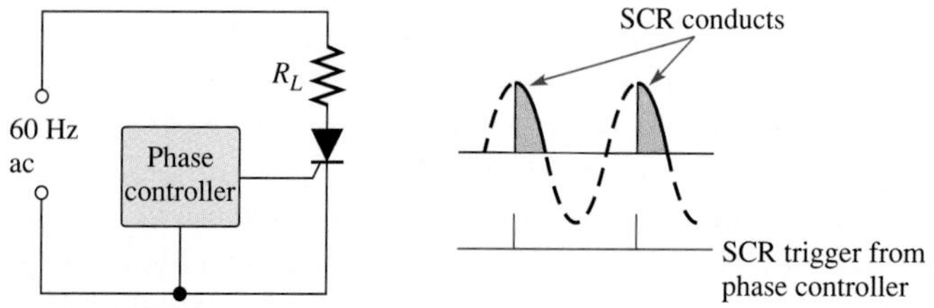

FIGURE 18–49
SCR controls average power to load.

Triacs

The **triac** is also known as a *bidirectional triode thyristor*. It is equivalent to two SCRs connected to allow current in either direction. The SCRs share a common gate. The symbol for a triac is shown in Figure 18–50.

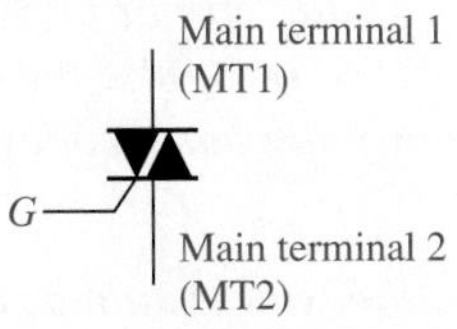

FIGURE 18–50
Triac symbol.

Applications Like the SCR, triacs are also used to control average power to a load by the method of phase control. The triac can be triggered such that the ac power is supplied to the load for a controlled portion of each half-cycle. During each positive half-cycle of the ac, the triac is off for a certain interval, called the *delay angle* (measured in degrees). Then it is triggered on and conducts current through the load for the remaining portion of the positive half-cycle, called the *conduction angle*. Similar action occurs on the negative half-cycle except that, of course, current is conducted in the opposite direction through the load. Figure 18–51 illustrates this action.

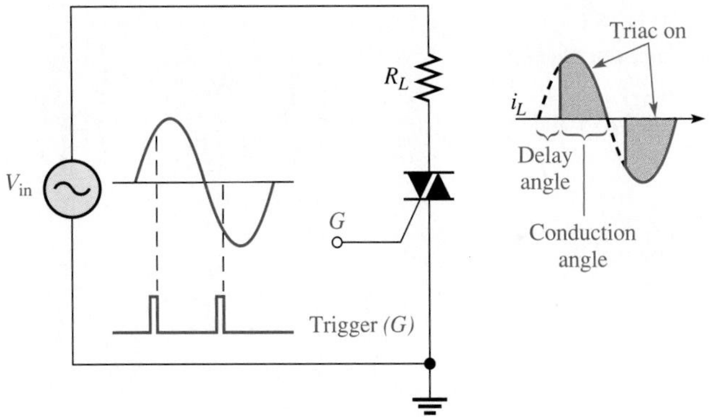

FIGURE 18–51
Basic triac phase control.

Diacs

The **diac** is a bidirectional device that does not have a gate. It conducts current in either direction when a sufficient voltage, called the *breakover potential*, is reached across the two terminals. The symbol for a diac is shown in Figure 18–52.

FIGURE 18–52
Diac symbol.

SECTION 18–11 REVIEW

1. What does "SCR" stand for?
2. Explain how an SCR basically works.
3. Name a typical application of a triac.
4. A diac is gate-controlled (T or F).

18–12 ■ TRANSISTOR PACKAGES AND TERMINAL IDENTIFICATION

Transistors are available in a wide range of package types for various applications. Those with mounting studs or heat sinks are power transistors. Low-power and medium-power transistors are usually found in smaller metal or plastic cases. Still another package classification is for high-frequency devices. As a technician, you often must check a transistor, either in the circuit or out of the circuit, for possible failure. First, of course, you must be able to recognize the transistor by its physical appearance. Also, you must be able to identify the base, collector, and emitter leads (or gate, drain, and source).

After completing this section, you should be able to

- **Recognize transistor packages and identify the terminals**

Transistors are found in a range of shapes and sizes. Figure 18–53 shows a few of the most typical case styles. Pin identification and construction diagrams are shown in Figure 18–54 for several types of transistor packages. Generally, the heat sink mounting or stud on power transistors is the collector terminal. On the metal "top hat" cases, the tab is closest to the emitter (or source) lead, and the collector (or drain) is often connected to the case.

FIGURE 18–53
Some common package configurations (copyright of Motorola, Inc. Used by permission.).

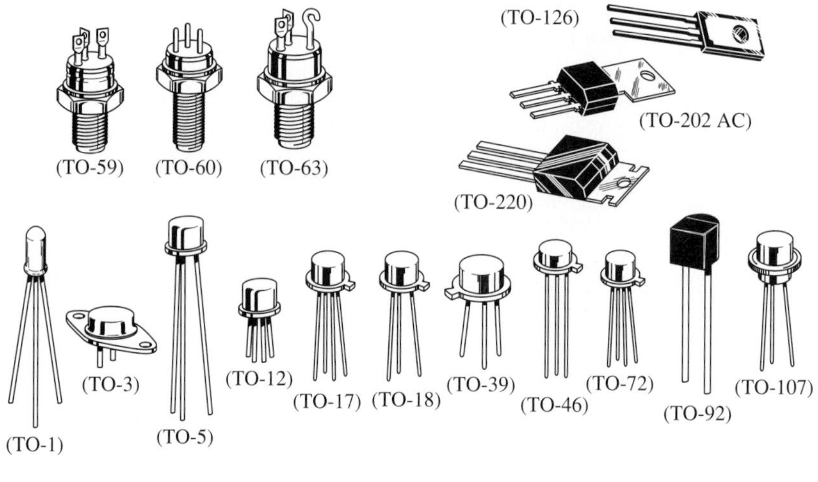

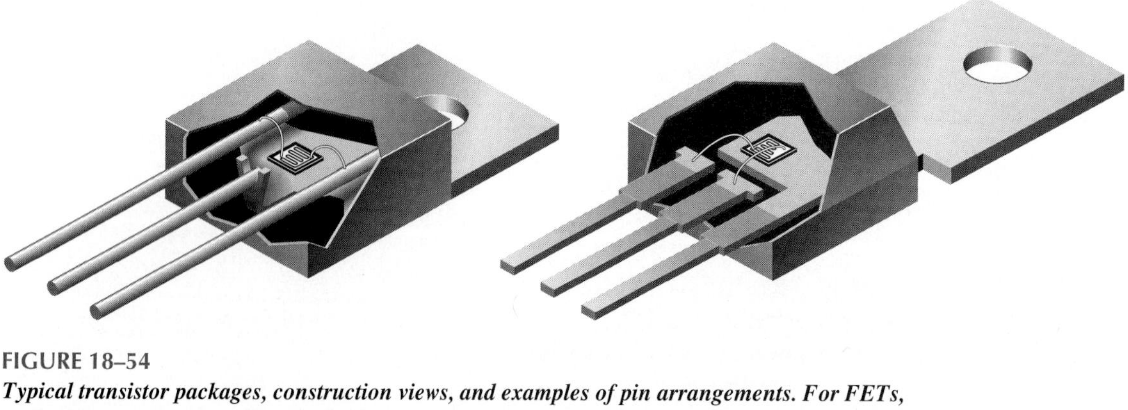

FIGURE 18–54
Typical transistor packages, construction views, and examples of pin arrangements. For FETs, replace base with gate, collector with drain, and emitter with source (copyright of Motorola, Inc. Used by permission.).

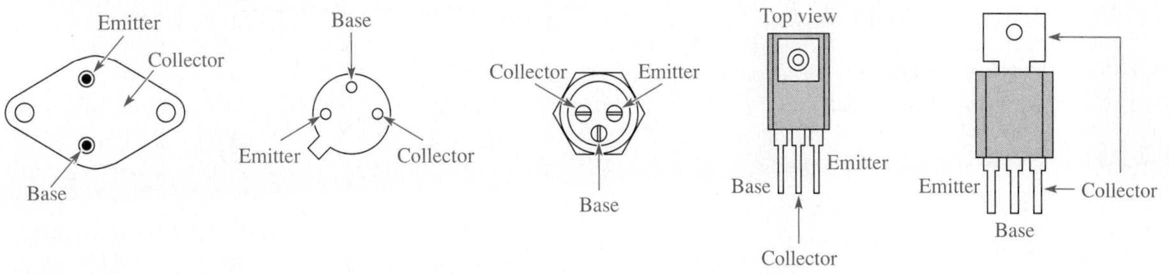

FIGURE 18–54 ***Continued***

SECTION 18–12 REVIEW

1. Identify the leads on the bipolar transistors in Figure 18–55.

FIGURE 18–55

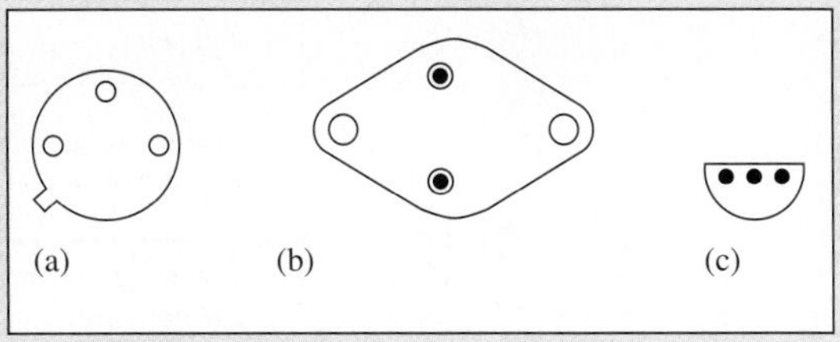

18–13 ■ TROUBLESHOOTING

As you already know, a critical skill in electronics is the ability to identify a circuit malfunction and to isolate the failure to a single component if possible. In this section, the basics of troubleshooting transistor bias circuits and testing individual transistors are covered.

After completing this section, you should be able to

- **Troubleshoot transistors**
  - □ Troubleshoot a transistor bias circuit
  - □ Test transistor in-circuit and out-of-circuit
  - □ Discuss leakage measurement
  - □ Discuss gain measurement

Troubleshooting a Transistor Bias Circuit

Several things can cause a failure in a transistor bias circuit. The most common faults are open base or collector resistor, loss of base or collector bias voltage, open transistor junctions, and open contacts or interconnections on the printed circuit board. Figure 18–56 illustrates a basic four-point check that will help isolate one of these faults in the circuit. The term *floating* refers to a point that is effectively disconnected from a "solid" voltage or ground. A very small, erratically fluctuating voltage is sometimes observed.

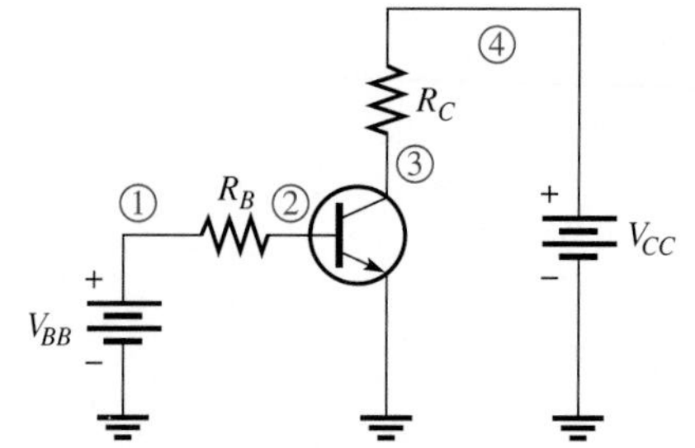

FIGURE 18–56
Test point troubleshooting guide for a basic transistor bias circuit.

Symptom	Test point voltages 1	2	3	4
A. No V_{BB} Source shorted	0 V	0 V	V_{CC}	V_{CC}
Source open	Floating	Floating	V_{CC}	V_{CC}
B. No V_{CC} Source shorted	V_{BB}	0.7 V	0 V	0 V
Source open	V_{BB}	0.7 V	Floating	Floating
C. R_{BN} open	V_{BB}	Floating	V_{CC}	V_{CC}
D. R_C open	V_{BB}	0.7 V	Floating	V_{CC}
E. Base-to-emitter junction open	V_{BB}	V_{BB}	V_{CC}	V_{CC}
F. Base-to-collector junction open	V_{BB}	0.7 V	V_{CC}	V_{CC}
G. Low β	V_{BB}	0.7 V	Higher than normal	V_{CC}

EXAMPLE 18–12 From the voltage measurements in Figure 18–57, what is the most likely problem?

FIGURE 18–57

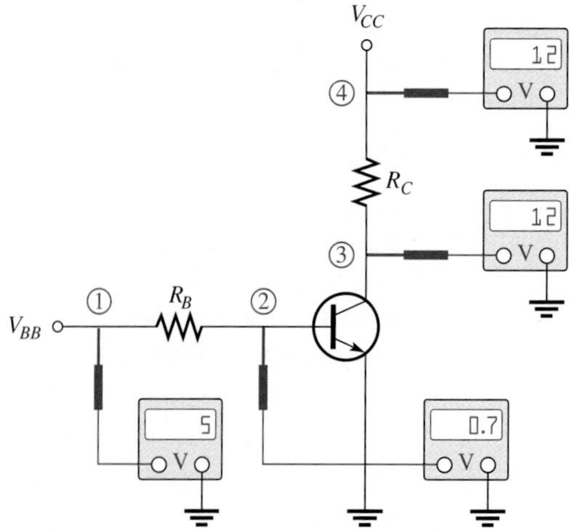

Solution Since $V_{CE} = 12$ V, the transistor is obviously in cutoff. The V_{BB} measurement appears OK and the voltage at the base is correct. Therefore, the collector-base junction of the transistor must be open. Replace the transistor and check the voltages again.

Related Exercise What is the voltage at point 2 if the *BE* junction opens in Figure 18–57?

Transistor Testing

An individual transistor can be tested either in-circuit or out-of-circuit. For example, let's say that an amplifier on a particular printed circuit (PC) board has malfunctioned. You perform a check and find that the symptoms are the same as Item F in the table of Figure 18–56, indicating an open from base to collector. This can mean either a bad transistor or an open in an external circuit connection. Your next step is to find out which one.

Good troubleshooting practice dictates that you do not remove a component from a circuit board unless you are reasonably sure that it is bad or you simply cannot isolate the problem down to a single component. When components are removed, there is a risk of damage to the PC board contacts and traces.

The next step is to do an in-circuit check of the transistor using a transistor tester similar to the ones shown in Figure 18–58. The three clip-leads are connected to the transistor terminals and the tester gives a GOOD/BAD indication.

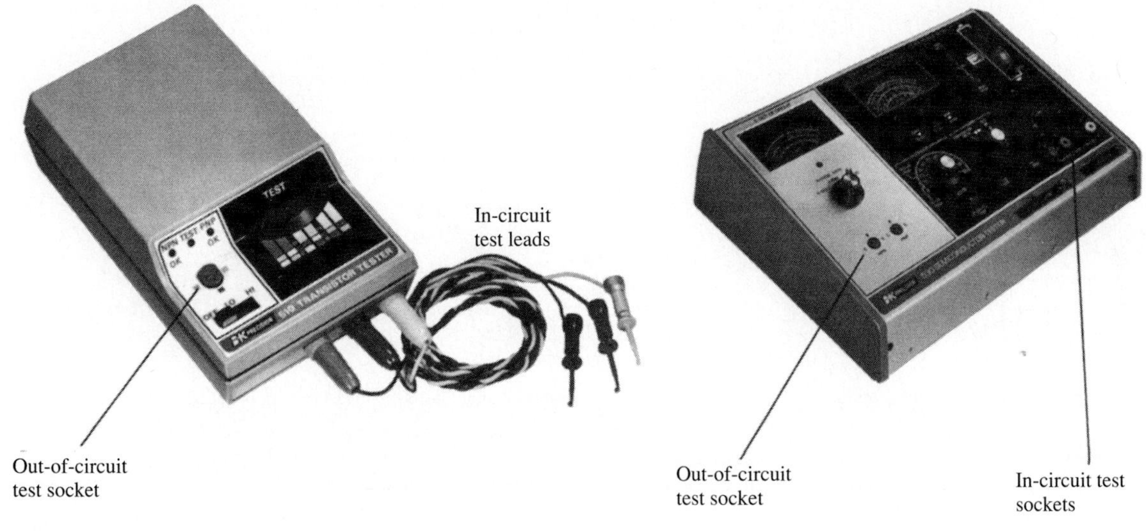

FIGURE 18–58
Transistor testers (courtesy of B & K Precision).

Case 1

If the transistor tests BAD, it should be carefully removed and replaced by a known good one. An out-of-circuit check of the replacement device is usually a good idea, just to make sure it is OK. The transistor is plugged into the socket on the transistor tester for out-of-circuit tests.

Case 2

If the transistor tests GOOD in-circuit, examine the circuit board for a poor connection at the collector pad or for a break in the connecting trace. A poor solder joint often results in an open or a highly resistive contact. The physical point at which you actually measure the voltage is very important in this case. If you measure on the collector side of an external open, you will get the indication listed in Item D of the table in Figure 18–56. If you

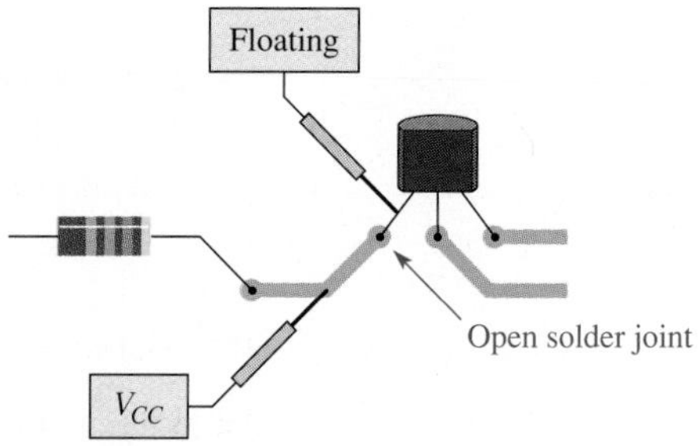

FIGURE 18–59
The indication of an open, when it is in the external circuit, depends on where you measure.

measure on the side of the external open closest to R_C, you will get the indication in Item F of the table. This situation is illustrated in Figure 18–59.

Importance of Point-of-Measurement in Troubleshooting

It is interesting to look back and notice that if you had taken the initial measurement on the transistor lead itself and the open were external to the transistor, you would have measured V_{CC}. This would have indicated a bad transistor even before the tester was used. This simple concept emphasizes the importance of point-of-measurement in certain troubleshooting situations as further illustrated in Figure 18–60 for the case just discussed. This is also valid for opens at the base and emitter of a transistor.

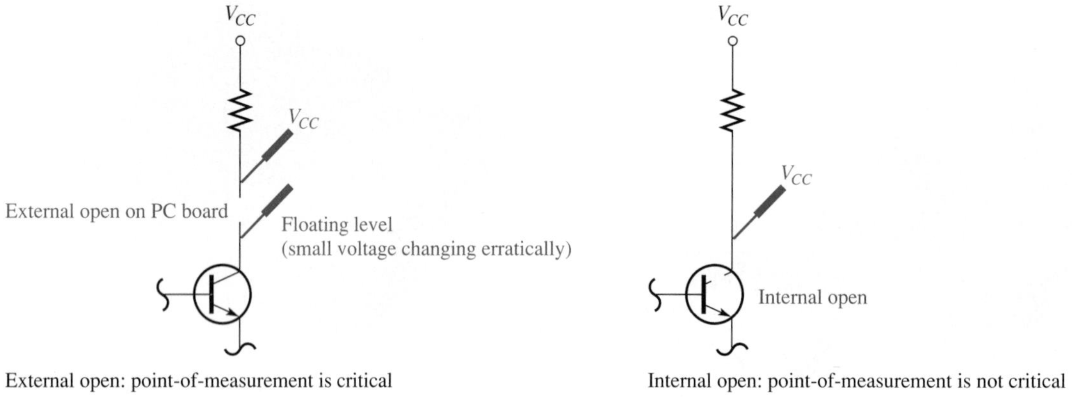

FIGURE 18–60
Importance of the point-of-measurement for certain open failures in transistor circuits.

EXAMPLE 18–13 What fault do the measurements in Figure 18–61 indicate?

FIGURE 18–61

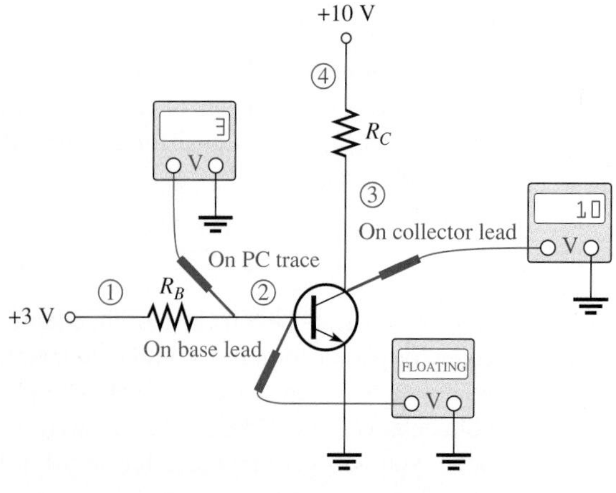

Solution The transistor is in cutoff, as indicated by the 10 V measurement on the collector lead. The base bias voltage of 3 V appears on the PC board contacts but not on the transistor lead. This indicates that there is an open external to the transistor between the two measured points. Check the solder joint at the base contact on the PC board. If the open were internal, there would be 3 V on the base lead.

Related Exercise If the meter in Figure 18–61 that now reads 3 V indicates a floating point, what is the most likely fault?

Out-of-Circuit Multimeter Test

If a transistor tester is not available, an analog multimeter can be used to perform a basic diode check on the transistor junctions for opens or shorts. For this test, a transistor can be viewed as two back-to-back diodes as shown in Figure 18–62. The *BC* junction is one diode and the *BE* junction is the other. Also some DMMs have a transistor test feature.

(a) Both junctions should read approximately 0.7 V when forward-biased

(b) Both junctions should read open (ideally) when reverse-biased

FIGURE 18–62
The transistor is viewed as two diodes for a junction check with a multimeter.

Leakage Measurement

Very small leakage currents exist in all transistors and in most cases are small enough to neglect (usually nA). When a transistor is connected as shown in Figure 18–63(a) with the base open ($I_B = 0$), it is in cutoff. Ideally, $I_C = 0$; but actually there is a small current from collector to emitter, as mentioned earlier, called I_{CEO} (collector-to-emitter current with base open). This leakage current is usually in the nA range for silicon. A faulty transistor will often have excessive leakage current and can be checked in a transistor tester, which connects an ammeter as shown in part (a). Another leakage current in transistors is the reverse collector-to-base current, I_{CBO}. This is measured with the emitter open, as shown in Figure 18–63(b). If it is excessive, a shorted collector-base junction is likely.

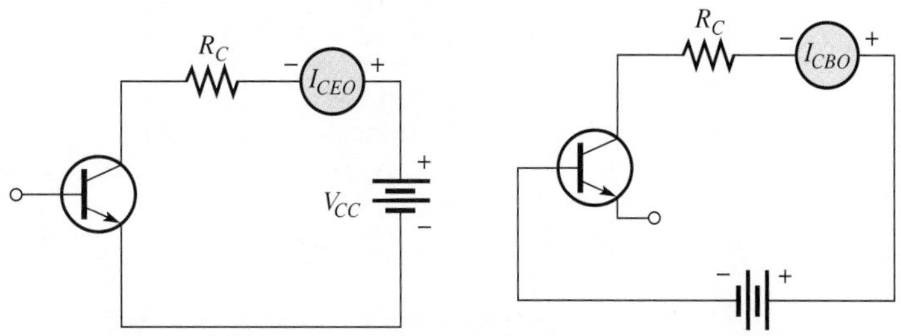

(a) Circuit for I_{CEO} test

(b) Circuit for I_{CBO} test

FIGURE 18–63
Leakage current test circuits.

Gain Measurement

In addition to leakage tests, the typical transistor tester also checks the β_{dc}. A known value of I_B is applied and the resulting I_C is measured. The reading will indicate the value of the I_C/I_B ratio, although in some units only a relative indication is given. Most testers provide for an in-circuit β_{dc} check, so that a suspected device does not have to be removed from the circuit for testing.

Curve Tracers

The curve tracer is an oscilloscope type of instrument that can display transistor characteristics such as a family of collector curves. In addition to the measurement and display of various transistor characteristics, diode curves can also be displayed, as well as the curves for other devices.

SECTION 18–13 REVIEW

1. If a transistor on a circuit board is suspected of being faulty, what should you do first?
2. In a transistor bias circuit, such as the one in Figure 18–56, what happens if R_B opens?
3. In a circuit such as the one in Figure 18–56, what are the base and collector voltages if there is an open between the emitter and ground?

18–14 ■ APPLICATION ASSIGNMENT

As stated in the chapter opener, you have been assigned to work on an industrial temperature-control system. The system has several parts, but in this assignment you are concerned with the transistor circuit that monitors the temperature changes in the fluid in a tank and provides a proportional output that is used to precisely control the temperature. In this application, the transistor is used as a dc amplifier.

Description of the System

The system in Figure 18–64 is an example of a process control system that uses the principle of closed-loop feedback to maintain a substance at a contant temperature of

FIGURE 18–64
Block diagram of the industrial temperature control system.

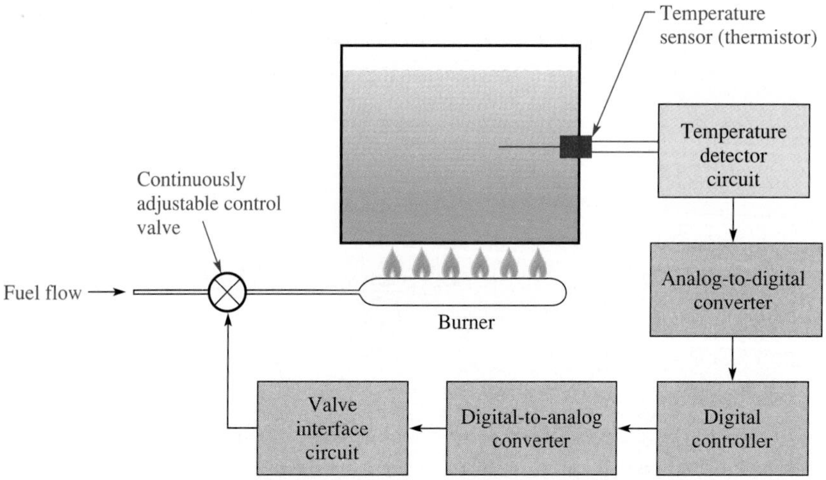

50°C ± 1°C. The temperature sensor in the tank is a **thermistor,** which is a temperature-sensitive resistor whose resistance has a negative temperature coefficient. The thermistor forms part of the bias circuit for the transistor detector.

A small change in temperature causes the resistance of the thermistor to change, which results in a proportional change in the output voltage from the transistor detector. Since thermistors are typically nonlinear devices, the purpose of the digital portions of the system is to precisely compensate for the nonlinear characteristic of the temperature detector in order to provide a continuous linear adjustment of the fuel flow to the burner to offset any small shift in temperature away from the desired value.

The temperature detector board (highlighted in color) that you are concerned with in this application assignment is obviously only a small part of the total system. It is interesting, however, to see how the variations in a transistor's bias point can be used in a situation such as this. Although the digital portions of the system are not covered, they show that typical systems have various types of elements.

The temperature detector circuit board and the thermistor probe that mounts into the side of the tank are shown in Figure 18–65. The thermistor symbol and the case configuration of the NPN transistor are included.

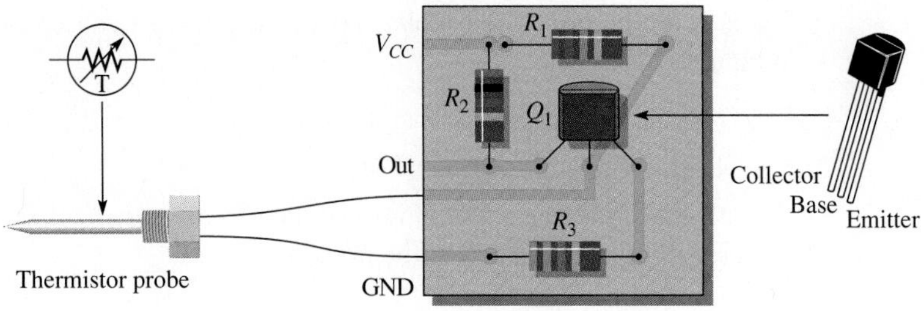

FIGURE 18–65
The temperature detector circuit.

Step 1: Relate the PC Board to the Schematic

Develop a schematic of the circuit board in Figure 18–65 including the thermistor. Arrange the schematic into a form that is familiar to you and identify the type of biasing used.

Step 2: Analyze the Circuit

With a dc supply of +15 V, analyze the temperature detector circuit to determine the output voltage at 50°C and at 1°C above and below 50°C. Use 0.7 V for the base-emitter voltage. Assume $\beta_{dc} = 100$ for the transistor. Since accuracy on the graph is difficult to achieve for this small range, assume that the thermistor resistances for the controlled temperature range are as follows:

At $T = 50$°C, $R = 2.75$ kΩ

At $T = 49$°C, $R = 3.1$ kΩ

At $T = 51$°C, $R = 2.5$ kΩ

Will your values change significantly if you take into account the 1 MΩ input resistance of the analog-to-digital converter?

Step 3: Check Output over Temperature Range

With the circuit in a controlled temperature environment, determine the output voltage over a range from 30°C to 110°C at 20° intervals. Use the graph in Figure 18–66.

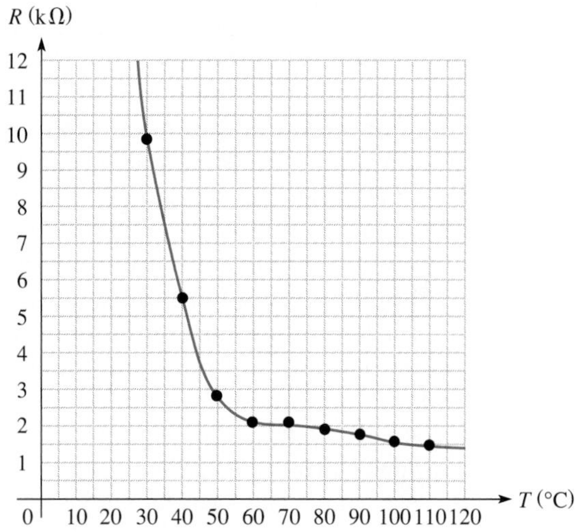

FIGURE 18–66
Graph of thermistor resistance versus temperature.

Step 4: Troubleshoot the Circuit Board

For each of the following problems state the probable cause or causes in each case (Refer to Figure 18–65):

1. V_{CE} is approximately 0.1 V and V_C is 3.8 V.
2. Collector of Q1 remains at 15 V.
3. For each of the above conditions, if there is more than one possible fault, describe how you will go about isolating the problem.

SECTION 18–14 REVIEW

1. What is the basic purpose of the detector circuit board in the process control system?
2. In Step 2, does the transistor go into saturation and/or cutoff?
3. Calculate the output voltage at 40°C.
4. Calculate the output voltage at 60°C.

■ SUMMARY

- A bipolar junction transistor (BJT) consists of three regions: emitter, base, and collector.
- The three regions of a BJT are separated by two PN junctions.
- The two types of bipolar transistor are the NPN and the PNP.
- The term *bipolar* refers to two types of current: electron current and hole current.
- A field-effect transistor (FET) has three terminals: source, drain, and gate.
- A junction field-effect transistor (JFET) operates with a reverse-biased gate-to-source PN junction.
- JFETs have very high input resistance due to the reverse-biased gate-source junction.
- JFET current flows between the drain and the source through a channel whose width is controlled by the amount of reverse bias on the gate-source junction.
- The two types of JFETs are N-channel and P-channel.
- Metal-oxide semiconductor field-effect transistors (MOSFETs) differ from JFETs in that the gate of a MOSFET is insulated from the channel.
- A depletion-enhancement MOSFET (D-MOSFET) can operate with a positive, negative, or zero gate-to-source voltage.
- The D-MOSFET has a physical channel between the drain and the source.

- An enhancement-only MOSFET (E-MOSFET) can operate only when the gate-to-source voltage exceeds a threshold value.
- The enhancement-only MOSFET has no physical channel.
- Transistors are used as either amplifying devices or switching devices.
- Transistor and thyristor symbols are shown in Figure 18–67.

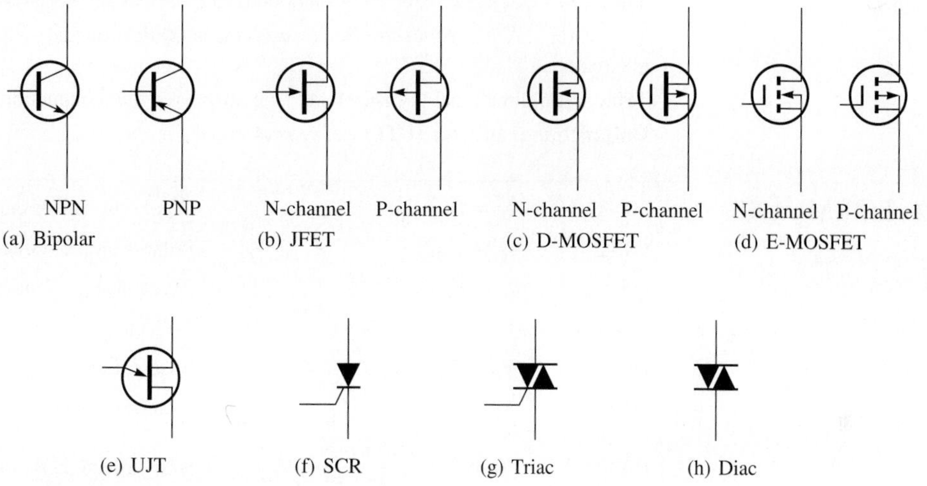

FIGURE 18–67
Transistor and thyristor symbols.

■ GLOSSARY

Alpha (α) The ratio of collector current to emitter in a bipolar junction transistor.

Amplification The process of producing a larger voltage, current, or power using a smaller input signal as a "pattern."

Amplifier An electronic circuit having the capability of amplification and designed specifically for that purpose.

Base One of the semiconductor regions in a bipolar junction transistor.

Beta (*β*) The ratio of collector current to base current in a bipolar junction transistor.

Bipolar Characterized by two PN junctions.

Bipolar junction transistor (BJT) A transistor constructed with three doped semiconductor regions separated by two PN junctions.

Collector One of the semiconductor regions in a BJT.

Cutoff The nonconducting state of a transistor.

Diac A semiconductor device that can conduct current in either of two directions when properly activated.

Drain One of the three terminals of a field-effect transistor.

Emitter One of the three semiconductor regions in a BJT.

Field-effect transistor (FET) A type of transistor that uses an induced electric field to control current.

Gain The amount of amplification.

Gate One of the three terminals of an FET.

Junction field-effect transistor (JFET) A type of FET that operates with a reverse-biased junction to control current in a channel.

MOSFET Metal-oxide semiconductor field-effect transistor.

Pinch-off voltage The value of the drain-to-source voltage of an FET at which the drain current becomes constant when the gate-to-source voltage is zero.

***Q*-point** The dc operating (bias) point of an amplifier.

Saturation The state of a BJT in which the collector has reached a maximum and is independent of the base current.

Silicon-controlled rectifier (SCR) A device that can be triggered on to conduct current in one direction.

Source One of the three terminals of an FET.

Switch An electrical or electronic device for opening and closing a current path.

Thermister A temperature-sensitive resistor whose resistance has a negative temperature coefficient.

Thyristor A class of four-layer semiconductor devices.

Transistor A semiconductor device used for amplification and switching applications in electronic circuits.

Triac A bidirectional thyristor that can be triggered into conduction.

Unijunction transistor (UJT) A type of transistor consisting of an emitter and two bases.

■ FORMULAS

(18–1)	$I_E = I_C + I_B$	Bipolar transistor currents
(18–2)	$I_C = \alpha_{dc} I_E$	Relationship of collector and emitter currents
(18–3)	$I_C = \beta_{dc} I_B$	Relationship of collector and base currents
(18–4)	$V_C = V_{CC} - I_C R_C$	Collector voltage
(18–5)	$V_B = V_E + V_{BE}$	Base voltage
(18–6)	$R_{IN} \cong \beta_{dc} R_E$	Input resistance at base
(18–7)	$V_B = \left(\frac{R_2 \parallel R_{IN}}{R_1 + R_2 \parallel R_{IN}}\right) V_{CC}$	Base voltage with voltage-divider bias
(18–8)	$V_B \cong \left(\frac{R_2}{R_1 + R_2}\right) V_{CC}$	Approximate base voltage ($R_{IN} >> R_2$)
(18–9)	$V_E = V_B - 0.7\text{ V}$	Emitter voltage
(18–10)	$\beta_{ac} = \frac{I_c}{I_b}$	ac beta
(18–11)	$A_v \cong \frac{R_C}{R_E}$	Voltage gain
(18–12)	$V_{CE(cutoff)} \cong V_{CE}$	V_{CE} at cutoff
(18–13)	$I_{C(sat)} \cong \frac{V_{CC}}{R_C}$	Collector saturation current
(18–14)	$I_{B(min)} = \frac{I_{C(sat)}}{\beta_{dc}}$	Minimum base current for saturation
(18–15)	$I_C = \frac{P_{D(max)}}{V_{CE}}$	I_C for maximum V_{CE}
(18–16)	$V_{CE} = \frac{P_{D(max)}}{I_C}$	V_{CE} for maximum I_C
(18–17)	$R_{IN} = \left\lvert \frac{V_{GS}}{I_{GSS}} \right\rvert$	JFET input resistance
(18–18)	$V_{GS} = -I_D R_S$	Self-bias voltage for N-channel JFET
(18–19)	$V_{GS} = +I_D R_S$	Self-bias voltage for P-channel JFET
(18–20)	$V_D = V_{DD} - I_D R_S$	Drain voltage
(18–21)	$V_{DS} = V_{DD} - I_D(R_D + R_S)$	Drain-to-source voltage
(18–22)	$V_{DS} = V_{DD} - I_{DSS} R_D$	Drain-to-source voltage D-MOSFET
(18–23)	$V_{GS} = \left(\frac{R_2}{R_1 + R_2}\right) V_{DD}$	Gate-to-source voltage E-MOSFET
(18–24)	$V_{DS} = V_{DD} - I_D R_D$	Drain-to-source voltage E-MOSFET

■ SELF-TEST

1. The N-type regions in an NPN bipolar junction transistor are
   (a) collector and base (b) collector and emitter
   (c) base and emitter (d) collector, base, and emitter
2. The N-region in a PNP transistor is the
   (a) base (b) collector (c) emitter (d) case
3. For normal operation of a NPN transistor, the base must be
   (a) disconnected (b) negative with respect to the emitter
   (c) positive with respect to the emitter (d) positive with respect to the collector
4. The three currents in a BJT are
   (a) forward, reverse, neutral (b) drain, source, gate
   (c) alpha, beta, and sigma (d) base, emitter, and collector
5. Beta (β) is the ratio of
   (a) collector current to emitter current (b) collector current to base current
   (c) emitter current to base current (d) output voltage to input voltage
6. Alpha (α) is the ratio of
   (a) collector current to emitter current (b) collector current to base current
   (c) emitter current to base current (d) output voltage to input voltage
7. If the beta of a certain transistor is 30 and the base current is 1 mA, the collector current is
   (a) 0.33 mA (b) 1 mA (c) 30 mA (d) unknown
8. If the base current is increased,
   (a) the collector current increases and the emitter current decreases
   (b) the collector current decreases and the emitter current decreases
   (c) the collector current increases and the emitter current does not change
   (d) the collector current increases and the emitter current increases
9. When an N-channel JFET is biased for conduction, the gate is
   (a) positive with respect to the source (b) negative with respect to the source
   (c) at the same voltage as the source (d) at the same voltage as the drain
10. When the gate-to-source voltage of an N-channel JFET is increased, the drain current
    (a) decreases (b) increases (c) stays constant (d) becomes zero
11. When a negative gate-to-source voltage is applied to an N-channel MOSFET, it operates in the
    (a) cutoff state (b) saturated state (c) enhancement mode (d) depletion mode
12. A UJT consists of
    (a) two emitters (b) a collector and two emitters
    (c) an emitter and two bases (d) an emitter, base, and collector
13. An SCR is like a rectifier diode except that
    (a) it can be triggered into conduction by a voltage at the gate
    (b) it can conduct current in both directions
    (c) it can handle more power
    (d) it has four terminals
14. A triac is a device that
    (a) can be triggered into conduction (b) can conduct current in both directions
    (c) both (a) and (b) (d) none of the above

■ PROBLEMS

SECTION 18–1 Bipolar Junction Transistors (BJTs)

1. What is the exact value of I_C for $I_E = 5.34$ mA and $I_B = 475\ \mu$A?
2. What is the α_{dc} when $I_C = 8.23$ mA and $I_E = 8.69$ mA?
3. A certain transistor has an $I_C = 25$ mA and an $I_B = 200\ \mu$A. Determine the β_{dc}.
4. In a certain transistor circuit, the base current is 2% of the 30 mA emitter current. Determine the approximate collector current.

5. Find I_B, I_E, and I_C in Figure 18–68 given that $\alpha_{dc} = 0.98$ and $\beta_{dc} = 49$.
6. Determine the terminal voltages of each transistor with respect to ground for each circuit in Figure 18–69. Also determine V_{CE}, V_{BE}, and V_{BC}. $\beta_{dc} = 25$.

FIGURE 18–68

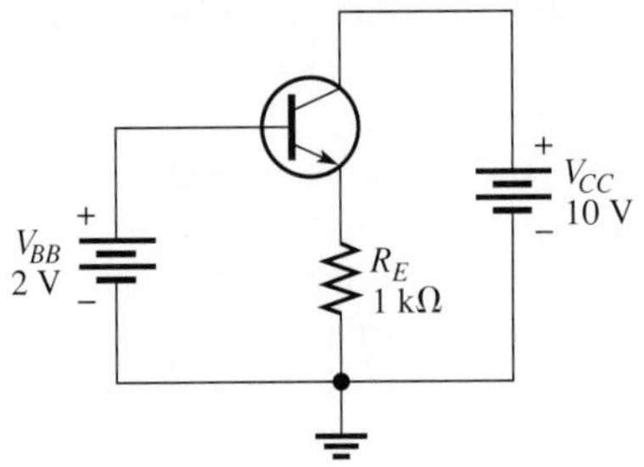

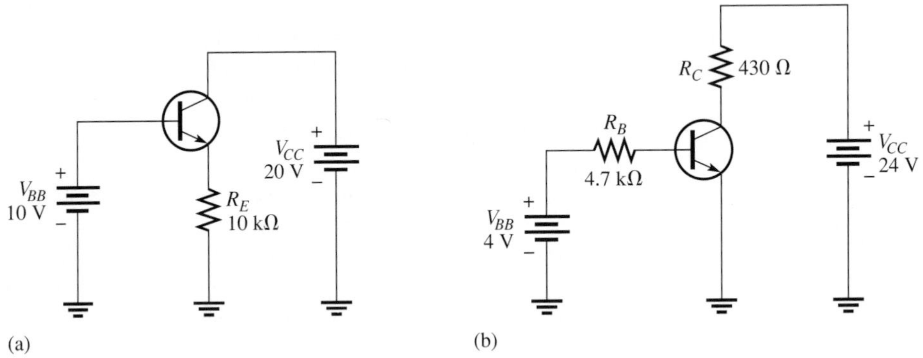

FIGURE 18–69

SECTION 18–2 Voltage-Divider Bias

7. Determine I_B, I_C, and V_C in Figure 18–70.
8. For the circuit in Figure 18–71, find V_B, V_E, I_E, I_C, and V_C.
9. In Figure 18–71, what is V_{CE}? What are the Q-point coordinates?

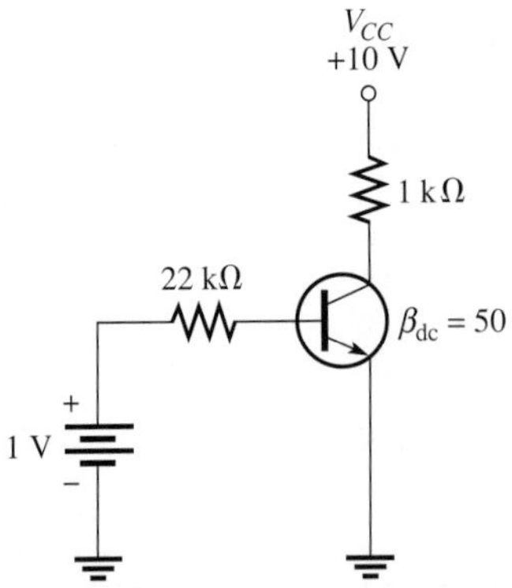

FIGURE 18–70

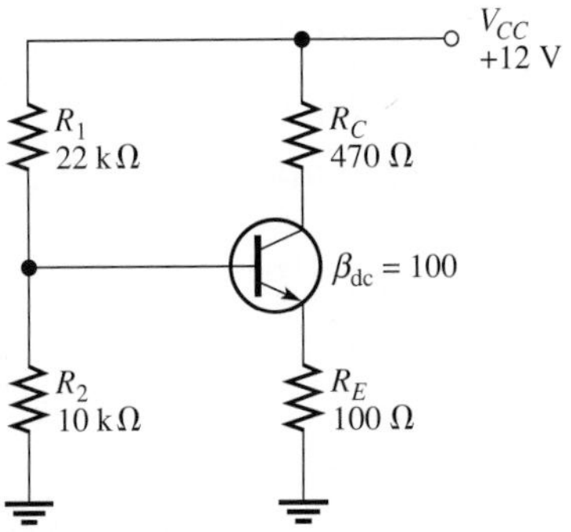

FIGURE 18–71

SECTION 18–3 The Bipolar Transistor As an Amplifier

10. A transistor amplifier has a voltage gain of 50. What is the output voltage when the input voltage is 100 mV?
11. To achieve an output of 10 V with an input of 300 mV, what voltage gain is required?
12. A 50 mV signal is applied to the base of a properly biased transistor with $R_E = 100\ \Omega$ and $R_C = 500\ \Omega$. Determine the signal voltage at the collector.

SECTION 18–4 The Bipolar Transistor As a Switch

13. Determine $I_{C(\text{sat})}$ for the transistor in Figure 18–72. What is the value of I_B necessary to produce saturation? What minimum value of V_{IN} is necessary for saturation?

14. The transistor in Figure 18–73 has a β_{dc} of 50. Determine the value of R_B required to ensure saturation when V_{IN} is 5 V. What must V_{IN} be to cut off the transistor?

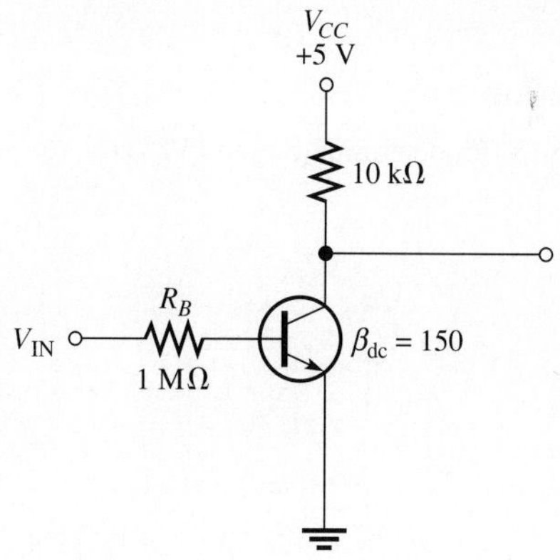

FIGURE 18–72

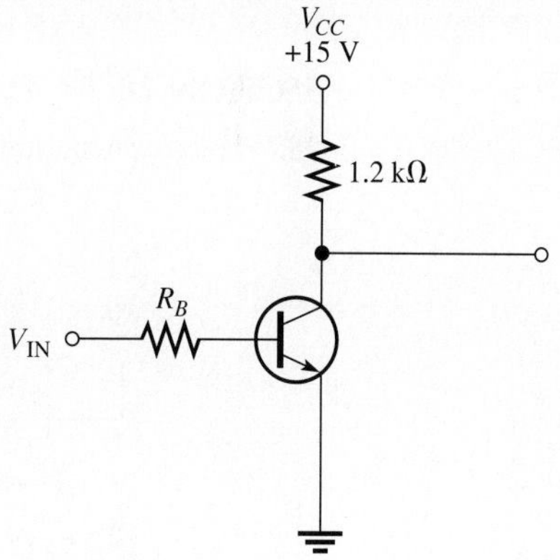

FIGURE 18–73

SECTION 18–5 BJT Parameters and Ratings

15. If the β_{dc} in Figure 18–69(a) changes from 100 to 150 due to a temperature increase, what is the change in collector current?

16. A certain transistor is to be operated at a collector current of 50 mA. How high can V_{CE} go without exceeding a $P_{D(\max)}$ of 1.2 W?

SECTION 18–6 The Junction Field-Effect Transistor (JFET)

17. The V_{GS} of a P-channel JFET is increased from 1 V to 3 V.

(a) Does the depletion region narrow or widen?

(b) Does the resistance of the channel increase or decrease?

18. Why must the gate-to-source voltage of an N-channel JFET always be either 0 V or negative?

SECTION 18–7 JFET Characteristics

19. A JFET has a specified pinch-off voltage of –5 V. When $V_{GS} = 0$, what is V_{DS}?

20. A certain JFET is biased such that $V_{GS} = -2$ V. What is the value of V_{DS} if V_P is specified to be –6 V?

21. A certain JFET data sheet gives $V_{GS(\text{off})} = -8$ V and $I_{DSS} = 10$ mA. Determine the value of V_{DS} at which pinch-off begins when $V_{GS} = -4$ V.

22. A certain P-channel JFET has a $V_{GS(\text{off})} = 6$ V. What is I_D when $V_{GS} = 8$ V?

23. The JFET in Figure 18–74 has a $V_{GS(\text{off})} = -4$ V. Assume that you increase the supply voltage, V_{DD}, beginning at 0 until the ammeter reaches a steady value. What does the voltmeter read at this point?

FIGURE 18–74

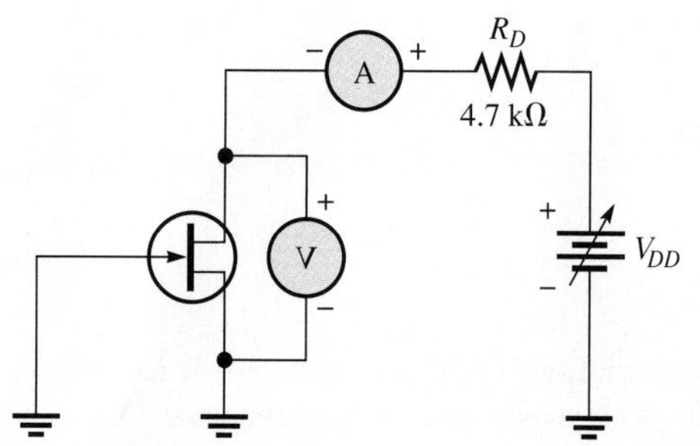

SECTION 18–8 The Metal-Oxide Semiconductor FET (MOSFET)

24. Sketch the schematic symbols for N-channel and P-channel D-MOSFETs and E-MOSFETs. Label the terminals.

25. Explain why both types of MOSFETs have an extremely high input resistance at the gate.

26. In what mode is an N-channel D-MOSFET operating with a positive V_{GS}?

27. A certain E-MOSFET has a $V_{GS(\text{thresh})} = 3$ V. What is the minimum V_{GS} for the device to turn on?

SECTION 18–9 FET Biasing

28. For each circuit in Figure 18–75, determine V_{DS} and V_{GS}.

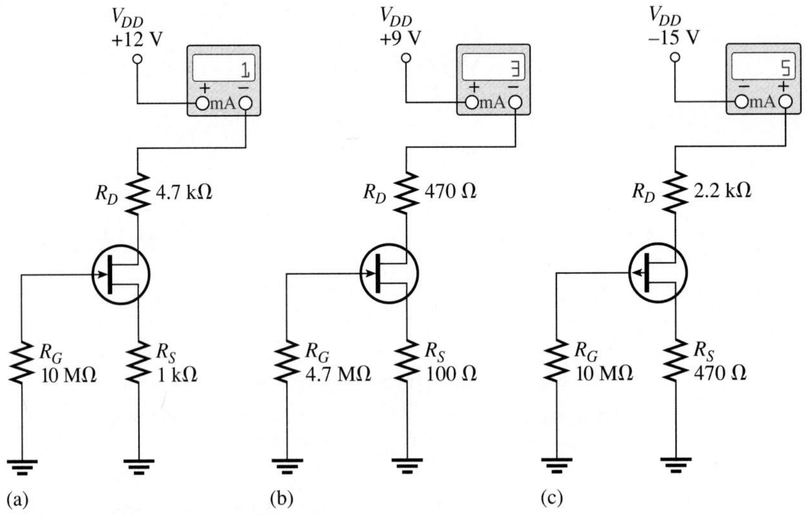

FIGURE 18–75

29. Determine in which mode (depletion or enhancement) each D-MOSFET in Figure 18–76 is biased.

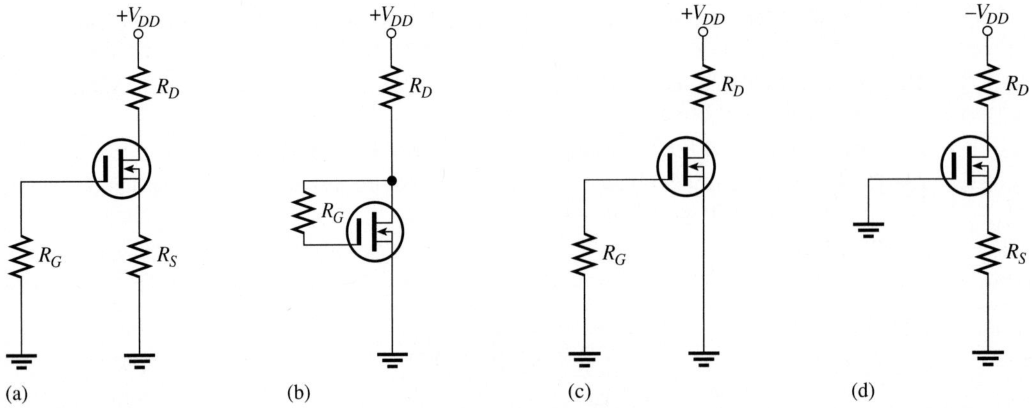

FIGURE 18–76

30. Each E-MOSFET in Figure 18–77 has a $V_{GS(\text{thresh})}$ of +5 V or −5 V, depending on whether it is an N-channel or a P-channel device. Determine whether each MOSFET is *on* or *off*.

FIGURE 18–77

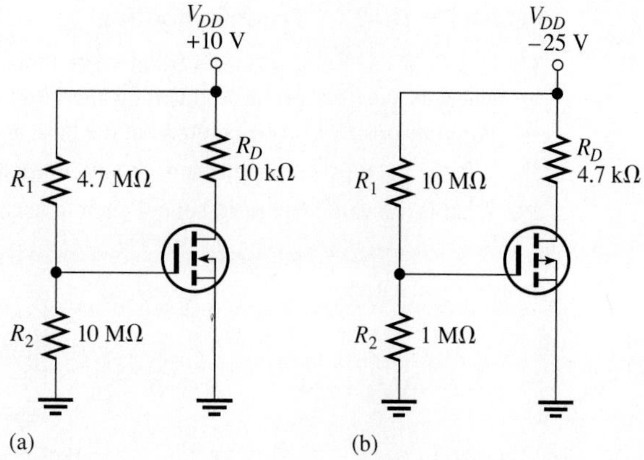

SECTION 18–10 Unijunction Transistors (UJTs)

31. In a UJT, if R_{BB} is 5 kΩ and R_{B1} is 3 kΩ, what is η?

32. What is the charging time constant in the oscillator circuit of Figure 18–78?

33. How long will the capacitor in Figure 18–78 initially charge if the peak voltage of the UJT is 6.3 V?

34. If the valley voltage of the UJT in Figure 18–78 is 2.23 V, what is the peak-to-peak value of the voltage across the capacitor, if the peak voltage is the same as in Problem 33? Sketch its general shape.

FIGURE 18–78

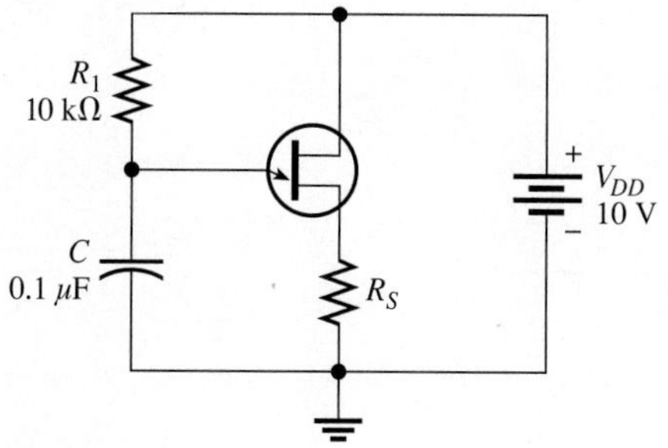

SECTION 18–11 Thyristors

35. Assume that the holding current for a particular SCR is 10 mA. What is the maximum value of R in Figure 18–79 necessary to keep the SCR in conduction once it is turned on? Neglect the drop across the SCR.

FIGURE 18–79

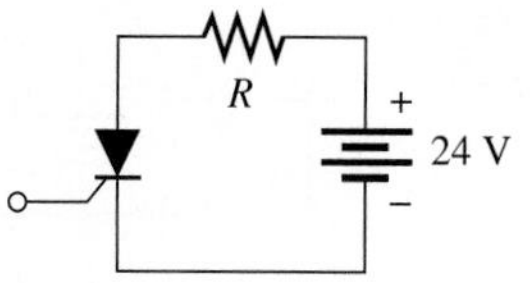

36. Repeat Problem 35 for a holding current of 500 μA.

SECTION 18–13 Troubleshooting

37. In an out-of-circuit test of a good NPN transistor, what should an analog ohmmeter indicate when its positive probe is touching the emitter and the negative probe is touching the base? When its positive probe is touching the base and the negative probe is touching the emitter?

38. What is the most likely problem, if any, in each circuit of Figure 18–80? Assume a β_{dc} of 75.

39. What is the value of the dc beta of each transistor in Figure 18–81?

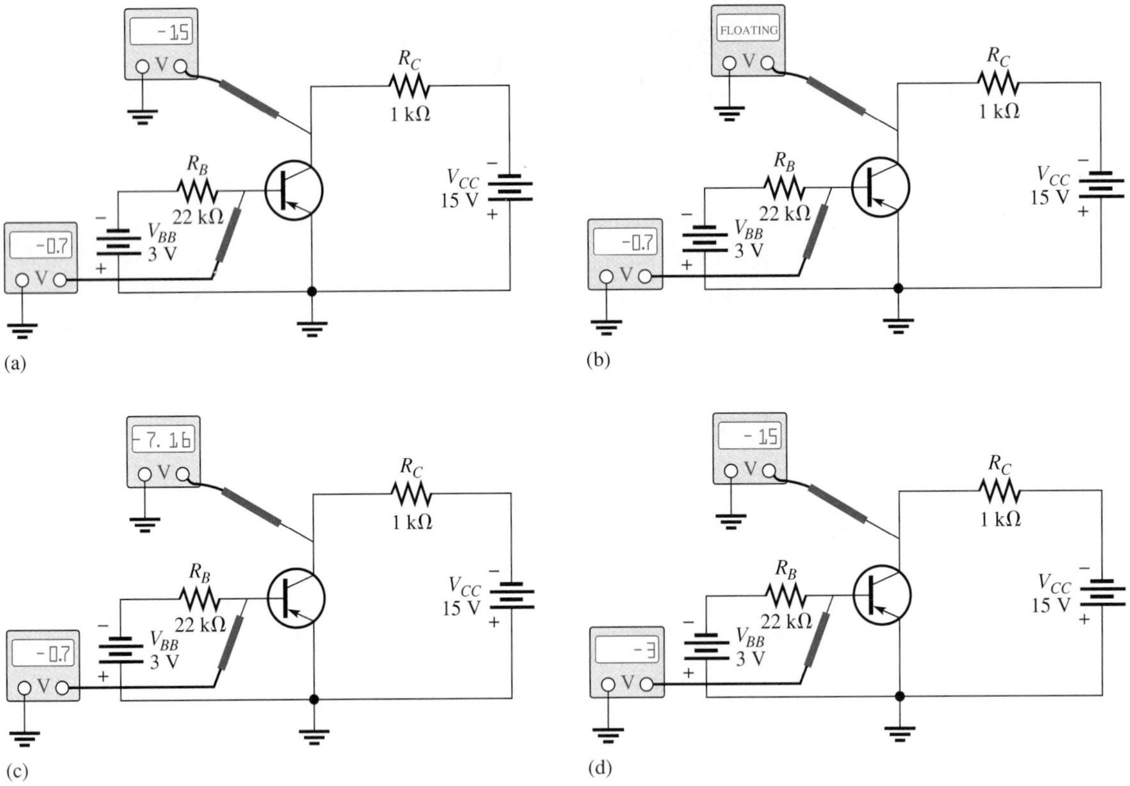

FIGURE 18–80

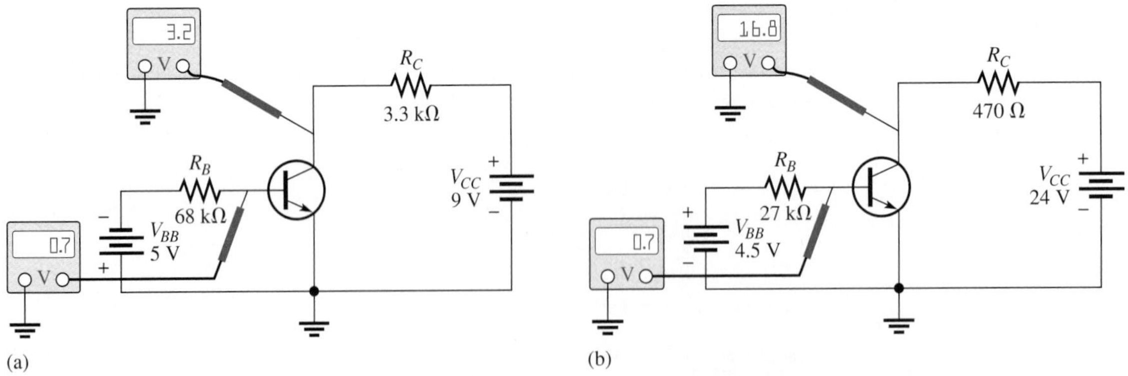

FIGURE 18–81

ANSWERS TO SECTION REVIEWS

Section 18–1

1. The three transistor terminals are emitter, base, and collector.
2. Forward-reverse bias means that the base-emitter junction is forward-biased and the base-collector junction is reverse-biased.
3. β_{dc} is direct current gain.
4. $I_C = \beta_{dc} I_B = 1$ mA

Section 18–2

1. One source is required for voltage-divider bias.
2. $V_B = (4.49\text{ k}\Omega/26.49\text{ k}\Omega)\ 30\text{ V} = 5.08\text{ V}$
3. $I_C \cong I_E = V_E/R_E = 5.15\text{ V}/1\text{ k}\Omega = 5.15\text{ mA}$

Section 18–3

1. Amplification is the process of increasing the amplitude of an electrical signal.
2. The Q point is the dc operating point.
3. $A_v = R_C/R_E = 21.4$

Section 18–4

1. A switching transistor operates in saturation and cutoff.
2. I_C is maximum at saturation.
3. $I_C \cong 0$ A at cutoff.
4. Saturation occurs when the base-emitter is forward-biased, and there is sufficient base current.
5. $V_{CE} = V_{CC}$ at cutoff.

Section 18–5

1. β_{dc} increases with temperature.
2. β_{dc} increases with I_C to a certain value, and then decreases.
3. $I_C = P_{D(max)}/V_{CE} = 40$ mA

Section 18–6

1. The three JFET terminals are drain, source, and gate.
2. A N-channel JFET requires a negative V_{GS}.
3. I_D is controlled by V_{GS}.

Section 18–7

1. $V_P = 7$ V
2. I_D decreases when V_{GS} is increased negatively.
3. $V_{GS} = +3$ V

Section 18–8

1. Depletion-enhancement MOSFET; enhancement only MOSFET. The D-MOSFET has a physical channel; the E-MOSFET does not.
2. Yes, there is I_{DS} when $V_{GS} = 0$ V.
3. No, there is no I_{DS} when $V_{GS} = 0$ V.

Section 18–9

1. A P-channel JFET must have a positive V_{GS}.
2. $V_{GS} = -I_D R_S = -8$ V
3. For $V_{GS} = 0$ V, $I_D = I_{DSS}$.
4. $V_{GS} > 2$ V

Section 18–10

1. UJT stands for unijunction transistor.
2. The terminals of a UJT are emitter, base 1, and base 2.
3. The intrinsic standoff ratio is R_{B1}/R_{BB}.

Section 18–11

1. SCR stands for silicon-controlled rectifier.
2. A positive voltage on the gate turns the SCR on, and the SCR conducts current from anode cathode. If the current drops below the holding value, the SCR turns off.
3. A triac can be used in a lamp dimmer.
4. False

Section 18–12

1. See Figure 18–82.

FIGURE 18–82

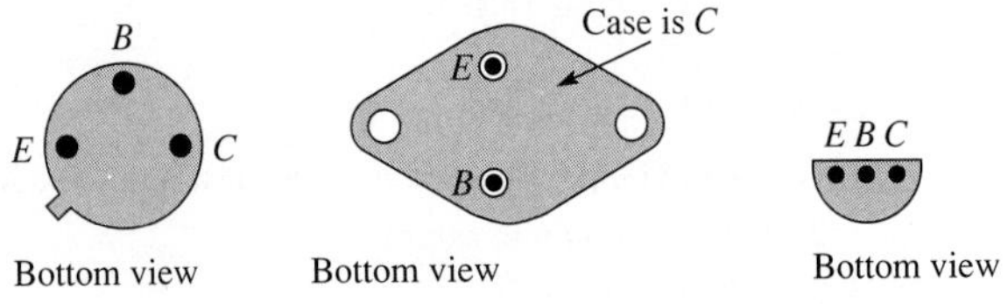

Section 18–13

1. First test the transistor in-circuit.
2. If R_B opens, the transistor is in cutoff.
3. The base voltage is V_{BB} and the collector voltage is V_{CC}.

Section 18–14

1. The detector circuit senses a change in temperature and produces a proportional change in output voltage.
2. No.
3. 9.09 V
4. 13.5 V

■ ANSWERS TO RELATED EXERCISES FOR EXAMPLES

18–1 $I_B = 241\ \mu A$; $I_C = 21.7$ mA; $I_E = 21.9$ mA; $V_{CE} = 4.23$ V; $V_{CB} = 3.53$ V

18–2 $V_B = 8.77$ V; $V_E = 8.07$ V; $V_C = 21.9$ V; $V_{CE} = 13.9$ V; $I_B = 80.7\ \mu A$; $I_E = 8.07$ mA; $I_C = 8.07$ mA

18–3 Along the horizontal axis

18–4 An 850 mV peak sine wave riding on a 6.16 V dc level

18–5 78.4 μA

18–6 $V_{CC(max)} = 44.6$ V; $V_{CE(max)}$ is exceeded first.

18–7 $I_D = 12$ mA

18–8 80,000 MΩ

18–9 $V_{DS} = 2$ V; $V_{GS} = -3.12$ V

18–10 6.8 V

18–11 2.13 mA

18–12 5 V

18–13 R_B open

19

AMPLIFIERS AND OSCILLATORS

As you learned in the previous chapter, the biasing of a transistor is purely a dc operation. The purpose of biasing, however, is to establish an operating point (Q point) about which variations in current and voltage can occur in response to an ac input signal.

When very small signal voltages must be amplified, such as from an antenna in a receiver, variations about the Q point of an amplifier are relatively small.

Amplifiers designed to handle these small ac signals are called *small-signal amplifiers.* When large swings or variations in voltage and current about the Q point are required for power amplification, large signal amplifiers are used. An example is the power amplifier that drives the speakers in a stereo system.

Regardless of whether an amplifier is in the small-signal or large-signal category, it will be in one of three configurations of amplifier circuits: common-emitter, common-collector, and common-base. For FETs the configurations are common-source, common-drain, and common-gate. Also, for any of the configurations there are three basic modes of operation possible—class A, class B, and class C—which are covered in this chapter.

Oscillators are also introduced in this chapter. An oscillator is a circuit that produces a sustained sine wave output without an input signal. Oscillators operate on the principle of positive feedback.

■ CHAPTER OBJECTIVES

- □ Analyze a common-emitter amplifier
- □ Analyze a common-collector amplifier
- □ Analyze a common-base amplifier
- □ Analyze three types of FET amplifiers
- □ Analyze a multistage amplifier
- □ Explain class A amplifier operation
- □ Analyze class B amplifiers
- □ Explain the basic operation of a class C amplifier
- □ Discuss the theory and analyze the operation of several types of oscillators
- □ Troubleshoot amplifier circuits

■ APPLICATION ASSIGNMENT

Your boss has asked you to check out and troubleshoot an audio preamplifier board that is thought to be malfunctioning. This circuit board consists of transistor amplifiers used in a superheterodyne AM receiver being manufactured by your company. The circuit amplifies the audio signal before it goes to the power amplifier that drives the speaker. After studying this chapter, you should be able to complete the application assignment in Section 19–11.

19–1 ■ COMMON-EMITTER AMPLIFIERS

The common-emitter (CE) is a type of bipolar amplifier configuration in which the emitter is at ac ground. The other two types of amplifier configuration, the common-collector and common-base, are covered in the following sections.

After completing this section, you should be able to

- **Analyze a common-emitter amplifier**
  - ☐ Identify a common-emitter configuration
  - ☐ Explain how a bypass capacitor increases voltage gain
  - ☐ Calculate voltage gain
  - ☐ Discuss phase inversion
  - ☐ Calculate input resistance
  - ☐ Calculate current gain
  - ☐ Calculate power gain

Figure 19–1 shows a typical **common-emitter (CE)** amplifier. The one shown has voltage-divider bias, although other types of bias methods are possible. C_1 and C_2 are coupling capacitors used to pass the signal into and out of the amplifier such that the source or load will not affect the dc bias voltages. C_3 is a bypass capacitor that shorts the emitter signal voltage (ac) to ground without disturbing the dc emitter voltage. Because of the bypass capacitor, the emitter is at signal ground (but not dc ground), thus making the circuit a common-emitter amplifier. The purpose of the bypass capacitor is to increase the signal voltage **gain.** (The reason why this increase occurs will be discussed shortly.) Notice that the input signal is applied to the base, and the output signal is taken off the collector. All capacitors are assumed to have a reactance of approximately zero at the signal frequency.

FIGURE 19–1
Typical common-emitter (CE) amplifier.

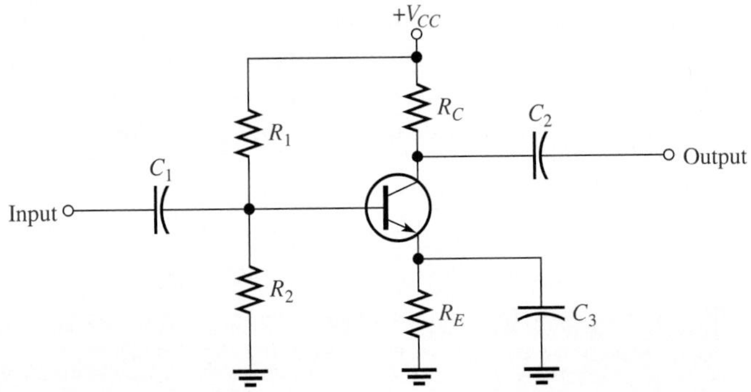

A Bypass Capacitor Increases Voltage Gain

The bypass capacitor shorts the signal around the emitter resistor, R_E, in order to increase the voltage gain. To understand why, let's consider the amplifier without the bypass capacitor and see what the voltage gain is. The CE amplifier with the bypass capacitor removed is shown in Figure 19–2.

As before, lowercase subscripts indicate signal (ac) voltages and signal (alternating) currents. The **voltage gain** of the amplifier is V_{out}/V_{in}. The output signal voltage is

$$V_{out} = I_c R_C$$

The signal voltage at the base is approximately equal to

$$V_b \cong V_{in} \cong I_e(r_e + R_E)$$

where r_e is the internal emitter resistance of the transistor. The voltage gain A_v can now be expressed as

$$A_v = \frac{V_{out}}{V_{in}} = \frac{I_c R_C}{I_e(r_e + R_E)}$$

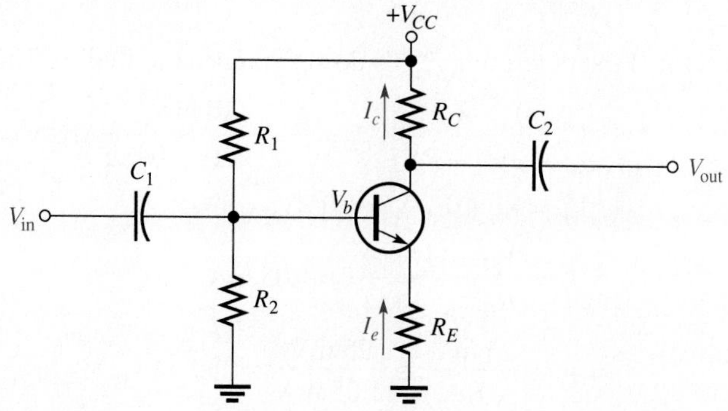

FIGURE 19–2
CE amplifier with bypass capacitor removed.

Since $I_c \cong I_e$, the currents cancel and the gain is the ratio of the resistances:

$$A_v = \frac{R_C}{r_e + R_E} \tag{19–1}$$

Keep in mind that this formula is for the CE configuration without the bypass capacitor. If R_E is much greater than r_e, then $A_v \cong R_C/R_E$.

If the bypass capacitor is connected across R_E, it effectively shorts the signal to ground, leaving only r_e in the emitter. Thus, the voltage gain of the CE amplifier with the bypass capacitor shorting R_E is

$$A_v = \frac{R_C}{r_e} \tag{19–2}$$

Now, r_e is a very important transistor parameter because it determines the voltage gain of a CE amplifier in conjunction with R_C. A formula for estimating r_e is given without derivation in the following equation:

$$r_e \cong \frac{25 \text{ mV}}{I_E} \tag{19–3}$$

EXAMPLE 19–1 Determine the voltage gain of the amplifier in Figure 19–3 both with and without a bypass capacitor.

FIGURE 19–3

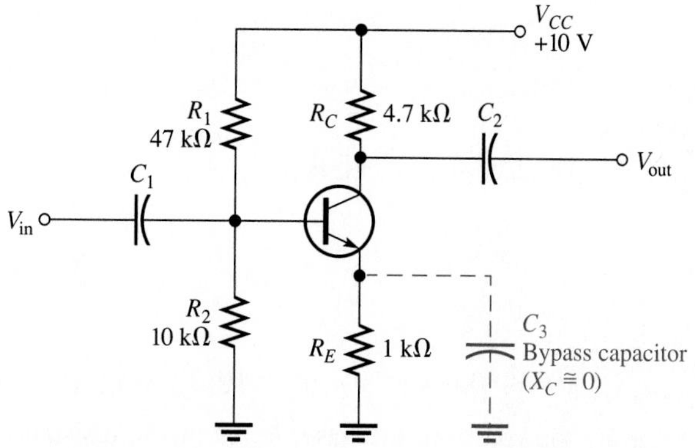

Solution First determine r_e. To do so, you need to find I_E. Thus,

$$V_B \cong \left(\frac{R_2}{R_1 + R_2}\right)V_{CC} = \left(\frac{10\text{ k}\Omega}{47\text{ k}\Omega + 10\text{ k}\Omega}\right)10\text{ V} = 1.75\text{ V}$$

$$V_E = V_B - 0.7\text{ V} = 1.05\text{ V}$$

$$I_E = \frac{V_E}{R_E} = \frac{1.05\text{ V}}{1\text{ k}\Omega} = 1.05\text{ mA}$$

$$r_e \cong \frac{25\text{ mV}}{I_E} \cong \frac{25\text{ mV}}{1.05\text{ mA}} = 23.8\ \Omega$$

The voltage gain without a bypass capacitor is

$$A_v = \frac{R_C}{r_e + R_E} = \frac{4.7\text{ k}\Omega}{1023.8\ \Omega} = 4.59$$

The voltage gain with the bypass capacitor installed is

$$A_v = \frac{R_C}{r_e} = \frac{4.7\text{ k}\Omega}{23.8\ \Omega} = 197$$

As you can see, the voltage gain is greatly increased by the addition of the bypass capacitor. In terms of decibels (dB), the voltage gain is

$$A_v = 20\log(197) = 45.9\text{ dB}$$

Related Exercise What is the voltage gain with the bypass capacitor if $R_C = 5.6\text{ k}\Omega$?

Phase Inversion

As we discussed in Chapter 18, the output voltage at the collector is 180° out of phase with the input voltage at the base. Therefore, the CE amplifier is characterized by a phase inversion between the input and the output. As mentioned before, this inversion is sometimes indicated by a negative voltage gain.

AC Input Resistance

The dc input resistance (R_{IN}), viewed from the base of the transistor, was developed in Section 18–2. The input resistance "seen" by the signal at the base is derived in a similar manner when the emitter resistor is bypassed to ground.

$$R_{in} = \frac{V_b}{I_b}$$

$$V_b = I_e r_e$$

$$I_e \cong \beta_{ac} I_b$$

$$R_{in} \cong \frac{\beta_{ac} I_b r_e}{I_b}$$

The I_b terms cancel, leaving

$$R_{in} \cong \beta_{ac} r_e \qquad \textbf{(19–4)}$$

Total Input Resistance to a CE Amplifier

Viewed from the base, R_{in} is the ac resistance. The actual resistance seen by the source includes that of bias resistors. We will now develop an expression for the total input resistance. The concept of ac ground was mentioned earlier. At this point it needs some additional explanation because it is important in the development of the formula for total input resistance, $R_{in(tot)}$.

You have already seen that the bypass capacitor effectively makes the emitter appear as ground to the ac signal, because the X_C of the capacitor is nearly zero at the signal frequency. Of course, to a dc signal the capacitor looks like an open and thus does not affect the dc emitter voltage.

In addition to seeing ground through the bypass capacitor, the signal also sees ground through the dc supply voltage source, V_{CC}. It does so because there is zero signal voltage at the V_{CC} terminal. Thus, the $+V_{CC}$ terminal effectively acts as ac ground. As a result, the two bias resistors, R_1 and R_2, appear in parallel to the ac input, because one end of R_2 goes to actual ground and one end of R_1 goes to ac ground (V_{CC} terminal). Also, R_{in} at the base appears in parallel with $R_1 \parallel R_2$. This situation is illustrated in Figure 19–4.

FIGURE 19–4
Total input resistance.

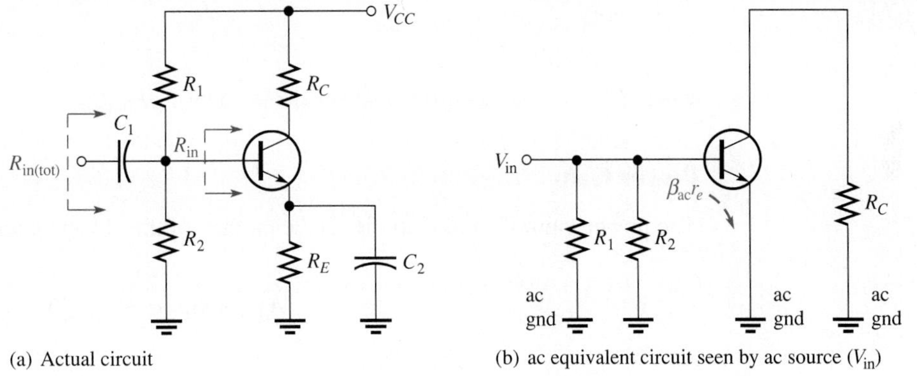

The expression for the total input resistance to the CE amplifier as seen by the ac source is as follows:

$$R_{in(tot)} = R_1 \parallel R_2 \parallel R_{in} \tag{19–5}$$

R_C has no effect because of the reverse-biased, base-collector junction.

EXAMPLE 19–2 Determine the total input resistance seen by the signal source in the CE amplifier in Figure 19–5. $\beta_{ac} = 150$.

FIGURE 19–5

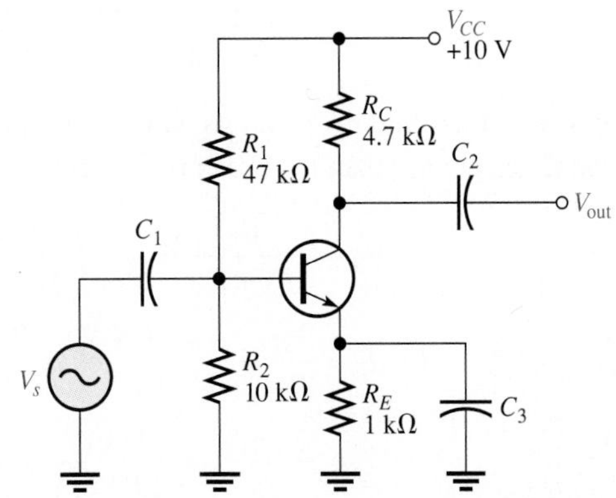

Solution In Example 19–1, you found r_e for the same circuit. Thus,

$$R_{in} = \beta_{ac} r_e = 150(23.8\ \Omega) = 3.57\ k\Omega$$
$$R_{in(tot)} = R_1 \parallel R_2 \parallel R_{in} = 47\ k\Omega \parallel 10\ k\Omega \parallel 3.57\ k\Omega = 2.49\ k\Omega$$

Related Exercise What is the total input resistance seen by the signal source in Figure 19–5 if C_3 is removed?

Current Gain

The signal **current gain** of a CE amplifier is

$$A_i = \frac{I_c}{I_s} \quad \textbf{(19–6)}$$

where I_s is the source current and is calculated by $V_s/R_{in(tot)}$.

Power Gain

The **power gain** of a CE amplifier is the product of the voltage gain and the current gain:

$$A_p = A_v A_i \quad \textbf{(19–7)}$$

EXAMPLE 19–3 Determine the voltage gain, current gain, and power gain for the CE amplifier in Figure 19–6. $\beta_{ac} = 100$.

FIGURE 19–6

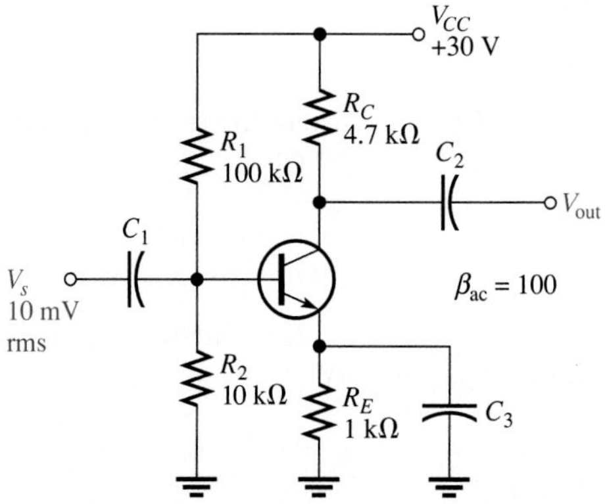

Solution First find r_e. To do so, you must find I_E. Begin by calculating V_B. Since R_{IN} is ten times greater than R_2, it can be neglected. Thus,

$$V_B \cong \left(\frac{R_2}{R_1 + R_2}\right)V_{CC} = \left(\frac{10\ k\Omega}{110\ k\Omega}\right)30\ V = 2.73\ V$$

$$I_E = \frac{V_E}{R_E} = \frac{V_B - 0.7\ V}{R_E} = \frac{2.03\ V}{1\ k\Omega} = 2.03\ mA$$

$$r_e \cong \frac{25\ mV}{I_E} = \frac{25\ mV}{2.03\ mA} = 12.3\ \Omega$$

The ac voltage gain is

$$A_v = \frac{R_C}{r_e} = \frac{4.7\ \text{k}\Omega}{12.3\ \Omega} = 382$$

Determine the signal current gain by first finding $R_{in(tot)}$ to get I_s:

$$R_{in(tot)} = R_1 \parallel R_2 \parallel \beta_{ac} r_e = 100\ \text{k}\Omega \parallel 10\ \text{k}\Omega \parallel 1.23\ \text{k}\Omega = 1.08\ \text{k}\Omega$$

$$I_s = \frac{V_s}{R_{in(tot)}} = \frac{10\ \text{mV}}{1.08\ \text{k}\Omega} = 9.26\ \mu\text{A}$$

Next determine I_c:

$$I_c = \frac{V_{out}}{R_C} = \frac{A_v V_s}{R_C} = \frac{(382)(10\ \text{mV})}{4.7\ \text{k}\Omega} = 813\ \mu\text{A}$$

$$A_i = \frac{I_c}{I_s} = \frac{813\ \mu\text{A}}{9.26\ \mu\text{A}} = 87.8$$

The power gain is

$$A_p = A_v A_i = (382)(87.8) = 33{,}540$$

The voltage gain and the power gain in decibels are as follows:

$$A_v = 20 \log(382) = 51.6\ \text{dB}$$

$$A_p = 10 \log(33{,}540) = 45.3\ \text{dB}$$

Related Exercise Determine A_v, A_i, and A_p in this example when R_{IN} is taken into account.

SECTION 19–1 REVIEW

1. What is the purpose of the bypass capacitor in a CE amplifier?
2. How is the voltage gain of a CE amplifier determined?
3. If A_v is 50 and A_i is 200, what is the power gain of a CE amplifier?

19–2 ■ COMMON-COLLECTOR AMPLIFIERS

The common-collector (CC) amplifier, commonly referred to as an emitter-follower, is the second of the three basic amplifier configurations. The input is applied to the base and the output is at the emitter. There is no collector resistor. The voltage gain of a CC amplifier is approximately 1, and its main advantage is its high input resistance.

After completing this section, you should be able to

- **Analyze a common-collector amplifier**
  - ☐ Calculate voltage gain
  - ☐ Calculate input resistance
  - ☐ Calculate current gain
  - ☐ Calculate power gain
  - ☐ Describe the darlington pair configuration

Figure 19–7 shows a **common-collector (CC)** circuit with a voltage-divider bias. Notice that the input is applied to the base and the output is taken off the emitter.

FIGURE 19–7
Typical emitter-follower (common-collector, CC) amplifier.

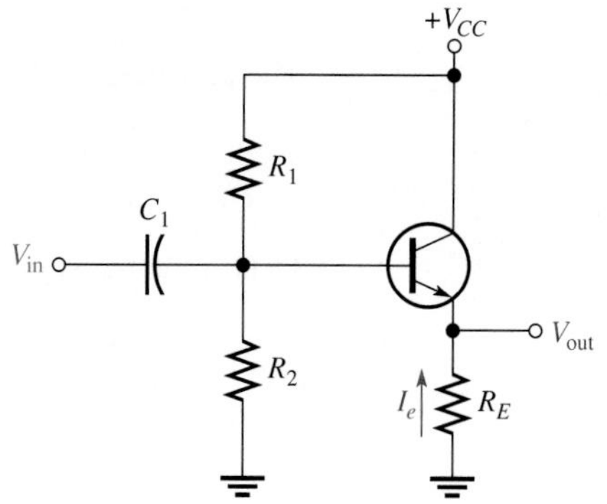

Voltage Gain

As in all amplifiers, the voltage gain in a CC amplifier is $A_v = V_{out}/V_{in}$. For the emitter-follower, V_{out} is I_eR_E, and V_{in} is $I_e(r_e + R_E)$. Therefore, the gain is $I_eR_E/I_e(r_e + R_E)$. The currents cancel, and the gain expression simplifies to

$$A_v = \frac{R_E}{r_e + R_E} \quad \textbf{(19–8)}$$

It is important to notice here that *the gain is always less than 1*. If r_e is much less than R_E, then a good approximation is $A_v \cong 1$.

Since the output voltage is the emitter voltage, it is in phase with the base or the input voltage. As a result, and because the voltage gain is close to 1, the output voltage follows the input voltage—thus the term **emitter-follower.**

Input Resistance

The emitter-follower is characterized by a high input resistance, which makes it a very useful circuit. Because of the high input resistance, the emitter-follower can be used as a buffer to minimize loading effects when one circuit is driving another.

The derivation of the input resistance viewed from the base is similar to that for the CE amplifier. In this case, however, the emitter resistor is not bypassed:

$$R_{in} = \frac{V_b}{I_b} = \frac{I_e(r_e + R_E)}{I_b} \cong \frac{\beta_{ac} I_b(r_e + R_E)}{I_b} = \beta_{ac}(r_e + R_E)$$

If R_E is at least ten times larger than r_e, then the input resistance at the base is

$$R_{in} \cong \beta_{ac} R_E \quad \textbf{(19–9)}$$

In Figure 19–7, the bias resistors appear to the input signal to be in parallel with R_{in}, just as in the voltage-divider biased CE amplifier. The total ac input resistance is

$$R_{in(tot)} = R_1 \parallel R_2 \parallel R_{in} \quad \textbf{(19–10)}$$

Since R_{in} can be made large with the proper selection of R_E, a much higher input resistance results for this configuration than for the CE circuit.

Current Gain

The signal current gain for the emitter-follower is I_e/I_s where I_s is the signal current and can be calculated as $V_s/R_{in(tot)}$. If the bias resistors are large enough to be neglected so that $I_s = I_b$, then the current gain of the amplifier is equal to the current gain of the transistor, β_{ac}. Of course, the same was also true for the CE amplifier. β_{ac} is the maximum achievable current gain in both types of amplifiers.

$$A_i = \frac{I_e}{I_s} \tag{19–11}$$

Power Gain

The power gain is the product of the voltage gain and the current gain. For the emitter-follower, the power gain is approximately equal to the current gain because the voltage gain is approximately 1:

$$A_p \cong A_i \tag{19–12}$$

EXAMPLE 19–4 Determine the input resistance of the emitter-follower in Figure 19–8. Also find the voltage gain, current gain, and power gain.

FIGURE 19–8

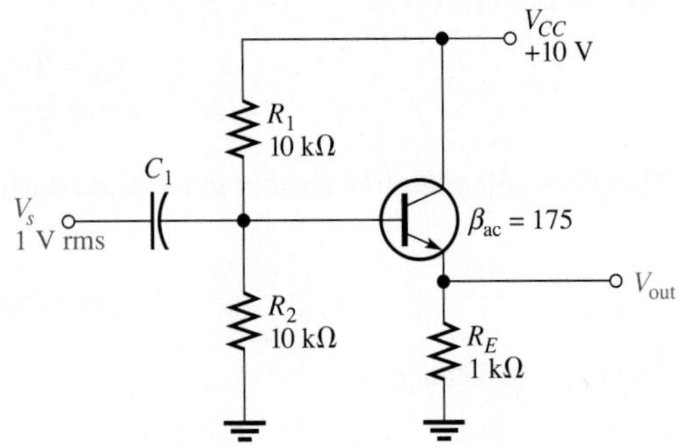

Solution The approximate input resistance viewed from the base is

$$R_{in} \cong \beta_{ac} R_E = (175)(1\text{ k}\Omega) = 175\text{ k}\Omega$$

The total input resistance is

$$R_{in(tot)} = R_1 \parallel R_2 \parallel R_{in} = 10\text{ k}\Omega \parallel 10\text{ k}\Omega \parallel 175\text{ k}\Omega = 4.86\text{ k}\Omega$$

The voltage gain is, neglecting r_e,

$$A_v \cong 1$$

The current gain is

$$A_i = \frac{I_e}{I_s}$$

$$I_e = \frac{V_e}{R_E} = \frac{A_v V_b}{R_E} \cong \frac{1\text{ V}}{1\text{ k}\Omega} = 1\text{ mA}$$

$$I_s = \frac{V_s}{R_{in(tot)}} = \frac{1\text{ V}}{4.86\text{ k}\Omega} = 206\ \mu\text{A}$$

$$A_i = \frac{1\text{ mA}}{206\ \mu\text{A}} = 4.85$$

The power gain is

$$A_p \cong A_i = 4.85$$

Related Exercise If R_E in Figure 19–8 is decreased to 820 Ω, what is the power gain?

The Darlington Pair

As you have seen, β is a major factor is determining the input resistance. The β of the transistor limits the maximum achievable input resistance you can get from a given emitter-follower circuit.

One way to boost input resistance is to use a **darlington pair,** as shown in Figure 19–9. The collectors of two transistors are connected together, and the emitter of the first drives the base of the second. This configuration achieves β multiplication as shown in the following steps. The emitter current of the first transistor is

$$I_{e1} \cong \beta_1 I_{b1}$$

This emitter current becomes the base current for the second transistor, producing a second emitter current of

$$I_{e2} \cong \beta_2 I_{e1}$$
$$I_{e2} \cong \beta_1 \beta_2 I_{b1}$$

Therefore, the effective current gain of the darlington pair is

$$\beta = \beta_1 \beta_2 \quad \textbf{(19–13)}$$

The input resistance is

$$R_{\text{in}} = \beta_1 \beta_2 R_E \quad \textbf{(19–14)}$$

FIGURE 19–9
Darlington pair.

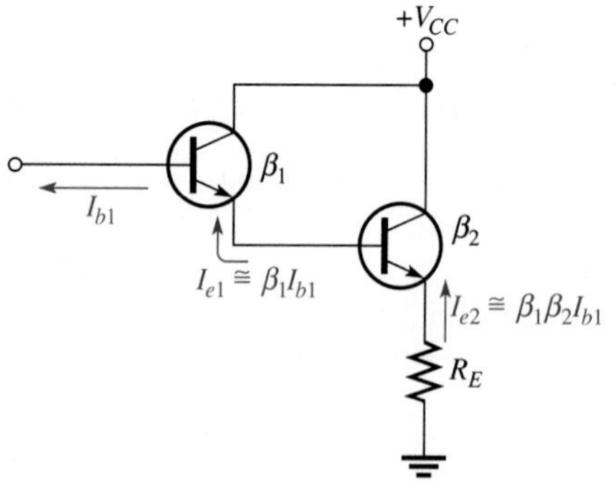

SECTION 19–2 REVIEW

1. What is a common-collector amplifier called?
2. What is its ideal maximum voltage gain of a CC amplifier?
3. What is the most important characteristic of the CC amplifier?

19–3 ▪ COMMON-BASE AMPLIFIERS

The third basic amplifier configuration is the common-base (CB). It provides high voltage gain with unity current gain. Since it has a low input resistance, the CB amplifier is the most appropriate type for certain high-frequency applications where sources tend to have very low output resistances.

After completing this section, you should be able to

- **Analyze a common-base amplifier**
  - ☐ Calculate voltage gain
  - ☐ Calculate input resistance
  - ☐ Calculate current gain
  - ☐ Calculate power gain
  - ☐ Compare the three basic amplifier configurations: CE, CC, and CB

A typical **common-base (CB)** circuit is pictured in Figure 19–10. The base is at signal (ac) ground, and the input is applied to the emitter. The output is taken off the collector.

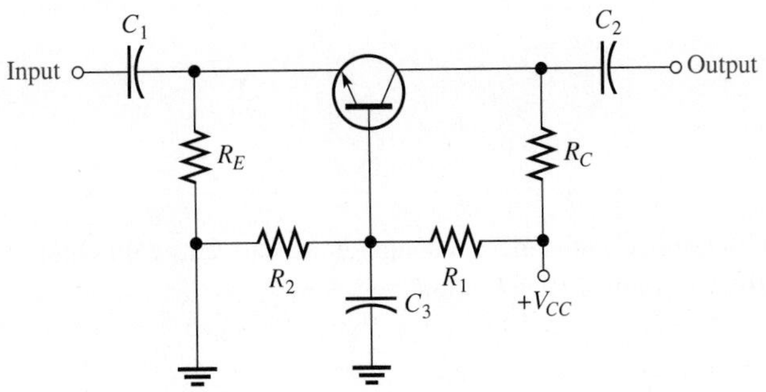

FIGURE 19–10
Typical common-base (CB) amplifier.

Voltage Gain

The input voltage is the emitter voltage V_e. The output voltage is the collector voltage V_c. With this in mind, we develop the voltage gain formula as follows:

$$A_v = \frac{V_c}{V_e} = \frac{I_c R_C}{I_e r_e} \cong \frac{I_e R_C}{I_e r_e}$$

$$A_v = \frac{R_C}{r_e} \qquad \textbf{(19–15)}$$

Notice that the gain expression is the same as that for the CE amplifier (when R_E is bypassed).

Input Resistance

The resistance viewed from the emitter appears to the input signal as follows:

$$R_{\text{in}} = \frac{V_{\text{in}}}{I_{\text{in}}} = \frac{V_e}{I_e} = \frac{I_e r_e}{I_e}$$

$$R_{\text{in}} = r_e \qquad \textbf{(19–16)}$$

Viewed from the source, R_E appears in parallel with R_{in}. However, r_e is normally so small compared to R_E that Equation (19–16) is also valid for the total input resistance, $R_{in(tot)}$.

Current Gain

The current gain is the output current I_c divided by the input current I_e. Since $I_c \cong I_e$, the signal current gain is approximately 1:

$$A_i \cong 1 \tag{19–17}$$

Power Gain

Since the current gain is approximately 1 for the CB amplifier, the power gain is approximately equal to the voltage gain:

$$A_p \cong A_v \tag{19–18}$$

EXAMPLE 19–5 Find the input resistance, voltage gain, current gain, and power gain for the CB amplifier in Figure 19–11.

FIGURE 19–11

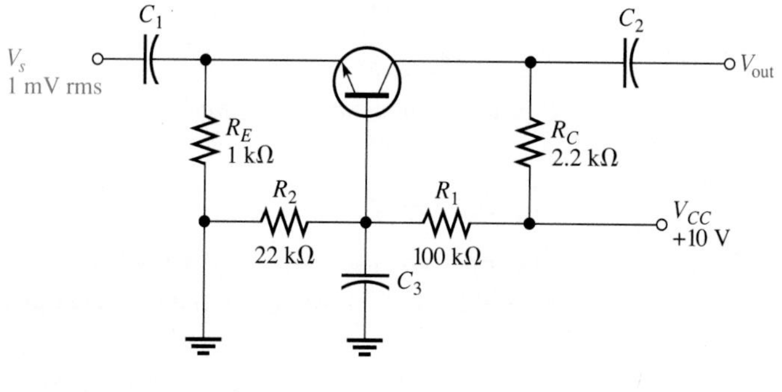

Solution First find I_E so that you can determine r_e. Then $R_{in} = r_e$. Thus,

$$V_B \cong \left(\frac{R_2}{R_1 + R_2}\right)V_{CC} = \left(\frac{22\ \text{k}\Omega}{122\ \text{k}\Omega}\right)10\ \text{V} = 1.80\ \text{V}$$

$$V_E = V_B - 0.7\ \text{V} = 1.80\ \text{V} - 0.7\ \text{V} = 1.10\ \text{V}$$

$$I_E = \frac{V_E}{R_E} = \frac{1.10\ \text{V}}{1\ \text{k}\Omega} = 1.10\ \text{mA}$$

$$R_{in} = r_e \cong \frac{25\ \text{mV}}{1.10\ \text{mA}} = 22.7\ \Omega$$

The signal voltage gain is

$$A_v = \frac{R_C}{r_e} = \frac{2.2\ \text{k}\Omega}{22.7\ \Omega} = 96.9$$

Thus,

$$A_i \cong 1$$
$$A_p \cong 96.9$$

Related Exercise If R_E is increased to 1.5 kΩ in Figure 19–11, what is the voltage gain?

Summary

Table 19–1 summarizes the important characteristics of each of the three amplifier configurations. Also, relative values are indicated for general comparison of the amplifiers.

TABLE 19–1

Comparison of amplifier configurations. The current gains and the input resistance are the maximum achievable values, with the bias resistors neglected.

	CE	CC	CB
Voltage gain, A_v	High R_C/r_e	Low $\cong 1$	High R_C/r_e
Current gain, $A_{i(max)}$	High β_{ac}	High β_{ac}	Low $\cong 1$
Power gain, A_p	Very high A_iA_v	High $\cong A_i$	High $\cong A_v$
Input resistance, $R_{in(max)}$	Low $\beta_{ac}r_e$	High $\beta_{ac}R_E$	Very low r_e

SECTION 19–3 REVIEW

1. Can the same voltage gain be achieved with a CB as with a CE amplifier?
2. Is the input resistance of a CB amplifier very low or very high?

19–4 ▪ FET AMPLIFIERS

Field-effect transistors, both JFETs and MOSFETs, can be used as amplifiers in any of three circuit configurations similar to those for the bipolar transistor. The FET configurations are common-source, common-drain, and common-gate. These are similar to the bipolar configurations of common-emitter, common-collector, and common-base, respectively.

After completing this section, you should be able to

- **Analyze three types of FET amplifiers**
  - ☐ Calculate the transconductance of an FET
  - ☐ Analyze a common-source amplifier
  - ☐ Analyze a common-drain amplifier
  - ☐ Analyze a common-gate amplifier
  - ☐ Compare the three FET amplifier configurations

Transconductance of an FET

Recall that in a bipolar transistor, the base current controls the collector current, and the relationship between these two currents is expressed as $I_c = \beta_{ac} I_b$. In an FET, the gate voltage controls the drain current. An important FET parameter is the **transconductance, g_m,** which is defined as

$$g_m = \frac{I_d}{V_{gs}} \tag{19–19}$$

The transconductance is one factor that determines the voltage gain of an FET amplifier. On data sheets, the transconductance is sometimes called the *forward transadmittance* and is designated y_{fs}.

Common-Source (CS) Amplifiers

A self-biased N-channel JFET with an ac source capacitively coupled to the gate is shown in Figure 19–12. The resistor R_G serves two purposes: (1) It keeps the gate at approximately 0 V dc (because I_{GSS} is extremely small), and (2) its large value (usually several megohms) prevents loading of the ac signal source. The bias voltage is created by the drop across R_S. The bypass capacitor, C_3, keeps the source of the FET effectively at ac ground.

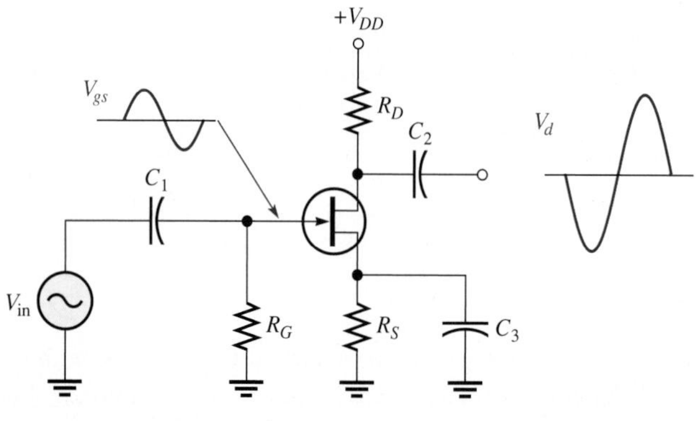

FIGURE 19–12
JFET common-source amplifier.

The signal voltage causes the gate-to-source voltage to swing above and below its Q-point value, causing a swing in drain current. As the drain current increases, the voltage drop across R_D also increases, causing the drain voltage (with respect to ground) to decrease.

The drain current swings above and below its Q-point value in phase with the gate-to-source voltage. The drain-to-source voltage swings above and below its Q-point value 180° out of phase with the gate-to-source voltage, as illustrated in Figure 19–12.

D-MOSFET A zero-biased N-channel D-MOSFET with an ac source capacitively coupled to the gate is shown in Figure 19–13. The gate is at approximately 0 V dc and the source terminal is at ground, thus making $V_{GS} = 0$ V.

The signal voltage causes V_{gs} to swing above and below its 0 value, producing a swing in I_d. The negative swing in V_{gs} produces the depletion mode, and I_d decreases. The positive swing in V_{gs} produces the enhancement mode, and I_d increases.

FIGURE 19–13
Zero-biased D-MOSFET common-source amplifier.

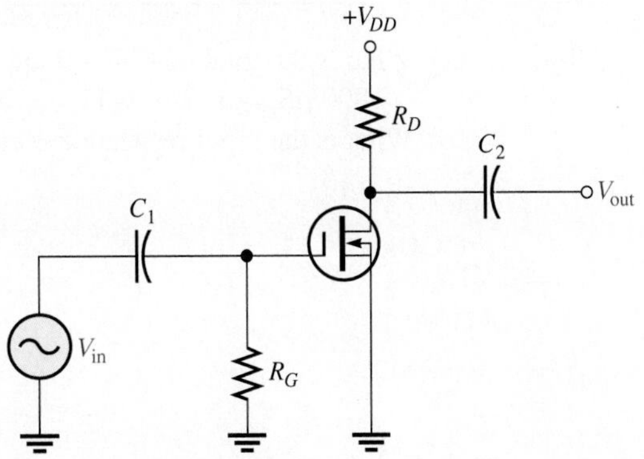

E-MOSFET Figure 19–14 shows a voltage divider-biased, N-channel E-MOSFET with an ac signal source capacitively coupled to the gate. The gate is biased with a positive voltage such that $V_{GS} > V_{GS(thresh)}$.

FIGURE 19–14
Common-source E-MOSFET amplifier with voltage-divider bias.

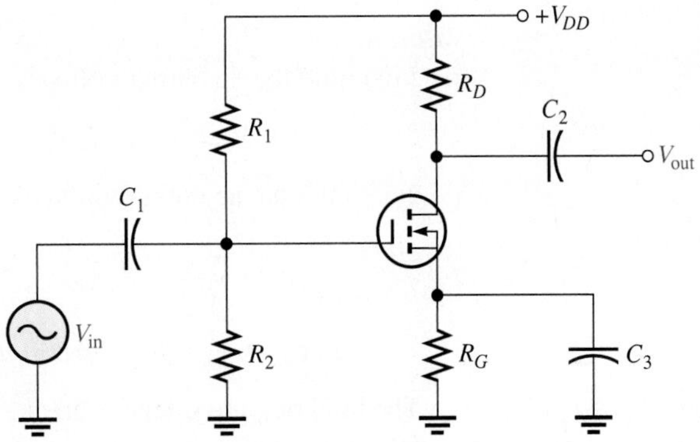

As with the JFET and D-MOSFET, the signal voltage produces a swing in V_{gs} above and below its Q-point value. This swing, in turn, causes a swing in I_d. Operation is entirely in the enhancement mode.

Voltage Gain Voltage gain, A_v, of an amplifier always equals V_{out}/V_{in}. In the case of the CS amplifier, V_{in} is equal to V_{gs}, and V_{out} is equal to the signal voltage developed across R_D, which is I_dR_D. Thus,

$$A_v = \frac{I_dR_D}{V_{gs}}$$

Since $g_m = I_d/V_{gs}$, the common-source voltage gain is

$$A_v = g_mR_D \qquad \textbf{(19–20)}$$

Input Resistance Because the input to a CS amplifier is at the gate, the input resistance is extremely high. Ideally, it approaches infinity and can be neglected. As you know, the high input resistance is produced by the reverse-biased PN junction in a JFET and by the insulated gate structure in a MOSFET.

The actual input resistance seen by the signal source is the gate-to-ground resistor R_G in parallel with the FET's input resistance, V_{GS}/I_{GSS}. The reverse leakage current I_{GSS} is typically given on the data sheet for a specific value of V_{GS} so that the input resistance of the device can be calculated.

EXAMPLE 19–6 (a) What is the total output voltage (dc + ac) of the amplifier in Figure 19–15? The g_m is 1800 μS, I_D is 2 mA, $V_{GS(\text{off})}$ is −3.5 V, and I_{GSS} is 15 nA.
(b) What is the input resistance seen by the signal source?

FIGURE 19–15

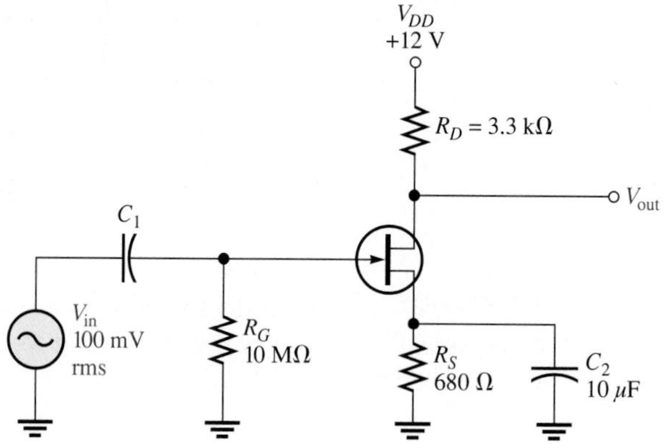

Solution

(a) First, find the dc output voltage:

$$V_D = V_{DD} - I_D R_D = 12\text{ V} - (2\text{ mA})(3.3\text{ k}\Omega) = 5.4\text{ V}$$

Next, find the ac output voltage by using the gain formula:

$$A_v = \frac{V_{out}}{V_{in}} = g_m R_D$$

$$V_{out} = g_m R_D V_{in} = (1800\ \mu\text{S})(3.3\text{ k}\Omega)(100\text{ mV}) = 594\text{ mV rms}$$

The total output voltage is an ac signal with a peak-to-peak value of 594 mV × 2.828 = 1.67 V, riding on a dc level of 5.4 V.

(b) The input resistance is determined as follows (since $V_G = 0$ V):

$$V_{GS} = I_D R_S = (2\text{ mA})(680\ \Omega) = 1.36\text{ V}$$

The input resistance at the gate of the JFET is

$$R_{IN(gate)} = \frac{V_{GS}}{I_{GSS}} = \frac{1.36\text{ V}}{15\text{ nA}} = 91\text{ M}\Omega$$

The input resistance seen by the signal source is

$$R_{in} = R_G \parallel R_{IN(gate)} = 10\text{ M}\Omega \parallel 91\text{ M}\Omega = 9.0\text{ M}\Omega$$

Related Exercise What is the total output voltage in the amplifier of Figure 19–15 if V_{DD} is changed to 15 V? Assume the other parameters are the same.

Common-Drain (CD) Amplifier

A common-drain JFET amplifier is shown in Figure 19–16 with voltages indicated. Self-biasing is used in this circuit. The input signal is applied to the gate through a coupling capacitor, and the output is at the source terminal. There is no drain resistor. This circuit, of course, is analogous to the bipolar emitter-follower and is sometimes called a *source-follower*.

FIGURE 19–16
JFET common-drain amplifier (source-follower).

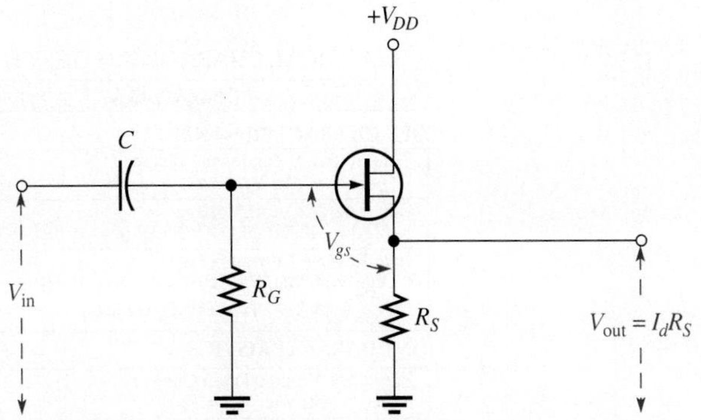

Voltage Gain As in all amplifiers, the voltage gain is $A_v = V_{out}/V_{in}$. For the source-follower, V_{out} is I_dR_S and V_{in} is $V_{gs} + I_dR_S$, as shown in Figure 19–16. Therefore, the gate-to-source voltage gain is $I_dR_S/(V_{gs} + I_dR_S)$. Substituting $I_d = g_mV_{gs}$ into the expression gives the following result:

$$A_v = \frac{g_mV_{gs}R_S}{V_{gs} + g_mV_{gs}R_S}$$

Canceling V_{gs}, we get

$$A_v = \frac{g_mR_S}{1 + g_mR_S} \qquad \textbf{(19–21)}$$

Notice here that *the gain is always slightly less than 1*. If $g_mR_S >> 1$, then a good approximation is $A_v \cong 1$. Since the output voltage is at the source, it is in phase with the gate (input) voltage.

Input Resistance Because the input signal is applied to the gate, the input resistance seen by the input signal source is extremely high, just as in the CS amplifier configuration. The gate resistor R_G, in parallel with the input resistance looking in at the gate, is the total input resistance.

EXAMPLE 19–7

(a) Determine the voltage gain of the amplifier in Figure 19–17(a) using the data sheet information in Figure 19–17(b) on page 834.

(b) Also determine the input resistance. Assume minimum data sheet values where available.

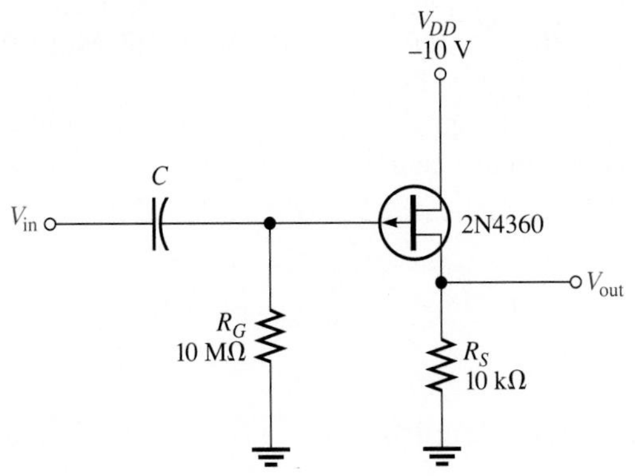

(a)

FIGURE 19–17

*ELECTRICAL CHARACTERISTICS (T_A = 25°C unless otherwise noted)

Characteristic	Symbol	Min	Max	Unit
OFF CHARACTERISTICS				
Gate-Source Breakdown Voltage (I_G = 10 μAdc, V_{DS} = 0)	$V_{(BR)GSS}$	20	–	Vdc
Gate-Source Cutoff Voltage (V_{DS} = –10 Vdc, I_D = 1.0 μAdc)	$V_{GS(off)}$	0.7	10	Vdc
Gate Reverse Current	I_{GSS}			
(V_{GS} = 15 Vdc, V_{DS} = 0)		–	10	nAdc
(V_{GS} = 15 Vdc, V_{DS} = 0, T_A = 65°C)		–	0.5	μAdc
ON CHARACTERISTICS				
Zero-Gate Voltage Drain Current (Note 1) (V_{DS} = –10 Vdc, V_{GS} = 0)	I_{DSS}	3.0	30	mAdc
Gate-Source Breakdown Voltage (V_{DS} = –10 Vdc, I_D = 0.3 mAdc)	V_{GS}	0.4	9.0	Vdc
SMALL SIGNAL CHARACTERISTICS				
Drain-Source "ON" Resistance (V_{GS} = 0, I_D = 0, f = 1.0 kHz)	$r_{ds(on)}$	–	700	Ohms
Forward Transadmittance (Note !) (V_{DS} = –10 Vdc, V_{GS} = 0, f = 1.0 kHz)	$\|y_{fs}\|$	2000	8000	μmhos
Forward Transconductance (V_{DS} = –10 Vdc, V_{GS} = 0, f = 1.0 MHz)	$Re(y_{fs})$	1500	–	μmhos
Output Admittance (V_{DS} = –10 Vdc, V_{GS} = 0, f = 1.0 kHz)	$\|y_{os}\|$	–	100	μmhos
Input Capacitance (V_{DS} = –10 Vdc, V_{GS} = 0, f = 1.0 MHz)	C_{iss}	–	20	pF
Reverse Transfer Capacitance (V_{DS} = –10 Vdc, V_{GS} = 0, f = 1.0 MHz)	C_{rss}	–	5.0	pF
Common-Source Noise Figure (V_{DS} = –10 Vdc, I_D = 1.0 mAdc, R_G = 1.0 Megohm, f = 100 Hz)	NF	–	5.0	dB
Equivalent Short-Circuit Input Noise Voltage (V_{DS} = –10 Vdc, I_D = 1.0 mAdc, f = 100 Hz, BW = 15 Hz)	E_n	–	0.19	$\mu V/\sqrt{Hz}$

*Indicates JEDEC Registered Data.

Note 1: Pulse Test: Pulse Width ≤ 630 ms, Duty Cycle ≤ 10%.

(b)

FIGURE 19–17 ***Continued***

Solution

(a) From the data sheet, $g_m = y_{fs} = 2000\ \mu S$ minimum. The gain is

$$A_v \cong \frac{g_m R_S}{1 + g_m R_S} = \frac{(2000\ \mu S)(10\ k\Omega)}{1 + (2000\ \mu S)(10\ k\Omega)} = 0.952$$

(b) From the data sheet, $I_{GSS} = 10$ nA at $V_{GS} = 15$ V. Therefore,

$$R_{IN(gate)} = \frac{15\ V}{10\ nA} = 1500\ M\Omega$$

$$R_{IN} = R_G \parallel R_{IN(gate)} = 10\ M\Omega \parallel 1500\ M\Omega = 9.93\ M\Omega$$

Related Exercise If the g_m of the JFET in the source-follower of Figure 19–17 is doubled, what is the voltage gain?

Common-Gate (CG) Amplifier

A typical common-gate amplifier is shown in Figure 19–18. The gate is effectively at ac ground because of the bypass capacitor C_3. The input signal is applied at the source terminal through C_1. The output is coupled through C_2 from the drain terminal.

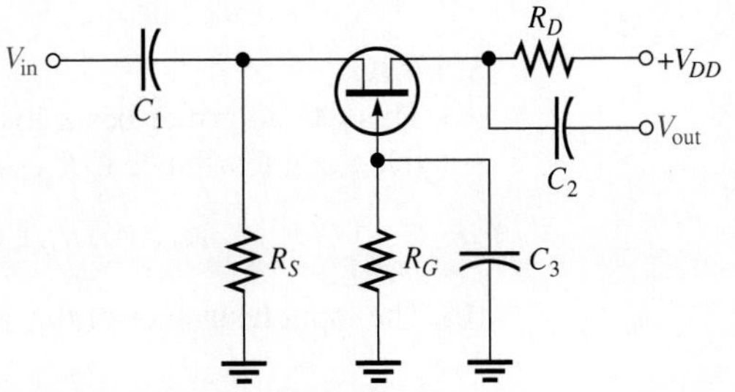

FIGURE 19–18
JFET Common-gate amplifier.

Voltage Gain The voltage gain from source to drain is developed as follows:

$$A_v = \frac{V_{out}}{V_{in}} = \frac{V_d}{V_{gs}} = \frac{I_d R_D}{V_{gs}} = \frac{g_m V_{gs} R_D}{V_{gs}}$$

$$A_v = g_m R_D \qquad \textbf{(19–22)}$$

Notice that the gain expression is the same as for the CS JFET amplifier.

Input Resistance As you have seen, both the CS and the CD configurations have extremely high input resistances because the gate is the input terminal. In contrast, the common-gate configuration has a low input resistance, as shown in the following steps.

First, the input current (source current) is equal to the drain current:

$$I_{in} = I_d = g_m V_{gs}$$

The input voltage equals V_{gs}:

$$V_{in} = V_{gs}$$

The input resistance at the source terminal is, therefore,

$$R_{in(source)} = \frac{V_{in}}{I_{in}} = \frac{V_{gs}}{g_m V_{gs}}$$

$$R_{in(source)} = \frac{1}{g_m} \qquad \textbf{(19–23)}$$

EXAMPLE 19–8

(a) Determine the voltage gain of the amplifier in Figure 19–19.
(b) Determine the input resistance.

FIGURE 19–19

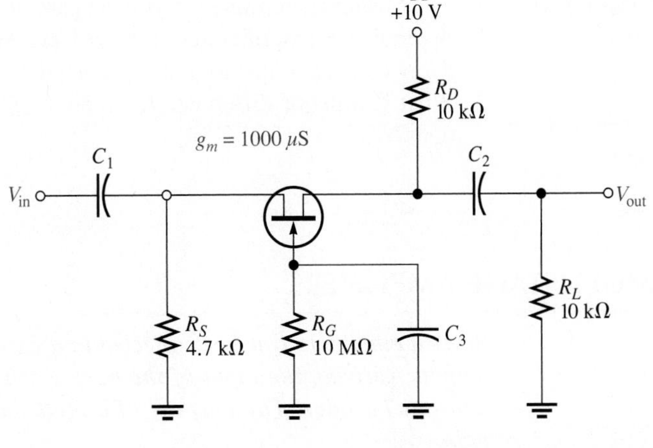

Solution

(a) This CG amplifier has a load resistor effectively in parallel with R_D, so the effective drain resistance is $R_D \parallel R_L$ and the gain is

$$A_v = g_m(R_D \parallel R_L) = (1000\ \mu\text{S})(10\ \text{k}\Omega \parallel 10\ \text{k}\Omega) = 5$$

(b) The input resistance at the source terminal is

$$R_{in(source)} = \frac{1}{g_m} = \frac{1}{1000\ \mu\text{S}} = 1\ \text{k}\Omega$$

The signal source actually sees R_S in parallel with $R_{in(source)}$, so the total input resistance is

$$R_{in(tot)} = R_{in(source)} \parallel R_S = 1\ \text{k}\Omega \parallel 4.7\ \text{k}\Omega = 825\ \Omega$$

Related Exercise How much does the input resistance change in Figure 19–19 if R_S is changed to 10 kΩ?

Summary

A summary of the gain and input resistance characteristics for the three FET amplifier configurations is given in Table 19–2.

TABLE 19–2
Comparison of FET amplifier configurations.

	CS	CD	CG
Voltage gain, A_v	g_mR_D	$\dfrac{g_mR_S}{1 + g_mR_S}$	g_mR_D
Input resistance, R_{in}	$\left(\dfrac{V_{GS}}{I_{GSS}}\right) \parallel R_G$	$\left(\dfrac{V_{GS}}{I_{GSS}}\right) \parallel R_G$	$\left(\dfrac{1}{g_m}\right) \parallel R_G$

SECTION 19–4 REVIEW

1. What factors determine the voltage gain of a CS FET amplifier?
2. A certain CS amplifier has an $R_D = 1$ kΩ. When a load resistance of 1 kΩ is capacitively coupled to the drain, how much does the gain change?
3. What is a major difference between a CG amplifier and the other two configurations?

19–5 ■ MULTISTAGE AMPLIFIERS

Several amplifiers can be connected in a cascaded arrangement with the output of one amplifier driving the input of the next. Each amplifier in the cascaded arrangement is known as a stage. The purpose of a multistage arrangement is to increase the overall gain.

After completing this section, you should be able to

- **Analyze a multistage amplifier**
  - ☐ Determine the multistage voltage gain
  - ☐ Convert the voltage gain to decibels
  - ☐ Determine the effects of loading on the gain of each stage and on the overall gain

Multistage Gain

The overall gain A_v' of cascaded amplifiers as in Figure 19–20 is the product of the individual gains:

$$A_v' = A_{v1}A_{v2}A_{v3} \cdots A_{vn} \quad \textbf{(19–24)}$$

where n is the number of stages.

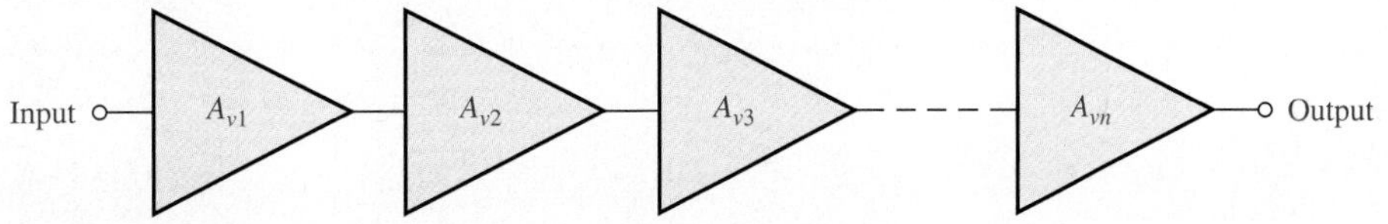

FIGURE 19–20
Cascaded amplifiers.

Decibel Voltage Gain

Amplifier voltage gain is often expressed in **decibels (dB)** as follows:

$$A_v \text{ (dB)} = 20 \log A_v \quad \textbf{(19–25)}$$

This formula is particularly useful in multistage systems because the overall dB voltage gain is the *sum* of the individual dB gains:

$$A_v' \text{ (dB)} = A_{v1} \text{ (dB)} + A_{v2} \text{ (dB)} + \cdots + A_{vn} \text{ (dB)} \quad \textbf{(19–26)}$$

EXAMPLE 19–9 A given cascaded amplifier arrangement has the following voltage gains: $A_{v1} = 10$, $A_{v2} = 15$, and $A_{v3} = 20$. What is the overall gain? Also express each gain in decibels and determine the total decibel voltage gain.

Solution

$$A_v' = A_{v1}A_{v2}A_{v3} = (10)(15)(20) = 3000$$
$$A_{v1} \text{ (dB)} = 20 \log 10 = 20 \text{ dB}$$
$$A_{v2} \text{ (dB)} = 20 \log 15 = 23.5 \text{ dB}$$
$$A_{v3} \text{ (dB)} = 20 \log 20 = 26.0 \text{ dB}$$
$$A_v' \text{ (dB)} = 20 \text{ dB} + 23.5 \text{ dB} + 26.0 \text{ dB} = 69.5 \text{ dB}$$

Related Exercise In a certain multistage amplifier, the individual stages have the following voltage gains: $A_{v1} = 25$, $A_{v2} = 5$, and $A_{v3} = 12$. What is the overall gain? Express each stage gain in dB and determine the total dB gain.

Multistage Analysis

We will use the two-stage amplifier in Figure 19–21 to illustrate multistage analysis. Notice that both stages are identical CE amplifiers with the output of the first stage capacitively coupled to the input of the second stage. Capacitive coupling prevents the dc bias of one stage from affecting that of the other. Also notice that the transistors are designated Q_1 and Q_2.

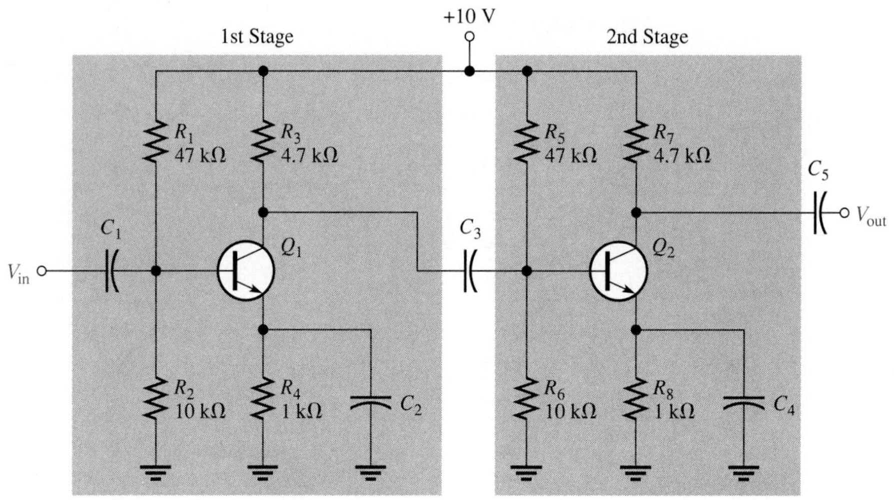

FIGURE 19–21
Two-stage common-emitter amplifier.

Loading Effects In determining the gain of the first stage, we must consider the loading effect of the second stage. Because the coupling capacitor C_3 appears as a short to the signal frequency, the total input resistance of the second stage presents an ac load to the first stage.

Looking from the collector of Q_1, the two biasing resistors, R_5 and R_6, appear in parallel with the input resistance at the base of Q_2. In other words, the signal at the collector of Q_1 "sees" the collector resistor R_3 and R_5, R_6, and $R_{in(base2)}$ of the second stage all in parallel to ac ground. Thus, the effective ac collector resistance of Q_1 is the total of all these in parallel, as Figure 19–22 illustrates.

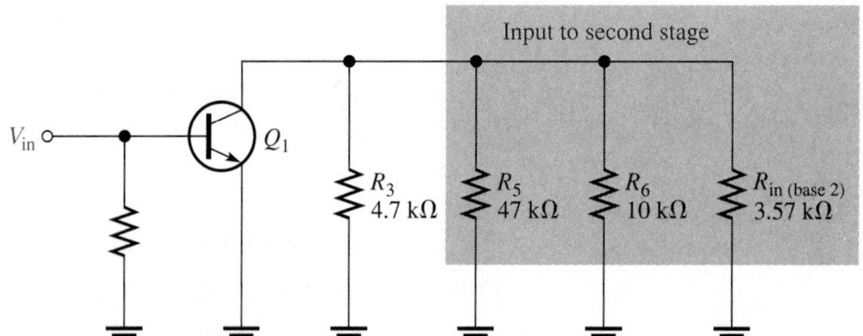

FIGURE 19–22
AC equivalent of first stage in Figure 19–21, showing loading from second stage.

The voltage gain of the first stage is reduced by the loading of the second stage, because the effective ac collector resistance of the first stage is less than the actual value of its collector resistor, R_3. Remember that $A_v = R_C/r_e$ for an unloaded amplifier.

Voltage Gain of the First Stage The ac collector resistance of the first stage, as shown in Figure 19–21, is

$$R_{c1} = R_3 \parallel R_5 \parallel R_6 \parallel R_{in(base2)}$$

Keep in mind that lowercase subscripts denote ac quantities such as for R_c.

You can verify that $I_E = 1.05$ mA, $r_e = 23.8\ \Omega$, and $R_{in(base2)} = 3.57\ \text{k}\Omega$. The effective ac collector resistance of the first stage is as follows:

$$R_c = 4.7\ \text{k}\Omega \parallel 47\ \text{k}\Omega \parallel 10\ \text{k}\Omega \parallel 3.57\ \text{k}\Omega = 1.63\ \text{k}\Omega$$

Therefore, the base-to-collector voltage gain of the first stage is

$$A_v = \frac{R_c}{r_e} = \frac{1.63\ \text{k}\Omega}{23.8\ \Omega} = 68.5$$

Voltage Gain of the Second Stage The second stage, as shown in Figure 19–21, has no load resistor, so the ac collector resistance is R_7, and the gain is

$$A_v = \frac{R_7}{r_e} = \frac{4.7\ \text{k}\Omega}{23.8\ \Omega} = 198$$

Compare this to the gain of the first stage, and notice how much the second-stage loading reduced the gain of the first stage.

Overall Voltage Gain The overall amplifier gain is

$$A_v' = A_{v1}A_{v2} = (68.5)(198) = 13{,}563$$

If an input signal of, say, 100 μV, is applied to the first stage and if the attenuation of the input base circuit is neglected, an output from the second stage of (100 μV)(13,563) ≅ 1.36 V will result. The overall gain can be expressed in decibels as follows:

$$A_v'\ (\text{dB}) = 20\log(13{,}563) = 82.6\ \text{dB}$$

SECTION 19–5 REVIEW

1. What does the term *stage* mean?
2. How is the overall gain of a multistage amplifier determined?
3. Express a voltage gain of 500 in decibels.

19–6 ■ CLASS A OPERATION

When an amplifier, whether it is one of the three configurations of bipolar or an FET type, is biased such that it always operates in the linear region where the output signal is an amplified replica of the input signal, it is a class A amplifier. The discussion and formulas in the previous sections apply to class A operation.

After completing this section, you should be able to

- **Explain class A amplifier operation**
  - ☐ Define *Q* point
  - ☐ Use a load line to analyze class A operation
  - ☐ Calculate power gain of a common-emitter class A amplifier
  - ☐ Determine dc quiescent power
  - ☐ Determine output power
  - ☐ Determine maximum efficiency of class A amplifiers

When the output signal takes up only a small percentage of the total load line excursion, the amplifier is a small-signal amplifier. When the output signal is larger and approaches the limits of the load line, the amplifier is a large-signal type. Amplifiers are typically operated as large-signal devices when power amplification is the major objective. Figure 19–23 illustrates **class A** operation of an inverting amplifier.

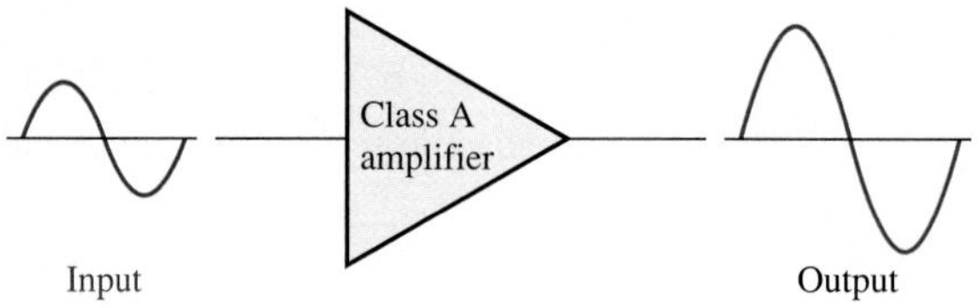

FIGURE 19–23
Class A operation.

The *Q* Point Is Centered for Maximum Output Signal

When the dc operating point (*Q* point) is at the center of the load line, a maximum class A signal can be obtained. You can see this concept by examining the graph of the load line for a given amplifier in Figure 19–24(a). This graph shows the load line with the *Q* point at its center. The collector current can vary from its *Q*-point value, I_{CQ}, up to its saturation value, $I_{C(sat)}$, and down to its cutoff value of zero. Likewise, the collector-to-emitter voltage can swing from its *Q*-point value, V_{CEQ}, up to its cutoff value, $V_{CE(cutoff)}$, and down to its saturation value of near zero. This operation is indicated in Figure 19–24(b).

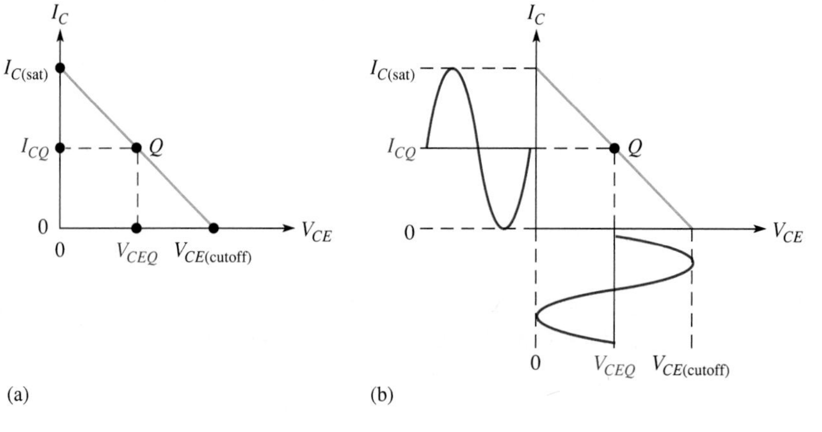

FIGURE 19–24
Maximum class A output occurs when the Q point is centered on the load line.

The peak value of the collector current is I_{CQ}, and the peak value of the collector-to-emitter voltage is V_{CEQ} in this case. This signal is the maximum that we can obtain from the class A amplifier. Actually, we cannot quite reach saturation or cutoff, so the practical maximum is slightly less.

A Noncentered *Q* Point Limits Output Swing

If the *Q* point is not centered, the output signal is limited. Figure 19–25 shows a load line with the *Q* point moved away from center toward cutoff. The output variation is limited by cutoff in this case. The collector current can only swing down to near zero and an equal amount above I_{CQ}. The collector-to-emitter voltage can only swing up to its cutoff value and an equal amount below V_{CEQ}. This situation is illustrated in Figure 19–25(a). If the amplifier is driven any further than this, it will "clip" at cutoff, as shown in Figure 19–25(b).

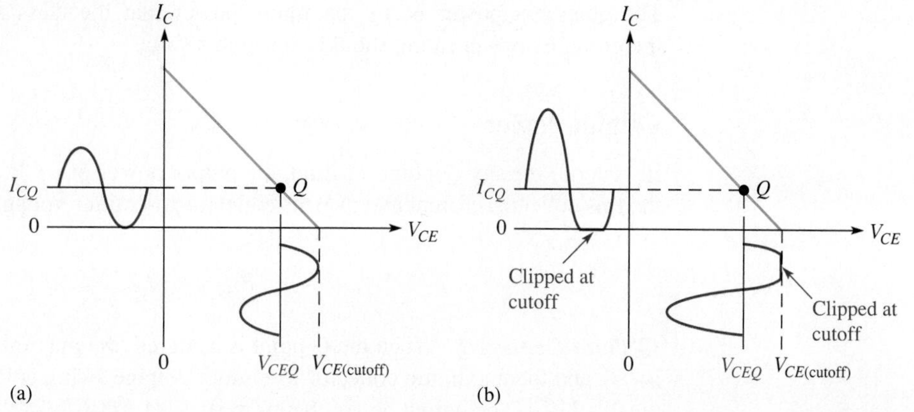

FIGURE 19–25
***Q* point closer to cutoff.**

Figure 19–26 shows a load line with the Q point moved away from center toward saturation. In this case, the output variation is limited by saturation. The collector current can only swing up to near saturation and an equal amount below I_{CQ}. The collector-to-emitter voltage can only swing down to its saturation value and an equal amount above V_{CEQ}. This situation is illustrated in Figure 19–26(a). If the amplifier is driven any further, it will "clip" at saturation, as shown in Figure 19–26(b).

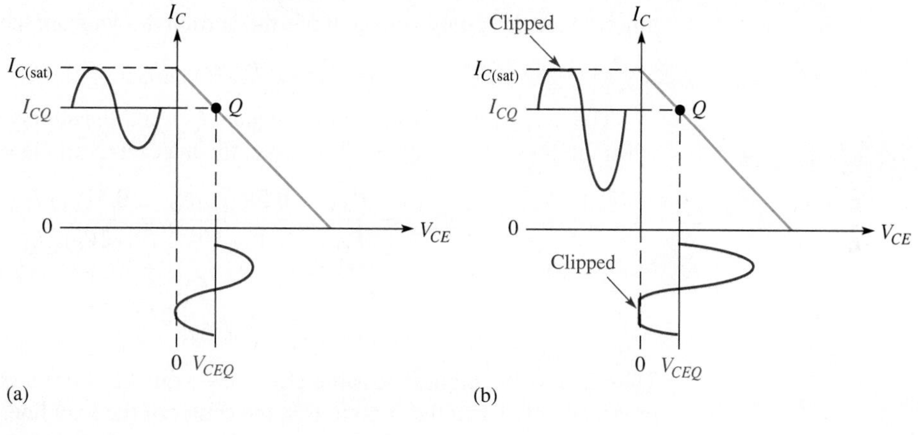

FIGURE 19–26
***Q* point closer to saturation.**

Power Gain

The main purpose of a large-signal amplifier is to achieve power gain. If we assume that the large-signal current gain A_i is approximately equal to β_{dc}, then the power gain for a common-emitter amplifier is

$$A_p \cong \beta_{\text{dc}} A_v \qquad \textbf{(19–27)}$$

DC Quiescent Power

The power dissipation of a transistor with no signal input is the product of its Q-point current and voltage:

$$P_{DQ} = I_{CQ} V_{CEQ} \qquad \textbf{(19–28)}$$

The quiescent power is the maximum power that the class A transistor must handle; therefore, its power rating should exceed this value.

Output Power

In general, for any Q-point location, the output power of a CE amplifier is the product of the rms collector current and the rms collector-to-emitter voltage:

$$P_{\text{out}} = V_{ce}I_c \qquad \textbf{(19–29)}$$

Q Point Centered When the Q point is centered, the maximum collector current swing is I_{CQ}, and the maximum collector-to-emitter voltage swing is V_{CEQ}, as was shown in Figure 19–24(b). The output power therefore is

$$P_{\text{out}} = (0.707V_{CEQ})(0.707I_{CQ})$$

$$P_{\text{out(max)}} = 0.5V_{CEQ}I_{CQ} \qquad \textbf{(19–30)}$$

This is the maximum ac output power from a class A amplifier under signal conditions. Notice that it is one-half the quiescent power dissipation.

Efficiency

Efficiency of an amplifier is the ratio of ac output power to dc input power. The dc input power is the dc supply voltage times the current drawn from the supply.

$$P_{\text{dc}} = V_{CC}I_{CC}$$

The average supply current I_{CC} equals I_{CQ}, and the supply voltage V_{CC} is twice V_{CEQ} when the Q point is centered. Therefore, the maximum efficiency is

$$\text{eff}_{\text{max}} = \frac{P_{\text{out}}}{P_{\text{dc}}} = \frac{0.5V_{CEQ}I_{CQ}}{V_{CC}I_{CC}} = \frac{0.5V_{CEQ}I_{CQ}}{2V_{CEQ}I_Q} = \frac{0.5}{2}$$

$$\text{eff}_{\text{max}} = 0.25 \qquad \textbf{(19–31)}$$

Thus 25% is the highest possible efficiency available from a class A amplifier and is approached only when the Q point is at the center of the load line.

EXAMPLE 19–10 Determine the following values for a class A amplifier operated with a centered Q point with $I_{CQ} = 50$ mA and $V_{CEQ} = 7.5$ V:
(a) minimum transistor power rating **(b)** ac output power **(c)** efficiency

Solution

(a) The maximum power that the transistor must be able to handle is the minimum rating that you would use. Thus,

$$P_{DQ} = I_{CQ}V_{CEQ} = (50\text{ mA})(7.5\text{ V}) = 375\text{ mW}$$

(b) $P_{\text{out}} = 0.5V_{CEQ}I_{CQ} = 0.5(50\text{ mA})(7.5\text{ V}) = 188\text{ mW}$

(c) Since the Q point is centered, the efficiency is at its maximum possible value of 25%.

Related Exercise Explain what happens to the output voltage of the class A amplifier in this example if the Q point is shifted to $V_{CE} = 3$ V.

SECTION 19–6 REVIEW

1. What is the optimum Q-point location for class A amplifiers?
2. What is the maximum efficiency of a class A amplifier?
3. A certain amplifier has a centered Q point of $I_{CQ} = 10$ mA and $V_{CEQ} = 7$ V. What is the maximum ac output power?

19–7 ▪ CLASS B PUSH-PULL AMPLIFIER OPERATION

When an amplifier is biased such that it operates in the linear region for 180° of the input cycle and is in cutoff for 180°, it is a class B amplifier. The primary advantage of a class B amplifier over a class A is that the class B is more efficient; you can get more output power for a given amount of input power. A disadvantage of class B is that it is more difficult to implement the circuit in order to get a linear reproduction of the input waveform. As you will see in this section, the term* push-pull *refers to a common type of class B amplifier circuit in which the input wave shape is approximately reproduced at the output.

After completing this section, you should be able to

- **Analyze class B amplifiers**
  - ☐ Explain class B operation
  - ☐ Discuss the meaning of push-pull amplifier operation
  - ☐ Define crossover distortion
  - ☐ Describe how a class B push-pull amplifier is biased
  - ☐ Determine the maximum output power
  - ☐ Determine the efficiency of a class B amplifier

Class B amplifier operation is illustrated in Figure 19–27, where the output waveform is shown relative to the input.

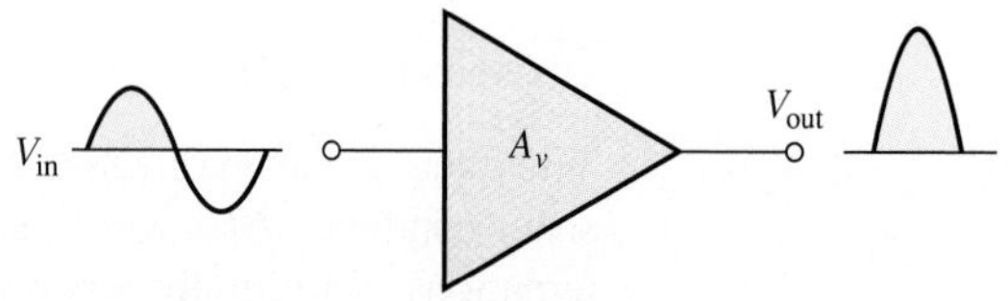

FIGURE 19–27
Class B amplifier (noninverting).

The Q Point Is at Cutoff

The class B amplifier is biased at cutoff so that $I_{CQ} = 0$ and $V_{CEQ} = V_{CE(cutoff)}$. It is brought out of cutoff and operates in its linear region when the input signal drives it into conduction. This is illustrated in Figure 19–28 with an emitter-follower circuit.

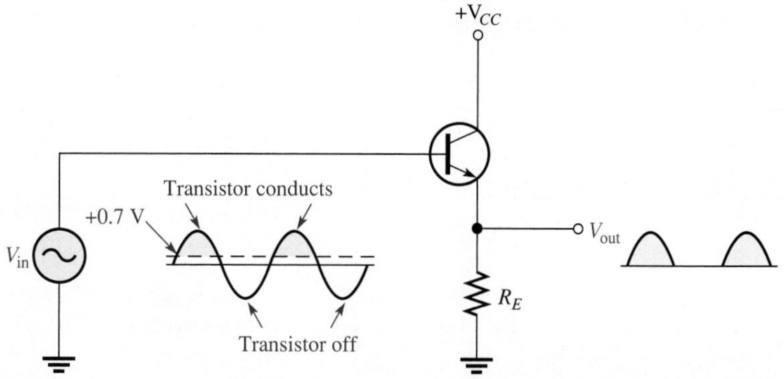

FIGURE 19–28
Common-collector class B amplifier.

Push-Pull Operation

Figure 19–29 shows one type of push-pull class B amplifier using two emitter-followers. This is a complementary amplifier because one emitter-follower uses an NPN transistor and the other a PNP, which conducts on opposite alternations of the input cycle. Notice that there is no dc base bias voltage ($V_B = 0$). Thus, only the signal voltage drives the transistors into conduction. Q_1 conducts during the positive half of the input cycle, and Q_2 conducts during the negative half.

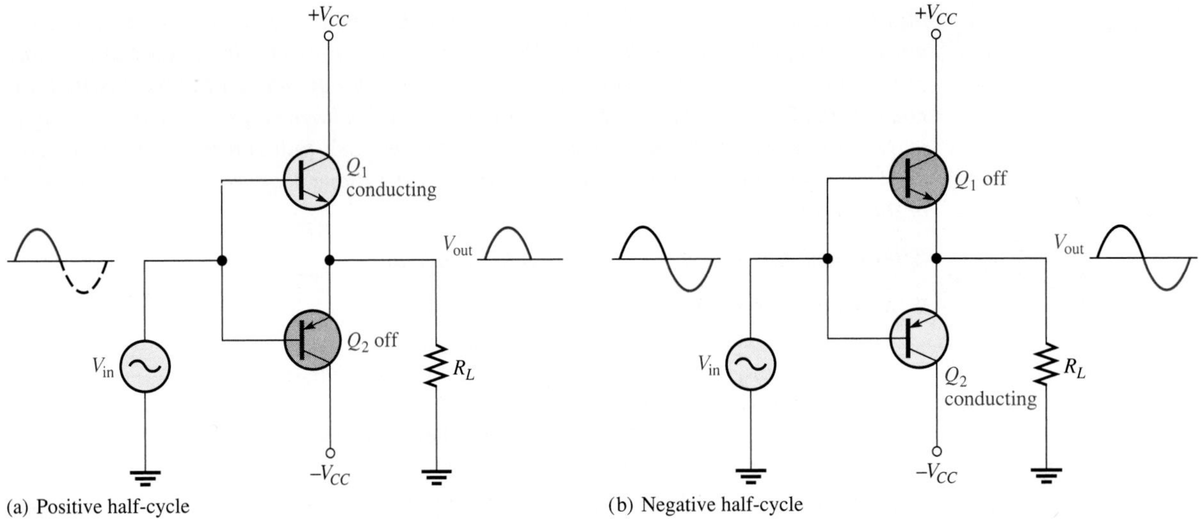

FIGURE 19–29
Class B push-pull operation.

Crossover Distortion

When the dc base voltage is zero, the input signal voltage must exceed V_{BE} before a transistor conducts. As a result, there is a time interval between the positive and negative alternations of the input when neither transistor is conducting, as shown in Figure 19–30. The resulting distortion in the output waveform is quite common and is called *crossover distortion.*

FIGURE 19–30
Illustration of crossover distortion in a class B push-pull amplifier.

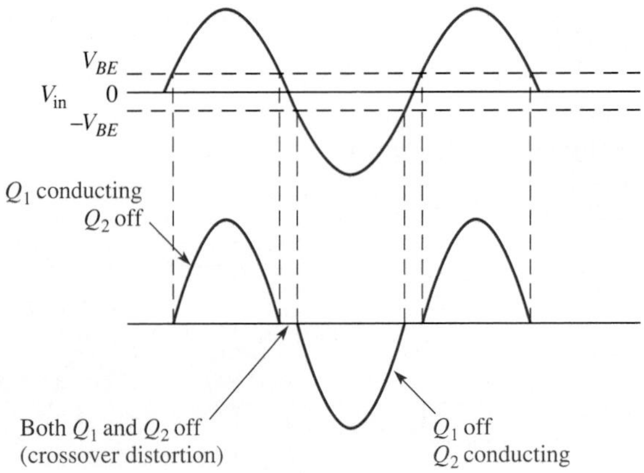

Biasing the Push-Pull Amplifier

To eliminate crossover distortion, both transistors in the push-pull arrangement must be biased slightly above cutoff when there is no signal. This can be done with a voltage divider and diode arrangement, as shown in Figure 19–31. When the diode characteristics of D_1 and D_2 are closely matched to the characteristics of the transistor base-emitter junctions, a stable bias is maintained.

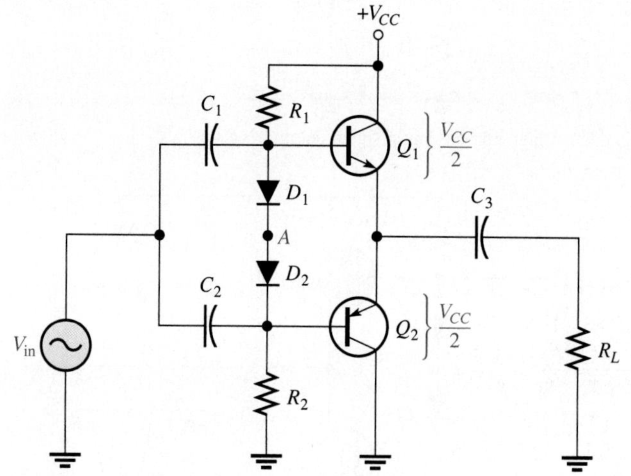

FIGURE 19–31
Biasing the push-pull amplifier to eliminate crossover distortion.

Since R_1 and R_2 are of equal value, the voltage with respect to ground at point A between the two diodes is $V_{CC}/2$. Assuming that both diodes and both transistors are identical, the drop across D_1 equals the V_{BE} of Q_1, and the drop across D_2 equals the V_{BE} of Q_2. As a result, the voltage at the emitters is also $V_{CC}/2$, therefore, $V_{CEQ1} = V_{CEQ2} = V_{CC}/2$, as indicated. Because both transistors are biased near cutoff, $I_{CQ} \cong 0$.

AC Operation

Under maximum conditions, transistors Q_1 and Q_2 are alternately driven from near cutoff to near saturation. During the positive alternation of the input signal, the Q_1 emitter is driven from its Q-point value of $V_{CC}/2$ to near V_{CC}, producing a positive peak voltage approximately equal to V_{CEQ}. At the same time, the Q_1 current swings from its Q-point value near zero to near-saturation value, as shown in Figure 19–32(a).

During the negative alternation of the input signal, the Q_2 emitter is driven from its Q-point value of $V_{CC}/2$ to near zero, producing a negative peak voltage approximately equal to V_{CEQ}. Also, the Q_2 current swings from near zero to near-saturation value, as shown in Figure 19–32(b).

Because the peak voltage across each transistor is V_{CEQ}, the ac saturation current is

$$I_{c(sat)} = \frac{V_{CEQ}}{R_L} \qquad \textbf{(19–32)}$$

Since $I_e \cong I_c$ and the output current is the emitter current, the peak output current is also V_{CEQ}/R_L.

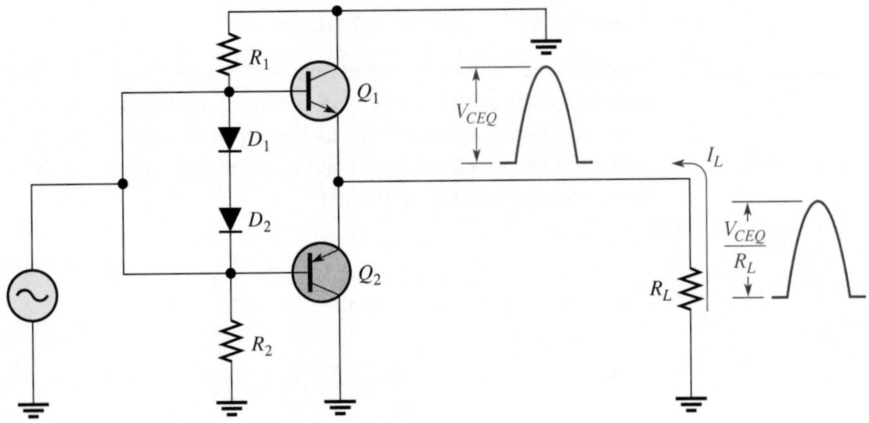

(a) Q_1 conducting with maximum signal output

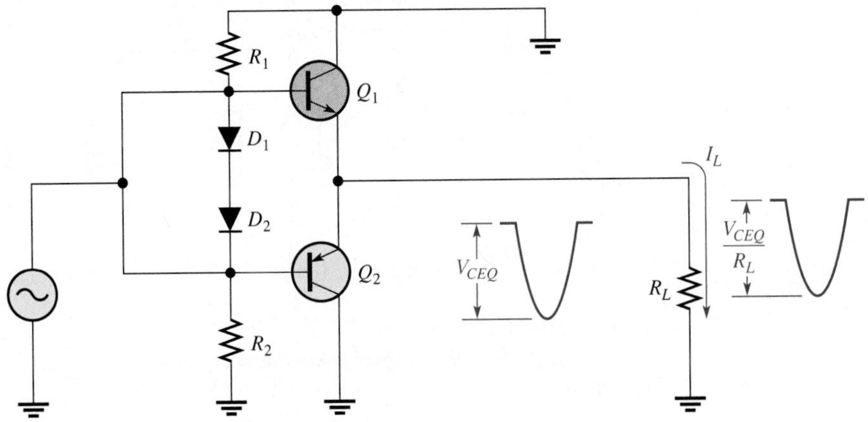

(b) Q_2 conducting with maximum signal output

FIGURE 19–32
AC push-pull operation. Capacitors are assumed to be shorts at the signal frequency, and the dc source is at ac ground.

EXAMPLE 19–11 Determine the maximum peak values for the output voltage and current in Figure 19–33.

FIGURE 19–33

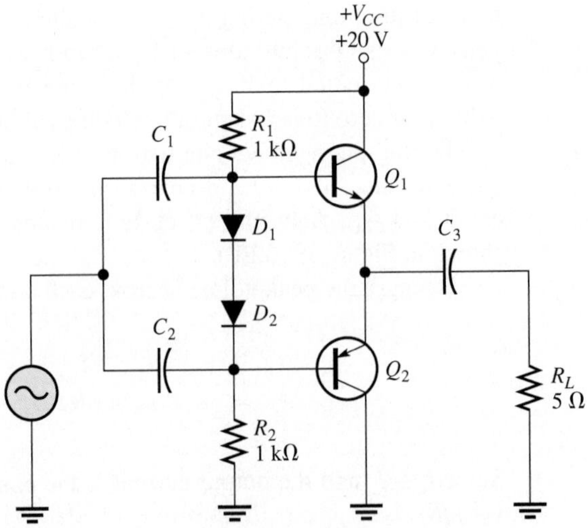

Solution The maximum peak output voltage is

$$V_{p(\text{out})} \cong V_{CEQ} = \frac{V_{CC}}{2} = \frac{20\text{ V}}{2} = 10\text{ V}$$

The maximum peak output current is

$$I_{p(\text{out})} \cong I_{c(\text{sat})} = \frac{V_{CEQ}}{R_L} = \frac{10\text{ V}}{5\ \Omega} = 2\text{ A}$$

Related Exercise Find the maximum peak values for the output voltage and current in Figure 19–33 if V_{CC} is lowered to 15 V and the load resistor is changed to 10 Ω.

Maximum Output Power

It has been shown that the maximum peak output current is approximately $I_{c(\text{sat})}$, and the maximum peak output voltage is approximately V_{CEQ}. The maximum average output power therefore is

$$P_{\text{out}} = V_{\text{rms(out)}}I_{\text{rms(out)}}$$

Since

$$V_{\text{rms(out)}} = 0.707V_{p(\text{out})} = 0.707V_{CEQ}$$

and

$$I_{\text{rms(out)}} = 0.707I_{p(\text{out})} = 0.707I_{c(\text{sat})}$$

then

$$P_{\text{out}} = 0.5V_{CEQ}I_{c(\text{sat})}$$

Substituting $V_{CC}/2$ for V_{CEQ},

$$P_{\text{out(max)}} = 0.25V_{CC}I_{c(\text{sat})} \quad \textbf{(19–33)}$$

Input Power

The input power comes from the V_{CC} supply and is

$$P_{\text{dc}} = V_{CC}I_{CC}$$

Since each transistor draws current for a half-cycle, the current is a half-wave signal with an average value of

$$I_{CC} = \frac{I_{c(\text{sat})}}{\pi}$$

Thus,

$$P_{\text{dc}} = \frac{V_{CC}I_{c(\text{sat})}}{\pi} \quad \textbf{(19–34)}$$

Efficiency

The great advantage of push-pull class B amplifiers over class A is a much higher efficiency. This advantage usually overrides the difficulty of biasing the class B push-pull amplifier to eliminate crossover distortion.

The efficiency is again defined as the ratio of ac output power to dc input power:

$$\text{eff} = \frac{P_{out}}{P_{dc}}$$

The maximum efficiency for a class B amplifier is designated eff_{max} and is developed as follows, starting with Equation (19–33):

$$P_{out(max)} = 0.25V_{CC}I_{c(sat)}$$

$$\text{eff}_{max} = \frac{P_{out}}{P_{dc}} = \frac{0.25V_{CC}I_{c(sat)}}{V_{CC}I_{c(sat)}/\pi} = 0.25\pi$$

$$\text{eff}_{max} = 0.785 \qquad \textbf{(19–35)}$$

Therefore, the maximum efficiency is 78.5%. Recall that the maximum efficiency for class A is 0.25 (25%).

EXAMPLE 19–12 Find the maximum ac output power and the dc input power of the amplifier in Figure 19–33 of Example 19–11.

Solution In Example 19–11, $I_{c(sat)}$ was found to be 2 A. Thus,

$$P_{out(max)} = 0.25V_{CC}I_{c(sat)} = 0.25(20\text{ V})(2\text{ A}) = 10\text{ W}$$

$$P_{dc} = \frac{V_{CC}I_{c(sat)}}{\pi} = \frac{(20\text{ V})(2\text{ A})}{\pi} = 12.7\text{ W}$$

Related Exercise What is the maximum ac output power in Figure 19–33 if R_L = 8 Ω?

SECTION 19–7 REVIEW

1. Where is the Q point for a class B amplifier?
2. What causes crossover distortion?
3. What is the maximum efficiency of a push-pull class B amplifier?
4. Explain the purpose of the push-pull configuration for class B.

19–8 ■ CLASS C OPERATION

Class C amplifiers are biased so that conduction occurs for much less than 180°. Class C amplifiers are more efficient than either class A or push-pull class B. Thus, more output power can be obtained from class C operation. Because the output waveform is severely distorted, class C amplifiers are normally limited to applications as tuned amplifiers at radio frequencies (RF).

After completing this section, you should be able to

- **Explain the basic operation of a class C amplifier**
  - ☐ Discuss power in a class C amplifier
  - ☐ Discuss a tuned amplifier
  - ☐ Determine maximum output power
  - ☐ Determine class C efficiency

Basic Operation

Class C amplifier operation is illustrated in Figure 19–34. A basic common-emitter class C amplifier with a resistive load is shown in Figure 19–35(a). It is biased below cutoff with the $-V_{BB}$ supply. The ac source voltage has a peak value that is slightly greater than $V_{BB} + V_{BE}$ so that the base voltage exceeds the barrier potential of the base-emitter junction for a short time near the positive peak of each cycle, as illustrated in Figure 19–35(b). During this short interval, the transistor is turned on.

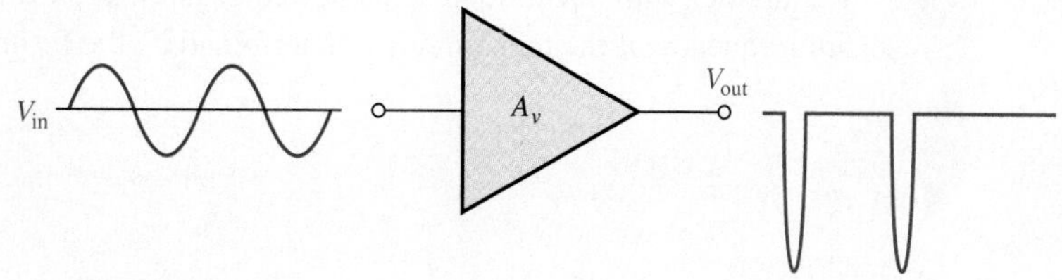

FIGURE 19–34
Class C amplifier (inverting).

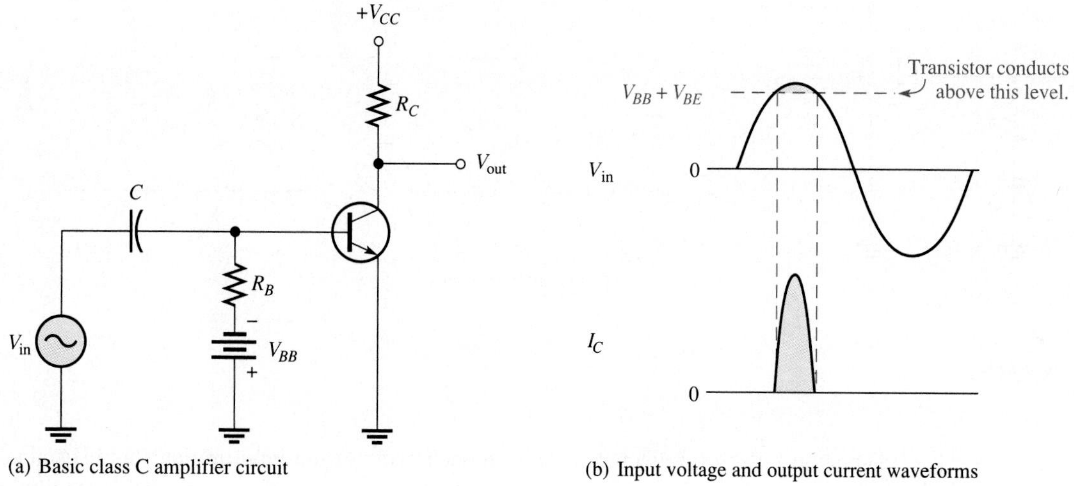

FIGURE 19–35
Class C operation.

When the entire ac load line is used, the maximum collector current is approximately $I_{C(sat)}$, and the minimum collector voltage is approximately $V_{CE(sat)}$.

Power Dissipation

The power dissipation of the transistor in a class C amplifier is low because it is on for only a small percentage of the input cycle. Figure 19–36 shows the collector current pulses. The time between the pulses is the period (T) of the ac input voltage.

FIGURE 19–36
Collector current pulses in a class C amplifier.

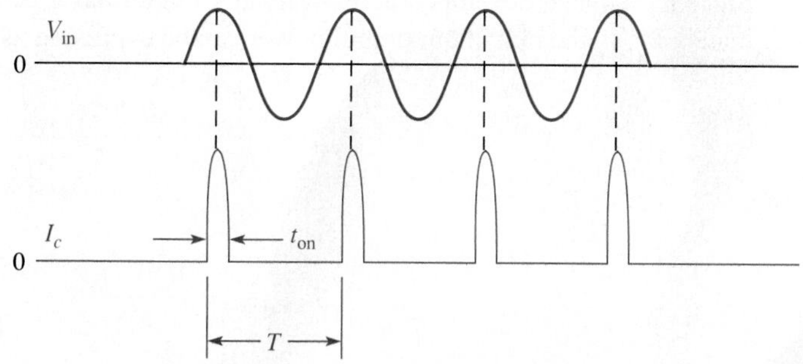

The transistor is on for a short time, t_{on}, and off for the rest of the input cycle. Since the power dissipation averaged over the entire cycle depends on the ratio of t_{on} to T and on the power dissipation during t_{on}, it is typically very low.

Tuned Operation

Because the collector voltage (output) is not a replica of the input, the resistively loaded class C amplifier is of no value in linear applications. Therefore, it is necessary to use a class C amplifier with a parallel resonant circuit (tank), as shown in Figure 19–37(a). The resonant frequency of the tank circuit is determined by the formula $f_r = 1/(2\pi\sqrt{LC})$.

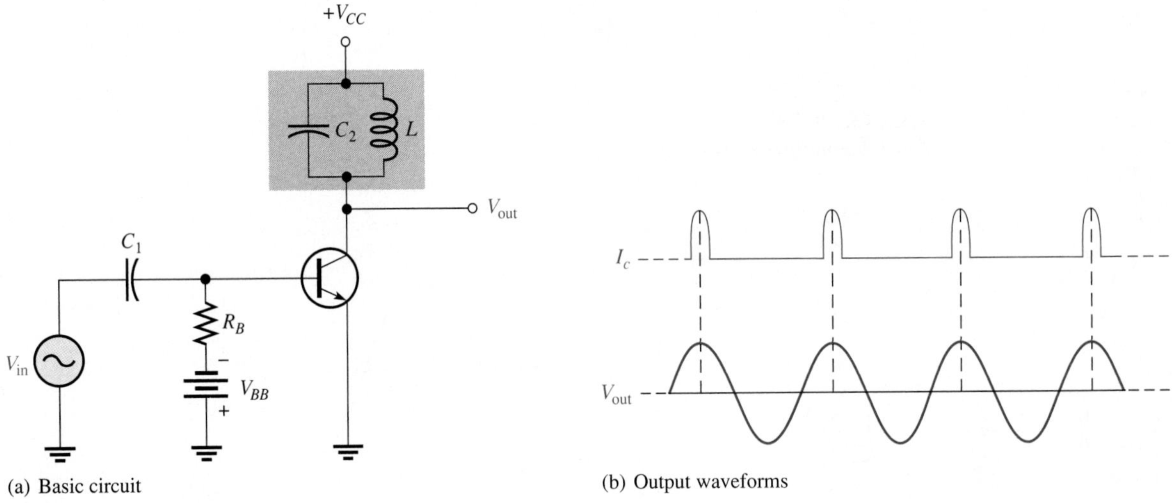

FIGURE 19–37
Tuned class C amplifier.

The short pulse of collector current on each cycle of the input initiates and sustains the oscillation of the tank circuit so that an output sine wave voltage is produced, as illustrated in Figure 19–37(b).

The amplitude of each successive cycle of the oscillation would be less than that of the previous cycle because of energy loss in the resistance of the tank circuit, as shown in Figure 19–38(a), and the oscillation would eventually die out. However, the regular recurrences of the collector current pulse re-energizes the resonant circuit and sustains the oscillations at a constant amplitude. When the tank circuit is tuned to the frequency of the input signal, re-energizing occurs on each cycle of the tank voltage, as shown in Figure 19–38(b).

Maximum Output Power

Since the voltage developed across the tank circuit has a peak-to-peak value of approximately $2V_{CC}$, the maximum output power can be expressed as

$$P_{out} = \frac{V_{rms}^2}{R_c} = \frac{(0.707V_{CC})^2}{R_c}$$

$$P_{out(max)} = \frac{0.5V_{CC}^2}{R_c} \tag{19–36}$$

where R_c is the equivalent parallel resistance of the collector tank circuit and represents the parallel combination of the coil resistance and the load resistance. It usually has a low value.

The total power that must be supplied to the amplifier is

$$P_T = P_{out} + P_{D(avg)}$$

Therefore, the efficiency is

$$\text{eff} = \frac{P_{out}}{P_{out} + P_{D(avg)}} \qquad \textbf{(19–37)}$$

When $P_{out} >> P_{D(avg)}$, the class C efficiency closely approaches 100%.

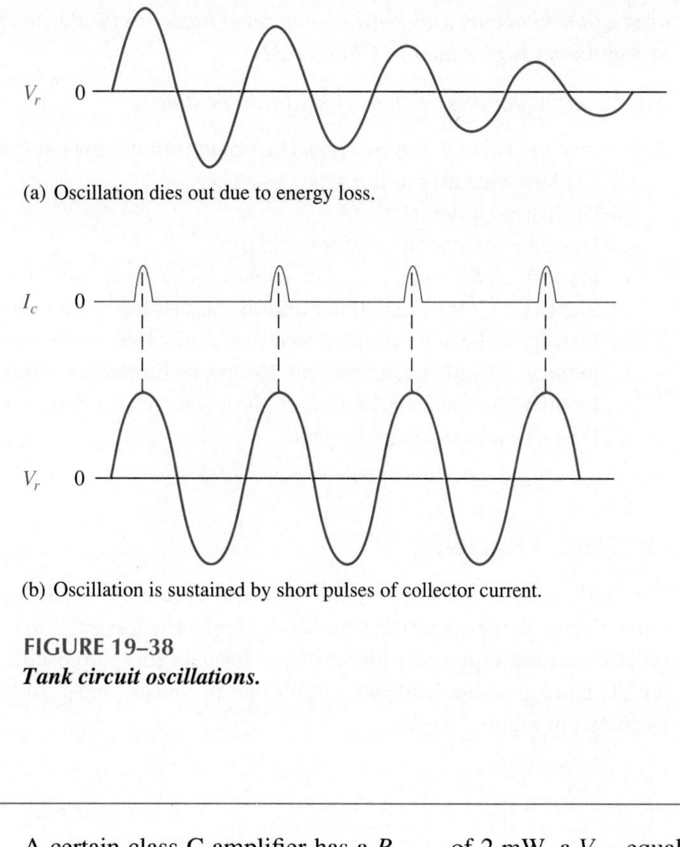

(a) Oscillation dies out due to energy loss.

(b) Oscillation is sustained by short pulses of collector current.

FIGURE 19–38
Tank circuit oscillations.

EXAMPLE 19–13 A certain class C amplifier has a $P_{D(avg)}$ of 2 mW, a V_{CC} equal to 24 V, and an R_c of 100 Ω. Determine the efficiency.

Solution

$$P_{out(max)} = \frac{0.5V^2_{CC}}{R_c} = \frac{0.5(24\text{ V})^2}{100\ \Omega} = 2.88\text{ W}$$

Therefore,

$$\text{eff} = \frac{P_{out}}{P_{out} + P_{D(avg)}} = \frac{2.88\text{ W}}{2.88\text{ W} + 2\text{ mW}} = 0.999$$

or

$$\%\text{eff} = 99.9\%$$

Related Exercise What happens to the efficiency of the amplifier if R_c is increased?

SECTION 19–8 REVIEW

1. How is a class C amplifier normally biased?
2. What is the purpose of the tuned circuit in a class C amplifier?
3. A certain class C amplifier has an average power dissipation of 100 mW and an output power of 1 W. What is its efficiency?

19–9 ■ OSCILLATORS

An oscillator is a circuit that produces a repetitive waveform on its output with only the dc supply voltage as an input. A repetitive input signal is not required. The output voltage can be either sinusoidal or nonsinusoidal, depending on the type of oscillator. Generally, oscillator operation is based on the principle of positive feedback. In this section, we will examine this concept, look at the general conditions required for oscillation to occur, and introduce several basic oscillator circuits. Additional coverage of oscillators is provided in Chapter 21.

After completing this section, you should be able to

- **Discuss the theory and analyze the operation of several types of oscillators**
  - ☐ Explain what an oscillator is
  - ☐ Discuss positive feedback
  - ☐ Describe the conditions for oscillation
  - ☐ Identify an *RC* oscillator and discuss its basic operation
  - ☐ Identify a Colpitts oscillator and discuss its basic operation
  - ☐ Identify a Hartley oscillator and discuss its basic operation
  - ☐ Identify a Clapp oscillator and discuss its basic operation
  - ☐ Identify a crystal oscillator and discuss its basic operation
  - ☐ Describe a basic quartz crystal

Oscillator Principles

The basic oscillator concept is illustrated in Figure 19–39. Essentially, an **oscillator** converts electrical energy in the form of dc to electrical energy in the form of ac. A basic oscillator consists of a transistor amplifier for gain (op-amp oscillators are covered in Chapter 21) and a positive feedback circuit that produces phase shift and provides attenuation, as shown in Figure 19–40.

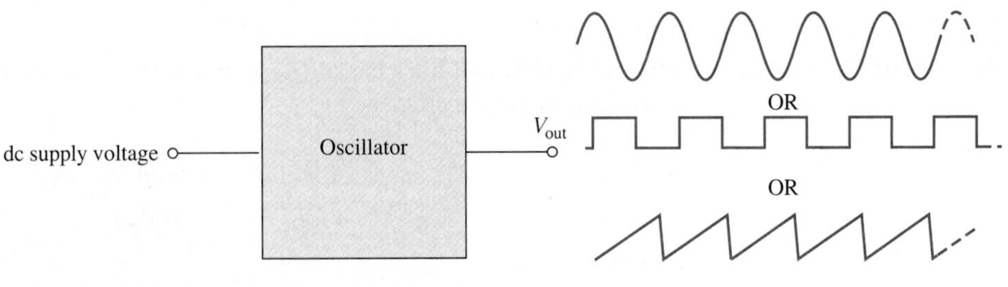

FIGURE 19–39
The basic oscillator concept showing three possible types of output waveforms.

Positive Feedback **Positive feedback** is characterized by the condition wherein a portion of the output voltage of an amplifier is fed back to the input with no net phase shift, resulting in a reinforcement of the output signal. This basic idea is illustrated in Figure 19–41. As you can see, the in-phase **feedback** voltage is amplified to produce the output

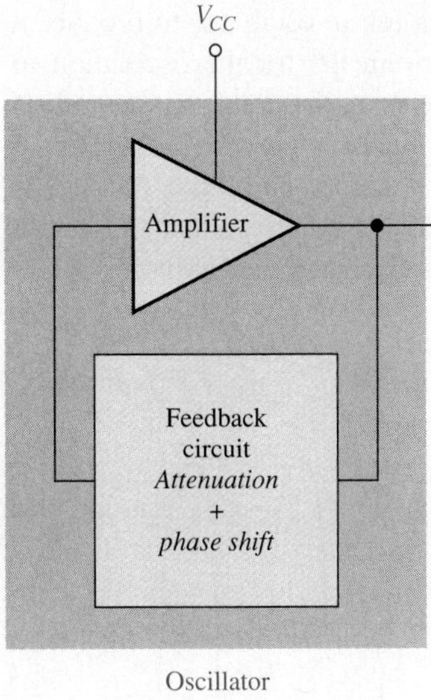

FIGURE 19–40
Basic elements of an oscillator.

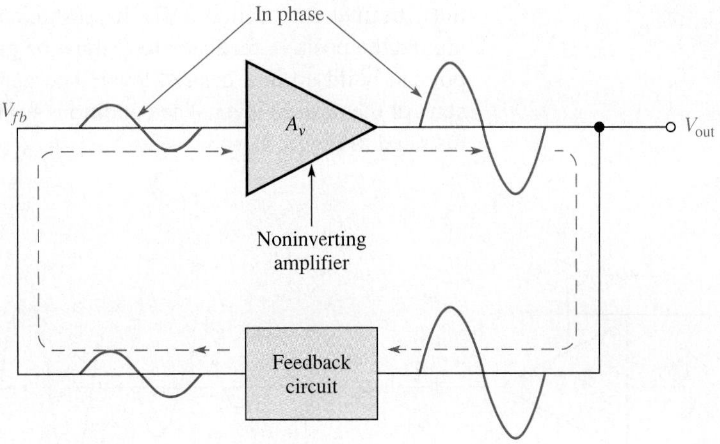

FIGURE 19–41
Positive feedback produces oscillation.

voltage, which in turn produces the feedback voltage. That is, a loop is created in which the signal sustains itself and a continuous sine wave output is produced. This phenomenon is called *oscillation*.

Conditions for Oscillation Two conditions are required for a sustained state of oscillation:

1. The phase shift around the feedback loop must be 0°.
2. The voltage gain around the closed feedback loop must equal 1 (**unity gain**).

The voltage gain around the closed feedback loop (A_{cl}) is the product of the amplifier gain (A_v) and the attenuation of the feedback circuit (B).

$$A_{cl} = A_v B \tag{19–38}$$

For example, if the amplifier has a gain of 100, the feedback circuit must have an attenuation of 0.01 to make the loop gain equal to 1 ($A_v B = 100 \times 0.01 = 1$). These conditions for oscillation are illustrated in Figure 19–42.

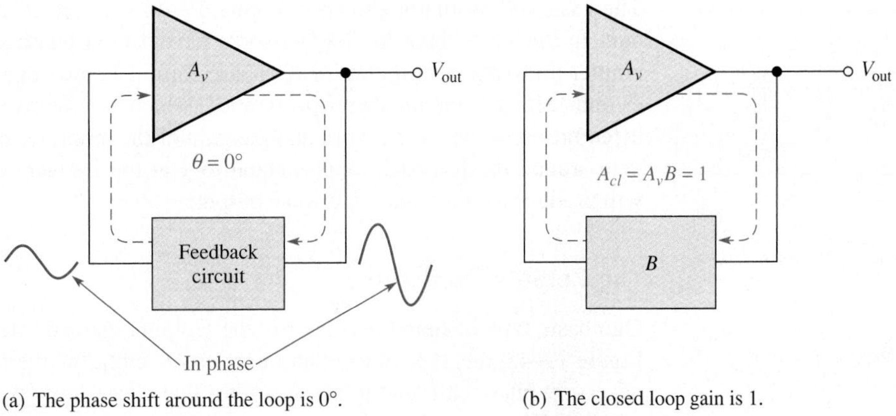

FIGURE 19–42
Conditions for oscillation.

Start-Up Conditions So far, we have seen what it takes for an oscillator to produce a continuous sine wave output. Now we examine the requirements for the oscillation to start when the dc supply voltage is turned on. As you have seen, the unity-gain condition must be met for oscillation to be sustained. For oscillation to begin, the voltage gain around the positive feedback loop must be greater than 1 so that the amplitude of the output can build up to a desired level. The gain must then decrease to 1 so that the output stays at the desired level. The conditions for both starting and sustaining oscillation are illustrated in Figure 19–43.

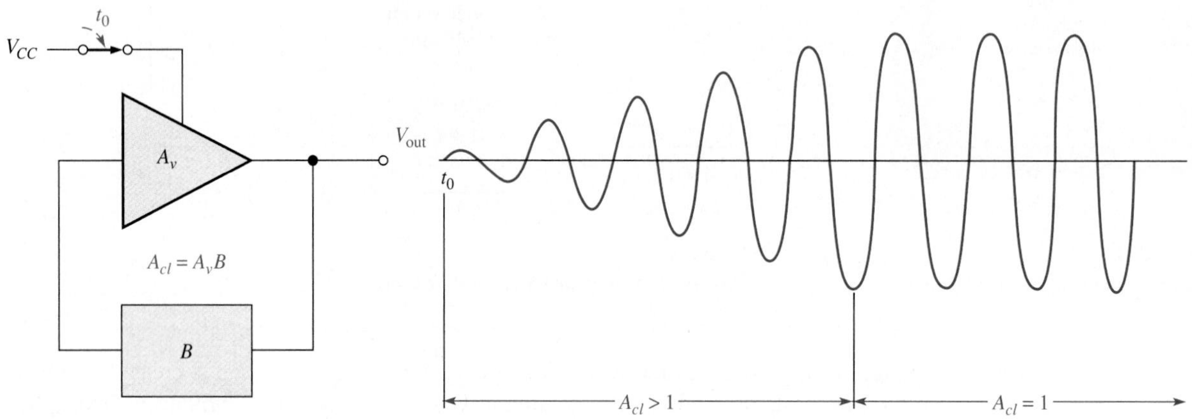

FIGURE 19–43
When oscillation starts at t_o, the condition $A_{cl} > 1$ causes the sinusoidal output voltage amplitude to build up to a desired level, where A_{cl} decreases to 1 and maintains the desired amplitude.

A question that normally arises is this: If the oscillator is off (no dc voltage) and there is no output voltage, how does a feedback signal originate to start the positive feedback build-up process? Initially, when power is turned on, a small positive feedback voltage develops from thermally produced broad-band noise in the resistors or other components or from turn-on transients. The feedback circuit permits only a voltage with a frequency equal to the selected oscillation frequency to appear in-phase on the amplifier's input. This initial feedback voltage is amplified and continually reinforced, resulting in a buildup of the output voltage as previously discussed.

The *RC* Oscillator

The basic *RC* oscillator shown in Figure 19–44 uses an *RC* network as its feedback circuit. In this case, three *RC* lag networks have a total phase shift of 180°. The common-emitter transistor contributes a 180° phase shift. The total phase shift through the amplifier and feedback circuit therefore is 360°, which is effectively 0° (no phase shift). The **attenuation** of the *RC* network and the gain of the amplifier must be such that the overall gain around the feedback loop is equal to 1 at the frequency of oscillation. This circuit will produce a continuous sine wave output.

The Colpitts Oscillator

One basic type of tuned oscillator is the Colpitts, named after its inventor. As shown in Figure 19–45, this type of oscillator uses an *LC* circuit in the feedback loop to provide the necessary phase shift and to act as a filter that passes only the specified frequency of oscillation. The approximate frequency of oscillation is established by the values of C_1, C_2, and L according to the following familiar formula:

$$f_r \cong \frac{1}{2\pi\sqrt{LC_T}} \tag{19–39}$$

Because the capacitors effectively appear in series around the tank circuit, the total capacitance is

$$C_T = \frac{C_1 C_2}{C_1 + C_2} \tag{19–40}$$

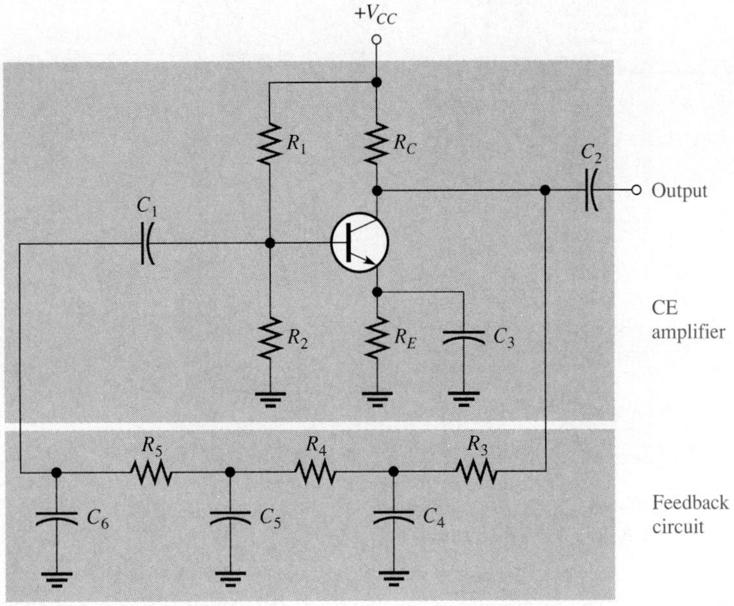

FIGURE 19–44
Basic RC oscillator.

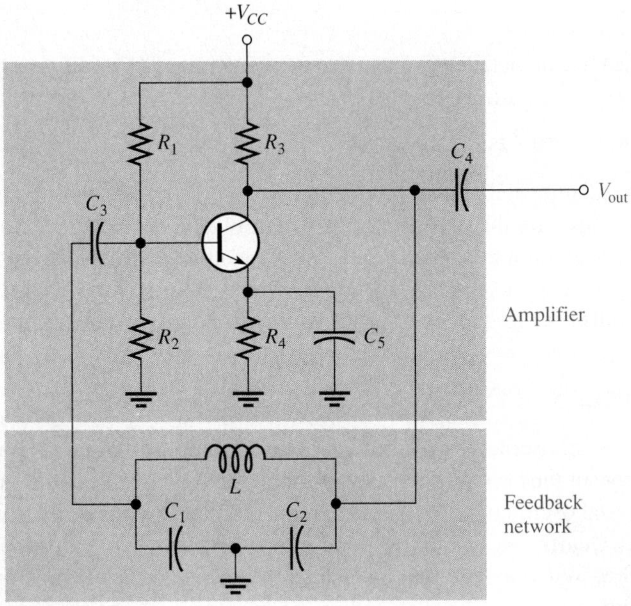

FIGURE 19–45
A basic Colpitts oscillator.

The Hartley Oscillator

Another basic type of oscillator circuit is the Hartley, which is similar to the Colpitts except that the feedback network consists of two inductors and one capacitor, as shown in Figure 19–46.

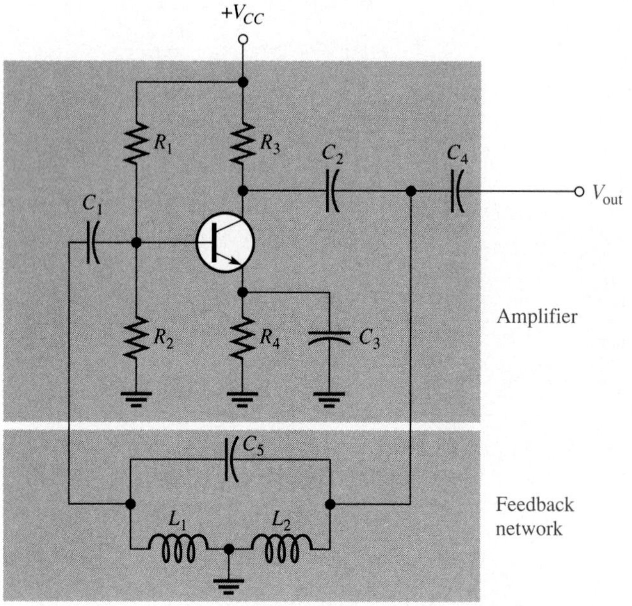

FIGURE 19–46
A basic Hartley oscillator.

The frequency of oscillation for the Hartley oscillator is

$$f_r \cong \frac{1}{2\pi\sqrt{L_T C_5}} \qquad \textbf{(19–41)}$$

The total inductance is the series combination of L_1 and L_2.

The Clapp Oscillator

The Clapp oscillator is similar to the Colpitts except that there is an additional capacitor in series with the inductor, as shown in Figure 19–47. Capacitors C_1 and C_2 can be selected for optimum feedback, and C_3 can be adjusted to obtain the desired frequency of oscillation. Also, a capacitor having a negative temperature coefficient can be used for C_3 to stabilize the frequency of oscillation when there are temperature changes.

The Crystal Oscillator

A crystal oscillator is essentially a tuned-circuit oscillator that uses a quartz crystal as the resonant tank circuit. Other types of crystals can be used, but quartz is the most prevalent. Crystal oscillators offer greater frequency stability than other types.

Quartz is a substance found in nature that exhibits a property called the *piezoelectric effect.* When a changing mechanical stress is applied across the crystal to cause it to vibrate, a voltage is developed at the frequency of the mechanical vibration. Conversely, when an ac voltage is applied across the crystal, it vibrates at the frequency of the applied voltage.

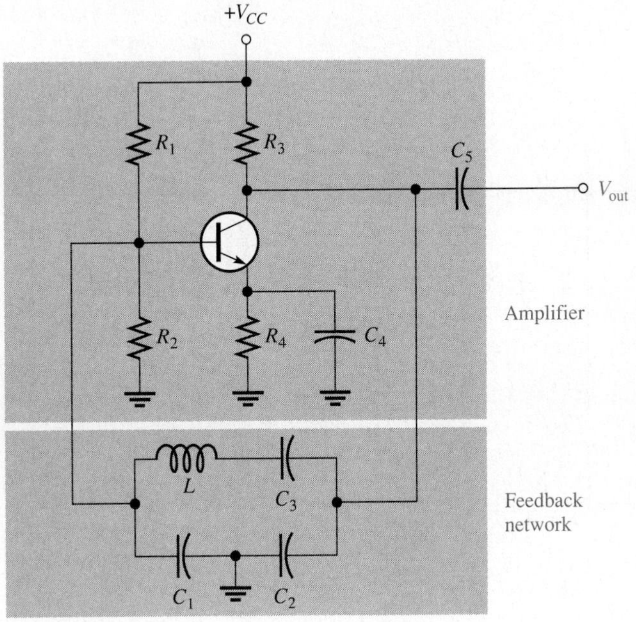

FIGURE 19–47
A basic Clapp oscillator.

The symbol for a crystal is shown in Figure 19–48(a), the electrical equivalent is shown in part (b), and a typical mounted crystal is shown in part (c). In construction, a slab of quartz is mounted as shown in Figure 19–48(d).

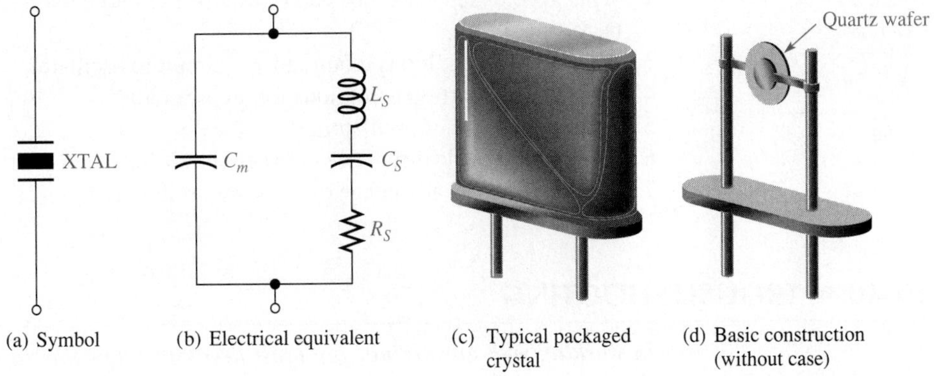

FIGURE 19–48
Quartz crystal.

Series resonance occurs in the crystal when the reactances in the series branch are equal. Parallel resonance occurs, at a higher frequency, when the reactance of L_S equals the reactance of C_m.

A crystal oscillator using the crystal as a series resonant tank circuit is shown in Figure 19–49(a). The impedance of the crystal is *minimum* at the series resonance, thus providing maximum feedback.

The capacitor C_c is a tuning capacitor used to fine-tune the frequency. A modified Colpitts configuration, shown in Figure 19–49(b), uses the crystal in its parallel resonant mode. The impedance of the crystal is maximum at parallel resonance, thus developing the maximum voltage across both C_1 and C_2. The voltage across C_2 is fed back to the input.

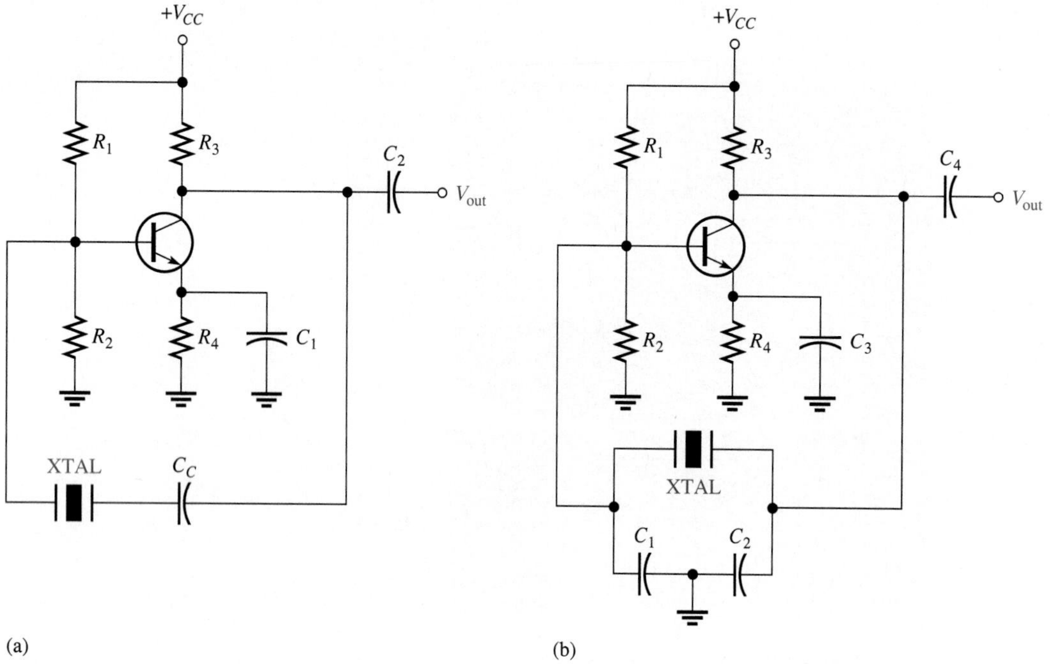

FIGURE 19–49
Basic crystal oscillators.

SECTION 19–9 REVIEW

1. What is an oscillator?
2. What type of feedback does an oscillator use and what is the purpose of the feedback circuit?
3. What are the conditions required for a circuit to oscillate?
4. What are the start-up conditions for an oscillator?
5. Name four types of oscillators.
6. Describe the basic difference between the Colpitts and Hartley oscillators.
7. What is the main advantage of a crystal oscillator?

19–10 ■ TROUBLESHOOTING

In working with any circuit, you must first know how it is supposed to work before you can troubleshoot it for a failure. The two-stage capacitively coupled amplifier discussed in Section 19–5 is used to illustrate a typical troubleshooting procedure.

After completing this section, you should be able to

- **Troubleshoot amplifier circuits**
  - □ Describe the process of signal tracing

The proper signal levels and dc voltage levels for the capacitively coupled two-stage amplifier are shown in Figure 19–50.

Troubleshooting Procedure

A basic procedure for troubleshooting called *signal tracing,* which is usually done with an oscilloscope, is illustrated in Figure 19–51 using the two-stage amplifier as an example. This general procedure can be expanded to any number of stages.

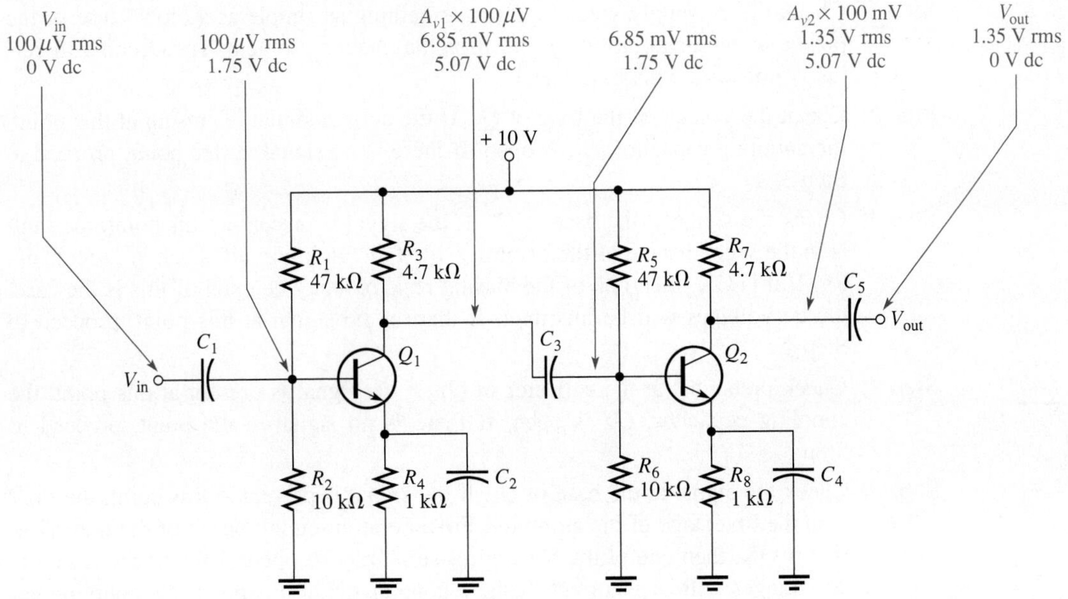

FIGURE 19–50
Two-stage amplifier with proper ac and dc voltage levels indicated.

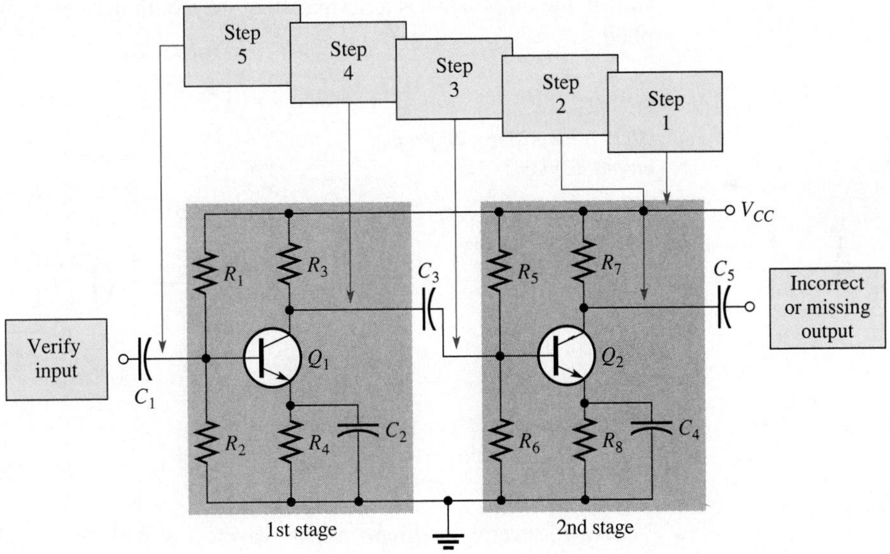

FIGURE 19–51
Basic troubleshooting procedure for a two-stage amplifier with no output signal.

We will begin by assuming that the final output signal of the amplifier has been determined to be missing. We will also assume that there is an input signal and that this has been verified. When troubleshooting, you generally start at the point where the signal is missing and work back point-by-point toward the input until a correct voltage is found. The fault then lies somewhere between the point of the first good voltage check and the missing or incorrect voltage point.

Before you begin checking the voltages, it is usually a good idea to visually check the circuit board or assembly for obvious problems such as broken or poor connections, solder splashes, wire clippings, or burned components.

Step 1: Check the dc supply voltage. Often something as simple as a blown fuse or the power switch being in the *off* position may be the problem. The circuit will certainly not work without power.

Step 2: Check the voltage at the base of Q_2. If the correct signal is present at this point, the coupling capacitor, C_5, is open. If there is no signal at this point, proceed to Step 3.

Step 3: Check the signal at the base of Q_2. If the signal is present at this point, the fault is in the second stage of the amplifier. First, do an in-circuit check of the transistor. If it is OK, then one of the biasing resistors may be open; if this is the case, the dc voltages will be incorrect. If there is no signal at this point, proceed to Step 4.

Step 4: Check the signal at the collector of Q_1. If the signal is present at this point, the coupling capacitor, C_3, is open. If there is no signal at the point, proceed to Step 5.

Step 5: Check the signal at the base of Q_1. If the signal is present at this point, the fault is in the first stage of the amplifier. First, do an in-circuit check of the transistor. If it is OK, then one of the biasing resistors may be open; if this is the case, the dc voltages will be incorrect. If there is no signal at this point, the coupling capacitor, C_1, is open because we started out by verifying that there is a correct signal at the input.

As another example of isolating a component failure in a circuit, we will use a class A amplifier with the output monitored by an oscilloscope, as shown in Figure 19–52. As shown, the amplifier has a normal sine wave output when a sinusoidal input signal is applied.

FIGURE 19–52
Class A amplifier with proper output display.

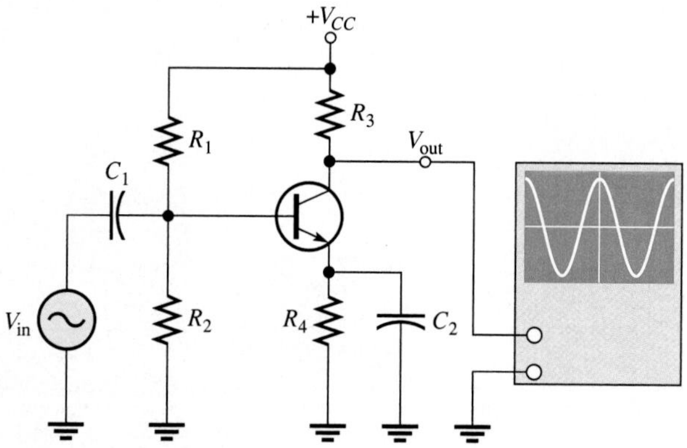

Now, several incorrect output waveforms will be considered and the most likely causes discussed. In Figure 19–53(a), the scope displays a dc level equal to the dc supply voltage, indicating that the transistor is in cutoff. The two possible causes of this condition are (1) the transistor is open from collector to emitter, or (2) R_4 is open, preventing collector and emitter current.

In Figure 19–53(b), the scope displays a dc level at the collector approximately equal to the emitter voltage. The two possible causes of this indication are (1) the transistor is shorted from collector to emitter, or (2) R_2 is open, causing the transistor to be biased in saturation. In the second case, a sufficiently large input signal can bring the transistor out of saturation on its negative peaks, resulting in short pulses on the output.

In Figure 19–53(c), the scope displays an output waveform that is clipped at cutoff. Possible causes of this indication are (1) the Q point has shifted down due to a drastic out-of-tolerance change in a resistor value, or (2) R_1 is open, biasing the transistor in cutoff. In the second case, the input signal is sufficient to bring it out of cutoff for a small portion of the cycle.

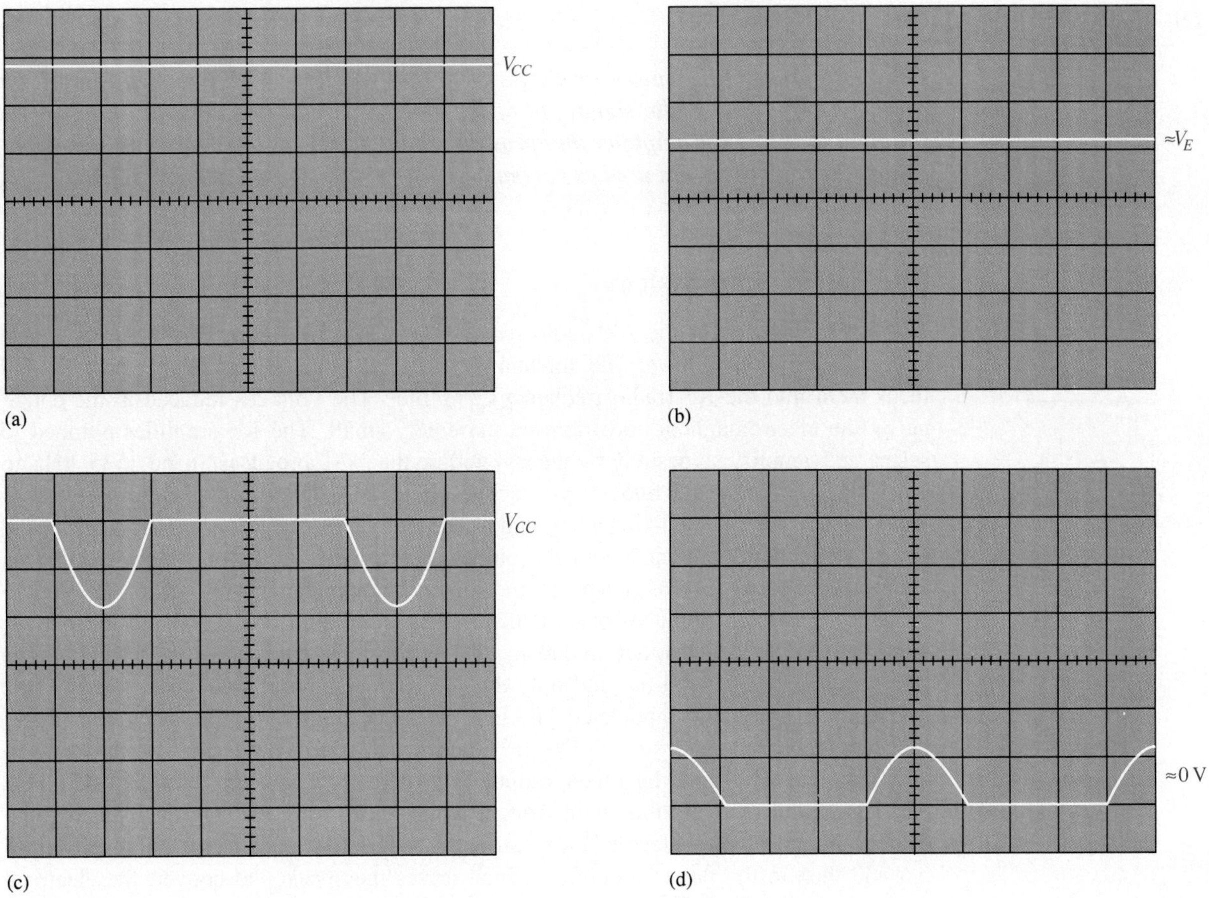

FIGURE 19–53
Oscilloscope displays of output voltage for the amplifier in Figure 19–52, illustrating several types of failures.

In Figure 19–53(d), the scope displays an output waveform that is clipped at saturation. Again, it is possible that a resistance change has caused a drastic shift in the Q point up toward saturation, or R_2 is open, causing the transistor to be biased in saturation, and the input signal is bringing it out of saturation for a small portion of the cycle.

SECTION 19–10 REVIEW

1. If C_4 in Figure 19–51 were open, how would the output signal be affected? How would the dc level at the collector of Q_2 be affected?
2. If R_5 in Figure 19–51 were open, how would the output signal be affected?
3. If the coupling capacitor C_3 in Figure 19–51 shorted out, would any of the dc voltages in the amplifier be changed? If so, which ones?
4. Assume that the base-emitter junction of Q_2 in Figure 19–50 shorts.
   (a) Will the ac signal at the base of Q_2 change? In what way?
   (b) Will the dc level at the base of Q_2 change? In what way?
5. What would you check for if you noticed clipping at both peaks of the output waveform in Figure 19–50?
6. A significant loss of gain in the amplifier of Figure 19–52 would most likely be caused by what type of failure?

19–11 ■ APPLICATION ASSIGNMENT

The audio preamplifier in the receiver system that you have been assigned to work on accepts a very small audio signal out of the detector circuit and amplifies it to a level for input to the power amplifier that provides audio power to the speaker. In this application, the transistors are used as common-emitter small-signal linear amplifiers.

Description of the System

The block diagram for the AM radio receiver is shown in Figure 19–54. A very basic system description follows. The antenna picks up all radiated signals that pass by and feeds them into the RF (radio frequency) amplifier. The voltages induced in the antenna by the electromagnetic radiation are extremely small. The RF amplifier is tuned to select and amplify a desired frequency within the AM broadcast band (535 kHz to 1605 kHz). Since it is a frequency-selective circuit, the RF amplifier eliminates essentially all but the selected frequency band. The output of the RF amplifier goes to the mixer where it is combined with the output of the local oscillator which is 455 kHz above the selected RF frequency. In the mixer, a nonlinear process called *heterodyning* takes place and produces one frequency that is the sum of the RF and local oscillator frequencies and another frequency that is the difference (always 455 kHz). The sum frequency is filtered out and only the 455 kHz difference frequency is used. This frequency is amplitude modulated just like the much higher RF frequency and, therefore, contains the audio signal. The IF (intermediate frequency) amplifier is tuned to 455 kHz and amplifies the mixer output. The detector takes the amplified 455 kHz AM signal and recovers the audio from it while eliminating the intermediate frequency. The output of the detector is a small audio signal that goes to the audio preamplifier and then to the power amplifier, which drives the speaker to convert the electrical audio signal into sound.

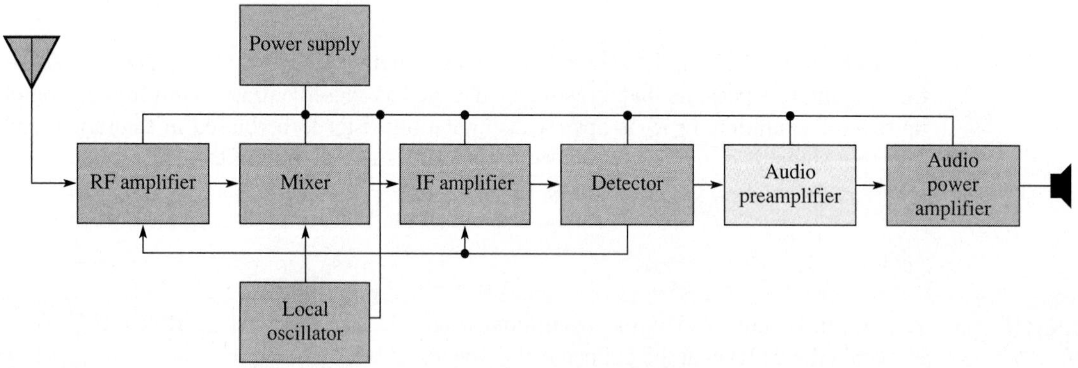

FIGURE 19–54
Block diagram of AM radio receiver.

Step 1: Relate the PC Board to the Schematic

Notice that there are no component labels on the PC board shown in Figure 19–55. Label the components using the schematic in Figure 19–56 as a guide. That is, find and label R_1, R_2, and so forth on the PC board and make the board and schematic agree. Figure 19–55 may be copied for this purpose.

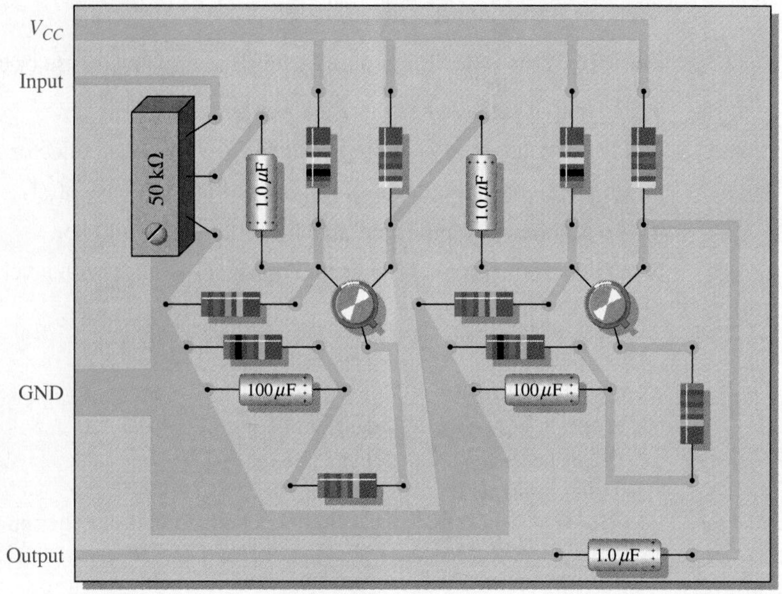

FIGURE 19–55
Audio preamplifier circuit board.

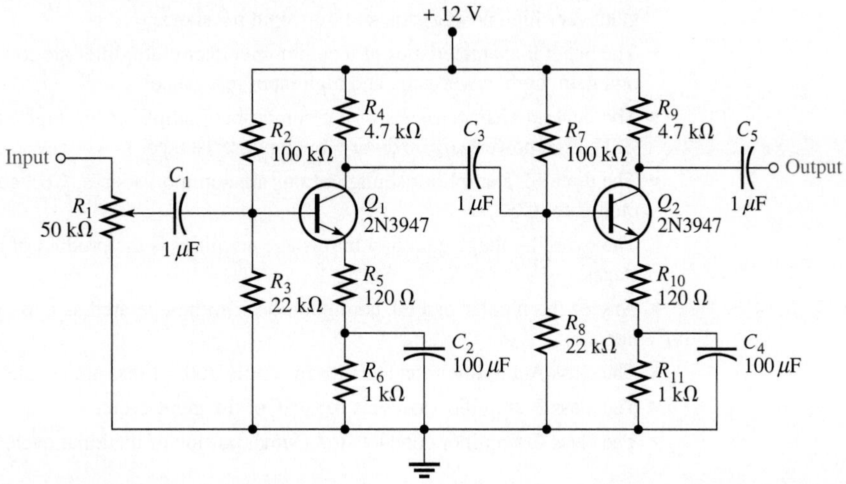

FIGURE 19–56
Audio preamplifier schematic.

Step 2: Analyze the Amplifier

Assume $\beta_{ac(min)} = 175$.

1. Calculate all of the dc voltages in the audio preamp.
2. Determine the voltage gain of each stage and the overall voltage gain of the two-stage amplifier. Don't forget to take into account the loading effect of the input resistance of the second stage. Assume that the second stage is unloaded and all capacitive reactances are zero.
3. Specify what you would expect to happen as the frequency is reduced to a very low value. Why?
4. Compute the voltage gain for the case of a 15 kΩ load resistor connected to the output of the second stage.

Step 3: Troubleshoot the Circuit

For each of the following problems state the probable cause or causes in each case.

1. No output signal when there is a verified input signal.
2. Proper voltages at base of Q_2, but signal voltage at collector is less than it should be.
3. Proper signal at collector of Q_1, but no signal at base of Q_2.
4. Amplitude of output signal much less than it should be.
5. Collector of Q_2 at + 12 V dc and no signal voltage with a verified signal at the base.

SECTION 19–11 REVIEW

1. Explain why R_1 is in the circuit of Figure 19–56.
2. What happens if C_2 opens?
3. What happens if C_3 opens?
4. How can you reduce the gain of each stage without changing the dc voltages?

▪ SUMMARY

- The three bipolar transistor amplifier configurations are common-emitter (CE), common-collector (CC), and common-base (CB).
- The general characteristics of a common-emitter amplifier are high voltage gain, high current gain, very high power gain, and low input resistance.
- The general characteristics of a common-collector amplifier are low voltage gain (≤ 1), high current gain, high power gain, and high input resistance.
- The general characteristics of a common-base amplifier are high voltage gain, low current gain (≤ 1), high power gain, and very low input resistance.
- The three FET amplifier configurations are common-source (CS), common-drain (CD), and common-gate (CG).
- The overall voltage gain of a multistage amplifier is the product of the gains of all the individual stages.
- Any of the bipolar or FET configurations can be operated as class A, class B, or class C amplifiers.
- The class A amplifier conducts for the entire 360° of the input cycle.
- The class B amplifier conducts for 180° of the input cycle.
- The class C amplifier conducts for a small portion of the input cycle.

▪ GLOSSARY

Attenuation A gain less than unity.

Class A A category of amplifier circuit that conducts for the entire input cycle and produces an output signal that is a replica of the input signal in terms of its waveshape.

Class B A category of amplifier circuit that conducts for half of the input cycle.

Class C A category of amplifier circuit that conducts for a very small portion of the input cycle.

Common-base (CB) A BJT amplifier configuration in which the base is the common (grounded) terminal.

Common-collector (CC) A BJT amplifier configuration in which the collector is the common (grounded) terminal.

Common-emitter (CE) A BJT amplifier configuration in which the emitter is the common (grounded) terminal.

Current gain The ratio of output current to input current.

Darlington pair A two-transistor arrangement that produces a multiplication of current gain.

Decibel (dB) The unit of logarithmic expression of a ratio, such as a power ratio or a voltage ratio.

Emitter-follower A popular term for a common-collector amplifier.

Feedback The process of returning a portion of a circuit's output signal to the input in such a way as to create certain specified operating conditions.

Gain The amount by which an electrical signal is increased or decreased; the ratio of output to input.

Oscillator An electronic circuit consisting of an amplifier and a phase-shift network connected in a feedback loop that produces a time-varying output signal without an external input signal using positive feedback.

Positive feedback The return of a portion of the output signal to the input such that it is in phase with the input signal.

Power gain The ratio of output power to input power; the product of voltage gain and current gain.

Push-pull A type of class B amplifier in which one output transistor conducts for one half-cycle and the other conducts for the other half-cycle.

Transconductance (g_m) The ratio of drain current to gate-to-source voltage in an FET.

Unity gain A gain of 1.

Voltage gain The ratio of output voltage to input voltage.

■ FORMULAS

(19–1)	$A_v = \frac{R_C}{r_e + R_E}$	CE voltage gain (unbypassed)
(19–2)	$A_v = \frac{R_C}{r_e}$	CE voltage gain (bypassed)
(19–3)	$r_e \cong \frac{25\text{ mV}}{I_E}$	Internal emitter resistance
(19–4)	$R_{in} \cong \beta_{ac} r_e$	CE input resistance
(19–5)	$R_{in(tot)} = R_1 \parallel R_2 \parallel R_{in}$	CE total input resistance
(19–6)	$A_i = \frac{I_c}{I_s}$	CE current gain
(19–7)	$A_p = A_v A_i$	CE power gain
(19–8)	$A_v = \frac{R_E}{r_e + R_E}$	CC voltage gain
(19–9)	$R_{in} \cong \beta_{ac} R_E$	CC input resistance
(19–10)	$R_{in(tot)} = R_1 \parallel R_2 \parallel R_{in}$	CC total input resistance
(19–11)	$A_i = \frac{I_e}{I_s}$	CC current gain
(19–12)	$A_p \cong A_i$	CC power gain
(19–13)	$\beta = \beta_1 \beta_2$	Beta for a darlington pair
(19–14)	$R_{in} = \beta_1 \beta_2 R_E$	Darlington input resistance
(19–15)	$A_v = \frac{R_C}{r_e}$	CB voltage gain
(19–16)	$R_{in} = r_e$	CB input resistance
(19–17)	$A_i \cong 1$	CB current gain
(19–18)	$A_p \cong A_v$	CB power gain
(19–19)	$g_m = \frac{I_d}{V_{gs}}$	FET transconductance
(19–20)	$A_v = g_m R_D$	CS voltage gain
(19–21)	$A_v = \frac{g_m R_S}{1 + g_m R_S}$	CD voltage gain

(19–22)	$A_v = g_m R_D$	CG voltage gain
(19–23)	$R_{in(source)} = \dfrac{1}{g_m}$	CG input resistance
(19–24)	$A'_v = A_{v1}A_{v2}A_{v3} \cdots A_{vn}$	Multistage gain
(19–25)	$A_v\,(\text{dB}) = 20 \log A_v$	Voltage gain in dB
(19–26)	$A'_v\,(\text{dB}) = A_{v1}\,(\text{dB}) + A_{v2}\,(\text{dB}) + \cdots + A_{vn}\,(\text{dB})$	Multistage dB gain
(19–27)	$A_p \cong \beta_{dc} A_v$	CE large-signal power gain
(19–28)	$P_{DQ} = I_{CQ} V_{CEQ}$	Transistor power dissipation
(19–29)	$P_{out} = V_{ce} I_c$	CE output power
(19–30)	$P_{out(max)} = 0.5 V_{CEQ} I_{CQ}$	CE output power with centered Q point (maximum)
(19–31)	$\text{eff}_{max} = 0.25$	Class A maximum efficiency
(19–32)	$I_{c(sat)} = \dfrac{V_{CEQ}}{R_L}$	Class B ac saturation current
(19–33)	$P_{out(max)} = 0.25 V_{CC} I_{c(sat)}$	Class B output power (maximum)
(19–34)	$P_{dc} = \dfrac{V_{CC} I_{c(sat)}}{\pi}$	Class B input power
(19–35)	$\text{eff}_{max} = 0.785$	Class B maximum efficiency
(19–36)	$P_{out(max)} = \dfrac{0.5 V_{CC}^2}{R_c}$	Class C output power (maximum)
(19–37)	$\text{eff} = \dfrac{P_{out}}{P_{out} + P_{D(avg)}}$	Class C efficiency
(19–38)	$A_{cl} = A_v B$	Closed loop gain
(19–39)	$f_r \cong \dfrac{1}{2\pi\sqrt{LC_T}}$	Colpitts frequency of oscillation
(19–40)	$C_T = \dfrac{C_1 C_2}{C_1 + C_2}$	Colpitts total feedback capacitance
(19–41)	$f_r \cong \dfrac{1}{2\pi\sqrt{L_T C_5}}$	Hartley frequency of oscillation

▪ SELF-TEST

1. In a common-emitter (CE) amplifier, the capacitor from emitter to ground is called the
(a) coupling capacitor **(b)** decoupling capacitor
(c) bypass capacitor **(d)** tuning capacitor

2. If the capacitor from emitter to ground in a CE amplifier is removed, the voltage gain
(a) increases **(b)** decreases **(c)** is not affected **(d)** becomes erratic

3. When the collector resistor in a CE amplifier is increased in value, the voltage gain
(a) increases **(b)** decreases **(c)** is not affected **(d)** becomes erratic

4. The input resistance of a CE amplifier is affected by
(a) α and r_e **(b)** β and r_e **(c)** R_c and r_e **(d)** R_e, r_e, and β

5. The output signal of a CE amplifier is always
(a) in phase with the input signal **(b)** out of phase with the input signal
(c) larger than the input signal **(d)** equal to the input signal

6. The output signal of a common-collector amplifier is always
(a) in phase with the input signal **(b)** out of phase with the input signal
(c) larger than the input signal **(d)** exactly equal to the input signal

7. The largest *theoretical* voltage gain obtainable with a CC amplifier is
   (a) 100 (b) 10 (c) 1 (d) dependent on β
8. Using a darlington pair has the advantage of
   (a) increasing the overall voltage gain (b) less cost
   (c) decreasing the input resistance (d) increasing the overall β
9. Compared to CE and CC amplifiers, the common-base (CB) amplifier has
   (a) a lower input resistance (b) a much larger voltage gain
   (c) a larger current gain (d) a higher input resistance
10. In terms of higher power gain, the amplifier of choice is
    (a) CB (b) CC (c) CE (d) any of these
11. The most efficient amplifier configuration is
    (a) class A (b) class B (c) class C
12. When an FET with a lower transconductance is substituted into an FET amplifier circuit,
    (a) the voltage gain increases (b) the voltage gain decreases
    (c) the input resistance decreases (d) nothing changes
13. When amplifiers are cascaded,
    (a) the gain of each amplifier is increased (b) each amplifier has to work less
    (c) a lower supply voltage is required (d) the overall gain is increased
14. If three amplifiers, each with a voltage gain of 30, are connected in a multistage arrangement, the overall voltage gain is
    (a) 27,000 (b) 90 (c) 10 (d) 30
15. In a class A amplifier, the output signal is
    (a) distorted (b) clipped
    (c) the same shape as the input (d) smaller in amplitude than the input
16. Oscillators operate on the principle of
    (a) signal feedthrough (b) positive feedback (c) negative feedback (d) attenuation

▪ PROBLEMS

SECTION 19–1 Common-Emitter Amplifiers

1. Determine the voltage gain for Figure 19–57.
2. Determine each of the dc voltages, V_B, V_C, and V_E, with respect to ground in Figure 19–57.
3. Determine the following dc values for the amplifier in Figure 19–58:
   (a) V_B (b) V_E (c) I_E (d) I_C (e) V_C (f) V_{CE}

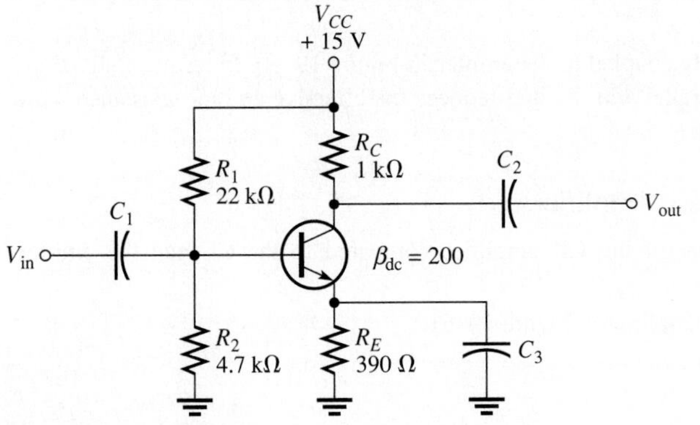

FIGURE 19–57

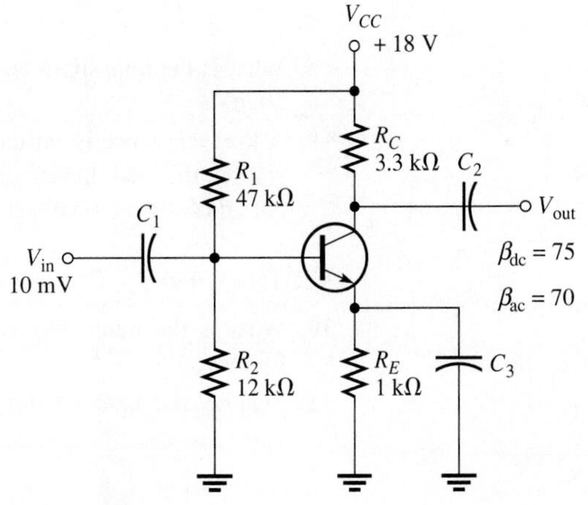

FIGURE 19–58

4. Determine the following ac values for the amplifier in Figure 19–58:
   (a) R_{in} (b) $R_{in(tot)}$ (c) A_v (d) A_i (e) A_p
5. The amplifier in Figure 19–59 has a variable gain control, using a 100 Ω potentiometer for R_E with the wiper ac grounded. As the potentiometer is adjusted, more or less of R_E is bypassed to ground, thus varying the gain. The total R_E remains constant to dc, keeping the bias fixed. Determine the maximum and minimum gains for this amplifier.

FIGURE 19–59

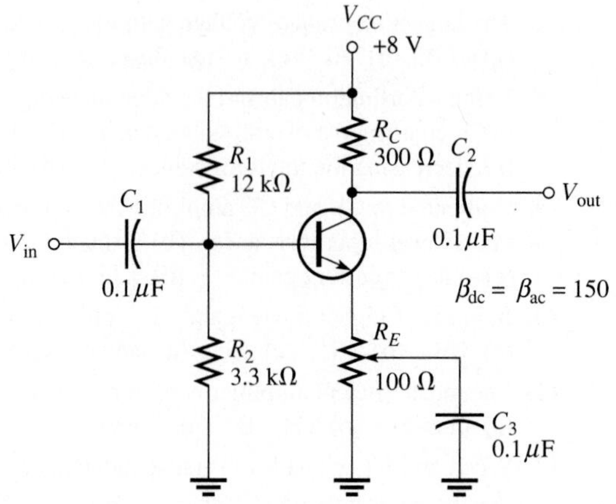

6. If a load resistance of 600 Ω is placed on the output of the amplifier in Figure 19–59, what is the maximum gain?

SECTION 19–2 Common-Collector Amplifiers

7. Determine the *exact* voltage gain for the emitter-follower in Figure 19–60.

FIGURE 19–60

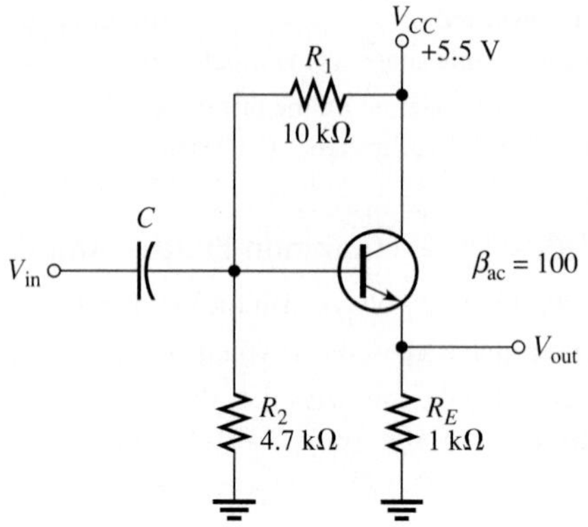

8. What is the total input resistance in Figure 19–60? What is the dc output voltage when V_{in} = 10 mV?

9. A load resistance is capacitively coupled to the emitter in Figure 19–60. In terms of signal operation, the load appears in parallel with R_E and reduces the effective emitter resistance. How does this affect the voltage gain?

SECTION 19–3 Common-Base Amplifiers

10. What is the main disadvantage of the CB amplifier compared to the CE and the emitter-follower?

11. Find R_{in}, A_v, A_i, and A_p for the amplifier in Figure 19–61.

FIGURE 19–61

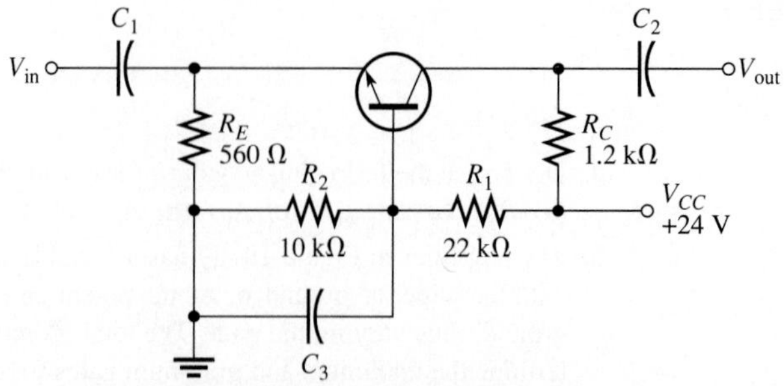

SECTION 19–4 FET Amplifiers

12. Determine the voltage gain of each CS amplifier in Figure 19–62.

FIGURE 19–62

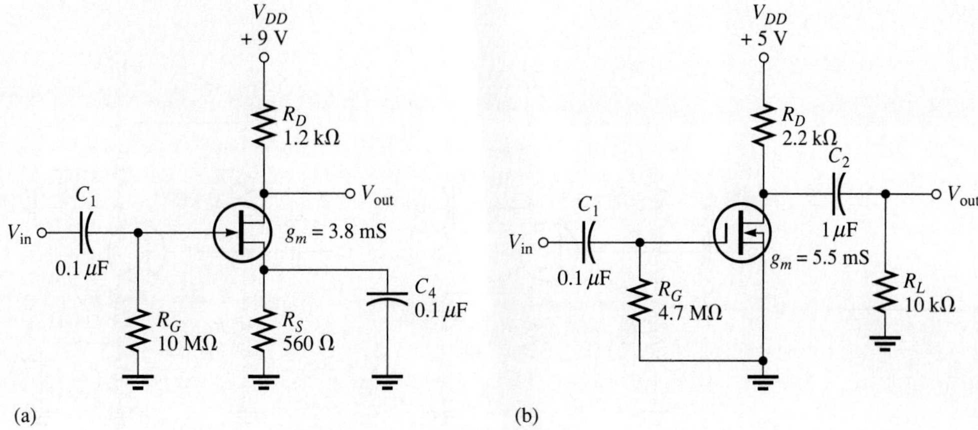

13. Find the gain of each amplifier in Figure 19–63.

14. Determine the gain of each amplifier in Figure 19–63 when a 10 kΩ load is capacitively coupled from source to ground.

FIGURE 19–63

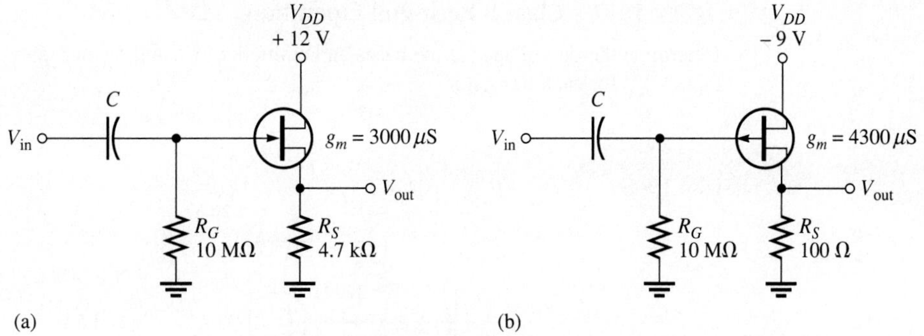

SECTION 19–5 Multistage Amplifiers

15. Each of three cascaded amplifier stages has a decibel voltage gain of 10. What is the overall decibel voltage gain? What is the actual overall voltage gain?

16. For the two-stage, capacitively coupled amplifier in Figure 19–64, find the following values:

(a) Voltage gain of each stage

(b) Overall voltage gain

(c) Express the gains found above in decibels

FIGURE 19–64

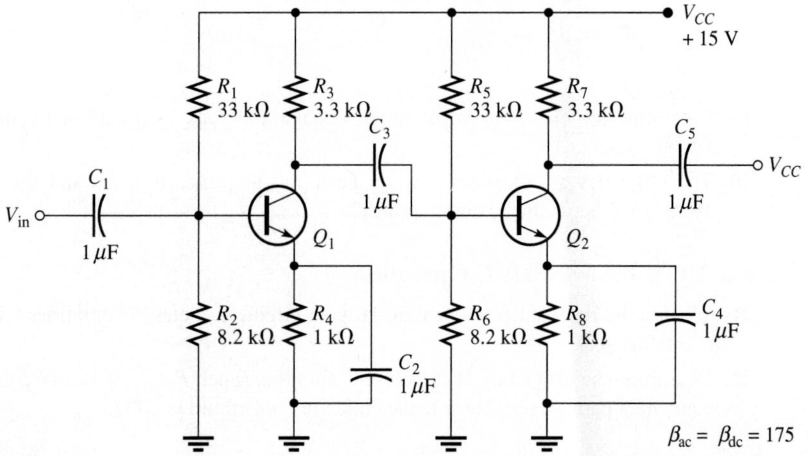

SECTION 19–6 Class A Operation

17. Determine the minimum power rating for each of the transistors in Figure 19–65.

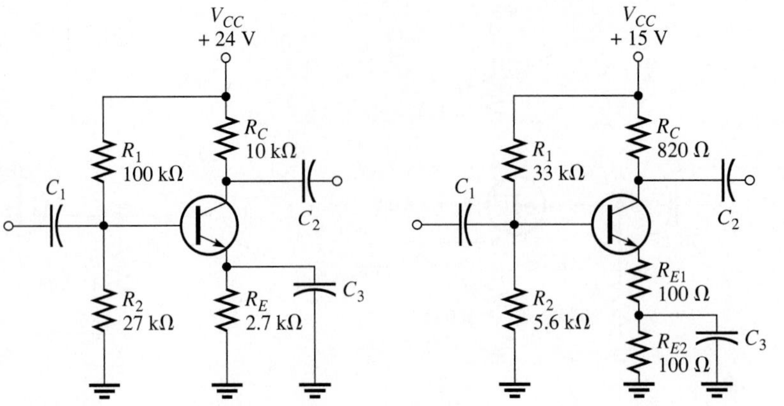

(a) $\beta_{ac} = \beta_{dc} = 90$; $r_e = 10\ \Omega$ (b) $\beta_{ac} = \beta_{dc} = 175$; $r_e = 1.2\ \Omega$

FIGURE 19–65

SECTION 19–7 Class B Push-Pull Operation

18. Determine the dc voltages at the bases and emitters of Q_1 and Q_2 in Figure 19–66. Also determine V_{CEQ} for each transistor.

FIGURE 19–66

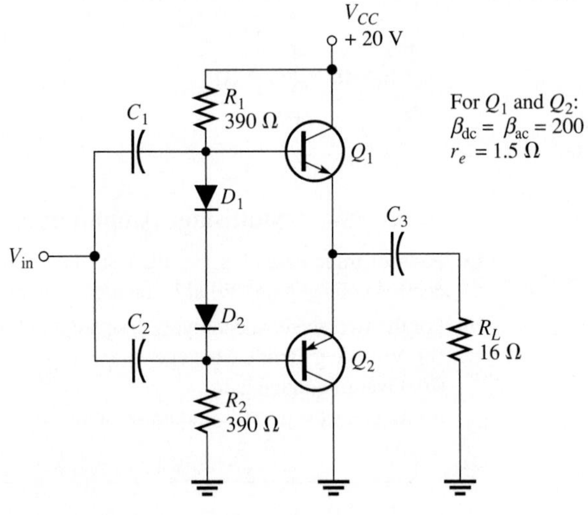

19. Determine the maximum peak output voltage and peak load current for the circuit in Figure 19–66.

20. The efficiency of a certain class B push-pull amplifier is 0.71, and the dc input power is 16.25 W. What is the ac output power?

SECTION 19–8 Class C Operation

21. What is the resonant frequency of the tank circuit in a class C amplifier with $L = 10$ mH and $C = 0.001\ \mu$F?

22. Determine the efficiency of the class C amplifier when $P_{D(avg)} = 10$ mW, $V_{CC} = 15$ V, and the equivalent parallel resistance in the collector tank circuit is 50 Ω.

SECTION 19–9 Oscillators

23. If the voltage gain of the amplifier portion of an oscillator is 75, what must be the attenuation of the feedback circuit to sustain the oscillation?

24. Generally describe the change required to the oscillator of Problem 23 in order for oscillation to begin when the power is initially turned on.

25. Calculate the frequency of oscillation for each circuit in Figure 19–67, and identify each type of oscillator.

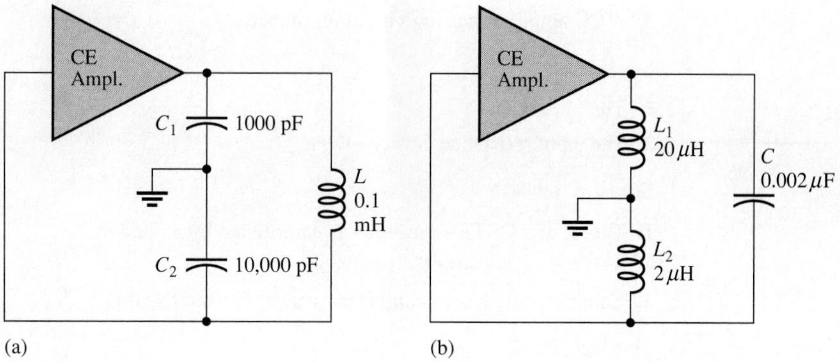

FIGURE 19–67

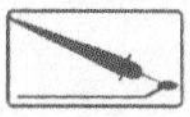

SECTION 19–10 Troubleshooting

26. Assume that the coupling capacitor C_3 is shorted in Figure 19–21. What dc voltage will appear at the collector of Q_1? $\beta_{dc} = 125$.

27. Assume that R_5 opens in Figure 19–21. Will Q_2 be in cutoff or in conduction? What dc voltage will you observe at the Q_2 collector?

28. Refer to Figure 19–64 and determine the general effect of each of the following failures:
(a) C_2 opens **(b)** C_3 opens
(c) C_4 opens **(d)** C_2 shorts
(e) Base-collector junction of Q_1 opens **(f)** Base-emitter of Q_2 opens

29. Assume that you must troubleshoot the amplifier in Figure 19–64. Set up a table of test point values, input, output, and all transistor terminals that include both dc and rms values that you expect to observe when a 300 Ω test signal source with a 25 μV rms output is used.

30. What symptom(s) would indicate each of the following failures under signal conditions in Figure 19–68?
(a) Q_1 open from drain to source **(b)** R_3 open **(c)** C_2 shorted
(d) C_3 shorted **(e)** Q_2 open from drain to source

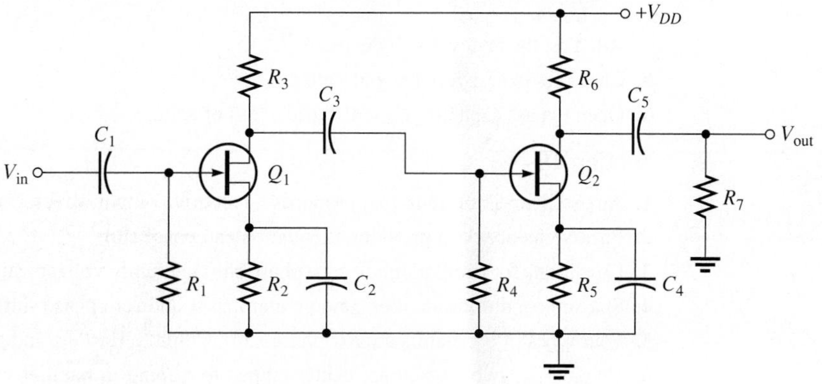

FIGURE 19–68

ANSWERS TO SECTION REVIEWS

Section 19–1

1. The bypass capacitor increases voltage gain.
2. The voltage gain of a CE amplifier is the ratio of collector resistance to total emitter resistance.
3. $A_p = A_v A_i = 10{,}000$

Section 19–2

1. A common-collector amplifier is an emitter-follower.
2. $A_{v(max)} = 1$
3. A CC amplifier has high input resistance.

Section 19–3

1. Yes
2. The input resistance is very low.

Section 19–4

1. The A_v of a CS FET amplifier is determined by g_m and R_d.
2. A_v is halved because R_d is halved.
3. Common-gate has low input resistance; CS and CD do not.

Section 19–5

1. A stage is one amplifier in a cascaded arrangement.
2. The overall gain is the product of individual gains.
3. $20 \log(500) = 54$ dB

Section 19–6

1. The Q-point is centered on the load line.
2. $\text{Eff}_{(max)} = 25\%$
3. $P_{out(max)} = 0.5 V_{CEQ} I_{CEQ} = 35$ mW

Section 19–7

1. The Q-point is at cutoff.
2. The barrier potential of the base-emitter junction causes crossover distortion.
3. $\text{Eff}_{max} = 78.5\%$
4. The push-pull configuration reproduces both positive and negative alternations of the input signal with greater efficiency.

Section 19–8

1. A Class C amplifier is biased in cutoff.
2. The tuned circuit produces a sine wave output.
3. $P_{out}/(P_{out} + P_{D(avg)}) = 90.9\%$
4. **(a)** The ac signal will disappear because it is shorted to ground through the base-emitter junction and C_4.
   (b) Yes, the dc level will decrease.
5. Check for excess input signal voltage.
6. Open bypass capacitor C_2 will cause a loss of gain.

Section 19–9

1. An oscillator is a circuit that produces a repetitive output waveform with no input signal.
2. Positive feedback; it provides attenuation and phase shift.
3. Conditions for oscillation are zero phase shift and unity voltage gain around the closed loop.
4. Start-up conditions are loop gain greater than 1 and zero phase shift.
5. Four types of oscillators are RC phase-shift, Colpitts, Hartley, and crystal.
6. Colpitts use two capacitors, center-tapped to ground in parallel with an inductor. Hartley uses two coils, center-tapped to ground in parallel with a capacitor.
7. A crystal oscillator has a greater frequency stability.

Section 19–10

1. If C_4 opens, the gain drops. The dc level would not be affected.
2. Q_2 would be biased in cutoff.
3. The collector voltage of Q_1 and the base voltage of Q_2 would change. A change in V_{B2} will also cause V_{E2}, I_{E2}, and V_{C2} to change.

Section 19–11

1. R_1 adjusts the volume by increasing or decreasing the input voltage.
2. The gain of the first stage decreases if C_2 opens.
3. There will be no output if C_3 opens because the signal is not coupled through to the second stage.
4. The gain can be reduced by partially bypassing the emitter resistors.

■ ANSWERS TO RELATED EXERCISES FOR EXAMPLES

19–1 235

19–2 7.83 kΩ

19–3 $A_v = 338$; $A_i = 77.6$; $A_p = 26{,}228$

19–4 5.89

19–5 64.5

19–6 4.21 V peak-to-peak signal riding on a dc level of 8.4 V

19–7 0.976

19–8 R_{in} increases 16 Ω to 385 Ω

19–9 $A_v' = 1500$, A_{v1} (dB) = 28 dB, A_{v2} (dB) = 14 dB, A_{v3} (dB) = 21.6 dB, A_v' (dB) = 63.5 dB

19–10 The peak unclipped output voltage decreases to 3 V.

19–11 7.5 V; 750 mA

19–12 6.25 W

19–13 Efficiency decreases.

20

OPERATIONAL AMPLIFIERS (OP-AMPS)

In the previous chapters of this book, a number of electronic devices were introduced. These devices, such as the diode and the transistor, are individually packaged and are connected in a circuit with other individual devices to form a functional unit. Individually packaged devices are referred to as *discrete components.*

We will now introduce the topic of linear integrated circuits (ICs), in which many transistors, diodes, resistors, and capacitors are fabricated on a single silicon chip and packaged in a single case to form an operational amplifier. The manufacturing process for ICs is complex and beyond the scope of this coverage.

In our study of ICs, we will treat the entire circuit as a single device. That is, we will be more concerned with what the circuit does from an external point of view and will be less concerned about the internal, component-level operation.

■ CHAPTER OBJECTIVES

- ☐ Discuss the basic op-amp
- ☐ Explain the basic operation of a differential amplifier
- ☐ Interpret op-amp data sheets
- ☐ Analyze the noninverting, voltage-follower, and inverting op-amp configurations
- ☐ Analyze the effects of negative feedback on the three basic op-amp configurations
- ☐ Troubleshoot op-amp circuits

■ APPLICATION ASSIGNMENT

As an electronics technician in a chemical laboratory, you are assigned responsibility for the maintenance and troubleshooting of a spectrophotometer system that is used to analyze chemical solutions to determine their contents. The system combines electronic, mechanical, optical technology to accomplish its task. The electronics portion uses a photocell and an op-amp to detect the wavelength of light passing through the solution and to amplify the resulting signal prior to sending it to a processor for analysis and display. After studying this chapter, you should be able to complete the application assignment in Section 20–7.

20–1 ▪ INTRODUCTION TO OPERATIONAL AMPLIFIERS

Early operational amplifiers (op-amps) were used primarily to perform mathematical operations such as addition, subtraction, integration, and differentiation—hence the term operational. These early devices were constructed with vacuum tubes and worked with high voltages. Today's op-amps are linear integrated circuits (ICs) that use relatively low dc supply voltages and are very reliable and inexpensive.

After completing this section, you should be able to

- **Discuss the basic op-amp**
  - ☐ Recognize an op-amp symbol
  - ☐ Identify terminals on op-amp packages
  - ☐ Describe an ideal op-amp
  - ☐ Describe a practical op-amp

Symbol and Terminals

The standard op-amp symbol is shown in Figure 20–1(a). It has two input terminals, the inverting input (−) and the noninverting input (+), and one output terminal. The typical **operational amplifier** requires two dc supply voltages, one positive and the other negative, as shown in Figure 20–1(b). Usually these dc voltage terminals are left off the schematic symbol for simplicity, but they are always understood to be there. Typical integrated circuit packages and a pin diagram are shown in parts (c) and (d) respectively.

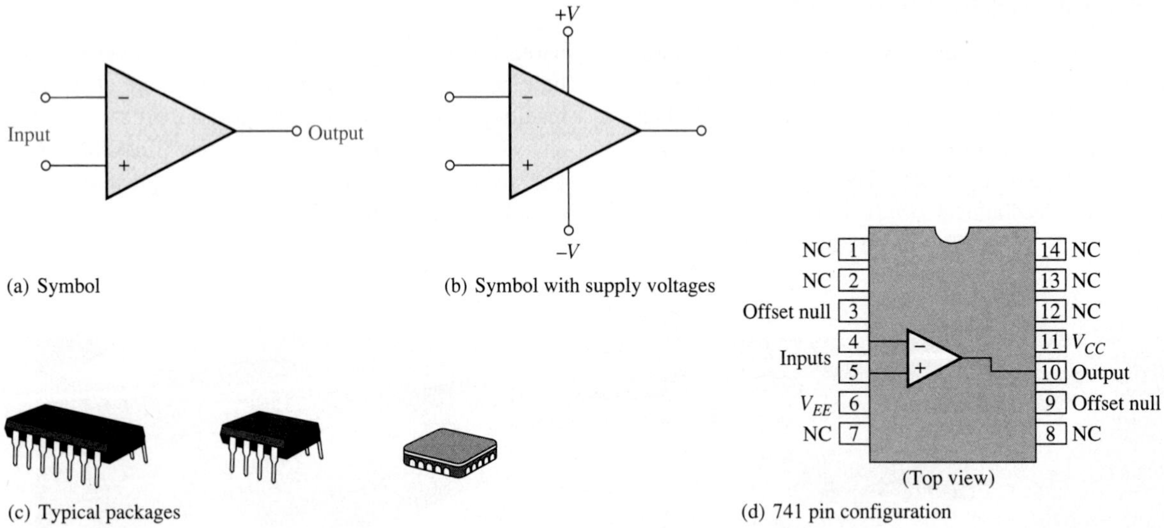

FIGURE 20–1
Op-amp symbols and packages.

The Ideal Op-Amp

To illustrate what an op-amp is, we will consider its *ideal* characteristics. A practical op-amp, of course, falls short of these ideal standards, but it is much easier to understand and analyze the device from an ideal point of view.

First, the ideal op-amp has infinite voltage gain and infinite bandwidth. Also, it has an infinite input impedance (open), so that it does not draw any power from the driving source. Finally, it has a 0 (zero) output impedance. These characteristics are illustrated in Figure 20–2. The input voltage V_{in} appears between the two input terminals, and the output voltage is A_vV_{in}, as indicated by the symbol for internal voltage source. The concept

FIGURE 20–2
Ideal op-amp representation.

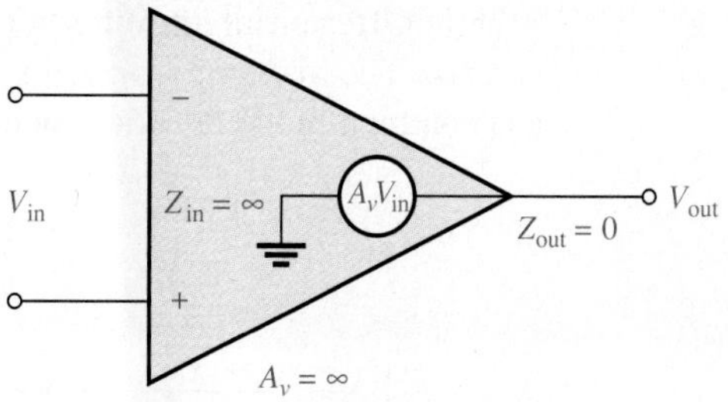

of infinite input impedance is a particularly valuable analysis tool for the various op-amp configurations, which will be covered later.

The Practical Op-Amp

Although modern IC op-amps approach parameter values that can be treated as ideal in many cases, the ideal device cannot be made.

Any device has limitations, and the IC op-amp is no exception. Op-amps have both voltage and current limitations. Peak-to-peak output voltage, for example, is usually limited to slightly less than the two supply voltages. Output current is also limited by internal restrictions such as power dissipation and component ratings.

Characteristics of a practical op-amp are high voltage gain, high input impedance, low output impedance, and wide bandwidth, as illustrated in Figure 20–3.

FIGURE 20–3
Practical op-amp representation.

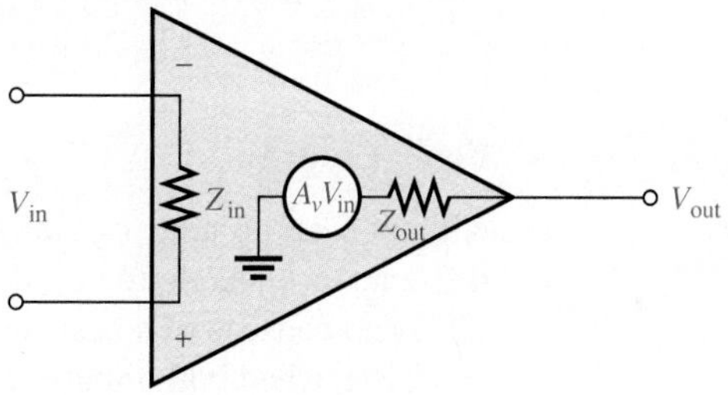

SECTION 20–1 REVIEW

1. Sketch the symbol for an op-amp.
2. Describe the ideal op-amp.
3. Describe some of the characteristics of a practical op-amp.

20–2 ■ THE DIFFERENTIAL AMPLIFIER

The op-amp, in its basic form, typically consists of two or more differential amplifier stages. Because the differential amplifier (diff-amp) is fundamental to the op-amp's internal operation, it is useful to spend some time in acquiring a basic understanding of this type of circuit.

After completing this section, you should be able to

- **Explain the basic operation of a differential amplifier**
  - □ Describe single-ended input operation
  - □ Describe differential input operation
  - □ Describe common-mode input operation
  - □ Define common-mode rejection ratio (CMRR)
  - □ Describe how diff-amps are used to implement an op-amp

A basic **differential amplifier** circuit is shown in Figure 20–4(a), and its block symbol in part (b). The diff-amp stages that make up part of the op-amp provide high voltage gain and common-mode rejection (defined later).

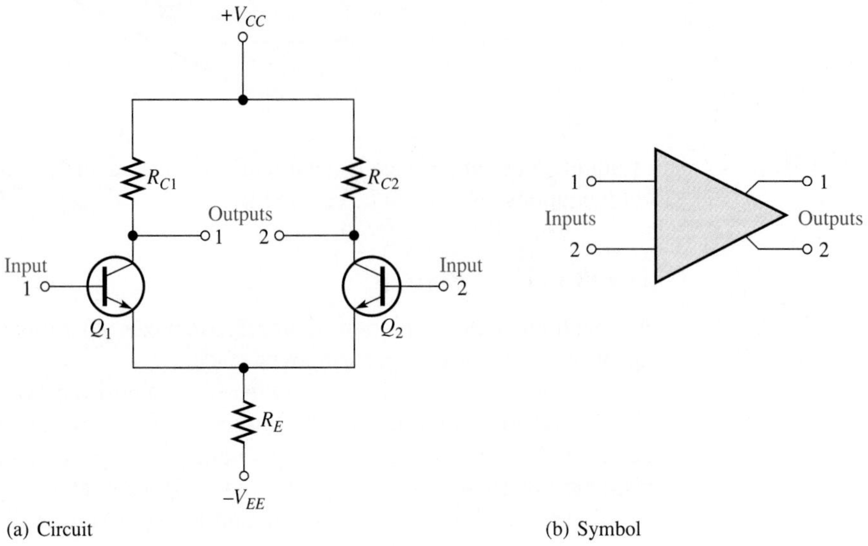

FIGURE 20–4
Basic differential amplifier.

Basic Operation

Although an op-amp typically has more than one diff-amp stage, we will use a single diff-amp to illustrate the basic operation. The following discussion is in relation to Figure 20–5 and consists of a basic dc analysis of the diff-amp's operation.

First, when both inputs are grounded (0 V), the emitters are at –0.7 V, as indicated in Figure 20–5(a). It is assumed that the transistors are identically matched by careful process control during manufacturing so that their dc emitter currents are the same with no input signal. Thus,

$$I_{E1} = I_{E2}$$

Since both emitter currents combine through R_E,

$$I_{E1} = I_{E2} = \frac{I_{RE}}{2}$$

where

$$I_{RE} = \frac{V_E - V_{EE}}{R_E}$$

Based on the approximation that $I_C \cong I_E$, it can be stated that

$$I_{C1} = I_{C2} \cong \frac{I_{RE}}{2}$$

Since both collector currents and both collector resistors are equal (when the input voltage is zero),

$$V_{C1} = V_{C2} = V_{CC} - I_{C1}R_{C1}$$

This condition is illustrated in Figure 20–5(a).

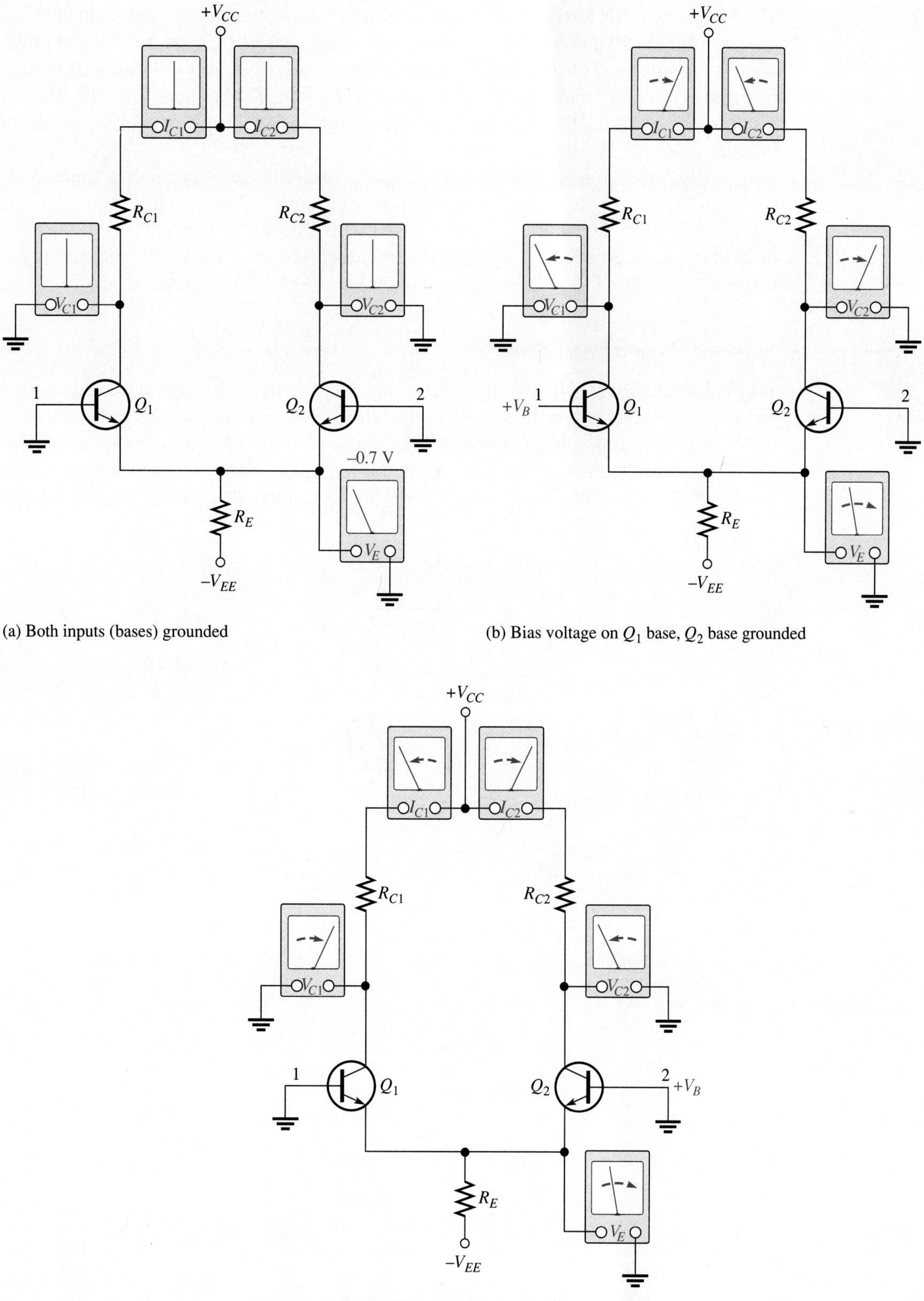

(a) Both inputs (bases) grounded

(b) Bias voltage on Q_1 base, Q_2 base grounded

(c) Bias voltage on Q_2 base, Q_1 base grounded

FIGURE 20–5

Basic operation of a differential amplifier. Ground is 0 V.

Next, input 2 is left grounded, and a positive bias voltage is applied to input 1, as shown in Figure 20–5(b). The positive voltage on the base of Q_1 increases I_{C1} and raises the emitter voltage. This action reduces the forward bias (V_{BE}) of Q_2 because its base is held at 0 V (ground), thus causing I_{C2} to decrease, as indicated in Figure 20–5(b). The net result is that the increase in I_{C1} causes a decrease in V_{C1}, and the decrease in I_{C2} causes an increase in V_{C2}, as shown.

Finally, input 1 is grounded and a positive bias voltage is applied to input 2, as shown in Figure 20–5(c). The positive bias voltage causes Q_2 to conduct more, thus increasing I_{C2}. Also, the emitter voltage is raised. This reduces the forward bias of Q_1 since its base is held at ground, and causes I_{C1} to decrease. The result is that the increase in I_{C2} produces a decrease in V_{C2}, and the decrease in I_{C1} causes V_{C1} to increase as shown.

Modes of Signal Operation

Single-Ended Input When a diff-amp is operated in this mode, one input is grounded and the signal voltage is applied only to the other input, as shown in Figure 20–6(a) and (b). When the signal voltage is applied to input 1, as in part (a), an inverted, amplified signal voltage appears at output 1 as shown. Also, a signal voltage appears in phase at the emitter of Q_1. Since the emitters of Q_1 and Q_2 are common, this emitter signal becomes

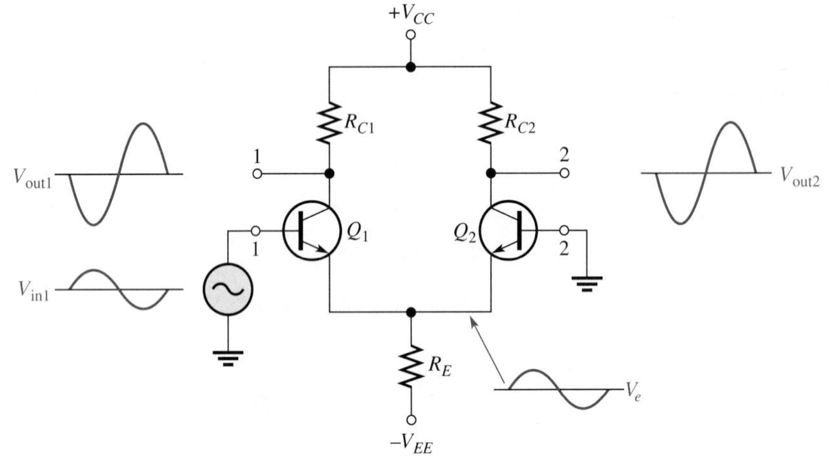

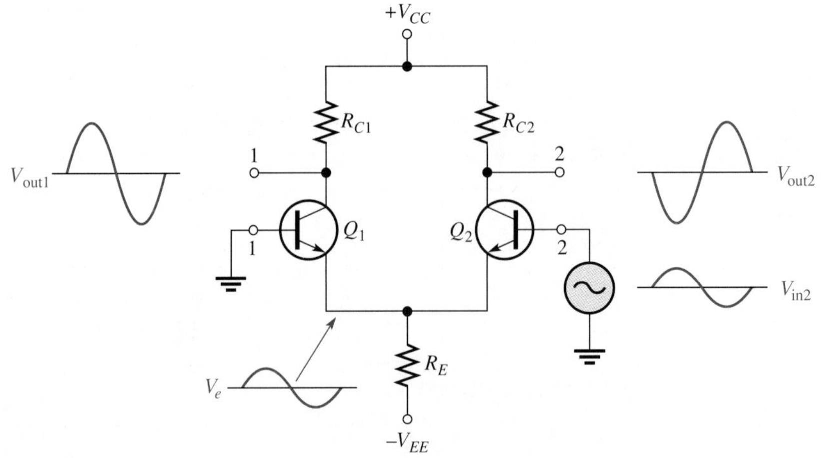

FIGURE 20–6
Single-ended input operation of a differential amplifier.

an input to Q_2 which functions as a common-base amplifier. The signal is amplified by Q_2 and appears, noninverted, at output 2. When the signal is applied to input 2 with input 1 grounded, as in Figure 20–6(b), an inverted, amplified signal voltage appears at output 2. In this situation, Q_1 acts as a common-base amplifier, and a noninverted, amplified signal appears at output 1.

Differential Input In this mode, two signals of opposite polarity (out-of-phase) are applied to the inputs, as shown in Figure 20–7(a). This type of operation is also referred to as *double-ended*. Each input affects the outputs, as you will see in the following discussion.

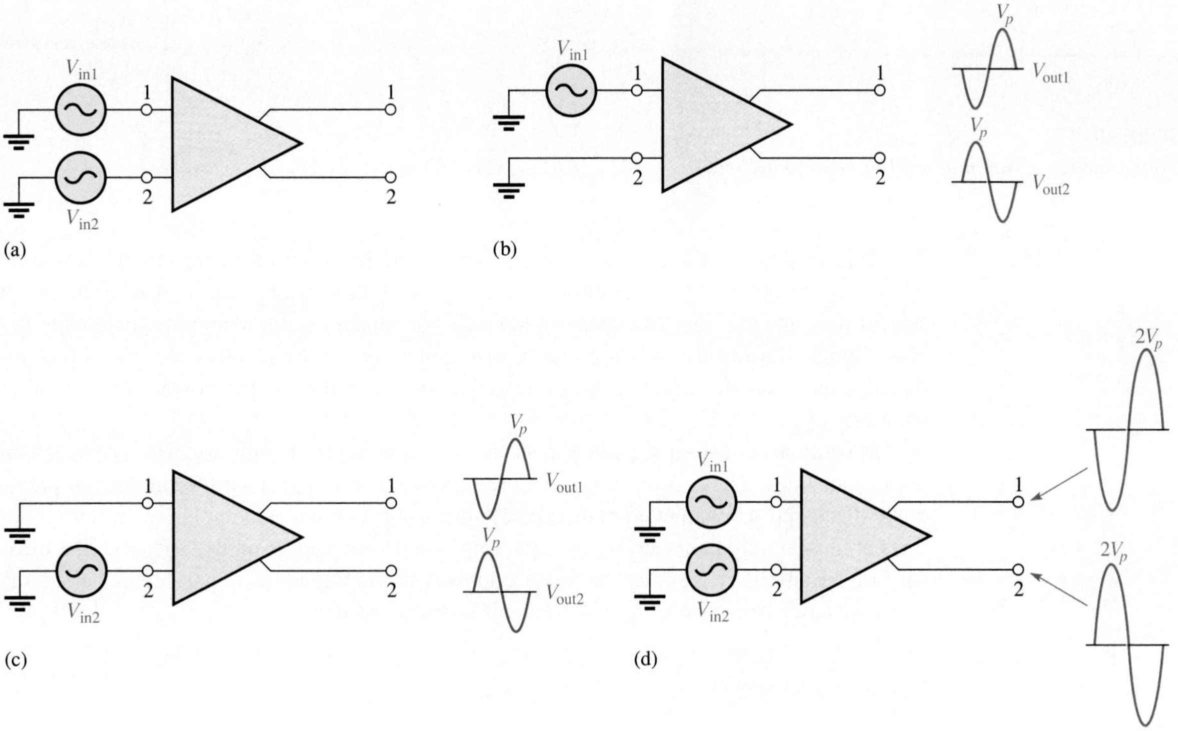

FIGURE 20–7
Differential operation of a differential amplifier.

Figure 20–7(b) shows the output signals due to the signal on input 1 acting alone as a single-ended input. Figure 20–7(c) shows the output signals due to the signal on input 2 acting alone as a single-ended input. In parts (b) and (c), notice that the signals on output 1 are of the same polarity. The same is also true for output 2. By superimposing both output 1 signals and both output 2 signals, we get the total differential operation, as pictured in Figure 20–7(d).

Common-Mode Input We will illustrate one of the most important aspects of the operation of a diff-amp by considering the case where two signal voltages of the same phase, frequency, and amplitude are applied to the two inputs, as shown in Figure 20–8(a). Again, by considering each input signal as acting alone, you can understand the basic operation.

Figure 20–8(b) shows the output signals due to the signal only on input 1, and part (c) shows the output signals due to the signal on only input 2. Notice in parts (b) and (c) that the signals on output 1 are of the opposite polarity, and so are those on output 2. When the input signals are applied to both inputs, the outputs are superimposed and they cancel, resulting in a near zero output voltage, as shown in Figure 20–8(d).

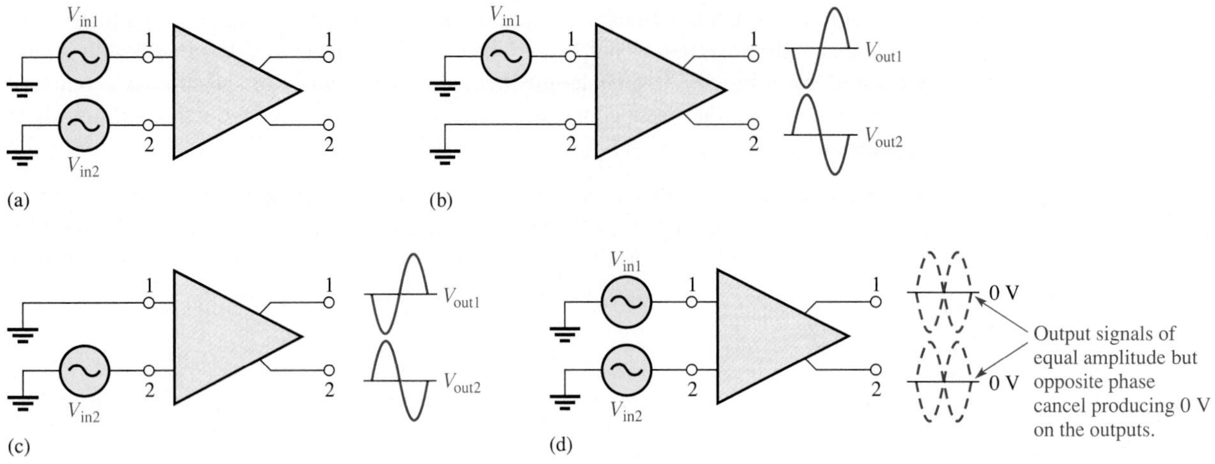

FIGURE 20–8
Common-mode operation of a differential amplifier.

This action is called *common-mode rejection.* Its importance lies in the situation where an unwanted signal appears commonly on both diff-amp inputs. Common-mode rejection means that this unwanted signal will not appear on the outputs to distort the desired signal. **Common-mode signals** (noise) generally are the result of the pick-up of radiated energy on the input lines, or adjacent lines, or the 60 Hz power line, or other sources.

In summary, desired signals appear on only one input or with opposite polarities on both input lines. These desired signals are amplified and appear on the outputs as previously discussed. Unwanted signals (noise) appearing with the same polarity on both input lines are essentially canceled by the diff-amp and do not appear on the outputs. The measure of an amplifier's ability to reject common-mode signals is a parameter called the *common-mode rejection ratio (CMRR)* and is discussed next.

Common-Mode Gain

Ideally, a diff-amp provides a very high gain for desired signals (single-ended or differential), and zero gain for common-mode signals. Practical diff-amps, however, do exhibit a very small common-mode gain (usually much less than one), while providing a high differential voltage gain (usually several thousand). The higher the differential gain with respect to the common-mode gain, the better the performance of the diff-amp in terms of rejection of common-mode signals. This suggests that a good measure of the diff-amp's performance in rejecting unwanted common-mode signals is the ratio of the differential gain, $A_{v(d)}$, to the common-mode gain, A_{cm}. This ratio is called the **common-mode rejection ratio, CMRR.**

$$\text{CMRR} = \frac{A_{v(d)}}{A_{cm}} \qquad \textbf{(20–1)}$$

The higher the CMRR, the better, as you can see. A very high value of CMRR means that the differential gain $A_{v(d)}$ is high and the common-mode gain A_{cm} is low.

The CMRR is often expressed in decibels (dB) as

$$\text{CMRR} = 20\log\left(\frac{A_{v(d)}}{A_{cm}}\right) \qquad \textbf{(20–2)}$$

EXAMPLE 20–1 A certain diff-amp has a differential voltage gain of 2000 and a common-mode gain of 0.2. Determine the CMRR and express it in decibels.

Solution $A_{v(d)} = 2000$ and $A_{cm} = 0.2$. Therefore,

$$\text{CMRR} = \frac{A_{v(d)}}{A_{cm}} = \frac{2000}{0.2} = 10{,}000$$

In decibels,

$$\text{CMRR} = 20\log(10{,}000) = 80 \text{ dB}$$

Related Exercise Determine the CMRR and express it in dB for an amplifier with a differential voltage gain of 8500 and a common-mode gain of 0.25.

A CMRR of 10,000, for example, means that the desired input signal (differential) is amplified 10,000 times more than the unwanted noise (common-mode). So, as an example, if the amplitudes of the differential input signal and the common-mode noise are equal, the desired signal will appear on the output 10,000 times greater in amplitude than the noise. Thus, the noise or interference has been essentially eliminated.

The following example illustrates further the idea of common-mode rejection and the general signal operation of the diff-amp.

EXAMPLE 20–2 The diff-amp shown in Figure 20–9 has a differential voltage gain of 2500 and a CMRR of 30,000. In part (a), a single-ended input signal of 500 μV rms is applied. At the same time a 1 V, 60 Hz common-mode interference signal appears on both inputs as a result of radiated pick-up from the ac power system. In part (b), differential input signals of 500 μV rms each are applied to the inputs. The common-mode interference is the same as in part (a).

(a) Determine the common-mode gain.
(b) Express the CMRR in dB.
(c) Determine the rms output signal for parts (a) and (b).
(d) Determine the rms interference voltage on the output.

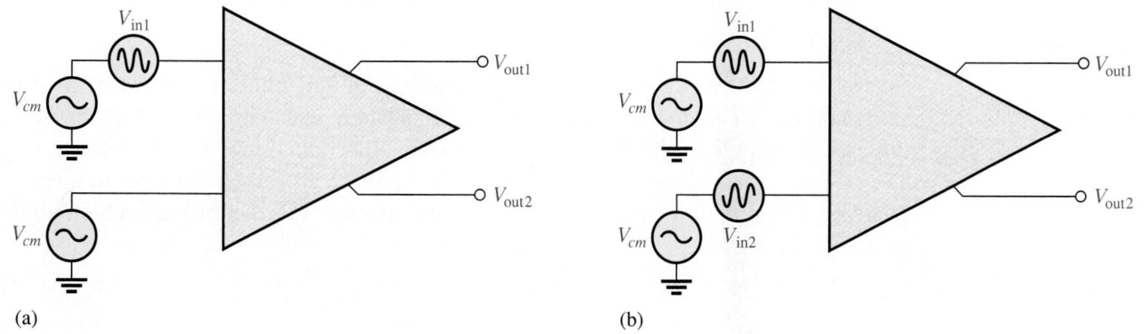

FIGURE 20–9

Solution

(a) $\text{CMRR} = \dfrac{A_{v(d)}}{A_{cm}}$. Therefore,

$$A_{cm} = \frac{A_{v(d)}}{\text{CMRR}} = \frac{2500}{30{,}000} = 0.083$$

(b) $\text{CMRR} = 20\log(30{,}000) = 89.5 \text{ dB}$

(c) In Figure 20–9(a), the differential input voltage, $V_{in(d)}$, is the difference between the voltage on input 1 and that on input 2. Since input 2 is grounded, its voltage is zero. Therefore,

$$V_{in(d)} = V_{in1} - V_{in2} = 500\ \mu\text{V} - 0\text{ V} = 500\ \mu\text{V}$$

The output signal voltage in this case is taken at output 1:

$$V_{out1} = A_{v(d)}V_{in(d)} = (2500)(500\ \mu\text{V}) = 1.25\text{ V rms}$$

In Figure 20–9(b), the differential input voltage is the difference between the two opposite-polarity, 500 μV signals:

$$V_{in(d)} = V_{in1} - V_{in2} = 500\ \mu\text{V} - (-500\ \mu\text{V}) = 1000\ \mu\text{V} = 1\text{ mV}$$

The output signal voltage is

$$V_{out1} = A_{v(d)}V_{in(d)} = (2500)(1\text{ mV}) = 2.5\text{ V rms}$$

This shows that a differential input (two opposite-polarity signals) results in a gain that is double that for a single-ended input.

(d) The common-mode input is 1 V rms. The common-mode gain A_{cm} is 0.083. The interference voltage on the output therefore is

$$A_{cm} = \frac{V_{out(cm)}}{V_{in(cm)}}$$

$$V_{out(cm)} = A_{cm}V_{in(cm)} = 0.083(1\text{ V}) = 83\text{ mV}$$

Related Exercise The amplifier in Figure 20–9 has a differential voltage gain of 4200 and a CMRR of 25,000. For the same single-ended and differential input signals as described in the example: **(a)** Find A_{cm}. **(b)** Express the CMRR in dB. **(c)** Determine the rms output signal for parts (a) and (b) of the figure. **(d)** Determine the rms interference (common-mode) voltage appearing on the output.

A Simple Op-Amp Arrangement

Figure 20–10 shows two diff-amp stages and an emitter-follower connected to form a simple op-amp. The first stage can be used with a single-ended or a differential input. The differential outputs of the first stage feed into the differential inputs of the second stage. The output of the second stage is single-ended to drive an emitter-follower to achieve a relatively low output impedance. Both differential stages together provide a high voltage gain and a high CMRR.

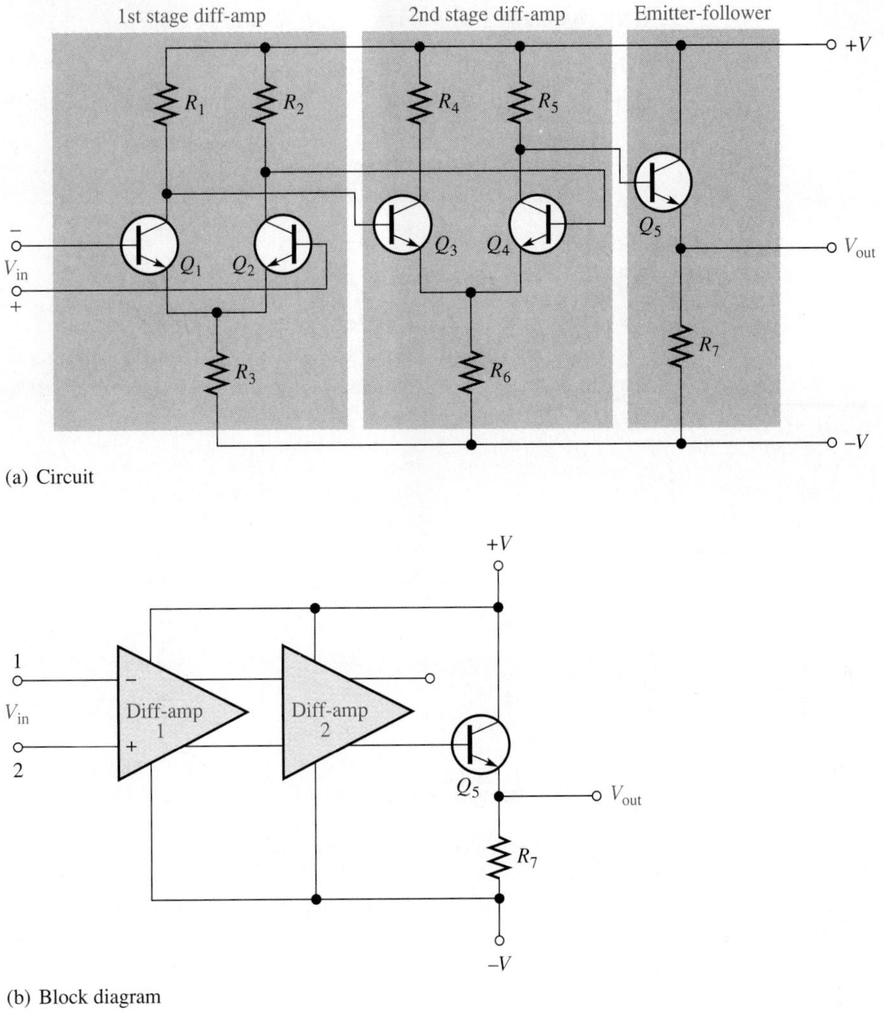

FIGURE 20–10
Internal circuitry of a simple op-amp.

SECTION 20–2 REVIEW

1. Distinguish between differential and single-ended inputs.
2. Define *common-mode rejection.*
3. For a given value of differential gain, does a higher CMRR result in a higher or lower common-mode gain?

20–3 ■ OP-AMP DATA SHEET PARAMETERS

In this section, several important op-amp parameters are defined. These are the input offset voltage, the input offset voltage drift with temperature, the input bias current, the input impedance, the input offset current, the output impedance, the common-mode range, the open-loop voltage gain, the common-mode rejection ratio, the slew rate, and the frequency response. Also four popular IC op-amps are compared in terms of these parameters.

After completing this section, you should be able to

- **Interpret op-amp data sheets**
  - ☐ Define all the parameters listed in the introductory paragraph
  - ☐ Determine the CMRR of an op-amp
  - ☐ Compare the parameters of four types of IC op-amps

Input Offset Voltage

The ideal op-amp produces zero volts out for zero volts in. In a practical op-amp, however, a small dc voltage appears at the output when no differential input voltage is applied. Its primary cause is a slight mismatch of the base-emitter voltages of the differential input stage, as illustrated in Figure 20–11(a).

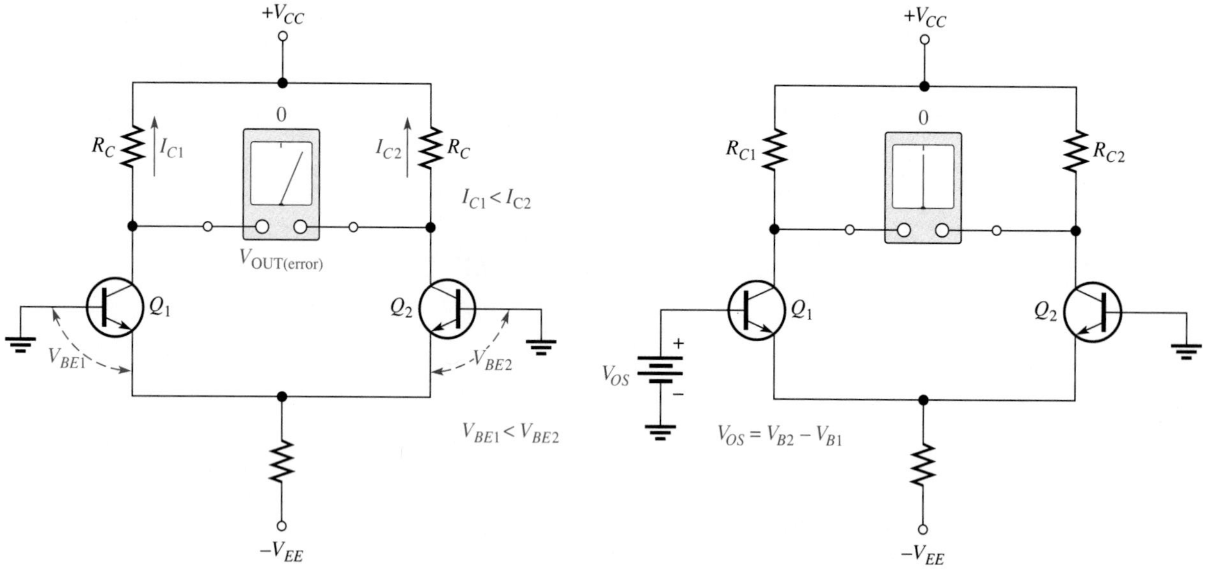

(a) A V_{BE} mismatch causes a small output error voltage.

(b) The input offset voltage is the difference in the voltage between the inputs that is necessary to eliminate the output error voltage (makes $V_{OUT} = 0$)

FIGURE 20–11
Input offset voltage, V_{OS}.

The output voltage of the differential input stage is expressed as

$$V_{OUT} = I_{C2}R_C - I_{C1}R_C \qquad \textbf{(20–3)}$$

A small difference in the base-emitter voltages of Q_1 and Q_2 causes a small difference in the collector currents. This results in a nonzero value of V_{OUT}. (The collector resistors are equal.)

As specified on an op-amp data sheet, the *input offset voltage, V_{OS}*, is the differential dc voltage required between the inputs to force the differential output to zero volts, as demonstrated in Figure 20–11(b). Typical values of input offset voltage are in the range of 2 mV or less. In the ideal case, it is 0 V.

Input Offset Voltage Drift with Temperature

The *input offset voltage drift* is a parameter related to V_{OS} that specifies how much change occurs in the input offset voltage for each degree change in temperature. Typical values range anywhere from about 5 μV per degree Celsius to about 50 μV per degree Celsius. Usually, an op-amp with a higher nominal value of input offset voltage exhibits a higher drift.

Input Bias Current

You have seen that the input terminals of a diff-amp are the transistor bases and, therefore, the input currents are the base currents.

The *input bias current* is the direct current required by the inputs of the amplifier to properly operate the first stage. By definition, the input bias current is the average of both input currents and is calculated as follows:

$$I_{BIAS} = \frac{I_1 + I_2}{2} \tag{20–4}$$

The concept of input bias current is illustrated in Figure 20–12.

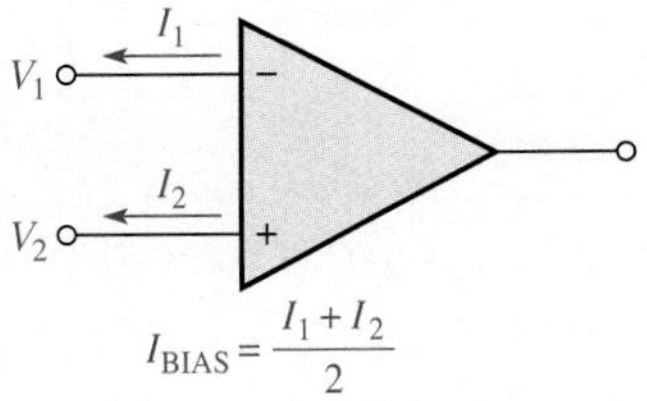

FIGURE 20–12
Input bias current is the average of the two op-amp input currents.

Input Impedance

Two basic ways of specifying the input impedance of an op-amp are the differential and the common mode. The *differential input impedance* is the total resistance between the inverting and the noninverting inputs, as illustrated in Figure 20–13(a). Differential impedance is measured by determining the change in bias current for a given change in differential input voltage. The *common-mode input impedance* is measured from the inputs to ground and is depicted in Figure 20–13(b).

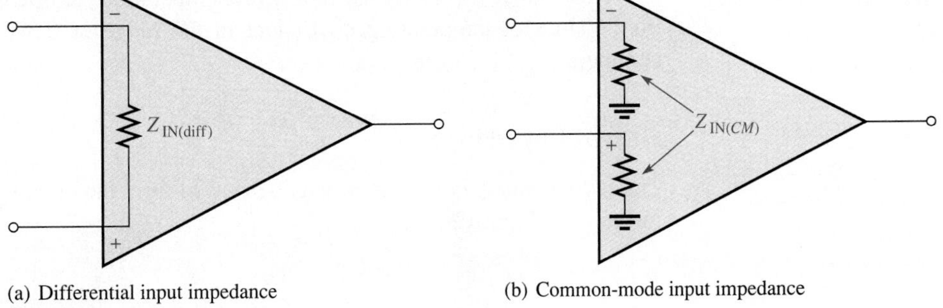

FIGURE 20–13
Op-amp input impedance.

Input Offset Current

Ideally, the two input bias currents are equal, and thus their difference is zero. In a practical op-amp, however, the bias currents are not exactly equal.

The *input offset current* is the difference of the input bias currents, expressed as

$$I_{OS} = |I_1 - I_2| \quad \textbf{(20–5)}$$

Actual magnitudes of offset current are usually at least an order of magnitude (ten times) less than the bias current. In many applications the offset current can be neglected. However, high-gain, high-input impedance amplifiers should have as little I_{OS} as possible, because the difference in currents through large input resistances develops a substantial offset voltage, as shown in Figure 20–14.

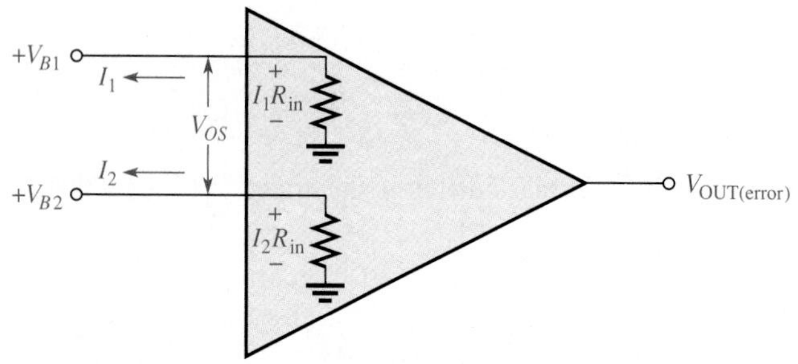

FIGURE 20–14
Effect of input offset current.

The offset voltage developed by the input offset current is

$$V_{OS} = I_1R_{in} - I_2R_{in} = (I_1 - I_2)R_{in}$$

$$V_{OS} = I_{OS}R_{in} \quad \textbf{(20–6)}$$

The error created by I_{OS} is amplified by the gain A_v of the op-amp and appears in the output as

$$V_{OUT(error)} = A_vI_{OS}R_{in} \quad \textbf{(20–7)}$$

The change in offset current with temperature is often an important consideration. Values of temperature coefficient in the range of 0.5 nA per degree Celsius are common.

Output Impedance

The *output impedance* is the resistance viewed from the output terminal of the op-amp, as indicated in Figure 20–15.

FIGURE 20–15
Op-amp output impedance.

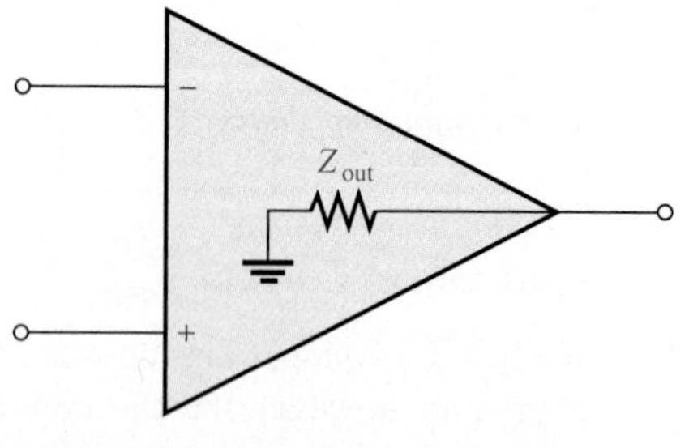

Common-Mode Range

All op-amps have limitations on the range of voltages over which they will operate. The *common-mode range* is the range of input voltages which, when applied to both inputs, will not cause clipping or other output distortion. Many op-amps have common-mode ranges of ±10 V with dc supply voltages of ±15 V.

Open-Loop Voltage Gain, A_{ol}

The *open-loop voltage gain* is the gain of the op-amp without any external feedback from output to input. A good op-amp has a very high **open-loop gain;** 50,000 to 200,000 is typical.

Common-Mode Rejection Ratio

The *common-mode rejection ratio (CMRR),* as discussed in conjunction with the diff-amp, is a measure of an op-amp's ability to reject common-mode signals. An infinite value of CMRR means that the output is zero when the same signal is applied to both inputs (common-mode).

An infinite CMRR is never achieved in practice, but a good op-amp does have a very high value of CMRR. As previously discussed, common-mode signals are undesired interference voltages such as 60 Hz power-supply ripple and noise voltages due to pickup of radiated energy. A high CMRR enables the op-amp to virtually eliminate these interference signals from the output.

The accepted definition of CMRR for an op-amp is the open-loop gain (A_{ol}) divided by the common-mode gain:

$$\text{CMRR} = \frac{A_{ol}}{A_{cm}} \qquad \textbf{(20–8)}$$

It is commonly expressed in decibels as follows:

$$\text{CMRR} = 20 \log\left(\frac{A_{ol}}{A_{cm}}\right) \qquad \textbf{(20–9)}$$

EXAMPLE 20–3 A certain op-amp has an open-loop gain of 100,000 and a common-mode gain of 0.25. Determine the CMRR and express it in decibels.

Solution

$$\text{CMRR} = \frac{A_{ol}}{A_{cm}} = \frac{100{,}000}{0.25} = 400{,}000$$

$$\text{CMRR} = 20 \log(400{,}000) = 112 \text{ dB}$$

Related Exercise If a particular op-amp has a CMRR of 90 dB and a common-mode gain of 0.4, what is the open-loop gain?

Slew Rate

The maximum rate of change of the output voltage in response to a step input voltage is the *slew rate* of an op-amp. The slew rate is dependent upon the frequency response of the amplifier stages within the op-amp.

Slew rate is measured with an op-amp connected as shown in Figure 20–16(a). This particular op-amp connection is a unity-gain, noninverting configuration that will be discussed later. It gives a worst-case (slowest) slew rate. As shown in part (b), a pulse is applied to the input, and the ideal output voltage is measured. The width of the input pulse must be sufficient to allow the output to "slew" from its lower limit to its upper limit, as shown.

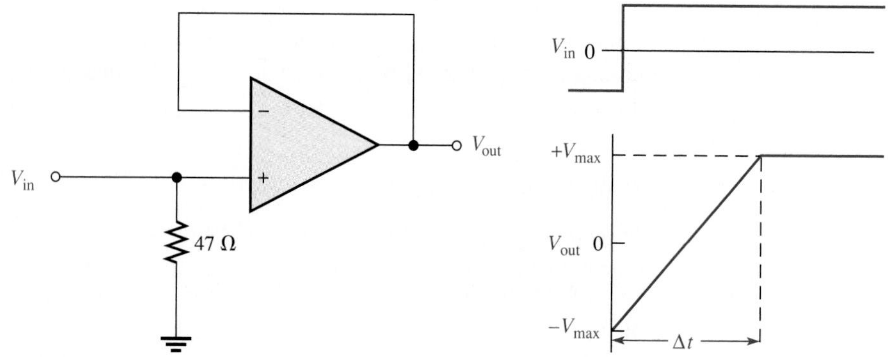

(a) Test circuit

(b) Step input voltage and the resulting output voltage

FIGURE 20–16
Measurement of slew rate.

As you can see, a certain time interval, Δt, is required for the output voltage to go from its lower limit $-V_{max}$ to its upper limit $+V_{max}$, once the input step is applied. The slew rate is expressed as

$$\text{Slew rate} = \frac{\Delta V_{out}}{\Delta t} \qquad \textbf{(20–10)}$$

where $\Delta V_{out} = +V_{max} - (-V_{max})$. The unit of slew rate is volts per microsecond (V/μs).

EXAMPLE 20–4 The output voltage of a certain op-amp appears as shown in Figure 20–17 in response to a step input. Determine the slew rate.

FIGURE 20–17

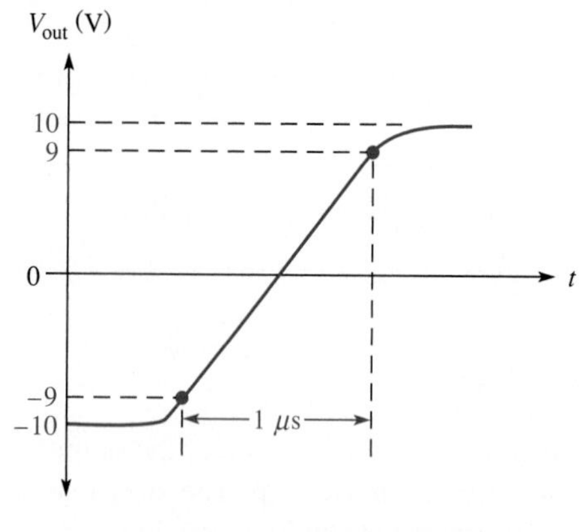

Solution The output goes from the lower to the upper limit in 1 μs. Since this response is not ideal, the limits are taken at the 90% points, as indicated. Thus, the upper limit is +9 V and the lower limit is −9 V.

The slew rate is

$$\frac{\Delta V}{\Delta t} = \frac{+9\text{ V} - (-9\text{ V})}{1\ \mu\text{s}} = 18\text{ V}/\mu\text{s}$$

Related Exercise When a pulse is applied to an op-amp, the output voltage goes from −8 V to +7 V in 0.75 μs. What is the slew rate?

Frequency Response

The internal amplifier stages that make up an op-amp have voltage gains limited by junction capacitances. Although the diff-amps used in op-amps are somewhat different from the basic amplifiers discussed earlier, the same principles apply. An op-amp has no internal coupling capacitors, however; therefore, the low frequency response extends down to dc.

Comparison of Op-Amp Parameters

Table 20–1 provides a comparison of values of some of the parameters just described for four popular IC op-amps. Any values not listed were not given on the manufacturer's data sheet. All values are typical at 25°C.

TABLE 20–1

	Op-Amp Type			
PARAMETER	741C	LM101A	LM108	LM118
Input offset voltage	1 mV	1 mV	0.7 mV	2 mV
Input bias current	80 nA	120 nA	0.8 nA	120 nA
Input offset current	20 nA	40 nA	0.05 nA	6 nA
Input impedance	2 MΩ	800 kΩ	70 MΩ	3 MΩ
Output impedance	75 Ω	—	—	—
Open-loop gain	200,000	160,000	300,000	200,000
Slew rate	0.5 V/μs	—	—	70 V/μs
CMRR	90 dB	90 dB	100 dB	100 dB

Other Features

Most available op-amps have two very important features: short-circuit protection and no latch-up. Short-circuit protection keeps the circuit from being damaged if the output becomes shorted, and the no-latch-up feature prevents the op-amp from hanging up in one output state (high or low voltage level) under certain input conditions.

SECTION 20–3 REVIEW

1. List ten or more op-amp parameters.
2. List two parameters, not including frequency response, that are frequency dependent.

20–4 ■ OP-AMP CONFIGURATIONS WITH NEGATIVE FEEDBACK

In this section we will discuss several basic ways in which an op-amp can be connected using negative feedback to stabilize the gain and increase frequency response. The extremely high open-loop gain of an op-amp creates an unstable situation because a small noise voltage on the input can be amplified to a point where the amplifier is driven out of its linear region. Also, unwanted oscillations can occur. In addition, the open-loop gain parameter of an op-amp can vary greatly from one device to the next. Negative feedback takes a portion of the output and applies it back out-of-phase with the input, creating an effective reduction in gain. This closed-loop gain is usually much less than the open-loop gain and independent of it.

After completing this section, you should be able to

- **Analyze the noninverting, voltage-follower, and inverting op-amp configurations**
  - ☐ Determine the closed-loop gain of the noninverting configuration
  - ☐ Determine the closed-loop gain of the voltage-follower configuration
  - ☐ Determine the closed-loop gain of the inverting amplifier configuration

Noninverting Amplifier

An op-amp connected in a **closed-loop** configuration as a noninverting amplifier with a controlled amount of voltage gain is shown in Figure 20–18.

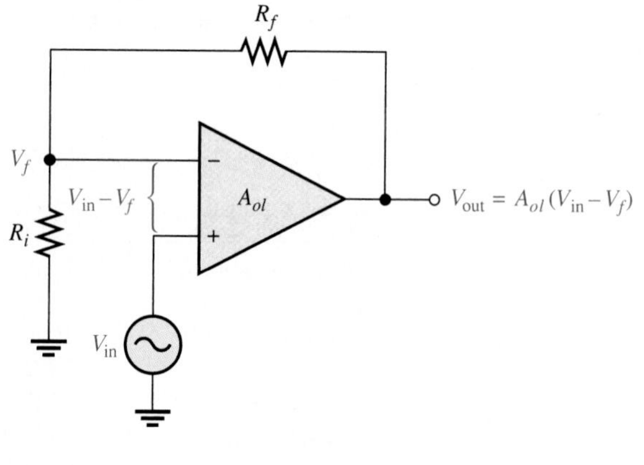

FIGURE 20–18
Noninverting amplifier showing differiential input $V_{in} - V_f$.

The input signal is applied to the noninverting input. The output is applied back to the inverting input through the feedback network formed by R_i and R_f, creating **negative feedback** as follows.

R_i and R_f form a voltage-divider network that reduces the output V_{out} and connects the reduced voltage V_f to the inverting input. The feedback voltage is expressed as

$$V_f = \left(\frac{R_i}{R_i + R_f}\right)V_{out} \qquad \textbf{(20–11)}$$

The difference of the input voltage V_{in} and the feedback voltage V_f is the differential input to the op-amp, as shown in Figure 20–18. This differential voltage is amplified by the open-loop gain of the op-amp (A_{ol}) and produces an output voltage expressed as

$$V_{out} = A_{ol}(V_{in} - V_f) \qquad \textbf{(20–12)}$$

Letting $R_i/(R_i + R_f) = B$ and then substituting BV_{out} for V_f in Equation (20–12), we get the following algebraic steps:

$$V_{out} = A_{ol}(V_{in} - BV_{out})$$
$$V_{out} = A_{ol}V_{in} - A_{ol}BV_{out}$$
$$V_{out} + A_{ol}BV_{out} = A_{ol}V_{in}$$
$$V_{out}(1 + A_{ol}B) = A_{ol}V_{in}$$

Since the total voltage gain of the amplifier in Figure 20–18 is V_{out}/V_{in}, it can be expressed as

$$\frac{V_{out}}{V_{in}} = \frac{A_{ol}}{1 + A_{ol}B}$$

The product $A_{ol}B$ is usually much greater than 1, so the equation simplifies to

$$\frac{V_{out}}{V_{in}} \cong \frac{A_{ol}}{A_{ol}B}$$

Since the closed-loop gain of the noninverting (NI) amplifier is

$$A_{cl(NI)} = \frac{V_{out}}{V_{in}}$$

then

$$A_{cl(NI)} \cong \frac{1}{B} \qquad \textbf{(20–13)}$$

or

$$A_{cl(NI)} = 1 + \frac{R_f}{R_i} \qquad \textbf{(20–14)}$$

Equation (20–13) shows that the **closed-loop gain,** $A_{cl(NI)}$, of the noninverting amplifier is the reciprocal of the attenuation (B) of the feedback network (voltage divider). It is very interesting to note that the closed-loop gain is not at all dependent on the op-amp's open-loop gain under the condition $A_{ol}B >> 1$. The closed-loop gain can be set by selecting values of R_i and R_f.

EXAMPLE 20–5 Determine the gain of the amplifier in Figure 20–19. The open-loop voltage gain is 100,000.

FIGURE 20–19

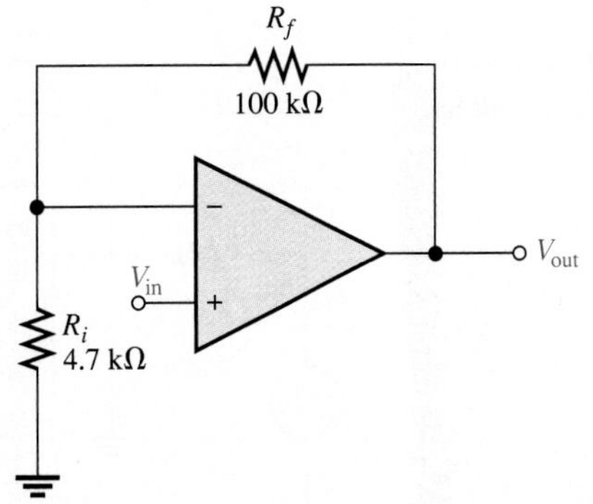

Solution This is a noninverting op-amp configuration. Therefore, the closed-loop gain is

$$A_{cl(\mathrm{NI})} = 1 + \frac{R_f}{R_i} = 1 + \frac{100\ \mathrm{k\Omega}}{4.7\ \mathrm{k\Omega}} = 22.3$$

Related Exercise If the open-loop gain of the amplifier in Figure 20–19 is 150,000 and R_f is increased to 150 kΩ, determine the closed-loop gain.

Voltage-Follower

The voltage-follower configuration is a special case of the noninverting amplifier in which all of the output voltage is fed back to the inverting input, as shown in Figure 20–20. As you can see, the straight feedback connection has a voltage gain of approximately one. The closed-loop voltage gain of a noninverting amplifier is $1/B$, as previously derived. Since $B = 1$, the closed-loop gain of the voltage-follower (VF) is

$$A_{cl(\mathrm{VF})} = 1 \qquad \textbf{(20–15)}$$

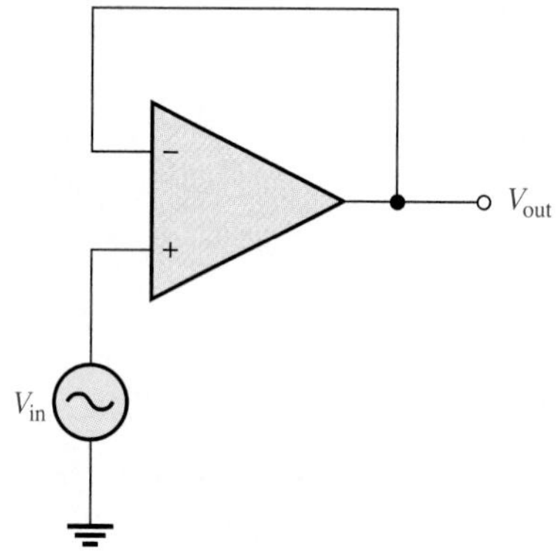

FIGURE 20–20
Op-amp voltage-follower.

The most important features of the voltage-follower configuration are its very high input impedance and its very low output impedance. These features make it a nearly ideal buffer amplifier for interfacing high-impedance sources and low-impedance loads.

Inverting Amplifier

An op-amp connected as an inverting amplifier with a controlled amount of voltage gain is shown in Figure 20–21. The input signal is applied through a series input resistor R_i to

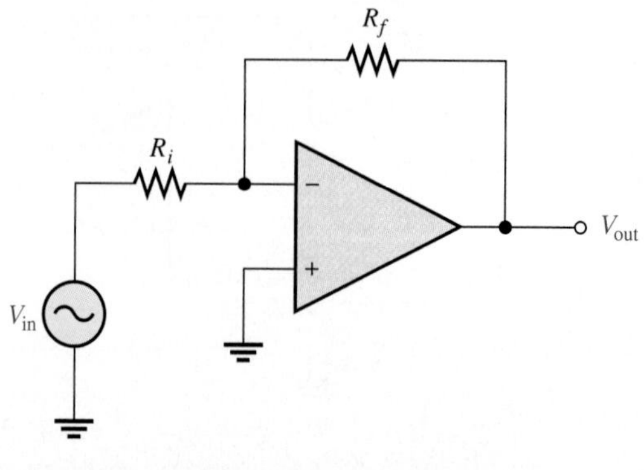

FIGURE 20–21
Inverting amplifier.

the inverting input. Also, the output is fed back through R_f to the same input. The noninverting input is grounded.

At this point, the ideal op-amp parameters mentioned earlier are very useful in simplifying the analysis of this circuit. In particular, the concept of infinite input impedance is of great value. An infinite input impedance implies zero current to the inverting input. If there is zero current through the input impedance, then there must be no voltage drop between the inverting and noninverting inputs. That is, the voltage at the inverting (–) input is zero because the other input (+) is grounded. This zero voltage at the inverting input terminal is referred to as *virtual ground* and is illustrated in Figure 20–22(a).

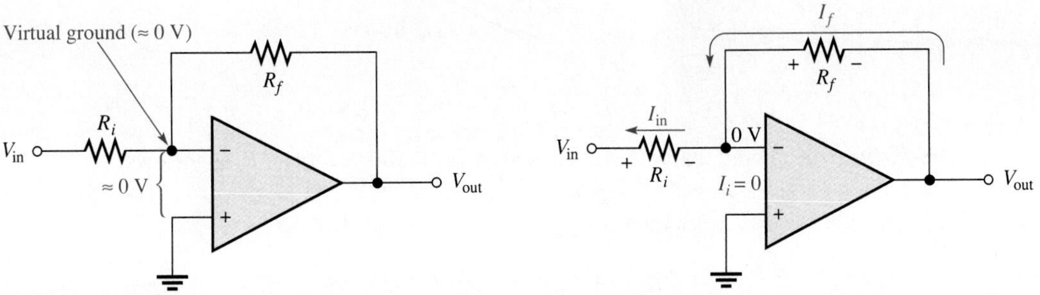

(a) Virtual ground

(b) $I_{in} = I_f$ and $I_i = 0$

FIGURE 20–22
Virtual ground concept and development of closed-loop voltage gain for the inverting amplifier.

Since there is no current into the inverting input, the current through R_i and the current through R_f are equal, as shown in Figure 20–22(b).

$$I_{in} = I_f$$

The voltage across R_i equals V_{in} because of virtual ground on the other side of the resistor. Therefore,

$$I_{in} = \frac{V_{in}}{R_i}$$

Also, the voltage across R_f equals $-V_{out}$ because of virtual ground, and therefore

$$I_f = \frac{-V_{out}}{R_f}$$

Since $I_f = I_{in}$,

$$\frac{-V_{out}}{R_f} = \frac{V_{in}}{R_i}$$

Rearranging the terms,

$$\frac{V_{out}}{V_{in}} = -\frac{R_f}{R_i}$$

Of course, V_{out}/V_{in} is the overall gain of the inverting (I) amplifier.

$$A_{cl(\mathrm{I})} = -\frac{R_f}{R_i} \qquad \textbf{(20–16)}$$

Equation (20–16) shows that the closed-loop voltage gain $A_{cl(\mathrm{I})}$ of the inverting amplifier is the ratio of the feedback resistance R_f to the resistance R_i. *The closed-loop gain is independent of the op-amp's internal open-loop gain.* Thus, the negative feedback stabilizes the voltage gain. The negative sign indicates inversion.

EXAMPLE 20–6 Given the op-amp configuration in Figure 20–23, determine the value of R_f required to produce a closed-loop voltage gain with an absolute value of 100.

FIGURE 20–23

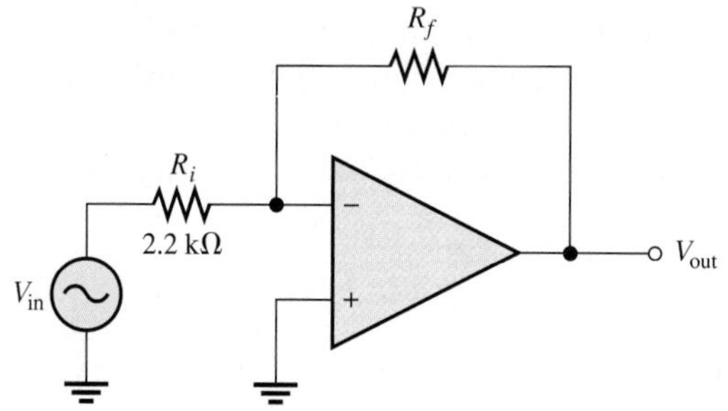

Solution Knowing that $R_i = 2.2\ \text{k}\Omega$ and $|A_{cl(\text{I})}| = 100$, calculate R_f as follows:

$$|A_{cl(\text{I})}| = \frac{R_f}{R_i}$$

$$R_f = |A_{cl(\text{I})}|R_i = (100)(2.2\ \text{k}\Omega) = 220\ \text{k}\Omega$$

Related Exercise If R_i is changed to 2.7 kΩ in Figure 20–23, what value of R_f is required to produce a closed-loop gain of −25?

SECTION 20–4 REVIEW

1. What is the main purpose of negative feedback?
2. The closed-loop voltage gain of each of the op-amp configurations discussed is dependent on the internal open-loop voltage gain of the op-amp (T or F).
3. The attenuation (B) of the negative feedback network of a noninverting op-amp configuration is 0.02. What is the closed-loop gain of the amplifier?

20–5 ▪ EFFECTS OF NEGATIVE FEEDBACK ON OP-AMP IMPEDANCES

Data sheets for IC op-amps usually give values for the input impedance (Z_{in}) and the output impedance (Z_{out}). These values are for the op-amp without any external feedback connections. Each of the negative feedback configurations discussed in the preceding section produces an overall input and output impedance that is different from the op-amp's internal Z_{in} and Z_{out} (except for the inverting amplifier).

After completing this section, you should be able to

■ **Analyze the effects of negative feedback on the three basic op-amp configurations.**

- ☐ Determine the input and output impedances for the noninverting amplifier
- ☐ Determine the input and output impedances for the voltage-follower
- ☐ Determine the input and output impedances for the inverting amplifier

Input and Output Impedances of the Noninverting Amplifier

The input impedance of the noninverting amplifier configuration, $Z_{in(NI)}$, shown in Figure 20–24, is greater than the internal input impedance of the op-amp itself (without feedback) by a factor of $1 + A_{ol}B$, as expressed by the following equation:

$$Z_{in(NI)} = (1 + A_{ol}B)Z_{in} \tag{20–17}$$

FIGURE 20–24
Noninverting amplifier.

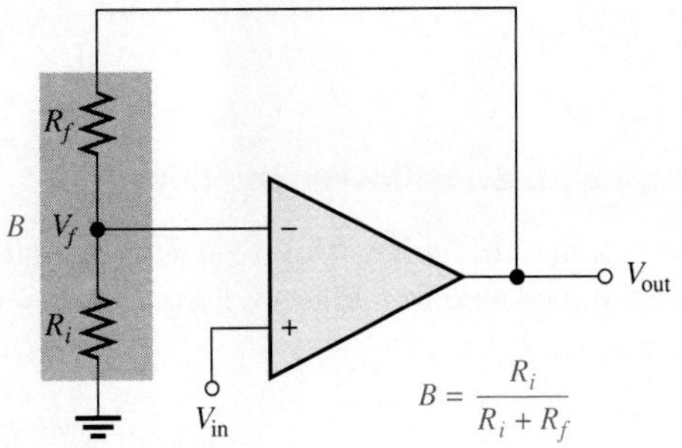

The output impedance with the negative feedback, $Z_{out(NI)}$, is less than the op-amp output impedance by a factor of $1/(1 + A_{ol}B)$, expressed as follows:

$$Z_{out(NI)} = \frac{Z_{out}}{1 + A_{ol}B} \tag{20–18}$$

In summary, the negative feedback in a noninverting configuration increases the input impedance and decreases the output impedance.

EXAMPLE 20–7

(a) Determine the input and output impedances of the amplifier in Figure 20–25. The op-amp data sheet gives $Z_{in} = 2\ \text{M}\Omega$, $Z_{out} = 75\ \Omega$, and $A_{ol} = 200{,}000$.
(b) Find the closed-loop voltage gain.

FIGURE 20–25

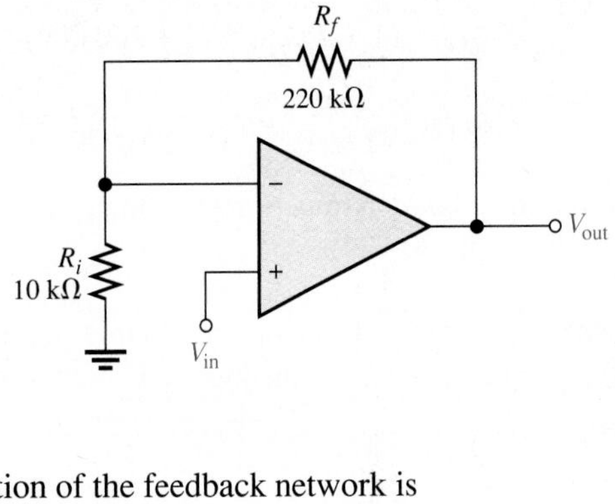

Solution

(a) The attenuation of the feedback network is

$$B = \frac{R_i}{R_i + R_f} = \frac{10\ \text{k}\Omega}{230\ \text{k}\Omega} = 0.043$$

$$Z_{in(NI)} = (1 + A_{ol}B)Z_{in} = [1 + (200{,}000)(0.043)]2\ \text{M}\Omega$$
$$= (1 + 8600)2\ \text{M}\Omega = 17.2 \times 10^3\ \text{M}\Omega = 17.2\ \text{G}\Omega$$

$$Z_{out(NI)} = \frac{Z_{out}}{1 + A_{ol}B} = \frac{75\ \Omega}{1 + 8600} = 0.0087\ \Omega = 8.7\ \text{m}\Omega$$

(b) $A_{cl(\text{NI})} = \dfrac{1}{B} = \dfrac{1}{0.043} = 23.3$

Related Exercise

(a) Determine the input and output impedances in Figure 20–25 for op-amp data sheet values of $Z_{\text{in}} = 3.5\ \text{M}\Omega$, $Z_{\text{out}} = 82\ \Omega$, and $A_{ol} = 135{,}000$.

(b) Find $A_{cl(\text{NI})}$.

Voltage-Follower Impedances

Since the voltage-follower is a special case of the noninverting configuration, the same impedance formulas are used with $B = 1$:

$$Z_{\text{in(VF)}} = (1 + A_{ol})Z_{\text{in}} \qquad \textbf{(20–19)}$$

$$Z_{\text{out(VF)}} = \frac{Z_{\text{out}}}{1 + A_{ol}} \qquad \textbf{(20–20)}$$

As you can see, the voltage-follower input impedance is greater for a given A_{ol} and Z_{in} than for the noninverting configuration with the voltage divider feedback network. Also, its output impedance is much smaller because B is normally much less than 1 for a noninverting configuration.

EXAMPLE 20–8 The same op-amp in Example 20–7 is used in a voltage-follower configuration. Determine the input and output impedance.

Solution Since $B = 1$,

$$Z_{\text{in(VF)}} = (1 + A_{ol})Z_{\text{in}} = (1 + 200{,}000)2\ \text{M}\Omega \cong 400 \times 10^3\ \text{M}\Omega = 400\ \text{G}\Omega$$

$$Z_{\text{out(VF)}} = \frac{Z_{\text{out}}}{1 + A_{ol}} = \frac{75\ \Omega}{1 + 200{,}000} = 0.00038\ \Omega = 380\ \mu\Omega$$

Notice that $Z_{\text{in(VF)}}$ is much greater than $Z_{\text{in(NI)}}$, and $Z_{\text{out(VF)}}$ is much less than $Z_{\text{out(NI)}}$ from Example 20–7.

Related Exercise If the op-amp in this example is replaced with one having a higher open-loop gain, how are the input and output impedances affected?

Input and Output Impedances of the Inverting Amplifier

For the inverting amplifier configuration shown in Figure 20–26, the input impedance, $Z_{\text{in(I)}}$, approximately equals the external input resistance, R_i:

$$Z_{\text{in(I)}} \cong R_i \qquad \textbf{(20–21)}$$

FIGURE 20–26
Inverting amplifier.

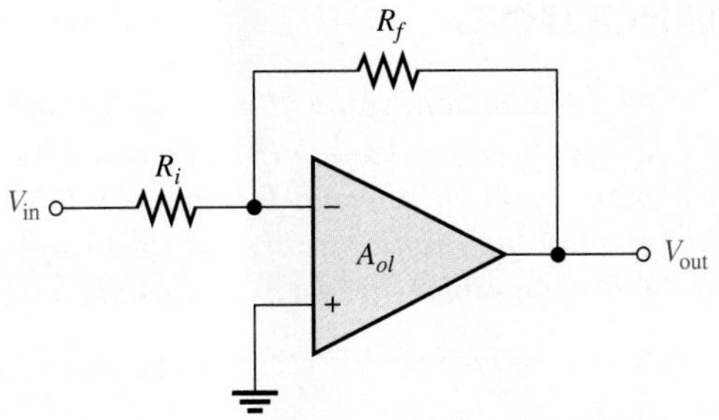

The output impedance, $Z_{out(I)}$, approximately equals the internal output impedance of the op-amp, Z_{out}:

$$Z_{out(I)} \cong Z_{out} \tag{20–22}$$

EXAMPLE 20–9 The op-amp in Figure 20–27 has the following parameters: $A_{ol} = 50{,}000$, $Z_{in} = 4\ \text{M}\Omega$, and $Z_{out} = 50\ \Omega$.

(a) Find the values of the input and output impedances.
(b) Determine the closed-loop voltage gain.

FIGURE 20–27

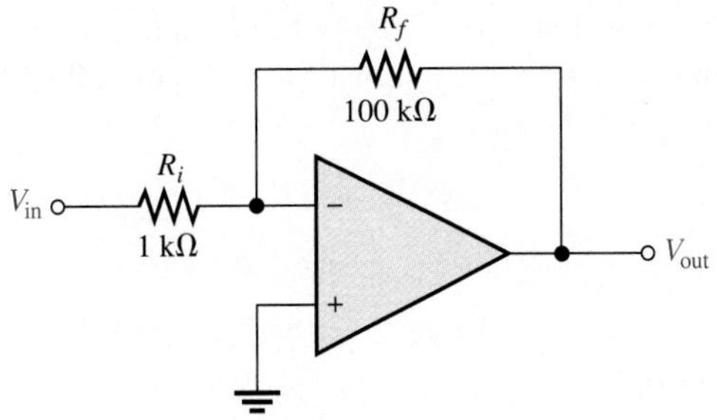

Solution

(a) $Z_{in(I)} \cong R_i = 1\ \text{k}\Omega$

$Z_{out(I)} \cong Z_{out} = 50\ \Omega$

(b) $A_{cl(I)} = -\dfrac{R_f}{R_i} = -\dfrac{100\ \text{k}\Omega}{1\ \text{k}\Omega} = -100$

Related Exercise Determine $Z_{in(I)}$, $Z_{out(I)}$, and A_{cl} in Figure 20–27 for the following values: $A_{ol} = 100{,}000$; $Z_{in} = 5\ \text{M}\Omega$; $Z_{out} = 75\ \Omega$; $R_i = 560\ \Omega$, $R_f = 82\ \text{k}\Omega$.

SECTION 20–5 REVIEW

1. How does the input impedance of a noninverting amplifier configuration compare to the input impedance of the op-amp itself?
2. Connecting an op-amp in a voltage-follower configuration (increases, decreases) the input impedance.
3. Given that $R_f = 100\ \text{k}\Omega$, $R_i = 2\ \text{k}\Omega$, $A_{ol} = 120{,}000$, $Z_{in} = 2\ \text{M}\Omega$, and $Z_{out} = 60\ \Omega$, determine $Z_{in(I)}$ and $Z_{out(I)}$ for an inverting amplifier configuration.

20–6 ▪ TROUBLESHOOTING

As a technician, you will no doubt encounter situations in which an op-amp or its associated circuitry has malfunctioned. The op-amp is a complex integrated circuit with many types of internal failures possible. However, since you cannot troubleshoot the op-amp internally, you treat it as a single device with only a few connections to it. If it fails, you replace it just as you would a resistor, capacitor, or transistor.

After completing this section, you should be able to

- **Troubleshoot op-amp circuits**
  - □ Compensate for input offset voltage
  - □ Troubleshoot for faulty op-amps, open resistors, and incorrect offset compensation

In the basic op-amp configurations, there are only a few external components that can fail. These are the feedback resistor, the input resistor, and the potentiometer used for offset voltage compensation. Also, of course, the op-amp itself can fail or there can be faulty contacts in the circuit. We will now examine the three basic configurations for possible faults and the associated symptoms.

Input Offset Voltage Compensation

Most integrated circuit op-amps provide a means of compensating for offset voltage. This is usually done by connecting an external potentiometer to designated pins on the IC package, as illustrated in Figure 20–28(a) and (b) for a 741 op-amp. The two terminals are labelled *offset null.* With no input, the potentiometer is simply adjusted until the output voltage reads 0, as shown in Figure 20–28(c).

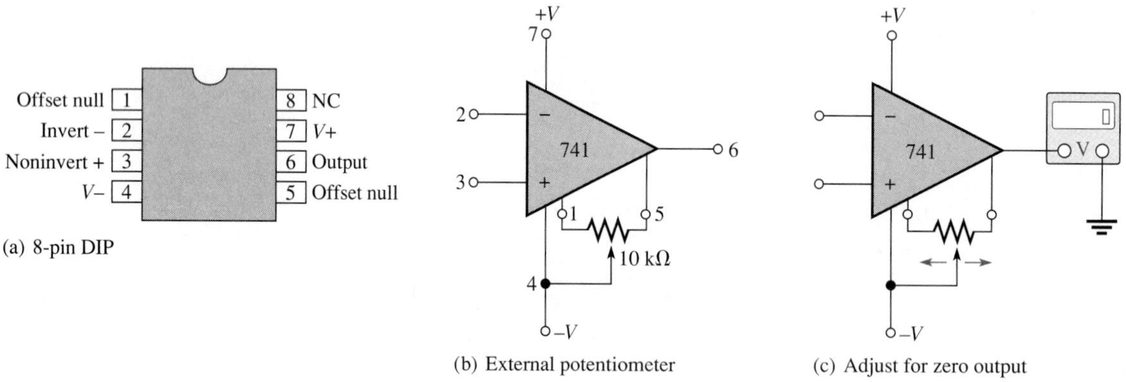

FIGURE 20–28
Input offset voltage compensation for a 741.

Faults in the Noninverting Amplifier

The first thing to do when you suspect a faulty circuit is to check for the proper supply voltage and ground. Having done that, several other possible faults are as follows.

Open Feedback Resistor If the feedback resistor, R_f, in Figure 20–29 opens, the op-amp is operating with its very high open-loop gain, which causes the input signal to drive the device into nonlinear operation and results in a severely clipped output signal as shown in part (a).

FIGURE 20–29
Faults in the noninverting amplifier.

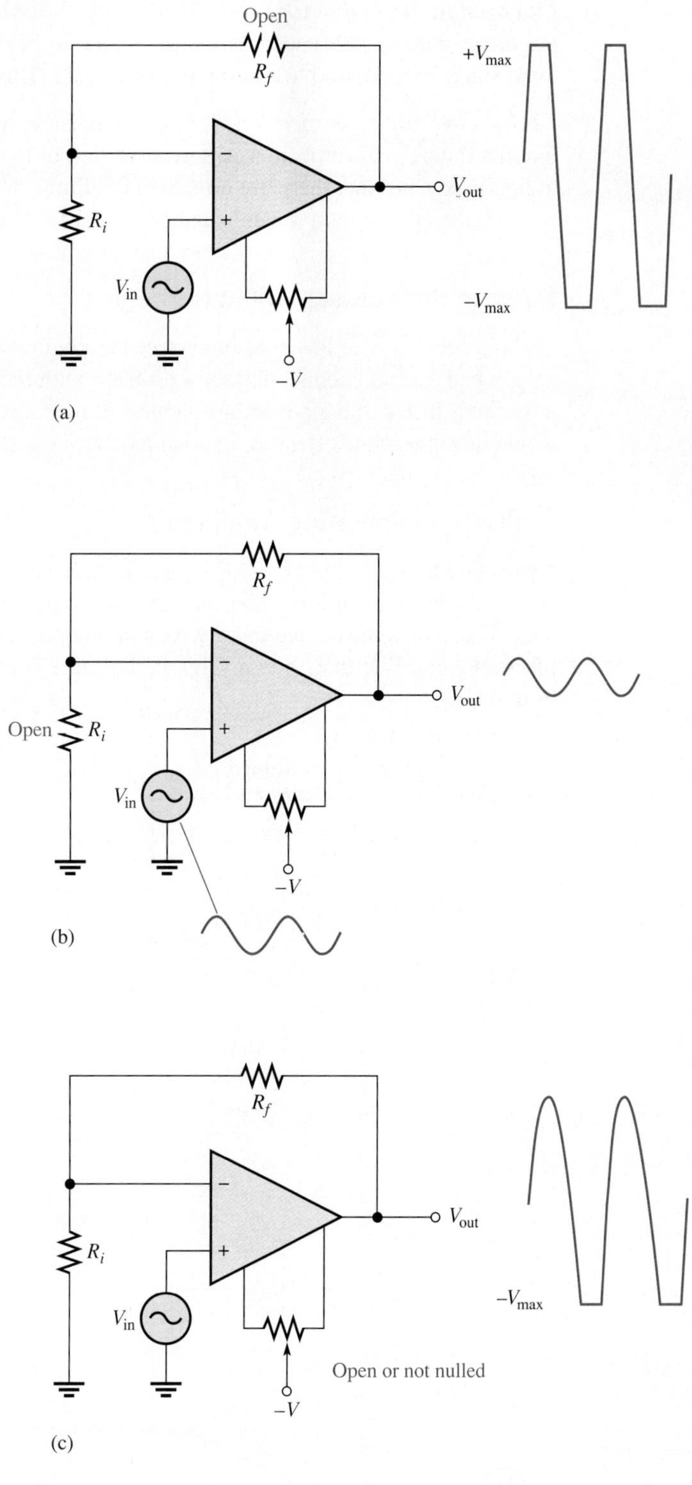

Open Input Resistor In this case, we still have a closed-loop configuration. But, since R_i is open and effectively equal to infinity (∞), the closed-loop gain from Equation (20–14) is

$$A_{cl(\text{NI})} = 1 + \frac{R_f}{R_i} = 1 + \frac{R_f}{\infty} = 1 + 0 = 1$$

This shows that the amplifier acts like a voltage-follower. You would observe an output signal that is the same as the input, as indicated in Figure 20–29(b).

Open or Incorrectly Adjusted Offset Null Potentiometer In this situation, the output offset voltage will cause the output signal to begin clipping on only one peak as the input signal is increased to a sufficient amplitude. This is indicated in Figure 20–29(c).

Faulty Op-Amp As mentioned, many things can happen to an op-amp. In general, an internal failure will result in a loss or distortion of the output signal. The best approach is to first make sure that there are no external failures or faulty conditions. If everything else is good, then the op-amp must be bad.

Faults in the Voltage-Follower

The voltage-follower is a special case of the noninverting amplifier. Except for a bad op-amp, a bad external connection, or a problem with the offset null potentiometer, about the only thing that can happen in a voltage-follower circuit is an open feedback loop. This would have the same effect as an open feedback resistor as previously discussed.

Faults in the Inverting Amplifier

Open Feedback Resistor If R_f opens as indicated in Figure 20–30(a), the input signal still feeds through the input resistor and is amplified by the high open-loop gain of the op-amp. This forces the device to be driven into nonlinear operation, and you will see an output something like that shown. This is the same result as in the noninverting configuration.

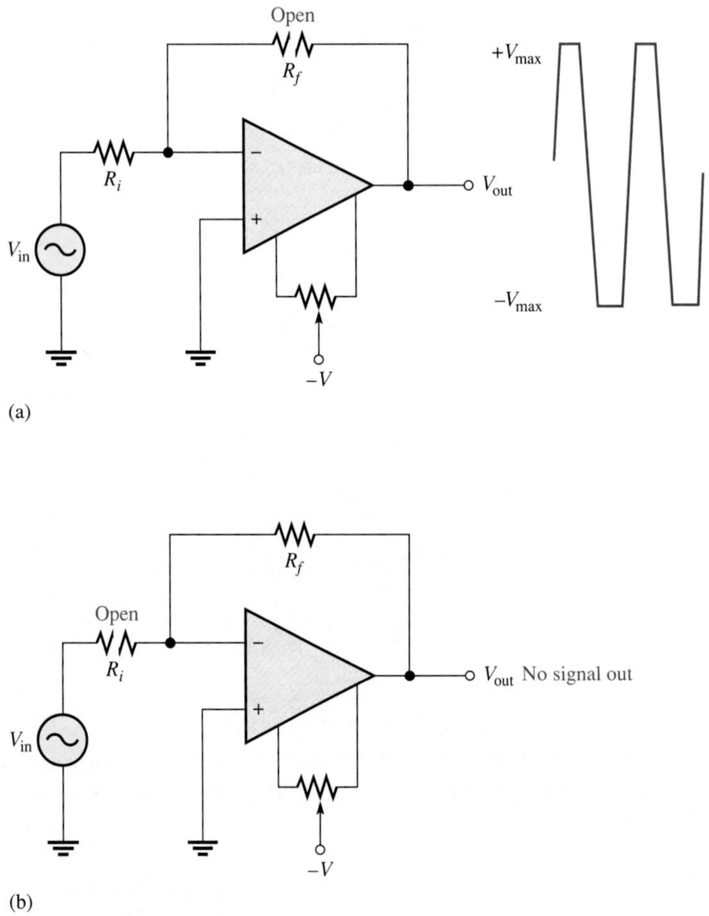

FIGURE 20–30
Faults in the inverting amplifier.

Open Input Resistor This prevents the input signal from getting to the op-amp input, so there will be no output signal, as indicated in Figure 20–30(b).

Failures in the op-amp itself or the offset null potentiometer have the same effects as previously discussed for the noninverting amplifier.

SECTION 20–6 REVIEW

1. If you notice that the op-amp output signal is beginning to clip on one peak as you increase the input signal, what should you check?
2. If there is no op-amp output signal when there is a verified input signal, what would you suspect as being faulty?

20–7 ■ APPLICATION ASSIGNMENT

As an electronics technician in a chemical lab, you have been assigned to work on a spectrophotometer system which combines light optics with electronics to analyze the chemical makeup of various solutions. This type of system is common in chemical and medical laboratories as well as many other areas. It is another example of mixed systems in which electronic circuits interface with other types of systems, such as mechanical and optical, to accomplish a specific function.

Description of the System

The light source in Figure 20–31 produces a beam of visible light containing a wide spectrum of wavelengths. Each component wavelength in the beam of light is refracted at a different angle by the prism as indicated. Depending on the angle of the platform as set

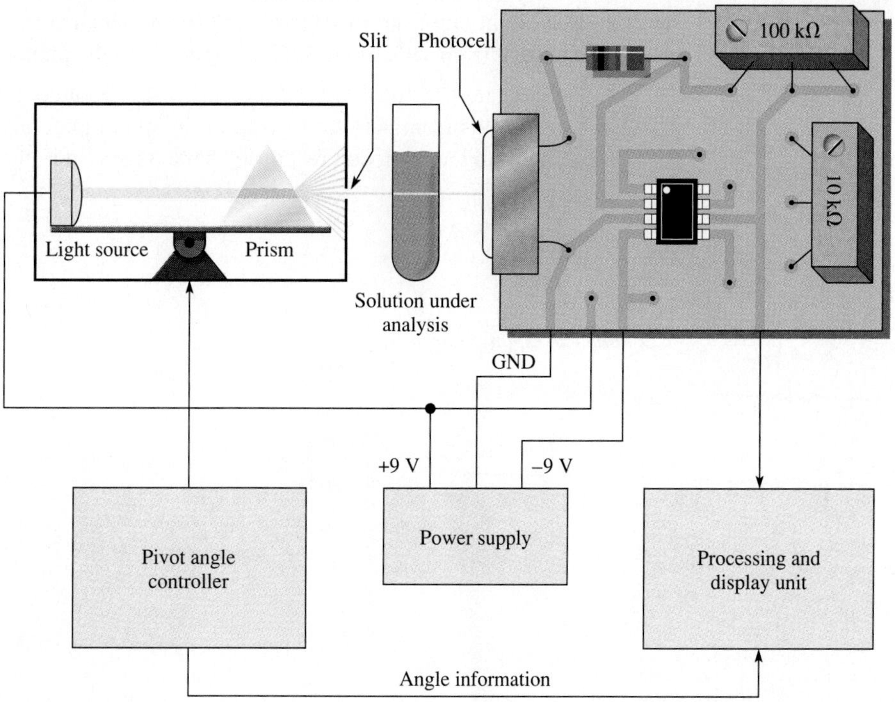

FIGURE 20–31
Spectrophotometer system.

by the pivot angle controller, a certain wavelength passes through the narrow slit and is transmitted through the solution under analysis. By precisely pivoting the light source and prism, a selected wavelength can be transmitted. Every chemical and compound absorbs different wavelengths of light in different ways, so the resulting light coming through the solution has a unique "signature" that can be used to define the chemicals in the solution.

The photocell on the circuit board, produces a voltage that is proportional to the amount of light and wavelength. The op-amp circuit amplifies the photovoltaic cell output and sends the resulting signal to the processing and display unit where the type of chemical(s) in the solution is identified. This is usually a microprocessor-based digital system. Although these other system blocks are interesting, our focus in this section is the photocell and op-amp circuit board.

Step 1: Relate the PC Board to a Schematic

Develop a complete schematic by carefully following the conductive traces on the PC board shown in Figure 20–31 to see how the components are interconnected. The middle pin on the potentiometer is the variable. Some of the interconnecting traces are on the reverse side of the board, but if you are familiar with basic op-amp configurations, you should have no trouble figuring out the connections. Refer to Figure 20–28 for the pin layout of the 741 op-amp. This op-amp is housed in a surface-mount SO-8 package.

Step 2: Analyze the Circuit

1. Determine the resistance value to which the feedback rheostat must be adjusted for a voltage gain of 10.
2. Assume the maximum linear output of the op-amp is 1 V less than the supply voltage. Determine the voltage gain required and the value to which the feedback resistance must be set to achieve the maximum linear output. The system light source produces constant light output from 400 nm to 700 nm, which is approximately the full range of visible light from violet to red. The voltage from the photocell is 0.5 V at 700 nm.
3. Using the gain previously found, determine the op-amp output voltage over the range of wavelengths from 400 nm to 700 nm in 50 nm intervals and plot a graph of the results. Refer to the photocell response characteristic in Figure 20–32.

FIGURE 20–32
Photocell response curve.

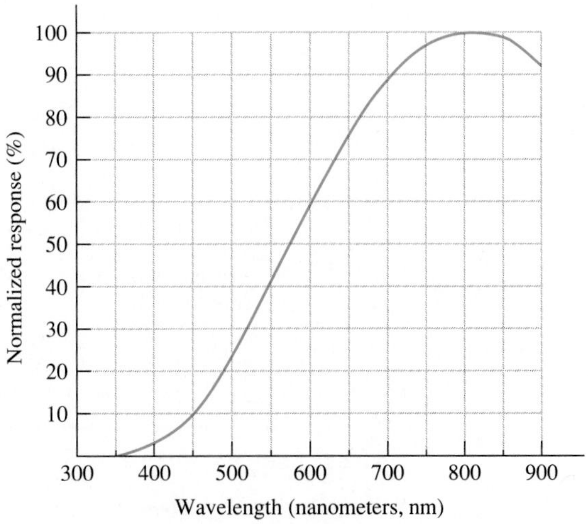

Step 3: Troubleshoot the Circuit

For each of the following problems state the probable cause or causes:

1. Zero voltage at the op-amp output. List three possible causes.
2. Output of op-amp stays at approximately –8 V.
3. A small dc voltage on the op-amp output under no-light conditions.

SECTION 20–7 REVIEW

1. What is the purpose of the 100 kΩ potentiometer on the circuit board?
2. What is the purpose of the 10 kΩ potentiometer?
3. Explain why the light source and prism must be pivoted.

■ SUMMARY

- The op-amp has three terminals not including power and ground: inverting input (–), noninverting input (+), and output.
- Most op-amps require both a positive and a negative dc supply voltage.
- The ideal (perfect) op-amp has infinite input impedance, zero output impedance, infinite open-loop voltage gain, infinite bandwidth, and infinite CMRR.
- A good practical op-amp has high input impedance, low output impedance, high open-loop voltage gain, and wide bandwidth.
- A diff-amp is normally used for the input stage of an op-amp.
- A differential input voltage appears between the inverting and noninverting inputs of a diff-amp.
- A single-ended input voltage appears between one input and ground (with the other inputs grounded).
- A differential output voltage appears between two output terminals of a diff-amp.
- A single-ended output voltage appears between the output and ground of a diff-amp.
- Common mode occurs when equal, in-phase voltages are applied to both input terminals.
- Input offset voltage produces an output error voltage (with no input voltage).
- Input bias current also produces an output error voltage (with no input voltage).
- Input offset current is the difference between the two bias currents.
- Open-loop voltage gain is the gain of the op-amp with no external feedback connections.
- Slew rate is the rate (in volts per microsecond) that the output voltage of an op-amp can change in response to a step input.
- There are three basic op-amp configurations: inverting, noninverting, and voltage-follower.
- All op-amp configurations (except comparators, covered in Chapter 21) employ negative feedback. Negative feedback occurs when a portion of the output voltage is connected back to the inverting input such that it subtracts from the input voltage, thus reducing the voltage gain but increasing the stability and bandwidth.
- A noninverting amplifier configuration has a higher input impedance and a lower output impedance than the op-amp itself (without feedback).
- An inverting amplifier configuration has an input impedance approximately equal to the input resistor R_i and an output impedance approximately equal to the internal output impedance of the op-amp itself.
- The voltage-follower has the highest input impedance and the lowest output impedance of the three configurations.

■ GLOSSARY

Closed loop An op-amp configuration in which the output is connected back to the input.

Closed loop gain (A_{cl}) The overall voltage gain of an op-amp with feedback.

Common-mode rejection ratio (CMRR) A measure of an op-amp's ability to reject signals that appear the same on both inputs; the ratio of open-loop gain to common-mode gain.

Common-mode signals Signals that appear the same on both inputs of an op-amp.

Differential amplifier An amplifier that produces an output proportional to the difference of two inputs.

Negative feedback The return of a portion of the output signal to the input such that it is out of phase with the input signal.

Open-loop gain (A_{ol}) The internal voltage gain of an op-amp without feedback.

Operational amplifier (op-amp) A special type of amplifier exhibiting very high gain, very high input impedance, very low output impedance, and good rejection of common-mode signals.

▪ FORMULAS

(20–1)	$\text{CMRR} = \dfrac{A_{v(d)}}{A_{cm}}$	Common-mode rejection ratio (diff-amp)
(20–2)	$\text{CMRR} = 20 \log\left(\dfrac{A_{v(d)}}{A_{cm}}\right)$	Common-mode rejection ratio (dB)
(20–3)	$V_{\text{OUT}} = I_{C2}R_C - I_{C1}R_C$	Differential output
(20–4)	$I_{\text{BIAS}} = \dfrac{I_1 + I_2}{2}$	Input bias current
(20–5)	$I_{OS} = \lvert I_1 - I_2 \rvert$	Input offset current
(20–6)	$V_{OS} = I_{OS}R_{\text{in}}$	Offset voltage
(20–7)	$V_{\text{OUT(error)}} = A_v I_{OS} R_{\text{in}}$	Output error voltage
(20–8)	$\text{CMRR} = \dfrac{A_{ol}}{A_{cm}}$	Common-mode rejection ratio (op-amp)
(20–9)	$\text{CMRR} = 20 \log\left(\dfrac{A_{ol}}{A_{cm}}\right)$	Common-mode rejection ratio (dB)
(20–10)	$\text{Slew rate} = \dfrac{\Delta V_{\text{out}}}{\Delta t}$	Slew rate
(20–11)	$V_f = \left(\dfrac{R_i}{R_i + R_f}\right)V_{\text{out}}$	Feedback voltage (noninverting)
(20–12)	$V_{\text{out}} = A_{ol}(V_{\text{in}} - V_f)$	Output voltage (noninverting)
(20–13)	$A_{cl(\text{NI})} \cong \dfrac{1}{B}$	Voltage gain (noninverting)
(20–14)	$A_{cl(\text{NI})} = \dfrac{R_f}{R_i}$	Voltage gain (noninverting)
(20–15)	$A_{cl(\text{VF})} = 1$	Voltage gain (voltage-follower)
(20–16)	$A_{cl(\text{I})} = -\dfrac{R_f}{R_i}$	Voltage gain (inverting)
(20–17)	$Z_{\text{in(NI)}} = (1 + A_{ol}B)Z_{\text{in}}$	Input impedance (noninverting)
(20–18)	$Z_{\text{out(NI)}} = \dfrac{Z_{\text{out}}}{1 + A_{ol}B}$	Output impedance (noninverting)
(20–19)	$Z_{\text{in(VF)}} = (1 + A_{ol})Z_{\text{in}}$	Input impedance (voltage-follower)
(20–20)	$Z_{\text{out(VF)}} = \dfrac{Z_{\text{out}}}{1 + A_{ol}}$	Output impedance (voltage-follower)
(20–21)	$Z_{\text{in(I)}} \cong R_i$	Input impedance (inverting)
(20–22)	$Z_{\text{out(I)}} \cong Z_{\text{out}}$	Output impedance (inverting)

■ SELF-TEST

1. Which characteristic does not necessarily apply to an op-amp:
(a) high gain **(b)** low power **(c)** high input impedance **(d)** low output impedance

2. In selecting an op-amp suppose you have several choices. Of the CMRR values listed, the most desirable is
(a) 10 dB **(b)** 20 dB **(c)** 50 dB **(d)** 100 dB

3. The output voltage of a particular op-amp increases 8 V in 12 μs in response to a step voltage on the input. The slew rate is
(a) 0.667 V/μs **(b)** 1.5 V/μs **(c)** 96 V/μs **(d)** 0.75 V/μs

4. A noninverting op-amp configuration has an R_i of 1 kΩ and an R_f of 100 kΩ. If V_{out} is 5 V, the value of V_f is
(a) 50 mV **(b)** 49.5 mV **(c)** 495 mV **(d)** 500 mV

5. In the amplifier described in Question 4, the value of B is
(a) 0.01 **(b)** 0.1 **(c)** 0.0099 **(d)** 101

6. The closed loop gain of the amplifier in Question 4 is
(a) 0.0099 **(b)** 1 **(c)** 99 **(d)** 101

7. One characteristic of a voltage-follower is
(a) $A_{cl} > 1$ **(b)** inversion **(c)** high Z_{out} **(d)** noninversion

8. An inverting amplifier has the following circuit values: R_f = 220 kΩ, R_i = 2.2 kΩ, and A_{ol} = 25,000. The closed-loop gain is
(a) −100 **(b)** −0.01 **(c)** 100 **(d)** −250

9. If you know an op-amp's open-loop gain and nothing else, you can determine the closed-loop gain of
(a) an inverting amplifier **(b)** a noninverting amplifier **(c)** a voltage-follower
(d) none of the amplifier configurations without additional information

10. The feedback attenuation of a voltage-follower is
(a) unity **(b)** less than unity **(c)** greater than unity **(d)** variable

11. The value of B in a certain noninverting amplifier is 0.025. The closed-loop gain is
(a) unity **(b)** 40 **(c)** 0.025 **(d)** undeterminable

12. The highest possible input impedance is achieved with the
(a) inverting amplifier **(b)** noninverting amplifier
(c) differential amplifier **(d)** voltage-follower

■ PROBLEMS

SECTION 20–1 Introduction to Operational Amplifiers

1. Compare a practical op-amp to the ideal.

2. Two IC op-amps are available to you. Their characteristics are as follows:
Op-amp 1: Z_{in} = 5 MΩ, Z_{out} = 100 Ω, A_{ol} = 50,000
Op-amp 2: Z_{in} = 10 MΩ, Z_{out} = 75 Ω, A_{ol} = 150,000
Choose the one you think is generally more desirable.

SECTION 20–2 The Differential Amplifier

3. Identify the type of input and output configuration for each diff-amp in Figure 20–33.

FIGURE 20–33

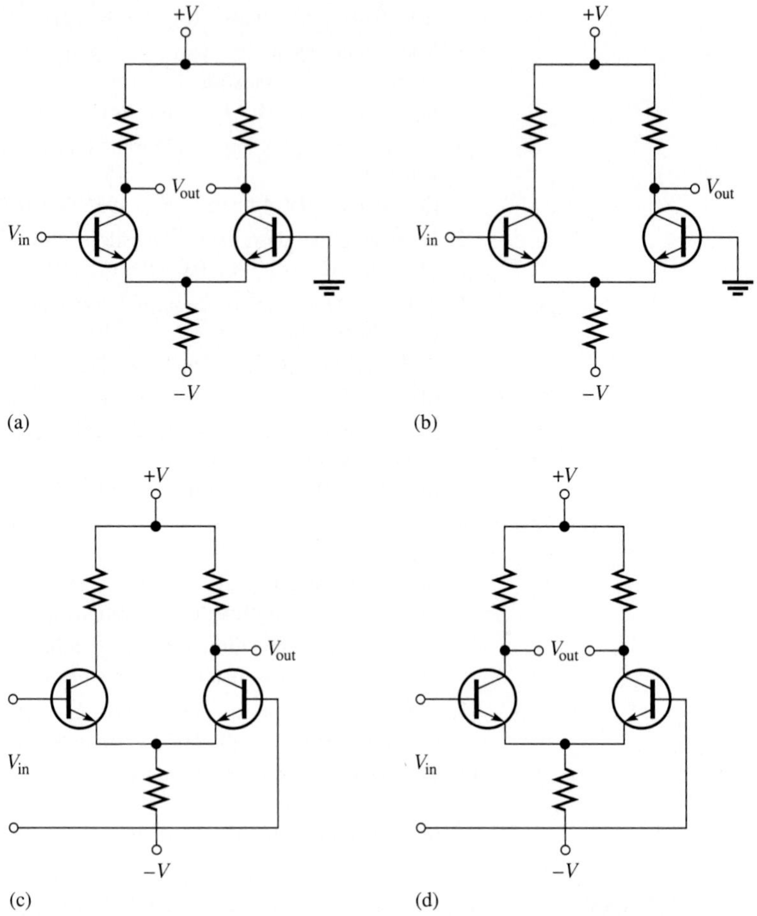

4. The dc base voltages in Figure 20–34 are zero. Using your knowledge of transistor analysis, determine the dc differential output voltage. Assume that Q_1 has an $\alpha = 0.98$ and Q_2 has an $\alpha = 0.975$.

FIGURE 20–34

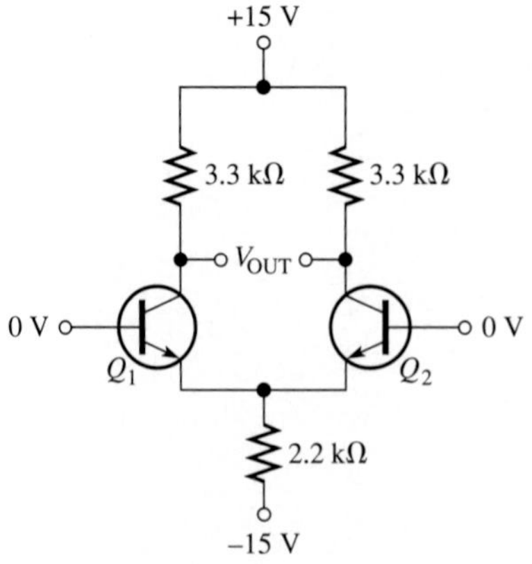

SECTION 20–3 Op-Amp Data Sheet Parameters

5. Determine the bias current, I_{BIAS}, given that the input currents to an op-amp are 8.3 μA and 7.9 μA.

6. Distinguish between *input bias current* and *input offset current,* and then calculate the input offset current in Problem 5.

7. A certain op-amp has a CMRR of 250,000. Convert this to decibels.

8. The open-loop gain of a certain op-amp is 175,000. Its common-mode gain is 0.18. Determine the CMRR in decibels.

9. An op-amp data sheet specifies a CMRR of 300,000 and an A_{ol} of 90,000. What is the common-mode gain?

10. Figure 20–35 shows the output voltage of an op-amp in response to a step input. What is the slew rate?

11. How long does it take the output voltage of an op-amp to go from −10 V to + 10 V if the slew rate is 0.5 V/μs?

FIGURE 20–35

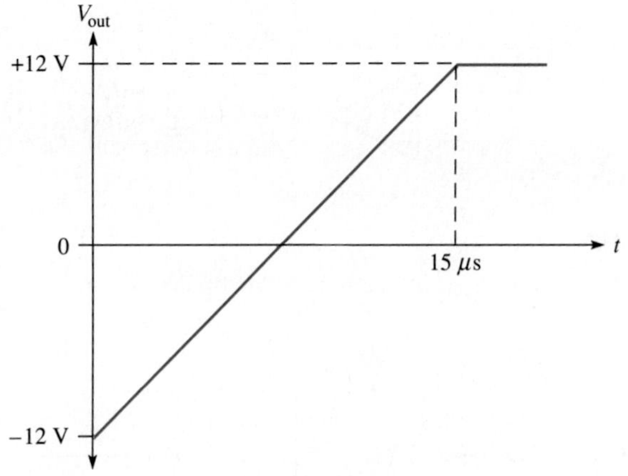

SECTION 20–4 Op-Amp Configurations with Negative Feedback

12. Identify each of the op-amp configurations in Figure 20–36.

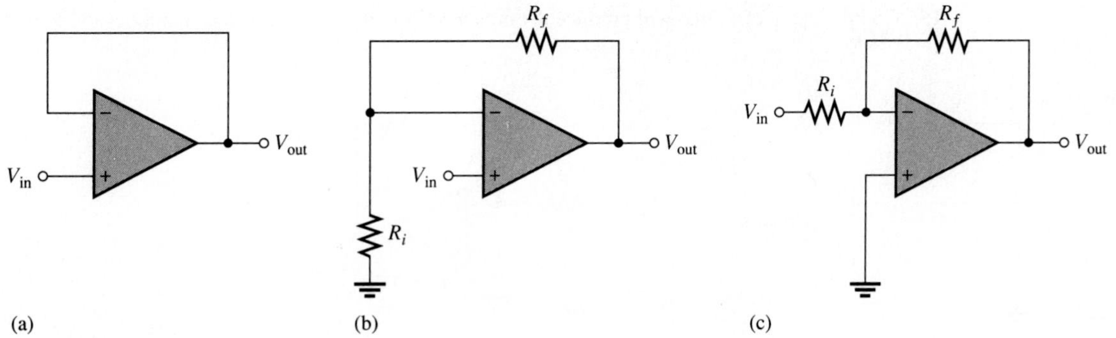

FIGURE 20–36

13. For the amplifier in Figure 20–37, determine the following:
(a) $A_{cl(NI)}$ **(b)** V_{out} **(c)** V_f

FIGURE 20–37

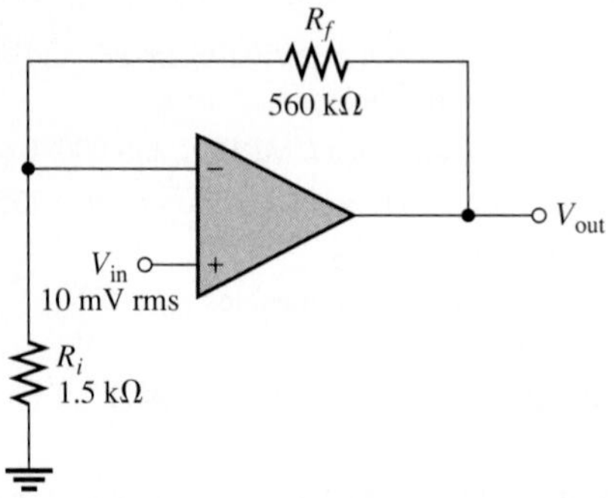

14. Determine the closed-loop gain of each amplifier in Figure 20–38.

FIGURE 20–38

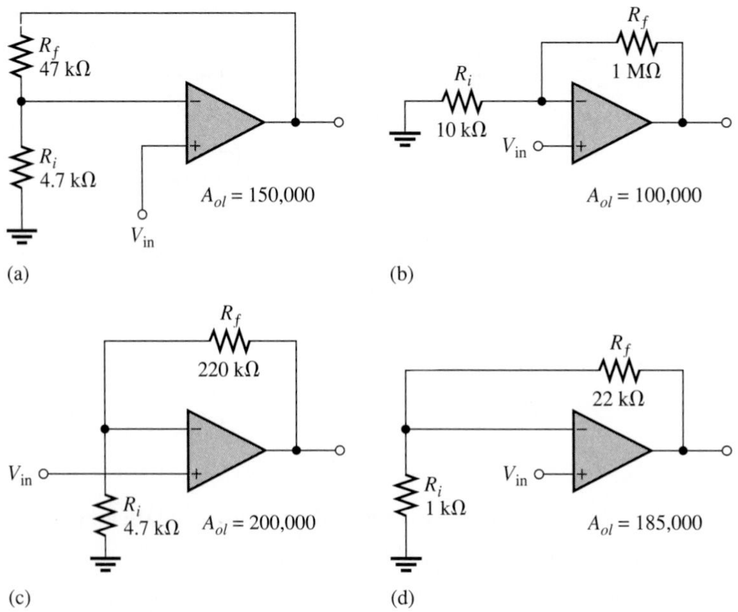

15. Find the value of R_f that will produce the indicated closed-loop gain in each amplifier in Figure 20–39.

FIGURE 20–39

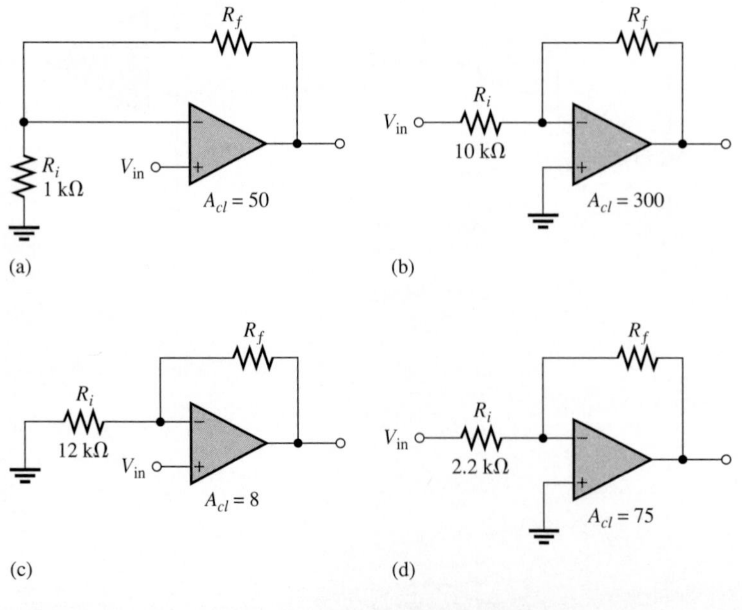

16. Find the gain of each amplifier in Figure 20–40.

17. If a signal voltage of 10 mV rms is applied to each amplifier in Figure 20–40, what are the output voltages and what is their phase relationship with inputs?

FIGURE 20–40

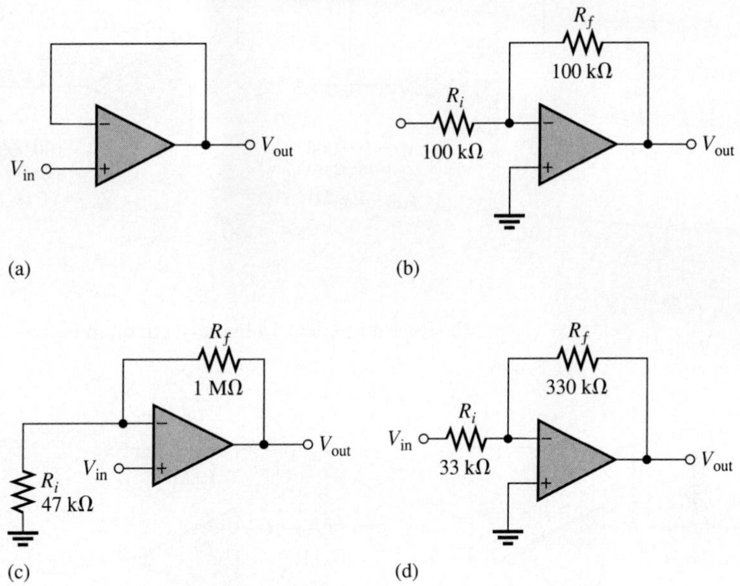

18. Determine the approximate values for each of the following quantities in Figure 20–41:
(a) I_{in} **(b)** I_f **(c)** V_{out} **(d)** closed-loop gain

FIGURE 20–41

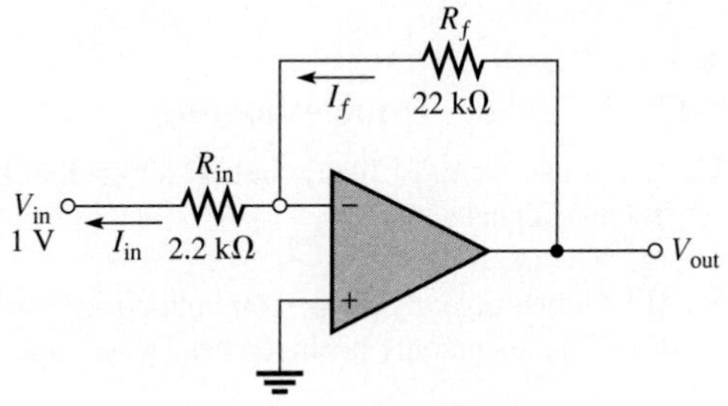

SECTION 20–5 Effects of Negative Feedback on Op-Amp Impedances

19. Determine the input and output impedances for each amplifier configuration in Figure 20–42.

FIGURE 20–42

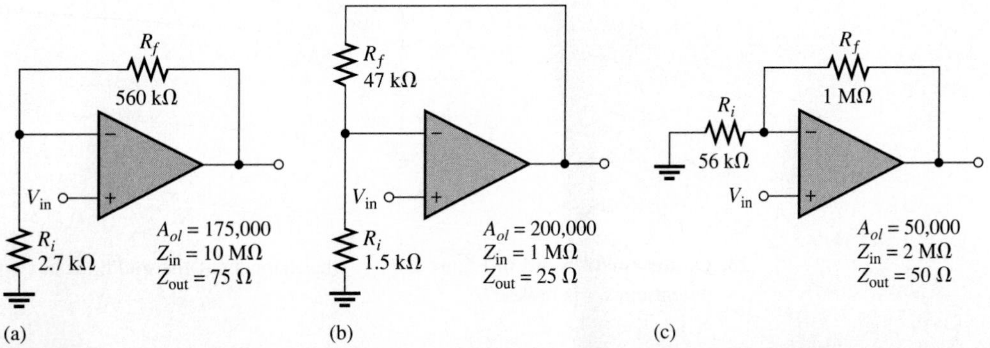

20. Repeat Problem 19 for each circuit in Figure 20–43.

FIGURE 20–43

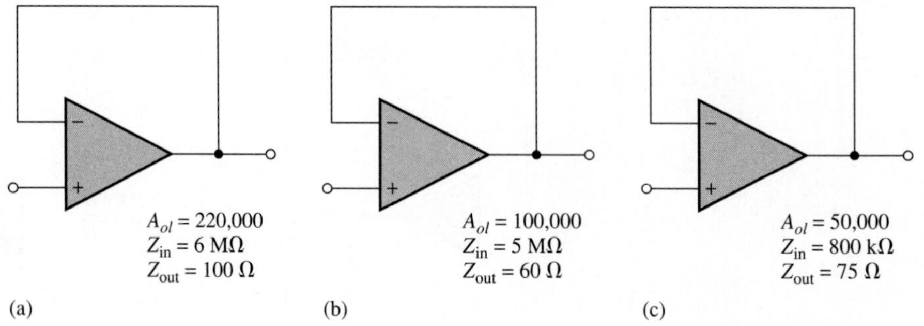

21. Repeat Problem 19 for each circuit in Figure 20–44.

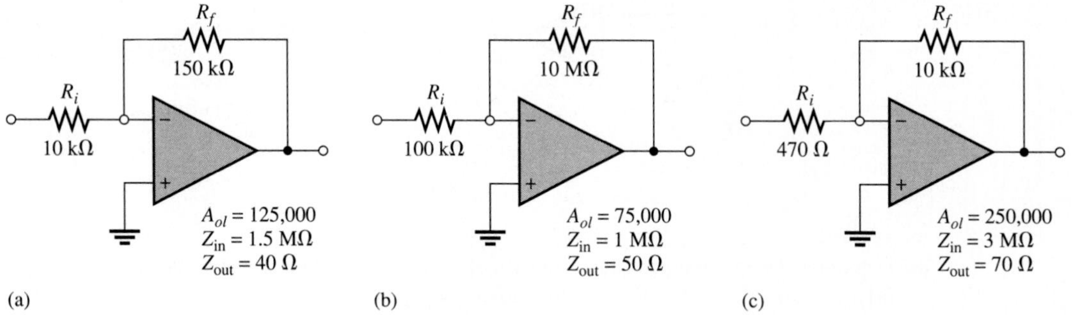

FIGURE 20–44

SECTION 20–6 Troubleshooting

22. Determine the most likely fault(s) for each of the following symptoms in Figure 20–45 with a 100 mV signal applied.

(a) No output signal.

(b) Output severely clipped on both positive and negative swings.

(c) Clipping on only positive peaks when input signal is increased to a certain point.

FIGURE 20–45

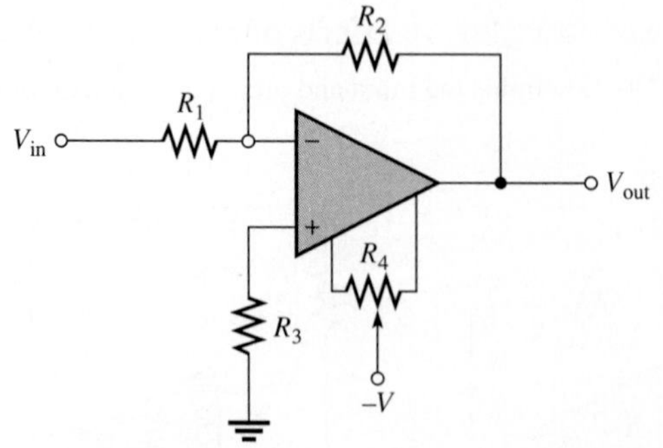

23. On the circuit board in Figure 20–31, what happens if the middle lead (wiper) of the 100 kΩ potentiometer is broken?

ANSWERS TO SECTION REVIEWS

Section 20–1

1. See Figure 20–46.

FIGURE 20–46

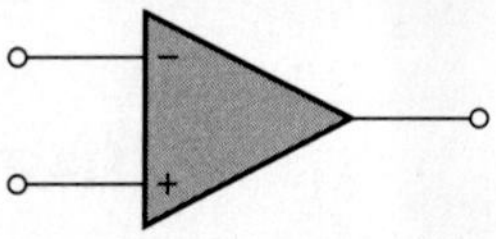

2. Infinite Z_{in}, zero Z_{out}, infinite voltage gain, infinite bandwidth
3. High Z_{in}, low Z_{out}, high voltage gain, wide bandwidth

Section 20–2

1. Differential input is between two input terminals; single-ended input is from one input terminal to ground (with the other inputs grounded).
2. Common-mode rejection is the ability of an op-amp to reject common-mode signals.
3. A higher CMRR results in a lower A_{cm}.

Section 20–3

1. Op-amp parameters are input bias current, input offset voltage, drift, input offset current, input impedance, output impedance, common-mode range, CMRR, open-loop voltage gain, slew rate, and frequency response.
2. Slew rate and gain are frequency-dependent parameters.

Section 20–4

1. Negative feedback stabilizes gain.
2. False
3. $A_{cl} = 1/B = 50$

Section 20–5

1. Z_{in} of the noninverting amplifier is higher than that of the op-amp itself.
2. The voltage-follower configuration increases Z_{in}.
3. $Z_{in(I)} = 2\ k\Omega$; $Z_{out(I)} = 60\ \Omega$

Section 20–6

1. Check the output null adjustment.
2. The op-amp

Section 20–7

1. The 100 kΩ potentiometer is the feedback resistor.
2. The 10 kΩ potentiometer is for nulling the output.
3. The light source and the prism must be pivoted to allow different wavelengths of light to pass through the slit.

ANSWERS TO RELATED EXERCISES FOR EXAMPLES

20–1 34,000; 90.6 dB

20–2 **(a)** 0.168 **(b)** 88 dB **(c)** 2.1 V rms, 4.2 V rms **(d)** 0.168 V

20–3 12,649

20–4 20 V/μs

20–5 32.9

20–6 67.5 kΩ

20–7 **(a)** 20.6×10^3 MΩ, 14 mΩ **(b)** 23

20–8 Z_{in} increases, Z_{out} decreases.

20–9 $Z_{in(I)} = 560\ \Omega$; $Z_{out(I)} = 75\ \Omega$; $A_{cl} = -146$

21

BASIC APPLICATIONS OF OP-AMPS

Op-amps are used in such a wide variety of applications that it is impossible to cover all of them in one chapter or even in one book. Therefore, in this chapter we examine some of the more fundamental applications to illustrate the versatility of the op-amp.

■ CHAPTER OBJECTIVES

- □ Explain the basic operation of a comparator circuit
- □ Analyze summing amplifiers, averaging amplifiers, and scaling amplifiers
- □ Explain the operation of op-amp integrators and differentiators
- □ Discuss the operation of several types of op-amp oscillators
- □ Recognize and evaluate basic op-amp filters
- □ Describe the operation of basic series and shunt voltage regulators

■ APPLICATION ASSIGNMENT

You are the owner of an electronics repair service that handles all kinds of consumer electronic equipment. In this particular case, a dead FM stereo receiver system has come into your shop. After a preliminary check, you decide that the dual-polarity power supply that provides ±12 V to all the op-amps in both channels of the receiver is faulty. The power supply uses positive and negative integrated circuit voltage regulators. After studying this chapter, you should be able to complete the application assignment in Section 21–7.

21–1 ■ COMPARATORS

Operational amplifiers are often used to compare the amplitude of one voltage with another. In this application the op-amp is used in the open-loop configuration, with the input voltage on one input and a reference voltage on the other.

After completing this section, you should be able to

- **Explain the basic operation of a comparator circuit**
  - □ Discuss zero-level detection
  - □ Discuss nonzero-level detection

Zero-Level Detection

A basic application of the op-amp as a **comparator** is in determining when an input voltage exceeds a certain level. Figure 21–1(a) shows a zero-level detector. Notice that the inverting (−) input is grounded and the input signal voltage is applied to the noninverting (+) input. Because of the high open-loop voltage gain, a very small difference voltage between the two inputs drives the amplifier into saturation, causing the output voltage to go to its limit.

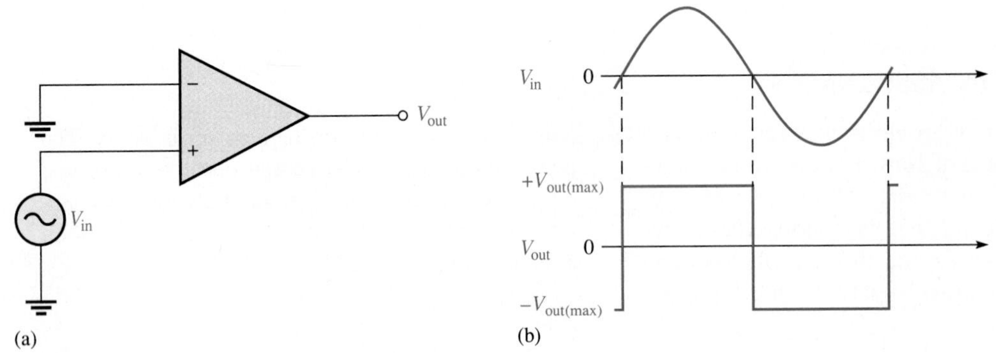

FIGURE 21–1
The op-amp as a zero-level detector.

For example, consider an op-amp having A_{ol} = 100,000. A voltage difference of only 0.25 mV between the inputs could produce an output voltage of (0.25 mV)(100,000) = 25 V if the op-amp were capable. However, since most op-amps have output voltage limitations of less than ±15 V, the device would be driven into saturation.

Figure 21–1(b) shows the result of a sine wave input voltage applied to the noninverting input of the zero-level detector. When the sine wave is negative, the output is at its maximum negative level. When the sine wave crosses 0, the amplifier is driven to its opposite state and the output goes to its maximum positive level, as shown.

As you can see, the zero-level detector can be used as a squaring circuit to produce a square wave from a sine wave.

Nonzero-Level Detection

The zero-level detector in Figure 21–1 can be modified to detect voltages other than zero by connecting a fixed reference voltage, as shown in Figure 21–2(a). A more practical arrangement is shown in part (b) using a voltage divider to set the reference voltage as follows:

$$V_{REF} = \frac{R_2}{R_1 + R_2}(+V) \quad (21\text{–}1)$$

where $+V$ is the positive op-amp supply voltage.

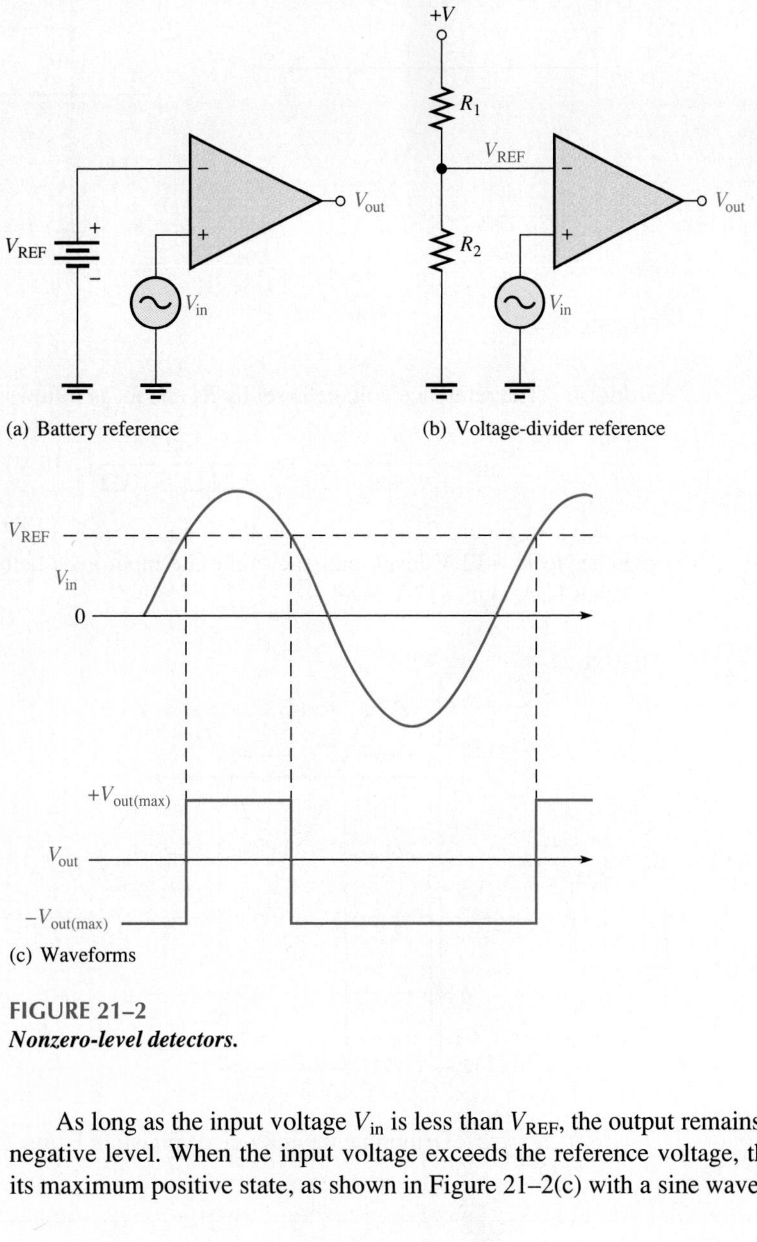

FIGURE 21–2
Nonzero-level detectors.

As long as the input voltage V_{in} is less than V_{REF}, the output remains at the maximum negative level. When the input voltage exceeds the reference voltage, the output goes to its maximum positive state, as shown in Figure 21–2(c) with a sine wave input voltage.

EXAMPLE 21–1

The input signal in Figure 21–3(a) is applied to the comparator circuit in part (b). Make a sketch of the output showing its proper relationship to the input signal. Assume that the maximum output levels are ±12 V.

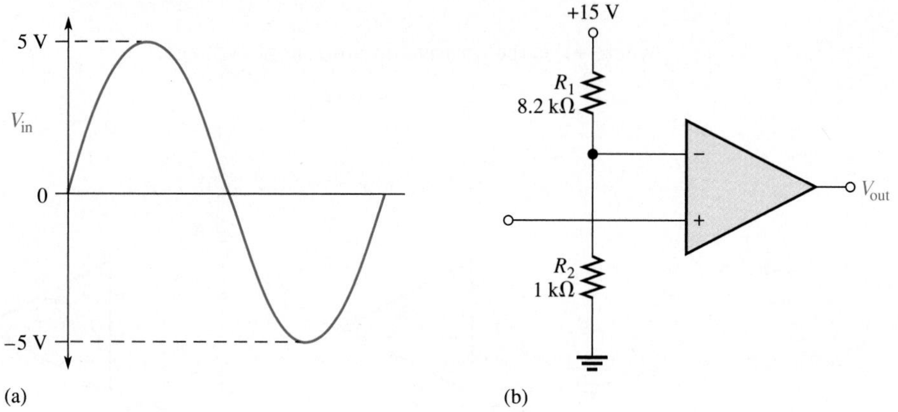

FIGURE 21–3

Solution The reference voltage is set by R_1 and R_2 as follows:

$$V_{REF} = \frac{R_2}{R_1 + R_2}(+V) = \frac{1\ \text{k}\Omega}{8.2\ \text{k}\Omega + 1\ \text{k}\Omega}(+15\ \text{V}) = 1.63\ \text{V}$$

As shown in Figure 21–4, each time the input exceeds +1.63 V, the output voltage switches to its +12 V level, and each time the input goes below +1.63 V, the output switches back to its −12 V level.

FIGURE 21–4

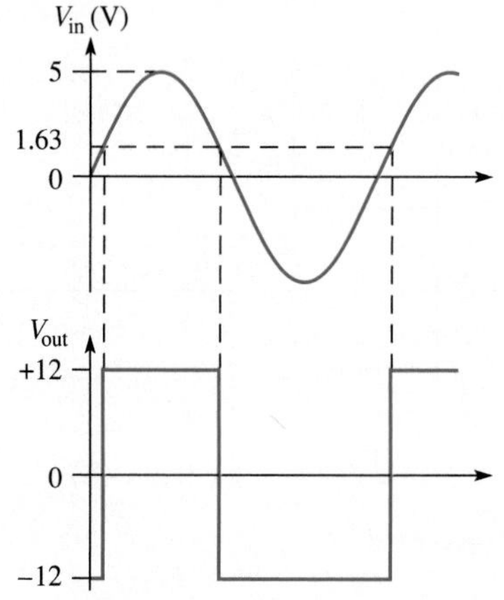

Related Exercise Determine the reference voltage in Figure 21–3 if $R_1 = 22\ \text{k}\Omega$ and $R_2 = 3.3\ \text{k}\Omega$.

SECTION 21–1 REVIEW

1. What is the reference voltage for the comparator in Figure 21–5?
2. Sketch the output waveform for Figure 21–5 when a sine wave with a 5 V peak is applied to the input.

FIGURE 21–5

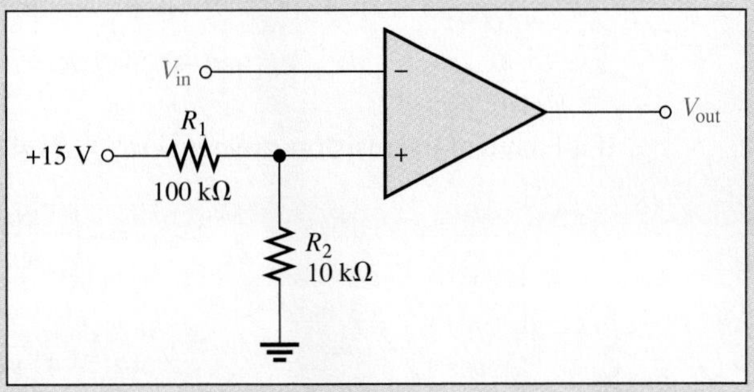

21–2 ▪ SUMMING AMPLIFIERS

The summing amplifier is a variation of the inverting op-amp configuration covered in Chapter 20. The summing amplifier has two or more inputs, and its output voltage is proportional to the negative of the algebraic sum of its input voltages. In this section, you will see how a summing amplifier works, and you will learn about the averaging amplifier and the scaling amplifier, which are variations of the basic summing amplifier.

After completing this section, you should be able to

- **Analyze summing amplifiers, averaging amplifiers, and scaling amplifiers**
  - ☐ Calculate summing amplifier output voltage for given inputs for both unity gain and nonunity gain conditions
  - ☐ Calculate the output voltage for an averaging amplifier
  - ☐ Calculate the output voltage for a scaling adder
  - ☐ Explain how a scaling adder can be used in a digital-to-analog converter

A two-input summing amplifier is shown in Figure 21–6, but any number of inputs can be used. The operation of the circuit and derivation of the output expression are as follows: Two voltages, V_{IN1} and V_{IN2}, are applied to the inputs and produce currents I_1 and I_2, as shown.

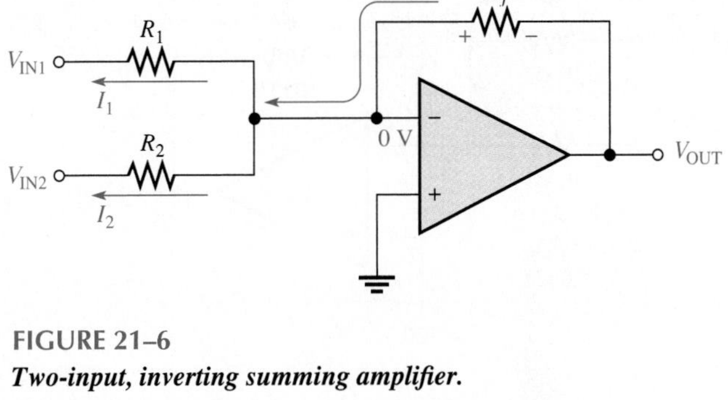

FIGURE 21–6
Two-input, inverting summing amplifier.

Using the concepts of infinite input impedance and virtual ground that you learned about earlier, you can see that the inverting input of the op-amp is approximately 0 V, and there is no current from the input. Therefore, the total current, which is the sum of I_1 and I_2, is through R_f, as indicated in Figure 21–6.

$$I_T = I_1 + I_2$$

Since $V_{OUT} = -I_T R_f$, the following steps apply:

$$V_{OUT} = -(I_1 + I_2)R_f = -\left(\frac{V_{IN1}}{R_1} + \frac{V_{IN2}}{R_2}\right)R_f$$

If all three of the resistors are equal ($R_1 = R_2 = R_f = R$), then

$$V_{OUT} = -\left(\frac{V_{IN1}}{R} + \frac{V_{IN2}}{R}\right)R$$

$$V_{OUT} = -(V_{IN1} + V_{IN2}) \qquad \textbf{(21–2)}$$

Equation (21–2) shows that the output voltage is the sum of the two input voltages. A general expression is given in Equation (21–3) for a summing amplifier with n inputs, as shown in Figure 21–7 where all the resistors are equal.

$$V_{OUT} = -(V_{IN1} + V_{IN2} + V_{IN3} + \cdots + V_{INn}) \qquad \textbf{(21–3)}$$

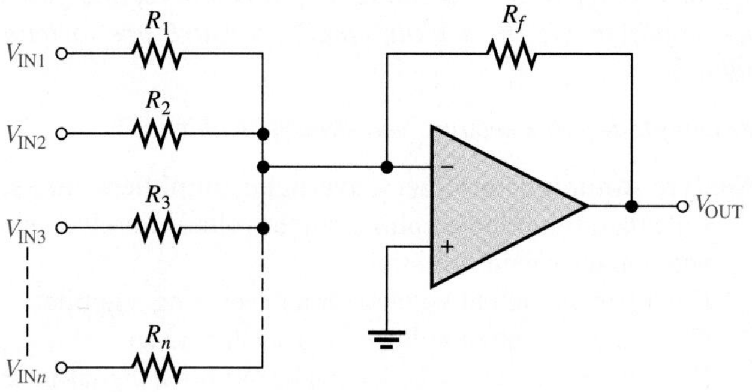

FIGURE 21–7
Summing amplifier with n inputs.

EXAMPLE 21–2 Determine the output voltage in Figure 21–8.

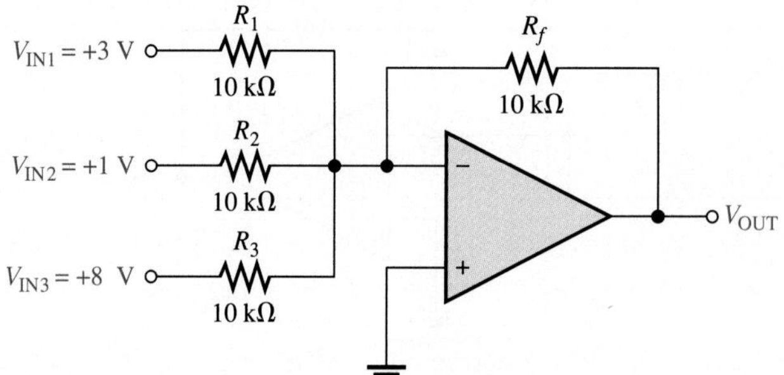

FIGURE 21–8

Solution $V_{OUT} = -(V_{IN1} + V_{IN2} + V_{IN3}) = -(3\text{ V} + 1\text{ V} + 8\text{ V}) = -12\text{ V}$

Related Exercise If a fourth input of +0.5 V is added to Figure 21–8 with a 10 kΩ resistor, what is the output voltage?

Summing Amplifier with Gain Greater Than Unity

When R_f is larger than the input resistors, the amplifier has a gain of R_f/R where R is the value of each input resistor. The general expression for the output is

$$V_{OUT} = -\frac{R_f}{R}(V_{IN1} + V_{IN2} + \cdots + V_{INn}) \qquad \textbf{(21–4)}$$

As you can see, the output is the sum of all the input voltages multiplied by a constant determined by the ratio R_f/R.

EXAMPLE 21–3 Determine the output voltage for the summing amplifier in Figure 21–9.

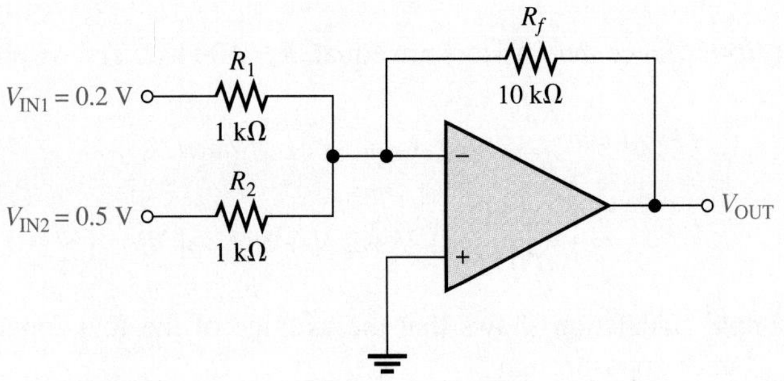

FIGURE 21–9

Solution $R_f = 10\text{ k}\Omega$ and $R = 1\text{ k}\Omega$. Therefore,

$$V_{OUT} = -\frac{R_f}{R}(V_{IN1} + V_{IN2}) = -\frac{10\text{ k}\Omega}{1\text{ k}\Omega}(0.2\text{ V} + 0.5\text{ V}) = -10(0.7\text{ V}) = -7\text{ V}$$

Related Exercise Determine the output voltage in Figure 21–9 if the two input resistors are 2.2 kΩ and the feedback resistor is 18 kΩ.

Averaging Amplifier

A summing amplifier can produce the mathematical average of the input voltages by setting the ratio R_f/R equal to the reciprocal of the number of inputs. You know that the average of several numbers is obtained by first adding the numbers and then dividing by the quantity of numbers you have. Examination of Equation (21–4) and a little thought will convince you that a summing amplifier will do the same. The next example will illustrate.

EXAMPLE 21–4 Show that the amplifier in Figure 21–10 produces an output whose magnitude is the average of the input voltages.

FIGURE 21–10

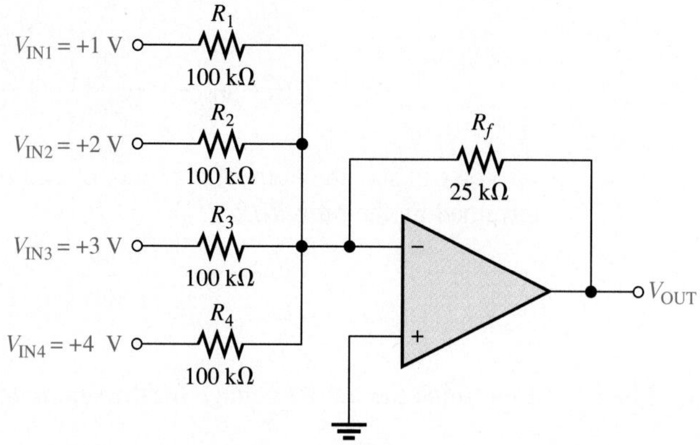

Solution Since the resistors are equal, $R = 100\text{ k}\Omega$. The output voltage is

$$V_{OUT} = -\frac{R_f}{R}(V_{IN1} + V_{IN2} + V_{IN3} + V_{IN4})$$

$$= -\frac{25\text{ k}\Omega}{100\text{ k}\Omega}(1\text{ V} + 2\text{ V} + 3\text{ V} + 4\text{ V}) = -\frac{1}{4}(10\text{ V}) = -2.5\text{ V}$$

A simple calculation shows that the average of the four input values is the same as V_{OUT} but of opposite sign:

$$\frac{(1\text{ V} + 2\text{ V} + 3\text{ V} + 4\text{ V})}{4} = \frac{10\text{ V}}{4} = 2.5\text{ V}$$

Related Exercise Specify the changes required in the averaging amplifier in Figure 21–10 in order to handle five inputs.

Scaling Adder

A different weight can be assigned to each input of a summing amplifier by simply adjusting the values of the input resistors. As you have seen, the output voltage can be expressed as

$$V_{OUT} = -\left(\frac{R_f}{R_1}V_{IN1} + \frac{R_f}{R_2}V_{IN2} + \cdots + \frac{R_f}{R_n}V_{INn}\right) \quad \textbf{(21–5)}$$

The weight of a particular input is set by the ratio of R_f to the input resistance R. For example, if an input voltage is to have a weight of 1, then $R_f = R$. Or, if a weight of 0.5 is required, $R_f = 0.5R$. The smaller the value of R relative to R_f, the greater the weight.

EXAMPLE 21–5 For the scaling adder in Figure 21–11 determine the weight of each input voltage and find the output voltage.

FIGURE 21–11

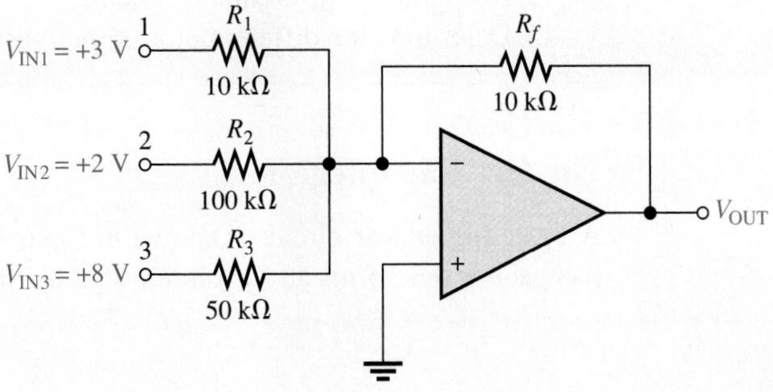

Solution

Weight of input 1: $\dfrac{R_f}{R_1} = \dfrac{10\text{ k}\Omega}{10\text{ k}\Omega} = 1$

Weight of input 2: $\dfrac{R_f}{R_2} = \dfrac{10\text{ k}\Omega}{100\text{ k}\Omega} = 0.1$

Weight of input 3: $\dfrac{R_f}{R_3} = \dfrac{10\text{ k}\Omega}{50\text{ k}\Omega} = 0.2$

The output voltage is

$$V_{OUT} = -\left(\frac{R_f}{R_1}V_{IN1} + \frac{R_f}{R_2}V_{IN2} + \frac{R_f}{R_3}V_{IN3}\right)$$
$$= -[1(3\text{ V}) + 0.1(2\text{ V}) + 0.2(8\text{ V})]$$
$$= -(3\text{ V} + 0.2\text{ V} + 1.6\text{ V}) = -4.8\text{ V}$$

Related Exercise Determine the weight of each input voltage in Figure 21–11 if $R_1 = 22$ kΩ, $R_2 = 82$ kΩ, $R_3 = 56$ kΩ, and $R_f = 10$ kΩ. Also find V_{OUT}.

SECTION 21–2 REVIEW

1. Define *summing point.*
2. What is the value of R_f/R for a five-input averaging amplifier?
3. A certain scaling adder has two inputs, one having twice the weight of the other. If the value of the lowest input resistor is 10 kΩ, what is the value of the other input resistor?

21–3 ▪ THE INTEGRATOR AND DIFFERENTIATOR

An op-amp integrator simulates mathematical integration, which is basically a summing process that determines the total area under the curve of a function. An op-amp differentiator simulates mathematical differentiation, which is a process of determining the instantaneous rate of change of a function. It is not necessary for you to understand mathematical integration or differentiation, at this point, in order to learn how an integrator and differentiator work.

After completing this section, you should be able to

- **Explain the operation of op-amp integrators and differentiators**
  - ☐ Recognize an integrator circuit
  - ☐ Determine the rate of change of the integrator output voltage
  - ☐ Recognize a differentiator circuit
  - ☐ Determine the differentiator output voltage

The Op-Amp Integrator

A basic **integrator** circuit is shown in Figure 21–12. Notice that the feedback element is a capacitor that forms an *RC* circuit with the input resistor.

FIGURE 21–12
An op-amp integrator.

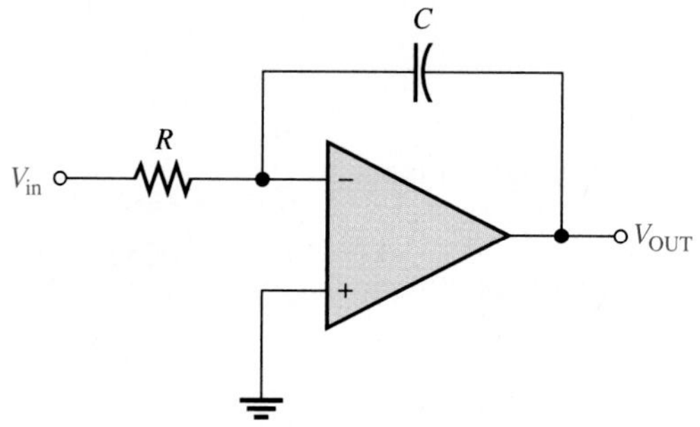

How a Capacitor Charges To understand how the integrator works, it is important to review how a capacitor charges. Recall that the charge Q on a capacitor is proportional to the charging current and the time.

$$Q = I_C t$$

Also, in terms of the voltage, the charge on a capacitor is

$$Q = CV_C$$

From these two relationships, the capacitor voltage can be expressed as

$$V_C = \left(\frac{I_C}{C}\right)t$$

This expression has the form of an equation for a straight line beginning at zero with a constant slope of I_C/C. (Remember from algebra that the general formula for a straight line is $y = mx + b$. In this case, $y = V_C$, $m = I_C/C$, $x = t$, and $b = 0$).

Recall that the capacitor voltage in a simple *RC* circuit is not linear but is exponential. This is because the charging current continuously decreases as the capacitor charges and causes the rate of change of the voltage to continuously decrease. The key thing about using an op-amp with an *RC* circuit to form an integrator is that the capacitor's charging current is made constant, thus producing a straight-line (linear) voltage rather than an exponential voltage. Now let's see why this is true.

In Figure 21–13, the inverting input of the op-amp is at virtual ground (0 V), so the voltage across R_i equals V_{in}. Therefore, the input current is

$$I_{in} = \frac{V_{in}}{R_i}$$

If V_{in} is a constant voltage, then I_{in} is also a constant because the inverting input always remains at 0 V, keeping a constant voltage across R_i. Because of the very high

FIGURE 21–13
Currents in an integrator.

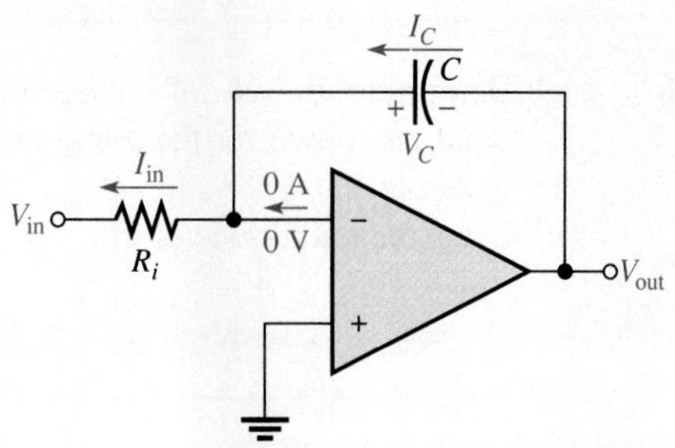

input impedance of the op-amp, there is negligible current from the inverting input. This makes all of the input current flow through the capacitor, as indicated in Figure 21–13, so

$$I_C = I_{in}$$

The Capacitor Voltage Since I_{in} is constant, so is I_C. The constant I_C charges the capacitor linearly and produces a linear voltage across C. The positive side of the capacitor is held at 0 V by the virtual ground of the op-amp. The voltage on the negative side of the capacitor decreases linearly from zero as the capacitor charges, as shown in Figure 21–14. This voltage is called a *negative ramp.*

The Output Voltage V_{out} is the same as the voltage on the negative side of the capacitor. When a constant input voltage in the form of a step or pulse (a pulse has a constant amplitude when high) is applied, the output ramp decreases negatively until the op-amp saturates at its maximum negative level. This is indicated in Figure 21–15.

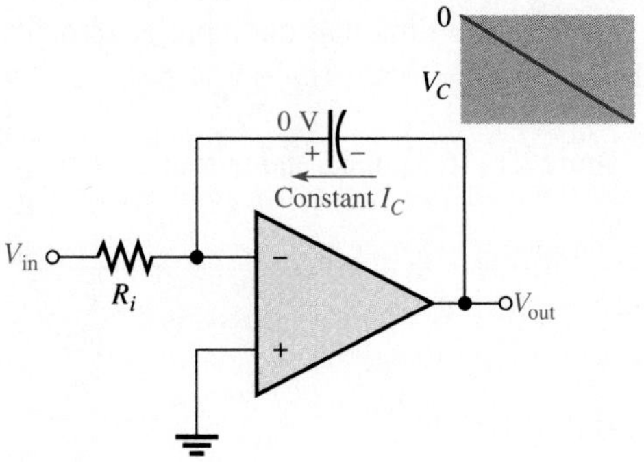

FIGURE 21–14
A linear ramp voltage is produced across C by the constant charging current.

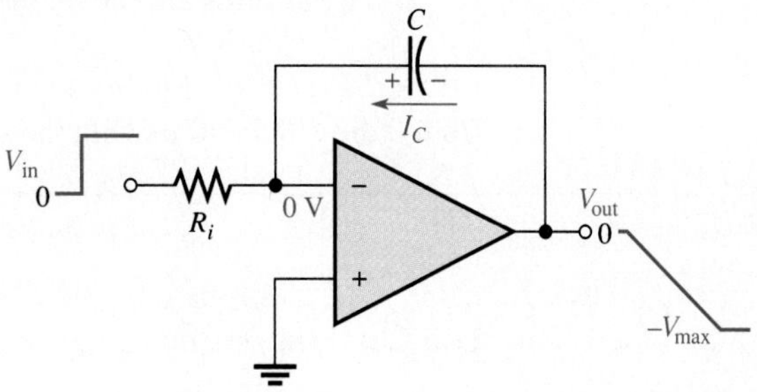

FIGURE 21–15
A constant input voltage produces a ramp on the output.

Rate of Change of the Output The rate at which the capacitor charges, and therefore the slope of the output ramp, is set by the ratio I_C/C, as you have seen. Since $I_C = V_{in}/R_i$, the rate of change or slope of the integrator's output voltage is

$$\frac{\Delta V_{out}}{\Delta t} = -\frac{V_{in}}{R_i C} \qquad \textbf{(21–6)}$$

EXAMPLE 21–6

(a) Determine the rate of change of the output voltage in response to a single pulse input, as shown for the integrator in Figure 21–16(a). The output voltage is initially zero.

(b) Draw the output waveform.

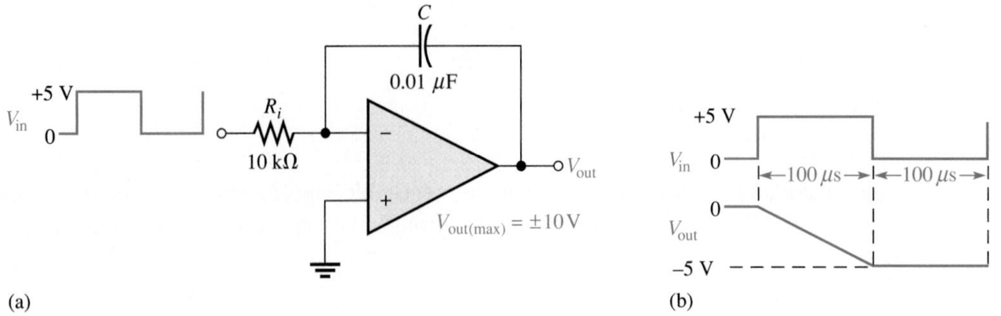

FIGURE 21–16

Solution

(a) The rate of change of the output voltage is

$$\frac{\Delta V_{out}}{\Delta t} = -\frac{V_{in}}{R_iC} = -\frac{5\text{ V}}{(10\text{ k}\Omega)(0.01\ \mu\text{F})} = -50\text{ kV/s} = -50\text{ mV}/\mu\text{s}$$

(b) The rate of change was found to be –50 mV/μs in part (a). When the input is at +5 V, the output is a negative-going ramp. When the input is at 0 V, the output is a constant level. In 100 μs, the voltage decreases.

$$\Delta V_{out} = (-50\text{ mV}/\mu\text{s})(100\ \mu\text{s}) = -5\text{ V}$$

Therefore, the negative-going ramp reaches –5 V at the end of the pulse. The output voltage then remains constant at –5 V for the time that the input is zero. The waveforms are shown in Figure 21–16(b).

Related Exercise Modify the integrator in Figure 21–16 to make the output change from 0 to –5 V in 50 μs with the same input.

The Op-Amp Differentiator

A basic **differentiator** is shown in Figure 21–17. Notice how the placement of the capacitor and resistor differ from the integrator. The capacitor is now the input element. A differentiator produces an output that is proportional to the rate of change of the input voltage.

FIGURE 21–17
An op-amp differentiator.

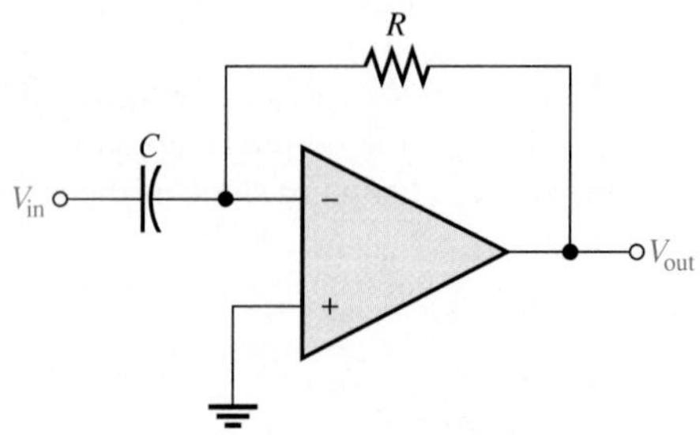

To see how the differentiator works, we will apply a positive-going ramp voltage to the input as indicated in Figure 21–18. In this case, $I_C = I_{in}$ and the voltage across the capacitor is equal to V_{in} at all times ($V_C = V_{in}$) because of virtual ground on the inverting input.

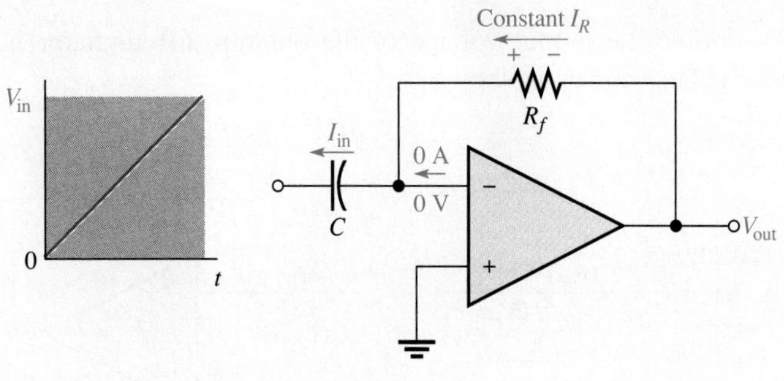

FIGURE 21–18
A differentiator with a ramp input.

From the basic formula, $V_C = (I_C/C)t$, we get

$$I_C = \left(\frac{V_C}{t}\right)C$$

Since the current from the inverting input is negligible, $I_R = I_C$. Both currents are constant because the slope of the capacitor voltage (V_C/t) is constant. The output voltage is also constant and equal to the voltage across R_f because one side of the feedback resistor is always 0 V (virtual ground).

$$V_{out} = I_R R_f = I_C R_f$$

$$V_{out} = \left(\frac{V_C}{t}\right)R_f C \qquad \textbf{(21–7)}$$

The output is negative when the input is a positive-going ramp and positive when the input is a negative-going ramp as illustrated in Figure 21–19. During the positive slope of the input, the capacitor is charging from the input source and the constant current through the feedback resistor is in the direction shown. During the negative slope of the input, the current is in the opposite direction because the capacitor is discharging.

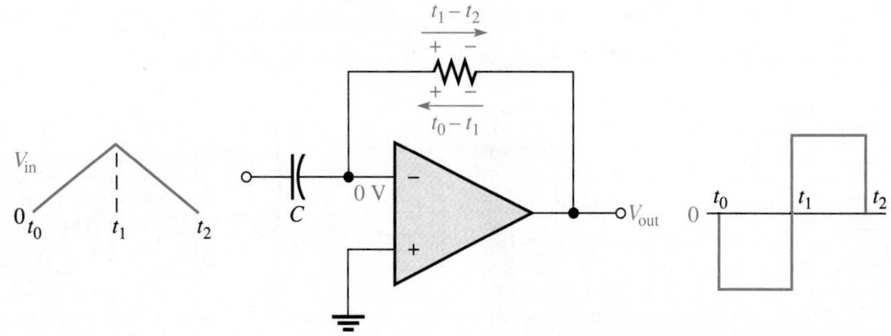

FIGURE 21–19
Output of a differentiator with a series of positive and negative ramps (triangle wave) on the input.

Notice in Equation (21–7) that the term V_C/t is the slope of the input. If the slope increases, V_{out} increases. If the slope decreases, V_{out} decreases. So, the output voltage is proportional to the slope (rate of change) of the input. The constant of proportionality is the time constant, R_fC.

EXAMPLE 21–7 Determine the output voltage of the op-amp differentiator in Figure 21–20 for the triangular-wave input shown.

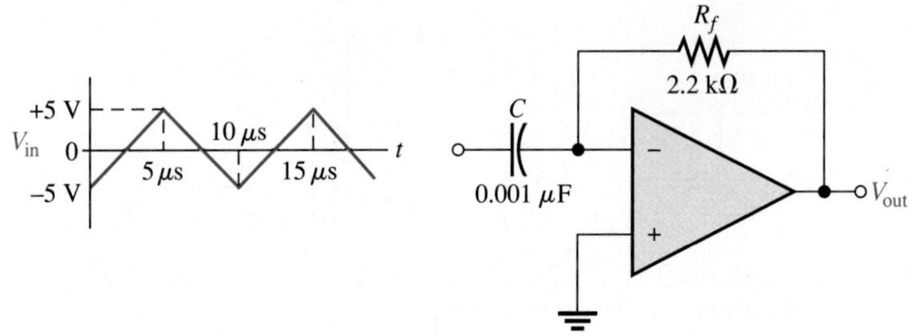

FIGURE 21–20

Solution Starting at $t = 0$, the input voltage is a positive-going ramp ranging from –5 V to +5 V (a +10 V change) in 5 μs. Then it changes to a negative-going ramp ranging from +5 V to –5 V (a –10 V change) in 5 μs.

The time constant is

$$R_fC = (2.2\text{ k}\Omega)(0.001\ \mu\text{F}) = 2.2\ \mu\text{s}$$

The slope or rate of change (V_C/t) of the positive-going ramp is determined, and the output voltage is calculated as follows:

$$\frac{V_C}{t} = \frac{10\text{ V}}{5\ \mu\text{s}} = 2\text{ V}/\mu\text{s}$$

$$V_{out} = (2\text{ V}/\mu\text{s})2.2\ \mu\text{s} = +4.4\text{ V}$$

Likewise, the slope of the negative-going ramp is –2 V/μs, and the output voltage is calculated as follows:

$$V_{out} = (-2\text{ V}/\mu\text{s})2.2\ \mu\text{s} = -4.4\text{ V}$$

Finally, the output voltage waveform is graphed relative to the input as shown in Figure 21–21.

FIGURE 21–21

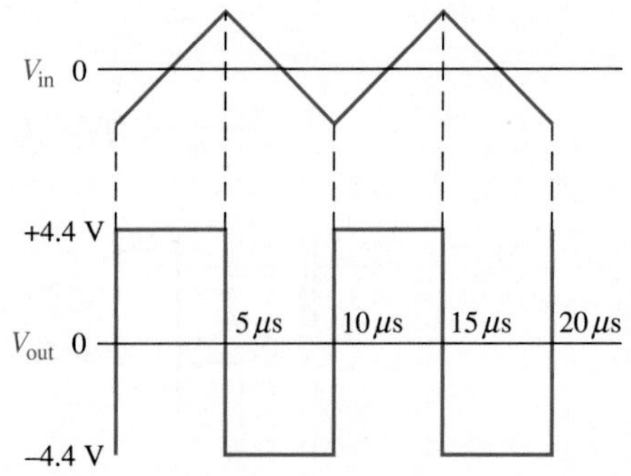

Related Exercise What would the output voltage be if the feedback resistor in Figure 21–20 is changed to 3.3 kΩ?

SECTION 21–3 REVIEW

1. What is the feedback element in an op-amp integrator?
2. For a constant input voltage to an integrator, why is the voltage across the capacitor linear?
3. What is the feedback element in an op-amp differentiator?
4. How is the output of a differentiator related to the input?

21–4 ▪ OSCILLATORS

Oscillators were introduced in Chapter 19 and the principles of operation were discussed. Also, several types of oscillator circuits using discrete transistors were covered. In this section, several types of oscillators implemented with op-amps are introduced.

After completing this section, you should be able to

- **Discuss the operation of several types of op-amp oscillators**
  - □ Identify a Wien-bridge oscillator and analyze its operation
  - □ Identify a triangular-wave oscillator and analyze its operation
  - □ Identify a voltage-controlled oscillator (VCO) and analyze its operation
  - □ Identify a relaxation oscillator and analyze its operation

The Wien-Bridge Oscillator

One type of sine wave oscillator is the *Wien-bridge oscillator*. A fundamental part of the Wien-bridge oscillator is a lead-lag network like that shown in Figure 21–22(a). R_1 and C_1 together form the lag portion of the network; R_2 and C_2 form the lead portion. The operation of this circuit is as follows. At lower frequencies, the lead network dominates due to the high reactance of C_2. As the frequency increases, X_{C2} decreases, thus allowing the output voltage to increase. At some specified frequency, the response of the lag network takes over, and the decreasing value of X_{C1} causes the output voltage to decrease.

So, we have a response curve like that shown in Figure 21–22(b) where the output voltage peaks at a frequency f_r. At this point, the attenuation (V_{out}/V_{in}) of the network is ⅓ if $R_1 = R_2$ and $X_{C1} = X_{C2}$ as stated by the following equation:

$$\frac{V_{out}}{V_{in}} = \frac{1}{3} \qquad \textbf{(21–8)}$$

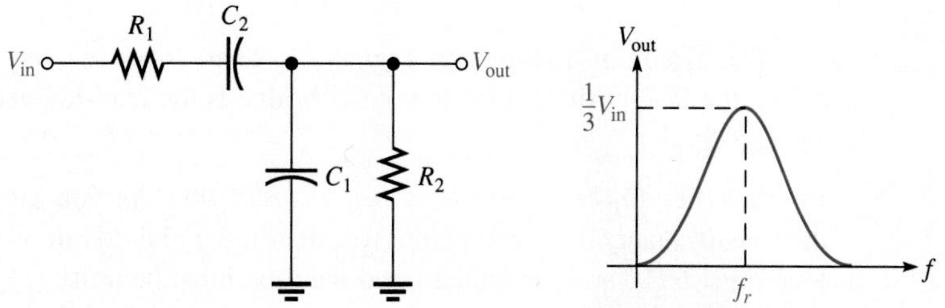

(a) Network (b) Response curve

FIGURE 21–22
A lead-lag network.

The formula for the resonant frequency is

$$f_r = \frac{1}{2\pi RC} \tag{21–9}$$

To summarize, the lead-lag network has a resonant frequency, f_r, at which the phase shift through the network is 0° and the attenuation is 1/3. Below f_r, the lead network dominates and the output leads the input. Above f_r, the lag network dominates and the output lags the input.

The lead-lag network is used in the positive feedback loop of an op-amp, as shown in Figure 21–23(a). A voltage divider is used in the negative feedback loop.

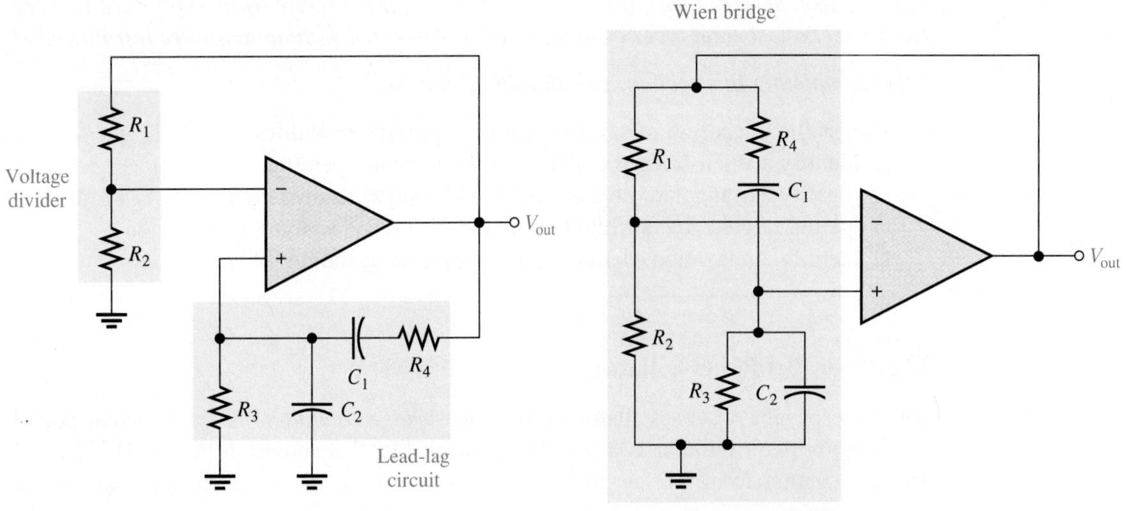

(a)

(b) Bridge network combines voltage divider and lead-lag circuit

FIGURE 21–23
Two ways to draw the schematic of a Wien-bridge oscillator.

The Basic Circuit The Wien-bridge oscillator circuit can be viewed as a noninverting amplifier configuration with the input signal fed back from the output through the lead-lag network. Recall that the closed-loop gain of the amplifier is determined by the voltage divider.

$$A_{cl} = \frac{1}{B} = \frac{1}{R_2/(R_1+R_2)} = \frac{R_1+R_2}{R_2}$$

The circuit is redrawn in Figure 21–23(b) to show that the op-amp is connected across the Wien bridge. One leg of the bridge is the lead-lag network, and the other is the voltage divider.

Positive Feedback Conditions for Oscillation As you know, for the circuit to produce a sustained sine wave output (oscillate), the phase shift around the positive feedback loop must be 0° and the gain around the loop must be unity (1). The 0° phase-shift condition is met when the frequency is f_r, because the phase shift through the lead-lag network is 0° and there is no inversion from the noninverting input (+) of the op-amp to the output. This is shown in Figure 21–24(a).

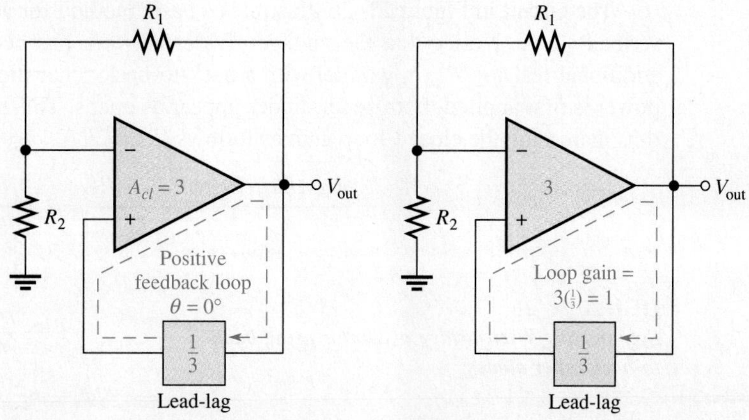

(a) The phase shift around the loop is 0°. (b) The voltage gain around the loop is 1.

FIGURE 21–24
Conditions for oscillation in the Wien-bridge circuit.

The unity-gain condition in the feedback loop is met when

$$A_{cl} = 3$$

This offsets the ⅓ attenuation of the lead-lag network, thus making the gain around the positive feedback loop equal to 1, as depicted in Figure 21–24(b). To achieve a closed-loop gain of 3,

$$R_1 = 2R_2$$

Then

$$A_{cl} = \frac{R_1 + R_2}{R_2} = \frac{2R_2 + R_2}{R_2} = \frac{3R_2}{R_2} = 3$$

Start-Up Conditions Initially, the closed-loop gain of the amplifier must be more than 1 ($A_{cl} > 3$) until the output signal builds up to a desired level. The gain must then decrease to 1 so that the output signal stays at the desired level. This is illustrated in Figure 21–25.

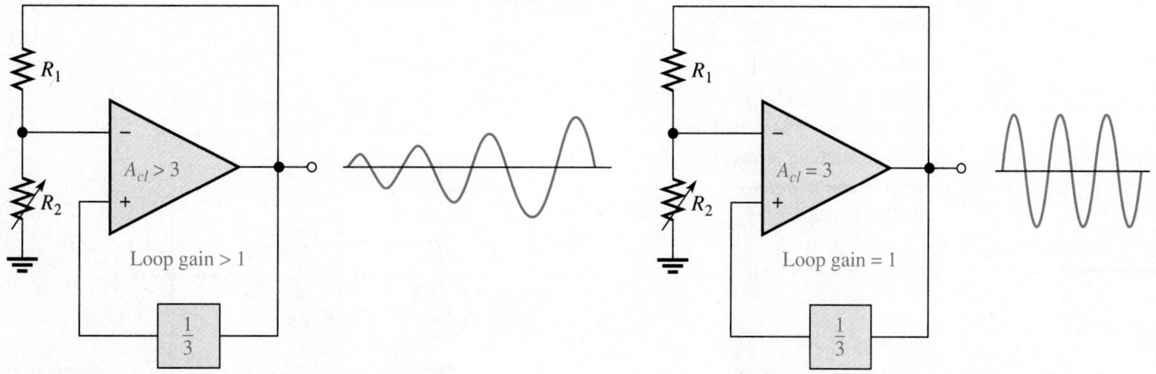

(a) Loop gain greater than 1 causes output to build up. (b) Loop gain of 1 causes a sustained constant output.

FIGURE 21–25
Oscillator start-up conditions.

The circuit in Figure 21–26 illustrates a basic method for achieving the condition described above. Notice that the voltage-divider network has been modified to include an additional resistor, R_3, in parallel with a back-to-back zener diode arrangement. When dc power is first applied, both zener diodes appear as opens. This places R_3 in series with R_1, thus increasing the closed-loop gain as follows ($R_1 = 2R_2$).

$$A_{cl} = \frac{R_1 + R_2 + R_3}{R_2} = \frac{3R_2 + R_3}{R_2} = 3 + \frac{R_3}{R_2}$$

FIGURE 21–26
Self-starting Wien-bridge oscillator using back-to-back zener diodes.

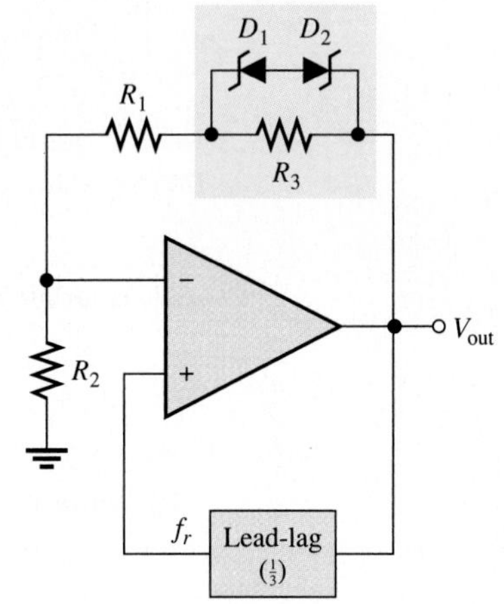

Initially when the power is turned on, a small positive feedback signal develops from noise or turn-on transients. The lead-lag network permits only a signal with a frequency equal to f_r to appear in-phase on the noninverting input. This feedback signal is amplified and continually reinforced, resulting in a buildup of the output voltage. When the output signal reaches the zener breakdown voltage, the zeners conduct and effectively short out R_3. This lowers the amplifier's closed-loop gain to 3. At this point the output signal levels off and the oscillation is sustained. (Incidentally, the frequency of oscillation can be adjusted by using gang-tuned capacitors in the lead-lag network.)

Another method sometimes used to ensure self-starting employs a tungsten lamp in the voltage divider, as shown in Figure 21–27. When the power is first turned on, the resistance of the lamp is lower than its nominal value. This keeps the negative feedback small and makes the closed-loop gain of the amplifier greater than 3. As the output voltage builds up, the voltage across the tungsten lamp—and thus its current—increases. As a result, the lamp resistance increases until it reaches a value equal to one-half the feedback resistance. At this point the closed-loop gain is 3, and the output is sustained at a constant level.

FIGURE 21–27
Self-starting Wien-bridge oscillator using a tungsten lamp in the negative feedback loop.

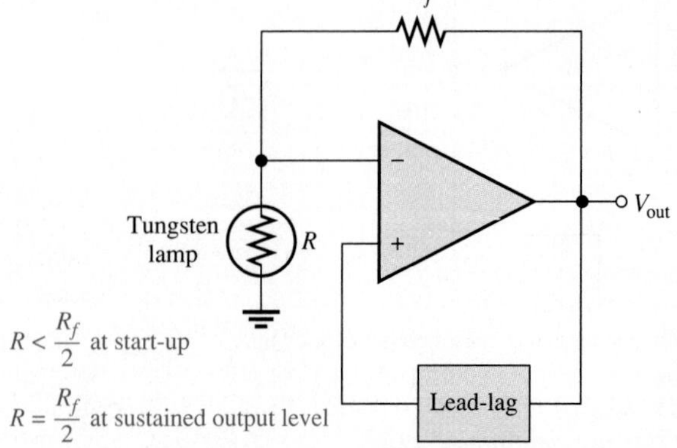

EXAMPLE 21–8 Determine the frequency of oscillation for the Wien-bridge oscillator in Figure 21–28. Also, verify that oscillations will start and then continue when the output signal reaches 5.4 V.

FIGURE 21–28

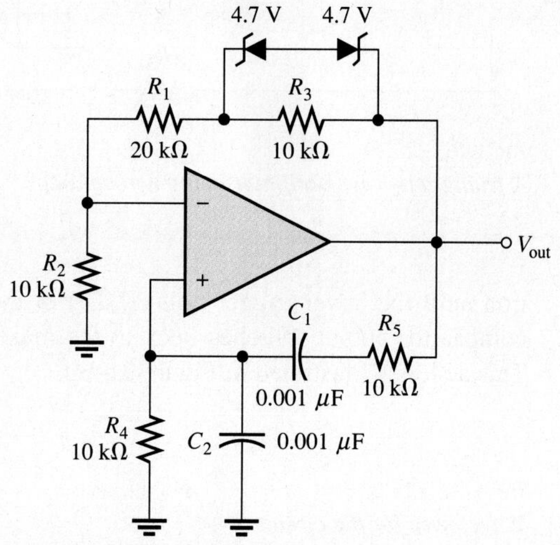

Solution For the lead-lag network, $R_4 = R_5 = R = 10\ \text{k}\Omega$ and $C_1 = C_2 = C = 0.001\ \mu\text{F}$. The frequency is

$$f_r = \frac{1}{2\pi RC} = \frac{1}{2\pi(10\ \text{k}\Omega)(0.001\ \mu\text{F})} = 15.9\ \text{kHz}$$

Initially, the closed-loop gain is

$$A_{cl} = \frac{R_1 + R_2 + R_3}{R_2} = \frac{40\ \text{k}\Omega}{10\ \text{k}\Omega} = 4$$

Since $A_{cl} > 3$, the start-up condition is met.

When the output reaches 5.4 V (4.7 V + 0.7 V), the zeners conduct (their forward resistance is assumed small, compared to 10 kΩ), and the unity closed-loop gain is reached. Thus, oscillation is sustained.

$$A_{cl} = \frac{R_1 + R_2}{R_2} = \frac{30\ \text{k}\Omega}{10\ \text{k}\Omega} = 3$$

Related Exercise What change is required in the oscillator in Figure 21–28 to produce an output with an amplitude of 6.8 V?

A Triangular-Wave Oscillator

One practical implementation of a triangular-wave oscillator utilizes an op-amp comparator as shown in Figure 21–29. The operation is as follows. To begin, assume that the output voltage of the comparator is at its maximum negative level. This output is connected to the inverting input of the integrator through R_1, producing a positive-going ramp on the output of the integrator. When the ramp voltage reaches the upper trigger point (UTP), the comparator switches to its maximum positive level. This positive level causes the integrator ramp to change to a negative-going direction. The ramp continues in this direc-

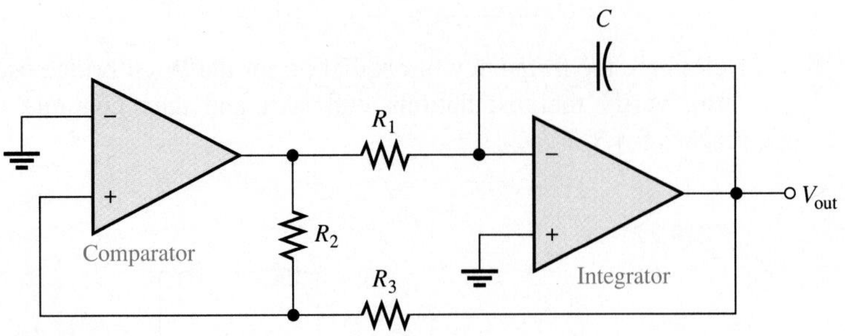

FIGURE 21–29
A triangular-wave oscillator using two op-amps.

tion until the lower trigger point (LTP) of the comparator is reached. At this point, the comparator output switches back to the maximum negative level and the cycle repeats. This action is illustrated in Figure 21–30.

FIGURE 21–30
Waveforms for the circuit in Figure 21–29.

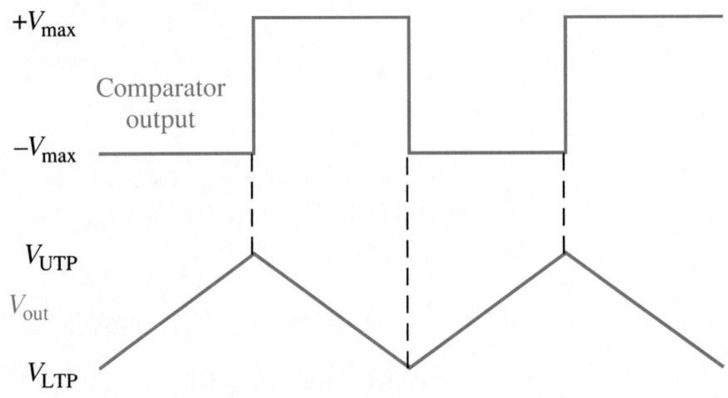

Since the comparator produces a square-wave output, the circuit in Figure 21–29 can be used as both a triangular-wave oscillator and a square-wave oscillator. Devices of this type are commonly known as *function generators* because they produce more than one output function. The output amplitude of the square wave is set by the output swing of the comparator, and the resistors R_2 and R_3 set the amplitude of the triangular output by establishing the UTP and LTP voltages according to the following formulas:

$$V_{UTP} = +V_{max}\left(\frac{R_3}{R_2}\right) \tag{21–10}$$

$$V_{LTP} = -V_{max}\left(\frac{R_3}{R_2}\right) \tag{21–11}$$

where the comparator output levels, $+V_{max}$ and $-V_{max}$, are equal. The frequency of both waveforms depends on the R_1C time constant as well as the amplitude-setting resistors, R_2 and R_3. By varying R_1, the frequency of oscillation can be adjusted without changing the output amplitude.

$$f = \frac{1}{4R_1C}\left(\frac{R_2}{R_3}\right) \tag{21–12}$$

EXAMPLE 21–9 Determine the frequency of the circuit in Figure 21–31. To what value must R_1 be changed to make the frequency 20 kHz?

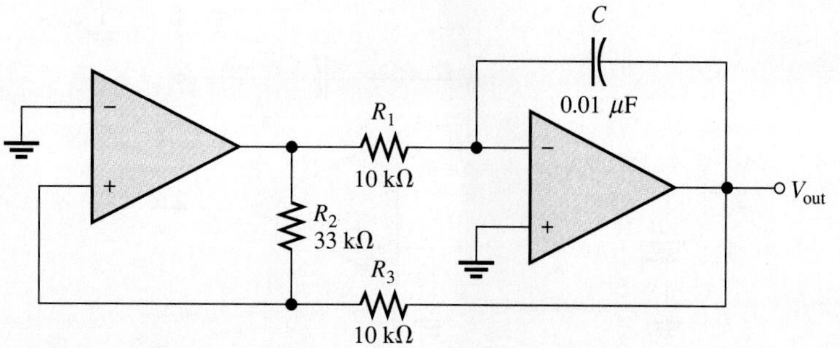

FIGURE 21–31

Solution

$$f = \frac{1}{4R_1C}\left(\frac{R_2}{R_3}\right) = \left(\frac{1}{4(10\text{ k}\Omega)(0.01\ \mu\text{F})}\right)\left(\frac{33\text{ k}\Omega}{10\text{ k}\Omega}\right) = 8.25\text{ kHz}$$

To make $f = 20$ kHz,

$$R_1 = \frac{1}{4fC}\left(\frac{R_2}{R_3}\right) = \left(\frac{1}{4(20\text{ kHz})(0.01\ \mu\text{F})}\right)\left(\frac{33\text{ k}\Omega}{10\text{ k}\Omega}\right) = 4.13\text{ k}\Omega$$

Related Exercise What is the amplitude of the triangular wave in Figure 21–31 if the comparator output is ±10 V?

A Voltage-Controlled Sawtooth Oscillator (VCO)

The voltage-controlled oscillator (VCO) is an oscillator whose frequency can be changed by a variable dc control voltage. VCOs can be either sinusoidal or nonsinusoidal. One way to build a voltage-controlled sawtooth oscillator is with an op-amp integrator that uses a switching device (PUT) in parallel with the feedback capacitor to terminate each ramp at a prescribed level and effectively "reset" the circuit. Figure 21–32(a) shows the implementation.

The PUT is a programmable unijunction transistor with an anode, a cathode, and a gate terminal. The gate is always biased positively with respect to the cathode. When the anode voltage exceeds the gate voltage by approximately 0.7 V, the PUT turns on and acts as a forward-biased diode. When the anode voltage falls below this level, the PUT turns off. Also, the current must be above the holding value to maintain conduction.

The operation of the sawtooth oscillator begins when the negative dc input voltage, $-V_{IN}$, produces a positive-going ramp on the output. During the time that the ramp is increasing, the circuit acts as a regular integrator. The PUT triggers on when the output ramp (at the anode) exceeds the gate voltage by 0.7 V. The gate is set to the approximate desired sawtooth peak voltage. When the PUT turns on, the capacitor rapidly discharges, as shown in Figure 21–32(b). The capacitor does not discharge completely to zero because of the PUT's forward voltage, V_F. Discharge continues until the PUT current falls below the holding value. At this point, the PUT turns off and the capacitor begins to charge again, thus generating a new output ramp. The cycle continually repeats, and the resulting output is a repetitive sawtooth waveform, as shown. The sawtooth amplitude and period can be adjusted by varying the PUT gate voltage.

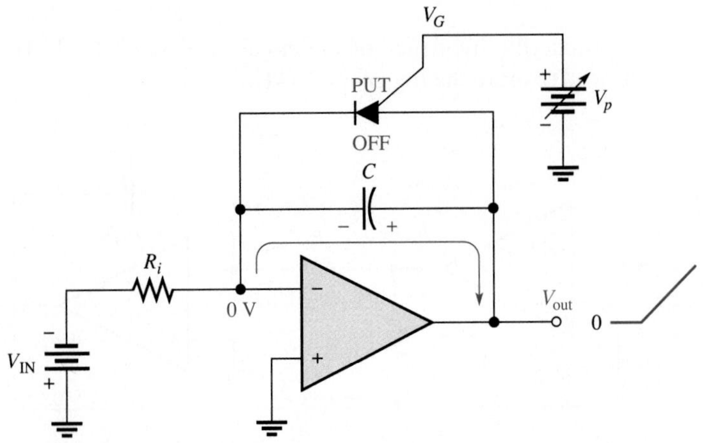

(a) Initially, the capacitor charges, the output ramp begins, and the PUT is off.

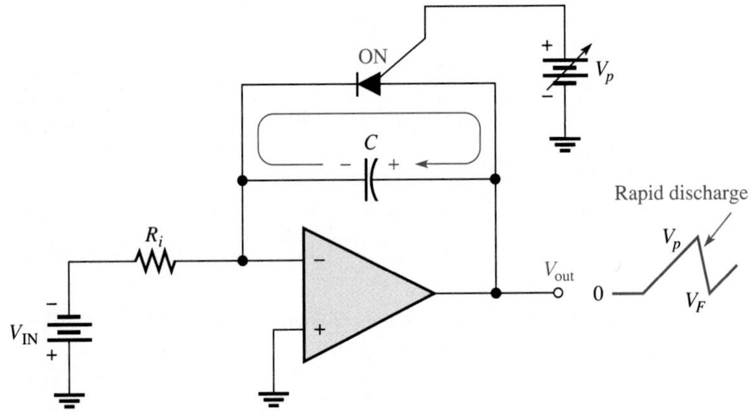

(b) The capacitor rapidly discharges when the PUT momentarily turns on.

FIGURE 21–32
Voltage-controlled sawtooth oscillator operation.

The frequency is determined by the R_iC time constant of the integrator and the peak voltage set by the PUT. Recall that the charging rate of the capacitor is V_{IN}/R_iC. The time it takes the capacitor to charge from V_F to V_p is the period, T, of the sawtooth (neglecting the rapid discharge time).

$$T = \frac{V_p - V_F}{|V_{IN}|/R_iC} \qquad \textbf{(21–13)}$$

From $f = 1/T$,

$$f = \frac{|V_{IN}|}{R_iC}\left(\frac{1}{V_p - V_F}\right) \qquad \textbf{(21–14)}$$

EXAMPLE 21–10

(a) Find the amplitude and frequency of the sawtooth output in Figure 21–33. Assume that the forward PUT voltage V_F is approximately 1 V.

(b) Sketch the output waveform.

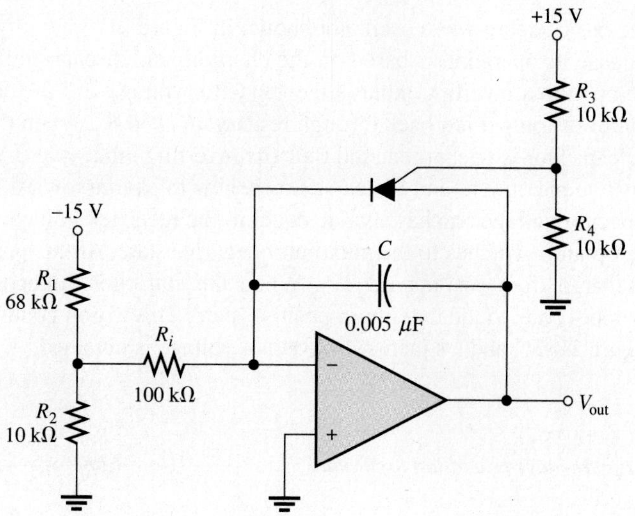

FIGURE 21–33

Solution

(a) First, find the gate voltage in order to establish the approximate voltage at which the PUT turns on.

$$V_G = \frac{R_4}{R_3 + R_4}(+V) = \frac{10\text{ k}\Omega}{20\text{ k}\Omega}(15\text{ V}) = 7.5\text{ V}$$

This voltage plus 0.7 V by which the PUT anode must exceed the gate voltage sets the approximate maximum peak value of the sawtooth output.

$$V_p = 7.5\text{ V} + 0.7\text{ V} = 8.2\text{ V}$$

The minimum peak value (low point) is

$$V_F \cong 1\text{ V}$$

The period and frequency are determined as follows:

$$V_{IN} = \frac{R_2}{R_1 + R_2}(-V) = \frac{10\text{ k}\Omega}{78\text{ k}\Omega}(-15\text{ V}) = -1.92\text{ V}$$

$$T = \frac{V_p - V_F}{|V_{IN}|/R_iC} = \frac{8.2\text{ V} - 1\text{ V}}{1.92\text{ V}/(100\text{ k}\Omega)(0.005\ \mu\text{F})} = 1.88\text{ ms}$$

$$f = \frac{1}{1.88\text{ ms}} = 532\text{ Hz}$$

(b) The output waveform is shown in Figure 21–34.

FIGURE 21–34
Output of the circuit in Figure 21–33.

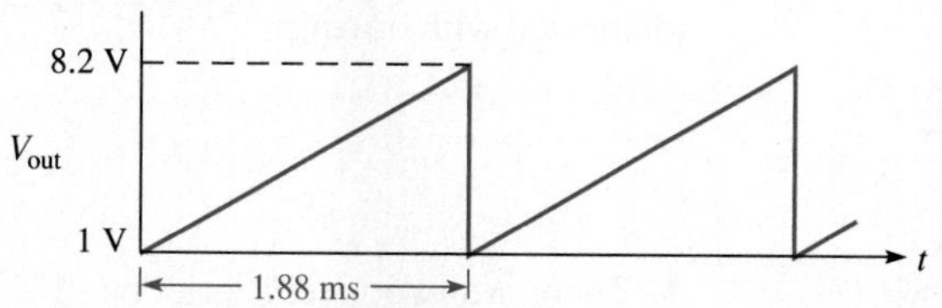

Related Exercise If R_i is changed to 56 kΩ in Figure 21–33, what is the frequency?

A Square-Wave Relaxation Oscillator

The basic square-wave oscillator shown in Figure 21–35 is a type of relaxation oscillator because its operation is based on the charging and discharging of a capacitor. Notice that the op-amp's inverting input is the capacitor voltage and the noninverting input is a portion of the output fed back through resistors R_2 and R_3. When the circuit is first turned on, the capacitor is uncharged, and thus the inverting input is at 0 V. This makes the output a positive maximum, and the capacitor begins to charge toward V_{out} through R_1. When the capacitor voltage reaches a value equal to the feedback voltage on the noninverting input, the op-amp switches to the maximum negative state. At this point, the capacitor begins to discharge from $+V_f$ toward $-V_f$. When the capacitor voltage reaches $-V_f$, the op-amp switches back to the maximum positive state. This action continues to repeat, as shown in Figure 21–36, and a square-wave output voltage is obtained.

FIGURE 21–35
A square-wave relaxation oscillator.

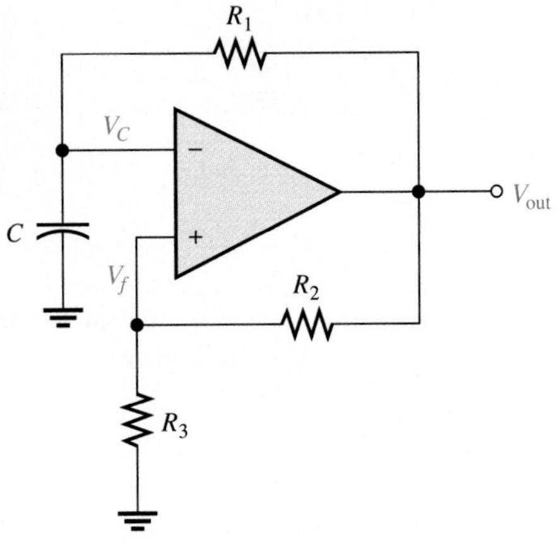

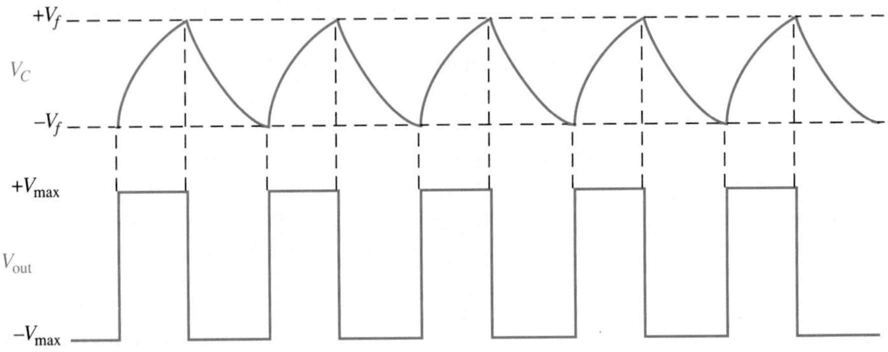

FIGURE 21–36
Waveforms for the square-wave relaxation oscillator.

The Hartley, Colpitts, and Clapp oscillators discussed in Chapter 19 can also be implemented with op-amps.

SECTION 21–4 REVIEW

1. There are two feedback loops in a Wien-bridge oscillator. What is the purpose of each?
2. What is a VCO, and basically, what does it do?
3. Upon what principle does a relaxation oscillator operate?

21–5 ■ ACTIVE FILTERS

Filters are usually categorized by the manner in which the output voltage varies with the frequency of the input voltage. The categories of active filters that we will examine in this section are low-pass, high-pass, and band-pass.

After completing this section, you should be able to

- **Recognize and evaluate basic op-amp filters**
  - ☐ Evaluate single-pole and two-pole low-pass filters
  - ☐ Evaluate single-pole and two-pole high-pass filters
  - ☐ Determine the resonant frequency of a certain type of band-pass filter

Low-Pass Active Filters

Figure 21–37 shows a basic **active filter** and its response curve. Notice that the input circuit is a single low-pass *RC* network, and unity gain is provided by the op-amp with a negative feedback loop. Simply stated, this is a voltage-follower with an *RC* filter between the input signal and the noninverting input.

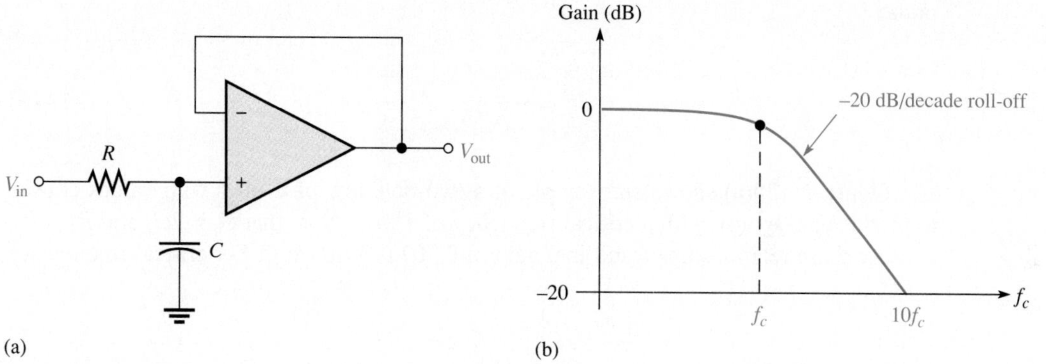

FIGURE 21–37
Single-pole, active low-pass filter and response curve.

The voltage at the noninverting input, *V*+, is as follows:

$$V+ = \left(\frac{X_C}{\sqrt{R^2 + X_C^2}}\right)V_{in}$$

Since the gain of the op-amp is 1, the output voltage is equal to *V*+:

$$V_{out} = \left(\frac{X_C}{\sqrt{R^2 + X_C^2}}\right)V_{in} \qquad \textbf{(21–15)}$$

A filter with one *RC* network that produces a –20 dB/decade **roll-off** beginning at f_c is said to be a *single-pole* or *first-order filter*. The term "–20 dB/decade" means that the voltage gain decreases by ten times (–20 dB) when the frequency increases by ten times (**decade**).

Low-Pass Two-pole Filters There are several types of active filters and they can have varying numbers of **poles,** but we will use a two-pole filter to illustrate. Figure 21–38(a) shows a two-pole (second-order) low-pass filter. Since each *RC* network in a filter is considered to be one-pole, the two-pole filter uses two *RC* networks to produce a roll-off rate

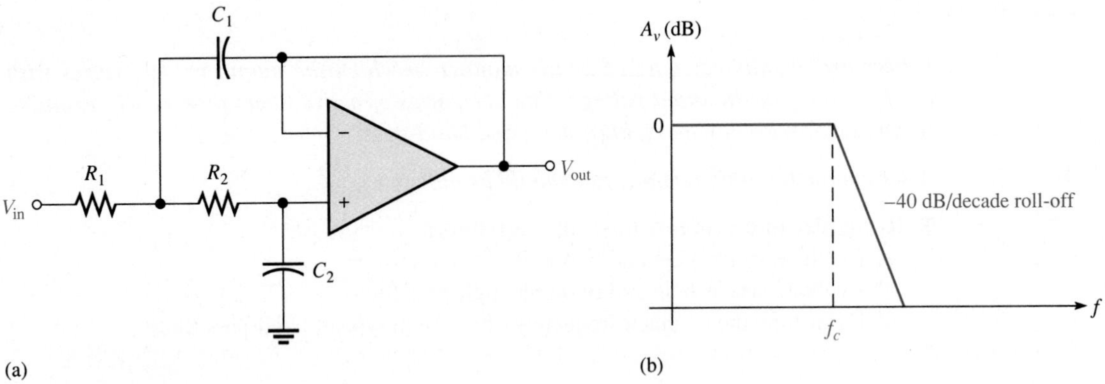

FIGURE 21–38
Two-pole, active low-pass filter and its ideal response curve.

of –40 dB/decade, as indicated in Figure 21–38(b). The active filter in Figure 21–38 has unity gain up to near f_c because the op-amp is connected as a voltage-follower.

One of the RC networks in Figure 21–38(a) is formed by R_1 and C_1, and the other by R_2 and C_2. The critical frequency of this filter can be calculated using the following formula:

$$f_c = \frac{1}{2\pi\sqrt{R_1 R_2 C_1 C_2}} \tag{21–16}$$

Figure 21–39(a) shows an example of a two-pole low-pass filter with values chosen to produce a response with a critical frequency of 1 kHz. Note that $C_1 = 2C_2$ and $R_1 = R_2$, because these relationships result in a gain of 0.707 (–3 dB) at f_c. For critical frequencies

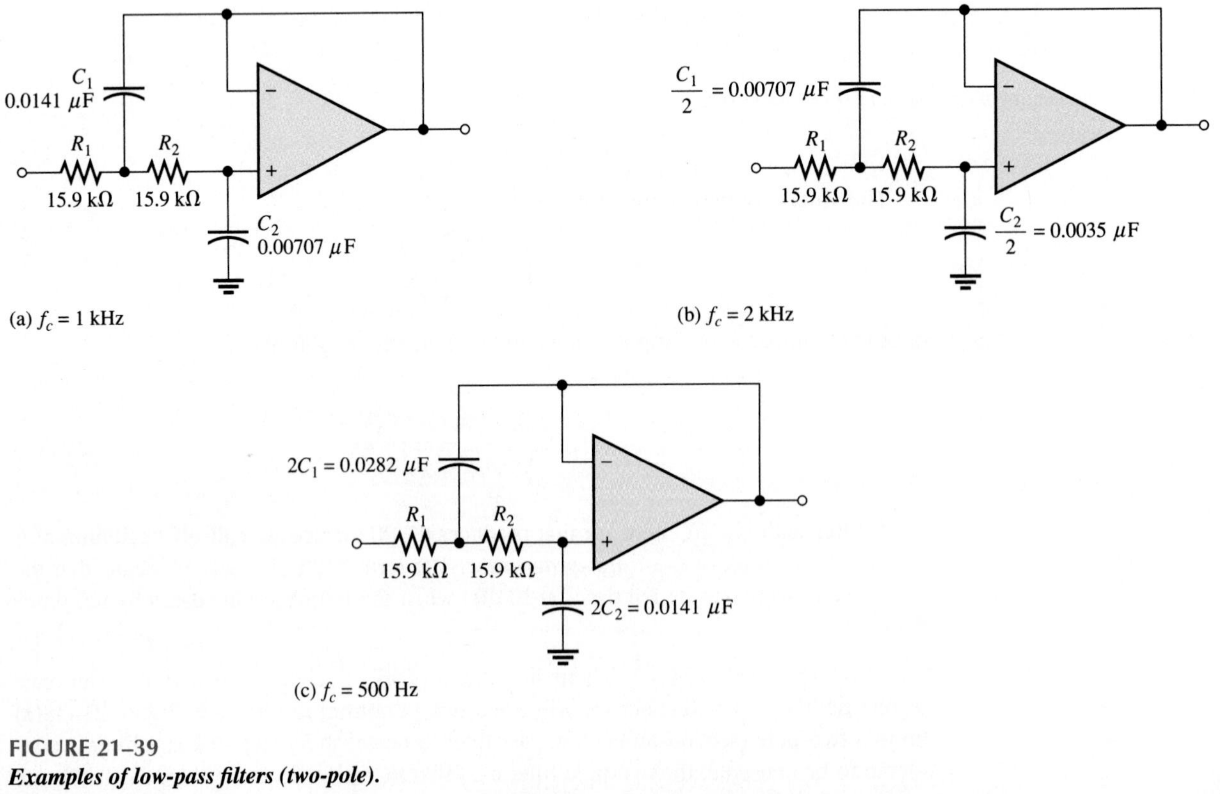

FIGURE 21–39
Examples of low-pass filters (two-pole).

other than 1 kHz, the capacitance values can be scaled inversely with the frequency. For example, as shown in Figure 21–39(b) and (c), to get a 2 kHz filter, halve the values of C_1 and C_2; for a 500 Hz filter, double the values.

EXAMPLE 21–11 Calculate the capacitance values required to produce a 3 kHz critical frequency in the low-pass filter of Figure 21–40.

FIGURE 21–40

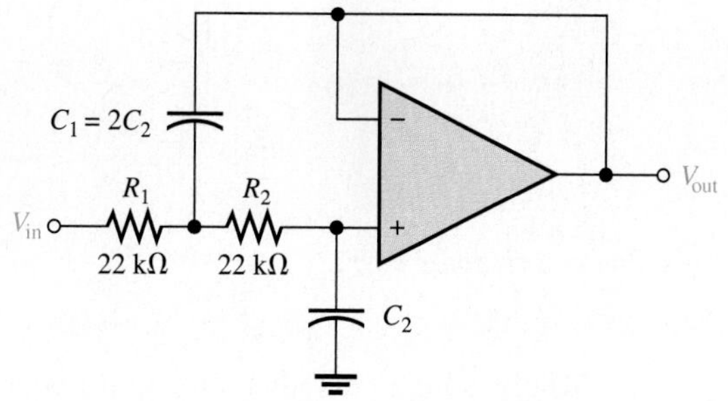

Solution The resistor values have already been set at 22 kΩ each. Since these differ from the 1 kHz reference filter, you cannot use the scaling method to get the capacitance values. Use Equation (21–16):

$$f_c = \frac{1}{2\pi\sqrt{R_1R_2C_1C_2}}$$

Then square both sides:

$$f_c^2 = \frac{1}{4\pi^2R_1R_2C_1C_2}$$

Since $C_1 = 2C_2$ and $R_1 = R_2 = R$,

$$f_c^2 = \frac{1}{4\pi^2R^2(2C_2^2)}$$

Solve for C_2, and then determine C_1 as follows:

$$C_2^2 = \frac{1}{8\pi^2R^2f_c^2}$$

$$C_2 = \frac{1}{\sqrt{2}\,2\pi Rf_c} = \frac{0.707}{2\pi Rf_c} = \frac{0.707}{2\pi(22\text{ k}\Omega)(3\text{ kHz})} = 0.0017\ \mu\text{F}$$

$$C_1 = 2C_2 = 2(0.0017\ \mu\text{F}) = 0.0034\ \mu\text{F}$$

Related Exercise Determine f_c in Figure 21–40 for $R_1 = R_2 = 27$ kΩ, $C_1 = 0.001\ \mu$F, and $C_2 = 500$ pF.

High-Pass Active Filters

In Figure 21–41(a), a high-pass active filter with a 20 dB/decade roll-off is shown. Notice that the input circuit is a single high-pass *RC* network and that unity gain is provided by the op-amp with negative feedback. The response curve is shown in Figure 21–41(b).

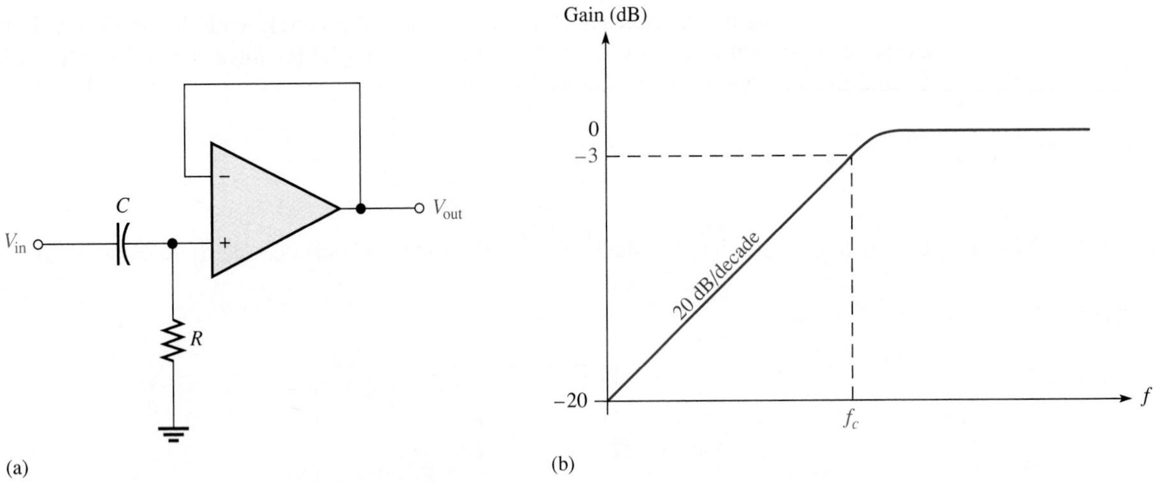

FIGURE 21–41
Single-pole, active high-pass filter and response curve.

Ideally, a high-pass filter passes all frequencies above f_c without limit, as indicated in Figure 21–42(a). In practice, of course, such is not the case. All op-amps inherently have internal *RC* networks that limit the amplifier's response at high frequencies. Such is the case with the active high-pass filter. There is an upper frequency limit to its response, which essentially makes this type of filter a wide band-pass filter rather than a true high-pass filter, as indicated in Figure 21–42(b). In many applications, the internal high-frequency cutoff is so much greater than the filter's f_c that the internal high-frequency cutoff can be neglected.

FIGURE 21–42
High-pass filter response.

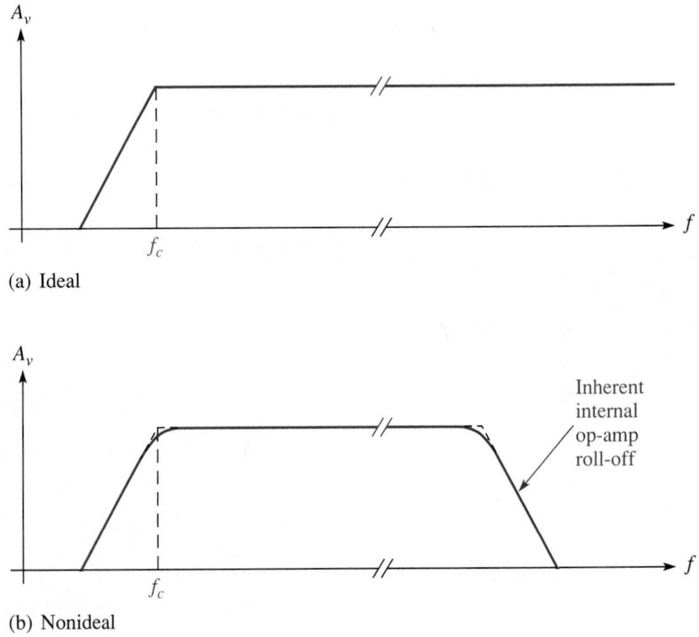

The voltage at the noninverting input is as follows:

$$V+ = \left(\frac{R}{\sqrt{R^2 + X_C^2}} \right) V_{in}$$

Since the op-amp is connected as a voltage-follower with unity gain, the output voltage is the same as $V+$:

$$V_{out} = \left(\frac{R}{\sqrt{R^2 + X_C^2}}\right)V_{in} \tag{21–17}$$

If the internal critical frequencies of the op-amp are assumed to be much greater than the desired f_c of the filter, the gain will roll off at 20 dB/decade as shown in Figure 21–42(b). This is a single-pole filter because it has one RC network.

High-Pass Two-Pole Filters Figure 21–43 shows a two-pole active high-pass filter. Notice that it is identical to the corresponding loss-pass type, except for the positions of the resistors and capacitors. This filter has a roll-off rate of 40 dB/decade below f_c, and the critical frequency is the same as for the low-pass filter given in Equation (21–16).

Figure 21–44 shows a two-pole high-pass filter with values chosen to produce a response with a critical frequency of 1 kHz. Note that $R_2 = 2R_1$ and $C_1 = C_2$ because these relationships result in a gain of 0.707 (–3 dB) at f_c. For frequencies other than 1 kHz, the resistance values can be scaled inversely, as was done with the capacitors in the low-pass case.

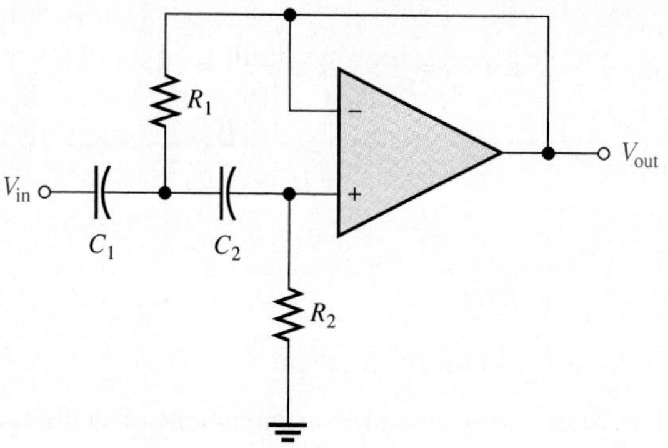

FIGURE 21–43
Two-pole, active high-pass filter.

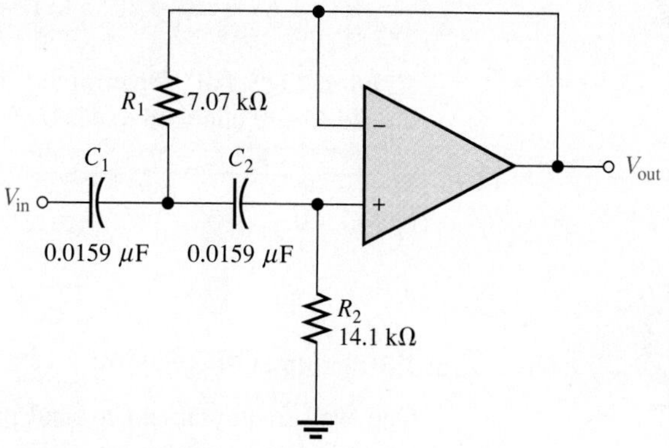

FIGURE 21–44
Two-pole, high-pass filter (f_c = 1 kHz).

EXAMPLE 21–12 For the filter of Figure 21–45, calculate the resistance values required to produce a critical frequency of 5.5 kHz.

FIGURE 21–45

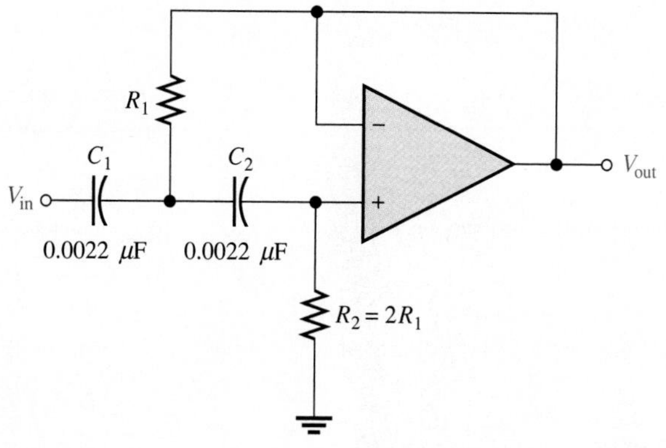

Solution The capacitor values have been preselected to be 0.0022 μF each. Since these differ from the 1 kHz reference filter, you cannot use the scaling method to get the resistor values. Use Equation (21–16) and square both sides.

$$f_c = \frac{1}{2\pi\sqrt{R_1R_2C_1C_2}}$$

$$f_c^2 = \frac{1}{4\pi^2R_1R_2C_1C_2}$$

Since $R_2 = 2R_1$ and $C_1 = C_2 = C$,

$$f_c^2 = \frac{1}{4\pi^2(2R_1^2)C^2}$$

Solve for R_1, and then determine R_2 as follows:

$$R_1^2 = \frac{1}{8\pi^2C^2f_c^2}$$

$$R_1 = \frac{1}{\sqrt{2}\,2\pi Cf_c} = \frac{0.707}{2\pi Cf_c} = \frac{0.707}{2\pi(0.0022\ \mu\text{F})(5.5\ \text{kHz})} = 9.3\ \text{k}\Omega$$

$$R_2 = 2R_1 = 2(9.3\ \text{k}\Omega) = 18.6\ \text{k}\Omega$$

Related Exercise Determine f_c in Figure 21–45 for $R_1 = 9.3$ kΩ and $R_2 = 18.6$ kΩ if C_1 and C_2 are changed to 4700 pF.

Band-Pass Filter Using a High-Pass/Low-Pass Combination

One way to implement a band-pass filter is to use a cascaded arrangement of a high-pass filter followed by a low-pass filter, as shown in Figure 21–46(a). Each of the filters shown is a two-pole configuration so that the roll-off rates of the response curve are ±40 dB/decade, as indicated in the composite response curve of part (b). The critical frequency of each filter is chosen so that the response curves overlap, as indicated. The critical frequency of the high-pass filter is lower than that of the low-pass filter.

The lower frequency, f_{c1}, of the pass band is set by the critical frequency of the high-pass filter. The upper frequency, f_{c2}, of the pass band is the critical frequency of the low-pass filter. Ideally, the center frequency, f_r, of the pass band is the geometric average of f_{c1} and f_{c2}. The following formulas express the three frequencies of the band-pass filter in Figure 21–46:

$$f_{c1} = \frac{1}{\sqrt{2\pi R_1R_2C_1C_2}} \tag{21–18}$$

$$f_{c2} = \frac{1}{\sqrt{2\pi R_3R_4C_3C_4}} \tag{21–19}$$

$$f_r = \sqrt{f_{c1}f_{c2}} \tag{21–20}$$

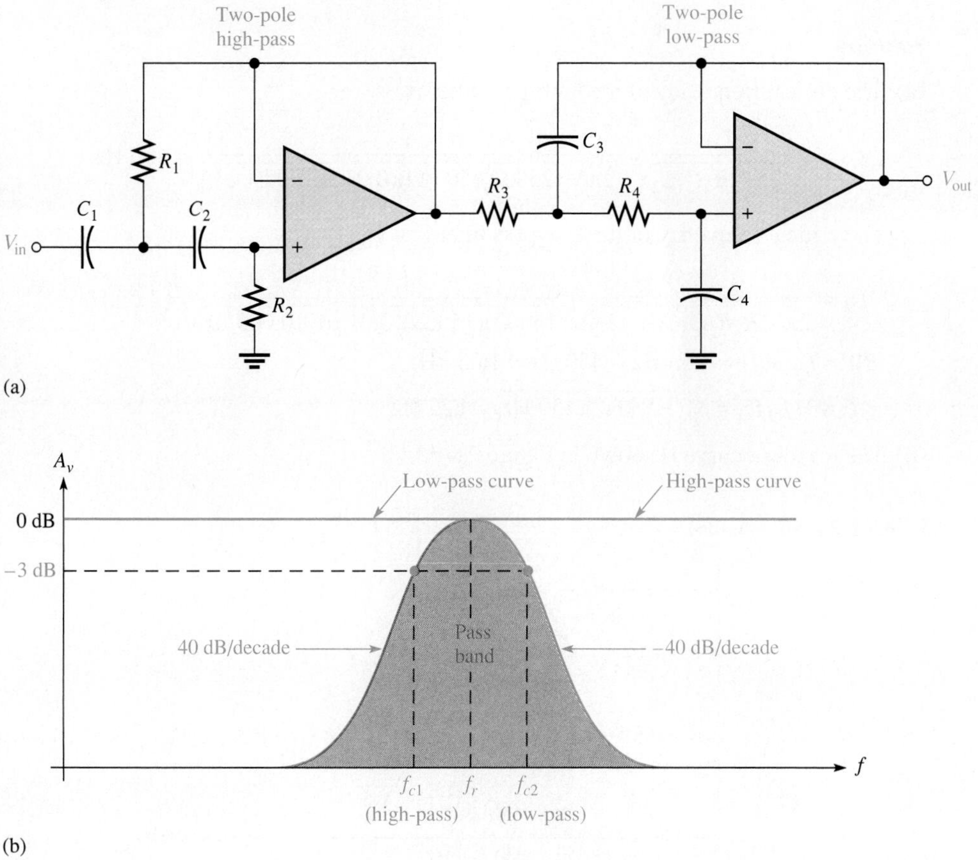

FIGURE 21–46

Band-pass filter formed by combining two-pole, high-pass filter with two-pole, low-pass filter. (The order in which the filters are cascaded does not matter.)

EXAMPLE 21–13 **(a)** Determine the bandwidth and center frequency for the filter in Figure 21–47.
(b) Sketch the response curve.

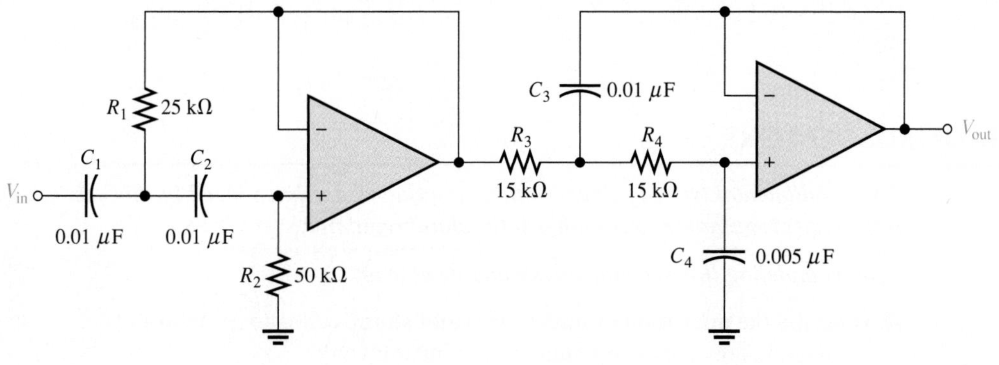

FIGURE 21–47

Solution

(a) The critical frequency of the high-pass filter is

$$f_{c1} = \frac{1}{\sqrt{2\pi R_1 R_2 C_1 C_2}} = \frac{1}{2\pi\sqrt{(25\text{ k}\Omega)(50\text{ k}\Omega)(0.01\ \mu\text{F})(0.01\ \mu\text{F})}} = 450\text{ Hz}$$

The critical frequency of the low-pass filter is

$$f_{c2} = \frac{1}{2\pi\sqrt{R_3 R_4 C_3 C_4}} = \frac{1}{2\pi\sqrt{(15\text{ k}\Omega)(15\text{ k}\Omega)(0.01\ \mu\text{F})(0.005\ \mu\text{F})}} = 1.5\text{ kHz}$$

$$BW = f_{c2} - f_{c1} = 1.5\text{ kHz} - 450\text{ Hz} = 1.05\text{ kHz}$$

$$f_r = \sqrt{f_{c1} f_{c2}} = \sqrt{(1.5\text{ kHz})(450\text{ Hz})} = 822\text{ Hz}$$

(b) The response curve is shown in Figure 21–48.

FIGURE 21–48

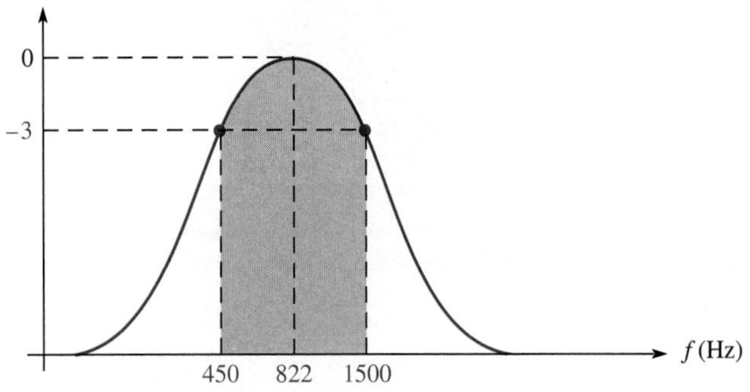

Related Exercise Describe how you can increase the bandwidth of the filter in Figure 21–47 without changing f_{c1}.

SECTION 21–5 REVIEW

1. In terms of circuit components, what does the term *pole* refer to?
2. What type of response characterizes the single-pole low-pass filter?
3. How does a high-pass filter differ in its implementation from a corresponding low-pass version?
4. If the resistance values of a high-pass filter are doubled, what happens to the critical frequency?

21–6 ■ VOLTAGE REGULATORS

Two fundamental types of linear voltage regulators are introduced in this section. One is the series regulator and the other is the shunt regulator.

After completing this section, you should be able to

- **Describe the operation of basic series and shunt voltage regulators**
  - □ Explain how a basic op-amp series regulator works
  - □ Discuss short circuit and overload protection
  - □ Explain how a basic op-amp shunt regulator works

Basic Series Regulator

A simple representation of a series linear regulator is shown in Figure 21–49(a), and the basic components are shown in the block diagram in part (b). Notice that the control element is in series with the load between input and output. The output sample circuit senses a change in the output voltage. The error detector compares the sample voltage with a reference voltage and causes the control element to compensate in order to maintain a constant output voltage.

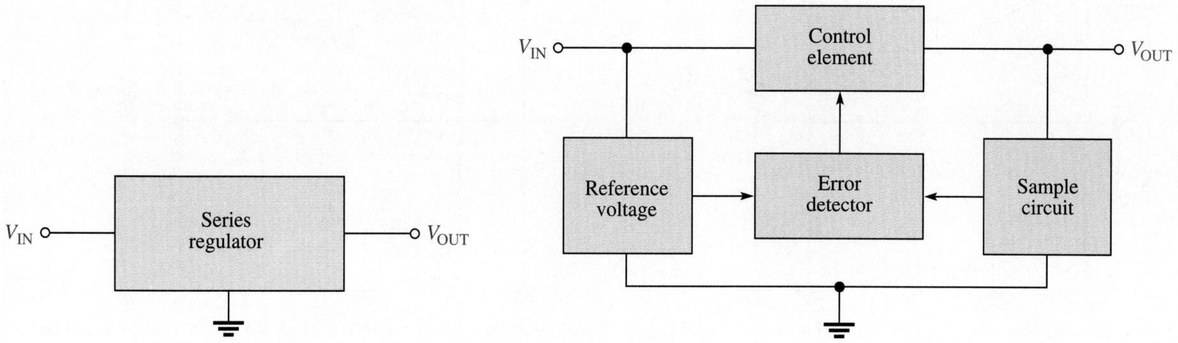

(a) (b)

FIGURE 21–49
Block diagram of simple, three-terminal, series voltage regulator.

Regulating Action A basic op-amp series regulator circuit is shown in Figure 21–50. Its operation is illustrated in Figure 21–51 and is as follows: The resistive voltage divider formed by R_2 and R_3 senses any change in the output voltage. When the output tries to decrease, as shown in part (a), because of a decrease in V_{IN} or because of an increase in I_L (decrease in R_L), a proportional voltage decrease is applied to the op-amp's inverting input by the voltage divider. Since the zener diode holds the other op-amp input at a nearly constant reference voltage V_{REF}, a small difference voltage (error voltage) is developed across the op-amp's inputs. This difference voltage is amplified, and the op-amp's output voltage increases. This increase is applied to the base of Q_1, causing the emitter voltage V_{OUT} to increase until the voltage to the inverting input again equals the reference (zener) voltage. This action offsets the attempted decrease in output voltage, thus keeping it nearly constant, as shown in part (b). The power transistor Q_1 is often used with a heat sink because it must handle all of the load current.

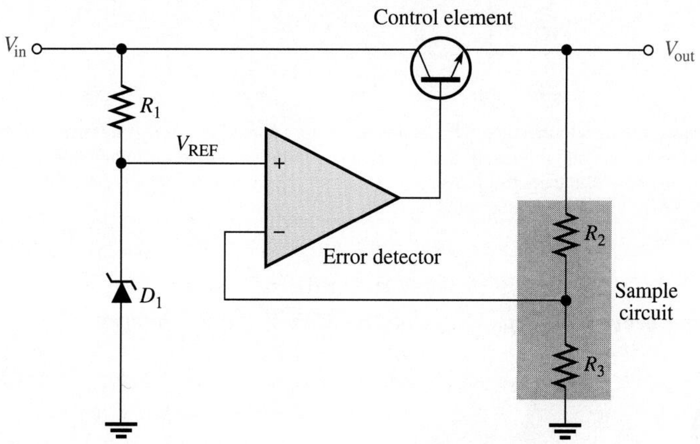

FIGURE 21–50
Basic op-amp series regulator.

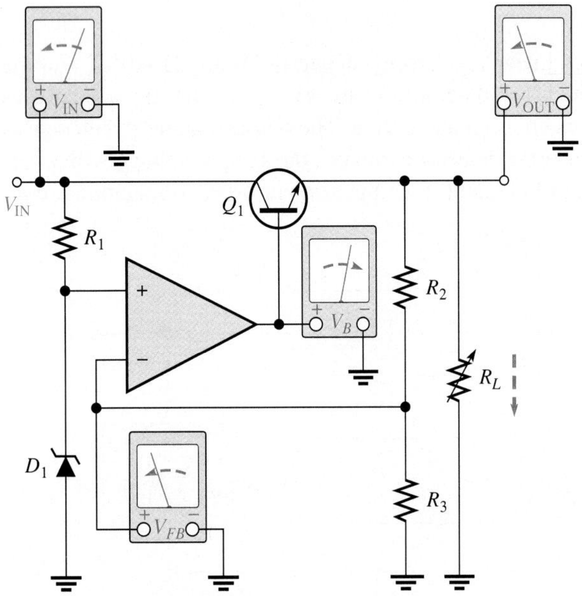

(a) When V_{IN} or R_L decreases, V_{OUT} attempts to decrease, V_{FB} also attempts to decrease, and as a result, V_B attempts to increase, thus compensating for the attempted decrease in V_{OUT} by increasing the Q_1 emitter voltage. Changes in V_{OUT} are exaggerated for illustration.

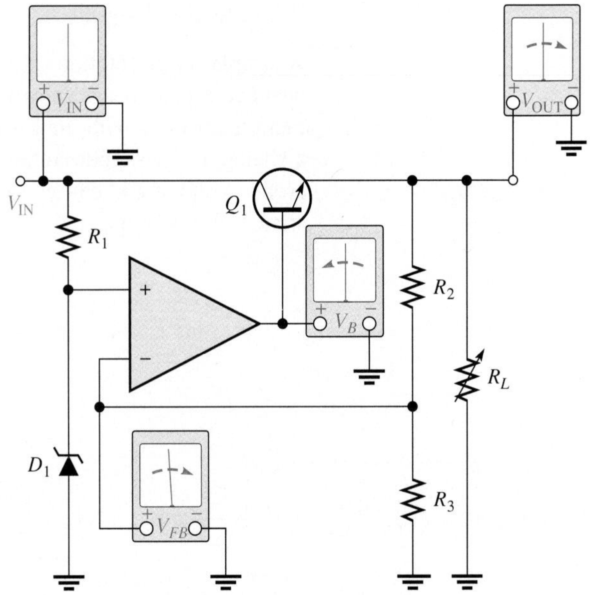

(b) When V_{IN} (or R_L) stabilizes at its new lower value, the voltages return to their original values, thus keeping V_{OUT} constant as a result of the negative feedback.

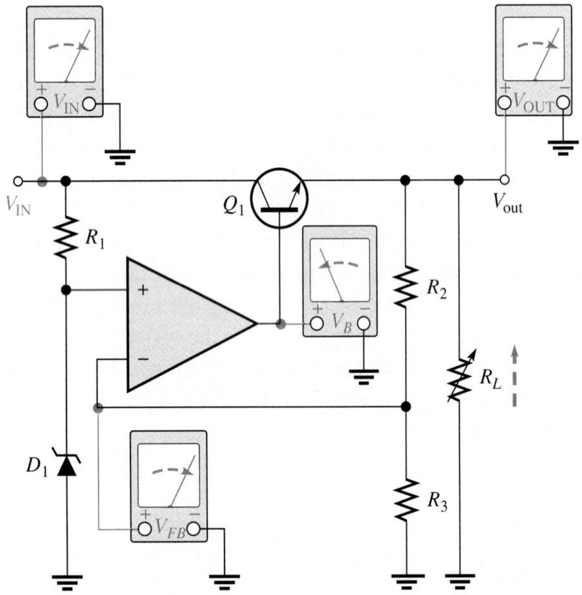

(c) When V_{IN} or R_L increases, V_{OUT} attempts to increase. The feedback voltage, V_{FB}, also attempts to increase, and, as a result, the op-amp's output voltage, V_B, applied to the base of the control transistor, Q_1, attempts to decrease, thus compensating for the attempted increase in V_{OUT} by decreasing the Q_1 emitter voltage.

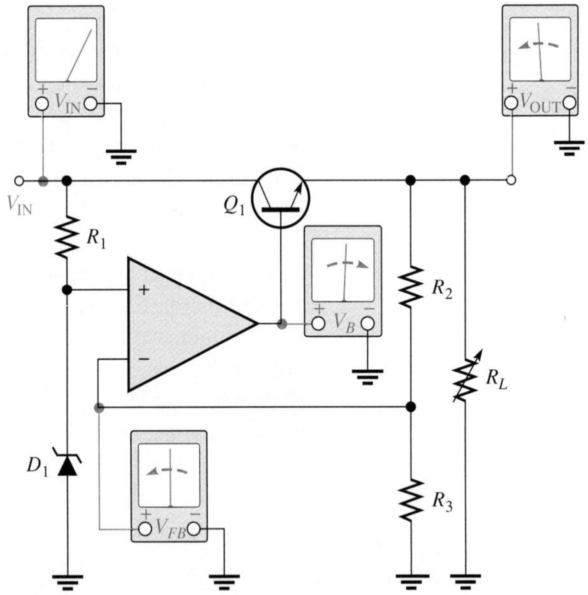

(d) When V_{IN} (or R_L) stabilizes at its new higher value, the voltages return to their original values, thus keeping V_{OUT} constant as a result of the negative feedback.

FIGURE 21–51

Illustration of series regulator action that keeps V_{OUT} constant when V_{IN} or R_L changes.

The opposite action occurs when the output tries to increase, as illustrated in Figure 21–51(c) and (d). Percent regulation was discussed in Chapter 17.

The op-amp in Figure 21–50 is actually connected as a noninverting amplifier in which the reference voltage V_{REF} is the input at the noninverting terminal, and the R_2/R_3 voltage divider forms the negative feedback network. The closed-loop voltage gain is

$$A_{cl} = 1 + \frac{R_2}{R_3} \quad \textbf{(21–21)}$$

Therefore, the regulated output voltage (neglecting the base-emitter voltage of Q_1) is

$$V_{OUT} = \left(1 + \frac{R_2}{R_3}\right)V_{REF} \quad \textbf{(21–22)}$$

From this analysis you can see that the output voltage is determined by the zener voltage and the resistors R_2 and R_3. It is relatively independent of the input voltage, and therefore, regulation is achieved (as long as the input voltage and load current are within specified limits).

EXAMPLE 21–14 Determine the output voltage for the regulator in Figure 21–52.

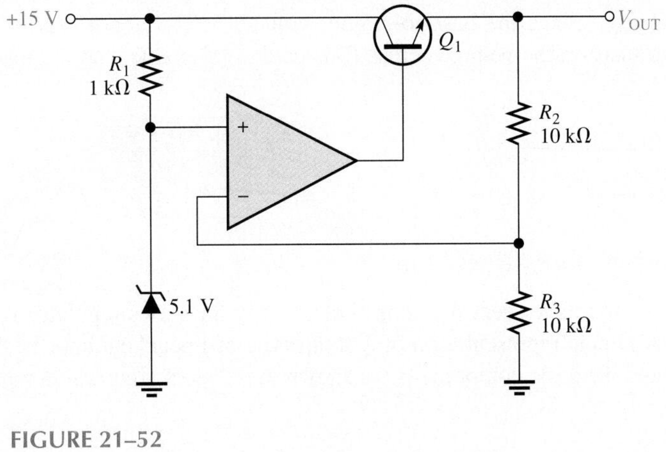

FIGURE 21–52

Solution $V_{REF} = 5.1$ V. Therefore,

$$V_{OUT} = \left(1 + \frac{R_2}{R_3}\right)V_{REF} = \left(1 + \frac{10\text{ k}\Omega}{10\text{ k}\Omega}\right)5.1\text{ V} = (2)5.1\text{ V} = 10.2\text{ V}$$

Related Exercise The following changes are made in the circuit of Figure 21–52: A 3.3 V zener replaces the 5.1 V zener, $R_1 = 1.8$ kΩ, $R_2 = 22$ kΩ, and $R_3 = 18$ kΩ. What is the output voltage?

Short-Circuit or Overload Protection If an excessive amount of load current is drawn, the series-pass transistor can be quickly damaged or destroyed. Most regulators employ some type of protection from excess current in the form of a current-limiting mechanism.

Figure 21–53 shows one method of current limiting to prevent overloads called *constant-current limiting.* The current-limiting circuit consists of transistor Q_2 and resistor R_4.

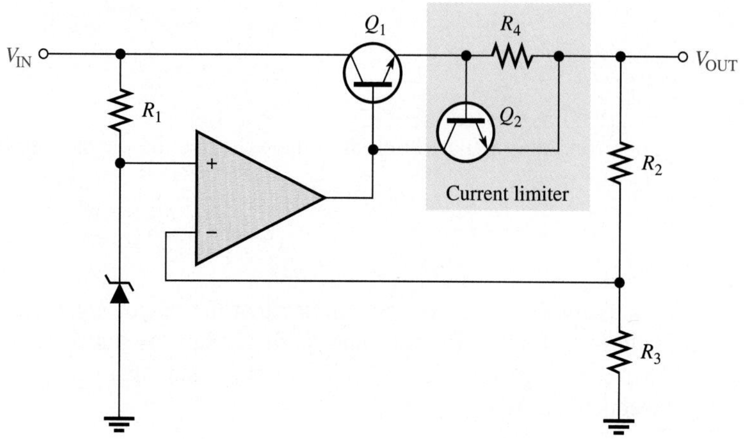

FIGURE 21–53
Series regulator with constant-current limiting.

The load current through R_4 creates a voltage from base to emitter of Q_2. When I_L reaches a predetermined maximum value, the voltage drop across R_4 is sufficient to forward-bias the base-emitter junction of Q_2, thus causing it to conduct. Enough Q_1 base current is diverted into the collector of Q_2 so that I_L is limited to its maximum value $I_{L(max)}$. Since the base-to-emitter voltage of Q_2 cannot exceed about 0.7 V for a silicon transistor, the voltage across R_4 is held to this value, and the load current is limited to

$$I_{L(max)} = \frac{0.7\ V}{R_4} \tag{21–23}$$

Basic Shunt Regulator

As you have seen, the control element in the series regulator is the series-pass transistor. A simple representation of a shunt type of linear regulator is shown in Figure 21–54(a), and the basic components are shown in the block diagram in part (b).

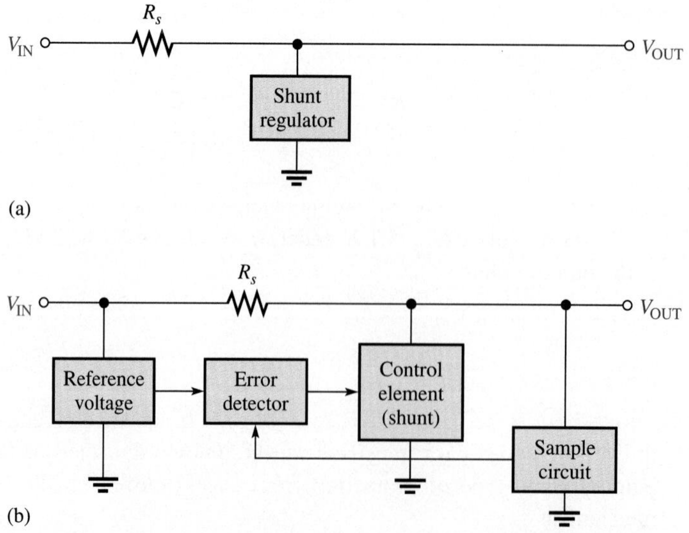

FIGURE 21–54
Block diagrams of simple, three-terminal shunt regulator.

In the basic shunt regulator, the control element is a series resistor (R_1) and a transistor (Q_1) in parallel with the load, as shown in Figure 21–55. The operation of the circuit is similar to that of the series regulator, except that regulation is achieved by controlling the current through the parallel transistor Q_1.

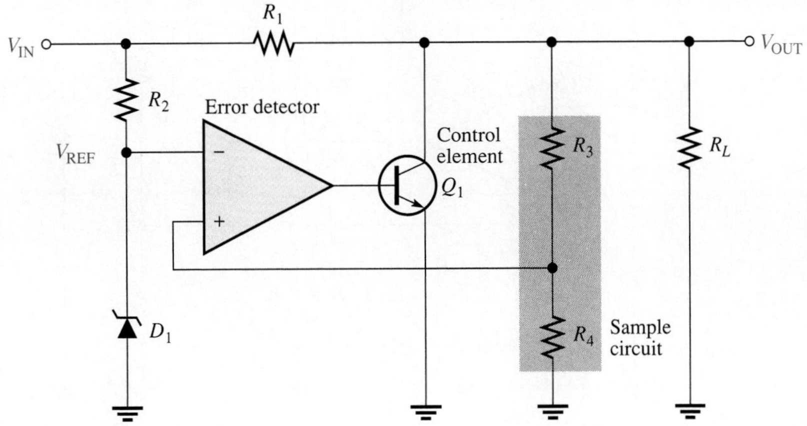

FIGURE 21–55
Basic op-amp shunt regulator.

When the output voltage tries to decrease due to a change in input voltage as shown in Figure 21–56(a), load current, or temperature, the attempted decrease is sensed by R_3 and R_4 and applied to the op-amp's noninverting input. The resulting difference in voltage reduces the op-amp's output, driving Q_1 less, thus reducing its collector current (shunt current) and increasing its internal collector-to-emitter resistance, r_{ce}. Since r_{ce} acts as a voltage divider with R_1, this action offsets the attempted decrease in V_{OUT} and maintains it at an almost constant level.

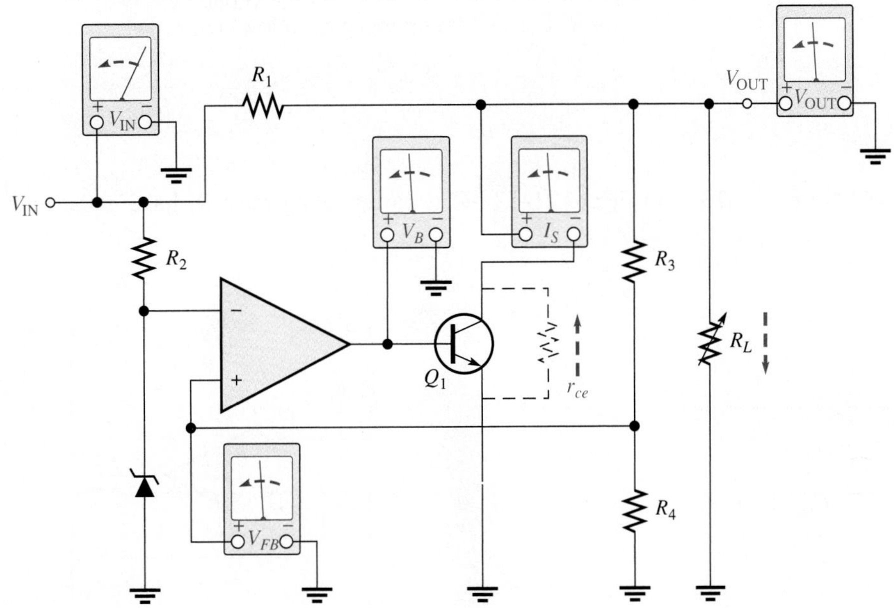

(a) Initial response to a decrease in V_{IN} or R_L

FIGURE 21–56
Sequence of responses when V_{IN} or R_L decreases.

FIGURE 21–56
Continued

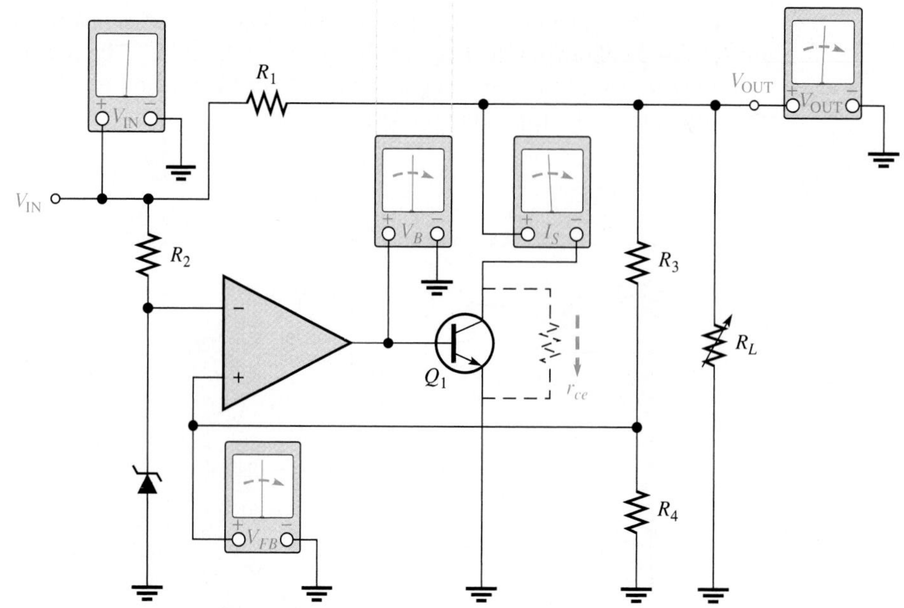

(b) V_{OUT} held constant by the feedback action.

The opposite action occurs when the output tries to increase, as shown in Figure 21–56(b). With I_L and V_{OUT} constant, a change in the input voltage produces a change in shunt current (I_S) as follows:

$$\Delta I_S = \frac{\Delta V_{IN}}{R_1}$$

With a constant V_{IN} and V_{OUT}, a change in load current causes an opposite change in shunt current:

$$\Delta I_S = -\Delta I_L$$

This formula says that if I_L increases, I_S decreases, and vice versa.

The shunt regulator is less efficient than the series type but offers inherent short-circuit protection. If the output is shorted ($V_{OUT} = 0$), the load current is limited by the series resistor R_1 to a maximum value as follows ($I_S = 0$):

$$I_{L(max)} = \frac{V_{IN}}{R_1}$$

EXAMPLE 21–15

In Figure 21–57, what power rating must R_1 have if the maximum input voltage is 12.5 V?

FIGURE 21–57

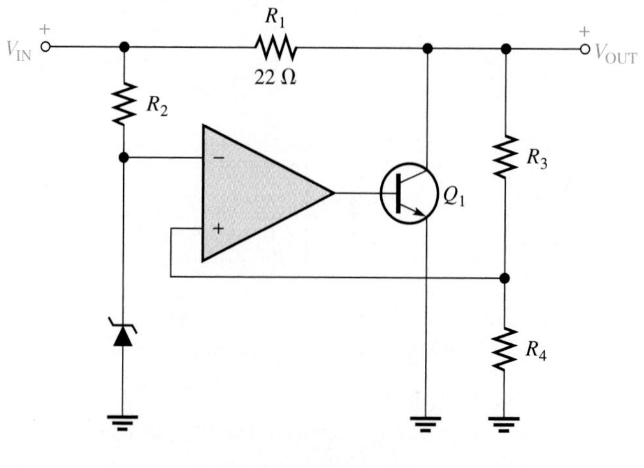

Solution The worst-case power dissipation in R_1 occurs when the output is short-circuited. $V_{OUT} = 0$, and when $V_{IN} = 12.5$ V, the voltage dropped across R_1 is

$$V_{IN} - V_{OUT} = 12.5 \text{ V}$$

The power dissipation in R_1 is

$$P_{R1} = \frac{V_{R1}^2}{R_1} = \frac{(12.5 \text{ V})^2}{22\ \Omega} = 7.1 \text{ W}$$

Therefore, a resistor of at least 10 W should be used.

Related Exercise In Figure 21–57, R_1 is changed to 33 Ω. What must be the power rating of R_1 if the maximum input voltage is 24 V?

SECTION 21–6 REVIEW

1. How does the control element in a shunt regulator differ from that in a series regulator?
2. Name one advantage and one disadvantage of a shunt regulator over a series type.

21–7 ■ APPLICATION ASSIGNMENT

In this application assignment, you will focus on the regulated power supply which provides the FM stereo receiver with dual polarity dc voltages. The op-amps in the channel separation circuits and the audio amplifiers operate from ±12 V. Both positive and negative voltage regulators are used to regulate the rectified and filtered voltages from a bridge rectifier.

This power supply utilizes a full-wave bridge **rectifier** with both the positive and negative rectified voltages taken off the bridge at the appropriate points and filtered by electrolytic capacitors. Integrated circuit voltage regulators (7812 and 7912) provide regulation for the positive and negative voltages.

Step 1: Relate the PC Board to a Schematic

Develop a schematic for the power supply in Figure 21–58. Add any missing labels and include the IC pin numbers. The rectifier diodes are 1N4001s, the filter capacitors C1 and

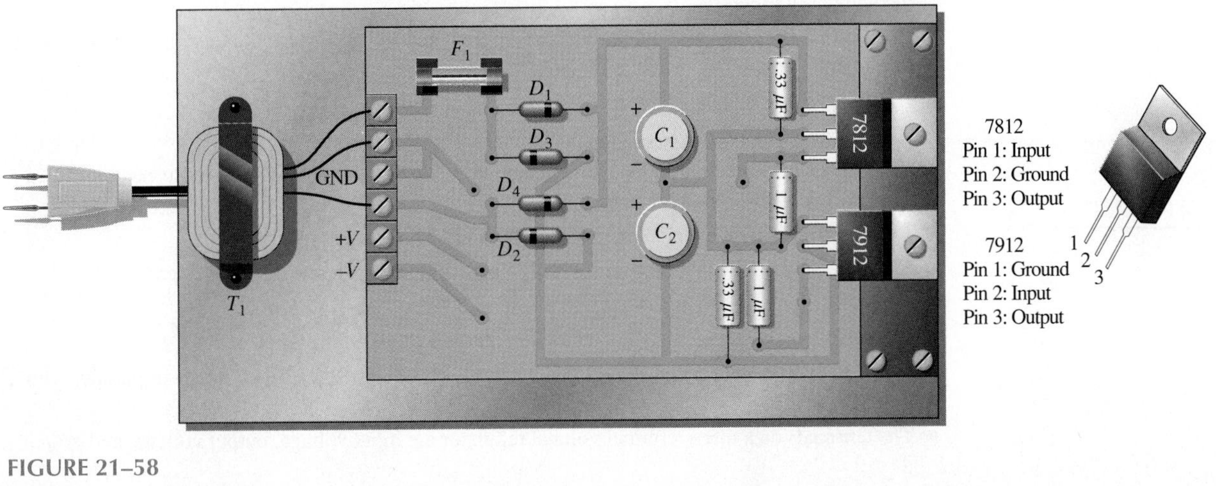

FIGURE 21–58

C2 are 100 μF, and the transformer has a turns ratio of 5:1. Determine the backside PC board connections as you develop the schematic.

Step 2: Analyze the Power Supply Circuits

1. Determine the approximate voltage with respect to ground at each of the four "corners" of the bridge. State whether each voltage is ac or dc.
2. Calculate the peak inverse voltage of the rectifier diodes.
3. Show the voltage waveform across D_1 for a full cycle of the ac input.

Step 3: Troubleshoot the Power Supply

State the probable cause or causes for the following:

1. Both positive and negative output voltages are zero.
2. Positive output voltage is zero and the negative output voltage is −12 V.
3. Negative output voltage is zero and the positive output voltage is +12 V.
4. Radical voltage fluctuations on output of positive regulator.

Indicate the voltages you should measure at the four corners of the diode bridge for the following faults:

1. Diode D_1 open.
2. Capacitor C_2 open.

SECTION 21–7 REVIEW

1. What should be the rating of the power supply fuse?
2. What purpose do the 0.33 μF capacitors serve?
3. Which regulator provides the negative voltage?

▪ SUMMARY

- In an op-amp comparator, when the input voltage exceeds a specified reference voltage, the output changes state.
- The output voltage of a summing amplifier is proportional to the sum of the input voltages.
- An averaging amplifier is a summing amplifier with a closed-loop gain equal to the reciprocal of the number of inputs.
- In a scaling adder, a different weight can be assigned to each input, thus making the input contribute more or contribute less to the output.
- The integral of a step is a ramp.
- The derivative of a ramp is a step.
- In a Wien-bridge oscillator, the closed-loop gain must be equal to 3 in order to have unity gain around the positive feedback loop.
- In filter terminology, a single RC network is called a *pole*.
- Each pole in a filter causes the output to roll off (decrease) at a rate of 20 dB/decade.
- Two-pole filters roll off at a rate of 40 dB/decade.
- In a series voltage regulator, the control element is a transistor in series with the load.
- In a shunt voltage regulator, the control element is a transistor (or zener diode) in parallel with a load.
- The terminals on a three-terminal voltage regulator are input voltage, output voltage, and ground.

▪ GLOSSARY

Active filter A frequency-selective circuit consisting of active devices such as transistors or op-amps combined with reactive (*RC*) circuits.

Comparator A circuit which compares two input voltages and produces an output in either of two states indicating the greater or less than relationship of the inputs.

Decade A tenfold change in the value of a quantity. When a quantity becomes ten times less or ten times greater, it has changed a decade.

Differentiator A circuit that produces an output which can approximate the instantaneous rate of change of the input function.

Integrator A circuit that produces an output which can approximate the area under the curve of the input function.

Pole In practical terms, a single *RC* network in a filter or amplifier that causes the response to change at a 20 dB/decade rate above or below a certain frequency.

Rectifier An electronic circuit that converts ac into pulsating dc.

Roll-off The decrease in the response of a filter below or above a critical frequency.

▪ FORMULAS

(21–1)	$V_{REF} = \frac{R_2}{R_1 + R_2}(+V)$	Comparator reference
(21–2)	$V_{OUT} = -(V_{IN1} + V_{IN2})$	Two-input adder
(21–3)	$V_{OUT} = -(V_{IN1} + V_{IN2} + V_{IN3} + \cdots + V_{INn})$	n-input adder
(21–4)	$V_{OUT} = -\frac{R_f}{R}(V_{IN1} + V_{IN2} + \cdots + V_{INn})$	Adder with gain
(21–5)	$V_{OUT} = -\left(\frac{R_f}{R_1}V_{IN1} + \frac{R_f}{R_2}V_{IN2} + \cdots + \frac{R_f}{R_n}V_{INn}\right)$	Adder with gain
(21–6)	$\frac{\Delta V_{out}}{\Delta t} = -\frac{V_{IN}}{RC}$	Rate of change in integrator
(21–7)	$V_{OUT} = \left(\frac{V_C}{t}\right)R_f C$	Differentiator output with ramp input
(21–8)	$\frac{V_{out}}{V_{in}} = \frac{1}{3}$	Lead-lag attenuation at f_r
(21–9)	$f_r = \frac{1}{2\pi RC}$	Lead-lag resonant frequency
(21–10)	$V_{UTP} = +V_{max}\left(\frac{R_3}{R_2}\right)$	Upper trigger point, triangle oscillator
(21–11)	$V_{LTP} = -V_{max}\left(\frac{R_3}{R_2}\right)$	Lower trigger point, triangle oscillator
(21–12)	$f = \frac{1}{4R_1C}\left(\frac{R_2}{R_3}\right)$	Frequency of oscillation, triangle oscillator
(21–13)	$T = \frac{V_p - V_F}{\lvert V_{IN}\rvert/RC}$	Sawtooth period, VCO
(21–14)	$f = \frac{\lvert V_{IN}\rvert}{RC}\left(\frac{1}{V_p - V_F}\right)$	Sawtooth frequency, VCO
(21–15)	$V_{out} = \left(\frac{X_C}{\sqrt{R^2 + X_C^2}}\right)V_{in}$	Output of one-pole filter (low-pass only)
(21–16)	$f_c = \frac{1}{2\pi\sqrt{R_1R_2C_1C_2}}$	Critical frequency of two-pole filter (low-pass)

(21–17)	$V_{out} = \left(\dfrac{R}{\sqrt{R^2 + X_C^2}}\right)V_{in}$	Output of one-pole filter (high-pass only)
(21–18)	$f_{c1} = \dfrac{1}{2\pi\sqrt{R_1R_2C_1C_2}}$	Lower critical frequency
(21–19)	$f_{c2} = \dfrac{1}{2\pi\sqrt{R_3R_4C_3C_4}}$	Upper critical frequency
(21–20)	$f_r = \sqrt{f_{c1}f_{c2}}$	Center frequency
(21–21)	$A_{cl} = 1 + \dfrac{R_2}{R_3}$	Closed-loop voltage gain, voltage regulator
(21–22)	$V_{OUT} = \left(1 + \dfrac{R_2}{R_3}\right)V_{REF}$	Regulator output
(21–23)	$I_{L(max)} = \dfrac{0.7\text{ V}}{R_4}$	For constant-current limiting

SELF-TEST

1. The purpose of a comparator is to
(a) amplify an input voltage
(b) detect the occurrence of a changing input voltage
(c) produce a change in output when an input voltage equals a reference voltage
(d) maintain a constant output when the dc input voltage changes

2. To use a comparator for zero-level detection, the inverting input is connected to
(a) ground **(b)** the dc supply voltage
(c) a positive reference voltage **(d)** a negative reference voltage

3. In a 5 V level detector circuit,
(a) the noninverting input is connected to +5 V
(b) the inverting input is connected to +5 V
(c) the input signal is limited to a 5 V peak value
(d) the input signal must be riding on a +5 dc level

4. In a certain four-input summing amplifier, all the input resistors are 2.2 kΩ and the feedback resistor is 2.2 kΩ. If all the input voltages are 2 V, the output voltage is
(a) 2 V **(b)** 10 V **(c)** 2.2 V **(d)** 8 V

5. The gain of the amplifier in Question 4 is
(a) 1 **(b)** 2.2 **(c)** 4 **(d)** unknown

6. To convert a summing amplifier to an averaging amplifier,
(a) all input resistors must be a different value
(b) the ratio R_f/R must equal to the reciprocal of the number of inputs
(c) the ratio R_f/R must equal to the number of inputs
(d) both (a) and (b)

7. In a scaling adder,
(a) the input resistors are all the same value
(b) the input resistors are all different values
(c) the input resistors have values depending on the assigned weight of each input
(d) the ratio R_f/R must be the same for each input

8. The feedback path in an op-amp integrator consists of
(a) a resistor **(b)** a capacitor
(c) a resistor and a capacitor in series **(d)** a resistor and a capacitor in parallel

9. The feedback path in an op-amp differentiator consists of
(a) a resistor **(b)** a capacitor
(c) a resistor and a capacitor in series **(d)** a resistor and a capacitor in parallel

10. The op-amp comparator circuit uses
(a) positive feedback **(b)** negative feedback **(c)** regenerative feedback **(d)** no feedback

11. Unity gain and zero phase shift around the feedback loop are conditions that describe
(a) an active filter **(b)** a comparator **(c)** an oscillator **(d)** an integrator or differentiator

12. The input frequency of a single-pole, low-pass active filter increases from 1.5 kHz to 150 kHz. If the critical frequency is 1.5 kHz, the gain decreases by
(a) 3 dB **(b)** 20 dB **(c)** 40 dB **(d)** 60 dB

■ PROBLEMS

SECTION 21–1 Comparators

1. Determine the output level (maximum positive or maximum negative) for each comparator in Figure 21–59.

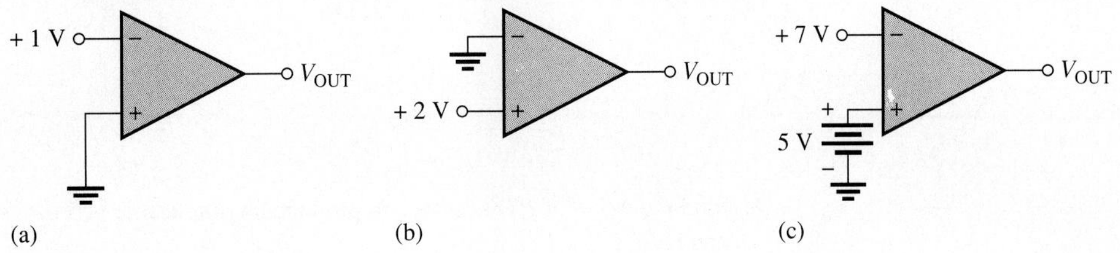

FIGURE 21–59

2. A certain op-amp has an open-loop gain of 80,000. The maximum saturated output levels of this particular device are ±12 V when the dc supply voltages are ±15 V. If a differential voltage of 0.15 mV rms is applied between the inputs, what is the peak-to-peak value of the output?

3. Sketch the output voltage waveform for each circuit in Figure 21–60 with respect to the input. Show voltage levels.

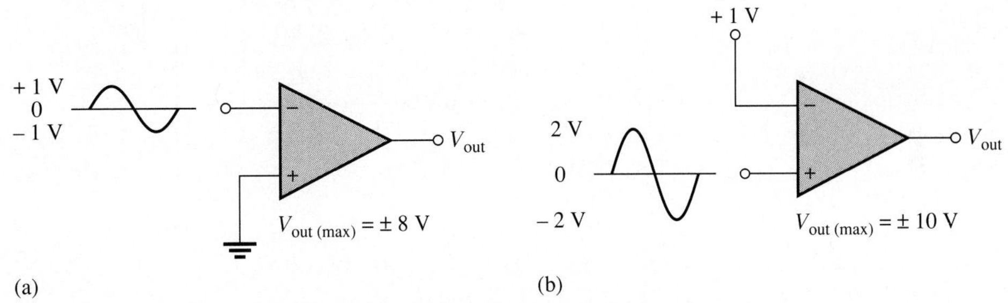

FIGURE 21–60

SECTION 21–2 Summing Amplifiers

4. Determine the output voltage for each circuit in Figure 21–61.

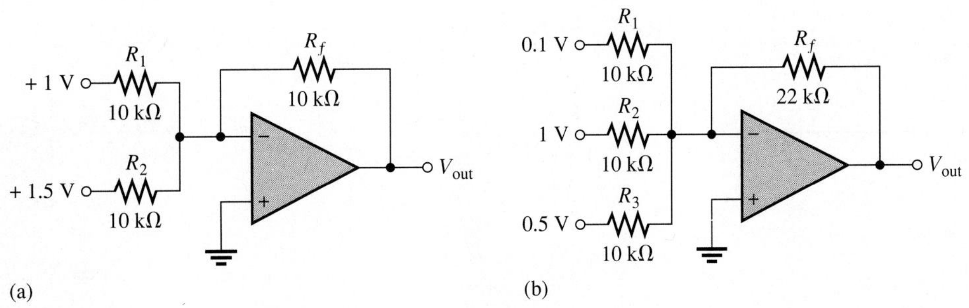

FIGURE 21–61

5. Determine the following in Figure 21–62:
(a) V_{R1} and V_{R2} (b) current through R_f (c) V_{OUT}

FIGURE 21–62

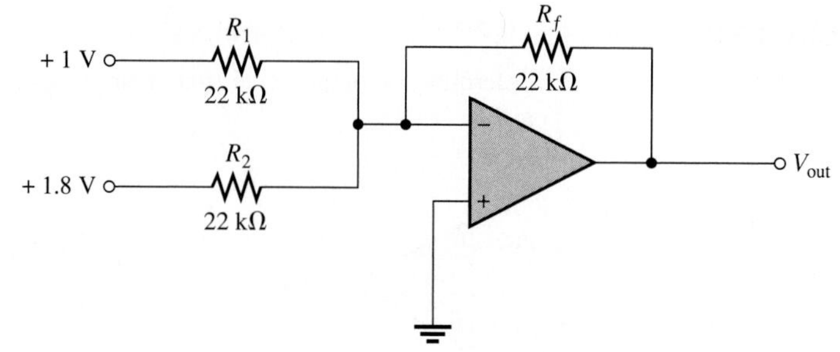

6. Find the value of R_f necessary to produce an output that is 5 times the sum of the inputs in Figure 21–62.

7. Find the output voltage when the input voltages shown in Figure 21–63 are applied to the scaling adder. What is the current through R_f?

FIGURE 21–63

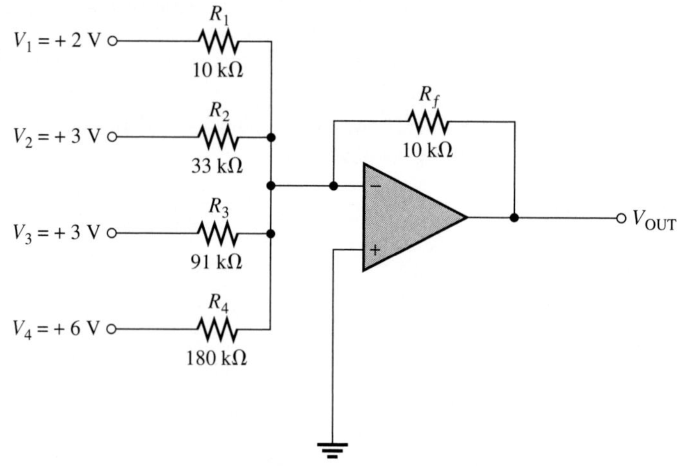

8. Determine the values of the input resistors required in a six-input scaling adder so that the lowest weighted input is 1 and each successive input has a weight *twice* the previous one. Use $R_f = 100\text{ k}\Omega$.

SECTION 21–3 Integrators and Differentiators

9. Determine the rate of change of the output voltage in response to the step input to the integrator in Figure 21–64.

FIGURE 21–64

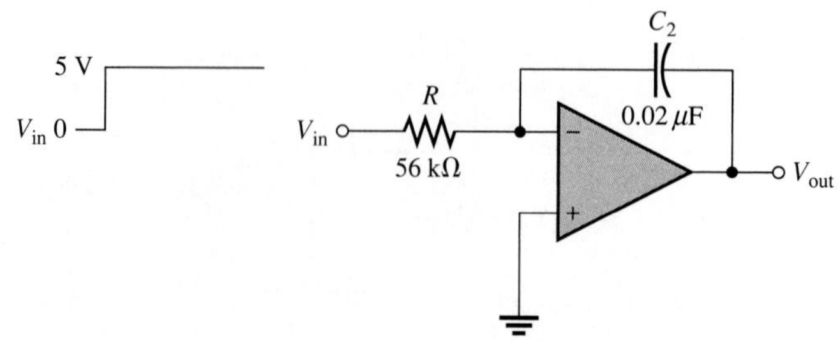

10. A triangular waveform is applied to the input of the circuit in Figure 21–65 as shown. Determine what the output should be, and sketch its waveform in relation to the input.

FIGURE 21–65

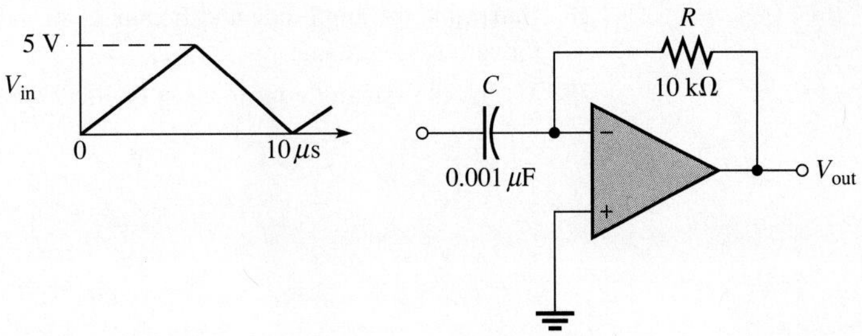

SECTION 21–4 Oscillators

11. A certain lead-lag network has a resonant frequency of 3.5 kHz. What is the rms output voltage if an input signal with a frequency equal to f_r and with an rms value of 2.2 V is applied to the input?

12. Calculate the resonant frequency of a lead-lag network with the following values: $R_1 = R_2 =$ 6.2 kΩ, and $C_1 = C_2 = 0.02$ μF.

13. Determine the necessary value of R_2 in Figure 21–66 so that the circuit will oscillate. Neglect the forward resistance of the zener diodes.

14. Explain the purpose of R_3 in Figure 21–66.

15. Find the frequency of oscillation for the Wien-bridge oscillator in Figure 21–66.

FIGURE 21–66

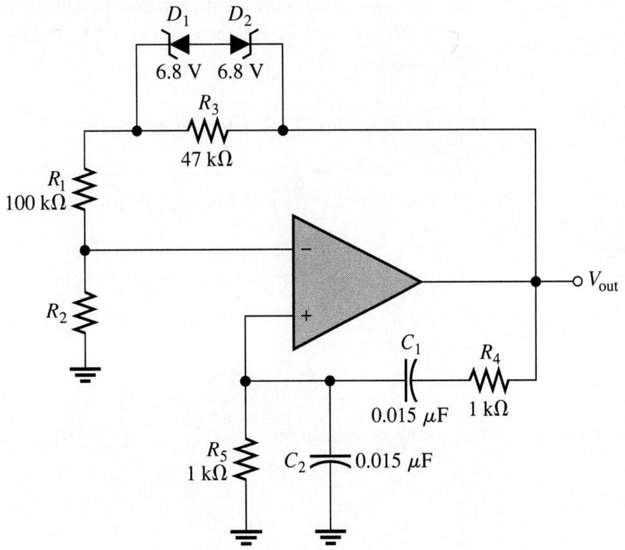

16. What type of signal does the circuit in Figure 21–67 produce? Determine the frequency of the output.

17. Show how to change the frequency of oscillation in Figure 21–67 to 10 kHz.

FIGURE 21–67

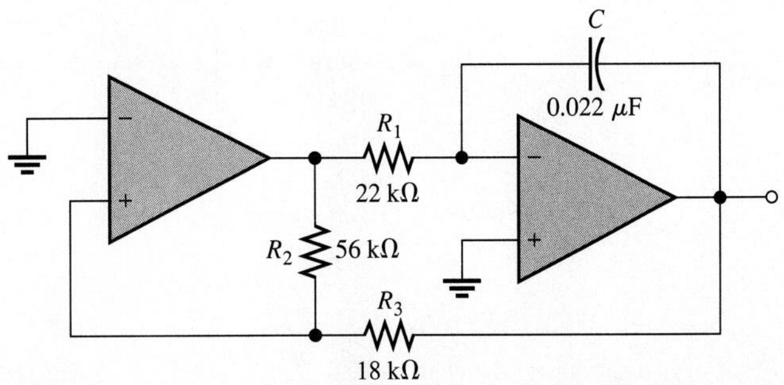

18. Determine the amplitude and frequency of the output voltage in Figure 21–68. Use 1 V as the forward PUT voltage.

19. Modify the sawtooth oscillator in Figure 21–68 so that its peak-to-peak output is 4 V.

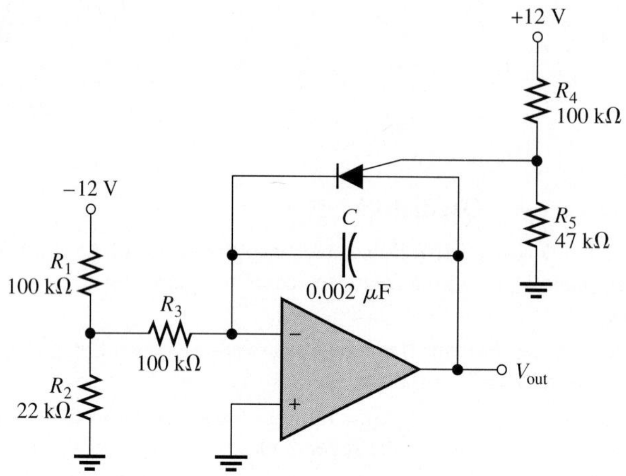

FIGURE 21–68

20. A certain sawtooth oscillator has the following parameter values: $V_{IN} = 3$ V, $R = 4.7$ kΩ, $C = 0.001$ μF, and V_F for the PUT is 1.2 V. Determine its peak-to-peak output voltage if the period is 10 μs.

SECTION 21–5 Active Filters

21. Determine the number of poles in each active filter in Figure 21–69, and identify its type.

22. Calculate the critical frequencies for the filters in Figure 21–69.

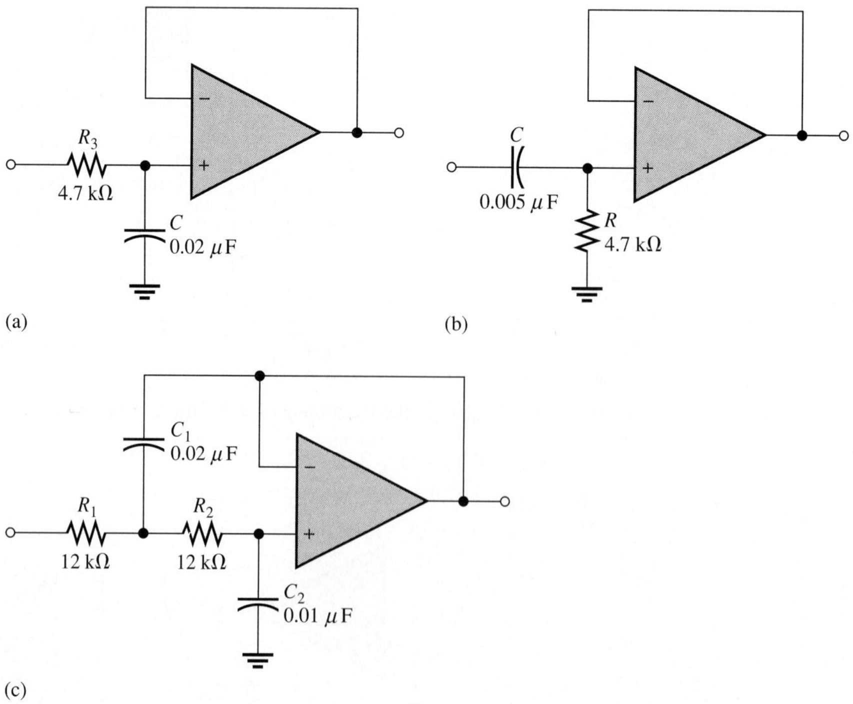

FIGURE 21–69

23. Determine the bandwidth and center frequency of each filter in Figure 21–70.

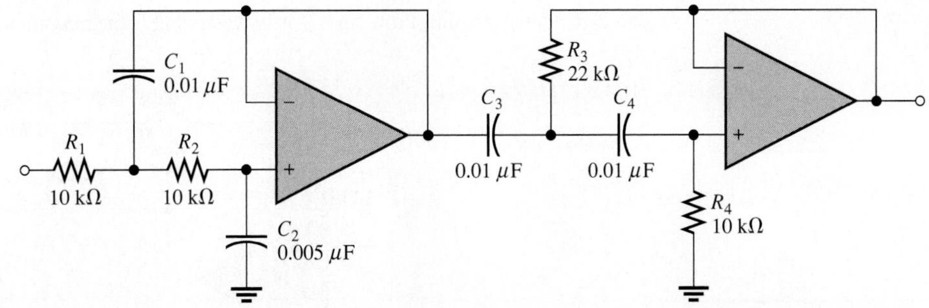

(a)

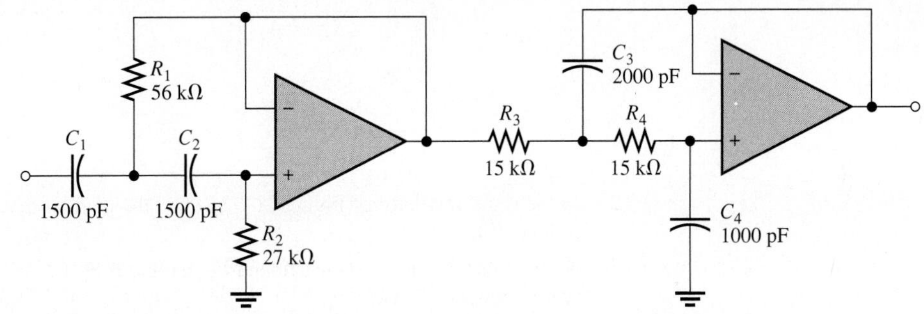

(b)

FIGURE 21–70

SECTION 21–6 Voltage Regulators

24. Determine the output voltage for the series regulator in Figure 21–71.

25. If R_3 in Figure 21–71 is doubled, what happens to the output voltage?

26. If the zener voltage is 2.7 V instead of 2 V in Figure 21–71, what is the output voltage?

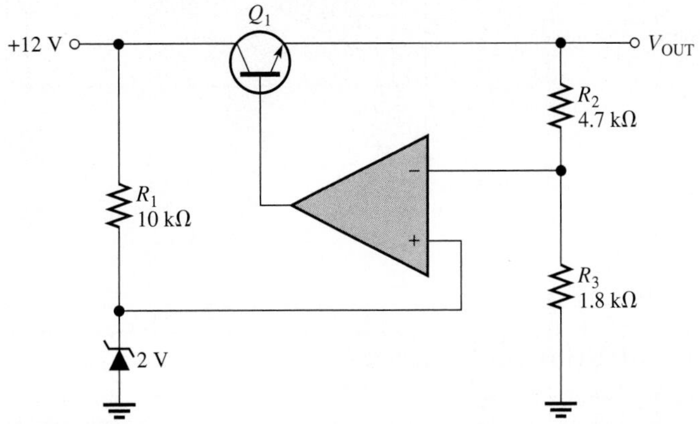

FIGURE 21–71

27. A series voltage regulator with constant current limiting is shown in Figure 21–72. Determine the value of R_4 if the load current is to be limited to a maximum value of 250 mA. What power rating must R_4 have?

28. If R_4 (determined in Problem 27) is halved, what is the maximum load current?

FIGURE 21–72

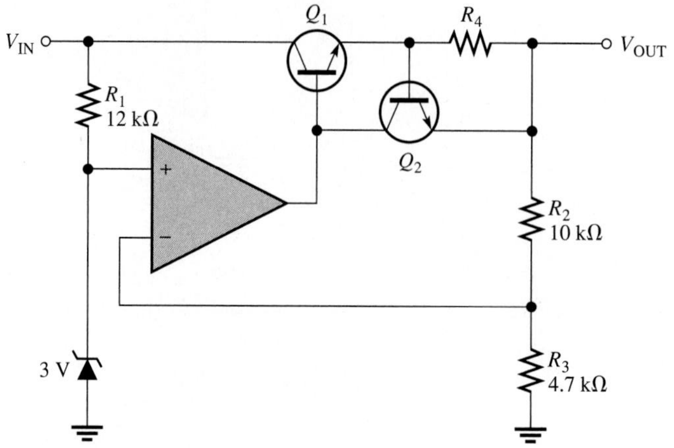

29. In the shunt regulator of Figure 21–73, when the load current increases, does Q_1 conduct more or less? Why?

30. Assume that I_L remains constant and V_{IN} increases by 1 V in Figure 21–73. What is the change in the collector current of Q_1?

31. With a constant input voltage of 18 V, the load resistance in Figure 21–73 is varied from 1 kΩ to 1.2 kΩ. Neglecting any change in output voltage, how much does the shunt current through Q_1 change?

FIGURE 21–73

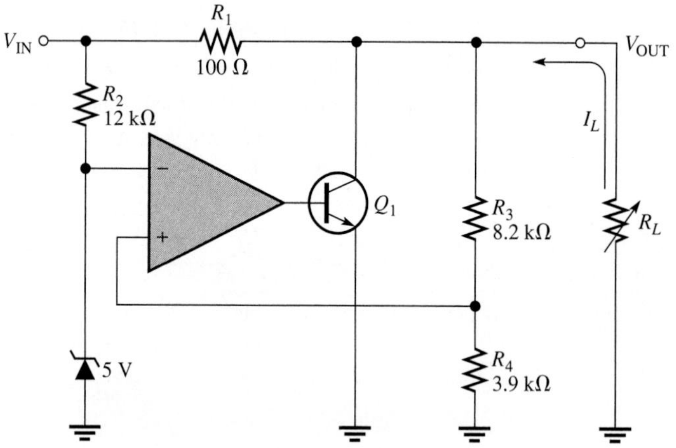

■ ANSWERS TO SECTION REVIEWS

Section 21–1

1. (10 kΩ/110 kΩ)15 V = 1.36 V

2. See Figure 21–74.

FIGURE 21–74

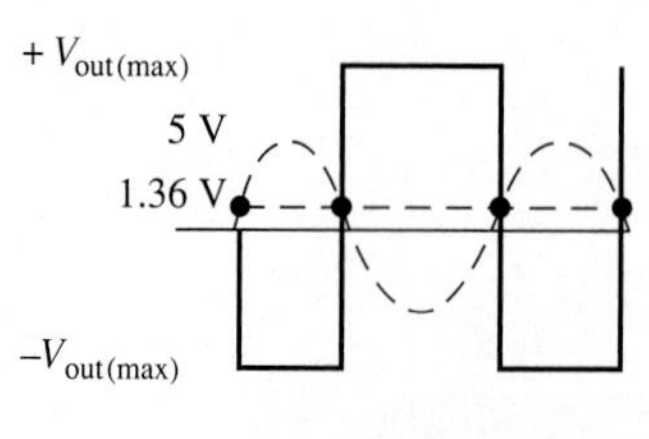

Section 21–2

1. The summing point is the terminal of the op-amp where the input resistors are commonly connected.
2. $R_f/R = 1/5$
3. $2(10\ k\Omega) = 20\ k\Omega$

Section 21–3

1. The feedback element in an integrator is a capacitor.
2. The capacitor voltage is linear because the capacitor current is constant.
3. The feedback element in a differentiator is a resistor.
4. The output of a differentiator is proportional to the rate of change of the input.

Section 21–4

1. The negative feedback loop sets the closed-loop gain; the positive feedback loop sets the frequency.
2. A VCO exhibits a frequency that can be varied with a dc control voltage.
3. The basis of a relaxation oscillator is the charging and discharging of a capacitor.

Section 21–5

1. A pole is a single *RC* network.
2. A single-pole low-pass is flat from dc to the critical frequency.
3. The *R* and *C* positions are interchanged.
4. The critical frequency is halved.

Section 21–6

1. In a shunt regulator, the control element is in parallel with the load rather than in series with the load.
2. A shunt regulator has inherent current limiting, but it is less efficient than a series regulator because part of the load current must be bypassed through the control element.

Section 21–7

1. The fuse rating should be 1 A.
2. These are optional capacitors that prevent oscillations.
3. The 7912 is a negative-voltage regulator.

▪ ANSWERS TO RELATED EXERCISES FOR EXAMPLES

21–1 1.96 V

21–2 –12.5 V

21–3 –5.73 V

21–4 Add a 100 kΩ input resistor and change R_f to 20 kΩ.

21–5 0.45; 0.12; 0.18; $V_{OUT} = -3.03$ V

21–6 Change *C* to 5000 pF.

21–7 Same waveform but with amplitude of 6.6 V

21–8 Use 6.1 V zeners.

21–9 6.06 V peak-to-peak

21–10 952 Hz

21–11 8.34 kHz

21–12 2.57 kHz

21–13 Increase f_{c2} by reducing the resistor and/or capacitor values.

21–14 7.33 V

21–15 17.5 W

A TABLE OF STANDARD RESISTOR VALUES

Resistance Tolerance (± %)

0.1% 0.25% 0.5%	1%	2% 5%	10%	0.1% 0.25% 0.5%	1%	2% 5%	10%	0.1% 0.25% 0.5%	1%	2% 5%	10%	0.1% 0.25% 0.5%	1%	2% 5%	10%	0.1% 0.25% 0.5%	1%	2% 5%	10%	0.1% 0.25% 0.5%	1%	2% 5%	10%
10.0	10.0	10	10	14.7	14.7	—	—	21.5	21.5	—	—	31.6	31.6	—	—	46.4	46.4	—	—	68.1	68.1	68	68
10.1	—	—	—	14.9	—	—	—	21.8	—	—	—	32.0	—	—	—	47.0	—	47	47	69.0	—	—	—
10.2	10.2	—	—	15.0	15.0	15	15	22.1	22.1	22	22	32.4	32.4	—	—	47.5	47.5	—	—	69.8	69.8	—	—
10.4	—	—	—	15.2	—	—	—	22.3	—	—	—	32.8	—	—	—	48.1	—	—	—	70.6	—	—	—
10.5	10.5	—	—	15.4	15.4	—	—	22.6	22.6	—	—	33.2	33.2	33	33	48.7	48.7	—	—	71.5	71.5	—	—
10.6	—	—	—	15.6	—	—	—	22.9	—	—	—	33.6	—	—	—	49.3	—	—	—	72.3	—	—	—
10.7	10.7	—	—	15.8	15.8	—	—	23.2	23.2	—	—	34.0	34.0	—	—	49.9	49.9	—	—	73.2	73.2	—	—
10.9	—	—	—	16.0	—	16	—	23.4	—	—	—	34.4	—	—	—	50.5	—	—	—	74.1	—	—	—
11.0	11.0	11	—	16.2	16.2	—	—	23.7	23.7	—	—	34.8	34.8	—	—	51.1	51.1	51	—	75.0	75.0	75	—
11.1	—	—	—	16.4	—	—	—	24.0	—	24	—	35.2	—	—	—	51.7	—	—	—	75.9	—	—	—
11.3	11.3	—	—	16.5	16.5	—	—	24.3	24.3	—	—	35.7	35.7	—	—	52.3	52.3	—	—	76.8	76.8	—	—
11.4	—	—	—	16.7	—	—	—	24.6	—	—	—	36.1	—	36	—	53.0	—	—	—	77.7	—	—	—
11.5	11.5	—	—	16.9	16.9	—	—	24.9	24.9	—	—	36.5	36.5	—	—	53.6	53.6	—	—	78.7	78.7	—	—
11.7	—	—	—	17.2	—	—	—	25.2	—	—	—	37.0	—	—	—	54.2	—	—	—	79.6	—	—	—
11.8	11.8	—	—	17.4	17.4	—	—	25.5	25.5	—	—	37.4	37.4	—	—	54.9	54.9	—	—	80.6	80.6	—	—
12.0	—	12	12	17.6	—	—	—	25.8	—	—	—	37.9	—	—	—	56.2	—	—	—	81.6	—	—	—
12.1	12.1	—	—	17.8	17.8	—	—	26.1	26.1	—	—	38.3	38.3	—	—	56.6	56.6	56	56	82.5	82.5	82	82
12.3	—	—	—	18.0	—	18	18	26.4	—	—	—	38.8	—	—	—	56.9	—	—	—	83.5	—	—	—
12.4	12.4	—	—	18.2	18.2	—	—	26.7	26.7	—	—	39.2	39.2	39	39	57.6	57.6	—	—	84.5	84.5	—	—
12.6	—	—	—	18.4	—	—	—	27.1	—	27	27	39.7	—	—	—	58.3	—	—	—	85.6	—	—	—
12.7	12.7	—	—	18.7	18.7	—	—	27.4	27.4	—	—	40.2	40.2	—	—	59.0	59.0	—	—	86.6	86.6	—	—
12.9	—	—	—	18.9	—	—	—	27.7	—	—	—	40.7	—	—	—	59.7	—	—	—	87.6	—	—	—
13.0	13.0	13	—	19.1	19.1	—	—	28.0	28.0	—	—	41.2	41.2	—	—	60.4	60.4	—	—	88.7	88.7	—	—
13.2	—	—	—	19.3	—	—	—	28.4	—	—	—	41.7	—	—	—	61.2	—	—	—	89.8	—	—	—
13.3	13.3	—	—	19.6	19.6	—	—	28.7	28.7	—	—	42.2	42.2	—	—	61.9	61.9	62	—	90.9	90.9	91	—
13.5	—	—	—	19.8	—	—	—	29.1	—	—	—	42.7	—	—	—	62.6	—	—	—	92.0	—	—	—
13.7	13.7	—	—	20.0	20.0	20	—	29.4	29.4	—	—	43.2	43.2	43	—	63.4	63.4	—	—	93.1	93.1	—	—
13.8	—	—	—	20.3	—	—	—	29.8	—	—	—	43.7	—	—	—	64.2	—	—	—	94.2	—	—	—
14.0	14.0	—	—	20.5	20.5	—	—	30.1	30.1	30	—	44.2	44.2	—	—	64.9	64.9	—	—	95.3	95.3	—	—
14.2	—	—	—	20.8	—	—	—	30.5	—	—	—	44.8	—	—	—	65.7	—	—	—	96.5	—	—	—
14.3	14.3	—	—	21.0	21.0	—	—	30.9	30.9	—	—	45.3	45.3	—	—	66.5	66.5	—	—	97.6	97.6	—	—
14.5	—	—	—	21.3	—	—	—	31.2	—	—	—	45.9	—	—	—	67.3	—	—	—	98.8	—	—	—

NOTE: These values are generally available in multiples of 0.1, 1, 10, 100, 1 k, and 1 M.

B BATTERIES

Batteries are an important source of dc voltage. They are available in two basic categories: the wet cell and the dry cell. A battery generally is made up of several individual cells.

A cell consists basically of two electrodes immersed in an electrolyte. A voltage is developed between the electrodes as a result of the chemical action between the electrodes and the electrolyte. The electrodes typically are two dissimilar metals, and the electrolyte is a chemical solution or substance.

Wet Cells and Dry Cells

In a wet cell, the electrolyte is in liquid form, as you might guess. The automotive battery is a common type of wet-cell battery. In a dry cell, the electrolyte is typically in a paste-like form. There are many more varieties of dry-cell batteries than wet-cell batteries.

Primary and Secondary Cells

Primary cells cannot be recharged and thus are used once and then discarded. The 1.5 V carbon-zinc or alkaline flashlight battery is an example of a primary cell.

Secondary cells are rechargable and can be reused many times. Examples are nickel-cadmium (Ni-Cad) cells and lead-acid automotive cells.

Charging Rate

The charging rate for secondary cells specifies the amount of reverse current recommended over a given period of time. For example, the charging rate might be specified as 100 mA for 10 h. The same battery could be charged at a faster equivalent rate of, say, 200 mA for 5 h or at a slower equivalent rate of 50 mA for 20 h.

Charge/Discharge Cycle

When a battery charges and is then discharged to below its useful output voltage, it has gone through one charge/discharge cycle. The total number of charge/discharge cycles that a battery can go through before it becomes no longer usable is called the *cycle life*. The cycle life of a typical rechargable battery can be several hundred cycles.

Capacity

The capacity of any type of battery specifies how much current the battery can deliver to a load over a certain period of time before its output voltage decreases to an unusable level. Capacity is expressed in ampere-hours (Ah). For example, if a certain battery has a capacity of 50 Ah, it can deliver 50 A for 1 h or 100 A for ½ h, 25 A for 2 h, and so on.

TYPES OF BATTERIES

Lead-Acid Cell

The positive electrode of a lead-acid cell is lead peroxide (PbO_2), and the negative electrode is spongy lead (Pb). The electrolyte is sulfuric acid (H_2SO_4) in water. Thus, the lead-acid cell is classified as a wet cell.

Two positive hydrogen ions ($2H^+$) and one negative sulfate ion (SO_4^{--}) are formed when the sulfuric acid ionizes in the water. Lead ions (Pb^{++}) from both electrodes displace the hydrogen ions in the electrolyte solution. When the lead ion from the spongy lead electrode enters the solution, it combines with a sulfate ion (SO_4^{--}) to form lead sulfate ($PbSO_4$), and it leaves two excess electrons on the electrode.

When a lead ion from the lead peroxide electrode enters the solution, it also leaves two excess electrons on the electrode and forms lead sulfate in the solution. However, because this electrode is lead peroxide, two free oxygen atoms are created when a lead atom leaves and enters the solution as a lead ion. These two oxygen atoms take four electrons from the lead peroxide electrode and become oxygen ions (O^{--}). This process creates a deficiency of two electrons on this electrode (there were initially two excess electrons).

The two oxygen ions ($2O^{--}$) combine in the solution with four hydrogen ions ($4H^+$) to produce two molecules of water ($2H_2O$). This process dilutes the electrolyte over a period of time. Also, there is a buildup of lead sulfate on the electrodes. These two factors result in a reduction in the voltage produced by the cell and necessitate periodic recharging.

As you have seen, for each departing lead ion, there is an excess of two electrons on the spongy lead electrode, and there is a deficiency of two electrons on the lead peroxide electrode. Therefore, the lead peroxide electrode is positive, and the spongy lead electrode is negative. This chemical reaction is pictured in Figure B–1.

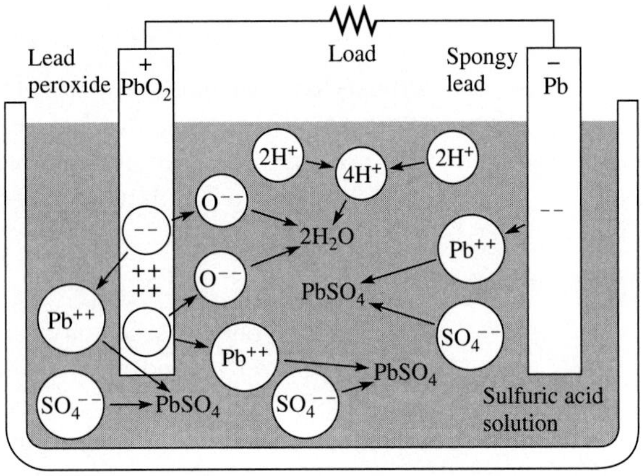

FIGURE B–1
Chemical reaction in a discharging lead-acid cell.

As mentioned, the dilution of the electrolyte by the formation of water and lead sulfate requires that the lead-acid cell be recharged to reverse the chemical process. The cell is recharged by connection of an external voltage source to the electrodes, as shown in Figure B–2. The formula for the chemical reaction in a lead-acid cell is as follows:

$$Pb + PbO_2 + 2H_2SO_4 \longrightarrow 2PbSO_4 + 2H_2O$$

Carbon-Zinc Cell

A typical carbon-zinc dry cell is illustrated in Figure B–3. The zinc container or can is dissolved by the electrolyte. As a result of this reaction, an excess of electrons accumulates on the container, making it the negative electrode. The hydrogen ions in the electrolyte take electrons from the carbon rod, making it the positive electrode. Hydrogen gas is formed near the carbon electrode, but his gas is eliminated by reaction with manganese dioxide (called a *depolarizing agent*). This depolarization prevents bursting of the container due to gas formation. Because the chemical reaction is not reversible, the carbon-zinc cell is a primary cell.

Several common types of dry cells are summarized in Table B–1.

FIGURE B–2
Recharging a lead-acid cell.

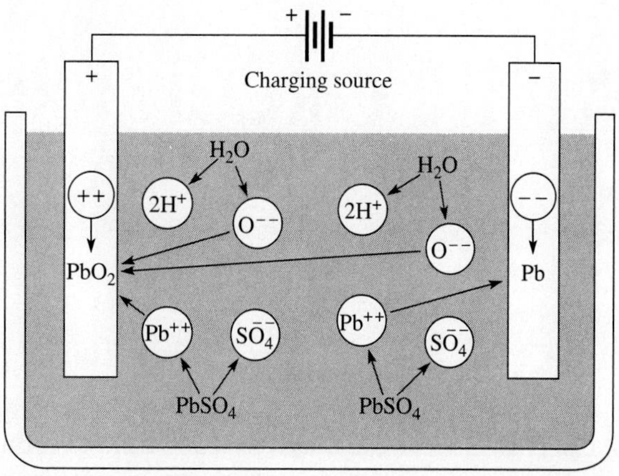

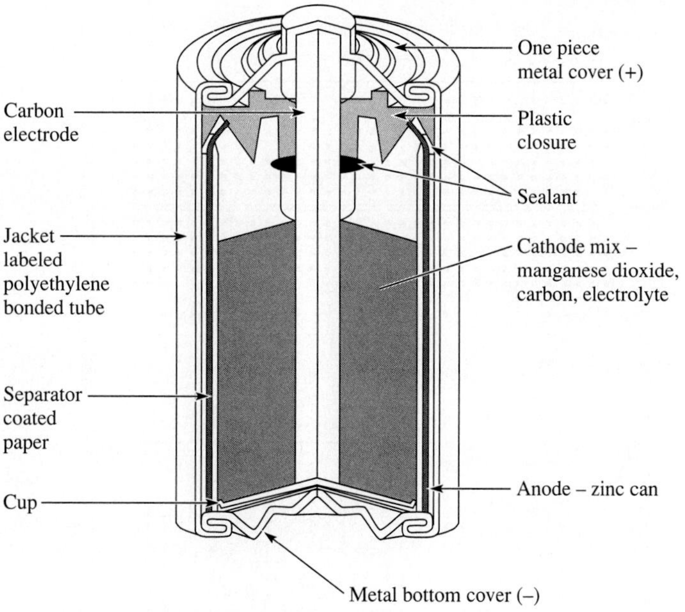

FIGURE B–3
Cutaway view of a typical carbon-zinc cell.

TABLE B–1
Types of battery cells

TYPE	+ ELECTRODE	–ELECTRODE	ELECTROLYTE	VOLTS	COMMENTS
Carbon-zinc	Carbon	Zinc	Ammonium and zinc chloride	1.5	Dry, primary
Lead-acid	Lead peroxide	Spongy lead	Sulfuric acid	2.0	Wet, secondary
Manganese-alkaline	Manganese dioxide	Zinc	Potassium hydroxide	1.5	Dry, primary or secondary
Mercury	Zinc	Mercuric oxide	Potassium hydroxide	1.3	Dry, primary
Nickel-cadmium	Nickel	Cadmium hydroxide	Potassium hydroxide	1.25	Dry, secondary
Nickel-iron (Edison cell)	Nickel oxide	Iron	Potassium hydroxide	1.36	Wet, secondary

C CIRCUIT BREADBOARDING AND MEASUREMENTS

Breadboarding a Circuit

The term *breadboarding* refers to the process of installing components on a circuit board and interconnecting them to form a specified circuit. One common type of circuit board used for constructing circuits in the laboratory is shown in Figure C–1. It consists of rows of small sockets into which component leads and jumper wires are inserted.

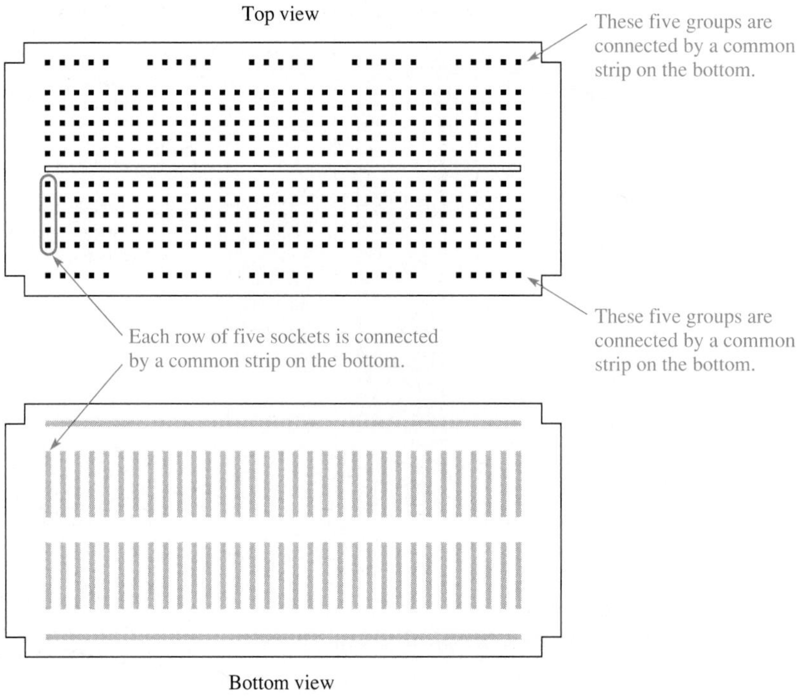

FIGURE C–1
A typical circuit board used for breadboarding.

In this particular configuration, all five sockets in each row are connected together and are effectively one electrical point as shown in the bottom view. All the sockets arranged on the outer edges of the board are typically connected together as shown. These two strips of sockets are normally used to distribute the supply voltages and ground to the circuit.

As examples of breadboarding a circuit, Figure C–2 shows how both simple series and parallel resistive circuits might be connected on the circuit board.

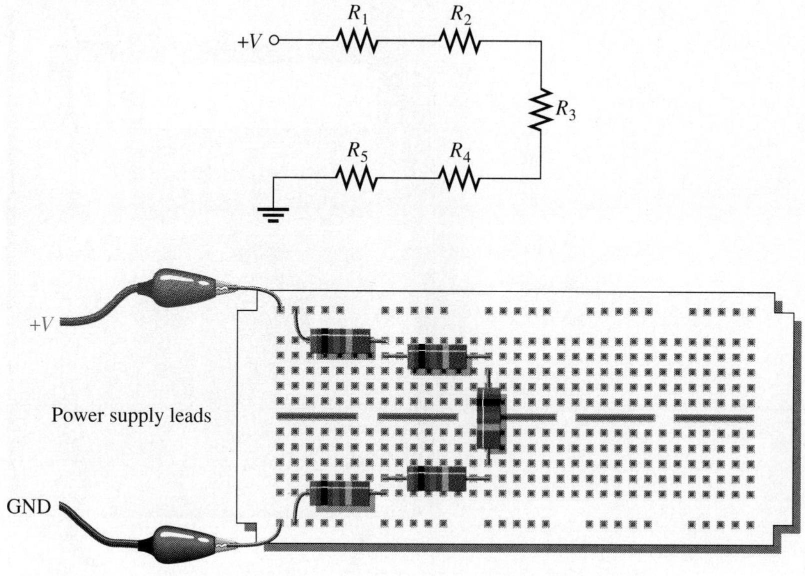

(a) Series

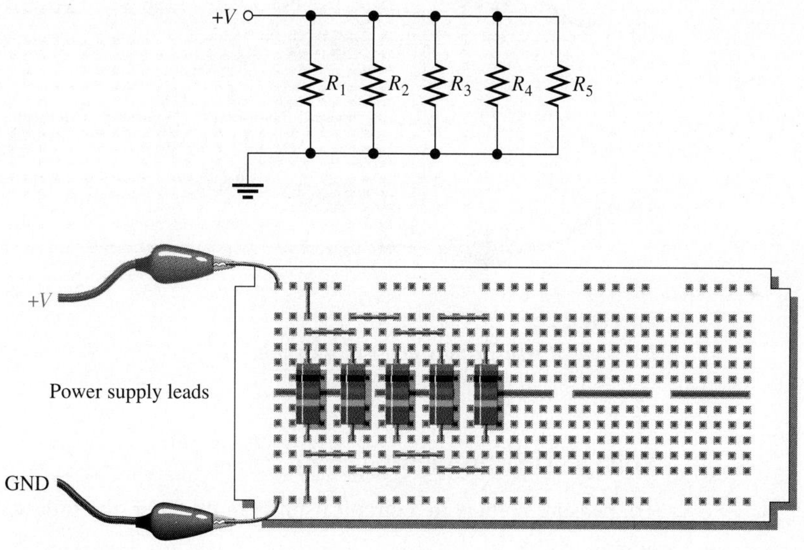

(b) Parallel

FIGURE C–2
Examples of breadboarding a circuit.

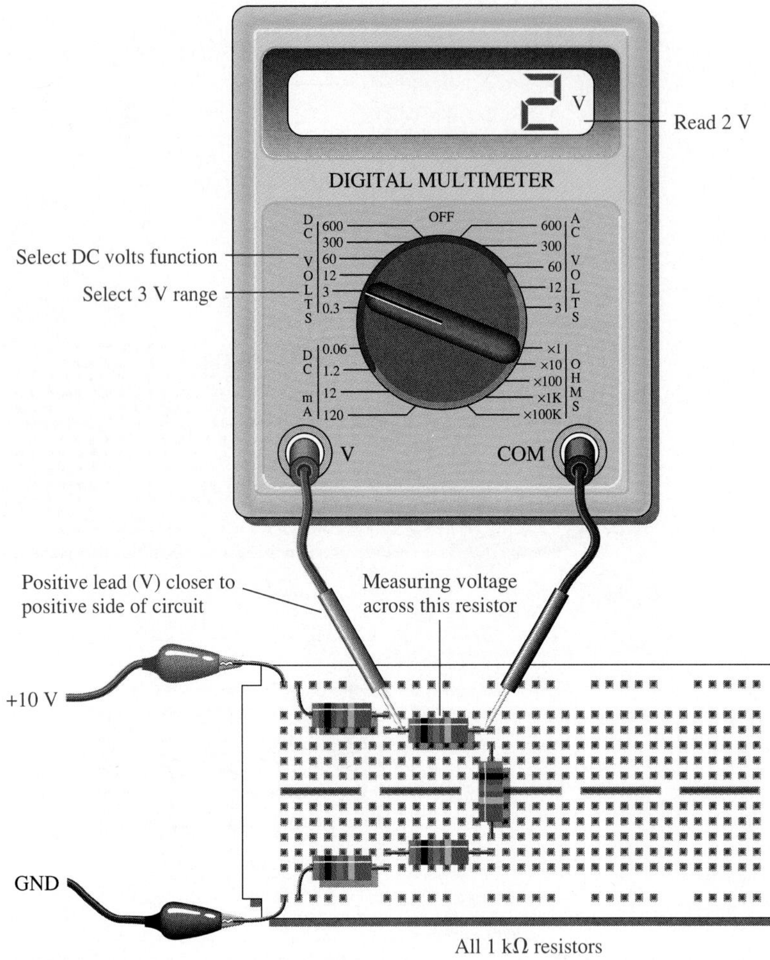

FIGURE C–3
Example of measuring voltage in a circuit.

Voltage Measurement

To measure voltage in a circuit using a multimeter, the following general steps should be taken:

1. Insert the red meter lead into the connector labeled for *voltage* (V) on the meter and insert the black meter lead into the connector labeled *common* (COM).
2. Set the function switch to the *volts* position (select either dc or ac).
3. Set the range switch (the range switch and function switch are often the same) to the lowest setting that is higher than the voltage you expect to measure.
4. Select the two points in the circuit between which you want to measure the voltage.
5. Place the red meter lead on the point closest to the positive side of the circuit and the black lead on the point closest to the negative or ground side of the circuit as shown in Figure C–3.
6. Read the voltage.

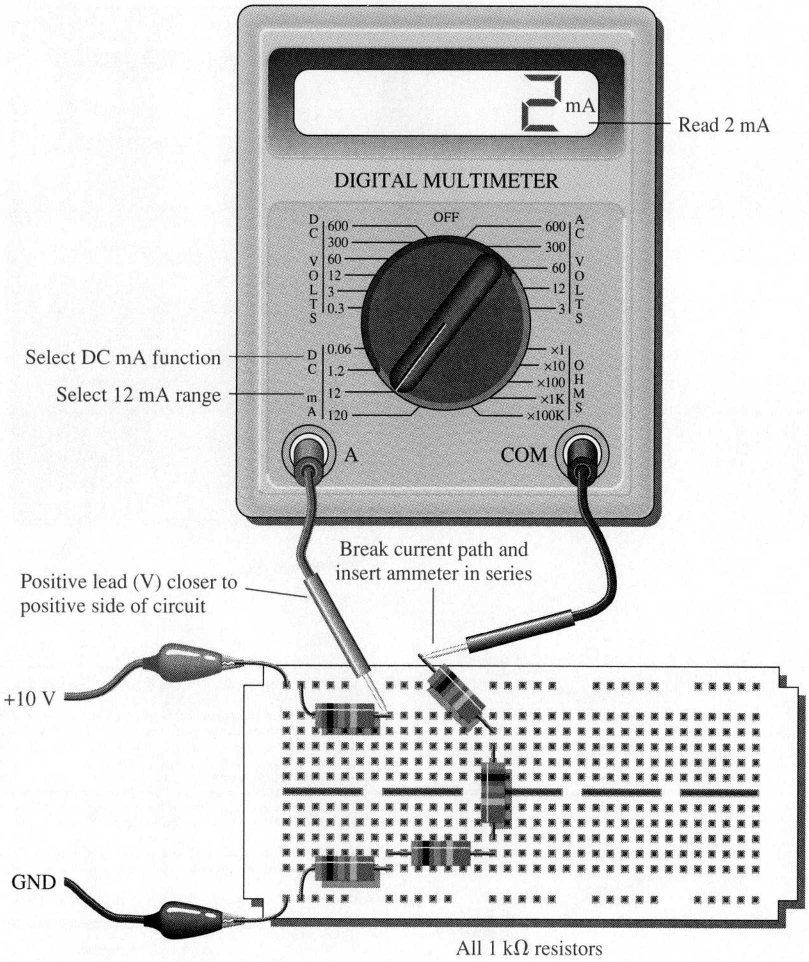

FIGURE C–4
Example of measuring current in a circuit.

Current Measurement

To measure current in a circuit using a multimeter, the following general steps are taken:

1. Insert the red meter lead into the connector labeled for *current* (A or mA) on the meter and insert the black meter lead into the connector labeled *common* (COM).
2. Set the function switch to the *amps* position (select either dc or ac).
3. Set the range switch to the lowest setting that is higher than the current you expect to measure.
4. Select the point in the circuit at which you want to measure the current.
5. Turn off the power and break the circuit at the selected point by simply lifting a component lead or one end of a jumper wire from the socket. This permits the ammeter to be placed in series with the current path.
6. Insert the ammeter in the break by connecting the red lead to the point closest to the positive side of the circuit and the black lead to the point closest to the negative or ground side of the circuit as shown in Figure C–4.
7. Turn the power back on and read the current.

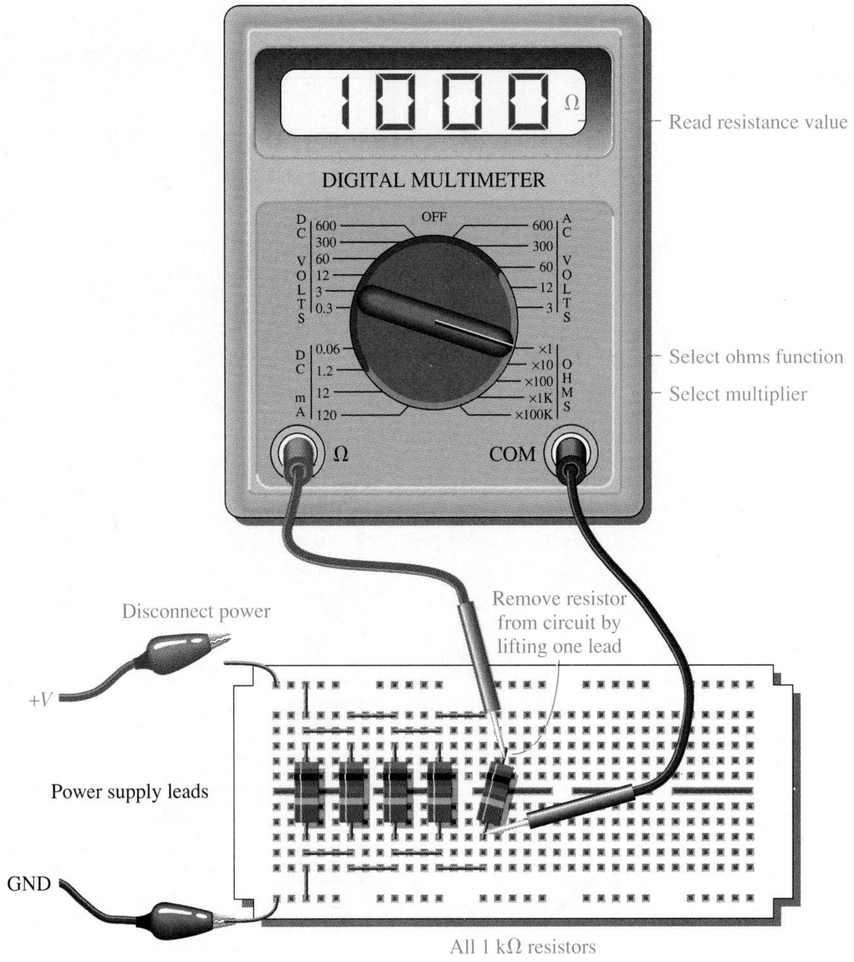

FIGURE C–5
Example of measuring resistance in a circuit.

Resistance Measurement

To measure resistance in a circuit using a multimeter, the following general steps are taken:

1. Insert one meter lead into the connector labeled for *ohms* (Ω) on the meter and the other lead into the connector labeled *common* (COM). Zero the ohmmeter if necessary.
2. Set the function switch to the *ohms* (Ω) position.
3. Select an appropriate multiplier setting (×1, ×10, etc.).
4. Disconnect power from the circuit.
5. Remove the resistor to be measured or lift one lead so that it is disconnected from the rest of the circuit. If you wish to measure total resistance, none of the components has to be removed.
6. Connect the ohmmeter leads across the selected resistor as shown in Figure C–5. Polarity does not matter.
7. Read the resistance and reconnect the resistor and the power.

D CAPACITOR COLOR CODING

Some capacitors have color-coded designations. The color code used for capacitors is basically the same as that used for resistors. Some variations occur in tolerance designation. The basic color codes are shown in Table D–1, and some typical color-coded capacitors are illustrated in Figure D–1.

TABLE D–1
Typical composite color codes for capacitors (picofarads)

COLOR	DIGIT	MULTIPLIER	TOLERANCE
Black	0	1	20%
Brown	1	10	1%
Red	2	100	2%
Orange	3	1000	3%
Yellow	4	10000	
Green	5	100000	5% (EIA)
Blue	6	1000000	
Violet	7		
Gray	8		
White	9		
Gold		0.1	5% (JAN)
Silver		0.01	10%
No color			20%

NOTE: EIA stands for Electronic Industries Association, and JAN stands for Joint Army-Navy, a military standard.

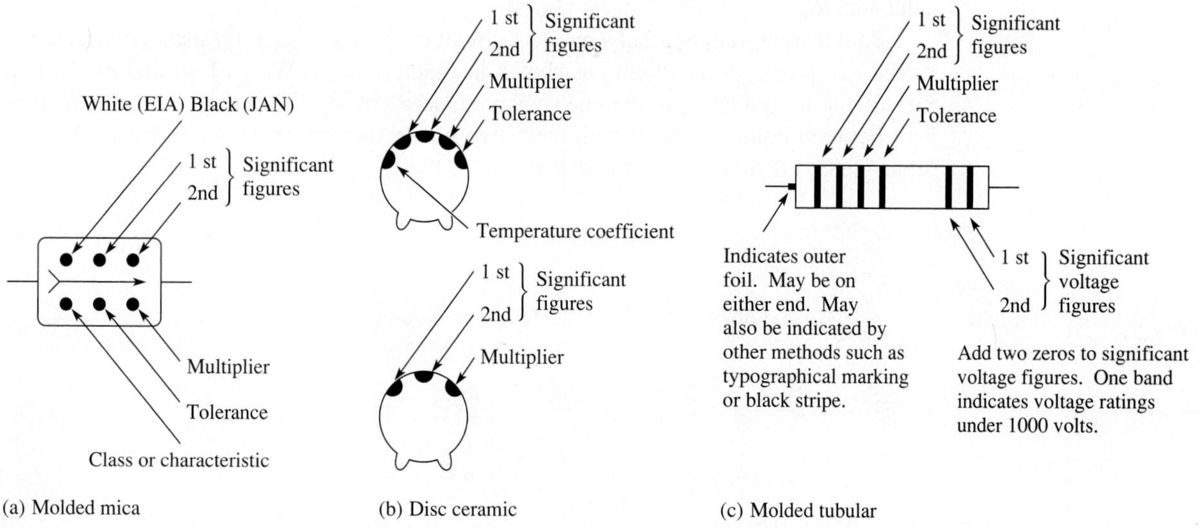

FIGURE D–1
Typical color-coded capacitors.

E NORTON'S THEOREM, MILLMAN'S THEOREM, AND Δ-TO-Y/Y-TO-Δ NETWORK CONVERSIONS

NORTON'S THEOREM

Like Thevenin's theorem, Norton's theorem provides a method of reducing a more complex circuit to a simpler form. The basic difference is that Norton's theorem gives an equivalent current source in parallel with an equivalent resistance. The form of Norton's equivalent circuit is shown in Figure E–1. Regardless of how complex the original circuit is, it can always be reduced to this equivalent form. The equivalent current source is designated I_N, and the equivalent resistance, R_N.

FIGURE E–1
The form of Norton's equivalent circuit.

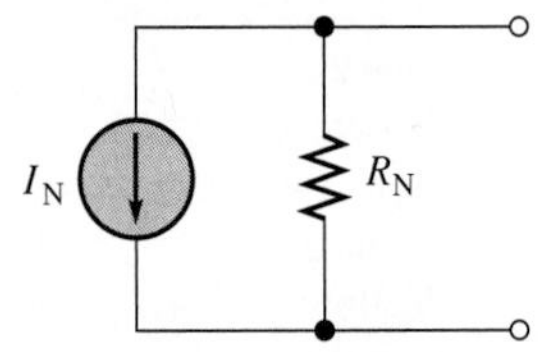

To apply Norton's theorem, you must know how to find the two quantities I_N and R_N. Once you know them for a given circuit, simply connect them in parallel to get the complete Norton circuit.

Norton's Equivalent Current (I_N)

As stated, I_N is one part of the complete Norton equivalent circuit; R_N is the other part. I_N is defined to be the short circuit current between two points in a circuit. Any component connected between these two points effectively "sees" a current source of value I_N in parallel with R_N.

To illustrate, suppose that a resistive circuit of some kind has a resistor connected between two points in the circuit, as shown in Figure E–2(a). We wish to find the Norton circuit that is equivalent to the one shown as "seen" by R_L. To find I_N, calculate the current between points A and B with these two points shorted, as shown in Figure E–2(b). Example E–1 demonstrates how to find I_N.

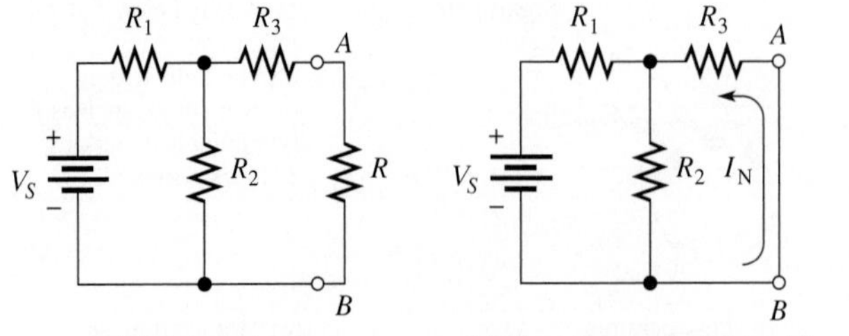

FIGURE E–2
Determining the Norton equivalent current, I_N.

EXAMPLE E–1 Determine I_N for the circuit within the shaded area in Figure E–3(a).

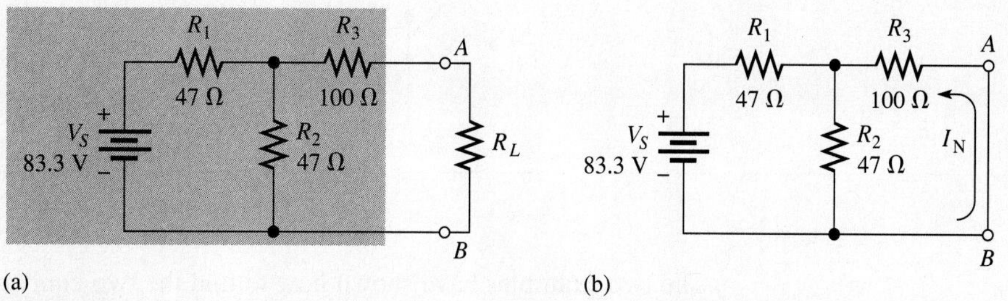

FIGURE E–3

Solution Short terminals A and B as shown in Figure E–3(b). I_N is the current through the short and is calculated as follows: First, the total resistance seen by the voltage source is

$$R_T = R_1 + \frac{R_2 R_3}{R_2 + R_3} = 47\ \Omega + \frac{(47\ \Omega)(100\ \Omega)}{147\ \Omega} = 79\ \Omega$$

The total current from the source is

$$I_T = \frac{V_S}{R_T} = \frac{83.3\ \text{V}}{79\ \Omega} = 1.05\ \text{A}$$

Now apply the current-divider formula to find I_N (the current through the short):

$$I_N = \left(\frac{R_2}{R_2 + R_3}\right) I_T = \left(\frac{47\ \Omega}{147\ \Omega}\right) 1.05\ \text{A} = 336\ \text{mA}$$

This is the value for the equivalent Norton current source.

Norton's Equivalent Resistance (R_N)

We define R_N in the same way as R_{TH}: it is the total resistance appearing between two terminals in a given circuit with all sources replaced by their internal resistances. Example E–2 demonstrates how to find R_N.

EXAMPLE E–2 Find R_N for the circuit within the shaded area of Figure E–3 (see Example E–1).

Solution First reduce V_S to zero by shorting it, as shown in Figure E–4.

Looking in at terminals A and B, you can see that the parallel combination of R_1 and R_2 is in series with R_3. Thus,

$$R_N = R_3 + \frac{R_1}{2} = 100\ \Omega + \frac{47\ \Omega}{2} = 124\ \Omega$$

FIGURE E–4

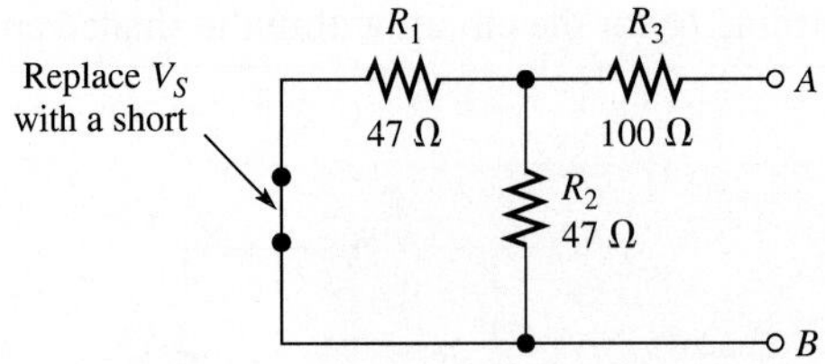

The two examples have shown how to find the two equivalent components of a Norton equivalent circuit, I_N and R_N. Keep in mind that these values can be found for any linear circuit. Once these are known, they must be connected in parallel to form the Norton equivalent circuit, as illustrated in Example E–3.

EXAMPLE E–3

Draw the complete Norton circuit for the original circuit in Figure E–3 (Example E–1).

Solution In Examples E–1 and E–2 you found that $I_N = 336$ mA and $R_N = 124\ \Omega$. The Norton equivalent circuit is shown in Figure E–5.

FIGURE E–5

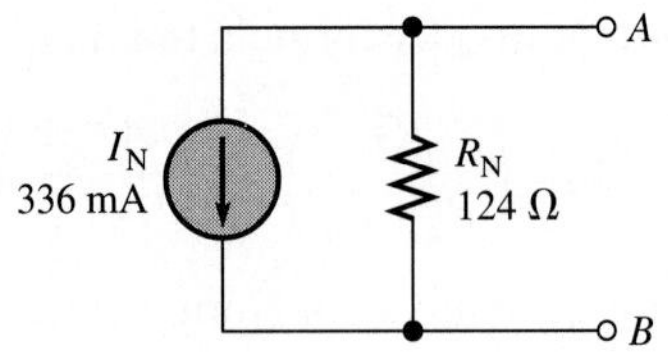

Summary of Norton's Theorem

Any load resistor connected between the terminals of a Norton equivalent circuit will have the same current through it and the same voltage across it as if it were connected to the terminals of the original circuit. A summary of steps for theoretically applying Norton's theorem is as follows:

1. Short the two terminals between which you want to find the Norton equivalent circuit.
2. Determine the current (I_N) through the shorted terminals.
3. Determine the resistance (R_N) between the two terminals (opened) with all voltage sources shorted and all current sources opened ($R_N = R_{TH}$).
4. Connect I_N and R_N in parallel to produce the complete Norton equivalent for the original circuit.

Norton's equivalent circuit can also be derived from Thevenin's equivalent circuit by use of the source conversion method.

MILLMAN'S THEOREM

Millman's theorem permits any number of parallel voltage sources to be reduced to a single equivalent voltage source. It simplifies finding the voltage across or current through a load. Millman's theorem gives the same results as Thevenin's theorem for the special case of parallel voltage sources. A conversion by Millman's theorem is illustrated in Figure E–6.

FIGURE E–6
Reduction of parallel voltage sources to a single equivalent voltage source.

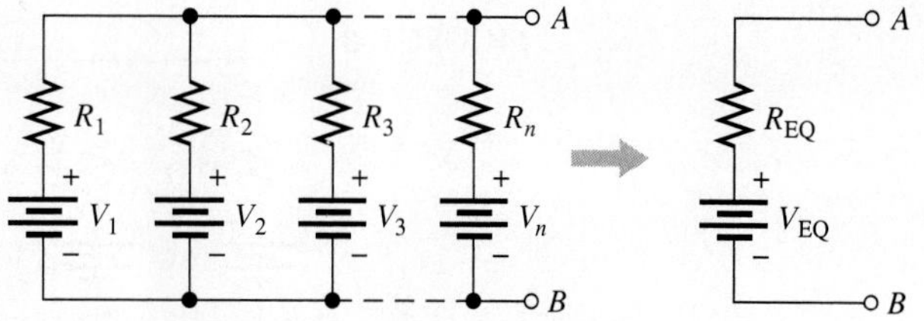

Millman's Equivalent Voltage (V_{EQ}) and Equivalent Resistance (R_{EQ})

Millman's theorem gives us a formula for calculating the equivalent voltage, V_{EQ}. To find V_{EQ}, convert each of the parallel voltage sources into the current sources, as shown in Figure E–7.

In Figure E–7(b), the total current from the parallel current sources is

$$I_T = I_1 + I_2 + I_3 + \cdots + I_n$$

The total conductance between terminals A and B is

$$G_T = G_1 + G_2 + G_3 + \cdots + G_n$$

where $G_T = 1/R_T$, $G_1 = 1/R_1$, and so on. Remember, the current sources are effectively open. Therefore, by Millman's theorem, the equivalent resistance, R_{EQ}, is the total resistance R_T.

$$R_{EQ} = \frac{1}{G_T} = \frac{1}{(1/R_1) + (1/R_2) + (1/R_3) + \cdots + (1/R_n)} \qquad \textbf{(E–1)}$$

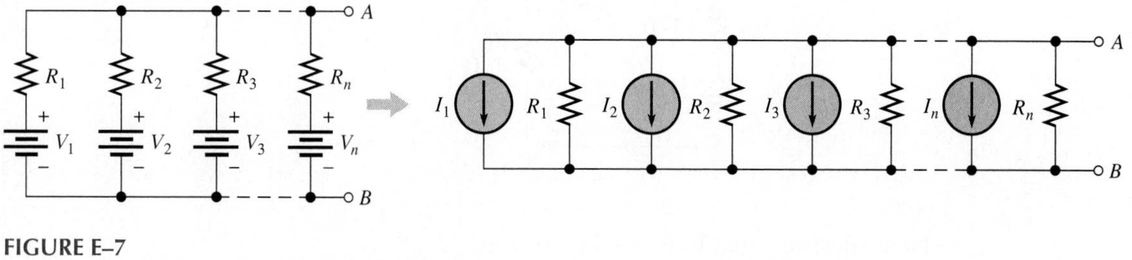

FIGURE E–7
Parallel voltage sources converted to current sources.

By Millman's theorem, the equivalent voltage is $I_T R_{EQ}$, where I_T is expressed as follows:

$$I_T = I_1 + I_2 + I_3 + \cdots + I_n = \frac{V_1}{R_1} + \frac{V_2}{R_2} + \frac{V_3}{R_3} + \cdots + \frac{V_n}{R_n}$$

The following is the formula for the equivalent voltage:

$$V_{EQ} = \frac{(V_1/R_1) + (V_2/R_2) + (V_3/R_3) + \cdots + (V_n/R_n)}{(1/R_1) + (1/R_2) + (1/R_3) + \cdots + (1/R_n)} \qquad \textbf{(E–2)}$$

Equations (E–1) and (E–2) are the two Millman formulas. The equivalent voltage source has a polarity such that the total current through a load will flow in the same direction as in the original circuit.

EXAMPLE E–4 Use Millman's theorem to find the voltage across R_L and the current through R_L in Figure E–8.

FIGURE E–8

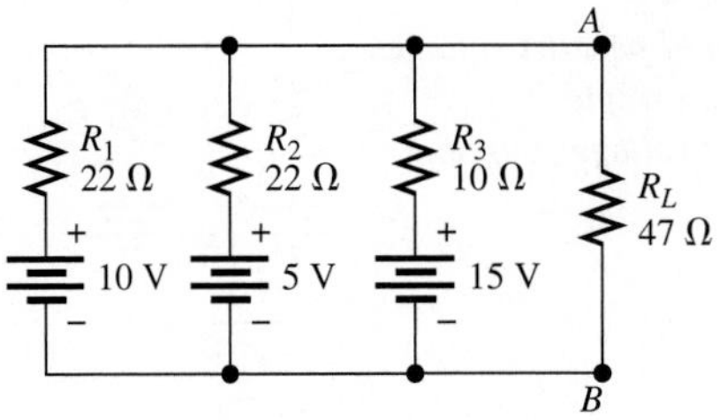

Solution Apply Millman's theorem as follows:

$$R_{EQ} = \frac{1}{(1/R_1) + (1/R_2) + (1/R_3)}$$

$$= \frac{1}{(1/22\ \Omega) + (1/22\ \Omega) + (1/10\ \Omega)} = \frac{1}{0.19} = 5.24\ \Omega$$

$$V_{EQ} = \frac{(V_1/R_2) + (V_2/R_2) + (V_3/R_3)}{(1/R_1) + (1/R_2) + (1/R_3)}$$

$$= \frac{(10\ \text{V}/22\ \Omega) + (5\ \text{V}/22\ \Omega) + (15\ \text{V}/10\ \Omega)}{(1/22\ \Omega) + (1/22\ \Omega) + (1/10\ \Omega)} = \frac{2.18\ \text{A}}{0.19\ \text{S}} = 11.5\ \text{V}$$

The single equivalent voltage source is shown in Figure E–9.

FIGURE E–9

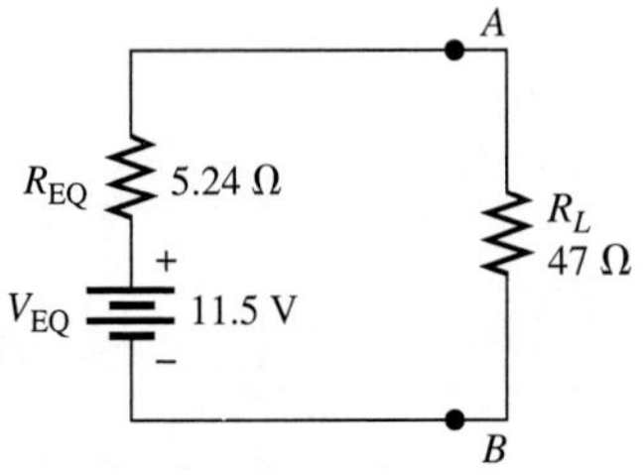

Now calculate I_L and V_L for the load resistor:

$$I_L = \frac{V_{EQ}}{R_{EQ} + R_L} = \frac{11.5\ \text{V}}{52.2\ \Omega} = 220\ \text{mA}$$

$$V_L = I_L R_L = (220\ \text{mA})(47\ \Omega) = 10.3\ \text{V}$$

DELTA-WYE (Δ-TO-Y) AND WYE-DELTA (Y-TO-Δ) NETWORK CONVERSIONS

A resistive delta (Δ) network has the form shown in Figure E–10(a). A wye (Y) network is shown in Figure E–10(b). Notice that letter subscripts are used to designate resistors in the delta network and that numerical subscripts are used to designate resistors in the wye.

FIGURE E–10
Delta and wye networks.

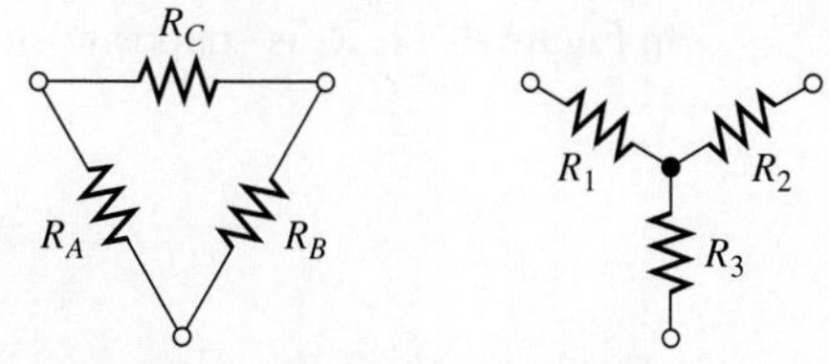

(a) Delta (b) Wye

Conversion between these two forms of circuits is sometimes helpful in areas such as bridge analysis and three-phase power systems. In this section, the conversion formulas and rules for remembering them are given.

Δ-to-Y Conversion

It is convenient to think of the wye positioned within the delta, as shown in Figure E–11. To convert from delta to wye, we need R_1, R_2, and R_3 in terms of R_A, R_B, and R_C. The conversion rule is as follows: *Each resistor in the wye is equal to the product of the resistors in two adjacent delta branches, divided by the sum of all three delta resistors.*

FIGURE E–11
"Y within Δ" aid for conversion formulas.

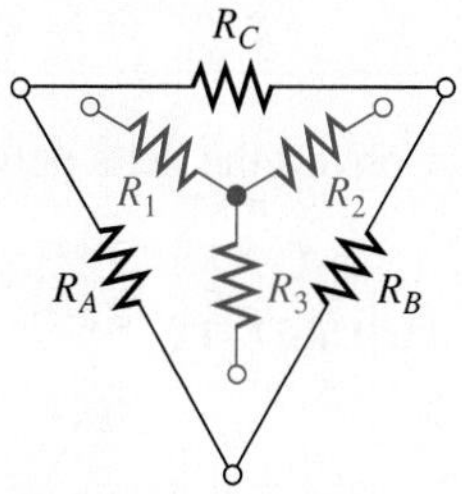

In Figure E–11, R_A and R_C are "adjacent" to R_1:

$$R_1 = \frac{R_A R_C}{R_A + R_B + R_C} \quad \text{(E–3)}$$

Also, R_B and R_C are "adjacent" to R_2:

$$R_2 = \frac{R_B R_C}{R_A + R_B + R_C} \quad \text{(E–4)}$$

and R_A and R_B are "adjacent" to R_3:

$$R_3 = \frac{R_A R_B}{R_A + R_B + R_C} \quad \text{(E–5)}$$

Y-to-Δ Conversion

To convert from wye to delta, we need R_A, R_B, and R_C in terms of R_1, R_2 and R_3. The conversion rule is as follows: *Each resistor in the delta is equal to the sum of all possible products of wye resistors taken two at a time, divided by the opposite wye resistor.*

In Figure E–11, R_2 is "opposite" to R_A:

$$R_A = \frac{R_1R_2 + R_1R_3 + R_2R_3}{R_2} \qquad \text{(E–6)}$$

Also, R_1 is "opposite" to R_B:

$$R_B = \frac{R_1R_2 + R_1R_3 + R_2R_3}{R_1} \qquad \text{(E–7)}$$

and R_3 is "opposite" to R_C:

$$R_C = \frac{R_1R_2 + R_1R_3 + R_2R_3}{R_3} \qquad \text{(E–8)}$$

The following two examples illustrate conversion between these two forms of circuits.

EXAMPLE E–5

Convert the delta network in Figure E–12 to a wye network.

FIGURE E–12

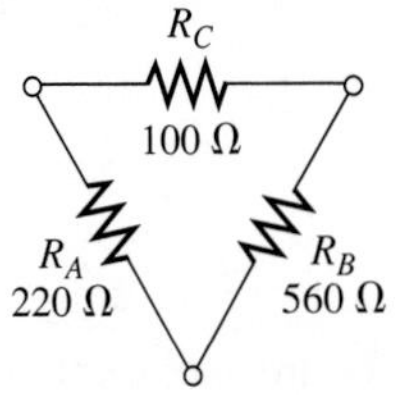

Solution Use Equations (E–3), (E–4), and (E–5):

$$R_1 = \frac{R_AR_C}{R_A + R_B + R_C} = \frac{(220\ \Omega)(100\ \Omega)}{220\ \Omega + 560\ \Omega + 100\ \Omega} = 25\ \Omega$$

$$R_2 = \frac{R_BR_C}{R_A + R_B + R_C} = \frac{(560\ \Omega)(100\ \Omega)}{880\ \Omega} = 63.6\ \Omega$$

$$R_3 = \frac{R_AR_B}{R_A + R_B + R_C} = \frac{(220\ \Omega)(560\ \Omega)}{880\ \Omega} = 140\ \Omega$$

The resulting wye network is shown in Figure E–13.

FIGURE E–13

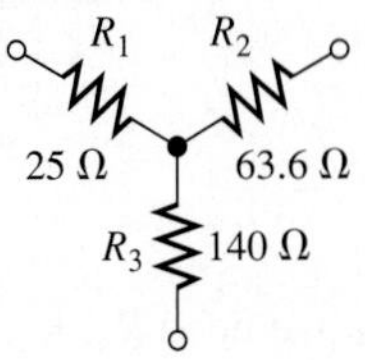

EXAMPLE E–6 Convert the wye network in Figure E–14 to a delta network.

FIGURE E–14

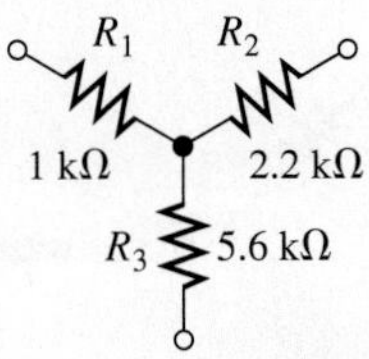

Solution Use Equations (E–6), (E–7), and (E–8):

$$R_A = \frac{R_1R_2 + R_1R_3 + R_2R_3}{R_2}$$

$$= \frac{(1\ \text{k}\Omega)(2.2\ \text{k}\Omega) + (1\ \text{k}\Omega)(5.6\ \text{k}\Omega) + (2.2\ \text{k}\Omega)(5.6\ \text{k}\Omega)}{2.2\ \text{k}\Omega} = 9.15\ \text{k}\Omega$$

$$R_B = \frac{R_1R_2 + R_1R_3 + R_2R_3}{R_1}$$

$$= \frac{(1\ \text{k}\Omega)(2.2\ \text{k}\Omega) + (1\ \text{k}\Omega)(5.6\ \text{k}\Omega) + (2.2\ \text{k}\Omega)(5.6\ \text{k}\Omega)}{1\ \text{k}\Omega} = 20.1\ \text{k}\Omega$$

$$R_C = \frac{R_1R_2 + R_1R_3 + R_2R_3}{R_3}$$

$$= \frac{(1\ \text{k}\Omega)(2.2\ \text{k}\Omega) + (1\ \text{k}\Omega)(5.6\ \text{k}\Omega) + (2.2\ \text{k}\Omega)(5.6\ \text{k}\Omega)}{5.6\ \text{k}\Omega} = 3.59\ \text{k}\Omega$$

The resulting delta network is shown in Figure E–15.

FIGURE E–15

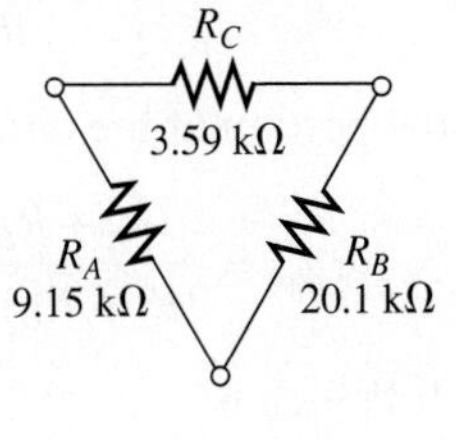

Application of Δ-to-Y Conversion to the Simplification of a Bridge Circuits

In Chapter 6, you saw how Thevenin's theorem can be used to simplify a bridge circuit. Now you will see how Δ-to-Y conversion can be used for conversion of a bridge circuit to a series parallel form for easier analysis.

Figure E–16 illustrates how the delta (Δ) formed by R_A, R_B, and R_C can be converted to a wye (Y), thus creating an equivalent series-parallel circuit. Equations (E–3), (E–4), and (E–5) are used in this conversion.

In a bridge circuit, the load is connected across points C and D. In Figure E–16(a), R_C represents the load resistor. When voltage is applied across points A and B, the voltage

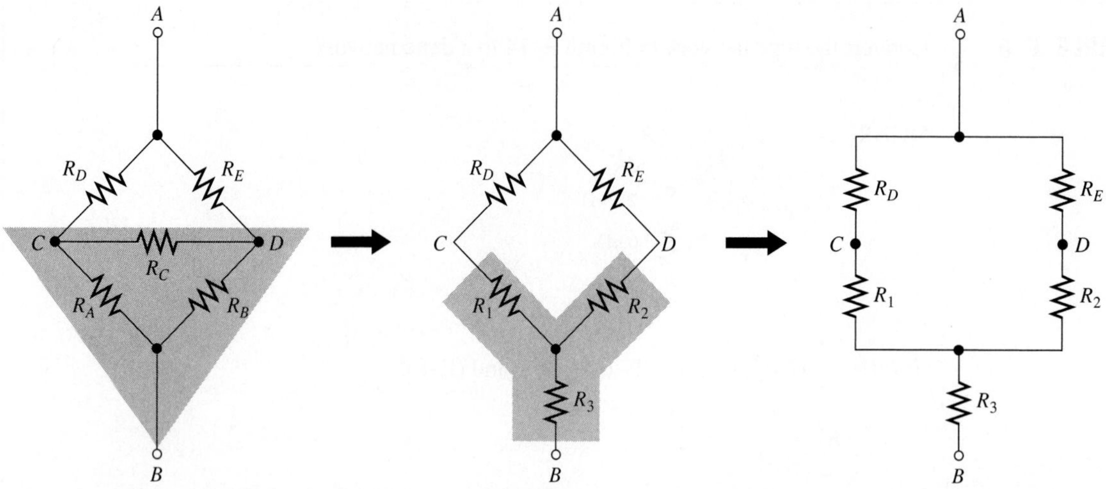

(a) R_A, R_B, and R_C form a delta.

(b) R_1, R_2, and R_3 form an equivalent wye.

(c) Part (b) redrawn as a series-parallel circuit.

FIGURE E–16
Conversion of a bridge circuit to a series-parallel configuration.

from C to D (V_{CD}) can be determined using the equivalent series-parallel circuit in Figure E–16(c) as follows. The total resistance from point A to point B is

$$R_T = \frac{(R_1 + R_D)(R_2 + R_E)}{(R_1 + R_D) + (R_2 + R_E)} + R_3$$

Then

$$I_T = \frac{V_{AB}}{R_T}$$

The resistance of the parallel portion of the circuit in Figure E–16(c) is

$$R_{T(\mathrm{p})} = \frac{(R_1 + R_D)(R_2 + R_E)}{(R_1 + R_D) + (R_2 + R_E)}$$

The current through the left branch is

$$I_{AC} = \left(\frac{R_{T(\mathrm{p})}}{R_1 + R_D}\right) I_T$$

The current through the right branch is

$$I_{AD} = \left(\frac{R_{T(\mathrm{p})}}{R_2 + R_E}\right) I_T$$

The voltage at point C with respect to point A is

$$V_{CA} = V_A - I_{AC} R_D$$

The voltage at point D with respect to point A is

$$V_{DA} = V_A - I_{AD} R_E$$

The voltage from point C to point D is

$$V_{CD} = V_{CA} - V_{DA} = (V_A - I_{AC}R_D) - (V_A - I_{AD}R_E)$$
$$= I_{AD}R_E - I_{AC}R_D$$

V_{CD} is the voltage across the load (R_C) in the bridge circuit of Figure E–16(a). The current through the load can be found by Ohm's law:

$$I_C = \frac{V_{CD}}{R_C}$$

EXAMPLE E–7 Determine the load voltage and the load current in the bridge circuit in Figure E–17. Notice that the resistors are labeled for convenient conversion using the previous formulas. R_C is the load resistor.

FIGURE E–17

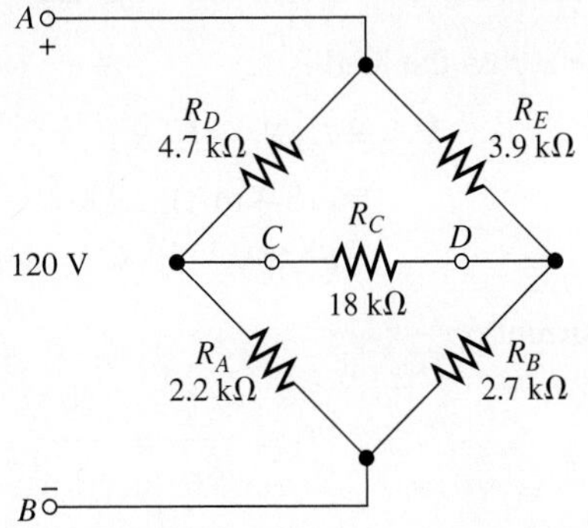

Solution First convert the delta formed by R_A, R_B, and R_C to a wye:

$$R_1 = \frac{R_A R_C}{R_A + R_B + R_C} = \frac{(2.2\text{ k}\Omega)(18\text{ k}\Omega)}{2.2\text{ k}\Omega + 2.7\text{ k}\Omega + 18\text{ k}\Omega} = 1.73\text{ k}\Omega$$

$$R_2 = \frac{R_B R_C}{R_A + R_B + R_C} = \frac{(2.7\text{ k}\Omega)(18\text{ k}\Omega)}{22.9\text{ k}\Omega} = 2.12\text{ k}\Omega$$

$$R_3 = \frac{R_A R_B}{R_A + R_B + R_C} = \frac{(2.2\text{ k}\Omega)(2.7\text{ k}\Omega)}{22.9\text{ k}\Omega} = 259\ \Omega$$

The resulting equivalent series-parallel circuit is shown in Figure E–18.

FIGURE E–18

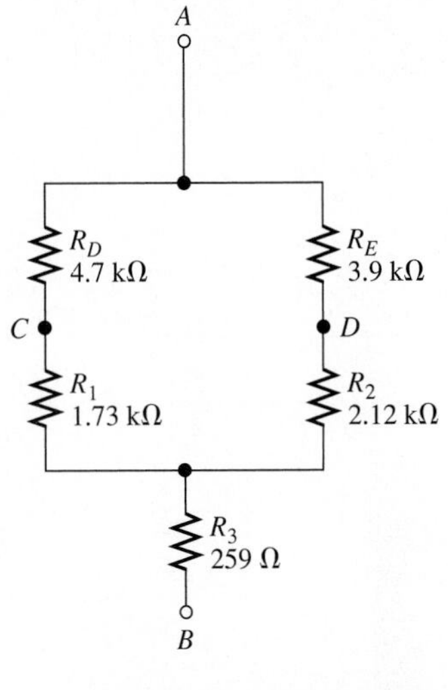

Now determine R_T and the branch currents in Figure E–18.

$$R_T = \frac{(R_1 + R_D)(R_2 + R_E)}{(R_1 + R_D) + (R_2 + R_E)} + R_3 = \frac{(6.43\ \text{k}\Omega)(6.02\ \text{k}\Omega)}{6.43\ \text{k}\Omega + 6.02\ \text{k}\Omega} + 259\ \Omega = 3.37\ \text{k}\Omega$$

$$I_T = \frac{V_{AB}}{R_T} = \frac{120\ \text{V}}{3.37\ \text{k}\Omega} = 35.6\ \text{mA}$$

The total resistance of the parallel part of the circuit, $R_{T(\text{p})}$, is 3.11 kΩ.

$$I_{AC} = \left(\frac{R_{T(\text{p})}}{R_1 + R_D}\right) I_T = \left(\frac{3.11\ \text{k}\Omega}{1.73\ \text{k}\Omega + 4.7\ \text{k}\Omega}\right) 35.6\ \text{mA} = 17.2\ \text{mA}$$

$$I_{AD} = \left(\frac{R_{T(\text{p})}}{R_2 + R_E}\right) I_T = \left(\frac{3.11\ \text{k}\Omega}{2.12\ \text{k}\Omega + 3.9\ \text{k}\Omega}\right) 35.6\ \text{mA} = 18.4\ \text{mA}$$

The voltage across the load is

$$\begin{aligned} V_{CD} &= I_{AD} R_E - I_{AC} R_D \\ &= (18.4\ \text{mA})(3.9\ \text{k}\Omega) - (17.2\ \text{mA})(4.7\ \text{k}\Omega) \\ &= 71.8\ \text{V} - 80.8\ \text{V} = -9\ \text{V} \end{aligned}$$

The load current is

$$I_C = \frac{V_{CD}}{R_C} = \frac{-9\ \text{V}}{18\ \text{k}\Omega} = -500\ \mu\text{A}$$

F DEVICES DATA SHEETS

MOTOROLA

1N4001 thru 1N4007

Designers▲ Data Sheet

"SURMETIC"▲ RECTIFIERS

. . . subminiature size, axial lead mounted rectifiers for general-purpose low-power applications.

Designers Data for "Worst Case" Conditions

The Designers▲ Data Sheets permit the design of most circuits entirely from the information presented. Limit curves – representing boundaries on device characteristics – are given to facilitate "worst case" design.

LEAD MOUNTED
SILICON RECTIFIERS

50-1000 VOLTS
DIFFUSED JUNCTION

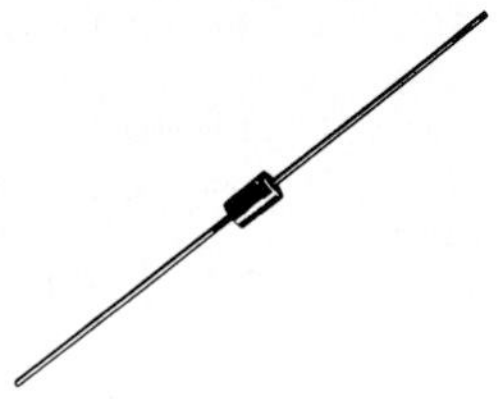

*MAXIMUM RATINGS

Rating	Symbol	1N4001	1N4002	1N4003	1N4004	1N4005	1N4006	1N4007	Unit
Peak Repetitive Reverse Voltage Working Peak Reverse Voltage DC Blocking Voltage	V_{RRM} V_{RWM} V_R	50	100	200	400	600	800	1000	Volts
Non-Repetitive Peak Reverse Voltage (halfwave, single phase, 60 Hz)	V_{RSM}	60	120	240	480	720	1000	1200	Volts
RMS Reverse Voltage	$V_{R(RMS)}$	35	70	140	280	420	560	700	Volts
Average Rectified Forward Current (single phase, resistive load, 60 Hz, see Figure 8, T_A = 75°C)	I_O	1.0							Amp
Non-Repetitive Peak Surge Current (surge applied at rated load conditions, see Figure 2)	I_{FSM}	30 (for 1 cycle)							Amp
Operating and Storage Junction Temperature Range	T_J, T_{stg}	-65 to +175							°C

*ELECTRICAL CHARACTERISTICS

Characteristic and Conditions	Symbol	Typ	Max	Unit
Maximum Instantaneous Forward Voltage Drop (i_F = 1.0 Amp, T_J = 25°C) Figure 1	v_F	0.93	1.1	Volts
Maximum Full-Cycle Average Forward Voltage Drop (I_O = 1.0 Amp, T_L = 75°C, 1 inch leads)	$V_{F(AV)}$	–	0.8	Volts
Maximum Reverse Current (rated dc voltage)	I_R			μA
T_J = 25°C		0.05	10	
T_J = 100°C		1.0	50	
Maximum Full-Cycle Average Reverse Current (I_O = 1.0 Amp, T_L = 75°C, 1 inch leads	$I_{R(AV)}$	–	30	μA

*Indicates JEDEC Registered Data.

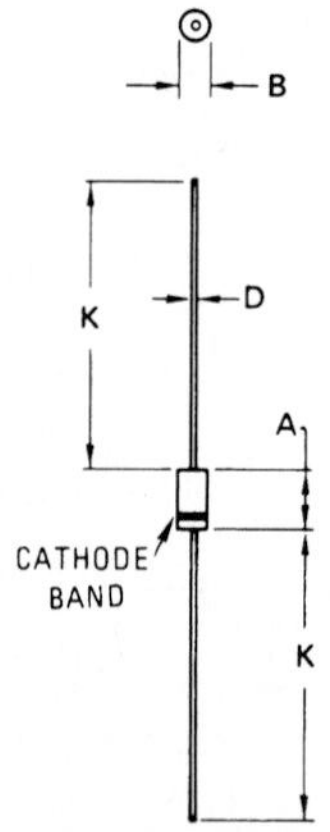

	MILLIMETERS		INCHES	
DIM	MIN	MAX	MIN	MAX
A	5.97	6.60	0.235	0.260
B	2.79	3.05	0.110	0.120
D	0.76	0.86	0.030	0.034
K	27.94	–	1.100	–

CASE 59-04
Does Not Conform to DO-41 Outline.

MECHANICAL CHARACTERISTICS

CASE: Transfer Molded Plastic
MAXIMUM LEAD TEMPERATURE FOR SOLDERING PURPOSES: 350°C, 3/8" from case for 10 seconds at 5 lbs. tension
FINISH: All external surfaces are corrosion-resistant, leads are readily solderable
POLARITY: Cathode indicated by color band
WEIGHT: 0.40 Grams (approximately)

▲Trademark of Motorola Inc.

MOTOROLA

1N4728, A thru 1N4764, A

Designers▲ Data Sheet

ONE WATT HERMETICALLY SEALED GLASS SILICON ZENER DIODES

- Complete Voltage Range – 2.4 to 100 Volts
- DO-41 Package – Smaller than Conventional DO-7 Package
- Double Slug Type Construction
- Metallurgically Bonded Construction
- Nitride Passivated Die

Designer's Data for "Worst Case" Conditions

The Designers▲ Data sheets permit the design of most circuits entirely from the information presented. Limit curves – representing boundaries on device characteristics – are given to facilitate "worst case" design.

1.0 WATT

ZENER REGULATOR DIODES

3.3–100 VOLTS

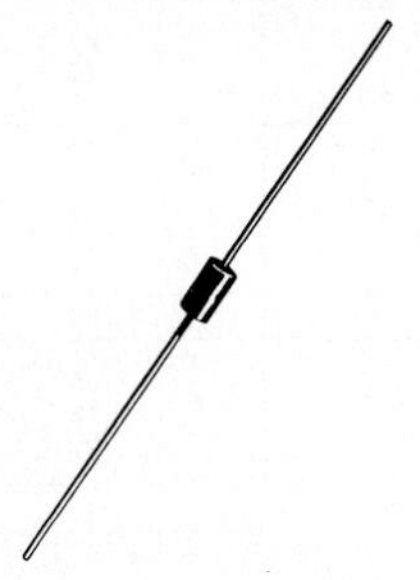

*MAXIMUM RATINGS

Rating	Symbol	Value	Unit
DC Power Dissipation @ $T_A = 50^oC$ Derate above 50^oC	P_D	1.0 6.67	Watt $mW/^oC$
Operating and Storage Junction Temperature Range	T_J, T_{stg}	-65 to +200	^oC

MECHANICAL CHARACTERISTICS

CASE: Double slug type, hermetically sealed glass

MAXIMUM LEAD TEMPERATURE FOR SOLDERING PURPOSES: 230°C, 1/16" from case for 10 seconds

FINISH: All external surfaces are corrosion resistant with readily solderable leads.

POLARITY: Cathode indicated by color band. When operated in zener mode, cathode will be positive with respect to anode.

MOUNTING POSITION: Any

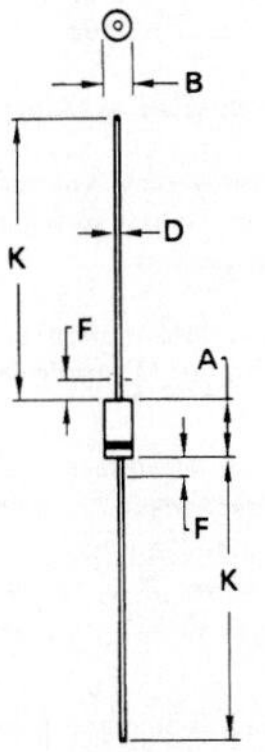

NOTE:
1. POLARITY DENOTED BY CATHODE BAND
2. LEAD DIAMETER NOT CONTROLLED WITHIN "F" DIMENSION.

DIM	MILLIMETERS MIN	MILLIMETERS MAX	INCHES MIN	INCHES MAX
A	4.07	5.20	0.160	0.205
B	2.04	2.71	0.080	0.107
D	0.71	0.86	0.028	0.034
F	–	1.27	–	0.050
K	27.94	–	1.100	–

All JEDEC dimensions and notes apply.

CASE 59-03 (DO-41)

FIGURE 1 – POWER TEMPERATURE DERATING CURVE

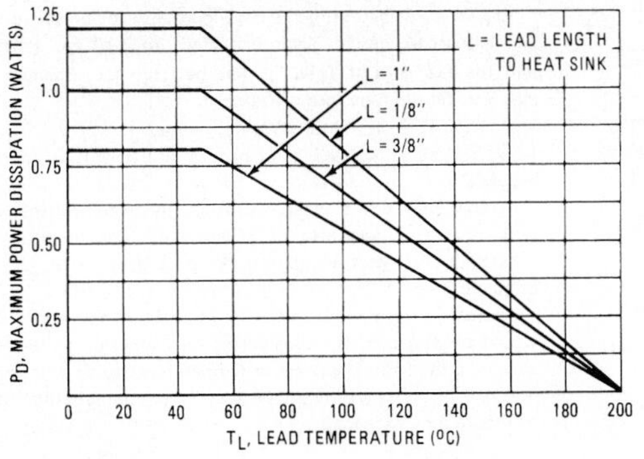

*Indicates JEDEC Registered Data
▲Trademark of Motorola Inc.

*ELECTRICAL CHARACTERISTICS (T_A = 25°C unless otherwise noted) V_F = 1.2 V max, I_F = 200 mA for all types.

JEDEC Type No. (Note 1)	Nominal Zener Voltage V_Z @ I_{ZT} Volts (Notes 2 and 3)	Test Current I_{ZT} mA	Maximum Zener Impedance (Note 4)			Leakage Current		Surge Current @ T_A = 25°C i_r – mA (Note 5)
			Z_{ZT} @ I_{ZT} Ohms	Z_{ZK} @ I_{ZK} Ohms	I_{ZK} mA	I_R μA Max	V_R Volts	
1N4728	3.3	76	10	400	1.0	100	1.0	1380
1N4729	3.6	69	10	400	1.0	100	1.0	1260
1N4730	3.9	64	9.0	400	1.0	50	1.0	1190
1N4731	4.3	58	9.0	400	1.0	10	1.0	1070
1N4732	4.7	53	8.0	500	1.0	10	1.0	970
1N4733	5.1	49	7.0	550	1.0	10	1.0	890
1N4734	5.6	45	5.0	600	1.0	10	2.0	810
1N4735	6.2	41	2.0	700	1.0	10	3.0	730
1N4736	6.8	37	3.5	700	1.0	10	4.0	660
1N4737	7.5	34	4.0	700	0.5	10	5.0	605
1N4738	8.2	31	4.5	700	0.5	10	6.0	550
1N4739	9.1	28	5.0	700	0.5	10	7.0	500
1N4740	10	25	7.0	700	0.25	10	7.6	454
1N4741	11	23	8.0	700	0.25	5.0	8.4	414
1N4742	12	21	9.0	700	0.25	5.0	9.1	380
1N4743	13	19	10	700	0.25	5.0	9.9	344
1N4744	15	17	14	700	0.25	5.0	11.4	304
1N4745	16	15.5	16	700	0.25	5.0	12.2	285
1N4746	18	14	20	750	0.25	5.0	13.7	250
1N4747	20	12.5	22	750	0.25	5.0	15.2	225
1N4748	22	11.5	23	750	0.25	5.0	16.7	205
1N4749	24	10.5	25	750	0.25	5.0	18.2	190
1N4750	27	9.5	35	750	0.25	5.0	20.6	170
1N4751	30	8.5	40	1000	0.25	5.0	22.8	150
1N4752	33	7.5	45	1000	0.25	5.0	25.1	135
1N4753	36	7.0	50	1000	0.25	5.0	27.4	125
1N4754	39	6.5	60	1000	0.25	5.0	29.7	115
1N4755	43	6.0	70	1500	0.25	5.0	32.7	110
1N4756	47	5.5	80	1500	0.25	5.0	35.8	95
1N4757	51	5.0	95	1500	0.25	5.0	38.8	90
1N4758	56	4.5	110	2000	0.25	5.0	42.6	80
1N4759	62	4.0	125	2000	0.25	5.0	47.1	70
1N4760	68	3.7	150	2000	0.25	5.0	51.7	65
1N4761	75	3.3	175	2000	0.25	5.0	56.0	60
1N4762	82	3.0	200	3000	0.25	5.0	62.2	55
1N4763	91	2.8	250	3000	0.25	5.0	69.2	50
1N4764	100	2.5	350	3000	0.25	5.0	76.0	45

*Indicates JEDEC Registered Data.

NOTE 1 – Tolerance and Type Number Designation. The JEDEC type numbers listed have a standard tolerance on the nominal zener voltage of ±10%. A standard tolerance of ±5% on individual units is also available and is indicated by suffixing "A" to the standard type number.

NOTE 2 – Specials Available Include:

A. Nominal zener voltages between the voltages shown and tighter voltage tolerances,

B. Matched sets.

For detailed information on price, availability, and delivery, contact your nearest Motorola representative.

NOTE 3 – Zener Voltage (V_Z) Measurement. Motorola guarantees the zener voltage when measured at 90 seconds while maintaining the lead temperature (T_L) at 30°C ± 1°C, 3/8" from the diode body.

NOTE 4 – Zener Impedance (Z_Z) Derivation. The zener impedance is derived from the 60 cycle ac voltage, which results when an ac current having an rms value equal to 10% of the dc zener current (I_{ZT} or I_{ZK}) is superimposed on I_{ZT} or I_{ZK}.

NOTE 5 – Surge Current (i_r) Non-Repetitive. The rating listed in the electrical characteristics table is maximum peak, non-repetitive, reverse surge current of 1/2 square wave or equivalent sine wave pulse of 1/120 second duration superimposed on the test current, I_{ZT}, per JEDEC registration; however, actual device capability is as described in Figures 4 and 5.

APPLICATION NOTE

Since the actual voltage available from a given zener diode is temperature dependent, it is necessary to determine junction temperature under any set of operating conditions in order to calculate its value. The following procedure is recommended:

Lead Temperature, T_L, should be determined from

$$T_L = \theta_{LA} P_D + T_A$$

θ_{LA} is the lead-to-ambient thermal resistance (°C/W) and P_D is the power dissipation. The value for θ_{LA} will vary and depends on the device mounting method. θ_{LA} is generally 30 to 40°C/W for the various clips and tie points in common use and for printed circuit board wiring.

The temperature of the lead can also be measured using a thermocouple placed on the lead as close as possible to the tie point. The thermal mass connected to the tie point is normally large enough so that it will not significantly respond to heat surges generated in the diode as a result of pulsed operation once steady-state conditions are achieved. Using the measured value of T_L, the junction temperature may be determined by:

$$T_J = T_L + \Delta T_{JL}.$$

ΔT_{JL} is the increase in junction temperature above the lead temperature and may be found as follows:

$$\Delta T_{JL} = \theta_{JL} P_D$$

θ_{JL} may be determined from Figure 3 for dc power conditions. For worst-case design, using expected limits of I_Z, limits of P_D and the extremes of $T_J(\Delta T_J)$ may be estimated. Changes in voltage, V_Z, can then be found from:

$$\Delta V = \theta_{VZ} \Delta T_J$$

θ_{VZ}, the zener voltage temperature coefficient, is found from Figure 2.

Under high power-pulse operation, the zener voltage will vary with time and may also be affected significantly by the zener resistance. For best regulation, keep current excursions as low as possible.

Surge limitations are given in Figure 5. They are lower than would be expected by considering only junction temperature, as current crowding effects cause temperatures to be extremely high in small spots resulting in device degradation should the limits of Figure 5 be exceeded.

MOTOROLA

2N4877

MEDIUM-POWER NPN SILICON TRANSISTOR

. . . designed for switching and wide band amplifier applications.

- Low Collector-Emitter Saturation Voltage — $V_{CE(sat)}$ = 1.0 Vdc (Max) @ I_C = 4.0 Amp
- DC Current Gain Specified to 4 Amperes
- Excellent Safe Operating Area
- Packaged in the Compact TO-39 Case for Critical Space-Limited Applications.

4 AMPERE POWER TRANSISTOR

NPN SILICON
60 VOLTS
10 WATTS

* MAXIMUM RATINGS

Rating	Symbol	Value	Unit
Collector-Emitter Voltage	V_{CEO}	60	Vdc
Collector-Base Voltage	V_{CB}	70	Vdc
Emitter-Base Voltage	V_{EB}	5.0	Vdc
Collector Current — Continuous	I_C	4.0	Adc
Base Current	I_B	1.0	Adc
Total Device Dissipation @ T_C = 25°C Derate above 25°C	P_D	10 57.2	Watts mW/°C
Operating and Storage Junction Temperature Range	T_J, T_{stg}	-65 to +200	°C

*Indicates JEDEC Registered Data

THERMAL CHARACTERISTICS

Characteristic	Symbol	Max	Unit
Thermal Resistance, Junction to Case	θ_{JC}	17.5	°C/W

FIGURE 1 — POWER-TEMPERATURE DERATING CURVE

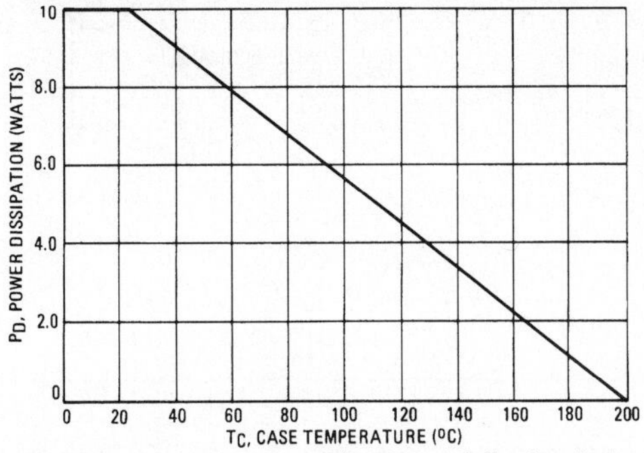

Safe Area Curves are indicated by Figure 2. All limits are applicable and must be observed.

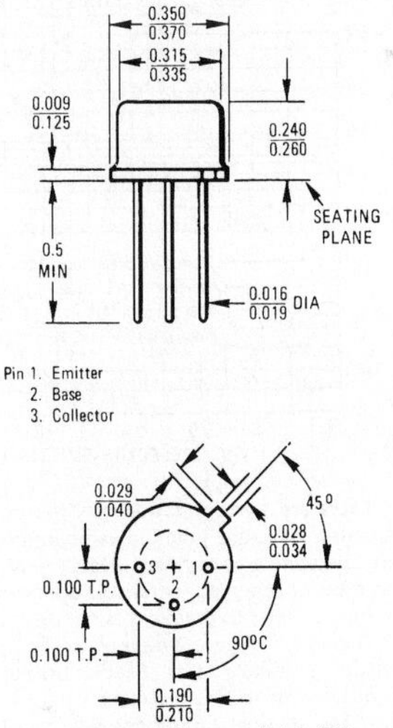

To convert inches to millimeters multiply by 25.4.
All JEDEC TO-39 dimensions and notes apply.

CASE 79 (1)
TO-39

***ELECTRICAL CHARACTERISTICS** ($T_C = 25^oC$ unless otherwise noted)

Characteristic	Symbol	Min	Max	Unit
OFF CHARACTERISTICS				
Collector-Emitter Sustaining Voltage (1) (I_C = 200 mAdc, I_B = 0)	$V_{CEO(sus)}$	60	–	Vdc
Collector Cutoff Current (V_{CE} = 70 Vdc, $V_{EB(off)}$ = 1.5 Vdc)	I_{CEX}	–	100	μAdc
(V_{CE} = 70 Vdc, $V_{EB(off)}$ = 1.5 Vdc, $T_C = 100^oC$)		–	1.0	mAdc
Collector Cutoff Current (V_{CB} = 70 Vdc, I_E = 0)	I_{CBO}	–	100	μAdc
Emitter Cutoff Current (V_{BE} = 5.0 Vdc, I_C = 0)	I_{EBO}	–	100	μAdc
ON CHARACTERISTICS(1)				
DC Current Gain (I_C = 1.0 Adc, V_{CE} = 2.0 Vdc)	h_{FE}	30	–	–
(I_C = 4.0 Adc, V_{CE} = 2.0 Vdc)		20	100	
Collector-Emitter Saturation Voltage (I_C = 4.0 Adc, I_B = 0.4 Adc)	$V_{CE(sat)}$	–	1.0	Vdc
Base-Emitter Saturation Voltage (I_C = 4.0 Adc, I_B = 0.4 Adc)	$V_{BE(sat)}$	–	1.8	Vdc
DYNAMIC CHARACTERISTICS				
Current-Gain-Bandwidth Product (I_C = 0.25 Adc, V_{CE} = 10 Vdc, f = 1.0 MHz)	f_T	4.0	–	MHz
(I_C = 0.25 Adc, V_{CE} = 10 Vdc, f = 10 MHz)**		30	–	
SWITCHING CHARACTERISTICS				
Rise Time (V_{CC} = 25 Vdc, I_C = 4.0 Adc, I_{B1} = 0.4 Adc)	t_r	–	100	ns
Storage Time (V_{CC} = 25 Vdc, I_C = 4.0 Adc, $I_{B1} = I_{B2}$ = 0.4 Adc)	t_s	–	1.5	μs
Fall Time	t_f	–	500	ns

*Indicates JEDEC Registered Data.
**Motorola guarantees this value in addition to JEDEC Registered Data.
Note 1: Pulse Test: Pulse Width ≤ 300 μs, Duty Cycle ≤ 2.0%.

FIGURE 2 – ACTIVE-REGION SAFE OPERATING AREA

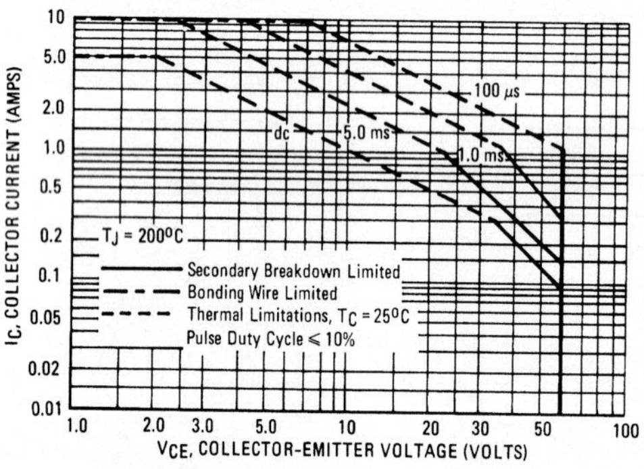

FIGURE 3 – SWITCHING TIME TEST CIRCUIT

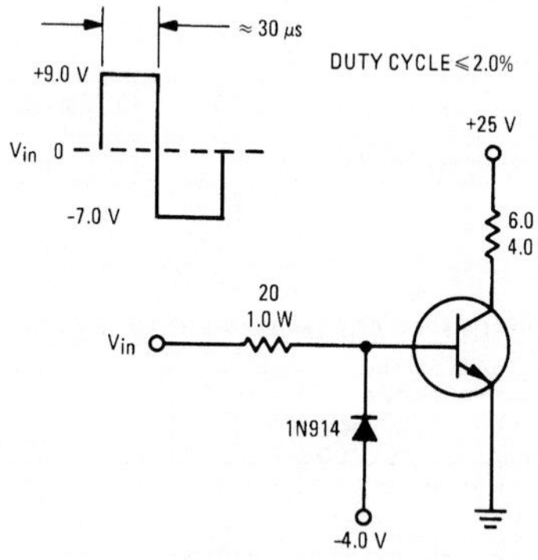

There are two limitations on the power handling ability of a transistor: average junction temperature and second breakdown. Safe operating area curves indicate $I_C - V_{CE}$ limits of the transistor that must be observed for reliable operation; i.e., the transistor must not be subjected to greater dissipation than the curves indicate.

The data of Figure 2 is based on $T_{J(pk)} = 200^oC$; T_C is variable depending on conditions. Second breakdown pulse limits are valid for duty cycles to 10% provided $T_{J(pk)} \leq 200^oC$. At high case temperatures, thermal limitations will reduce the power that can be handled to values less than the limitations imposed by second breakdown. (See AN-415)

MOTOROLA
Semiconductors
BOX 20912 • PHOENIX, ARIZONA 85036

2N1595 thru 2N1599

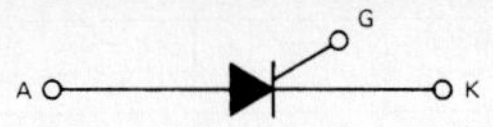

SILICON CONTROLLED RECTIFIERS

1.6 AMPERE RMS
50 thru 400 VOLTS

REVERSE BLOCKING TRIODE THYRISTORS

These devices are glassivated planar construction designed for gating operation in mA/μA signal or detection circuits.

- Low-Level Gate Characteristics — I_{GT} = 10 mA (Max) @ 25°C
- Low Holding Current — I_H = 5.0 mA (Typ) @ 25°C
- Glass-to-Metal Bond for Maximum Hermetic Seal

***MAXIMUM RATINGS** (T_J = 125°C unless otherwise noted).

Rating	Symbol	Value	Unit
Repetitive Peak Reverse Blocking Voltage	V_{RRM}		Volts
2N1595		50	
2N1596		100	
2N1597		200	
2N1598		300	
2N1599		400	
Repetitive Peak Forward Blocking Voltage	V_{DRM}		Volts
2N1595		50	
2N1596		100	
2N1597		200	
2N1598		300	
2N1599		400	
RMS On-State Current (All Conduction Angles)	$I_{T(RMS)}$	1.6	Amps
Peak Non-Repetitive Surge Current (One Cycle, 60 Hz, T_J = -65 to +125°C)	I_{TSM}	15	Amps
Peak Gate Power	P_{GM}	0.1	Watt
Average Gate Power	$P_{G(AV)}$	0.01	Watt
Peak Gate Current	I_{GM}	0.1	Amp
Peak Gate Voltage — Forward	V_{GFM}	10	Volts
Reverse	V_{GRM}	10	
Operating Junction Temperature Range	T_J	-65 to +125	°C
Storage Temperature Range	T_{stg}	-65 to +150	°C

*Indicates JEDEC Registered Data.

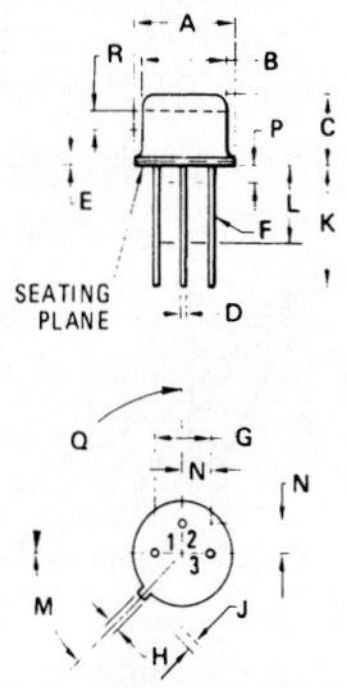

STYLE 3:
PIN 1. CATHODE
2. GATE
3. ANODE (CONNECTED TO CASE)

DIM	MILLIMETERS MIN	MILLIMETERS MAX	INCHES MIN	INCHES MAX
A	8.89	9.40	0.350	0.370
B	8.00	8.51	0.315	0.335
C	6.10	6.60	0.240	0.260
D	0.406	0.533	0.016	0.021
E	0.229	3.18	0.009	0.125
F	0.406	0.483	0.016	0.019
G	4.83	5.33	0.190	0.210
H	0.711	0.864	0.028	0.034
J	0.737	1.02	0.029	0.040
K	12.70	–	0.500	–
L	6.35	–	0.250	–
M	45° NOM		45° NOM	
P	–	1.27	–	0.050
Q	90° NOM		90° NOM	
R	2.54	–	0.100	–

All JEDEC dimensions and notes apply.

CASE 79-02
TO-39

2N1595 thru 2N1599

ELECTRICAL CHARACTERISTICS (T_C = 25°C unless otherwise noted).

Characteristic	Symbol	Min	Typ	Max	Unit
*Peak Reverse Blocking Current (Rated V_{RRM}, T_J = 125°C)	I_{RRM}	–	–	1000	μA
*Peak Forward Blocking Current (Rated V_{DRM}, T_J = 125°C)	I_{DRM}	–	–	1000	μA
*Peak On-State Voltage (I_F = 1.0 Adc, Pulsed, 1.0 ms (Max), Duty Cycle ≤ 1%)	V_{TM}		1.1	2.0	Volts
*Gate Trigger Current (V_{AK} = 6.0 V, R_L = 12 Ohms)	I_{GT}	–	2.0	10	mA
*Gate Trigger Voltage	V_{GT}				Volts
(V_{AK} = 6.0 V, R_L = 12 Ohms)		–	0.7	3.0	
(V_{AK} = 6.0 V, R_L = 12 Ohms, T_J = 125°C)		0.2	–	–	
Reverse Gate Current (V_{GK} = 10 V)	I_{GR}	–	17	–	mA
Holding Current (V_{AK} = 12 V)	I_H	–	5.0	–	mA
Turn-On Time	t_{gt}				μs
(I_{GT} = 10 mA, I_F = 1.0 A)		–	0.8	–	
(I_{GT} = 20 mA, I_F = 1.0 A)		–	0.6	–	
Turn-Off Time (I_F = 1.0 A, I_R = 1.0 A, dv/dt = 20 V/μs, T_J = 125°C)	t_q	–	10	–	μs

*Indicates JEDEC Registered Data.

CURRENT DERATING

FIGURE 1 – CASE TEMPERATURE REFERENCE

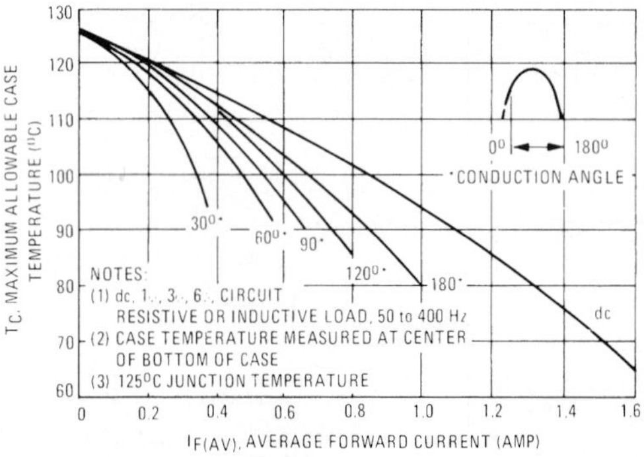

FIGURE 2 – AMBIENT TEMPERATURE REFERENCE

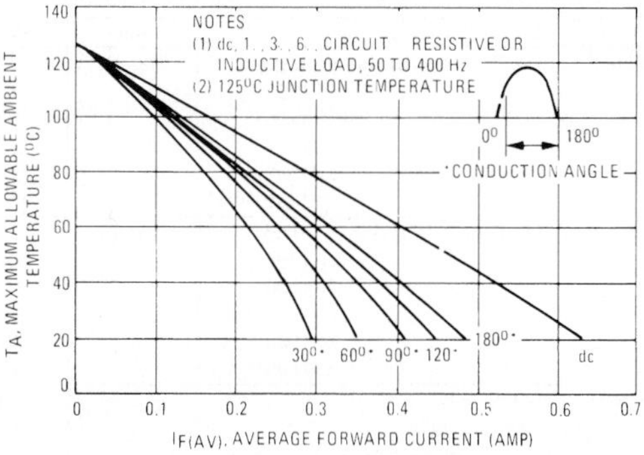

10137 PRINTED IN USA (8-77) MPS 10M

MOTOROLA Semiconductors
BOX 20912 • PHOENIX, ARIZONA 85036

MC1741, MC1741C MC1741N, MC1741NC

INTERNALLY COMPENSATED, HIGH PERFORMANCE OPERATIONAL AMPLIFIERS

. . . designed for use as a summing amplifier, integrator, or amplifier with operating characteristics as a function of the external feedback components.

- No Frequency Compensation Required
- Short-Circuit Protection
- Offset Voltage Null Capability
- Wide Common-Mode and Differential Voltage Ranges
- Low-Power Consumption
- No Latch Up
- Low Noise Selections Offered – N Suffix

OPERATIONAL AMPLIFIER
SILICON MONOLITHIC
INTEGRATED CIRCUIT

G SUFFIX
METAL PACKAGE
CASE 601

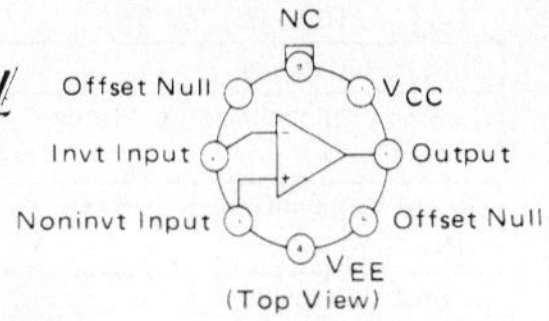

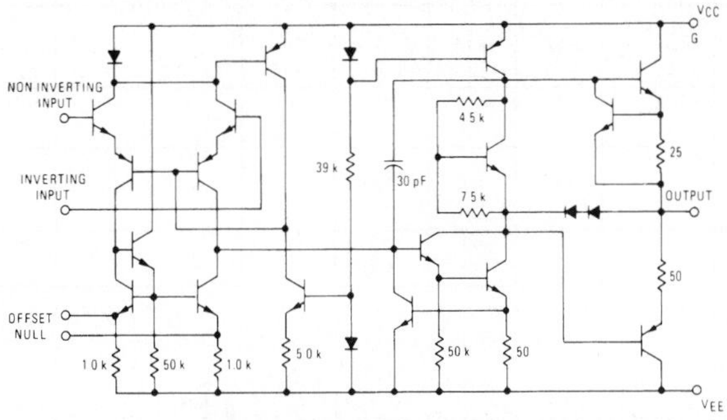

P1 SUFFIX
PLASTIC PACKAGE
CASE 626
(MC1741C,MC1741NC)

U SUFFIX
CERAMIC PACKAGE
CASE 693

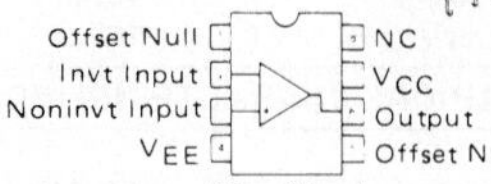

L SUFFIX
CERAMIC PACKAGE
CASE 632
TO-116

P2 SUFFIX
PLASTIC PACKAGE
CASE 646
(MC1741C,MC1741NC)

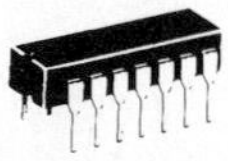

F SUFFIX
CERAMIC PACKAGE
CASE 606-04
TO-91

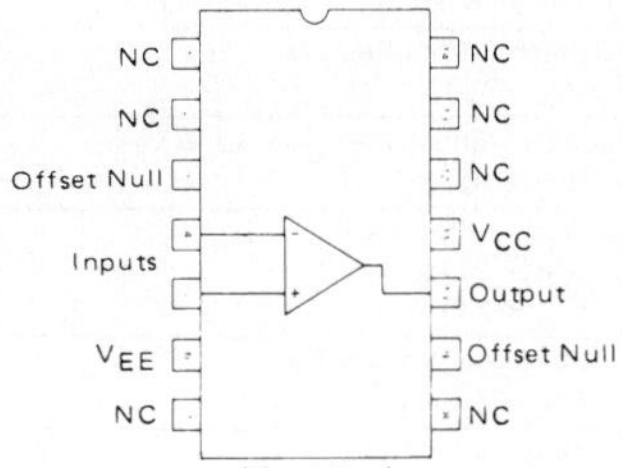

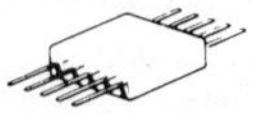

MAXIMUM RATINGS (T_A = +25°C unless otherwise noted)

Rating	Symbol	MC1741C	MC1741	Unit
Power Supply Voltage	V_{CC} V_{EE}	+18 -18	+22 -22	Vdc Vdc
Input Differential Voltage	V_{ID}	±30		Volts
Input Common Mode Voltage (Note 1)	V_{ICM}	±15		Volts
Output Short Circuit Duration (Note 2)	t_S	Continuous		
Operating Ambient Temperature Range	T_A	0 to +70	-55 to +125	°C
Storage Temperature Range Metal, Flat and Ceramic Packages Plastic Packages	T_{stg}	 -65 to +150 -55 to +125		°C

Note 1. For supply voltages less than ± 15 V, the absolute maximum input voltage is equal to the supply voltage.
Note 2. Supply voltage equal to or less than 15 V.

EQUIVALENT CIRCUIT SCHEMATIC

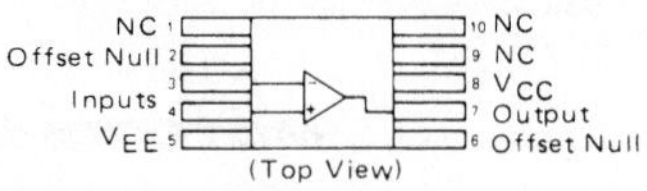

DS 9123 R3

MC1741 • MC1741C • MC1741N • MC1741NC

ELECTRICAL CHARACTERISTICS (V_{CC} = 15 V, V_{EE} = 15 V, T_A = 25°C unless otherwise noted).

Characteristic	Symbol	MC1741			MC1741C			Unit
		Min	Typ	Max	Min	Typ	Max	
Input Offset Voltage ($R_S \leqslant$ 10 k)	V_{IO}	–	1.0	5.0	–	2.0	6.0	mV
Input Offset Current	I_{IO}	–	20	200	–	20	200	nA
Input Bias Current	I_{IB}	–	80	500	–	80	500	nA
Input Resistance	r_i	0.3	2.0	–	0.3	2.0	–	MΩ
Input Capacitance	C_i	–	1.4	–	–	1.4	–	pF
Offset Voltage Adjustment Range	V_{IOR}	–	±15	–	–	±15	–	mV
Common Mode Input Voltage Range	V_{ICR}	±12	±13	–	±12	±13	–	V
Large Signal Voltage Gain (V_O = ±10 V, $R_L \geqslant$ 2.0 k)	A_v	50	200	–	20	200	–	V/mV
Output Resistance	r_o	–	75	–	–	75	–	Ω
Common Mode Rejection Ratio ($R_S \leqslant$ 10 k)	CMRR	70	90	–	70	90	–	dB
Supply Voltage Rejection Ratio ($R_S \leqslant$ 10 k)	PSRR	–	30	150	–	30	150	μV/V
Output Voltage Swing	V_O							V
($R_L \geqslant$ 10 k)		±12	±14	–	±12	±14	–	
($R_L \geqslant$ 2 k)		±10	±13	–	±10	±13	–	
Output Short-Circuit Current	I_{OS}	–	20	–	–	20	–	mA
Supply Current	I_D	–	1.7	2.8	–	1.7	2.8	mA
Power Consumption	P_C	–	50	85	–	50	85	mW
Transient Response (Unity Gain – Non-Inverting)								
(V_I = 20 mV, $R_L \geqslant$ 2 k, $C_L \leqslant$ 100 pF) Rise Time	t_{TLH}	–	0.3	–	–	0.3	–	μs
(V_I = 20 mV, $R_L \geqslant$ 2 k, $C_L \leqslant$ 100 pF) Overshoot	os	–	15	–	–	15	–	%
(V_I = 10 V, $R_L \geqslant$ 2 k, $C_L \leqslant$ 100 pF) Slew Rate	SR	–	0.5	–	–	0.5	–	V/μs

ELECTRICAL CHARACTERISTICS (V_{CC} = 15 V, V_{EE} = 15 V, T_A = *T_{high} to T_{low} unless otherwise noted.)

Characteristic	Symbol	MC1741			MC1741C			Unit
		Min	Typ	Max	Min	Typ	Max	
Input Offset Voltage ($R_S \leqslant$ 10 kΩ)	V_{IO}	–	1.0	6.0	–	–	7.5	mV
Input Offset Current	I_{IO}							nA
(T_A = 125°C)		–	7.0	200	–	–	–	
(T_A = -55°C)		–	85	500	–	–	–	
(T_A = 0°C to +70°C)		–	–	–	–	–	300	
Input Bias Current	I_{IB}							nA
(T_A = 125°C)		–	30	500	–	–	–	
(T_A = -55°C)		–	300	1500	–	–	–	
(T_A = 0°C to +70°C)		–	–	–	–	–	800	
Common Mode Input Voltage Range	V_{ICR}	±12	±13	–	–	–	–	V
Common Mode Rejection Ratio ($R_S \leqslant$ 10 k)	CMRR	70	90	–	–	–	–	dB
Supply Voltage Rejection Ratio ($R_S \leqslant$ 10 k)	PSRR	–	30	150	–	–	–	μV/V
Output Voltage Swing	V_O							V
($R_L \geqslant$ 10 k)		±12	±14	–	–	–	–	
($R_L \geqslant$ 2 k)		±10	±13	–	±10	±13	–	
Large Signal Voltage Gain ($R_L \geqslant$ 2 k, V_{out} = ±10 V)	A_v	25	–	–	15	–	–	V/mV
Supply Currents	I_D							mA
(T_A = 125°C)		–	1.5	2.5	–	–	–	
(T_A = -55°C)		–	2.0	3.3	–	–	–	
Power Consumption (T_A = +125°C)	P_C	–	45	75	–	–	–	mW
(T_A = -55°C)		–	60	100	–	–	–	

*T_{high} = 125°C for MC1741 and 70°C for MC1741C
T_{low} = -55°C for MC1741 and 0°C for MC1741C

ANSWERS TO SELF-TESTS

Chapter 1

1. (c)	**2.** (c)	**3.** (c)	**4.** (d)	**5.** (b)	**6.** (b)	**7.** (a)	**8.** (d)
9. (d)	**10.** (d)	**11.** (c)	**12.** (c)				

Chapter 2

1. (b)	**2.** (a)	**3.** (c)	**4.** (c)	**5.** (b)	**6.** (b)	**7.** (c)	**8.** (d)
9. (b)	**10.** (b)	**11.** (e)	**12.** (b)	**13.** (b)	**14.** (c)		

Chapter 3

1. (b)	**2.** (c)	**3.** (b)	**4.** (d)	**5.** (a)	**6.** (d)	**7.** (b)	**8.** (c)
9. (d)	**10.** (a)	**11.** (a)	**12.** (d)	**13.** (c)	**14.** (a)	**15.** (c)	

Chapter 4

1. (a)	**2.** (d)	**3.** (b)	**4.** (d)	**5.** (d)	**6.** (a)	**7.** (b)	**8.** (c)
9. (b)	**10.** (c)	**11.** (a)	**12.** (d)	**13.** (d)	**14.** (d)		

Chapter 5

1. (b)	**2.** (c)	**3.** (b)	**4.** (d)	**5.** (a)	**6.** (c)	**7.** (c)	**8.** (c)
9. (d)	**10.** (b)	**11.** (a)	**12.** (d)	**13.** (c)	**14.** (b)		

Chapter 6

1. (e)	**2.** (c)	**3.** (c)	**4.** (c)	**5.** (b)	**6.** (a)	**7.** (b)	**8.** (b)
9. (d)	**10.** (b)	**11.** (b)	**12.** (c)	**13.** (d)	**14.** (a)	**15.** (d)	

Chapter 7

1. (b)	**2.** (c)	**3.** (a)	**4.** (d)	**5.** (b)	**6.** (c)	**7.** (a)	**8.** (d)
9. (c)	**10.** (c)	**11.** (b)	**12.** (c)				

Chapter 8

1. (b)	**2.** (b)	**3.** (c)	**4.** (b)	**5.** (d)	**6.** (a)	**7.** (a)	**8.** (a)
9. (c)	**10.** (b)	**11.** (a)	**12.** (d)	**13.** (c)	**14.** (b)	**15.** (d)	

Chapter 9

1. (g)	**2.** (b)	**3.** (c)	**4.** (d)	**5.** (a)	**6.** (d)	**7.** (a)	**8.** (f)
9. (c)	**10.** (b)	**11.** (d)	**12.** (a)	**13.** (b)	**14.** (a)	**15.** (b)	**16.** (d)

Chapter 10

1. (c)	**2.** (d)	**3.** (c)	**4.** (b)	**5.** (d)	**6.** (a)	**7.** (d)	**8.** (c)
9. (b)	**10.** (a)	**11.** (d)	**12.** (b)				

Chapter 11

1. (b)	**2.** (c)	**3.** (d)	**4.** (a)	**5.** (b)	**6.** (c)	**7.** (d)	**8.** (b)
9. (a)	**10.** (c)	**11.** (d)	**12.** (c)	**13.** (a)	**14.** (c)		

Chapter 12

1. (c)	**2.** (b)	**3.** (b)	**4.** (a)	**5.** (d)	**6.** (b)	**7.** (a)	**8.** (d)
9. (c)	**10.** (b)	**11.** (d)	**12.** (d)	**13.** (c)	**14.** (b)	**15.** (d)	

Chapter 13

1. (f) **2.** (b) **3.** (a) **4.** (d) **5.** (a) **6.** (d) **7.** (b) **8.** (b)
9. (a) **10.** (d) **11.** (c) **12.** (d) **13.** (b) **14.** (c) **15.** (a)

Chapter 14

1. (a) **2.** (c) **3.** (b) **4.** (c) **5.** (d) **6.** (c) **7.** (a) **8.** (b)
9. (d) **10.** (a) **11.** (b) **12.** (c)

Chapter 15

1. (b) **2.** (c) **3.** (a) **4.** (d) **5.** (a) **6.** (a) **7.** (a) **8.** (c)
9. (b) **10.** (d)

Chapter 16

1. (b) **2.** (c) **3.** (d) **4.** (c) **5.** (b) **6.** (a) **7.** (b) **8.** (d)
9. (c) **10.** (a) **11.** (d) **12.** (c) **13.** (a) **14.** (d) **15.** (f) **16.** (a)
17. (d) **18.** (b)

Chapter 17

1. (c) **2.** (b) **3.** (a) **4.** (a) **5.** (c) **6.** (d) **7.** (b) **8.** (d)
9. (a) **10.** (d) **11.** (b) **12.** (a) **13.** (c) **14.** (a) **15.** (c) **16.** (d)
17. (b) **18.** (a)

Chapter 18

1. (b) **2.** (a) **3.** (c) **4.** (d) **5.** (b) **6.** (a) **7.** (c) **8.** (d)
9. (b) **10.** (a) **11.** (d) **12.** (c) **13.** (a) **14.** (c)

Chapter 19

1. (c) **2.** (b) **3.** (a) **4.** (d) **5.** (b) **6.** (a) **7.** (c) **8.** (d)
9. (a) **10.** (c) **11.** (c) **12.** (b) **13.** (d) **14.** (a) **15.** (c) **16.** (b)

Chapter 20

1. (b) **2.** (d) **3.** (a) **4.** (b) **5.** (c) **6.** (d) **7.** (d) **8.** (a)
9. (c) **10.** (a) **11.** (b) **12.** (d)

Chapter 21

1. (c) **2.** (a) **3.** (b) **4.** (d) **5.** (a) **6.** (b) **7.** (c) **8.** (b)
9. (a) **10.** (d) **11.** (c) **12.** (c)

ANSWERS TO ODD-NUMBERED PROBLEMS

Chapter 1

1. (a) 3×10^3 (b) 75×10^3 (c) 2×10^6
3. (a) $8.4 \times 10^3 = 0.84 \times 10^4 = 0.084 \times 10^5$ (b) $99 \times 10^3 = 9.9 \times 10^4 = 0.99 \times 10^5$
   (c) $200 \times 10^3 = 20 \times 10^4 = 2 \times 10^5$
5. (a) 0.0000025 (b) 5000 (c) 0.39
7. (a) 126×10^6 (b) 5.00085×10^3 (c) 6060×10^{-9}
9. (a) 20×10^8 (b) 36×10^{14} (c) 15.4×10^{-15}
11. (a) 2370×10^{-6} (b) 18.56×10^{-6}
    (c) $0.00574389 \times 10^{-6}$ (d) $100{,}000{,}000{,}000 \times 10^{-6}$
13. (a) 12.25×10^{14} (b) 5×10^3 (c) 10.575×10^{-6} (d) 2×10^{10}
15. (a) 3 μF (b) 3.3 MΩ (c) 350 nA
17. (a) 5000 μA (b) 3.2 mW (c) 5 MV (d) 10,000 kV

Chapter 2

1. 80×10^{12} C
3. (a) 10 V (b) 2.5 V (c) 4 V
5. 20 V
7. (a) 75 A (b) 20 A (c) 2.5 A
9. 2 s
11. (1) 6800 Ω ± 10% (2) 33 Ω ± 10% (3) 47,000 Ω ± 20%
13. (a) Red, violet, brown (b) 2—330 Ω, 4—2.2 kΩ, 12—56 kΩ, 6—100 kΩ, 1—39 kΩ
15. Current flows through lamp 2.
17. Ammeter in series with resistors, with its negative terminal to the negative terminal of source and its positive terminal to one side of R_1. Voltmeter placed across (in parallel with) the source (negative to negative, positive to positive).
19. Position 1: V1 = 0, V2 = V_S
    Position 2: V1 = V_S, V2 = 0
21. (a) 0.25 V (b) 250 V
23. (a) 200 Ω (b) 150 MΩ (c) 4500 Ω
25. 33.3 V
27. AWG #27
29. Circuit (b)
31. One ammeter in series with battery. One ammeter in series with each resistor (seven total).
33. See Figure P–1.

FIGURE P–1

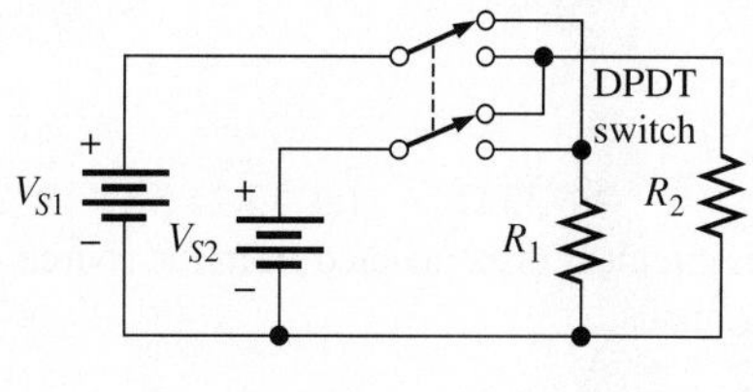

Chapter 3

1. **(a)** 3 A **(b)** 0.2 A **(c)** 1.5 A

3. 15 mA

5. **(a)** 3.33 mA **(b)** 550 μA **(c)** 588 μA **(d)** 500 mA **(e)** 6.60 mA

7. **(a)** 2.50 mA **(b)** 2.27 mA **(c)** 8.33 mA

9. **(a)** 10 mV **(b)** 1.65 V **(c)** 15 kV **(d)** 3.52 V
(e) 0.25 V **(f)** 750 kV **(g)** 8.5 kV **(h)** 3.75 mV

11. **(a)** 81 V **(b)** 500 V **(c)** 125 V

13. **(a)** 2 kΩ **(b)** 3.5 kΩ **(c)** 2 kΩ **(d)** 100 kΩ **(e)** 1 MΩ

15. **(a)** 4 Ω **(b)** 3 kΩ **(c)** 200 kΩ

17. 417 mW

19. **(a)** 1 MW **(b)** 3 MW **(c)** 150 MW **(d)** 8.7 MW

21. **(a)** 2,000,000 μW **(b)** 500 μW **(c)** 250 μW **(d)** 6.67 μW

23. 16.5 mW

25. 1.18 kW

27. 6 W

29. 25 Ω

31. 0.00186 kWh

33. 1 W

35. **(a)** positive at top **(b)** positive at bottom **(c)** positive at right

37. 36 Ah

39. 13.5 mA

41. 4.25 W

43. 150 Ω

45. $V = 0$ V, $I = 0$ A; $V = 10$ V, $I = 100$ mA; $V = 20$ V, $I = 200$ mA; $V = 30$ V, $I = 300$ mA; $V = 40$ V, $I = 400$ mA; $V = 50$ V, $I = 500$ mA; $V = 60$ V, $I = 600$ mA; $V = 70$ V, $I = 700$ mA; $V = 80$ V, $I = 800$ mA; $V = 90$ V, $I = 900$ mA; $V = 100$ V, $I = 1$ A
The graph is a straight line. There is a linear relationship between V and I.

47. $R_1 = 0.5\ \Omega$; $R_2 = 1\ \Omega$; $R_3 = 2\ \Omega$

49. 10 V; 30 V

51. $I_{MAX} = 3.83$ mA; $I_{MIN} = 3.46$ mA

53. 216 kWh

55. 12 W

57. 2 A

Chapter 4

1. See Figure P–2.

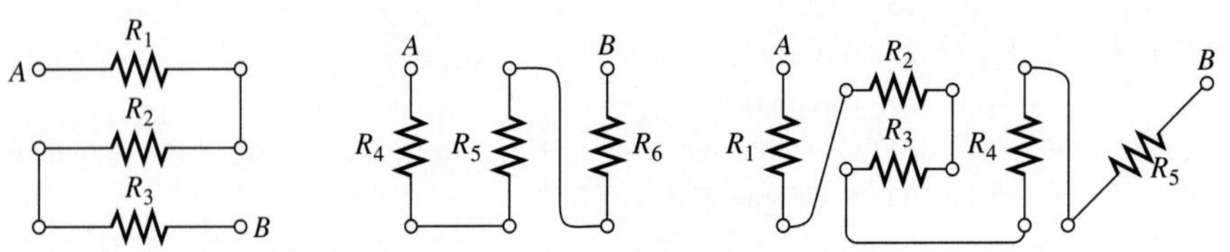

FIGURE P–2

3. 0.1 A

5. 138 Ω

7. **(a)** 7.9 kΩ **(b)** 33 Ω **(c)** 13.24 MΩ
The series circuit is disconnected from the source, and the ohmmeter is connected across the circuit terminals.

9. 1126 Ω

11. **(a)** 170 kΩ **(b)** 50 Ω **(c)** 12.4 kΩ **(d)** 1.97 kΩ

13. **(a)** 625 μA **(b)** 4.26 μA. The ammeter is connected in series.

15. 355 mA

17. 14 V

19. 26 V

21. **(a)** $V_2 = 6.8$ V **(b)** $V_R = 8$ V, $V_{2R} = 16$ V, $V_{3R} = 24$ V, $V_{4R} = 32$ V
The voltmeter is connected across (in parallel with) each resistor for which the voltage is unknown.

23. **(a)** 3.84 V **(b)** 6.77 V

25. $V_{5.6k\Omega} = 10$ V; $V_{1k\Omega} = 1.79$ V; $V_{560\Omega} = 1$ V; $V_{10k\Omega} = 17.9$ V

27. 55 mW

29. Measure V_A and V_B with respect to ground; then $V_{R2} = V_A - V_B$.

31. **(a)** R_4 open **(b)** R_4 and R_5 shorted

33. 780 Ω

35. $V_{AF} = 100$ V; $V_{BF} = 77.2$ V; $V_{CF} = 66.9$ V; $V_{DF} = 18.2$ V; $V_{EF} = 5.81$ V; $V_F = 0$ V

37. 500 Ω

39. **(a)** 19.1 mA **(b)** 45.8 V **(c)** $R(\frac{1}{8}$ W$) = 343$ Ω, $R(\frac{1}{4}$ W$) = 686$ Ω, $R(\frac{1}{2}$ W$) = 1372$ Ω

41. See Figure P–3.

FIGURE P–3

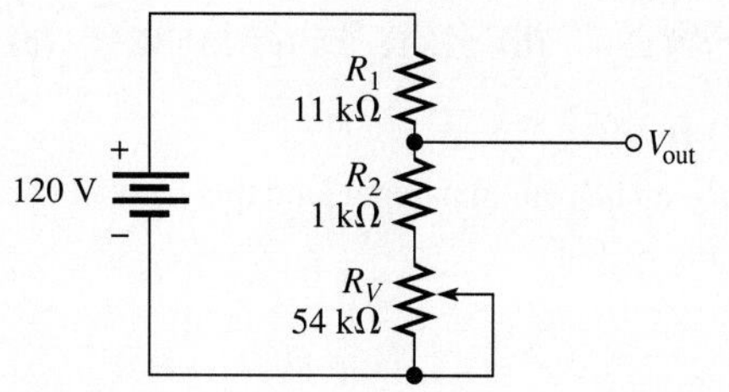

43. $R_1 + R_7 + R_8 + R_{10} = 4.23$ kΩ; $R_2 + R_4 + R_6 + R_{11} = 23.6$ kΩ;
$R_3 + R_5 + R_9 + R_{12} = 19.9$ kΩ

45. A: 5.45 mA; B: 6.06 mA; C: 7.95 mA; D: 12 mA

47. A: $V_1 = 6.03$ V, $V_2 = 3.35$ V, $V_3 = 2.75$ V, $V_4 = 1.88$ V, $V_5 = 4.0$ V;
B: $V_1 = 6.71$ V, $V_2 = 3.73$ V, $V_3 = 3.06$ V, $V_5 = 4.5$ V;
C: $V_1 = 8.1$ V, $V_2 = 4.5$ V, $V_5 = 5.4$ V;
D: $V_1 = 10.8$ V, $V_5 = 7.2$ V

49. Yes, R_3 and R_5 are shorted.

51. **(a)** R_{11} has burned open due to excessive power. **(b)** Replace R_{11} (10 kΩ). **(c)** 338 V

Chapter 5

1. See Figure P–4.

FIGURE P–4

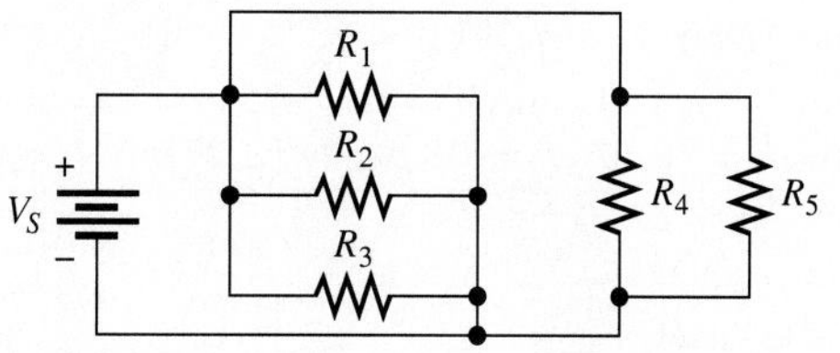

3. 12 V; 5 mA

5. 1350 mA

7. $I_2 = I_3 = 7.5$ mA. An ammeter in series with each resistor in each branch.

9. **(a)** 359 Ω **(b)** 25.6 Ω **(c)** 819 Ω **(d)** 996 Ω

11. 2 kΩ

13. **(a)** 909 mA **(b)** 76 mA

15. Circuit (a)

17. $I_1 = 2.19$ A; $I_2 = 811$ mA

19. 200 mW

21. 682 mA; 4.09 A

23. The 1 kΩ resistor is open.

25. $R_2 = 25\ \Omega$; $R_3 = 100\ \Omega$; $R_4 = 12.5\ \Omega$

27. $I_R = 4.8$ A; $I_{2R} = 2.4$ A; $I_{3R} = 1.6$ A; $I_{4R} = 1.2$ A

29. **(a)** $R_1 = 100\ \Omega$, $R_2 = 200\ \Omega$, $I_2 = 50$ mA
(b) $I_1 = 125$ mA, $I_2 = 74.9$ mA, $R_1 = 79.9\ \Omega$, $R_2 = 134$ mA, $V_s = 10$ V
(c) $I_1 = 253$ mA, $I_2 = 147$ mA, $I_3 = 100$ mA, $R_1 = 395\ \Omega$

31. 53.7 Ω

33. 1.08 A

35. R_1, R_2, R_5, R_9, R_{10}, and R_{12}; $R_T = 2.57$ MΩ
R_4, R_6, R_7, and R_8; $R_T = 2.77$ MΩ
R_3 and R_{11}; $R_T = 2.13$ MΩ

37. $R_2 = 750\ \Omega$; $R_4 = 423\ \Omega$

39. The 4.7 kΩ resistor is open.

41. **(a)** One of the resistors has opened due to excessive power dissipation. **(b)** 30 V
(c) Replace the 1.8 kΩ resistor.

43. **(a)** 940 Ω **(b)** 518 Ω **(c)** 518 Ω **(d)** 422 Ω

Chapter 6

1. R_2, R_3, and R_4 are in parallel, and this parallel combination is in series with both R_1 and R_5.

3. See Figure P–5.

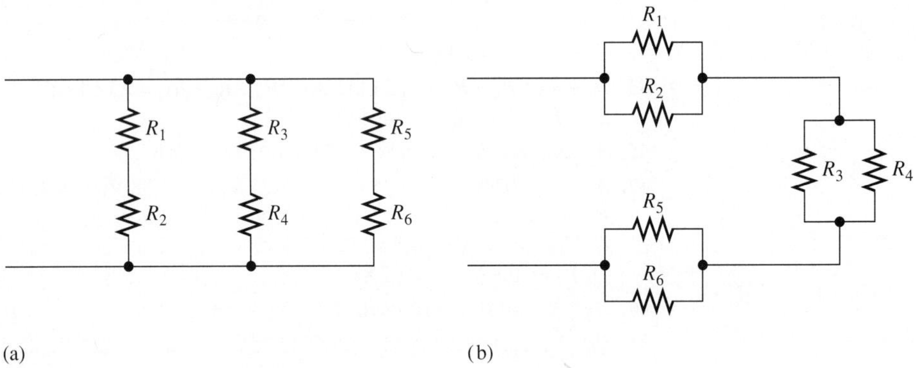

FIGURE P–5

5. 2003 Ω

7. **(a)** 128 Ω **(b)** 791 Ω

9. **(a)** $I_1 = I_4 = 11.7$ mA, $I_2 = I_3 = 5.85$ mA, $V_1 = 655$ mV, $V_2 = V_3 = 585$ mV, $V_4 = 257$ mV
(b) $I_1 = 3.8$ mA, $I_2 = 618\ \mu$A, $I_3 = 1.27$ mA, $I_4 = 1.91$ mA, $V_1 = 2.58$ V,
$V_2 = V_3 = V_4 = 420$ mV

11. 7.5 V unloaded; 7.29 V loaded

13. 56 kΩ load

15. 1000 V

17. **(a)** 0.5 V range **(b)** Approximately 1 mV

19. 360 Ω

21. 21 mA

23. $R_{TH} = 18$ kΩ; $V_{TH} = 2.7$ V

25. No, the meter should read 4.39 V. The 680 Ω is open.

27. The 7.62 V and 5.24 V readings are incorrect, indicating that the 3.3 kΩ resistor is open.

29. **(a)** $V_1 = -10$ V, all others 0 V **(b)** $V_1 = -2.33$ V, $V_4 = -7.67$ V, $V_2 = -2.33$ V, $V_3 = 0$ V
(c) $V_1 = -2.33$ V, $V_4 = -7.67$ V, $V_2 = V_3 = 0$ V **(d)** $V_1 = -10$ V, all others 0 V

31. See Figure P–6.

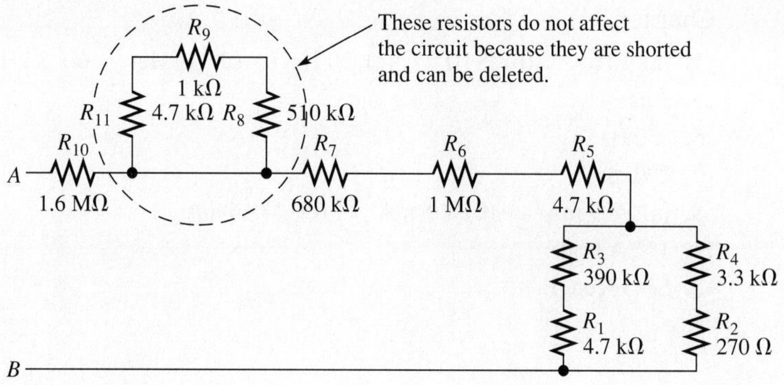

FIGURE P–6

33. $R_T = 5.76$ kΩ; $V_A = 3.3$ V; $V_B = 1.7$ V; $V_C = 850$ mV

35. $V_1 = 1.61$ V; $V_2 = 6.7$ V; $V_3 = 1.77$ V; $V_4 = 3.4$ V; $V_5 = 387$ mV; $V_6 = 2.63$ V;
$V_7 = 387$ mV; $V_8 = 1.77$ V; $V_9 = 1.61$ V

37. 110 Ω

39. $R_1 = 180$ Ω; $R_2 = 60$ Ω. Output across R_2.

41. 842 μA

43. 11.7 V

45. See Figure P–7.

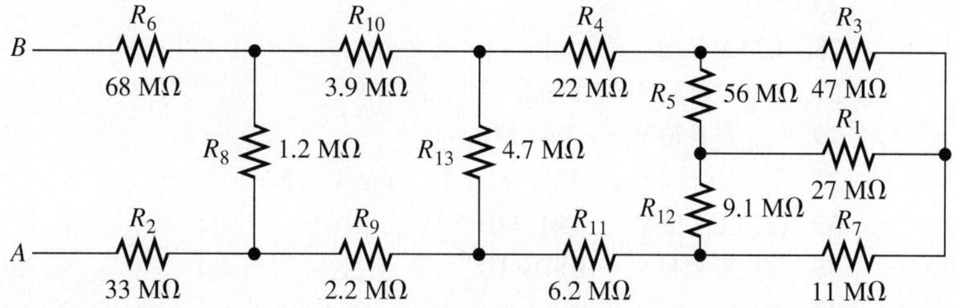

FIGURE P–7

47. Pos 1: $V_1 = 81.0$ V, $V_2 = 54.0$ V, $V_3 = 27.0$ V;
Pos 2: $V_1 = 86.2$ V, $V_2 = 52.4$ V, $V_3 = 26.2$ V;
Pos 3: $V_1 = 89.0$ V, $V_2 = 58.0$ V, $V_3 = 27.0$ V

49. One of the 12 kΩ resistors is open.

51. The 2.2 kΩ resistor is open.

53. $V_A = 0$ V; $V_B = 11.1$ V

Chapter 7

1. Decrease

3. 37.5 μWb

5. 597

7. 150 At

9. **(a)** Electromagnetic force **(b)** Spring force

11. Electromagnetic force

13. Change the current.

15. Material A

17. 1 mA

19. To electrically connect the loop to the external circuit

21. The output voltage appears as the solid curve in Figure 7–38 with a frequency of 120 Hz and a peak of 10 V.

Chapter 8

1. **(a)** 1 Hz **(b)** 5 Hz **(c)** 20 Hz **(d)** 1 kHz **(e)** 2 kHz **(f)** 100 kHz

3. 2 μs

5. 250 Hz

7. 200 rps

9. **(a)** 7.07 mA **(b)** 4.5 mA **(c)** 14.14 mA

11. 15°, A leads.

13. See Figure P–8.

FIGURE P–8

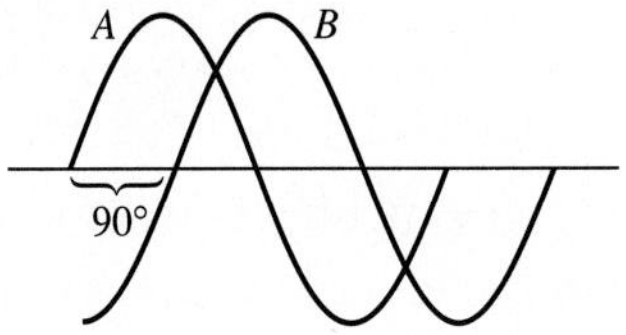

15. **(a)** 22.5° **(b)** 60° **(c)** 90° **(d)** 108° **(e)** 216° **(f)** 324°

17. **(a)** 57.4 mA **(b)** 99.6 mA **(c)** −17.4 mA
(d) −57.4 mA **(e)** −99.6 mA **(f)** 0 mA

19. 30°: 13.0 V; 45°: 14.5 V; 90°: 13.0 V; 180°: −7.5 V; 200°: −11.5 V; 300°: −7.5 V

21. **(a)** 7.07 mA **(b)** 6.37 mA **(c)** 10 mA **(d)** 20 mA **(e)** 10 mA

23. 7.38 V

25. 4.24 V

27. $V_p = 0.6$ V; $T = 500$ ms

29. $V_{rms} = 1.56$ V; $f = 1.47$ MHz

31. $t_r \cong 3.5$ ms; $t_f \cong 3.5$ ms; $t_W \cong 12.5$ ms; $V = 5$ V

33. **(a)** −0.375 V **(b)** 3.01 V

35. **(a)** 50 kHz **(b)** 10 Hz

37. 25 kHz

39. $I_{max} = 3.5$ A; $V_{avg} = 200$ V; See Figure P–9.

FIGURE P–9

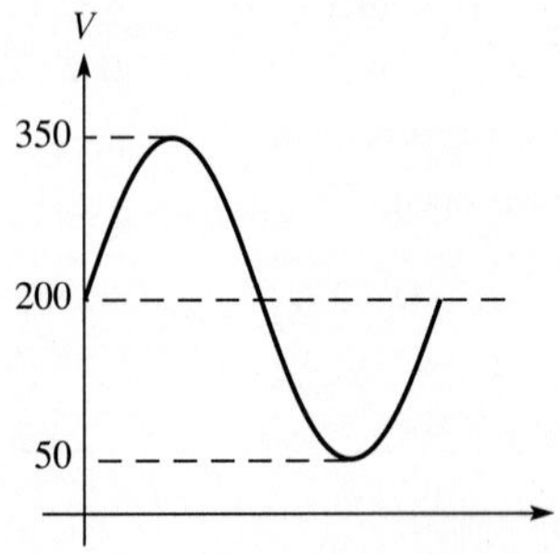

41. **(a)** 2.5 **(b)** 3.54 V **(c)** 12.5 kHz

43. See Figure P–10.

45. $V_p = 1.89$ V; $f = 2$ Hz

FIGURE P–10

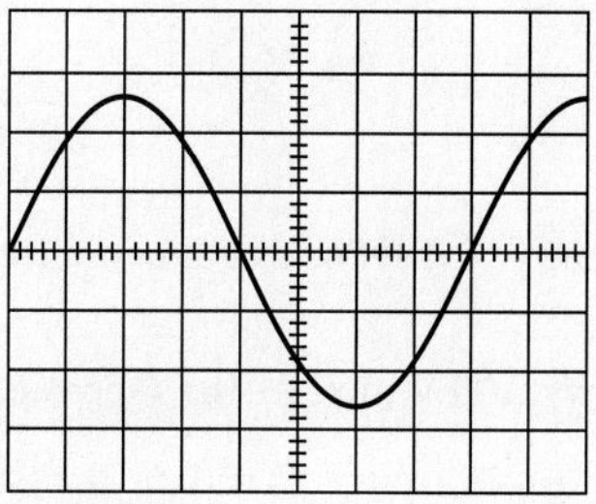

Chapter 9

1. **(a)** 5 μF **(b)** 1 μC **(c)** 10 V
3. **(a)** 0.001 μF **(b)** 0.0035 μF **(c)** 0.00025 μF
5. 2 μF
7. 88.5 pF
9. A 12.5 pF increase
11. Ceramic
13. **(a)** 0.02 μF **(b)** 0.047 μF **(c)** 0.001 μF **(d)** 220 pF **(e)** 100,000 pF
15. **(a)** Encapsulation **(b)** Dielectric (ceramic disk) **(c)** Plate (metal disk) **(d)** Leads
17. **(a)** 0.67 μF **(b)** 69 pF **(c)** 2.7 μF
19. $V_1 = 2$ V; $V_2 = 10$ V; $V_3 = 5$ V; $V_4 = 1$ V
21. **(a)** 2.5 μF **(b)** 717 pF **(c)** 1.6 μF
23. **(a)** 100 μs **(b)** 500 μs **(c)** 23.5 μs **(d)** 15 ms
25. **(a)** 9.48 V **(b)** 13.0 V **(c)** 14.3 V **(d)** 14.7 V **(e)** 14.9 V
27. **(a)** 2.72 V **(b)** 5.90 V **(c)** 11.7 V
29. **(a)** 339 kΩ **(b)** 13.6 kΩ **(c)** 677 Ω **(d)** 33.9 Ω
31. **(a)** 31.8 Ω **(b)** 111 kΩ **(c)** 49.7 Ω
33. 200 Ω
35. $P_{true} = 0$ W; $P_r = 3.39$ mVAR
37. 0 Ω
39. Shorted capacitor
41. 3.18 ms
43. 2.94 μs
45. **(a)** Charges to 3.32 V in 10 ms, then discharges to 0 V in 215 ms.
(b) Charges to 3.32 V in 10 ms, then discharges to 2.85 V in 5 ms, then charges toward 20 V.
47. 0.0052 μF
49. $V_1 = 7.25$ V; $V_2 = 2.76$ V; $V_3 = 0.79$ V; $V_4 = 1.98$ V; $V_5 = 1.19$ V; $V_6 = 0.79$ V

Chapter 10

1. **(a)** 1000 mH **(b)** 0.25 mH **(c)** 0.01 mH **(d)** 0.5 mH
3. 3536 turns
5. 50 mJ
7. 155 μH
9. 7.14 μH
11. **(a)** 4.33 H **(b)** 50 mH **(c)** 0.57 μH
13. **(a)** 1 μs **(b)** 2.13 μs **(c)** 2 μs
15. **(a)** 5.52 V **(b)** 2.03 V **(c)** 0.75 V **(d)** 0.27 V **(e)** 0.1 V
17. **(a)** 136 kΩ **(b)** 1.57 kΩ **(c)** 179 mΩ
19. $I_T = 10.1$ mA; $I_{L2} = 6.7$ mA; $I_{L3} = 3.35$ mA
21. 101 VAR
23. **(a)** Infinite resistance **(b)** Zero resistance **(c)** Lower R_W
25. 2.86 mA
27. 26.1 mA

Chapter 11

1. 1.5 μH

3. 3

5. **(a)** Positive at top **(b)** Positive at bottom **(c)** Positive at bottom

7. 500 turns

9. **(a)** Same polarity, 100 V rms **(b)** Opposite polarity, 100 V rms

11. 240 V

13. 33.3 mA

15. 27.3 Ω

17. 5.75 mA

19. 0.5

21. 5 kΩ

23. No

25. 94.5 W

27. 0.98

29. 25 kVA

31. Secondary 1: 2; Secondary 2: 0.5; Secondary 3: 0.25

33. **(a)** 48 V **(b)** 25 V

35. Excessive primary current is drawn, potentially burning out the source and/or the transformer, unless the primary is protected by a fuse.

37. **(a)** $V_{L1} = 35$ V, $I_{L1} = 2.92$ A, $V_{L2} = 15$ V, $I_{L2} = 1.5$ A **(b)** 28.9 Ω

39. **(a)** 20 V **(b)** 10 V

41. 0.0145 (69.2:1)

Chapter 12

1. 8 kHz; 8 kHz

3. **(a)** 288 Ω **(b)** 1209 Ω

5. **(a)** 798 kΩ **(b)** 169 kΩ **(c)** 97.3 kΩ **(d)** 64.4 kΩ

7. **(a)** 34.7 mA **(b)** 4.14 mA

9. $I_T = 12$ mA; $V_{C(0.1)} = 1.27$ V; $V_{C(0.2)} = 0.637$ V; $V_R = 0.6$ V; $\theta = 72.5°$ (V_s lagging I_T)

11. 807 Ω; −36.1°

13. 353 Ω; 62°

15. 261 Ω; −79.8°

17. $I_{C1} = 118$ mA; $I_{C2} = 50.3$ mA; $I_{R1} = 36.4$ mA; $I_{R2} = 44.4$ mA; $I_T = 187$ mA; $\theta = 64.4°$ (V_s lagging I_T)

19. **(a)** 3.9 kΩ **(b)** 20 μA **(c)** 15.7 μA **(d)** 25.6 μA **(e)** 38.1° (V_s lagging I_T)

21. $V_{C1} = 8.77$ V; $V_{C2} = 3.24$ V; $V_{C3} = 3.24$ V; $V_{R1} = 2.1$ V; $V_{R2} = 1.14$ V

23. $I_T = 82.8$ mA; $I_{C2} = 15.3$ mA; $I_{C3} = 67.5$ mA; $I_{R1} = I_{R2} = 6.35$ mA

25. 4.03 VA

27. 0.91

29. **(a)** 0.056° **(b)** 5.60° **(c)** 44.4° **(d)** 84.2°

31. **(a)** 90° **(b)** 86.4° **(c)** 57.8° **(d)** 9.04°

33. See Figure P–11.

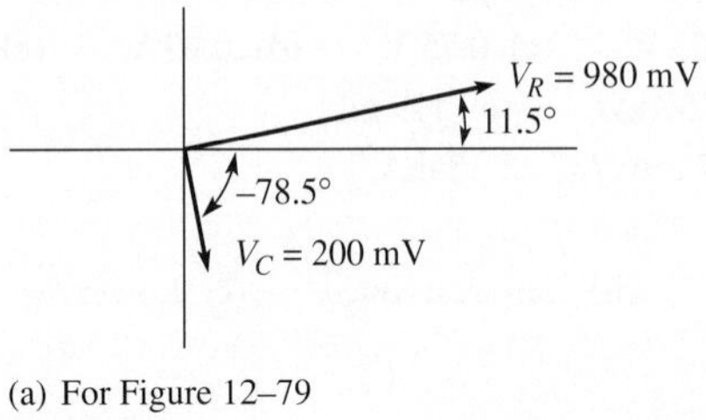

(a) For Figure 12–79

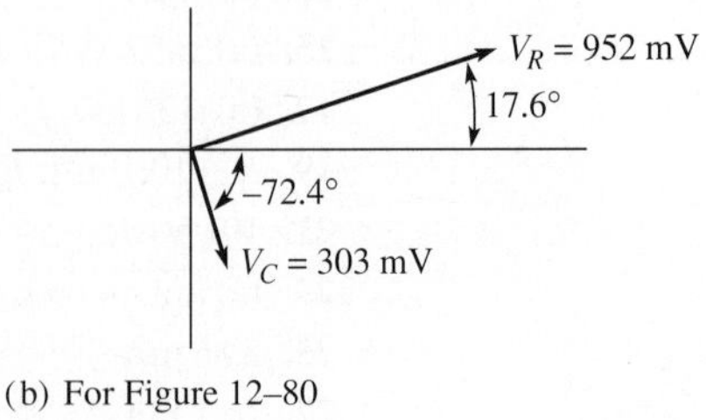

(b) For Figure 12–80

FIGURE P–11

35. Figure 12–79: 1.02 kHz; Figure 12–80: 1.59 kHz

37. No leakage: $V_{out} = 3.21$ V, $\theta = -18.7°$; Leakage: $V_{out} = 2.83$ V, $\theta = -33.3°$

39. **(a)** 0 V **(b)** 0.3 V **(c)** 0.5 V **(d)** 0 V

41. **(a)** $I_{LA} = 4.4$ A; $I_{LB} = 3.06$ A **(b)** $P_{rA} = 509$ VAR; $P_{rB} = 211$ VAR
(c) $P_{true(A)} = 823$ W; $P_{true(B)} = 641$ W **(d)** $P_{aA} = 968$ VA; $P_{aB} = 674$ VA
(e) Load A

43. 11.4 kΩ

45. $P_r = 1.32$ kVAR; $P_a = 2$ kVA

47. 0.1 μF

Chapter 13

1. 15 kHz

3. **(a)** 112 Ω **(b)** 1.8 kΩ

5. **(a)** 17.4 Ω **(b)** 64 Ω **(c)** 127 Ω **(d)** 251 Ω

7. **(a)** 89.3 mA **(b)** 2.78 mA

9. 38.7°

11. See Figure P–12.

13. 7.69 Ω

FIGURE P–12

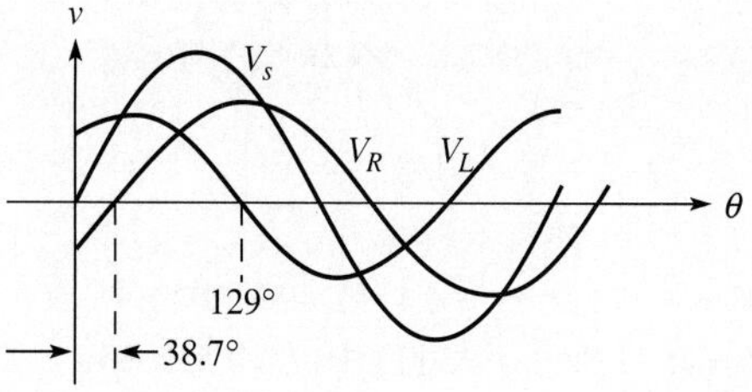

15. 2.39 kHz

17. **(a)** 274 Ω **(b)** 89.3 mA **(c)** 159 mA **(d)** 183 mA **(e)** 60.7° (I_T lagging V_s)

19. $V_{R1} = 7.92$ V; $V_{R2} = V_L = 20.8$ V

21. $I_T = 36$ mA; $I_L = 33.2$ mA; $I_{R2} = 13.9$ mA

23. 1.30 W; 1.04 VAR

25. $PF = 0.386$; $P_{true} = 347$ mW; $P_r = 692$ mVAR; $P_a = 900$ mVA

27. See Figure P–13.

29. See Figure P–14.

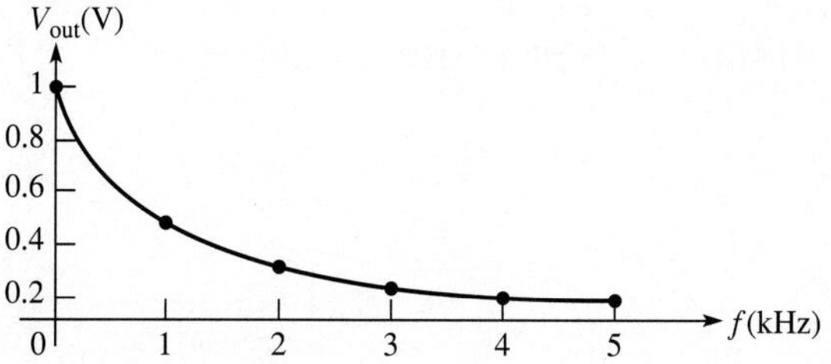

FIGURE P–13

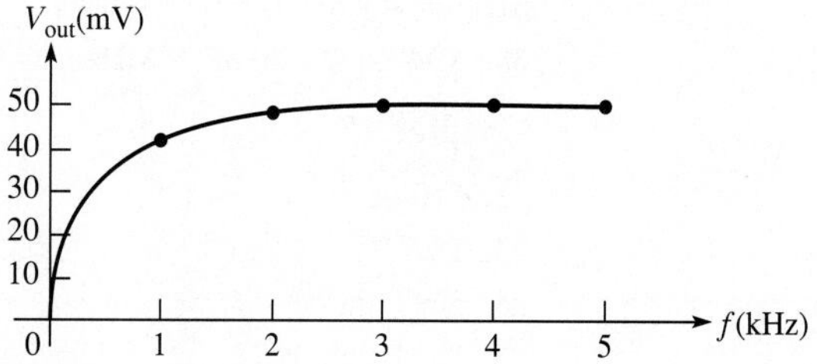

FIGURE P–14

31. $V_{R1} = V_{L1} = 18$ V; $V_{R2} = V_{R3} = V_{L2} = 0$ V

33. 5.57 V

35. 88.7 mA

37. **(a)** 406 mA **(b)** 228 mA **(c)** 333 mA **(d)** 336 mA

39. 0.133

41. See Figure P–15.

FIGURE P–15

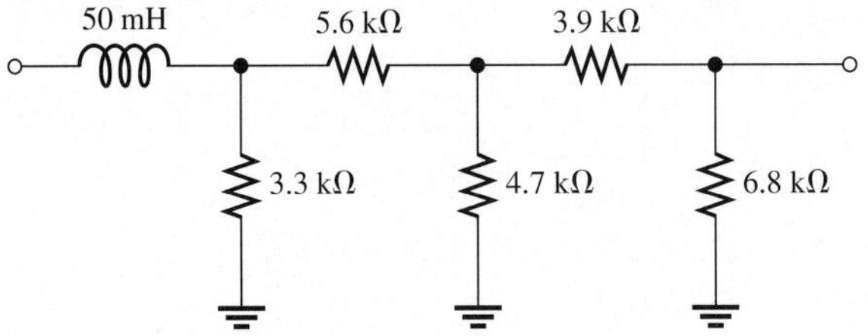

Chapter 14

1. 480 Ω, 88.8° (V_s lagging I); 480 Ω capacitive

3. Impedance increases.

5. See Figure P–16.

FIGURE P–16

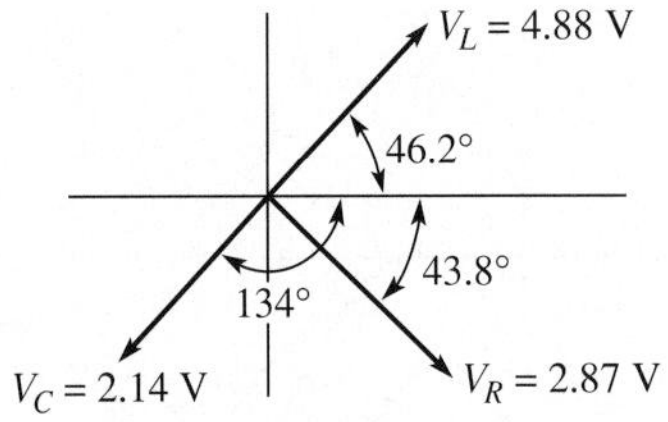

7. $X_L = 4.47$ kΩ; $X_C = 4.47$ kΩ, Z = 20 Ω; $I = 600$ mA

9. $f_r = 459$ kHz; $f_{C1} = 360$ kHz; $f_{C2} = 559$ kHz

11. **(a)** 14.5 kHz **(b)** 25.2 kHz

13. **(a)** $f_r = 339$ kHz, $BW = 239$ kHz **(b)** $f_r = 10.4$ kHz, $BW = 2.61$ kHz

15. Capacitive, $X_C < X_L$.

17. 758 Ω

19. 531 MΩ; 104 kHz

21. 62.5 Hz

23. 1.38 W

25. 200 Hz

27. $I_{R1} = I_C = 2.11$ mA; $I_{L1} = 1.33$ mA; $I_{L2} = 667$ μA; $I_{R2} = 667$ μA;
$V_{R1} = 6.96$ V; $V_C = 2.11$ V; $V_{L1} = V_{L2} = V_{R2} = 6.67$ V

29. $I_{R1} = I_{L1} = 41.5$ mA; $I_C = I_{L2} = 105$ mA; $I_T = 77.0$ mA

31. $L = 989$ μH; $C = 0.064$ μF

33. 8 MHz: $C = 40$ pF; 9 MHz: $C = 31$ pF; 10 MHz: $C = 25$ pF; 11 MHz: $C = 21$ pF

Chapter 15

1. 110 μs

3. 12.6 V

5. See Figure P–17.

7. See Figure P–18.

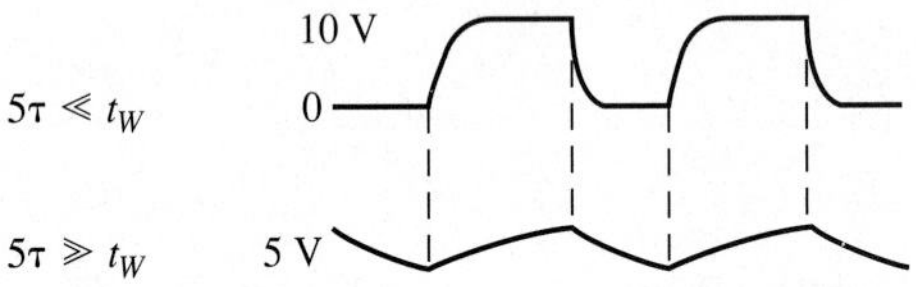

FIGURE P–17

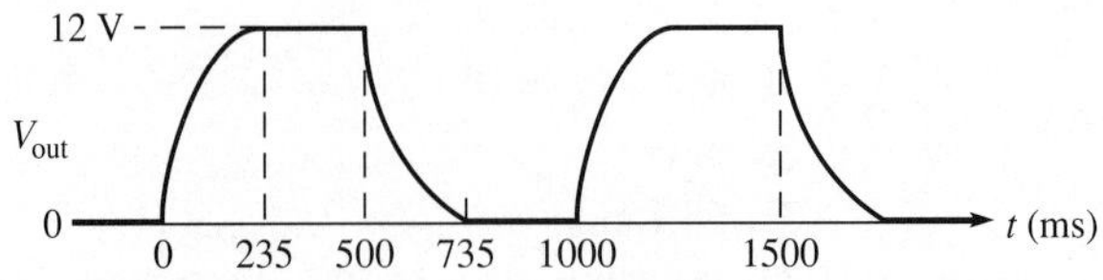

FIGURE P–18

9. 15 V dc level with a very small charge/discharge fluctuation

11. Exchange positions of *R* and *C*. See Figure P–19 for the output voltage.
$5\tau = 5$ ms (repeating Problem 6)

FIGURE P–19

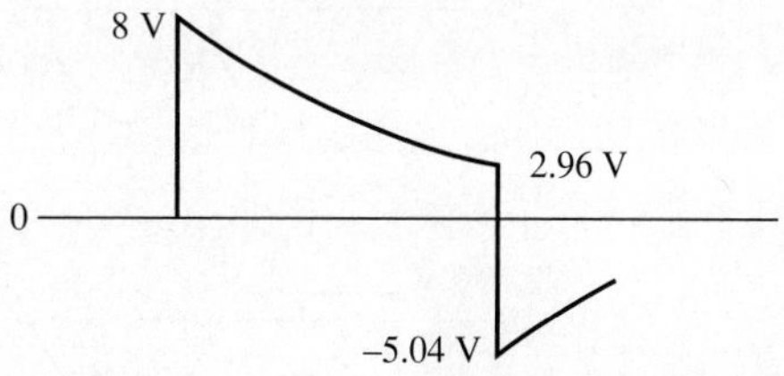

13. Approximately the same shape as input, but with an average value of 0 V

15. See Figure P–20.

FIGURE P–20

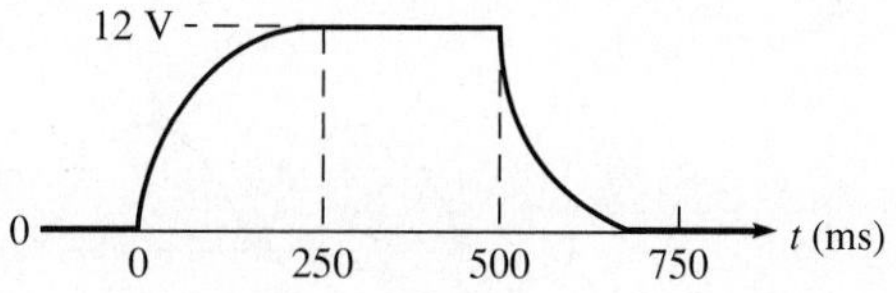

17. See Figure P–21.

FIGURE P–21

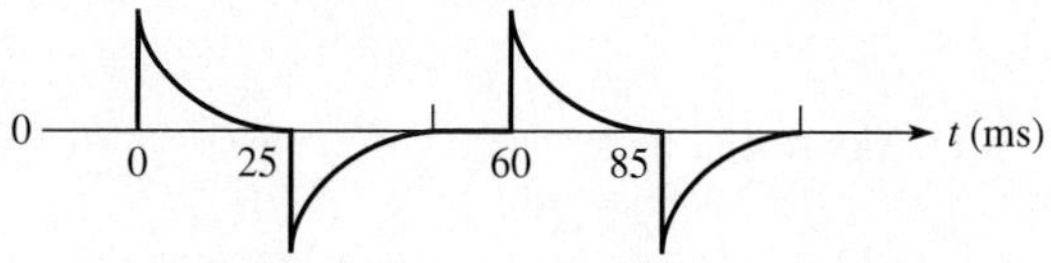

19. A 6 V dc level

21. **(b)** No fault **(c)** Leaky capacitor **(d)** Open *C* or shorted *R*

23. **(a)** 25 ms **(b)** See Figure P–22.

FIGURE P–22

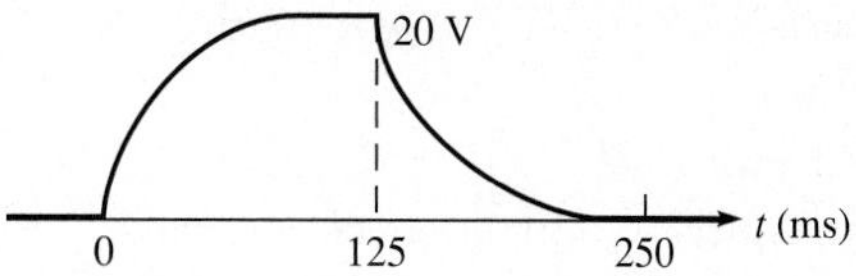

25. 1.44 s

Chapter 16

No problem set.

Chapter 17

1. 63.7 V

3. Yes

5. 47.8 V

7. 173 V

9. 78.5 V

11. See Figure P–23.

FIGURE P–23

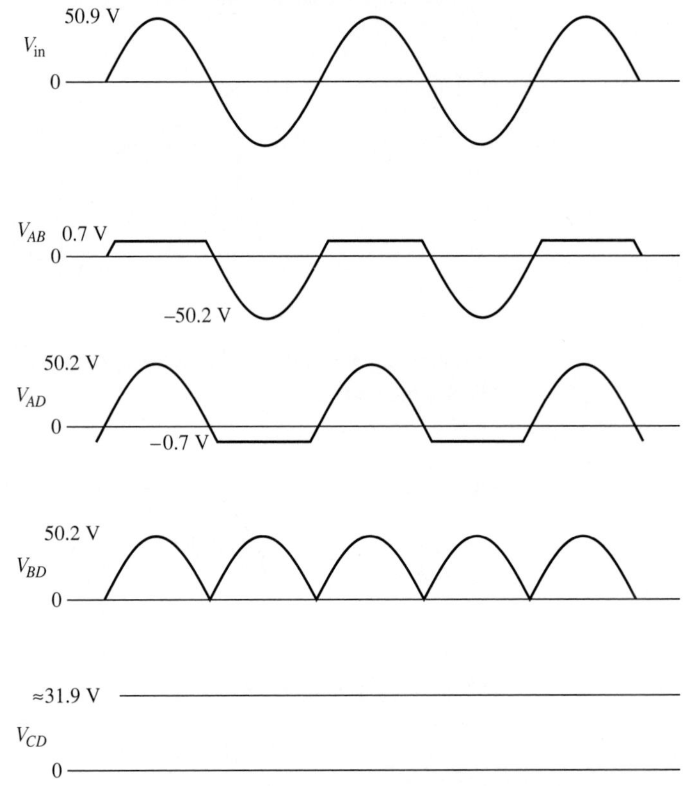

13. **(a)** A sine wave with a positive peak at +0.7 V, a negative peak at −7.3 V, and a dc value of −3.3 V
(b) A sine wave with a positive peak at +29.3 V, a negative peak at −0.7 V, and a dc value of +14.3 V

15. 15.0 V

17. $I_{L(\text{min})} = 0$ A; $I_{L(\text{max})} = 28$ mA

19. 12%; 8%

21. A decrease of 10 pF

23. 26 pF each

25. Increase

27. 50 V

29. 62.5 mΩ

31. **(a)** Correct **(b)** Incorrect, open diode **(c)** Correct **(d)** Incorrect, open diode

33. **(a)** Blown fuse, replace **(b)** Open transformer winding or connection
(c) Open transformer winding or connection **(d)** Some primary windings shorted
(e) Some secondary windings shorted **(f)** C_1 open **(g)** C_1 leaky
(h) A diode open **(i)** C_2 shorted

Chapter 18

1. 4.87 mA

3. 125

5. $I_B = 25.9\ \mu$A; $I_E = 1.3$ mA; $I_C = 1.27$ mA

7. $I_B = 13.6\ \mu$A; $I_C = 682\ \mu$A; $V_C = 9.32$ V

9. $V_{CE} = 3.45$ V; Q point: $I_C = 15$ mA, $V_{CE} = 3.45$ V

11. 33.3

13. 0.5 mA; 3.33 μA; 4.03 V

15. 3 μA

17. **(a)** Narrows **(b)** Increases

19. 5 V

21. 4 V

23. 4 V

25. The gate is insulated from the channel by an SiO_2 layer.

27. 3 V

29. **(a)** Depletion **(b)** Enhancement **(c)** Zero bias **(d)** Depletion

31. 0.6

33. 1 ms

35. 2.4 kΩ

37. Very high resistance; very low resistance

39. **(a)** 36 **(b)** 109

Chapter 19

1. 199

3. **(a)** 3.25 V **(b)** 2.55 V **(c)** 2.55 mA **(d)** ≅ 2.55 mA **(e)** 9.59 V **(f)** 7.04 V

5. $A_{v(max)} = 92.3$; $A_{v(min)} = 2.91$

7. 0.978

9. A_v is reduced slightly.

11. $R_{in} = 2.07\ \Omega$; $A_v = 580$; $A_i \cong 1$; $A_p = 580$

13. **(a)** 0.934 **(b)** 0.301

15. 30 dB; 31.6

17. **(a)** 7.71 mW **(b)** 53.6 mW

19. 10 V; 625 mA

21. 50.3 kHz

23. 1/75

25. **(a)** 528 kHz, Colpitts **(b)** 759 kHz, Hartley

27. Cutoff, 10 V

29.

DC VALUES							
V_{B1}	V_{B2}	V_{E1}	V_{E2}	V_{C1}	V_{C2}	V_{IN}	V_{OUT}
2.88 V	2.88 V	2.18 V	2.18 V	7.85 V	7.85 V	0 V	0 V

AC VALUES (RMS)							
V_{b1}	V_{b2}	V_{e1}	V_{e2}	V_{c1}	V_{c2}	V_{in}	V_{out}
20.9 μV	1.96 mV	0 V	0 V	1.96 mV	594 mV	20.9 μV	594 mV

Chapter 20

1. *Practical op-amp:* High open-loop gain, high input impedance, low output impedance, large bandwidth, high CMRR
Ideal op-amp: Infinite open-loop gain, infinite input impedance, zero output impedance, infinite bandwidth, infinite CMRR

3. **(a)** Single-ended input, differential output **(b)** Single-ended input, single-ended output **(c)** Differential input, single-ended output **(d)** Differential input, differential output

5. 8.1 μA

7. 108 dB
9. 0.3
11. 40 μs
13. **(a)** 375 **(b)** 3.75 V **(c)** 10 mV
15. **(a)** 49 kΩ **(b)** 3 MΩ **(c)** 84 kΩ **(d)** 165 kΩ
17. **(a)** 10 mV, in-phase **(b)** −10 mV, 180° out-of-phase
(c) 223 mV, in-phase **(d)** −100 mV, 180° out-of-phase
19. **(a)** $Z_{in(NI)} = 8410$ MΩ, $Z_{out(NI)} = 89$ mΩ **(b)** $Z_{in(NI)} = 6201$ mΩ, $Z_{out(NI)} = 4$ mΩ
(c) $Z_{in(NI)} = 5302$ mΩ, $Z_{out(NI)} = 19$ mΩ
21. **(a)** $Z_{in(I)} = 10$ kΩ, $Z_{out(I)} = 40$ Ω **(b)** $Z_{in(I)} = 100$ kΩ, $Z_{out(I)} = 50$ Ω
(c) $Z_{in(I)} = 470$ Ω, $Z_{out(I)} = 70$ Ω
23. A_{cl} increases to 100.

Chapter 21

1. **(a)** Maximum negative **(b)** Maximum positive **(c)** Maximum negative
3. See Figure P–24.

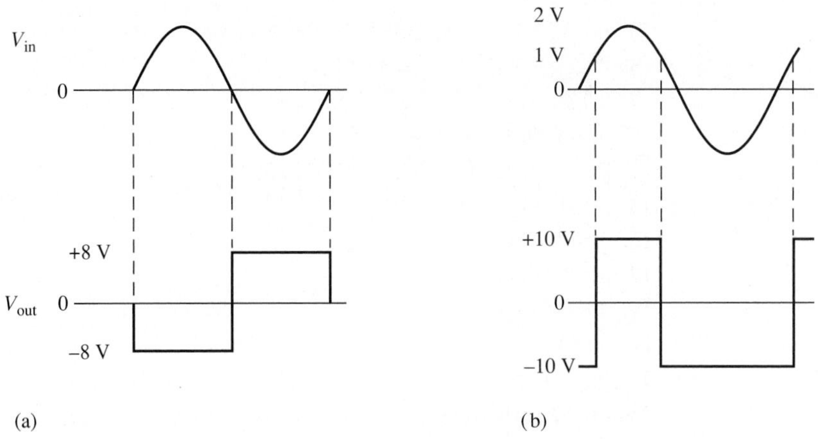

FIGURE P–24

5. **(a)** $V_{R1} = 1$ V, $V_{R2} = 1.8$ V **(b)** 127 μA **(c)** −2.8 V
7. −3.57 V; 357 μA
9. −4.46 mV/μs
11. 733 mV
13. 50 kΩ
15. 10.6 kHz
17. Change R_1 to 3.54 kΩ.
19. Change R_4 to 65.8 kΩ.
21. **(a)** 1 pole, low pass **(b)** 1 pole, high pass **(c)** 2 poles, band pass
23. **(a)** $BW = 1.18$ kHz; $f_r = 1.55$ kHz **(b)** $BW = 4.77$ kHz; $f_r = 4.52$ kHz
25. Output voltage decreases.
27. 2.8 Ω; > 175 mW
29. Q_1 conducts more.
31. Increases by 2.6 mA

GLOSSARY

Active filter A frequency-selective circuit consisting of active devices such as transistors or op-amps combined with reactive (RC) circuits.

Admittance (Y) A measure of the ability of a reactive circuit to permit current; the reciprocal of impedance. The unit is siemens (S).

Alpha (α) The ratio of collector current to emitter in a bipolar junction transistor.

Alternating current (ac) Current that reverses direction in response to a change in source voltage polarity.

Ammeter An electrical instrument used to measure current.

Ampere (A or amp) The unit of electrical current.

Ampere-hour rating A number given in ampere-hours determined by multiplying the current (A) times the length of time (h) a battery can deliver that current to a load.

Ampere-turn The unit of magnetomotive force (mmf).

Amplification The process of producing a larger voltage, current, or power using a smaller input signal as a pattern.

Amplifier An electronic circuit having the capability of amplification and designed specifically for that purpose.

Amplitude The maximum value of a voltage or current.

Anode The more positive terminal of a diode or other electronic device.

Apparent power (P_a) The product of the voltage times the current, expressed in volt-amperes (VA). In a purely resistive circuit it is the same as the true power.

Apparent power rating The method of rating transformers in which the power capability is expressed in volt-amperes (VA).

Atom The smallest particle of an element possessing the unique characteristics of that element.

Atomic number The number of electrons in a neutral atom.

Atomic weight The number of protons and neutrons in the nucleus of an atom.

Attenuation A gain less than unity.

Autotransformer A transformer in which the primary and secondary windings are in a single winding.

Average value The average of a sine wave over one half-cycle. It is 0.637 times the peak value.

AWG American wire gage. A standardization based on wire diameter.

Band-pass filter A filter that passes frequencies within a certain range and blocks all others.

Band-stop filter A filter that blocks frequencies within a certain range.

Bandwidth (BW) The characteristic of certain electronic circuits that specifies the usable range of frequencies for which signals pass from input to output without significant reduction in amplitude.

Barrier potential The inherent voltage across the depletion layer of a PN junction.

Base One of the semiconductor regions in a bipolar junction transistor.

Baseline The normal level of a pulse waveform; the voltage level in the absence of a pulse.

Battery An energy source that uses a chemical reaction to convert chemical energy into electrical energy.

Beta (β) The ratio of collector current to base current in a bipolar junction transistor.

Bias The application of dc voltage to a diode or other electronic device to produce a desired mode of operation.

Bipolar Characterized by two PN junctions.

Bipolar junction transistor (BJT) A transistor constructed with three doped semiconductor regions separated by two PN junctions.

Bleeder current The current left after the total load current is subtracted from the total current into the circuit.

Branch One current path in a parallel circuit.

Capacitance The ability of a capacitor to store electrical charge.

Capacitive reactance The opposition of a capacitor to sinusoidal current. The unit is the ohm.

Capacitive susceptance (B_C) The ability of a capacitor to permit current; the reciprocal of capacitive reactance. The unit is the siemens (S).

Capacitor An electrical device consisting of two conductive plates separated by an insulating material and possessing the property of capacitance.

Cathode The more negative terminal of a diode or other electronic device.

Cathode-ray tube (CRT) A vacuum tube device containing an electron gun that emits a narrow focused beam of electrons onto a phosphor-coated screen.

Center tap (CT) A connection at the midpoint of the secondary winding of a transformer.

Charge An electrical property of matter that exists because of an excess or a deficiency of electrons. Charge can be either positive or negative.

Choke An inductor. The term is used more commonly in connection with inductors used to block or choke off high frequencies.

Circuit An interconnection of electrical components designed to produce a desired result. A basic circuit consists of a source, a load, and an interconnecting current path.

Circuit breaker A resettable protective device used for interrupting excessive current in an electric circuit.

Circuit ground A method of grounding whereby a large conducive area on a printed circuit board or the metal chassis that houses the assembly is used as the common or reference point; also called *chassis ground.*

Circular mil (CM) The unit of the cross-sectional area of a wire.

Clamper A circuit that adds a dc level to an ac signal; a dc restorer.

Class A A category of amplifier circuit that conducts for the entire input cycle and produces an output signal that is a replica of the input signal in terms of its waveshape.

Class B A category of amplifier circuit that conducts for half of the input cycle.

Class C A category of amplifier circuit that conducts for a very small portion of the input cycle.

Clipper A circuit that removes part of a waveform above or below a specified level; a limiter.

Closed circuit A circuit with a complete current path.

Closed loop An op-amp configuration in which the output is connected back to the input.

Closed loop gain (A_{cl}) The overall voltage gain of an op-amp with feedback.

Coefficient of coupling (k) A constant associated with transformers that is the ratio of secondary flux to primary flux. The ideal value of 1 indicates that all the flux in the primary winding is coupled into the secondary winding.

Coil A common term for an inductor.

Collector One of the semiconductor regions in a BJT.

Common base (CB) A BJT amplifier configuration in which the base is the common (grounded) terminal.

Common-collector (CC) A BJT amplifier configuration in which the collector is the common (grounded) terminal.

Common-emitter (CE) A BJT amplifier configuration in which the emitter is the common (grounded) terminal.

Common-mode rejection ratio (CMRR) A measure of an op-amp's ability to reject signals that appear the same on both inputs; the ratio of open-loop gain to common-mode gain.

Common-mode signals Signals that appear the same on both inputs of an op-amp.

Comparator A circuit which compares two input voltages and produces an output in either of two states indicating the greater or less than relationship of the inputs.

Conductance (*G*) The ability of a circuit to allow current; the reciprocal of resistance. The unit is the siemens (S).

Conductor A material in which electrical current can flow with relative ease. An example is copper.

Core The structure around which the winding of an inductor is formed. The core material influences the electromagnetic characteristics of the inductor.

Coulomb (C) The unit of electrical charge.

Coulomb's law A physical law that states a force exists between two charged bodies that is directly proportional to the product of the two charges and inversely proportional to the square of the distance between them.

Covalent Related to the bonding of two or more atoms by the interaction of their valence electrons.

CRT See *Cathode-ray tube*.

Crystal The pattern or arrangement of atoms forming a solid material.

Current The rate of flow of electrons.

Current gain The ratio of output current to input current.

Cutoff The nonconducting state of a transistor.

Cutoff frequency (f_c) The frequency at which the output voltage of a filter is 70.7% of the maximum output voltage.

Cycle One repetition of a periodic waveform.

Darlington pair A two-transistor arrangement that produces a multiplication of current gain.

DC component The average value of a pulse waveform.

Decade A tenfold change in the value of a quantity. When a quantity becomes ten times less or ten times greater, it has changed a decade.

Decibel (dB) The unit of logarithmic expression of a ratio, such as a power ratio or a voltage ratio.

Degree The unit of angular measure corresponding to 1/360 of a complete revolution.

Derivative The instantaneous rate of change of a function determined mathematically.

Diac A semiconductor device that can conduct current in either of two directions when properly activated.

Dielectric The insulating material between the plates of a capacitor.

Dielectric constant A measure of the ability of a dielectric material to establish an electric field.

Dielectric strength A measure of the ability of a dielectric material to withstand voltage.

Differential amplifier An amplifier that produces an output proportional to the difference of two inputs.

Differentiator A circuit that produces an output which can approximate the instantaneous rate of change of the input function.

Diode An electronic device that permits current in only one direction.

Discrete device An individual electrical or electronic component that must be used in combination with other components to form a complete functional circuit.

Doping The process of imparting impurities to an intrinsic semiconductor material in order to control its conduction characteristics.

Drain One of the three terminals of a field-effect transistor.

Duty cycle A characteristic of a pulse waveform that indicates the percentage of time that a pulse is present during a cycle; the ratio of pulse width to period.

Earth ground A method of grounding whereby one side of a power line is neutralized by connecting it to a water pipe or a metal rod driven into the ground.

Effective value A measure of the heating effect of a sine wave; also known as the *rms* (root mean square) *value*.

Efficiency The ratio of the output power to the input power, expressed as a ratio.

Electrical Related to the use of electrical voltage and current to achieve desired results.

Electrical isolation The condition that exists when two coils are magnetically linked but have no electrical connection between them.

Electromagnetic field A formation of a group of magnetic lines of force surrounding a conductor created by electrical current in the conductor.

Electromagnetic induction The phenomenon or process by which a voltage is produced in a conductor when there is relative motion between the conductor and a magnetic or electromagnetic field.

Electron The basic particle of negative electrical charge in matter.

Electronic Related to the movement and control of free electrons in semiconductors or vacuum devices.

Element One of the unique substances that make up the known universe. Each element is characterized by a unique atomic structure.

Emitter One of the three semiconductor regions in a BJT.

Emitter-follower A popular term for a common-collector amplifier.

Energy The fundamental ability to do work and produce action or results. The unit is the joule (J).

Falling edge The negative-going transition of a pulse.

Fall time (t_f) The time interval required for a pulse to change from 90% to 10% of its full amplitude.

Farad (F) The unit of capacitance.

Faraday's law A law stating that the voltage induced across a coil of wire equals the number of turns in the coil times the rate of change of the magnetic flux.

Feedback The process of returning a portion of a circuit's output signal to the input in such a way as to create certain specified operating conditions.

Field-effect transistor (FET) A type of transistor that uses an induced electric field to control current.

Filter A type of electrical circuit that passes certain frequencies and rejects all others.

Free electron A valence electron that has broken away from its parent atom and is free to move from atom to atom within the atomic structure of a material.

Frequency A measure of the rate of change of a periodic function; the number of cycles completed in 1 s. The unit of frequency is the hertz.

Frequency response The way the output voltage of a circuit varies as the input frequency is changed.

Full-wave rectifier A circuit that converts an alternating sine wave into a pulsating dc consisting of both halves of a sine wave for each input cycle.

Fundamental frequency The repetition rate of a waveform.

Fuse A protective device that burns open when excessive current flows in a circuit.

Gain The amount by which an electrical signal is increased or decreased; the ratio of output to input; the amount of amplification.

Gate One of the three terminals of an FET.

Generator An energy source that produces electrical signals.

Germanium A semiconductor material.

Ground In electronic circuits, the common or reference point.

Half-power frequency The frequency at which the output of a filter is 70.7% of maximum.

Half-wave rectifier A circuit that converts an alternating sine wave into a pulsating dc consisting of one-half of a sine wave for each input cycle.

Harmonics The frequencies contained in a composite waveform, which are integer multiples of the repetition frequency (fundamental).

Henry (H) The unit of inductance.

Hertz (Hz) The unit of frequency. One hertz equals one cycle per second.

High-pass filter A certain type of filter in which higher frequencies are passed and lower frequencies are rejected.

Hole The absence of an electron.

Hypotenuse The longest side of a right triangle.

Hysteresis A lagging in the values of resulting magnetization in a magnetic material due to a changing magnetizing force.

Impedance (Z) The total opposition to sinusoidal current expressed in ohms.

Impedance matching A technique used to match a load resistance to an internal source resistance in order to achieve a maximum transfer of power.

Induced voltage Voltage produced as a result of a changing magnetic field.

Inductance (L) The property of an inductor that produces an opposition to any change in current.

Inductive reactance (X_L) The opposition of an inductor to sinusoidal current. The unit is ohms.

Inductive susceptance (B_L) The reciprocal of inductive reactance. The unit is siemens.

Inductor An electrical device formed by a wire wound around a core having the property of inductance and the capability to store energy in its electromagnetic field; also known as a coil or a choke.

Input The voltage, current, or power applied to an electrical circuit to produce a desired result.

Instantaneous value The voltage or current value of a waveform at a given instant in time.

Insulator A material that does not allow current under normal conditions.

Integral The area under the curve of a function, determined mathematically.

Integrator A circuit that produces an output which can approximate the area under the curve of the input function.

Interface To make the output of one type of circuit compatible with the input of another so that they can operate properly together.

Intrinsic semiconductor A pure material with relatively few free electrons.

Ion An atom that has gained or lost a valence electron resulting in a net positive or negative charge.

Ionization The removal or addition of an electron from or to an atom so that the resulting atom (ion) has a net positive or negative charge.

Joule (J) The unit of energy.

Junction A point at which two or more branches are connected together.

Junction field-effect transistor (JFET) A type of FET that operates with a reverse-biased junction to control current in a channel.

Kilowatt-hour (kWh) A common unit of energy used mainly by utility companies.

Kirchhoff's current law A law stating that the total current into a junction equals the total current out of the junction.

Kirchhoff's voltage law A law stating that (1) the sum of the voltage drops around a closed loop equals the source voltage or (2) the sum of all the voltages (drops and sources) around a closed loop is zero.

Lag To be behind; describes a condition of the phase or time relationship of waveforms in which one waveform is behind the other in phase or time.

Lead To be ahead; describes a condition of the phase or time relationship of waveforms in which one waveform is ahead of the other in phase or time; also, a wire or cable connection to a device or instrument.

Leading edge The first step or transition of a pulse.

Lenz's law A physical law that states when the current through a coil changes, an induced voltage is created in a direction to oppose the change in current. The current cannot change instantaneously.

Light-emitting diode (LED) A type of diode that emits light when forward current flows.

Linear Characterized by a straight-line relationship.

Line regulation The percent change in output voltage for a given change in line (input) voltage.

Load An element (resistor or other component) connected across the output terminals of a circuit that draws current from the circuit.

Load regulation The percent change in output voltage for a given change in load current.

Low-pass filter A certain type of filter in which lower frequencies are passed and higher frequencies are rejected.

Magnetic coupling The magnetic connection between two coils as a result of the changing magnetic flux lines of one coil cutting through the second coil.

Magnetic flux The lines of force between the north and south poles of a permanent magnet or an electromagnet.

Magnetic flux density The number of lines of force per unit area in a magnetic field.

Magnetizing force The amount of mmf per unit length of magnetic material.

Magnetomotive force The force that produces a magnetic field.

Magnitude The value of a quantity, such as the number of volts of voltage or the number of amperes of current.

Maximum power transfer theorem A theorem that states the maximum power is transferred from a source to a load when the load resistance equals the internal source resistance.

MOSFET Metal-oxide semiconductor field-effect transistor.

Multimeter An instrument that measures voltage, current, and resistance.

Mutual inductance (L_M) The inductance between two separate coils, such as a transformer.

Negative feedback The return of a portion of the output signal to the input such that it is out of phase with the input signal.

Neutron An atomic particle having no electrical charge.

Ohm (Ω) The unit of resistance.

Ohmmeter An instrument for measuring resistance.

Ohm's law A law stating that current is directly proportional to voltage and inversely proportional to resistance.

Open circuit A circuit in which the current path is broken.

Open-loop gain (A_{ol}) The internal voltage gain of an op-amp without feedback.

Operational amplifier (op-amp) A special type of amplifier exhibiting very high open-loop gain, very high input impedance, very low output impedance, and good rejection of common-mode signals.

Oscillator An electronic circuit consisting of an amplifier and a phase-shift network connected in a positive feedback loop that produces a time-varying output signal without an external input signal using positive feedback.

Oscilloscope A measurement instrument that displays signal waveforms on a screen.

Parallel The relationship in electric circuits in which two or more current paths are connected between the same two points.

Parallel resonance In a parallel *RLC* circuit, the condition where the impedance is maximum and the reactances are equal.

Pass band The range of frequencies passed by a filter.

Peak inverse voltage (PIV) The maximum value of reverse voltage which occurs at the peak of the input cycle when the diode is reversed-biased.

Peak value The voltage or current value of a waveform at its maximum positive or negative points.

Peak-to-peak value The voltage or current value of a waveform measured from its minimum to its maximum points.

Period The time interval of one complete cycle of a given sine wave or any periodic waveform.

Periodic Characterized by a repetition at fixed time intervals.

Permeability The measure of ease with which a magnetic field can be established in a material.

Phase The relative displacement of a time-varying waveform in terms of its occurrence with respect to a reference.

Phasor A representation of a sine wave in terms of both magnitude and phase angle.

Photoconductive cell A type of variable resistor that is light-sensitive.

Photodiode A diode whose reverse resistance changes with incident light.

Pinch-off voltage The value of the drain-to-source voltage of an FET at which the drain current becomes constant when the gate-to-source voltage is zero.

PN junction The boundary between N-type and P-type materials.

Pole In practical terms, a single *RC* network in a filter or amplifier that causes the response to change at a 20 dB/decade rate above or below a certain frequency.

Positive feedback The return of a portion of the output signal to the input such that it is in phase with the input signal.

Potentiometer A three-terminal variable resistor.

Power The rate of energy consumption.

Power factor The relationship between volt-amperes and true power or watts. Volt-amperes multiplied by the power factor equals true power.

Power gain The ratio of output power to input power; the product of voltage gain and current gain.

Power rating The maximum amount of power that a resistor can dissipate without being damaged by excessive heat buildup.

Power supply An electronic instrument that produces voltage, current, and power from the ac power line or batteries in a form suitable for use in various applications to power electronic equipment.

Primary winding The input winding of a transformer; also called *primary*.

Proton A positively charged atomic particle.

Pulse A type of waveform that consists of two equal and opposite steps in voltage or current separated by a time interval.

Pulse response The reaction of a circuit to a given input.

Pulse width The time interval between the opposite steps of an ideal pulse. For a nonideal pulse, the time between the 50% points on the leading and trailing edges.

Push-pull A type of class B amplifier in which one output transistor conducts for one half-cycle and the other conducts for the other half-cycle.

***Q*-point** The dc operating (bias) point of an amplifier.

Quality factor (*Q*) The ratio of reactive power to true power in a coil or a resonant circuit.

Radian A unit of angular measurement. There are 2π radians in one complete revolution. One radian equals 57.3°.

Ramp A type of waveform characterized by a linear increase or decrease in voltage or current.

Reactive power (P_r) The rate of which energy is stored and alternately returned to the source by a capacitor or inductor. The unit is the VAR.

Recombination The process of a free electron falling into a hole in the valence band of an atom.

Rectifier An electronic circuit that converts ac into pulsating dc; one part of a power supply.

Reflected load The secondary load as it appears to the source in the primary circuit.

Reflected resistance The resistance in the secondary circuit reflected into the primary circuit.

Regulator An electronic circuit that maintains an essentially constant output voltage with a changing input voltage or load.

Relay An electromagnetically controlled mechanical device in which electrical contacts are open or closed by a magnetizing current.

Reluctance The opposition to the establishment of a magnetic field.

Resistance Opposition to current. The unit is the ohm (Ω).

Resistor An electrical component designed specifically to provide resistance.

Resolution The smallest increment of a quantity that a meter can measure.

Resonant frequency The frequency at which a resonant condition occurs in a series or parallel *RLC* circuit.

Retentivity The ability of a material, once magnetized, to maintain a magnetized state without the presence of a magnetizing force.

Reverse breakdown The condition of a diode in which excessive reverse-bias voltage causes a rapid buildup of reverse current.

Rheostat A two-terminal variable resistor.

Right triangle A triangle with a 90° angle.

Ripple voltage The variation in the dc voltage on the output of a filtered rectifier caused by the slight charging and discharging action of the filter capacitor.

Rise time (t_r) The time interval required for a pulse to change from 10% to 90% of its amplitude.

Rising edge The positive-going transition of a pulse.

Roll-off The decrease in the response of a filter below or above a critical frequency.

Root mean square (rms) The value of a sine wave that indicates its heating effect, also known as the *effective value*. It is equal to 0.707 times the peak value.

Saturation The state of a BJT in which the collector current has reached a maximum and is independent of the base current.

Sawtooth wave A type of electrical waveform composed of ramps; a special case of a triangular wave.

Schematic A symbolized diagram of an electrical or electronic circuit.

Secondary winding The output winding of a transformer; also called *secondary.*

Selectivity A measure of how well a filter passes certain frequencies and rejects others. The narrower the bandwidth, the greater the selectivity.

Semiconductor A material that has a conductance value between that of a conductor and that of an insulator. Silicon and germanium are examples.

Sensitivity factor The ohms-per-volt rating of a voltmeter.

Series In an electrical circuit, a relationship of components in which the components are connected such that they provide a single current path between two points.

Series resonance In a series *RLC* circuit, the condition where the impedance is minimum and the reactances are equal.

Shell The orbit in which an electron revolves.

Short circuit A zero or abnormally low resistance between two points; usually an inadvertent condition.

Silicon A semiconductor material used in diodes and transistors.

Silicon-controlled rectifier (SCR) A device that can be triggered on to conduct current in one direction.

Solenoid An electromagnetically controlled device in which the mechanical movement of a shaft or plunger is activated by a magnetizing current.

Source Any device that produces energy; one of the three terminals of an FET.

Steady state The equilibrium condition of a circuit that occurs after an initial transient time.

Step-down transformer A transformer in which the secondary voltage is less than the primary voltage.

Step-up transformer A transformer in which the secondary voltage is greater than the primary voltage.

Stop band The range of frequencies between the upper and lower cutoff points.

Superposition A method for analyzing circuits with two or more sources by examining the effects of each source by itself and then combining the effects.

Switch An electrical or electronic device for opening and closing a current path.

Tank circuit A parallel resonant circuit.

Tapered Nonlinear, such as a tapered potentiometer.

Temperature coefficient A constant specifying the amount of change in the value of a quantity for a given change in temperature.

Terminal An external contact point on an electronic device.

Terminal equivalency A condition that occurs when two circuits produce the same load voltage and load current where the same value of load resistance is connected to either circuit.

Tesla The unit of flux density.

Termistor A type of variable resistor that is temperature-sensitive.

Thevenin's theorem A circuit theorem that provides for reducing any resistive circuit to a single equivalent voltage source in series with an equivalent resistance.

Thyristor A class of four-layer semiconductor devices.

Time constant A fixed time interval set by *R*,*C*, and/or *L* values that determines the time response of a circuit.

Tolerance The limits of variation in the value of a component.

Trailing edge The second step or transition of a pulse.

Transconductance (g_m) The ratio of drain current to gate-to-source voltage in an FET.

Transformer A device formed by two or more windings that are magnetically coupled to each other and providing a transfer of power electromagnetically from one winding to another.

Transient A temporary passing condition in a circuit; a sudden or temporary change in circuit conditions.

Transient time An interval equal to approximately five time constants.

Transistor A semiconductor device used for amplification and switching applications in electronic circuits.

Triac A bidirectional thyristor that can be triggered into conduction.

Triangular wave A type of electrical waveform that consists of two ramps.

Trigger The activating mechanism of some electronic devices or instruments.

Trimmer A small variable capacitor.

Turns ratio The ratio of turns in the secondary winding to turns in the primary winding.

Unijunction transistor (UJT) A type of transistor consisting of an emitter and two bases.

Unit gain A gain of 1.

Valence Related to the outer shell or orbit of an atom.

Valence electron An electron that is present in the outermost shell of an atom.

Varactor A diode that is used as a voltage-variable capacitor.

Volt The unit of voltage or electromotive force.

Voltage The amount of energy available to move a certain number of electrons from one point to another in an electrical circuit.

Voltage drop The potential difference in voltage between two points when there is a drop in energy level due to reactance.

Voltage gain The ratio of output voltage to input voltage.

Voltage regulation The process of maintaining an essentially constant output voltage over variations in input voltage or load.

Volt-ampere reactive (VAR) The unit of reactive power.

Voltmeter An instrument used to measure voltage.

Watt (W) The unit of power.

Watt's law A law that states the relationships of power to current, voltage, and resistance.

Waveform The pattern of variations of a voltage or current showing how the quantity changes with time.

Weber The unit of magnetic flux.

Winding The loops of wire in an inductor.

Wiper The sliding contact in a potentiometer.

Zener diode A type of diode that operates in reverse breakdown (called zener breakdown) to provide voltage regulation.

INDEX

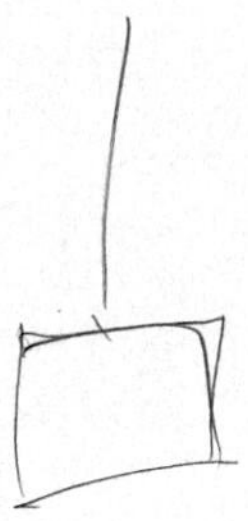